DATE DUE FOR RETURN

This book may be recalled before the above date.

BIOLOGICAL AND MEDICAL PHYSICS, BIOMEDICAL ENGINEERING

BIOLOGICAL AND MEDICAL PHYSICS, BIOMEDICAL ENGINEERING

The fields of biological and medical physics and biomedical engineering are broad, multidisciplinary and dynamic. They lie at the crossroads of frontier research in physics, biology, chemistry, and medicine. The Biological and Medical Physics, Biomedical Engineering Series is intended to be comprehensive, covering a broad range of topics important to the study of the physical, chemical and biological sciences. Its goal is to provide scientists and engineers with textbooks, monographs, and reference works to address the growing need for information.

Books in the series emphasize established and emergent areas of science including molecular, membrane, and mathematical biophysics; photosynthetic energy harvesting and conversion; information processing; physical principles of genetics; sensory communications; automata networks, neural networks, and cellular automata. Equally important will be coverage of applied aspects of biological and medical physics and biomedical engineering such as molecular electronic components and devices, biosensors, medicine, imaging, physical principles of renewable energy production, advanced prostheses, and environmental control and engineering.

Irving P. Herman

Physics of the Human Body

With 571 Figures and 135 Tables

Professor Dr. Irving P. Herman
Department of Applied Physics and Applied Mathematics
Columbia University
MC 4701
500 W. 120th Street, New York, NY 10027, USA
E-mail: iph1@columbia.edu

Library of Congress Control Number: 2006933935

ISSN 1618-7210

ISBN-10 3-540-29603-4 Springer Berlin Heidelberg New York

ISBN-13 978-3-540-29603-4 Springer Berlin Heidelberg New York

Springer is a part of Springer Science+Business Media

springer.com

Cover concept by eStudio Calamar Steinen

Typesetting by the author and SPi using a Springer LaTeX package
Final layout: SPi India
Cover production: *design & production* GmbH, Heidelberg

Printed on acid-free paper SPIN 11355717 57/3100/SPI - 5 4 3 2 1 0

This book is dedicated to Daniel, Jonathan, and Janet,

and to Sandy Koufax

Preface

Physics explains everything from the beginning to the end of any complete description of the human body. Such a comprehensive discussion should begin with the basic structure of matter, as explained by quantum mechanics – the physics at small dimensions, and end with the mechanics of human motion, the energetics of metabolism, the fluid dynamics of blood flow through vessels, the mechanisms for speaking and hearing, and the optical imaging system we call the eye. All of required combinations of atoms to form the complex molecules and organs of organisms that live and reproduce can be explained by quantum mechanics; however, such explanations can get pretty complex. The fields of chemistry and biology have been developed, in part, to explain the gap between the extremes – the microphysics and macrophysics of organisms such as the human body.

This book focuses mostly on the macrophysics end of the human body. We will assume that atoms form molecules that form cells that form organs. We want to understand the physics of human organs and of humans themselves. We will apply and somewhat extend freshman level physics to see how the body works. In addition to applying physical concepts to the body, we will try to understand the body from a viewpoint that is more numerical than is often adopted in biological and medical presentations.

One way to characterize this text is by saying what it is and what it is not. It is certainly about the physics of the human body. It is not about human anatomy, although we will need to use some basic anatomical concepts. It is not about human physiology, although it can be called a book about the physics of physiology. It is not a monograph in biomedical engineering per se, although about half of this volume concerns biomechanics, one important area in biomedical engineering. Medical physics is more closely related to health physics, the use of ionizing radiation, imaging, and instrumentation than to the macrophysics of the body. Biophysics concerns how physics can be used to study biology and focuses much more on the molecular basis and the cellular basis than will we (see Appendix E). One could say that the physics of the human body is synonymous with understanding the human machine.

Our goal is to understand physical issues concerning the human body, in part by solving problems to further this understanding. The focus is not at all on learning and memorizing medical terminology. Still some very basic concepts in anatomy and physiology will be introduced and used. Several of the many excellent general anatomy and physiology texts are cited at the end of the chapter [11, 16, 21, 22, 23, 24, 25, 26, 27, 29].

One theme that runs throughout this text is developing and then using simple and subsequently more refined models of the macrophysics of the human body [7, 13, 15]. Physicists tend to model concepts in as simple terms as possible at first. For example, to zero order a physicist would model a cow as a sphere. (This is sometimes used as part of a joke.) We will get a bit more complex here, but not much more. Another theme is to address issues in human biology quantitatively that are often addressed only qualitatively. The call for more quantitative thinking in physiology by Burton in *Physiology by Numbers* [5] is much appreciated by the author. In addition, we will present real physiological data and tie them with quantitative analysis and modeling.

If there is an applied force, energy, fluid flow, a light ray, an electric current, or an electric or magnetic field associated with the body, we will call it physics and we will analyze it. We will tend to avoid topics that delve into more chemistry and biology issues, but will briefly address physical chemistry issues involving concentration gradients and such, as they relate to fluid exchange in capillaries and conduction in nerves. Although we emphasize the physics of the body over the instrumentation used to make physical measurements on the body and probe body function, such instrumentation is addressed as needed.

Our intent is to use basic physics and not to teach it, particularly from scratch. Many chapters include a brief review of the physics principles needed in that chapter and subsequent chapters. Some topics are developed a bit further, and some even a bit further – and these are identified as advanced topics. More detailed overviews are given for topics seldom covered in detail during a two-semester physics course, such as fluids (Chap. 7), acoustics (Chap. 10), and optics (Chap. 11) and for areas used in several contexts, such as harmonic motion (Chap. 3). Some differential and integral calculus is used. (Partial differentiation is used sparingly, and mostly in sections labeled as advanced topics.) A brief review of the solutions to the simple differential equations used here is presented in Appendix C to help students with a limited background in calculus.

We will start with a comparison of medical and physics-type terminology in Chap. 1. The first chapter also includes a discussion of the "standard" human and introduces the concept of scaling relations. We can group the topics in subsequent 11 chapters into four areas in human body physics. (1) In Chaps. 2–5, the mechanics of the static body (Chap. 2) and the body in motion (Chap. 3) are analyzed and are then linked to the mechanical properties of the materials of the body (Chap. 4) and the body's motors: muscles (Chap. 5); these topics can be characterized as *Locomotion on Land*. (2) The second topic, *Energetics of the Body Metabolism*, is discussed in Chap. 6 and is needed to understand the discussions of body locomotion and function that precede and follow it. (3) Chapters 7–9 cover the *Locomotion of Humans in Fluids (other than on land) and the Motion of Fluids in Humans*. Chapter 7 overviews the physics of fluids and addresses locomotion in water (swimming) and in air – above ground (at least, the prospect for human flying). Chapters 8 and 9 respectively cover the fluidics of blood (cardiovascular system) and air (respiratory system) in the body. (4) Chapter 10 explores the acoustics of sound waves in speaking and hearing. The optics of eyes and vision are investigated in Chap. 11. Basic electrical properties of the body are developed in Chap. 12, along with a brief description of the magnetic properties of the body. So these three chapters respectively address sound, electromagnetic, and electrical waves, which we can collectively call *Waves and Signals*. (The electromagnetic nature of light waves is not discussed in Chap. 11.)

Chapter 13 examines how the body automatically uses the basic engineering principle of feedback and control in regulating all aspects of function. The physics of sensation of three of the five senses are described: hearing, seeing, and touch – the last briefly in Chap. 2. Some connection is made between the physics of sensation, biochemistry of sensation, and perception (psychophysics) in Chap. 1. The sense of taste and smell are purely chemical, with little basis in physics (other than the chemistry of the molecular interactions in each being clear applications of physics), and are not covered – except for a brief discussion of the electrical properties of the taste and smell sensory neurons in Chap. 12. The emphasis throughout is on how physics can explain the functioning of the body under normal and unusual circumstances. We will concern ourselves with the human body with its common body coverings: footware to minimize stress during movement (Chap. 4), clothes to regulate heat loss (Chap. 6), and corrective lenses to improve vision (Chap. 11).

The chapters are set more to address specific areas in physics rather than specific parts or systems in the body. It is difficult to construct chapters with clean divisions because different areas of physics are needed to understand many components of the body. For example, to understand the physics of the heart, you need to address its role in circulation (Chap. 8), the action of muscles (Chap. 5, which is more focused on skeletal muscle than the fairly similar cardiac muscle), and the electrical signals generated by the heart (Chap. 12).

This text concludes with five appendices. Appendix A overviews symbols and units, and references tables of units presented in the chapters. Appendix B lists the figures and tables that describe the main features of human anatomical and anthropometric information, which are used throughout this text. The types of differential equations used in the text are reviewed in Appendix C. These same differential equations are used throughout the text in mechanical, fluid flow, and electrical models; the connections between these models are made in Appendix D. Appendix E attempts to define the field of biophysics, and connects the contents of this text with this field.

This text has been developed from the author's lecture notes developed for the course *Physics of the Human Body*, which is a "professional-level" restricted elective course he developed taken mostly by first and second year undergraduates in the Columbia University Fu Foundation School of Engineering and Applied Science. This course was designed so it could be taken by all first year students in their second term (in conjunction with second-term physics and calculus). The author usually covers Chaps. 1–10 in some detail and Chaps. 11–13 in less detail in a full semester.

Courses at different levels, including mid-level and upper-level undergraduate courses, can be taught by purposely including or excluding more detailed and advanced topics in the text and problems. Depending of the level of desired depth, material in about half to all the chapters can be covered in one term.

This text can also be used as a companion volume in introductory physics courses, and assist premedical undergraduates in learning and reviewing physics. It can also serve as a text in introductory biomedical engineering or medical materials courses. Medical students interested in a more quantitative approach to physiology and those doing medical research may also appreciate the approaches adopted here.

Many problems are presented at the end of each chapter, ranging from simple to more advanced problems (the latter are denoted as such). Several problems have multiple parts, and only a few of these parts can be assigned. Answers to selected problems are given after the appendices.

Usually SI (MKS, m-kg-s) units are used; when more convenient, other metric units, including CGS (cm-g-s) units and mixed metric units are used. English FPS (ft-lb-s) units are sometimes purposely used to make a connection to the real world (at least in countries such as the USA and UK). For example, it would be strange to hear a baseball announcer say, "This pitcher is really throwing some heat. The radar gun clocked his last pitch at 43.8 m/s (or 158 km/h)", as opposed to 98 mph. It would be stranger to hear a football (i.e., American football) announcer say, "They have first (down) and 9.144 to go", meaning 9.144 m instead of 10 yd. Similarly, it would be strange to discuss the physics of the body in these sports, such as in throwing a baseball, in any but the usual units. Angles are given in radians, except when using degrees gives a more physical picture.

Several excellent texts cover material that overlaps topics covered here, each with a different focus. They are magnificent resources in their own right. *Physics of the Body* by Cameron, Skofronick, and Grant [6] spans most of the topics in this book and provides excellent physical insight. It is at a level of physics that is lower than that used here and derives and presents fewer of the equations necessary for a more rigorous treatment, but it provides a very good basic background in human physiology for nonexperts. In a way, the emphasis of *The Human Machine* by Alexander [2] coincides with ours, but, again, the explanations are more qualitative. The mode of physical thinking it presents is impressive. *Physics with Examples from Medicine and Biology* by

Benedek and Villars [3] is a series of three comprehensive introductory physics texts in which excellent examples and problems have been chosen concerning the physics of the body. The authors have taken several fairly complex topics and have made them utterly understandable. Many other first-year general physics texts commonly used nowadays have several examples and chapter problems dealing with the body. *Intermediate Physics for Medicine and Biology* by Hobbie [14] is a more advanced text that emphasizes both physics and physical chemistry. *Medical Physics and Biomedical Engineering* by Brown et al. [4] is a bit more advanced and focuses also on classic areas in medical physics, such as radioactivity and instrumentation. Many of the illustrative problems concerning human biology and related topics have been collected in the beautiful books: *Biomedical Applications of Introductory Physics* by Tuszynski and Dixon [28], *Physics in Biology and Medicine* by Davidovits [9], *Biophysics Problems: A Textbook with Answers* by Maróti, Berkes, and T́olgyesi [17], *Physics for the Biological Sciences: A Topical Approach to Biophysical Concepts* by Hallett, Stinson, and Speight [12], and *Topics in Classical Biophysics* by Metcalf [18]. Many of the issues in exercise physiology, such as the metabolism during sporting activities, are described in elementary terms in *Fox's Physiological Basis for Exercise and Sport* by Foss and Keteyian [10] and *Physiology of Sport and Exercise* by Wilmore and Costill [30]. *Basic Biomechanics of the Musculoskeletal System*, edited by Nordin and Frankel [20] is a comprehensive and clear overview of the biomechanics of structures, joints, and motion. The applications of physics at a more molecular and cellular level, more in the classical domain of biophysics, are described in *Biophysics: An Introduction*, by Cotterill [8] and *Biological Physics: Energy, Information* by Nelson [19]. The more general application of physics to animals is addressed in the exciting and very comprehensive book *Zoological Physics: Quantitative Models, Body Design, Actions and Physical Limitations in Animals* by Ahlborn [1]. All of these texts are highly recommended for more details. They, along with the anatomy and physiology texts cited earlier, have contributed to the preparation of this text.

The author thanks the many people who have made valuable comments contributing to this book, including Marlene Arbo, Gerard Ateshian, Sarbajit Benerjee, Alex Breskin, Bill Burdick, Yi-Ting Chiang, Kevin Costa, Ted Ducas, Yossi Goffer, Daniel Herman, Jonathan Herman, Steven Heymsfield, Jeffrey Holmes, Mark Langill, Barclay Morrison III, Elizabeth Olson, Thomas Pedersen, Harry Radousky, Paul Sajda, Michael Sheetz, and Samuel Sia. He would also like to thank the Columbia University Library system.

This author began writing this text when he was a Lady Davis Scholar on sabbatical at Hebrew University in Jerusalem as a guest of Uri Banin, and he gratefully acknowledges this support.

New York, NY, November 2006 *Irving P. Herman*

Contents

1

Terminology, the Standard Human, and Scaling

Several concepts will appear throughout our discussion of the human body: medical terminology, the characteristics of a "typical" human, and how body properties and responses scale with parameters. Much of the problem we have in comprehending specialists in any field is in understanding their jargon, and not in understanding their ideas. This is particularly true for medicine. Much of medical jargon of interest to us is the terminology used in anatomy, and much of that in anatomy relates to directions and positions. To make things clearer for people who think in more physics-type terms, we will relate some of the anatomical coordinate systems used in medicine to coordinate systems that would be used by physicists to describe any physical system. We will also extend this terminology to describe the degrees of freedom of rotational motion about the joints needed for human motion. In all of our discussions we will examine a typical human. To be able to do this, we will define and characterize the concept of a standard human. The final concept in this introductory chapter will be that of scaling relationships. We will examine how the properties of a standard human scale with body mass and how the perception level of our senses varies with the level of external stimulus.

1.1 Anatomical Terminology

The first series of anatomical "coordinate systems" relate to direction, and the first set of these we encounter is right vs. left. With the xyz coordinate system of the body shown in Fig. 1.1, we see that *right* means $y < 0$ and *left* means $y > 0$. Right and left, as well as all other anatomical terms, are always from the "patient's" point of view. This was made perfectly clear to the author during a visit to his son's ophthalmologist. When he tried to discuss what he thought was his son's right eye, it was pointed out to the author in no uncertain terms that he was really referring to the patient's left eye and that he was doing so in an improper manner. Case closed! (Stages in theaters have a similar convention, with stage left and stage right referring to the left and right sides of an actor on stage facing the audience. This was evident in

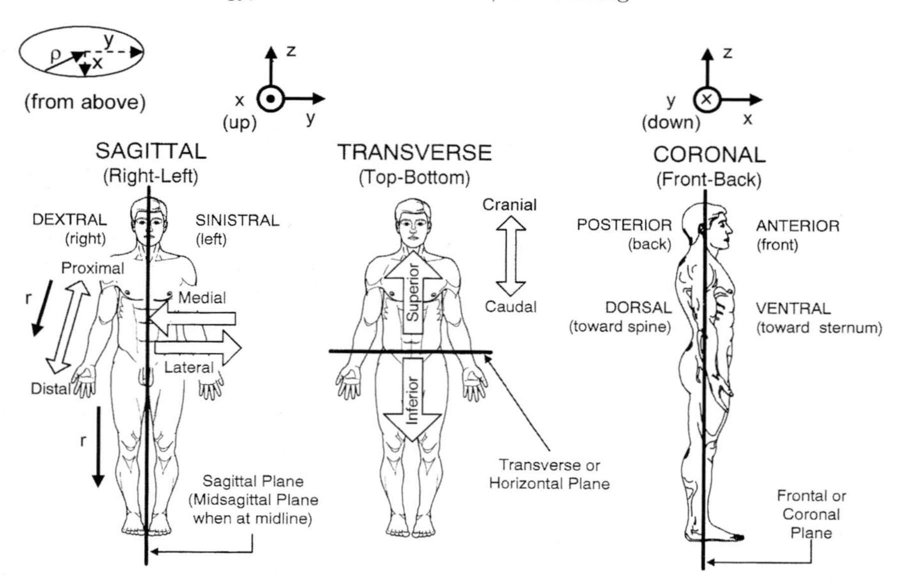

Fig. 1.1. Directions, orientations, and planes used to describe the body in anatomy, along with common coordinate systems described in the text. We will assume both terms in the following pairs mean the same: superior/cranial, inferior/caudal, anterior/ventral, and posterior/dorsal, even though there may be fine distinctions in what they mean, as is depicted here. (From [43], with additions. Used with permission)

a funny scene in the movie **Tootsie** when a stagehand was told to focus on the right side of the face of Dorothy Michaels, aka Michael Dorsey, aka Dustin Hoffman – and Dorothy heard this and then turned her (i.e., his) head so the camera would be focusing on the left side of her (i.e., his) face. A comical debate then ensued concerning whose "right" was correct, that of a person on stage or one facing the stage.)

The second direction is *superior* (or *cranial*), which means towards the head or above, i.e., to larger z. *Inferior* (or *caudal* (kaw'-dul)) means away from the head, i.e., to smaller z – in an algebraic sense, so more and more inferior means smaller positive numbers and then more highly negative values of z. (This is relative to a defined $z = 0$ plane. We could choose to define the origin of the coordinate system at the center of mass of the body.) So, the head is superior to the feet, which are inferior to the head. After supplying the body with oxygen, blood returns to the heart through two major veins, the superior and inferior vena cava (vee'-nuh cave'-uh), which collect blood from above and below the heart, respectively. (As you see, words that the author has trouble pronouncing are also presented more or less phonetically, with an apostrophe after the accented syllable.)

Anterior (or *ventral*) means towards or from the front of the body, i.e., to larger x. *Posterior* (or *dorsal*) means towards or from the back, corresponding to smaller algebraic x. The nose is anterior to the ears, which is posterior to the nose.

There is another pair of terms that relate to the y coordinate, specifically to its magnitude. *Medial* means nearer the midline of the body, i.e., towards smaller $|y|$. *Lateral* means further from the midline, i.e., towards larger $|y|$.

Other anatomical terms require other types of coordinate systems. One set describes the distance from the point of attachment of any of the two arms and two legs from the trunk. Figure 1.1 depicts this with the coordinate r, where $r = 0$ at the trunk. r is never negative. *Proximal* means near the point of attachment, i.e., to smaller r. *Distal* means further from the point of attachment, or larger r.

The last series of directional terms relates to the local surface of the body. This can be depicted by the coordinate ρ (inset in Fig. 1.1), which is related to x and y in an $x - y$ plane. $\rho = 0$ on the local surface of the body. *Superficial* means towards or on the surface of the body, or to smaller ρ. *Deep* means away from the surface, or towards larger ρ.

These directional terms can refer to any locality of the body. Regional terms designate a specific region in the body (Tables 1.1 and 1.2). This is illustrated by an example we will use several times later. The region between the shoulder and elbow joints is called the brachium (brae'-kee-um). The adjective used to describe this region in anatomical terms is brachial (brae'-kee-al). The muscles in our arms that we usually call the biceps are really the brachial biceps or biceps brachii, while our triceps are really our brachial triceps or triceps brachii. The terms biceps and triceps refer to any muscles with two or three points of origin, respectively (as we will see) – and not necessarily to those in our arms.

The final set of terms describes two-dimensional planes, cuts or sections of the body. They are illustrated in Fig. 1.1. A *transverse* or *horizontal* section separates the body into superior and inferior sections. Such planes have constant z. *Sagittal* sections separate the body into right and left sections, and are planes with constant y. The *midsagittal* section is special; it occurs at the midline and is a plane with $y = 0$. The *frontal* or *coronal* section separates the body into anterior and posterior portions, as described by planes with constant x.

Much of our outright confusion concerning medical descriptions is alleviated with the knowledge of these three categories of anatomical terminology. There is actually a fourth set of anatomical terms that relates to types of motion. These are discussed in Sect. 1.2.

1.2 Motion in the Human Machine

Anatomical terms refer to the body locally whether it is at rest or in motion. Since we are also concerned with how we move, we need to address human motion [32]. We will describe how we move by examining the *degrees of freedom*

Table 1.1. Anatomical terms in anterior regions

anatomical term	common term
abdominal	abdomen
antebrachial	forearm
axilliary	armpit
brachial	upper arm
buccal	cheek
carpal	wrist
cephalic	head
cervical	neck
coxal	hip
crural	front of leg
digital	finger or toe
frontal	forehead
iguinal	groin
lingual	tongue
mammary	breast
mental	chin
nasal	nose
oral	mouth
palmar	palm
pedal	foot
sternal	breastbone
tarsal	ankle
thoracic	chest
umbilical	navel

of our motion and the means for providing such motion by our joints. We will see that our arms and legs are constructed in a very clever manner. Because joints involve motion between bones, we will need to refer to the anatomy of the skeletal system, as in Fig. 1.2.

Table 1.2. Anatomical terms in posterior regions

anatomical term	common term
acromial	top of shoulder
femoral	thigh
gluteal	buttock
occipital	back of head
plantar	sole of foot
popliteal	back of knee
sacral	between hips
sural	back of leg
vertebral	spinal column

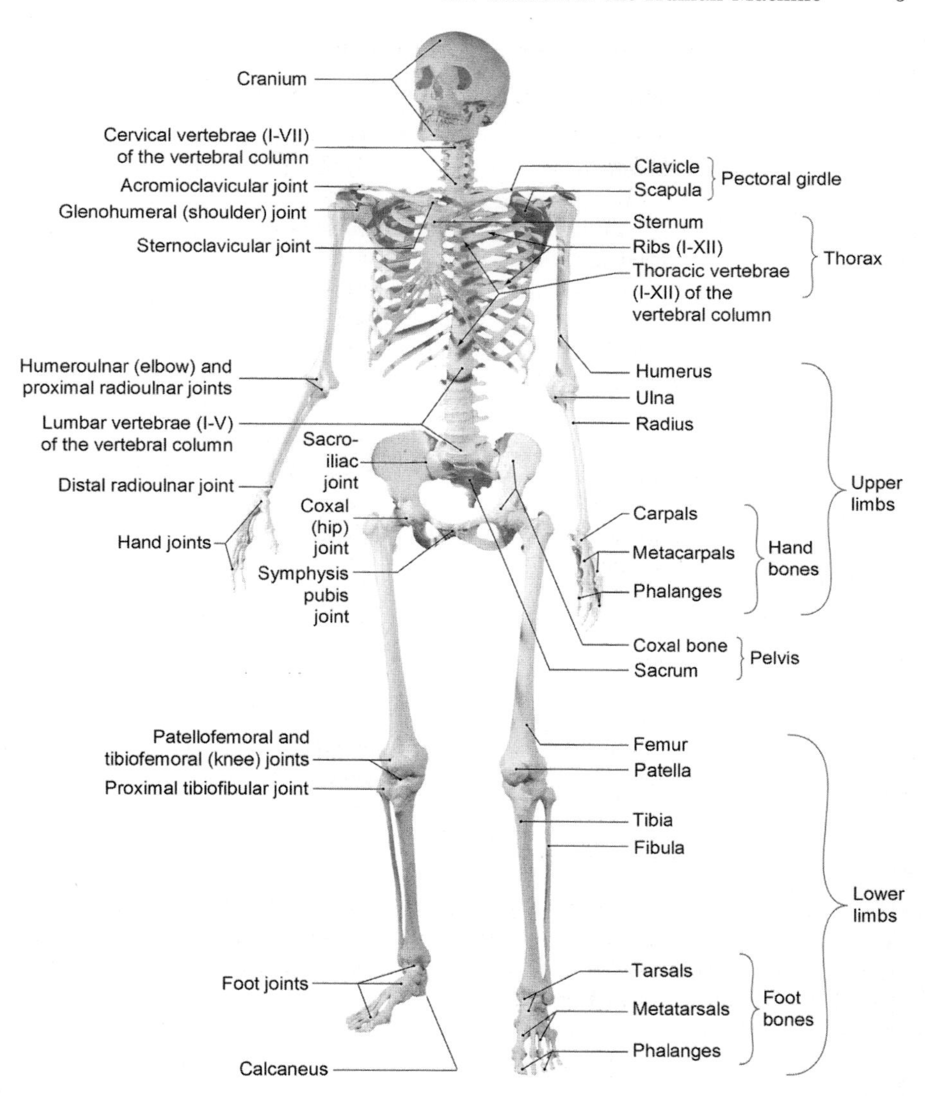

Fig. 1.2. Anatomy of the skeletal system, anterior view, with major bones and joints listed. (From [59])

Think of a degree of freedom (DOF) of motion as a coordinate needed to describe that type of motion. If you want to relocate an object, you are generally interested in changing its center of mass and its angular orientation. You may want to change its center of mass from an (x, y, z) of $(0, 0, 0)$ to (a, b, c). Because three coordinates are needed to describe this change, there are three "translational" degrees of freedom. Similarly, you can change the angular orientation of the object about the x, y, and z axes, by changing the angles this object can be rotated about these three axes: θ_x, θ_y, and θ_z, respectively. So, there are also three rotational degrees of freedom. (Sometimes, these three independent rotations are defined differently, by the three Eulerian angles, which will not be introduced here.)

These six (three plus three) degrees of freedom are independent of each other. Keeping your fingers rigid as a fist, you should be able to change independently either the x, y, z, θ_x, θ_y, and θ_z of your fist by moving your arms in different ways. You should try to change the x, y, and z of your fist, while keeping θ_x, θ_y, and θ_z fixed. Also, try changing the θ_x, θ_y, and θ_z of your fist, while keeping its x, y, and z constant.

We would like each of our arms and legs to have these six degrees of freedom. *How does the body do it?* It does it with joints, also known as *articulations.* Two types of articulations, fibrous (bones joined by connective tissue) and cartilaginous (bones joined by cartilage) joints, can bend only very little. There is a joint cavity between the articulating bones in synovial joints. Only these synovial joints have the large degree of angular motion needed for motion. As seen in Fig. 1.3, in synovial joints cartilage on the ends of opposing bones are contained in a sac containing synovial fluid. The coefficient of friction in such joints is lower than any joints made by mankind. (More on this later.)

There are several types of synovial joints in the body, each with either one, two, or three degrees of angular motion. Each has an analog with physical objects, as seen in Fig. 1.4. For example, a common door hinge is a model of one degree of angular freedom. Universal joints, which connect each axle to a wheel in a car, have two angular degrees of freedom. A ball-and-socket joint has three independent degrees of angular motion. The water faucet in a shower is a ball-and-socket joint. The balls and sockets in these joints are spherical. Condyloid or ellipsoidal joints are ball-and-socket joints with ellipsoidal balls and sockets. They have only two degrees of freedom because rotation is not possible about the axis emanating from the balls. A saddle joint, which looks like two saddles meshing into one another, also has two degrees of angular motion. Other examples are shown in Fig. 1.4.

Now back to our limbs. Consider a leg with rigid toes. The upper leg bone (femur) is connected to the hip as a ball-and-socket joint (three DOFs) (as in the song "Dry Bones" aka "Them Bones" in which "The hip bone is connected to the thigh bones, ..." The knee is a hinge (one DOF). The ankle is a saddle joint (two DOFs). This means that each leg has six degrees of angular motion, as needed for complete location of the foot. Of course, several of these degrees of freedom have only limited angular motion.

Now consider each arm, with all fingers rigid. The upper arm (humerus) fits into the shoulder as a ball-and-socket joint (three DOFs). The elbow is a hinge (one DOF). The wrist is an ellipsoidal joint (two DOFs). That makes

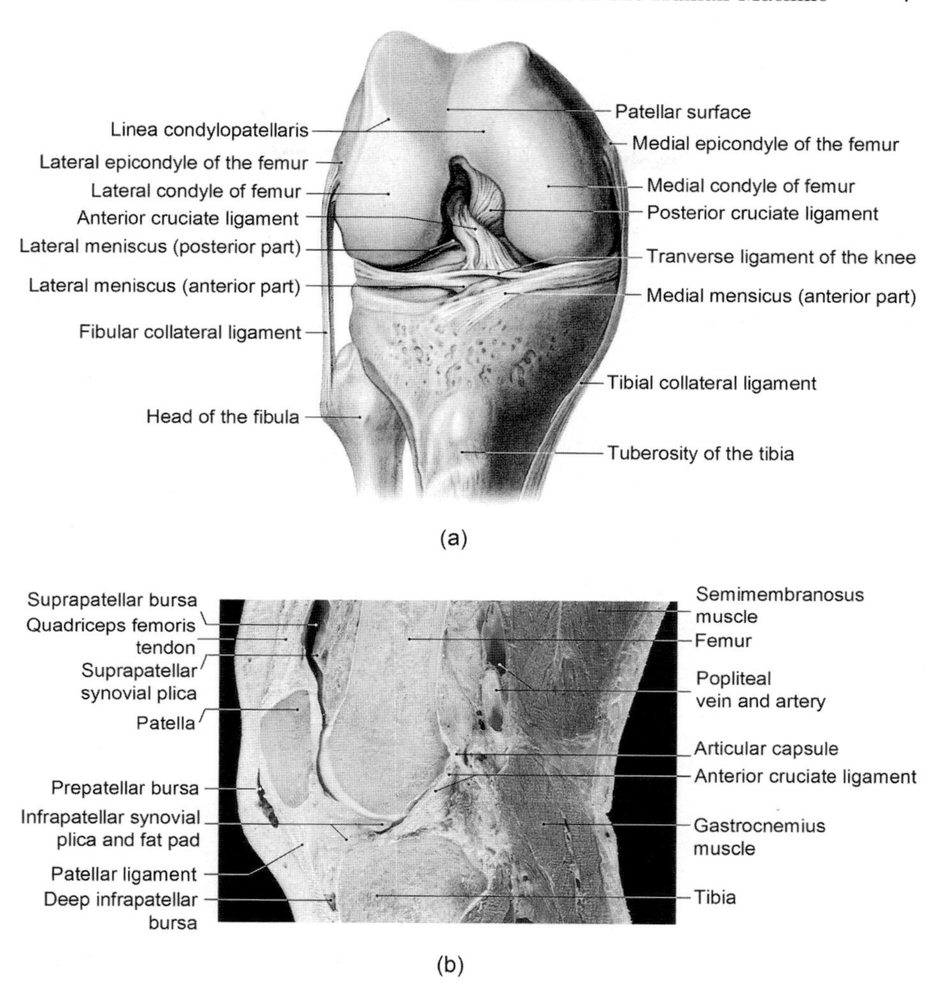

Fig. 1.3. The right knee synovial joint, with **(a)** anterior view with the kneecap (patella) removed and **(b)** in sagittal section (photo). Also see Fig. 3.2e. (From [59])

six DOFs. The leg has these six DOFs, but the arm has one additional DOF, for a total of seven. This additional DOF is the screwdriver type motion of the radius rolling on the ulna (Figs. 1.2, 2.7, and 2.8), which is a pivot with 1 DOF. With only six DOFs you would be able to move your hand to a given x, y, z, $\theta_x, \theta_y, \theta_z$ position in only one way. With the additional DOF you can do it in many ways, as is seen for the person sitting in a chair in Fig. 1.5. There are many more degrees of freedom available in the hand, which enable the complex operations we perform, such as holding a ball. Figure 1.6 shows the bones of the hand, and the associated articulations and degrees of freedom associated with the motion of each finger.

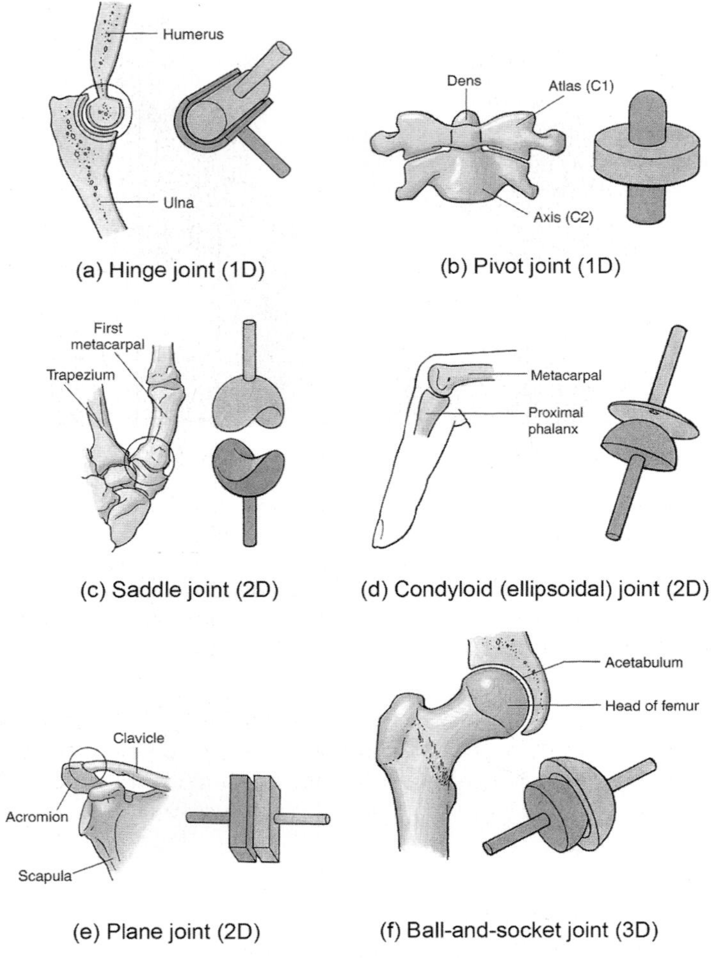

Fig. 1.4. Six types of synovial joints, including a: **(a)** hinge joint (1D joint), as in the elbow joint for flexion and extension, **(b)** pivot joint (1D joint), as in the atlantoaxial joint in the spinal cord for rotation, **(c)** saddle joint (2D), which is both concave and convex where the bones articulate, as in the joint between the first metacarpal and the trapezium in the hand, **(d)** condyloid or ellipsoidal joint (2D), as in the metacarpophalangeal (knuckle) joint between the metacarpal and proximal phalanx for flexion and extension, abduction and adduction, and circumduction, **(e)** plane joint (2D), as in the acromioclavicular joint in the shoulder for gliding or sliding, and **(f)** ball-and-socket joint (3D), as in the hip joint (and the shoulder joint) for flexion and extension, abduction and adduction, and medial and lateral rotation. See Figs. 1.9 and 1.10 for definitions of the terms describing the types of motion about joints and the diagrams in Fig. 1.11 for more information about synovial joints. (From [49]. Used with permission)

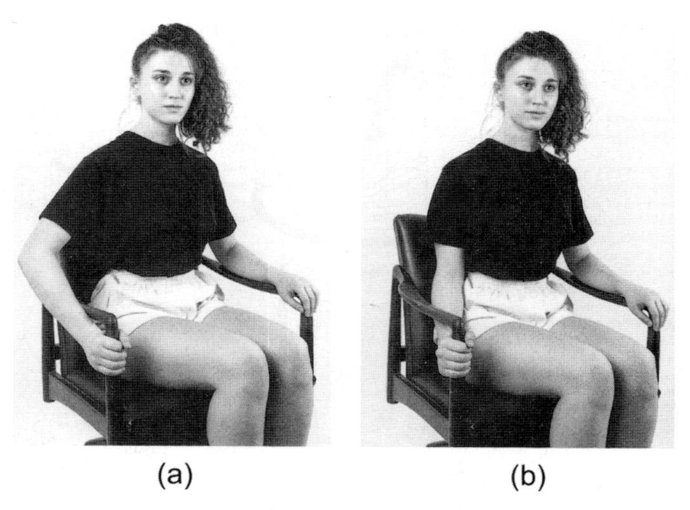

Fig. 1.5. Nonunique way of positioning the right arm. This is demonstrated by grasping the armrest while sitting, with the six coordinates of the hand (three for position and three for angle) being the same in both arm positions. This is possible because the arm can use its seven degrees of freedom to determine these six coordinates. (From [32]. Copyright 1992 Columbia University Press. Reprinted with the permission of the press)

We can also see why it is clever and good engineering that the knee hinge divides the leg into two nearly equal sections and the elbow hinge divides the arm into two nearly equal sections. In the two-dimensional world of Fig. 1.7 this enables a greater area (volume for 3D) to be covered than with unequal sections.

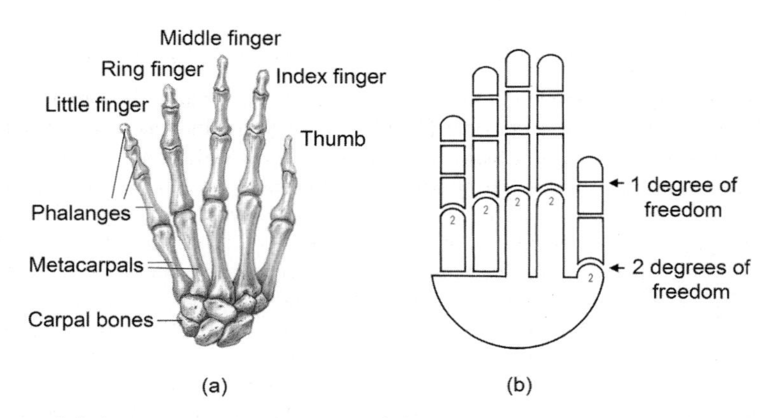

Fig. 1.6. (a) Anatomy of the hand and **(b)** the degrees of freedom of the hand and fingers, with joints (spaces) having one (spaces with flat terminations) or two (curved terminations, with a "2" below the joint) degrees of freedom. (From [32]. Copyright 1992 Columbia University Press. Reprinted with the permission of the press)

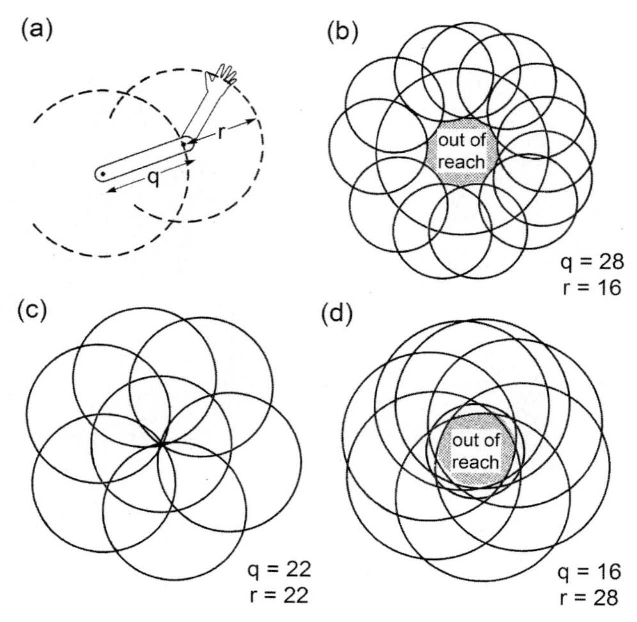

Fig. 1.7. Range of hand motion in two dimensions for different lengths of the upper and lower arms. (From [32]. Copyright 1992 Columbia University Press. Reprinted with the permission of the press)

In preparation for our discussion of statics and motion of the body, we should consider the building blocks of human motion. There are four types of components: *bones*, *ligaments*, *muscles*, and *tendons*. Each has a very different function and mechanical properties. Bones are often lined with hyaline (high'u-lun) articular cartilage at the synovial joints. Ligaments hold bones together. Muscles, in particular skeletal muscles, are the motors that move the bones about the joints. (There is also cardiac muscle – the heart – and smooth muscle – of the digestive and other organs.) Tendons connect muscles to bones. Muscles are connected at *points of origin and insertion* via tendons; the points of insertion are where the "action" is. Figure 1.8 shows several of the larger muscles in the body, along with some of the tendons.

Muscles work by contraction only, i.e., only by getting shorter. Consequently, to be able to move your arms one way and then back in the opposite direction, you need pairs of muscles on the same body part for each opposing motion. Such opposing pairs, known as "antagonists," are very common in the body.

We now return to our brief review of terminology, this time to describe the angular motion of joints. It is not surprising that these come in opposing pairs (Figs. 1.9 and 1.10) as supplied by antagonist muscles. When the angle of a 1D hinge, such as the elbow, increases it is called *extension* and when it decreases it is *flexion*. When you rotate your leg away from the midline of your body, it

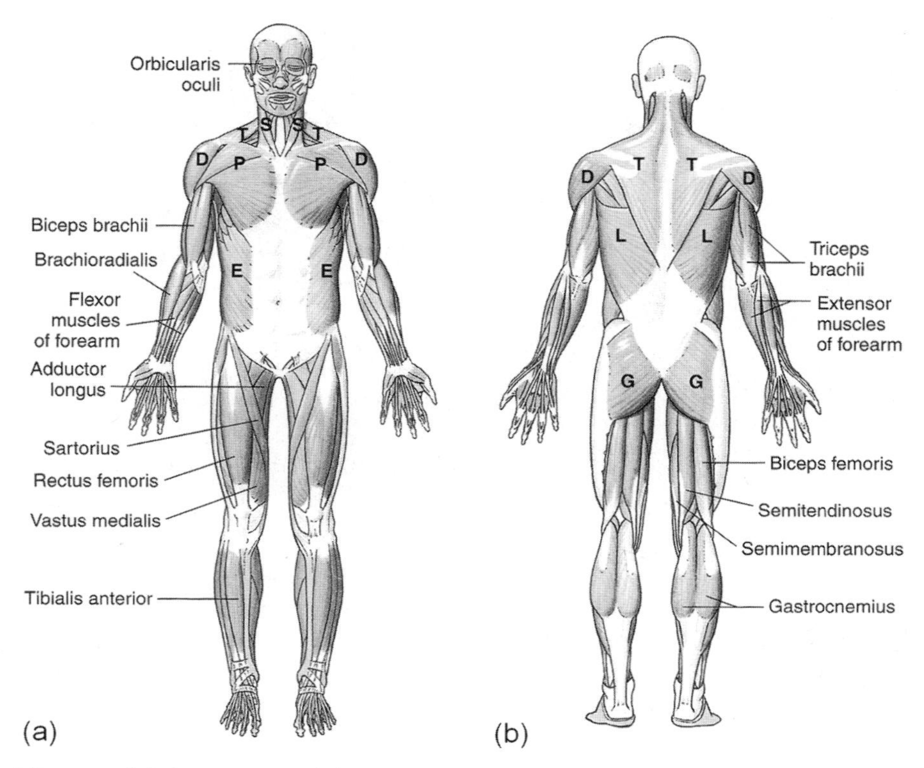

Fig. 1.8. **(a)** Anterior and **(b)** posterior views of some of the larger skeletal muscles in the body. Several muscles are labeled: S, sternocleidomastoid; T, trapezius; D, deltoid;, P, pectoralis major; E, external oblique; L, latissimus dorsi; G, gluteus maximus. In **(b)**, the broad-banded tendon extending from the gastrocnemius and soleus (deep to the gastrocnemius, not shown) muscles to the ankle (calcaneus) is the calcaneal (or Achilles) tendon. (From [49]. Used with permission)

is *abduction*, and when you bring is closer to the midline, it is *adduction*. When you rotate a body part about its long axis it is called *rotation*. The screwdriver motion in the arm is *pronation* (a front facing hand rotates towards the body) or *supination* (away from the body), and so supination is the motion of a right hand screwing in a right-handed screw (clockwise looking from the shoulder distally) and pronation is that of a right hand unscrewing a right-handed screw (counterclockwise looking from the shoulder distally). Examples of the rotation axes for the synovial joints used in these opposing motions are given in Fig. 1.11.

One example of opposing motion is the motion of the arm (Fig. 1.12). The biceps brachii have two points of origin and are inserted on the radius (as shown in Fig. 2.10 below). When they contract, the radius undergoes flexion about the pivot point in the elbow. The triceps brachii have three points of origin, and a point of insertion on the ulna. They are relaxed during flexion.

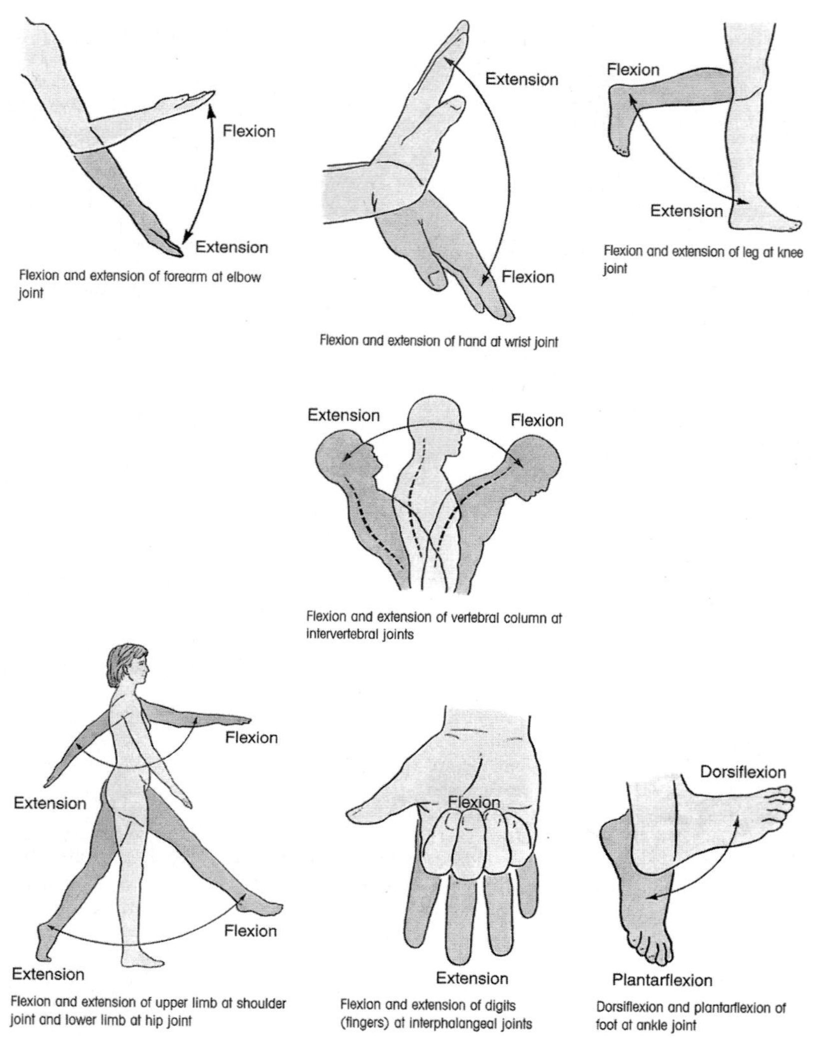

Fig. 1.9. Several antagonistic motions allowed by synovial joints. See other motions in Fig. 1.10. (From [49]. Used with permission)

During extension they contract, while the biceps brachii are relaxed. This is an example of a lever system about a pivot point. (This is really a pivot axis normal to the plane of the arm, as is illustrated in Fig. 1.11a for a hinge joint.)

A second place where there is such opposing motion is the eye. The three types of opposite motion in each eye (monocular rotations) are shown in Fig. 1.13. During *adduction* the eye turns in to the midline, while during *abduction* it turns out. The eyeball can also undergo *elevation* (eye rotating

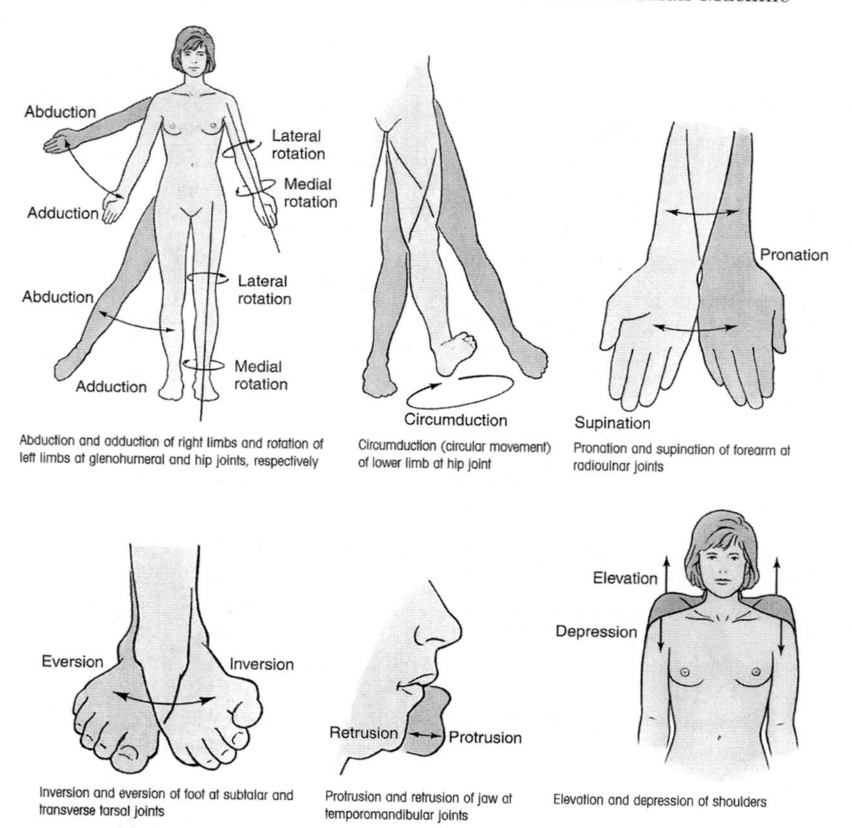

Fig. 1.10. More antagonistic motions allowed by synovial joints. See other motions in Fig. 1.9. (From [49]. Used with permission)

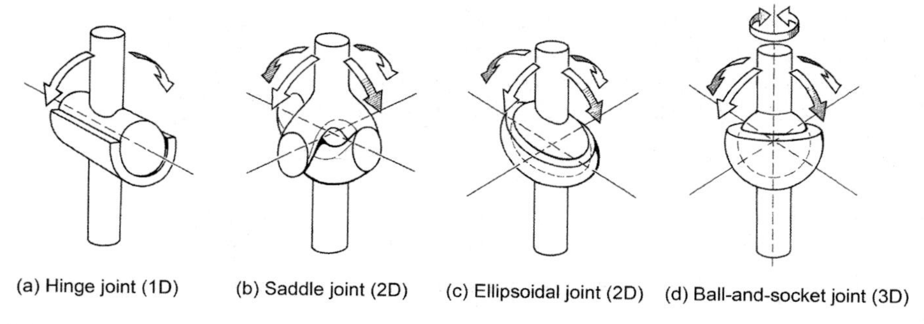

Fig. 1.11. Rotation axes for four types of synovial joints are shown for each depicted rotation direction: **(a)** one axis for a hinge joint (1D joint), **(b)** two axes for a saddle joint (2D), **(c)** two axes for an ellipsoidal joint (2D), and **(d)** three axes for a ball-and-socket joint (3D). (From [54])

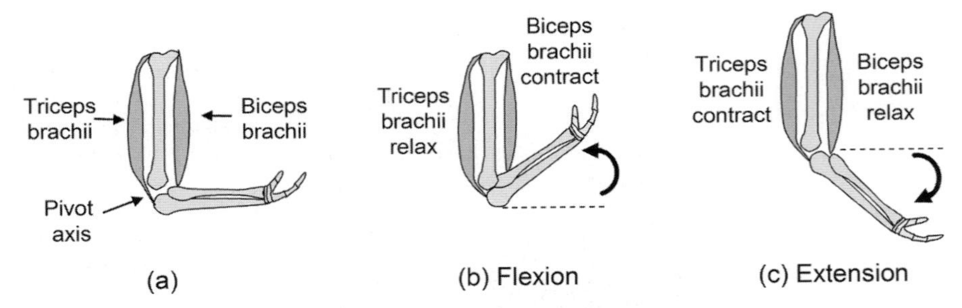

Fig. 1.12. Opposing motions of the lower arm with antagonist muscles, with flexion by contraction of the biceps brachii and extension by the contraction of the triceps brachii. The axis of rotation is seen in Fig. 1.11a

upward, or *supraduction*) or *depression* (eyeball rotating downward, or *infraduction*). Less common is the rotation of the eyeball about an axis normal to the iris, in opposing *intorsion* (*incycloduction*) or *extorsion* (*excycloduction*) motions. There are three pairs of opposing muscles per eye, each attached to the skull behind the eye, that control these motions (Fig. 1.14, Table 1.3). However, of these three pairs, only one is cleanly associated with only one of these pairs of opposing motions. Adduction occurs with the contraction of the medial rectus muscle, while abduction occurs when the lateral rectus contracts. The primary action of the superior rectus is elevation, while that of the opposing inferior rectus is depression. The primary action of the superior oblique is also depression, while that of the opposing inferior oblique is also elevation. These last two pairs of muscles have secondary actions in adduction/abduction and intorsion/extorsion that depend on the position of the eye. Binocular vision requires coordinated motion of the three opposing muscle pairs in both eyes, as described in Table 1.4.

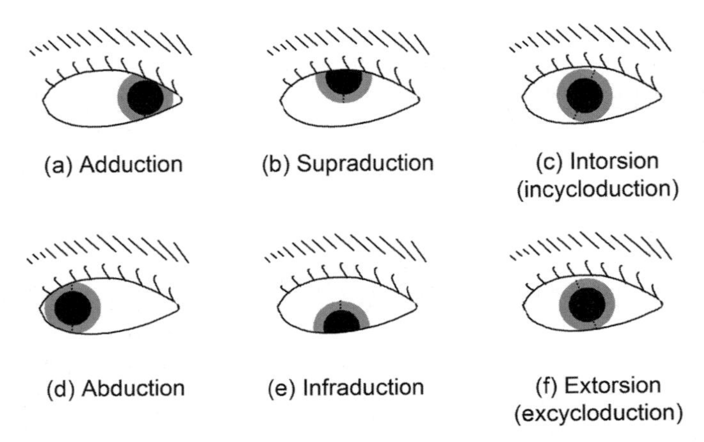

Fig. 1.13. Rotations of the right eye. A dashed line has been added across the iris to help view the rotations. (Based on [60])

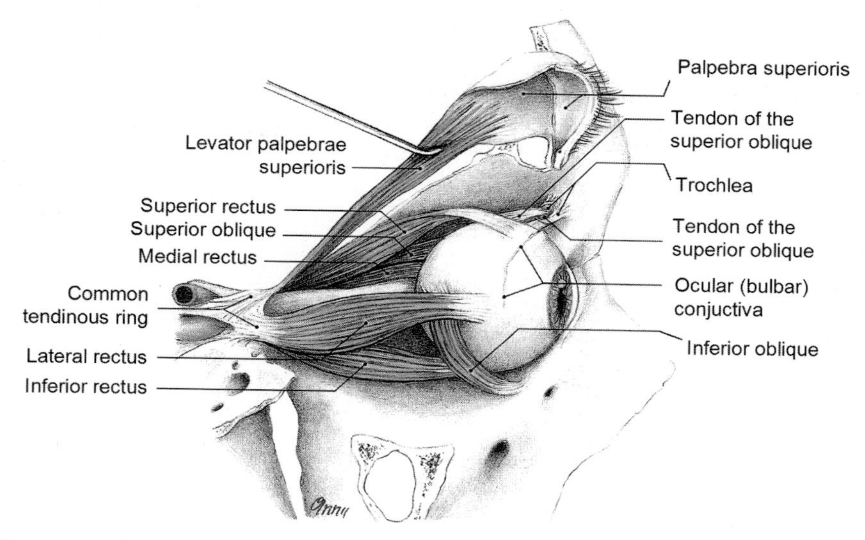

Fig. 1.14. Ocular muscles, with the eyelid (palpebra) pulled up as shown. The tendon of the superior oblique muscle (marked in two regions) passes through the trochlea loop. (From [59])

Table 1.3. Ocular muscle functions. (Based on [60])

muscle	primary action	secondary action
lateral rectus	abduction	none
medial rectus	adduction	none
superior rectus	elevation	adduction, intorsion
inferior rectus	depression	abduction, extorsion
superior oblique	depression	intorsion, abduction
inferior oblique	elevation	extorsion, abduction

Table 1.4. Muscle combinations of both eyes for gaze directions. (Based on [60])

direction of gaze	right eye muscle	left eye muscle
eyes up, right	superior rectus	inferior oblique
eyes right	lateral rectus	medial rectus
eyes down, right	inferior rectus	superior oblique
eyes down, left	superior oblique	inferior rectus
eyes left	medial rectus	lateral rectus
eyes up, left	inferior oblique	superior rectus

1.3 The Standard Human

We will often, but not always, model humans assuming numerical values for mass, height, etc. of a "standard" human, a 70 kg man with parameters similar to those in Table 1.5.

We will need details of human anatomy in some cases, and these are provided now and in subsequent chapters as needed. We will also need to use the

Table 1.5. A description of the "Standard Man". (Using data from [37, 44])

age	30 yr
height	1.72 m (5 ft 8 in)
mass	70 kg
weight	690 N (154 lb)
surface area	1.85 m^2
body core temperature	37.0°C
body skin temperature	34.0°C
heat capacity	0.83 kcal/kg-°C (3.5 kJ/kg-°C)
basal metabolic rate	70 kcal/h (1,680 kcal/day, 38 kcal/m^2-h, 44 W/m^2)
body fat	15%
subcutaneous fat layer	5 mm
body fluids volume	51 L
body fluids composition	53% intracellular; 40% interstitial, lymph; 7% plasma
heart rate	65 beats/min
blood volume	5.2 L
blood hematocrit	0.43
cardiac output (at rest)	5.0 L/min
cardiac output (in general)	3.0 + 8 × O_2 consumption (in L/min) L/min
systolic blood pressure	120 mmHg (16.0 kPa)
diastolic blood pressure	80 mmHg (10.7 kPa)
breathing rate	15/min
O_2 consumption	0.26 L/min
CO_2 production	0.21 L/min
total lung capacity	6.0 L
vital capacity	4.8 L
tidal volume	0.5 L
lung dead space	0.15 L
lung mass transfer area	90 m^2
mechanical work efficiency	0–25%

There are wide variations about these typical values for body parameters. Also, these values are different for different regions; the ones in the table typify American males in the mid-1970s. Values for women are different than for men; for example, their typical heights and weights are lower and their percentage of body fat is higher.

Table 1.6. Body segment lengths. Also see Fig. 1.15. (Using data from [63])

segment	segment length[a]/ body height H
head height	0.130
neck height	0.052
shoulder width	0.259
upper arm	0.186
lower arm	0.146
hand	0.108
shoulder width	0.259
chest width	0.174
hip width/leg separation	0.191
upper leg (thigh)	0.245
lower leg (calf)	0.246
ankle to bottom of foot	0.039
foot breadth	0.055
foot length	0.152

[a]Unless otherwise specified.

findings of *anthropometry*, which involves the measurement of the size, weight, and proportions of the human body. Of particular use will be anthropometric data, such as those in Table 1.6 and Fig. 1.15, which provide the lengths of different anatomical segments of the "average" body as a fraction of the body height H.

Table 1.7 gives the masses (or weights) of different anatomical parts of the body as fractions of total body mass m_{b} (or equivalently, total body weight

Table 1.7. Masses and mass densities of body segments. (Using data from [63])

segment	segment mass/ total body mass m_{b}	mass density (g/cm^3)
hand	0.006	1.16
forearm	0.016	1.13
upper arm	0.028	1.07
forearm and hand	0.022	1.14
total arm	0.050	1.11
foot	0.0145	1.10
lower leg (calf)	0.0465	1.09
upper leg (thigh)	0.100	1.05
foot and lower leg	0.061	1.09
total leg	0.161	1.06
head and neck	0.081	1.11
trunk	0.497	1.03

Table 1.10. Range of joint mobility for opposing movements, with mean and standard deviation (SD) in degrees. (Using data from [39], as from [33, 61])

opposing movements	mean	SD
shoulder flexion/extension	188/61	12/14
shoulder abduction/adduction	134/48	17/9
shoulder medial/lateral rotation	97/34	22/13
elbow flexion	142	10
forearm supination/pronation	113/77	22/24
wrist flexion/extension	90/99	12/13
wrist abduction/adduction	27/47	9/7
hip flexion	113	13
hip abduction/adduction	53/31	12/12
hip medial/lateral rotation (prone)	39/34	10/10
hip medial/lateral rotation (sitting)	31/30	9/9
knee flexion (prone) – voluntary, arm assist	125,144	10,9
knee flexion – voluntary (standing), forced (kneeling)	113,159	13,9
knee medial/lateral rotation (sitting)	35/43	12/12
ankle flexion/extension	35/38	7/12
foot inversion/eversion	24/23	9/7

The subjects were college-age males. Also see Fig. 1.16.

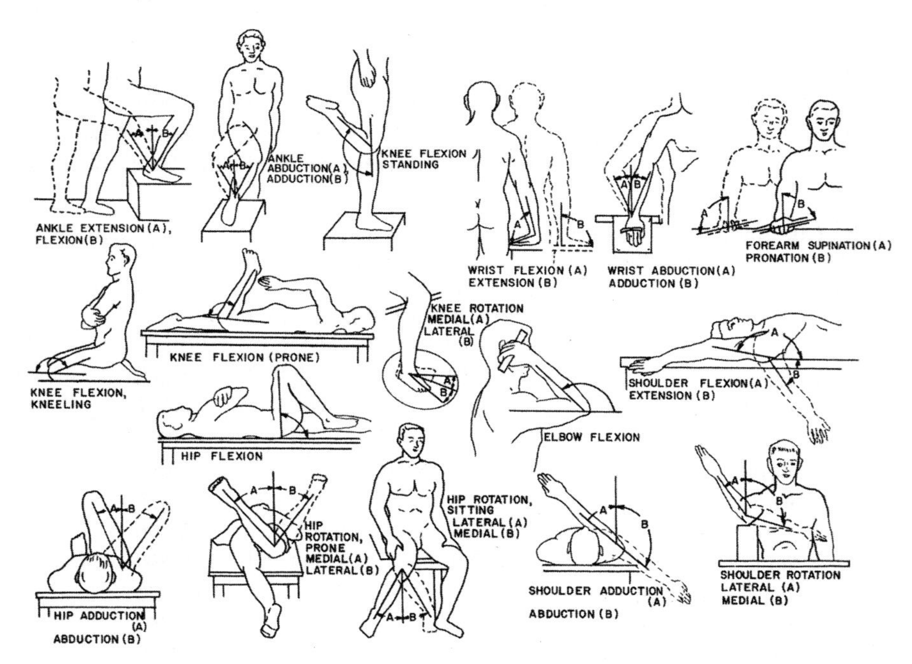

Fig. 1.16. Postures used for Table 1.10, for range of opposing motions. (From [38]. Reprinted with permission of Wiley. Also see [33, 61])

with the standard deviations about these values. (For normal or Gaussian distributions with an average, A, and standard deviation, SD, about 68% of all values are between A − SD and A + SD.) Three degrees of freedom are given for the shoulder and hip, two for the wrist and the foot (listed separately as foot and ankle), and one each for the elbow and forearm. The knee, as idealized above, has one DOF, but two are listed here: the flexion in a 1 D hinge and also some rotation of the upper and lower leg about the knee.

Table 1.11 gives the mass and volumes of different systems and parts of the body. The components of a typical human cell are given in Table 1.12. Although, most of our discussion will not concern these components of a cell,

Table 1.11. Mass and volume of the organs of the human body. (Using data from [42])

fluid, tissue, organ, or system	total mass (g)	total volume (cm^3)
adult male body	70,000	60,000
muscle	30,000	23,000
fat	10,500	12,000
skin	2,000	1,800
subcutaneous tissue	4,100	3,700
skeleton	10,000	6,875
gastrointestinal track	2,000	1,800
contents (chyme/feces)	∼2,000	∼2,000
blood vessels	1,800	1,700
contents (blood)	5,600	5,400
liver	1,650	1,470
brain	1,400	1,350
lungs (2)	825	775
contents (air)	∼7.7	∼6,000
heart	330	300
chamber volume	–	450
kidneys (2)	300	270
urinary bladder	150	140
contents (urine)	∼500	∼500
digestive fluids	∼150	∼150
pancreas	110	100
salivary glands (6)	50	48
synovial fluid	∼50	∼50
teeth (32)	42	14
eyes (2)	30	27
hair (average haircut)	21	16
gall bladder	7	7
contents (bile)	∼50	∼50
fingernails and toenails (20)	1.1	0.9

Table 1.12. Estimated gross molecular contents of a typical 20-μm human cell. (Using data from [42])

molecule	mass	molecular weight (amu, daltons)	number of molecules	number of molecular entities
water	65	18	1.74×10^{14}	1
other inorganic	1.5	55	1.31×10^{12}	20
lipid (fat)	12	700	8.4×10^{11}	50
other organic	0.4	250	7.7×10^{10}	∼200
protein	20	50,000	1.9×10^{10}	∼5,000
RNA	1.0	1×10^{6}	5×10^{7}	–
DNA	0.1	1×10^{11}	46	–

this listing is instructive because it provides a quantitative assessment of these components.

So far we have considered only the physical specifications of a typical person. In subsequent chapters, we will address the physical aspects of how a typical person responds to a wide range of physical conditions, for healthy and unhealthy people, and under normal and extreme conditions. The prospect of space exploration motivated extensive studies of how people respond to a wide range of extreme physical conditions, such as extreme pressures, temperatures, linear and rotary accelerations, collisions, vibrations, weightlessness, and sound [50].

We will also see that many processes can be described in terms of characteristic times or distances, such as the time needed for a muscle activation to decay or a molecule to diffuse in a cell. There are also more general characteristic times within the human body. Your heart beats and you breathe roughly once every second. Your blood flows throughout your body roughly once every minute, and each ATP molecule (the molecule which is the ultimate form of energy usage in your body) is used and then regenerated roughly once every minute.

1.4 Scaling Relationships

In human and animal biology there are diverse ranges of properties that scale with either a property, such as mass or length, or a physical or chemical input. We summarize them here; they relate to many concepts developed and used in subsequent chapters.

1.4.1 Allometric Rules

Some properties scale with body mass in a fairly predictable way, and are characterized by scaling relationships called *allometric rules*. For a property

Table 1.13. Allometric parameters (1.1) for mammals. (Using data from [31, 55])

parameter	a	α
basal metabolic rate (BMR), in W	4.1	0.75
body surface area, in m^2	0.11	0.65
brain mass in man, in kg	0.085	0.66
brain mass in nonprimates, in kg	0.01	0.7
breathing rate, in Hz	0.892	−0.26
energy cost of running, in J/m-kg	7	−0.33
energy cost of swimming, in J/m-kg	0.6	−0.33
effective lung volume, in m^2	5.67×10^{-5}	1.03
heart beat rate, in Hz	4.02	−0.25
heart mass, in kg	5.8×10^{-3}	0.97
lifetime, in y	11.89	0.20
muscle mass, in kg	0.45	1.0
skeletal mass (terrestrial), in kg	0.068	1.08
speed of flying, in m/s	15	0.167
speed of walking, in m/s	0.5	0.167

f for animals with body mass m_b (in kg), an allometric relation has the form

$$f(\text{in a given set of units}) = am_b^{\alpha}. \tag{1.1}$$

Technically, the relationship is allometric if $\alpha \neq 1$. Some examples are given in Table 1.13. By the way, *allometric* means "by a different measure" from the Greek *alloios*, which means "different" – so how body height scales with body mass is an allometric relationship. *Isometric* means "by the same measure" – so how leg mass scales with body mass is an example of an isometric relationship. For a delightful discussion on allometry and scaling see [46]. Other equally intriguing discussions have been presented in [31, 35, 36, 45, 47, 55, 56].

These relationships can hold for many species of a given type, such as land-based mammals, etc. Some are also valid within a species, such as for man – and as such would be called anthropometric relationships. Sometimes they apply only to adults in a species, and not across all age groups. See Problems 1.54 and 1.55 for more on this.

The "predicted" values from a scaling relation would be the expected average values only if the parameters for all species follow the relation *exactly* – and there is no reason why this must be so. Moreover, there is always a spread, or dispersion, about these average values. Some of these allometric relationships are empirical, and others can be derived, or at least rationalized, as we will see for Kleiber's Law of basal (6.19) (i.e., minimum) metabolic rates (BMRs) in Chap. 6.

One obvious example of such scaling is that the legs of bigger mammals tend to be wider in proportion to their overall linear dimension L (and mass m_b) than those for smaller mammals, and this is reflected in the larger ratio of the width of their long skeletal bones, w_{bone}, to the length, L_{bone}, for larger m_{b}. Such bones need to support their body weight or mass, which are proportional to their volume $\sim L^3$ because the mass densities are very nearly the same for all mammals. Such long bones in the different mammals can withstand approximately the same force per unit cross-sectional area, which means that they have the same ultimate compressive stress or UCS. (The UCS is a measure of fracture conditions that will be discussed more in Chap. 4.). If this maximum force is a fixed multiple of the weight, so it is $\propto L^3$, and the cross-sectional area is $\sim w_{\text{bone}}^2$, then $w^2 \propto L^3$ or $w \propto L^{3/2}$. Because the bone length $L_{\text{bone}} \propto L$, we see that $w/L_{\text{bone}} \propto L^{1/2} \propto m_{\text{b}}^{1/6}$. Body mass m_{b} increases by 10^4 from rat to elephant, and so this ratio increases by $10^{2/3} \sim 4$. This means that the body shapes and the bones themselves are not geometrically similar.

In some cases more accurate scaling relations necessitate the use of parameters in addition to the body mass. One example is the scaling of the surface area of a person A (in m^2), which is empirically seen to depend on height H (in m), as well as body mass m_{b} (in kg):

$$A = 0.202 m_{\text{b}}^{0.425} H^{0.725}. \tag{1.2}$$

Accurate scaling relationships often involve the index W_{b}/H^p (or equivalently m_{b}/H^p) with p ranging from 1.0 to 3.0. (Sometimes the reciprocal or square root or cube root of this parameter is used. Complicating matters further, different sets of units are commonly used for these parameters.) One such index is the specific stature or ponderal index $S = H/m_{\text{b}}^{1/3}$ (for which $p = 3$), which is used in the Harris–Benedict versions of Kleiber's Law specialized for people (6.30–6.31). It is also used in the expression for the average human density

$$\rho\ (\text{in kg/L or g/cm}^3) = 0.69 + 0.9S \tag{1.3}$$

where within S the units of H are m and those of m_{b} are kg. Another scaling parameter is Quételet's index or the body mass index (BMI), $Q = m_{\text{b}}/H^2$ (for which $p = 2.0$). It is often considered the best index for epidemiological studies. For example, with m_{b} in kg and H in m, the average fat content of the body increases with this index:

$$\text{Men:}\quad \text{fat (\% of body weight)} = 1.28Q - 10.1 \tag{1.4}$$

$$\text{Women:}\quad \text{fat (\% of body weight)} = 1.48Q - 7.0. \tag{1.5}$$

Normal or ideal body fat is 14–20% in men and 21–27% in women. Not coincidentally, mortality increases as this index Q increases.

Some properties do not scale with mass. From mouse to elephant, a mass range of over 10,000, the maximum jumping height (of the center of mass) of most every mammal is within a factor of 2 of 2/3 m. Similarly, the maximum running speed of most mammals is within a factor of 2 of 7 m/s (15 mph).

Table 1.14. Size-independent dimensionless groups in mammals. (Using data from [46])

parameter	a	α
breathing flow rate/blood flow rate^a	2.0	0.00
mass of blood/mass of heart	8.3	0.01
time for 50% of growth/lifespan in captivity	0.03	0.05
gestation period/lifespan in captivity	0.015	0.05
breathing cycle/lifespan in captivity	3×10^{-9}	0.06
cardiac cycle/lifespan in captivity	6.8×10^{-10}	0.05
half-life of drug^b/lifespan in captivity	0.95×10^{-5}	0.01

The value of a in (1.1) is that for a 1 kg mammal. Also see Chaps. 8 and 9.
^{a}(tidal volume/breath time)/(heart stroke volume/pulse time).
^{b}Methotrexate.

The main reason behind these two relationships is that the force a muscle can exert is proportional to its cross-sectional area and therefore it varies as the square of the characteristic linear dimension, such as height H. This is explored in Problems 3.28, 3.29, and 3.41 in Chap. 3. The dimensionless ratios of some physical and physiological properties are noteworthy in that they are essentially independent of size (and mass), as is seen in Table 1.14.

1.4.2 Scaling in the Senses

A very different type of scaling is exhibited by Stevens' Law, which characterizes how the perceived strength P of a sense varies with the intensity of a stimulus S for a given sensation. This scaling is

$$P = K(S - S_0)^n \tag{1.6}$$

above a threshold S_0 [57, 58]. As seen in Table 1.15, sometimes the perception of a sense, called the *psychoperception*, is sublinear with the strength of the

Table 1.15. Exponent n for perceived strength (P) of a stimulus (S) above a threshold S_0, with $P = K(S - S_0)^n$ in Steven's Law. (Using data from [57, 58])

psychoperception	n	stimulus
brightness	0.33, 0.5	5° target, point source – dark adapted eye
loudness	0.54, 0.60	monoaural, binaural
smell	0.55, 0.60	coffee odor, heptane
vibration	0.6, 0.95	250 Hz, 60 Hz – on finger
taste	0.8, 1.3, 1.3	saccharine, sucrose, salt
temperature	1.0, 1.6	cold, warm – on arm
pressure on palm	1.1	static force on skin
heaviness	1.45	lifted weights
electric shock	3.5	60 Hz through fingers

stimulus and sometime it is superlinear. Three of the senses involve very significant physical aspects: hearing (Chap. 10, loudness) and seeing (Chap. 11), which we will examine in detail, and touch (Chap. 2, vibration, temperature, pressure on palm, heaviness; Chap. 12, electric shock), which we will study in less detail. These three also have very important chemical and biological origins. The senses of taste and smell are essentially solely chemical and biological in basis and will not be discussed any further, except for a brief discussion of the electrical properties of the membranes of the taste and smell sensory neurons in Chap. 12. For these three other senses, we will concentrate mostly on the physical input and the beginning of the sensation process (detection). The nonlinearities inherent in Stevens' Law are due in part to the detection process. The final parts of detection process are the generation and transmission of neural signals sent to the brain – which we will cover – and the processing in the brain – which we will cover only briefly. There are approximately 12 orders of magnitude sensitivity in hearing and vision.

1.5 Summary

Much about the body can be understood by learning the terminology of directions and local regions in the body. Much about the motion of the body can be explained by examining the rotation about bones about joints. In analyzing the physics of the body, reference can be made to anthropometric data on body parts for a standard human. Many phenomena concerning anatomy and physiology can be characterized and understood by using scaling relations.

Problems

Body Terminology

1.1. (a) Is the heart superior or inferior to the large intestine?
(b) Is the large intestine superior or inferior to the heart?

1.2. (a) Is the navel posterior or anterior to the spine?
(b) Is the spine posterior or anterior to the navel?

1.3. Is the nose lateral or medial to the ears?

1.4. Are the eyes lateral or medial to the nose?

1.5. Is the foot proximal or distal to the knee?

1.6. Is the elbow proximal or distal to the wrist?

1.7. Is the skeleton superficial or deep to the skin?

1.8. The blind spot in the eye retina is said to be nasal to the fovea (center of the retina). What does this mean?

1.9. What would you expect the term *cephalid* to mean? What would be an equivalent term?

1.10. Which is the anterior part of the heart in Fig. 8.7? Is this a superior or inferior view of the heart?

1.11. Consider the directional terms *ipsilateral* and *contralateral.* One means on the same side of the body, while the other means on opposite sides of the body. Which is which?

1.12. The directional term *intermediate* means "in between." What is intermediate between the upper and lower legs?

1.13. Encephalitis is the inflammation, i.e., "itis", of what?

1.14. *Presbyopia* refers to disorders in vision due to old age, such as lack of accommodation in the crystalline lens (see Chap. 11). *Presbycusis* refers to old age-related auditory impairments (see Chap. 10.) What parts of these two terms mean old age, vision, and hearing?

1.15. The three tiny bones in the middle ear, the malleus, incus, and stapes are interconnected by the incudomallear articulation and the incudostapedial joint. Describe the origin of the names of these connections.

1.16. The quadriceps muscles in the upper leg attach to the kneecap (patella) through the quadriceps tendon. The kneecap is connected to the tibia by connective tissue that is sometimes called the patellar tendon and sometimes the patellar ligament. Explain why both designations have merit and why neither designation completely describes the linkage perfectly well by itself.

1.17. Consider the drawing of the hand skeleton and the schematic of a hand showing joints with one or (labeled 2) two degrees of freedom in Fig. 1.6.
(a) How many degrees of freedom does each hand have? (Ignore the wrist joint.)
(b) Do we need so many degrees of freedom? Why? (There is no right or wrong answer to this part. Just think about what a human hand should be able to do (in clutching, etc.) and try to express your conclusions in terms of degrees of freedom.)

1.18. Estimate the angle of each of the joints in the hand for each of the following functions [52]. (Define the angle of each joint as shown in the left hand in Fig. 1.6b to be 0°. Define rotations into the paper and clockwise motions in the plane of the paper as being positive.)

(a) lifting a pail (a hook grip)
(b) holding a cigarette (a scissors grip)

(c) lifting a coaster (a five-jaw chuck)
(d) holding a pencil (a three-jaw chuck)
(e) threading a needle (a two-jaw pad-to-pad chuck)
(f) turning a key (a two-jaw pad-to-side chuck)
(g) holding a hammer (a squeeze grip)
(h) opening a jar (a disc grip)
(i) holding a ball (a spherical grip)

1.19. We said that the seven DOFs available for arm motion enabled nonunique positioning of the hand, but analogous nonunique positioning of the foot is not possible because the leg has only six DOFs. Use Table 1.10 to explain why this is not exactly correct.

1.20. Use Table 1.3 to show that the coordinated eye motions in Table 1.4 use the muscles listed for primary motion.

1.21. Consider a limb, of length L, composed of upper and lower limbs with respective lengths r_1 and r_2, with $L = r_1 + r_2$. There is a total range of motion in the angles the upper limb makes with the torso and the lower limb makes with the upper limb. Assume motion only in two dimensions (see Fig. 1.7.)
(a) What area is subtended by the end of the lower limb (hand or foot) when $r_1 = r_2$?
(b) What area is subtended by the end of the lower limb when $r_1 > r_2$? What fraction of that in (a) is this?
(c) What area is subtended by the end of the lower limb when $r_1 < r_2$? What fraction of that in (a) is this?

1.22. Redo Problem 1.21 in three dimensions, finding the volume subtended by the end of the lower limb in each case.

The Standard Human

1.23. Qualitatively explain the differences of density in Table 1.7 in the different segments of the body. The average densities of blood, bone, muscle, fat, and air (in the lungs) can be determined from Table 1.11.

1.24. (a) Use Table 1.7 to determine the average density of the body.
(b) Use this to determine the average volume of a 70 kg body.
Your answers will be a bit different from the rough volume estimate given in Table 1.11.

1.25. (a) Calculate the range of segment masses alternatively using Tables 1.7 and 1.16, for each type of segment listed in the latter table, for people with masses in the range 40–100 kg.
(b) Give several reasons why these ranges seem to be different.

Table 1.16. An alternative set of relations of weights of body segments (all in lb). (Using data from [48], which used unpublished data by [40])

segment	segment weight
head	$0.028W_{\mathrm{b}} + 6.354$
trunk	$0.552W_{\mathrm{b}} - 6.417$
upper arms	$0.059W_{\mathrm{b}} + 0.862$
forearms	$0.026W_{\mathrm{b}} + 0.85$
hands	$0.009W_{\mathrm{b}} + 0.53$
upper legs	$0.239W_{\mathrm{b}} - 4.844$
lower legs	$0.067W_{\mathrm{b}} + 2.846$
feet	$0.016W_{\mathrm{b}} + 1.826$

1.26. (a) Show that (1.3) becomes

$$\rho(\text{in kg/L or g/cm}^3) = 0.69 + 0.0297S\,, \tag{1.7}$$

when $S = H/W_{\mathrm{b}}^{1/3}$ is expressed with H in inches and W_{b} in lb.
(b) Show that the average density for an adult of height 5 ft 10 in and weight 170 lb is 1.065 g/cm^3, with $S = 12.64$.
(c) Show that the average density for an adult of height 1.78 m and mass 77.3 kg is 1.066 g/cm^3, with $S = 0.418$.

1.27. What percentage of body mass is fat, skin, the skeleton, blood, liver, the brain, the lungs, heart, kidneys, and eyes?

1.28. Use Table 1.11 to determine the mass density of blood, skin, the lungs, the air in the lungs, fat, liver, hair, eyes, and blood vessels.

1.29. In modeling heat loss in Chap. 6, a typical man is modeled as a cylinder that is 1.65 m high with a 0.234 m diameter. If the human density is 1.1 g/cm^3, what is the mass (in kg) and weight (in N and lb) of this man?

1.30. (a) If a man has a mass of 70 kg and an average density of 1.1 g/cm^3, find the man's volume.
(b) If this man is modeled as a sphere, find his radius and diameter.
(c) If this man is 1.72 m high, and is modeled as a right circular cylinder, find the radius and diameter of this cylinder.
(d) Now model a man of this height and mass as a rectangular solid with square cross-section, and find the length of the square.
(e) Repeat this for a constant rectangular cross-section, and determine the sizes if the long and short rectangle dimensions have a ratio of either 2:1, 3:1, or 4:1.
(f) In each above case calculate his surface area and compare it to that predicted by (1.2) for a 1.72-m tall man. Which of the above models seems best?

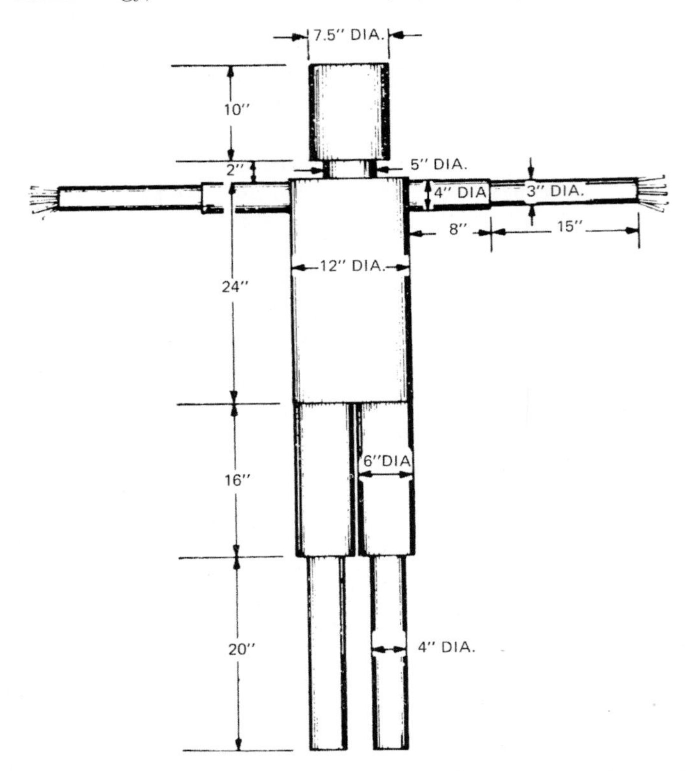

Fig. 1.17. Cylindrical model of a man used in studies of convective cooling. (From [34], adapted from [51])

1.31. The cylindrical model of a man in Fig. 1.17 was once used in studies of convective cooling. What are the volume, mass, and exposed surface area (including the bottom of the lower limb) for this person? Assume each finger is 3.5 in long and has a diameter of 0.875 in and that the mass density of all components is $1.05\,\mathrm{g/cm^3}$.

1.32. How much heavier is someone with a totally full stomach, small intestine, large intestine and rectum, than when each system is empty? Assume the mass density of the contents is $1\,\mathrm{g/cm^3}$. Express your answer in mass (kg) and weight (N and lb). (Use Table 7.4.)

1.33. Use the anthropometric data to determine the average cross-sectional area and diameter of an arm and a leg of a 70 kg man. Assume the cross section of each is circular.

1.34. Compare the surface area of the standard man given in Table 1.5, alternatively as predicted by (1.1) and (1.2). Use the data given in Table 1.13.

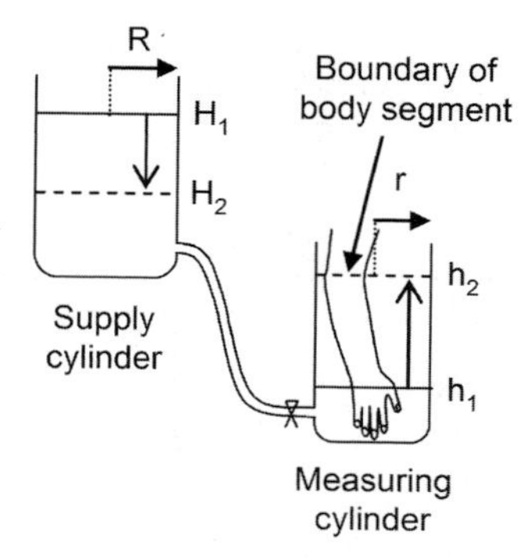

Fig. 1.18. Immersion technique for measuring the volume of various body segments, with the solid lines denoting the initial water level and the dashed lines the final water level. (Based on [48].) For Problem 1.37

1.35. Compare the surface area of a 50 kg, 5 ft 5 in woman, alternatively as predicted by (1.1) and (1.2). Use the data given in Table 1.13.

1.36. For an adult, the average fractional surface area is 9% for the head, 9% for each upper limb, 18% both for front and back of the torso, and 18% for each lower limb. (The remaining 1% is for genitalia.)
(a) This is used to estimate the fraction of damaged area in burn victims. It is known as the "Rule of Nines." Why?
(b) Use the data given in Table 1.13 to determine the average surface area for each of these parts of the body for the standard man.

1.37. In the system depicted in Fig. 1.18, a body segment is put in the measuring cylinder and the valve is opened to allow flow of water up to the "beginning" of the body segment (giving the "1" heights). The valve is then opened until water flows into the measuring cylinder to the "end" of the body segment (giving the "2" heights). Explain how this can be used to measure the volume of the body segment.

1.38. (a) Calculate Quételet's index (also known as the BMI) and the specific stature (also known as the ponderal index) of a person of average density ρ modeled as a cube of length L.
(b) How do these change if the person has the same overall mass, but is modeled as a rectangular solid of height H, width $0.20H$, and depth $0.15H$?

1.39. (a) A person with mass M is modeled as a rectangular solid of height H, width $0.25H$, and depth $0.25H$. The person loses weight, maintaining the same mass density, and then has a width $0.20H$ and depth $0.15H$. Calculate Quételet's index and the specific stature of the person before and after the weight loss.
(b) Would you expect the mass density of the person to change during the weight loss? If so, how would you expect it to change?

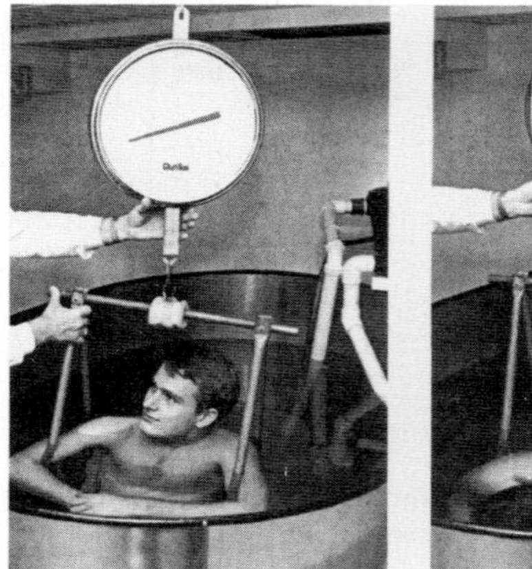
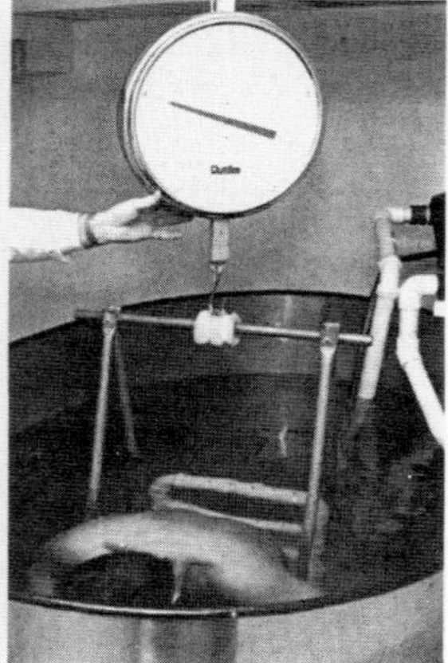

Fig. 1.19. Determining human body fat content and density by weighing a person in water. (Photo by Clifton Boutelle, News and Information Service, Bowling Green State University. Used with permission of Brad Phalin. Also see [41].) For Problem 1.40

1.40. You can determine the density and percentage of fat in people by weighing them underwater, as in Fig. 1.19. Data for two men with the same height and mass, but with different underwater masses are given in Table 1.17.
(a) Why are the volumes given as listed?
(b) What assumption has been made about the relative densities of fat and the average of the rest of the body?
(c) Is the value assumed for the density for the rest of the body reasonable? Why or why not?

Table 1.17. Comparison of the density and fat percentage for two men with the same height and mass, but different underwater masses. (Using data from [62].) For Problem 1.40

parameter	man A	man B
height, m (in)	1.88(74)	1.88(74)
mass, kg (lb)	93(205)	93(205)
underwater mass, kg	5.00	3.50
volume, L	88.0	89.5
$\text{volume}_{\text{corrected}}$[a], L	86.5	88.0
body density[b], g/cm^3	1.075	1.057
relative fat, %	10.4	18.4
fat mass, kg (lb)	9.7(21.4)	17.1(37.7)
fat-free mass, kg (lb)	83.3(183.6)	75.9(167.3)

[a]The volume is corrected for the water density, intestinal gas volume, and residual lung volume.
[b]The body density is the mass-corrected volume. Relative fat (in %) = 100(4.95/(body density) − 4.50).

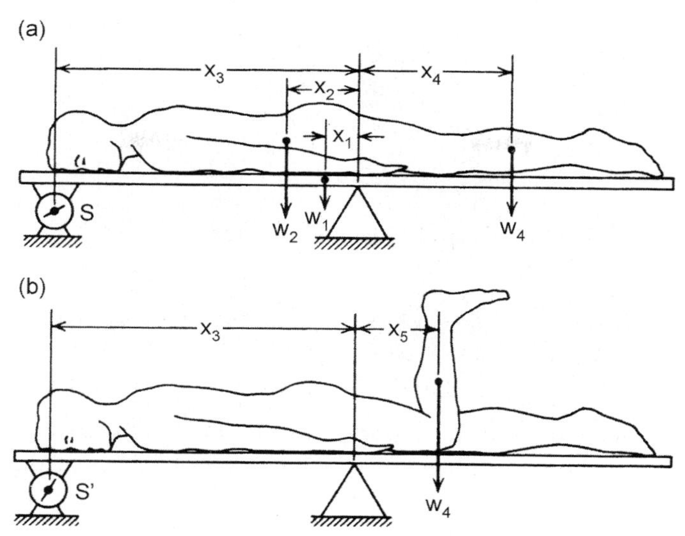

Fig. 1.20. In vivo estimation of (**a**) body center of mass and (**b**) mass of a distal segment, for Problems 1.42 and 1.43. (From [63]. Reprinted with permission of John Wiley & Sons)

(d) Table 1.17 uses the Siri formula for the percentage of body fat: 100(4.95/(body density) − 4.50). How do the results for the percentage and mass of body fat differ using the Brozek formula: 100(4.57/(body density) − 4.142) for the two cases in the table? (Both formulas may, in fact, give values for body fat percentage that are too high.)

1.41. Explain why in Problem 1.40 you can either measure the weight of the water displaced by the body or the weight of the body when it is completely submerged [41].

1.42. You can measure the location of the anatomical center of mass of the body using the arrangement in Fig. 1.20a. The weight (w_1) and location of the mass (x_1) of the balance board are known along with the body weight w_2. The location of the body center of mass relative to the pivot point is x_2. The distance from the pivot to the scale is x_3. With the body center of mass to the left of the pivot point there is a measurable force S on the scale (under the head). Show that

$$x_2 = \frac{Sx_3 - w_1 x_1}{w_2}. \tag{1.8}$$

1.43. You can determine the weight of the lower part of a limb (w_4) using the same balance board as in Problem 1.42, using Fig. 1.20b. The center of mass of the limb changes from x_4 to x_5 relative to the pivot point when the limb is set vertically; concomitantly the scale reading changes from S to S'.

Show that

$$w_4 = \frac{(S' - S)x_3}{(x_4 - x_5)} . \tag{1.9}$$

The location of the center of mass of the limb relative the joint near the trunk is assumed to be known. To determine the weight of the entire limb the subject should be lying on his or her back and the entire limb is flexed to a right angle.

1.44. (a) Determine the goal mass (in kg, and find the weight in lb) to achieve 10% fat for the two men described in Table 1.17, by using the fat-free mass. (b) How much fat mass (and weight) must be lost by each to attain this goal?

1.45. The normalized distances of the segment center of mass from the proximal and distal ends in Table 1.8 always sum to 1. Is this a coincidence, a trivial point, or significant? Why?

Allometry and Scaling

1.46. Determine the parameters for a 70 kg person for each set of allometric relation parameters in Table 1.13. How do they compare with similar parameters listed in Tables 1.5 and 1.11?

1.47. Derive the allometric laws for the percentages of the total body mass residing in the brain, heart, muscle, and skeletal mass for mammals (such as humans).

1.48. Compare the prediction of the fat in a standard man using Quetelet's index, with those listed in Table 1.11.

1.49. Compare the % body fat in:
(a) a male and female who are both 5 ft 6 in, 140 lb.
(b) males who are 6 ft 2 in and 5 ft 8 in tall, both weighing 190 lb.

1.50. For a 70 kg person living 70 years, determine the person's total lifetime
(a) number of heart beats
(b) number of breaths
(c) energy consumed
(d) energy consumed per unit mass.

1.51. Does it make sense that the ratio of the volumetric flow rates in the respiratory and circulatory systems in mammals (first entry line in Table 1.14) is essentially independent of mammal mass? Why?

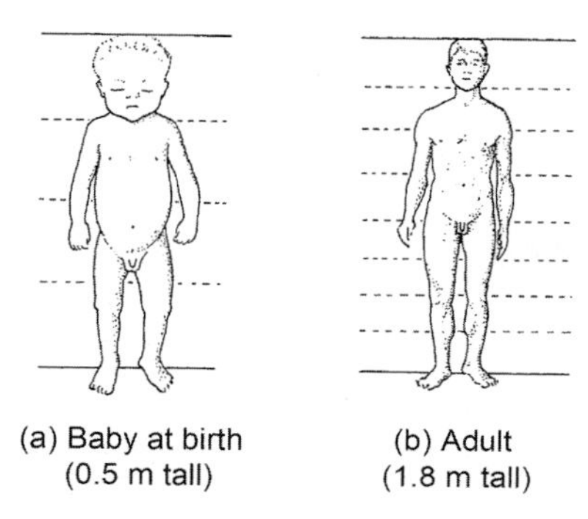

Fig. 1.21. Human development, showing the change in body shape from birth to adulthood, for Problem 1.54. (From [54])

1.52. Use Table 1.14 to find the allometry parameters for the ratio of the respiratory and cardiac rates (both in 1/s).

1.53. Use Table 1.14 to find the allometry parameters for the ratio of the volumes per breath (tidal volume) and per heart beat (heart stroke volume).

1.54. Use Fig. 1.21 to comment on whether the same mass-dependent-only allometric rules should be used within a species from birth to adulthood.

1.55. (a) Arm length scales as the body height to the 1.0 power for people older that 9 months and to the 1.2 power for those younger. At 9 months of age, a male is 61 cm tall and has an arm length of 23 cm. When that male was 0.42 yr old he had a height of 30 cm and when he will be 25.75 yr old he will have a height of 190 cm. In each what is their expected arm length, alternatively using the age-correct and age-incorrect scaling exponents?
(b) Is this an example of allometric or isometric scaling?

1.56. Scaling arguments can also be used to understand some general trends.
(a) If the linear dimension of an object is L, show that its surface area varies as L^2, its volume as L^3, and its surface to volume ratio as $1/L$, by using a sphere (diameter L) and a cube (length L) as examples.
(b) An animal loses heat by loss at the surface, so its rate of losing heat varies as its surface area, whereas its metabolic rate varies as its volume. In cold environments this loss of heat can be devastating. Do scaling arguments suggest animals would be bigger or smaller in cold climates?
(c) A cell receives oxygen and nutrients across its membrane to supply the entire volume of the cell. Do scaling arguments suggest that limitations in supplying oxygen and nutrients place a lower limit or upper limit in the size of cells? Why?

1.57. The strength of bones varies as their cross-sectional area, as we will see in Chap. 4. We have seen that this suggests how the diameter of a long bone scales with its length.
(a) Does this scaling relation mean that smaller creatures have thinner or thicker bones than bigger creatures assuming the same strength criterion?
(b) Does this "static" argument imply a limitation on how small or how large an animal can be?

1.58. (a) The work an animal of dimension L needs to propel itself a distance equal to its dimension is the needed force – which is proportional to its mass – times its dimension. Show this work scales as L^4.
(b) This force must be supplied by muscles, and the work done by the muscles is this force times the distance the muscles can contract; this distance scales as the length of the muscles, which in turn scales with L. In Chap. 5 we will see that the force exerted by a muscle is proportional to its cross-sectional area. If the lateral dimension of the muscle also scales as L, show that the maximum work that can be done by the muscle scales as L^3.
(c) For work done by muscles to scale as fast as that needed for locomotion, how must the lateral dimensions of muscles vary?
(d) Do these "dynamic" arguments limit how small or how large an animal can be [47]?

2

Statics of the Body

The study of the force balance of an object at rest is called "statics." Moreover, the study of very slow motion can usually be treated as a series of static conditions – as if there were no motion; this is called "quasistatics." After reviewing the conditions for static equilibrium in three dimensions, we will examine the useful simplification to two dimensions, examples of which can often be characterized as one of the three types of levers. We will then apply these equilibrium conditions to the lower arm, hip, and the spine (lower back). Statics is one important area in biomechanics [75, 82, 86, 94].

2.1 Review of Forces, Torques, and Equilibrium

Each force $\mathbf{F}$ can be resolved into components in the x, y, and z directions (F_x, F_y, F_z). In a static condition the sum of the forces $\mathbf{F}$ in each the x, y, and z directions is zero:

$$\sum F_x = 0, \qquad \sum F_y = 0, \qquad \sum F_z = 0. \tag{2.1}$$

The speed of the center of mass of the object in each direction is then constant, and will usually be assumed to be zero here. These forces can be in balance either for the entire body or for any part of the body.

Similarly, each torque τ can be resolved into components in the x, y, and z directions (τ_x, τ_y, τ_z). In a static condition, the torques τ about the x-, y-, and z-axis also each sum to zero for the entire body and for any body part:

$$\sum \tau_x = 0, \qquad \sum \tau_y = 0, \qquad \sum \tau_z = 0. \tag{2.2}$$

The speed of angular rotation of the object about each axis is then constant, and will usually be assumed to be zero here.

What actually is a torque? Forces describe changes in linear motion – which means changes in velocities, while torques describe how these same forces

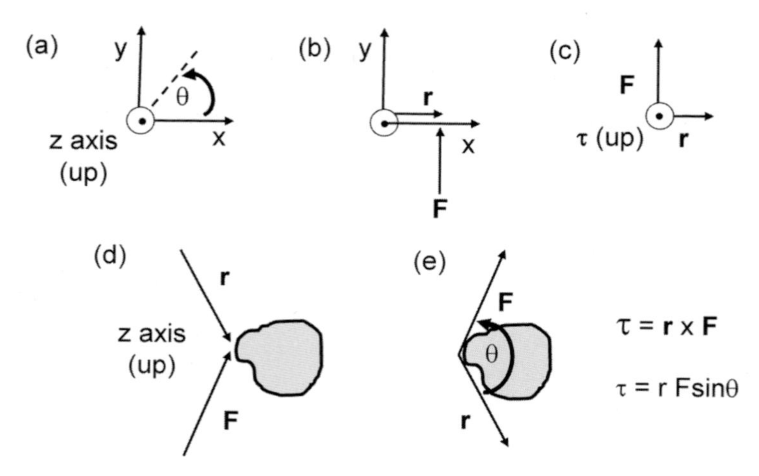

Fig. 2.1. Torques and relevant axes

can change angular motion – which means changes in angular velocities. The diagram in Fig. 2.1b shows that a force $\mathbf{F}$ applied in the positive y direction (with component F) a distance $+r$ from the z-axis, leads to a torque about the z-axis τ_y of magnitude rF. This leads to motion in the counterclockwise direction, caused by an angular acceleration that increases the angle θ. This is defined as a positive torque about this axis. A negative torque would occur, for example, if the force were applied in the negative y direction. This would lead to motion in the clockwise direction, caused by an angular acceleration that decreases the (signed) angle θ.

In general, the torque (vector τ) about any axis is defined as the vector cross product between the distance vector from that axis to the point where the force is applied $\mathbf{r}$ and the force vector $\mathbf{F}$ (Fig. 2.1e)

$$\tau = \mathbf{r} \times \mathbf{F}. \tag{2.3}$$

(You do not need to understand or use this vector cross product, just the results that are given below.)

Because vectors can be translated anywhere, things may be clearer if we move both $\mathbf{r}$ and $\mathbf{F}$, as in Fig. 2.1b, d, so they originate from where the force is applied, as in Fig. 2.1c, e, respectively. We will call the angle from the $\mathbf{r}$ vector to the $\mathbf{F}$ vector θ. The torque τ_z about the upward axis is

$$\tau_z = rF \sin\theta, \tag{2.4}$$

where r is the magnitude of vector $\mathbf{r}$ (the distance from the axis to the point where the force is applied) and F is the magnitude of vector $\mathbf{F}$. For $0° < \theta < 180°$ (or $0 < \theta < \pi$ in radians), $\sin\theta$ is positive and the torque is positive (Fig. 2.2a), while for $180° < \theta < 360°$ ($\pi < \theta < 2\pi$), $\sin\theta$ is negative and the torque is negative (Fig. 2.2b). When $\theta = 90°(= \pi/2)$, $\sin\theta = 1$ and the torque

(a)
F
θ
r
sinθ > 0
τ > 0
ccw
d²θ/dt² > 0
(b)
F
θ
r
sinθ < 0
τ < 0
cw
d²θ/dt² < 0

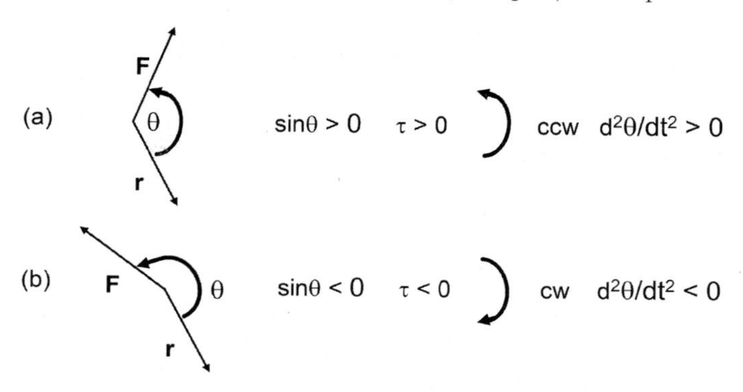

Fig. 2.2. Direction of torques, showing (**a**) positive and (**b**) negative torques

is rF, as above. When $\mathbf{r}$ and $\mathbf{F}$ are either parallel ($\theta = 0°(= 0)$) or antiparallel ($\theta = 180°(= \pi)$), the torque is zero.

Clearly, only the component of the $\mathbf{r}$ normal to the $\mathbf{F}$, which we will call r', contributes to the torque action. In fact, as Fig. 2.3 proves, $\tau_z = r'F$. Equivalently, only the component of $\mathbf{F}$ normal to $\mathbf{r}$, i.e., F', contributes to the torque action, and $\tau_z = rF'$. As we will see, sometimes information is provided where these normal components of displacement or force are provided, and the torques can be calculated without explicitly determining the angle between the displacement and force vectors. Consequently,

$$\tau_z = \mathbf{rF} \sin\theta = r'F = rF'. \tag{2.5}$$

This is true for any axis. The axis can be chosen cleverly for a particular problem to simplify analysis.

In linear motion, a force leads to an acceleration $\mathbf{a} = \mathrm{d}^2\mathbf{r}/\mathrm{d}t^2$, which is equivalent to a change in velocity $\mathbf{v} = \mathrm{d}\mathbf{r}/\mathrm{d}t$ (the magnitude of which is the speed v) or momentum $\mathbf{p} = m\,\mathrm{d}\mathbf{v}/\mathrm{d}t$, by

$$\mathbf{F} = m\mathbf{a} = m\frac{\mathrm{d}\mathbf{v}}{\mathrm{d}t} = \frac{\mathrm{d}\mathbf{p}}{\mathrm{d}t}, \tag{2.6}$$

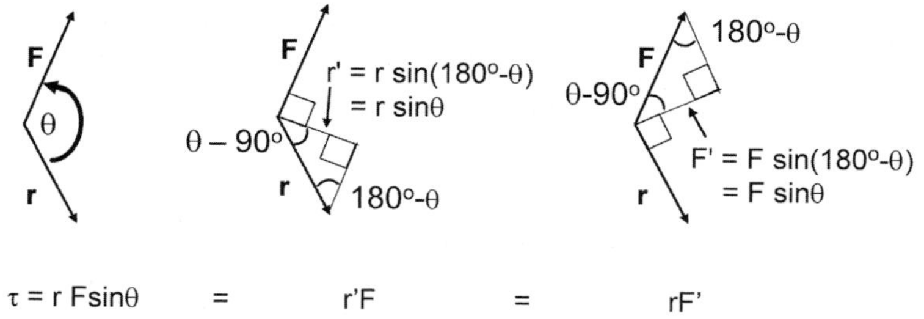

F
θ
r
F
r' = r sin(180°-θ)
= r sinθ
θ – 90°
r
180°-θ
180°-θ
F
θ-90°
F' = F sin(180°-θ)
= F sinθ
r
τ = r Fsinθ = r'F = rF'

Fig. 2.3. Determining torques from using components of the displacement and force vectors that are normal to the force and displacement vectors, respectively where m is the mass of the object. Similarly, a torque leads to an analogous change in the angle θ and angular frequency $\Omega = \mathrm{d}\theta/\mathrm{d}t$, and orbital angular momentum L

$$\tau = I\frac{\mathrm{d}\Omega}{\mathrm{d}t} = \frac{\mathrm{d}L}{\mathrm{d}t}, \tag{2.7}$$

where I is the moment of inertia and $L = I\Omega$. In static situations the sum of the forces and torques on the object is zero, so the right-hand side of (2.7) equals zero.

By the way, what we are defining as torques are indeed commonly called "torques" in connection to the rotational and twisting motions of objects, as in this chapter and Chap. 3, but are instead called "moments" in connection to the bending of objects, as in Chap. 4.

2.2 Statics: Motion in One Plane and Levers

Many problems involve motion in one plane, say the xy plane – for which z is a constant. For example, the motion of knees and elbows is in one plane. Some problems involving motion of the leg about the hip can be treated in these two dimensions. The six equations in (2.1) and (2.2) then reduce to three equations:

$$\sum F_x = 0, \qquad \sum F_y = 0, \qquad \sum \tau_z = 0. \tag{2.8}$$

(We will adopt this xyz coordinate system because it is conventionally used in two-dimensional problems, even though it differs from the coordinate system convention we adopted for the body in Fig. 1.1.)

These types of problems can be classified as one of the three types of levers (Fig. 2.4). There are examples in the body of each. They can be described by how a weight W and a force M, provided by a muscle, act on a solid object, say a bone resting on a fulcrum; this represents an articular joint. The weight can include that of parts of the body as well as external weights. The weight and muscle act at distances d_{W} and d_{M} from the joint. For each type of lever the total torque is zero when

$$Md_{\mathrm{M}} = Wd_{\mathrm{W}}, \tag{2.9}$$

so

$$M = \frac{d_{\mathrm{W}}}{d_{\mathrm{M}}}W. \tag{2.10}$$

The relative directions of the forces and the relative distances of the weight and muscle forces from the joint are different for each type of lever.

In a first class lever, the weight and muscle act on opposite sides of the fulcrum and are in the same direction (Fig. 2.4a). This is the least common type of lever in the body. Using the x, y coordinate system shown, there are

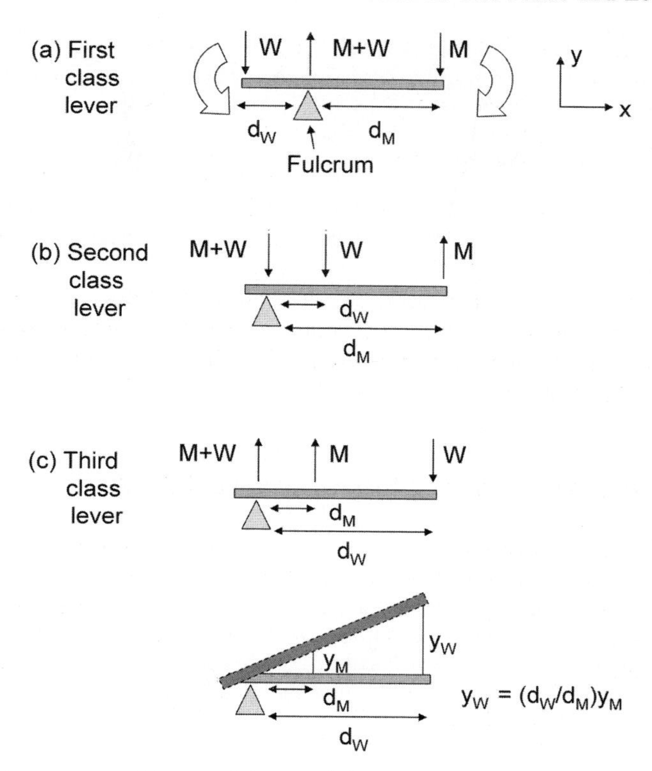

Fig. 2.4. Three types of levers, (**a**) first, (**b**) second, and (**c**) third class levers. The large increase in the distance the weight moves over the change in muscle length in the third class lever is also seen in part (**c**)

clearly no forces in the x direction so the first equation in (2.8) is automatically satisfied. Since the weight and muscle both act in the same direction – downward – force balance in the y direction requires that the fulcrum provides an upward force of $W + M$. Balancing torques in the z direction requires a choice of a z-axis. Any axis normal to the xy plane can be chosen. The simplest one is an axis at the fulcrum. The weight provides a torque of Wd_{W}, while the muscle provides a torque of $-Md_{\mathrm{M}}$. The signs are consistent with the above discussion. The fulcrum provides no torque about this axis because the distance from the fulcrum to the axis is zero. So

$$\sum \tau_z = Wd_{\mathrm{W}} - Md_{\mathrm{M}} = 0. \tag{2.11}$$

This leads to (2.9), which tells how large the muscle force must be to maintain equilibrium. If the muscle cannot provide this large of a force, there can be no static condition. (Example: The lead ball is too heavy to hold up.) If the muscle provides more than this force, there is motion. (Example: The baseball is being thrown, as we will see later.) In these two cases, $\sum \tau_z$ in (2.11) is not zero.

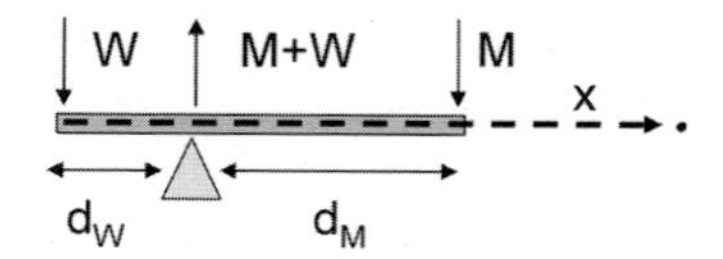

Fig. 2.5. Displacing the axis for calculating torques to the right of the weight by a distance x, as shown for a first class lever. For the axis chosen at the fulcrum $x = d_W$. The axis can be laterally displaced anywhere, to the left or right (as shown) of the lever, above or below it, or in it

It may seem that we cheated by choosing the axis at the fulcrum. Actually, we could have chosen the axis anyway in the xy plane. To prove this let us choose the axis anywhere along the bone, say a distance x to the right of the weight (Fig. 2.5). The torques provided by the weight, fulcrum, and muscle are now Wx, $(W + M)(d_W - x)$, and $-M(d_W + d_M - x)$, respectively. Balance requires

$$\sum \tau_z = Wx + (W + M)(d_W - x) - M(d_W + d_M - x) = 0, \qquad (2.12)$$

which reduces to (2.11) again.

One type of the first class lever is a seesaw or teeter totter. A second type is the head atop the spinal cord, where the weight of the head is balanced by the downward effective force of the muscles (Fig. 2.6a). In a third example, the

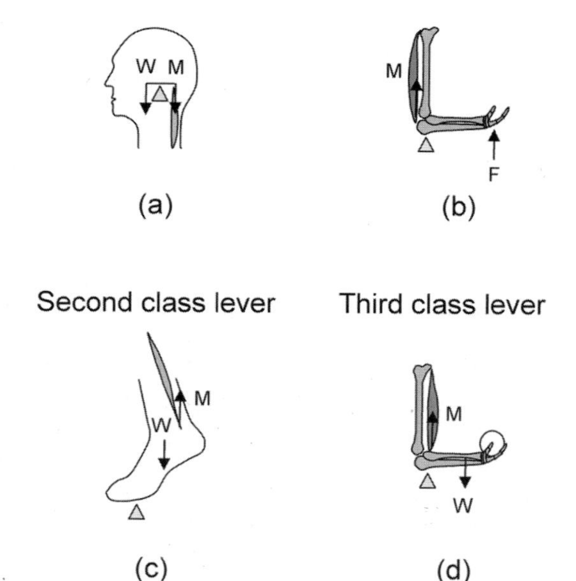

Fig. 2.6. Examples of first (**a**, **b**), second (**c**), and third (**d**) class levers in the body

triceps brachii pull on the ulna about the elbow pivot balanced by the forces on the forearm. With the upper arm down, the triceps brachii can balance an upward force pushing the hand up (Fig. 2.6b). (With the upper arm pointed up, the triceps brachii can also balance the hand holding a weight.) Because $d_{\mathrm{M}} \ll d_{\mathrm{W}}$ here, the force that the triceps brachii needs to exert is much greater than the forces exerted at the hand.

In a second class lever, the muscle and weight act on the same side of the fulcrum, and the weight is nearer to the fulcrum, so $M < W$ (Fig. 2.4b). This type of lever is the second most common in the body. One example is standing on tiptoes: the rotation of the foot about the toes (the fulcrum), which would be caused by the weight of the foot, is balanced by the muscle force transmitted by the Achilles tendon (Fig. 2.6c). Another example is pushing down with the triceps brachii.

In third class levers, the muscle and weight are again on the same side of the fulcrum, but now the muscle is nearer to the fulcrum than the weight (Fig. 2.4c). This is the most common example in the body. Because often $d_{\mathrm{M}} \ll d_{\mathrm{W}}$, we see that $M = (d_{\mathrm{W}}/d_{\mathrm{M}})W \gg W$. This arrangement means that very large forces must be exerted by the muscles because of this $d_{\mathrm{W}}/d_{\mathrm{M}}$ amplification, which seems to be a big disadvantage (and is literally a mechanical disadvantage). However, something else is gained in this tradeoff in design. As seen in the Fig. 2.4c, when the bone rotates a given angle, causing a vertical displacement y_{M} at the muscle, there is an amplification of the distance traveled at the position of the weight by $d_{\mathrm{W}}/d_{\mathrm{M}}$. As we will see in Chap. 5, muscles are able to contract only a small fraction of their length – which amounts to at most several cm in many muscles. The length of the biceps is about 25 cm, and the maximum contraction is by ~7–8 cm. With this amplification, the weight can now move much more than this. One example of a third class lever is the balancing of the lower arm by the biceps brachii inserted on the radius (Fig. 2.6d). Another is holding a weight with an outstretched arm.

2.3 Statics in the Body

We will examine the planar forces in the static equilibrium of the lower arm, at the hip, and in the back. In analyzing the lower arm, we will choose successively more complex and realistic models. We will see that the forces in the hip and back are quite large, much more than one would expect, and explains why people often have problems in these parts of the body, problems that can lead to hip replacements and life-long lower back pain. The approach for these problems is the same. We consider all elements in one plane and examine the forces in the (as defined) x and y directions and the torque in the z direction. Some of the approaches of [65] and [86] are followed.

2.3.1 The Lower Arm

We will examine the equilibrium of the forearm balanced by the contraction of the biceps brachii inserted on the forearm long bone called the radius; this is a continuation of the discussion of third class levers. The relevant bones are shown in Figs. 2.7 and 2.8. In equilibrium, the biceps brachii force counters the potential rotation about the elbow joint by the weight held in the hand (Fig. 2.9a). We will examine this example for different models, using Fig. 1.15 and Tables 1.6 and 1.7 to provide anthropometric information. The forearm is $0.146H$ long and the hand length is $0.108H$, where H is the body height, so the weight held in the hand is about $(0.146 + 0.108/2)H = 0.2H$ from the pivot. (The ball is in the middle of the hand.)

Case 1

The biceps brachii insert about 4 cm from the pivot axis. Say there is a weight W_{W} held in the hand, which is $d_{\mathrm{W}} = 36$ cm from the pivot. (With $H = 180\,\mathrm{cm}$, $0.2H = 36\,\mathrm{cm}$.) Therefore $M = (d_{\mathrm{W}}/d_{\mathrm{M}})W_{\mathrm{W}} = (36\,\mathrm{cm}/4\,\mathrm{cm})W_{\mathrm{W}} = 9W_{\mathrm{W}}$. So for a weight of 100 N the muscle must provide a force of 900 N for balance. Here N stands for the MKS/SI unit of newtons. Since $1\,\mathrm{N} \simeq 0.225\,\mathrm{lb}$, equivalently, a 22 lb weight is balanced by 200 lb of force exerted by the biceps brachii (Fig. 2.9b).

We have made several assumptions and approximations in this example without explicitly stating them. It is always good to start with simple models. It is equally important to understand exactly what assumptions and approximations are being made. Then, the model can be made more realistic. Here, we have assumed that the forearm and upper arm make a 90° angle. We have also neglected the mass of the forearm.

Case 2

Now let us improve the model by including the weight of the forearm W_{F} (Fig. 2.9c). This is about $0.022W_{\mathrm{b}}$, (where W_{b} is the body weight) (Table 1.7). For a 70 kg (700 N, 160 lb) person, this is ≈15 N (3.4 lb). We can treat the effect of the weight of the forearm as if it were acting at its center of mass, which is approximately in the middle of the forearm, $d_{\mathrm{F}} = 0.146H/2 = 13\,\mathrm{cm}$ from the pivot:

$$\sum \tau_z = M d_{\mathrm{M}} - W_{\mathrm{W}} d_{\mathrm{W}} - W_{\mathrm{F}} d_{\mathrm{F}} = 0 \tag{2.13}$$

$$M d_{\mathrm{M}} = W_{\mathrm{W}} d_{\mathrm{W}} + W_{\mathrm{F}} d_{\mathrm{F}} \tag{2.14}$$

$$M = \frac{d_{\mathrm{W}}}{d_{\mathrm{M}}} W_{\mathrm{W}} + \frac{d_{\mathrm{F}}}{d_{\mathrm{M}}} W_{\mathrm{F}}. \tag{2.15}$$

The ratio $d_{\mathrm{F}}/d_{\mathrm{M}}$ $(= 13\,\mathrm{cm}/4\,\mathrm{cm})$, so now $M = 9W_{\mathrm{W}} + 3.25W_{\mathrm{F}}$ and the muscle force required to maintain equilibrium has increased to $900\,\mathrm{N} + 3.25\,(15\,\mathrm{N}) = 950\,\mathrm{N}$ (210 lb).

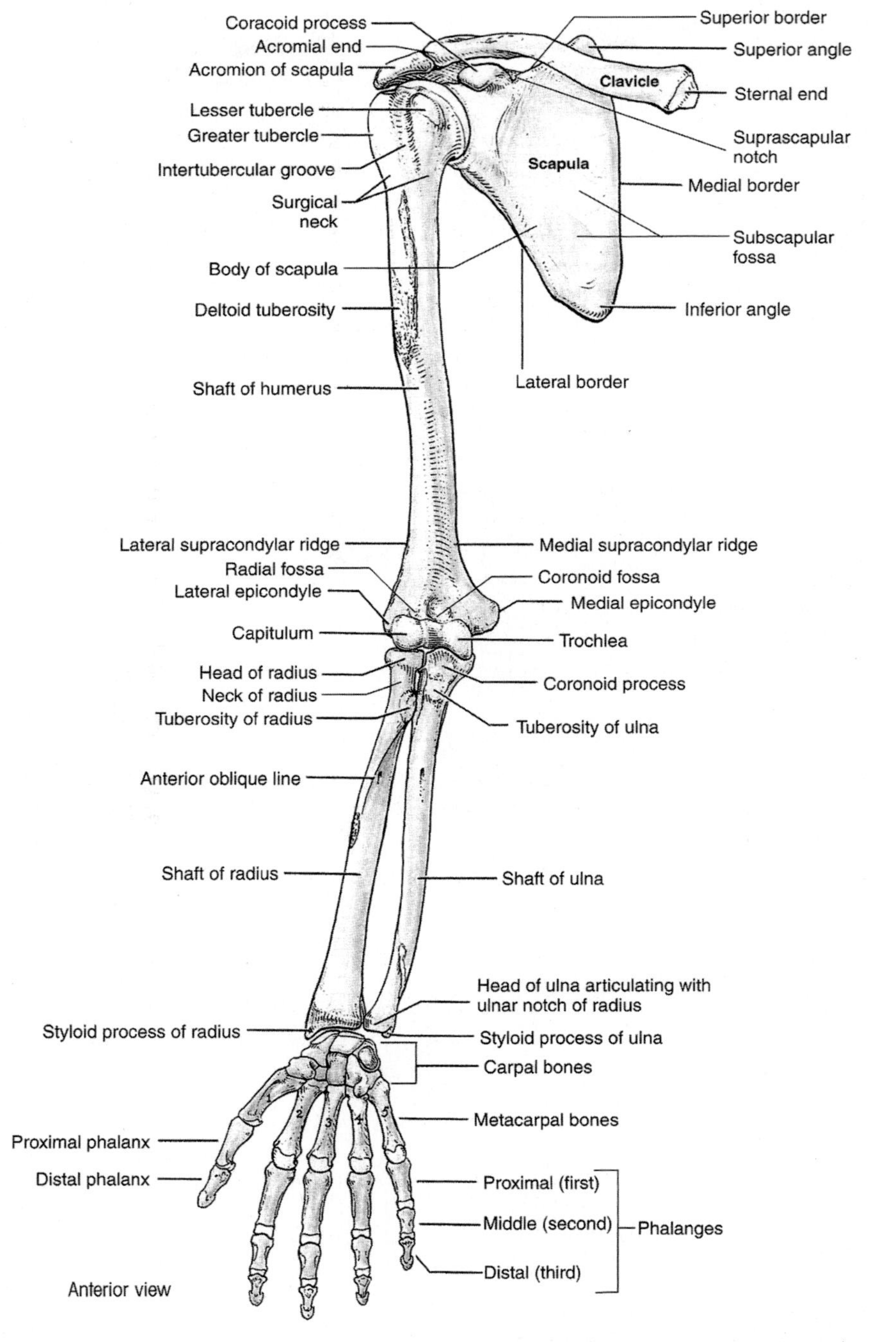

Fig. 2.7. Bones of the arm, anterior view. (From [78]. Used with permission)

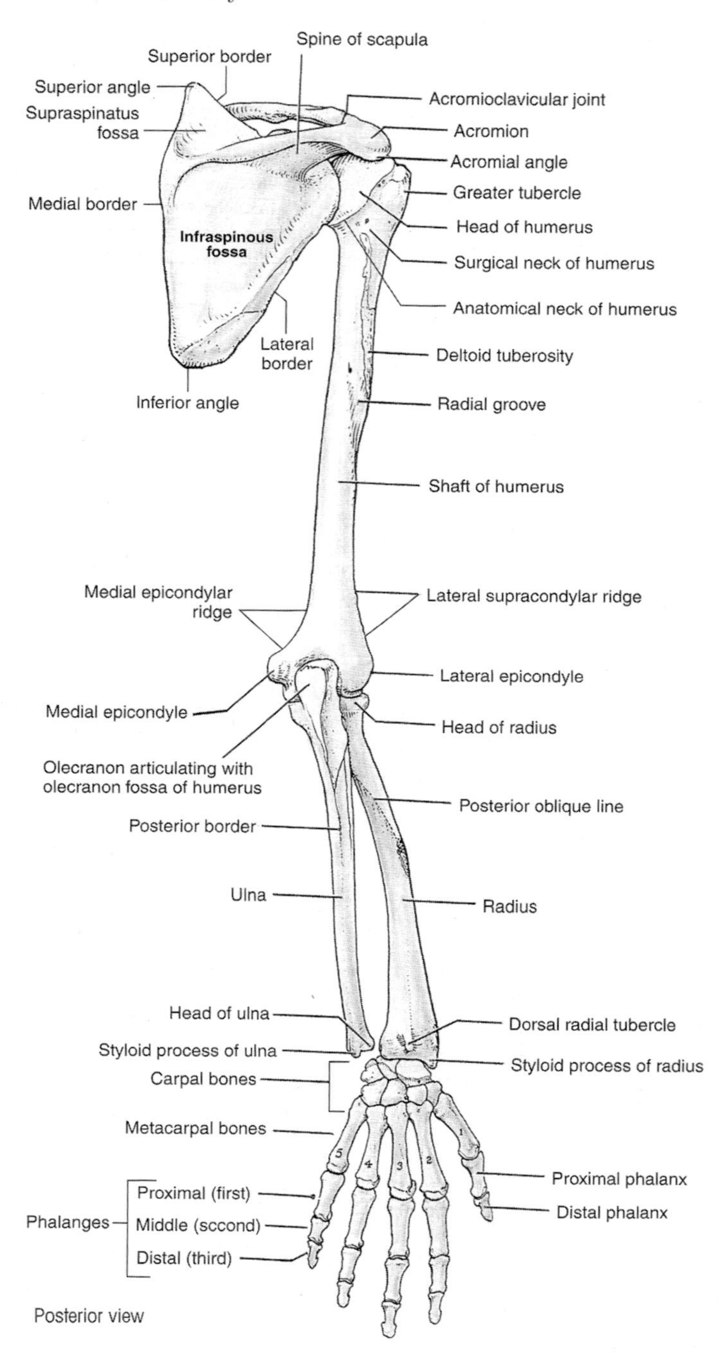

Fig. 2.8. Bones of the arm, posterior view. (From [78]. Used with permission)

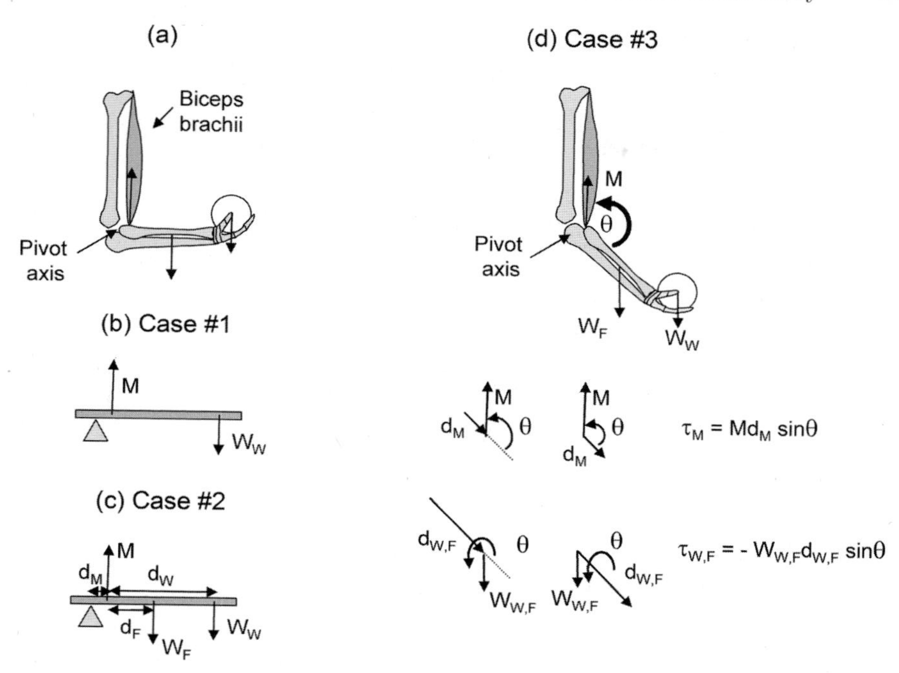

Fig. 2.9. (**a**) Weight held in the hand, showing the biceps brachii muscles. (**b**–**d**) Forces for the equilibrium of a weight held in the hand for Cases 1–3

Case 3

What happens if we no longer assume that the forearm and upper arm make a 90° angle? Let us keep the upper arm vertical and let the forearm make an angle θ, which can range over 142° (Table 1.10). The force due to the muscle is then still vertical, and those due to the weights of the forearm and ball are, of course, downward. From Fig. 2.9d we see that the torque caused by each of these three forces is multiplied by $\sin\theta$. Now

$$\sum \tau_z = Md_{\mathrm{M}} \sin\theta - W_{\mathrm{W}} d_{\mathrm{W}} \sin\theta - W_{\mathrm{F}} d_{\mathrm{F}} \sin\theta = 0 \tag{2.16}$$

and we arrive at the same result that

$$Md_{\mathrm{M}} = W_{\mathrm{W}} d_{\mathrm{W}} + W_{\mathrm{F}} d_{\mathrm{F}}. \tag{2.17}$$

Actually, we made additional assumptions in this example that we will re-examine later. The distance from the pivot where the biceps brachii insert on the radius really changes with θ (Fig. 3.42). Also, while this analysis suggests that the muscle force M required for equilibrium is the same for all angles, there is a subtlety in this result. Equation (2.16) gives the muscle force needed to maintain equilibrium. As we will see in Chap. 5, muscles can exert forces up to a maximum value. If the M from (2.17) can be achieved, then there can be equilibrium; if it cannot, then the static condition cannot be achieved. The maximum force that a muscle can exert depends on its length, which, from Fig. 2.9, is clearly a function of θ. So the M in (2.17) may be achievable at some angles (nearer 90°, where the maximum force turns out to be greatest) and not at others.

Case 4

The biceps brachii are not the only muscles used to flex the elbow. What happens if we also include the contributions of these other muscles? Figure 2.10 shows that the biceps brachii, the brachialis, and the brachioradialis all contribute to this flexing. Assuming that $\theta = 90°$ (which may not be a good assumption for each muscle), (2.13) is modified to

$$\sum \tau_z = M_1 d_{\mathrm{M}_1} + M_2 d_{\mathrm{M}_2} + M_3 d_{\mathrm{M}_3} - W_{\mathrm{W}} d_{\mathrm{W}} - W_{\mathrm{F}} d_{\mathrm{F}} = 0 \tag{2.18}$$

$$M_1 d_{\mathrm{M}_1} + M_2 d_{\mathrm{M}_2} + M_3 d_{\mathrm{M}_3} = W_{\mathrm{W}} d_{\mathrm{W}} + W_{\mathrm{F}} d_{\mathrm{F}} \tag{2.19}$$

where M_1, M_2, and M_3 represent the forces exerted by the three muscles M_i, respectively. If the physiological cross-sectional areas of the three muscles are A_1, A_2, and A_3, respectively (which we usually call PCA), and the muscle force for each can be assumed to be proportional to this area (which is a pretty good assumption), then $M_i = kA_i$, for $i = 1, 2, 3$. (We will see

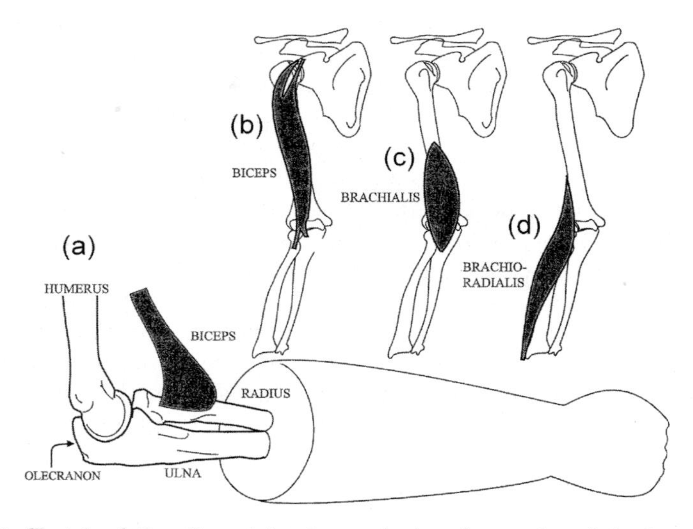

Fig. 2.10. Sketch of the elbow joint for analyzing the statics of the lower arm for Case 4, with the three muscles, the biceps (brachii), brachialis, and brachioradialis, shown in (**b**–**d**). (From [76])

Table 2.1. Data for the three elbow muscles used for flexion. (Using data from [76, 95])

muscle	moment arm d_i (cm)	physiological cross-section (PCA) (cm^2)
biceps (muscle 1)	4.6	4.6
brachialis (muscle 2)	3.4	7.0
brachioradialis (muscle 3)	7.5	1.5

that this is a good assumption with k reaching a maximum of $\sim$40 N/cm^2 or so.) So,

$$kA_1 d_{\text{M}_1} + kA_2 d_{\text{M}_2} + kA_3 d_{\text{M}_3} = W_{\text{W}} d_{\text{W}} + W_{\text{F}} d_{\text{F}} \tag{2.20}$$

$$k = \frac{W_{\text{W}} d_{\text{W}} + W_{\text{F}} d_{\text{F}}}{A_1 d_{\text{M}_1} + A_2 d_{\text{M}_2} + A_3 d_{\text{M}_3}} \tag{2.21}$$

$$M_1 = kA_1 = A_1 \frac{W_{\text{W}} d_{\text{W}} + W_{\text{F}} d_{\text{F}}}{A_1 d_{\text{M}_1} + A_2 d_{\text{M}_2} + A_3 d_{\text{M}_3}} \tag{2.22}$$

$$M_2 = kA_2 = A_2 \frac{W_{\text{W}} d_{\text{W}} + W_{\text{F}} d_{\text{F}}}{A_1 d_{\text{M}_1} + A_2 d_{\text{M}_2} + A_3 d_{\text{M}_3}} \tag{2.23}$$

$$M_3 = kA_3 = A_3 \frac{W_{\text{W}} d_{\text{W}} + W_{\text{F}} d_{\text{F}}}{A_1 d_{\text{M}_1} + A_2 d_{\text{M}_2} + A_3 d_{\text{M}_3}}. \tag{2.24}$$

Using the parameters from Table 2.1, we get $M_1 = 262$ N (biceps), $M_2 = 399$ N (brachialis), and $M_3 = 85$ N (brachioradialis) when we generalize Case 2. This compares to the $M_1 = 696$ N that we would obtain for Case 2 with the biceps alone, using $d_1 = 4.6$ cm (instead of the 4 cm used before, which led to 800 N). The total muscle force is 746 N, which is greater than 696 N because the brachialis has a relatively small moment arm.

Life is a bit more complex than this result suggests because we assumed that k has the same value for each muscle. Really $M_i = k_i A_i$, and all the k_i's need not be the same, as long as k_i is less than the maximum that can be exerted by the muscles. Unfortunately, if all the k_is are not assumed to be equal, we do not have enough information to solve this problem uniquely as posed. The body may solve the indeterminate nature of this problem (with more variables than conditions) by minimizing energy or optimizing the force distribution (shifting the load from one muscle to another) to rest specific muscles or to keep the weight balanced better (so it will not tip in the hand).

2.3.2 Hip Problems

The hip (pelvis) is not a single bone, but several bones that are fused together (Figs. 2.11 and 2.12). The pelvis is composed of the pelvic girdle and two

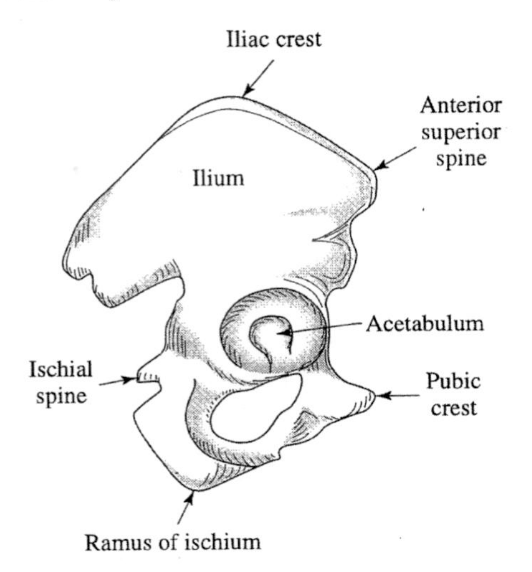

Fig. 2.11. Right hip bone in adult. (From [65])

parts of the spinal cord, the sacrum and coccyx. The pelvic girdle itself is composed of the right and left coxal (or hip) bones. Each coxal (hip) bone is composed of three bones: the ilium (at the top = superior), pubis (bottom front = inferior, anterior), and ischium (iss-kee'-um) (bottom back = inferior, posterior). The acetabulum (a-si-tab-yoo'-lum) is the socket area where the femur of the leg (Fig. 2.13) is attached ("hip joint"). Actually, the head of the femur is in this socket, and is maintained there by the muscles attached at the greater trochanter. These muscles are collectively called the hip abductor muscles. (The hip abductor muscles are not the only one attached at the greater trochanter (see below), but they are the ones that contribute to the force needed for the equilibrium condition in this problem.)

First we will determine the force on the head of the femur and in the hip abductor muscles while the subject is standing on one leg, say the right

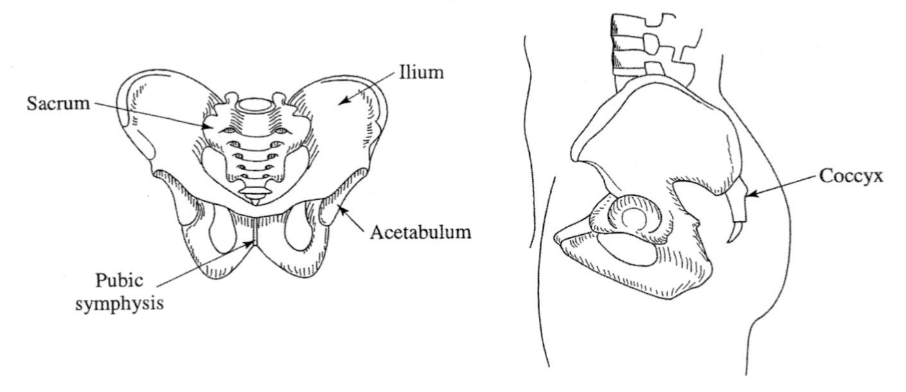

Fig. 2.12. Front and side views of the hip. (From [65])

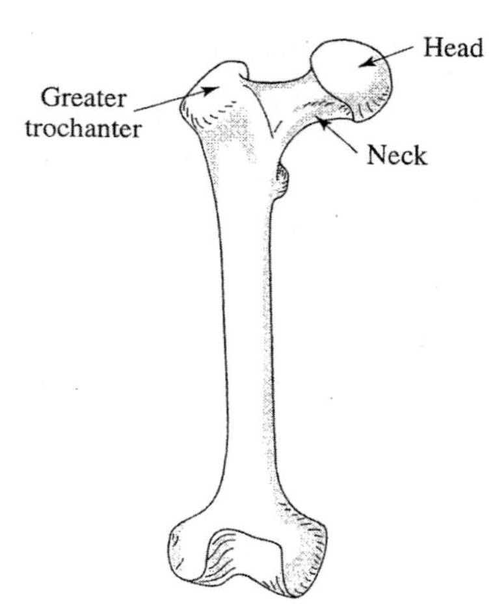

Fig. 2.13. Anterior view of right femur. (From [65])

leg. This is actually a good model for determining these forces during slow walking. The upper and lower leg and the foot are treated as a rigid body. This problem is solved in two steps. First, the forces in the whole body are analyzed and then the rigid leg is treated as a free rigid body, whose only interaction with the rest of the body will be the normal force from the hip [65, 85]. The bones in the leg are shown in Figs. 2.14 and 2.15.

Total Body Equilibrium

There are only two forces on the body. The body weight W_{b} acts downward, and as if it all originated at the center of mass of the body, which is in the midline in the hip (Fig. 2.16). The foot feels an upward normal force from the floor of magnitude N. There are no forces in the x direction, and these two forces in the y direction must balance in equilibrium, so $N = W$. In equilibrium the body cannot start to rotate, so the torques are zero. It is clear from the Fig. 2.16 that this occurs when the foot is directly below the hip, in the midline. If we choose the pivot axis at the center of mass, the torque from the center of mass is zero because the distance term (from the axis to the center of mass) is zero and the torque from the normal force is zero because the normal force is antiparallel to the distance vector ($\theta = 180°$). (You can prove for yourself that the total toque is zero for any other axis normal to the xy plane.)

Equilibrium of the Individual Body Component

There are four external forces on the leg (Figs. 2.16 and 2.17):

(a) $\mathbf{N}$ is the normal force on the leg from the floor, and we know that $N = W_{\mathrm{b}}$.

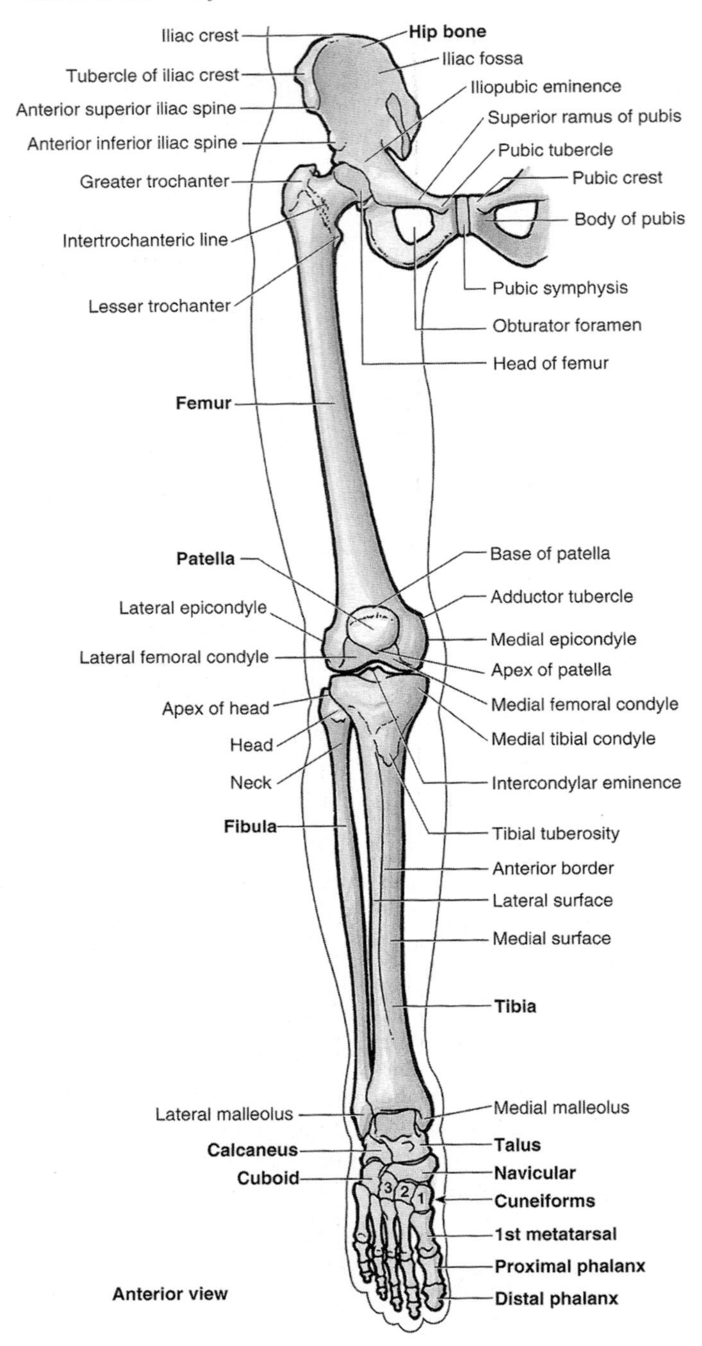

Fig. 2.14. Bones of the leg and hip, anterior view, with names of bones in bold. (From [79]. Used with permission)

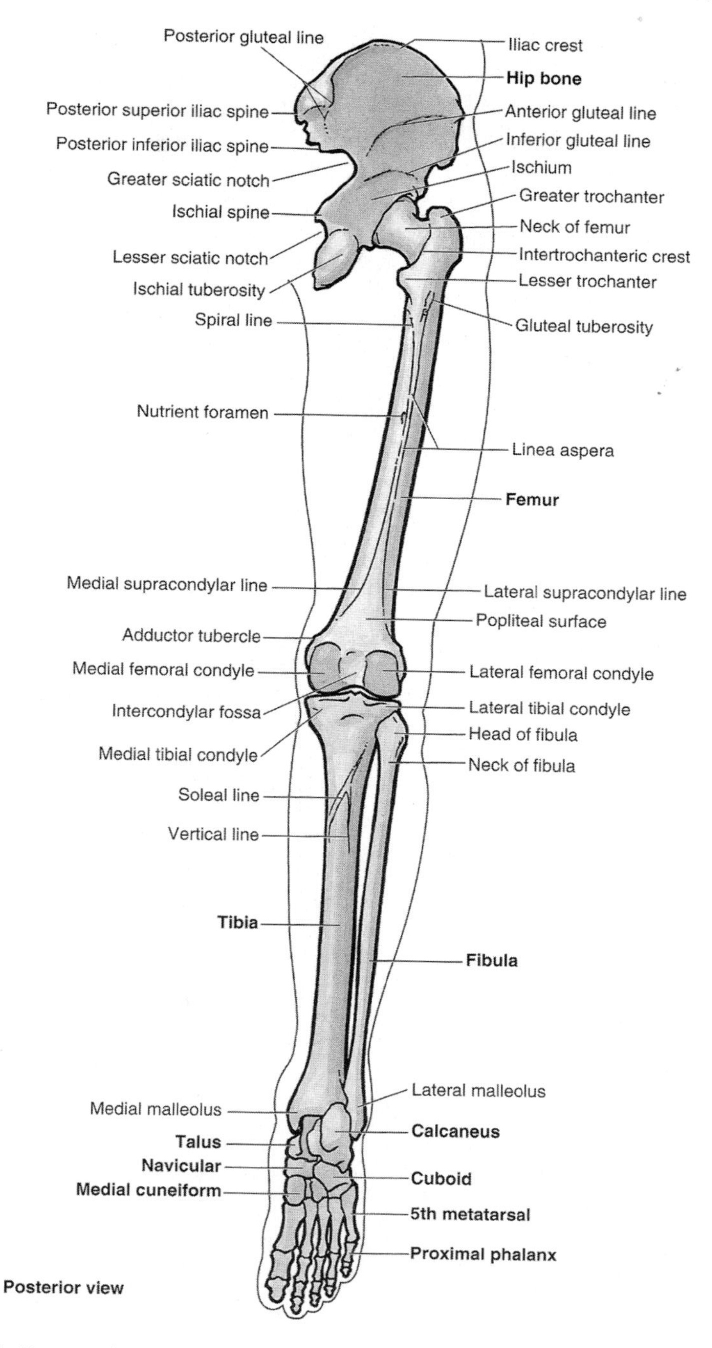

Fig. 2.15. Bones of the leg and hip, posterior view, with names of bones in bold. (From [79]. Used with permission)

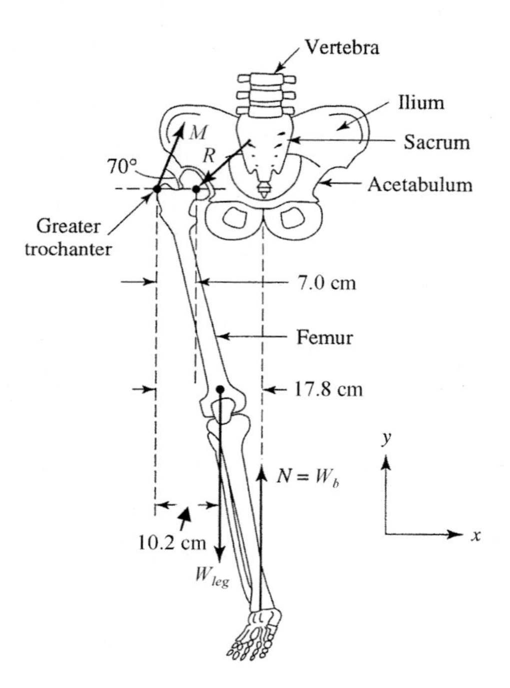

Fig. 2.16. Anatomical diagram of the leg and hip for someone standing on one leg, or during slow walking, showing the forces on them and relevant dimensions, including the force exerted on the head of the femur by the acetabulum R and the net force exerted by the hip abductor muscles. (From [65])

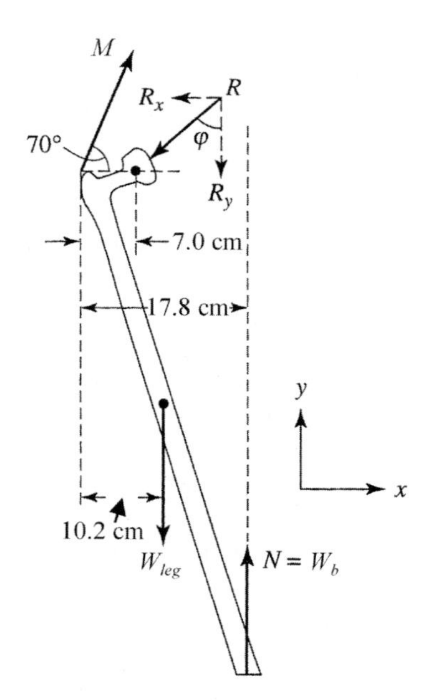

Fig. 2.17. Force diagram for a leg for someone standing on one foot, using Fig. 2.16. (From [65])

(b) W_{leg} is the weight of the leg. Table 1.7 shows that this is about $0.16W_{\text{b}}$. It acts as if it were applied at the center of mass of the leg, which is approximately halfway down the leg.
(c) $\mathbf{R}$ is the reaction force on the leg from the hip, and it is normal to the hip socket. We will define the x and y components R_x and R_y so they are positive (Fig. 2.17). Equivalently, we could define the magnitude of R and the angle it makes with the vertical, φ.
(d) $\mathbf{M}$ is the force (of magnitude M) due to the hip abductor muscles. There are actually three muscles involved here: the tensor fascia (fash-ee'-uh) latae (la-tuh) (see Fig. 3.2a), gluteus (gloo'-tee-us) minimus (see Fig. 3.3c), and the gluteus medius (see Fig. 3.3a, b). (The gluteus maximus muscle is what the author is sitting on as he is typing this.) These three muscles have a mass ratio of about 1:2:4 and, as we will see, this is roughly the ratio of the forces each can exert. The hip abductor muscle structure we consider is a composite of the three muscles. It has been shown that the effective action of this system is $\simeq 70°$ to the horizontal, acting on the greater trochanter.

We have three equations with three unknowns: R_x, R_y, and M. The two force equations are

$$\sum F_x = M\cos 70° - R_x = 0, \tag{2.25}$$

$$\sum F_y = M\sin 70° - R_y - 0.16W_{\text{b}} + W_{\text{b}} = 0, \tag{2.26}$$

where W_{leg} has been replaced by $0.16W_{\text{b}}$.

We will choose the rotation axis to emanate from the center of the head of the femur because the reaction force from the acetabulum passes through this point. This makes the analysis easier, but, of course, the solution would be the same if we chose any other parallel axis. The relevant distances of interest are shown in the diagram (obtained from anatomical dimensions and geometry), as needed for torque analysis.

(a) The component of the distance vector perpendicular to the normal force (r') is 10.8 cm, so the normal force causes a torque of $(10.8\,\text{cm})W_{\text{b}}$. This is a positive torque because the normal force induces a counter clockwise rotation about the chosen z-axis (see Fig. 2.2).
(b) The component of the distance vector normal to the force of the weight of the leg is 3.2 cm and this force tends to induce a clockwise rotation, so it contributes a torque of $-(3.2\,\text{cm})W_{\text{leg}} = -(3.2\,\text{cm})(0.16W_{\text{b}}) = -(0.5\,\text{cm})W_{\text{b}}$.
(c) With the choice of the axis, the torque from the reaction force from the hip is zero, because the distance vector and normal force are antiparallel.
(d) The component of the force from the hip abductor muscles normal to the horizontal distance vector (of magnitude 7.0 cm) is $M\sin 70°$. Since this

causes a clockwise rotation, the torque is $-(7.0\,\text{cm})M\sin 70°$. So we see that

$$\sum \tau_z = (10.8\,\text{cm})W_\text{b} - (3.2\,\text{cm})(0.16W_\text{b}) + 0 - (7.0\,\text{cm})M\sin 70° = 0, \tag{2.27}$$

$$M = \frac{10.8 - 0.5}{7.0\sin 70°}W_\text{b} = 1.57W_\text{b}. \tag{2.28}$$

We see that torque provided by the hip abductor muscles is needed to counter the torques from the normal force from the floor and the weight of the leg. This normal force torque is much more important than that due to the leg, because of the greater magnitude of the force and the larger moment arm.

Using this value for the muscle force, the force balance in the x direction gives $R_x = M\cos 70° = 0.54W_\text{b}$. From the balance in the y direction, $R_y = M\sin 70° + 0.84W_\text{b} = 2.31W_\text{b}$. The magnitude of $R = (R_x^2 + R_y^2)^{1/2} = 2.37W_\text{b}$, and $\tan\theta = R_x/R_y = 0.54/2.31 = 0.23$, so $\theta = 13°$.

Because $M \simeq 1.6W_\text{b}$ and $R \simeq 2.4W_\text{b}$, for $m_\text{b} = 90\,\text{kg}$ we have $W_\text{b} = 880\,\text{N}$ (200 lb), and so $M \simeq 1{,}400\,\text{N}$ (320 lb) and $R \simeq 2{,}100\,\text{N}$ (470 lb). The origin of hip problems is clear: The force from the hip is much greater than the body weight because of the large moment arms.

We next examine a variation of this problem. The person now uses a cane to provide support on the left side while standing on his or her right leg (Fig. 2.18). As shown in Fig. 2.19, the cane is 30.5 cm (1 ft) from the body midline. It is supported by and pushed down by the left arm or shoulder. Consequently, there is a normal force N_c from the floor. We assume $N_\text{c} = W_\text{b}/6$. We will see that this has two immediate consequences. The right foot is no longer directly in the body midline but is displaced a distance L to the right (in the reference of the body) and the normal force felt by the right foot N_f is no longer the body weight.

Whole body equilibrium gives

$$N_\text{f} + N_\text{c} - W_\text{b} = N_\text{f} + W_\text{b}/6 - W_\text{b} = 0 \tag{2.29}$$

or $N_\text{f} = 5W_\text{b}/6$. Using the same axis as before, the torque balance is

$$\sum \tau_z = (30.5\,\text{cm})N_\text{c} - L(N_\text{f}) = (30.5\,\text{cm})(W_\text{b}/6) - L(5W_\text{b}/6) = 0 \tag{2.30}$$

or $L = (N_c/N_\text{f})30.5\,\text{cm} = (1/5)30.5\,\text{cm} = 6.1\,\text{cm}$.

For the same leg as in Fig. 2.17, the corresponding distances are different because the angle of the leg is now different (as obtained from anatomical dimensions and the new geometry) (Fig. 2.20). We will examine the whole leg equilibrium again. The cane is not explicitly involved, but implicitly through the changes in the leg position and the load borne by the leg.

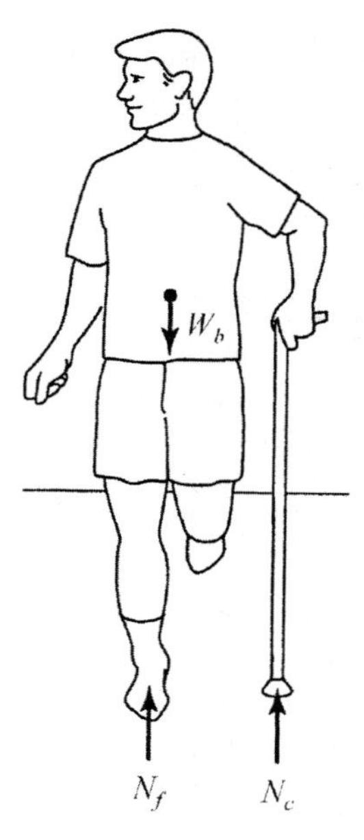

Fig. 2.18. Forces on entire person for someone walking with a cane. (From [65])

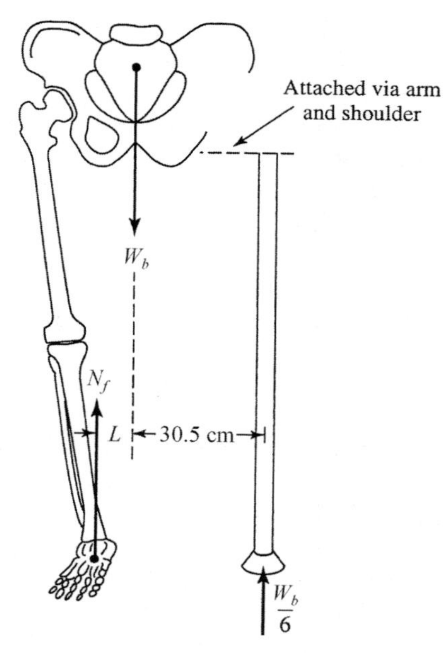

Fig. 2.19. Force diagram for a person using a cane for some support. (From [65])

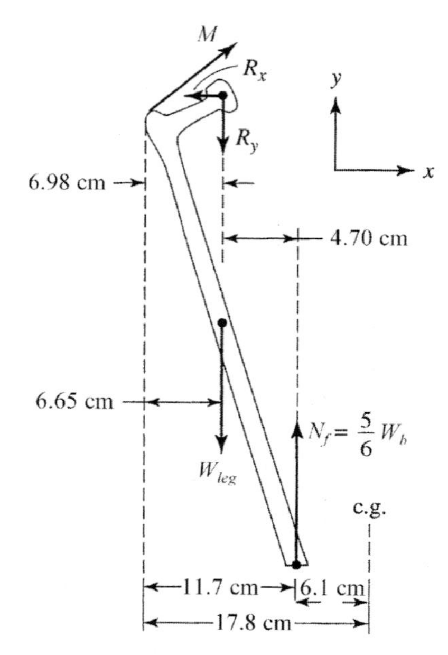

Fig. 2.20. Free-body force diagram of the leg for someone walking with a cane. Note that the center of mass of the leg is now 0.33 cm to the left of the vertical from the center of the head of the femur, whereas without the cane it was 3.2 cm to the right of the vertical. Consequently, the leg center of mass is 6.65 cm from the vertical line from the greater trochanter, whereas the center of the head of the femur is 6.98 cm from it. (From [65])

Now

$$\sum F_x = M \cos 70° - R_x = 0, \tag{2.31}$$

$$\sum F_y = M \sin 70° - R_y - 0.16W_\mathrm{b} + (5/6)W_\mathrm{b} = 0, \tag{2.32}$$

$$\sum \tau_z = +(4.7\ \mathrm{cm})(5/6)W_\mathrm{b} + (0.33\ \mathrm{cm})(0.16W_\mathrm{b}) + 0 - (6.98\ \mathrm{cm})M \sin 70° = 0. \tag{2.33}$$

The first equation is unchanged. (We are assuming that the effective angle of the hip abductor muscles with the x-axis is still 70°, even though this is no longer rigorously true.) The only change in the second equation is the smaller normal force on the leg. There are two changes in the torque equation. The first term, due to the normal force from the floor, is much smaller due to the change in the moment arm and the normal force of the leg – the former effect being much larger than the latter. The second term, due to the weight of the leg, is much smaller in magnitude and is now a positive torque instead of a negative torque, because the center of mass of the leg is now to the left of the vertical drawn down from the axis (Figs. 2.2 and 2.20).

Table 2.2. Analytic estimates of peak hip forces. (Using data from [82])

activity	magnitude/body weight, W_b
walking	4.8–5.5
walking slowly with/without a cane	2.2/3.4
stair ascending/climbing	7.2–7.4
stair descending	7.1
chair raising	3.3

Now $M = 0.61W_\mathrm{b}$, $R_x = 0.21W_\mathrm{b}$ and $R_y = 1.24W_\mathrm{b}$, with $R = 1.26W_\mathrm{b}$ and $\varphi = 9.5°$. With W_b still 880 N (200 lb), the muscle force M is now 540 N (120 lb) (instead of 1,400 N (320 lb) without the cane) and the reaction force at the hip R is now 1,100 N (250 lb) (instead of 2,100 N (470 lb)). This is a very big effect, considering that only about 145 N (35 lb) ($\approx W_\mathrm{b}/6$) is resting on the cane. By far the major consequence of the cane is the change of the moment arm of the normal force from the floor from 10.8 to 4.7 cm, because the foot shifted to the right by 6.1 cm.

Another variation of this situation is presented in Problems 2.8 and 2.9. Instead of using a cane with the left hand, you will examine the consequences of carrying a weight on the left side. This greatly increases the moment arms, thereby increasing M and R. Table 2.2 shows estimates for peak hip forces for several activities from more detailed analyses.

Excessive hip forces for extended periods of time can thin cartilage and cause other damage; this leads to pain during walking that can sometimes only be alleviated with a hip replacement. Total hip replacements entail replacing the ball of the femur and the acetabulum. The femoral component is a highly polished ball of a high-strength alloy, such as cobalt–chromium or titanium, with a step that is placed in the canal of the femur. It is often fixed with an acrylic plastic, such as poly(methyl methacrylate), PMMA, "bone cement." The acetabulum component is a socket made from ultrahigh molecular weight polyethylene, and may have a metal backing.

2.3.3 Statics of Other Synovial Joints

Other synovial joints can be examined in similar ways by using models of the muscles and tendons. They are briefly outlined here; see [86] for more details.

Shoulder

The anatomy and musculature of the shoulder are shown in Figs. 1.2 and 1.8. Figure 2.21 shows a model of the static equilibrium of a horizontal arm that holds a weight by means of the deltoid muscle. This weight and muscle

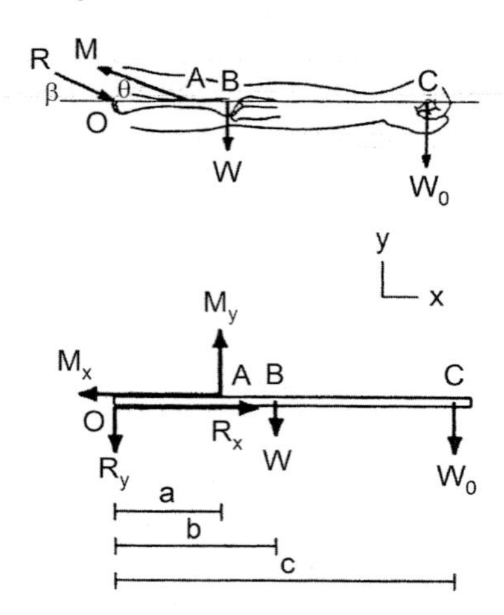

Fig. 2.21. Forces on the arm and shoulder, when the arm is abducted to the horizontal position and the hand holds a weight, along with the force diagram. (From [86])

are in equilibrium with the weight of the arm and the reaction force on the joint. Problems 2.15–2.17 address this with and without a weight held in the hand. For reasonable parameters, the component of the muscle force in the x direction is about $4\times$ that in the y direction, suggesting that holding the arm horizontally is not stable – as we all know.

Dislocation of the shoulder (glenohumeral) joint is common because it is shallow. The large degree of rotational motion of the head of the humerus about the articulating surface of the glenoid fossa enables this large motion, albeit with little stability. The shoulder joint angles range by $\sim 249^\circ$ during flexion/extension, $\sim 182^\circ$ in abduction/adduction, and $\sim 131^\circ$ for inward/outward rotation (Table 1.10). Fracture of the humerus is also relatively common.

Knee

The knee joint is really two joints (Fig. 1.3). The tibiofemoral joint is located between the medial and lateral condyles of the femur and tibia, which are separated by cartilaginous regions called menisci. (A *condyle* is the rounded prominence at the end of a bone, often at an articulation joint.) The second is the patellofemoral joint between the kneecap (patella) and the anterior end of the femoral condyles. Muscle control of the knee is mostly through the quadriceps and hamstring muscles. The quadriceps attach to the quadriceps tendon, which attaches to the kneecap, which attaches to the patellar tendon, which attaches to the tibia.

The forces on the static lower leg loaded with an ankle weight, as during exercise (Fig. 2.22), are shown in Figs. 2.23 and 2.24. The forces shown are

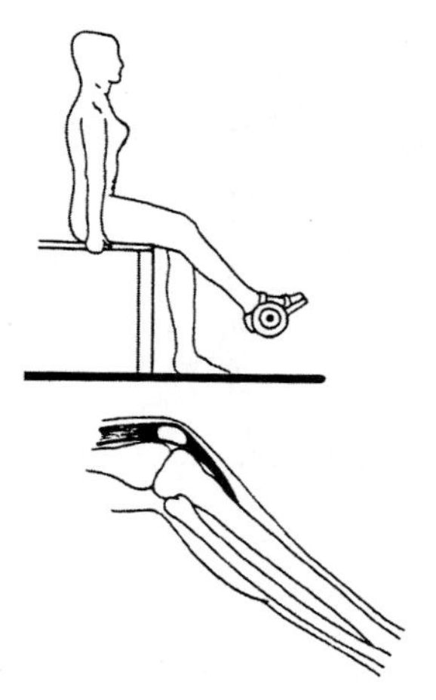

Fig. 2.22. Exercising muscles near and at the knee. (From [86])

due to this added weight, the weight of the lower leg, the quadriceps muscle force transmitted by the patellar tendon **M** (of magnitude M), and the joint reaction force **R** (of magnitude R), while the angle between the horizontal and the leg is β. In equilibrium, the muscle force and the x and y components

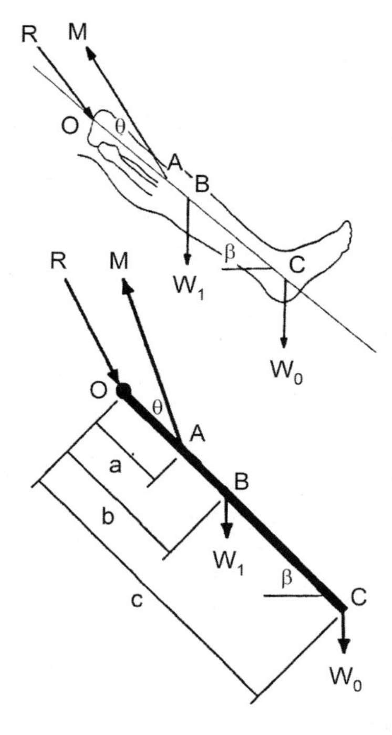

Fig. 2.23. Forces on the lower leg, while exercising the muscle around the knee. (From [86])

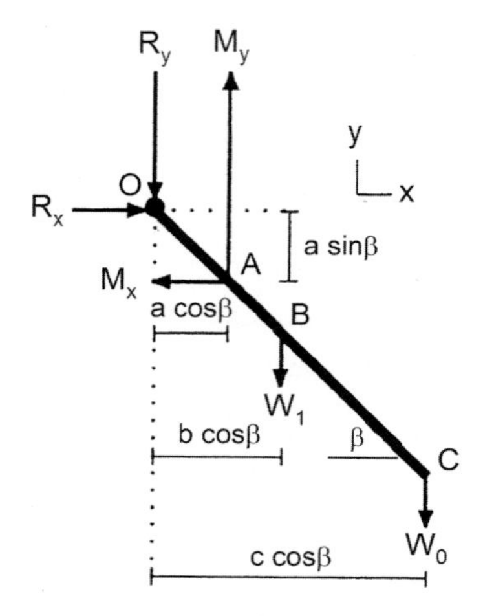

Fig. 2.24. Resolution of the forces on the lower leg in Fig. 2.23. (From [86])

of the joint force are

$$M = \frac{(bW_1 + cW_0)\cos\beta}{a\sin\theta}, \tag{2.34}$$

$$R_x = M\cos(\theta+\beta), \tag{2.35}$$

$$R_y = M\sin(\theta+\beta) - W_0 - W_1. \tag{2.36}$$

These forces are very large [84, 86]. For $a = 12\,\mathrm{cm}$, $b = 22\,\mathrm{cm}$, $c = 50\,\mathrm{cm}$, $W_1 = 150\,\mathrm{N}$, $W_0 = 100\,\mathrm{N}$, $\theta = 15°$ and $\beta = 45°$, we see that the muscle force $M = 1{,}381\,\mathrm{N}$ and the joint force $R = 1{,}171\,\mathrm{N}$. This is examined further in Problem 2.18. Problems 2.21–2.24 address a related condition, that of the crouching position, as occurs during ascending and descending stairs or jumping.

One function of the kneecap is to increase the moment arm (Fig. 2.25). We can analyze the equilibrium of the kneecap at the patellofemoral joint between the reaction force on the kneecap from the anterior end of the femoral condyles, the patellar tendon and the quadriceps tendon. This is shown in Fig. 2.26. The compressive force applied on the kneecap is

$$F_{\mathrm{P}} = \frac{\cos\gamma - \cos\alpha}{\cos\phi} M \tag{2.37}$$

at an angle

$$\phi = \arctan\left(\frac{\sin\alpha - \sin\gamma}{\cos\gamma - \cos\alpha}\right). \tag{2.38}$$

Ankle

We now examine the equilibrium of the foot. The anatomy of the ankle and foot are depicted in Figs. 2.14 and 2.15 (also see Figs. 1.8 and 3.4). Figures 2.27

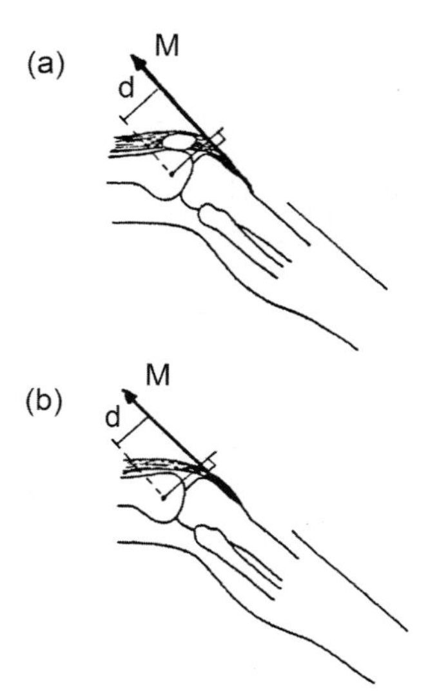

Fig. 2.25. The presence of the kneecap (patella) increases the moment arm in the lever. (From [86])

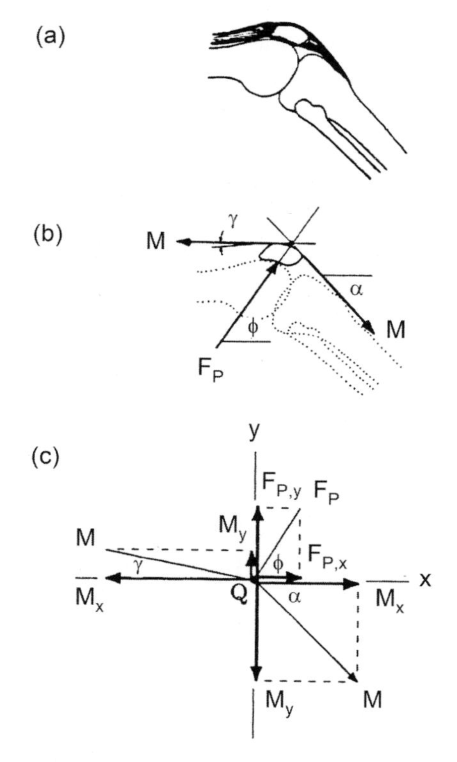

Fig. 2.26. Force diagram of the kneecap (patella) in equilibrium. (From [86])

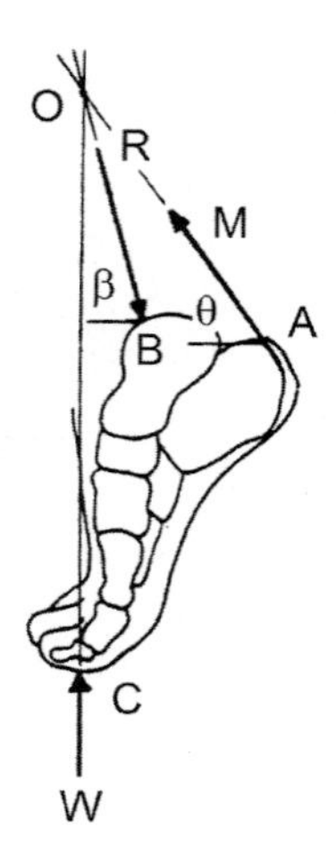

Fig. 2.27. Force diagram of the foot on tiptoe, showing that they form a concurrent system. (From [86])

and 2.28 show the forces when someone stands on tiptoes on one foot. The reaction force on the talus bone of the foot is in balance with the normal force from the floor (equal to the body weight) and the muscle force transmitted by the Achilles tendon on the calcaneus (heel). (The mass of the foot itself is neglected here.) The muscle and reaction forces are

$$M = W_{\mathrm{b}} \frac{\cos\beta}{\cos(\theta+\beta)}, \tag{2.39}$$

$$R = W_{\mathrm{b}} \frac{\cos\theta}{\cos(\theta+\beta)}. \tag{2.40}$$

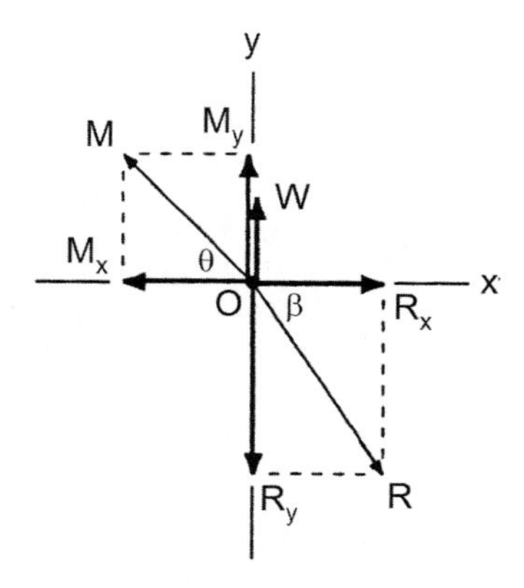

Fig. 2.28. Components of the forces acting on a foot on tiptoe. (From [86])

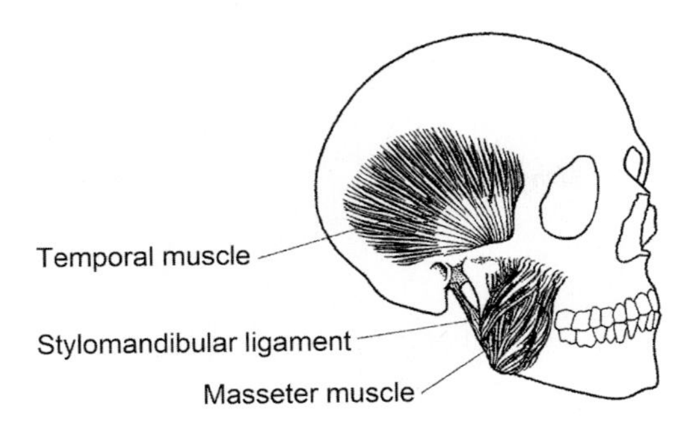

Fig. 2.29. The masseter and temporal muscles in the jaw, about the temporomandibular joint. (From [88])

These are both much greater than the body weight $W_{\rm b}$. For $\theta = 45°$ and $\beta = 60°$, we see that $M = 1.93W_{\rm b}$ and $R = 2.73W_{\rm b}$.

Jaws and Teeth

Forces on teeth arise from several sources [72]. Figure 2.29 shows how the masseter muscles provide the force in the lever system involved in chewing and biting. Lever models can be used to examine the quasistatics of chewing and biting. (See Problem 2.27.)

Orthodontics is the practical application of biomechanics to move teeth using forces applied by *appliances*, such as wires, brackets, and elastics [81, 87, 90]. Each tooth has a center of mass, but since teeth are not free bodies – they are restrained by the periodontium – a more useful position in the tooth is defined, the *center of resistance*. This is the balance point for the tooth. Figure 2.30 shows how forces and torques (moments) applied to the crown of a tooth, can be designed to create a lateral force at the center of resistance, but no torque about it. Appliances can affect several teeth, such as the intrusion

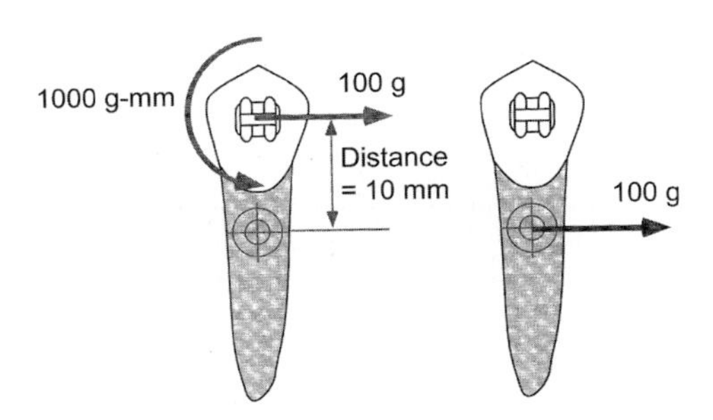

Fig. 2.30. In orthodontics forces and torques are applied to the crown (*left*), leading to forces (and in this case no torques) about the center of resistance. (Reprinted from [81]. Used with permission of Elsevier)

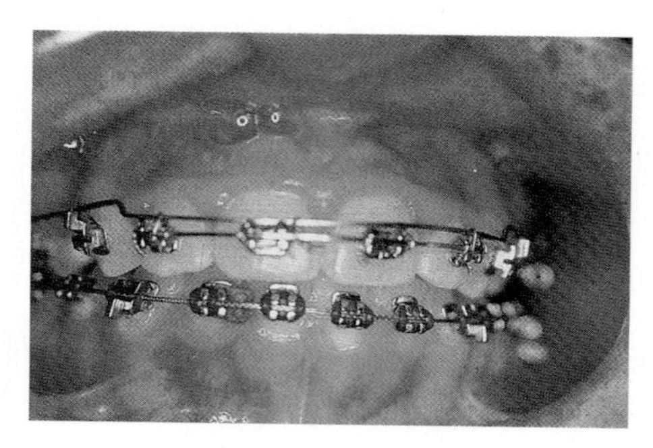

Fig. 2.31. Photo of an intrusion arch used in orthodontics. (Reprinted from [81]. Used with permission of Elsevier)

arch shown in Fig. 2.31, which leads to the application of forces and torques shown in Fig. 2.32.

2.3.4 Lower Back Problems

Most people eventually develop chronic lower back problems. The best way to avoid such a persistent and annoying problem is to try to avoid situations that might trigger your first back problem, such as bending over and lifting heavy objects, or sleeping on very hard beds. We will examine a simple statics model that will show why you should never bend and lift. (You should use the muscles in your legs to lift and not those in your back.) We will examine the force on the fifth lumbar vertebra when you bend and lift.

The spinal cord consists of 33 vertebra with 26 bones, and is classified into five sections. Starting from the top (superior) in Fig. 2.33, there are seven cervical, twelve thoracic – which rhymes with "Jurassic" – and five lumbar vertebra, and then five fused vertebra in the sacrum (sae'krum) and four fused vertebra in the coccyx (koak'-sis) (tail bone). Figure 2.34 shows two lumbar vertebra with the central regions of each, the centrum, separated by the intervertebral disc; more detail about the vertebrae is shown in Fig. 2.35. Note

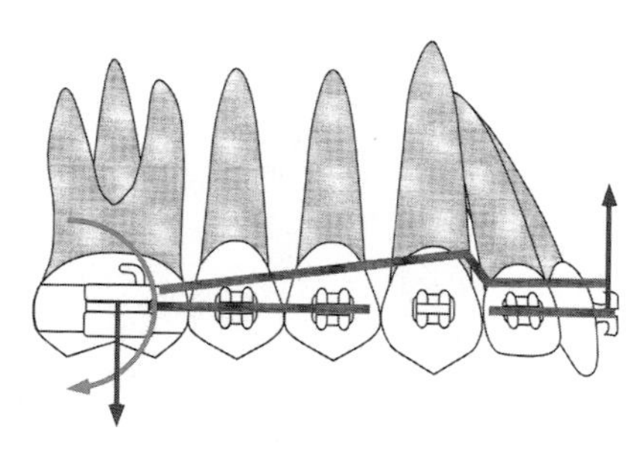

Fig. 2.32. Forces and torques applied by the intrusion arch in Fig. 2.31. (Reprinted from [81]. Used with permission of Elsevier)

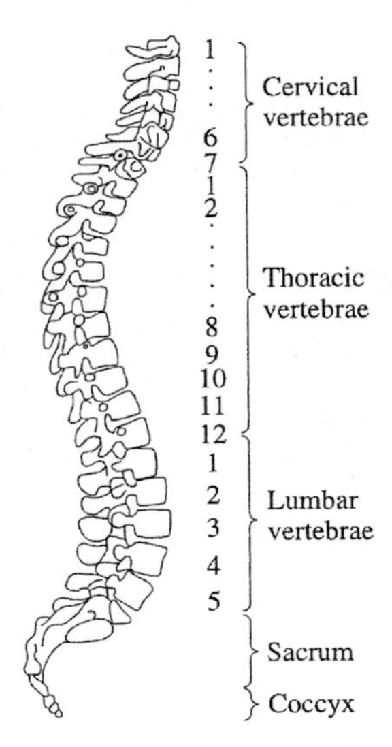

Fig. 2.33. The vertebral column (spine). (From [65].) The thoracic and sacral curves are primary curves, while the cervical and lumbar curves are secondary curves

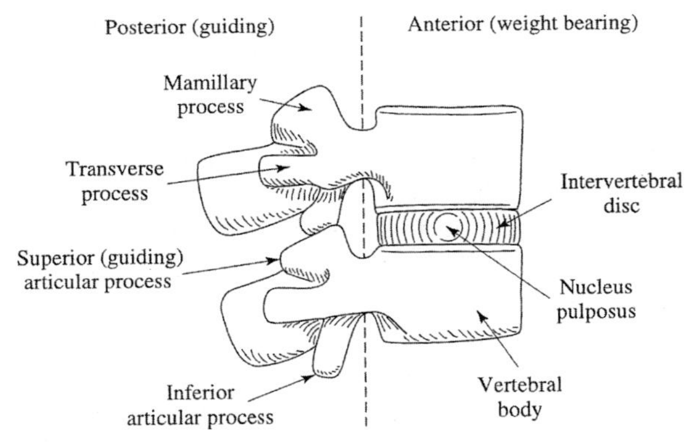

Fig. 2.34. Side view of two vertebrae separated by a vertebral disc. (From [65])

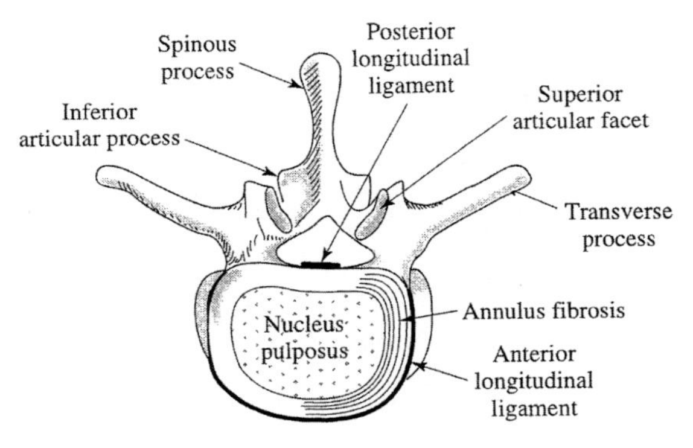

Fig. 2.35. Vertebra viewed from above. (From [65])

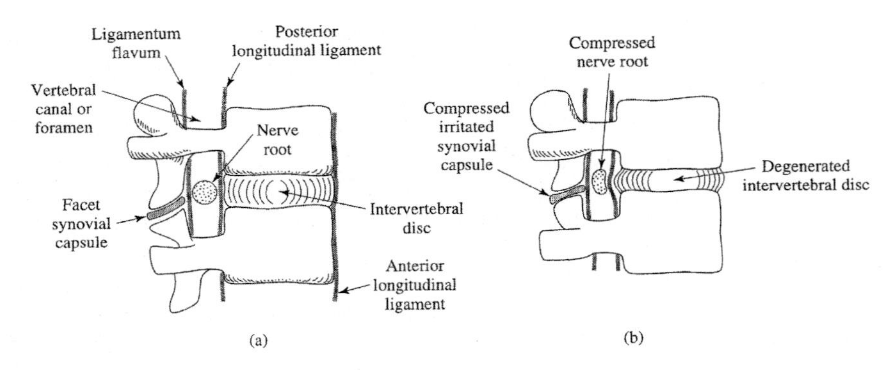

Fig. 2.36. Cross-section of two vertebrae and a vertebral disc with nerve for (**a**) a normal intervertebral disc and (**b**) one that has degenerated and is compressing the nerve root. (From [65])

the spinal nerve root (in Fig. 2.36) (pain region) and the spinous processes. The distinct vertebrae become successively larger down the spinal cord, because of the additional load they bear. This combination of vertebrae and intervertebral discs provides flexibility in the spinal cord, but also causes potential problems.

The spinal cord is not straight; each section is curved. At birth, only the thoracic and sacral curves are developed. These *primary curves* are in the same direction and lead to the "fetal position." At three months, the cervical curve develops, so the baby can hold his/her head up. When the baby learns to stand and walk, the lumbar curve develops. These *secondary curves* have curvature opposite to that of the primary curves (Fig. 2.33). Figure 2.37 shows the lumbosacral angle between the fifth lumbar vertebra and the sacrum.

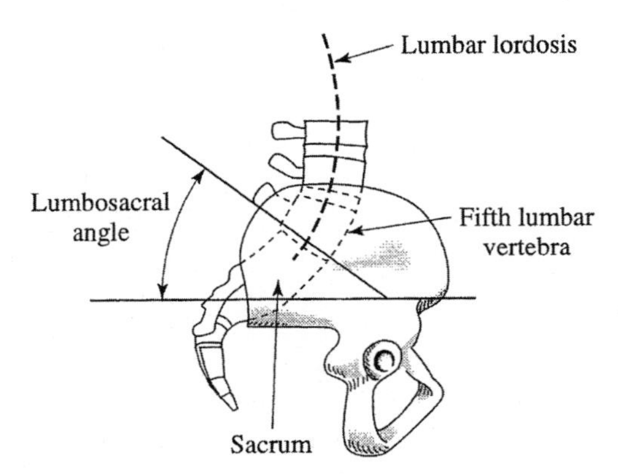

Fig. 2.37. The lumbosacral angle is defined as that between the horizontal and the top surface of the sacrum. (From [65])

Deviations in the angle from ~30° can lead to lower back pain. We will model the spinal cord as a rigid bar even though this description of the spinal cord curves would suggest a more complex model.

There is a series of massive muscles from the posterior parts of the iliac crest and sacrum to the skull (occipital bone), called the erector spinae (spy'nee) (or sacrospinalis (sae-kro-spy-na'-lis)) (Figs. 2.38 and 2.39). They will be modeled as a single muscle inserted 2/3 from the center of mass of the head and arms, at a 12° angle (Fig. 2.40).

Consider the spinal cord at an angle θ to the horizontal (Fig. 2.40); initially we will take $\theta = 30°$, corresponding to a deep bend. It is hinged at the lumbosacral disc, just below the fifth lumbar vertebra and the sacrum (Figs. 2.37 and 2.40). We will choose the axis for torque analysis right there. There is a reaction force **R** (of magnitude R) from the sacrum with components R_x and R_y. The weight of the trunk (above the hips, excluding arms and head) W_1 acts half way down the spinal cord. The weight of the arms, head, and any object lifted, W_2, act at the top of the spinal cord. These are shown in Fig. 2.40 along with the erector spinae force **M** (of magnitude M). Using Table 1.7, it is reasonable to approximate $W_1 = 0.4W_\mathrm{b}$ and $W_2 = 0.2W_\mathrm{b}$, with nothing being lifted.

Figure 2.40 shows that **M** acts at an angle that is $\theta - 12°$ relative to the horizontal; for our first example $\theta = 30°$ and so this angle is 18°. The force balances are

$$\sum F_x = R_x - M\cos 18° = 0, \tag{2.41}$$

$$\sum F_y = R_y - M\sin 18° - 0.4W_\mathrm{b} - 0.2W_\mathrm{b} = 0. \tag{2.42}$$

With this choice of rotation axis, the torque due to the reaction force is zero. The component of **M** normal to the spinal cord is $F' = M\sin 12°$, leading to a torque $(2L/3)(\sin 12°)(M)$. The fraction of each weight force normal to the spinal cord is $\theta = \cos 30°$. Torque balance requires

$$\sum \tau_z = \frac{2L}{3}\sin 12°(M) - \frac{L}{2}\cos 30°(0.4W_\mathrm{b}) - L\cos 30°(0.2W_\mathrm{b}) = 0. \tag{2.43}$$

The torque equation gives $M = 2.5W_\mathrm{b}$ or 2,200 N (500 lb) for the 880 N (200 lb) body weight of a 90 kg body mass. The reaction force parameters are $R_x = M\cos 18° = 2.38W_\mathrm{b}$, $R_y = 1.37W_\mathrm{b}$, $\phi = \arctan(R_y/R_x) = 30°$ (which is the angle the reaction force makes with the horizontal), and $R = 2.74W_\mathrm{b}$, which is 2,400 N (540 lb) here. Clearly, the muscle forces and reaction force on the lower spinal cord are much larger than the body weight. The moment arms of the weights are about the same as that of the muscle. However, the direction of weight forces lead to large torques at large bending angles, while the direction of the muscle force does not lead to a large torque to balance the weights – at any bending angle.

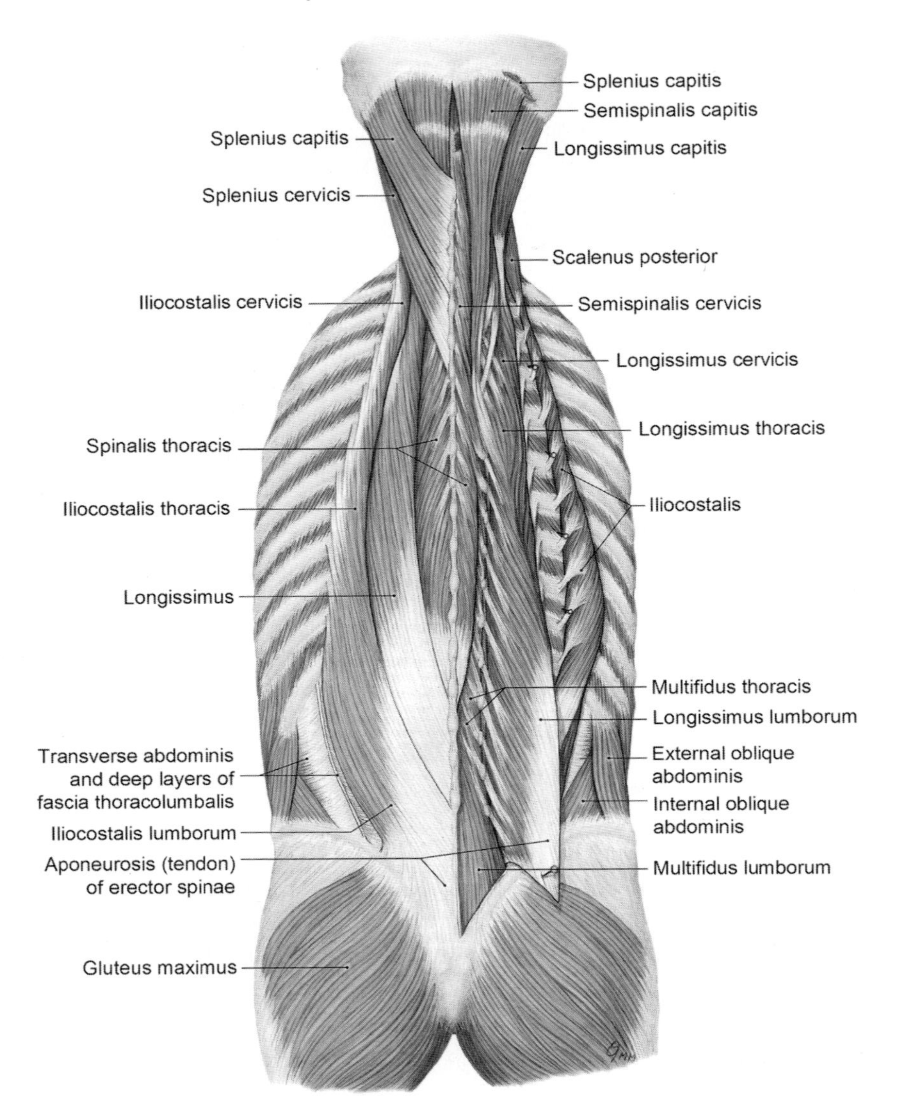

Fig. 2.38. Intermediate (*left*) and deep (*right*) layers of back muscles – showing the erector spinae muscles. The erector spinae consists of lateral columns (the iliocostalis lumborum, thoracis, and cervicis muscles), intermediate columns (the longissimus thoracis, cervicis, and capitis muscles), and a medial column (spinalis thoracis). (From [93])

Matters are even worse if you hold a weight in your arms; this simulates bending and lifting an object of this weight. All this does is to increase W_2. Let us increase W_2 by $0.2W_\mathrm{b}$ (180 N (40 lb) in our example) to $0.4W_\mathrm{b}$. Now

$$\sum F_x = R_x - M\cos 18^\circ = 0, \tag{2.44}$$

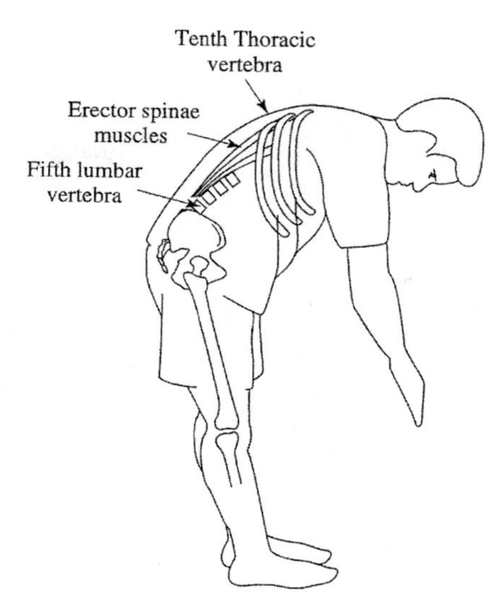

Fig. 2.39. Diagram of the erector spinae muscles used to control the trunk when bending. (From [65])

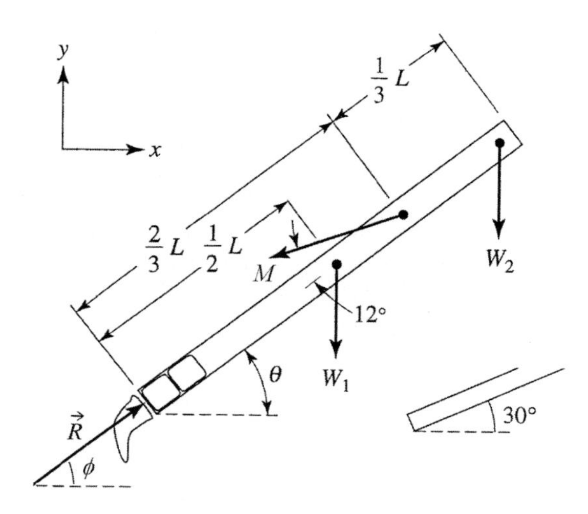

Fig. 2.40. Free-body diagram of the vertebral column while bending, with the spine modeled as a straight bar at an angle θ to the horizontal, which we will take to be $\sim 30^{\circ}$ – a bit steeper bend than is depicted here and shown in the inset. The angle of **R** to the horizontal is ϕ. With nothing being lifted, we will take $W_1 = 0.4W_{\rm b}$ and $W_2 = 0.2W_{\rm b}$. (From [65])

Table 2.3. Forces in the body during bending and lifting

θ	M (no lifting)	M (extra $0.2W_b$)	R (no lifting)	R (extra $0.2W_b$)
30°	2.50	3.74	2.74	4.07
60°	1.44	2.16	1.93	2.81
80°	0.50	0.75	1.08	1.53
90°	0	0	0.60	0.80

All forces are in units of the body weight. For a body mass of 90 kg, multiply each number by 880 N (200 lb)

$$\sum F_y = R_y - M \sin 18° - 0.4W_b - 0.4W_b = 0, \tag{2.45}$$

$$\sum \tau_z = \frac{2L}{3} \sin 12°(M) - \frac{L}{2} \cos 30°(0.4W_b) - L \cos 30°(0.4W_b) = 0. \tag{2.46}$$

The force the erector spinae muscles need to exert increases to $3.74W_b$ or 3,300 N (740 lb). The muscle must exert an additional 1,100 N (250 lb) to balance only an additional 180 N (40 lb). We see that $R_x = 3.56W_b$, $R_y = 1.96W_b$, and $R = 4.07W_b$. This is 3,600 N (810 lb), an additional 1,200 N (270 lb) of reaction force on the fifth lumbar vertebra.

Table 2.3 shows these forces for several bending angles. We can see why bending itself, and bending and lifting can lead to problems with the back muscles and the lower vertebra discs.

What does this mean for the lumbosacral (intervertebral) disc? Let us consider our initial example of bending by 60° from the upright position, to 30° from the horizontal, without any lifting. The reaction force on this disc is 2,400 N (540 lb). This force pushes down on the top and up on the bottom of this cylindrical disc. We are assuming a load that is normal to the axis, which is not exactly correct at these angles. Figure 2.41 shows how much the height of the disc decreases (fractionally) with this type of load. The weight corresponding to 250 kg is 2,400 N (540 lb), so by using this figure a 20% contraction is expected parallel to the spine. If the disc is a cylinder of radius r and height H, then its volume is $\pi r^2 H$. If the material in the disc is incompressible, its volume will not change with this compression and if H decreases by 20%, r will increase by about 10%. The bulging disc can press against the nerve, as seen in Fig. 2.36b. With lifting, the reaction force goes up more, the compression of the disc increases, and there is more bulging and more irritation of the nerves. (We have compared a 245 kg (2,400 N, 540 lb) load on the disc to a situation with no force on this disc. In the upright position, there is 0.6× the body mass (and weight), 54 kg (530 N, 120 lb) already supported by this disc due to body weight, so the load is really changing from 530 N (120 lb) to 2,400 N (540 lb). (How does this affect our numerical results and conclusions?) Also the loading is not uniform when you lift or bend, and there can be tension and compression and bulging due to this, as in Fig. 2.42.)

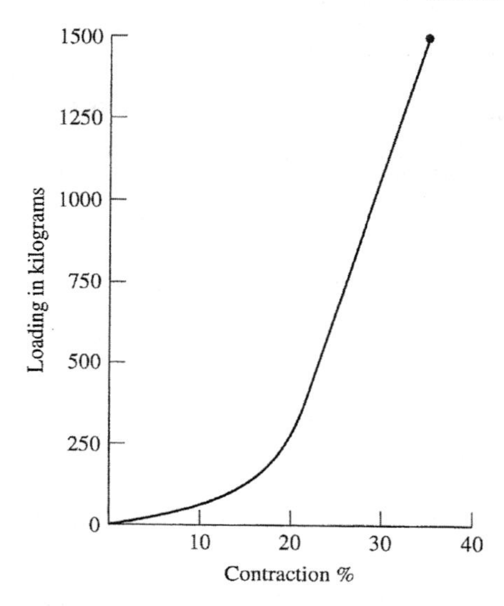

Fig. 2.41. Loading of wet lumbar vertebral discs of persons 40–59 years of age vs. percent compression. For the loading in N, the ordinate scale needs to be multiplied by 9.8. (From [65]. Based on [96])

This teaches us two things. (1) Understanding the mechanical properties of the parts of the body is essential to understand the implications of forces on the body. (2) You should never bend and lift. Problems 2.33–2.35 examine the torques during lifting of objects of different sizes, with different upper body positions (as above), and with different techniques.

These lower back forces depend not only on spine angle and load, but on knee bending during lifting and where the load is positioned relative to the body. During lying the lower back forces are 20–50% of those while standing upright relaxed (Table 2.4), and reach values over double the weight and over

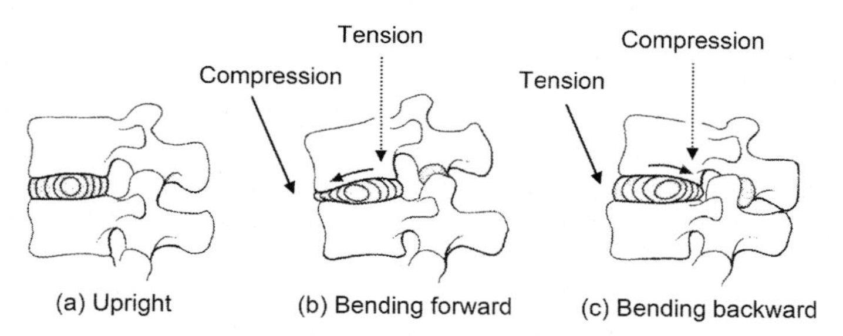

Fig. 2.42. Disc compression for a person who is (**a**) vertical, (**b**) bending forward, and (**c**) bending backward. Bending forward leads to disc compression anteriorly and tension posteriorly, bulging on the compressive side, and the shifting of the disc nucleus posteriorly. (From [88])

Table 2.4. Values of intradiscal pressure for different positions and exercise, relative to that during relaxed standing. (Using data from [80, 83, 92])

position or activity	%
still	
lying supine	20
side-lying	24
lying prone	22
lying prone, extended back, supporting elbows	50
relaxed standing	100
standing, bent forward	220
sitting relaxed, no back rest	92
sitting actively straightening back	110
sitting with maximum flexion (bent forward)	166
sitting bent forward, thigh supporting the elbows	86
sitting slouched in a chair	54
motion	
standing up from chair	220
walking barefoot or in tennis shoes	106–130
jogging with shoes	70–180
climbing stairs, one at a time	100–140
climbing stairs, two at a time	60–240
walking down stairs, one at a time	76–120
walking down stairs, two at a time	60–180
lifting	
lifting 20 kg, no bent knees	460
lifting 20 kg, bent knees, weight near body	340
holding 20 kg near body	220
holding 20 kg, 60 cm from chest	360

3× the torso weight during fast walking (Fig. 2.43). Table 2.4 shows that the pressure between vertebral discs for people sitting is minimized when they slouch, so when your parents tell you to stop slouching you can respond that you are trying to minimize intradisc pressure for long-term care of your spinal cord.

How can you maintain your back and relieve lower back pain if you have relatively minor lower back damage? Stretch your back muscles. Always bend your knees and use your leg muscles (and not your back) when you lift. Make sure that you maintain the curvature of your lower back at all times. You should sit on chairs with lower back support (see Fig. 2.44 and [91]), but when you must sit on flat back chairs, you may want to use a rolled-up towel or a pillow at the bottom of your back to help maintain the curvature. When you sleep, maintain one of two positions: on your back with your knees bent

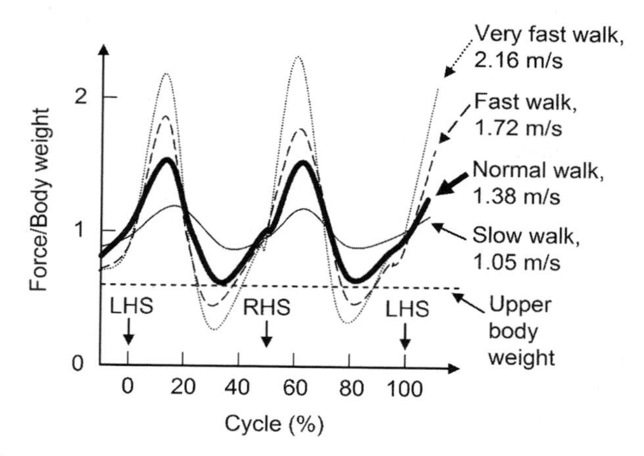

Fig. 2.43. Axial load on the disc between the L3 and L4 vertebra while walking at different speeds. LHS and RHS are left and right heel strike, respectively. (Based on [68, 83])

and feet pulled up (bent hips and knees) or the fetal position; never sleep on your front. (One reason for this is that in the supine position with straight legs, the vertebral portion of the psoas muscle puts a load on the lumbar spine, whereas this muscle relaxes and this load decreases when the hips and knees are bent and supported [83].) Only sleep on beds that allow your back

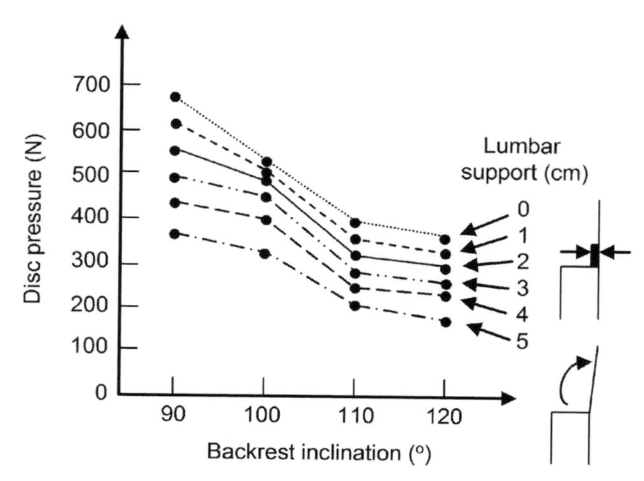

Fig. 2.44. The pressure on the third lumbar disc is decreased with backward backrest inclination and with lumbar support. Also, it is increased with support in the thoracic region, which is not shown here. Chairs with some backward inclination and lumbar support provide needed support, while those with an upright flat back or that curve toward the body can cause painful pressure. (Based on [64, 69])

Fig. 2.45. Hyperextension exercises recommended (at the Tientzin Hospital in China) to strengthen the back of lumbago patients (i.e., those with mild to severe pain or discomfort in the lower back); each is performed with the patient's back kept hollow. Several other exercises are also recommended for those with lower back pain. If you have lower back pain, please consult your physician before attempting any of these exercises. (From [70])

to maintain its natural curvature; beds should not be too firm or too soft. Some hyperextension exercises that can strengthen muscles to help your back are shown in Fig. 2.45. (Guess, who has a lower back problem?)

2.3.5 Three-Force Rule

Consider the foot–lower leg combination shown in Fig. 2.46a, which is a model for a person walking upstairs. There are three forces acting on this isolated

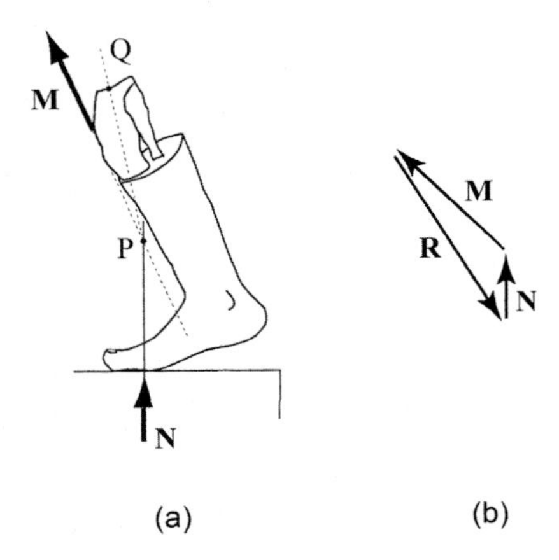

Fig. 2.46. (a) Illustrating the three-force rule on the free lower leg, with reaction force **N**, patellar tendon force **M**, point where these extended vector lines meet P, and the point where the joint force acts on the tibia Q. (b) Vector diagram of three-force rule with **R** being the joint reaction force at Q. (From [76])

system. The normal force **N** acts vertically at the front of the foot and for slow stair climbing has a magnitude equal to the body weight W_b. This normal force tries to rotate the lower leg clockwise about the knee. The patellar tendon transmits the force of the thigh quadriceps muscle **M**, as shown, and tries to counter this rotation. The direction of this force is along this tendon and the direction of this tendon is known. There is also a reaction force **R** exerted by the femoral condyles on the top of the tibia, acting as shown at point Q. In this two-dimensional problem, there are three equations (two for force balance and one for torque balance) and three unknowns: the magnitudes of **M** and **R**, and the direction of **R**. As such, this problem can be solved in the straightforward way illustrated earlier. However, there is a graphical procedure called the three-force rule that offers some additional insight and, of course, the same answers as the straightforward method. (For simplicity, we are ignoring the mass of the leg and foot here.)

Extending the **N** and **M** vectors, we see that they intersect at point P. If we choose the torque rotation axis perpendicular to this point, we see that these two forces contribute nothing to the torque since their **r** and **F** vectors are, respectively, antiparallel and parallel to each other. Since in static equilibrium (or the quasistatic situation here) the net torque is zero, the reaction force vector **R** must also pass through P when it is extended, and so **R** is parallel to the QP line segment. Because the directions of **M** and **R** are now known, we can place them at the head and tail of the **N**, respectively, as shown in Fig. 2.46b (always with the head of one vector to the tail of the other, as in adding vectors). The lengths of both vectors are now determined by lengthening them until they hit each other. Then the sum of the force in the x and y directions is zero (because of the closed triangle), and we have obtained all of the needed information.

By the way, the reaction forces at the knee are $\sim$3–4W_b during walking [82] and are much greater during stair climbing.

2.3.6 Multisegment Modeling

Several parts of the body are often important in modeling the body, both in static situations and for those involving motion. As an example, Fig. 2.47 shows how an anatomical model of leg can be modeled by three segments, describing the upper leg, lower leg, and the foot. Each segment is labeled by its mass (m), center of mass (dot), and moment of inertia (I). In this example there are also three joints (freely rotating hinges or pins), between the hip and upper leg, the upper leg and lower leg (the knee), and the lower leg and foot (the ankle), each denoted by an open circle. The forces by muscles and normal forces at the joints and other places (such as the floor) can be added to this link segment model, and the forces on torques on the entire object can be analyzed or those on each segment can be analyzed by itself. (This is

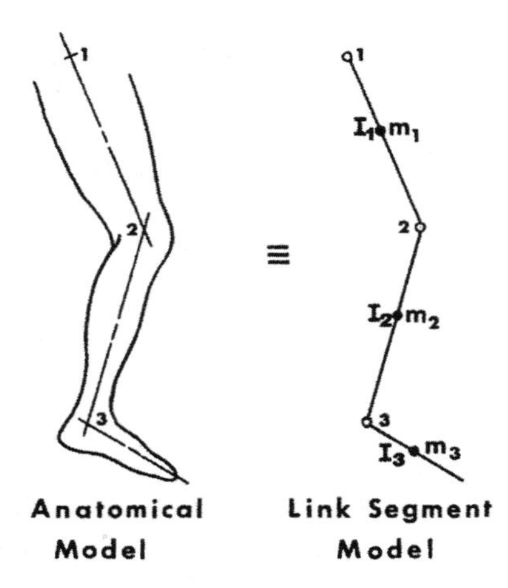

Fig. 2.47. Relationship between an anatomical model of the leg and a link segment model of the upper leg, lower leg, and foot. (From [95]. Reprinted with permission of Wiley)

somewhat similar to examining the whole body and then the leg by itself in the above analysis of the hip.)

In analyzing each segment in Fig. 2.47 by itself, you arrive at a free body diagram for each segment, as in Fig. 2.48. In this example, the reaction forces

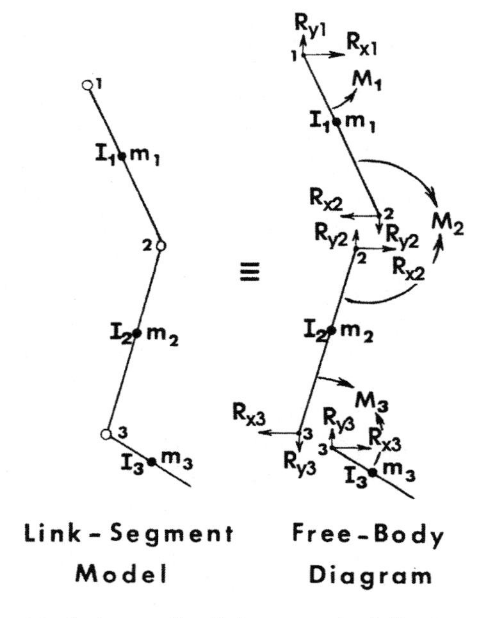

Fig. 2.48. Relationship between the link segment of the leg with a free-body diagram, with individual upper leg, lower leg, and foot. (From [95]. Reprinted with permission of Wiley)

at every body joint are shown (**R**), and at each particular joint the forces are equal in magnitude and opposite in direction due to Newton's Third Law, such as $\mathbf{R_2}$ shown for the knee. The forces due to each muscle (or effective set of muscles) could be added, acting with equal magnitudes but opposite directions at the origin and point of insertion (again Newton's Third Law). (This is examined in Problem 2.37.) In the free-body diagram shown in Fig. 2.48, the total torque (which is called moment M in the figure) about each joint is shown.

2.4 The Sense of Touch

The body also "feels" force. This is the sense of touch. The *somatic senses*, include the sense of touch by the skin (cutaneous sensations), the sense of position of the limbs (proprioception), and the sense of movement of the limbs (kinesthesis). There are sensors in the skin for tactile perception, perception of temperature, and the perception of pain [73].

There are four types of tactile receptors in glabrous (hair-free) skin (Fig. 2.49), each sensitive to stresses and displacements at different frequencies. Merkel receptors (or disks), located near the border between the epidermis (outermost skin layer) and dermis (the underlying layer), are most sensitive to pressure disturbances which vary in the 0.3–3 Hz range. The other receptors are deeper within the dermis. Meissner corpuscles sense light

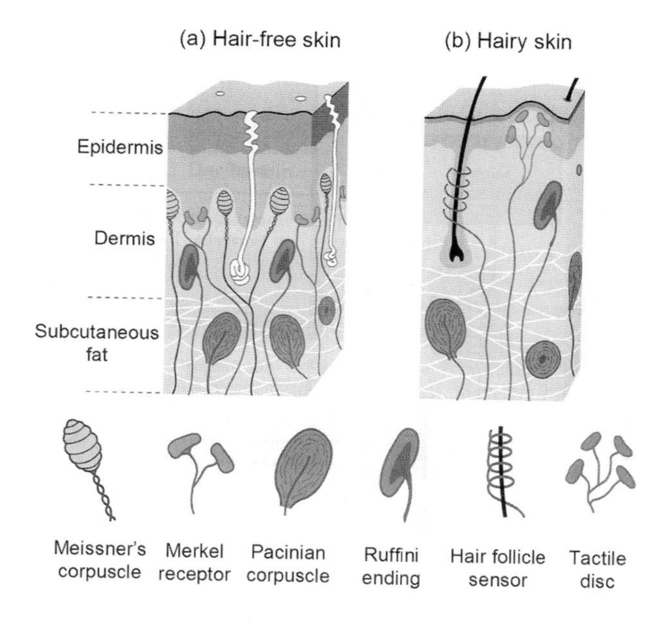

Fig. 2.49. Cross-section of (**a**) hair-free and (**b**) hairy skin, showing tactile sensors. (From [89])

tapping, which is characterized by 3–40 Hz variations, and sensed as flutter. Ruffini cylinders (or endings) sense faster vibrations, 15–400 Hz, such as those caused by stretching of the skin or joint movement, and sensed as buzzing. Pacinian corpuscles sense a range of rapid vibrations, from 10 to 500 Hz. Motion of hair also stimulates these receptors.

The spatial density of mechanical receptors on the skin varies from $25/\text{cm}^2$ at the tip of the tongue to $0.02/\text{cm}^2$ on the back [66]. The movement of hair on hairy skin contributes to the sense of touch, along with sensing by these four receptors. The mechanism for converting these impulses into electrical signals is discussed in Chap. 12.

The skin also has two types of receptors that sense temperature. Warm fiber thermoreceptors increase their firing rate only when it is warmer and cold fibers only when it is colder; they are not sensitive to mechanical stimulation. Warm fibers respond from 30 to 48°C, and best at about 44°C, while cold fibers respond from 20 to 45°C, and best at 30°C. There are on the order of 5–10 thermal receptors per cm^2 over most of the skin. This sense of temperature is important in the body's drive for temperature control (Chap. 13).

Nociceptors in the skin sense excessive pressure, extreme temperature, and corrosive chemicals, each which can damage the skin. Some pain receptors transmit signals very rapidly to the brain, on myelinated axons in neurons with conduction speeds up to 30 m/s (see Chap. 12). Some pain receptors indicate persistent pain and have signals that are transmitted on very slow unmyelinated axons with speeds of 2 m/s and slower.

As characterized by Stevens' Law ((1.6), Table 1.15), the variation of the perceived stimulus is slightly sublinear with stimulus strength for vibration, somewhat superlinear for sensing pressure on the palm and the heaviness of lifted objects, linear for coldness, and superlinear for warmth.

A problem in Chap. 7 (Problem 7.14, Fig. 7.24) addresses how to use the sense of touch to estimate the internal pressure in an elastic vessel. This is an example of *palpation*, which is the use of a physician's hands to examine parts of the body during a medical examination. (A physician also taps the body in the diagnostic known as *percussion*. A third physical diagnostic is listening to sounds emanating from the body, which is called *auscultation*; it is discussed in Chap. 10.)

2.5 Diversion into the Units of Force and Pressure

2.5.1 Force

We should be clear on the correct units of force, as in Table 2.5. In the English (or FPS) system it is pounds (lb). In the metric MKS-SI system it is newtons (N) and in the CGS systems it is dynes. Most technical work is in MKS-SI; however, much medical work is done using the English system (in the US). We

Table 2.5. Units of force

$1\,\mathrm{N} = 1\,\mathrm{kg\text{-}m/s^2}$
$1\,\mathrm{N} = 10^5$ dyne
$1\,\mathrm{N} = 0.225\,\mathrm{lb}$ ($\sim 2/9\,\mathrm{lb}$)
$1\,\mathrm{lb} = 4.45\,\mathrm{N}$

will usually use lb or N, with lb used when referring to body weights. (Also see Appendix A.)

One kilogram (kg) is a mass (m) that at sea level (on the planet Earth) has a weight of 9.8 N (or 2.2 lb). This is a consequence of gravitational acceleration on earth and Newton's Second Law of motion

$$F = mg, \tag{2.47}$$

where $g = 9.8\,\mathrm{m/s^2}$. "g" is the acceleration of any freely falling object due to gravity. In CGS units, a mass of 1 g has a weight of 980.7 dynes because $g = 980.7\,\mathrm{cm/s^2}$. In the English system $g = 32.2\,\mathrm{ft/s^2}$. The unit of mass in the English system is the slug; it is rarely used in the US. One slug has a weight of 32.2 lb. (Sometimes in the biomechanics literature the loading mass is given, and it is possible for the unit "kg" to be mistakenly referred to as a force. What is really meant is the weight of an object with a mass of 1 kg. To lessen confusion, this should be expressed as 1 kg (force). Even better, when 1 kg (force) is seen it should be replaced by 9.8 N ($\sim$10 N).)

2.5.2 Pressure

We will encounter the concept of pressure later. Pressure (P) is simply a force per unit area. (More generally, a force per unit area is a stress (Table 2.6). Pressure or more precisely hydrostatic pressure is a stress that is the same in all directions.) The units of pressure (or stress σ) are the units of force divided by the units of area. Because work (or energy) is force $\times$ distance, the units of pressure are the same as those of energy/volume. (Also see Appendix A.)

The standard MKS-SI unit of pressure is a pascal (Pa), with $1\,\mathrm{Pa} = 1\,\mathrm{N/m^2}$. We will find that often the numbers we will encounter are simpler in units of $\mathrm{N/mm^2}$, which are the same as the units MPa. Common units in English units are pounds per square inch (psi).

These units are independent of the planet we happen to be on. Some common units of pressure are specific to Earth (at sea level, at 0°C temperature). A column of a liquid of mass density ρ and height h exerts a pressure

$$P = \rho g h. \tag{2.48}$$

This is the same as (2.47), after both sides have been divided by the column area A, because $P = F/A$ and $m = \rho \times \text{volume} = \rho A h$. At sea level on

Table 2.6. Units of pressure

$1\,\mathrm{N/m^2} = 1\,\mathrm{Pa} = 9.87 \times 10^{-6}\,\mathrm{atm.} = 0.0075\,\mathrm{mmHg} = 0.102\,\mathrm{mmH_2O}$
$1\,\mathrm{N/mm^2} = 10^6\,\mathrm{N/m^2} = 10^6\,\mathrm{Pa} = 1\,\mathrm{MPa} = 145\,\mathrm{psi} = 9.87\,\mathrm{atm.} \simeq 10\,\mathrm{atm.}$
$1\,\mathrm{psi} = 0.0069\,\mathrm{N/mm^2} = 6{,}894.8\,\mathrm{N/m^2}$ (or Pa) $= 1/14.7\,\mathrm{atm.} = 0.068\,\mathrm{atm.}$
$1\,\mathrm{bar} = 10^5\,\mathrm{N/m^2}$ (or Pa) $= 0.1\,\mathrm{N/mm^2} \simeq 1\,\mathrm{atm.}$
$1\,\mathrm{atm.} = 1.013 \times 10^5\,\mathrm{Pa} = 1.013\,\mathrm{bar} = 0.103\,\mathrm{N/mm^2}$ (or MPa) $= 14.7\,\mathrm{psi}$
$= 760\,\mathrm{mmHg} = 29.9\,\mathrm{inchHg} = 1{,}033\,\mathrm{cmH_2O} = 407\,\mathrm{inchH_2O}$
$1\,\mathrm{mmHg} = 0.00132\,\mathrm{atm.} = 133\,\mathrm{N/m^2} = 13.6\,\mathrm{mmH_2O}$
$1\,\mathrm{mmH_2O} = 0.1\,\mathrm{cmH_2O} = 9.68 \times 10^{-5}\,\mathrm{atm.} = 9.81\,\mathrm{N/m^2} = 0.0735\,\mathrm{mmHg}$
$1\,\mathrm{kg\ (force)/cm^2} = 9.8\,\mathrm{N/cm^2}$

We will often use MPa ($= \mathrm{N/mm^2}$), but occasionally use other units, such as mmHg when discussing blood pressure

Earth (0°C) the air pressure is 1 atmosphere (1 atm.), which is the pressure exerted by the air column above it. The same pressure is exerted by 760 mm of Hg (mercury) or 1,033 cm of water. The units of mmHg and $\mathrm{cmH_2O}$ (or inches of H_2O) are very commonly used even though they really refer to h in (2.48) and not pressure. (In a calculation, they need to be multiplied by $\rho_{\mathrm{Hg}} = 13.6\,\mathrm{g/cm^3}$ or $\rho_{\mathrm{H_2O}} = 1.0\,\mathrm{g/cm^3}$ and then by g. Note that $\rho_{\mathrm{blood}} = 1.0\,\mathrm{g/cm^3}$.) In the US, air pressure is commonly expressed in weather reports in units of inches of Hg, as in "the air pressure is 29.8 in and dropping." Blood pressure is commonly reported in mmHg, as in 120/80, which means that the systolic and diastolic pressures are, respectively, 120 mmHg and 80 mmHg. Air pressures in the body, such as in the lungs, are sometimes expressed in terms of cm or inches of water, because it is a smaller and much more convenient unit. Pressure is often referenced to atmospheric pressure, so a blood pressure of 120 mmHg really means an absolute pressure that is 120 mmHg above atmospheric pressure (with 1 atm. = 760 mmHg), and as such is called a gauge pressure.

It is an amazing coincidence that the Earth-based unit of atmospheric pressure (1 atm.) is within 1% of the Earth-independent MKS-SI-based unit of a "bar" ($1\,\mathrm{bar} = 10^5\,\mathrm{N/m^2}$ (or Pa)), with 1 atm. = 1.013 bar.

2.6 Summary

Force and torque balance can often be analyzed in terms of levers, such as for the lower arm. The equilibrium of a part of the body can be analyzed by examining the forces due to gravity, muscles, and reaction forces in the whole body and that body part separately. Muscle and internal reaction forces can exceed the body weight, as was seen in analyzing the forces in the hip and lower back, and this can have serious consequences.

Problems

Statics of the Arm and Levers

2.1. A person holds a weight with her arm extended horizontally, using her deltoid muscles to balance the weight about the shoulder joint. What type of lever is this? Sketch and label this lever.

2.2. A person holds himself up in a pushup position. Consider one of the person's arms, with the pivot point being the hand on the ground. Are the biceps brachii or triceps brachii involved? What type of lever is this? Sketch and label this lever.

2.3. For each of the following, identify the type of lever, and show in a sketch the locations of the applied force, the fulcrum, and the load being applied:
(a) cutting with a pair of scissors
(b) lifting a wheelbarrow
(c) picking up something with a pair of tweezers.

2.4. Analyze the force balance in the x and y directions and the torque balance in the z-direction for the lower arm, with a vertical upper arm and a horizontal lower arm (length 35 cm) (Fig. 2.9). As with Case 1 ignore the forearm weight and other muscles. Now the effect of the reaction force $\mathbf{N}$ on the joint (with components N_x and N_y) is explicitly included (because it contributes no torque it did not have to be included in the earlier Case 1 analysis) and assume the biceps are attached 4 cm from the pivot at an angle θ to the lower arm. With $\theta = 75^\circ$, find the muscle force M and the magnitude of the reaction force in terms of the weight W and the angle of this reaction force relative to the x (horizontal) axis.

Statics of the Hip and Leg

2.5. Using Fig. 1.15, what can you say about the height of the person whose leg is depicted in Fig. 2.17?

2.6. Show how the dimensions given in Fig. 2.16 change to those in Fig. 2.20 when the person holds a cane.

2.7. In examining the hip forces during the equilibrium of a man standing on one foot, we assumed that the effective angle of the hip abductor muscles to the x-axis was the same, 70°, without or with a cane. Would using more realistic (and different) values of this angle without or with a cane, significantly affect the conclusions concerning the effect of using a cane?

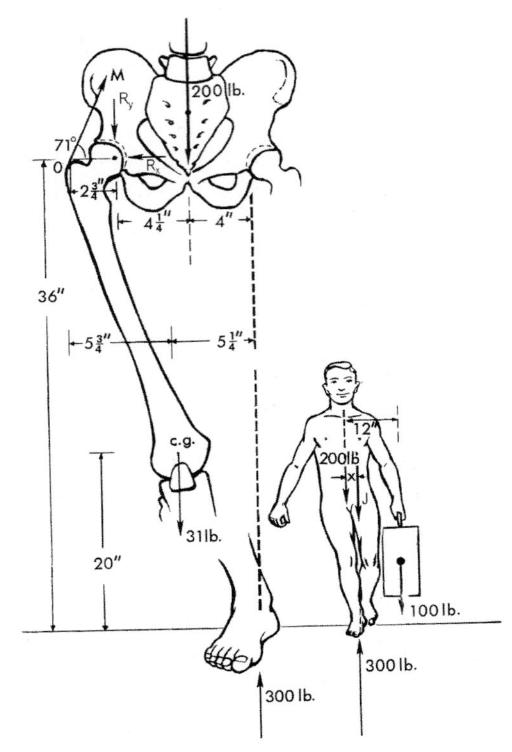

Fig. 2.50. Forces on hip and femoral head while standing on one leg and lifting a weight with the opposite hand. (Reprinted from [75]. Used with permission of Elsevier.) For Problem 2.8

2.8. A 200 lb man stands on his right foot while carrying a 100 lb bag in his left hand. The center of mass of the bag is 12 in from his center of mass (see Fig. 2.50).
(a) Show that the placement of the foot (as shown) leads to no net torque in the body.
(b) Find the force (its magnitude and direction) on the head of the support femur and the force in the hip abductor muscle by examining the right leg.
(c) Compare your answers in parts (a) and (b) with what was found for the man holding no mass – without and with a cane (for $W_b = 880\,\mathrm{N} = 200\,\mathrm{lb}$). Are the forces here greater than for a (200 lb + 100 lb =) 300 lb man (with no cane). Why? (The muscle angle and leg mass are only trivially different for the problem given here and those analyzed above.)

2.9. Redraw Fig. 2.50, changing all distances and forces into metric units and then do Problem 2.8.

2.10. Calculate the force on the hip abductor muscles for a person standing symmetrically on two feet, as a function of foot separation. For what position is this force zero?

2.11. In the arabesque position an initially upright gymnast kicks one leg backward and upward while keeping it straight (mass m_{leg}), pushes her torso and head forward (mass $m_{\mathrm{torso+head}}$), and propels her arms backward and upward while keeping them straight (each m_{arm}). They, respectively, make acute angles θ_{leg}, $\theta_{\mathrm{torso+head}}$, and θ_{arms}, to the horizontal. The centers of mass of the extended leg and vertical balancing leg are $x_{\mathrm{extended\ leg}}$ and $x_{\mathrm{balancing\ leg}}$ behind the vertical from the center of mass (in the midsagittal plane), and that of her upper body (torso/head/arms combination) is $x_{\mathrm{upper\ body}}$ in front of this vertical (so all of these distances are defined as positive). In achieving this arabesque position her center of mass drops vertically from height y_{before} to y_{after}.
(a) Draw a diagram showing the gymnast before and during this maneuver.
(b) Find the equilibrium condition in terms of these masses and distances. (Hint: Analyze the torques about her center of mass in the arabesque position. You can ignore the contribution from her arms. Why? You may not need all of the information that is presented.)
(c) Assume the gymnast is 1.49 m (4 ft 11 in) tall and has a mass of 38 kg (weight 84 lb) and her free leg and arms make a 30° angle with the horizontal. What angle does her upper body make with the horizontal? (Assume the anthropometric relations for a standard human.)

2.12. Redo Problem 2.11c if the free leg of the gymnast is horizontal and her arms are vertical. (This is the cheerleading position.)

2.13. (a) Calculate the torque of the diver of mass m_{b} about the axis through her toes (normal to a sagittal plane) when she is on a diving board and leaning over and about to dive, so the vertical axis through her center of mass is a distance x in front of her toes.
(b) Now say that the diver has a height H and that her body is proportioned as per the data given in Chap. 1. Calculate the torque in terms of H, with her body straight with arms stretched parallel to her torso. Assume that her body can be straight (and so everything in her body can be approximated as being in one plane) and ignore the change in position because she is on her toes.
(c) Redo this if her arms are instead along her sides. How does this torque differ from that in part (b)?

2.14. The split Russel traction device is used to stabilize the leg, as depicted in Fig. 2.51, along with the relevant force diagram [86]. The leg is stabilized by two weights, W_1 and W_2, attached to the leg by two cables. The leg and cast have a combined weight of $W_1 = 300\,\mathrm{N}$ and a center of mass 2/3 of the way from the left, as shown. The cable for W_2 makes an angle $\beta = 45°$ with

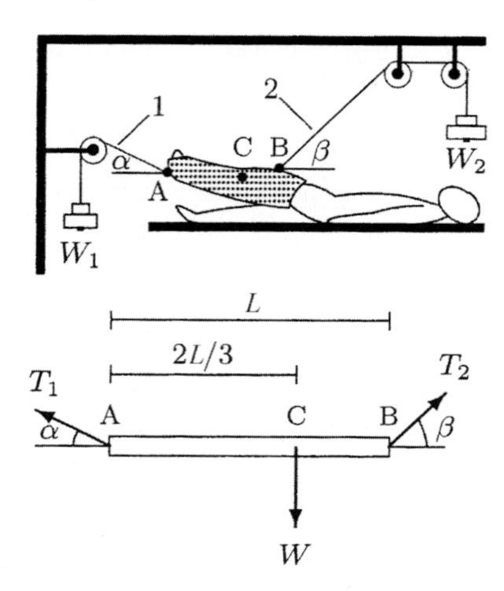

Fig. 2.51. The split Russel traction device. (From [86].) For Problem 2.14

the horizontal. For equilibrium, find the tension in the cables T_1 and T_2 and angle the cable for W_1 makes with the horizontal, α.

Statics of the Shoulder, Knee, Ankle, and Jaw

2.15. You are able to hold your arm in an outstretched position because of the deltoid muscle (Fig. 2.52). The force diagram for this is shown in Fig. 2.53. Use the three equilibrium conditions to determine the tension T in the deltoid muscle needed to achieve this equilibrium, and the vertical and horizontal components of the force exerted on the scapula (shoulder blade) on the humerus. Assume the weight of the humerus is $mg = 8\,\text{lb}$ and the deltoid muscle make an angle of $\alpha = 17°$ to the humerus. (From [65].)

2.16. Solve the more general shoulder problem with a weight in the hand, depicted in Fig. 2.21, by finding M, R, and β. Now evaluate the x and y components of the muscle force, the magnitude of the joint reaction force, and its angle for the following parameters: $a = 15\,\text{cm}$, $b = 30\,\text{cm}$, $c = 60\,\text{cm}$, $\theta = 15°$, $W = 40\,\text{N}$, and $W_0 = 60\,\text{N}$.

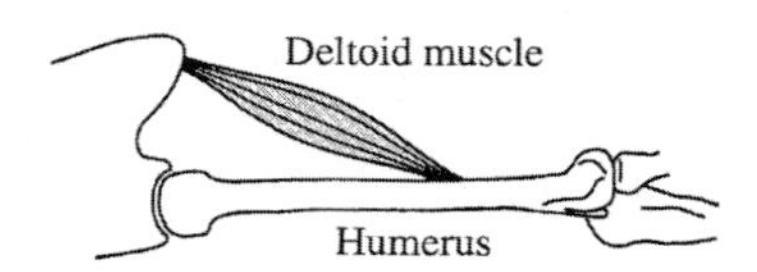

Fig. 2.52. Deltoid muscle during lifting with an outstretched arm. (From [65].) For Problem 2.15

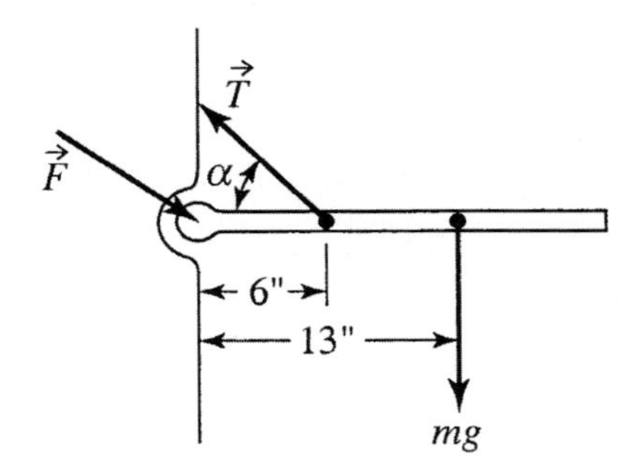

Fig. 2.53. Force diagram for the deltoid muscle and reaction forces during lifting with an outstretched arm. (This is a simpler version of Fig. 2.21.) (From [65].) For Problem 2.15

2.17. (a) A gymnast of mass m_b suspends himself on the rings with his body upright and straight arms that are horizontal with which he clutches the rings. Each ring is suspended by a rope with tension T that makes an acute angle θ with his arms, and the rings are separated by a distance d. Solve for the T and θ. Assume symmetry.
(b) If the gymnast weighs 600 N, $\theta = 75°$, and $d = 1.8$ m, find T. (From [74].)

2.18. Derive (2.34)–(2.36) for the equilibrium of the lower leg with an ankle weight.

2.19. Determine the angle between the leg and the reaction force at the knee for the conditions given in the text for Fig. 2.23.

2.20. Analyze how the patellar tendon and reaction forces depend on the ankle weight (for a fixed leg weight) and leg angle for Fig. 2.23. During exercise, what are the advantages of varying this weight vs. this angle?

2.21. Consider the equilibrium of the foot during crouching for a 200-lb person, with the force through the Achilles tendon, the reaction force of the tibia, and the normal force from the floor in balance, as in Fig. 2.54 – neglecting the weight of the foot for simplicity. Take the angle $\alpha = 38°$.
(a) Why is the normal force from the floor 100 lb?
(b) Find the magnitude of the Achilles tendon tension T and the magnitude and direction of the reaction force $\mathbf{F}$.

2.22. The topic of Problem 2.21 is similar to the discussion of forces on the foot in Chap. 3, where we will assume that all forces are parallel or antiparallel to each other and normal to a bar, as in a lever. Is this totally valid? Why or why not?

2.23. Redo Problem 2.21, now including the mass of the foot. If the distance from the bottom of the tibia (where the normal force emanates) is 4 in and the center of mass of the foot is halfway between it and the ground. Use the data for the mass of the foot in Chap. 1.

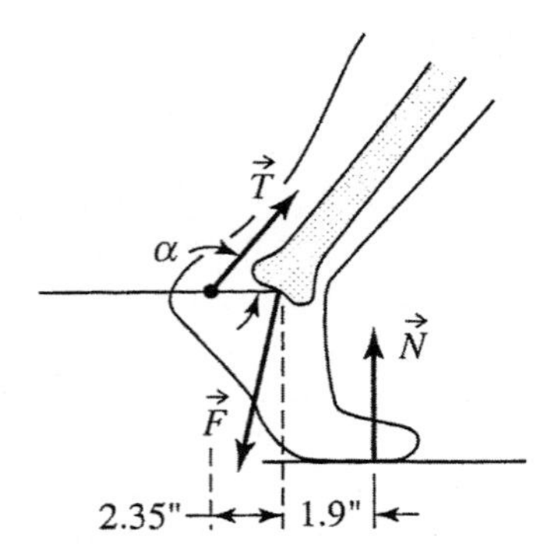

Fig. 2.54. Forces on the foot during crouching. (From [65].) For Problem 2.21

2.24. In the crouching position, the lower leg is held in equilibrium through the action of the patellar ligament, which is attached to the upper tibia and runs over the kneecap. As depicted in Fig. 2.55, the forces acting on the lower leg are **N**, **R**, and **T**. If the lower leg is in equilibrium, determine the magnitude of the tension **T** in the patellar ligament, and the direction and magnitude of **R**. Assume that the tension acts at a point directly below the point of action of **R**. Take the normal force equal to 100 lb (half the body weight), the weight of the leg W_{leg} as 20 lb, and the angle $\alpha = 40°$ (for the leg at a 45° angle). (From [65].)

2.25. Derive (2.37) and (2.38) for the equilibrium of the kneecap.

2.26. Derive (2.39) and (2.40) for the equilibrium of the foot.

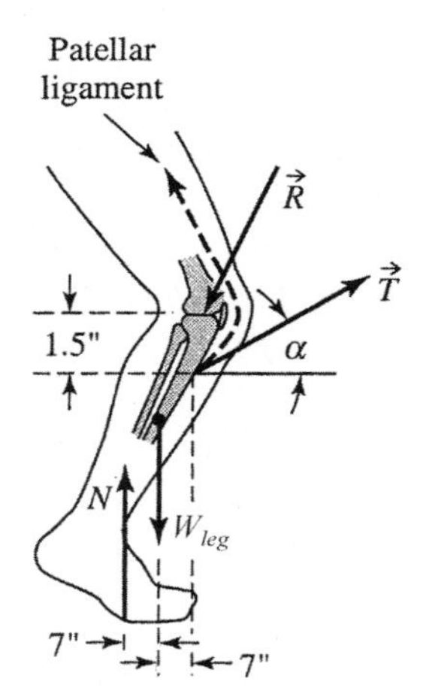

Fig. 2.55. Forces on the lower leg during crouching. (From [65].) For Problem 2.24

2.27. Laterally from the midline in either jaw, we have eight teeth: two incisors, one canine, two bicuspids (or premolars), and three molars (the last molar being a wisdom tooth). (This tooth order can be remembered from its acronym: ICBM – which itself is easy to remember because it is also the acronym for Intercontinental Ballistic Missiles.) The lateral distances from the temporomandibular joint to the insertion of the masseter muscles, the first bicuspids, and the central incisors, are $0.4L$, L, and $1.2L$, respectively.
(a) What type of lever is involved in biting and chewing with the masseter muscles?
(b) For biting in equilibrium with a masseter muscle force of 1,625 N, show that the force on the first bicuspid of 650 N, assuming there is no force on the central incisors [67]. Draw a force diagram for this.
(c) Under these conditions, show that the needed counter force on the central incisors is 540 N, now assuming no force on the first bicuspid. Draw a force diagram for this.

2.28. When you bite an apple with your incisors only, you exert a force of 650 N on it. When you bite an apple with your bicuspids only, you exert a force of 540 N on it. Find the force per unit area (which is called the stress) on the apple for both cases if the effective contact areas of the incisors and bicuspids are $5\,\text{mm}^2$ and $1\,\text{mm}^2$, respectively.

Statics of the Back

2.29. In analyzing bending, we assumed that the weight of the trunk (above the hips, excluding arms and head) is $W_1 = 0.4W_\text{b}$ and the weight of the arms and head is $W_2 = 0.2W_\text{b}$. Is this reasonable. Why?

2.30. Consider a woman of height 1.6 m and mass 50 kg.
(a) Calculate the reaction force on her lower vertebrae and the force in her erector spinae muscle when she is either upright or bent at 60° (and consequently 30° to the horizontal).
(b) Recalculate these forces when she is pregnant. Assume that during pregnancy the mass of her torso increases by 15 kg, but the center of mass of the torso is the same.
(c) The forces in part (b) are equivalent to those for the same nonpregnant woman who lifts a weight of what mass?

2.31. Describe the designs of the back of a chair that could lead to pain in the lumbar vertebrae and those that would give good lumbar support.

2.32. We showed that when the force on the lumbosacral (intervertebral) disc increases from 0 to 2,400 N, the disc height H decreases by 20% and the disc radius r increases by about 10%. When it is recognized that the load on the disc for a vertical person is really 530 N (and not 0), how do the disc dimensions really change when the person bends to an angle of 30° (and then to a load of 550 lb)?

2.33. *Why is it more difficult to lift bulky objects?* A person lifts a package of mass 20 kg in front of her so the back of the package touches her abdomen. The horizontal distance from the person's lumbar-sacral disc to the front of her abdomen in 20 cm. Calculate the bending moments (in N-m) about the center of mass of her disc caused by the lifted loads, assuming the package is alternatively 20 or 40 cm deep [83]. Draw force diagrams for these two cases. The other dimensions of the packages are the same and they both have uniform density. How does this show that the size of the lifted object affects the load on the lumbar spine?

2.34. *Why is it better to standard erect when you hold an object?* A person holds a 20 kg object while either standing erect or bending over. The mass of the person above his lumbar-sacral disc (his torso) is 45 kg. When upright, the center of mass of the torso is (horizontally) 2 cm in front of his disc and that of the object is 30 cm in front of his disc. When bent, the center of mass of the torso is 25 cm in front of his disc and that of the object is 40 cm in front of his disc. Draw force diagrams for these two cases. Calculate the bending moments (in N-m) about the center of mass of his disc caused by holding this load while either being upright or bent over [83]. How does this show that bending when lifting an object affects the load on the lumbar spine?

2.35. *Why is it best to lift an object with bent legs and the object very close to you?* A person lifts a 20 kg object while either bending over with legs straight, with bent knees and the object near to her body, or with bent knees and the object far from her body. The mass of the person above her lumbar-sacral disc (her torso) is 45 kg. When bent over with straight legs, the center of mass of her torso is (horizontally) 25 cm in front of her disc and that of the object is 40 cm in front of her disc. When bent over with bent knees and the object near her body, the center of mass of her torso is (horizontally) 18 cm in front of her disc and that of the object is 35 cm in front of her disc. When bent over with bent knees and the object far from her body, the center of mass of her torso is (horizontally) 25 cm in front of her disc and that of the object is 50 cm in front of her disc. Draw force diagrams for these three cases. Calculate the bending moments (in N-m) about the center of mass of her disc for these three lifting methods [83]. Which position is the worst? How does this show that the position when lifting an object affects the load on the lumbar spine?

2.36. One position during shoveling snow or soil is shown in Fig. 2.56. Assume this is an equilibrium position.
(a) If the shovel and contents have a mass of 10 kg, with a center of mass 1 m from the lumbar vertebra, find the moment about that vertebra.
(b) If the back muscles are 5 cm behind the center of the disc, find the magnitude and direction of the muscle force needed for equilibrium.

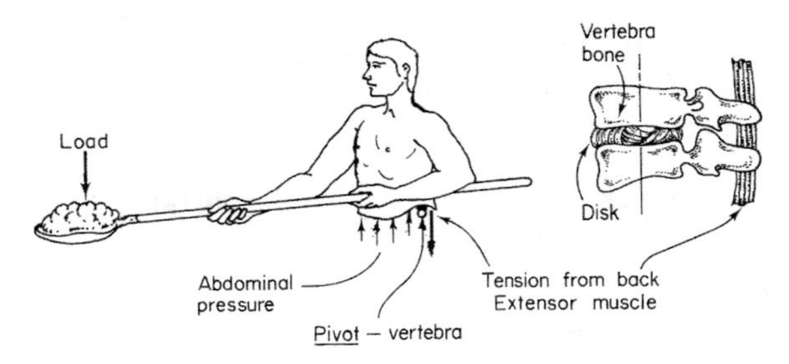

Fig. 2.56. Shoveling. (From [71].) For Problem 2.36

(c) Find the force on the intervertebral disc.
(d) If the abdominal muscles of the person are strong and can provide some upward force, would that help relieve stress to the back muscles and the disc? Why?

Multisegment Modeling

2.37. (a) Sketch an (in-plane) multisegment model of the leg showing the forces on the upper leg, lower leg, and ankle – using the resultant forces in Fig. 2.57. Show the center of mass gravity forces on each segment, along with the normal forces (at each body joint and with the floor).
(b) Label all distances and angles needed to analyze the in-plane forces and torques. For each segment, label the distances starting from the proximal end. Label each angle between muscle and bone; use the angles as shown, with the acute angle when possible.
(c) Write the equilibrium force balance and torque equation for each of the segments.

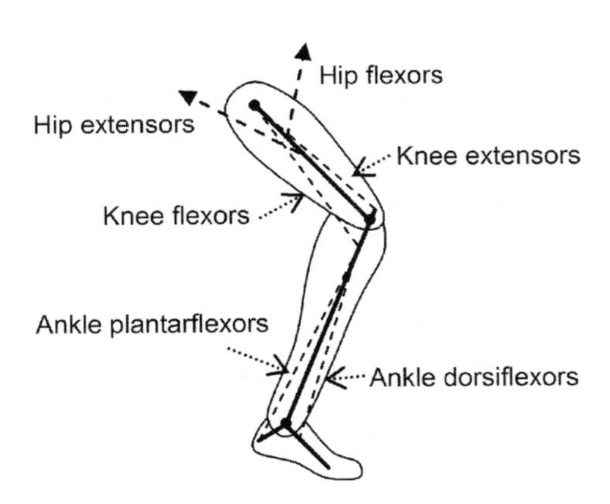

Fig. 2.57. Idealized resultant forces for six muscle groups in the leg. (Based on [77].) For Problem 2.37

(d) For each of these equilibrium conditions, sum the equations – such as the torque equations – for the upper leg, lower leg, and foot. Show that the three resulting balance equations are the correct equations for the entire leg.

Sense of Touch

2.38. (a) A 50 kg person stands on her fingertips. Assuming each finger makes a $1\,\mathrm{cm}^2$ contact area with the ground, find the pressure on each finger tip. (b) Which tactile sensors are sensing this pressure?

3

Motion

We now discuss human motion. We start with the absence of motion, standing, and then progress to walking and running. Another type of motion is jumping, and we will examine the vertical jump, high jump, long jump, and pole vault. We will develop a model of throwing a ball and then examine collisions of the human body with other objects and collisions caused by the body. In each case we will ask: *What does the body do? What can the body do? How does this translate into performance?* In many instances we will focus on the body itself, but in some we will need to consider the system of the body plus other objects such as running shoes, the floor surface, a ball, or a baseball bat. We will delay our discussion of swimming until the discussion of fluids in Chap. 7.

In this chapter we will follow (1) the motion itself – the kinematics of motion, (2) the forces causing this motion and the ensuing motion – the dynamics or kinetics, (3) the muscles that provide these forces, and (4) the energetics of the motion. Both linear and rotational motion can be important. The discussion of pendulums in the section on walking includes a brief review of the harmonic oscillator and rotational kinematics and kinetics. In Chap. 5 we will address how muscles work.

Human locomotion is a very important area in biomechanics. Many of the motions we will describe are analyzed in the wonderful books by Alexander [99] and Hay [127].

3.1 Kinematics and Musculature

Following the motion in walking, running, cycling, and jumping means tracking the angles the thigh makes with the hip, the lower leg with the thigh, and the foot with the lower leg, as shown in Fig. 3.1 [111]. (Some conventions for defining these angles are somewhat different than in this figure.) The thigh angle should be that with respect to the trunk, because the trunk usually

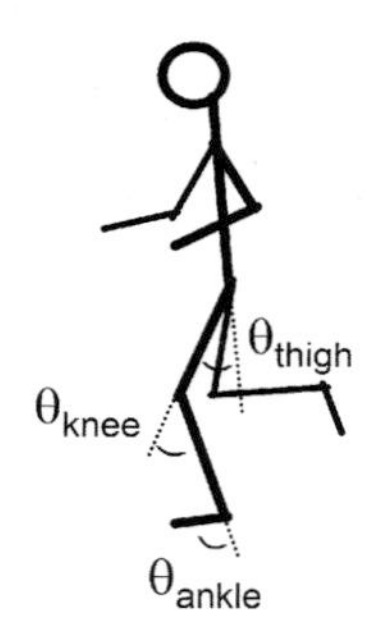

Fig. 3.1. One set of conventions for hip (thigh), knee (lower leg), and ankle (foot) angles. Other conventions are also used, such as with the thigh angle being relative to the vertical and the ankle angle defined as 90° less than here. (Based on [150])

tilts forward during running; this differs from the angle from the vertical (which is sometimes given instead). These three angles correspond to rotation in the sagittal plane of Fig. 3.1 about the hip (coxofemoral), knee, and ankle (talocrural) joints. Increasing θ_{thigh} is abduction, decreasing it is adduction (as in Figs. 1.9 and 1.10). Increasing θ_{knee} is flexion, decreasing it is extension. Increasing θ_{ankle} is dorsiflexion (flexion), decreasing it is plantarflexion (extension).

The hip has three degrees of freedom of rotation, as does a ball-and-socket joint (Fig. 1.4). This sagittal plane flexion/extension motion about the mediolateral axis of the hip in Fig. 3.1 can be larger in amplitude than the other two rotations (adduction–abduction about an anteroposterior axis and internal–external rotation about the longitudinal axis of the femur) (Table 1.10). While these two rotations are not negligible, we can ignore them here. The knee sagittal plane flexion/extension motion about the mediolateral axis is the one angular degree of freedom of the knee described in Fig. 1.4 as a "hinge." The knee joint actually has three degrees of freedom, but the other two (internal–external rotation and adduction–abduction (varus-valgus)) are much less important because of soft tissue and bony constraints. Ankle motion of interest here is restricted to extension (plantarflexion) and flexion (dorsiflexion). These three major sagittal plane rotations of hip, knee, and ankle constitute a three-segment model of the leg, in contrast to the one segment models of levers examined in Chap. 2. The head, arms, and trunk also move relative to the rest of the body during the walking and running cycles.

Kinematics involves tracking these angles (here only in the sagittal plane) vs. time and vs. each other during a walking or running cycle. These three angles are changed by forces controlled by the different sets of muscles outlined in Tables 3.1–3.3. Several of them are depicted in Figs. 1.8 and 3.2–3.4. (See [170] for more details.)

Table 3.1. Muscles that move the thigh (upper leg) (θ_{thigh}, femur relative to hip). (See [153] and [170] for more details)

location, muscle	action	approximate origin/insertion
pelvic girdle		
psoas major[a]	flexes thigh	ilium/upper femur
iliacus[a]	flexes thigh	ilium/upper femur
anterior thigh		
pectineus	flexes, adducts thigh	pubis/upper femur
adductor longus	adducts, rotates, flexes thigh	pubis/upper femur
adductor brevis	adducts thigh	pubis/upper femur
adductor magnus	adducts, rotates thigh	pubis, ischium/upper femur
gracilis	adducts thigh, flexes lower leg	pubis/upper tibia
lateral thigh		
tensor fasciae latae	aids thigh flex, abduction, rotation	ilium/upper tibia
posterior thigh		
gluteus maximus	extends, rotates thigh; stabilizes knee	hip/top femur
gluteus medius	abducts, rotates thigh	ilium/GT femur
gluteus minimus	abducts, rotates thigh	ilium/GT femur
piriformis	rotates, abducts flexed thigh	sacrum, ilium/GT femur
obturator internus	rotates, abducts flexed thigh	obdurator foramen/GT femur
obturator externus	rotates thigh	obdurator foramen/upper femur
quadratus femoris	rotates thigh	ischium/GT femur

GT is the greater trochanter of the femur
[a]The iliopsoas consists of the psoas major, ilicacus, and psoas minor.

3.2 Standing

3.2.1 Stability

Stability is essential during standing, as well as during any type of motion. We will examine the overall stability of the body and then local stability.

First consider the body as a rigid mass. The criterion for *overall stability* during standing is for the center of mass to be over the area spanned by the feet (Figs. 3.5 and 3.6). Otherwise there would be torques that are not balanced, and a "rigid" human would be unstable and topple over. Let us examine the torques about a rotation axis emanating from the center of mass. We see in Fig. 3.5 that when the center of mass is above this area spanned by the feet, the right foot causes a negative torque and the left foot causes a positive torque, and they cancel. (The torque due to the weight of the center

Table 3.2. Muscles that move the lower leg ($\theta_{\rm knee}$, lower leg relative to thigh)

location, muscle	action	approximate origin/insertion
anterior		
rectus femoris[a]	extends leg, flexes thigh	hip/patellar tendon
vastus lateralis[a]	extends leg	upper femur/patellar tendon
vastus medialis[a]	extends leg	upper femur/patellar tendon
vastus intermedius[a]	extends leg	upper femur/patellar tendon
sartorius	flexes leg; flexes, rotates thigh	ilium/upper tibia
gracilis (Table 3.1)		
posterior		
biceps femoris[b]	flexes leg, extends thigh	ischium, upper femur/upper fibula, tibia
semitendinosus[b]	flexes leg, extends thigh	ischium/upper tibia
semimembranosus[b]	flexes leg, extends thigh	ischium/upper tibia

[a]The quadriceps femoris consists of the rectus femoris, vastus lateralis, vastus medialis, and vastus intermedius.
[b]The hamstrings consist of the biceps femoris, semitendinosus, and semimembranosus. (See [153] and [170] for more details.)

of mass about this axis is zero.) When the center of mass is to the left of the foot area both torques are negative, and when it is to the right both torques are positive. The torques cannot balance in either case, and there is instability. Try it! (Stand and try to lean over.) (The term center of gravity is also often used in stability analysis; technically it refers to the center of mass in the direction of gravity.) Figure 3.7 suggests different stability conditions for an upright person with mass fairly well balanced in front and back to one with unbalanced mass, such as one with a "beer belly" or a woman who is pregnant.

Such overall stability presumes a rigid body and consequently rigid joints. Such stable joints indicate *local stability.* The design of the human knee is an example of good human design. Four ligaments provide much of the needed lateral and cross support.

The knee is the junction of the femur in the upper leg, and the tibia and fibula bones in the lower leg (Fig. 1.3, right knee). The *tibial (or medial) collateral ligament* connects the femur and tibia and the *fibular (or lateral) collateral ligament* connects the femur and the fibula. These collateral ligaments prevent left/right sliding and provide overall leg tautness. They are slightly posterior, so they are taut during extension, and can produce the straight leg needed during standing.

These two ligaments are not sufficient for stability. There are two cruciate ligaments that cross each other, and prevent twisting and forward/backward sliding of the knee. They cross in between the *lateral and medial menisci,*

Table 3.3. Muscles that move the foot (θ_{ankle}, foot relative to lower leg) and toe. (See [153] and [170] for more details)

location, muscle	action	approximate origin/insertion
anterior		
tibialis anterior	dorsiflexes, inverts foot	upper tibia/ first metatarsal
extensor hallucis longus	extends great toe, dorsiflexes foot	mid fibula/great toe
extensor digitorum longus	extends toes, dorsiflexes foot	tibia, fibula/ small toe phalanges
peroneus tertius	dorsiflexes, everts foot	lower fibula/metatarsal V
peroneus longus	plantar flexes, everts foot	fibula/metatarsal I
peroneus brevus	plantar flexes, everts foot	fibula/metatarsal V
posterior		
gastrocnemius[a]	plantar flexes foot, flexes leg	lower femur /calcaneal tendon
soleus[a]	plantar flexes foot	fibula, tibia/ calcaneal tendon
plantaris	plantar flexion of foot/ flexion of leg	femur/calcaneal tendon
popliteus	rotates leg	lower femur/upper tibia
tibialis posterior	inverts foot, plantar flexion of foot	tibia, fibula/foot
flexor hallucis longus	flexes great toe, plantar flexes foot[b]	fibula/ phalanx of great toe
flexor digitorum longus	flexes toes, plantar flexes foot	tibia/ small toe distal phalanges

The calcaneal (Achilles) tendon attaches to the calcaneus (heel).
[a]The triceps surae consists of the lateral and medial heads of the gastrocnemius and the soleus.
[b]Active in takeoff and tip-toeing.

which are two curved half sockets of cartilage on the tibia. The *posterior cruciate ligament* connects the medial side of the femur to a more lateral part of the tibia. The *anterior cruciate ligament* connects a lateral part of the femur to a more medial part of the tibia. The anterior cruciate ligament is so-named because it attaches to the tibia at a position anterior to that of the posterior cruciate ligament (which is attached more posteriorly).

Local stability of the knee is also provided by the oblique ligaments, as well as others. The kneecap (patella), attached below to the tibia by the patellar ligament and above to the tendon of the quadriceps femoris muscle, also provides stability. Other functions of the kneecap are to lessen the compressive

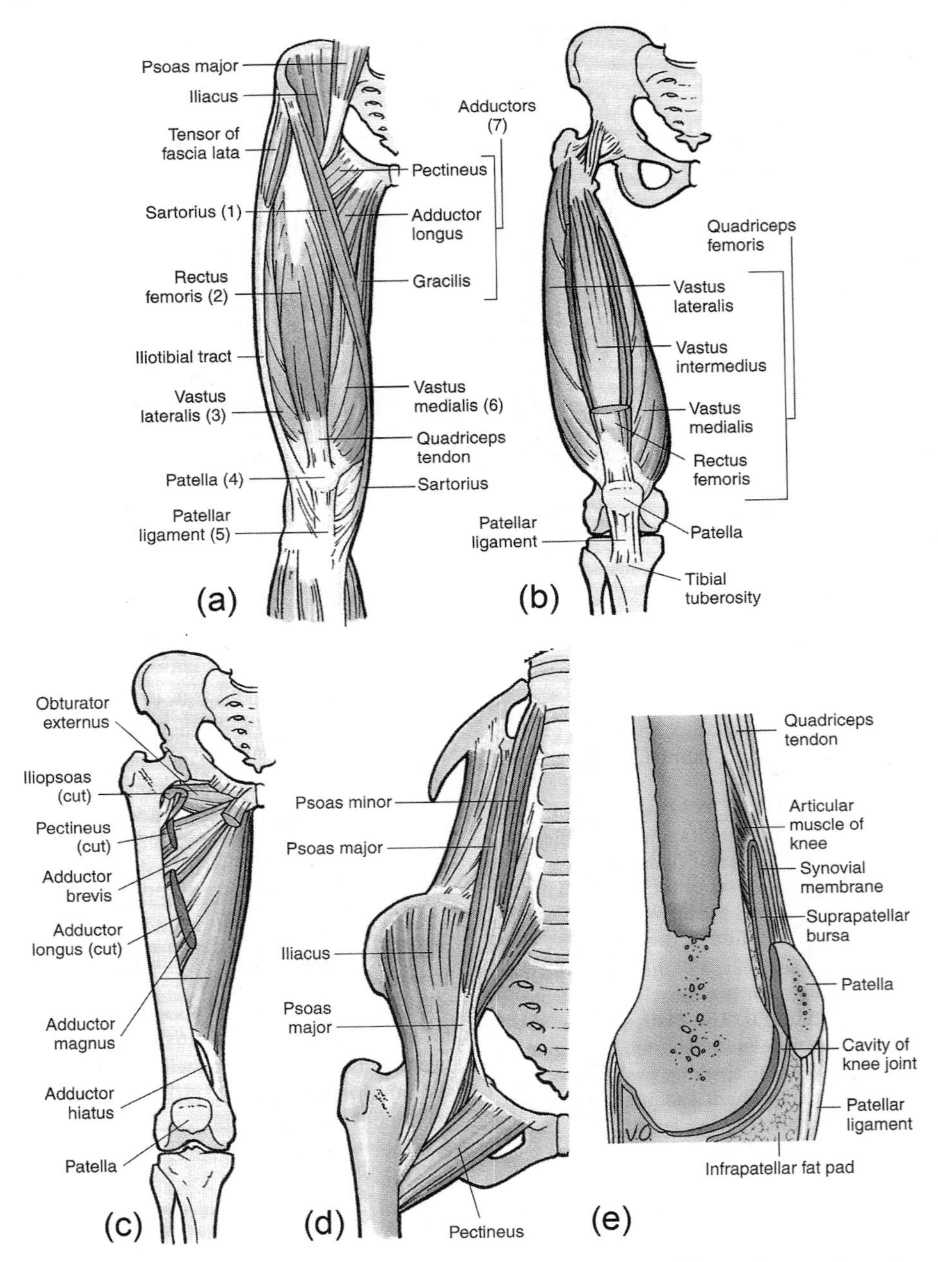

Fig. 3.2. (**a**) Anterior and medial muscles of the thigh, with (**b**) quadriceps femoris, with much of the rectus femoris removed, (**c**) deeper muscles of the medial thigh muscles, (**d**) the iliopsoas (psoas major and iliacus) and pectineus muscles, and (**e**) articular muscles of the knee. (From [152]. Used with permission)

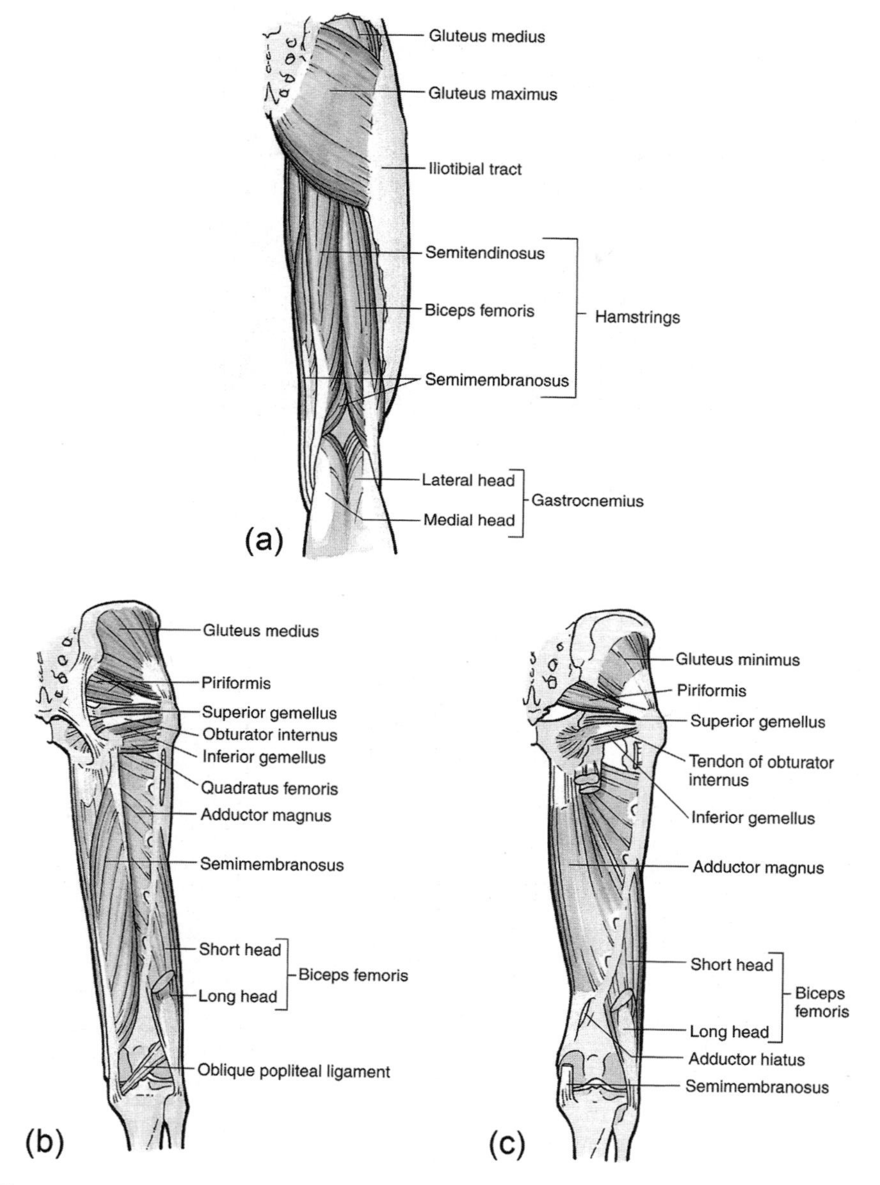

Fig. 3.3. Posterior thigh and gluteal region muscles, showing (**a**) superficial muscles, (**b**) deeper muscles, and (**c**) even deeper muscles. (From [152]. Used with permission)

stress on the femur by increasing the area of contact between the patellar tendon and the femur and to aid knee extension by lengthening the lever arm of the quadriceps muscle via an anterior displacement of the quadriceps tendon (Fig. 2.25).

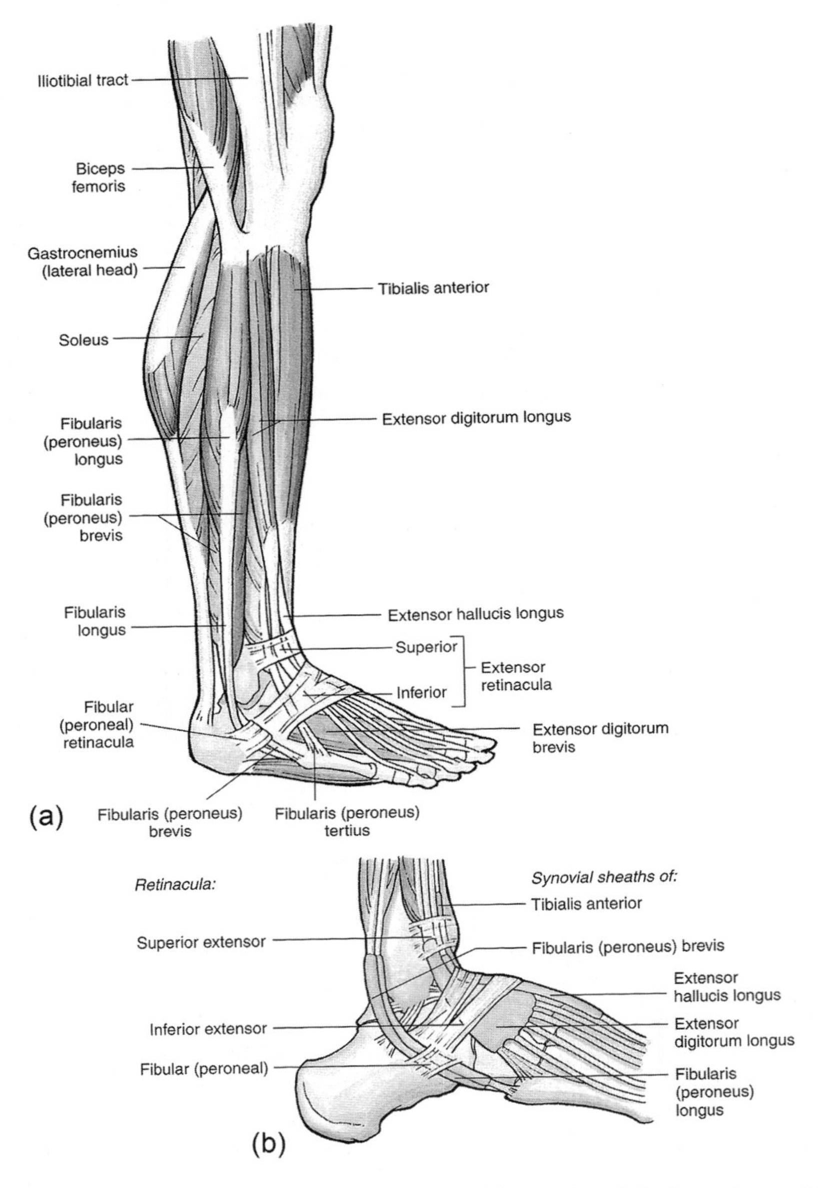

Fig. 3.4. Lateral views of the right leg, showing (**a**) muscles of the lower leg and foot and (**b**) the tendons extending into the foot. (From [152]. Used with permission)

3.2.2 Forces on the Feet

Consider a person weighing 700 N (160 lb). During standing, each foot of the person must support 350 N. The total cross-sectional area of the two feet is about 350 cm^2, so the average force/area or pressure on the feet is about

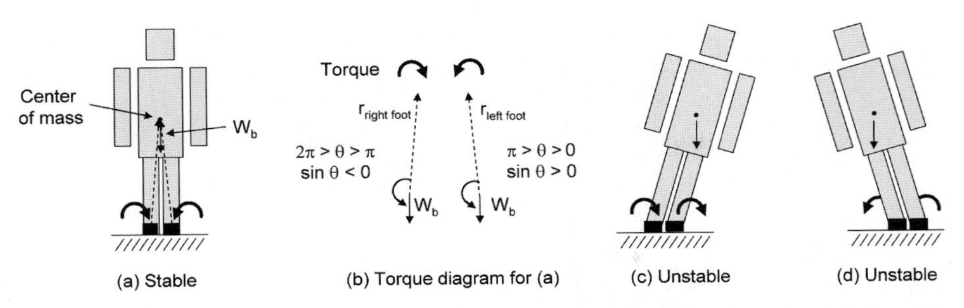

Fig. 3.5. Stability during standing for a model person (**a**) standing upright and with (**b**) torque diagrams, (**c**) leaning to her left, and (**d**) leaning to her right

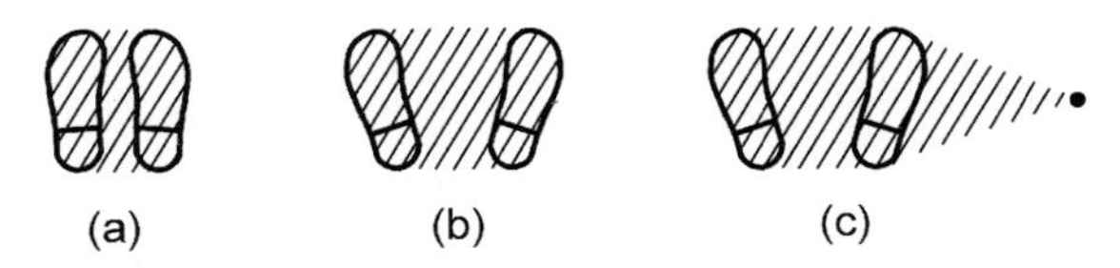

Fig. 3.6. A standing person is stable when her center of mass is over the cross-hatched region spanned by her feet, as shown here with her feet (**a**) together, (**b**) apart, (**c**) and apart, with stability also provided by a cane or crutch. (Based on [110])

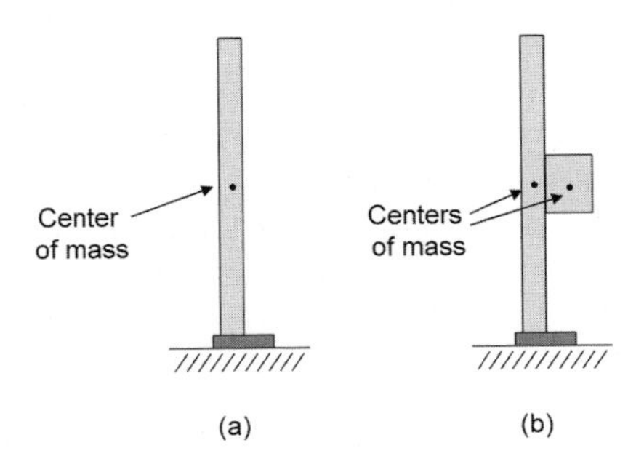

Fig. 3.7. Simple models of people facing to the right: (**a**) an upright person with mass fairly well balanced in front and back, with the person's center of mass shown – which is normally 58% of the person's height over the soles of his feet, and (**b**) one with unbalanced mass, such as a person with a beer belly or a woman who is pregnant, with separate centers of mass shown for the main body and the additional mass (which are approximately, but not exactly, at the same height). Clearly, the overall center of mass of the person in (**b**) is displaced to the right relative to that in (**a**). To prevent this being in front of the balls of the feet, the person often contorts his position, which can lead to muscle strain and a bad back. (See [110] for more details)

Fig. 3.8. Walking, with arrows showing the directions of the forces on the feet. (From [99]. Copyright 1992 Columbia University Press. Reprinted with the permission of the press)

$2\,N/cm^2$. However, not all of the foot touches the ground. Most of the contact is at the ball and heel of each foot. Because of this much smaller area, the pressure on these more limited points of contact is much higher, about $10\,N/cm^2$ during standing on both feet.

The peak-forces on the feet are much higher during walking. During walking there is usually only one foot on the ground, so the force on that foot is twice that with both feet on the ground and consequently the pressure is twice that during standing. Also, the normal forces involved in braking and forward propulsion during walking can be twice those during standing, which are only those needed to balance gravity. Overall, this leads to a peak force of about 1,400 N per foot, which translates to a pressure of about $40\,N/cm^2$. Because the whole foot is not flat on the ground during most phases of walking, the contact area is less than during standing and the peak-pressure during walking can be about $60\,N/cm^2$. Clearly, these forces are larger during running. Force plots are routinely measured during motion by using pressure sensors. (One such force plot is shown later in Fig. 3.18.)

3.3 Walking

3.3.1 Kinematics

The photographs in Fig. 3.8 show the stages of walking during a step by the right foot. The stages of walking are diagrammed in Fig. 3.9. The hip, knee, and ankle sagittal plane angles are usually referenced to the time the foot touches the ground (foot strike, FS – or heel strike/contact) and when the toe

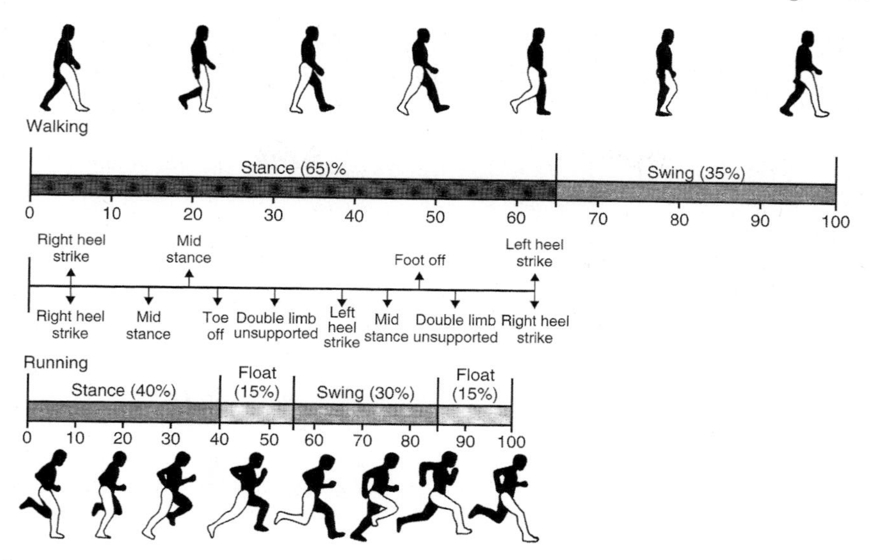

Fig. 3.9. Walking and running gait sequences. For walking, from foot strike (or heel strike) to toe off (or foot off) is the stance phase, with subdivisions sometimes called: foot flat or loading, midstance, terminal stance, and then preswing. From toe off to foot strike is the swing phase, with subdivisions called: initial swing, midswing, and then terminal swing. (From [166]. Used with permission)

leaves the ground (toe off, TO or foot off) (Fig. 3.9). From foot strike to toe off is the stance phase (foot flat or loading, midstance, terminal stance, and then preswing). For walking, from toe off to foot strike is the swing (initial swing, midswing, and then terminal swing). The stance occurs during 60–65% of the cycle for each leg and the swing phase occurs during 35–40% of the gait cycle. There are two feet on the ground for the first and last ∼10% of the stance phase for walking, just after foot strike and just before toe off. Both feet are never off the ground at the same time. The evolution of the three motion angles during a walking cycle is shown in the top row of Fig. 3.10.

At contact the upper leg (hip) is flexed by about 30° (relative to the vertical in this analysis), then extends to about 10° extension, and during preswing and then most of swing it flexes to 35° (Fig. 3.10). At contact the lower leg (knee) is nearly extended (∼5°), flexes to 20° during midstance, extends again, flexes to 40° during preswing, peaks at 60–70° flexion at midswing, and then extends again. At contact, the foot (ankle) is slightly plantarflexed (0–5°), extends a bit more to ∼7°, dorsiflexes as the leg rotates forward over the planted foot to 15°, dorsiflexes to 15° for propulsion near the end of stance; after toe off, it dorsiflexes to get toe clearance and plantarflexes to prepare for contact again.

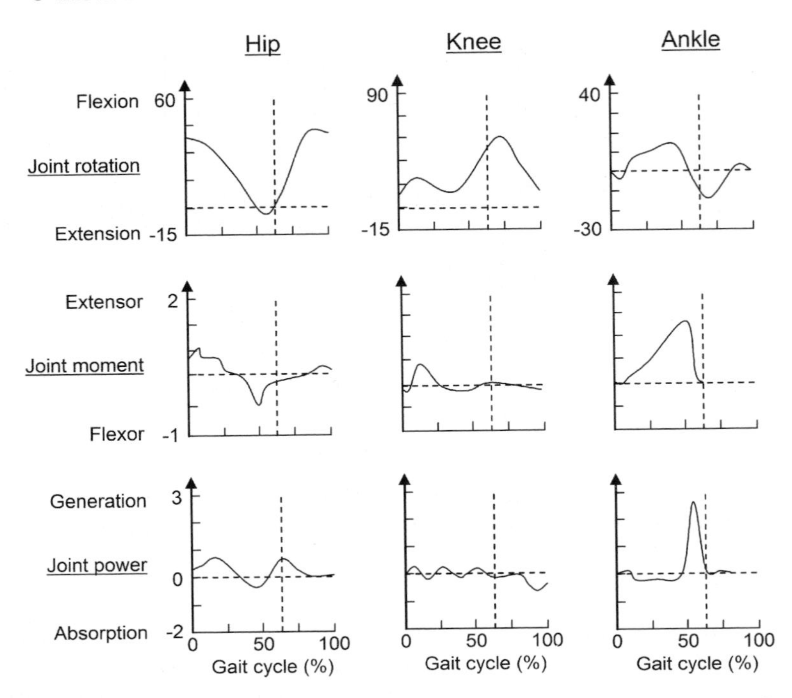

Fig. 3.10. Average values for sagittal plane joint rotation (in °), moment (torque) per mass (in N-m/kg body mass), and power per mass (in W/kg body mass) for the hip, knee, and ankle during a step in walking. (Based on [121]. Also see [104])

3.3.2 Muscular Action

Each flexion/extension action described in the kinematics is caused by the several flexor/extensor muscles for that joint. The muscular activity at each phase of the walking cycle is shown for the three sagittal joints in Figs. 3.11 and 3.12. Tables 3.1–3.3 describe these muscle groups [153, 165].

First, we describe which muscle groups are important during the stance phase and then the swing phase during walking. The ankle dorsiflexors help to lower the leg to the ground at heel strike (the initial contact, and the first part of stance) by eccentric muscle contraction (see Chap. 5). The hip extensors act after contact in early stance to continue this deceleration by reversing the forward swing of the leg, while the intrinsic muscles and long tendons of the foot preserve the shape of the arch of the foot. The knee extensors help the person accept the body weight during loading and then stabilize the knee during midstance. The ankle plantarflexors decelerate mass and control dorsiflexion up to mid-stance, while the hip abductors stabilize the pelvis. In terminal stance (heel off) and then preswing (toe off), the body is accelerated by the ankle plantarflexors by concentric muscle contraction (see Chap. 5) and then by the long flexors of the toes, while the pelvis is still stabilized, and the thigh is decelerated by eccentric contraction in preswing.

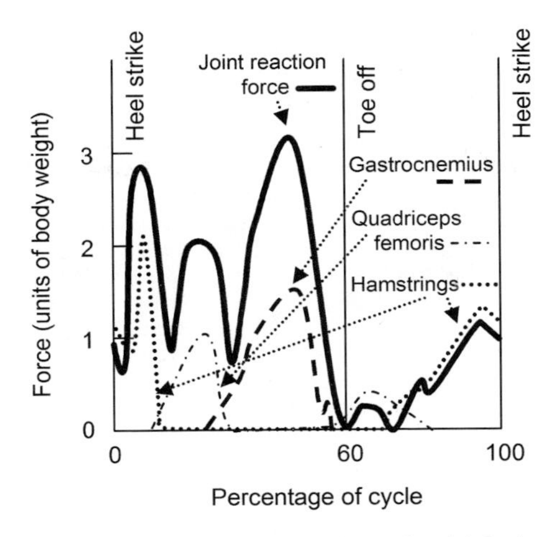

Fig. 3.11. Joint reaction forces transmitted through tibial plateau during one gait normal walking (*solid line*), along with the muscles forces – averaged for 12 subjects. (Based on [154] and [160])

In initial or preswing to midswing, the hip flexors advance the thigh by concentric contraction, the ankle dorsiflexors enable foot clearance from initial swing to midswing, and in terminal swing the hip extensors decelerate the thigh by eccentric contraction, the knee-flexors decelerate the lower leg by eccentric contraction, the ankle dorsiflexors help position the foot, and the knee extensors extend the knee to prepare for foot contact.

Alternatively, we could analyze this motion by describing when each muscle group is activated. The hip extensors act after contact in early stance and the hip flexors then advance the limb during preswing to midswing; from stance to swing the hip adductors and hamstrings are activated. The quadriceps muscles are activated before contact, then resist extensive knee flexion (by eccentric muscle contractions) during stance to avoid buckling due to gravity, and then extend the knee (by concentric muscle contractions) through early midstance as the leg rotates over the planted foot; they also control hip flexion prior to stance. The dorsiflexors fire during the swing phase to achieve foot clearance and during contact to control the foot placement. The plantar flexing soleus and gastrocnemius muscles fire during stance to control the motion of the leg over the foot and then to propel the foot and body forward. Figure 3.10 shows the joint moment and power generated during the gait cycle.

3.3.3 Friction

Is friction good or bad for walking? It is both.

First, what is friction? On a microscopic scale it is a complex set of interactions between two surfaces and any medium in between them and it is

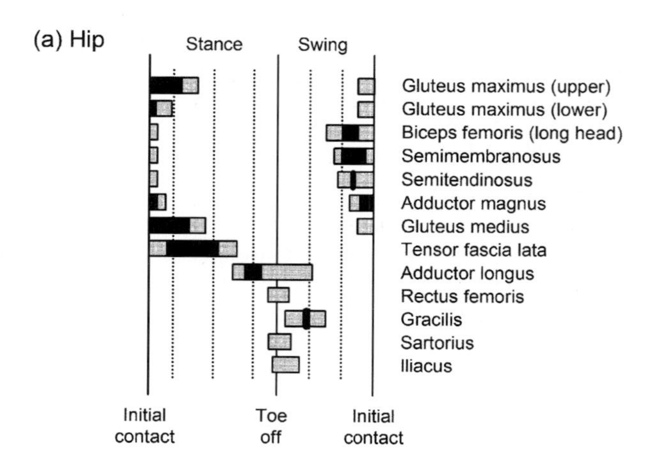
(a) Hip
Stance
Swing
Gluteus maximus (upper)
Gluteus maximus (lower)
Biceps femoris (long head)
Semimembranosus
Semitendinosus
Adductor magnus
Gluteus medius
Tensor fascia lata
Adductor longus
Rectus femoris
Gracilis
Sartorius
Iliacus
Initial contact
Toe off
Initial contact

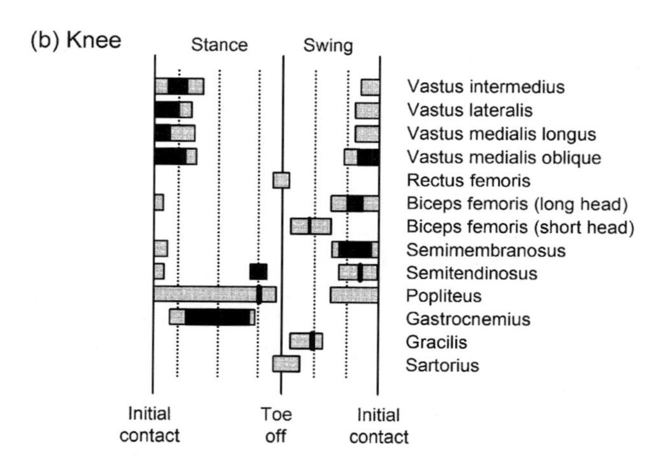
(b) Knee
Stance
Swing
Vastus intermedius
Vastus lateralis
Vastus medialis longus
Vastus medialis oblique
Rectus femoris
Biceps femoris (long head)
Biceps femoris (short head)
Semimembranosus
Semitendinosus
Popliteus
Gastrocnemius
Gracilis
Sartorius
Initial contact
Toe off
Initial contact

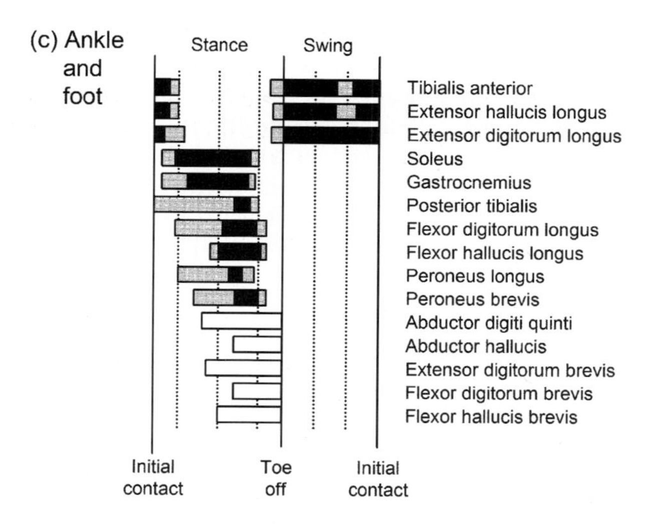
(c) Ankle and foot
Stance
Swing
Tibialis anterior
Extensor hallucis longus
Extensor digitorum longus
Soleus
Gastrocnemius
Posterior tibialis
Flexor digitorum longus
Flexor hallucis longus
Peroneus longus
Peroneus brevis
Abductor digiti quinti
Abductor hallucis
Extensor digitorum brevis
Flexor digitorum brevis
Flexor hallucis brevis
Initial contact
Toe off
Initial contact

Table 3.4. Static and kinetic (when noted) coefficient of friction, μ. (Using data from [109, 110, 159])

common objects and the body	
rubber (tire) on dry (wet) concrete road, static	1.0
brake material on cast ion, dry; static	0.4
brake material on cast ion, with mineral oil; static	0.1
graphite on steel, static	0.1
steel on unlubricated steel	0.7
steel on lubricated steel or ice	0.15
teflon on teflon (or steel)	0.04
ice on ice	0.1
ice on ice, 4 m/s, 0°C, kinetic	0.02
wood on wood	0.25–0.50
articular cartilage in the human knee, kinetic	0.005–0.02
articular cartilage in the human hip, kinetic	0.01–0.04
athletic equipment (kinetic)	
skates on ice	0.003–0.007
skis on snow	0.05–0.20
tennis balls on wood	0.25
basketball shoes on clean (dusty) wooden floor	1.0–1.2 (0.3–0.6)
cleated shoes on astroturf	1.2–1.7
jogging shoes on felt carpet, clay, asphalt	0.9–1.1, 0.3–0.5, 0.6–0.8

not easily modeled. On a macroscopic scale it is modeled using a coefficient of friction μ. An object of weight $W_b = m_b g$ feels a normal force $N = m_b g$ (see Fig. 3.13). Initially the object is static and the friction interaction is characterized by a static coefficient of friction μ_s (Table 3.4). Consider a lateral force on that object F. For low forces, the frictional force is equal in magnitude and opposite in direction to this force and the object does not move. The object will move only if this force exceeds $F_f = \mu_s N$. With the object moving, the net lateral force on the object is $F - F_f$, where now $F_f = \mu_k N$. μ_k is the kinetic, dynamic, or translational coefficient of friction, which can be different from μ_s.

←

Fig. 3.12. Electromyographic (EMG) activity of the muscles of the (**a**) hip, (**b**) knee, and (**c**) ankle and foot during walking in a healthy person. *Dotted* regions represent activation <20% of maximum voluntary contractions, while *black* regions represent >20% activation. *White* regions in (**c**) for the intrinsic muscles of the foot show at least some level of activation. The four phases of stance are, in time sequence, loading, midstance, terminal stance, and preswing; the three phases of swing are initial swing, midswing, and terminal swing. (Also see Fig. 3.9.) (Based on [104] and [162])

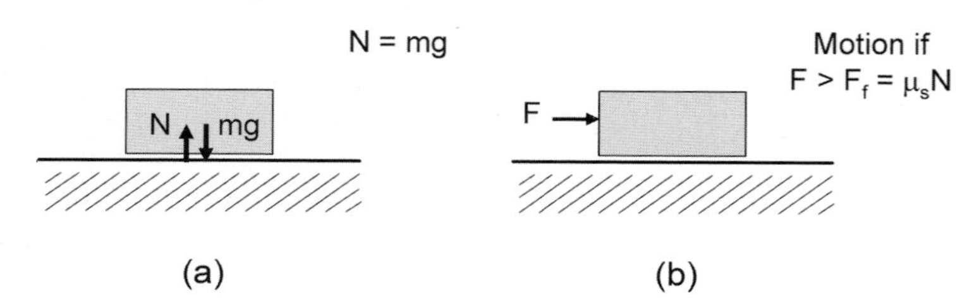

Fig. 3.13. (**a**) Static equilibrium of an object on a table and (**b**) lateral force required to overcome static friction to enable motion

Internal friction is usually troublesome. Friction in the knee joints during walking is bad. In fact, it is terrible. The knee is a synovial joint where the cartilage on the femur and tibia, and the surrounding synovial fluid contained in a sac, provide a very low coefficient of kinetic friction $\mu_{\mathrm{k}} = 0.003$–0.01. (This is smaller than most – if not all – man-made systems. See Table 3.4.) Such synovial joints are found in the knees and hips, which have relatively thick $\geq$5 mm cartilage, and in the finger joints, which have relatively thin cartilage layers, $\sim$1–2 mm. Lubrication at such articulating surfaces can occur by fluid-film lubrication, in which the surfaces are relatively far apart and are separated by synovial fluid; boundary lubrication, in which the surfaces are very close together (1–100 nm) and the lubrication is provided by the lubricating glycoprotein on the surfaces (Fig. 3.14a); or by a mixed lubrication, in which there is a mixture of both (Fig. 3.14b). Synovial joints with fluid-film lubrication have an extremely low coefficient of friction of $\simeq$0.02, whereas boundary-lubricated surfaces typically have coefficients of friction two orders of magnitude higher than those that are fluid-film lubricated. (Articular

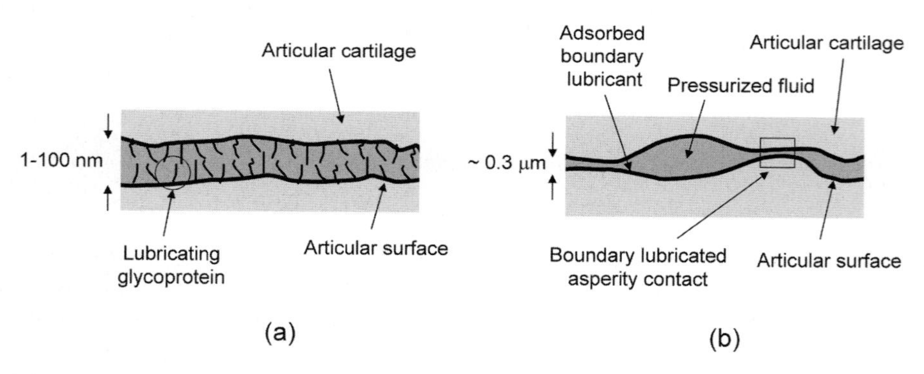

Fig. 3.14. In synovial joints (**a**) boundary layer lubrication of the articular cartilage surfaces for relatively flat and nearby surfaces and (**b**) mixed lubrication at articular cartilage, showing boundary lubrication when the separation is on the order of the surface roughness and fluid-film lubrication in areas of more widely separated surfaces. (Based on [155])

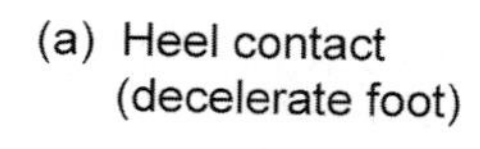

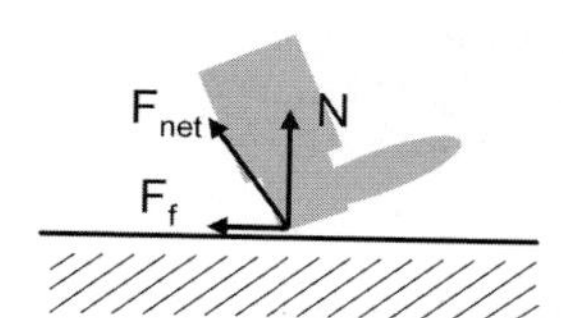

(b) Toe off
(accelerate foot)

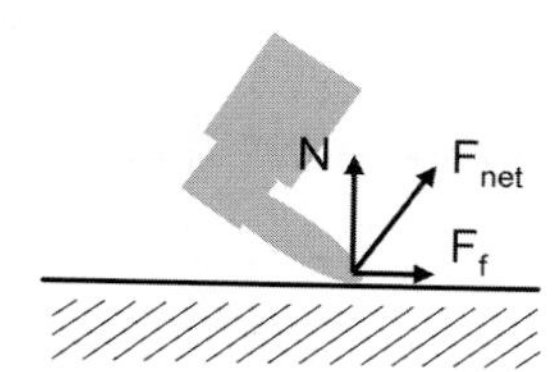

Fig. 3.15. (**a**) Heel contact stage and decelerating the foot and (**b**) toe-off stage and accelerating the foot

cartilage is a fairly complex and dynamic material, containing synovial fluid, etc. (Fig. 4.51).)

Loss of the synovial fluid from traumatic injury, loss of cartilage from long-term wear or injury, and excessive bone growth in the joint region – as in osteoarthritis – produce major pain during attempted motion. (The synovial fluid about the author's knee left the containing sac during a pantomime in junior high school, as a result of a collision of his knee with a door. Until his leg was stabilized by a cast, his severe pain was lessened during locomotion by his walking backward, with no rotation in the afflicted joint.)

External friction can be necessary. Without friction we could not walk or run. We would merely slip. When the heel of the foot touches the ground during walking, friction from the ground must slow it and then stop the forward motion of the foot (Fig. 3.15). The fraction of a second or so later when the ball of the foot is about to leave the ground, friction from the ground helps propel the body forward. This initial deceleration and subsequent acceleration of the foot typically requires forces of about $0.15mg$ and so a static coefficient of friction $\mu_\mathrm{s} > 0.15$ is necessary to walk. Figure 3.16 shows how this required minimum coefficient of friction increases with walking speed; it is about 0.17 for a typical walking speed of 3 mph (miles per hour) (a 20 minute mile). Figure 3.17 shows how it varies during the course of a step, and how it increases when pushing different loads [115]. Table 3.4 lists the kinetic coefficients of friction for several cases. Obviously, walking on ice is difficult. (For other reasons, walking on thin ice is also ill-advised.)

3.3.4 Energetics

Our motion during walking is surprisingly and deceptively simple when we analyze the kinetic and potential energy of the body during a walking cycle. Let us choose the horizontal direction of walking as the x direction and the vertical direction as the z direction. The kinetic energy KE of an object is

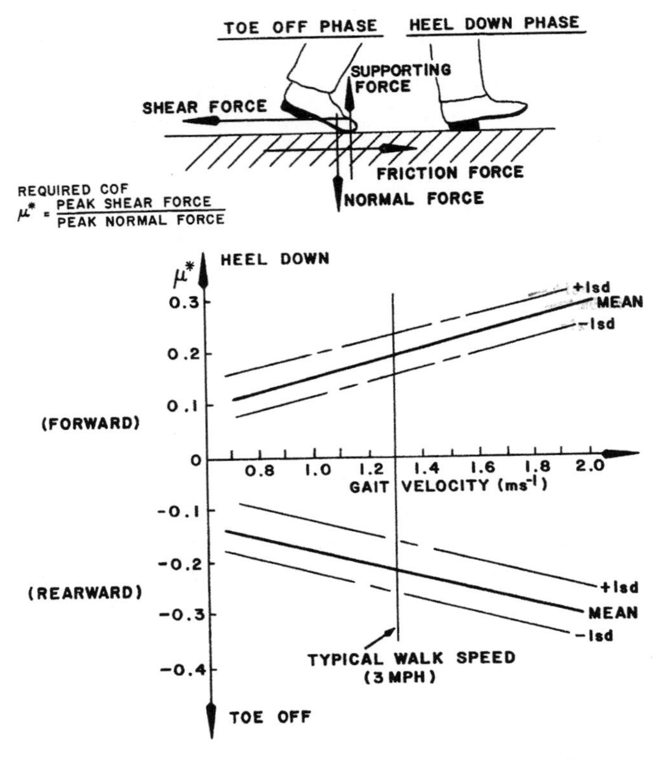

Fig. 3.16. Coefficient of friction required during walking at different speeds. (From [114]. Reprinted with permission of Wiley)

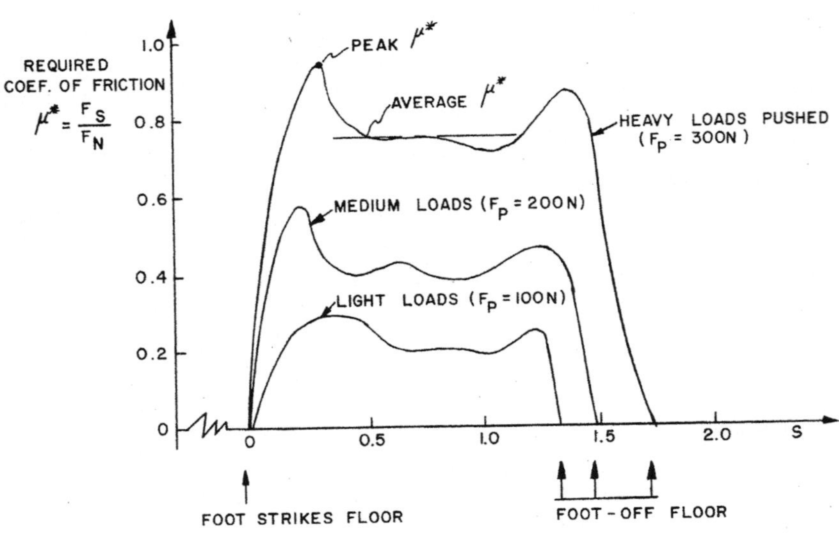

Fig. 3.17. Dynamic coefficient of friction required to push a weighted cart with different loads. (From [113]. Reprinted with permission of Wiley)

$m_b v^2/2$, where m_b is the body mass and v is the magnitude of the velocity, the speed, of the center of mass. This total KE can be separated into that due to motion in the horizontal and vertical directions, as

$$\mathrm{KE} = \mathrm{KE_H} + \mathrm{KE_V}, \tag{3.1}$$

$$\mathrm{KE_H} = \frac{1}{2} m_b v_x^2 = \frac{1}{2} m_b \left(\frac{\mathrm{d}x}{\mathrm{d}t}\right)^2, \tag{3.2}$$

$$\mathrm{KE_V} = \frac{1}{2} m_b v_z^2 = \frac{1}{2} m_b \left(\frac{\mathrm{d}z}{\mathrm{d}t}\right)^2. \tag{3.3}$$

The potential energy is

$$\mathrm{PE} = m_b g z_{\mathrm{CM}}, \tag{3.4}$$

where z_{CM} is the vertical position of the center of mass of the body.

When we describe a system where energy is not supplied, say by an external driving force, or lost, such as to friction, the total energy E of this system is constant, where

$$E = \mathrm{KE} + \mathrm{PE}. \tag{3.5}$$

We would not expect KE+PE to be constant during walking, but it is almost constant.

The photographs in Fig. 3.8 show the stages of walking during a step by the right foot. In stage (a), the right foot has just decelerated, much of the heel of the foot is on the ground, and the foot is in front of the body's center of mass. In stage (b), the right foot is squarely on the ground and under the center of mass. In stage (c), the right foot is behind the center of mass, mostly the ball is in contact with the ground, and forward propulsion is beginning. In stage (d), the front of the right foot is propelling the body forward, while the heel of the left foot has made contact with the ground and has begun to decelerate. We will examine the horizontal and vertical forces (F_{H} and F_{V}) and energies at each of these stages (Fig. 3.18).

The horizontal forces are due to friction. During stage (a) in Fig. 3.18, F_{H} slows the right foot, and during stages (c) and (d) it accelerates it. Because the horizontal speed of the center of mass of the whole body decreases when the right foot decelerates and it increases when the foot accelerates, during stage (a) $\mathrm{KE_H}$ decreases and during stages (c) and (d) it increases.

The center of mass (CM) of the body is changing during this step. The right leg is becoming more vertical and straighter from stage (a) to stage (b), so the center of mass (z_{CM}) is rising and PE is increasing. From stages (b) to (c) to (d), the right leg is becoming less vertical and is also bending, so the center of mass is falling and PE is decreasing. After stage (d) the left leg gets straighter and PE starts to increase again.

Because changes in the height of the center of mass must be the result of vertical forces, we see that from stages (a) to (b), z_{CM} is rising, which must

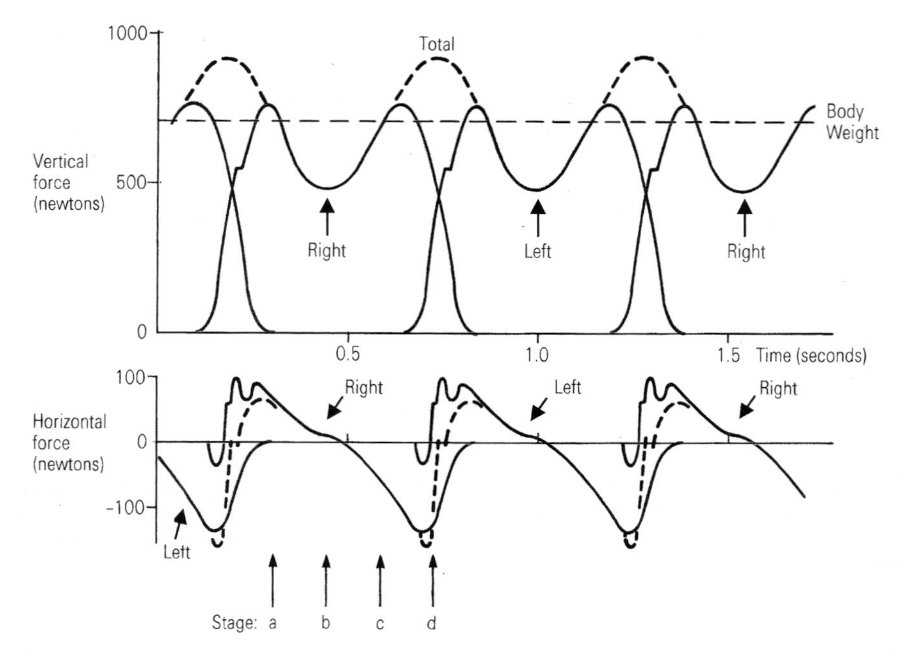

Fig. 3.18. Vertical and horizontal forces on feet during walking. (From [99]. Copyright 1992 Columbia University Press. Reprinted with the permission of the press)

mean that there is a net positive vertical force and $F_\mathrm{V} > m_\mathrm{b}g$. From stages (b) to (c), z_CM is falling, which must mean that there is a net negative vertical force and $F_\mathrm{V} < m_\mathrm{b}g$.

So we see that from stage (a) to (b), the center of mass is rising and PE is increasing, while the right foot is decelerating and $\mathrm{KE_H}$ is decreasing. During stage (c), the center of mass is falling and PE is decreasing, while the right foot is accelerating and $\mathrm{KE_H}$ is increasing. We see that PE and $\mathrm{KE_H}$ are out of phase, one is decreasing, while the other in increasing, and vice versa. This also occurs in harmonic motion, such as that for a mass on a spring or a pendulum.

Figure 3.19 shows that during a step PE + $\mathrm{KE_V}$ and $\mathrm{KE_H}$ are out of phase. ($\mathrm{KE_V}$ also changes during these stages, as a result of the motion of the center of mass z_CM; however, it changes relatively little.) More surprisingly, their sum PE + $\mathrm{KE_V}$ + $\mathrm{KE_H}$ is fairly constant during the step. Again, these two observations are reminiscent of harmonic motion, during which the PE and KE are out of phase and their sum does not change during a cycle. This exchange of kinetic and potential energy also occurs in a pendulum and for a rolling egg, as in Fig. 3.20, and this observation has led some to compare the leg during walking to a pendulum, by what is known as the ballistic or pendulum model of walking. Before pursuing this analogy further, we will review harmonic and pendulum motion.

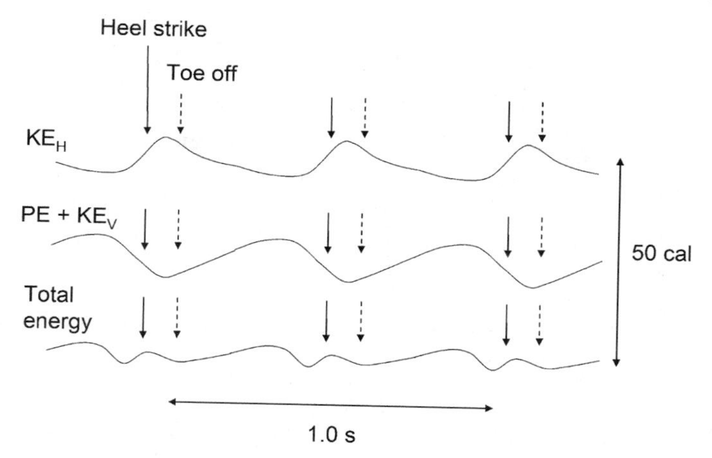

Fig. 3.19. Changes in mechanical energy during walking from force plates, with heel strike and toe off shown, for (from top to bottom curve) forward kinetic energy, gravitational potential energy plus vertical kinetic energy, and total energy. (Based on [112] and [146])

3.3.5 Review of Harmonic Motion, Pendulums, and Moments of Inertia

Simple Harmonic Oscillator

The position x from equilibrium for a body of mass m attached by a spring with spring constant k (Fig. 3.21) is determined from Newton's second law $F = ma$, where F is the restoring force on the mass due to the spring and

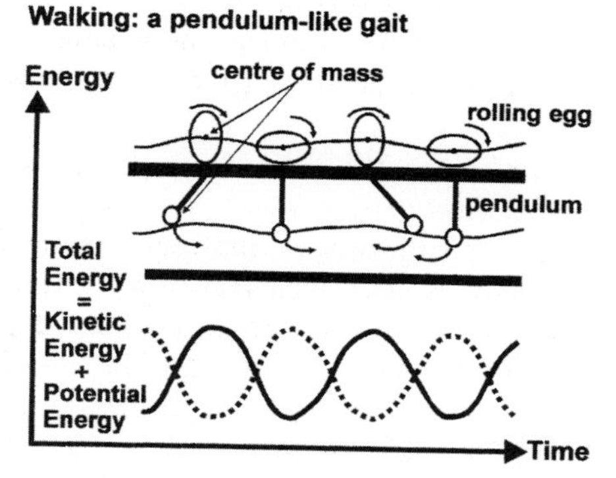

Fig. 3.20. The cyclic exchange of kinetic and potential energy in a pendulum and a rolling egg, with constant total energy, is similar to that in walking. (From [151]. Used with permission)

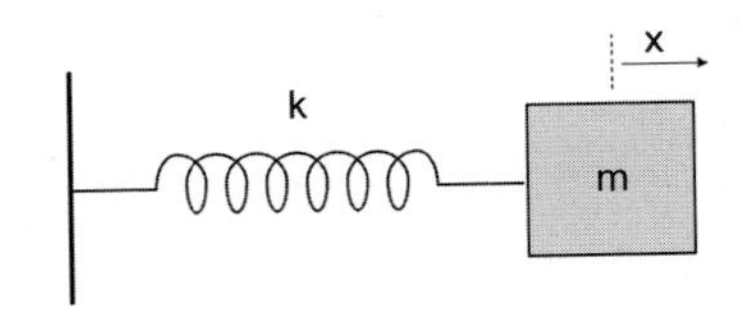

Fig. 3.21. Simple harmonic oscillator, with an object with mass m attached to a spring with spring constant k

$a = \mathrm{d}^2x/\mathrm{d}t^2$ is the acceleration of the mass. So

$$m\frac{\mathrm{d}^2x}{\mathrm{d}t^2} = -kx \tag{3.6}$$

$$\frac{\mathrm{d}^2x}{\mathrm{d}t^2} = -\frac{k}{m}x = -\omega^2 x, \tag{3.7}$$

where $\omega = (k/m)^{1/2}$ is the resonant frequency of this harmonic oscillator. The solution can be written as

$$x = A\cos(\omega t + \phi), \tag{3.8}$$

where A is an arbitrary amplitude and ϕ is a phase (which is not very significant here). (See Appendix C for more information about the solution.)

The kinetic energy during this motion is

$$\mathrm{KE} = \frac{1}{2}mv^2 = \frac{1}{2}m\left(\frac{\mathrm{d}x}{\mathrm{d}t}\right)^2 = \frac{1}{2}mA^2\omega^2\sin^2(\omega t + \phi), \tag{3.9}$$

because $\mathrm{d}x/\mathrm{d}t = -A\omega\sin(\omega t + \phi)$. The potential energy is

$$\mathrm{PE} = \frac{1}{2}kx^2 = \frac{1}{2}kA^2\cos^2(\omega t + \phi) = \frac{1}{2}mA^2\omega^2\cos^2(\omega t + \phi) \tag{3.10}$$

using $\omega = (k/m)^{1/2}$. The PE and KE are clearly 90° out of phase during this oscillatory motion. The total energy,

$$E = \mathrm{KE} + \mathrm{PE} = \frac{1}{2}mA^2\omega^2\sin^2(\omega t + \phi) + \frac{1}{2}mA^2\omega^2\cos^2(\omega t + \phi) = \frac{1}{2}mA^2\omega^2, \tag{3.11}$$

is constant during this motion.

The radial frequency ω has units of radians per second. The cycle frequency f is $\omega/2\pi$, which has units of cycles per second (cps) or Hz (Hertz). This means that a complete cycle (period) occurs in a time $T_{\mathrm{period}} = 1/f = 2\pi/\omega$, during which time ωt changes by 2π rad.

Mass on a Pendulum

We will use the pendulum to model walking. In a simple pendulum, a ball of mass m is at the end of a plumb (string) of length L (Fig. 3.22a). For now

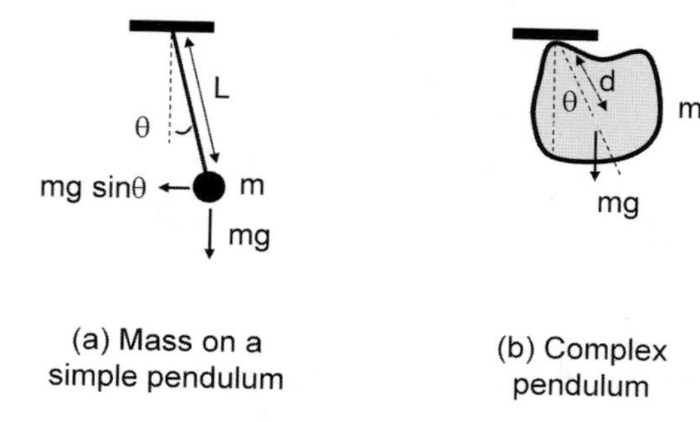

Fig. 3.22. (**a**) Mass on a simple pendulum and (**b**) complex pendulum

we will assume that the plumb is massless. Assume that the plumb is at an angle θ from the vertical. The position of the ball along the arc of the plumb is $L\theta$. The speed of the ball normal to the plumb is $v = L\mathrm{d}\theta/\mathrm{d}t$, and the acceleration of the ball normal to the plumb is $L\mathrm{d}^2\theta/\mathrm{d}t^2$. (For small angles, $|\theta| \ll 1$, the lateral coordinate is $x \simeq L\theta$, and so $\mathrm{d}^2x/\mathrm{d}t^2 \simeq L\mathrm{d}^2\theta/\mathrm{d}t^2$.) From the diagram, the tension T in the string is balanced by $mg\cos\theta$, and there is a net acceleration of the ball $mg\sin\theta$ that tends to decrease the magnitude of θ. Newton's Second Law $F = ma$, can be written as (with $ma = F$)

$$mL\frac{\mathrm{d}^2\theta}{\mathrm{d}t^2} = -mg\sin\theta. \tag{3.12}$$

For small angle motion ($|\theta| \ll 1$), so $\sin\theta \approx \theta$ and

$$mL\frac{\mathrm{d}^2\theta}{\mathrm{d}t^2} = -mg\theta \tag{3.13}$$

$$\frac{\mathrm{d}^2\theta}{\mathrm{d}t^2} = -\frac{g}{L}\theta = -\omega^2\theta. \tag{3.14}$$

Physically and mathematically this is similar to the simple harmonic oscillator, now with a resonant frequency $\omega = (g/L)^{1/2}$ and solution

$$\theta(t) = B\cos(\omega t + \beta), \tag{3.15}$$

where B is an arbitrary amplitude and β is a phase (which again is not very significant here). (See Appendix C for more information about the solution.)

The kinetic energy during this motion is

$$\mathrm{KE} = \frac{1}{2}mv^2 = \frac{1}{2}m\left(L\frac{\mathrm{d}\theta}{\mathrm{d}t}\right)^2 = \frac{1}{2}mB^2L^2\omega^2\sin^2(\omega t + \beta), \tag{3.16}$$

because $v = L\mathrm{d}\theta/\mathrm{d}t = -BL\omega\sin(\omega t + \beta)$. The potential energy is

$$\mathrm{PE} = mgL(1 - \cos\theta) \approx \frac{1}{2}mgL\theta^2, \tag{3.17}$$

because for small θ ($|\theta| \ll 1$), $1 - \cos\theta \approx 1 - \theta^2/2$, and so

$$\mathrm{PE} \approx \frac{1}{2} mgLB\cos^2(\omega t + \beta) = \frac{1}{2} mB^2L^2\omega^2\cos^2(\omega t + \beta) \tag{3.18}$$

using $\omega = (g/L)^{1/2}$. Again, the PE and KE are out of phase during this oscillatory motion. As with the harmonic oscillator, the total energy

$$\begin{aligned} E &= \mathrm{KE} + \mathrm{PE} \\ &= \frac{1}{2} mB^2L^2\omega^2\sin^2(\omega t + \beta) + \frac{1}{2} mB^2L^2\omega^2\cos^2(\omega t + \beta) && (3.19) \\ &= \frac{1}{2} mB^2L^2\omega^2 && (3.20) \end{aligned}$$

is constant during the motion of the pendulum.

During walking the leg can be modeled as a pendulum because of this interplay between the KE and PE. We will use this pendulum (ballistic) model to determine the oscillation frequency of the leg $\omega = (g/L)^{1/2}$ and from this we will obtain the walking speed. Before doing this, we need to modify this simple pendulum model because the leg cannot be approximated by a mass (foot) at the end of a massless string (upper and lower legs). (We will also use this refined model in our analysis of throwing a ball.)

More Complex Pendulum

If we multiply both sides of (3.12) by L, we get

$$mL^2\frac{\mathrm{d}^2\theta}{\mathrm{d}t^2} = -mgL\sin\theta. \tag{3.21}$$

The right side is actually the net torque τ acting on the ball of the simple pendulum about an axis at the top of the string, where the magnitude of the distance vector $r = L$, the force on the ball is mg, and the angle from this $\mathbf{r}$ to $\mathbf{F}$ is $360° - \theta$, according to our convention (Fig. 3.22b). With $\sin(360° - \theta) = -\sin\theta$,

$$\tau = mgL\sin(360° - \theta) = -mgL\sin\theta = mL^2\frac{\mathrm{d}^2\theta}{\mathrm{d}t^2}. \tag{3.22}$$

The mL^2 term is the *moment of inertia* I of the ball about the pivot axis, which is at the hip. More generally, the moment of inertia of a complex object can be written as the sum of contributions from distinct masses m_i each a distance R_i from the axis

$$I = \sum_i m_i R_i^2 \tag{3.23}$$

or as an integral of mass density $\rho(\mathbf{r})$ at positions $\mathbf{r}$ that are a distance R from the axis, integrated over the volume V

$$I = \int \rho(\mathbf{r})R^2\mathrm{d}V. \tag{3.24}$$

For these complex objects with moment of inertia I, total mass m, and center of mass a distance d from the axis

$$\tau = -mgd \sin\theta = I\frac{\mathrm{d}^2\theta}{\mathrm{d}t^2} \tag{3.25}$$

and with the small angle approximation $\sin\theta \approx \theta$

$$\frac{\mathrm{d}^2\theta}{\mathrm{d}t^2} = -\frac{mgd}{I}\sin\theta \approx -\frac{mgd}{I}\,\theta. \tag{3.26}$$

The resonant frequency is now $\omega = (mgd/I)^{1/2}$. (See Appendix C for more information about the solution.)

Also, the angular momentum of the pendulum $L = I\Omega$ with $\Omega = \mathrm{d}\theta/\mathrm{d}t$, and $\tau = \mathrm{d}L/\mathrm{d}t$.

More on Moments of Inertia

Consider an object of mass m with moment of inertia I about a given axis through its center of mass. You can obtain its moment of inertia about any axis laterally displaced a distance q from this axis by using the parallel axis theorem

$$I_{\text{about displaced axis}} = I_{\text{about cm}} + mq^2. \tag{3.27}$$

This *parallel axis theorem* is illustrated in Fig. 3.23a, and can be proved from (3.24).

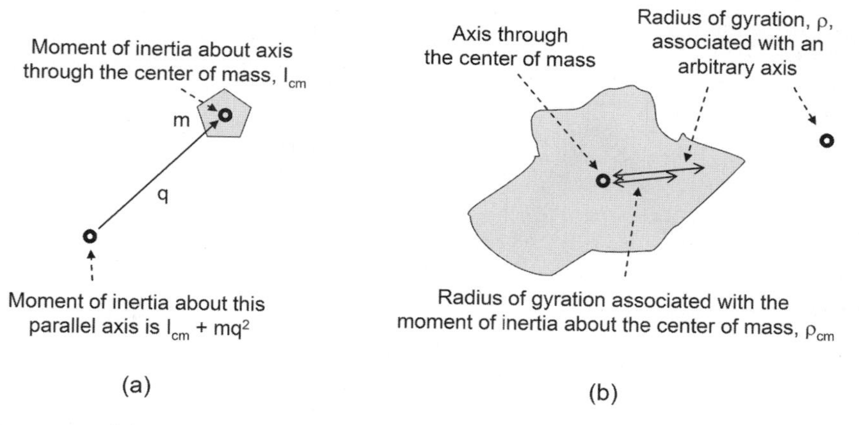

Fig. 3.23. (**a**) The parallel axis theorem is illustrated for an object of mass m for an axis about the center of mass normal to the page and about an arbitrary parallel axis, and (**b**) the radius of gyration is schematically illustrated for moments of inertia about the center of mass and an arbitrary axis

One simple way of providing the moment or inertia of any object is by giving its *radius of gyration*, ρ, which is defined by

$$I = m\rho^2. \tag{3.28}$$

This means the moment of inertia of the object would be the same if all of the mass were distributed a distance ρ from the axis. One example of this is the mass m distributed on a circle of radius ρ about the axis. In another example, two points with mass $m/2$ are placed on either side of the axis. Of course, the radius of gyration of a ball on a string is the length of the string, but in this case the center of mass is not on the rotation axis. Figure 3.23b illustrates the radius of gyration for an axis through the center of mass and about an arbitrary axis. The average radii of gyration for body segments are given in Table 1.9.

3.3.6 Ballistic (or Pendulum) Model of Walking

The ballistic model says that during the swing cycle your leg is like a pendulum with radial frequency ω. This means that a complete cycle (period) occurs in a time $T_{\text{period}} = 2\pi/\omega$. One forward swing of the leg corresponds to a half cycle, which takes $T_{\text{half period}} = \pi/\omega$. For a leg of length L_{leg} swinging through an arc of $\Delta\theta$, the step length is $\sim L_{\text{leg}}(\Delta\theta)$ for small $\Delta\theta$ (Fig. 3.24a), so the stepping or walking speed v is this step length divided by $T_{\text{half period}}$ or

$$v = \frac{\omega}{\pi} L_{\text{leg}}(\Delta\theta). \tag{3.29}$$

For ball on a string $\omega = (g/L_{\text{leg}})^{1/2}$, so $T_{\text{half period}} = \pi(L_{\text{leg}}/g)^{1/2}$. For a body height $H = 1.8$ m, Fig. 1.15 and Table 1.6 show that the leg length is $L_{\text{leg}} = 0.53H = 0.95$ m. With $g = 9.8\,\text{m/s}^2$, $T_{\text{half period}} = 0.98$ s.

How does this compare to real walking? The speed for a slow walk is about $0.5\,\text{m/s} = 1.1$ mph. (Test this!) With a step length of 0.3 m, the time for a step

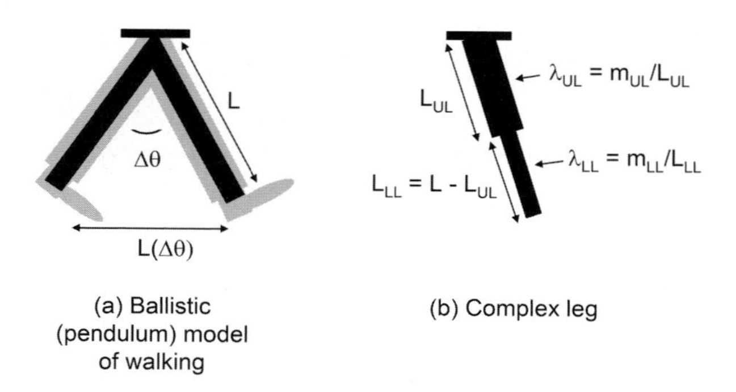

Fig. 3.24. (**a**) Ballistic (pendulum) model of walking, with simple leg of uniform linear mass density, and with (**b**) complex leg with upper and lower legs with different linear mass densities

is $T_{\text{half period}} \simeq 0.60$ s. For a fast walk the speed is about 2.0 m/s = 4.5 mph and the step length is larger, 0.7 m, and so the step time is about 0.35 s. The model predictions are fair for the slow walk with this ball on a string model (0.98 s model result compared to 0.6 s actual time). Fast walking cannot be modeled as a free pendulum; it is clearly a forced pendulum. You can feel this in your legs as you change from a slow, leisurely pace to a fast, vigorous pace.

Can we improve the model of slow walking by using a more refined model for the mass distribution of the leg? For a leg of length L with a uniform linear density (and the same thickness throughout, which is not exactly true) and mass m_{leg}, as suggested by Fig. 3.24a, we need to use the expression from the more general, refined model (3.26) with $\omega = (m_{\text{leg}}gd/I)^{1/2}$ and $T_{\text{half period}} = \pi(I/m_{\text{leg}}gd)^{1/2}$. The distance of the center of mass from the axis is $d = L_{\text{leg}}/2$. The moment of inertia $I = m_{\text{leg}}L_{\text{leg}}^2/3$. This is obtained by integrating over a uniform linear mass density (mass per unit length) $\lambda = m_{\text{leg}}/L_{\text{leg}}$ from $R = 0$ to L_{leg} in (3.24):

$$I = \int \rho(\mathbf{r})R^2\mathrm{d}V = \int_0^{L_{\text{leg}}} \lambda R^2\mathrm{d}R = \frac{m_{\text{leg}}}{L_{\text{leg}}}\frac{L_{\text{leg}}^3}{3} = \frac{1}{3}m_{\text{leg}}L_{\text{leg}}^2. \tag{3.30}$$

(With a uniform cross sectional area A, $\rho(\mathbf{r}) = \lambda/A$ and $dV = A\ \mathrm{d}R$, so $\rho(\mathbf{r})\mathrm{d}V = \lambda\ \mathrm{d}R$.) Then $\omega = (m_{\text{leg}}gd/I)^{1/2} = (m_{\text{leg}}g(L_{\text{leg}}/2)/(m_{\text{leg}}L_{\text{leg}}^2/3))^{1/2} = (3g/2L_{\text{leg}})^{1/2}$ and $T_{\text{half period}} = \pi(2L_{\text{leg}}/3g)^{1/2} = 0.80$ s, again for $L_{\text{leg}} = 0.95$ m. This prediction is fairly close to the estimated 0.60 s step time, and agreement is much better than for the ball on string model. We should remember that this is just a model; energy recovery is not perfect, and ~70%, for walking (Fig. 3.25).

What happens if we refine the model a bit more? The leg does not have a uniform linear mass density; the upper leg (thigh) is heavier per unit length than the lower leg (with the foot) (Fig. 3.24b). The entire leg has a mass $m_{\text{leg}} = 0.161m_{\text{b}}$ and length $L_{\text{leg}} = 0.530H$. Tables 1.6 and 1.7 and Fig. 1.15 show that the upper leg has a mass $m_{\text{u,leg}} = 0.10m_{\text{b}} = 0.621m_{\text{leg}}$ and has a length $L_{\text{u,leg}} = 0.245H = 0.462L_{\text{leg}}$ (H = body height), while the lower leg and foot have a mass $m_{\text{l,leg}} = 0.061m_{\text{b}} = 0.379m_{\text{leg}}$ and a length $L_{\text{l,leg}} = 0.285H = 0.538L_{\text{leg}}$.

The center of mass distance d from the hip joint and the moment of inertia about the hip are needed. The center of mass of two objects is in general given by

$$r_{\text{CM}} = \frac{m_1r_1 + m_2r_2}{m_1 + m_2}, \tag{3.31}$$

which here is

$$d = \frac{m_{\text{u,leg}}(L_{\text{u,leg}}/2) + m_{\text{l,leg}}(L_{\text{u,leg}} + L_{\text{l,leg}}/2)}{m_{\text{u,leg}} + m_{\text{l,leg}}} \tag{3.32}$$

$$= \frac{(0.621m_{\text{leg}})(0.231L_{\text{leg}}) + (0.379m_{\text{leg}})(0.731L_{\text{leg}})}{m_{\text{leg}}} = 0.421L_{\text{leg}}. \tag{3.33}$$

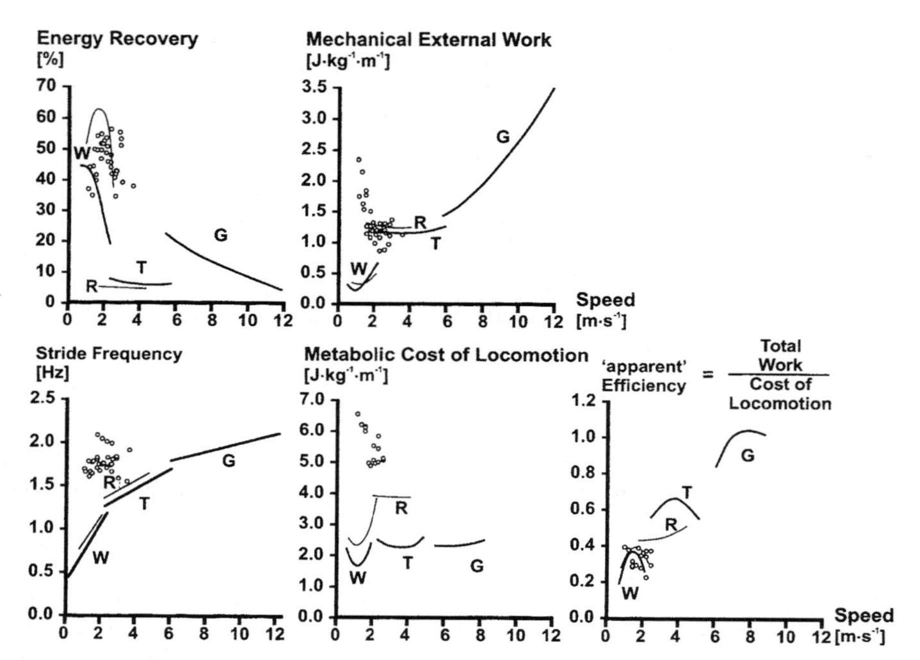

Fig. 3.25. Gait parameters as a function of speed for humans (*thin curves*) and horses (*thick curves*), for different gaits: walking (W), running (R), trotting (T), galloping (G), and human skipping (*open circles*). (From [151]. Used with permission. Also see [123])

The moment of inertia is (Fig. 3.24b)

$$I = \int_0^{L_{\mathrm{u,leg}}} \lambda R^2 \mathrm{d}R + \int_{L_{\mathrm{u,leg}}}^{L_{\mathrm{leg}}} \lambda R^2 \mathrm{d}R = \frac{1}{3}\lambda_{\mathrm{u,leg}} L_{\mathrm{u,leg}}^3 + \frac{1}{3}\lambda_{\mathrm{l,leg}}(L_{\mathrm{leg}}^3 - L_{\mathrm{u,leg}}^3), \tag{3.34}$$

where $\lambda_{\mathrm{u,leg}} = m_{\mathrm{u,leg}}/L_{\mathrm{u,leg}} = 0.62 m_{\mathrm{leg}}/0.46 L_{\mathrm{leg}} = 1.34 m_{\mathrm{leg}}/L_{\mathrm{leg}}$ and $\lambda_{\mathrm{l,leg}} = m_{\mathrm{l,leg}}/L_{\mathrm{l,leg}} = 0.38 m_{\mathrm{leg}}/0.54 L_{\mathrm{leg}} = 0.70 m_{\mathrm{leg}}/L_{\mathrm{leg}}$. With $L_{\mathrm{u,leg}} = 0.46 L_{\mathrm{leg}}$, we see that $I = 0.256 m_{\mathrm{leg}} L_{\mathrm{leg}}^2$. (Problem 3.17 compares these models of the moments on inertia with measured data, using the parallel axis theorem and the radii of gyration in Table 1.9.)

Then $\omega = (m_{\mathrm{leg}} g d/I)^{1/2} = (m_{\mathrm{leg}} g (0.421 L_{\mathrm{leg}})/(0.256 m_{\mathrm{leg}} L_{\mathrm{leg}}^2))^{1/2} = (1.64 g/L_{\mathrm{leg}})^{1/2}$. We see that $T_{\text{half period}} = \pi(0.61 L_{\mathrm{leg}}/g)^{1/2} = 0.76$ s, which is 4.5% faster than the uniform density model, and not a great improvement over it.

3.3.7 Inverted Pendulum Model

The leg is like a pendulum during the swing cycle for that leg. During stance it is like an inverted pendulum. The center of mass travels on a circular arc

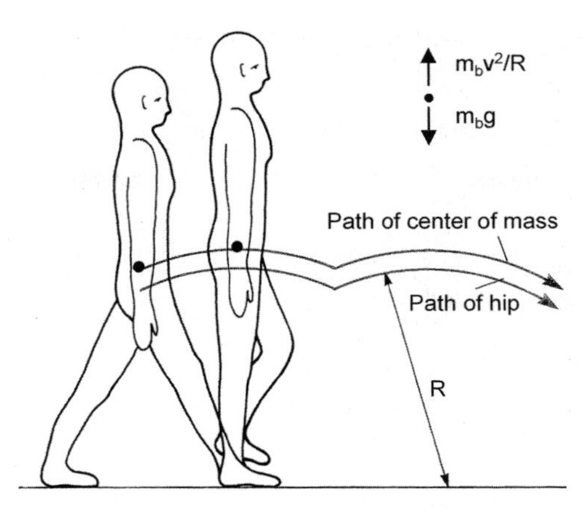

Fig. 3.26. Takeoff during fast walking, in the inverted pendulum model. (From [99]. Copyright 1992 Columbia University Press. Reprinted with the permission of the press)

of radius R during stance (Fig. 3.26). To maintain the circular motion, a centripetal force with magnitude $m_b v^2/R$ is needed, where v is approximately the walking speed. (This is the same as an upward "centrifugal force" in a frame that is rotating with the leg.) When this exceeds $m_b g$, the center of mass will not come down. This sets a maximum walking speed of $v_{max} = (gR)^{1/2} \sim 3.0\,\mathrm{m/s}$ for $R = 0.9$ m; at faster speeds people have to move in a different way: jogging, running, etc. On the moon ($g_{Moon} = 0.17 g_{Earth}$) and Mars ($g_{Mars} = 0.4 g_{Earth}$) this limit is smaller, 1.2 m/s and 1.9 m/s, respectively. This low upper limit for the walking speed on the moon explains why astronauts had to resort to hopping on the moon to move fast; this is the real "moonwalking." Children (on earth) have smaller R and consequently have to start running at lower speeds than adults (and sometimes need to run to keep up with adults). Adults usually start running at speeds even slower than the 3.0 m/s estimate, about 2.0 m/s, possibly in an effort by the body to minimize the metabolic energy cost per unit distance traveled. (See Fig. 3.25.) Speeds near 4 m/s are attainable in race walking because the racers distort their hips at midstance in a way that flattens the trajectory of the center of mass – thereby increasing R.

3.4 Running

Running is not just fast walking; it is qualitatively different. During walking, each foot is on the ground for more than half the time, and sometimes both are on the ground at the same time. Both feet are never off the ground at the same time. During running, each foot is on the ground for less than half the time, and – unlike walking – sometimes neither foot is on the ground. This

Fig. 3.27. Stages of a running stride, with the arrows indicating the directions of the forces on the feet. (From [99]. Copyright 1992 Columbia University Press. Reprinted with the permission of the press)

distinction can have serious implications. Jane Saville, an Australian race walker, was in first place 200 m from the finish line in the 20 km race walking competition in the 2000 Summer Olympics. As photographs later confirmed, both of her feet were off the ground at the same time and she was disqualified. (Following standard practice, she had received two prior cautions before her red card disqualification.) She later won the bronze medal in this event in the 2004 Summer Olympics.

3.4.1 Kinematics

Figure 3.27 shows snapshots of different stages of running. One difference between walking and running is that the leg becomes almost straight during walking at midstance, but it never becomes straight during running. Another difference is that during walking the stance is longer than the swing, while the reverse is true for running. As seen in Fig. 3.9, during running the stance occurs during 40% of the cycle after foot strike, and the foot is off the ground for 60% of the cycle from toe off to the next foot strike. The leg floats for the first 15% of the cycle after toe off, swings forward during the next 30%, and then floats for 15% until foot strike.

The changes of sagittal joint angles are qualitatively the same as in walking (Fig. 3.10 top), but with differences, such as with the knee angle. Figure 3.28 shows a representative example of how the locus of these angles varies with respect to each other during a gait cycle. For a running speed of 3.4 m/s ($\sim$8 minute mile), the thigh angle varies from 39° (flexion) to $-21°$ (extension) for a range of 60°. At footstrike the thigh is about 25° from the vertical because the hip is flexed; any initial angle change after FS is relatively small. During the extension prior to footstrike the knee angle is 13° and increases to 41° during cushioning flexion (for a range of 27° during cushioning flexion), and

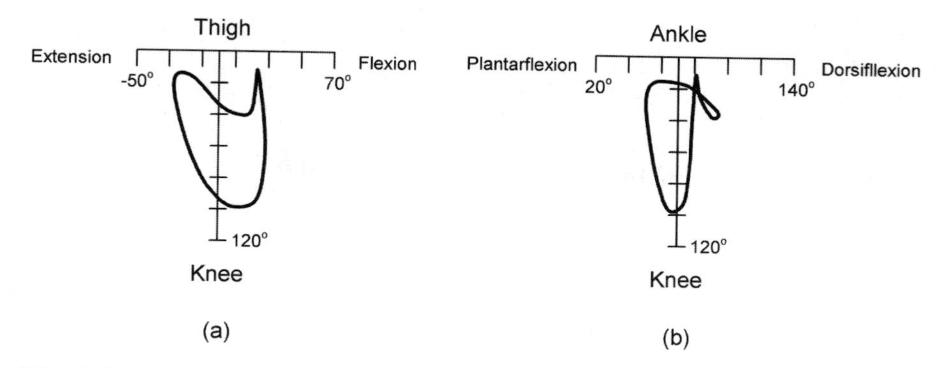

Fig. 3.28. Mean locus of (**a**) thigh angle vs. knee angle and (**b**) ankle angle vs. knee angle during running at 3.57 m/s, using joint angles as defined in Fig. 3.1 (Based on [150] and [175])

then decreases to 15° during extension during the propulsive phase. At swing phase flexion the angle is 104°, for a total knee range of 90°. For the ankle, the dorsiflexion prior to footstrike is 91°, that at the stance phase is 112°, and that at plantar flexion is 64°, for a total range of ankle motion of 48°. These values change with running speed and with grade for uphill and downhill running. Table 3.5 shows typical loci of the thigh and ankle angles vs. the knee angle during a stride.

Table 3.5. Maximum thigh, knee, and ankles angles (in degrees) for running at different speeds and grades on a treadmill. (Using data from [150])

	level			ramp		
speed (m/s)	3.4	4.2	5.0	3.4	3.4	3.4
grade	0%	0%	0%	−20%	0%	20%
thigh						
flexion	39.0	41.0	46.6	25.8	39.0	54.5
extension	−20.6	−23.1	−28.0	−18.4	−20.6	−18.7
knee						
extension before footstrike	13.4	16.7	15.5	4.0	13.4	37.7
cushioning flexion	40.6	39.4	42.9	41.2	40.6	46.3
propulsion phase extension	15.1	16.8	14.2	32.1	15.1	13.0
swing phase flexion	103.9	108.1	117.3	99.6	103.9	113.2
ankle						
dorsiflexion before footstrike	91.3	91.8	92.5	–	91.3	97.6
stance phase dorsiflexion	112.4	108.7	110.8	103.5	112.4	116.4
plantar flexion	64.2	58.7	57.1	76.4	64.2	59.4

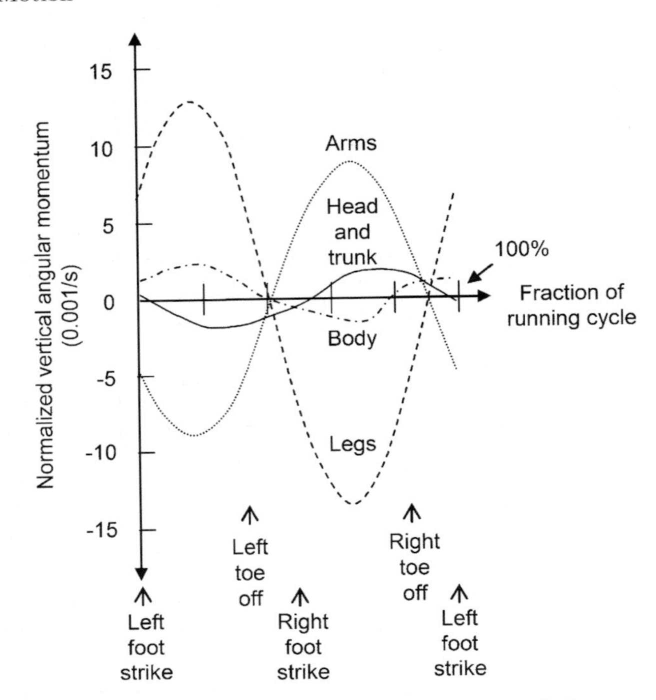

Fig. 3.29. Normalized mean vertical angular momenta of the arms, head-plus-trunk, legs, and whole body during a running cycle, at 4.5 m/s. The normalized vertical angular momentum is obtained by dividing the vertical angular momentum (in kg-m^2/s) by body mass (in kg) and the square of the runner's standing height (height in m). (Based on [128] and [129])

Your legs are not the only things that move during running and fast walking (other than your center of mass). Your arms swing back and forth and your torso rotates, both 180° out of phase with your legs. Why? Consider what would happen if this did not happen. With every stride by your right leg you would need to gain much positive angular momentum about the vertical axis, while with every stride with your left leg you would need to acquire negative angular momentum, as is evident in Fig. 3.29. This is just like the rotor in a washing machine. The motion in your arms and torso cancels ~90% of these changes, as seen in this figure.

3.4.2 Muscular Action

Each flexion/extension action described by kinematics is caused by the flexor/extensor muscles of that joint. The muscular activity at each phase of the cycle is qualitatively similar to that shown for the three sagittal joints for walking in Figs. 3.10–3.12. Tables 3.1–3.3 describe these muscle groups. Figure 3.30 shows muscle activity during running.

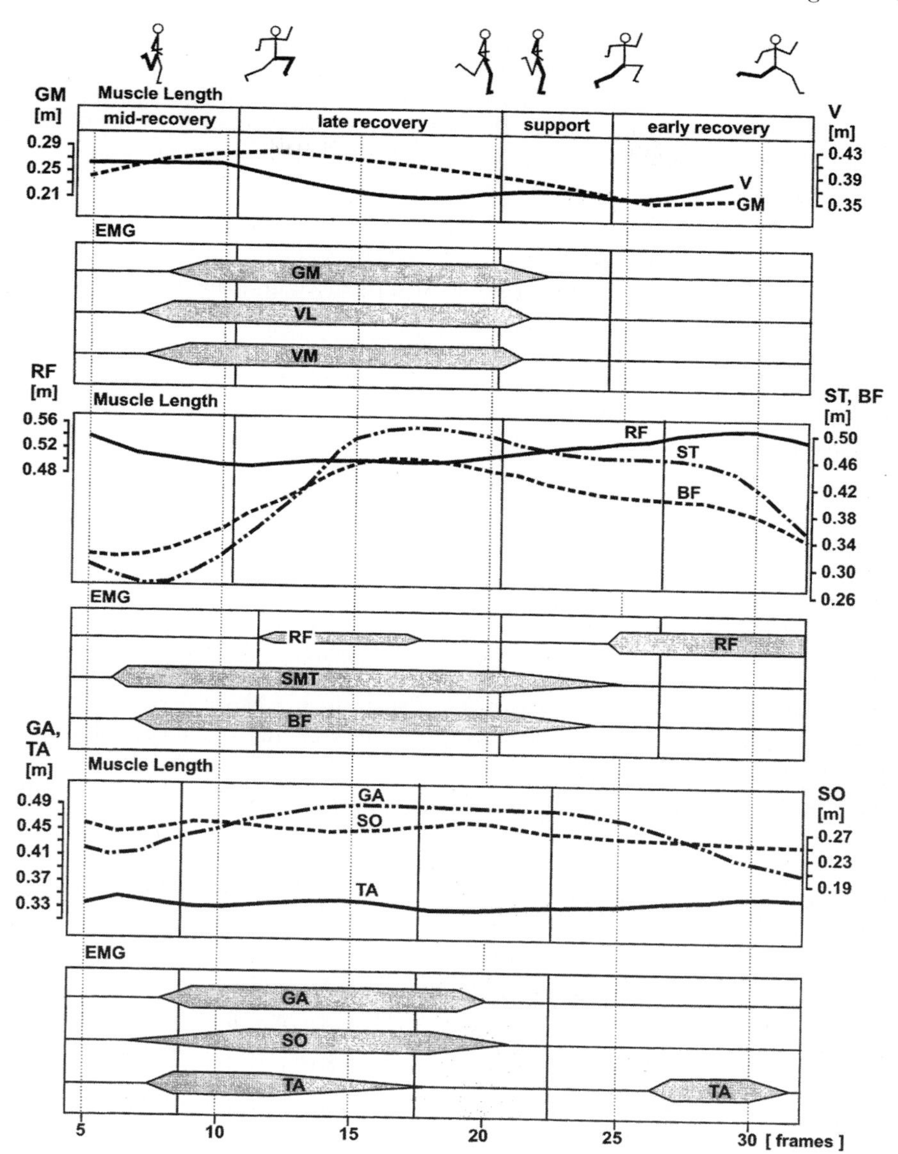

Fig. 3.30. Muscle length and electromyographic (EMG) activity vs. time during one running cycle of the right leg, for the gluteus maximus (GM), vastus lateralis (VL), biceps femoris (BF), vastus medialis (VM), combined VL and VM (V), rectus femoris (RF), semitendinosus (ST), combined semimembranosus (SM) and ST (SMT), gastrocnemius (GA), soleus (SO), and tibialis anterior (TA). (From [126]. Used with permission). Also see [145]

3.4.3 Energetics

Figure 3.27 shows that the body slows down during stage (a) when the right foot is planted on the ground, and it accelerates during stage (c) as the right foot propels the body forward, as in walking. This means that the horizontal kinetic energy

$$\mathrm{KE_H}(t = \text{stage b}) < \mathrm{KE_H}(t = \text{stage d}), \tag{3.35}$$

because stage (b) is right after (a) and stage (d) is right after (c). During stage (d) both feet are off the ground, the body is in the air, and the body center of mass and potential energy are relatively high. In contrast, during stage (b) the right foot is on the ground, the right leg is bent (while it is straight during walking) and so the center of mass and potential energy are relatively low. Consequently,

$$\mathrm{PE}(t = \text{stage b}) < \mathrm{PE}(t = \text{stage d}) \tag{3.36}$$

and so

$$\mathrm{KE_H}(t = \text{stage b}) + \mathrm{PE}(t = \text{stage b}) \ll \mathrm{KE_H}(t = \text{stage d}) + \mathrm{PE}(t = \text{stage d}). \tag{3.37}$$

We see that during running $\mathrm{KE_H}$ and PE (or really $\mathrm{PE+KE_V}$, with $\mathrm{KE_V}$ being relatively small) are not out of phase and their sum changes much during each stride (see Fig. 3.31). The pendulum model cannot be applied to running at all [99].

Not all of the lessened KE and PE during stage (b) is lost; some is saved within the body as stretched tendons and ligaments, as in a pogo stick. We

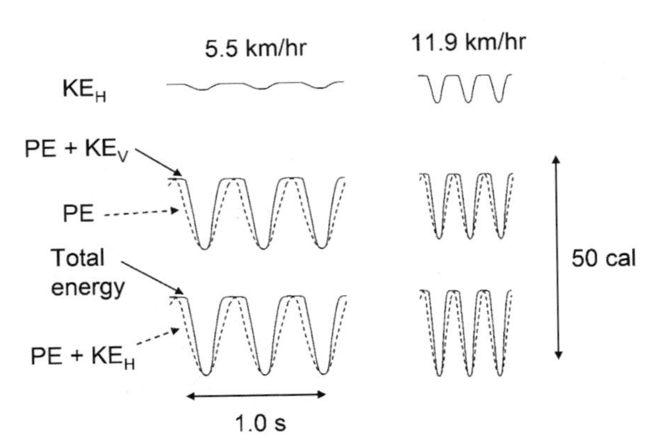

Fig. 3.31. Changes in mechanical energy during running at two different speeds from force plates, for (from top to bottom curve) horizontal kinetic energy; gravitational potential energy plus vertical kinetic energy and gravitational potential energy; and total energy and gravitational potential energy plus horizontal kinetic energy. (Based on [112] and [146])

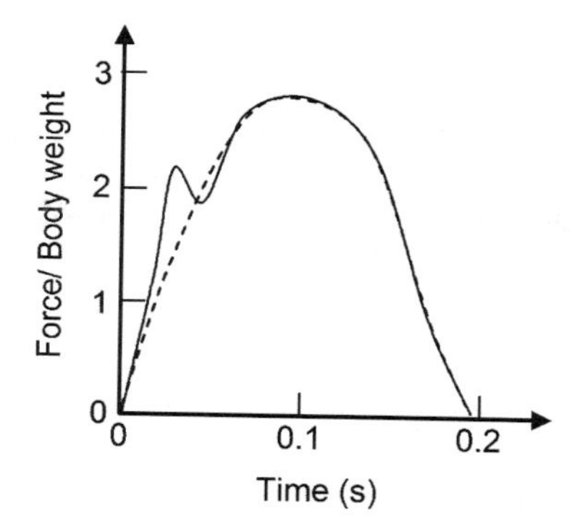

Fig. 3.32. Vertical force exerted on the ground during a running stride (*solid line*, with *dashed line* approximation). This was calculated from force plates and was averaged for five-runners running at 4.5 m/s. (Based on [99])

will examine this now and revisit it again later after we have learned a bit more about the elastic properties of parts of the body in Chap. 4.

Figure 3.32 shows a force plot of the vertical normal forces on the foot on the ground during a run at 4.5 m/s, which corresponds to a competitive speed during a marathon (2 h 37 min over 42.2 km or 26 miles, 385 yd). The peak force in this example is $2.7m_{\mathrm{b}}g = 2.7\ W_{\mathrm{b}}$. For a 70 kg person (700 N, 160 lb), this is $\sim$1,900 N. During sprinting, this force can increase to $3.6m_{\mathrm{b}}g$. (These forces are estimated later in this chapter in the discussion of collisions.)

The torque balance in the quasistatic condition when this normal force peaks at 1,900 N is depicted in Fig. 3.33, which is similar to the second class lever discussed in Chap. 2. In equilibrium, the Achilles tendon exerts an

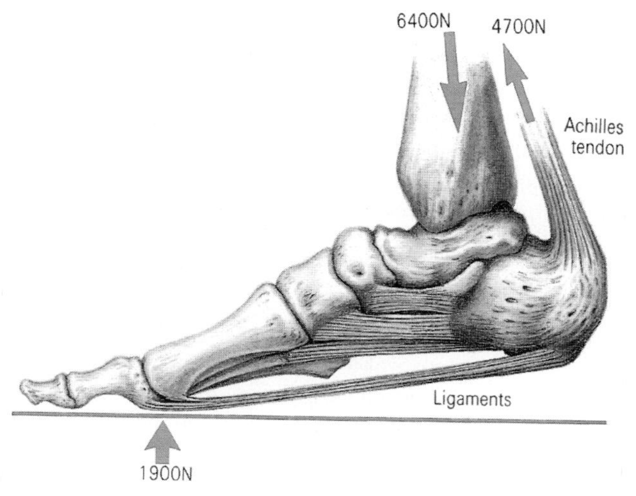

Fig. 3.33. Peak forces on foot during running stride. (From [99]. Copyright 1992 Columbia University Press. Reprinted with the permission of the press)

upward force of 4,700 N and there is a 6,400 N reaction force down on the foot from the rest of the body, because 4,700 N $\sim$ 1,900 N (120 mm/47 mm).

The Achilles tendon is also known as the calcaneal (kal-kane'-ee-ul) tendon. It attaches the triceps surae (sur-ee'), which includes the soleus and the medial and lateral heads of the gastrocnemius (gas-trok-nee'-mee-us) muscles (Figs. 1.8 and 3.4, Table 3.3), to the calcaneus (kal-cane'-ee-us), also known as the heel.

During this marathon-type run, there is a loss of about 100 J of kinetic energy each time the foot touches the ground and decelerates to a stop. Not all of it is lost. Some of it is stored in the body, and a portion of that is recoverable, mostly from body components that store the energy like springs. To maintain a steady average running speed, the portion that cannot be recovered must be supplied by the body in the acceleration phase of the stride. *How much does the body slow down during this step?* For given mass and speed, just calculate the change in speed when 100 J of kinetic energy is lost. (See Problem 3.22.)

Approximately 35 J of this 100 J is stored in stretching the Achilles tendon. We will see in the discussion about mechanical properties in Chap. 4 how this value is consistent with the forces exerted on the Achilles tendon. Of this 35 J, about 93% is recoverable during the acceleration phase, so this tendon acts almost like a friction-free spring. During the step, the foot arch flattens and it takes about 17 J to do this. About 80% of this is recoverable. Some energy is also lost to the quadriceps tendon over the kneecap. Much of the remainder is lost to muscles acting as brakes, and this is not recoverable. Overall, about 50 J is recoverable from the body's springs and about 50 J must be supplied by the muscles during the acceleration phase. The notion that 50% of the energy is recoverable energy is not universally accepted; Fig. 3.25 suggests that only about 5% is recoverable. The materials properties of the body are clearly significant – for analyzing the recoverable energy storage and the needed braking-action of muscles. We will examine both soon. The long bones in the leg can also store energy, but we will see that they store little energy. There are also questions about the efficiency and timing of the energy return, and potential confusions between the loss of kinetic energy during a step with the much larger ($\sim$5×) metabolic energy needed to replace this lost kinetic energy.

Good running shoes lessen the impact in each step. *Can they also help by storing energy as springs, thereby lessening the amount of energy needed in each step?* It is not clear that this occurs. Good running shoes are squeezed up to at most 10 mm in the shoe sole under the forefoot upon impact. They can store about 7 J, of which about 54–66% could be returned, so it would seem that they can help a bit. However, by cushioning the forces felt by the feet, significantly less recoverable energy is stored by the body. Even more cushioning would produce instability during stance and take off. Furthermore, even if much energy could be stored elastically in shoes, it is not clear if much of it would be recoverable. Stored energy has to be released at the right time, over the right duration (frequency), and at the right position [159]. The resonance frequencies of unloaded shoes are $\sim$100–200 Hz and loaded shoes likely 20–30 Hz. Because any stored energy in the heel needs to be released 100–200 ms after foot contact and it is really released 50–100 ms (∼1/20 Hz) after contact, the timing does not seem right. Also, if energy is stored in one part of the shoe because of impact (say the heel) it does little good if the extra "kick" is needed in the front of the foot for forward propulsion. Sprinters use running shoes with essentially no padding during races because their body returns more energy without the extra padding. During practice, they wear shoes with padding to limit the wear and tear on their legs, i.e., the stress on their tendons and so on.

A good running track can return up ∼12 J per step (Table 3.8, later). Such a track has a stiffness (spring constant) about 3× the lower leg stiffness (80,000 N/m), and can increase running speeds by ∼2%. Such an increase in speed decreases 100 m sprint times by ∼0.2 s and could decrease marathon times by ∼80 s. (Of course, marathons are not run on such a track.) These tracks can be tuned to have resonant frequencies of ∼2 Hz, which matches the times needed for the recovery of the stored energy. This impact of the step is a type of collision that will be discussed later in this chapter. This is also analyzed in a problem at the end of Chap. 4 (Problem 4.39).

Still, more of the energy lost per step is recoverable in walking than during running, perhaps somewhat over 70% (Fig. 3.25). Kangaroos recover about 50% during each hop.

Accelerating to Sprint Speed

The average speed for world class runners at the 100 and 200 m distances is about the same, ≃10.2 m/s. (See Table 6.34 later.) It is slower at longer distances, because the stores of energy that activate muscles at the fastest rate are depleted after 200 m (Chap. 6). It is also slower at shorter distances, because it takes several strides to achieve peak speeds, starting from a dead stop. We now model this acceleration process by examining the energetics after each stride [144]. The stride is step (c) in Fig. 3.27, with one foot on the ground propelling the body forward. The body consists of the other leg, which has mass m_{leg}, and the body above the legs (upper body), which has a mass $m_{\mathrm{b}} - 2m_{\mathrm{leg}}$, where m_{b} is the total body mass. (In this model we are ignoring the other stages: free flight, the slowing down of the body during touchdown, etc., and we are effectively incorporating all into this step (c).)

During the first stride, the foot on the ground, say the right foot, propels the upper body and the left leg to a speed with a force F as they move a distance L, and they attain a speed of v_1 at the end of the stride. The work done FL equals the kinetic energy in the upper body, $(m_{\mathrm{b}} - 2m_{\mathrm{leg}})v_1^2/2$, and that in the left leg, $m_{\mathrm{leg}}v_1^2/2$; the right foot is still on the ground, so the kinetic energy of the right leg is small and can be neglected. Therefore, after the first stride

$$FL = \frac{1}{2}(m_{\mathrm{b}} - 2m_{\mathrm{leg}})v_1^2 + \frac{1}{2}m_{\mathrm{leg}}v_1^2 = \frac{1}{2}(m_{\mathrm{b}} - m_{\mathrm{leg}})v_1^2 \tag{3.38}$$

and

$$v_1^2 = \frac{2FL}{m_{\rm b} - m_{\rm leg}}. \tag{3.39}$$

In the second stride the left foot is on the ground, does work FL, and accelerates the upper body from v_1 to v_2 and the right leg from 0 to v_2, so

$$\begin{aligned} FL &= \frac{1}{2}(m_{\rm b} - 2m_{\rm leg})(v_2^2 - v_1^2) + \frac{1}{2}m_{\rm leg}v_2^2 \\ &= \frac{1}{2}(m_{\rm b} - m_{\rm leg})v_2^2 - \frac{1}{2}(m_{\rm b} - 2m_{\rm leg})v_1^2 \end{aligned} \tag{3.40}$$

and

$$v_2^2 = \frac{1}{m_{\rm b} - m_{\rm leg}}(2FL + (m_{\rm b} - 2m_{\rm leg})v_1^2). \tag{3.41}$$

Using (3.39) for v_1^2

$$v_2^2 = \frac{2FL}{m_{\rm b} - m_{\rm leg}}\left(1 + \frac{m_{\rm b} - 2m_{\rm leg}}{m_{\rm b} - m_{\rm leg}}\right). \tag{3.42}$$

Similarly, after the third stride, with the right foot on the ground,

$$v_3^2 = \frac{1}{m_{\rm b} - m_{\rm leg}}(2FL + (m_{\rm b} - 2m_{\rm leg})v_2^2) \tag{3.43}$$

and using (3.42)

$$v_3^2 = \frac{2FL}{m_{\rm b} - m_{\rm leg}}\left(1 + \frac{m_{\rm b} - 2m_{\rm leg}}{m_{\rm b} - m_{\rm leg}} + \left(\frac{m_{\rm b} - 2m_{\rm leg}}{m_{\rm b} - m_{\rm leg}}\right)^2\right). \tag{3.44}$$

After n strides, v_n is obviously given by

$$\begin{aligned} v_n^2 = {} & \frac{2FL}{m_{\rm b} - m_{\rm leg}} \\ & \times \left(1 + \frac{m_{\rm b} - 2m_{\rm leg}}{m_{\rm b} - m_{\rm leg}} + \left(\frac{m_{\rm b} - 2m_{\rm leg}}{m_{\rm b} - m_{\rm leg}}\right)^2 + \cdots + \left(\frac{m_{\rm b} - 2m_{\rm leg}}{m_{\rm b} - m_{\rm leg}}\right)^n\right). \end{aligned} \tag{3.45}$$

The terms in the brackets form a geometric series $1 + x + x^2 + \cdots + x^n = (1 - x^n)/(1 - x)$ for $0 < x < 1$, so

$$v_n^2 = \frac{2FL}{m_{\rm b} - m_{\rm leg}} \frac{1 - \left(\frac{m_{\rm b}-2m_{\rm leg}}{m_{\rm b}-m_{\rm leg}}\right)^n}{1 - \left(\frac{m_{\rm b}-2m_{\rm leg}}{m_{\rm b}-m_{\rm leg}}\right)} = \frac{2FL}{m_{\rm leg}}\left(1 - \left(\frac{m_{\rm b} - 2m_{\rm leg}}{m_{\rm b} - m_{\rm leg}}\right)^n\right). \tag{3.46}$$

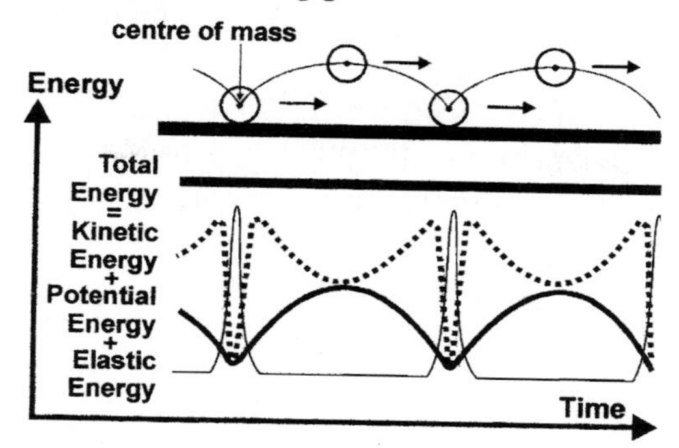

Fig. 3.34. Bouncing ball model of running, showing that the sum of the ball kinetic energy (*dotted line*), gravitational potential energy (*thick line*), and elastic potential energy (*thin line*) is constant (*very thick line*) during bounces. (From [151]. Used with permission)

The final speed is (with $n \to \infty$)

$$v_{\text{final}} = \sqrt{\frac{2FL}{m_{\text{leg}}}}. \tag{3.47}$$

Table 1.7 shows that $m_{\text{leg}} = 0.161 m_{\text{b}}$, so a 70 kg sprinter, with $F = 560\,\text{N}$ and $L = 1.0\,\text{m}$ stride, attains a final speed of 10.0 m/s.

3.4.4 Bouncing Ball/Pogo Stick Model

Much as walking can be modeled as a pendulum, running can be modeled using mechanical analogies. Figure 3.34 shows the bouncing ball model. This is fairly equivalent to the even simpler pogo stick model, which does not have the complication of rotational ball motion. When the ball or stick is at its highest point, the gravitational PE is maximized, the KE is zero, and there is no stored *elastic energy* EE in the ball due to compression of the ball or in the pogo stick due to compression of the spring (which is also a form of potential energy). When it is about to touch the ground, the PE is almost minimized, the KE is almost maximized and the EE is still zero. During the elastic collision with the ground, the EE increases while the KE goes to zero, and then all of the EE is recovered, initially mostly as KE. The total energy TE = KE + PE + EE is always constant [151]. There is exchange between the sum of the kinetic and gravitational potential energies with the elastic energy in a bouncing ball.

There is an analogy between running and this model, but it is not very strong. The PE and KE are both maximized during stage (d) in Fig. 3.25, while EE is zero. During stage (b), PE and KE are both minimized, and

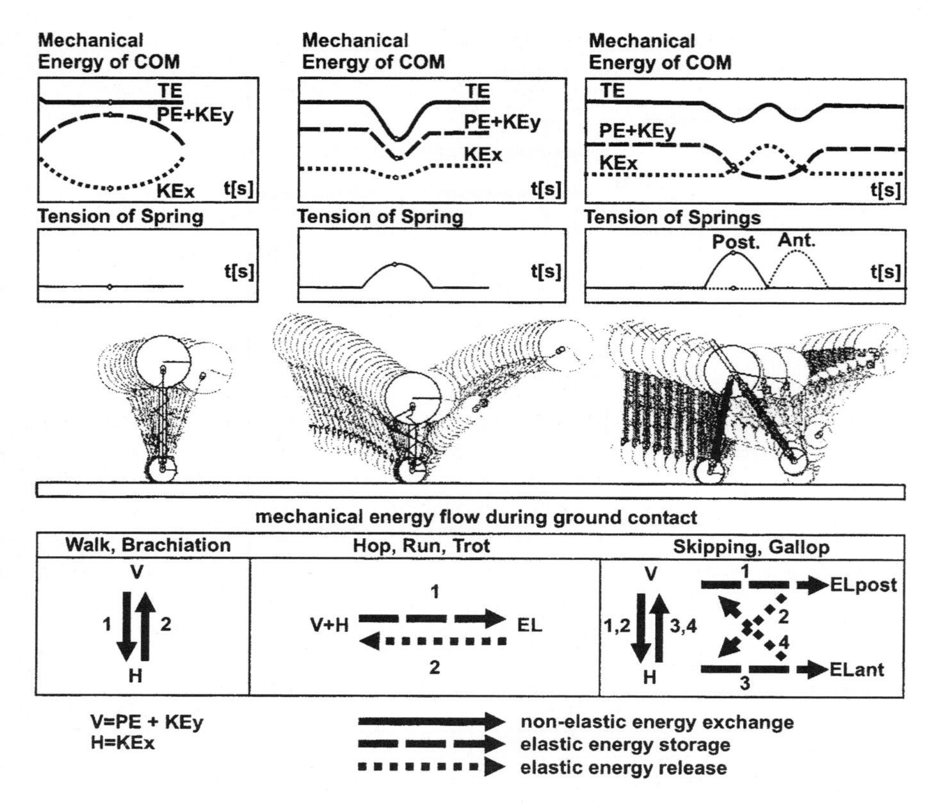

Fig. 3.35. Three mechanical models of locomotion, the (inverted) pendulum (left, for walking), single pogo stick (middle, for hopping and running), and double pogo stick (right, for skipping), with tension in pogo stick spring and the modes of energy exchange. (From [151]. Used with permission)

presumably there is EE stored in the tendons, foot arch, etc. Consequently, the phase of KE is different from that in the model. However, the analogy is fine for the phases of PE + KE = TE − EE and of EE. Unlike walking, there is relatively little stored energy returned in running (Fig. 3.25); it is not very elastic.

Figure 3.35 shows the pogo stick model for running and a double pogo stick model for skipping. Skipping is a third, and fairly uncommon gait for bipeds. It differs from walking in that it has a significant flight phase, and from running in that both feet can be on the ground (Table 3.6). Skipping is energetically very efficient: about 50% of the energy is recoverable (Fig. 3.25). These models for biped locomotion can also be used to model quadruped locomotion (Fig. 3.35). Whereas there are two main types of locomotion in bipeds (walking, running), there are three for quadrupeds (walking, trotting, galloping). The types of quadruped motion are described in Table 3.6.

Table 3.6. Comparison of motion in bipeds and quadrupeds

biped type	foot sequence[a]	features
walk	R, L, R, L, etc.	no flight phase
run	R, flight, L, flight, etc.	
skip		
right unilateral skipping	L, R, flight; L, R, flight; etc.	left unilateral is R, L, flight, etc.
bilateral skipping	L, R, flight; R, L, flight; etc.	
quadruped type (biped analog)	**foot sequence[b]**	**features**
walk (walk)	FL, HR, FR, HL	stride: up to 3 hooves on ground
trot (run)	two running bipeds w/50% phase lag	rack is trot with 0% phase lag
gallop (none)		
slow gallop or canter	HL, HR, FL simul.; FR flight	right cantor; left switches R and L
transverse gallop (fast run)	HL, HR flight; FL, FR flight	right transverse
rotary gallop (fast run)	HL, HR flight; FR, FL flight	clockwise or counterclockwise

[a]R right foot, L left foot.
[b]FL: fore left, HR: hind right, FR: fore right, HL: hind left. (From [151].)

3.5 Jumping

We will examine the vertical jump and pole vault here and the high jump and long jump in the problems at the end of this chapter (Problems 3.34–3.37, and 3.38–3.40, respectively). In the last three athletic endeavors, kinetic energy from running is converted into potential energy. The vertical jump is a half-collision with the floor. (Here the half-collision being the second half of a collision with the floor.) Other types of jumping, as on trampolines and in others common in gymnastics, involve full collisions with surfaces; how much energy these surfaces return to the jumper becomes important. This is addressed in the discussion of collisions.

3.5.1 Vertical Jump

How high can you jump? It depends on how fast you can take off. The four stick diagrams in Fig. 3.36 show the four stages of the vertical jump.

First the person stands upright. The center of mass is about 1.0 m from the ground, 5 cm above the hip joints. Figure 3.36b shows the body after

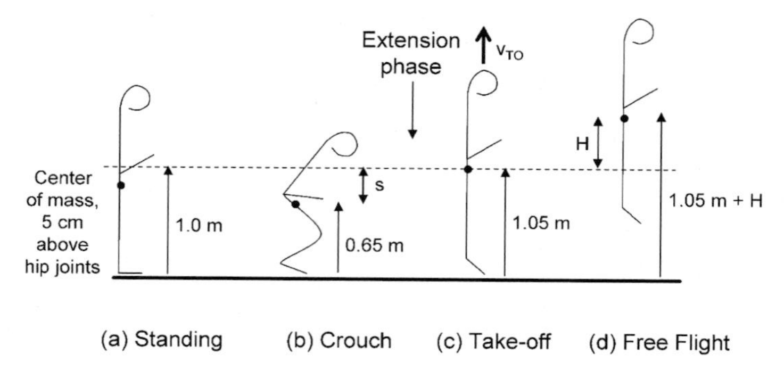

Fig. 3.36. Stick diagram of a vertical jump, including initial standing in (**a**), crouching in (**b**) and then extension to take-off in (**c**) and free flight in (**d**). (Based on [161])

the *crouch*, after which the center of mass is about 0.65 m above the ground. The *extension* phase is next, which ends in *takeoff* shown in the third diagram. At takeoff (Fig. 3.36c) the center of mass is about 1.05 m above the ground, a bit higher than in Fig. 3.36a since the person is on his or hers toes. The center of mass rises by a distance s during extension, and attains a vertical speed of v_{TO} at takeoff. After takeoff the person is in *free flight* for a distance H and then the vertical speed v is zero, as shown in Fig. 3.36d.

How high can you jump? The kinetic energy at takeoff is all converted into potential energy at the apex of free flight. Therefore

$$\frac{1}{2}m_{\mathrm{b}}v_{\mathrm{TO}}^2 + m_{\mathrm{b}}g(1.05\ \mathrm{m}) = 0 + m_{\mathrm{b}}g(1.05\ \mathrm{m} + H) \tag{3.48}$$

$$\frac{1}{2}m_{\mathrm{b}}v_{\mathrm{TO}}^2 = m_{\mathrm{b}}gH, \tag{3.49}$$

so $H = v_{\mathrm{TO}}^2/2g$ and the faster the takeoff speed the higher the jump.

During extension both legs generate a normal force $N(t)$ that leads to a net vertical force $F_{\mathrm{V}}(t) = N(t) - m_{\mathrm{b}}g$. Figure 3.37 shows this normal force in measurements made on several men in the early 1930s at Columbia University [161]. This vertical force increases and then decreases, and while it is applied the center of mass accelerates upward and travels a distance s. The work W done on the center of mass during this phase is

$$W = \int_0^s F_{\mathrm{V}}(t)\mathrm{d}z. \tag{3.50}$$

(Note that to perform this integration $F_{\mathrm{V}}(t)$ needs to be converted to a function of z.) This gets converted into the vertical kinetic energy so

$$\int_0^s F_{\mathrm{V}}(t)\mathrm{d}z = \frac{1}{2}m_{\mathrm{b}}v_{\mathrm{TO}}^2 = m_{\mathrm{b}}gH \tag{3.51}$$

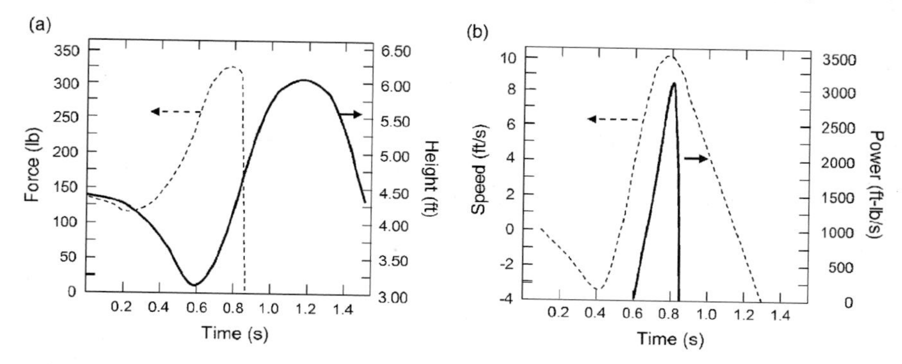

Fig. 3.37. Kinematic and dynamic data during a vertical jump, including (**a**) reaction force from the ground (left scale) and height of the center of mass (right scale), and (**b**) vertical speed of the center of mass (left scale) and applied power (right scale, this power is the product of the force in (**a**) and speed in (**b**)). (Based on [161])

and

$$H = \frac{\int_0^s F_V(t)\mathrm{d}z}{m_b g}. \tag{3.52}$$

For $W_b = m_b g = 140\,\mathrm{lb}$ (64 kg, 620 N), an average normal force during extension $\langle N(t)\rangle$ of 300 lb (Fig. 3.37), and $s = 1.4\,\mathrm{ft}$ (0.43 m), we find

$$H = \frac{\int_0^s F_V(t)\mathrm{d}z}{m_b g} = \frac{(N(t) - m_b g)s}{m_b g} \tag{3.53}$$

$$= \frac{(300\,\mathrm{lb} - 140\,\mathrm{lb})1.4\,\mathrm{ft}}{140\,\mathrm{lb}} = 1.6\text{ ft }(0.49\text{ m}). \tag{3.54}$$

How long is takeoff? Let us assume that the vertical acceleration is constant and equal to a during extension. (This would be true only if the vertical normal force were constant – and it is not.) From $F_V = m_b a$

$$a = \frac{N(t) - m_b g}{m_b}. \tag{3.55}$$

If the duration of takeoff is τ

$$v_{TO} = a\tau \tag{3.56}$$

and using

$$s = \frac{1}{2}a\tau^2 \tag{3.57}$$

we see

$$\tau^2 = 2\left(\frac{m_b g}{N(t) - m_b g}\right)\left(\frac{s}{g}\right) = 2\left(\frac{140\text{ lb}}{160\text{ lb}}\right)\left(\frac{1.4\text{ ft}}{32.2\text{ ft/s}^2}\right) \tag{3.58}$$

because $\langle N(t)\rangle - m_\mathrm{b}g = 300\,\mathrm{lb} - 140\,\mathrm{lb} = 160\,\mathrm{lb}$ (710 N). The acceleration occurs over a time $\tau = 0.28$ s.

How are the jump height and takeoff time related? Dividing (3.56) by (3.57) eliminates a to give

$$v_\mathrm{TO} = \frac{2s}{\tau}, \tag{3.59}$$

so

$$H = \frac{v_\mathrm{TO}^2}{2g} = \frac{(2s/\tau)^2}{2g} = \frac{2s^2}{g\tau^2}. \tag{3.60}$$

If you can decrease the extension time, you can jump much higher! We will see that extension times are shorter in people with larger fractions of fast-twitch muscles in their legs. However, there can be a trade-off with shorter extension times (Chap. 5). When muscles achieve steady-state motion, they develop less tension the faster they contract, which is much less than the isometric (fixed length) force that they can generate. Measurements of force on the floor using force plates have shown that while the peak force on a representative runner's foot might be as high as 1,710 N, the force on each foot during his high jump turns out to be much less, 715 N [99]. This surprising observation is explained by the faster muscle contraction during jumping. Average forces are also expected to be lower in the high jump because muscles develop less tension when they are much longer or shorter than their resting length. Also, the momentum arm about the knee joint is smaller for the bent knee in the squatting position of a vertical jump.

Figure 3.38 shows a model of two variations of the vertical jump, which includes muscle activation, tendon stretching, and body response [101]. The squat jump starts in a squatting position. Initially, as the muscles contract there is only slow movement of the knees and consequently much tendon stretching. Due to fast contraction, muscle forces in the squat jump are smaller

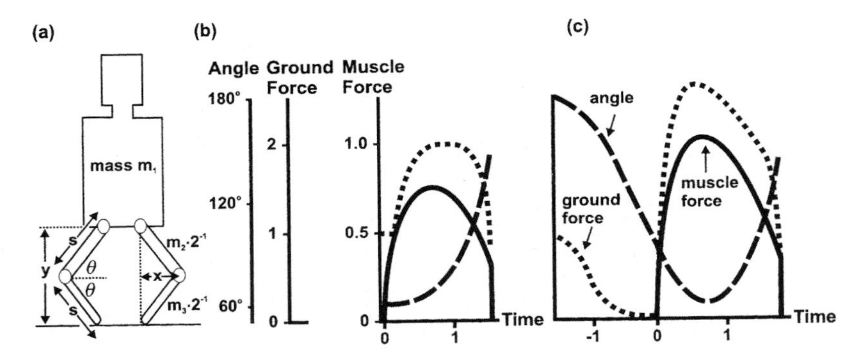

Fig. 3.38. (**a**) A mechanical model of vertical jumping, with (**b**) a simulated jump and (**c**) a simulated countermovement jump, with muscle force, ground reaction force, and angle plotted vs. time in (**b**) and (**c**). (From [101], as from [100]. Used with permission)

than (0.2×) the isometric level. Just before liftoff these muscle forces and tendon recoil produce forces 0.4× the isometric level. The countermovement jump starts with an upright person first falling freely to the starting position of the squat and then jumping upward. The muscle forces initially generated are larger than the isometric values because the muscles are initially braking (eccentric contraction, Chap. 5), and this leads to even more tendon stretching and upward force generated, and a higher elevation jump than the squat jump. This countermovement jump is higher also because muscles are stretched immediately before they are contracted, leading to greater forces during this contraction (as in the later phases of this jump). (This effect is not included in the Fig. 3.38 model.)

3.5.2 Pole Vault

The pole vault is not really a jump, but a propulsion. Still, it is fairly similar to the long jump in that horizontal kinetic energy is converted to a propulsion. More specifically, the pole vaulter's horizontal kinetic energy is mostly converted into elastic potential energy stored in the pole, which is then mostly converted to gravitational potential energy of the pole vaulter (Fig. 3.39). A fast running pole vaulter can propel above and beyond the bar if the two "mostly"s in the previous sentence are really "essentially all"s. This happens with good techniques and good poles. To propel over the bar there has to be some remnant horizontal kinetic energy, but this necessary amount is minimal.

Figure 3.39 shows the relevant heights: the height of the center of mass at takeoff (the takeoff height, H_1), the maximum increase in the height of the

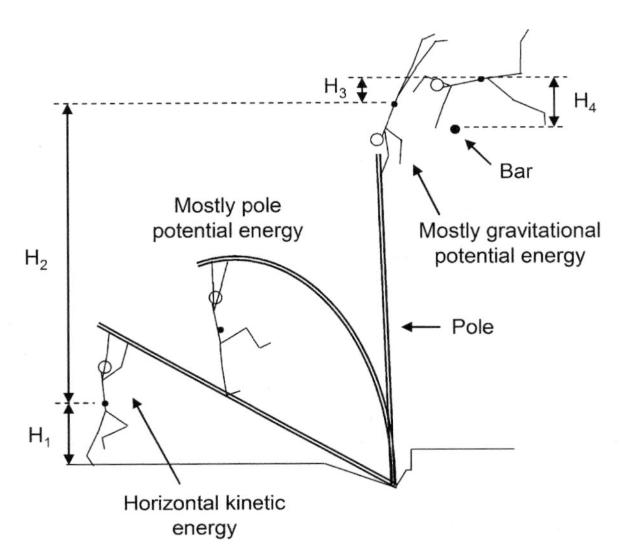

Fig. 3.39. Schematic of the heights involved in the pole vault. (Based on [127])

pole vaulter's center of mass while on the pole (the swing height, H_2), the further increase in height due to vertical kinetic energy after release of the pole (the flight height, H_3), and maximum height of the center of mass above the cross bar (the clearance height, H_4 – which could be negative). The takeoff angle of world-class pole vaulters using fiberglass poles is $\simeq$13–15° (relative to the vertical).

The ideal case of total conversion of one type of energy to another gives a surprisingly good prediction of actual pole vaulting performance. Initially the pole vaulter accelerates to a speed $v_{\mathrm{H,i}}$ and has a center of mass h_{CM} above the ground (H_1 in Fig. 3.39). During running h_{CM} is approximately 0.9 m, corresponding to the slightly crouched position during running. The goal is to convert the initial $E_{\mathrm{H,i}} = mv_{\mathrm{H,i}}^2/2$ into the potential energy. Over the bar, the pole vaulter has a center of mass $h_{\mathrm{bar}} + h_{\mathrm{min}}$, where h_{bar} is the height of the bar and h_{min} (H_4 in Fig. 3.39) is the distance the center of mass of the vaulter needs to be above the bar without touching it, which we will take as $\sim$0.1 m. This corresponds to an increase in the height of the center of mass by $h_{\mathrm{bar}} + h_{\mathrm{min}} - h_{\mathrm{CM}}$. A good pole vaulter can achieve a speed $v_{\mathrm{H,i}}$ $\sim$9.5 m/s. (World-class sprinters in a 100 m race can attain speeds of 10.2 m/s.)

If the initial kinetic energy is converted into potential energy over the bar:

$$\frac{1}{2}mv_{\mathrm{H,i}}^2 = mg(h_{\mathrm{bar}} + h_{\mathrm{min}} - h_{\mathrm{CM}}) \tag{3.61}$$

$$h_{\mathrm{bar}} = h_{\mathrm{CM}} - h_{\mathrm{min}} + \frac{v_{\mathrm{H,i}}^2}{2g}. \tag{3.62}$$

With $g = 9.8\,\mathrm{m/s^2}$, we see that $h_{\mathrm{bar}} = 5.4$ m. In a good pole vault $h_{\mathrm{bar}} \sim 5.4$ m; the world record is $\sim$6.14 m (in 2006). (See Problem 3.33.) Of course, the pole vaulter is still moving forward over the bar and so v_{H} does not actually become zero. One feature that is not included in this analysis is the athlete extending his or her hands just before releasing the pole to improve performance.

The pole is very elastic. It stores and returns energy to the pole vaulter much more efficiently than do human body components during jumping and running. There is an optimal pole stiffness that maximizes performance [171]. If the pole is too stiff, it straightens before the athlete is at the maximum height and the pole pushes the athlete horizontally away from the bar. If the pole is not stiff enough, it straightens too slowly and the athlete passes the bar before attaining the highest position.

3.6 Throwing a Ball

The biomechanics of throwing a ball are quite complex. It involves the coordinated motion of many muscles about several body joints, as is also true for running, jogging, etc. We will model throwing a ball. In doing so, we will greatly simplify the problem and subsequently reassess many of our initial

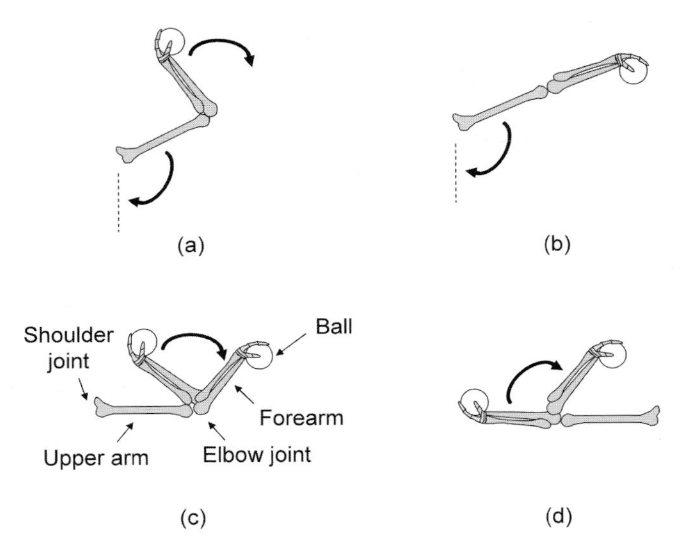

Fig. 3.40. Models of throwing a ball with (**a**) motion of the shoulder and elbow (extension), (**b**) motion of the shoulder only with a fixed arm, (**c**) extension of the elbow and a stationary shoulder, and (**d**) flexion of the elbow and a stationary shoulder

assumptions. The first assumption is that only the arm is involved in throwing. This is a major assumption.

The diagrams in Fig. 3.40 shows four models of throwing a ball overhand. One can consider throwing by:

(a) The entire motion with motion in the shoulder and elbow joints (Fig. 3.40a).
(b) Motion of a stiff arm (elbow fixed) – motion only in the shoulder joint, using the deltoid muscles (Fig. 3.40b).
(c) Motion of only the forearm (lower arm) (shoulder fixed) – motion only in the elbow joint using the extensor triceps brachii (Fig. 3.40c).
(d) Motion of only the forearm (shoulder fixed) – motion only in the elbow joint using the flexor biceps brachii (Fig. 3.40d).

Similar diagrams can be drawn showing several analogous ways of throwing a ball underhand, which involve these same (a), (b), and (d) models.

Each of these models represents a fairly unnatural motion for throwing (try them!), but they represent good first-order models. Model (a) involves two-body joints, and consequently is a two-segment model. Models (b)–(d) involve one-body joint and are one-segment models. More sophisticated models require a multisegment model.

We will analyze throwing with the single-segment model (d) where the forearm moves due to the flexor biceps brachii contraction, which causes the forearm to rotate about the elbow joint. This can be a model of either overhand or underhand throwing. Although it is very simple, this model has several

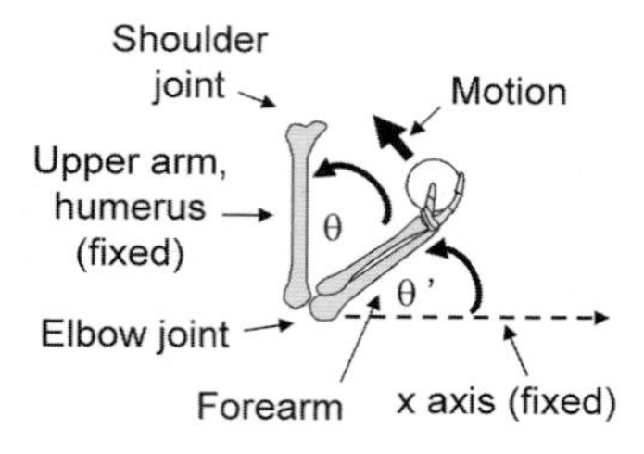

Fig. 3.41. Angles involved in throwing a ball, include the angles of the rotating forearm relative to the x-axis (θ') and the fixed upper arm (θ)

complexities and we will analyze each one. Our goal is to see how fast we can throw a ball, specifically a baseball. We will first assume that the biceps brachii are the only muscles used to flex the forearms; later we will examine other muscles that can contribute to form an "effective biceps" muscle.

We have analyzed the torque in this system already, in Chap. 2. As seen in Fig. 2.9d (Case 3)

$$\sum \tau_z = Md_{\mathrm{M}} \sin\theta - W_{\mathrm{F}} d_{\mathrm{F}} \sin\theta - W_{\mathrm{B}} d_{\mathrm{B}} \sin\theta$$
$$= (Md_{\mathrm{M}} - W_{\mathrm{F}} d_{\mathrm{F}} - W_{\mathrm{B}} d_{\mathrm{B}}) \sin\theta, \quad (3.63)$$

with the weight in the hand now called a ball (B). For simplicity, we first assume that the humerus is vertical. In a static situation the torques add up to zero. Here they do not. For our desired arm motion, the sum must be positive.

Figure 3.41 explains a small subtlety concerning the defined angle θ, which is the angle the moving forearm makes with the fixed humerus (upper arm). The forearm makes an angle θ' with the fixed x-axis. In relating the net torques to the second derivative of the angle, as in (3.63), the dynamical variable is actually what we have called θ' here, so

$$\sum \tau_z = I \frac{\mathrm{d}^2\theta'}{\mathrm{d}t^2}. \quad (3.64)$$

Because $\theta + \theta' = \pi/2$ $(= 90°)$, a fixed quantity, we see that $\mathrm{d}\theta/\mathrm{d}t + \mathrm{d}\theta'/\mathrm{d}t = 0$ and $\mathrm{d}^2\theta/\mathrm{d}t^2 + \mathrm{d}^2\theta'/\mathrm{d}t^2 = 0$, so

$$\frac{\mathrm{d}^2\theta'}{\mathrm{d}t^2} = -\frac{\mathrm{d}^2\theta}{\mathrm{d}t^2} \quad (3.65)$$

and from (3.25)

$$\sum \tau_z = -I \frac{\mathrm{d}^2\theta}{\mathrm{d}t^2} \quad (3.66)$$

or

$$\frac{\mathrm{d}^2\theta}{\mathrm{d}t^2} = -\frac{\sum \tau_z}{I}. \quad (3.67)$$

Now we need to evaluate the moment of inertia I and the torques. The moment of inertia about the elbow pivot equals the sum of the components due to the arm and the ball. Using (3.23) and (3.30) this is

$$I = \frac{1}{3} m_{\mathrm{F}} L^2 + m_{\mathrm{B}} L^2, \tag{3.68}$$

where L is the length of the forearm ($0.146H$) plus half the hand ($0.108H/2$), which is $L \sim 0.2H \sim 36$ cm. The weight of the forearm (with the whole hand) $\sim 0.022 W_{\mathrm{b}}$, so $m_{\mathrm{f}} \sim 2$ kg for a 90 kg player. (We are modeling a 6 ft tall (1.8 m), 200 lb (90 kg) major leaguer.) A regulation major league baseball must weigh between 5 and 5 $\frac{1}{4}$ oz (Official Baseball Rules (Major League Baseball), Rule 1.09), which corresponds to a mass of 0.146 kg. This gives $I = 1,053$ kg-cm^2. (These are mixed MKS/CGS units. This is often not the best practice, but is fine if we are careful.)

Evaluating the components of torque due to the forearm and ball is straightforward. Using $W = mg$,

$$W_{\mathrm{F}} d_{\mathrm{F}} = (2\ \mathrm{kg})(9.8\ \mathrm{m/s^2})(18\ \mathrm{cm}) = 353\ \mathrm{N\text{-}cm} \tag{3.69}$$

$$W_{\mathrm{B}} d_{\mathrm{B}} = (0.146\ \mathrm{kg})(9.8\ \mathrm{m/s^2})(36\ \mathrm{cm}) = 51.5\ \mathrm{N\text{-}cm}, \tag{3.70}$$

for a total of 404 N-cm.

In evaluating the torque due to the muscle, we will initially make some assumptions. (1) We assume that the distance from the point of biceps brachii insertion to the pivot axis, d_{M}, is 4 cm, independent of the arm angle θ. Figure 3.42 shows that $d_{\mathrm{M}} = d_{\mathrm{M}}(\theta)$ and so including this variation would change our final answer a bit, but it will not change our overall conclusions. (2) We assume a fixed value of the maximum muscle force M, to see how fast we can throw a ball. As we will see in Chap. 5, this is an assumption because (a) M depends on muscle length (l_{M}), which depends on θ, and (b) M depends on the muscle contraction speed $v_{\mathrm{M}} = \mathrm{d}l_{\mathrm{M}}/\mathrm{d}t$, which depends on $\mathrm{d}\theta/\mathrm{d}t$. We will revisit several of these assumptions in the problems in Chap. 5. For now we use the fact that most muscles exert a maximum force per unit cross-sectional area, which we will initially take as ~ 20 N/cm^2, a relatively modest value.

For biceps brachii with a 2 in diameter, the cross-sectional area is $\pi(1\ \mathrm{in})^2 = 20.3$ cm^2, so $M = 405$ N and $M d_{\mathrm{M}} = 1,620$ N-cm. So (3.63) becomes

$$\sum \tau_z = (1{,}620\ \mathrm{N\text{-}cm} - 404\ \mathrm{N\text{-}cm}) \sin\theta = (1{,}216\ \mathrm{N\text{-}cm}) \sin\theta \tag{3.71}$$

and

$$\frac{\mathrm{d}^2\theta}{\mathrm{d}t^2} = -\frac{\sum \tau_z}{I} = -\frac{(1{,}216\ \mathrm{N\text{-}cm}) \sin\theta}{1{,}053\ \mathrm{kg\text{-}cm^2}}. \tag{3.72}$$

Because $1\ \mathrm{N} = 1\ \mathrm{kg\text{-}m/s^2} = 100\ \mathrm{kg\text{-}cm/s^2}$,

$$\frac{\mathrm{d}^2\theta(t)}{\mathrm{d}t^2} = -116/\mathrm{s}^2 \sin\theta(t), \tag{3.73}$$

where we have expressed $\theta(t)$ as an explicit a function of time.

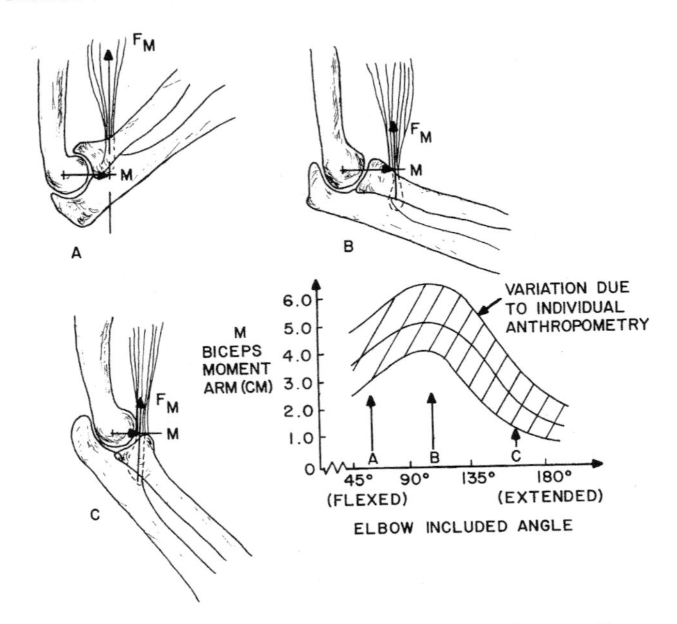

Fig. 3.42. Variation of the moment arm of biceps brachii vs. elbow angle. (From [113]. Reprinted with permission of Wiley)

We want to solve this equation to find the speed of the baseball at the end of the throwing motion. The speed of the ball at any time t is $v_{\mathrm{B}} = |L\mathrm{d}\theta(t)/\mathrm{d}t|$. Before examining the end of the motion, let us examine the conditions at the beginning of the throw, the so-called initial conditions. From the diagram in Fig. 3.43a, we see that at the beginning, $t = 0$, the arm is straight, so $\theta(t = 0) = \pi (= 180°)$. Because the ball is still, $\mathrm{d}\theta(t = 0)/\mathrm{d}t = 0$. At the end of the throwing motion the ball is released, say at t_{final}. This could occur at $\theta = 0$, so $\theta(t_{\mathrm{final}}) = 0$. In a normal throwing motion the release occurs before $\theta = 0$. In any case, if the release time is t_{final}, the ball leaves the hand with a speed $v_{\mathrm{B}} = |L\mathrm{d}\theta(t_{\mathrm{final}})/\mathrm{d}t|$.

We will restrict the range of motion for θ from $3\pi/4$ to $\pi/4$ (135° to 45°) (Fig. 3.43b) because this range involves the most efficient region for the muscles. So we see that

$$\theta(t = 0) = 3\pi/4 \qquad \mathrm{d}\theta(t = 0)/\mathrm{d}t = 0 \tag{3.74}$$

and

$$\theta(t = t_{\mathrm{final}}) = \pi/4 \qquad \mathrm{d}\theta(t = t_{\mathrm{final}})/\mathrm{d}t = ?, \tag{3.75}$$

where the last quantity will give us the final answer.

We can numerically integrate (3.73) to find the final speed; however, we will make some further approximations that will simplify the solution. We

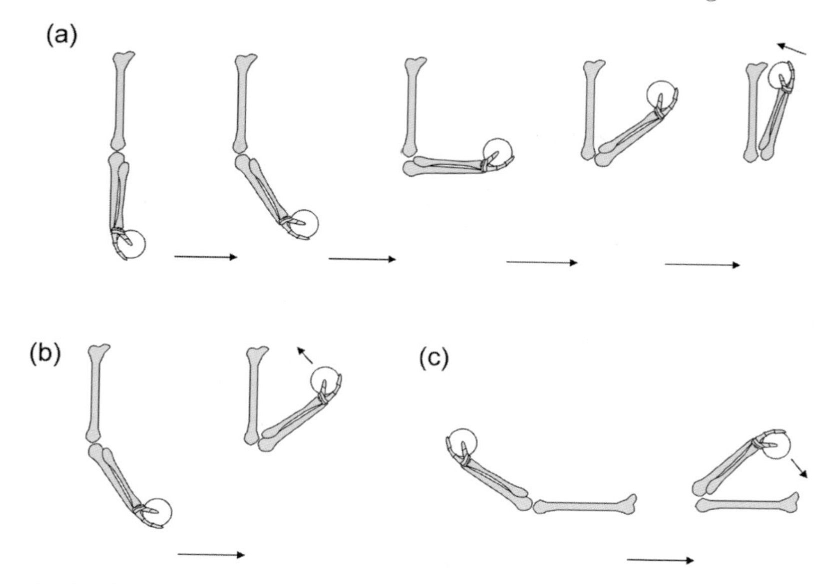

Fig. 3.43. Throwing models with (**a**) 180° rotation, (**b**) 90° rotation for vertical upper arm, and (**c**) 90° rotation for horizontal upper arm. The direction of the velocity vector of the released ball is shown in the last segment of each. Clearly, in this simple model a different position of the upper arm is needed to release the ball horizontally

approximate $\sin\theta(t)$ by a constant. Over the restricted angle range of interest, we replace $\sin\theta(t)$ by $\langle\sin\theta\rangle = 0.707$ ($= \sin\pi/4 = \sin 3\pi/4$). This will lead to a slight underestimate of the speed because $\sin\theta(t)$ varies from this value to 1.0. Now (3.73) becomes

$$\frac{\mathrm{d}^2\theta}{\mathrm{d}t^2} = -116/\mathrm{s}^2\langle\sin\theta\rangle = -81.7/\mathrm{s}^2 = -\alpha. \tag{3.76}$$

This is easily solved. (See Appendix C for more information about the solution.) After integrating both sides from time 0 to t

$$\frac{\mathrm{d}\theta}{\mathrm{d}t} = -\alpha t + c_1. \tag{3.77}$$

c_1 is determined from the initial condition that $\mathrm{d}\theta(t=0)/\mathrm{d}t = 0$, which gives $c_1 = 0$ and

$$\frac{\mathrm{d}\theta}{\mathrm{d}t} = -\alpha t. \tag{3.78}$$

This equation is integrated again to give

$$\theta(t) = -\frac{1}{2}\alpha t^2 + c_2. \tag{3.79}$$

Because $\theta(t=0) = 3\pi/4$, we find $c_2 = 3\pi/4$ and $\theta(t) = -\alpha t^2/2 + 3\pi/4$. At t_{final}, $\theta(t = t_{\mathrm{final}}) = \pi/4$, so

$$\pi/4 = -\frac{1}{2}\alpha t_{\mathrm{final}}^2 + \frac{3\pi}{4}, \tag{3.80}$$

which gives $t_{\mathrm{final}} = (\pi/\alpha)^{1/2}$. Using (3.78), $|\mathrm{d}\theta(t = t_{\mathrm{final}})/\mathrm{d}t| = |-\alpha t_{\mathrm{final}}| = (\pi\alpha)^{1/2}$. The speed of the ball when it is released by the hand is

$$v_{\mathrm{B}} = \left| L\frac{\mathrm{d}\theta(t = t_{\mathrm{final}})}{\mathrm{d}t} \right| = \sqrt{\pi\alpha}. \tag{3.81}$$

For $L = 36\,\mathrm{cm}$ this is $v_{\mathrm{B}} = 63.8(\alpha)^{1/2}$, in units of cm/s with α in units of $1/\mathrm{s}^2$.

Now we can evaluate several specific cases:

Case a: For these 2 in diameter biceps brachii with $\langle \sin\theta \rangle = 0.707$, we see that $\alpha = 81.7/\mathrm{s}^2$, $t_{\mathrm{final}} = 0.2\,\mathrm{s}$ and $v_{\mathrm{B}} = 577\,\mathrm{cm/s} = 5.77\,\mathrm{m/s} = 18.9\,\mathrm{ft/s} = 12.9\,\mathrm{mph}$. (Remember, 60 mph = 88 ft/s = 26.8 m/s = 96.6 km/h.) Because power major league baseball pitchers can throw a fastball at speeds approaching 100 mph, our calculated speed is very slow. Maybe we should expect a more modest speed of say ~70 mph, but our result is still much too slow. Let us reconsider our assumptions one by one.

Case b: Let us choose a larger value for $\langle \sin\theta \rangle$ say 1.0 (because of the smaller range of θ), so now we see that $\alpha = 115.5/\mathrm{s}^2$ and $t_{\mathrm{final}} = 0.165\,\mathrm{s}$. Now $v_{\mathrm{B}} = 6.86\,\mathrm{m/s} = 22.5\,\mathrm{ft/s} = 15.3\,\mathrm{mph}$. This is still too slow.

Case c: Now let us make our muscles bigger, so they have a 3 in diameter. Because the maximum muscle force is proportional to area, the torque provided by the muscles is now $1{,}620\,\mathrm{N\text{-}cm}\,\sin\theta \times 9/4 = 3{,}645\,\mathrm{N\text{-}cm}\,\sin\theta$. The total torque is now $(3{,}645 - 404\,\mathrm{N\text{-}cm})\sin\theta = (3{,}241\,\mathrm{N\text{-}cm})\sin\theta$. Still using $\langle \sin\theta \rangle = 1.0$ now, we find $\alpha = 308/\mathrm{s}^2$ and $t_{\mathrm{final}} = 0.10$ s. Now $v_{\mathrm{B}} = 11.2\,\mathrm{m/s} = 36.7\,\mathrm{ft/s} = 25.0\,\mathrm{mph}$. This is getting better, but the speed is still too slow.

Case d: The last thing we will do right now is to remove the effect of gravity. This amounts to ignoring the mass of the arm and ball. We have really been examining the situation in the diagram like Fig. 3.43b. If we were to throw it as in Fig. 3.43c, then gravity would slow the motion in the first half of the operation and would increase it in the second half, so to first-order let us ignore it. This means that we ignore the $-404\,\mathrm{N\text{-}cm}\,\sin\theta$ torque term in Case (c). Now $\alpha = 346/\mathrm{s}^2$, and we see that $t_{\mathrm{final}} = 0.094\,\mathrm{s}$ and $v_{\mathrm{B}} = 11.9\,\mathrm{m/s} = 38.9\,\mathrm{ft/s} = 26.5\,\mathrm{mph}$. This does not help much.

Something is clearly wrong. What? Let us examine some of our remaining assumptions:

1. Is the muscle force per unit area really $\sim 20\,\mathrm{N/cm}^2$? No. The maximum force per unit area is closer to 30–40 $\mathrm{N/cm}^2$. However, since the maximum occurs only for part of the throwing cycle, this assumption is not that bad. Using $30\,\mathrm{N/cm}^2$, would increase the speed by $(1.5)^{1/2}$ giving $\simeq 32.5$ mph for Case (d).

2. The biceps brachii are not the only muscles used to flex the elbow (Fig. 2.10, Case 4). Inclusion of the other muscles increases the torque by a factor of 2.6, which should increase the speed by $(2.6)^{1/2}$ to 52 mph. This is likely an overestimate given that a physiological 3 in diameter for the biceps brachii (only) is an overestimate. Let us estimate that this brings the speed up to 45 mph.
3. The underhand throwing assumption with elbow flexing does not limit us within the confines of the model. Still, in overhand throwing the elbow is extended, by the triceps brachii. For similar musculature and moments, elbow extension should still give the same upper throwing speed.
4. We assumed a simple one-segment model with motion of the elbow joint. We have ignored (a) the motion of the whole arm, with both the elbow and shoulder joints moving (two-segment model, Fig. 3.40a) and possibly also the wrist and (b) the use of back and leg muscles for propulsion (like a spring) in throwing (multisegment model). This is clearly seen for Sandy Koufax pitching in Fig. 3.44, and is why baseball pitchers do a lot of running and have very strong legs, especially power pitchers.

This last assumption is the real reason for the slow calculated speed. The physics is right. Our answer is the final speed for throwing a baseball in the manner described. The model is just much too simple to describe a real throwing motion. The effective muscle cross-section area is really much larger because many more muscles are involved and the moment arms of these additional torques are large.

Fig. 3.44. Sandy Koufax pitching in Game 1 of the 1963 World Series in Yankee Stadium on October 2, 1963. He and the Los Angeles Dodgers won Games 1 and 4, by the scores of 5-2 and 2-1, in a four-game sweep of the New York Yankees. A home run by Mickey Mantle was the sole run he gave up in his second complete-game victory over Whitey Ford. (Photo reprinted with the permission of the Los Angeles Dodgers)

Sandy Koufax has said, "Everybody who performs an athletic event of any kind is a system of levers...." He continued his description of the most critical point of the pitcher's windup, "The front leg is charged with stopping the torso. When the torso stops, the arm catches. ... So the arm now develops more speed than it had when it was simply moving with the body." He continued his description of the pitching motion, "It's a two-armed catapult. You try to get the front half out as far as you can. All power pitchers do that. When the front leg stops, the upper body catches, the arm straightens, and you see a straight line. Basically you try to work as much as you can with leverage and weight and energy transfer...." [140]

A more general multisegment model is obviously needed. Motion of six joints is possible: ankle, knee, hip, shoulder, elbow, and wrist. If each leads to the same forward motion of the ball at the end of the pitch (they do), the speed from each can be added to get the final speed of the ball. The ankle, knee, and hip all lead to forward propulsion, such as the first step of a sprint. Running 100 yd in 10 s translates to 30 fps or ≈20 mph. It takes several steps to reach this steady-state speed. We estimate that the ankle, knee, and hip can add 10 mph. The time required for the upper arm to rotate about the shoulder and this rotation angle are about the same as those for the lower arm about the elbow, so the elbow and shoulder joints both contribute about 45 mph. The wrist may contribute a little to the speed (but much to ball rotation, as is described in the problems). Altogether, the final speed is 100 mph, which makes sense for a power pitcher.

The mechanics of the pitching motion are clearer when viewing the six phases of baseball pitching in Fig. 3.45 [158, 176]. After the windup, the

Fig. 3.45. The mechanics of the pitching motion. The six phases of baseball pitching are shown from windup (**a**) to stride (**b**), arm cocking (**c**), arm acceleration (**d**–**f**), arm deceleration (**g**), and follow-through (**h**). (This photo was provided by the American Sports Medicine Institute)

pitcher is at the balance point, which we will call $t = 0$ s. The pitcher strides and makes foot contact at ~0.50 s. The pelvis rotates transferring energy from the legs to the upper body (maximum rotation speed of ~550–700°/s) and this causes the upper trunk to rotate (~1,100–1,350°/s). The arm straightens as the elbow extends (~2,200–2,800°/s) and the shoulder rotates (~6,900–9,800°/s), and then the ball is released at ~0.64 s, which is ~0.14 s after foot contact. Then the front leg straightens to help decelerate the body. We see that the measured acceleration times are consistent with our simple calculations, ~0.1 s. We should not be surprised that if we included the contribution of the even faster rotation of the shoulder to that of our calculated elbow extension, we could have predicted a ~100 mph fastball. The torques applied to the forearm from the elbow and the upper arm from the shoulder during a pitch are shown in Figs. 3.46 and 3.47 (along with the range of variation). These values are consistent with those in our simple model.

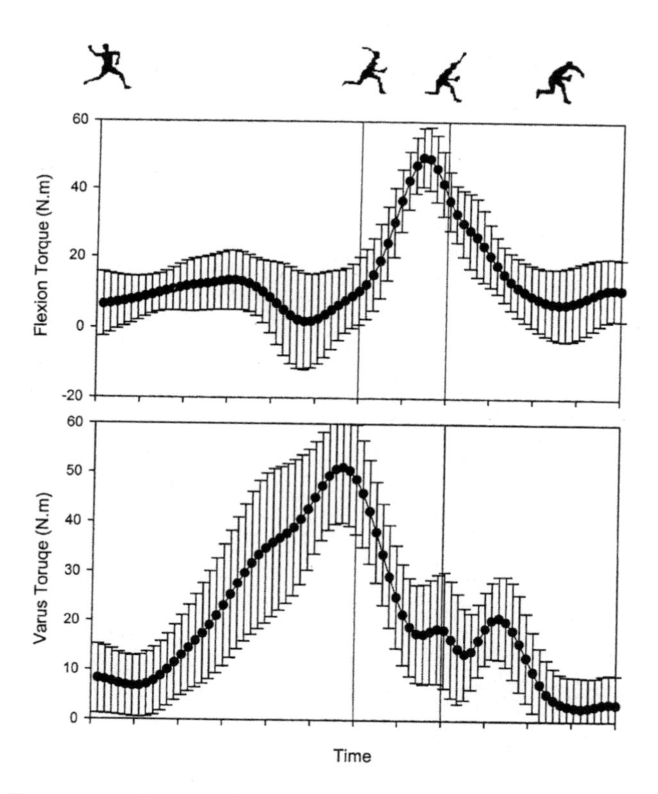

Fig. 3.46. Torques applied to the forearm at the elbow during a baseball pitch, with the time scale relative to the duration of the pitch. (From [176]. With the kind permission of Springer Science and Business Media)

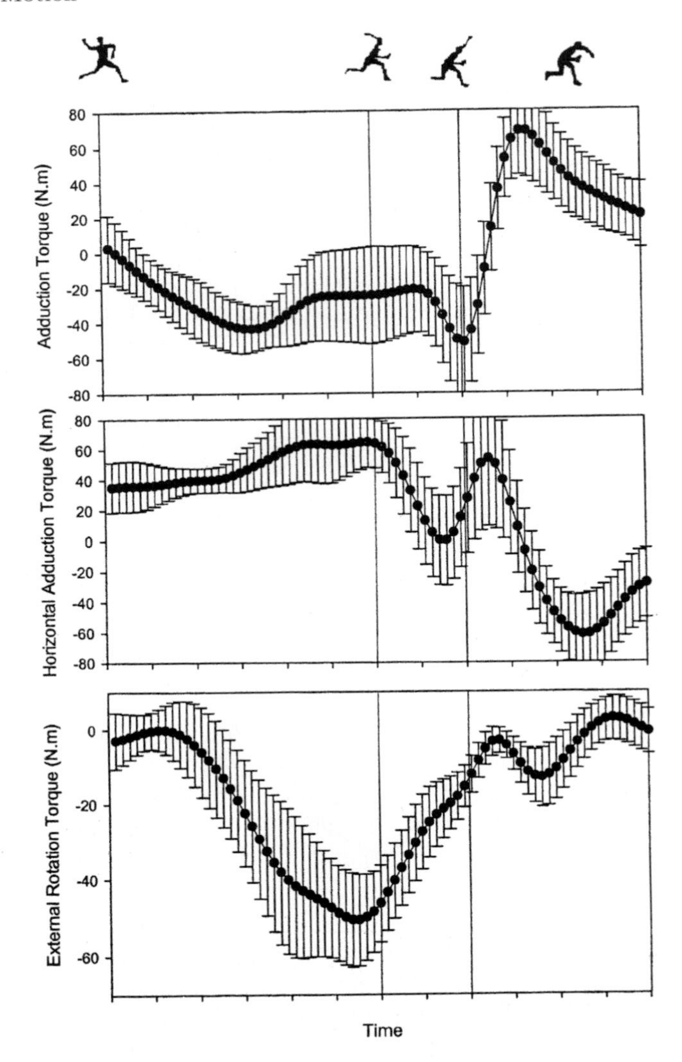

Fig. 3.47. Torques applied to the upper arm at the shoulder during a baseball pitch, with the time scale relative to the duration of the pitch. (From [176]. With the kind permission of Springer Science and Business Media)

3.6.1 Throwing a Spinning Ball

The rotation of a thrown (or batted) baseball affects its motion. The ball thrown by a pitcher rotates due to wrist rotation and arm motion, such as the screwdriver action of the radius rolling on the ulna (described in Chap. 1). The pitcher tries to maximize this rotation speed (ω, in rad/s) when throwing a curve ball and minimize it when throwing a knuckleball (as is seen in Table 3.9

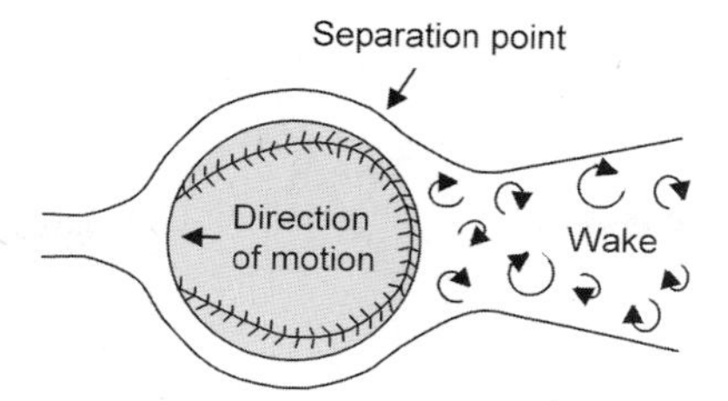

Fig. 3.48. In the rest frame of the ball (i.e., moving along in a frame with the ball), the flow of air around a baseball. The ball is moving to the left, so the air is moving to the right in the drawing. The air flows along the ball and detaches at the separation point to form a wake of chaotic, swirling flow. (Based on [173])

in the problem section). A spinning ball feels the *Magnus force* (or the lift force, which is described in Chap. 7) due to this rotation, which changes its path from that of a rotation-free ball. When a nonspinning ball travels in air, the air travels symmetrically on the top and bottom and it leaves a wake in its path. As sketched in Fig. 3.48 in the rest frame of the ball, the air streamlines separate from the ball symmetrically and this wake is symmetrical and there is no force normal to the direction of the ball. For the spinning baseball in the photograph in the rest frame of the ball in Fig. 3.49, the wake is noncentered because the air flows farther along the "top" (where the upper surface of the rotating ball is moving in the same direction as the airstream) than along the bottom (where the lower surface of the rotating ball is moving in the opposite direction as the airstream). The wake is pushing air "downward" and there is an equal and opposite reaction that pushes the ball "upward" according to Newton's Third Law. Figure 3.50 shows that the direction of this Magnus

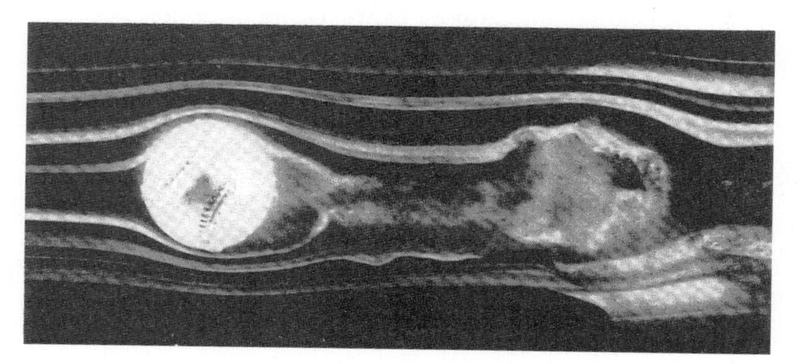

Fig. 3.49. A baseball rotating clockwise in a wind tunnel, with air moving to the right. (Smoke has been added to make the air motion visible.) The streamline detaches later on the top – where the upper surface of the ball moves in the same direction as the streaming air – leading to a downward wake and a net upward force on the ball. (From F.N.M. Brown, courtesy of the University of Notre Dame Hessert Laboratory)

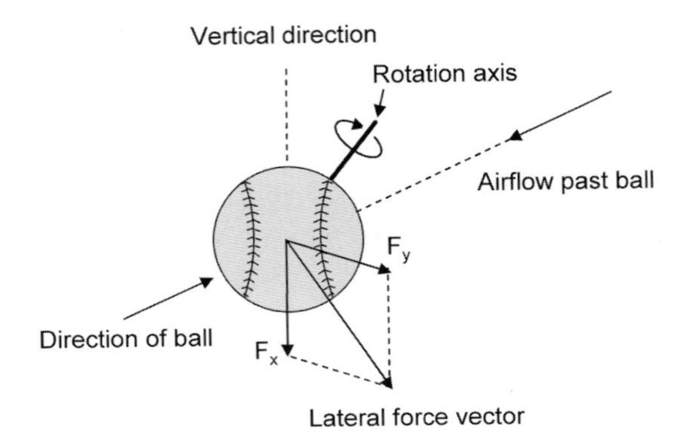

Fig. 3.50. Direction of the Magnus force. (Based on [173])

force is normal to the rotation axis vector of the ball (whose direction is to you if you see it rotating counterclockwise) and the ball velocity vector, in the direction of their cross-product. (If these two vectors point in the x and y directions, respectively, the force will be in the z direction.)

The Magnus force explains why thrown balls do not always follow the same trajectories as expected by gravity and drag, i.e., they can curve, such as with thrown curve balls and batted balls hit to right and left field that slice to the respective foul line. Balls thrown with topspin drop faster than expected by gravity alone and those with backspin drop slower. Balls thrown by a right-handed pitcher normally spin counterclockwise, when viewed from above, and curve away from a right-handed hitter. Balls thrown with clockwise spin (curve ball from a left-hander or screwball from a right-hander) curve to a right-handed hitter. Balls rotating at other angles have motion both vertically and horizontally, aside from the drop by gravity. This is discussed more in [97] and [173], and in Problems 3.54–3.59. Problem 3.60 addresses throwing knuckleballs and scuffed balls.

3.6.2 Power Generated During a Throw

How much power is exerted during a throw, even the quite slow throw using elbow motion only? In linear motion with a constant force exerted F, the work done in moving the amount a distance Δx is

$$W = F\Delta x \tag{3.82}$$

and the power needed to do this in a time Δt is

$$P_{\text{power}} = Fv, \tag{3.83}$$

where the linear speed $v = \Delta x/\Delta t$.

For rotational motion the analogous expressions are

$$W = \tau \Delta\theta \tag{3.84}$$

$$P_{\text{power}} = \tau \Omega, \tag{3.85}$$

for an angular motion of $\Delta\theta$ in a time Δt with rotational speed $\Omega = \Delta\theta/\Delta t$. This is consistent with the linear results, as is seen by examining the power needed to exert a torque in a circular motion with a radius r. Because $\tau = rF$ and $v = \Omega r$,

$$P_{\text{power}} = \tau\Omega = (rF)\Omega = (\Omega r)F = Fv. \tag{3.86}$$

For multijoint models, the net torque and angular speed about each joint need to be considered and so

$$P_{\text{power}} = \sum_k \tau_k \Omega_k. \tag{3.87}$$

The torque generated in the case with the 3 in diameter muscles is 3,645 N-cm (Case c, 25.0 mph), and it has been assumed to be constant during the throw. The average angular speed is $\Omega_{\text{av}} = \Delta\theta/\Delta t$ where $\Delta\theta = 3\pi/4 - \pi/4 = \pi/2$ and $\Delta t = t_{\text{final}} = 0.10\,\text{s}$, so $\Omega_{\text{av}} = 15.7$ rad/s. The angular speed increases linearly with time during this motion (from (3.78)) from 0 at the beginning to the peak value at the end $\Omega_{\text{peak}} = 31.4$ rad/s.

Using (3.85), the average power generated is:

$$P_{\text{power, av}} = (3{,}645\,\text{N-cm})(15.7\,\text{rad/s}) = 57{,}000\,\text{N-cm/s} = 570\text{ W} = 0.76\text{ hp}. \tag{3.88}$$

We have made use of 1 W = 1 N-m/s and 1 horsepower (hp) = 746 W. The peak power

$$P_{\text{power peak}} = (3{,}645\,\text{N-cm})(31.4\,\text{rad/s}) = 1{,}140\text{ W} = 1.5\,\text{hp}, \tag{3.89}$$

which seems large, even for this slow throw. (For the 2 in diameter biceps, the average and peak powers are 178 W and 356 W, respectively.)

3.7 Other Types of Motion

References [99, 127, 159] describe many other types of motion. One such motion is bicycling. Figures 3.51–3.53 show the joint angles during a cycling crank cycle, along with changes in muscle-tendon length and positive (contracting muscles) and negative (extending muscles) work by muscles during this cycle [127]. We will analyze swimming and potential human flight in Chap. 7 when we discuss drag and lift in fluids.

Angular kinematics and kinetics of the entire body are very important in several sports, such as several gymnastics events, diving, and ice skating (see

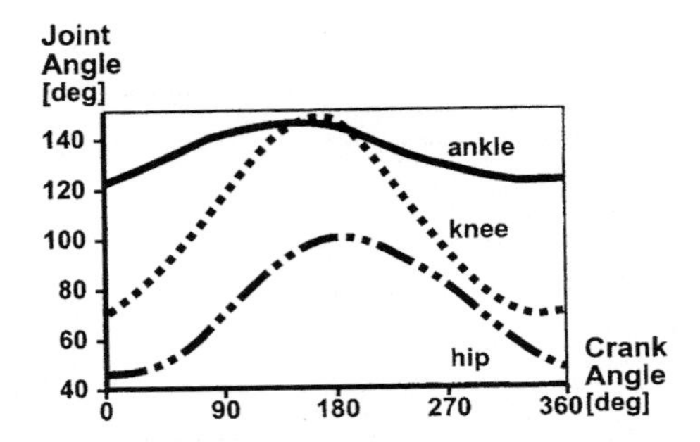

Fig. 3.51. Joint angles in the right leg during a crank cycle in bicycling. The hip angle is that between the pelvis and the upper leg (thigh) (180° with full extension), the knee angle is that between the lower leg (shank, calf) and upper leg (thigh) (180° with full extension), and the ankle angle is that between the lower leg and a line from the ankle joint axis to the pedal spindle axis. (From [126], as from [131]. Used with permission)

[127]). (In several activities, such as in throwing a ball, the angular motion of a few body segments dominate.) Figure 3.54 shows the moment of inertia of the body during diving and in high-bar events. You can clearly see that you can decrease your moment of inertia greatly during a dive and concomitantly increase your rotation rate, as seen in Fig. 3.55, because angular moment is constant when no external torques are applied. This control of rotation rate is important in ice skating, and is analyzed in Problem 3.42. During an athletic maneuver in flight total angular momentum of the body is constant but the angular momentum of different body segments can change, as is seen by the transfer of the angular momentum from the upper body to the legs during a piked front dive in Fig. 3.56. By appropriate motion of your arm, you replace some of this somersaulting angular to twisting angular during a dive. For

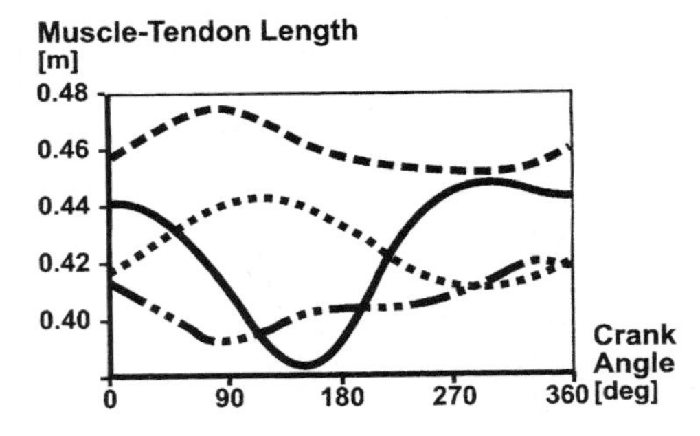

Fig. 3.52. Changes in the muscle/tendon lengths in biarticulate muscles (that span two joints) during a crank cycle in bicycling, for the rectus femoris (*dashes*), biceps femoris (*dots*), semimembranosus (*solid line*), and gastrocnemius (*dashes and dots*). (From [126], as from [131]. Used with permission)

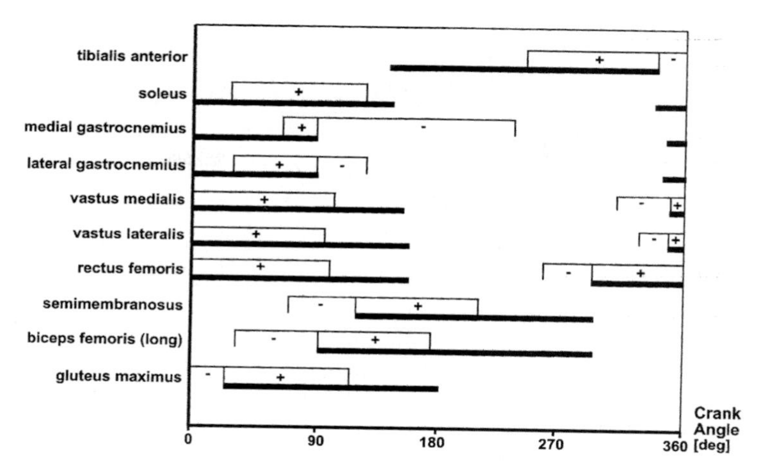

Fig. 3.53. Crank angles corresponding to positive work (contracting muscles) and negative work (extending muscles) for ten muscles during one crank cycle in bicycling. (From [126], as from [131]. Used with permission)

example, if your arms are initially extended above your head during a front dive and you swing one arm sideways bringing it alongside your body, your body will rotate in the opposite direction to conserve total angular momentum, in a side-somersaulting rotation.

3.8 Collisions of the Human Body

So far we have examined how people begin to move or continue to move. Now we will investigate what happens when motion is stopped very fast, in a collision. We humans tend to break down when we collide with objects. We can see the reasons for this by determining how fast given body parts decelerate during such collisions. We will also use this analysis to estimate the forces on a

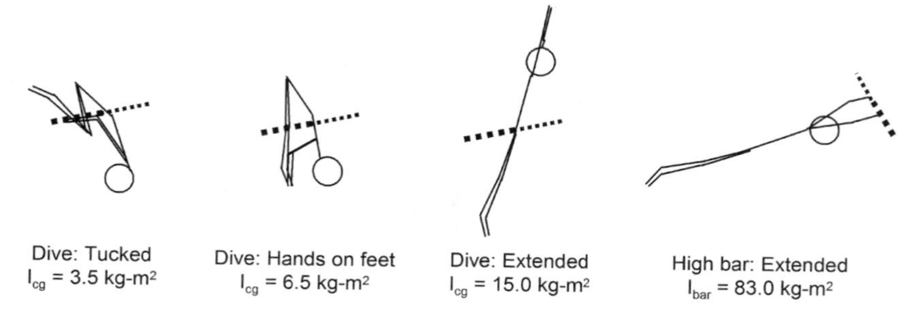

Fig. 3.54. Moments of inertia during diving positions (about the axis through center of mass) and in the high bar gymnastic event (about the bar). (Based on [127])

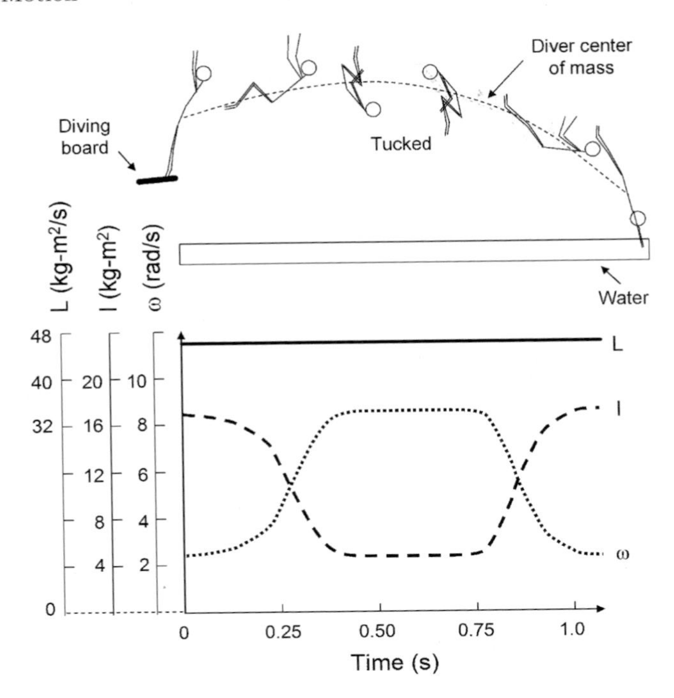

Fig. 3.55. Interplay between angular momentum, moment of inertia, and angular velocity during a tucked backward one-and-one half dive. In this backward dive the body is initially straight, curls up with arms around the tucked and bent legs by the knees, and becomes straight again before impact. The diver gains enough vertical momentum from the springboard to have enough time to complete this maneuver before impact. A sketch of the motion of the diver's center of mass is also shown. (Based on [127])

foot upon impact during running and the return of energy during jumping. In these examples the person is directly involved in the collision, for example in the collision of a person's head with the car dashboard in a car crash or with the opponent's fist in boxing. In some instances the person causes a collision, such as by hitting a baseball with a bat.

3.8.1 Kinematics of a Collision

We need to learn how humans react to rapid decelerations. During a constant acceleration a in a time t, the velocity of an object changes from the initial value v_i to the final value

$$v_\mathrm{f} = v_\mathrm{i} + at \tag{3.90}$$

and the distance traveled by the object during that time is

$$\Delta x = v_\mathrm{i} t + \frac{1}{2} a t^2. \tag{3.91}$$

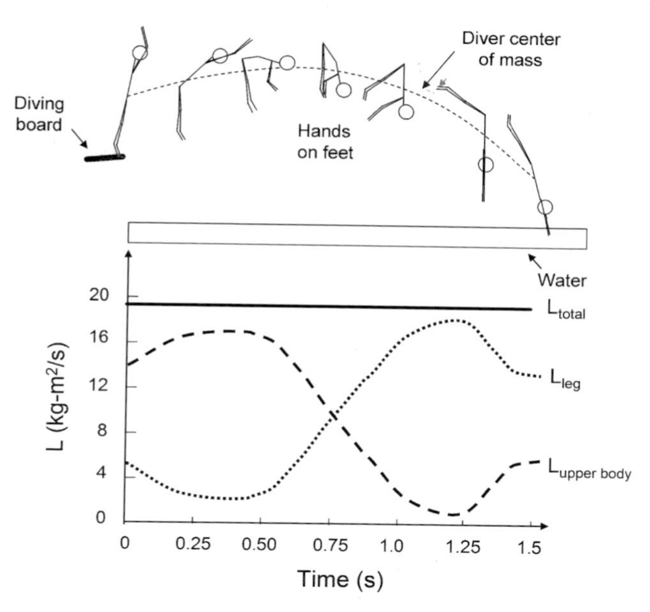

Fig. 3.56. Interplay between angular momentum localized first in the upper body and then in the legs during a piked front dive. In this forward dive the body is initially straight, bends over at the waist with the hands holding on to the ankles of the straight legs, and then becomes straight again before impact. The diver gains sufficient vertical momentum from the springboard to have enough time to complete this maneuver before impact. A sketch of the motion of the diver's center of mass is also shown. (Based on [127])

Say a mass m is traveling at a speed v, which we will assume is positive. If it decelerates at a uniform rate to speed $v_\mathrm{f} = 0$ in a time $t = t_\mathrm{coll}$, during the collision the acceleration $a_\mathrm{coll} = \Delta v/\Delta t = -v_\mathrm{i}/t_\mathrm{coll}$. The analysis will be clearer if we define the positive quantity $a_\mathrm{decel} = -a_\mathrm{coll} = v_\mathrm{i}/t_\mathrm{coll} > 0$. The distance the mass travels during that collision is

$$\Delta x = v_\mathrm{i} t_\mathrm{coll} + \frac{1}{2} a_\mathrm{coll} t_\mathrm{coll}^2 = v_\mathrm{i} t_\mathrm{coll} + \frac{1}{2}\left(\frac{-v_\mathrm{i}}{t_\mathrm{coll}}\right) t_\mathrm{coll}^2 \tag{3.92}$$

$$= \frac{1}{2} v_\mathrm{i} t_\mathrm{coll} = \frac{1}{2} a_\mathrm{decel} t_\mathrm{coll}^2 = \frac{v_\mathrm{i}^2}{2 a_\mathrm{decel}}, \tag{3.93}$$

where $v_\mathrm{i} = a_\mathrm{decel} t_\mathrm{coll}$ has been used.

The magnitude of the force felt by the object is:

$$|F| = \left| m \frac{\Delta v}{\Delta t} \right| = \frac{m v_\mathrm{i}}{t_\mathrm{coll}}. \tag{3.94}$$

The force per unit area A on the collision region is

$$P = \frac{|F|}{A} = \frac{m|\Delta v/\Delta t|}{A} = \frac{m v_\mathrm{i}}{A t_\mathrm{coll}}. \tag{3.95}$$

This is also called a pressure or a stress. Such decelerations to zero speed are totally inelastic. In totally elastic or partially elastic collisions, the objects bounce back and the momentum transfer and the resulting forces are greater.

Partially Elastic Collisions

So far we have implicitly assumed that the object is colliding with an infinitely massive body. Also we have not described the elasticity of the collision quantitatively.

In a collision, linear momentum and energy are conserved. However, if the collision is not elastic, kinetic energy is not conserved and heat is produced. Let us consider two objects with mass m_1 and m_2 that move with velocities in the x direction v_1 and v_2 before the collision and v_1' and v_2' after the collision. For motion to more positive x, $v > 0$, and to more negative x, $v < 0$. Conservation of linear momentum gives

$$m_1 v_1 + m_2 v_2 = m_1 v_1' + m_2 v_2'. \tag{3.96}$$

The fraction of the speed returned to a body as a result of a collision is characterized by the coefficient of restitution e (or COR).

$$e = -\frac{v_1' - v_2'}{v_1 - v_2}. \tag{3.97}$$

For a stationary massive object 2, we see that $e = v_1'/v_1$ and so it is 1 for a totally elastic collision, for which $v_1' = -v_1$, and 0 for a totally inelastic collision, for which $v_1' = 0$. Table 3.7 shows typical values for balls. The coefficient of restitution typically decreases with speed.

Table 3.7. Coefficient of restitution (e) for balls. (Using data from [97, 127, 171, 173])

ball	surface	e	speeds
"super ball"	hardwood floor	0.89	slow
golf	floor	0.83–0.89	slow
tennis	racket	0.76–0.88	slow
racket	floor	0.74–0.88	slow
basketball	hardwood floor	0.76	slow
volleyball	hardwood floor	0.74	slow
soccer	floor	0.69–0.80	slow
squash	plywood	0.48–0.60	slow
baseball	ash boards on concrete	0.563	58 mph
	wood	0.588	25 mph
	wood	0.584	18 mph
	wood	0.46	89 mph
	typical wood bat	0.55	typical pitch
softball	hardwood floor	0.31	slow

Inserting (3.97) into (3.96) gives

$$v_1' = \frac{(m_1 - em_2)v_1 + (m_2 + em_2)v_2}{m_1 + m_2} \tag{3.98}$$

and

$$v_2' = \frac{(m_1 - em_1)v_1 + (m_2 + em_1)v_2}{m_1 + m_2}. \tag{3.99}$$

3.8.2 Consequences of Collisions

Falls and Crashes

What are the results of a collision? We will examine two types of consequences: potentially fatal collisions due to brain trauma and the breakage of body parts [156, 169].

Let us examine the forces on a head during a collision. Consider a head of 3 kg mass moving at 1 m/s that hits a steel beam and stops in 0.01 s (= 10 ms). From (3.93), this corresponds to a distance of 5 mm. The deceleration felt by the head is $a_{\text{decel}} = (1\,\text{m/s})/(0.01\,\text{s}) = 100\,\text{m/s}^2 \simeq 10g$, which is very large. The force felt by the head is $ma_{\text{decel}} = 3\,\text{kg} \times 100\,\text{m/s}^2 = 300\,\text{N}$ (~70 lb) [110]. We will see the consequences of this soon. Years ago, before seat belts were standard in cars, car dashboards were made of hard metal instead of the softer foam-based materials used now. Imagine the head injuries from the rapid deceleration of heads on these dashboards during car crashes!

Now let us examine the forces in our legs when we jump from a ledge barefoot, a ledge that is only $h = 1\,\text{m}$ high. This may not seem to be a dare-devil activity, but if we land stiff-legged and on one leg, the consequences are startling. The deceleration during this type of landing occurs during the compression of the padding on the bottom of our foot.

The gravitational potential energy before the jump is mgh, which equals the kinetic energy before impact with the ground, $mv_{\text{i}}^2/2$, where v_i is the initial speed in the collision. We find

$$v_{\text{i}} = \sqrt{2gh} = \sqrt{2 \times 9.8\ \text{m/s}^2 \times 1\ \text{m}} \simeq \sqrt{20\ \text{m}^2/\text{s}^2} = 4.5\ \text{m/s}. \tag{3.100}$$

If the foot padding is compressed during impact by $\Delta x = 1\,\text{cm} = 10^{-2}\,\text{m}$, then $t_{\text{coll}} \sim 5\,\text{ms}$.

Using (3.92) and (3.93), the deceleration is

$$a_{\text{decel}} = \frac{v_{\text{i}}^2}{2\Delta x} = \frac{1}{2}\frac{20\ \text{m}^2/\text{s}^2}{10^{-2}\ \text{m}} = 10^3\ \text{m/s}^2 = 100g, \tag{3.101}$$

which is very high. For a $m_{\text{b}} = 70\,\text{kg}$ person, this corresponds to a force of $m_{\text{b}}a_{\text{decel}} = (70\,\text{kg})(10^3\,\text{m/s}^2) = 7 \times 10^4\,\text{N}$ on one foot.

This is a compressive force, and it has its greatest effect on the thinnest bones absorbing the impact. The tibia near the ankle has a radius $r \sim 1\,\text{cm}$

and a cross-sectional area $A \sim 3\,\mathrm{cm}^2 = 3 \times 10^{-4}\,\mathrm{m}^2$. During impact the force per unit area in this region of the tibia is

$$\frac{F}{A} = \frac{7 \times 10^4\ \mathrm{N}}{3 \times 10^{-4}\ \mathrm{m}^2} = 2.3 \times 10^8\ \mathrm{N/m}^2 = 230\ \mathrm{N/mm}^2. \qquad (3.102)$$

Upon compression bones typical break when subjected to a force per area above $\sim 1.7 \times 10^8\,\mathrm{N/m}^2 = 170\,\mathrm{N/mm}^2 = 170\,\mathrm{MPa}$. This damage threshold is called the ultimate compressive stress (UCS, Chap. 4). The stresses on the tibia exceed this limit, so jumping from 1 m and landing on one stiff leg will break your leg. (Please do not try this!) This also shows us why we need to learn more about the mechanical properties of parts of the body. Problem 3.68 considers the gentler case of using leg muscles to cushion the collision.

These two examples, along with (3.92) and (3.93), suggest how to decrease the impact of a collision. Because the stress during the collision is $mv_\mathrm{i}/At_\mathrm{coll}$, the effect of the impact can be lessened by increasing t_coll and/or the area of contact A. The collision time can be increased by (a) using better padding at the site of the collision (such as by wearing better sneakers during running), (b) increasing the duration of the impact by adjusting your body during actual impact (as by letting your knees bend during the impact after a jump – the muscles will then bear part of the impact of the collision), or (c) rolling over (as at the end of a parachute jump). Rolling over at the end of a parachute jump also serves to increase the impact area A.

Obviously, damage can occur during falls and collisions. *However, when are such crashes survivable?* It depends on what part of the body makes contact, what type of surface it hits, and the deceleration time. In whole body impact, your fate depends on what part of the body makes contact: the back (supine position), front (prone), or side (lateral). There are some guidelines gleaned from past (unfortunate) experience. You can barely survive lateral falls at 8 m/s and supine falls at 12 m/s. You can survive foot first falls at 12 m/s on concrete, 15 m/s on soil, and 35 m/s on water. You can survive head first falls at 9 m/s on concrete/ice [110]. (Do not try any of these!) Figure 3.57 shows that human collisions are usually not survivable for decelerations of 175–200 g and greater.

What is actually occurring in such collisions? Consider an injury to the head. In direct trauma, the head hits an object and the skull can fracture or the brain, which continues moving forward when the skull has been decelerated, hits the interior of the skull, leading to bruising (a *contusion*) or bleeding (a *hemorrhage*). In addition to this primary impact (the *coup*), the brain can bounce off the skull and hit the back of the skull (the *contrecoup*). Some think that this coup/contracoup injury is instead caused by negative gauge pressure that occurs in the part of the brain opposite the impact side (in the contracoup region), which leads to cavitation and this causes the injury. Severe injury can result when the brain is subject to large accelerations even without impact (*indirect trauma*): in inertial injuries due to rotational

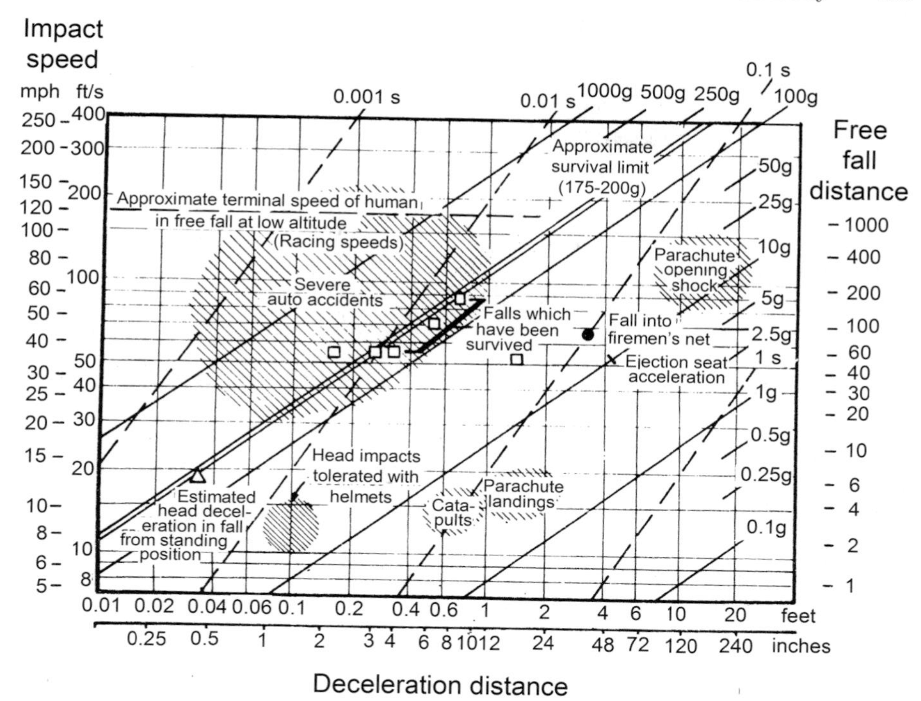

Fig. 3.57. Survivability of collisions. The free fall distance is that for the indicated impact speed, with allowance made for air resistance of the human body near sea level. The band shown for 175–200g separates the approximate survival and nonsurvival regions. (From [169])

accelerations and decelerations and in *Shaken Baby Syndrome* nerve cell axons are damaged by stretching (*diffuse axonal injury*) [106]. (*Whiplash* is a neck injury that can be caused in such collisions.)

The gravity of such injuries is qualitatively described by the abbreviated injury scale (AIS). This is a guide to the severity of injury to a body component or organ by its threat to life. It ranges from minor injury (AIS = 1), moderate injury (2), serious injury (3), severe injury (4), critical injury (5), and to unsurvivable injury (6).

There are semiquantitative guidelines for the severity and likely fatality of head injuries that have been determined from the records of past accidents [119, 156]. Using data from experiments on cadavers and animals, the Wayne State University Concussion Tolerance Curve was developed [124, 125, 143], which Gadd converted into an index [120]. The Gadd severity index (GSI) is defined as

$$\mathrm{GSI} = \int \left(\frac{a_{\mathrm{decel}}}{g} \right)^{2.5} \mathrm{d}t, \tag{3.103}$$

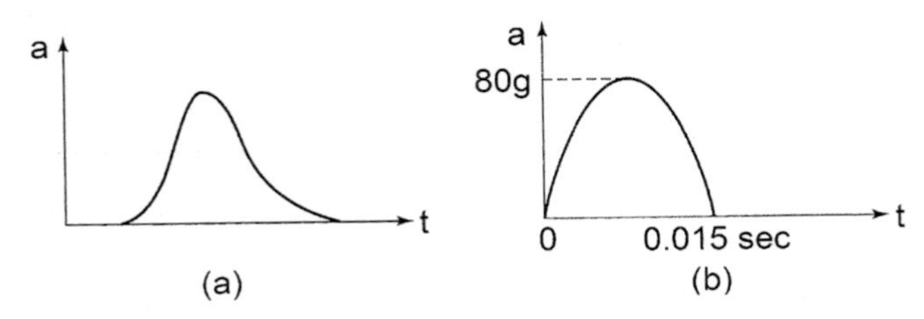

Fig. 3.58. Acceleration vs. time in a collision, (**a**) in general, (**b**) of the head of a dummy hit by a heavyweight boxer. (From [105])

expressed in seconds. This is integrated over the entire collision (a frontal blow to the cranium), as in Fig. 3.58a. a_{decel} is expressed in units of g and t is in s. If the deceleration is constant during the course of the collision, the integral is the product of this constant deceleration-based term and the collision time

$$\mathrm{GSI} = \left(\frac{\Delta v}{g t_{\mathrm{coll}}}\right)^{2.5} t_{\mathrm{coll}}. \tag{3.104}$$

This helps determine the likelihood of skull fracture or concussion for frontal collisions. When the GSI reaches ∼1,000 (in units of seconds), there is a 50% chance of fatality. If the GSI is much above 1,000, the collision is usually fatal. A GSI much below 1,000 can still indicate a severe injury. For example, when the GSI $\sim$ 400, there may be a mild concussion that could result in unconsciousness.

Versace recommended a modification of the GSI to remove potential long tails in the deceleration, which are probably unimportant. This truncated index is the head injury criterion (HIC)

$$\mathrm{HIC} = \max\left[\left\{\frac{1}{t_2 - t_1}\int_{t_1}^{t_2}\left(\frac{a_{\mathrm{decel}}}{g}\right)\mathrm{d}t\right\}^{2.5}(t_2 - t_1)\right], \tag{3.105}$$

expressed in seconds [172]. This is the maximum value of this expression between any two times t_1 and t_2 during the collision, separated by not more than 36 ms; in a more recent version (HIC-15) this time separation is 15 ms. One study has shown that the threshold for serious head injury is HIC = 1,000 (in units of seconds), for which the probability of death is about 7% (and is AIS 3). An index of HIC = 1,500 is the threshold for severe/critical injury, with a 26% probability of death (and is AIS 4–5), and this probability is about 100% for HIC = 3,200. Figure 3.59 plots the risk of serious brain injury (AIS $\geq$ 4) for HIC-15, which is the integral over a gaussian probability curve

$$R = \left(\frac{430}{\pi}\right)^{1/2}\int_0^{\mathrm{HIC-15}} \exp\left[-\{(x - 1,434)/430\}^2\right]\mathrm{d}x. \tag{3.106}$$

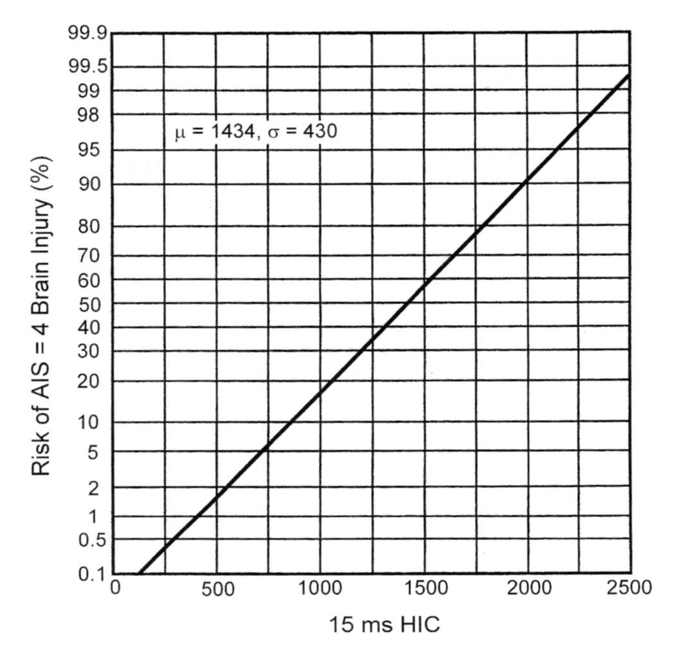

Fig. 3.59. Risk of AIS $\geq$ 4 brain injury as a function of 15-ms HIC for forehead impacts. (From [147])

(This integral is also known as the *error function*. See Appendix C.) The risk of serious brain injury is 5% for HIC-15 = 700 and 50% for HIC-15 $\sim$ 1,400–1,500 for this serious injury and for skull fracture. The maximum AIS from head injuries is plotted in Fig. 3.60. The risk of death is plotted in Fig. 3.61. The HIC is a better predictor of the seriousness of a trauma to the brain than is the GSI. Problems 3.70 and 3.71 explore the differences in the GSI and HIC indices for the same accident. The use of such indices to predict the outcomes of accidents is not universally accepted for several reasons. For example, the GSI and HIC assess the effects of linear (rectilinear) accelerations, but not rotational accelerations, which can also be very important. Other criteria have also been proposed [135].

Let us examine the consequences of the elasticity of the head in a car crash. If the head collides with a heavy hard object, the head will rebound elastically. We will also assume the object is much heavier than the head. In the cases we have considered so far we have assumed that $v_{\mathrm{f}} = 0$. Now we will consider $v_{\mathrm{f}} = -v_{\mathrm{i}}$, so $\Delta v = v_{\mathrm{i}} - v_{\mathrm{f}} = 2v_{\mathrm{i}}$ and $a_{\mathrm{decel}} = \Delta v/\Delta t = 2v_{\mathrm{i}}/t_{\mathrm{coll}}$. If this is constant during the collision,

$$\mathrm{GSI} = \left(\frac{2v_{\mathrm{i}}}{gt_{\mathrm{coll}}}\right)^{2.5} t_{\mathrm{coll}} \tag{3.107}$$

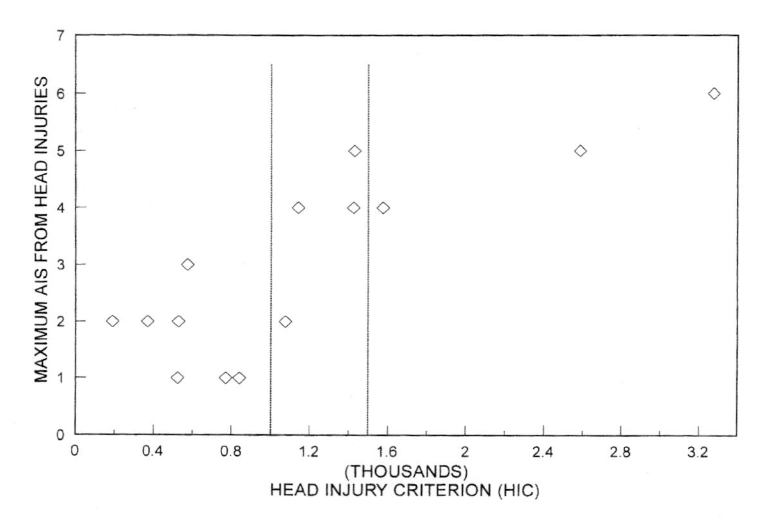

Fig. 3.60. Classification of case-studies of collisions leading to head injury, plotted as the maximum head injury (using the AIS scale) vs. the collision condition as quantified by the HIC. (From [167], as from [130])

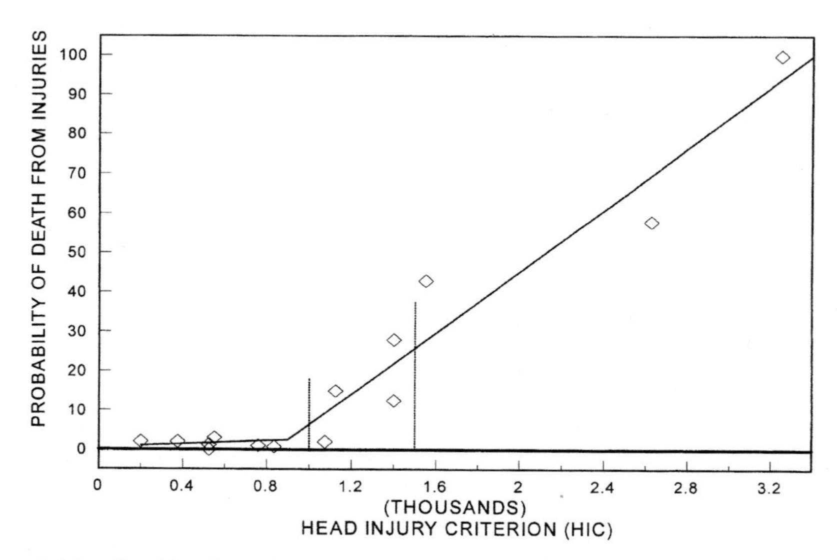

Fig. 3.61. Classification of case studies of collisions leading to death, plotted as the probability of death vs. the collision condition as quantified by the HIC. (From [167], as from [130])

for an elastic collision, compared to

$$\mathrm{GSI} = \left(\frac{v_\mathrm{i}}{g t_\mathrm{coll}}\right)^{2.5} t_\mathrm{coll} \tag{3.108}$$

for an inelastic collision. Say $v_\mathrm{i} = 50\,\mathrm{mph} = 73.3\,\mathrm{ft/s} = 22.3\,\mathrm{m/s} = 80.4\,\mathrm{km/h}$. Then t_coll can range from ∼2–50 ms. Let us assume a collision time of 10 ms, which gives GSI = 44,000 for a perfectly elastic collision, compared to 7,800 for a perfectly elastic collision. This indicates a definitely fatal collision in either case for this example.

Padding in car dashboards and air bags can help! Padding in bicycle and motorcycle helmets can help! The helmet padding helps in two ways. It cushions the blow to the head because the kinetic energy of the head is used to deform the padding. This decreases the deceleration rate of the head. Furthermore, the blow is spread over the inside area of the helmet, which is much larger than the typical collision regions of contact. The desired materials properties of the helmet shell and interior padding are discussed in Chap. 4. Clearly, the padding must absorb the kinetic energy of the head without returning it, as would happen if you used a spring-like material for the padding.

Boxing

The long-term decline of cognitive ability in boxers commonly results from repeated blows to the head and the resulting the "punch drunk" symptoms of *dementia pugilistica*. In heavyweight boxing, GSI ∼400 for a solid punch, which suggests a mild concussion consistent with a knockout punch. This is based on integrating the information in force plots (Fig. 3.58b). We can arrive at similar results from first principles.

We first ask: why do boxers wear boxing gloves? These gloves protect the one who throws the punch as well as the one who receives it. Boxing gloves decrease the deceleration a_decel of the puncher's fist by increasing t_coll, due to the thickness of the padding in the glove, and A, due to the large lateral dimensions of the boxing glove. (This large load area lessens the large forces on specific places on specific bones, thereby reducing the risk of fracture.) The deceleration is even smaller for yet another reason. The collision of the fist with head becomes much more inelastic (i.e., less elastic) with the glove. Without the glove, a_decel would be much larger and the boxer's hand would more easily break (Fig. 3.62) and the damage to the head of the opponent would be greater. The forces involved are tremendously large even with the boxing gloves. Let us examine what happens when a fist hits a head.

A good boxer uses his or her legs and whole body to deliver a punch, but for our model let us consider the motion of the fist and whole arm only, and what happens when it hits the opponent's head. Again, the other boxer's head is attached to his body (at least before the punch), but let us consider the

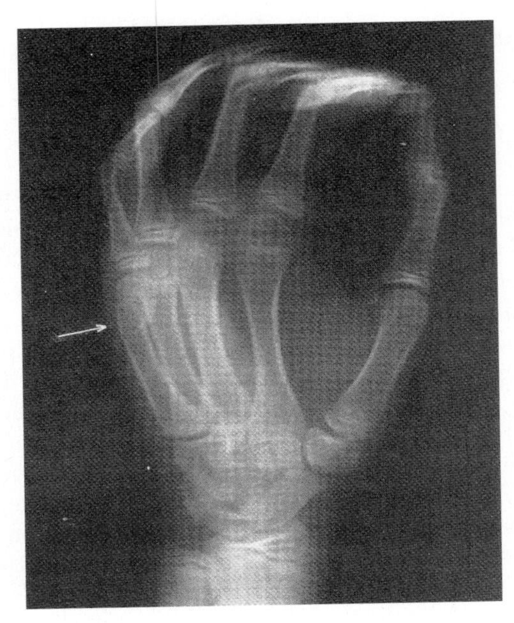

Fig. 3.62. X-ray of a boxer's fracture, which results when hitting with a closed fist. (Which bone is broken?) (From [141], as from [130])

effect of the punch on the head and neck only. The fist is initially moving at $v_{\mathrm{fist,i}}$ and if we ignore the motion of the shoulder the average speed of the arm is $\sim v_{\mathrm{fist,i}}/2$ (which we will use here). The mass of the arm (with fist) is $m_{\mathrm{arm}} = 0.050 m_{\mathrm{b}}$ (ignoring the mass of the 5 oz gloves and where again m_{b} is the body mass) (Table 1.7). The head (and neck) of the other boxer is $m_{\mathrm{head}} = 0.081 m_{\mathrm{b}}$.

The elasticity of the collision affects the analysis. In elastic collisions, the colliding bodies have different speeds after the collision, and the total linear momentum and kinetic energy are the same before and after the collision. In totally inelastic collisions, the colliding bodies have the same velocity after the collision, and total linear momentum is conserved. Total energy, but not merely kinetic energy, is still conserved. Most collisions are partially elastic and partially inelastic. (We will analyze the collision here without using the coefficient of restitution.)

Say the boxers are wearing gloves. Before the punch collision, the fist is moving at a speed $v_{\mathrm{fist,i}}$ and the linear momentum of the arm is $\sim m_{\mathrm{arm}}(v_{\mathrm{fist,i}}/2)$. The collision is inelastic, meaning that at the end of the punch the fist and head will both be moving at the same final speed v_{f} (Fig. 3.63a). The average speed of the arm is $v_{\mathrm{f}}/2$. The total momentum after the punch is $m_{\mathrm{arm}}(v_{\mathrm{f}}/2) + m_{\mathrm{head}} v_{\mathrm{f}}$, so

$$\frac{m_{\mathrm{arm}} v_{\mathrm{fist,i}}}{2} = \frac{m_{\mathrm{arm}} v_{\mathrm{f}}}{2} + m_{\mathrm{head}} v_{\mathrm{f}} \tag{3.109}$$

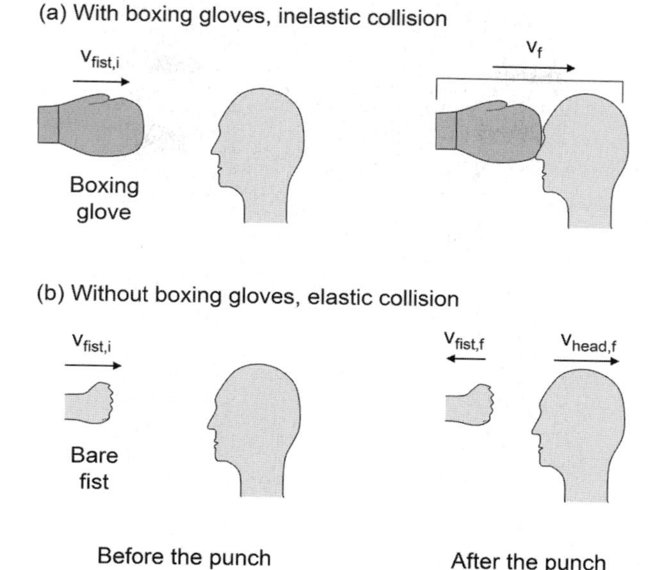

Fig. 3.63. Models of boxing, with (**a**) boxing gloves, leading to an inelastic collision with the head, and (**b**) bare fist, leading to an elastic collision with the head

and

$$v_\mathrm{f} = \frac{v_\mathrm{fist,i}}{1 + 2\,m_\mathrm{head}/m_\mathrm{arm}} = 0.236 v_\mathrm{fist,i}, \tag{3.110}$$

using the arm and head masses. This means that during contact the fist decelerates from $v_\mathrm{fist,i}$ to $0.236v_\mathrm{fist,i}$, while the head accelerates from 0 to $0.236v_\mathrm{fist,i}$. If the glove compresses by $\Delta x = 1\,\mathrm{cm}$ during the punch, the head accelerates at a rate $a_\mathrm{coll} = v_\mathrm{f}^2/2\Delta x = (0.236v_\mathrm{fist,i})^2/(2 \times 1\,\mathrm{cm})$. The collision time $t_\mathrm{coll} = v_\mathrm{f}/a_\mathrm{coll} = 2 \times 1\,\mathrm{cm}/0.236v_\mathrm{fist,i}$.

Our model for throwing a ball is also a good model for throwing a punch. Instead of the biceps brachii contracting causing θ to decrease during throwing a ball, the triceps brachii contract causing θ to increase. Ignoring the effect of the rest of the body, with similar assumptions we have $v_\mathrm{fist,i} \sim 30\,\mathrm{mph}$ ($13.4\,\mathrm{m/s}$) for 3 in diameter biceps. Assuming the boxers are wearing boxing gloves, the head of the opponent accelerates to $v_\mathrm{f} = 3.2\,\mathrm{m/s}$ at rate $a_\mathrm{coll} = 500\,\mathrm{m/s^2} = 51g$ in $t_\mathrm{coll} = 0.0063\,\mathrm{s}$. If we assume that (3.104) can be used for accelerations as well as decelerations, this leads to a severity index GSI = 117. This indicates damage, but not a fatal blow. With less padding in the glove, say leading to a deformation of $\Delta x = 0.5\,\mathrm{cm}$, the GSI increases to 332. With Δx again 1 cm and with $v_\mathrm{fist,i} = 45\,\mathrm{mph}$ ($20.1\,\mathrm{m/s}$, our faster estimate), we find $v_\mathrm{f} = 4.7\,\mathrm{m/s}$ and the GSI increases to 593, suggesting substantial damage. Boxers' heads clearly do get damaged even when they wear gloves. Because part of the rest of the body is involved in the punch, v_f will be larger than estimated here and the GSI will be larger in each case.

Now what happens if the boxers do not wear boxing gloves? Without gloves the collision of the fist and head is assumed elastic, with final speeds $v_{fist,f}$ and $v_{head,f}$ (Fig. 3.63b), and the collision distance is much shorter, say 1 mm. Both of these differences contribute to the more dire consequences of bare knuckle boxing, both for the puncher (broken hand) and the punched (more damage to the head). Using the same approximations for arm movement, along with conservation of momentum and kinetic energy, gives

$$\frac{m_{arm} v_{fist,i}}{2} = \frac{m_{arm} v_{fist,f}}{2} + m_{head} v_{head,f}, \tag{3.111}$$

$$\frac{m_{arm}}{2}\left(\frac{v_{fist,i}}{2}\right)^2 = \frac{m_{arm}}{2}\left(\frac{v_{fist,f}}{2}\right)^2 + \frac{m_{head}}{2} v_{head,f}^2, \tag{3.112}$$

and

$$v_{head,f} = \frac{v_{fist,i}}{1 + m_{head}/m_{arm}} = 0.382 v_{fist,i}, \tag{3.113}$$

which is 1.6× that with gloves. With $v_{fist,i} = 30$ mph and $\Delta x = 1$ mm, the head accelerates to $v_{head,f} = 5.1\,\mathrm{m/s}$ at a rate $a_{coll} = 13,100\,\mathrm{m/s^2} = 1,340g$ in $t_{coll} = 0.00039$ s and GSI = 25,600. Even though this is a bit of an overestimate since the collision is not perfectly elastic, it is clear that bare knuckle boxing can lead to death (as well as broken knuckles).

3.8.3 Hitting Balls

We will now examine collisions with objects such as baseballs [173] and tennis balls. The physics of hitting such objects combines motion – much as in throwing a baseball – and using a different object – bat or tennis racket – to hit it. The second part is a collision, much as in punching in boxing. However, the emphasis in boxing is the effect of the collision on the people delivering and receiving the punch. (As in many other activities in life, in boxing it is definitely better "to give than to receive.")

To examine what happens when a bat hits a baseball, we use (3.98) with 1 to denote the baseball and 2 to denote the bat to arrive at

$$v'_{ball} = \frac{(m_{ball} - e m_{bat}) v_{ball} + (m_{bat} + e m_{bat}) v_{bat}}{m_{ball} + m_{bat}}. \tag{3.114}$$

(In this and subsequent equations in this discussion, we will substitute all masses by their weights (because $W = mg$ and the g cancels out). This will be useful in using the weights of the ball and bat in the very common units of oz.)

To understand the plight of the batted ball we need to know the motion of the body, arms, and wrist during the swing (Fig. 3.64), the speed of the pitched ball, where the ball hits the bat (relative to its center of mass), and

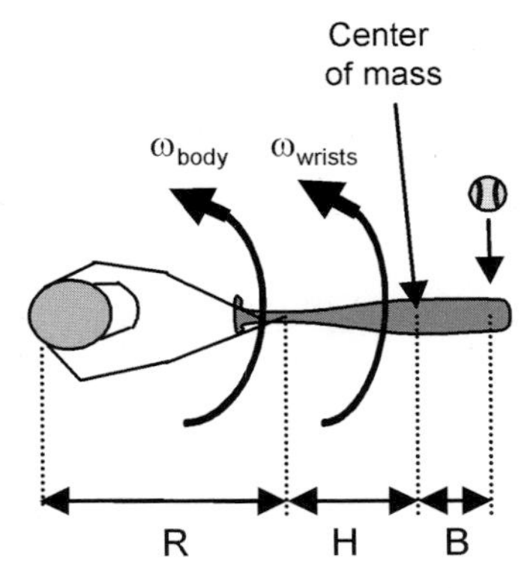

Fig. 3.64. Model of batting a ball. (Based on [173])

the weight of the bat. (Major league bats typically weigh approximately 32 oz; much heavier, 44–50 oz, bats were once routinely used by sluggers.) We know that $W_{\text{ball}} = 5.125$ oz and that the coefficient of restitution for the collision of a baseball with a wooden bat is 0.55.

To hit a home run in a major league park, the initial speed of the batted ball needs to be ∼100 mph, so it could clear a 15 ft tall fence that is ∼330 ft from home plate. (This includes the effect of drag. Also, note that it is best to hit the ball at an angle of 35° to maximize the distance traveled for the initial speed because of drag, which is smaller than 45° optimum angle for range with no drag. (See [97].)) To hit a ball this hard requires a bat speed of ≃50 mph at its center of mass or ≃58 mph a bit further out on the bat, at the "sweet spot," which is the most effect location for this collision (see [173]). We will now examine the different important aspects of this collision.

What is the optimal bat weight? In one model [103, 136, 157, 173] it is assumed that the optimal bat weight is one that requires the least energy input to achieve a given batted-ball speed. Minimizing the total initial kinetic energy of the ball and bat, and using that result in (3.114) gives

$$\frac{W_{\text{bat,ideal}}}{W_{\text{ball}}} = \frac{v'_{\text{ball}} - v_{\text{ball}}}{v'_{\text{ball}} + e v_{\text{ball}}}. \tag{3.115}$$

Using the typical pitch speed of $v_{\text{ball}} = -80$ mph and the batted ball speed to hit a home run $v'_{\text{ball}} = 110$ mph, the "optimized" bat weight is $W_{\text{bat}} = m_{\text{bat}} g =$ 15 oz, which is smaller than the actual weight of bats that are used by adults. (Is the assumption that the initial kinetic energy should be minimized really reasonable?)

Another way to determine the ideal bat weight combines a model with measured data. The fastest speed you can swing a bat has been observed to decrease with bat weight, typically as

$$v_{\text{bat}} = A - B W_{\text{bat}}. \tag{3.116}$$

Substituting (3.116) into (3.114) (and using weights instead of masses), and then differentiating the equation with respect to W_{bat} and setting the result equal to zero, gives the bat weight leading to the fastest batted ball

$$\frac{W_{\text{bat,ideal}}}{W_{\text{ball}}} = \sqrt{1 + (A - v_{\text{ball}})/(BW_{\text{ball}})} - 1. \tag{3.117}$$

In particular, for one slugger the bat speed relation (3.116) has been measured to be [103, 173]

$$v_{\text{bat}}(\text{in mph}) = 63 - 0.39W_{\text{bat}}(\text{in oz}), \tag{3.118}$$

so here $A = 63\,\text{mph}$ – the maximum bat speed for a massless bat – and $B = 0.39$. For $v_{\text{ball}} = -80\,\text{mph}$, the ideal bat weight is 38.5 oz, which is quite reasonable, and the batted ball speed is 95.1 mph. In fact, the batted ball speed varies very little for a broad range of bat weights.

Using multisegment models with more than one joint is important in analyzing statics and motion, such as for throwing a ball. Figure 3.64 shows that it is also very important in hitting balls. The body rotates at an angular speed ω_{body} due to the combined rotation about the hip, shoulders, and ankles and the wrists rotate at an angular speed ω_{wrists}. With the extension of the arms being R, the speed of the hands is $R\omega_{\text{body}}$. With the center of mass of the bat a distance H from the hands, the speed of the bat center of mass due to body rotation is $(R + H)\,\omega_{\text{body}}$. The speed of the center of mass of the bat due to wrist motion is $H\omega_{\text{wrists}}$. Therefore, before hitting the ball, the bat center of mass moves at the speed v_{bat}

$$v_{\text{bat}} = (R + H)\,\omega_{\text{body}} + H\omega_{\text{wrists}}. \tag{3.119}$$

If the ball hits the bat a distance H from its center of mass, then H is replaced by $H + B$, so

$$v_{\text{bat, not CM}} = (R + H + B)\,\omega_{\text{body}} + (H + B)\,\omega_{\text{wrists}}, \tag{3.120}$$

and this is related to the speed of the bat center of mass by

$$v_{\text{bat, not CM}} = v_{\text{bat}} + B(\omega_{\text{body}} + \omega_{\text{wrists}}). \tag{3.121}$$

This shows that the motion of the region where the bat hits the ball can be viewed as the rotation of the bat center of mass plus a rotation of this region about the bat center of mass at angular speed $\omega_{\text{bat}} = \omega_{\text{body}} + \omega_{\text{wrists}}$.

How does this change our analysis of the collision of the ball and bat? Conservation of linear momentum, (3.96), still holds for the ball and the bat center of mass. However, you need to use the bat speed at the actual point of impact before and after the collision in the coefficient of restitution equation, (3.97). Furthermore, there is a torque on the bat during the collision when the ball does not hit the center of mass of the bat. This is examined further in Problem 3.90.

3.8.4 Running

Your foot suffers a collision every time it touches the ground during walking, running, and landing after a jump. During a very slow step – a gentle collision – the forces on the foot slowly build up from 0 to $m_b g$. During running the forces on the foot reach a peak of ∼2.0–3.6 $m_b g$ (marathon running: ∼$2.7 m_b g$; sprinting: ∼3.6 $m_b g$). Does this make sense?

Equation (3.101) shows the deceleration during a collision is $a_{decel} = v_i^2/2\Delta x$. Here, v_i is the initial downward speed of the foot and Δx is the vertical braking distance of the foot. The downward speed is ∼1.5 m/s for a running speed of 5 m/s (5.4 min/mile). The braking distance is due to the compression of the fatty pad of the heel of the foot, which is ∼4 mm, plus that of the heel of the running shoe, ∼8 mm, for a total of ∼12 mm. The deceleration is 187.5 m/s^2 and the maximum force on the decelerating mass m is $19.1mg$. The approximate energy returned to the body as a consequence of this collision was addressed above and will be given in Table 3.8 later. Now let us estimate the forces involved in the downward deceleration.

If the entire body were being decelerated during the collision of the heel (the human body as a rigid body), the force on the heel would be due to the entire mass of the body, $19.1 m_b g$. That is excessive, so it appears that during the collision the entire body mass is not being decelerated at once by the heel. In the opposite extreme we can assume that only the foot is being decelerated. Because the mass of the foot is $0.0145 m_b$ (Table 1.7), the force on the heel would be $0.28 m_b g$ plus that from the slower rest of the body, $0.9855 m_b g$, for a total peak force of $1.26 m_b g$. This seems low. Following the lessons learned from "Goldilocks and the Three Bears" (if first too big and then too small, try somewhere in between), we should examine several intermediate cases. Let us say that during the collision the foot plus lower leg mass $= 0.061 m_b$ (or alternatively the foot plus lower leg plus thigh mass $= 0.161 m_b$) is being decelerated. Then the peak forces would be $1.165 m_b g + 0.939 m_b g = 2.104 m_b g$ (or $3.075 m_b g + 0.839 m_b g = 3.914 m_b g$ for the entire leg). Comparing this to the cited measurements suggests that during running the initial impact

Table 3.8. Energy return from elastic sports equipment. (Using data from [171])

equipment	k (1,000 N/m)	Δx_{max} (cm)	KE_{max} (J)	H_{max} (cm)
trampoline	5	80	1,600	230
tumbling floor	50	10	250	36
gymnastic floor	120	5	150	22
running track	240	1	12	2
gymnasium floor	400	0.5	5	1

Δx_{max} and KE_{max} are the maximum deformation and stored energy ($KE_{max} = k\Delta x_{max}^2/2$) and H_{max} is the maximum height of a $m_b = 70$ kg person with this energy, $H_{max} = KE_{max}/m_b g$.

of the collision is borne by your foot and lower leg, with some significant coupling to the thigh. Experience shows us that this is reasonable. Without some coupling to the thigh, your knees (patellar tendons, etc.) would not have the opportunity to hurt when you run.

Sliding in Baseball

The goal of sliding into second or third base in baseball is to touch the base and then maintain contact with it before being tagged by the fielder. By sliding, you are at ground level and therefore harder to tag. You also have a greater chance of maintaining contact with the base after "colliding" with it, without bouncing off of it. Even better, by sliding you decelerate and collide with the bag at a slower speed, so there is less chance of breaking your foot (or hands in the ill-advised head-first slide). However, because you are decelerating during the slide, it takes a longer time to reach the base, $\Delta t_{\mathrm{slide,extra}}$, and it is more likely you will be tagged out than if you had continued at the same initial speed v_{i} and then suddenly stopped at the base without overrunning it (and damaging your ligaments, etc.). You can (1) control "when" you start your slide – which really means the distance before the base where you start to slide, $\Delta x_{\mathrm{slide}}$, (2) try to control your (final) speed at the base, $v_{\mathrm{f,slide}}$ (the speed of the collision with the base), and (3) try to control the effective kinetic coefficient of friction in your deceleration μ_{slide}. (You slow down faster with more body area and weight on the ground and with your spikes dug in more. You decelerate slower on mud than on dry dirt.) Assuming a normal force equal to your body weight, the frictional force has a constant magnitude $\mu_{\mathrm{slide}} m_{\mathrm{b}} g$ and so $a_{\mathrm{coll}} = -a_{\mathrm{decel}} = -\mu_{\mathrm{slide}} g$.

Using (3.90), we see that $\Delta t_{\mathrm{slide}} = (v_{\mathrm{i}} - v_{\mathrm{f,slide}})/\mu_{\mathrm{slide}} g$, where $\Delta t_{\mathrm{slide}}$ is the duration of the slide. (Clearly, $\Delta t_{\mathrm{slide,extra}} = \Delta t_{\mathrm{slide}} - \Delta x_{\mathrm{slide}}/v_{\mathrm{i}}$.) Also, using (3.91), we see $\Delta x_{\mathrm{slide}} = v_{\mathrm{i}} \Delta t_{\mathrm{slide}} - 1/2 \mu_{\mathrm{slide}} g \Delta t_{\mathrm{slide}}{}^2$. This is examined more in Problems 3.95 and 3.96.

3.8.5 Jumping

Several forms of jumping involve full collisions with a surface. The spring constant (stiffness) of this surface greatly affects how much energy can be returned to the body. This is seen in Table 3.8, which shows different surfaces, and for each their force constant and maximum deformation, the maximum stored energy, and the maximum height a 70 kg person could attain with this maximum stored energy. This is obviously significant for jumping on trampolines and gym floors.

3.9 Sustained Acceleration

In the previous section we have seen that collisions involving the head can lead to damage from the direct impact and from accelerations induced by the collision. Sustained linear and rotational (or rotary) accelerations of the entire body are also generally important, and they are particulary significant for space flight and for fighter pilot excursions [117, 118].

We will define linear accelerations using the coordinate system in Fig. 1.1, with $+g_z$ corresponding to the normal gravity acceleration g we feel on earth, and consider the effect of these accelerations [116, 117]. For such positive vertical acceleration, it is difficult to raise oneself at $2.5g_z$; it is impossible to raise oneself and difficult to raise one's arms and legs, and vision dims after 3–4 s from 3–$4g_z$; and vision is blacked out after 5 s, hearing and consciousness is later lost, and convulsions are possible from 4.5–$6g_z$. As we will see in Chap. 7, one reason for these extreme responses is that blood pumped by heart cannot reach the brain. At $+1g_z$, the (systolic) arterial pressure is 120 mmHg at the heart, 96 mmHg at the head, and 170 mmHg at the foot. At $+5g_z$, this pressure is still 120 mmHg at the heart, but 0 mmHg at the base of the brain and 370 mmHg at the foot. This means your brain does not get blood (which means you become unconscious) and you need an extra 250 mmHg pressure to pump the blood in the veins in your legs and feet back to the heart. Even though your body compensates for this a bit (by increasing blood pressure, increasing the heart rate, decreasing the cardiac volumetric output, and so on), so you may be able to withstand $+5g_z$, the situation is still dire. Vision is affected at lower accelerations than for general brain function because blood flow to the retina has the barrier of the intraocular pressure, which is about 20 mmHg higher than the intracerebral pressure. (Vision may be affected in ways similar to *retinal dysfunction ischemia*.)

Negative vertical accelerations also lead to physiological consequences: there is severe facial congestion, throbbing headaches, and after 5 s blurring of vision from $-2g_z$ to $-3g_z$; and most subjects cannot even tolerate 5 s of $-5g_z$ because of bloodied eyes, nosebleeds, and hemorrhage.

We can tolerate forward $+g_x$ and backward $-g_x$ acceleration a bit better. A forward acceleration of $2g_x$ is tolerable for at least 24 h and $4g_x$ for at least 60 min. There is increased weight, abdominal pressure, and some difficulty in focusing vision from 2–$3g_x$; progressive chest tightness and pain, difficulty in breathing, blurring of vision from 3–$6g_x$; each symptom gets worse from 6–$9g_x$; arms cannot be lifted at $8g_x$; the head cannot be lifted at $9g_x$; breathing is difficult and peripheral vision is lost from 9–$12g_x$; and there is extreme difficulty in breathing and speaking and a recurrent complete loss of vision at $15g_x$. The effects of backward accelerations are similar, but there is none of the increased pressure on the chest that accompanies forward accelerations, so breathing is easier.

For lateral accelerations, $\pm g_y$, there is discomfort at $3g_y$ after 10 s, and external hemorrhage and severe headaches afterward for $5g_y$ for 14.5 s.

Rotational acceleration about the body x-axis is called *rolling*, about the y-axis *pitching*, and about the z-axis *yawing* or *spinning*, again using the coordinate system in Fig. 1.1. Sometimes the term *tumbling* is used instead of pitching and rolling. Tumbling (about the pitch axis) centered about the heart is safe for rotations rates of at least 120 rpm for about 3 s and for successively longer times at slower rates, such as 70 rpm for 400 s [117, 174].

3.10 Physics of Sports

This chapter has shown how many types of motion can be analyzed using simple physics. Not coincidentally we have also examined the physics of several sports, including those in track and field, baseball, and boxing. There are several books devoted to the physics of different sports. In most cases they emphasize the object of the sport and not how the human body generates the needed action. (They are nonetheless very interesting.) The excellent book by Hay [127] is one of several exceptions, in that it examines the biomechanics of many sports, including gymnastics, swimming, track and field, and several major team sports. The book edited by Hung and Pallis [133] examines the physics of golf, tennis, baseball, football and soccer, and basketball. Adair [97, 98] and Watts and Bahill [173] have analyzed baseball, concentrating on the motion of the ball during a pitch and after it has been hit. Gay [122] discusses the physics of football. Football is the epitome of collision "ball" sports. A player tries to advance the football. The goal of the defensive player is to collide with the ball carrier to stop his progress. The goal of most of the offensive players, other than the ball carrier, is to collide with the defensive players to stop them from colliding with the ball carrier. Jorgensen [134] has examined golf. A golf swing can be modeled as a double pendulum, one pendulum being your arms and the other the golf club. Lind and Sanders [142] have looked at the physics of skiing, including the skiing motion, snow, ski wax, and so on. The books by Laws and co-workers [137, 138, 139] examine a subject related to the physics of sports: the physics of dancing.

3.11 Summary

Motion of the body can be understood by analyzing the kinematics, dynamics, and energetics of the motion. Understanding body stability and the action of skeletal muscles is also important. Walking, running, jumping, and throwing a ball can be analyzed by using simple models, which can be refined to be more realistic. Similarly, collisions can be modeled to understand body motion and action, as in sports, and how the body is affected when it is involved in a crash.

Problems

Stability, Friction, and Human Moments of Inertia

3.1. Consider a person modeled as a one-dimensional being of height H and mass m_{b} in the two-dimensional world of Fig. 3.7a. The person has a constant mass per unit length and the body rests 25% from the back of her massless feet, which have length $0.152H$ (as from Table 1.6) and negligible height. To what angles can the person tilt in the forward and backward directions and still maintain overall balance and stability? (Consider potential rotations about the front and back of the feet, as in Fig. 3.7a.)

3.2. Can the pregnant woman, modeled in the two-dimensional rigid-figure world of Fig. 3.7b, maintain overall balance and stability? Assume the centers of mass of the woman without the womb and the womb itself are both $0.58H$ above her feet. The additional mass is $0.25m_{\mathrm{b}}$ and has a center of mass $0.1H$ in front of the rest of her body. (Assume the other information given about the feet in Problem 3.1.)

3.3. Refer to the gymnast in the arabesque position described in Problem 2.11. In attaining this position her center of mass descends but does not move laterally. How does this affect her overall stability?

3.4. (a) Why is a tripod stable?
(b) Explain why a moving six-legged creature (an insect) can always be stable if it lifts and moves three feet at a time, while keeping its other three feet on the ground [163].
(c) Explain how a four-legged animal (a horse) can move, albeit slowly, always maintaining stability, by keeping three of its feet on the ground at all times. Is this type of motion normal for a four-legged animal? Why?
(d) The two-legged motion of humans cannot take advantage of the stability of a tripod. Still, standing barefoot on one foot or both feet can sometimes approximate a tripod. Explain this and why such tripods are fairly unstable.

3.5. The normal force on the head of the femur in the knee is 300 N. Find the frictional force for a normal ($\mu = 0.003$) and arthritic ($\mu = 0.03$) joint.

3.6. The accelerating sprinter described in this chapter pushes off with 560 N force. If the net downward force on the foot on the ground is $3m_{\mathrm{b}}g$ for this 70 kg sprinter, what is the minimum coefficient of friction needed to prevent slipping? Under what conditions is this possible?

3.7. Use the data from Chap. 1 to show that the moments of inertia of a normally standing person about his or her center of mass are roughly the following, about these axes [127]:
(a) 1 kg-m^2 about a vertical axis.
(b) 11 kg-m^2 about the axis in the transverse plane that is normal to the vertical axis.
(c) 11 kg-m^2 about the axis in the mid-sagittal plane that is normal to the vertical axis.
(d) If a person curls up into a ball the moments of inertia about all three axes will be roughly the same. Estimate them.

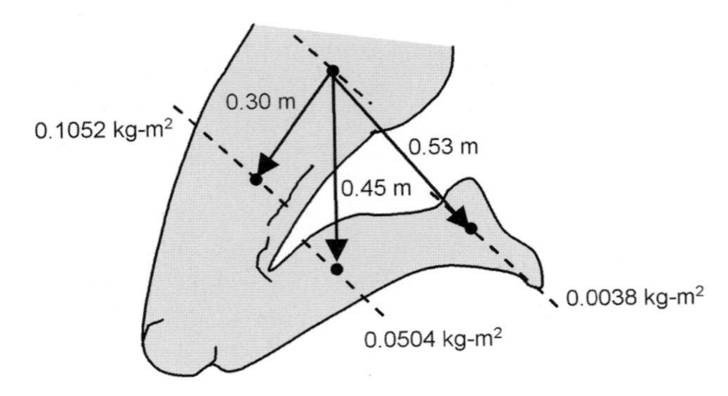

Fig. 3.65. Determination of the moment of inertia of a sprinter's leg about the hip axis. (Based on [127].) For Problem 3.10

3.8. Calculate the moments of inertia of a standard man about the three normal axes though his center of mass – described in Problem 3.7, using each of the models in Problem 1.30.

3.9. (a) For each limb and limb segment in Table 1.9, show that the sum of the squares of the radius of gyration about the center of mass and the distance of the proximal end from the center of mass equals the square of the radius of gyration about the proximal end. (All distances can remain normalized by the segment length.)
(b) Repeat this for the distal end.
(c) Why is this so?

3.10. Use the parallel axis theorem to calculate the moment of inertia of a sprinter's leg about the hip axis, as in (a) Fig. 3.65. The moments of inertia of the upper leg (thigh), lower leg (calf), and foot about their respective centers of mass are 0.1052, 0.504, and 0.0038 kg-m^2, these centers of mass are, respectively, 0.30, 0.45, and 0.53 m from the hip rotation axis, and these segments, respectively, have masses 7.21, 3.01, and 1.05 kg.

3.11. How do the leg joint angles defined in Fig. 3.1 differ from those in our discussion of walking and those defined in the caption to Fig. 3.51?

3.12. Relate the angles defined in Problem 2.37 for the multisegment model of Fig. 2.57 to the joint angles.

Walking and Running

3.13. Use Fig. 3.9 to determine the fraction of time that both legs, one leg, and neither leg is on the ground during:
(a) Walking
(b) Running.

3.14. (a) Use the pendulum model of walking to determine how fast someone walks. Every half period the person takes a step that corresponds to the arc length the foot (at the end of the leg of length L) that traverses 30° during the step (so the step length is $(\pi/6)L$). (This is approximately the maximum swing angle for a fast walk.)
(b) Calculate the pendulum frequency and this walking speed for a 2 m tall adult and a 1 m tall child. Do the ratios of their frequencies and speeds make physical sense? Explain. (You can use the model that assumes constant leg linear density.)

3.15. Show that it is reasonable that leg length L scales as $m_{\mathrm{b}}^{1/3}$, and therefore the walking speed scales as $m_{\mathrm{b}}^{1/6}$, as in Table 1.13.

3.16. Obtain the moment of inertia of a leg of mass m_{leg} and length L_{leg} with constant linear mass density by:
(a) Changing the integration limits in (3.30)
(b) Using the results of (3.30) and the parallel axis theorem.

3.17. Compare the moments of inertia of the whole leg as determined by the constant linear density model, (3.30) and the more refined model (result after (3.34)), with that determined from the total leg radius of gyration given in Table 1.9. (Use the parallel axis theorem.)

3.18. Estimate what fraction of the change in body angular momentum due to leg motion is canceled by swaying your arms out of phase with your legs when you walk. Assume standard parameters for your arms and legs, that they are straight and each has a uniform mass density per unit length:
(a) First assume that $\mathrm{d}\theta/\mathrm{d}t$ is equal in magnitude and opposite in sign for the legs and arms.
(b) Now assume that the forward displacements traversed by the foot and hand $L\mathrm{d}\theta(t)/\mathrm{d}t$ are equal in magnitude and opposite in sign, where L is alternately the leg or arm length.

3.19. Estimate the vertical angular momentum gained by your body with each stride (ignoring the motion of your arms and torso) for a 3.4 m/s running speed.

3.20. Each step in climbing a staircase can be modeled in two parts. First, one foot is placed on the next step and, second, that foot is used to propel the body up. Assume that the center of mass is raised an insignificant amount in the first part, so the whole change in the center of mass occurs in the second part of the step. Also, assume that the durations of both parts of the step are the same. Consider a staircase with 13 steps, each 20 cm high. It takes 3.6 s to go up the staircase "quickly" and 6.0 s to go up "slowly." Calculate the average total (not net) vertical force in the second part of the step for each case, in terms of the body weight W_{b}. (You can ignore horizontal motion.).

3.21. Determine the total range of motion of thigh, knee, and ankle for each example of running in Table 3.5.

3.22. (a) A 70 kg person runs at a speed of 4.50 m/s. If 100 J of kinetic energy is lost each time a foot touches the ground (Fig. 3.27), what is the speed after stage (a) (if it was 4.50 m/s just before stage (a)). (Ignore vertical motion and potential energy changes throughout this problem.)
(b) How much energy needs to be supplied by the body in the acceleration phase (stage (c)) to account for the loss when the foot earlier hit the ground, if the 93% of the 35 J which is stored in the Achilles tendon and the 80% of the 17 J which is stored in flattening the foot arch in stage (a) are both returned in stage (c) (and nothing else is)?
(c) If the runner takes 3 steps per second and the runner's muscles are 20% efficient in converting energy into the mechanical work of running, how much extra energy is used by the body per hour when it is running? Express your answer in kcal/h, where 1 kcal = 1×10^3 cal and 1 cal = 4.184 J.

3.23. After how many strides does the accelerating sprinter reach 90% of her final speed? What distance does this correspond to (in m)?

3.24. The final speed attained of the accelerating sprinter refers to the upper body and one leg. If the other leg can be considered to be still, what is the speed of the sprinter's center of mass after acceleration?

3.25. We assumed that the pushoff leg of the accelerating sprinter does not contribute any kinetic energy in (3.38)–(3.47). Let us say that the center of mass of that leg is moving at half of the speed of the upper body and other leg:
(a) Is this reasonable? Why?
(b) How would this change (3.38)–(3.47)?
(c) How would it change the numerical value of the calculated final speed?
(d) How would it change the number of strides needed to reach 90% of the final speed?

3.26. We assumed that the pushoff leg of the accelerating sprinter does not contribute any kinetic energy in (3.38)–(3.47). Let us say that this leg rotates while this foot is on the ground, so the top of it moves at the same speed as the rest of the body, but the bottom of it is still. Assume this leg is straight and of uniform linear density. What fraction of the kinetic energy of the leg is lost as it changes from a translating to a rotating leg?

3.27. Do parts (b)–(d) of Problem 3.25 for the case posed by Problem 3.26.

3.28. Show that the maximum (horizontal) running speed of mammals is expected to be roughly independent of mass by using the following approach [148, 168]. Assume that the maximum muscle force is proportional to its cross-sectional area, which is $\propto L^2$ for characteristic dimension L. The body mass

is proportional to the volume, and so $m_\mathrm{b} \propto L^3$. Proceed as follows:
(a) The work done by leg muscles in each stride is the muscle force $\times$ the muscle contraction distance. Show that this scales at L^3.
(b) The kinetic energy of that limb is $I(\mathrm{d}\theta/\mathrm{d}t)^2/2$, where I is the moment of inertia of the leg and $\mathrm{d}\theta/\mathrm{d}t$ is its angular speed, which is the running speed v divided by the leg length. Show that this energy scales as L^3v^2.
(c) By equating these, show that v is independent of L and therefore m_b.

3.29. Show that the maximum uphill running speed of mammals should decrease with linear dimension L, as $1/L$, and therefore with mass m_b as $1/m_\mathrm{b}^{1/3}$ [148, 168]. Do this by equating the power available from the muscles to the power required to work against gravity. (Hint: The power available from muscles is the work done by leg muscles per stride ($\propto L^3$, from Problem 3.28) divided by the time per stride (which is the leg length divided by the speed, and Problem 3.28 shows the speed is independent of L). The power needed to work against gravity is the body weight times this uphill running speed.)

Jumping

3.30. Neglect air resistance and muscle atrophy and assume the same upward normal reaction force as on earth:
(a) If someone can increase her center of mass by 0.7 m in a vertical jump on earth, how high could she jump on the moon?
(b) If an athlete can long jump 25 ft on earth, how far could she long jump on the moon? (Assume her takeoff angle is 30°.)

3.31. (a) During a vertical jump the center of mass of a 70 kg person is 0.65 m from the ground in the crouch phase and 1.05 m at takeoff. During the extension phase the average (total) force exerted by the floor (summed on both feet) is 1,600 N. (Remember that part of this counters the body weight and the remainder – the net vertical force – counters the forces due to muscles.) When necessary, assume that this force is constant during the extension phase. Find how high the center of mass rises during free vertical flight, the speed at takeoff, and the (temporal) duration of the extension phase.
(b) About 43% of the mass of the average person is muscle and essentially all of it is skeletal muscle mass. This average person in (a) does weight training, which increases his body mass by 10 kg. Assume all of it goes into muscle proportionately throughout the body and that the total vertical force is proportional to the total body muscle mass. How high does his center of mass rise now during a vertical jump?
(c) Repeat (b) (starting with the person in (a)) if the "weight training" instead leads to adding only 10 kg of muscle.
(d) Repeat (b) (starting with the person in (a)) if the "weight training" instead leads to adding only body fat – with no change in muscle mass.

(e) Find how high the center of mass rises in (a) if the person takes a wonder drug that decreases the extension phase by 0.05 s.
(f) On the moon, how high does the person in (a) rise in a vertical jump (from a hard floor)? Assume $g_{\text{Moon}} = g_{\text{Earth}}/6$ and that the muscles function exactly as on Earth.
(g) If the person in (a) were to jump from a rubber pad or sand instead of a hard floor the center of mass would rise less. Why?

3.32. Redo Problem 3.31b if the person now does not gain weight, but instead either increases his muscle mass fraction from 43% to 53% or decreases it to 33%.

3.33. The world record pole vault (6.14 m) exceeds that calculated here (5.4 m). Compare this world record to the calculation assuming world class speed, 10.2 m/s, and $h_{\text{min}} = 0.02$ m. Why are they different?

3.34. In a high jump the athlete takes a running start and then hurls himself over a horizontal pole. (In the older, straddle high jump (face down when over the bar), the athlete's center of mass is ∼150 mm over the bar, while in the newer Fosbury flop method (face up when over the bar), the center of mass is slightly under the bar. Assume here that the center of mass needs to clear the bar by 100 mm during a successful high jump (H_3 in Fig. 3.66)):
(a) If a 70 kg athlete's initial center of mass is 0.9 m high (H_1), how high can the bar be for a successful jump ($H_1 + H_2 - H_3$) if the athlete runs at 7.0 m/s and all of his initial kinetic energy (corresponding to motion in the horizontal direction) is converted into kinetic energy corresponding to motion in the vertical direction and then into potential energy?
(b) Since the world record high jump is 2.45 m (in 2006), how much of the initial kinetic energy could not have gone into kinetic energy associated with vertical motion?

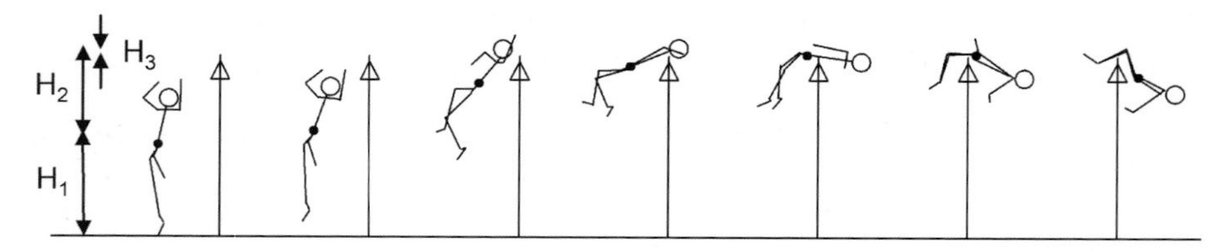

Fig. 3.66. Multisegment model of a high jump using the Fosbury flop method, showing that the center of mass (the *closed circle*) of the jumper is always below the bar (in this example). Heights during a high jump are also shown, with H_1 being the initial height of the center of mass, H_2 the maximum elevation of the center of mass, and H_3 the distance that the final center of mass is above the bar. The arrows are shown assuming the highest center of mass is above the bar, so H_3 would really be negative for a successful Fosbury flop. (Based on [127].) For Problems 3.34 and 3.35

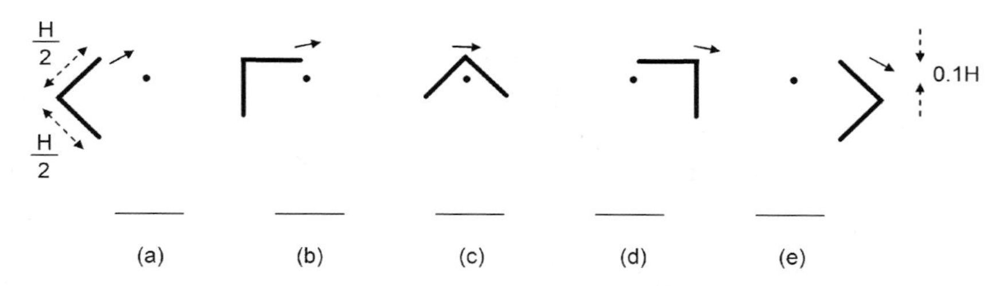

Fig. 3.67. Simple one-segment model of a high jump using the Fosbury flop method for a person with height H, with the body modeled as two segments of equal length, normal to each other. In (**a**)–(**e**), the top of the modeled body is always $0.1H$ above the bar (which is the dot). This is a reasonable assumption for (**c**), because the center of the person's chest is high enough for her to clear the bar. For Problems 3.36 and 3.37

(c) Measurements show that when 76 kg athletes run 6.7 m/s and high jump over a bar that is 2.0 m high, their horizontal speed over the bar is 4.2 m/s. How much energy is still not accounted for? What happened to it? (Note that the total kinetic energy (in two-dimensions, x and z) is the sum of that for horizontal and vertical motion $= mv_x^2/2 + mv_z^2/2$.)

3.35. In a high jump using the Fosbury flop method (face up when over the bar, as in Fig. 3.66), the center of mass can be slightly below the bar. Find H_3 if the athlete height is 1.96 m, $H_1 = 1.40$ m, $H_2 = 0.97$ m, and the height of the bar is 2.30 m (as for the jumper Dwight Stones, [127]).

3.36. For the high-jumper modeled with a two-segment model of the Fosbury flop in Fig. 3.67, calculate how much lower the center of mass of the jumper is than the bar for positions (a)–(e). How much lower is it for position (c)?

3.37. Assume the two segments of the Fosbury flop jumper in Fig. 3.67 are at an obtuse angle, rather than the right angle shown in the figure. (This is more realistic.) Also assume that the general orientation of the jumper is otherwise the same in each part of the figure. Find the maximum angle for which the center of mass of the jumper will always remain below the bar.

3.38. (a) The length of the long jump has three parts, the takeoff distance (L_1, the center of mass precedes the foot at takeoff, take as 0.24 m here), the flight distance (L_2, the distance the center of mass travels during flight), and the landing distance (L_3, the distance the heel lands in the sand in front of the center of mass, take as 0.53 m), as in Fig. 3.68. Calculate the length of a jump for a world-class male and female long jumps with speeds at takeoff of 9.8 and 8.6 m/s, respectively, and a takeoff angle of 20.0°.
(b) We have ignored drag here, as well as the difference in height of the center of mass at takeoff and landing. How are the results in (a) affected if the center of mass at takeoff is 60 cm above that during landing?

Fig. 3.68. Lengths during a hang-style long jump. (Based on [127].) For Problem 3.38

3.39. A 70 kg world-class long jumper accelerates to a speed of 10.5 m/s and then jumps the longest distance possible:
(a) If drag due to air resistance is neglected, show that this occurs at a 45° takeoff angle, and find the length of this longest possible long jump. (Assume that the long jump distance is the same as the horizontal distance traveled by the center of mass. (They can be slightly different because of the different take-off and landing arrangements of the body, as is addressed in Problem 3.38.))
(b) Compare this distance to the world record of 9.0 m, and qualitatively account for any differences between this record jump and your calculated value. (Consider that many long jumpers take off at an angle closer to 20°.)

3.40. If drag due to air resistance is neglected, a projectile, initially at ground level, travels farthest with a 45° takeoff angle. Including air resistance the takeoff angle is closer to ~35°. Given this, does it makes sense that many long jumpers take off at an angle closer to 20°?

3.41. Show that the maximum jumping height of the center of mass of mammals is expected to be roughly independent of mass. Do this by first showing the acceleration a due to muscles varies as $1/L$, for characteristic mammal linear dimension L. Then show that the launch speed, $v = \sqrt{2as}$ – where s is the vertical distance traveled during acceleration – is independent of L and so the height of the jump $v^2/2g$ is expected to be independent of mass. Use the same scaling rules for the maximum muscle force and body mass as in Problem 3.28. (How does s scale with L?)

Other Motions

3.42. (a) A figure skater, rotating at an angular frequency of f (in revolutions per s, which can be ~0.6 revolutions/s) with outstretched arms, rotates even faster after pulling in her arms. Why?
(b) The moment of inertia of this person changes by a factor α by this pulling action. What is the skater's new rotation speed?

(c) Estimate this factor. (You can use Tables 1.6–1.9 and Problems 1.30 and 1.31 for guidance.)

Throwing Balls and Other Objects

3.43. (a) Use the one-segment model in this chapter that predicts you can throw a baseball at 45 mph to determine how fast can you throw a football. A National Football League (NFL) football must weigh between 14 and 15 oz (and have a mass between 0.40 and 0.43 kg). Is this consistent with an NFL quarterback throwing a ball at 68 mph? Does this make sense?
(b) Would it make sense if you modified the one-segment ball throwing model so the baseball is released at 100 mph and then used it for throwing a football?

3.44. In the one-segment model for throwing a ball in this chapter, what should the angle of the upper arm be to the vertical so that the velocity vector of the released ball is horizontal (with a release at $\theta = \pi/4$)?

3.45. (a) Modify the one-segment model of throwing a baseball so that the whole arm is always straight and is rotated about the shoulder by the deltoid muscles.
(b) How fast is the ball released, assuming 3 in diameter deltoid muscles and the other assumptions in the final model in this chapter?

3.46. How do the torques we calculated in our pitching model compare to those in Figs. 3.46 and 3.47? Should they be comparable?

3.47. Estimate the relative final angular rotation speeds of the elbow and wrist in throwing a ball by using Fig. 3.45, and then estimate the relative contributions to the speed of the pitched ball from the motion of the lower arm and the wrist. Assume the ball is a distance of half of the length of the hand from the wrist joint.

3.48. (a) Show that a ball travels farthest when released at a 45° takeoff angle, and when it is released at speed v, it travels a distance of v^2/g. (This is an overestimate because air drag is omitted. When drag is included, the best angle is really 35°. Also, we are ignoring any difference in the heights of the release point and the ground. This is examined later.)
(b) How far does the baseball travel when thrown at 90 mph?
(c) If the ball is wet and weighs 7 oz instead of 5 oz, how far will it travel? Assume the ball leaves the hand with the same linear momentum as the dry ball does in part (b).

3.49. If the initial velocity vector component of a ball is $v_{x,0}$ in the forward direction and $v_{z,0}$ in the vertical direction, show that it travels a distance $2v_{x,0}v_{z,0}/g$ (which is called its range).

3.50. The farthest distance an object can be thrown is with a takeoff angle of 45°, ignoring drag, only if it is thrown from ground height. Otherwise this is just an approximation, as we will now see:
(a) If an object is thrown with an initial speed v and angle α, from a height $z = h$ show that at a later time: $x(t) = vt\cos\alpha$ and $z(t) = h + vt\sin\alpha - gt^2/2$.
(b) Show that it reaches the ground, $z(t) = 0$, after it has traveled a distance $x_{\text{final}} = (v^2\cos\alpha/g)(\sin\alpha + \sqrt{q^2 + \sin^2\alpha})$, where $q = \sqrt{2gh}/v$. (Note that $q^2 = mgh/(mv^2/2)$, which is the initial potential energy divided by the initial kinetic energy.)
(c) By setting $dx/d\alpha = 0$, show that the ball will travel farthest for an initial angle given by $\tan\alpha_{\text{max}} = 1/\sqrt{1+q^2}$ and it travels a distance $x_{\text{final,max}} = (v/g)\sqrt{v^2 + 2gh}$.

3.51. A baseball is thrown from a height of 6.5 ft from the ground with an initial speed of 90 mph. Ignoring drag, find how far it travels if:
(a) it is released at 45°, and you ignore the difference in the release height and the ground.
(b) it is released at 45°, and you take into account the difference in the release height and the ground (see Problem 3.50).
(c) it is released at the optimal angle for distance (which you also must find), and you take into account the difference in the release height and the ground (see Problem 3.50).

3.52. A baseball pitcher throws a ball at 90 mph releasing it 6 ft in front of the pitcher's rubber (the stripe on the pitcher's mound), at a height of 5 ft. The ball arrives at the plate, which is 60 ft 6 in from the pitcher's rubber at a height of 3 ft. (Ignore ball deceleration due to drag):
(a) How long does is take the ball to arrive?
(b) What angle does the initial velocity vector of the ball make with the horizontal?
(c) When the ball arrives at the plate, how much does it appear to have fallen relative to the path it would have taken with its initial velocity vector? (Ignore any effects due to ball rotation.)

3.53. A baseball pitcher throws a ball with initial velocity vector component $v_{x,0}$ in the forward direction and $v_{z,0}$ in the vertical direction. (Ignore ball deceleration due to drag):
(a) Show that after a time t, its coordinates are: $x = v_{x,0}t$ and $z = v_{z,0}t - gt^2/2$.
(b) After it travels a distance d in the x direction, show that the ball seems to have fallen a distance $gd^2/2v_{x,0}^2$ from its initial trajectory.
(c) Show that the ball always stays in the $y = 0$ plane.

3.54. A baseball is rotated by a half revolution by the pitcher during the last 0.04 s before release, due to a constant rotational acceleration due to wrist action and the rotation of the radius on the ulna. Show that the released ball spins at a rate of 1,500 revolutions/min.

Table 3.9. Speed and spinning of pitches by professional-level pitchers. (Using data from [173])

pitch	speed (mph)	typical spin rate (rpm)
fastball	85–95	1,600
slider	75–85	1,700
curve ball	70–80	1,900
change-up	60–70	1,500
knuckleball	60–70	25–50

The term rpm stands for revolutions per minute.

3.55. The pitcher tries to maximize the rotation speed of the ball (ω, in rad/s) when throwing a curve ball, as seen in Table 3.9. A spinning ball feels the Magnus force (or lift) due to this rotation, which changes its path from that of a rotation-free ball. This lift force has magnitude $= \pi \rho r^3 \omega v/2$ for air mass density ρ, ball radius r, spinning rate ω (in rad/s), and ball speed v. (From [173]):

(a) The Magnus force on a baseball has been determined from measurements to have magnitude (in lb) $= 6.4 \times 10^{-7} fv$, where f is in revolution/min and the speed v is in ft/s. Is this consistent with the magnitude of the lift force just given?

(b) How many revolutions do a fast fastball, a slow curve ball, and a medium-speed, medium-rotation-rate knuckleball undergo if each is traveling at a constant forward speed and travels 55 ft before it reaches home plate?

(c) Sketch a diagram showing the initial velocity vector of the ball, how the ball rotates, and the direction of the Magnus forces when a right-handed pitcher rotates the ball in the direction of the screwdriver motion and releases it so that the rotation axis is vertical.

(d) If the Magnus force is constant, how much has the ball moved laterally – relative to the expected trajectory with no spinning – when it reaches the plate. (Ignore changes to speed and direction due to drag and gravity, so assume the ball is moving forward and horizontally at 80 mph. Also assume that changes in the trajectory due to the Magnus force are so small that you can assume this trajectory in your calculation.) (To learn more about drag forces see Chap. 7.)

3.56. As in Problem 3.53, a baseball pitcher throws a ball with initial velocity vector component $v_{x,0}$ in the forward direction and $v_{z,0}$ in the vertical direction, but now it has additional constant forces in the vertical z direction $\alpha_z g$ and lateral y direction $\alpha_y g$ due to its rotation and the Magnus force:

(a) Show that after a time t, its coordinates are: $x = v_{x,0}t$, $y = \alpha_y g t^2/2$, and $z = v_{z,0}t - (1-\alpha_z)gt^2/2$.

(b) After it travels a distance d in the x direction, show that the ball seems to have fallen a distance $(1-\alpha_z)gd^2/2v_{x,0}^2$ and moved laterally a distance $\alpha_y g d^2/2v_{x,0}^2$ from its initial trajectory.

(c) Show that the ball always stays in the tilted plane $z = (v_{z,0}/v_{x,0})t - ((1-\alpha_z)/\alpha_y)y$.
(d) Explain why the ball seems to move laterally when $\alpha_y \neq 0$, it falls even faster than expected when $\alpha_z < 0$, and it falls slower than expected – "the rising fastball" – when $\alpha_z > 0$.
(Note that this additional force vector has been assumed to be constant in this simple model. More specifically, the Magnus force has been assumed to be that for a ball thrown with a constant velocity in the x direction. Changes in the force direction and magnitude due to changes in ball velocity due to gravity and the Magnus force itself – and the addition of drag, make the ball motion even more complex.)

3.57. (a) Show that a 80 mph baseball curve ball subjected to a horizontal force of 2.6 oz ($\alpha_y = 0.5$) curves laterally by 2.1 ft from its expected path (assuming it travels a distance of 60 ft).
(b) In what direction is the spin (given by the right-hand rule) if the ball veers to the right, as seen at the plate. (This is a curve ball as thrown by a right-handed pitcher and a screwball as thrown by a left-handed pitcher.)
(c) How fast is the ball rotating?

3.58. Can the Magnus force described in Problem 3.57 explain the "rising fastball" – for which the batter seems to think that the pitched fastball is rising? This could be possible if the ball spun fast enough in the correct direction to give the ball a lift force exceeding g. (This corresponds to $\alpha_z > 1$ in Problem 3.56.) Calculate it for a translational speed of 100 mph and a rotation speed of 2,300 rpm (typical of some excellent pitchers), and show that under the best of circumstances, the Magnus force can account for at most about $\sim 2g/3$ (or $\alpha_z \sim 2/3$ in Problem 3.56), and so a "rising fastball" is really a consequence of a hitter's perception and expectation [102, 173].

3.59. A right-handed pitcher has similar wrist action when throwing overhand and side arm. When throwing overhand (arm motion in a vertical plane) his ball drops very fast. When throwing side arm (arm motion in a horizontal plane), it moves to the left. When throwing at three-quarter overhand (arm motion in a plane that bisects these two planes), it moves a bit down and to the left. Why?

3.60. In throwing a knuckleball, the pitcher is not concerned with the effects of the Magnus force, but with the sideways force on a ball that is not a perfect sphere because of the stitches on the ball (Fig. 3.69). The pitcher purposely tries to minimize the baseball rotation rate when throwing a knuckleball (Table 3.9), because this force will average to zero for a ball with a moderate rotation rate – and one that is too small for a significant Magnus force. With a slow rotation rate the lateral motion will seem irregular. (This irregular motion also occurs for a ball that has been scuffed – accidentally or on purpose.)

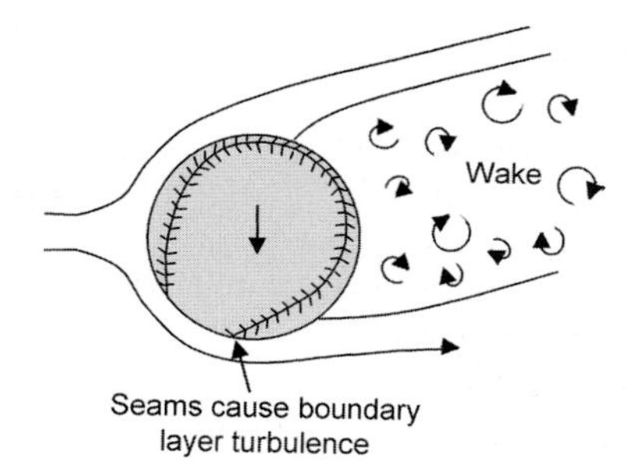

Fig. 3.69. The flow past a spinless ball can be asymmetric. In this example, the top is smooth and the stitches (or scuff mark) are on the bottom. The seams cause boundary layer turbulence, which delays airstream separation on the bottom surface. The wake moves upward and there is a downward force on the ball. (Based on [173])

The magnitude of this lift force is $C_{\mathrm{L}} A \rho v^2/2$, where C_{L} is the lift coefficient and A is the cross-sectional area, which is similar to (7.50) [173]:
(a) For a knuckle ball, this force is $2.16 \times 10^{-5} v^2$ (in lb) with v in ft/s. Show that this means that $C_L = 0.42$.
(b) For a scuffed ball, this force is $1 \times 10^{-5} v^2$ (with the same units). Show that this means that $C_L = 0.194$.
(c) Estimate how much a knuckleball can move laterally if it does not rotate at all? (If it does not rotate at all, its motion is predictable, and this is not desirable.)

3.61. Most basketball players shoot free throws from the free throw line in an overhand motion, but some professional players, notably Wilt Chamberlain and Rick Barry, have shot them underhanded. In both cases the basketball must go through an 18 in diameter horizontal hoop, with the hoop center 15 in. in front of the backboard. The hoop is 10 ft high on a backboard that is 15 ft from the free throw line. A basketball weighs 21 oz, has a circumference of 30 in, and therefore a diameter of 9.7 in.

One study has shown that for a release height of 7 ft, an error in release angle has the minimum affect on accuracy for release angles between 49° and 55°, and this corresponds to entry angles between 38° and 45°, both relative to the horizontal [127, 132]. More recent analysis suggest that the optimal release angle is between 51° and 56° with release speeds between 20.5 and 24.0 ft/s. Spin due to bending elbows and knees and snapping the wrist at the moment of release adds lifts due to the Magnus force [127].

Assume the ball goes through the middle of the hoop at a 41° angle to the horizontal (which is the entry angle). Assume that the 6 ft tall basketball player releases the ball with arms extended vertically for the overhand motion and with outstretched straight arms at shoulder level for the underhand motion. Sketch the motion and determine the initial speed and angle of release for both cases. (Hint: Use the anthropometric data in Chap. 1.)

3.62. A shot putter releases a 7.25 kg shot at a height 2.2 m from the ground. If the shot is released at the angle that maximizes the travel distance (see Problem 3.50), it travels 23.06 m. Find this angle and the initial shot speed [144]. (Ignore drag.)

3.63. (a) Find the initial kinetic energy and the maximum increase in the potential energy of the shot in Problem 3.62.
(b) If the putt is 20% efficient, how much energy is used in putting the shot (in J and kcal).
(c) If the putt is pushed from its resting point on the shoulder as the arm is extended, it is accelerated along a distance corresponding to the length of the whole arm, or about 0.7 m. During this acceleration phase (at this extension and release angle), what is the increase in potential energy and how does it compare to the energies in part (a)?
(d) What is the force exerted to accelerate the shot, if this force exerted during this extension phase is constant?
(e) If the muscles used to accelerate the shot can exert 20 N of force for each cm^2 of effective cross-sectional area, what is the net cross-sectional area of these muscles? If this were a muscle with a circular cross-section, what would its diameter be?

3.64. A hammer thrower releases a 7.25 kg hammer at a height 3.5 m from the ground. If the hammer is released at the angle needed so it can travel a maximum distance (see Problem 3.50), it travels 102 m. Find this angle and the initial hammer speed [144]. (Ignore drag.)

3.65. (a) Find the initial kinetic energy and the maximum increase in the potential energy of the hammer in Problem 3.64.
(b) If the throw is 20% efficient, how much energy is used in throwing the hammer (in J and kcal).

3.66. The forces exerted by muscles are proportional to their physiological cross-sectional area PCA. For each of the following cases, the result is roughly proportional to $(\text{PCA})^x$ (ignoring the force needed to counter the effects of gravity on the body, etc.). Find x in each case. (It should be either 0.5 or 1):
(a) The weight lifted by a weightlifter.
(b) The speed of a ball thrown by a pitcher.
(c) The farthest a ball can be thrown.
(d) How fast a bat can be swung. (Is this similar to case (b))?
(e) The speed of a ball after being hit by a bat. (Assume the pitcher and hitter have muscles proportional to PCA.)
(f) How far a ball can be hit.

Collisions of Humans

3.67. Show that the lines denoting specific deceleration times and decelerations in Fig. 3.57 are expected from simple kinematics, for constant deceleration rates.

3.68. Let us say as we land at the end of a jump we use the muscles in our knees to help cushion the collision. (Earlier in this chapter we considered a case where we ignored this potential cushioning.) The center of mass of a person of mass m_b falls a distance h before the person's feet make contact with the ground, and then the person's center of mass is lowered by a distance s and the reaction force from the ground F is used to cushion the collision:
(a) By using conservation of energy, show that the required force is $F = m_b g(1 + h/s)$.
(b) If we land on both feet and want to keep stresses to no more than 10% of the UCS, show that the maximum force felt by the feet during the collision should be no more than 10^4 N. (Use the analysis in the text for landing with stiff legs.)
(c) If we slow the fall by bending our knees by 0.5 m, what is the maximum height from which a 70 kg person can safely land? (Do not try this!!!)

3.69. A skater falls on ice [105]. Assume that his head hits the ice with a speed corresponding to free fall from a height of 6 ft.
(a) Assuming a constant force during impact, show the GSI for this collision is $\mathrm{GSI} = (t_f/\Delta t)^{2.5}\Delta t$, where t_f is the time of the free fall and Δt is the collision time.
(b) Calculate the value of the GSI for collision times $\Delta t = 1$, 2, 5, and 10 ms.
(c) Let $H = 6$ ft be the height of the free fall and Δs be the total compressional distance of the collision, which is the flattening experienced by the skin and/or protective padding. Show that the GSI can be expressed as $\mathrm{GSI} = (H/\Delta s)^{1.5}\sqrt{2H/g}$.
(d) Calculate the GSI for $\Delta s = 0.1$, 0.2, 0.5, and 1.0 cm.
(e) Identify the range of values of Δt in case (b) and Δs in case (d), for which serious injury can be expected.

3.70. Explore how the GSI and HIC-15 indices differ for constant deceleration, when:
(a) The magnitude of the deceleration is αg and it lasts for a time τ.
(b) $\alpha = 70$ and $\tau = 15$ ms.
(c) $\alpha = 40$ and $\tau = 60$ ms.
(d) What can you say about the weighting of parameters for each index and how each depends on the duration of the collision?

3.71. Let us explore how the GSI and HIC indices change when the rate of deceleration is not constant during the collision. Consider an initial speed v and a total collision time τ (<15 ms). It will be simpler if we express v as $\beta g \tau$,

where β is a constant. The deceleration $a(t)$ is expressed as $\alpha(t)g$. For each of these cases, first confirm that the speed decreases to 0 in a time τ and plot $\alpha(t)$ on the same graph. Then calculate general expressions for the GSI and HIC for each:
(a) $\alpha(t) = \beta$ is constant during the time τ.
(b) $\alpha(t) = 1.5\beta$ from time 0 to $\tau/2$ and 0.5β from time $\tau/2$ to τ.
(c) $\alpha(t) = 2(1-\tau/t)\beta$, so the deceleration decreases linearly from time 0 to τ.
(d) How do the GSI and HIC differ? Which is more sensitive to changes in deceleration?

3.72. In the text the GSI was calculated for an elastic and inelastic collisions of a head with $v_{\mathrm{i}} = 50$ mph and a collision time of 10 ms. For what collision times would the elastic collision be expected to be fatal and the inelastic collision expected to be survivable?

3.73. Modify (3.104), for a partially elastic collision, with a coefficient of restitution e, again with a very massive object.

3.74. (a) Calculate the GSI for the partially elastic collision of a head moving with $v_{\mathrm{i}} = 25$ mph and a collision time of 20 ms with a very massive object, for a general coefficient of restitution e, and then specifically for $e = 0$ (totally inelastic collision), 0.5, and 1.0 (totally elastic collision).
(b) Over what ranges of e will the collision definitely be fatal, likely be fatal, likely to not be fatal but likely result in a significant injury, and likely to lead to a relatively minor injury only?

3.75. (a) A pitcher throws a baseball at the head of an 80 kg batter (who has an average-sized head and who is not wearing a batting helmet). It hits his head at a speed of 90 mph at normal incidence and the collision is elastic. Consider the collision of the ball with the head only (ignore the rest of the body) and calculate the GSI assuming that during the collision the baseball decelerates at a constant rate, the baseball deforms its linear dimension (diameter) by 6%, and the head does not deform at all.
(b) What is the fate of the batter?

3.76. Some professional soccer players have the same type of loss of cognitive ability as do boxers, likely due to the repeated heading of soccer balls. What is the GSI when a 82 kg (180 lb) soccer player hits or redirects the ball with his/her head (a header)? Assume that the 430 g ball is moving at 50 mph and hits the head at a normal angle, so it bounces back on the original path. Say that the collision with the head is elastic and the ball is squeezed by 3 cm in the collision. Treat the collision as with the head (and not the rest of the body).

3.77. Figure 4.74 shows the deceleration and the square wave function approximation to this measured deceleration for a head impacting a helmet with initial speed 5.63 m/s:

(a) Does this square wave approximation stop the head in the time shown?
(b) Calculate the GSI and HIC-15 for the square wave function approximation.

3.78. In professional football, a linebacker and fullback, both weighing 245 lb, are racing toward each other running at 30 ft/s. (Actually, the linebacker is running to the fullback and the fullback is trying to run away from him.) They collide and then both decelerate at a constant rate and become stationary in 0.2 s:
(a) What is the deceleration of each? (Also express your answer in g.)
(b) What is the force on each during the collision?

3.79. (a) A big defensive lineman in professional football weighing 310 lb runs at a speed of 24 ft/s into a small quarterback weighing 189 lb, who is initially still. Using conservation of linear momentum, what are their speeds after this collision and in what directions are they moving? Assume all frictional forces during the collision can be neglected, so kinetic energy is also conserved during the collision.
(b) Alternatively, assume that the lineman holds on to the quarterback during and after the collision, i.e., "tackles" or "sacks" him. In this case, what fraction of the initial kinetic energy is lost? If the forces with the ground can be neglected, where did this lost kinetic energy go?

Hitting and Kicking Balls

3.80. A "super ball," basketball, volleyball, and softball are dropped from a height of 1.83 m (6 ft) onto a hardwood floor. How high do they bounce?

3.81. A baseball is found to bounce to a height of 0.46, 0.51, and 0.55 m from a height of 1.83 m (6 ft), when it has been previously cooled for 1 h in a freezer, left at room temperature, and heated for 15 min at 225°C, respectively. What are the coefficients of restitution for the balls in the three cases? [127]

3.82. How far can a batted baseball travel? Assume the ball speed and bat speed are both 90 mph, $e = 0.46$, and the bat is much heavier than the ball. (Go into the rest frame of the bat, analyze the collision, and then return back.) (Your answer will be an overestimate because drag is omitted.)

3.83. If the same torque is generated by a pitcher in throwing a ball and a batter in hitting a ball, would the ability to throw a 90 mph fastball mean you could swing a bat with a 60 mph speed of the center of mass of the bat? Assume the moment of inertia for throwing the ball is that of a straight arm about the shoulder and that for hitting a ball is the body with two outstretched arms and a fully extended bat (as in Fig. 3.64).

3.84. Derive (3.115) by using the method described preceding it in the text.

3.85. Use our simple torque model for throwing a ball to show how you would expect the final bat speed to depend on bat weight. Is the dependence used in the text, (3.116), reasonable in terms of its linearity and the magnitude of coefficients?

3.86. Derive (3.117) by using the method described preceding it in the text.

3.87. Use (3.114)–(3.118) to find the range of bat weights for which the batted ball speed for this example is within 2 mph of the maximum speed (obtained with the optimized bat weight).

3.88. For the example of optimizing the bat weight for a ball with speed −80 mph and a bat speed that decreases with bat weight, find the bat speed before and after the collision with the baseball.

3.89. Use (3.117) to determine the optimal bat weight and batted ball speed for a major leaguer whose bat speed varies as v_{bat} (in mph) $= 48 - 0.34W_{\text{bat}}$ (with W_{bat} in oz). Over what range of bat weights is the batted ball speed within 5% of this maximum speed?

3.90. Equation (3.114) gives the speed of the baseball after it collides with a bat at its center of mass. In this problem we show that the analysis is a bit different if it hits the bat elsewhere, as in Fig. 3.64. We will use notation similar to that in the chapter, so v'_{ball} is the speed of the ball after the collision and v'_{bat} is the speed of the bat center of mass after the collision. (Before the collision these variables are unprimed):

(a) Conservation of linear momentum, (3.96), still holds. Show this is now

$$m_{\text{ball}}v_{\text{ball}} + m_{\text{bat}}v_{\text{bat}} = m_{\text{ball}}v'_{\text{ball}} + m_{\text{bat}}v'_{\text{bat}} \,. \tag{3.122}$$

(b) The total angular momentum of the bat and ball is conserved in the collision. The rotational kinetic energy of the ball is ≃5–10% of its translational kinetic energy (show this), so we can neglect the rotation of the ball. Show that conservation of angular momentum gives

$$I_{\text{bat}}(\omega'_{\text{bat}} - \omega_{\text{bat}}) + Bm_{\text{ball}}(v'_{\text{ball}} - v_{\text{ball}}) = 0 \,, \tag{3.123}$$

where I_{bat} is the moment of inertia of the bat about its center of mass, and ω_{bat} and ω'_{bat} are the angular velocity of the bat before and after the collision, respectively.

(c) The coefficient of restitution equation, (3.97), now involves the speed of the bat at the point of impact. Show that it now becomes

$$e = -\frac{v'_{\text{ball}} - v_{\text{bat}} - B\omega'_{\text{bat}}}{v_{\text{ball}} - v_{\text{bat}} - B\omega_{\text{bat}}} \,. \tag{3.124}$$

(d) Solve these three equations to show that the speed of the ball after the collision is

$$v'_{\text{ball}} = \frac{(m_{\text{ball}} - e m_{\text{bat}} + m_{\text{ball}} m_{\text{bat}} B^2/I_{\text{bat}}) v_{\text{ball}} + m_{\text{bat}}(1+e)(v_{\text{bat}} + B\omega_{\text{bat}})}{m_{\text{ball}} + m_{\text{bat}} + m_{\text{ball}} m_{\text{bat}} B^2/I_{\text{bat}}} \tag{3.125}$$

(e) Show that this reduces to (3.114) when the ball hits the bat center of mass ($B = 0$).
For more details see [107, 108, 173].

3.91. Kicking a football, such as a "place kick," is a form of a collision – of the foot with the football:
(a) If the average force on the ball is 450 lb during the kick and the kick lasts for 8 ms, how fast does the ball move after the kick? Assume the football weighs 0.91 lb.
(b) How far does the ball travel (in yards) if its takeoff angle is 45° and drag is neglected?

3.92. A football player kicks a football at a takeoff angle of 45°; it lands 50 yd away (ignoring air resistance). Assume the person has a mass of 80 kg and height of 1.9 m (use Tables 1.6 and 1.7, and Fig. 1.15), and the mass of the football can be ignored during the kick. Also, assume that at the end of the kick, the kicker's foot is moving at the same speed as the football. Examine the kicking leg by itself, assuming that it has a constant mass per unit length, and find the average power generated by the leg during the kick in W and in hp, if the duration of the kick is 0.2 s. If the body can produce this mechanical motion with 15% efficiency, what average power does it need, in W and in hp, to achieve this motion?

3.93. Assume that after a soccer ball is hit, the ball moves with the same speed as the player's foot. For a soccer player with the dimensions of the average human in Chap. 1, how fast does the soccer ball move with thigh and shank (lower leg) motion as in Fig. 3.70.

3.94. Does Problem 3.93 assume that the collision is inelastic, partially elastic, or elastic? Is this a good assumption? If not, how would you correct the approach in that problem?

Sliding

3.95. A base runner has an initial speed of 9 m/s and slides to a stop over 4 m. What is the coefficient of friction during the slide?

3.96. (a) When you slide into a base at speed $v_{\text{slide,final}}$ your foot stops after crushing into the base a distance d. Find this constant deceleration and,

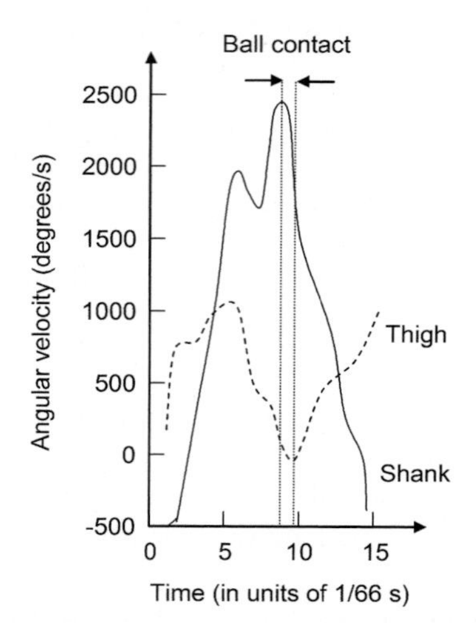

Fig. 3.70. The angular speed of the hip joint (thigh) and knee (shank) during kicking of a soccer ball. (The unusual units of 1/66 s are from the 66/s frame rate of the photographs taken by [164].) (Based on [164] and [149].) For Problem 3.93

assuming the decelerating force is transmitted to the whole body, this collision force F_{coll}.
(b) Assume this force causes a fracture when this force F_{coll} is distributed over a bone area A and this force per unit area F_{coll}/A exceeds the UCS of the bone. Find the threshold sliding speed $v_{\text{slide,fracture}}$ for fracture.
(c) For reasonable values of A and the ultimate compressive stress (UCS) (see Chap. 4), find $v_{\text{slide,fracture}}$.

4

Mechanical Properties of the Body

We now examine the mechanical properties of organs and components of the body. In subsequent chapters we will consider their other materials properties, specifically their thermal, electrical, and optical properties. We need to understand these mechanical properties to evaluate how body components function and to assess the impact of injuries. They are also essential in assessing the suitability of biomedical devices, such as hip replacements. Research and development teams are rightly concerned with how human bodies react to such prosthetic devices in a biochemical sense and whether they will "reject" the implants. They are equally concerned with how such implants match the other body components in a mechanical sense [184] (Problems 4.13 and 4.14, Fig. 4.76). For example, if they are softer than what they replace, they will wear out; if they are harder, there could be excessive wear on other body parts. In Chap. 3 we also saw how human motion is affected by the mechanical properties of objects outside the body, such as running shoes, floors, and vaulting poles.

Our goal in this chapter is to characterize the mechanical behavior of body components by using basic models that are routinely used in materials science and engineering. Once we have modeled the body component, we will use that model to understand the consequences of that modeled property, such as: For a given impact, will the bone break or just bend?

These mechanical properties all have a biological basis that is very complex and this will not be discussed here. Much of these details are still not understood well at all. We will assume that these properties have known averages among humans. There are distributions about these averages due to variations in our genes, gender, age, health, past injuries, and so on.

From a mechanical perspective, the different parts of the body can be classified in a variety of ways. For example, components can be either *passive* or *active*. Passive components, such as bones and tendons, respond to outside forces. Active elements, muscles, generate forces. This division is not perfect. Muscles are indeed active elements, but they also have some properties of passive components, and when they are modeled, the model must include both their active and passive properties. These passive elements and properties are discussed in this chapter. Active elements are discussed in Chap. 5.

The response of passive elements to applied stresses (forces/area) is by no means simple. Passive components can respond to forces in ways that are either independent or dependent of time. By this we mean that the component can respond to only currently applied forces or to both current forces and forces applied earlier.

The simplest type of passive response is *harmonic or Hookean* behavior, in which the properties of the material behave exactly like that of an ideal harmonic oscillator spring. Deformations are linear with the applied forces and stresses. The response is independent of time. All the potential energy stored in such media can be extracted. Bones and tendons are fairly well (but not perfectly) modeled as such elastic media. The elastic nature of tendons makes them very important in energy storage and retrieval during motion. Some materials systems, such as metal springs and bones, behave similarly under *tension* and *compression*. Others, such as cartilage and tendons, do not. (Why?)

No material is perfectly harmonic. Most materials deviate from perfectly harmonic behavior for large applied forces and large deformations. A material can deviate from a harmonic oscillator dependence with the deformation depending nonlinearly on force or stress, and yet this deformation can still be reversible. This means that the material returns to its initial state when the stress is removed both in the linear and nonlinear parts of this *elastic regime or region* (see Fig. 4.1). For even larger stresses, the material is no longer elastic because it undergoes *plastic deformation*, which is irreversible. This means that the material never returns to the same size or shape when the stress is removed. For even larger stresses, there is *fracture*. One glaring example is the fracture of bones.

Whereas elastic behavior is independent of history and enables total recovery of stored energy, this is not so in the opposite extreme of *viscous*

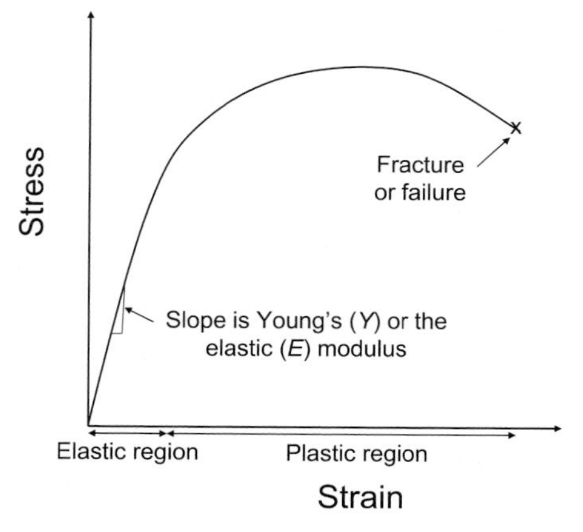

Fig. 4.1. General stress–strain relationship. For more detail, see Fig. 4.15

behavior, for which the response depends on the history of applied stresses and no energy is recoverable. Viscous materials dissipate energy; friction is one manifestation of viscous behavior. Most materials have properties that are in part elastic and in part viscous, and as such are *viscoelastic*. We will examine models describing such viscoelasticity.

We will need to distinguish between the *intensive* and *extensive* properties of the body component (or any other object). Let us say we were to examine a $100\,\mathrm{cm}^3$ ball of solid iron that has a 787 g mass. Obviously, the iron ball has a mass density of $787\,\mathrm{g}/100\,\mathrm{cm}^3 = 7.87\,\mathrm{g/cm}^3$. This property per unit volume is an *intensive* property. It does not depend on the size or shape of the ball and applies to any object composed of this type of iron. An *extensive* property of this ball is that it has a mass of 787 g; another is its $100\,\mathrm{cm}^3$ volume. Such extensive properties depend on the intensive property of the object and the size and shape of the object.

Why do some people's bones break more readily than others? There are several reasons: (a) They could have different intensive properties. For example, they could be more porous and concomitantly have lower damage thresholds – such as for those with *osteoporosis*, which is common in older people who have lost much calcium. (b) They could have different extensive properties, such as thinner bones. (c) They could have bad luck. Reason (a) is yet another illustration of why body materials are complex. They are *composite* materials, composed of different types of materials on a microscopic basis. Bone is a composite composed of calcium-based inorganic matter and organic matter.

We have actually seen the implications of several of these mechanical properties earlier. In analyzing running we saw that about a third of the kinetic energy lost each time the foot hits the ground goes into stretching the Achilles tendon, and that most of this energy is recoverable (which is nearly elastic behavior). In modeling throwing a ball, we neglected any friction about the elbow joint during the throwing motion. This followed our discussion of the very low coefficient of friction in synovial joints. In our discussion of collisions, we saw that the tibia can break if we jump stiff-legged from a height of only 1 m (which is fracture). Our model of throwing a ball used the force generated by the biceps brachii (which is an active element).

Interesting references for these materials properties include [182, 184, 185, 195, 199, 201, 210, 212, 220, 221, 223, 225]. Reference [228] examines quite extensively the materials of the body and materials used in medicine. Mechanical properties are given in [177, 190, 236].

4.1 Material Components of the Body

We will briefly characterize some of the major structural components of the body: bones, and several soft materials, such as ligaments, tendons, and cartilage, and then analyze their mechanical properties. More generally, there are four categories of tissues:

(1) *Epithelial tissue* covers the body and lines organs or secretes hormones. It has closely packed cells, little intercellular material, nerves, and no blood vessels (and so it is *avascular*).

(2) *Connective tissue* includes bone, cartilage, dense connective tissue (such as ligaments and tendons), loose connective tissue – such as "fat" – and blood and lymph vascular tissue. Most connective tissue has nerves and scattered cells in a background called a matrix. There are many blood vessels in bone and at the periphery of the menisci – and so they are highly *vascularized*, but tendons, ligaments, and (the bulk of) cartilage are not. The matrix consists of fibers and ground substances. The fibers include collagen fibers (made of the protein *collagen*) that are tough and flexible; elastic fibers (made of the protein *elastin*) that are strong and stretchable; and reticular, web-like fibers. The ground substance includes cell adhesion proteins to hold the tissue together and proteoglycans to provide firmness.

Epithelial membranes consist of epithelial and connective tissue. These line the body (skin (*cutaneous* membrane)), internal organs (*serous* membranes of the heart (*pericardium*), lungs (*pleura*), and abdominal structures (*peritoneum*)), cavities that open to the outside world (*mucous* membranes of the nasal cavity, and the respiratory, gastrointestinal, and urogenital tracts), and cavities at bone joints (*synovial* membranes).

(3) *Nervous tissue*, for body control, consists of *neurons* to transmit electrical signals and neuroglia (or glial cells) to support the neurons, by insulating them or anchoring them to blood vessels.

(4) *Muscle tissue* controls movement, and includes passive components (such as in the connective tissue) and active, motor-like components. Its structure and properties are detailed in Chap. 5.

The different fractions of the common building blocks in these components are shown in Fig. 4.2.

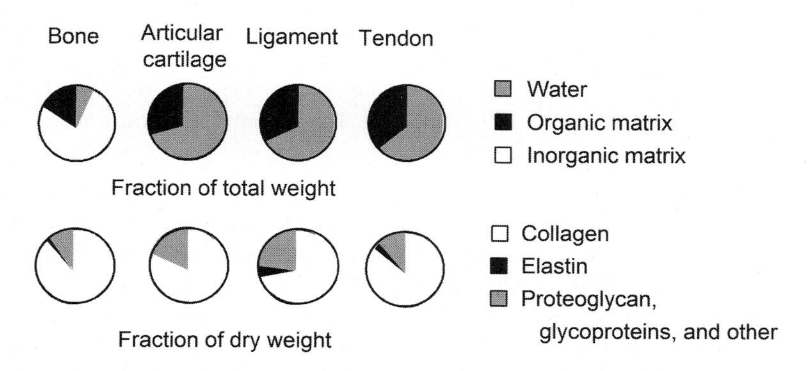

Fig. 4.2. Typical composition of several human musculoskeletal structural components by fractional total and dry weight. (Based on [180])

4.1.1 Bone

Bones provide a structural framework to attach muscles and organs, enable movement through the attachment of muscles, provide physical protection of organs (such as the skull for the brain and the rib cage for the lungs), store minerals (calcium and phosphorus) and some fats (in the yellow marrow), and produce red blood cells (in the red marrow). The stiff nature of bone clearly enables it to form a semirigid framework, enable motion (because how could muscles do their job with flexible bones?), and provide organ protection. We will see it also means that large bones can serve these functions and still be hollow and filled with the soft marrow. There are long bones, as in the arms and legs; short cube-like bones; flat bones, as in the skull and ribs; and irregularly shaped bones, as in the pelvis and vertebrae.

Bone is a complex composite material, with living and nonliving matter. The living matter includes the cells osteoblasts and osteoclasts, which, respectively, make new bone and resorb (erode) existing bone, and osteocytes, which are former osteoblasts buried in bone they have made. Excluding water, the nonliving matter of bone is 40% by weight (60% by volume) collagen and 60% by weight (40% by volume) calcium hydroxyappatite ($Ca_{10}(PO_4)_6(OH)_2$). The $\sim$5 nm $\times$ 5 nm $\times$ 40 nm rod or plate crystals with hexagonal symmetry of the ceramic-like calcium hydroxyappatite are bound by the elastomer-like collagen. The inorganic ceramic component gives compact bone its large strength (a large elastic constant Y) and a large ultimate compressive stress (UCS). The collagen component makes bone much more flexible than a ceramic and much more stable under tension and bending. If you let a turkey leg sit for 24 h in 1 M HCl it becomes very flexible because the ceramic crystals have been dissolved and all that remains is a collagen structure [186]. About 1% of the organic component is proteoglycans (mucopolysaccharides). About 25% of the volume of bone is water, $\sim$60% of which is bound to the collagen. Spongy (or trabecular) bone has voids with lateral dimensions of 50–500 μm.

Figure 4.3a shows the structure of a typical long bone, such as the femur. It has a long tubular shaft, the *diaphysis* (die-a'-phi-sus), which is a relatively thin shell of *compact, cortical, or dense bone* for strength. We will see later in this chapter that this type of hollow design maintains much of the strength of the corresponding solid structure, but with much less weight. At either end, the shaft broadens to form the *epiphyses* (e-pi-fi-sees'), where there is an overlayer of articular cartilage for lubrication and inside the bone, beneath the compact bone, is *trabecular, cancellous, or spongy bone*, which is a porous mesh of trabeculae (tra-bic'-you-lee) that can absorb shock. This porous bone is also found in the bones in the spinal column, where it provides some structural support and absorbs shock. Figure 4.3b shows that the layer of cortical bone in the shaft is thick and it becomes relatively very thin at the proximal end, where it surrounds the trabecular bone. There is bone marrow in the hollow shaft, the diaphysis. In short and irregular bones, spongy bone is

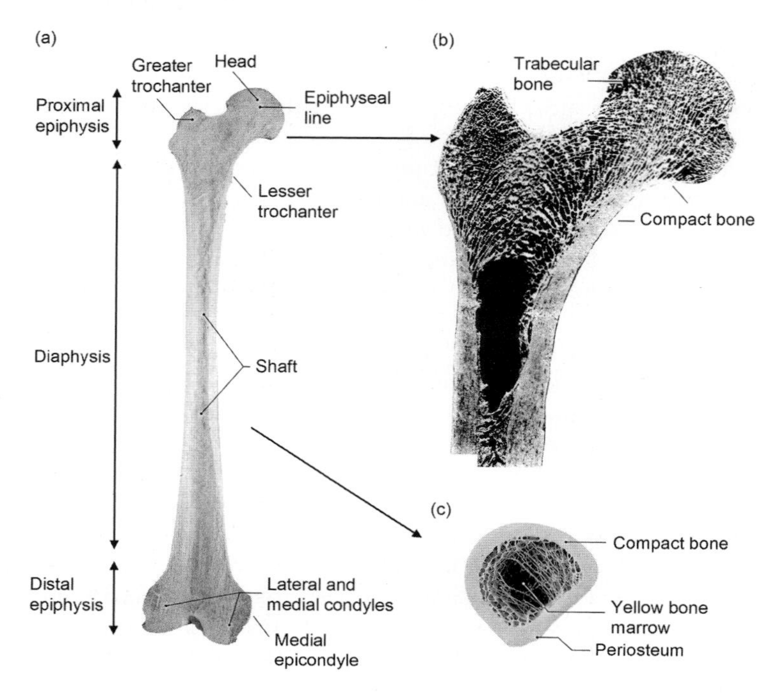

Fig. 4.3. Structure of a long bone, as exemplified by the femur, with a (**a**) schematic of the frontal section, (**b**) photo of the proximal epiphysis, and (**c**) schematic of the cross-section of the diaphysis. (**b**) is a photograph of a coronal section of the upper end of the femur of a 31-year-old male. The cut passes through the head, neck, greater trochanter, and part of the shaft, and is off-center between the middle and posterior thirds. The uniform sections are compact bone, while the meshed regions are trabecular bone. (From [233] (for (**a**), (**c**)) and [234] (for (**b**)))

encircled by a thin layer of compact bone, while in flat bones it is sandwiched by it.

4.1.2 Ligaments and Tendons

Ligaments and tendons are dense connective tissue with a dense network of fibers, with few cells and little ground substance. Ligaments are tough bands of fibrous connective tissue. They are 55–65% water and 35–45% dry matter, which consists of 70–80% collagen (mostly type I), 10–15% elastin, and a small amount, 1–3%, of proteoglycans. The collagen (Fig. 4.4) gives ligaments their high tensile strength. The collagen helices assemble into microfibrils (4 nm in diameter), which assemble into subfrils (20 nm in diameter), which assemble into fibrils (50–500 nm in diameter), and then into collagen fibers (100–300 μm in diameter) with fibroblast cells that synthesize the collagen.

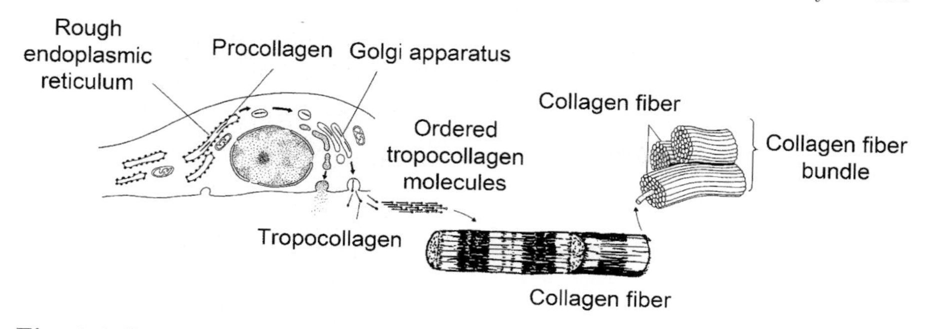

Fig. 4.4. Structure of collagen in fibers and bundles in tendons and ligaments, with ordered arrangement of collagen molecules in the microstructure. See Fig. 4.5 for more details about structure. (From [231])

The dry weight of tendons is 75–85% collagen (95% type I and 5% type III or V), < 3% elastin, and 1–2% proteoglycans. The structural hierarchy (Fig. 4.5) is like that of ligaments except they are arranged into packets called fascicles. Also, the bundles of collagen fibers are more parallel in tendons than in ligaments, as seen in Fig. 4.6.

In contrast, the dry matter of skin is 56–70% collagen (mostly type I), 5–10% elastin, and 2–4% proteoglycans.

In each of these soft materials, the collagen gives it tensile strength, while the elastin gives it elasticity, which is more important in ligaments than in tendons.

4.1.3 Cartilage

There are three types of cartilage: *Hyaline* (high'-uh-lun) *cartilage*, the most common in adults, is found in the ventral ends of ribs and covering the joint surfaces of bones. *Elastic cartilage* is more flexible, and is found in the external ear and eustachian tubes. *Fibrocartilage* occurs in the intervertebral disks.

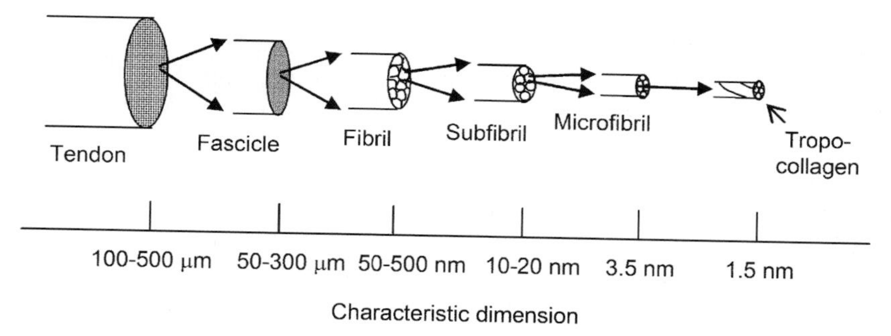

Fig. 4.5. Hierarchical structure of tendons. (Based on [180, 205])

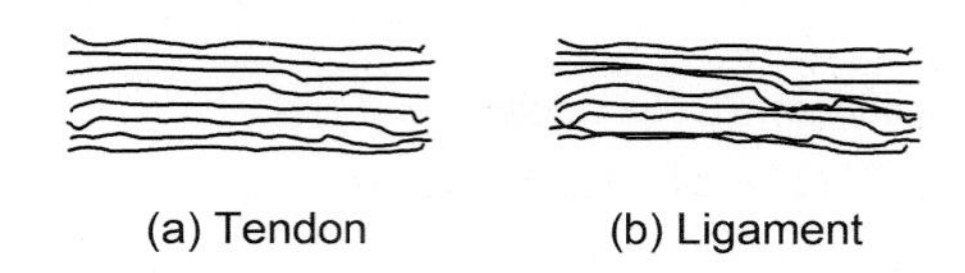

Fig. 4.6. Collagen fibers are (**a**) parallel in a tendon and (**b**) nearly parallel in a ligament. (Based on [222])

Cartilage that lines the bones in synovial joints (1–6 mm thick) is also called *articular cartilage*; it serves as a self-renewing, well-lubricated load bearing surface with wear prevention. It is most often hyaline cartilage, except in joints, such as the knee (the menisci), which contain fibrocartilaginous disks.

Articular cartilage is not meant to serve as a shock absorber to cushion forces or slow joint rotation [210], because it is so thin that it can absorb very little energy even though it is less stiff than cortical bone. It absorbs much less energy than muscles resisting joint rotation (eccentric contractions, Chap. 5) or the bones on either side of the joint (see Problem 4.12). About 30% of cartilage by mass is a solid matrix of collagen (40–70% of the dry mass, ∼80% type II collagen and several other types: V, VI, IX, X, and XI) and proteoglycan (15–40% of the dry mass) and 70% is water and inorganic salts (see the structure in Fig. 4.7). Chondrocyte cells that manufacture the cartilage organic material comprise less than 5–10% of the volume. Cartilage is viscoelastic because it is a very flexible, porous material (50 Å voids) with voids that are filled with water. The water dissipates energy as it flows through the voids under compression.

In tension the collagen of the solid phase carries most of the load, while in compression both the solid and liquid phases carry the load. The viscoelasticity of cartilage is controlled by the exudation of fluid through the pores in this biphasic material.

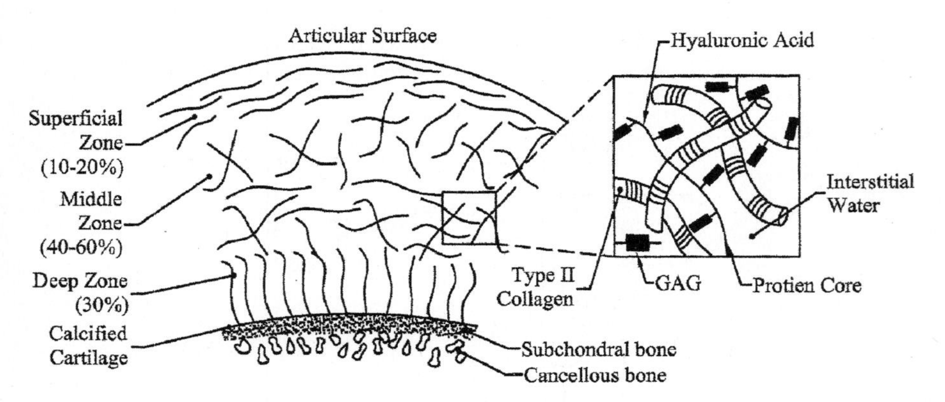

Fig. 4.7. Structure of articular cartilage, showing its inhomogeneity and solid–fluid constitution. The inset shows the local molecular organization of cartilage. (From [203])

4.2 Elastic Properties

4.2.1 Basic Stress–Strain Relationships

In the harmonic regime, elastic materials are modeled as perfect springs obeying Hooke's Law. This is usually expressed as

$$F = -kx, \tag{4.1}$$

where F is the force felt by an object attached to a spring, with spring constant k, when the spring is extended a distance x. When the spring is extended a distance x, say to the right, the attached body feels a restoring force kx to the left (Fig. 4.8).

In examining such Hookean materials we will need to alter this viewpoint a bit. There is a length of spring or material for which there is no restoring force. We will call this equilibrium length x_0. In (4.1), x is implicitly the deviation from this equilibrium length, the *deformation.* For reasons that will become clear soon, we prefer to refer x to this equilibrium length and so

$$F = -k(x - x_0). \tag{4.2}$$

Also, in studying problems with springs, we usually examine the effect of the spring forces on other masses. Here we are concerned with the effect of other forces on materials modeled as springs. Therefore we consider the force applied to the spring-like object, F_{applied}, which is the negative of the above

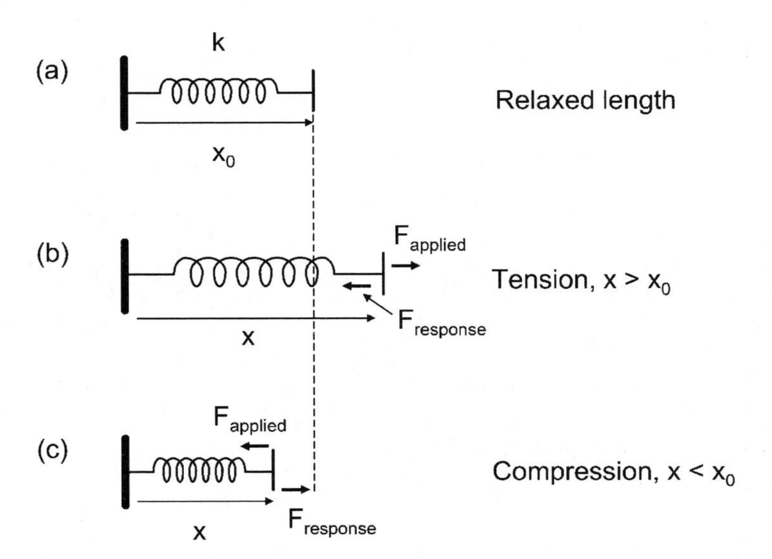

Fig. 4.8. Spring model of elastic materials, (**a**) relaxed, (**b**) under tension, (**c**) under compression. The text calls the material length L instead of x

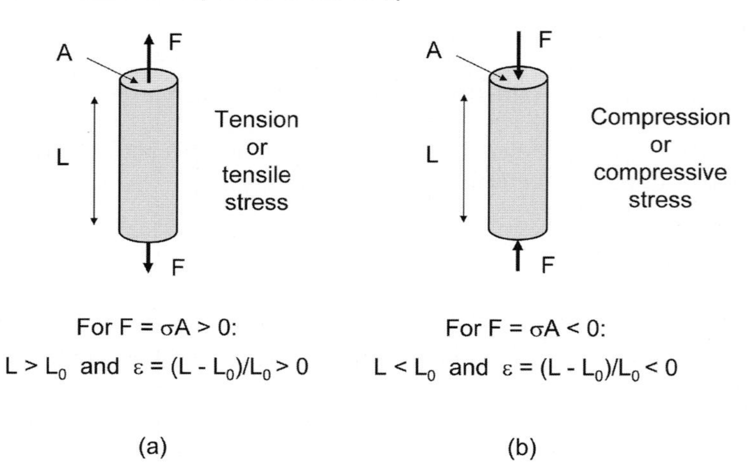

Fig. 4.9. Cylinder of relaxed length L_0 (**a**) under tension and (**b**) under compression

force F felt by the object attached to the spring, and is

$$F_{\mathrm{applied}} = k(x - x_0) = k(L - L_0). \tag{4.3}$$

We have also changed notation so that length of the material is L and its relaxed length is L_0.

When $L = L_0$ the material is relaxed. When there is a positive F_{applied} (Figs. 4.8b and 4.9a), the material is under *tension* and $L > L_0$. When there is a negative F_{applied} (Figs. 4.8c and 4.9b), the material is under *compression* and $L < L_0$.

Equation (4.3) represents the extensive properties of the material. While this is very important, we first want to examine the intensive properties of the material. If the object has a cross-sectional area A and length L, we can rewrite (4.3) as

$$\frac{F_{\mathrm{applied}}}{A} = \frac{kL_0}{A}\frac{L - L_0}{L_0}. \tag{4.4}$$

Each fraction represents an intensive parameter. The applied force/area, F_{applied}/A, is called the stress σ. The fractional increase in length, $(L-L_0)/L_0$, is called the strain (or the engineering strain) ϵ. ($L - L_0$ is the elongation.) The normalized spring constant, kL_0/A, is called either Young's modulus or the elastic modulus and is represented by Y (or E). This modulus is a fundamental intensive property of the material. Consequently,

$$\sigma = Y\epsilon. \tag{4.5}$$

This linear *constitutive* relationship describing this material is valid only for small strains. It is usually valid for $|\epsilon| \ll 1$, but the range of validity really depends on the type of material. We have ignored any change in cross-sectional

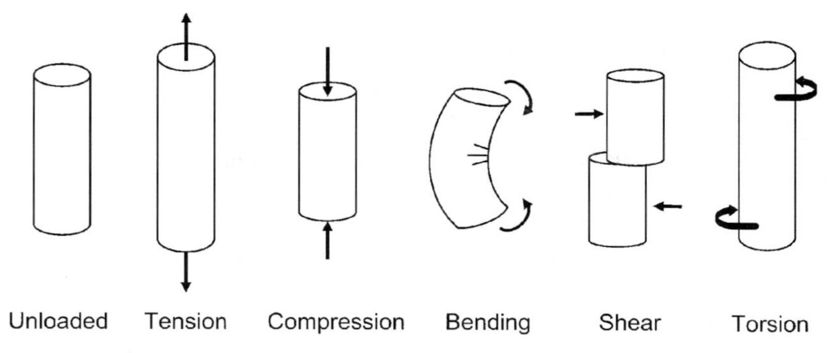

Fig. 4.10. Schematic of various loading modes. (Based on [222])

area. There is usually a change in A with a change in L (see below), which we will usually ignore here.

As seen in Fig. 4.8, tensile stress means $\sigma > 0$ and leads to a tensile strain $\epsilon > 0$. A compressive stress means $\sigma < 0$ and leads to a compressive strain $\epsilon < 0$. For such elastic materials in the *proportional (or harmonic or Hookean) regime* the stress–strain relation is linear, as is seen in Fig. 4.1. The units of stress σ and modulus Y are both those of force/area, such as N/m^2 (=1 Pa) or the more convenient unit of N/mm^2 (=1 MPa); we will usually use these last two equivalent units. Strain, ϵ, is unitless. Remember from Table 2.6 that $1\,\text{N/mm}^2 = 10^6\,\text{N/m}^2 = 1\,\text{MPa} = 145\,\text{psi}$.

4.2.2 Other Stress–Strain Relations

In addition to these linear relations between stress and strain, there are other types of deformations (Fig. 4.10). Figure 4.11a shows the geometry of shear deformations with force F and shear stress $\tau = F/A$. (τ is not torque here.) The response is the shear strain $\gamma = \tan\theta$, and for small deviations $\gamma \approx \theta$. The shear stress and strain are related by

$$\tau = G\gamma, \tag{4.6}$$

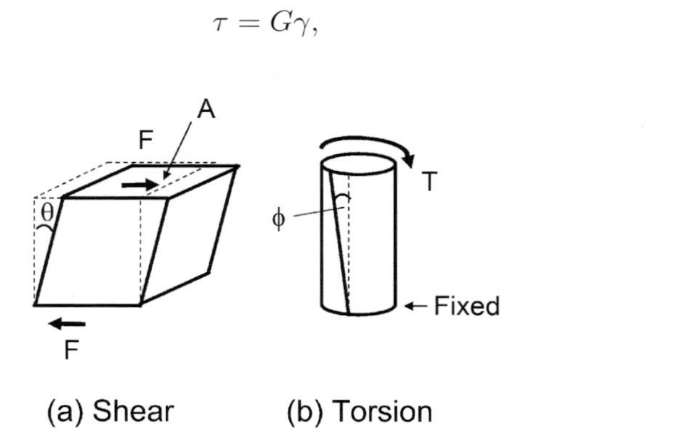

Fig. 4.11. Shear and torsion forces

where G is the shear modulus. This shear deformation is related to the torsion of the top of a cylinder, with the bottom fixed, as seen in Fig. 4.11b, where the torsion T is related to the deformation angle ϕ.

Let us consider the deformation of a cylinder with the long axis along the z-axis. We have already called the *axial strain* response in the z direction ϵ, but because stresses lead to strain deformations in different directions, we could be more specific (for the moment) and call it ϵ_z. We assumed earlier that the cross-sectional area of such a cylinder does not change under tension or compression, but it does to a certain extent. We will call the fractional strains in these lateral x and y directions – the *lateral or transverse strains* – ϵ_x and ϵ_y, respectively. (In more advanced discussions, these three x, y, and z components of strain are really referred to as ϵ_{xx}, ϵ_{yy}, and ϵ_{zz}.) For the linear deformation described above, symmetry implies that $\epsilon_x = \epsilon_y$. For a given material there is a relationship between these longitudinal and lateral strains provided by Poisson's ratio

$$\upsilon = -\frac{\epsilon_x}{\epsilon_z}. \tag{4.7}$$

For isotropic materials, the range of possible υ is $-1 < \upsilon < 0.5$, although materials with negative υ are not found in nature. For anisotropic materials, such as many materials in the body, υ can exceed 0.5. For metals and many engineering materials $\upsilon = 0.25$–0.35, but it tends to be higher for biological materials. For bone, υ ranges from 0.21 to 0.62 [212]. For tissues like those in the brain, $\upsilon \sim 0.5$.

After this deformation the new volume is the old one $\times$ $(1+\epsilon_x)(1+\epsilon_y)(1+\epsilon_z) \approx 1+\epsilon_x+\epsilon_y+\epsilon_z$, when each strain $\ll 1$ (Fig. 4.12). Using Poisson's ratio, the new volume is $1+(1-2\upsilon)\epsilon_z$ $\times$ the old volume and the fractional change in the volume is $(1-2\upsilon)\epsilon_z$. For example, with $\upsilon = 0.25$ this fractional volume change is $0.5\epsilon_z$. If $\upsilon = 0.5$, there is no volume change even with the change in

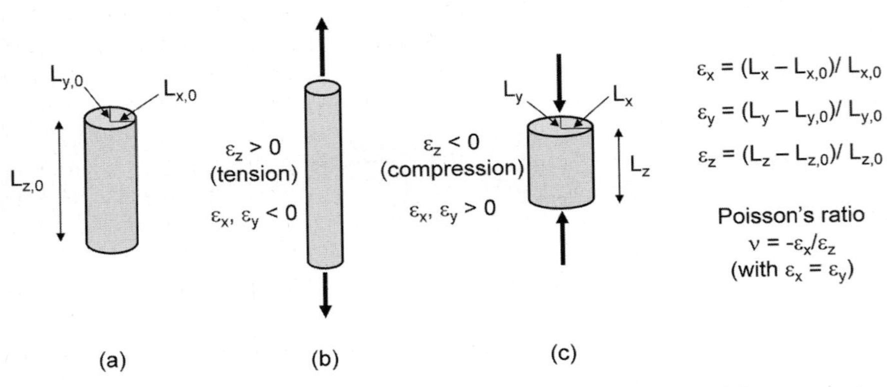

Fig. 4.12. Changes in lateral dimensions during (**b**) tension and (**c**) compression, as determined by Poisson's ratio, compared to those with no forces applied in (**a**)

shape. Similarly, after deformation the new cross-sectional area is the old one $\times\ (1+\epsilon_x)(1+\epsilon_y) \approx 1+\epsilon_x+\epsilon_y = 1-2\upsilon\epsilon_z$, and the fractional change in area is $-2\upsilon\epsilon_z$.

Like Y and G, υ is an intensive property of the material. For isotropic materials they are interrelated by

$$Y = 2G(1+\upsilon). \tag{4.8}$$

For example, if $\upsilon = 0.25$, the shear and elastic moduli are related by $G = 0.4Y$.

4.2.3 Bone Shortening

How much do our bones shorten under compression? We will assume that the relation $\sigma = Y\epsilon$ is valid until the stress reaches its maximum just before fracture occurs, which is called the *ultimate compressive stress*, UCS = 170 MPa for compact bone; this is a good approximation for this calculation. Then $\sigma = Y(L-L_0)/L_0$ and the bone shortens by

$$\Delta L = L - L_0 = \frac{\sigma L_0}{Y} \tag{4.9}$$

and fractionally by

$$\epsilon = \frac{\Delta L}{L_0} = \frac{\sigma}{Y}. \tag{4.10}$$

How much does the femur shorten when you stand on one foot? With no stress the femur is $L_0 = 0.5\,\text{m} = 500\,\text{mm}$ long. The body weight of 700 N (70 kg) is distributed over the femur cross-sectional area A = 370 mm^2, so $\sigma = 700\,\text{N}/370\,\text{mm}^2 = 2.1\,\text{N/mm}^2 = 2.1\,\text{MPa}$. The femur shortens by only $\Delta L = (\sigma/Y)L_0 = ((2.1\,\text{N/mm}^2)/(179\times 10^2\,\text{N/mm}^2))\,500\,\text{mm} = 0.06\,\text{mm}$. This corresponds to a strain of $\Delta L/L_0 = \sigma/Y = (2.1\,\text{N/mm}^2)/(179\times 10^2\,\text{N/mm}^2) = 0.01\%$. In units of microstrain (10^{-6} mm/mm), this is 100 microstrain (or $100\mu\epsilon$).

The maximum stress in compression is the UCS. *What is the strain at the UCS (assuming linear behavior)?* At the breaking limit the bone shortens by $\Delta L = (\text{UCS}/Y)L_0$, which is a fractional shortening of $\Delta L/L_0 =$ UCS/Y. The femur shortens by $((170\,\text{N/mm}^2)/(179\times 10^2\,\text{N/mm}^2))$ $500\,\text{mm} = (0.95\%)\,500\,\text{mm} = 5\,\text{mm}$ or 0.5 cm. This is a fractional decrease of $0.95\% \sim 1\%$, and a microstrain of $10{,}000\mu\epsilon$.

4.2.4 Energy Storage in Elastic Media

There are several essentially equivalent ways to determine the potential energy stored in elastic materials. From (4.1), in a spring the potential energy

(PE) is

$$\mathrm{PE} = -\int_0^x F\,\mathrm{d}x' = \int_0^x kx'\mathrm{d}x' = \frac{1}{2}kx^2. \quad (4.11)$$

Changing to coordinates relative to the equilibrium position and changing the length to L gives

$$\mathrm{PE} = \frac{1}{2}k(L - L_0)^2. \quad (4.12)$$

Because $Y = kL_0/A$, $k = YA/L_0$, and $\epsilon = (L - L_0)/L_0$, we see that $L - L_0 = \epsilon L_0$. Therefore

$$\mathrm{PE} = \frac{1}{2}\frac{YA}{L_0}(\epsilon L_0)^2 = \frac{1}{2}(Y\epsilon^2)(AL_0) = \frac{1}{2}Y\epsilon^2 V, \quad (4.13)$$

where the volume $V = AL_0$. Because $\sigma = Y\epsilon$, this can be expressed as

$$\mathrm{PE} = \frac{1}{2}\sigma\epsilon V = \frac{1}{2}Y\epsilon^2 V = \frac{1}{2}\frac{\sigma^2}{Y}V. \quad (4.14)$$

The potential energy per unit volume PE/V is an intensive quantity.

This is equivalent to integrating

$$W = \int_{L_0}^{L} F\,\mathrm{d}L' = \frac{1}{2}F_{\mathrm{applied}}(L - L_0) = \frac{1}{2}k(L - L_0)^2 \quad (4.15)$$

using $F_{\mathrm{applied}} = k(L - L_0)$ or

$$W = \int_{L_0}^{L} F\,\mathrm{d}L' = \int_0^{\epsilon} (\sigma A)\,\mathrm{d}(\epsilon' L_0) = V\int_0^{\epsilon} \sigma\,\mathrm{d}\epsilon' = \frac{1}{2}\sigma\epsilon V, \quad (4.16)$$

where the last integral equals the area under the curve in Fig. 4.13 and can be obtained by replacing σ by $Y\epsilon$ and integrating to get $Y(\epsilon^2/2) = \sigma\epsilon/2$.

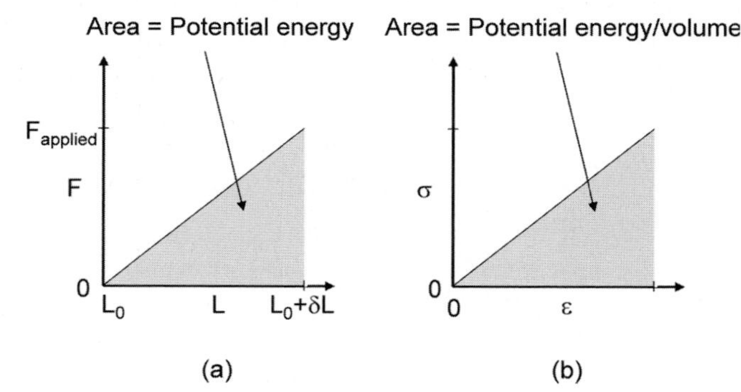

Fig. 4.13. Potential energy from area under (**a**) force–length and (**b**) stress–strain curves for a harmonic system

Designing Optimal Energy Storage Media

How can we design the best elastic storage medium for the body, such as would be desired for tendons? We would want (1) to store the maximum amount of potential energy for a given applied force F_{applied} and (2) the medium to withstand as large a F_{applied} as possible.

(1) The stored energy is

$$\mathrm{PE} = \frac{1}{2}\frac{\sigma^2}{Y}V = \left(\frac{(F_{\text{applied}}/A)^2}{2Y}\right)AL_0 = \frac{F_{\text{applied}}^2}{2Y}\frac{L_0}{A}, \tag{4.17}$$

so we would want to maximize the length L_0, minimize the cross-sectional area A, and minimize Y.

(2) However, to withstand a large F_{applied} we need to keep $\sigma = F_{\text{applied}}/A$ below the threshold for damage (which for tension is called the *ultimate tensile stress*, UTS), so we have a limit for how small we could make A to keep $\sigma \ll$ UTS. Also, there is a limit to how much the element can be lengthened $(L - L_0)$ for a large F_{applied}, given its motion requirements, such as that for a tendon. Because $L - L_0 = \epsilon L_0$, there are limits on how large both ϵ and L_0 can be. This sets a limit on the length L_0 and, because $\epsilon = \sigma/Y$, a limit on how small Y can be.

There is a tradeoff in the optimal values of L_0, A, and Y set by these two criteria. We want long and thin tendons with a small Y, but there are limits. In this design problem we also have to recognize that the medium, such as a tendon, is not perfectly harmonic or even elastic; all materials are really viscoelastic.

Energy Storage in Tendons and Long Bones

Let us return to the example of running in Fig. 3.33 [178]. The force on the Achilles tendon is 4,700 N. With a cross-sectional area of 89 mm^2, we see that $\sigma = 4{,}500\,\text{N}/89\,\text{mm}^2 = 53\,\text{N/mm}^2 = 53\,\text{MPa}$. Given the maximum stress for tendons, the UTS, is $\sim$100 N/mm^2 = 100 MPa, during running the stress in these tendons is not far from the damage threshold. It is not surprising that the Achilles tendons of athletes occasionally snap, either partially or totally.

Using the stress–strain relation shown in Fig. 4.14, this stress leads to a strain of 0.06 = 6%. The length of the Achilles tendon is $L_0 = 250$ mm, so this strain corresponds to the tendon lengthening by 15 mm and

$$\mathrm{PE} = \frac{1}{2}\sigma\epsilon V = \frac{1}{2}\sigma\epsilon A L_0 = \frac{1}{2}(53\,\text{N/mm}^2)(0.06)(89\,\text{mm}^2)(250\,\text{mm}) \tag{4.18}$$

$$= 35{,}000\,\text{N-mm} = 35\,\text{N-m} = 35\,\text{J}. \tag{4.19}$$

This is exactly the amount of energy we stated was being stored in the Achilles tendon during every step of a run.

How much energy is stored in the bones during this step? Let us examine the largest bone, the femur. We will use $L_0 = 0.5\,\text{m} = 500\,\text{mm}$ and

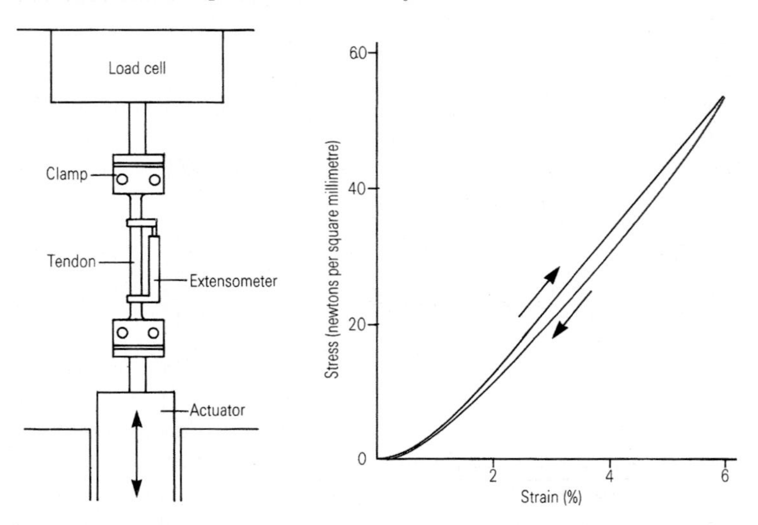

Fig. 4.14. Stress–strain (or force–length) for a human big toe flexor tendon, using the instrument on the left, with a 2-s-long stretch and recoil cycle. (From [178]. Copyright 1992 Columbia University Press. Reprinted with the permission of the Press)

$A = 330\,\text{mm}^2$, and so $V = 165{,}000\,\text{mm}^3$. Also $Y = 17{,}900\,\text{MPa} = 17{,}900\,\text{N/mm}^2$. The upward normal force in Fig. 3.33 is 6,400 N, which we will assume is transmitted all the way to the femur. The stress is $6{,}400\,\text{N}/330\,\text{mm}^2 = 19.4\,\text{N/mm}^2$ and

$$\text{PE} = \frac{1}{2}\frac{\sigma^2}{Y}V = \frac{1}{2}\frac{(19.4\,\text{N/mm}^2)^2}{17{,}900\,\text{N/mm}^2}\,165{,}000\,\text{mm}^3 \tag{4.20}$$

$$= 1{,}730\,\text{N-mm} = 1.73\,\text{N-m} \sim 2\,\text{J}. \tag{4.21}$$

If the same is stored in the tibia and fibula, then at most ∼3–4 J is stored in these long bones, which is a very small fraction of the 100 J kinetic energy lost per step.

4.3 Time-Independent Deviations in Hookean Materials

The Hookean (harmonic, linear) stress–strain relation is valid in tension and compression up to a limiting stress, corresponding to a strain $\ll 1$ that varies for different materials. Figure 4.15 shows a more realistic stress–strain relation. There is elastic Hookean behavior up to the point P, the *proportional limit*. The slope up to this stress is constant, the Young's modulus Y. The higher the Y, the *stiffer* or the less *compliant* the material (Fig. 4.16). At higher stresses, the stress–strain relation is nonlinear. Up to the *elastic limit*, denoted by E_L, the object returns to its initial length when the stress is removed and there

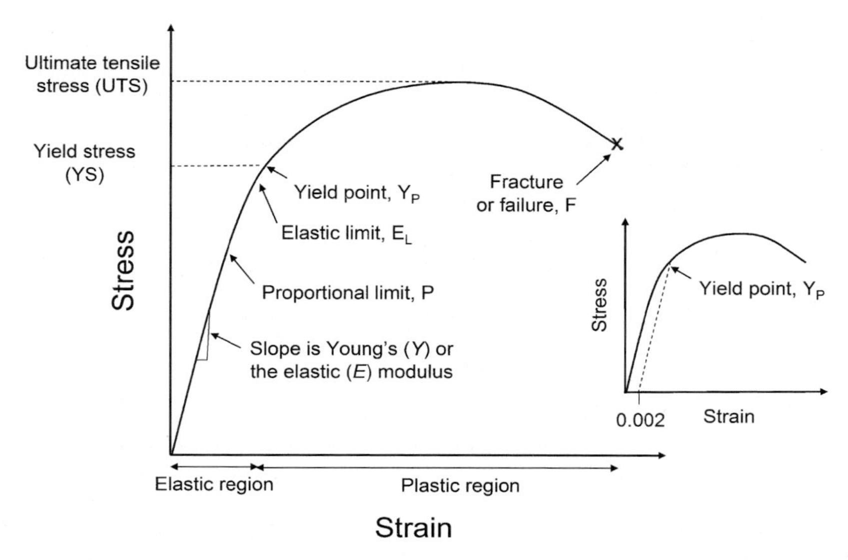

Fig. 4.15. General stress–strain relationship. The *engineering stress* is plotted here, which is the force divided by the initial area; it decreases after the UTS. The *true stress*, which is the force divided by the actual area increases after the UTS, due to the necking of the material. The inset shows the offset method to determine the yield point

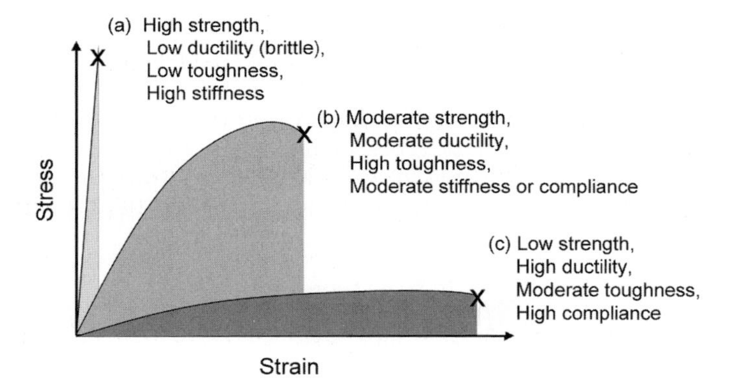

Fig. 4.16. Stress–strain curves of different types of materials with different levels of strength, ductility, and toughness. The engineering stress is plotted here. Strong materials fracture at very large ultimate tensile (or compressive) stress (UTS or UCS) (in Pa). Brittle materials have a small ultimate percent elongation (UPE) (unitless) and ductile materials have a large UPE. Tough materials can absorb much energy (when work is done on them) before they fracture, and so have a large work of fracture W_F (in J/m^2). Stiffer or less elastic materials have a larger Young's modulus Y (in Pa)

is no permanent deformation. In the linear and nonlinear elastic regimes, the stretched bonds relax totally and there is no rearrangement of atoms after the load is released.

For stresses beyond the elastic limit, there is *permanent or plastic deformation* and the length and shape of the object are different after the stress is removed. The *yield point or limit*, denoted by Y_P, is at a stress somewhat higher than the elastic limit; above it much elongation can occur without much increase in the load. (Some do not distinguish between the elastic limit and the yield point.) Because it is often difficult to determine, the yield point is usually estimated by the intersection of the stress–strain curve with a line parallel to the linear part of the stress–strain curve, but with an intercept set at a strain of 0.2% (or 0.002). This offset method is illustrated in the inset in Fig. 4.15. The yield point occurs at the yield stress (or strength), YS.

For tension, the material remains intact for larger stresses until the *ultimate tensile stress* (UTS), which is also called the *tensile strength* (TS) or, less commonly, the *tensile breaking strength* (TBS). The larger the breaking strength, the stronger is the material. Application of this stress leads to *fracture* at point F, which occurs at a strain called the *ultimate strain* or the *ultimate percent elongation* (UPE).

In Figs. 4.15 and 4.16, the actual type of stress being plotted is called the *engineering stress*. It is the force divided by the initial area, which is the area before any force is applied. Past the UTS, the engineering stress decreases as the material becomes narrower as it is pulled apart and the actual area becomes progressively smaller than this initial area, which is called "necking." (This narrowing is much, much more than that expected from the lateral strain, from Poisson's ratio.) The *true stress*, which is the force divided by the actual area, increases after the UTS, due to this necking.

Figure 4.17 shows that these stress–strain relations look qualitatively different for ceramics, metals, and elastomers because of the very different microscopic structures of these types of materials. Ceramics have a linear stress–strain relation with large slope Y. The fracture point appears only a little into the nonlinear elastic regime, and for smaller values of strain <0.1. To

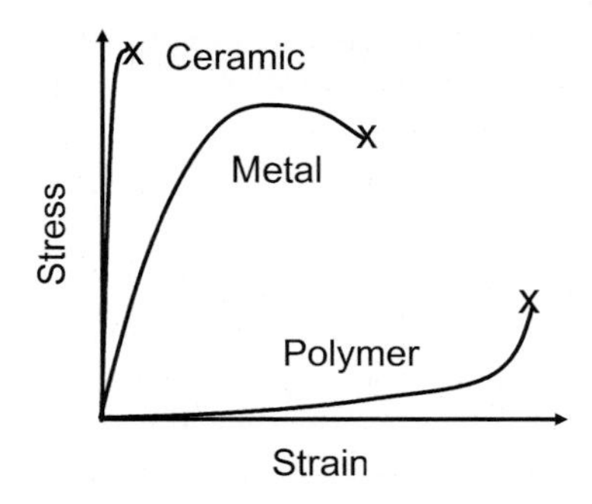

Fig. 4.17. Stress–strain curves for different types of materials under tension

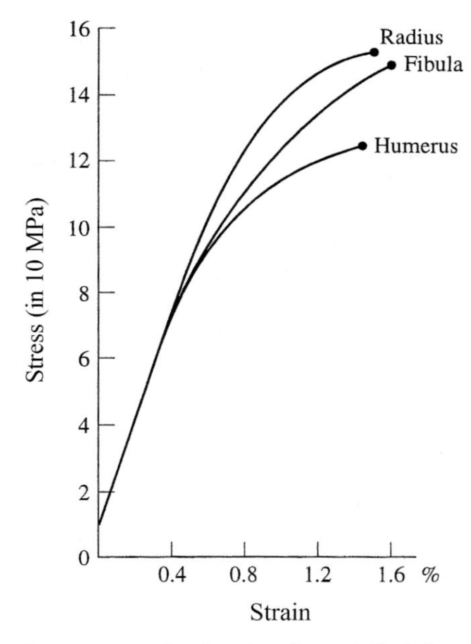

Fig. 4.18. Stress–strain curves under tension for wet limb bones of persons between 20 and 39 years of age. The *closed circles* are the fracture points. (From [182], based on [236])

first order, bone (Fig. 4.18) is like a ceramic. (It is actually more complicated than that, as we will see.) Metals have a smaller Y, a larger nonelastic and plastic regime, and a larger UPE ~ 50 (in %). Elastomers (rubber, polymers) distort greatly even with small stresses because in this regime long, tangled chain molecules are straightened out at low stress in this *toe region* (the region of positive curvature at low strain, as in Fig. 4.24). The stress–strain curve is not linear. We will examine this again later. It takes much larger stresses to increase strain further after all of the chains have been straightened, because now bonds must be stretched. These materials have a very large UPE, typically >1. Blood vessels are elastomers.

There are striking differences in the plastic deformation regimes of the curves in Fig. 4.17. *Ductile* materials, such as modeling clay, chewing gum, plastic, and most metals, have an extensive plastic deformation phase (metals, elastomers, Fig. 4.17). *Nonductile* or *brittle* materials, such as glass, ceramics (stone, brick, concrete, pottery), cast iron, bone, and teeth, have a limited or essentially no plastic phase (ceramics, Fig. 4.17). They break easily when they are dropped; cracks easily propagate in them. The bonding in ductile materials allows layers of atoms to slip or shear past each other, as in the bonding of metals. When thin rods of ductile materials are pulled at either end, they narrow in the center, forming a neck. In nonductile materials the covalent bonding is directional and does not permit this type of distortion.

Table 4.1. Mechanical properties of common materials. (Using data from [186, 201])

material	Y ($\times 10^3$ MPa = GPa)	UCS (MPa)	UTS (MPa)
hard steel	207	552	827
rubber	0.0010	–	2.1
nylon 66	1.2–2.9	–	59–83
gold	78	–	–
tungsten	411	–	–
granite	51.7	145	4.8
concrete	16.5	21	2.1
oak	10.0	59	117
fused quartz	73	–	69
diamond	965	–	–
porcelain	–	552	55
alumina (85% dense)	220	1,620	125
alumina (99.8% dense)	385	2,760	205
compact bone	17.9	170	120
trabecular bone	0.076	2.2	–

Typically, brittle materials have a small UPE and *ductile* materials have a large UPE.

Figures 4.15, 4.17, and 4.18 show the effects of tension. Under compression $\epsilon < 0$, and the stress–strain slope is the same Y for many conventional materials. For biological materials, like cartilage, they can be very different because of their complex nature. In cartilage, tension is resisted by the solid phase, while compression is resisted by the solid and liquid components. For ligaments and tendons, there is resistance to tension, but not to compression. For larger stresses the dependence is different even for many common nonbiological materials, and fracture occurs at the *ultimate compressive stress* (UCS), also called the *compressive strength* (CS) or the *compressive breaking strength* (CBS), which is different from the UTS in general (Table 4.1).

Table 4.1 gives the Y, UCS, and UTS for several types of materials. Note the very wide range of Y. Some ceramic-type materials, such as granite, porcelain, and concrete, can take much larger stresses in compression than in tension (UCS $\gg$ UTS). In others, UCS $<$ UTS. The two types of bones listed have a different porosity and very different properties. Compact bone, also known as cortical, or dense bone, has a large Young's modulus that is comparable to that of other strong materials (Fig. 4.18). It can withstand more stress in compression than in tension, but unlike the ceramics it has a fairly large UTS. Trabecular bone, also known as spongy or cancellous bone, is more porous and has a very small Y, almost as small as that of rubber (Fig. 4.19a).

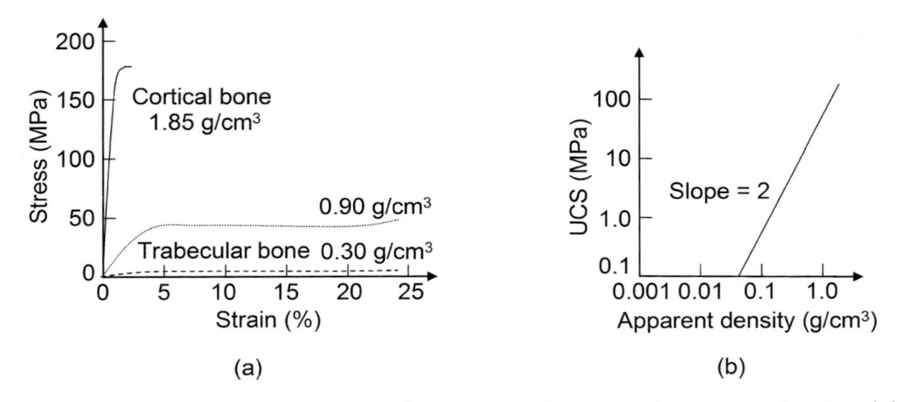

Fig. 4.19. Mechanical properties of bone as a function of apparent density. (**a**) Stress–strain of different densities of bones under compression. (**b**) UCS of trabecular bone vs. bone density. (Based on (**a**) [197, 206], and (**b**) [180, 206])

Typical stress–strain curves for structural materials in the body under tension are shown in Fig. 4.20. Yamada [236] has published extensive measurements of stress–strain relations for many components of the human body. Table 4.2 lists several elastic constants determined from these data. Figure 4.21 shows one series of these stress–strain relations, for different sections of the small intestine. Note that these curves are very nonlinear for a given stress, as are many soft human tissues; this is discussed more below. Furthermore, the tissue is very anisotropic. It stretches much more easily in the transverse direction than the longitudinal direction [183, 200, 216].

To first order, bones, teeth, and nails, all hard materials, have similar stress–strain curves that are ceramic-like. Tendons, cartilage, resting muscle, skin, arteries, and intestines all have more elastomer-like properties because they have much more collagen; they are really non-Hookean materials. This is

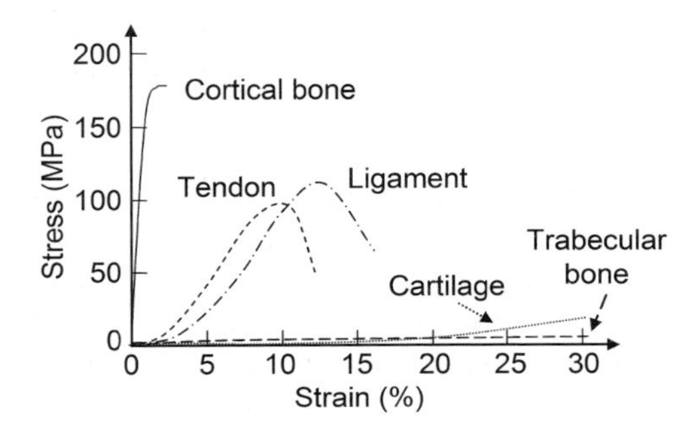

Fig. 4.20. Typical stress–strain of cortical and trabecular bone, cartilage, ligaments, and tendons under tensile loading. For some structural materials, the stress–strain relations are very different for different types of the material, such as for cartilage. (Based on [180, 236])

Table 4.2. Elastic properties of organs under tension (human, unless otherwise specified). (Determined using [236])

organ	UTS (MPa)	UPE (%)	Y (MPa)
hair (head)	197	40	12,000
dentin (wet teeth) (compression)	162	4.2	6,000
femoral compact bone (compression)	162	1.8	10,600
femoral compact bone	109	1.4	10,600
tendons (calcaneal =Achilles)	54	9.0	250
nail	18	14	160
nerves	13	18	10
intervertebral disc (compression)	11	32	6.0
skin (face)	3.8	58	0.3
vertebrae	3.5	0.8	410
elastic cartilage (external ear)	3.1	26	4.5
hyaline cartilage (synovial joints)	2.9	18	24
intervertebral disc	2.8	57	2.0
cardiac valves	2.5	15	1.0
ligaments (cattle)	2.1	130	0.5
gall bladder (rabbit)	2.1	53	0.05
umbilical cord	1.5	59	0.7
vena cava (longitudinal direction)	1.5	100	0.04
wet spongy bone (vertebrae)	1.2	0.6	200
coronary arteries	1.1	64	0.1
large intestine (longitudinal direction)	0.69	117	0.02
esophagus (longitudinal direction)	0.60	73	0.03
stomach (longitudinal direction)	0.56	93	0.015
small intestine (longitudinal direction)	0.56	43	0.2
skeletal muscle (rectus abdominis)	0.11	61	0.02
cardiac muscle	0.11	64	0.08
liver (rabbit)	0.024	46	0.02

The Young's modulus is given in the low strain limit.

also seen in Fig. 4.20. Each of these body materials has viscoelastic properties, as we will address later.

Compact bone in different long bones in the human body has slightly different properties (Table 4.6). These properties can be *anisotropic* (Fig. 4.22, Table 4.3), meaning that the properties are different along different directions. For example, this is true of bone and the esophagus, which are composed of very different materials. Many biological materials are anisotropic, as are many common materials, such as wood due it is grain structure. Some materials are fairly *isotropic*. Several elastic properties vary with age. These properties also change with density, which is a main reason why people with *osteoporosis* often fracture bones during a fall. Figure 4.19b shows the UCS decreases roughly as the square of bone density. (These changes with age

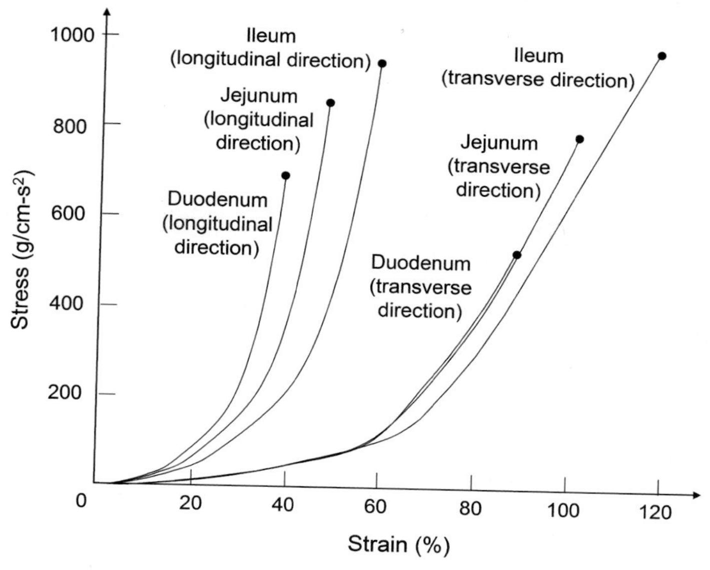

Fig. 4.21. Stress–strain curves for material in different sections of the small intestine of persons from 20 to 29 years of age, under tension in the longitudinal and transverse directions. The *closed circles* are the fracture points. (Based on [236])

are, in part, linked to such changes in density.) Figure 4.23 shows that the mechanical properties of soft tissues, in this case the anterior cruciate ligament (ACL) in the knee, also depend on age, as well as direction (also see Fig. 4.21.)

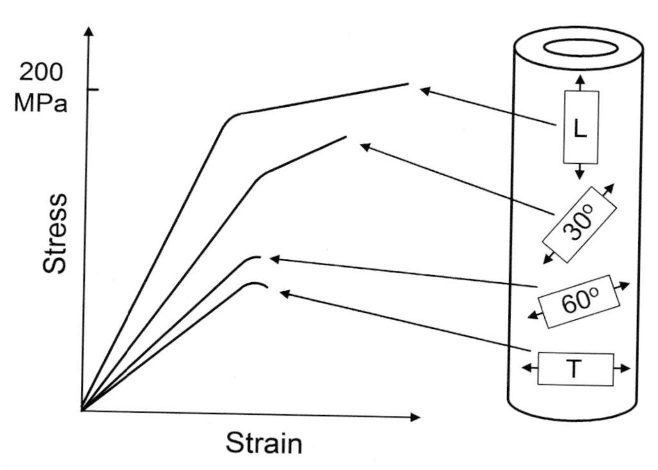

Fig. 4.22. Anisotropic properties of cortical bone specimens from a human femoral shaft tested under tension. Each curve ends at its point of failure. (Based on [196, 197])

Table 4.3. Mechanical properties of human cortical bone. (Using data from [210])

Young's modulus, Y (GPa)	
longitudinal	17.4
transverse	9.6
bending	14.8
shear modulus (GPa)	3.51
Poisson's ratio	0.39
yield stress (MPa)	
tensile – longitudinal	115
compressive – longitudinal	182
compressive – transverse	121
shear	54
ultimate stress (MPa)	
tensile – longitudinal	133
tensile – transverse	51
compressive – longitudinal	195
compressive – transverse	133
shear	69
bending	208.6
ultimate strain	
tensile – longitudinal	0.0293
tensile – transverse	0.0324
compressive – longitudinal	0.0220
compressive – transverse	0.0462
shear	0.33
bending	(0.0178 bovine)

We have seen that materials in the body are sometimes composed of different structures (i.e., they are composite materials), are anisotropic, and are sometimes layered. Moreover, a given material in a given organ or part of the body can also be very nonuniform. One example is seen in our teeth. Teeth

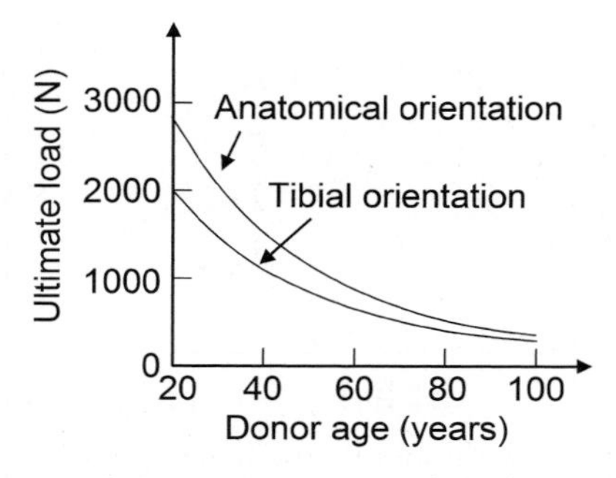

Fig. 4.23. Age variation of the ultimate load (UTS) of human anterior cruciate ligament (ACL) as a function of age and orientation. (Based on [180, 235])

are composed of pulp, which is mostly surrounded by dentin, which itself is overlayed by enamel. The enamel is very stiff and hard, and has very nonuniform properties [209]. Near the surface of the tooth (the occlusal surface), of say the second molar, the Young's modulus approaches 120 GPa, and it decreases to approximately 55 GPa near the enamel–dentine surface. It is also somewhat larger on the lingual (tongue) side than the buccal (cheek) side.

What is the strongest part of the body? If we were to define strength as the largest UTS, then of the body components in Table 4.2 it is not bone and not dentin in the teeth, but hair. (Of course, if we were to include tooth enamel, which is not in this table, it would beat out hair for this distinction. It is the hardest biological material in the body.)

4.3.1 Non-Hookean Materials

Figure 4.24 shows that many body materials cannot be modeled as Hookean springs, even for small stresses. This is typically true for many collagenous tissues, such as tendons, skin, mesentery (which are the folds attaching the intestines to the dorsal abdomen), the sclera, cartilage, and resting skeletal muscle. Experimentally, it is found for these materials that for larger strains

$$\frac{d\sigma}{d\epsilon} = \alpha(\sigma + \beta), \tag{4.22}$$

which is very different for materials such as bone for which $\sigma = Y\epsilon$, and so $d\sigma/d\epsilon = Y$ and is independent of stress.

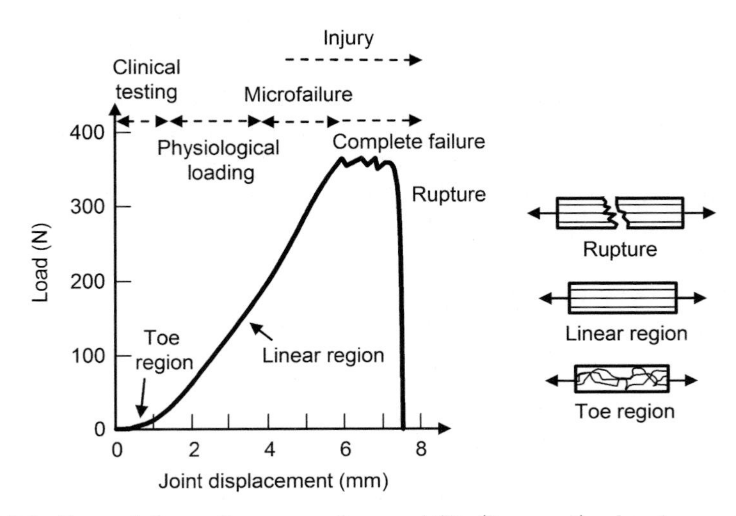

Fig. 4.24. Force–deformation curve for an ACL (ligament), showing regimes of clinical test loading, loads during physiological activity (toe and linear regions), and loads leading to microfailure and ultimate rupture and complete failure. (Based on [222, 224])

Equation (4.22) can be integrated after bringing the σ and ϵ terms to opposite sides of the equation

$$\frac{\mathrm{d}\sigma}{\sigma+\beta} = \alpha \,\mathrm{d}\epsilon \tag{4.23}$$

$$\ln(\sigma+\beta) = \alpha\epsilon + \gamma, \tag{4.24}$$

where γ is a constant. Exponentiating both sides gives

$$\sigma+\beta = \exp(\alpha\epsilon)\exp(\gamma) \qquad \text{and} \qquad \sigma = \mu\exp(\alpha\epsilon) - \beta, \tag{4.25}$$

where $\mu = \exp(\gamma)$. Because $\sigma(\epsilon=0)=0$, we see that $\beta=\mu$ and so

$$\sigma = \mu(\exp(\alpha\epsilon) - 1). \tag{4.26}$$

This is illustrated in the passive curves in Fig. 5.25. (See Appendix C for more information about this method of solution.)

Sometimes the Lagrangian strain $\lambda = L/L_0 = \epsilon + 1$ is defined, where L is the length and L_0 is the length with no stress. Equation (4.26) becomes

$$\sigma = \mu' \exp(\alpha\lambda) - \mu = \mu' \exp(\alpha L/L_0) - \mu, \tag{4.27}$$

with $\mu' = \mu \exp(-\alpha)$.

At larger strains this exponential form may not work well. In the *neo-Hookean* regime, finite strain, $E = \frac{1}{2}(\lambda^2 - 1)$, is defined, which for small deformations approaches the small-strain approximation, $\epsilon = \lambda - 1$. Soft, neo-Hookean materials tend to follow a linear relationship between stress and λ^2 (or E), and not a linear or exponential relationship between stress and λ (or ϵ). This neo-Hookean regime and other more general ways to define strain are described in Problems 4.20–4.22.

Such non-Hookean stress–strain curves are typical for materials with fibers. As mentioned earlier, there are large strains for small stresses where the tangled fibers are being aligned (in this toe regime), but much larger stresses are required to achieve much higher strains where the already-aligned fibers are being stretched (Fig. 4.25). Try this by stretching yarn. The fibers begin to tear at the UTS, corresponding to the load seen in Fig. 4.24.

We will now return to the deformation of Hookean materials, like bone. We will revisit the properties of these non-Hookean materials in the discussions of viscoelasticity and muscles.

4.4 Static Equilibrium of Deformable Bodies (Advanced Topic)

We now examine the deformation of bones under the action of forces in more detail. We have seen how they can be pulled (tension) and squeezed (compression); now we will see how they can bend. This analysis will help us understand

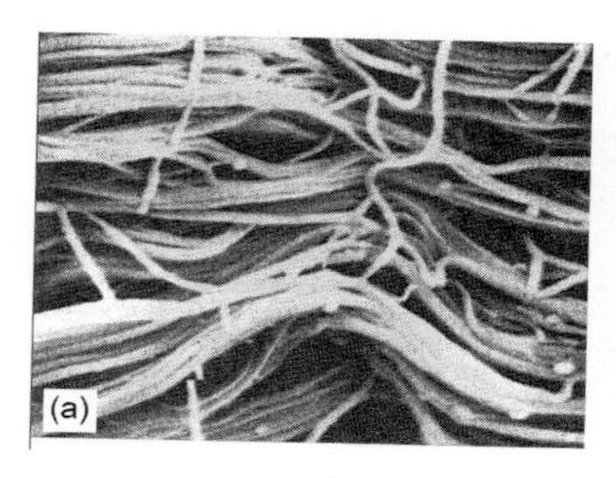

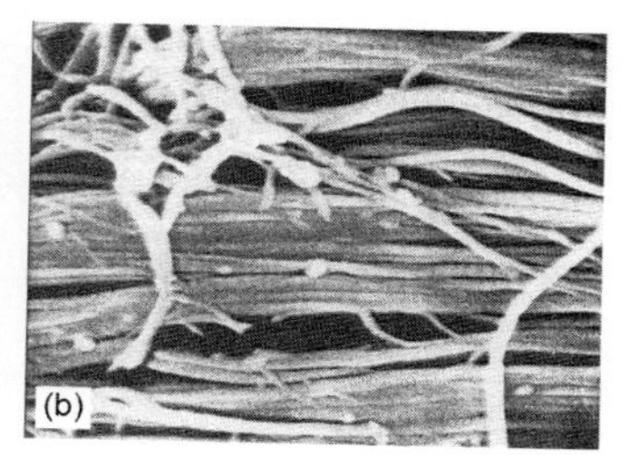

Fig. 4.25. Scanning electron micrographs ($10,000\times$) of (**a**) unloaded and (**b**) loaded collagen fibers from human knee ligaments, showing them straightening out under the tensile load. (From [207]. Used with permission)

how bones fracture when they are bent, such as during slipping and skiing accidents. We will also learn why long bones, like the femur, are strong even though they are hollow. As an added benefit, we will derive a scaling law that will help us understand some aspects of metabolism.

Physics classes usually describe the motion of point objects or more extended objects that never deform. However, no object is a point and objects do deform. Such extended objects are treated in great detail in mechanical and civil engineering curricula for obvious reasons. We will examine how such finite bodies bend to understand bone fracturing better, and will follow the treatment of [182]. The derivations in this section can be treated as a more advanced topic. They can be skipped and the final results can be used.

Let us consider the beam of length L shown in Fig. 4.26. It has a constant cross-section throughout its length; the cross-section need not be rectangular or circular. It is supported at both ends and a force F is applied to the center at the top as shown. We expect the beam to bend. For beams composed of most materials, we expect it to bend to a shape with a top surface that is somewhat cylindrical, and have a circular arc cross-section in the plane of the paper. If this were a rubber band, we would expect a more triangular deformation. Because we want to learn about bones, we anticipate some small degree of bending.

In this two-dimensional problem, in equilibrium $\sum F_x = 0$, $\sum F_y = 0$, and $\sum \tau_z = 0$ (2.8) for the entire beam. There are no forces in the x direction. We

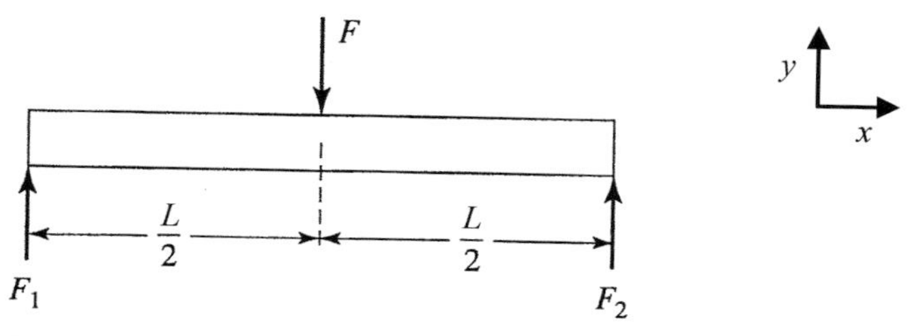

Fig. 4.26. Force diagram of a rectangular beam with a force applied to the middle. (From [182])

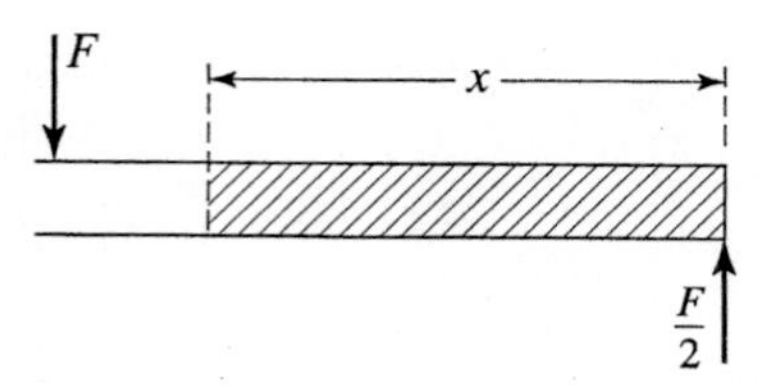

Fig. 4.27. External forces on a section of a beam. (From [182])

are assuming that in the equilibrium bent position, the amount of bending is small, so nothing interesting is happening in the x direction. The downward force F (which is the negative y direction) that is applied to the center of the beam is countered by the forces $F/2$ at the two supports, as is shown. The total torques on this beam are zero about any axis.

So far, this is how we treated the statics of rigid bodies in Chap. 2. Now let us examine the static equilibrium for only a part of the beam. As seen in Fig. 4.27, we consider the right side of the beam, from the right end to a distance x to the left of this end. The three equilibrium conditions also apply to this section, as well as to any other section. We can again ignore the x direction because there are no forces in that direction. For this example, with $x < L/2$ for now, there is apparently only one force acting on this piece of the beam, the upward force $F/2$ at the right support. This force also causes a torque and so

$$\sum F_y = \frac{1}{2}F \quad \text{and} \quad \sum \tau_z = \frac{1}{2}xF. \tag{4.28}$$

We have chosen the torque axis normal to the page at the left end of this portion of the beam (at a distance x from the right end). Because this portion of the beam is static, both terms must sum to zero. Something is wrong. What? We have excluded the force on this section from the other part of the beam. These internal forces must be $-F/2$ to balance the effect of the external force $F/2$ (Fig. 4.28). This "internal vertical force," often called the "internal shear force" or just the "shear force," supplied at the border with the other section is similar to the "normal" or reaction force felt by an isolated part of the body, as that on the leg from the hip. There must also be an internal torque applied by the other part of the beam equal to $-xF/2$. This "internal

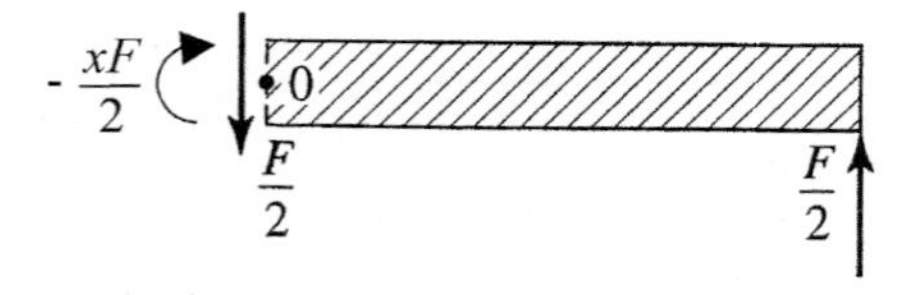

Fig. 4.28. Internal (*left*) and external (*right*) forces on an isolated section of the lined portion of the beam in Fig. 4.27. This is a free-body diagram of this portion. (From [182])

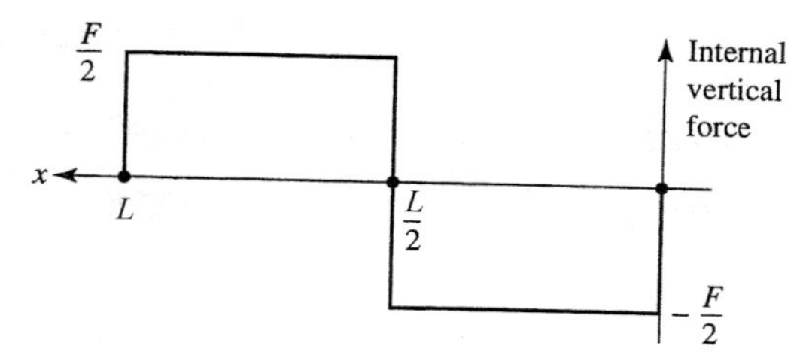

Fig. 4.29. Internal vertical force for a right-adjusted section in the beam. (From [182])

torque" is also called the "internal bending moment" or just the "bending moment."

For longer sections, with $L/2 < x < L$, the section feels the upward force at the right support $F/2$ and also the applied force $-F$ at the center. Both lead to torques. Excluding internal forces and torques

$$\sum F_y = -\frac{1}{2}F \qquad \text{and} \qquad \sum \tau_z = \frac{1}{2}(L-x)F \tag{4.29}$$

and so the internal force is $F/2$ and the internal torque is $-(L-x)F/2$. The internal vertical force and torque are plotted vs. x in Figs. 4.29 and 4.30. When we include these internal forces and torques, each portion of the beam is in static equilibrium. For now assume that $x < L/2$; extension to $L/2 < x < L$ is straightforward.

How do the internal torques arise? With the applied force, the beam deforms to that in Fig. 4.31 (in which the deformation is greatly exaggerated for a long bone). Clearly, the top portion is compressed and has a length $<L$, while the bottom portion is under tension and has a length $>L$. (This should become clearer if you take a spring or Slinky™ and bend it into a circular arc.) Somewhere in the middle (in the y direction) there is no compression or tension, so the length in this neutral axis is L. (The neutral axis is in the center (in the y direction) for symmetrical cross-sections.) For the top to be compressed there must be an internal force at the top in the $+x$ direction pushing into the section from the other portion (Fig. 4.32). Similarly, for the

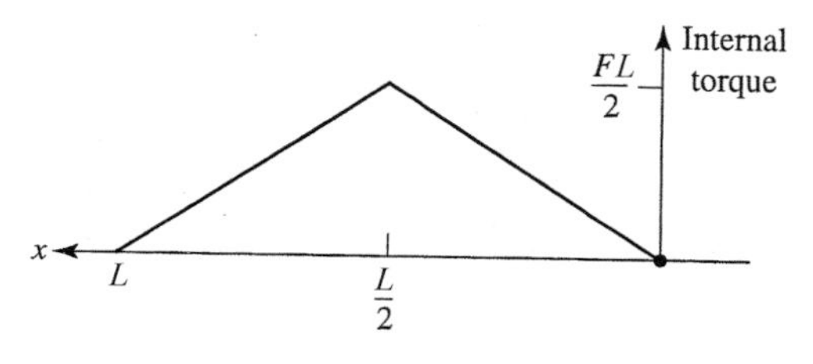

Fig. 4.30. Internal torque for a right-adjusted section in the beam. (From [182])

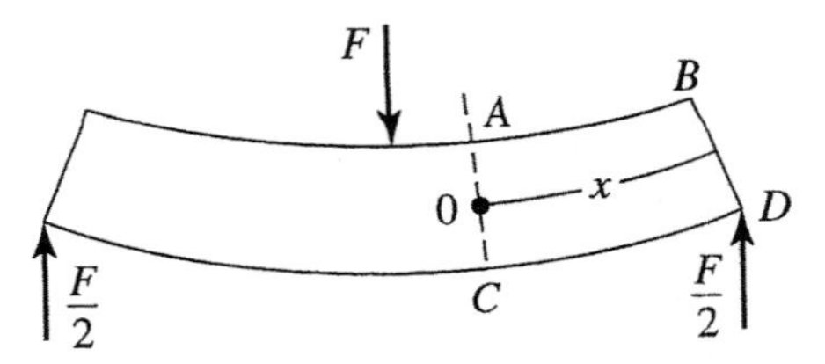

Fig. 4.31. Bending of a loaded beam. (From [182])

bottom to be under tension there must be an internal force at the bottom in the $-x$ direction, pulling into the section from the other portion. As shown in this figure, there is a smooth variation of this force from the top (called point A) to the bottom (point C), with it being zero at the neutral axis (point O). Clearly, the sum of these internal forces in the x direction must be zero.

Each of these internal forces causes a torque in the z direction, $-F_I y_I$, and each of these leads to a clockwise, or negative, torque about point O (Fig. 4.33). With forces to the right called positive, clearly $F_{\mathrm{A}} > 0$ and $F_{\mathrm{C}} < 0$, and with $y_{\mathrm{A}} > 0$ (measured upward from the neutral axis) and $y_{\mathrm{C}} < 0$, we find that the torque contributions from points A and C, $-F_{\mathrm{A}} y_{\mathrm{A}}$ and $-F_{\mathrm{C}} y_{\mathrm{C}}$, are equal in symmetrical situations. We can sum all of these internal torques to arrive at τ_{internal}, and then

$$\sum \tau_z = \tau_{\mathrm{internal}} + \frac{1}{2} F x = 0 \tag{4.30}$$

for static equilibrium.

What is the total internal torque? Consider a beam with arbitrary, but constant, cross-section, as shown in Fig. 4.34. The distance up from the neutral axis (with point O') is y, and there is a cross-section element with area $\mathrm{d}A$ at this position; $\mathrm{d}A = w(y)\mathrm{d}y$, where $w(y)$ is the width at y. There is a force acting on this area element at height y, which is

$$\mathrm{d}F(y) = \sigma(y)\mathrm{d}A(y). \tag{4.31}$$

For each element there is a torque

$$-y\,\mathrm{d}F = -y\sigma(y)\mathrm{d}A(y). \tag{4.32}$$

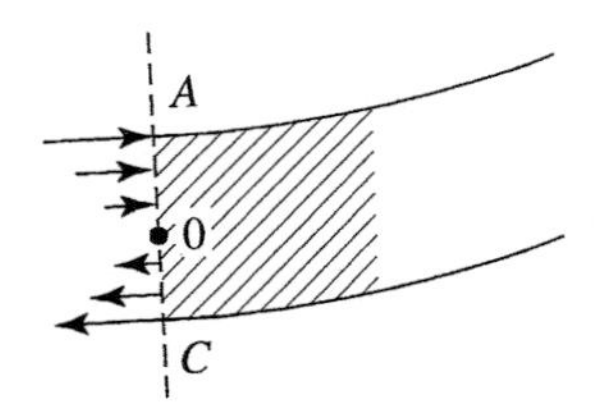

Fig. 4.32. Internal stresses in a bent beam. (From [182])

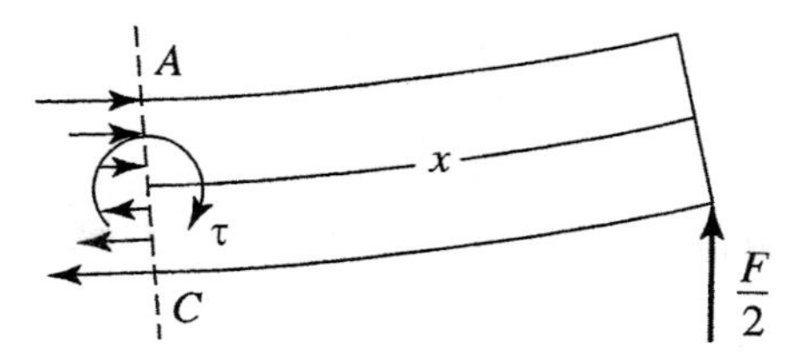

Fig. 4.33. Torques in a bent beam. (From [182])

So the total internal torque is

$$\tau_{\text{internal}} = -\int_{y_B}^{y_A} y\sigma(y)\mathrm{d}A(y) = -\frac{1}{2}Fx, \tag{4.33}$$

where $y_A = L'$ and $y_B = -L$ in Fig. 4.34. (The dA element includes the dy term.)

What is the distribution of σ? To first order, the beam deforms to a circular arc (Figs. 4.35 and 4.36) of radius R and angle α. At the midline neutral axis, where $y = 0$, we see that $L = R\alpha$ for $\alpha \ll 1$ and so $\alpha = L/R$. If the beam has a thickness in the y direction of d and the beam is symmetrical, then the top of the beam is $(R - d/2)\alpha$ long and the bottom is $(R + d/2)\alpha$ long. In general,

$$L(y) = (R - y)\alpha = (R - y)\frac{L}{R} = \left(1 - \frac{y}{R}\right)L, \tag{4.34}$$

so the elongation is $L(y) - L = -(y/R)L$ and the strain is

$$\epsilon(y) = -\frac{y}{R}, \tag{4.35}$$

where $1/R$ is the curvature.

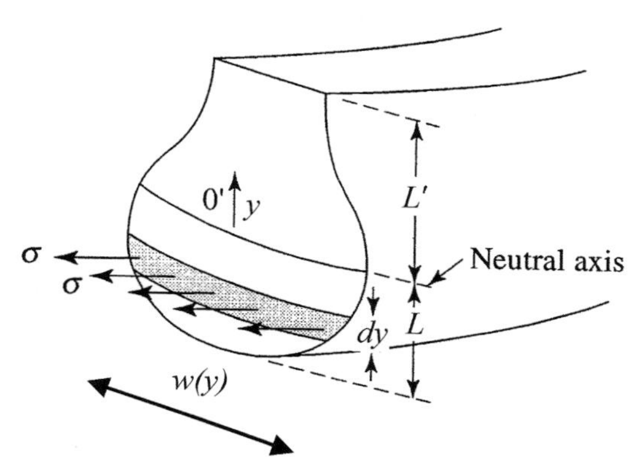

Fig. 4.34. Stress in a bent beam vs. position. An area element (for $y < 0$) is shown as a shaded region, with area d$A = w(y)$dy for width $w(y)$. (From [182])

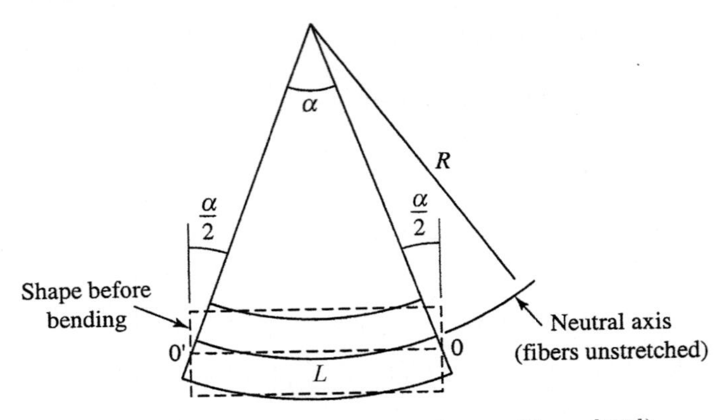

Fig. 4.35. Geometry of a bent beam. (From [182])

In the harmonic region, with $\sigma = Y\epsilon$, the stress would be expected to be $-Y(y/R)$. Given the direction of the forces shown in Fig. 4.34, the stress is defined to be positive for positive y, so

$$\sigma(y) = Y\frac{y}{R}. \tag{4.36}$$

4.4.1 Bending of a Beam (or Bone)

The total internal torque is

$$\tau_{\text{internal}} = -\int_{y_\text{B}}^{y_\text{A}} y\left(Y\frac{y}{R}\right) \mathrm{d}A(y) = -\frac{Y}{R}\int_{y_\text{B}}^{y_\text{A}} y^2 \mathrm{d}A(y) = -\frac{1}{2}Fx, \tag{4.37}$$

where $y_\text{A} = d/2$ and $y_\text{B} = -d/2$ for the symmetrical situation.

The *area moment of inertia* is defined as

$$I_\text{A} = \int_{y_\text{B}}^{y_\text{A}} y^2 \mathrm{d}A(y). \tag{4.38}$$

(This parameter is very different from the moment of inertia defined in (3.24). This one sums the squares of the distances from a plane, while the other and

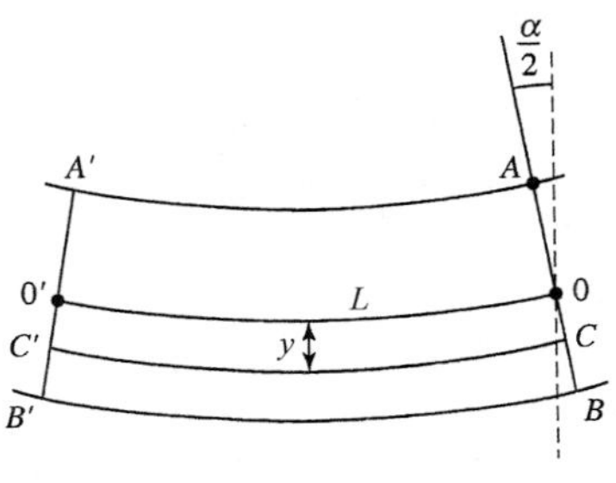

Fig. 4.36. Geometry of a bent beam in more detail. (From [182])

more usual one sums the squares of the distances from an axis.) Using this moment and the definition of the bending moment M_B due to applied forces ($M_\mathrm{B} = -Fx/2$ at equilibrium]), (4.38) is

$$M_\mathrm{B} = -\frac{Y}{R} I_\mathrm{A} \tag{4.39}$$

and the magnitude of the curvature is

$$\frac{1}{|R|} = \frac{|M_\mathrm{B}|}{YI_\mathrm{A}}. \tag{4.40}$$

Equations (4.39) and (4.40) interrelate four quantities (1) the applied forces, through M_B; (2) the intensive materials properties, through Y; (3) the physical deformation (response) of the beam due to the applied forces, through R; and (4) the shape of the object, through I_A.

For a given M_B and Y, when the area moment of inertia I_A is large there is little bending, while when I_A is small there is much bending. For example, the area moment of inertia for a rectangle of height h and width w is

$$I_\mathrm{A} = \int_{-h/2}^{h/2} y^2 (w\mathrm{d}y) = \frac{1}{12} wh^3, \tag{4.41}$$

where $\mathrm{d}A = (w)\mathrm{d}y$. Consider a $2\,\mathrm{cm} \times 6\,\mathrm{cm}$ rectangle arranged vertically (see Fig. 4.37) with $w = 2\,\mathrm{cm}$ and $h = 6\,\mathrm{cm}$. It has an $I_\mathrm{A} = 2\,\mathrm{cm} \times (6\,\mathrm{cm})^3/12 = 36\,\mathrm{cm}^4$. If this same rectangle were horizontal, then $w = 6\,\mathrm{cm}$ and $h = 2\,\mathrm{cm}$, and $I_\mathrm{A} = 6\,\mathrm{cm} \times (2\,\mathrm{cm})^3/12 = 4\,\mathrm{cm}^4$. For the same M_B and Y, the horizontal beam would bend 9× more. Try this with a yardstick!

The moment I_A is larger when the mass is distributed far from the central action, and there is less bending for a given bending moment when this occurs.

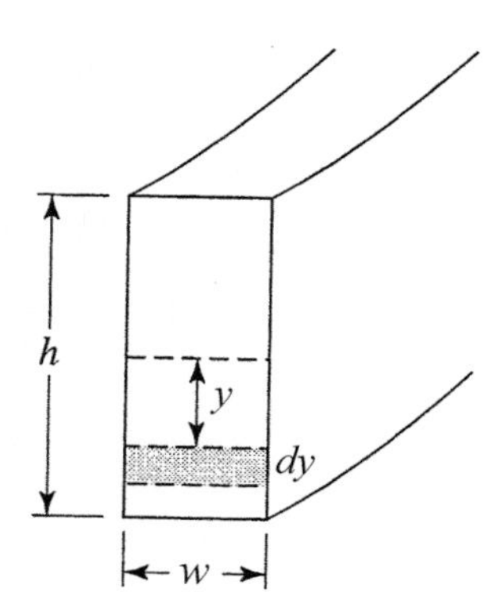

Fig. 4.37. Geometry for calculating the area moment of inertia for a rectangular beam. The same calculation can be used for the very different rectangular beam, with $h < w$. (From [182])

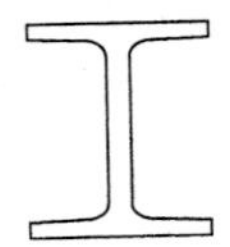

Fig. 4.38. An I-beam. (From [182])

This illustrates why "I beams" are used in construction instead of solid beams with the same overall rectangular cross-section (Fig. 4.38). The mass far from the neutral axis provides the resistance to bending, which is proportional to I_{A}, and the lack of material near the neutral axis lowers the weight of the beam.

Why Long Bones are Hollow

We now see why the long bones in the body can be hollow with much loss of weight and little loss of stiffness. The mass far from the neutral axis provides resistance to bending, while that near the neutral axis contributes little. Such hollow bones have sufficient resistance to bending, as well as larger resistance to bending per unit mass than do solid bones.

The area moment of inertia for a solid circular beam of radius a (Fig. 4.39a) is given from (4.38) $I_{\mathrm{A,solid}} = \int_{-a}^{a} y^2 \mathrm{d}A(y)$. Using Fig. 4.39b, we see that $y = a\sin\theta$, $\mathrm{d}y = a\cos\theta\ \mathrm{d}\theta$, and $w(y) = 2a\cos\theta$, so $\mathrm{d}A = 2a^2\cos^2\theta\,\mathrm{d}\theta$. Therefore

$$I_{\mathrm{A,solid}} = \int_{-\pi/2}^{\pi/2} (a\sin\theta)^2 (2a^2\cos^2\theta\,\mathrm{d}\theta) = 2a^4 \int_{-\pi/2}^{\pi/2} \sin^2\theta\cos^2\theta\,\mathrm{d}\theta = \frac{1}{4}\pi a^4, \tag{4.42}$$

because $\sin^2\theta\cos^2\theta = \sin^2\theta(1-\sin^2\theta) = \sin^2\theta - \sin^4\theta$, and $\int_{-\pi/2}^{\pi/2}\sin^2\theta\,\mathrm{d}\theta = \pi/2$ and $\int_{-\pi/2}^{\pi/2}\sin^4\theta\,\mathrm{d}\theta = 3\pi/8$. The mass of the solid circular beam with length L and mass per unit volume (mass density) ρ is $M_{\mathrm{solid}} = \rho\pi a^2 L$. Using

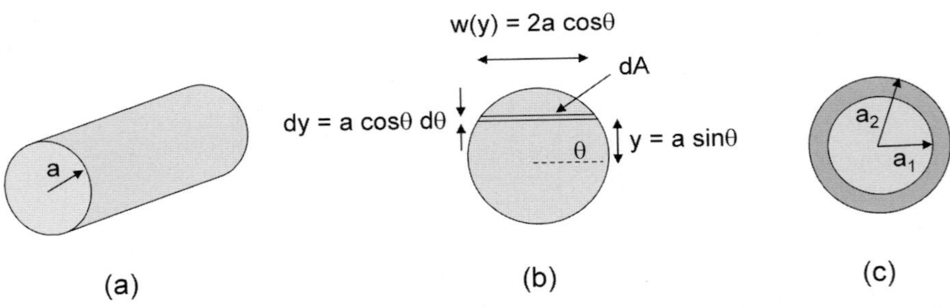

Fig. 4.39. (**a**) A solid circular beam; (**b**) determining the area moment of inertia for a solid circular cylinder beam; (**c**) a hollow circular beam

Table 4.4. Comparison of area moments of inertia and masses of hollow and solid circular beams

a_1/a_2	$I_{\text{A,hollow}}/I_{\text{A,solid}}$	$m_{\text{hollow}}/m_{\text{solid}}$	$(I_{\text{A,hollow}}/m_{\text{hollow}})/(I_{\text{A,solid}}/m_{\text{solid}})$
0	1.0	1.0	1.0
0.2	0.998	0.96	1.04
0.4	0.974	0.84	1.16
0.5	0.937	0.75	1.25
0.6	0.870	0.64	1.36
0.8	0.590	0.36	1.64
0.9	0.344	0.19	1.81

(4.42), it is clear that for a hollow circular beam with hollow radius a_1 and total radius a_2 (Fig. 4.39c),

$$I_{\text{A,hollow}} = \frac{\pi(a_2^4 - a_1^4)}{4} \tag{4.43}$$

and $m_{\text{hollow}} = \rho\pi(a_2^4 - a_1^4)L$. Table 4.4 shows that only 6% of the bending stiffness is lost with $a_1/a_2 = 0.5$, even though there is a 25% decrease in mass. For a beam of radius a and thin wall of thickness $w \ll a$ ($a_2 = a, a_1 = a - w$), we find that $I_{\text{A,hollow}} = \pi a^3 w$, $m_{\text{hollow}} = 2\rho\pi a w L$, and $I_{\text{A,hollow}}/m_{\text{hollow}} = a^2/2\rho L$. While the resistance to bending per unit mass increases as the beam (or bone) becomes more and more hollow, there is a limit to how much smaller I_{A} can become with smaller w before the beam can bend too much; it can also buckle (see below).

Bone Bending and Scaling Relationships

Let us consider a cantilever of length L that is firmly attached at the left and initially free at the right side (Fig. 4.40). A force F is applied downward at this free end. *How much does this end bend down?* For every section of length x (from the right), there is an applied moment $M_{\text{B}}(x) = F(L - x)$ (Figs. 4.41 and 4.42). Locally, at each x there is a curvature $1/R$, given by (4.40). The local curvature of any curve (in this case the beam) can be expressed by

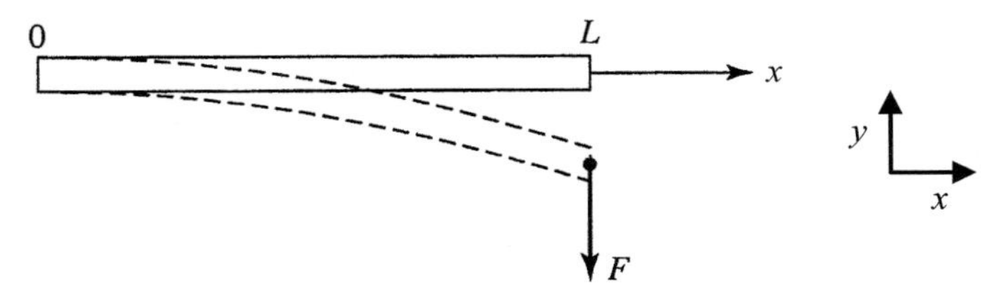

Fig. 4.40. Bending of a cantilever beam loaded at one end. (From [182])

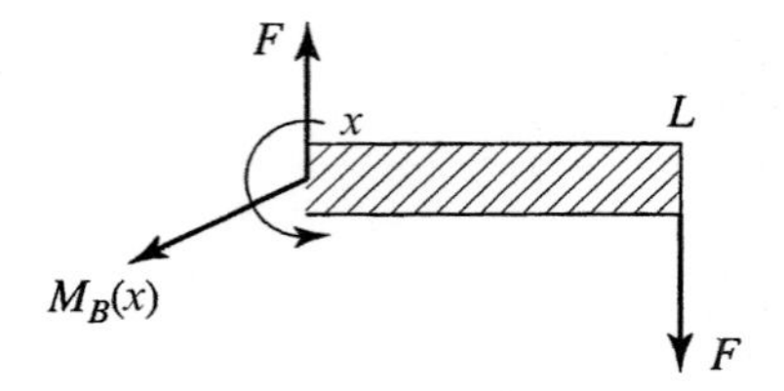

Fig. 4.41. Bending moment at end point x along the axis for the loaded cantilever beam. (From [182])

$\mathrm{d}^2y/\mathrm{d}x^2 = -1/R(x)$ and so

$$\frac{\mathrm{d}^2y}{\mathrm{d}x^2} = -\frac{F(L-x)}{YI_\mathrm{A}}. \tag{4.44}$$

At the wall ($x = 0$) the position is fixed, so $y = 0$ and $\mathrm{d}y/\mathrm{d}x = 0$ at $x = 0$.

Integrating (4.44) twice and applying these conditions gives the downward deflection at each x

$$y(x) = -\frac{F}{6YI_\mathrm{A}}((L-x)^3 + L^2(3x - L)). \tag{4.45}$$

(See Appendix C for more information about the solution.) At the end

$$y(L) = -\frac{FL^3}{3YI_\mathrm{A}}. \tag{4.46}$$

We will use this relation in the discussion about scaling in metabolism in Chap. 6.

4.5 Time-Dependent Deviations from Elastic Behavior: Viscoelasticity

So far we have asked how large of a force is needed to create a given strain or to break a bone. We have never asked whether it makes a difference if this force were applied quickly or slowly. (If we were to ask this, we would need

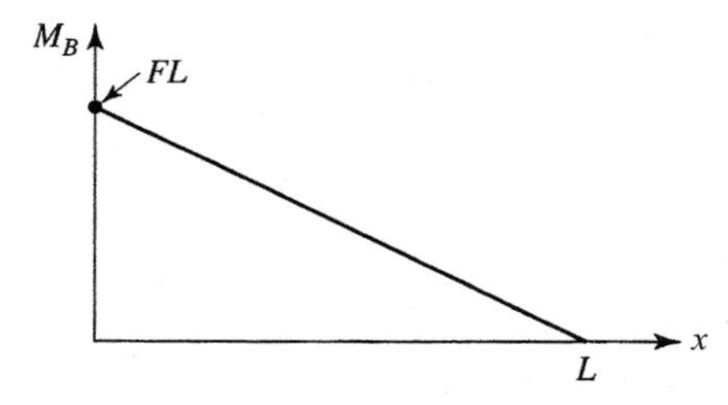

Fig. 4.42. Moment vs. x for the loaded cantilever beam. (From [182])

(a) k Spring $F_{applied} = kx$

(b) c Dashpot $F_{applied} = cv = c\overset{o}{x} = c\, dx/dt$

Fig. 4.43. (**a**) Ideal (or perfect) spring and (**b**) ideal (or perfect) dashpot. These are the two basic building blocks used in modeling the mechanical response of materials

to know if it were applied fast or slow relative to a defined time scale.) The responses of most materials inside or outside the body depend on these temporal dependences and on history, to some degree. This very important type of mechanical behavior is called *viscoelasticity*. Biological liquids and solids are usually viscoelastic, and this includes tendons, ligaments, cartilage, bone, and mucous. We will see how to model the viscoelasticity of body materials, and how it affects us – such as in fractures and collisions [195, 199].

Perfectly harmonic elastic behavior is modeled by a spring (Fig. 4.43a), with

$$F(t) = kx(t), \tag{4.47}$$

where F is now the applied force and x is the response, which is the displacement of the end of the spring. k is the spring constant. The force and displacement depend on the current state at the current time t and are independent of history.

Perfectly viscous behavior is modeled by a *dashpot* (Fig. 4.43b), with

$$F(t) = cv(t) = c\frac{\mathrm{d}x(t)}{\mathrm{d}t}, \tag{4.48}$$

where the response depends on the speed. c is a constant that describes damping due to viscosity. (In the biomedical engineering community this is sometimes called η.) An idealized dashpot is a piston moving in a cylinder, impeded by its movement in a viscous fluid. The displacement of the piston in the dashpot depends on its history! (The damping motion of a screen door closer is a dashpot.) This viscosity damping constant c describes the effects of viscosity for this macroscopic model and relates the force in the dashpot model to the speed of the piston in the viscous medium. It is related to, but is different from, the coefficient of viscosity η described in Chap. 7, which connects shear stress and the shear rate in a viscous fluid, as in (7.22) and (7.23).

We will combine these ideal springs and dashpots to arrive at models of realistic viscoelastic materials, and see how they respond to stimuli that vary with time. We can examine the extensive properties of applied forces and deformations of the material or the corresponding intensive properties of stress and strain.

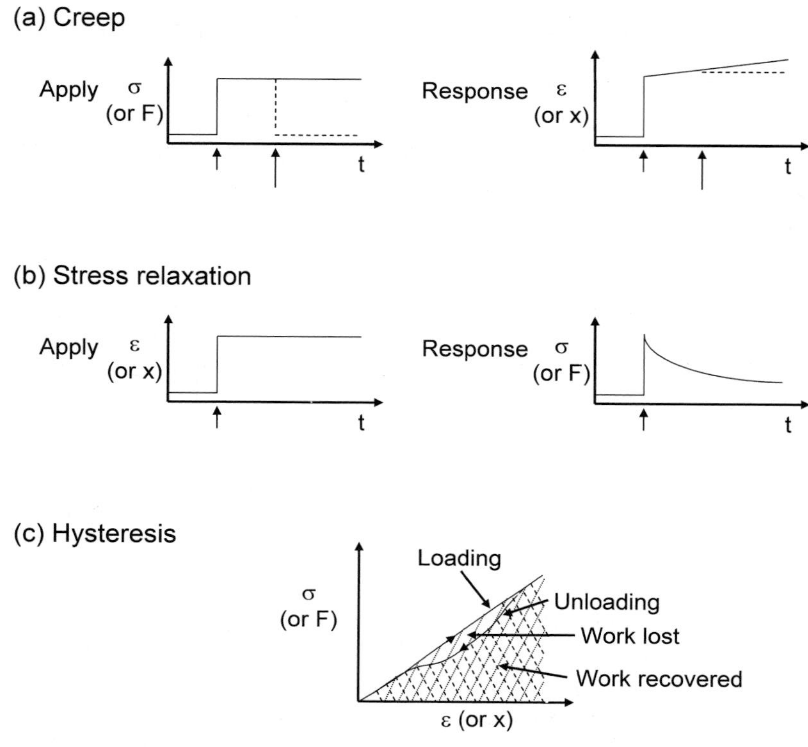

Fig. 4.44. General examples of (**a**) creep, (**b**) stress relaxation, and (**c**) hysteresis in viscoelastic systems. In (**a**) and (**b**) the stimulus is applied at the time of the *shorter arrow*. In (**a**) the possibility of removing the stimulus is also shown, at the time of the *longer arrow*, with the *dashed lines*. In (**c**) the recovered work is the area of the *cross-hatched region*, while the lost work is the area of the *lined region*. (More precisely, this is work per unit volume for stress σ and strain ϵ and work for force F and distortion x)

There are three interrelated manifestations of viscoelasticity (Fig. 4.44):

1. *Creep.* When a stress (or force) is applied and maintained, there is a strain (or deformation) in the medium that increases with time.
2. *Stress relaxation.* When a strain (or deformation) is applied and maintained, a stress (or force) is felt by the medium immediately, and it then relaxes in time.
3. *Hysteresis.* When stresses are applied and then released (forces loaded and unloaded), the stress–strain cycles are not reversible. Some, but not all, of the work done in the loading processes (during which the stress is increased) is recoverable in unloading (during which the stress is decreased).

Each of these effects can be observed and characterized by our models for *step function*, *impulse*, and *cyclic loading* (which are shown in the Fig. 4.45).

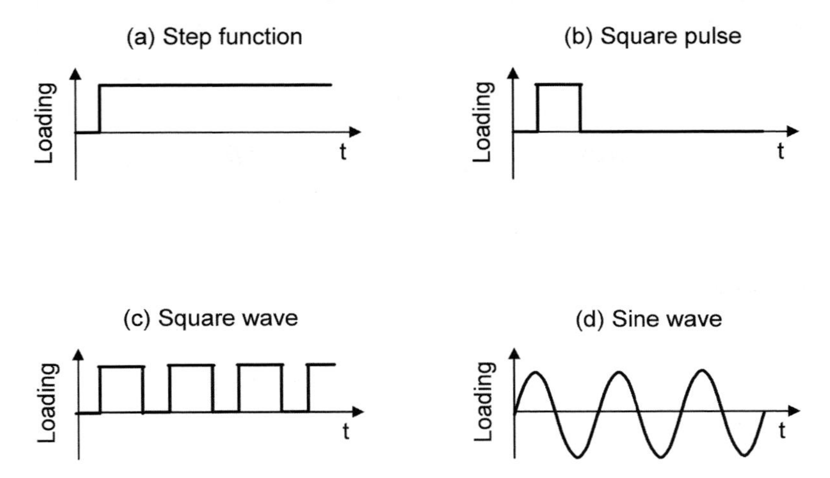

Fig. 4.45. Different types of loading protocols: (**a**) step function, (**b**) square pulse, and cyclic with (**c**) square pulses, or (**d**) sine waves

There are also outcomes other than those predicted by our models – such as ordinary and stress fractures – from long-term static loading and many cycles of loading.

One feature of viscoelasticity is that materials behave differently over different time scales. This is seen for one well-known viscoelastic material, Silly Putty^{TM}. When you throw silly putty against a wall, it bounces back like a ball. When you pull it, it stretches like putty. In the first example it behaves elastically. The time scale, the collision time with the wall, is short. It behaves in a viscous manner in the second example, because the time scale of pulling on it is long.

Figure 4.46 shows the stress–strain curves for bone when it is strained at different strain rates. When bone is strained slower, it develops less stress for the same applied strain. Figure 4.47 shows when stress is applied at slower rates, there is more strain for the same applied stress. Hysteresis in bone is shown in Fig. 4.48. Hysteresis and stress relaxation are shown for ligaments, tendons, and passive muscles in Figs. 4.49 and 4.50. The mechanism and the response of stress relaxation for cartilage are depicted in Fig. 4.51. This involves the exudation of fluid from the cartilage, which is tied to the lubrication of synovial joints, as shown in Fig. 3.14. The elastic modulus of cartilage increases from ~1 MPa for very slow rates of loading to 500 MPa for fast rates. Some mechanical properties of cartilage are plotted for different strain rates in Figs. 4.52 and 4.53. The modulus is ~0.70 MPa and Poisson's ratio = 0.10 for lateral femoral condyle cartilage.

Before developing models of viscoelastic materials, let us see how the perfect spring and dashpot components respond to idealized applications of stress and strain.

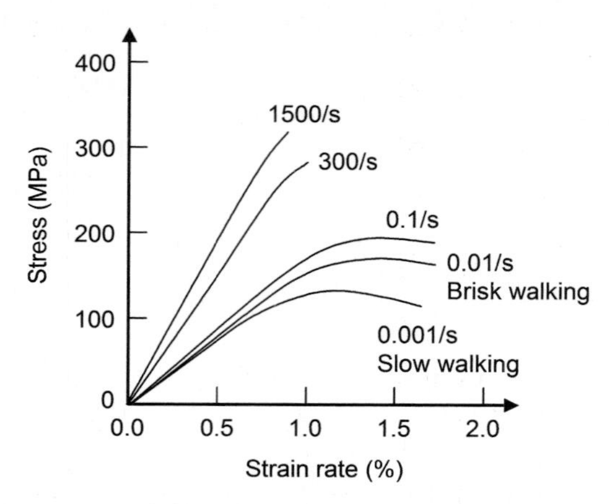

Fig. 4.46. Stress vs. strain for cortical bone for different strain rates, showing increased modulus and strength with increased strain rate. (Based on [197, 210, 211])

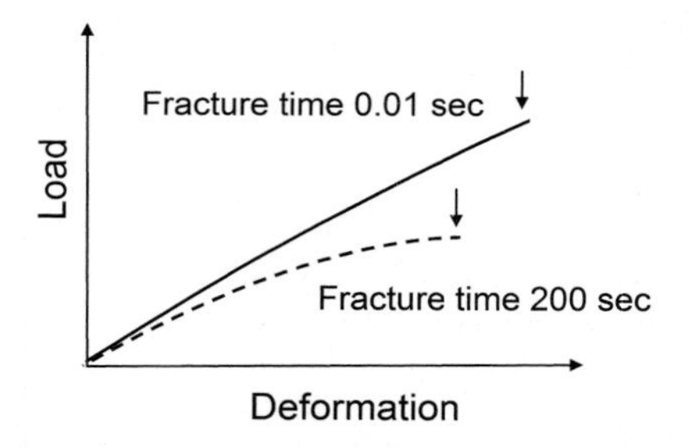

Fig. 4.47. Load (stress) vs. deformation (strain) for dog tibiae for different loading rates. The *arrow* shows the point of failure. At higher loading rates the load and the energy to failure are almost doubled, where energy is the area under the curve. (Based on [188, 229])

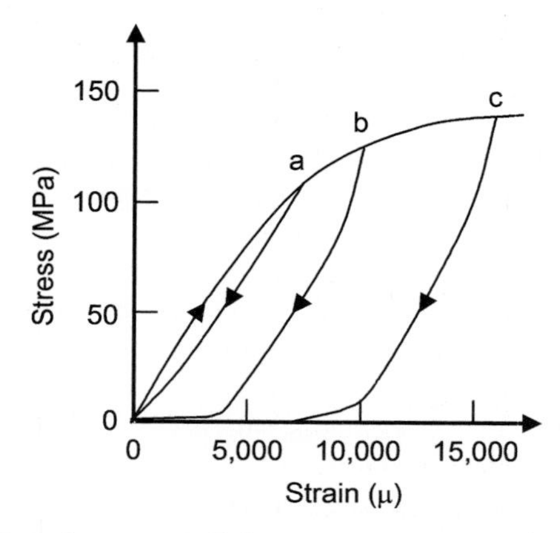

Fig. 4.48. Hysteresis in bone and shifting in the stress–strain curve with repeated loading (to a, b, c) and unloading. The units of strain are microstrain. (Based on [184, 191])

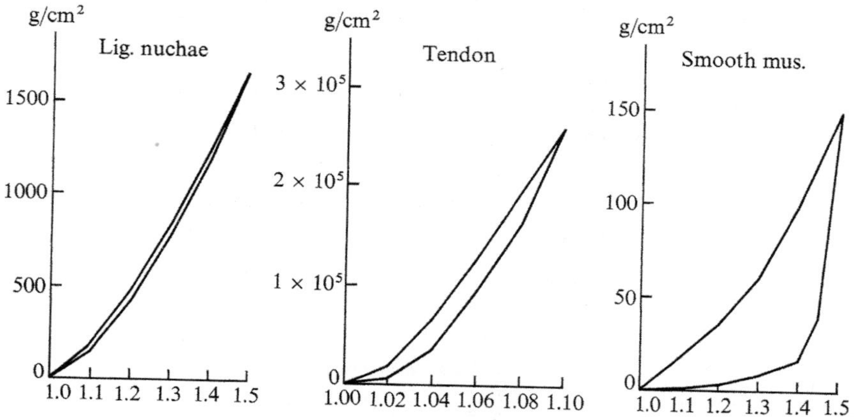

Fig. 4.49. Stress–strain hysteresis loop for nonvascular tissue: (**a**) the ligamentum nuchae (a ligament) (collagen denatured at 76°C, so it is mostly elastin), (**b**) tendon (mostly collagen), and (**c**) (passive) intestinal smooth muscle. The vertical axis units are those of stress when multiplied by g. (From [181, 198])

4.5.1 Perfect Spring

We use (4.47) to determine the response of a perfect spring to a stimulus. If we apply a force F_0 of any level, there will be an "instantaneous" deformation response of $x = F_0/k$. This creep response is seen in Fig. 4.54a for

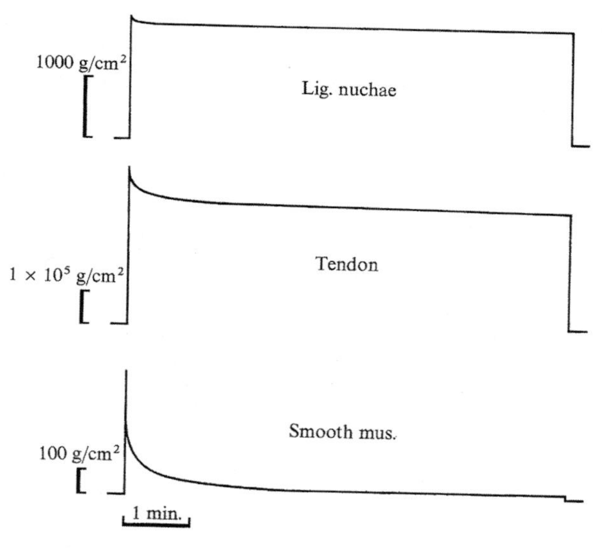

Fig. 4.50. Stress relaxation in nonvascular tissue: (**a**) the ligamentum nuchae (collagen denatured at 76°C, so it is mostly elastin), (**b**) tendon (mostly collagen), and (**c**) (passive) intestinal smooth muscle. The vertical axis units are those of stress when multiplied by g. (From [181, 198])

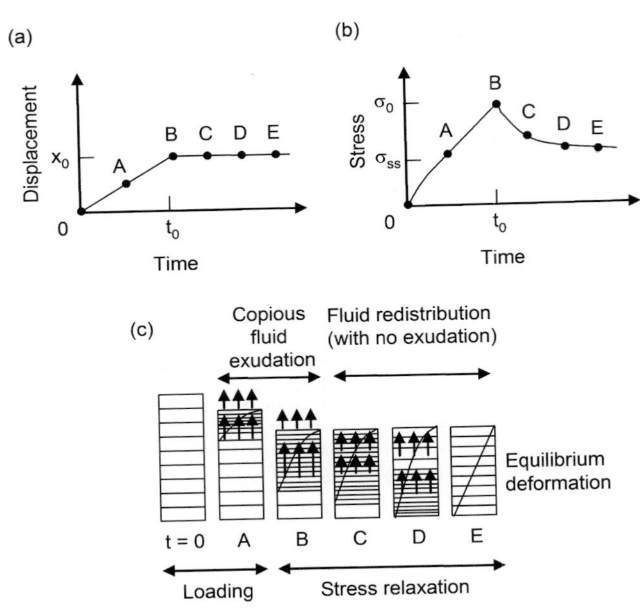

Fig. 4.51. (**a**) Controlled ramp deformation of cartilage from time 0 to t_0 and the (**b**) (viscoelastic) stress response, initially to σ_0, and later to the steady state value σ_{ss}, along with (**a**) and (**c**) physical model of the response. This response includes interstitial fluid flow (*arrows*) – initially out of and within the solid matrix and later only within the matrix – and also the deformation of the solid matrix of the cartilage. (Based on [202, 215, 217])

a step-like application of force. If we suddenly subject the material to a deformation x_0 (or a strain), with a step function ($\theta(t)$, as described below), there is an instantaneous step function response in the force (or stress) it feels (Fig. 4.54b).

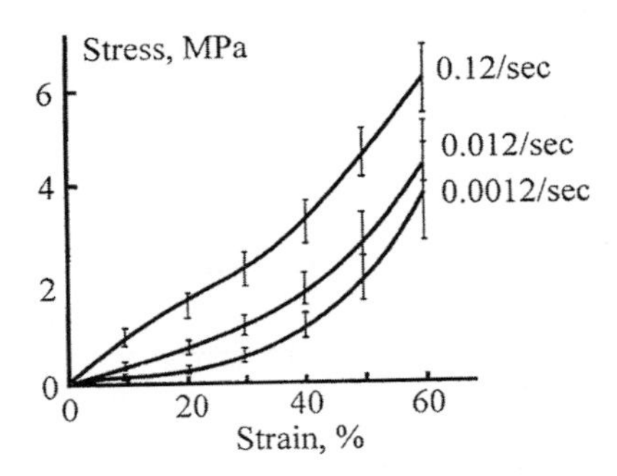

Fig. 4.52. Stress–strain for cartilage at different strain rates. (From [210], as from [232])

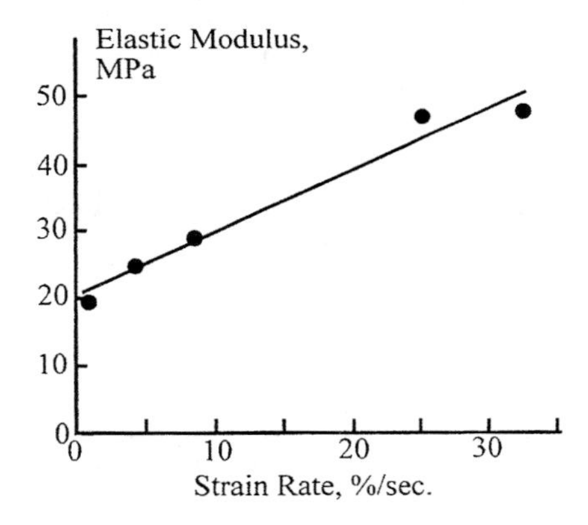

Fig. 4.53. Variation of compressive elastic modulus vs. strain rate in bovine articular cartilage. (From [210], as from [227])

4.5.2 Perfect Dashpot

Equation (4.48) describes the motion of a perfect dashpot with damping constant c. This characterization is often used to describe friction and other types of energy relaxation and dissipation. If we immediately apply a constant force F_0 at time $t = 0$, there is immediate motion with $v = \mathrm{d}x/\mathrm{d}t = F_0/c$. As seen in Fig. 4.55a, the dashpot piston is at position $(F_0/c)t$. This creep response stops suddenly when the force is removed, because $v = \mathrm{d}x/\mathrm{d}t$ immediately becomes zero. If we suddenly subject the material to a deformation x_0 (or a strain), with a step-like function, there is an immediate, very large, short-enduring force (described by a Dirac delta function response – see later) (Fig. 4.55b).

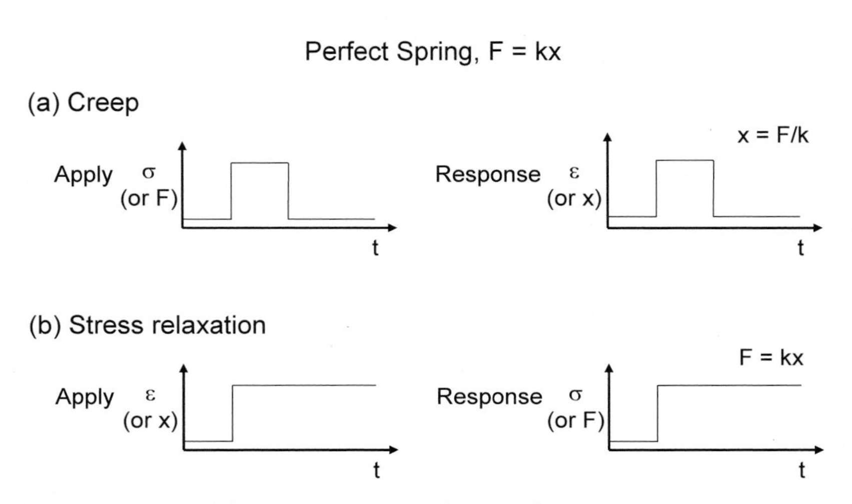

Fig. 4.54. Response by a perfect spring

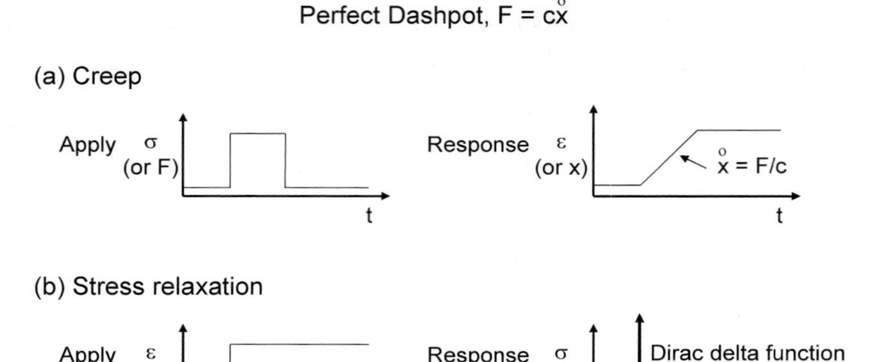

Fig. 4.55. Response by a perfect dashpot

With continued application of this deformation or strain, the force remains zero. Clearly, the responses of the dashpot and spring to applied stresses and applied strains are very different.

4.5.3 Simple Viscoelastic Models

Three models are commonly used to describe viscoelastic materials. Each combines these idealized springs and dashpots in different ways (see Fig. 4.56) [195, 199]. (a) A *Maxwell body* is a dashpot and spring in series. (b) A *Voigt body* is a dashpot and spring in parallel. (c) A *Kelvin body* is a dashpot and spring in series, which are in parallel with another spring. The Kelvin model is also called *the standard linear model.* The use of the terms "in series" and "in parallel" is similar to that in combining resistors in electrical systems, but the consequences are somewhat different.

The Maxwell and Voigt models are special cases of the Kelvin model, and consequently, the Kelvin model exhibits characteristic features of both simpler models. Why do not we always use the Kelvin model? Sometimes the use of the Kelvin model makes the analysis complex (at least at this level). Even more complex materials models consist of combinations of Kelvin systems.

What are the predictions for each model under the application of forces and deformations? We combine the forces (F_i) and deformations (x_i) of each individual unit to arrive at a constitutive relation that interconnects the applied force (F) and deformation (x) of the whole unit. We also express the total length of each object x_j^{T} as the sum of its equilibrium length x_j^{E} and its deformation from equilibrium x_j, $x_j^{\mathrm{T}} = x_j^{\mathrm{E}} + x_j$. Reference to an equilibrium length is important in describing springs, because $F = kx = k(x^{\mathrm{T}} - x^{\mathrm{E}})$.

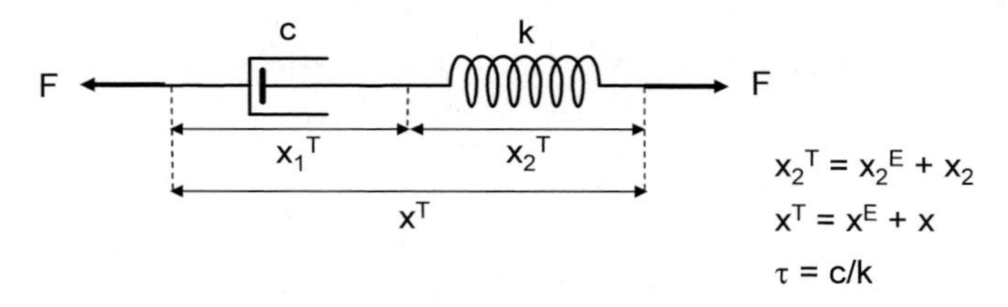

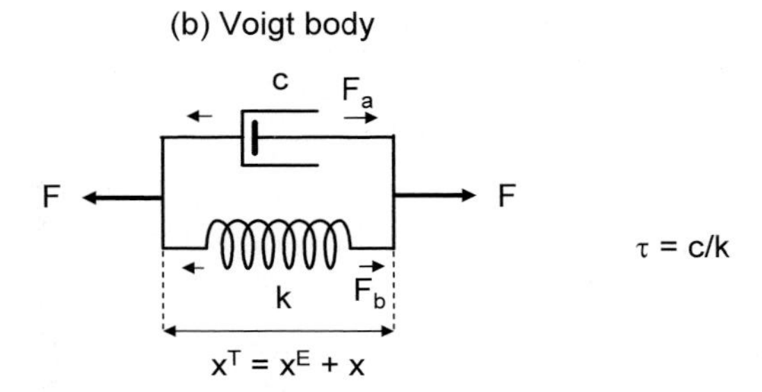

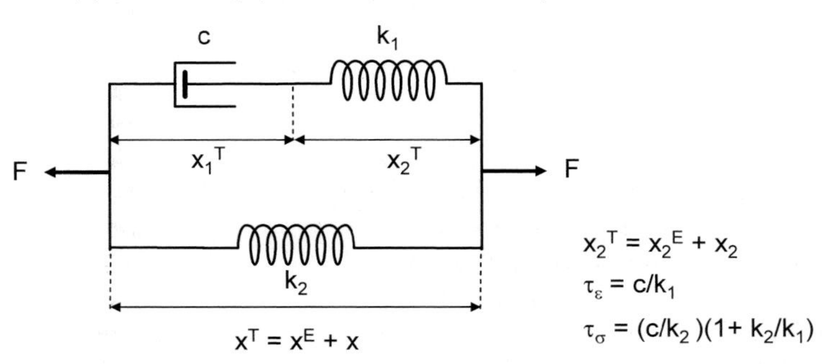

Fig. 4.56. Maxwell, Voigt, and Kelvin (standard linear model) mechanical models of viscoelasticity. The symbols for the springs and dashpots are the same as those used in Fig. 4.43. (Based on [198])

Maxwell Model

The force F applied at the end of a Maxwell body (Fig. 4.56a) is felt equally by the dashpot (unit "1") and spring (unit "2"), so $F = F_1 = F_2$. Therefore, for the dashpot

$$F_1 = F = c\,\frac{\mathrm{d}x_1}{\mathrm{d}t} \left(= c\,\frac{\mathrm{d}x_1^{\mathrm{T}}}{\mathrm{d}t}\right) \tag{4.49}$$

and for the spring

$$F_2 = F = kx_2 \ (= k(x_2^{\mathrm{T}} - x_2^{\mathrm{E}})). \tag{4.50}$$

The total length is $x^{\mathrm{T}} = x_1^{\mathrm{T}} + x_2^{\mathrm{T}}$, so $\mathrm{d}x^{\mathrm{T}}/\mathrm{d}t = \mathrm{d}x_1^{\mathrm{T}}/\mathrm{d}t + \mathrm{d}x_2^{\mathrm{T}}/\mathrm{d}t$. Because the equilibrium lengths do not vary with time ($\mathrm{d}x^{\mathrm{E}}/\mathrm{d}t = \mathrm{d}x_1^{\mathrm{E}}/\mathrm{d}t = \mathrm{d}x_2^{\mathrm{E}}/\mathrm{d}t = 0$), we find

$$\frac{\mathrm{d}x}{\mathrm{d}t} = \frac{\mathrm{d}x_1}{\mathrm{d}t} + \frac{\mathrm{d}x_2}{\mathrm{d}t}. \tag{4.51}$$

Using (4.49), we see that $\mathrm{d}x_1/\mathrm{d}t = F/c$. From (4.50), we see $x_2 = F/k$, and taking the first time derivative of both sides gives $\mathrm{d}x_2/\mathrm{d}t = (\mathrm{d}F/\mathrm{d}t)/k$. Using (4.51), we find

$$\frac{\mathrm{d}x}{\mathrm{d}t} = \frac{F}{c} + \frac{\mathrm{d}F/\mathrm{d}t}{k}. \tag{4.52}$$

This is the equation that relates the deformation $x(t)$ and force $F(t)$ for a Maxwell body.

To test creep, a force F_0 is suddenly applied at $t = 0$. There is no change in the displacement of the dashpot, so $x_1^{\mathrm{T}} = 0$ then. The spring immediately responds to give $x_2 = F_0/k$, so overall the initial condition is $x(0) = F(0)/k$ (for either creep or stress relaxation), and for $F(t = 0) = F_0$ it is $x(t = 0) = F_0/k$.

The sudden application of a constant force F_0 can be represented by $F(t) = F_0\theta(t)$, where $\theta(t)$ is the Heaviside step function (Fig. 4.57a), alluded to earlier, which is

$$\theta(t) = 0 \text{ for } t < 0; \quad = 0.5 \text{ at } t = 0; \quad = 1 \text{ for } t > 0. \tag{4.53}$$

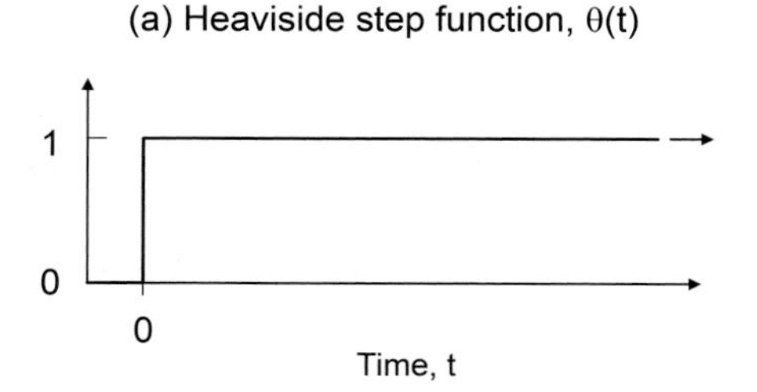

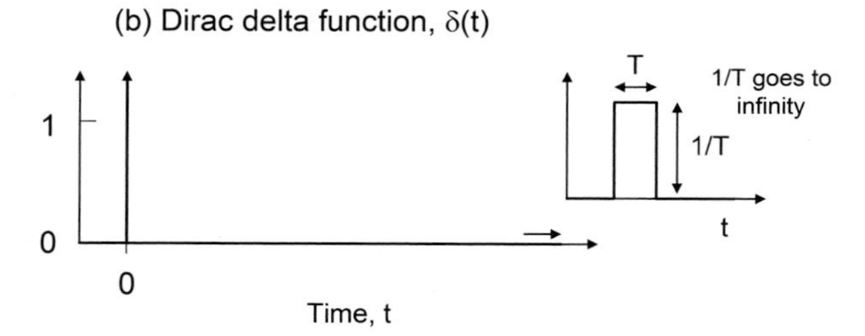

Fig. 4.57. (**a**) Heaviside step function $\theta(t)$ and (**b**) Dirac delta function $\delta(t)$. The area under the Dirac delta function remains 1, as the width $T \to 0$ and the height $1/T \to \infty$. If these functions start at a time other than $t = 0$, say at t_0, they are written as $\theta(t - t_0)$ and $\delta(t - t_0)$

The time derivative $d\theta(t)/dt$ is the Dirac delta function $\delta(t)$ (Fig. 4.57b) which is zero for all t except at $t = 0$, when it approaches infinity in such a way that its integral over time remains unity, as in

$$\delta(t) = 0 \text{ for } t < -T/2; \quad = 1/T \text{ for } -T/2 < t < T/2; \quad = 0 \text{ for } t > T/2 \tag{4.54}$$

in the limit that T goes to 0.

To test stress relaxation, a deformation x_0 is suddenly applied at $t = 0$. There is no change in the force of the dashpot, so $F_1^{\mathrm{T}} = 0$ then. A sudden application of a constant deformation x_0 can be represented by $x(t) = x_0\theta(t)$.

The response of the Maxwell body to the applied force $F(t) = F_0\theta(t)$ is

$$x(t) = F_0\left(\frac{1}{k} + \frac{t}{c}\right)\theta(t) \tag{4.55}$$

and to the deformation $x(t) = x_0\theta(t)$ it is

$$F(t) = kx_0 \exp(-(k/c)t\theta(t)). \tag{4.56}$$

These solutions can be proved by substitution in (4.52). (Also see Appendix C.) These results are plotted in Figs. 4.58a and 4.59a. (The plotted creep response is really that due to a sudden application of a constant force – say at time $t = 0$ – and then suddenly turning it off – say at time $t = T$ – and so the response to $F(t) = F(\theta(t) - \theta(t - T))$ is actually plotted for this and the other two models.)

In the creep experiment, there is an immediate spring-like response. Then the deformation increases (i.e., it creeps) linearly in time, as for the dashpot. When the force is removed, the deformation immediately decreases to the value determined by the spring component, and subsequently there is no more creep due to the dashpot. This is a simple linear combination of the responses seen for the individual elements in Figs. 4.54 and 4.55.

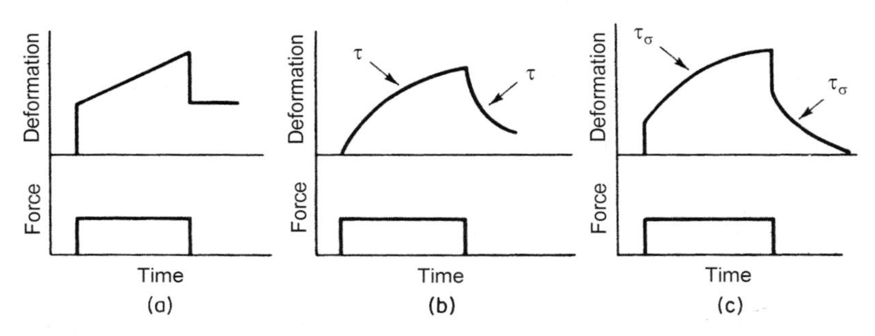

Fig. 4.58. Creep functions for the (**a**) Maxwell, (**b**) Voigt, and (**c**) Kelvin/linear standard models of viscoelasticity, with force loading and subsequent unloading. Characteristic relaxation times are shown. (From [198])

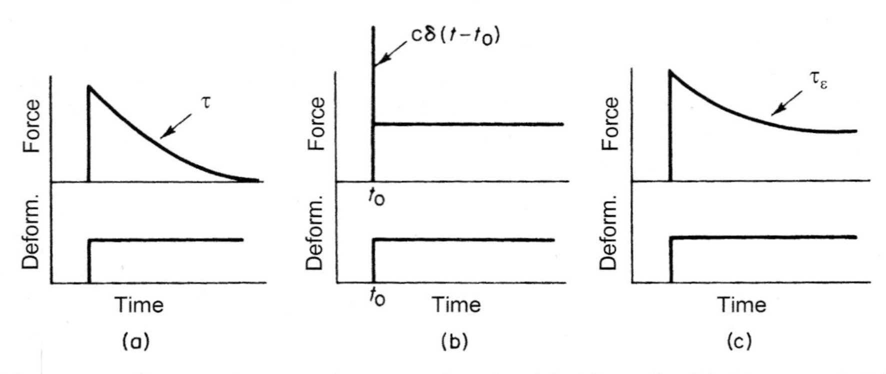

Fig. 4.59. Stress relaxation functions for the (**a**) Maxwell, (**b**) Voigt, and (**c**) Kelvin/linear standard models of viscoelasticity, with a step function deformation. Characteristic relaxation times are shown. (From [198])

In the stress relaxation experiment, there is an immediate force response due to the spring element, but this response decreases in an exponential manner, as $\exp(-t/\tau)$, due to the dashpot. The parameter $\tau = c/k$ is called a time constant; it has the units of seconds. This response is clearly not a mere linear combination of the responses for the individual elements.

Voigt Model

The combination of the elements is simple (Fig. 4.56b). The total force is the sum of the individual forces on each element $F = F_1 + F_2$, and the deformations of both elements are equal and they are equal to the whole $x = x_1 = x_2$. Because $F_1 = c\,\mathrm{d}x_1/\mathrm{d}t = c\,\mathrm{d}x/\mathrm{d}t$ and $F_2 = kx_2 = kx$,

$$F = c\frac{\mathrm{d}x}{\mathrm{d}t} + kx. \tag{4.57}$$

The initial condition is $x(t = 0) = 0$ for any applied F, because the dashpot prevents any immediate deformation.

The response of the Voigt body to the applied force $F(t) = F_0\theta(t)$ is

$$x(t) = \frac{F_0}{k}(1 - \exp(-(k/c)t))\theta(t) \tag{4.58}$$

and to the deformation $x(t) = x_0\theta(t)$ it is

$$F(t) = cx_0\delta(t) + kx_0\theta(t). \tag{4.59}$$

Again, these solutions can be proved by substitution in (4.57). (Also see Appendix C.) These results are plotted in Figs. 4.58b and 4.59b.

In the creep experiments, there is an exponential increase in creep, as $1 - \exp(-t/\tau)$, due to the dashpot – where again $\tau = c/k$. If the force is removed, this deformation decays to zero exponentially as $\exp(-t/\tau)$. This is qualitatively different from the predictions of the Maxwell model.

In the stress relaxation experiments, there is an immediate and temporary Dirac delta function increase in force, as seen for the dashpot alone, and then the response is the constant value expected from the spring alone. Again, this is qualitatively different from the predictions of the Maxwell model.

Kelvin Model (The "Standard" Linear Model)

In this model (Fig. 4.56c) the spring constant of the spring in series with the dashpot is called k_1; it was called k in the Maxwell model. The spring constant in parallel with the dashpot and spring in series is called k_2; the analogous constant in the Voigt model was also called k.

The length of the dashpot is $x_1^{\mathrm{T}} = x_1^{\mathrm{E}} + x_1$, while that of the top spring is $x_2^{\mathrm{T}} = x_2^{\mathrm{E}} + x_2$. The total length

$$x^{\mathrm{T}} = x_1^{\mathrm{T}} + x_2^{\mathrm{T}} \tag{4.60}$$

is also the length of the bottom spring. As with the Maxwell body, the same force

$$F_{\mathrm{a}} = c\,\frac{\mathrm{d}x_1}{\mathrm{d}t} = k_1 x_2 \tag{4.61}$$

is felt across the top dashpot and spring; the force

$$F_{\mathrm{b}} = k_2 x \tag{4.62}$$

is felt across the bottom spring. As with the Voigt model, the total force across the parallel elements is

$$F = F_{\mathrm{a}} + F_{\mathrm{b}}. \tag{4.63}$$

Our goal is to derive an equation that has only F, $\mathrm{d}F/\mathrm{d}t$, x, and $\mathrm{d}x/\mathrm{d}t$. Using (4.61), we see $\mathrm{d}x_1/\mathrm{d}t = F_{\mathrm{a}}/c$ and $x_2 = F_{\mathrm{a}}/k_1$. The time derivative of the second expression gives $\mathrm{d}x_2/\mathrm{d}t = (\mathrm{d}F_{\mathrm{a}}/\mathrm{d}t)/k_1$. Because $\mathrm{d}x_i^{\mathrm{E}}/\mathrm{d}t = 0$, from the first time derivative of (4.60), we have

$$\frac{\mathrm{d}x}{\mathrm{d}t} = \frac{\mathrm{d}x_1}{\mathrm{d}t} + \frac{\mathrm{d}x_2}{\mathrm{d}t} = \frac{F_{\mathrm{a}}}{c} + \frac{\mathrm{d}F_{\mathrm{a}}/\mathrm{d}t}{k_1}. \tag{4.64}$$

Using (4.62) and (4.63), we find $F_{\mathrm{a}} = F - F_{\mathrm{b}} = F - k_2 x$. The first time derivative of this is $\mathrm{d}F_{\mathrm{a}}/\mathrm{d}t = \mathrm{d}F/\mathrm{d}t - k_2\mathrm{d}x/\mathrm{d}t$. Using these in (4.64) gives

$$\frac{\mathrm{d}x}{\mathrm{d}t} = \frac{F - k_2 x}{c} + \frac{1}{k_1}\left(\frac{\mathrm{d}F}{\mathrm{d}t} - k_2\frac{\mathrm{d}x}{\mathrm{d}t}\right). \tag{4.65}$$

Collecting the force and deformation terms on opposite sides of the equation gives

$$F + \frac{c}{k_1}\frac{\mathrm{d}F}{\mathrm{d}t} = k_2 x + c\left(1 + \frac{k_2}{k_1}\right)\frac{\mathrm{d}x}{\mathrm{d}t}. \tag{4.66}$$

The second term on the left-hand side is absent in the Voigt model, while the first term on the right-hand side is absent in the Maxwell model. Factoring out k_2 gives

$$F + \frac{c}{k_1}\frac{\mathrm{d}F}{\mathrm{d}t} = k_2\left[x + \frac{c}{k_2}\left(1 + \frac{k_2}{k_1}\right)\frac{\mathrm{d}x}{\mathrm{d}t}\right]. \tag{4.67}$$

After introducing the time constants $\tau_\epsilon = c/k_1$ and $\tau_\sigma = (c/k_2)(1 + k_2/k_1) = c(1/k_1 + 1/k_2)$, this equation becomes

$$F + \tau_\epsilon\frac{\mathrm{d}F}{\mathrm{d}t} = k_2\left(x + \tau_\sigma\frac{\mathrm{d}x}{\mathrm{d}t}\right). \tag{4.68}$$

This tells us that the force terms relax with a time constant τ_ϵ, while the deformation terms relax with a time constant τ_σ. This is clear because the solution to $Q + \tau\,\mathrm{d}Q/\mathrm{d}t = 0$ is $Q(t) = Q(0)\exp(-t/\tau)$.

For a suddenly applied force or deformation, the initial condition is $\tau_\epsilon F(0) = k_2\tau_\sigma x(0)$. The response of the Kelvin body to the applied force $F(t) = F_0\theta(t)$ is

$$x(t) = \frac{F_0}{k_2}[1 - (1 - \frac{\tau_\epsilon}{\tau_\sigma})\exp(-t/\tau_\sigma)]\theta(t) \tag{4.69}$$

and to the deformation $x(t) = x_0\theta(t)$ it is

$$F(t) = k_2 x_0[1 - (1 - \frac{\tau_\sigma}{\tau_\epsilon})\exp(-t/\tau_\epsilon)]\theta(t). \tag{4.70}$$

Again, these solutions can be proved by substitution in (4.68). (Also see Appendix C.) These results are plotted in Figs. 4.58c and 4.59c.

In the creep experiments, there is an immediate increase due to the k_1 spring and then an exponential increase in creep, as $1-\exp(-t/\tau_\sigma)$, due to the dashpot. When the force is removed, this strain decays to zero exponentially as $\exp(-t/\tau_\sigma)$.

In the stress relaxation experiments, there is an immediate finite increase in force, and then the response relaxes as $\exp(-t/\tau_\epsilon)$ to a constant value.

These predictions incorporate features from both the Maxwell and Voigt models. The Kelvin model also cures the clear deficiencies in them, such as the unphysical Dirac delta function in the stress relaxation response in the Voigt model

4.6 Viscoelasticity in Bone

The stress in bone does not depend only on the current value of strain, but on how fast that strain was applied. Figure 4.46 shows that for a given strain the developed stress is larger when the strain is applied fast. Similarly, the

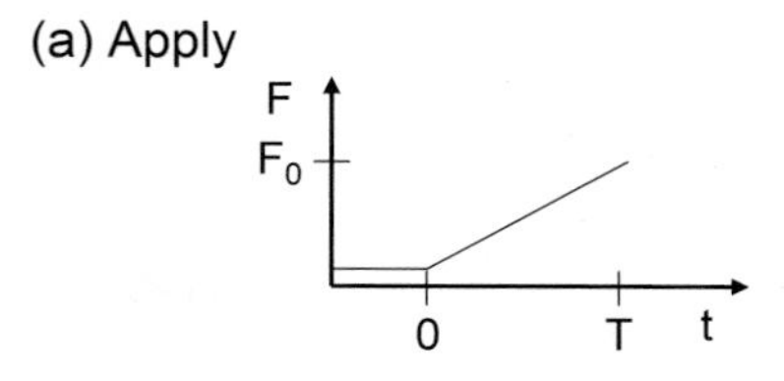

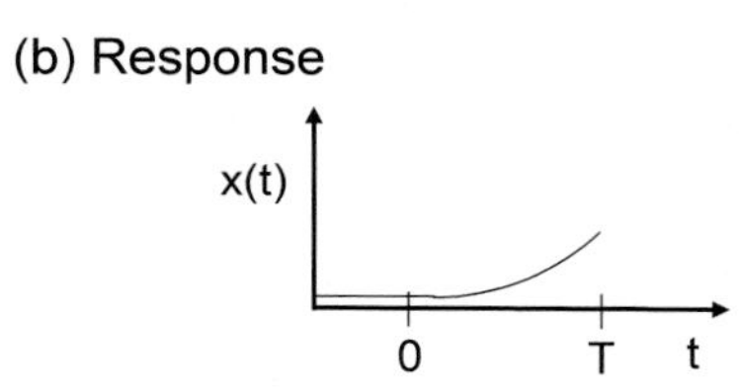

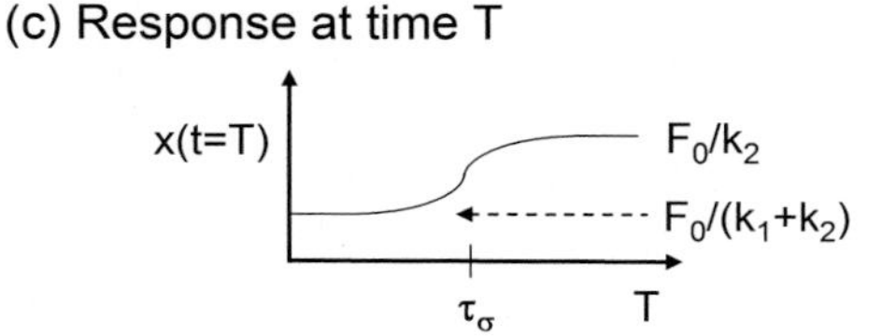

Fig. 4.60. The deformation (in (**b**)) resulting from a linearly increasing applied force (in (**a**)). The deformation from (**b**) at the end of the ramp, $t = T$, is plotted in (**c**) as a function of the ramp time T, which is referenced to τ_σ

strain in the bone depends not only on the current value of stress, but also on how fast that stress was applied. Figure 4.47 shows that for a given force load, the deformation is smaller when the load is applied fast.

We examine this second case quantitatively by using the Kelvin standard linear model. Let us apply a force F_0 in a linearly increasing manner over a time T. As seen in Fig. 4.60a, this means that $F = F_0(t/T)$ from $t = 0$ to $t = T$. We will determine the deformation $x(t)$, so we can obtain the deformation when the total force F_0 has been applied, $x(T)$ and see how $x(T)$ depends on T.

We use $F = F_0(t/T)$ and $\mathrm{d}F/\mathrm{d}t = F_0/T$ in (4.68) to get

$$k_2\left(x + \tau_\sigma \frac{\mathrm{d}x}{\mathrm{d}t}\right) = F + \tau_\epsilon \frac{\mathrm{d}F}{\mathrm{d}t} = F_0 \frac{t}{T} + \tau_\epsilon \frac{F_0}{T} \tag{4.71}$$

or

$$x + \tau_\sigma \frac{\mathrm{d}x}{\mathrm{d}t} = \frac{F_0}{k_2 T} t + \frac{\tau_\epsilon F_0}{k_2 T}, \tag{4.72}$$

with $x(t = 0) = 0$.

This equation has the form

$$x + a\frac{\mathrm{d}x}{\mathrm{d}t} = bt + e, \tag{4.73}$$

which has a solution

$$x(t) = bt + (e - ab)(1 - \exp(-t/a)), \tag{4.74}$$

that satisfies $x(t = 0) = 0$. The form of this solution can be verified by inserting it into the original equation. Such a substitution also gives a, b, and e, which leads to the final solution

$$x(t) = \frac{F_0}{k_2 T} t - \frac{cF_0}{k_2^2 T}(1 - \exp(-t/\tau_\sigma)). \tag{4.75}$$

(See Appendix C.) Clearly, this is valid only for $0 < t < T$. This solution can be checked in Problem 4.35, and is plotted in Fig. 4.60b. For $t \ll \tau_\sigma$, $x(t) \approx F_0(t/T)/(k_1 + k_2)$.

The deformation $x(T)$ at the end of the force ramp is

$$x(t = T) = \frac{F_0}{k_2} - \frac{cF_0}{k_2^2 T}(1 - \exp(-T/\tau_\sigma)), \tag{4.76}$$

which is plotted in Fig. 4.60c as a function of T. Applying the force quickly or slowly really means that the time T is either mach shorter or longer than the time constant, τ_σ. In these limits:

$$x(T) \to \frac{F_0}{k_1 + k_2} \text{ when } T \ll \tau_\sigma \text{ and } \to \frac{F_0}{k_2} \text{ when } T \gg \tau_\sigma. \tag{4.77}$$

This model agrees with the experimental observations that the deformation is less with faster loading (Fig. 4.47 for bone). More generally, the terms "fast" and "slow" are relative to a characteristic time constant, which in this case is τ_σ. The analogous stress relaxation experiment can be modeled in a similar way with the relevant time constant τ_ϵ, and is addressed in Problem 4.38.

4.7 Bone Fractures

Bones in the skeleton are designed for several properties and functions. Muscles create motion by swinging bones at articulations. If the bones are relatively stiff, the muscles are efficient in that when they contract they do not cause the bones to deform. If bones are stiff, they absorb relatively little energy before they fracture (see below), so more compliant bones resist fracture more and are lighter (and there is less mass for the body to lug around) – but they are less ideal for muscle action. Nature compromises as needed. In children, efficiency of motion is less important than resistance to fracture. The femoral bone is about 2/3 as stiff in children as in adults and requires about 50% more energy to break. The bones of the inner ear need to be stiff to transmit sound waves efficiently, but do not need to resist fracture because they bear no loads.

Table 4.5. Physical properties of different types of bone. (From [212]. Using data from [189])

property	femur	antler	bulla
Young's modulus (Y) (GPa)	13.5	7.4	31.3
ultimate bending stress (UBS) (GPa)	247	179	33
work of fracture (W_F) (J/m^2)	1,710	6,190	200
density (g/cm^3)	2.06	1.86	2.47

We now consider the *work of fracture* W_F, which is the amount of work that has to be performed on a material to break it. It is usually defined as the energy (J) needed for fracture per area (m^2). This can be estimated from the elastic energy stored using the stress–strain curves earlier in this chapter. Materials that have a higher work of fracture are *tougher* than those with a lower one (Fig. 4.16). Typical values are 1–10 J/m^2 for glass, $\sim$1,000 J/m^2 for nylon, $\sim$10,000 J/m^2 for wood, and 10^5–10^6 J/m^2 for bone. Materials with the same strength (UTS) are tougher (i.e., they require more energy to fracture) when they are less stiff (smaller Y), because the elastic energy stored per unit volume is $\mathrm{PE}/V = \sigma^2/2Y = (\mathrm{UTS})^2/2Y$ (4.17) in the linear stress–strain limit.

Table 4.5 shows that the mechanical properties of typical femurs, deer antlers, and tympanic bulla (which are bony capsule housings in the ear) are quite different, even though they have comparable densities. Femurs support weight during movement and need to be stiff (large Y), strong (large *ultimate bending stress*, UBS – see below), and tough (large W_F). Deer antlers need to be very tough with a very high work of fracture to avoid breakage in deer fights (and they are tougher than femurs), but they do not need to be really stiff or strong. Tympanic bulla house the middle/inner ear and keep out sounds other than those coming through the ear canal. This helps directional hearing, with sounds detected in each ear at different times. This acoustic separation is improved by increasing the ratio of Y for the bulla and water (see Chap. 10); Y for bulla is very high. The bulla do not need to be strong or tough. If forces were applied that would be large enough to break them, the person would be dead anyway.

4.7.1 Modes of Sudden Breaking of Bones

Let us revisit the example from earlier in this chapter, when we considered how much the femur shortens at the UCS = 170 MPa = 170 N/mm^2. With no stress applied, the femur is L_0 = 0.5 m = 500 mm long and has a cross-sectional area A = 370 mm^2. The UCS is reached when there is a force of (170 N/mm^2)(370 mm^2) = 56,000 N = 12,600 lb $\sim$6 tons on the femur. For a 70 kg person (700 N, 160 lb), this is 80× body weight. Because the maximum force on the hip bone and femur during walking is $\sim$2× body weight and during running it is $\sim 4\times$ body weight, we are fortunately well designed. There is a much larger overdesign protection in the long leg bones during running than in the Achilles tendon!

The potential energy available during a fall to the ground from standing is $m_{\mathrm{b}}g(\Delta h_{\mathrm{CM}})$. A 1.8 m tall person of mass 70 kg, has a center of mass 0.9 m above the ground. When this person falls, the center of mass decreases to 0.1 m and the available potential energy is $m_{\mathrm{b}}g(\Delta h_{\mathrm{CM}}) = (70\,\mathrm{kg})(9.8\,\mathrm{m/s^2})(0.8\,\mathrm{m}) \approx 550\,\mathrm{J}$.

How much energy is stored in the bones during this fall? Let us examine the largest bone, the femur. We use (4.17) with $L_0 = 0.5\,\mathrm{m} = 500\,\mathrm{mm}$ and $A = 330\,\mathrm{mm}^2$, and so $V = 165{,}000\,\mathrm{mm}^3$, and $Y = 17{,}900\,\mathrm{MPa} = 17{,}900\,\mathrm{N/mm^2}$. If the stress is either the UTS $= 122\,\mathrm{MPa} = 122\,\mathrm{N/mm^2}$ or UBS (ultimate bending stress, as described below) $= 170\,\mathrm{MPa} = 170\,\mathrm{N/mm^2}$ (Table 4.6), then respectively

$$\mathrm{PE} = \frac{(\mathrm{UTS})^2}{2Y}V = \frac{(122\,\mathrm{N/mm^2})^2}{2\times 17{,}900\,\mathrm{N/mm^2}}165{,}000\,\mathrm{mm}^3 \simeq 69\,\mathrm{J} \tag{4.78}$$

$$= \frac{(\mathrm{UBS})^2}{2Y}V = \frac{(170\,\mathrm{N/mm^2})^2}{2\times 17{,}900\,\mathrm{N/mm^2}}165{,}000\,\mathrm{mm}^3 \simeq 133\,\mathrm{J}. \tag{4.79}$$

The energy needed to break long bones is clearly a reasonable fraction of that available from the kinetic energy in common collisions, such as falls. If the available energy is distributed to several of the long bones, there is enough for sudden fracture. Our bones do not regularly break because most of the energy is absorbed by muscle contractions and the deformation of soft tissues. Loads normal to skin, fat, and muscles (and clothing) absorb energy upon compression and propagate stress waves in the body. Fascia, tendons, ligaments, joint capsules, and contracted muscles brace bones against bending by supporting part of the tensile forces and absorbing energy as they are stretched. In elderly people bones fracture more easily because their bones are weaker (because they are more porous, Fig. 4.19) and their tissues are less suited to absorb energy, which causes even more energy to be transmitted to the (already weaker) bones.

Bone fractures are determined by the mode of the applied loads and their orientations. Bones are strongest in compression, less strong in tension, and weakest in shear. Under some loading conditions there are tensile and shear or compressive and tensile loads at a given position. Bones usually break by shear (twisting) stresses or under tension, but not under compression because UTS $<$ UCS. Figure 4.61 shows crack formation in bent and twisted long bones. Under bending there is tension on one side and compression on the other. Because UTS $<$ UCS, the fracture starts at the side with tension. There are shear stresses at 45° to this load axis, but the tensile stress is larger on the left side, so the crack propagates normal to the bone axis. On the compressed side the shear stress (at 45°) is large and the compressive stress is $<$UCS, so cracking occurs at two 45° angles, leading to the *butterfly fragment*

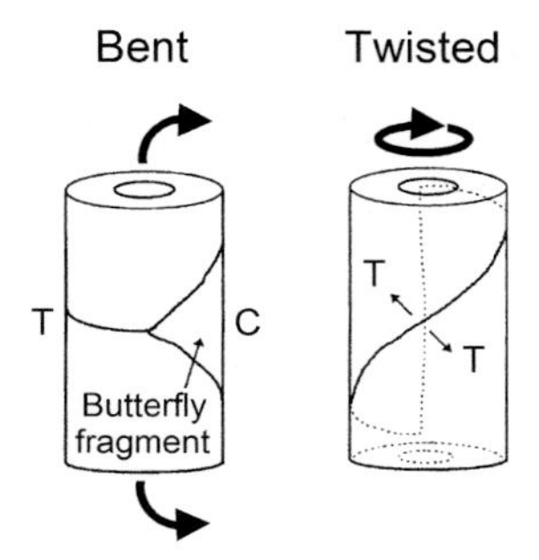

Fig. 4.61. Crack propagation in bent and twisted bone. T means tension and C means compression. (From [210])

seen in the figure. In twisting, the tensile stresses produce a spiral crack that winds around the bone and the bone breaks when the ends of the crack are connected by a longitudinal fissure [210].

Fracture can be due to direct blows, such as by blows to the soft tissue surrounding the bone or bullets, which break the bone in two (*noncomminuted*), at low energy – leading to a *transverse* fracture, or into many pieces (*comminuted*) at high energy (Fig. 4.62) [179, 208, 218]. Indirect blows, as in skiing, can lead to fractures that are *spiral*, *oblique*, transverse with a butterfly fragment, and so on (Fig. 4.62). The nightstick fracture of the ulna, shown in Fig. 4.63, is one type of low-energy, direct-blow injury, and it is transverse. Figure 4.64 depicts the classification of humerus fractures. (Analogous classifications exist for other bones as well.)

As with any collision, we can lessen the likelihood of bone breakage in falls by increasing the impact area and the collision time. Because bones are actually viscoelastic, they absorb shocks a bit, which lessens the chance of fracture.

Breaking of Bones by Bending

Let us say we have a bone of thickness d that is symmetrical in the y direction. *Will it break when bent by a force F like the one in Fig. 4.31?*

We use the analysis we developed earlier this chapter. Equation (4.36) tells us that $\sigma(y) = Y(y/R)$, so with $y = d/2$ on the top surface and $-d/2$ on the bottom, the maximum compressive and tensile stresses have magnitudes (Fig. 4.34)

$$|\sigma_{\mathrm{max,compression}}| = |\sigma_{\mathrm{max,tension}}| = Y\frac{d}{2R}. \tag{4.80}$$

We would expect that the bone will break if either $|\sigma_{\mathrm{max,compression}}| > \mathrm{UCS}$ or $|\sigma_{\mathrm{max,tension}}| > \mathrm{UTS}$. Because UCS = 170 MPa and UTS = 120 MPa, we would expect that the fracture will occur first in tension and consequently on the bottom surface for hard bone (Fig. 4.35). However, this fracture really

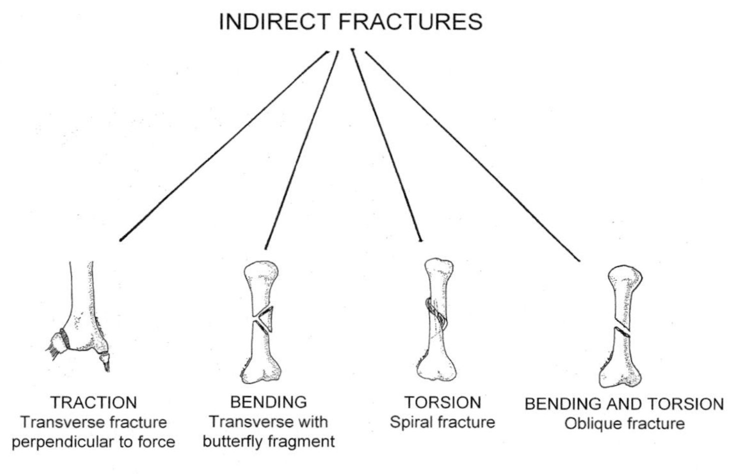

Fig. 4.62. Types of bone fractures resulting from different types of loading. (From [208])

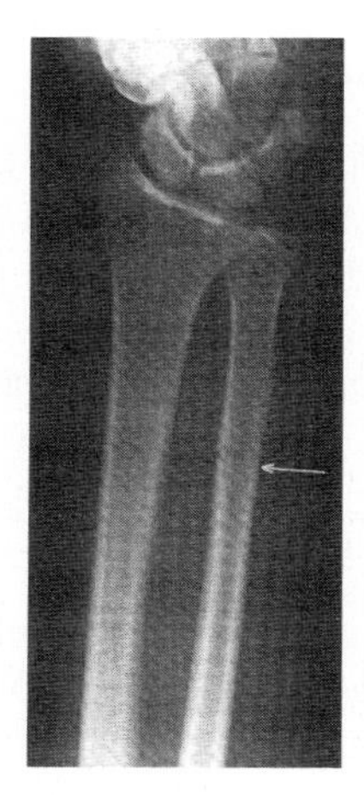

Fig. 4.63. X-ray of a nightstick fracture of the ulna bone. (From [208])

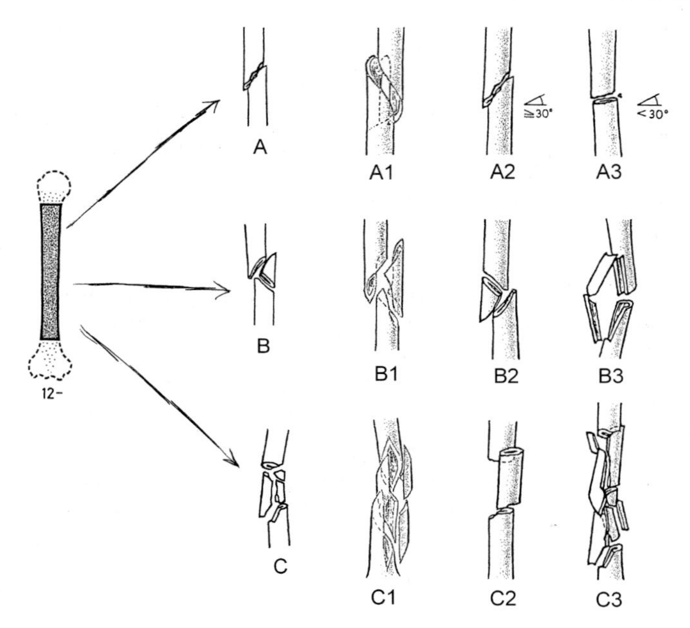

Fig. 4.64. AO-ASIF classification of humerus diaphysis fractures. A: Simple fractures; A1: spiral, A2: oblique (≥30°), A3: transverse (<30°); B: wedge fractures; B1: spiral wedge, B2: bending wedge, B3: fragmented wedge; C: complex fractures; C1: spiral, C2: segmental, C3: irregular. (From [208])

occurs at a slightly higher value called the *ultimate bending stress* (UBS). The UBS is higher than UTS for the long bones, as in seen in Table 4.6. This table also shows that the mechanical properties of the long bones in the leg and arm are very similar, but not identical.

Fracture occurs when

$$|\sigma_{\mathrm{max,bending}}| > Y\frac{d}{2R_{\mathrm{min}}} = \mathrm{UBS}. \tag{4.81}$$

Table 4.6. Properties of long bones. (From [182]. Using data from [236])

bone	Y ($\times 10^4$ MPa)	UTS ($\times 10^2$ MPa)	UBS ($\times 10^2$ MPa)
femur	1.72	1.21	2.08
tibia	1.80	1.40	2.13
fibula	1.85	1.46	2.16
humerus	1.71	1.22	2.11
radius	1.85	1.49	–
ulna	1.84	1.48	–

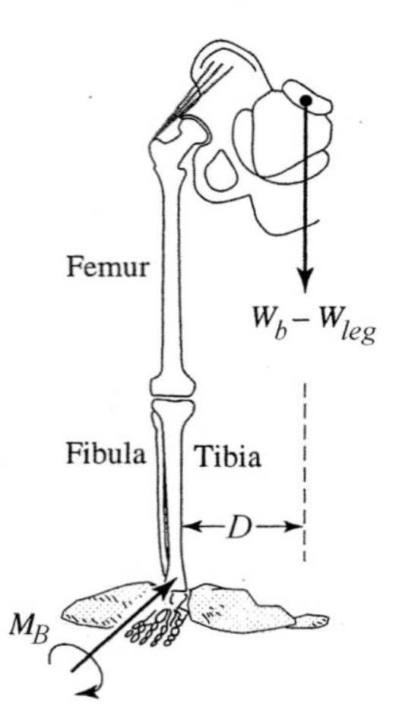

Fig. 4.65. Illustration of the origin of the bending moment in a person with a pinned ankle, as during falling. (From [182])

Using (4.40) for the curvature, $1/|R| = |M_\mathrm{B}|/YI_\mathrm{A}$, the bone breaks for bending moments

$$|M_\mathrm{B}| \geq \frac{2(\mathrm{UBS})I_\mathrm{A}}{d}. \tag{4.82}$$

Let us consider the example of one foot pinned at the ankle, while the other foot is slipping [182]. The pinned foot could be in a hole in frozen snow, pinned during a football tackle, or in a rigid ski boot. This situation is modeled in Fig. 4.65, where we see the force of the body (minus that of the leg) $W_\mathrm{b} - W_\mathrm{leg}$, creates a torque about the leg with a moment arm D of magnitude $D(W_\mathrm{b} - W_\mathrm{leg})$. D is the lateral distance of the midline of the body (center of mass) from the pinned leg, and so

$$|M_\mathrm{B}| = D(W_\mathrm{b} - W_\mathrm{leg}) \geq \frac{2(\mathrm{UBS})I_\mathrm{A}}{d}. \tag{4.83}$$

If the bone has a radius a, then $I_\mathrm{A} = \pi a^4/4$ (because we can ignore the hollow nature of the bone in this estimate) and using $d \sim 2a$, the bone breaks when the moment arm

$$D \geq \frac{\pi a^3}{4} \frac{\mathrm{UBS}}{W_\mathrm{b} - W_\mathrm{leg}}. \tag{4.84}$$

The tibia has its smallest cross-section about 1/3 of the way up from the ankle, where $a \sim 1$ cm. It is much thinner there than the humerus anywhere. The fibula is even narrower, but bears much less of the force than the tibia. For $W_\mathrm{b} - W_\mathrm{leg} = 640\,\mathrm{N}$ (145 lb) (for a 75 kg, 750 N, 170 lb person), this shows that fracture occurs when the midline of the body moves more than 25 cm from the pinned leg during the slip.

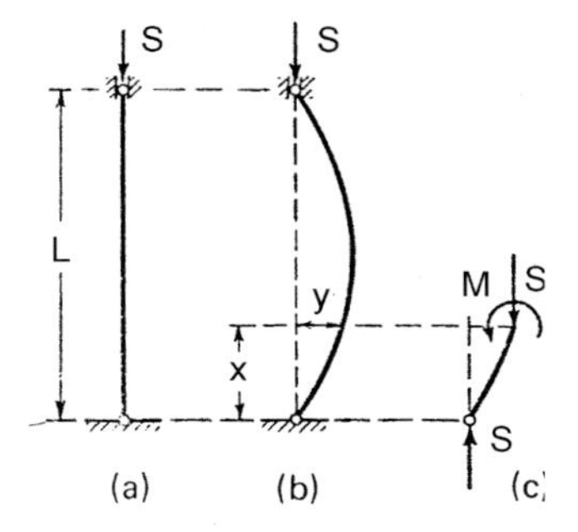

Fig. 4.66. (**a**) Before and (**b**) after Euler buckling of a bar or column. The bottom of the bar corresponds to $x = 0$ and $y = 0$. (**c**) A free body diagram of a part of the bar showing the external and internal forces and the moments acting on this column. (From [383])

Euler Buckling (Advanced Topic)

The occurrence of fractures depends on the ultimate strength, defects, and specifically how loads are applied. Another type of macroscopic failure is buckling. This *Euler buckling* can be demonstrated by pushing down on a drinking straw that is standing upright on a table. When long thin tubes are compressed, the middle bows to one side and collapses. This mode of failure is associated with the stiffness of the material, and not its strength. This is different from the bending that occurs when a bar along the x-axis is fixed at one end, and the free end is pushed by a force F in the y direction (normal to the x-axis), as with bending in (4.44)–(4.46) and Fig. 4.40; the moment there is the force along the y-axis $\times$ the moment arm along x. Here the force is actually applied along the y-axis.

Let us consider a bar or column of length L fixed at both ends ($x = 0$ and $x = L$) with a compressive load S applied along the x-axis (Fig. 4.66a). The moment M_{B} in (4.44) now becomes Sy because the y-axis is normal to the applied force, so the curvature, from (4.40) and (4.44), becomes

$$\frac{\mathrm{d}^2 y}{\mathrm{d}x^2} = -\frac{Sy}{YI_{\mathrm{A}}}, \tag{4.85}$$

where Y is Young's modulus and I_{A} is the area moment of inertia. So we see that

$$\frac{\mathrm{d}^2 y}{\mathrm{d}x^2} + \lambda^2 y = 0, \tag{4.86}$$

where we have used $\lambda^2 = S/YI_{\mathrm{A}}$. This has a solution

$$y(x) = A \sin \lambda x + B \cos \lambda x, \tag{4.87}$$

which can be proved by substituting this into the previous equation (see Appendix C). The constant $B = 0$ because $y = 0$ at $x = 0$ and so $y(x) = A \sin \lambda x$.

Because $y = 0$ at $x = L$, either $A = 0$ or $\sin\lambda L = 0$. The former boundary condition implies the bar will always be straight, while the latter allows for the possibility of buckling (with indeterminate and conceivably very large amplitude A). This latter condition is satisfied by $\lambda L = n\pi$, with $n = 1,2,3,\ldots$. This means $\lambda^2 = S/YI_\mathrm{A} = n^2\pi^2/L^2$, so $S = n^2\pi^2 YI_\mathrm{A}/L^2$. The lowest load that this buckling can occur at is the critical load S_c with $n = 1$

$$S_\mathrm{c} = \frac{\pi^2 YI_\mathrm{A}}{L^2}. \tag{4.88}$$

For a beam of radius a and thin wall of thickness $w \ll a$ ($a_2 = a, a_1 = a-w$), $I_\mathrm{A,hollow} = \pi a^3 w$, so $S_\mathrm{c} = \pi^3 Ya^3 w/L^2$. Consequently, although making the walls of bone progressively thinner does not hurt its resistance to bending per unit mass, it will buckle more easily.

4.7.2 Stress Fractures (Advanced Topic)

We have seen that bones can fracture when the stress on them suddenly exceeds a given failure limit. They can also fracture more gradually from damage from prolonged continuous stress (creep, as with sitting) or prolonged *cyclic* stress (*fatigue*, as with walking or running) (Fig. 4.67). When the rate of damage exceeds the rate of repair by the body (by *remodeling*), the bone fails as a result of a *stress fracture*.

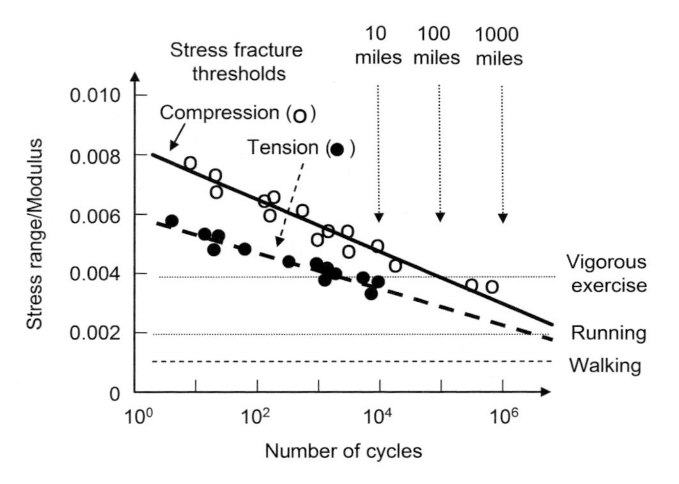

Fig. 4.67. Test strain ranges in compression and tension that lead to fatigue damage (stress fractures) in human cortical bone when applied for given number of cycles, referenced to the strains that simulate walking, running, and other strenuous exercise. There are approximately 5,000 cycles of testing (each corresponding to a step) in 10 miles (16 km) of running (which are the ranges of strain in either compressive or tensile loading experiments that simulate walking, running, and other strenuous exercise). (Based on [187, 197])

We will now assess the occurrence of such fractures by looking at the applied stress and the resulting strains. From earlier in this chapter, the microstrain in long bones is $\sim 10,000\mu\epsilon$ (1%) at the UCS. This is usually not reached. The peak functional microstrain in bones in most animals is between 2,000 and 3,000$\mu\epsilon$ at peak performance. Strains in thoroughbred horses are routinely 5,000–6,000$\mu\epsilon$ during racing. In humans, some studies indicate that the peak functional microstrains in the tibia, where stress fractures often occur, do not exceed 2,000$\mu\epsilon$, while others suggest that microstrains over 3,000$\mu\epsilon$ can occur during jumping; this explains why "shin splints" are not uncommon among basketball players.

The study of how a flaw or crack grows under stress and leads to catastrophic failure is called *fracture mechanics*. The derivation of relations of linear elastic fracture mechanics is beyond the level of this text (see [210, 214, 226, 230]). Nonetheless, we will present some results from this field to help us understand stress fractures better.

Let us consider a thin plate with an elliptical hole with minor and major radii a and b, as in Fig. 4.68. With stress s applied parallel to the minor axis, the stress is maximum at the semimajor axis end as shown, and has magnitude

$$\sigma = s\left(1 + \frac{2a}{b}\right). \tag{4.89}$$

(The value of stress far from the crack is the applied stress s.) As a becomes much greater than b, the ellipse becomes narrower and begins to look more like a crack; then this relation is no longer valid. The stress pattern can then be determined, and expressed as a function of the distance from this same point (the end of the major axis) r and the angle from this axis θ, as shown in Fig. 4.68. For a given r, the stress is a maximum for $\theta = 0°$ (which makes

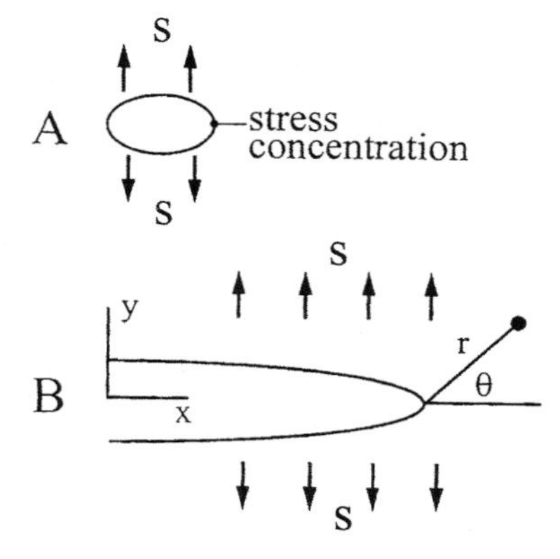

Fig. 4.68. An elliptical hole in a plate structure with stress tension s has much higher stress at the concentration point shown. (From [210])

Table 4.7. Fracture toughness of materials. (From [210])

material	K_c (MPa-m$^{1/2}$)
2024 aluminum	20–40
4330V steel	86–110
Ti–6Al–4V	106–123
concrete	0.23–1.43
Al_2O_3 ceramic	3.0–5.3
SiC ceramic	3.4
PMMA polymer	0.8–1.75
polycarbonate polymer	2.75–3.3
cortical bone	2.2–6.3

sense from symmetry) and it varies as

$$\sigma = s\sqrt{\frac{a}{2r}}. \tag{4.90}$$

The stress increases nearer and nearer the crack (as r becomes small), but it does not become infinite at the crack as this expression would suggest. The distance dependence can be brought to the left-hand side to obtain the stress intensity $\sigma(2r)^{1/2} = sa^{1/2}$. A stress intensity factor K is commonly defined, which is fairly similar to this stress intensity with the same (stress)(distance)$^{1/2}$ units, but it is more general:

$$K = Cs\sqrt{\pi a}. \tag{4.91}$$

C is a dimensionless constant that depends on the size and shape of the crack and object, and how the stress is loaded. If K exceeds a critical value, the fracture toughness K_c, the crack will propagate; the larger K is above this value, the faster the crack will propagate. If it is smaller, it will not propagate. K_c is an intensive property of the material (Table 4.7).

Figure 4.69 shows three ways stresses can be applied to cause crack propagation, to the right in each picture. In Mode I the load is tensile (and here

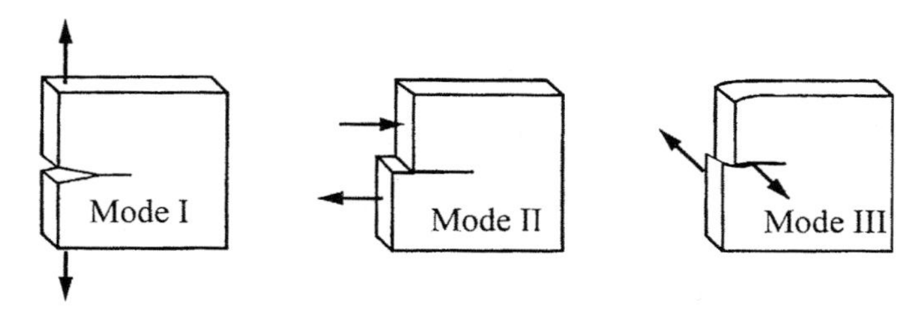

Fig. 4.69. Modes of cracking: I, the opening mode; II, the forward shear or sliding mode; III, the antiplane shear or tearing mode. (From [210])

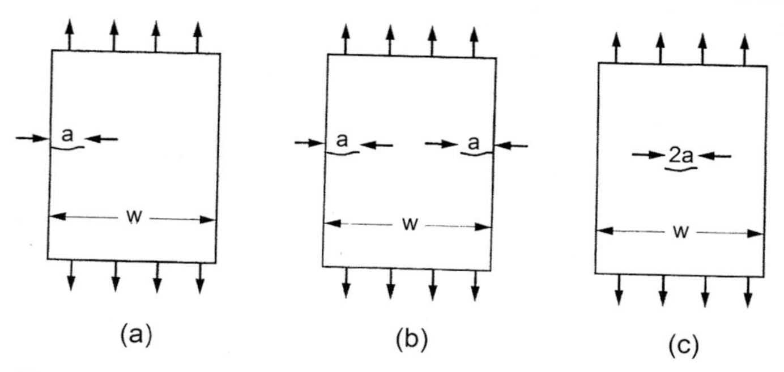

Fig. 4.70. Types of fracture for Mode I cracks, with (a)–(c) corresponding to (4.92)–(4.94), respectively. (Adapted from [210])

vertical) and perpendicular to the crack propagation direction, as in splitting a log lengthwise. In Mode II the load is shear, and parallel to the plane of the crack and the direction of crack propagation. In Mode III the shear load is perpendicular to the crack direction, as in tearing paper.

For Mode I cracks the constant C in (4.91) differs for different crack locations, such as those shown in Fig. 4.70. For a plate of width w under tension and a crack of length a on one edge, this constant is

$$C = \frac{0.752 + 2.02q/\pi + 0.37(1 - \sin(q/2))^3}{\cos(q/2)} \sqrt{\frac{2}{q} \tan(q/2)} \tag{4.92}$$

using $q = \pi a/w$.

For cracks of length a on both edges, it is

$$C = (1 + 0.122 \cos^4(q)) \sqrt{\frac{1}{q} \tan(q)}. \tag{4.93}$$

For a crack of length $2a$ in the center, it is

$$C = \left(1 - 0.10 \left(\frac{q}{\pi}\right)^2 + 0.96 \left(\frac{q}{\pi}\right)^4\right) \sqrt{\sec(q)}. \tag{4.94}$$

For cortical bone, $K_{\mathrm{c}} = 2.2$–$5.7\,\mathrm{MPa\text{-}m^{1/2}}$ for a Mode I transverse fracture of the tibia and a Mode I longitudinal fracture of the femur. It is 2.2–2.7 $\mathrm{MPa\text{-}m^{1/2}}$ for a Mode II fracture of the tibia. In each case the crack propagation is parallel to the long axis of the bone. Transverse propagation of cracks in long bones, perpendicular to the lamellar structure, causes the crack to turn along the long axis. Crack propagation in long bones is very anisotropic. The laminate structure of bones can stop or redirect a crack.

For a small crack in a large plate $q = \pi a/w \ll 1$, for which (4.92)–(4.94) each gives $C \sim 1$, and the critical condition is $K = K_{\mathrm{c}} \sim s(\pi a_{\mathrm{c}})^{1/2}$ or

$$a_{\mathrm{c}} \sim \frac{K_{\mathrm{c}}^2}{\pi s^2}. \tag{4.95}$$

As before, if the body weight of 700 N (70 kg) were distributed over the femur cross-sectional area 370 mm^2, the stress $s = 2.1\,\mathrm{N/mm^2}$. Because 1 MPa = 1 N/mm^2, a typical $K_c = 4\,\mathrm{MPa\text{-}m^{1/2}} = 4\,\mathrm{N/mm^2\text{-}m^{1/2}}$. This gives $a_c \sim 1$ m. Does it make sense that a crack in the femur would have to be 1 m long for the bone to spontaneously fracture – especially when we know the bone is shorter and much narrower than this? Yes, because we would expect (and hope) that normal people would not get stress fractures by standing on one leg. If a_c were very small ∼1 mm, then our bones would fracture with the slightest of flaws when we stood up. If the stress were 10× larger (corresponding to 10× body weight), the critical crack length would be 100× shorter or ∼1 cm. Also, note that our initial $a_c \sim 1$ m result violates the $q = \pi a/w \ll 1$ assumption we made that led to $C \sim 1$. We could have used the exact form(s) for C, but still would have obtained a large value for a_c.

The energy needed to break bonds in cracking comes from stored elastic energy. Cracks grow when the decrease in strain energy (from strain relief) $\mathrm{d}U/\mathrm{d}a$ (= G, *the strain energy release rate*) that occurs from the crack propagating a distance a exceeds or equals the energy or work $\mathrm{d}W$ needed to propagate the crack a distance a, which is $\mathrm{d}W/\mathrm{d}a$ (= *crack growth resistance* R). The strain energy release rate for Mode I cracks is [210]

$$G = \frac{\mathrm{d}U}{\mathrm{d}a} = \frac{\pi a s^2}{Y}. \tag{4.96}$$

for a crack of length a and stress s. Using (4.91),

$$G = \frac{K^2}{CY}. \tag{4.97}$$

There is much more understood about fracture, which we will not cover [201, 210, 230]. For example, the elastic model presented here ignores the plastic deformation that occurs very near the crack.

4.8 Common Sports Injuries

As we have seen, damage to bones, ligaments, muscles, etc. can result from collisions, excessive stress or strain, and from repeated use with moderately large stresses. These often lead to injuries in sports, including injuries to the following [204, 213]:

Head

1. *Concussions* are described in Chap. 3, and are common in boxing, football, and hockey. They also occur in baseball when pitchers successfully throw at batters' heads.

Shoulder

1. In a *separated shoulder* there is ligament damage that can occur from collisions in several sports. Ligament stretching is a first degree separation, a slight tear is a second degree injury, and a complete tear is a third degree injury.
2. In a *dislocated shoulder* the arm is out of the joint, which can result from collisions in several sports.
3. *Rotator cuff injuries* involve a strain or tear in the four muscles around the shoulder (supraspinatus, infraspinatus, subscapularis, and teres minor) that hold the humeral head into the scapula. They are not uncommon in activities requiring the arm to be moved over the head many times (leading to the overuse of the shoulder), as in baseball pitching, swimming, weightlifting, and racket sports.

Elbow

1. *Forehand tennis elbow (golfer's elbow, baseball elbow, suitcase elbow)* (*medial epicondylitis*) is due to forceful wrist flexion and pronation that can damage the tendons that attach to the medial epicondyle, and is common in tennis (when serving with topspin), pitching in baseball, and throwing a javelin.
2. *Backhand tennis elbow* (*lateral epicondylitis*) is caused when using the grasping and supination muscles. Damage occurs to the extensor tendons when the wrist is extended and to these muscles, such as during backhand returns in tennis.
3. A torn ulnar collateral ligament is common for baseball pitchers due to the overuse of the elbow. It is corrected by *Tommy John surgery* in which the torn ligament is replaced by the ulnar collateral ligament from the other elbow.

Hip

1. In a *hip flexor* there is damage to muscles around the hip.
2. In a *hip pointer* there is a bruise or fracture to the hip iliac crest (Figs. 2.14 and 2.15) and occurs in collisions in football and hockey.
3. *Avascular necrosis* is an injury due to collisions that results in a lack of blood supply to joint regions and their subsequent death. It is most common in the hips, but also is seen after collisions of the knees, shoulders, and ankles.

Legs

1. *Hamstring pulls* – as in: I pulled my "hammie" – are common in sports with much running, such as in track running and baseball, and occur in

simultaneous (eccentric) contractions of the quadriceps and hamstrings, when the hamstrings are $<60\%$ as strong as the quadriceps muscles.
2. *Shin splints* are muscle pulls, often found in running.

Knee

1. Increasingly common are sprains or tears to the *anterior cruciate ligament (ACL)* (Fig. 1.3). Injuries to the ACL are not uncommon in skiing, basketball, soccer, and football when the leg is contorted at the knee and this ligament is excessively elongated.
2. *Runner's knee* is pain behind and on either side of the kneecap (patella), due to the rubbing of the kneecap against the lateral condyle of the femur (cartilage), and can result from downhill running and walking downstairs. Soreness in the tendons above and below the knee, *patella tendinitis,* can occur from repetitive overloading due to jumping and running.

Foot

1. *Turf toe* is a bruise to the last joint in the toe, the metatarsal phalangeal joint. It can occur from jamming the toe into turf, as in football collisions.
2. *Metatarsal stress fractures* (in the toes) are common in running due to pushing off from the toes.
3. *Plantar fasciitis* is an injury to the plantar fascia under the arch of the foot, and is seen in long distance running, squash, tennis, and basketball. (The fascia are the surrounding soft tissues.)
4. Injuries to the *Achilles tendon* (Figs. 1.8 and 3.33), including tendinitis (inflammation) or tearing is common in many sports with repetitive overloading, as in speed running, squash, and tennis, due to excessive tendon elongation. We have seen that the stresses in this tendon during running are not that far below the UTS.

Spine and Back

1. In a *herniated disc* a vertebral disc (Fig. 2.36) is displaced and presses against nerves. *Lower back pain* can also result when muscles in the lower back become strained or when the ligaments interconnecting the lowermost five vertebral bones become sprained *(lumbar strain)*. Such injuries can result during weightlifting, moderate lifting using back muscles instead of leg muscles, and sitting or lying down in positions that do not permit your spine to assume its natural curvature.

Generally to Bones and Cartilage

1. *Stress fractures* of bones are slight fractures due to repeated stress, such as to the foot or shins after excessive running.

2. *In a compound fracture* the bone breaks through the skin, and this can occur from collisions in skiing and football.
3. *Fractured ribs* can result from collisions. (Strains or tearing of intercostal muscles between the ribs can result from awkward motions such as over-reaching.)

4.9 Avoiding Fractures and Other Injuries: Materials for Helmets

In this chapter we have discussed some of the consequences of collisions, i.e., bone breakage (and there are others, such as hematomas, etc.), and in Chap. 3 we briefly discussed how to lessen the effects of collisions by increasing the collision time and contact area. Helmets mitigate the effects of collisions of the head [219]. They consist of an outer shell and an interior liner. The shell transmits the impact load over the larger area of the liner, which absorbs most of the kinetic energy of the head.

To do this, the shell must be rigid (i.e., be very stiff to resist deformation), tough (i.e., have high bulk strength to limit fracture), and hard (i.e., have high surface strength to prevent penetrating injuries), so that a large area of impact of the head into the liner can be maintained, and it should be light. The shell is often made from fiber-reinforced plastics (fiberglass/resin composites) and thermoplastics (such as polycarbonate).

Liners must be capable of being compressed and absorbing energy at a force level low enough so the peak force and acceleration felt by the head are minimized and the collision time is maximized. Figure 4.71 shows three types of materials, with very different force/deformation (stress/strain) curves. Type A is a linear spring and type C is a more realistic material. Type B is an ideal helmet material because it deforms at a constant stress, which is low enough to be of value in collisions. Figure 4.72 shows three materials. The shaded area, which is the work done on the material, is the same under each curve.

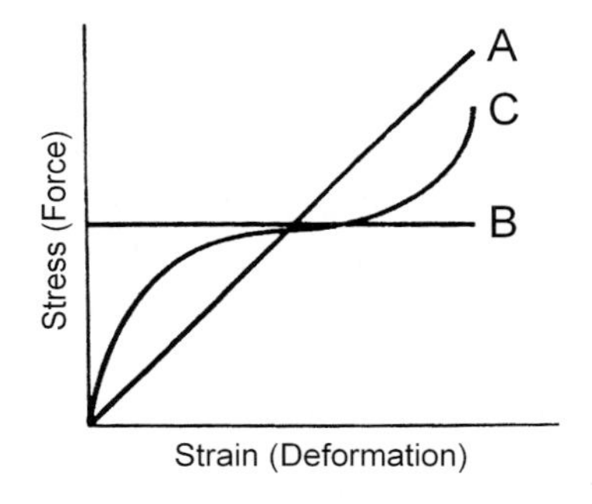

Fig. 4.71. Stress–strain curves for padding materials that could potentially be used in safety helmets. (From [219])

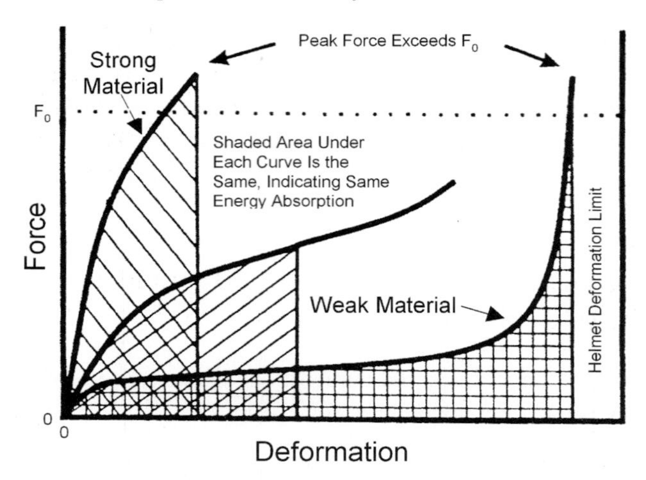

Fig. 4.72. Effect of the padding strength for the same energy absorbed. The peak force exceeds the maximum allowable force for the "strong" (or stiff) and "weak" (or compliant) materials. The weak material is crushed to just about its initial thickness and then becomes very stiff. (From [219])

For the stiff (large Y) and compliant (small Y) materials, the peak force is very high at the end of deformation, while for the intermediate material, it is much smaller. This makes it a better material for a liner. (Ideally, the force should be independent of deformation, as for material B in Fig. 4.71.) The stress should be relatively independent of the strain rate, so the liner would work well at high and low impact speeds. As shown in Fig. 4.73, the material

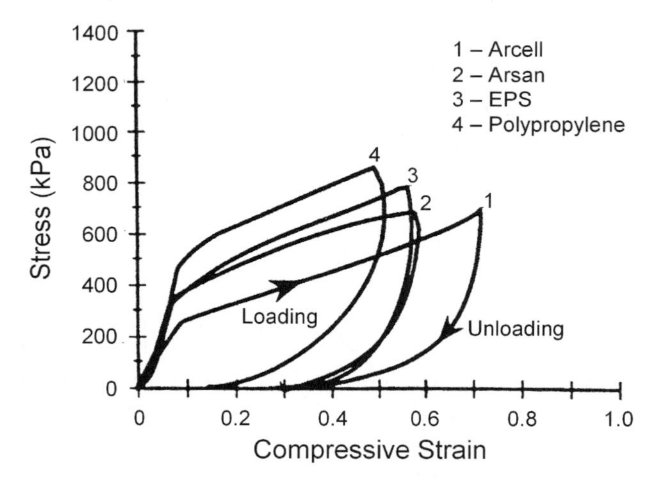

Fig. 4.73. Stress–strain for real padding materials: 1-Arcell, 2-Arsan, 3-EPS (expanded polystyrene), 4-Polypropylene. The energy absorbed is the area under the loading curve minus that under the unloading curve. (From [219])

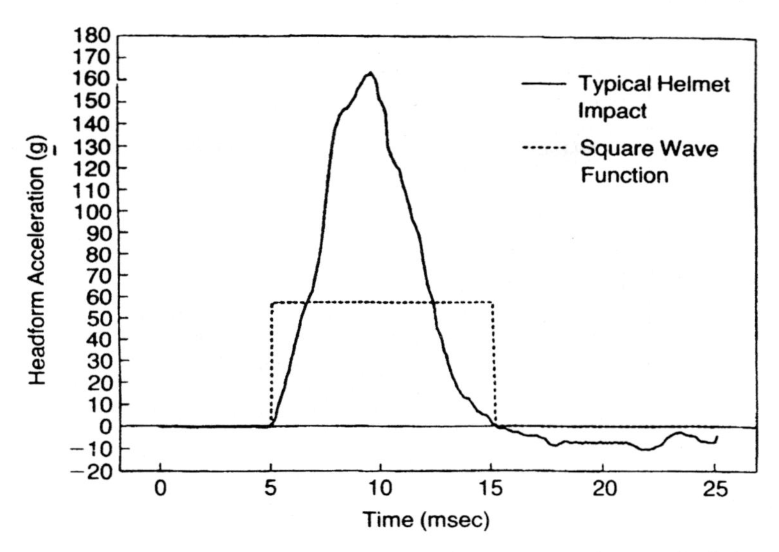

Fig. 4.74. Helmeted impact deceleration with initial speed of 5.63 m/s. (From [219])

should deform plastically and have a large stress–strain hysteresis loop. Then the liner material absorbs the energy of the head impact and does not transfer it back to the head (during the collision), as would a compliant, spring-like material. If the helmet is to be used over and over again, the deformation or strain remaining after a cycle should be minimal. The area of impact of the head on the padding should be maximized. The thickness of the padding should be increased as much as possible, subject to weight and bulkiness constraints, because of limitations on how much it can be compressed during the impact. Also, when the padding is fully crushed, it becomes very stiff, resulting in high forces. (The maximum designed compression is about 80%.) Energy absorbing liners are usually made of semirigid polyurethane foams or expanded polystyrene bead foams. (Spongy bone at the end of long bones should have similar properties.) The deceleration while wearing a good helmet is shown in Fig. 4.74.

4.10 Summary

Understanding the stress, strain, and fracture of body materials and parts is essential to explain the performance of the human machine under normal and extraordinary conditions. Time-independent material models, describing harmonic and non-harmonic elastic behavior, and time-dependent viscoelastic models can be used to characterize and understand the stress–strain relations of the body materials and components. The mechanical properties of the many parts of the body involved in structure, motion, and organ operation, are all very different; these properties depend on their composition, structure, and composite nature. The deformation of extended body parts, such as the bending of bones, and the mechanics of fracture are needed to analyze the body under extreme conditions that can lead to injury.

Problems

Stress and Strain

4.1. Determine the spring constant, k, in SI units for a solid cylinder of cortical bone of length 0.5 m, diameter 2 cm, and $Y = 17.4$ GPa.

4.2. An cylindrical spring of length 2 cm and diameter 3 mm has spring constant $k = 1.7 \times 10^5$ N/m.
(a) How much does it extend when a force of 100 N is applied to it?
(b) What is the strain?
(c) The spring is composed of a uniform material. Find its Young's modulus (in MPa).

4.3. Equation (4.8) shows how Young's modulus, the shear modulus, and Poisson's ratio are interrelated for an elastic isotropic material. The bulk modulus B is the negative of the pressure divided by the fractional change in volume caused by that pressure, and it can be related to any two of these three above parameters. Show that for isotropic harmonic materials $Y = 3B(1-2\upsilon)$. (When a stress σ is applied to compress a cylinder, the pressure is $\sigma/3$.)

4.4. Calculate the strain and change of length of the femur during a single step while running, using the data near (4.21).

4.5. As we will see in Chap. 11, we are able to see objects that are both near and far from us because of accommodation in the eye. This occurs because the shape of the crystalline lens in the eye changes when the force on it from the suspensory ligaments is changed. This is a fairly complex three-dimensional problem [192, 193, 194], which we will simplify. (It is actually a two-dimensional problem because of rotational symmetry, but it is still complex.)
(a) Use the simple one-dimensional model in this chapter to estimate how much stress is on the crystalline lens if it has a Young's modulus of 1×10^3 Pa (which is that for a 20 year old) and has a strain is 3%.
(b) If the total force on the lens is 0.002 N, determine the effective contact area (in mm^2).
(c) Determine the strain in the lens of someone who is 60 years old, if the stress is the same as in (a), but her Young's modulus has increased to 3×10^3 Pa.

4.6. Use the information in Table 4.2 to find the stress (in MPa) needed to stretch femoral compact bone, nails, nerves, skin, and coronary arteries to a strain of 0.01. (Assume harmonic behavior, with the low-strain value of Y.)

4.7. Use the information in Table 4.2 to find the strain resulting when femoral compact bone, nails, nerves, skin, and coronary arteries are subjected to a stress of 0.5 MPa. (Assume harmonic behavior, with the low-strain value of Y.)

4.8. Find the energy density for each case in Problem 4.6.

4.9. Find the energy density for each case in Problem 4.7.

4.10. Use the information for the fibula in Fig. 4.18 to:
(a) Calculate the maximum tension a bone with a cross-sectional area of 4 cm^2 could withstand just prior to fracture.
(b) Determine the elongation of a bone whose initial length is 0.35 m under the maximum tension from part (a).
(c) Calculate the stress on this bone if a tension force of 10^4 N were applied to it [186]. How much would this bone elongate?

4.11. Calculate the energy stored in parts (a) and (c) of Problem 4.10 for a bone that is 0.5 m long, always assuming that $\sigma = Y\epsilon$ (even until fracture).

4.12. Determine the relative amounts of strain energy absorption in the cartilage and bone of the proximal half of the tibia (Fig. 4.75) when it is loaded uniformly over the articular surface by a compressive force. Ignore the fibula and assume the tibia consists of three parts: a hollow cylindrical diaphyseal segment of cortical bone, a solid metaphyseal segment of cancellous (trabecular) bone, and a solid disk-shaped cartilage layer. The dimensions are $a = 10$ mm, $b = 30$ mm, $c = 50$ mm, $e = 4$ mm, $f = 70$ mm, and $g = 130$ mm. Assume each material is elastic, with a Young's modulus of 20,000 MPa for cortical bone,

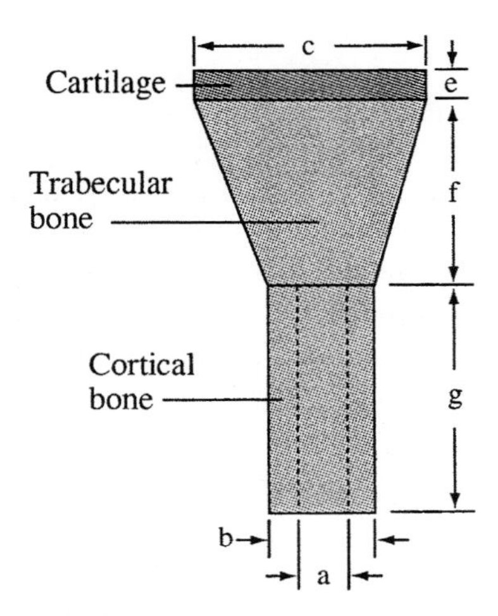

Fig. 4.75. Model for the proximal half of the human tibia. (From [210].) For Problem 4.12

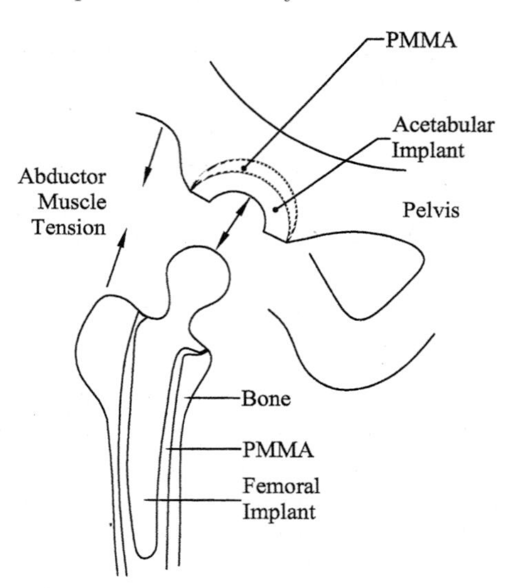

Fig. 4.76. Schematic of the components of a hip replacement. (From [203].) For Problem 4.14

200 MPa for cancellous bone, and either 20 MPa or 200 MPa for cartilage. Calculate the total strain energy in each segment of the model and the fraction of the total energy in each segment. Do this for both measured parameters for cartilage [210].

4.13. As a biomedical engineer you have been assigned to design a replacement for a femur (which consists of compact and trabecular bone). Using the materials (other than bone) listed in Table 4.1 (and other materials if you like), what materials would you use? (Would you want to use materials that match the properties of bones? Why?)

4.14. Figure 4.76 shows a diagram of parts of a hip replacement, with acetabular and femoral implants and PMMA (poly(methyl methacrylate)). Use the discussion in Chap. 2 to show that the lines of action of the loads on the femoral head and the long axis of the femur do not line up. Also, show that this means that there will bending and twisting moments on the implants, in addition to axial compression.

4.15. In the medieval torture device, the rack, the head is pulled apart from the feet.
(a) Is this compression or tension?
(b) Which body part is likely to break first?

4.16. Someone tells you that UPE $\times$ Y approximately equals the UTS for only a few organs and materials in Table 4.2, such as for femoral compact bone, but not for most of them. Explain why this is either true or false by using several specific examples. Why is this so?

4.17. Use the data in Fig. 4.19a to estimate a power-law relation between the Young's modulus for bone and its porosity.

4.18. Refer to Fig. 4.21.
(a) Why is it reasonable that the small intestine can stretch more and more easily in the transverse direction than the longitudinal direction?
(b) Is it reasonable that the small intestine can stretch significantly at low stress levels?
(c) How much should the radius of the small intestine be able to change under reasonable stresses and is the UPE large enough for this to occur?

4.19. (advanced problem) Two very simple models of composite material are shown in Fig. 4.77 [210]. In both cases there are two elastic materials, with respective Young's moduli Y_1 and Y_2 and volume fractions ρ_1 and ρ_2. In the Voigt composite model in (a) the materials are modeled as slabs in parallel so each material undergoes the same strain, while in the Reuss composite model in (b) the materials are modeled as slabs in series so each material bears the same stress. Show that the effective Young's modulus for the Voigt composite material is

$$Y_{\mathrm{c,Voigt}} = \rho_1 Y_1 + \rho_2 Y_2, \tag{4.98}$$

while for the Reuss composite model it is

$$Y_{\mathrm{c,Reuss}} = \frac{Y_1 Y_2}{(1-\rho_1)Y_1 + (1-\rho_2)Y_2}. \tag{4.99}$$

These models give the upper and lower limits to the Young's modulus of the actual composite material.

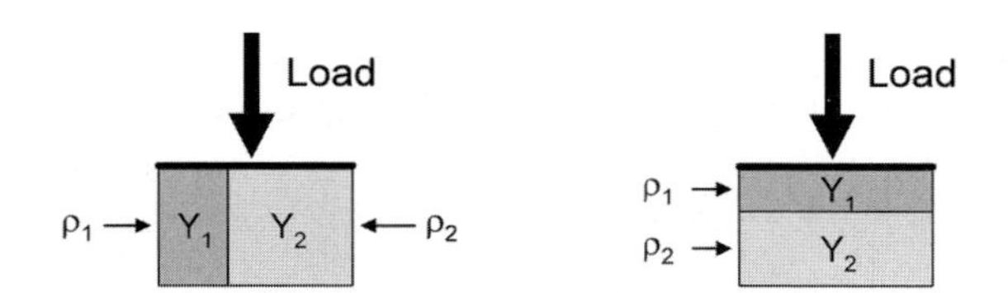

Fig. 4.77. (**a**) Voigt and (**b**) Reuss models of composite materials, with components of elastic modulus Y_1 and Y_2, with volume fractions ρ_1 and ρ_2. In the Voigt model both materials have the same strain, while in the Reuss model both materials are subjected to the same stress. Note that this Voigt model of composite materials is qualitatively different from the Voigt model used to model viscoelastic materials, which is simply called the Voigt model in this chapter. (Based on [210].) For Problem 4.19

4.20. In the text, strain, ϵ, was defined as the fractional deformation $(L - L_0)/L_0$, which equals $\lambda - 1$, where $\lambda = L/L_0$ is the Lagrangian strain (or the stretch ratio). This is quite common for small fractional deformations, so let us call it now $\epsilon_{\text{small}} = \lambda - 1$. More generally, for arbitrary fractional deformation, strain is defined as $\epsilon_{\text{general}} = \frac{1}{2}(\lambda^2 - 1)$. This is often called *finite strain* (and sometimes Green's strain) and is sometimes labeled as E.
(a) Calculate λ and compare ϵ_{small} and $\epsilon_{\text{general}}$ for $L = 2\,\text{cm}$ and $L_0 = 1\,\text{cm}$.
(b) Express $\epsilon_{\text{general}}$ in terms of ϵ_{small}, and vice versa.
(c) Show that for very small $\lambda - 1$, the finite strain, $\epsilon_{\text{general}}$, approaches the small-strain approximation, ϵ_{small}. Specifically compare them for $L = 1.01\,\text{cm}$ and $L_0 = 1.00\,\text{cm}$.
(d) What is the largest value of $\lambda - 1$ for which ϵ_{small} is within 10% of $\epsilon_{\text{general}}$?
(e) Say that the stress–strain relation can be written as $\sigma = Y\epsilon_{\text{general}}$. Find the relationship between σ and ϵ_{small} and sketch it. Compare it to the linear and exponential relations between σ and ϵ_{small}.

4.21. Another way used to define strain in general (Almansi's strain) is $e = \frac{1}{2}(1 - 1/\lambda^2)$. Repeat Problem 4.20, replacing $\epsilon_{\text{general}}$ by e.

4.22. Yet another way used to define strain in general ("true" strain) is as $\ln \lambda$. Repeat Problem 4.20, replacing $\epsilon_{\text{general}}$ by $\ln \lambda$.

Viscoelasticity

4.23. (a) The faster you try to open or close a screen door, the more resisting force you encounter. Does that mean you should use include a dashpot in the mechanical model of the door closer unit? Why?
(b) When the door is open and released, it returns to its initial position. Does that mean you should include a spring in the mechanical model of the door closer? Why?
(c) When the door is open and released, the length of the closer decreases linearly with time until it returns to its initial length. Do any of the three viscoelastic models (Maxwell, Voigt, Kelvin) model this behavior? If so, which? If not, which one comes closest?

4.24. A dashpot of length 3 cm is characterized by the constant $c = 2 \times 10^4\,\text{N-s/m}$. A constant force of 10 N is applied to it. Find its length and dx/dt after the force has been applied for 2 s.

4.25. Say that the constant c of a dashpot varies linearly with its cross-sectional area. Repeat Problem 4.24 for a dashpot that is smaller in all dimensions by a factor of 2.

4.26. Consider a function $f(t)$ that is 0 before time $-T/2$, then increases linearly in time until it becomes 1 at time $T/2$, and remains at 1 thereafter.

(a) Sketch it.
(b) What function does it become as T approaches 0?
(c) For arbitrary T, find and plot $\mathrm{d}f/\mathrm{d}t$.
(d) What does this derivative become as T approaches 0?

4.27. The text says that the time derivative of the step function $\mathrm{d}\theta(t)/\mathrm{d}t$ is the Dirac delta function $\delta(t)$. Show that the integral over the Dirac delta function is the step function.

4.28. Use the step function to compose the following functions:
(a) A function that is 0 for $t < -1$ s, 2 from $t = -1$ to 3 s, and 0 thereafter, which is a type of square pulse (Fig. 4.45b).
(b) A function that is 0 for $t < 0$ s, 1 from $t = 0$ to 3 s, -2 from $t = 3$ to 4 s, and 0 thereafter.
(c) A function that is 0 for $t < 0$ s, 1 from $t = 0$ to 1 s, 0 from $t = 1$ to 2 s, 1 from $t = 2$ to 3 s, and so on, which is a type of square wave (Fig. 4.45c).

4.29. Verify the solutions of the Maxwell model for applied force (4.55) and then for applied deformation (4.56), by inserting them into (4.52) – and verifying that you get an equality – (for $t > 0$ where the step function $\theta(t)$ can be replaced by 1 (unity)) and by checking initial conditions (i.e., solutions for $t = 0$).

4.30. Show that you obtain the solutions of the Maxwell model ((4.55) and (4.56)) by integrating (4.52) from early times to time t for the sudden application of force F_0 as described in the text. Hint: This involves the integral

$$x(t) = \int_{-\infty}^{t} \left(\frac{F(t')}{c} + \frac{\mathrm{d}F(t')/\mathrm{d}t'}{k} \right) \mathrm{d}t'. \tag{4.100}$$

4.31. Verify the solutions of the Voigt model for applied force (4.58) and then for applied deformation (4.59), by inserting them into (4.57) – and verifying that you get an equality – (for $t > 0$ where the step function $\theta(t)$ can be replaced by 1 (unity)) and by checking initial conditions (i.e., solutions for $t = 0$).

4.32. Verify the solutions of the Kelvin/standard linear solid model for applied force (4.69) and then for applied deformation (4.70), by inserting them into (4.68) – and verifying that you get an equality – (for $t > 0$ where the step function $\theta(t)$ can be replaced by 1 (unity)) and by checking initial conditions (i.e., solutions for $t = 0$).

4.33. What happens to the Kelvin model when either k_1 goes to 0 or k_2 goes to ∞, or when both occur? Why?

4.34. Show that the solution for a force $F(t) = F_0(\theta(t) - \theta(t-T))$ applied to a Maxwell model material is $x(t) = F_0(1/k + t/c)\theta(t) - F_0(1/k + (t-T)/c)\theta(t-T)$.

4.35. (a) Verify (4.75) by substitution into the Kelvin constitutive equation (4.68).
(b) Verify (4.77) – evaluated at the end of the force ramp. (Hint: Use the fact that $\exp(-x)$ is approximately $1 - x$ when $x \ll 1$. Remember that $\exp(-x)$ approaches 0 as x gets very large.)
(c) Now explain why this Kelvin solid model successfully describes the cited viscoelastic properties of bone (at least qualitatively). Carefully sketch $x(T)$ as a function of time from $t = 0$ to $T =$ many times τ.

4.36. Consider the deformation x resulting from a force that linearly increases from 0 to F_0 from $t = 0$ to T, as described by (4.75) and (4.77). The force is then maintained at F_0 for $t > T$. Use the Kelvin model to find x for $t > T$.

4.37. Repeat Problem 4.36 if instead at $t = T$ the force is very suddenly decreased to 0 and is maintained at that value.

4.38. Apply a deformation x_0 in a linearly increasing manner over a time T ($x = x_0(t/T)$ from $t = 0$ to $t = T$) to a material described by the Kelvin standard linear model. Determine the stress relaxation $F(t)$ from $t = 0$ to T and obtain the deformation when the total deformation x_0 has been applied, $F(T)$. Examine how $F(T)$ depends on T, and explain why this qualitatively agrees or disagrees with observations for bone in Fig. 4.46 with increasing strain rates.

4.39. (advanced problem) One way to model the phase during running when a foot is in contact with the ground is with a one dimensional model with the force of a mass on a muscle (which is modeled as a passive material – a Voigt material with elastic and viscous elements in parallel) acting on the track (which is modeled as an elastic element in series with the muscle). (The mass is that of the body minus that of the leg in contact with the ground. The force on the mass includes the effects of gravity and the downward acceleration.) This is mathematically equivalent to the Kelvin model of a material. Show that there is a range of stiffnesses for which the period of this damped oscillation is less than for a very hard track (very high stiffness, very low elasticity). (Is this consistent with the data in Table 3.6?) Because the running speed is thought to be inversely proportional to the time the foot is in contact with the track and the foot would be in contact for about a half of an oscillation period, this would mean that people could run faster on a track with the right elasticity – and this has been demonstrated.

Breaking

4.40. The goal is to find how much a tibia of length L bends before it breaks when a force F is applied to the center of it, as shown in Fig. 4.35.
(a) Show that the vertical deflection for a beam is $L^2/8R$, where R is the radius of curvature. (Define the vertical deflection as the maximum difference

in vertical position of the neutral axis, laterally across the beam. Assume that $R \gg L$.)
(b) Show that this maximum vertical deflection is $L^2(\mathrm{UBS})/4Yd$, by using (4.81). d is the thickness, which can be taken as the tibia diameter here.
(c) If the tibia is 440 mm long and its diameter is 20 mm, find this deflection. (Use the data in the tables such as Table 4.6.)
(d) Is your answer reasonable to you? Why?
(e) Quantitatively, how does this deflection compare to how much the tibia deforms laterally in length (i) at the top (compression) and bottom (tension) of the bone at the UBS (where the magnitude of the relevant strain is $d/2R_{\mathrm{min}}$) and (ii) before fracture in ordinary compression or tension?

4.41. Based on the discussions of bone breaking, below what porosity level (the fraction of maximum bone density) will bones break fairly easily.

4.42. (a) Estimate the critical load for Euler buckling of an adult femur. Assume the femur is solid.
(b) Do forces on the femur during exercise ever reach the levels needed for buckling?
(c) If your answer to (b) was no, how small would Y have to be for buckling to occur during exercise?
(d) If Y were proportional to porosity, how porous would the bone in the femur need to be, to be concerned about buckling?

4.43. Find the critical load to buckle a hair (on your head) (Euler buckling equation, (4.88)). Use Table 4.2 for Y, $L = 3$ mm, and thickness $= 0.02$ mm.

4.44. Replot (i.e., carefully sketch) Fig. 4.67 on linear–linear axes.

4.45. Use Fig. 4.67 to determine relationships for the number of cycles needed to create stress fractures in cortical bone for a given strain range, for both compressive and tensile loading. Also express these relations as the number of miles to failure as a function of the strain range.

5

Muscles

In this chapter we will discuss the motors of the body, which are the muscles. There are three types of muscles in the body (Fig. 5.1): (1) *Skeletal muscles* are involved in skeletal motion, lip motion, and in eyelid and eyeball motion. They consist of long cylindrical cells that appear striated, have many nuclei, and are under conscious control (*voluntary muscles*). Although skeletal muscles can be consciously controlled, some move without conscious effort, as in the diaphragm. (2) *Cardiac muscle* comprises most of the heart. These muscle cells are striated, have one centrally located nucleus, and often branch; when two cardiac cells meet they form an intercalated disc. (3) *Smooth muscles* line the walls of blood vessels and the digestive and urogenital organs, to help advance and control flow (see Figs. 7.21 and 8.14). They are elongated cells with tapered ends and no striation, have a single, centrally located nucleus, and are not under conscious control (*involuntary muscles*), as is also true for cardiac muscle. We will concentrate on skeletal muscles; many of the properties of cardiac and smooth muscles are similar. Some differences between cardiac and skeletal muscles are described in Chap. 8.

We will learn how muscles work and develop macroscopic, microscopic, and nanoscopic models of them [271, 272, 283]. We will also re-examine some of the assumptions we made in earlier chapters involving muscle forces. In the next chapter we will learn more about how energy is used in muscles.

5.1 Skeletal Muscles in the Body

Figure 1.8 shows many of the main skeletal muscles in the body.

Skeletal muscles account for ~43% of the typical body mass, or ~30 kg of a 70 kg person. At rest, they use ~18% of the body energy consumption rate (which is called the basal metabolic rate or BMR); they use much more during activities, such as motion. Of the total energy "burned" by skeletal muscle, only ~25% is used for work, and this is the muscular efficiency. The other

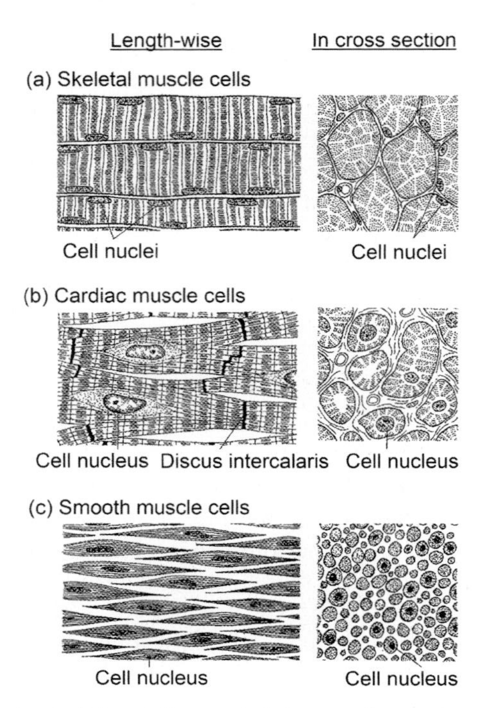

Fig. 5.1. **(a)** Skeletal, **(b)** cardiac, and **(c)** smooth muscle cells. (From [280])

$\sim$75% is released as heat. (This inefficiency of the body does have a positive purpose – it is a consistent and very important source of body heat.)

The maximum muscle force F_{M} or tension T that a muscle can develop is k_{M}PCA, where PCA is the physiological cross-sectional area of the muscle. The range of the maximum values of k_{M} is 20–100 N/cm^2 during isometric (constant length) conditions; the larger values are for muscles with the pinnate fiber structure that is described below. (See Fig. 5.38 and Problem 5.2.) In the quadriceps, the forces are up to $\sim$70 N/cm^2 during running and jumping and $\sim$100 N/cm^2 under isometric conditions. We will discuss this more later.

Many muscles that cross the hip, knee, and ankle joints are important in locomotion. The relative significance of these leg muscles depends on the locations of their points of origin and insertion, lengths, and their PCAs. Many of these muscles are seen in Figs. 3.2–3.4. (Analogous sets of muscles are involved in controlling our arms and hands, as seen in Figs. 5.26 and 5.27.) The relative PCAs are given in Tables 5.1–5.3. In general, longer muscles enable larger angles of rotation about joints. The larger the muscle PCA, the more the muscle strength. Muscles with smaller PCAs can be important in providing stability.

As is clear from Fig. 1.8 and Tables 5.1–5.3, many of these leg muscles pass over more than one joint. Figure 5.2 shows the three such major biarticulate

Table 5.1. Percent PCA of muscles crossing the hip joint. (From [292], data from [289])

muscle	%PCA
iliopsoas	9
sartorius	1
pectineus	1
rectus femoris	7
gluteus maximus	16
gluteus medius	12
gluteus minimus	6
adductor magnus	11
adductor longus	3
adductor brevis	3
tensor fasciae latae	1
biceps femoris (long)	6
semitendinosus	3
semimembranosus	8
piriformis	2
lateral rotators	13

(two joint) muscles in the leg. The forces exerted by the muscles on the bones at the points of origin and insertion are the same (Newton's Third Law), but the torques are different at the proximal and distal joints because the moment arms are different. As seen in Fig. 5.2, the gastrocnemius is a knee flexor and ankle plantarflexor (extensor), with the torque about the latter joint greater because of its larger moment arm (5 cm vs. 3.5 cm). During stance the net effect of this muscle is to cause the leg to rotate posteriorly and to prevent

Table 5.2. Percent PCA of muscles crossing the knee joint. (From [292], data from [289])

muscle	%PCA
gastrocnemius	19
biceps femoris (small)	3
biceps femoris (long)	7
semitendinosus	3
semimembranosus	10
vastus lateralis	20
vastus medialis	15
vastus intermedius	13
rectus femoris	8
sartorius	1
gracilis	1

Table 5.3. Percent PCA of muscles crossing the ankle joint. (From [292], data from [289])

muscle	%PCA
soleus	41
gastrocnemius	22
flexor hallucis longus	6
flexor digitorum longus	3
tibialis posterior	10
peroneus brevis	9
tiabialis anterior	5
extensor digitorum longus	3
extensor hallucis longus	1

the knee from collapsing. The hamstrings are extensors of the hip and flexors of the knee, with hip extension having twice the moment of knee flexion. During stance, this causes the thigh to rotate posteriorly and prevents the knee from collapsing. The rectus femoris of the quadriceps is a hip flexor and knee extensor, with a slightly larger moment about the hip. However, the major action of the quadriceps is knee extension because 84% of the quadriceps PCA is from other muscles, the uniarticulate knee extensors (and all of these muscles fire simultaneously). The net effect of these three major biarticulate

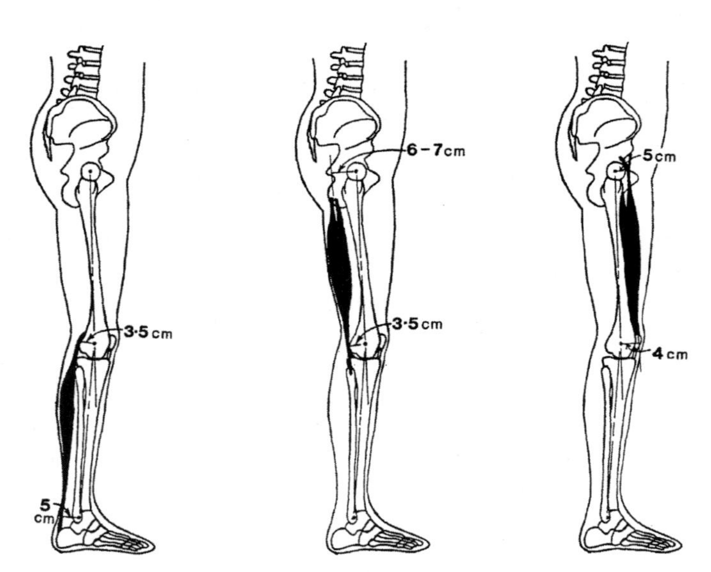

Fig. 5.2. Three major biarticulate muscles of the leg, from left to right the gastrocnemius, hamstrings, and rectus femoris, along with moment-arm lengths about the joints at their proximal and distal ends. (From [292]. Reprinted with permission of Wiley)

muscles is the extension of all three leg joints, and they help prevent against collapse due to gravity. Coordinated motion of these muscles, such as the lengthening of one of these muscles and shortening of another, is needed to achieve the full range of rotation of these joints because of the limited change of muscle length of any one of them [292].

5.1.1 Types of Muscle Activity

Muscles can be activated under a wide range of conditions. When the muscle length changes, the angle of the joint for this inserted muscle changes, and there is motion at the end of the bone emanating from the joint. Because there is motion, mechanical work is done. This can be called "dynamic" work. When the muscle length does not change, i.e., *isometric* conditions, there is no rotation of the joint and no mechanical work is done. This is important because tension is still supplied by the muscle to resist outside forces, as needed in holding objects or standing upright. Energy is still expended by the muscle (for such "static" work) to produce the tension that resists the load.

Mechanical work is performed with joint motion for nonisometric contractions. In *concentric* contractions, the muscle develops enough tension to overcome the load and the muscle length shortens, causing joint movement. In ascending stairs the quadriceps contract and the leg straightens as a result of this concentric contraction. In *eccentric* contractions, the muscle does not develop enough tension to overcome the load and the muscle length still lengthens, sometimes slowing joint movement (on purpose). In descending stairs the quadriceps fire but still extend during this essential controlled (eccentric) braking of knee flexion in fighting gravity. Therefore, the same flexor muscles that contract concentrically during flexion can contract eccentrically during extension to decelerate the extension. Concentric contractions are said to do *positive (mechanical) work*, while eccentric contractions do *negative work*.

During *isokinetic* contraction, the velocity of muscle shortening or lengthening, and consequently also the angular speed of the joint, are constant. During *isoinertial* contraction, the resistive load on the muscle – due to the gravity force, applied forces, etc. – is constant. During *isotonic* contraction, the tension is constant. (This is an idealized condition because the muscle tension changes with length.)

In elbow flexion the biceps brachii and brachialis muscles contract and they are the *agonists* or prime movers of the action. The brachioradialis is a *synergist* muscle in the motion; such synergist muscles assist the motion and sometimes add fine tuning. The triceps brachii are the *antagonists* of this action, and oppose the prime movers.

During walking and running the quadriceps act eccentrically during early stance to prevent the collapse of the knee angle, and then concentrically to extend the knee as the leg rotates over the foot in midstance (and this is more so in walking). The ankle plantarflexors fire eccentrically during stance to help

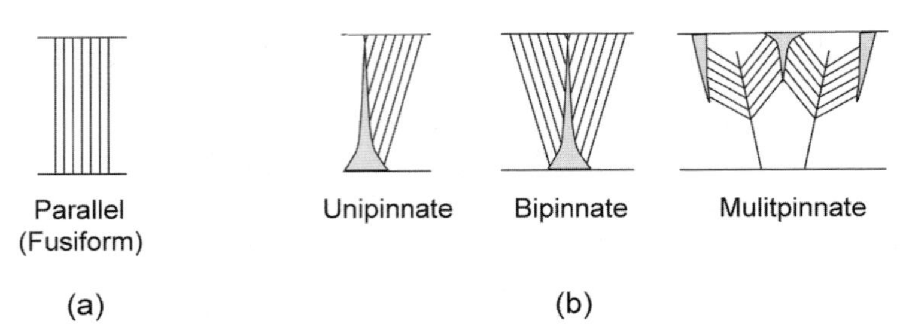

Fig. 5.3. Types of fiber arrangements in skeletal muscles, with **(a)** parallel and **(b)** pinnate fibers. (Based on [271])

advance the leg rotation over the foot controllably, and then concentrically to assist push off.

5.2 The Structure of Muscles

Muscles have a fiber structure with successive levels of fiber-like substructures called fasciculi (fuh-sik'-you-lie) (a single one is a fasciculus (fuh-sik'-you-lus)). On a macroscopic basis, these fibers are arranged in one of several ways. This is illustrated in Fig. 5.3.

In *fusiform* muscles the muscle fibers are parallel and they narrow and blend into tendons that attach to the skeleton. In *parallel* muscles the fibers are also parallel. They tend to be long, such as the sartorius – the longest muscle in the body, which spans from the hip to the tibia (Figs. 1.8 and 5.4). (The fibers are also parallel in the rectus abdominis (the "abs").) We will usually not differentiate between fusiform and parallel muscles because of their similar arrangement of muscle fibers. These muscles consist of many sarcomere components in series (see below), and as such they can become much shorter to produce much movement of bones for motion about joints, and they can do so quickly. In *unipinnate* muscles, parallel muscle fibers attach to tendons at an angle, such as with the flexor policis longus (poe-lee'-cis) in the lower hand/thumb and the extensor digitorium longus (di-gi-tor'-ee-um) in the lower leg/foot. In *bipinnate* muscles, such parallel muscle fibers attach to a tendon in two different directions, such as with the rectus femoris (fe-more'-is) in the thigh. These *pinnate* (or pennate or pinnation) structures resemble the structure of feathers and of some leaves on branches. The fibers attach on the central tendon at several angles in the *multipinnate* deltoid (shoulder) muscles. Many short muscle fibers can attach to such short tendons in pinnate muscles, leading to larger forces than for parallel muscles and more movement, but with less efficient use of the muscles forces (see below and Problem 5.18). The orbicularis oculi (or-bee-queue-lar'-is ok'-you-lie) about the eye are *circular* muscles, in which the muscle fibers are in a circle about the object. The pectoralis major (pec-tor-al'-is) in the upper chest (the "pects")

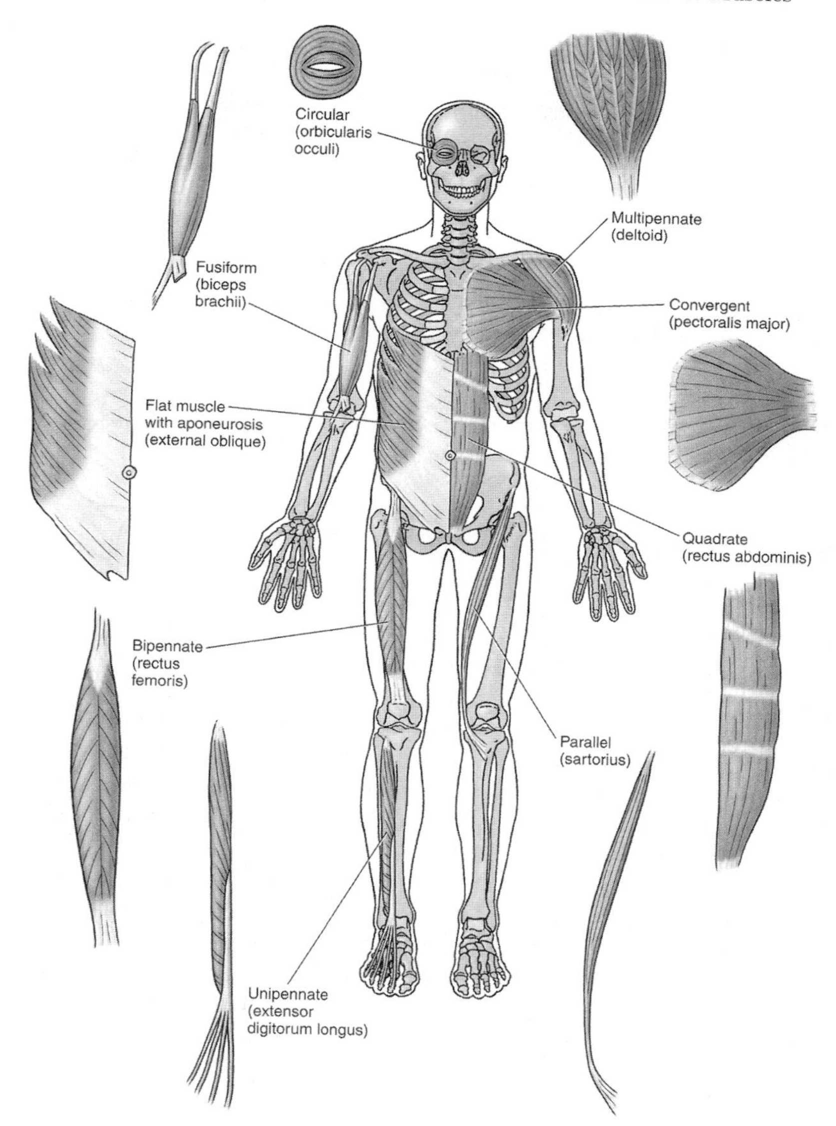

Fig. 5.4. Examples of muscle fiber arrangements in the body. (From [275]. Used with permission)

is a *convergent* muscle; such muscles are broad at their origin and narrow at insertion, leading to large forces near the insertion point. Examples of each of these muscle types in the body are shown in Fig. 5.4.

The way the forces from muscle fibers add to give the total force on the attaching tendon differs for these different muscle structures. In parallel,

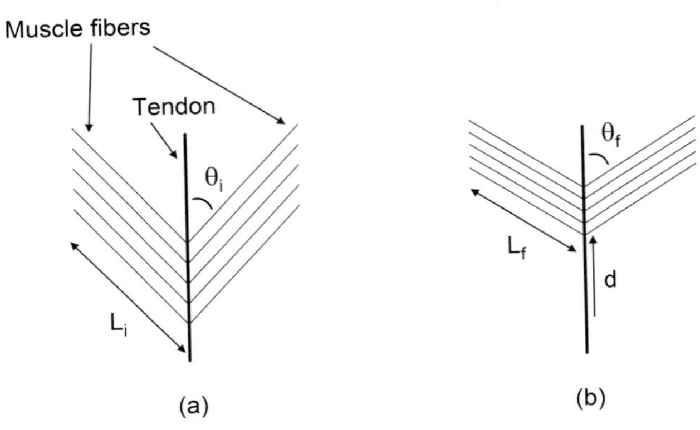

Fig. 5.5. Muscle fibers in pinnate muscles **(a)** before and **(b)** after fiber contraction with accompanying tendon movement

fusiform muscles, all the force of the fibers is transmitted to the tendon. In pinnate muscles, the fibers are attached to the tendon at an angle θ, and only $F\cos\theta$ of the force $\mathbf{F}$ of each fiber is effectively transmitted. While this is a distinct disadvantage of the pinnate design, it has other relative advantages. Because the geometry allows fibers to attach along part of the length of the tendon (Fig. 5.5), many more fibers can be attached to the tendon. Also, this geometry allows the central tendon to move a longer distance than in the fusiform scheme, so the bones attached to the tendon can move more. This overcomes the limited change in length in the muscles that can limit the range of angular motion of bones about joints. These and related issues are addressed in Problem 5.18. The mass, fiber length, PCA, and pinnation angle θ for several muscles are given in Table 5.4.

Table 5.4. Properties of some muscles. (From [292], data from [289])

muscle	mass (g)	fiber length (cm)	PCA (cm^2)	pinnation angle (°)
sartorius	75	38	1.9	0
biceps femoris (long)	150	9	15.8	0
semitendinosus	75	16	4.4	0
soleus	215	3.0	58	30
gastrocnemius	158	4.8	30	15
tibialis posterior	55	2.4	21	15
tibialis anterior	70	7.3	9.1	5
rectus femoris	90	6.8	12.5	5
vastus lateralis	210	6.7	30	5
vastus medialis	200	7.2	26	5
vastus intermedius	180	6.8	25	5

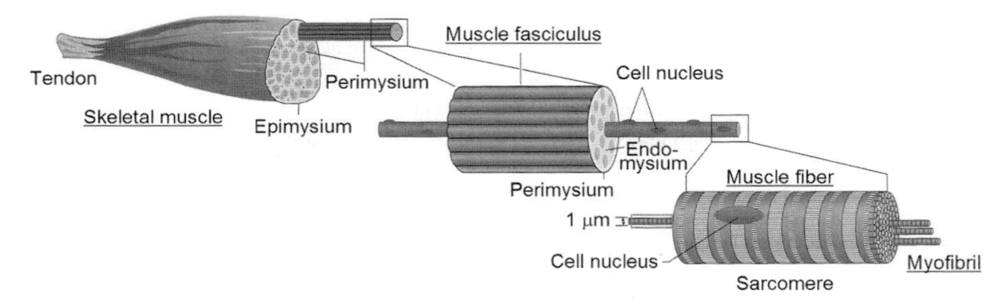

Fig. 5.6. Organization of skeletal muscles, down to the myofibril level. See Fig. 5.7 for the structure of the myofibril. (From [281])

The microscopic components of the fibers are illustrated in Figs. 5.6 and 5.7, where the muscle belly of a fusiform muscle is seen to be composed of many parallel *fasciculi.* Each fasiculus is composed of many parallel *muscle fibers* (or *muscle cells*). Each muscle fiber is composed of many parallel *myofibrils.* Each myofibril is composed of *myofilaments* arranged into ∼2–3 μm long units called *sarcomeres* (sar'-koe-meres). There are thick myofilaments with a serial arrangement of many *myosin* (my'-oh-sin) molecules and thin myofilaments with globules of *F-actin* (or *actin*) molecules that form twisting strands, that are surrounded by two other proteins: *tropomyosin* – which forms strands that twist about the actin strands – and *troponin-T* – which attaches at regular intervals to the actin and tropomyosin strands. The

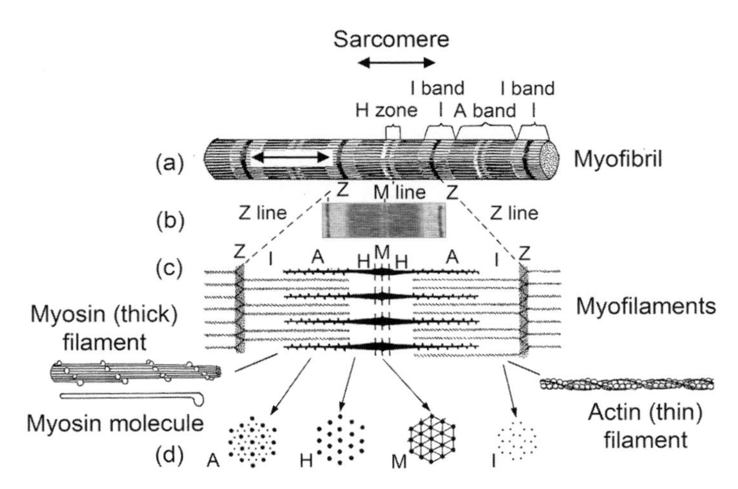

Fig. 5.7. **(a)** Structure and substructure of the myofibril shown in Fig. 5.6, with the banded structure of sarcomeres (whose periodicity is denoted by the arrows). **(b)** The scanning electron micrograph of skeletal muscle shows the structure of bands and lines in sarcomeres. This is associated with the thick and thin filaments in **(c)** from the longitudinal perspective of **(a)** and **(b)**, the transverse cross-section is seen in **(d)**. (From [280])

Table 5.5. Number of muscle fibers in human muscles. (Using data from [272])

muscle	number of muscle fibers
first lumbrical	10,250
external rectus	27,000
platysma	27,000
first dorsal interosseous	40,500
sartorius	128,150
brachioradialis	129,200
tiabialis anterior	271,350
medial gastrocnemius	1,033,000

interaction between the myosin and actin proteins on adjacent myofilaments is the fundamental, chemical-induced, force-producing interaction in the muscle, which involves the hydrolysis of ATP and the change in conformation of the myosin.

In the large limb of an adult the muscle fibers are ~50 μm in diameter. This diameter can double with weight training. Depending on the type of muscle, there are $\sim 10^4$–10^7 muscle fibers per muscle (Table 5.5). There are $\sim 1 \times 10^4$–1.7×10^5 sarcomeres per muscle fiber, depending on the type of muscle (Table 5.6).

The electron micrograph in Fig. 5.7 shows the banded myofilament structure that is also sketched in the figure. The A band is a wide dark, anisotropic region, while the I band is a wide light, isotropic region. There is a lighter H zone or band in the middle of the dark A band, and a darker M line in the middle of the H zone. The dark Z lines run through the light I bands.

Each sarcomere is bound between the adjacent Z lines. The dark A band in the center of each sarcomere consists of thick (myosin) myofilaments (or thick filaments), which are connected to each other in the central M line within the H zone. They are overlapped to some degree by the thin (actin) myofilaments (or thin filaments). There are thick filaments but no thin filaments in the

Table 5.6. Number of sarcomeres in human muscles. (Using data from [272])

muscle	number of sarcomeres per fiber ($\times 10^4$)		
	person I	person II	person III
tibialis posterior	1.1	1.5	0.8
soleus	1.4	–	–
medial gastrocnemius	1.6	1.5	1.5
semitendinosus	5.8	6.6	–
gracilis	8.1	9.3	8.4
sartorius	15.3	17.4	13.5

central H zone. The light I band regions next to the Z lines are the thin myofilaments in regions where they do not overlap the thick myofilaments; they are bound to each other at the Z line. This is shown schematically in Fig. 5.7.

5.3 Passive Muscles

A passive or resting muscle with no electrical stimulation has mechanical properties that need to be understood before exploring the active state of the muscle because they affect muscle performance during activation. Experimentally it is found that passive muscles are non-Hookean, as in (4.22). Equation (4.27) showed that for larger strains, stress and strain are related by

$$\sigma = \mu' \exp(\alpha L/L_0) - \mu, \tag{5.1}$$

with $\lambda = L/L_0 = \epsilon + 1$ being the Lagrangian strain, where L is the muscle length and L_0 is the relaxed length.

Using this, the force across a passive muscle is related to stress by

$$F_{\mathrm{M}} = \sigma(\mathrm{PCA}). \tag{5.2}$$

Consequently, the force needed to maintain a passive muscle at a length L is

$$F_{\mathrm{M}} = (\mathrm{PCA})\mu' \exp(\alpha L/L_0) - (\mathrm{PCA})\mu. \tag{5.3}$$

For large strains, more general relations need to be used for this neo-Hookean regime and for defining strains. Chapter 4 and Problems 4.20–4.22 explain this further.

5.4 Activating Muscles: Macroscopic View

Electrical stimuli lead to twitches in the muscles that temporarily increase the force exerted by them. These twitches are delayed by about 15 ms after the electrical stimulus. They peak ~40 ms later, and then decay to zero ~50 ms later (Fig. 5.8a, Table 5.7). The shape of a twitch can be modeled as

$$F(t) = F_0 \frac{t}{T} \exp(-t/T) \tag{5.4}$$

with twitch time T.

By increasing the frequency of these stimuli and consequently of the twitches, there is an increase in the force exerted by the muscle (Fig. 5.8 b,c). At a large enough frequency the twitches overlap to produce an almost steady level of force called *unfused tetanus,* and at an even higher frequency they produce a force that is constant in time called *tetanus* (Fig. 5.9). Roughly

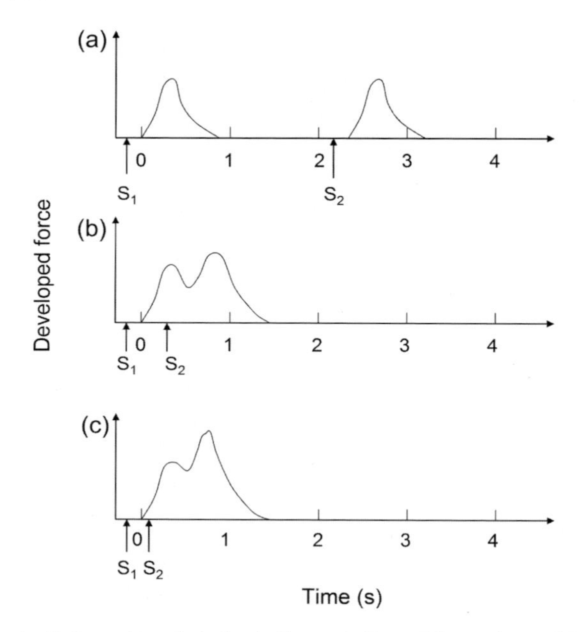

Fig. 5.8. Two twitches stimulated at S_1 and S_2, with various delays. (Based on [271])

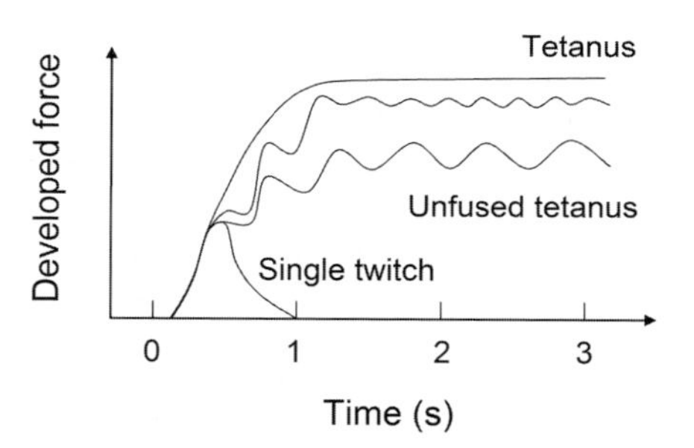

Fig. 5.9. A twitch, and then a series of twitches leading to unfused tetanus, and at an even higher frequency leading to tetanus. (Based on [271])

Table 5.7. Twitch time T, in ms. (Using data from [292])

muscle	typical mean time	range of times
triceps brachii	44.5	16–68
biceps brachii	52.0	16–85
tibialis anterior	58.0	38–80
soleus	74.0	52–100
medial gastrocnemius	79.0	40–110

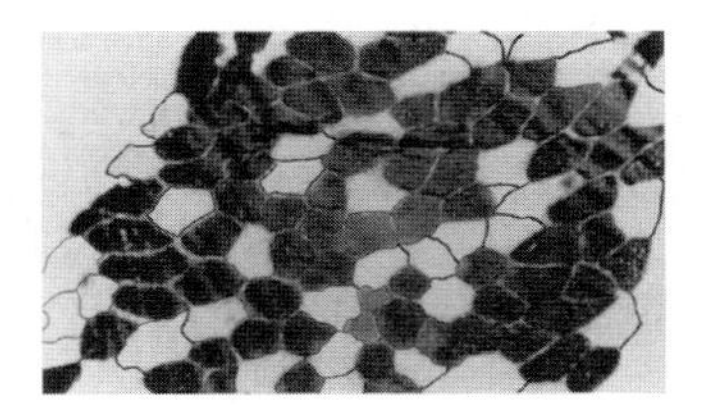

Fig. 5.10. Photomicrograph of stained muscle cells from a human vastus lateralis muscle, showing ST Type I (*dark*) and FT Type II (*lightly stained*) cells, with *fine lines* showing boundaries added. (From [250]. Used with permission)

50–60 electrical shocks per second are required to fully tetanize mammalian muscles at room temperature. This varies from about ∼30/s for the soleus muscle to ∼300/s for eye muscles. (Also see Fig. 5.11 below.)

There are three different types of muscle fibers, which differ in how fast they contract and their resistance to fatigue. They appear in different types of muscles in varying proportions (Fig. 5.10). For a given muscle type, their relative concentrations can be different in different people. Slow-twitch (ST) red fibers (Type I) have a long contraction time (∼110 ms) and are very resistant to fatigue because they are aerobic, i.e., they use oxygen to produce ATP. These fibers are red because they have blood to supply oxygen. Fast-twitch (FT) fibers generally can create more force than ST fibers, and reach a peak tension in less time (∼50 ms). FT red intermediate fibers (Type IIA) have a relatively short contraction time, an intermediate fatiguing rate, and aerobic generation of ATP. FT white fibers (Type IIB) have a short contraction time and fatigue quickly because they use anaerobic (i.e., no oxygen) processes to produce ATP. One of the reasons for the faster response of the fast twitch muscle is the larger neuron exciting it.

The "average" muscle has roughly 50% ST fibers and 25% red and white FT fibers. The contraction time of a muscle depends on the proportion of FT and ST fibers. Figure 5.11 shows the twitch response for three types of muscles.

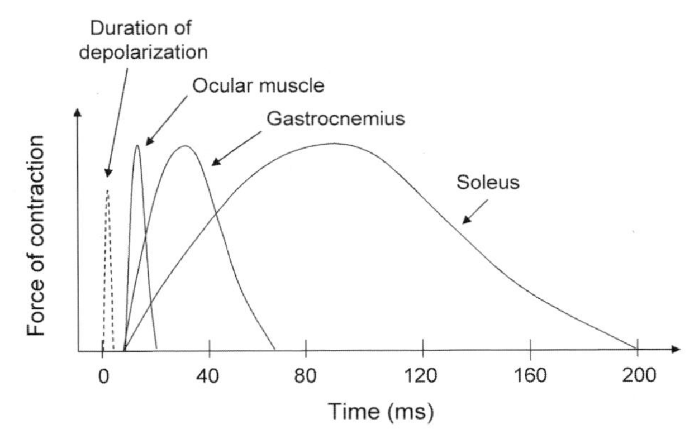

Fig. 5.11. Duration of isometric responses for different muscles with different FT and ST muscle fibers. (Based on [254])

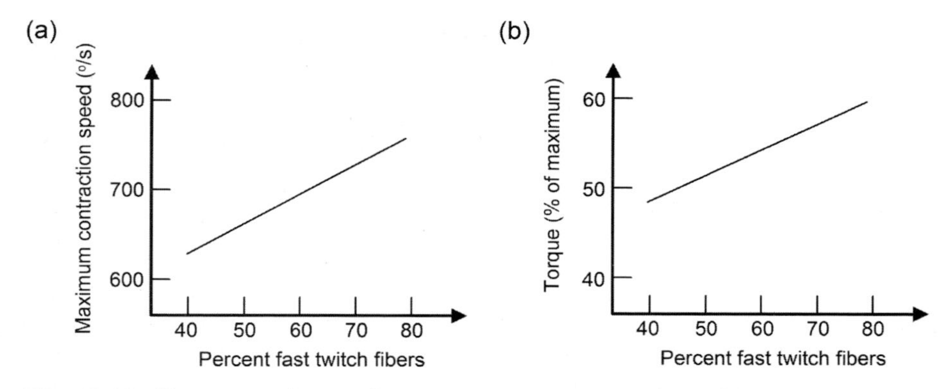

Fig. 5.12. Knee extension performance vs. percentage of FT fibers in the knee. (Based on [251] and [286])

The ocular muscles in the eye have mostly FT muscles and a response time of ~1/40 s. The gastrocnemius (gas-trok-nee'-mee-us) muscle has many FT muscles and a response time of ~1/15 s. The soleus (soh'-lee-us) muscle has many ST muscles and a response time of ~1/5 s.

All of this makes sense in terms of the body's needs. The eye muscles must be fast for our needed rapid eye reflex. The gastrocnemius and soleus muscles in the upper calf are both connected to the Achilles (calcaneal) tendon in the foot, which is connected to the calcaneus bone in the heel. The soleus muscle is a broad calf muscle that is deep relative to the medial and lateral heads of the gastrocnemius muscle (see Figs. 1.8b and 3.4). The soleus muscle is used more for standing and stability, for which endurance and resistance to fatigue are important and a fast response is not very important. The gastrocnemius muscle is used mostly for jumping and running, for which a fast response is necessary, even at the expense of endurance. The relative fraction of ST and FT muscles in the knee for trained athletes is depicted in Fig. 5.12 and Table 5.8. Clearly, athletes in sports that require endurance – such as marathon runners – have a higher fraction of ST muscle fibers in muscles used in running, while those in sports that require speed and strength – such as sprint runners and weight lifters – have a higher fraction of FT muscles in those muscles heavily used in these activities. There is a similar correlation for the cross-sectional areas of these different muscle fibers.

5.4.1 Mechanical Model of the Active State of Muscles

We now develop a macroscopic model of muscles in the active state due to electrical stimulation by using the type of spring–dashpot models we employed to characterize viscoelasticity in Chapter 4. This model includes springs and dashpots to account for the passive properties of the muscle and tension generators to characterize the active state. While it mathematically characterizes

Table 5.8. Percentages of ST and FT muscle fibers in selected muscles in male (M) and female (F) athletes, along with the cross-sectional areas of these muscle fibers. (From [291])

athlete	gender	muscle	%ST	%FT	ST area (μm^2)	FT area (μm^2)
sprint runners	M	gastrocnemius	24	76	5,878	6,034
	F	gastrocnemius	27	73	3,752	3,930
distance runners	M	gastrocnemius	79	21	8,342	6,485
	F	gastrocnemius	69	31	4,441	4,128
cyclists	M	vastus lateralis	57	43	6,333	6,116
	F	vastus lateralis	51	49	5,487	5,216
swimmers	M	posterior deltoid	67	33	–	–
weightlifters	M	gastrocnemius	44	56	5,060	8,910
	M	deltoid	53	47	5,010	8,450
triathletes	M	posterior deltoid	60	40	–	–
	M	vastus lateralis	63	37	–	–
	M	gastrocnemius	59	41	–	–
canoeists	M	posterior deltoid	71	29	4,920	7,040
shot-putters	M	gastrocnemius	38	62	6,367	6,441
nonathletes	M	vastus lateralis	47	53	4,722	4,709
	F	gastrocnemius	52	48	3,501	3,141

the mechanical features of skeletal muscles acting against loads quite well, the individual components of the model may or may not describe muscle properties microscopically. For example, the viscosity represented by the dashpot does not model the effect of the viscosity of the fluid in a muscle very well. The parallel elastic element may, in fact, be due to the sarcolemma, which is the outer membrane surrounding the muscle fiber. The highly elastic protein, titin (see Fig. 5.22 below), forms a net-like structure about the thick and thin filaments, and may also contribute to this element. The hinge regions of myosin may contribute to a series elastic element. (In cardiac muscle, it is believed that connective tissue is the major part of the parallel elastic element.)

Figure 5.13a shows a mechanical model with a unit composed of a tension generator $T_0(x,t)$ in parallel with a dashpot with viscosity c and a spring with spring constant k_{parallel}, that is in series with another spring with spring constant k_{series}. The total length of the muscle is $x^T = x^E + x$, where x^E is the equilibrium length and x is the displacement.

One can show that this model (which is also depicted in Fig. 5.14a) is mathematically equivalent to the model shown in Fig. 5.14b in which there is a unit with a spring in series with a tension generator in parallel with a dashpot, and this unit is in parallel with another spring. They are mathematically equivalent in the sense that combinations of the components give the same model predictions. The spring constants and viscosity coefficients in the two models are not equal to each other, but are related to each other (see Problem 5.9). This representation of the model in (a) is exactly the same as

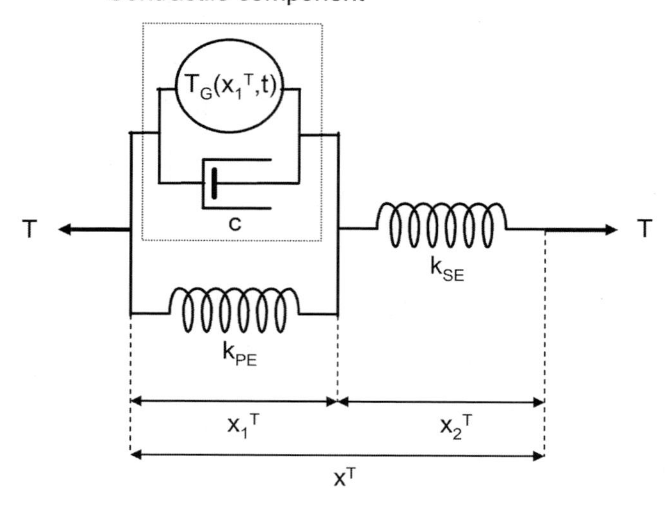

Fig. 5.13. Complete active state muscle model. (Based on [271])

the Kelvin/standard linear model, except that the dashpot is in parallel with the tension generator.

We will examine the state of the muscle by solving a slightly simpler model, with only the spring with k_{series} (now called k for simplicity) in Fig. 5.15. Without the tension generator, this reduces to the Maxwell model. The tension generator supplies a tension that could depend on muscle length x and time t, $T_{\mathrm{G}} = T_{\mathrm{G}}(x,t)$. We will also assume that the total length of the muscle does not change, i.e., isometric conditions. This means that $x^{\mathrm{T}} = x_1^{\mathrm{T}} + x_2^{\mathrm{T}}$ is constant, where x_1^{T} is the length of the dashpot/tension generator and x_2^{T} is the length of the spring. For isometric conditions, we know that $x(t) = 0$. Consequently, the tension generator $T_{\mathrm{G}} = T_{\mathrm{G}}(t)$. We will say this generator supplies the tension T_0 for specific durations of time.

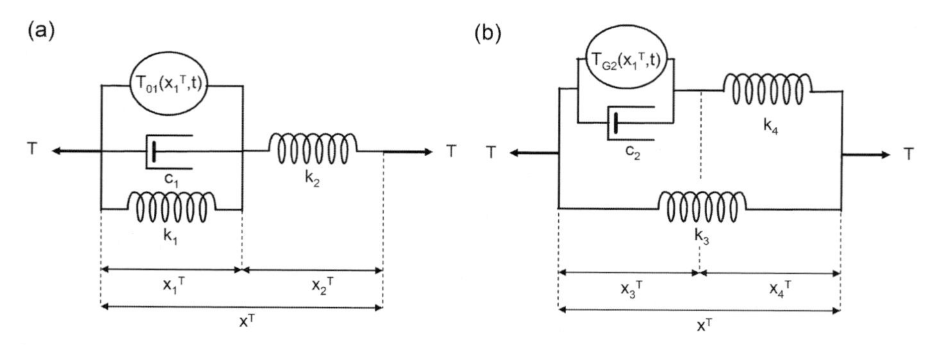

Fig. 5.14. Equivalent active state muscle models. The model in part **(a)** is the same as that in Fig. 5.13. (Based on [271])

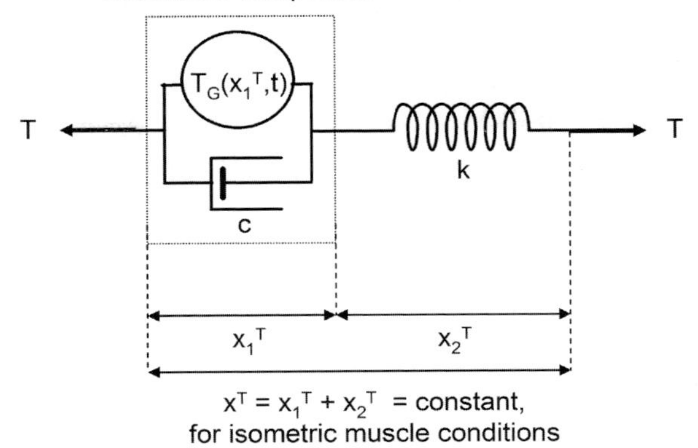

Fig. 5.15. Simpler active state muscle model. (Based on [271])

The total length of the dashpot/tension generator can be subdivided into the equilibrium length and the displacement $x_1^{\mathrm{T}} = x_1^{\mathrm{E}} + x_1$; for the spring we see that $x_2^{\mathrm{T}} = x_2^{\mathrm{E}} + x_2$. The force generated across the muscle is $T = T(t)$. This is equal to the tension across the spring

$$T = kx_2 \tag{5.5}$$

and is also equal to the sum of the tensions across the tension generator and dashpot, and so

$$T = T_{\mathrm{G}} + c\,\frac{\mathrm{d}x_1}{\mathrm{d}t}. \tag{5.6}$$

(Refer to the discussions of the Maxwell and Voigt models of viscoelasticity in Chap. 4 for a more detailed explanation.)

Because $x^{\mathrm{T}} = x_1^{\mathrm{T}} + x_2^{\mathrm{T}}$ is constant, we see that $\mathrm{d}x^{\mathrm{T}}/\mathrm{d}t = 0$. Also, because the equilibrium distances are constant

$$\frac{\mathrm{d}x}{\mathrm{d}t} = \frac{\mathrm{d}x_1}{\mathrm{d}t} + \frac{\mathrm{d}x_2}{\mathrm{d}t} = 0. \tag{5.7}$$

Equation (5.6) gives $\mathrm{d}x_1/\mathrm{d}t = (T - T_{\mathrm{G}})/c$ and the first derivative of (5.5) gives $\mathrm{d}x_2/\mathrm{d}t = (\mathrm{d}T/\mathrm{d}t)/k$, and so (5.7) becomes

$$\frac{T - T_{\mathrm{G}}}{c} + \frac{1}{k}\frac{\mathrm{d}T}{\mathrm{d}t} = 0 \tag{5.8}$$

or

$$\frac{\mathrm{d}T(t)}{\mathrm{d}t} + \frac{T(t)}{\tau} = \frac{T_{\mathrm{G}}(t)}{\tau}, \tag{5.9}$$

where $\tau = c/k$ is the relaxation time. $T(t)$ is a function of t driven by $T_{\mathrm{G}}(t)$.

The general solution to this with $T_G(t) = \alpha$ (a constant) for $t > 0$ and 0 for $t < 0$, with tension $T(0)$ at $t = 0$, is

$$T(t) = \alpha + (T(0) - \alpha)\exp(-t/\tau) = T(0)\exp(-t/\tau) + \alpha(1 - \exp(-t/\tau)). \quad (5.10)$$

This can be proved by substituting this in (5.9) and checking the solution at $t = 0$. (See Appendix C.)

Before the tension generator turns on, $T_G = 0$ and $T(0) = 0$. At $t = 0$, $T_G(t) = T_0 (= \alpha)$ for a period of time, during which time (5.10) is

$$T(t) = T_0(1 - \exp(-t/\tau)), \quad (5.11)$$

as shown in Fig. 5.16a.

If the tension generator turns off at time t_1, so $T_G(t) = 0$ for $t > t_1$, then the existing tension at t_1 is

$$T(t_1) = T_0(1 - \exp(-t_1/\tau)), \quad (5.12)$$

and (5.10) (now with $\alpha = 0$) shows that the tension then decays at a rate $\exp(-\Delta t/\tau)$, where $\Delta t = t - t_1$. Therefore, for $t > t_1$

$$T(t) = T_0(1 - \exp(-t_1/\tau))\exp(-(t - t_1)/\tau), \quad (5.13)$$

as is seen in Fig. 5.16b.

This is the model of a single twitch. We can consider sequences of two twitches separated in time one right after the other (see Fig. 5.16c). Say the activation is off for a time t_2 and is on again at $t_1 + t_2$ for a time t_1 (until $2t_1 + t_2$). Then the tension at the start of the second twitch is

$$T(t_1 + t_2) = T_0(1 - \exp(-t_1/\tau))\exp(-t_2/\tau). \quad (5.14)$$

For $t_1 + t_2 < t < 2t_1 + t_2$, the tension evolves as given by (5.10) with (5.14) used for the initial tension, $\alpha = T_0$, with t replaced by $t - (t_1 + t_2)$ on the right hand side of the equation (which is a new definition of the starting time). Therefore,

$$T(t) = T_0\{1 + [\exp(-t_2/\tau) - \exp(-(t_1 + t_2)/\tau) - 1]\exp(-(t - (t_1 + t_2))/\tau)\} \quad (5.15)$$

$$= T_0\{1 - \exp(-t/\tau) + \exp(-(t - t_1)/\tau) - \exp(-(t - (t_1 + t_2))/\tau)\}. \quad (5.16)$$

At the end of the second twitch $t = 2t_1 + t_2$, so

$$T(t) = T_0\{1 - \exp(-(2t_1 + t_2)/\tau) + \exp(-(t_1 + t_2)/\tau) - \exp(-t_1/\tau)\} \quad (5.17)$$

$$= T_0[1 - \exp(-t_1/\tau)][1 + \exp(-(t_1 + t_2)/\tau)], \quad (5.18)$$

which is $(1 + \exp(-(t_1 + t_2)/\tau))\times$ larger than at the end of the first twitch.

(a) Step function excitation

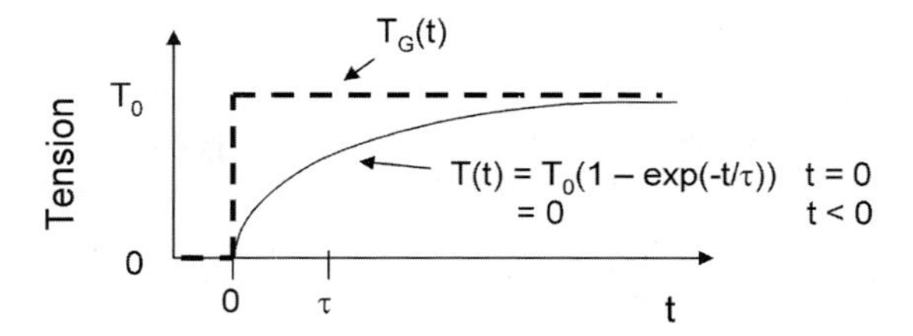

(b) Square pulse excitation

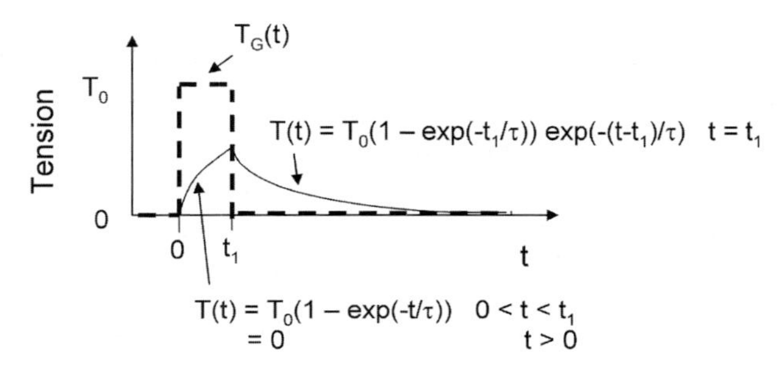

(c) Square wave excitation

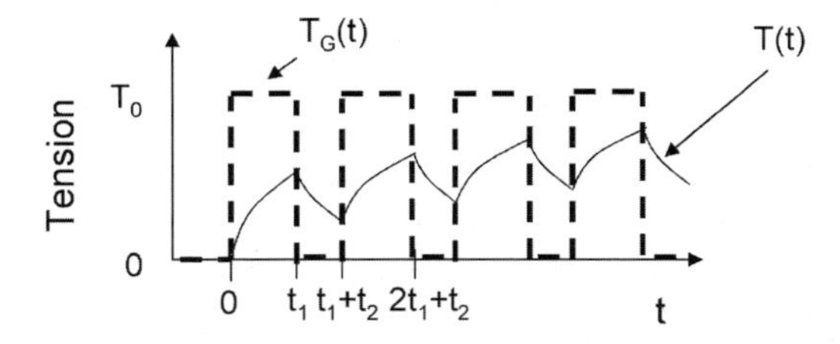

Fig. 5.16. Solutions of the active state muscle model in Fig. 5.15 for the tension with excitation by **(a)** a step function, **(b)** a square pulse, and **(c)** a square wave. The text derives the solution after two sequential square pulses at time $2t_1 + t_2$ in **(c)**. (Based on [271])

At the end of N such twitches, the tension is larger than at the end of the first twitch by a factor $1 + \exp(-(t_1 + t_2)/\tau) + \exp(-2(t_1 + t_2)/\tau) + \ldots + \exp(-N(t_1 + t_2)/\tau)$, which for large N approaches $1/[1 - \exp(-(t_1 + t_2)/\tau)]$. (The geometric sum $1+x+x^2+x^3+\ldots \rightarrow 1/(1-x)$ for $0 < x < 1$.) Therefore the total developed tension in the tetanized state is

$$T_{\text{tetanized}} = T_0 \, \frac{1 - \exp(-t_1/\tau)}{1 - \exp(-(t_1 + t_2)/\tau)}. \tag{5.19}$$

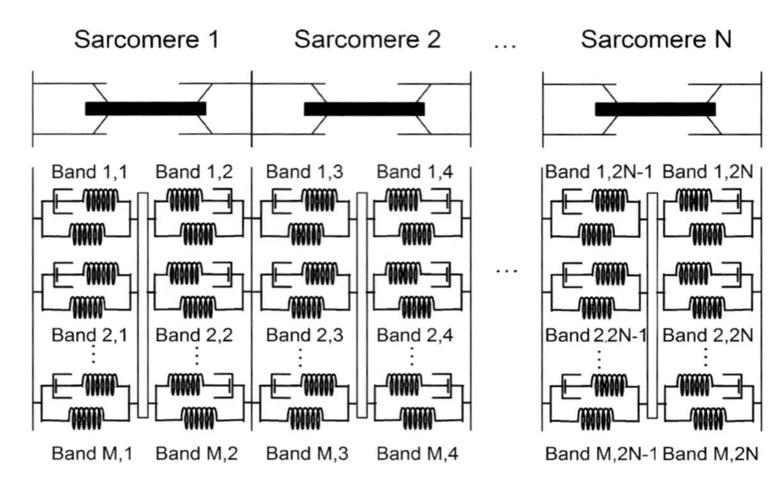

Fig. 5.17. Schematic diagram of a muscle fiber that consists of the simpler models. This more complete muscle model is built from a distributed network of N sarcomeres. (Based on [246] and [278])

If the pulses come right after each other, so $t_2 = 0$, this equation becomes $T_{\text{tetanized}} = T_0$, which makes sense.

Figure 5.17 shows a more complete mechanical model of a muscle, with individual sarcomeres modeled.

5.5 The Effect of Exercise

The PCA of a muscle increases with exercise. This increase is mostly due to increases in the cross-sectional area of individual muscle fibers, which is called *hypertrophy*. The formation of new fibers, *hyperplasia*, was once thought not to contribute to this increase at all, but now a few – but not all – studies have found evidence of it in humans. With continued training at a given load, hypertrophy continues until the muscle strength adjusts to the load. For a constant load, such as that provided by free weights, the maximum load that can be used during training (or lifting, etc.) over a large range of joint angles is that near the lower forces achieved by muscles that are shorter and longer than the optimal length (Fig. 5.23). This does not provide very satisfactory training because the load is well below the loads that could be overcome near the optimal muscle length. Some training systems vary the load with joint angle to train the muscle optimally for the full range of joint rotation. (Can training also be improved by exercising the muscle near its optimal length, with less rotation about the joint?)

In such resistance training a 1 RM load is defined as the highest load that can be moved, and it can be moved only once, a 10 RM load is one that can be moved at most 10 times, and so on. Optimizing the development of strength

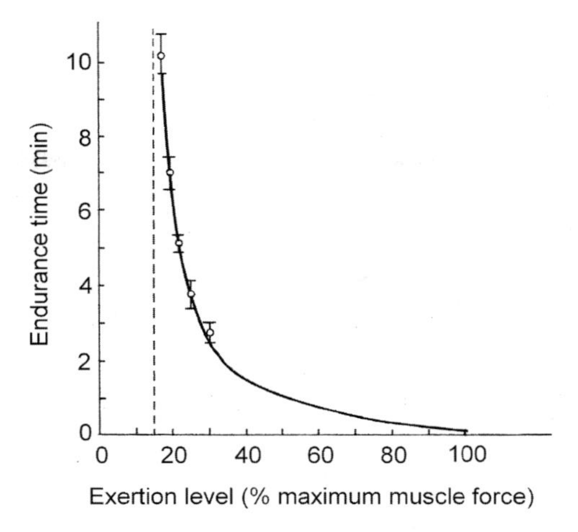

Fig. 5.18. Rohmert curve for static muscle endurance, with ±1 SD (standard devation) points shown. (From [243], adapted from [285]. Reprinted with permission of Wiley)

occurs with few repetitions at high resistance (6 RM or less). Optimizing the development of muscular endurance occurs with many repetitions at low resistance (20 RM or more). Optimizing muscle size occurs for loads between 6 RM and 12 RM, with short <90 s intervals between the (greater than three) repetition sets. For more details see [248, 268, 291].

5.5.1 Muscle Fatigue

The analyses of statics and motion in Chapts. 2 and 3 assumed muscle forces do not change with time, i.e., there is never muscle fatigue. The Rohmert curve in Fig. 5.18 characterizes the hyperbolic decrease in the endurance time of the voluntary control of a muscle with the exertion level, which is expressed as the % of maximum force level. Although not indicated in this figure, some fatigue is also expected below 15% exertion level. Metabolic changes in the muscle and impaired activation both contribute to muscle fatigue. Force can decline due to several metabolic factors that hinder the formation and detachment of crossbridges, including the formation of lactic acid, the increase in phosphate, and the decrease in the PCr (phosphocreatine) in the fibers (see Chap. 6); the depletion of ATP is not dominant. The increase in intramuscular pressure to that above the maximum blood pressure (systolic pressure, Chap. 8) restricts blood flow, which limits oxygen flow (i.e., ischemia) to the muscle and the removal of waste products. In some cases blood restriction begins at 10–20% of the maximum voluntary contraction level; some muscles are ischemic above 50% of this level. Impaired activation of muscles can be due to several factors, such as central fatigue – one's motivation – and peripheral fatigue – the failure of the neural control of muscles. Type I and IIA muscle fibers are less resistant to fatigue because they are surrounded by a rich network of capillaries. Muscles used to maintain posture, such as the soleus muscles in the lower leg, contain a large fraction of type I fibers. Muscles that have a short burst of extreme activity, such as the gastrocnemius muscle, have less need to be resistant to fatigue.

5.6 Coordination of Muscles

Many muscles are involved in most types of motion [256, 257, 258]. Sometimes they work together and are *agonists* or *protagonists* or help each other in other ways and are *synergists* (both mostly in phase) and sometimes they are *antagonists* (and out of phase). The discussion of the muscles involved in walking, running, and jumping in Chap. 3 (Figs. 3.2–3.4, 3.11, 3.12, and 3.30; Tables 3.1–3.3) and earlier in this chapter provides several examples of both protagonists (e.g., several flexors of the same joint) and antagonists (e.g., extensors/flexors of the same joint, etc.). Coordinated muscular activity during each phase of the walking cycle is shown for the three sagittal joints in Figs. 3.11 and 3.12.

Figure 5.19 shows the forces developed in the soleus and gastrocnemius muscles in the upper calf in a walking cat as a function of time during a step. They are both involved in this motion and need to work together. However, they serve different functions, the soleus (with more ST muscles) for stability and the gastrocnemius (with more FT muscles) for vertical force, and as such are not phased exactly the same. This is evident in Fig. 5.20, which plots how the soleus force evolves as a function of the gastrocnemius force, with time as an implicit variable. If they worked exactly in phase, the three sets of force plots in Fig. 5.19 would have the same shape. However, the force is developed in the gastrocnemius first during a step, and then a bit later the force in the soleus is developed. Also, as the speed of the walk increases, the force developed in the gastrocnemius increases relative to that developed in the soleus. The force plot in Fig. 5.21 shows that the tibialis anterior muscle (Fig. 3.4) and the soleus work out of phase during a step.

5.7 Active/Tetanized Muscles: Microscopic View

Unlike passive organ parts whose lengths change because of applied stresses, the lengths of muscles change and force is generated by the sliding of the thick filaments on the thin filaments, as seen in Figs. 5.22 and 5.23. Because the basic operating unit undergoing this motion is the sarcomere, which is $\sim 2\,\mu m$ long, these figures describe a microscopic view of muscle action.

The maximum tension force that can be developed by a muscle depends on several quantities, including the current muscle length, the current rate of

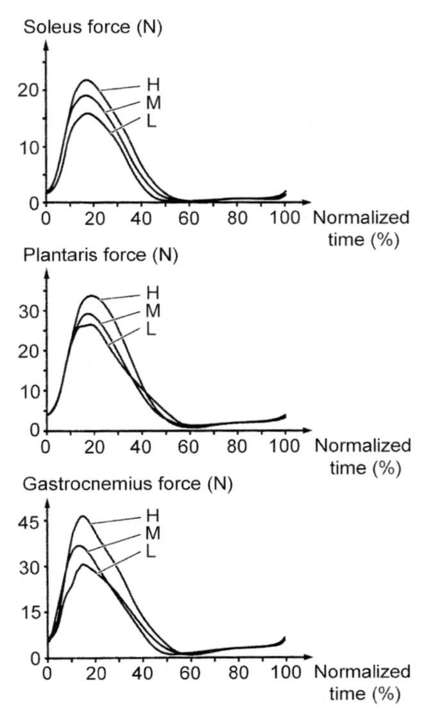

Fig. 5.19. Force vs. time curves for three synergistic muscles in the ankle plantar flexor group in a cat walking at 1.2 m/s, with mean (M), highest (H), and lowest (L) forces over 43 step cycles. (From [276]. Copyright Wiley. Reproduced with permission. Also see [258])

muscle contraction, and the physiological cross-sectional area (PCA) of the muscle. Under optimal conditions this force is $\sim$30 N/cm^2 for many muscles in most mammals.

The tension generated by a muscle when the two ends of the muscle are held fixed, called *isometric* conditions, is shown in Fig. 5.23. The maximum force generated by a muscle occurs at the optimal length L_0. (This occurs at a muscle length slightly longer than the initial or resting length L_i, prior to activation.) This maximum level of force decreases to half of this value when the muscle length is 70% (shortened muscle) or 130% (lengthened muscle) of this optimal length. It is almost zero at 170% of this resting length. These observations refer to both the total length of the actual muscle and the length of each $\sim$2 µm long sarcomere.

The physical basis of this tension is the extent of overlap of myosin and actin molecules on the thick and thin filaments. At the resting length the number of myosin/actin pair overlaps is at a maximum, and this number of overlapping pairs decreases when the sarcomere gets shorter or longer. These

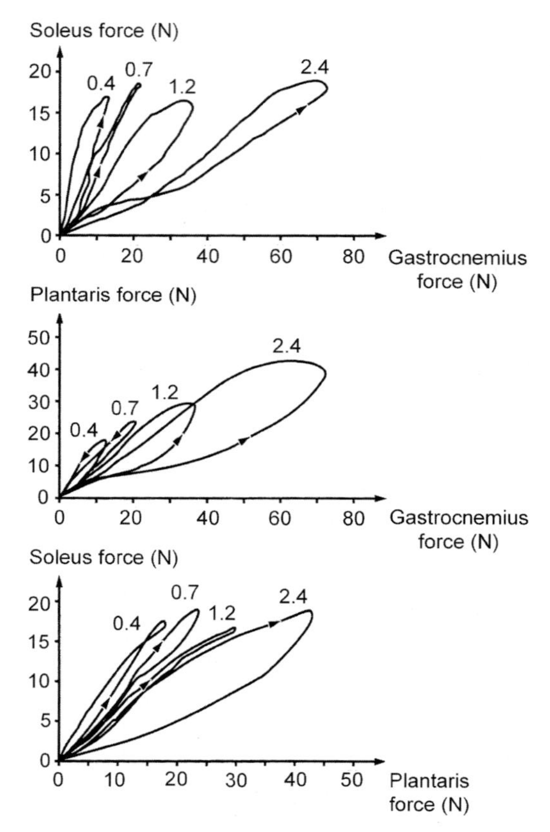

Fig. 5.20. Force vs. force curves for protagonist muscles, showing force sharing for three synergistic muscles in the ankle plantar flexor group in a cat walking at 0.4, 0.7, and 1.2 m/s and trotting at 2.4 m/s, averaged over 10 cycles. (From [276]. Copyright Wiley. Reproduced with permission. Also see [258])

interactions of these pairs of molecules are known as crossbridges between the thick and thin filaments.

One consequence of changing muscle length is the concomitant changing of joint angle (Problem 5.17). Consequently, the generated force and resulting torque also depend on the joint angle. Figure 5.24 shows the force vs. joint angle for several joints.

5.7.1 Total Muscle Tension

Figure 5.25 shows the total tetanized tension vs. muscle length. It clearly looks different than the response shown in Fig. 5.23. Also shown in Fig. 5.25 is the passive tension vs. muscle length. The contribution of the passive properties of muscle tissue cannot be ignored in modeling activated muscles. The passive

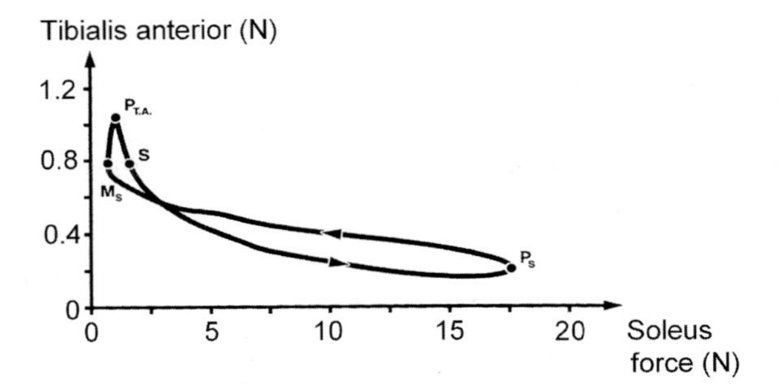

Fig. 5.21. Force vs. force curves for the antagonist pair of soleus and tibialis anterior muscles in a cat walking at 0.7 m/s. (From [276]. Copyright Wiley. Reproduced with permission)

component is given by (5.3), and it and the developed tension are added in Fig. 5.25 to give the total tension. In parallel, fusiform muscles, such as the sartorius muscle in the thigh, there are long muscle fibers with relatively little passive material. The maximum seen in the active part is still seen in the total muscle response (Fig. 5.25b). In pinnate muscles, such as the gastrocnemius muscle, there are short fibers and much connective tissue and so the passive material contribution is large and the maximum attributable to the developed portion alone is not seen (Fig. 5.25a).

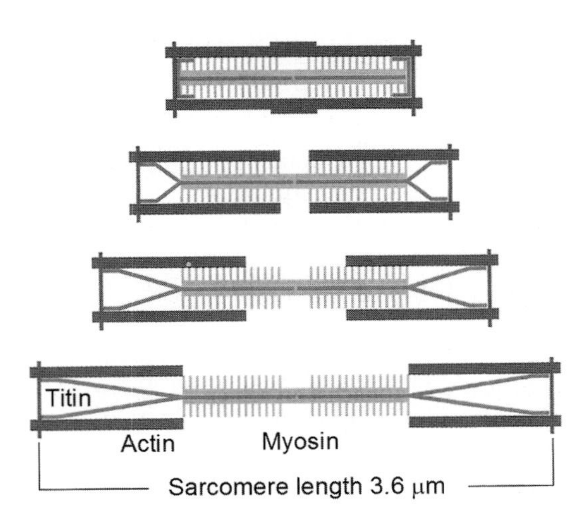

Fig. 5.22. Schematic of the sliding filament mechanism in sarcomeres, with relative sliding of the thick and thin filaments, for sarcomeres of increasing length: 1.6, 2.2, 2.9, and 3.6 μm (which roughly correspond to the sarcomeres highlighted in Fig. 5.23). Note that the elastic material titin is schematically shown to connect the thick filament to the structure; it plays an important role in the elastic properties of muscles. (From [281])

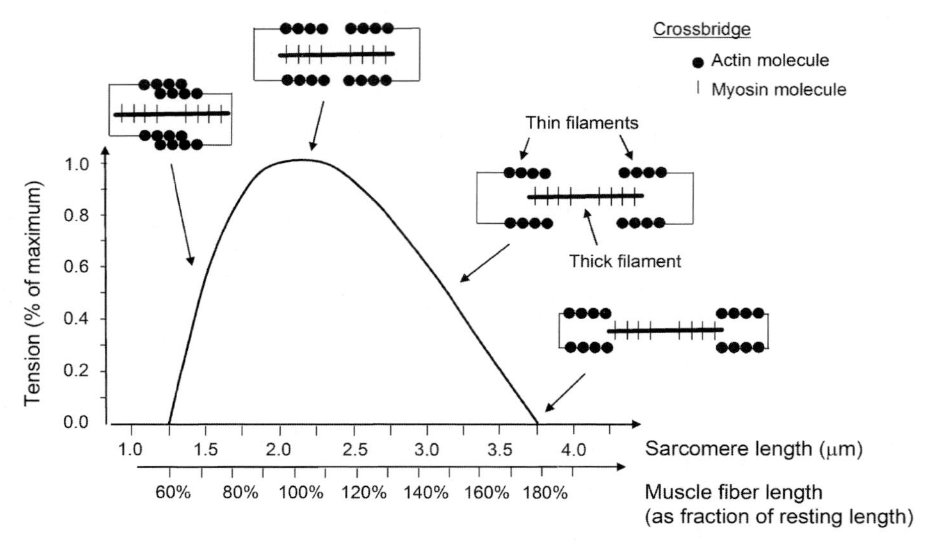

Fig. 5.23. Force vs. sarcomere or muscle length, with schematics of the variation of the overlap of the thick (myosin) and thin (actin) filaments for different sarcomere lengths. The sarcomere and total muscle lengths scale the same way. The crossbridge between an actin and myosin molecule is also shown. (Based on [254], [271], [280], and [284])

5.7.2 Everyday Proof of the Limited Range of Useful Muscle Length

The decrease in the tension that muscle is able to develop when it is very short or long can be seen in simple demonstrations [239].

For most of us, it is hard to do pull ups (chin ups) or push ups with our arms fully extended, because in this beginning position the muscles we need to use are much longer than their resting length. It is much easier to start these exercises half-way through the motion, with the muscles nearer their resting length.

Grip a ball (or pen) tightly in one fist. Have someone of comparable physical dimensions try to grasp the ball from your fist. Hopefully (and quite likely) he or she cannot. Now use your other hand to push the top of your clenched fist down at the wrist, to flex your wrist downward as much as possible without inflicting undue pain. Now have your colleague try to grasp the ball. It should be a relatively simple task. Why? Before you usually make a fist the relevant muscles are initially near their resting length, but when your wrist is flexed, the relevant muscles are much too long to generate much tension. What is the exact origin of this?

There are 20 muscles in the hand, with a total mass of about 91 g. The total hand itself has a mass of about 600 g. There are 20 muscles in the forearm,

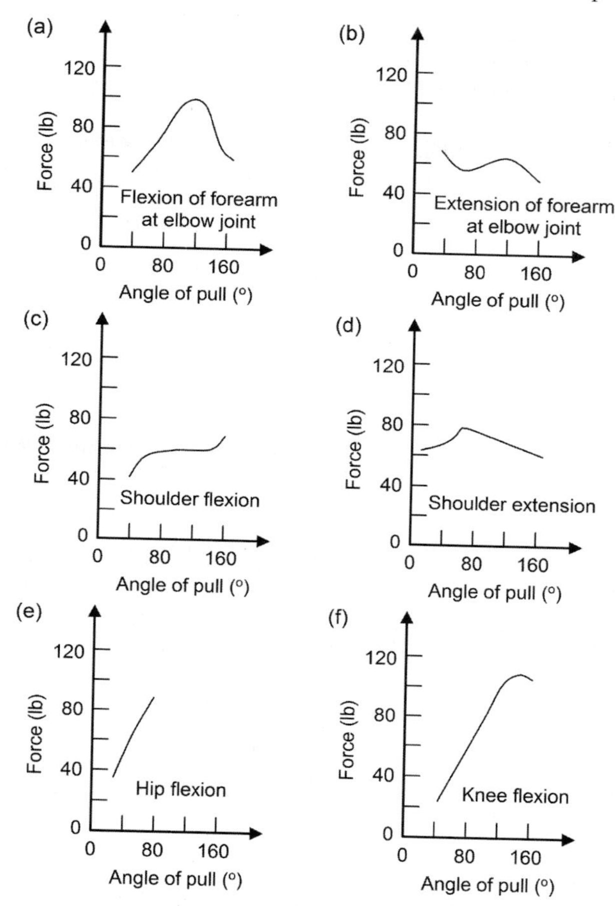

Fig. 5.24. Force vs. angle for elbow and shoulder flexion and extension, and hip and knee flexion. Only some of these plots look like the classic force vs. length locus (Fig. 5.23, converted to a plot vs. joint angle) for a single muscle fiber or sarcomere. (Based on [250])

with a mass of about 766 g. Nine of these, extensor and flexor muscles with a combined mass of 401 g, are connected to the hand via long tendons (Figs. 5.26 and 5.27). If these were physically in the hand, it would be very bulky.

Now back to holding the ball in a clenched fist. When the wrist is flexed (Fig. 5.28a), with no ball in place, the digital extensor muscles are already stretched, while the digital flexors are slackened (closed circles in Fig. 5.29 a,b). When the wrist is in this position and you try to clench an object, like a ball, the extensors need to be stretched even more and the flexors need to be slackened even more (arrows in Fig. 5.29 a,b). Both muscles enter regimes beyond the normal range of muscle length.

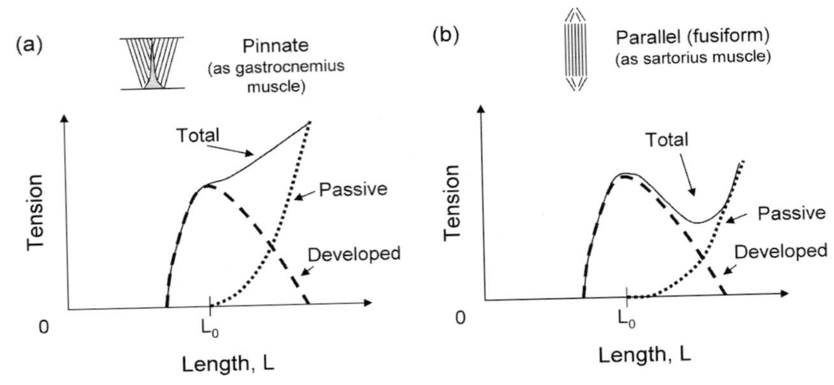

Fig. 5.25. Schematic of passive, developed (active), and total tension vs. length for pinnate and parallel muscles. (Based on [240] and [271]. Also see [237])

This type of knowledge does have practical applications! You can make an assassin with a knife in hand drop the knife by forcibly bending his or her wrist.

Similarly, you cannot straighten your fingers when your wrist is first hyperextended (hand back the other way). When the wrist is hyperextended (Fig. 5.28b), with no ball in place, the digital extensor muscles are slackened, while the digital flexors are stretched (closed circles in Fig. 5.29 c,d). When the wrist is in this position and you try to straighten your fingers, the extensors need to be slackened even more and the flexors need to be stretched even more (arrows in Fig. 5.29 c,d). Again, both muscles enter regimes beyond the normal range of muscle length.

5.8 Hill Force–Velocity Curve

The maximum tension T a muscle can develop also depends on how fast it contracts. An equivalent way of saying this is the speed at which a muscle can shorten depends on how much tension it must generate to overcome a load, such as a mass hanging at the end of it. The contraction speed is $v = |\mathrm{d}L/\mathrm{d}t|$, and it is measured with the muscle fixed at one end and with a fixed load on the other, i.e., isotonically. (Again, this fixed load can be a hanging mass.) The maximum tension and contraction speed are experimentally observed to follow the hyperbolic dependence in the Hill force–velocity curve (Fig. 5.30), with the muscle shortening faster with a smaller load mass. Note that muscles do work by contracting, and as such the term tension really refers to the tension felt by objects attached to them. (By the way, Archibald Vivian Hill, who formulated this force–velocity relation, shared the Nobel Prize in Physiology or Medicine in 1922 for his discovery relating to the production of heat in the muscle; much of his work concerning the force generated by muscles occurred after his Nobel-Prize-winning research.)

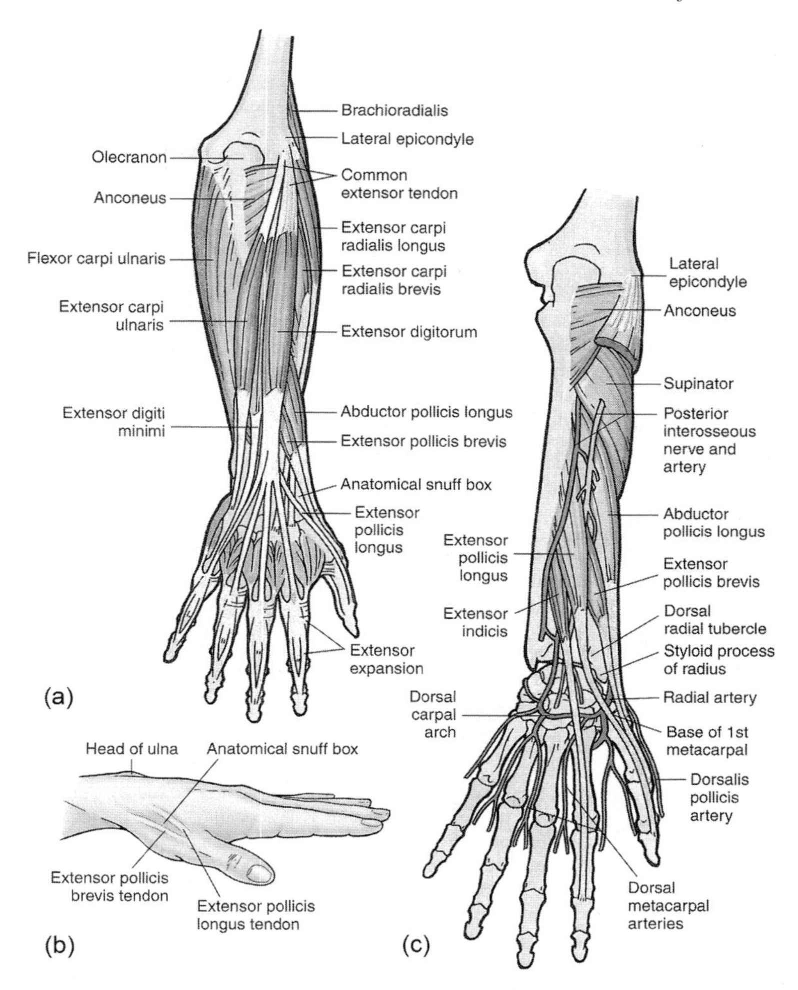

Fig. 5.26. Extensor muscles of the forearm, with **(a)** posterior view of superficial muscles, **(b)** selected features of the hand, and **(c)** deeper muscles, along with selected arteries. Tendons extending to the hand are also seen. (From [274]. Used with permission)

The maximum tension $T_{\max}$ that can be developed occurs for $v = 0$ (isometric conditions), which is the limit of very slow muscle contraction, $T(v = 0) = T_{\max}$ (sometimes called T_0). T decreases and becomes zero at $v_{\max}$. One form of the Hill force–velocity curve is

$$(T(v) + a)(v + b) = (T_{\max} + a)b \tag{5.20}$$

or

$$T(v) = \frac{bT_{\max} - av}{v + b}, \tag{5.21}$$

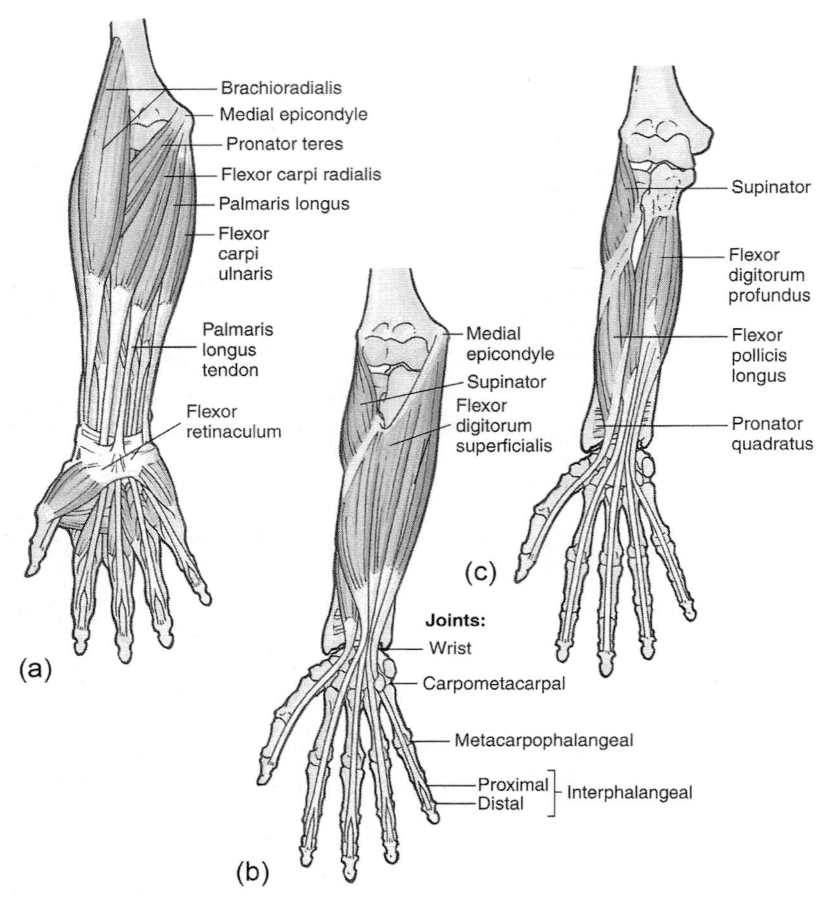

Fig. 5.27. Flexor muscles of the forearm, with anterior views of the first (most superficial) layer in **(a)**, the second layer in **(b)**, and the third and fourth layers in **(c)**. Tendons extending to the hand are also seen. (From [274]. Used with permission)

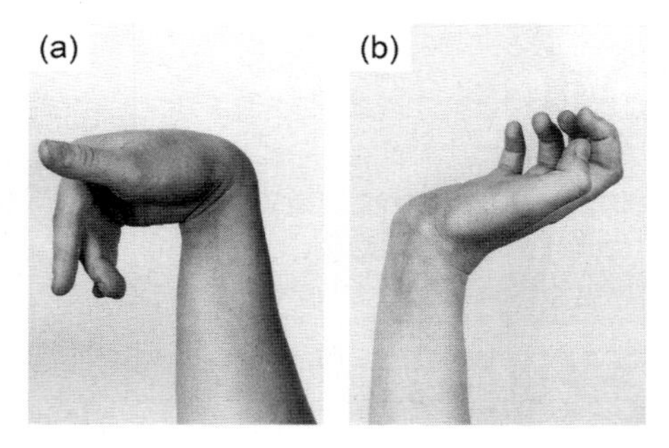

Fig. 5.28. It is difficult to clench your fist with a strongly flexed wrist **(a)** and straighten your fingers with a strongly hyperextended wrist **(b)**. (From [239]. Copyright 1992 Columbia University. Reprinted with the permission of the press)

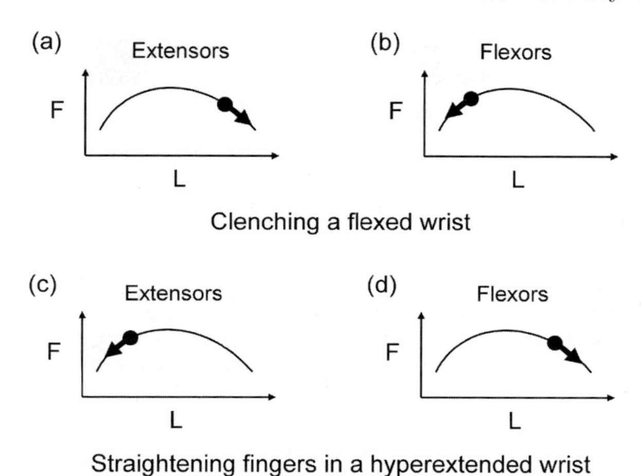

Fig. 5.29. Schematic of forces on extensors and flexors for clenching a flexed wrist and straightening fingers in a hyperextended wrist as shown by the closed circles. The arrows show the continued slackening or stretching described in the text

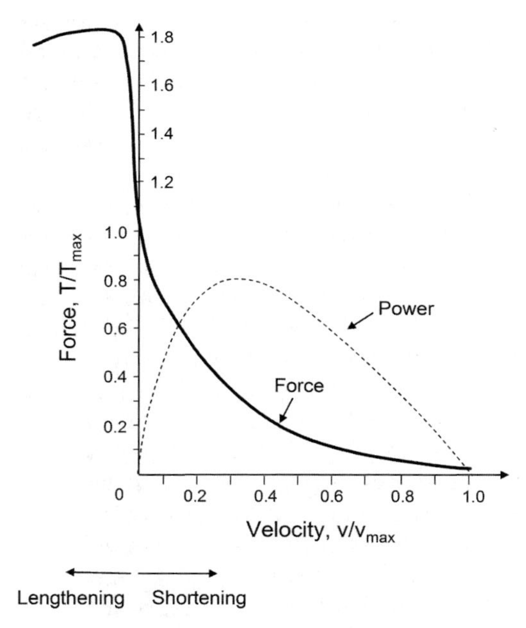

Fig. 5.30. Hill force–velocity curve. Both muscle shortening and lengthening are shown. Power = Force × velocity, is also shown, with arbitrary vertical units. (Based on [271])

Table 5.9. Parameters for the Hill force–velocity relationship for skinned muscles. (Using data from [270])

muscle	isometric force, P_0 (MPa)	maximum velocity, v_0 (FL/s)[a]	a/P_0
human soleus (15°C)[b]	0.145	0.52	0.037
human gastrocnemius (15°C)[b]	0.136	0.64	0.034
monkey soleus (15°C)[b]	0.146	0.7	0.044
monkey gastrocnemius (15°C)[b]	0.160	0.69	0.040
rabbit psoas (20°C)[c]	0.246	3.26	0.10
rat soleus (20°C)[c]	0.234	1.94	0.05

[a] FL is the fiber length.
[b] [290].
[c] [288].

which is plotted in Fig. 5.30. It is valid from $v = 0$ to $v_{\max}$. The right hand side of (5.20) represents the point on the hyperbola with maximum tension, with $v = 0$ and $T(v = 0) = T_{\max}$. We could have chosen the values at any other point on the hyperbola, such as that with minimum tension with $v = v_{\max}$ and $T(v = v_{\max}) = 0$, for which the right-hand side would have been $a(v_{\max} + b)$. Clearly these right-hand sides are equal, so $(T_{\max} + a)b = a(v_{\max} + b)$; this means $v_{\max} = (b/a)T_{\max}$, so there are only three independent parameters.

After rearranging terms, (5.21) can be expressed as

$$v' = \frac{1 - T'}{1 + T'/k}, \tag{5.22}$$

where $v' = v/v_{\max}$, $T' = T/T_{\max}$, and $k = a/T_{\max} = b/v_{\max}$. Typically, $0.15 < k < 0.25$. Table 5.9 gives the parameters in the force–velocity curve for several muscles.

We saw in (3.83) that the power needed to move an object is Fv, where F is the (constant) force on the object and v is the speed with which it is moved. Therefore the power generated by the muscle is

$$P(v) = T(v)v = v\frac{bT_{\max} - av}{v + b}. \tag{5.23}$$

Problem 5.21 shows that this is a peak when $v \sim 0.3v_{\max}$, for which $T \sim 0.3T_{\max}$. At this speed the power attains a peak value of $\sim 0.1T_{\max}v_{\max}$ (Fig. 5.30). Bicycles have gears that take advantage of this, keeping the muscle shortening velocity near the value where this peak power is achieved. This is analogous to shifting gears in cars traveling at different speeds so their engines can operate at an rpm near maximum torque.

Figure 5.31 shows experimental data for the torque that can be developed in knee extension vs. the speed of angular change [251]. (In this figure the

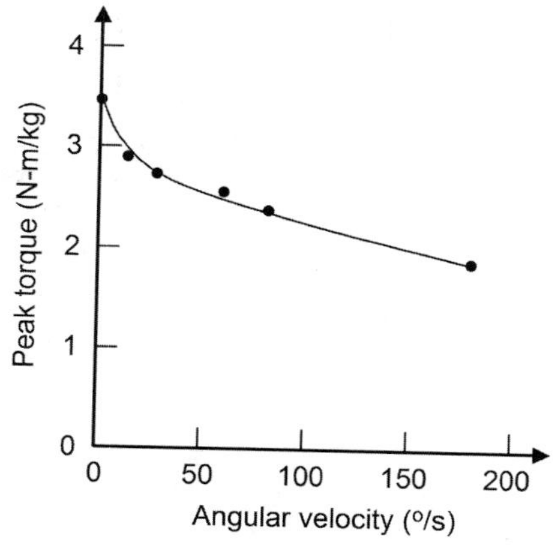

Fig. 5.31. Torque (per unit body mass) vs. knee angular velocity for knee extension. (Based on [251] and [286])

generated forces are proportional to this torque and the speed of muscle length change is proportional to this speed of angular change. This is analogous to interrelating muscle force vs. length with knee torque vs. angle.) Figure 5.12 shows that the maximum angular speed and torque depend on the fraction of FT muscles.

At lower levels of muscle excitation, there is less contractile force at each velocity, as is seen in Fig. 5.32. Also, muscles usually do not maintain a zero speed, as in Fig. 5.23, or a constant length, as in Fig. 5.30. The response of

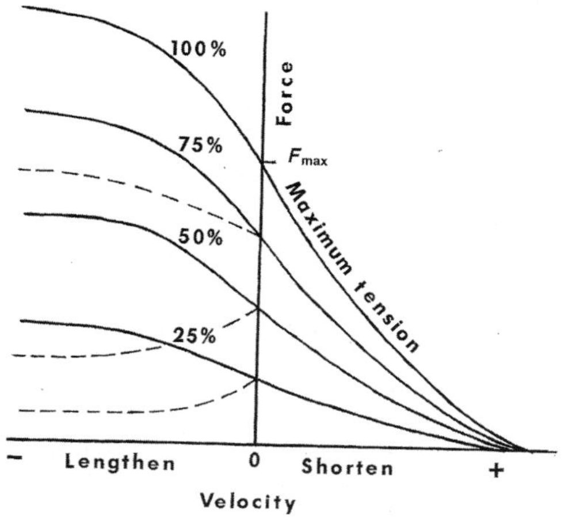

Fig. 5.32. Force–velocity curves at different levels of muscle activation, for a given length muscle. All shortening curves follow the Hill force–velocity curves, but for lengthening the curves are different for isotonic activity (*solid curves*) and isovelocity activity (*dashed curves*). (From [292]. Reprinted with permission of Wiley)

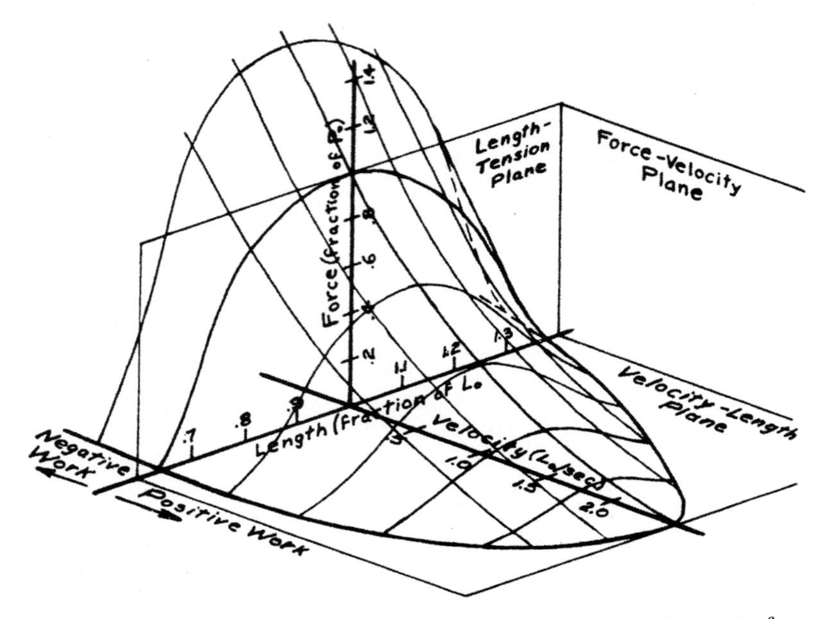

Fig. 5.33. Three-dimensional plot showing the tension of the active part of a muscle vs. velocity and length. This is for maximum muscle activation and excludes the passive muscle response. (From [292]. Reprinted with permission of Wiley)

a muscle really depends both on its length and the contraction velocity, as is shown in Fig. 5.33. In Hill's two-element model for a skeletal muscle, the contractile element, described by the Hill force–velocity curve, is in series with an elastic element [266].

The Hill force–velocity curve can be derived by considering the statistics (more precisely the statistical mechanics) of the crossbridges in the sliding filament model. The derivation of this Huxley model is beyond the level of this text; it can be found in [259, 262, 263, 264, 271].

So far we have considered muscle shortening from speeds 0 to v_{max}. This is important in doing positive work. However, Fig. 5.30 is seen to extend to the left of $v = 0$, where there is muscle lengthening ($\mathrm{d}L/\mathrm{d}t > 0$). The tension that is developed in this eccentric contraction regime is about $1.8T_{\mathrm{max}}$. In this regime the muscle is used for braking, not performing positive work. This application is used for braking motion (eccentric), resisting pulling, isometric (constant muscle length) activities, and keeping the body stable – as in standing. Remember that 100 J is lost in each step during running. Almost half of it is absorbed in the braking action of the leg muscles. When the muscles are not fully activated, this very large tension in braking always occurs for isotonic motions but not always for isovelocity modes, as seen in Fig. 5.32.

In weightlifting a heavier load can be raised if it is raised quickly, i.e., if it is "jerked," than if it is moved more slowly and smoothly. This is not really contradictory to the Hill force–velocity curve because the lifter is taking advantage of the higher muscle forces that are possible before the filaments can pick up speed. So, you can lift heavier weights by rapid lifting to shoulder height and then a jerk to full vertical extension – the "clean-and-jerk" – than by rapid lifting to shoulder height and then slow lifting to full vertical extension – the "press." (By the way, the press is no longer a competitive event because it is hard to distinguish from the clean-and-jerk.) A third weight lifting motion is the "snatch," in which the weight is raised in one continuous motion. The world records for the snatch and the clean-and-jerk are 138 kg and 182 kg, respectively for women ($m_b \geq 75$ kg, as of 2006) and 213 kg and 263 kg for men ($m_b \geq 105$ kg). (For more on this, see Problem 5.6.)

5.9 The Sliding Filament Model: Nanoscopic View

The body has several motor proteins that induce directed movement. Myosin (Figs. 5.7, 5.22, 5.23, and 5.34) does it in muscles [279]. Kinesin (kai-nee'sin) and dynein (die-nee-in) direct transport on microtubules in the body (Fig. 5.35) [253, 282, 287]. (Both are used in long-range transport in the body. Within the cell, kinesin is used for transport from the center of the cell to its edge, while dynein is used for transport from the edge of the cell to its

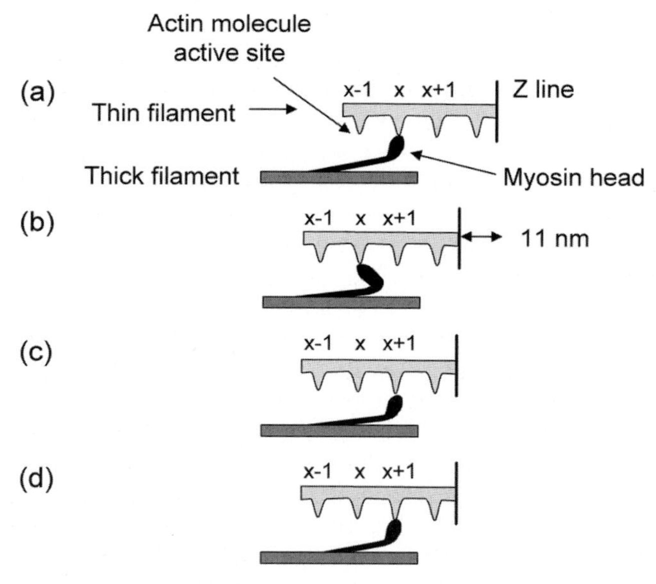

Fig. 5.34. Sliding of myofilaments, with **(a)** binding of the myosin head to an active actin site (site x), **(b)** the power stroke in which the myosin head hydrolyzes ATP and distorts, and moves the thin filament, **(c)** the myosin head detaches and relaxes, and **(d)** the myosin head reattaches to a new actin site ($x+1$) for a new cycle. (Based on [280] and [284])

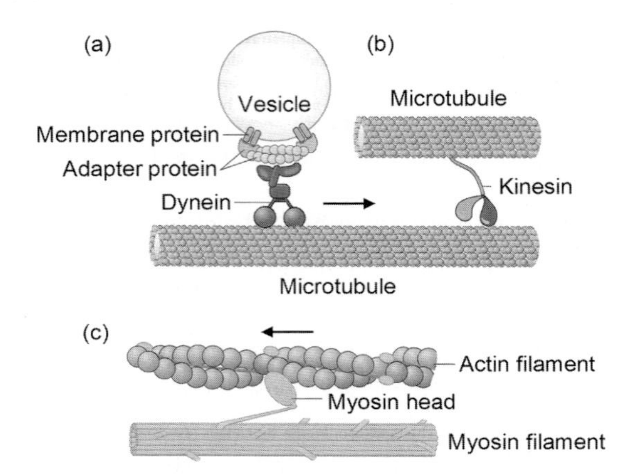

Fig. 5.35. Comparison of motor proteins, with **(a)** transport of a vesicle along a microtubule track by dynein, **(b)** transport of a microtubule on a microtubule track by kinesin, and **(c)** motion of the actin filament by a myosin filament. (From [281])

center.) Such direct transport is preferred to diffusion because it provides a directed motion and a motion that is faster than diffusion. As is described in more detail in Chap. 7, in one dimension diffusion leads to a slow gaussian-like, undirected spreading of the species over a distance $x \sim \sqrt{2D_{\text{diff}}t}$ in a time t, where D_{diff} is the diffusion coefficient.

Directed motion is necessary for muscle movement. Figure 5.34 shows the head of a myosin molecule from a thick filament attached to an actin molecule labeled x on a thin filament (Fig. 5.34a). This single coupling is a crossbridge. This stage in the interaction is called the cocked position. ATP is hydrolyzed to form ADP + inorganic phosphate + energy (see Chapter 6) and this energy is used to contort the myosin molecule, which is still bound to the same actin molecule, to a new configuration. This motion, along with the simultaneous action of other crossbridges on the same filaments, causes the thin filament to slide to the left. This is the power stroke stage. The myosin molecule detaches (the detach phase), relaxes, and then binds to the adjacent actin molecule labeled $x + 1$ (Fig. 5.34d) on the thin filament (the bind phase). *How many crossbridges are there in a muscle? How much force is generated per crossbridge?* We need to connect this nanoscopic crossbridge view with the macroscopic observation that muscles generate up to $\sim$30 N/cm^2.

Each muscle has 10^4–10^6 muscle fibers. Each fiber is 10–80 μm in diameter. Let us say 50 μm is typical so there are $\sim 1/(50\,\mu\text{m})^2 = 4 \times 10^4$ fibers/cm^2. Each fiber has several hundred to several thousand myofibrils, and let us say there are 2,000 of them. Each myofibril has about 1,500 myosin thick filaments and 3,000 actin thin filaments, and $\sim 1 \times 10^4$–1.7×10^5 sarcomeres. These filaments are arranged in an ordered manner as in Fig. 5.6a,b (lower). Each thick filament has 200 myosin molecules.

In a given sarcomere, all the crossbridges are in parallel, so the forces of each add for each filament. This is analogous to many people pulling on a rope in a tug-of-war or rowing together in a crew boat or a galley ship, where their combined forces add. All filaments in a sarcomere (myofibril) add in parallel and so these forces add. All myofibrils in a fiber add in parallel, so their forces add. All muscle fibers add in parallel and so their forces add.

Therefore, within the width of a half sarcomere there are: 4×10^4 fibers/cm^2 $\times$ 2,000 myofibrils/fiber $\times$ 1,500 thick filaments/myofibril $\times$ 200 myosin molecules/thick filament $\simeq 1 \times 10^{13}$ myosin molecules/cm^2. This means that the force developed across in each crossbridge is

$$\sim \frac{30\,\mathrm{N/cm}^2}{1 \times 10^{13}\,\text{myosin molecules/cm}^2} \sim 3 \times 10^{-12}\,\mathrm{N} = 3\,\mathrm{pN}. \tag{5.24}$$

This is also a conclusion of Huxley's sliding filament model of the Hill force–velocity curve. Direct measurement of the interaction of a single actin–myosin crossbridge using optical tweezers (which are focused laser traps used to confine these molecules) has shown that 3–4 pN is generated per crossbridge and the power stroke distance (Fig. 5.34b) – the relative motion of the filaments per ATP hydrolysis – is 11 nm (Fig. 5.36). This is truly a nanoscopic view of muscles. (In Problem 5.26 you can calculate the work by this motion and characterize it in terms of the fraction of the energy released in the hydrolysis of ATP, which is described in the next chapter.)

In examining the effect of the crossbridges in sarcomeres in exerting forces, we considered only the crossbridges in the sarcomeres in a cross-section of the muscle. These add to give the total force/area for the muscle. However, there are $\sim 1 \times 10^4$–1.7×10^5 sarcomeres along the length of each fiber in the muscle. These tens of thousands of sarcomeres do not increase the force exerted by the muscle, but they serve a quite important, though very different,

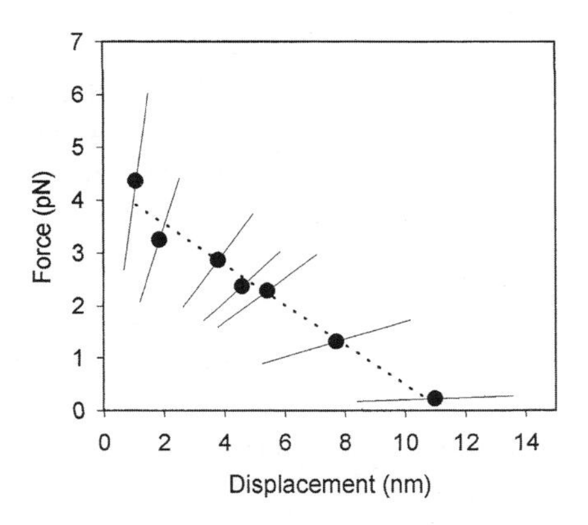

Fig. 5.36. Force vs. displacement curve of a single myosin molecule interacting with an actin molecule during a powerstroke, as measured by "optical tweezers." (From [247]. Used with permission of the Biophysical Society)

function. The necessary amount of rotation about joints can only occur if there is enough of a change in the muscle length (or tendon plus muscle length). Because the effective change in sarcomere and muscle length is only about ±30%, the change in total length is controlled by the resting muscle length, which depends on the number of sarcomeres along the length of the fibers. (This is related to the second advantage of the pinnated design of muscles.)

Our crossbridge thick/thin filament model of the sarcomere is fine so far, but real life is a bit more complicated. For example, it is geometrically more complicated. Two actin filaments are actually intertwined, producing a double helix and a tropomyosin strand runs down the grooves of this structure. (Tropomyosin is a regulatory protein, and this strand is different from the thick myosin strand.) Six actin filaments surround each myosin filament, and each actin filament is surrounded by three myosin filaments (Fig. 5.6). Moreover, this attachment/detachment cycle does not happen unless there are Ca^{2+} ions present to induce a topological change that causes the myosin head and actin to be physically close. These ions are present for a limited time after a nerve impulse arrives at the muscle. Also, for force to be exerted for useful motion, there must be some anchoring point and mechanism, and this is provided by the anchoring of each actin filament to a transverse structure by a titin molecule.

Furthermore, the model of concerted motion in which billions or trillions of myosin molecules all attach to actin, cock, then detach, and reattach in perfect unison seems highly improbable [260, 261]. It appears that instead of this concerted and rapid sequence of pull/wait, pull/wait, ... cycles, there are myosin molecules in different parts of the cycle at a given time – but all forming crossbridges pulling in the same direction. (This is like a boat with four oarsmen and eight oars, with, say only two oars in the water and propelling the boat at any one time.) There is a cycle time, τ_{cycle}, for a given crossbridge, during which time the myosin is attached to the actin for a time τ_{on} and is detached and inactive for a time τ_{off}, with $\tau_{\text{cycle}} = \tau_{\text{on}} + \tau_{\text{off}}$. During the on-time the crossbridge moves the attached phase via hydrolysis by the powerstroke distance or working distance δ, which is about 5.3 nm in vitro for skeletal muscle myosin II. (We will approximate this as 5 nm.) The duty ratio r_{duty} is the fraction of time that the crossbridge is attached:

$$r_{\text{duty}} = \frac{\tau_{\text{on}}}{\tau_{\text{on}} + \tau_{\text{off}}} = \frac{\tau_{\text{on}}}{\tau_{\text{cycle}}}. \tag{5.25}$$

For myosin τ_{on} is $\sim 10^{-3}$ s and r_{duty} is ~ 0.02, so a given myosin molecule is in contact with an actin filament for a relatively small fraction of the time.

For the motion of the strands to be continuous there have to be a minimum of N_{min} heads at different phases during the cycle time τ_{cycle}, where

$$N_{\text{min}} = \frac{1}{r_{\text{duty}}}. \tag{5.26}$$

With $N_{\min} \sim 1/0.02 = 50$, only about six of the myosin heads of the roughly 300 myosin crossbridges in a half a filament are attached and exerting a force at any given time.

During a complete crossbridge cycle, which is the time it takes to complete one ATP-hydrolysis cycle, the motor moves a distance Δ, where

$$\Delta = \frac{\delta}{r_{\text{duty}}}. \tag{5.27}$$

This is typically $\sim 5\,\text{nm} \times 50 = 250\,\text{nm}$ (for low loads).

The cycle time is related to the rate constant for the hydrolysis of an ATP molecule k_{ATPase}

$$\tau_{\text{cycle}} = \frac{1}{k_{\text{ATPase}}}. \tag{5.28}$$

Using $k_{\text{ATPase}} \sim 20/\text{s}$, we see that $\tau_{\text{cycle}} \sim 0.05\,\text{s}$ and this is consistent with $\tau_{\text{cycle}} = \tau_{\text{on}}/r_{\text{duty}} = 10^{-3}\ \text{s}/0.02$. The relative speed of the sliding actin and myosin filaments is given by the total displacement during the cycle time v, which is

$$v = k_{\text{ATPase}}\Delta = \frac{\Delta}{\tau_{\text{cycle}}} = \frac{\delta}{\tau_{\text{on}}}. \tag{5.29}$$

So we see that $v = 5\,\text{nm}/10^{-3}\,\text{s} = 5\,\mu\text{m/s}$.

In addition to δ and Δ, there is a third characteristic distance, d, which is the path distance or step size between consecutive binding sites, and is $\sim$36 nm for this crossbridge. This distance is related to Δ by $\Delta = nd$, where n is an integer. In the schematic of Fig. 5.37, $\delta \ll d$ and $\Delta = 3d$, and so $n = 3$. In the simpler model of concerted motion of actin–myosin crossbridges, it was implicitly assumed that these three characteristic distances were the same, and that r was almost 1. In contrast to the real $\ll 1$ duty cycle of myosin, in transport by conventional kinesin, the motor protein is just about always in contact with the tubule (and r is almost 1), with δ being about 8 nm, $d = 8\,\text{nm}$, and $\Delta = 16\,\text{nm}$.

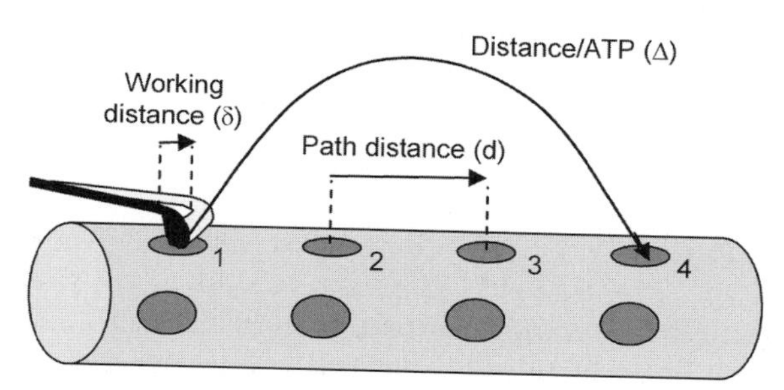

Fig. 5.37. Distances associated with a more sophisticated model of the actin–myosin crossbridge. This crossbridge moves parallel to the filament axis and skips two potential binding sites during a hydrolysis cycle. The duty ratio is low in this example. (Based on [261])

What is the maximum mechanical power provided by the legs during exercises like rapid step climbing? How is this provided by these crossbridges? Let us assume that the calf and thigh muscles both contract a distance of 3 cm during a step and calculate the energy needed to perform those needed contractions during a step time of 0.25 s, and therefore the necessary power. If these muscles both have an average total diameter of 10 cm and the effective diameter of a myosin filament is 20 nm, the number of myosin filaments in a cross-section of each muscle is about $(10\,\text{cm})^2/(20\,\text{nm})^2 = 2.5 \times 10^{13}$, and so there are about 5.0×10^{13} filaments in cross-section for both muscles. There are six chains of side chains per myosin filament (as seen in Fig. 5.6), so there are 3.0×10^{14} crossbridges in cross-section.

If the attachment/detachment of each crossbridge contracts the chain by 5.3 nm, because each muscle contracts by 3 cm the total number of crossbridge attachments/detachments during the motion is roughly $(3.0 \times 10^{14})(3\,\text{cm}/5\,\text{nm}) = 1.8 \times 10^{21}$. If each crossbridge motion requires the hydrolysis of one ATP molecule, which releases 14 kcal/mole $= 9.7 \times 10^{-20}$ J (as discussed in Chap. 6), the amount of energy needed is $\sim(1.8 \times 10^{21})(9.7 \times 10^{-20})\,\text{J} = 175\,\text{J}$. If this energy is used in 0.25 s, the power is $175\,\text{J}/0.25\,\text{s} = 700\,\text{W}$. This is consistent with the metabolic powers needed for vigorous activities such as step climbing (Problem 5.10 and Chap. 6). Note that part of the uncertainty in the final answer is the very rough estimation of the effective cross-sections of the muscles used.

5.10 Summary

The function of muscles can be understood by analyzing their substructure in terms of fibers, sarcomeres, and the filamentary structure of the sarcomeres. The forces developed by a muscle can be modeled to be a function of overall muscle length and contraction speed. Mechanical models, similar to those developed in Chap. 4, can be used to examine the consequences of the electrical stimulation of muscles. On a nanoscopic basis, forces developed by muscles can be understood and modeled in terms of the crossbridges between the myosin and actin filaments.

Problems

Muscles and Forces

5.1. There is an average normal vertical force of 1,710 N (2.7× body weight) on the one foot of an athlete in contact with the ground during running, but an average of only 715 N on each foot (and 1,430 N total vertical force with both feet = 2.3× body weight) when this same athlete does a vertical jump. Is this difference possible? Qualitatively why is it reasonable or unreasonable?

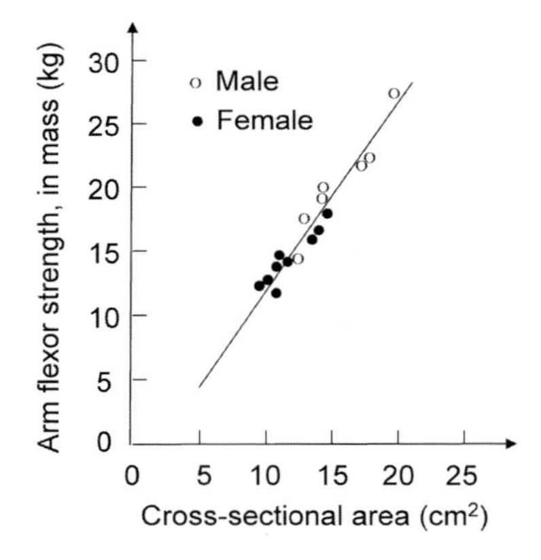

Fig. 5.38. The strength of the arm flexor muscles for males and females vs. muscle cross-sectional area. (Using data from [265].) For Problem 5.2

(Consider only how the ranges of muscle lengths in these activities contribute to these observations. References [238] and [239] suggest that the stimulation of the muscles during stretching that occurs immediately prior to contraction in running, but not in jumping, contributes to this observation. This is also consistent with the observation that down-and-up action prior to a high jump leads to higher jumps.)

5.2. (a) Do the data presented in Fig. 5.38 suggest that the strength of muscles varies linearly with their cross-sectional areas, as is assumed in the text?
(b) If so, what is the proportionality constant?
(c) How does it differ for males and females?

5.3. (a) Show that the physiological cross-sectional area (PCA) of a muscle for a parallel-fibered muscle is PCA $= (m/Ld)\,\mathrm{cm}^2$, where m is the mass of the muscle fibers, d is the density of muscles ($1.056\,\mathrm{g/cm}^3$), and L is the length of the muscle fibers.
(b) Repeat this for pinnate muscles, showing that PCA $= (m\cos\theta/Ld)\,\mathrm{cm}^2$, where θ is the pinnation angle (the angle between the long axis of the muscle and the fiber angle).

5.4. Check to see if the two relations in Problem 5.3 are consistent with the muscle data given in Table 5.4.

5.5. Let us estimate how much weight a person can lift above his head, with extended arms.
(a) Use the data for the 70 kg person in Chap. 1, and specifically the volume of skeletal muscle. Let us assume that all of these muscles are arranged as an upright right circular cylinder of height equal to the sum of the lengths of the person's legs, torso, and arms. Use the resulting cross-sectional area and

assume that skeletal muscle can exert a force of $20\,\mathrm{N/cm^2}$ to find the mass that the person can lift.
(b) Is this value reasonable? If not, describe how each of the following factors may or may not contribute to an unreasonable result: (i) His arms are narrower than his legs, so the cross-sectional area of muscle is not uniform and the limiting factor would be the weakest (i.e., narrowest) link. (ii) The muscles are not the length assumed and should be taken to be no longer than a fraction of the length of an arm or leg segment. Consequently, there are several muscles in series. (iii) Not all skeletal muscle in the body is designed to contribute to this lifting. (iv) Not all of the muscles involved are parallel fibered muscles. (v) The $20\,\mathrm{N/cm^2}$ value is inaccurate.
(c) Now re-evaluate the problem for female and male world-class weightlifters for their world-record lifted masses. (Use the data given in the text.) What fraction of their skeletal muscle cross-sectional area (assumed to be uniform here) effectively contributes to the lift?

5.6. Here are some data for world weightlifting records (in 2006 for clean-and-jerk) for people of different masses: for men, 168 kg lifted (by a 56 kg man), 182 kg (62 kg), 197 kg (69 kg), 210 kg (77 kg), 218 kg (85 kg), 232 kg (94 kg), and 242 kg (105 kg); and for women, 118 kg (by a 48 kg woman), 127 kg (53 kg), 139 kg (58 kg), 141 kg (63 kg), 157 kg (69 kg), and 159 (75 kg). How does the lifted weight vary with body weight? (Do this separately for men and women.) Does it follow a power law? Is it sublinear? How can you explain this dependence?

5.7. How can a measurement of force vs. muscle length be made with the muscle fixed at either end (isometric conditions), as in Fig. 5.23? Is there an inconsistency here?

5.8. Let us say that a fusiform muscle of length L and PCA is attached to bones by tendons of length $0.9L$ and cross-section area $f(\mathrm{PCA})$ ($f \ll 1$) on either end. This total length of $2.8L$ is kept fixed as the muscle contracts by 20% and develops a force corresponding to $15\,\mathrm{N/cm^2}$. Use the data in Table 4.2.
(a) Find the strain in the tendons. How does it compare to the UPE for tendons?
(b) Relate f to Y for the tendon.
(c) Using the value of Y from the table, compare the diameters of the muscle and tendons.
(d) Find the stress in the tendons. How does it compare to the UPE for tendons?

5.9. (advanced problem) Show that the two models in Fig. 5.14 are mathematically equivalent, with $c_1/(k_1+k_2) = c_2/k_4$; $k_1k_2/(k_1+k_2) = k_3$; $k_2 = k_3+k_4$; and $(k_2/(k_1+k_2))T_{01} = T_{02}$ [271].

Muscle Energy, Endurance, and Synergy

5.10. A 75 kg person climbs stairs to a height of 50 m in 53 s. Determine the person's metabolic rate during climbing, in W and kcal/h, assuming the process in 25% efficient. How does this relate to the estimates of energy used in such motion in the text? (See Chap. 6 for more on metabolism.)

5.11. Derive a hyperbolic expression that describes the Rohmert curve (Fig. 5.18) relating between the endurance time and exertion level of a muscle.

5.12. Figure 5.20 shows that as a cat walks faster his gastrocnemius develop more and more force, but his soleus muscles do not. Explain why the relative contribution of the gastrocnemius muscle increases with speed.

5.13. You repetitively extend your arm and then flex it (say as to touch your shoulder), with a period of 2 s. Sketch the forces in your biceps brachii and triceps brachii each vs. time, and both vs. each other (with time as an implicit parameter).

Muscle Tension vs. Length or Time

5.14. Why do you suppose that Fig. 3.52 plots the total length of the muscles and tendons?

5.15. Use the data in Fig. 3.52 to determine whether the major leg muscles used during bicycling are always near their optimal lengths (so they always exert nearly maximum forces) or do they become much longer or shorter than these optimal lengths during parts of the cycle (so they exert forces that are much less than the maxima)? Justify your answer.

5.16. In a weight room, you exercise your biceps brachii with a constant weight by trying to lift it (i.e., rotating your lower arm about your elbow). Sketch the maximum tension vs. length for your biceps brachii if
(a) The weight is so heavy that you can barely hold it as you try to lift it.
(b) The weight is so light that you can easily lift it.
For both cases, where is the weight on your sketch? Describe how much you are changing the length of your muscle during the "lift."

5.17. In the example of throwing a ball studied in Chap. 3, we assumed that the force developed by the biceps brachii did not depend on the angle between the upper and lower arm; this angle varied from 180–0° in the first statement of the problem and from 135–45° in the second. Use what we now know about how the contractile force of a muscle varies with its length to determine the validity of that second assumption. Assume that the muscle has a length of $L_0 = 12$ cm when resting at 90°, and that the distance from the elbow joint to the insertion point on the radius is 4 cm (see Fig. 5.39). (You should base your conclusions on calculations of the muscle length for various angles – at least for 0, 45, 90, 135, and 180°. Ignore the length of the tendons.)

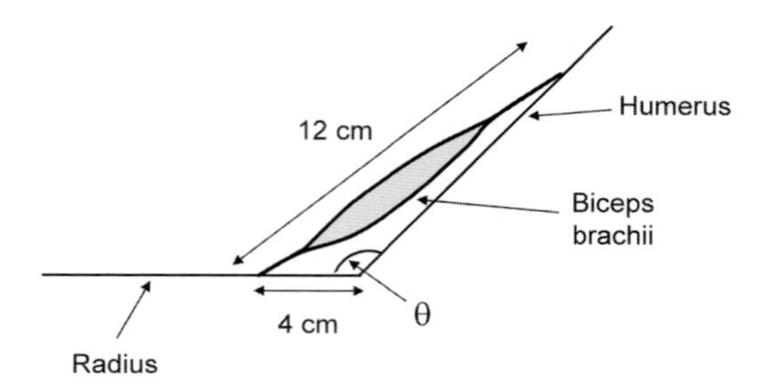

Fig. 5.39. Model of parallel fibered muscle. For Problem 5.17

5.18. In fusiform (or parallel) muscles all of the muscle contractile force is directed along the axis of the tendon and there is a relationship between changes in muscle and tendon length that depends on conditions (isometric contraction, etc.). In pinnate and bipinnate muscles these relationships are qualitatively different. Refer to Fig. 5.40, which depicts a bipinnate muscle in relatively more relaxed, initial (solid lines) and relatively more contracted, final (dashed lines) configurations.
(a) In the initial configuration, the tendon length (to the point where the muscle is inserted into the tendon, at point B) is 8.7 cm (= 2.9 cm + 5.8 cm in the figure) and each muscle fiber is 10 cm long, and is attached to the tendon at a 30° angle. Say that there are a total of N fibers attached to the tendon near this point (with an equal number of fibers on both sides) and that each fiber exerts a force F_{fiber}. Find the total force transmitted to the tendon F_{tendon}. (Is there "wasted" force here?)
(b) The muscle contracts so that the tendon length (to point C) is 2.9 cm, and the muscle/tendon angle is now 60°. Find the length of the muscle (see the dashed lines). During this contraction, which has gotten shorter more, the

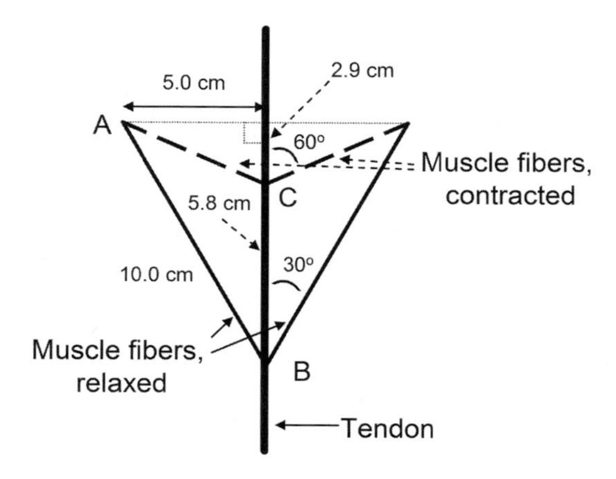

Fig. 5.40. Model of bipinnate fibered muscle. (Based on [272].) For Problem 5.18

muscle fibers or the tendon? Is this good or bad for muscular action? (Does this effect, along with the ability to attach more muscle fibers to this type of tendon counter the "wasted" force in part (a)?)
(c) Find the force transmitted to the tendon after the contraction in part (b) F_{tendon}, assuming that the contractile force of each muscle fiber is still F_{fiber}. How has the force transmitted to the tendon changed?
(d) Now consider the more general case, in which we acknowledge that the force exerted by a muscle fiber changes with length. Say that F_{fiber} peaks at a length $L = 8\,\text{cm}$, with $F_{\text{fiber}}(L) = F_{\max}(1-(L-8)^2/8)$. Sketch F_{tendon} as L contracts from 10 to 6 cm. Does it peak exactly at the length where $F_{\text{fiber}}(L)$ peaks? Why?

5.19. Let us model the decrease of muscle force F from its maximum value $F_{\max}$ at a length L_{peak} as being parabolic: $F(L) = F_{\max} - \Delta F((L - L_{\text{peak}})/L_{\text{dec}})^2$. Show that the work done by the muscle as it contracts from $L = L_{\text{peak}} + L_{\text{dec}}$ to $L = L_{\text{peak}} - L_{\text{dec}}$ is $W = W_0(1 - \Delta F/3F_{\max})$, where $W_0 = 2F_{\max}L_{\text{dec}}$ is the work that the muscle would do if the force did not change with length [273].

5.20. Sketch on one set of axes the six curves in Fig. 5.24, each as the % of maximum strength for that muscle group vs. the angle of pull (which is the joint angle). The ordinate axis should range from 0 to 100% (and of course 100% represents a different maximum force for each of the curves).

5.21. Determine the values of T and v that maximize the power output of a muscle described by the Hill force–velocity equation (5.20) and find the power output $P_{\max}$ of muscle. Take $a/T_{\max} = 0.25$; remember that $P = Tv$. (Hint: You should differentiate the expression for power with respect to v and set it equal to zero. One way to do this is to first solve for $T(v)$ from the Hill equation and express $P(v) = T(v)v$. The maximum power should turn out to be about $0.1T_{\max}v_{\max}$.)

5.22. In the example of throwing a ball studied in Chap. 3, we assumed that the force developed by the biceps brachii did not depend on how fast the elbow angle θ varied with time ($\text{d}\theta/\text{d}t$). In this problem we will examine two aspects of this assumption using the Hill force–velocity equation. Say $T_{\max} = (20\,\text{N/cm}^2)\text{PCA}$, where PCA is the physiological cross-sectional area of a muscle, and that $v_{\max} = 6L_0/\text{s}$ for muscles predominantly comprised of ST muscle fibers and $v_{\max} = 16L_0/\text{s}$ for muscles predominantly comprised of FT muscle fibers, where L_0 is the optimal fiber length of the muscle of interest. (Because muscles are often composed of combinations of FT and ST muscle fiber, the appropriate $v_{\max}$ is often in between these limiting values.) Also assume that $a/T_{\max} = b/v_{\max} = 0.25$ and use the results from Problem 5.21. For each part that follows, consider the case studied in Chap. 3 of 2 in diameter biceps brachii muscles, for which the calculated throwing speed is 17.8 mph (with no gravity and $\langle \sin\theta \rangle = 1$), and take $L_0 = 10\,\text{cm}$.

(a) Find a, b, and the maximum power (in W) that the muscle can generate, first assuming FT and then ST muscles.
(b) If the ball leaves the hand at a speed of x mph, geometry says that the speed of muscle contraction is smaller by the proportion of the distance of muscle insertion from the elbow pivot (4 cm) to the distance of the ball from the pivot (36 cm). How does this speed of muscle contraction, based on the throwing calculation, compare to the maximum muscle contraction speed (v_{max}) for FT and ST muscles? Also, how does it compare to the muscle contraction speed, for both types of muscles, at which the power generated by the muscle is maximized? (Ignore the response of the tendons. Can these be important?)
(c) The average and peak powers needed to be generated by the muscles to achieve these throwing speeds are 178 W and 356 W, respectively, for the 2 in diameter biceps using the kinematics calculations performed in Chap. 3. For both FT and ST muscles, calculate the maximum power that the muscle can generate and compare your answers with the values calculated using kinematics. Repeat this if you (incorrectly) assume that the force generated by the muscle is T_{max} independent of muscle contraction speed, so that the maximum muscle power would be $T_{max}v_{max}$ (which is clearly incorrect).
(d) Do the results in (b) and (c) cast doubt on the calculation in Chap. 3 (and why)? If so, does this totally invalidate the calculation or does it mean that after a certain muscle contraction speed is achieved the decrease in the muscle force must be included to improve the model. (Also, note that since the biceps brachii have much ST muscle fiber, the "FT" limit is not very realistic.)

Microscopic and Nanoscopic Processes in Muscles

5.23. (a) Estimate the time it takes Ca^{2+} ions to diffuse throughout a typical skeletal muscle cell if the cell has a diameter of 200 μm and the diffusion coefficient of the ions is 10^{-5} cm^2/s. Assume that the release of ions occurs just outside of the cell, due to a nerve pulse arriving there, and diffusion occurs in one transverse dimension of the cell.
(b) How does this time compare to the typical reaction times of muscles?
(c) The nerve signal actually activates the sarcoplasmic reticulum that runs transverse to the outer membrane of the cell, in the Z-line borders of each sarcomere, and the Ca^{2+} ions are released there; consequently, the maximum distance the ions need to travel to diffuse across the whole sarcomere is 2 μm. How long does this take and does this time seem more reasonable?

5.24. Estimate the number of sarcomeres in your biceps brachii: (a) along its length, (b) across its cross-section, and (c) in total.

5.25. When lifting a weight, a muscle goes from being 25% shorter than its resting length to 25% longer than it. If the resting length of the sarcomere is 2.5 μm, how many 11-nm crossbridge power strokes occur in each sarcomere during the lift?

5.26. Figure 5.36 is the force vs. displacement curve of a single myosin molecule interacting with an actin molecule during a powerstroke, as measured by "optical tweezers." Since work is the integral of force over distance, estimate the work done by the myosin molecule during one powerstroke. (Use the dotted line and find the area in the triangle.) If the energy available from ATP hydrolysis in a muscle cell is 10^{-19} J, calculate the efficiency of a myosin powerstroke in using the energy from ATP hydrolysis. This estimate uses the mean force. The efficiency is $\sim$1.75$\times$ larger if a less conservative estimate is made using the highest forces measured [247].

5.27. (a) If in a single actin–myosin crossbridge 4 pN is generated per crossbridge and the power stroke distance is 11 nm, as in Fig. 5.36, show that the mechanical work done is 22 pN–nm. Express this in J and eV (1 eV (electron volt) $= 1.6 \times 10^{-19}$ J) per crossbridge, and in kcal/mole (for a mole of crossbridges).
(b) If the hydrolysis of ATP releases 14 kcal/mole (as is seen in Chap. 6) and the hydrolysis of one ATP molecule activates one crossbridge, what fraction of the available energy is used in mechanical work?

6

Metabolism: Energy, Heat, Work, and Power of the Body

We cannot function without energy. The processes involved in the energy intake, storage, and use by the body are collectively called the *metabolism*; the discipline describing this area is sometimes called *bioenergetics.* More generally, metabolism is any energy usage by the body, and is the sum of all chemical processes performed by the cells in order to keep the body alive. For a complete picture we need to include input of food and oxygen to the body, energy storage, and loss of energy by the body through the loss of heat and work done by the body, as is shown in Fig. 6.1.

Metabolic processes can be divided into catabolic and anabolic reactions. In *catabolic* reactions complex molecules are broken into simple ones, for purposes such as energy usage. In *anabolic* reactions simple molecules are combined to form complex ones, for purposes such as energy storage.

The body uses food to (1) operate organs, (2) maintain a constant temperature by using some of the heat that is generated by operating the organs (while the rest is rejected), (3) do external work, and (4) build a stored energy supply (fat) for later needs. About 5–10% of the food energy intake is excreted in the feces and urine.

We will first consider the basics of the conservation of energy (thermodynamics). Then we will examine the energy content of food and the way it is stored in the body. We will see what the body's metabolic rate needs to be to perform tasks. (This is technically the catabolic rate.) We will then analyze how the body loses energy as heat. These steps are interrelated in a complex feedback and control mode that is discussed in Chap. 13.

For excellent general discussions about metabolism see [297, 298, 300, 301, 306, 321, 334, 340].

6.1 Conservation of Energy and Heat Flow

Let us briefly review some of the basics of the thermodynamics and heat flow physics that we will use in this discussion.

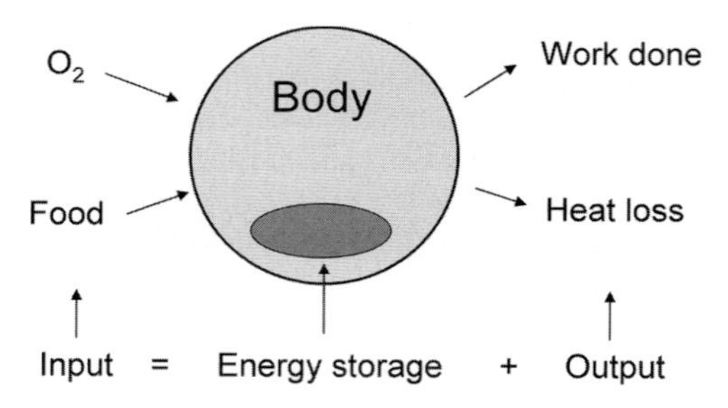

Fig. 6.1. Energy flow into and from the body

The First Law of Thermodynamics is essentially the conservation of energy in any process. In reference to the body, it can be stated as

$$\Delta U = Q - W, \tag{6.1}$$

where ΔU is the change in stored energy, Q is the heat flow to the body, and W is the mechanical work done by the body. The stored energy decreases, $\Delta U < 0$, when there is heat flow from the body, $Q < 0$, and work done by the body, $W > 0$. This type of work is purely mechanical in nature, such as in moving and lifting items. Heat flow includes heat production from the metabolism (Q_{met}) and heat loss (Q_{loss}) from radiation, convection, conduction, and evaporation. We can express $Q = Q_{\text{met}} + Q_{\text{loss}}$, where metabolic heat production is positive and a negative Q_{loss} indicates heat flow away from the body, so

$$\Delta U = Q_{\text{met}} + Q_{\text{loss}} - W. \tag{6.2}$$

Q_{met} is called the metabolic rate (MR).

Relationships in thermodynamics involve amounts of energies changing in a process at equilibrium and not those changing per unit time, i.e., the kinetics of that process, which involves the rates of energy changes or flows. The study of the metabolism usually involves rates and therefore

$$\frac{\mathrm{d}U}{\mathrm{d}t} = \frac{\mathrm{d}Q_{\text{met}}}{\mathrm{d}t} + \frac{\mathrm{d}Q_{\text{loss}}}{\mathrm{d}t} - \frac{\mathrm{d}W}{\mathrm{d}t} \tag{6.3}$$

is more appropriate. (We need to be careful about signs. The body increases its energy with terms such as $\mathrm{d}Q_{\text{met}}/\mathrm{d}t$ that are positive and loses it with terms such as $\mathrm{d}Q_{\text{loss}}/\mathrm{d}t$ that are negative. The amount of heat flowing from the body is $-\mathrm{d}Q_{\text{loss}}/\mathrm{d}t$, which is a positive quantity.)

All types of energy have the same units, including heat (often expressed in terms of calories) and work (often expressed in terms of joules). One important conversion between units is

$$1\,\text{calorie (cal)} = 4.184\,\text{joule (J)}. \tag{6.4}$$

1 kilocalorie (1 kcal = 1,000 cal) is sometimes called 1 Cal, which is also known as a food calorie. The energy content of food is always expressed in terms

Table 6.1. Units of power

1 watt (W) = 1 J/s
100 W = 1.43 kcal/min
1 horsepower (hp) = 746 W = 642 kcal/h
1 kcal/min = 69.7 W = 0.094 hp
1 kcal/h = 1.162 W

of these Cal (kcal) units. The relations between the various units of power (energy/time) are given in Table 6.1.

One important concept in the thermal physics of the body is the *heat capacity* C, which is the energy (or more specifically, the heat) required to raise the temperature T of an object by 1°C. The heat capacity per unit volume or mass is the *specific heat* c. The heat capacity is an *extensive property* of a given object, while the specific heat is an *intensive property* of a material. (This classification is analogous to that for mechanical properties described in Chap. 4.) If the specific heat is expressed per unit gram, it can be converted to that per unit volume by multiplying it by the mass density ρ. The heat capacity C is the specific heat (expressed per unit mass) $\times$ the total object mass m, so $C = mc$.

The temperature rise ΔT of an object with a heat flow Q to the body is

$$\Delta T = \frac{Q}{mc}. \tag{6.5}$$

For water, $c_{\text{water}} = 1.0\,\text{cal/g-}^\circ\text{C} = 1.0\,\text{kcal/kg-}^\circ\text{C}$. Even though the human body contains much water, the average specific heat of the body is a bit less, $c_{\text{b}} = 0.83\,\text{cal/g-}^\circ\text{C} = 0.83\,\text{kcal/kg-}^\circ\text{C}$. This means that it takes 83 kcal to raise the temperature of a 100 kg person by 1°C. This 83 kcal (83 food calories) is approximately the food energy content of a slice of bread. An obvious question arises: *If most of our metabolized energy becomes heat, why does not our body temperature increase by 1.0°C = 1.8°F each time we eat and metabolize a slice of bread?* We are very fortunate it does not. (The reason is heat loss by the body.)

The heat capacity of an object describes how its temperature changes with time due to heat flow to and from the object. The thermal conductivity K describes how the temperature varies (ΔT) spatially due to the heat flow between different regions that are separated by a distance Δx. (Conversely, it also describes how much heat flows due to this spatial variation in temperature.) This relation is

$$\frac{1}{A}\frac{\mathrm{d}Q}{\mathrm{d}t} = -K\frac{\mathrm{d}T}{\mathrm{d}x} \sim -K\frac{\Delta T}{\Delta x}. \tag{6.6}$$

The left-hand side is the amount of heat that flows per unit area A per unit time, and is also called the heat flux. The minus sign indicates that heat flows from hotter regions to colder regions. When there is a well-defined distance $d = \Delta x$ between two regions of different but uniform temperature, say due to the thickness of clothing or an air boundary layer, we can define a heat transfer coefficient per unit area $h = K/d$ and then

$$\frac{1}{A}\frac{\mathrm{d}Q}{\mathrm{d}t} = -h\,\Delta T. \tag{6.7}$$

Heat flow due to other mechanisms, such as due to radiation, can often be expressed in terms of (6.6) or (6.7).

One consequence of thermodynamics is that engines that convert chemical energy to heat and use that heat for mechanical work, so-called heat engines, have a limited efficiency to do such useful mechanical work. An ideal heat engine has a maximum efficiency of $\epsilon = 1 - (T_c/T_h)$ when it operates at a temperature T_h and rejects heat to a lower temperature T_c (both expressed in K). Humans operate internally at about T_h= 310 K and reject heat to a $T_c \simeq 293$ K ambient, so ϵ would be 5.5% if we were heat engines. This is much less than the ~25% efficiency of humans converting chemical energy into mechanical work. This is not a contradiction because we use the chemical energy directly to do mechanical work, as seen in Chap. 5, and do not produce heat in an intermediate step.

6.2 Energy Content of Body Fuel

There is some similarity between metabolic oxidation and combustion, even though the body does not "burn" its fuels in oxygen. It is useful to learn about the combustion of these fuels because combustion tells you the maximum amount of energy that is available from breaking and rearranging bonds. Metabolic oxidation is a bit less efficient, as we will see. The combustion energies (enthalpies) are obtained from "bomb" calorimetry in which the materials are burned in heavy-walled vessels. The resulting energies are useful and approximately correct even though the conditions of this bomb – its operation at constant volume and the actual "bomb" temperature – are different than those in the body. For example, many processes in the body occur at constant pressure and not constant volume.

Glucose. The metabolism of glucose is representative of that of carbohydrates. If we start with 1 mol of glucose (180 g, 6.02×10^{23} molecules) and oxidize it with 6 mol of oxygen molecules (192 g = 134.4 L at standard conditions, given there are 22.4 L/mol at 1 atm. and 0°C), there are 6 mol of carbon dioxide, 6 mol of water (108 g, 108 mL), and 686 kcal of energy produced. This can be represented by

$$C_6H_{12}O_6 + 6O_2 \rightarrow 6CO_2 + 6H_2O + 686\,\mathrm{kcal}. \tag{6.8}$$

The energy produced per mass of fuel is 686 kcal/180 g glucose = 3.80 kcal/g glucose. We also define a *calorific equivalent*, the energy produced per liter of oxygen consumed, which is 686 kcal/134.4 L O_2 = 5.5 kcal/L O_2 here.

We will see that in metabolic oxidation the body combines this 1 mol of glucose with 30–32 mol of ADP and 30–32 mol of the phosphate group P_i to form 30–32 mol of the energy storage molecule ATP.

Palmitic Acid. The oxidation of palmitic acid is representative of that of fatty acids. Oxidation proceeds by

$$CH_3(CH_2)_{14}COOH + 23O_2 \rightarrow 16CO_2 + 16H_2O + 2,397\,\text{kcal}, \qquad (6.9)$$

so 1 mol (256.4 g) of palmitic acid is burned by 23 mol (515.2 L) of oxygen to form 16 mol (358.4 L) each of carbon dioxide and water, and 2,397 kcal. The energy released is 2,397 kcal/256.4 g = 9.3 kcal/g and the calorific equivalent is 2,397 kcal/515.2 L O_2 = 4.7 kcal/L O_2. In metabolic oxidation the body uses this 1 mol of palmitic acid to combine 106 mol of both ADP and the phosphate group P_i to form 106 mol of ATP.

In comparing glucose $C_6H_{12}O_6$ with palmitic acid $C_{16}H_{32}O_2$, both have the same 1:2 ratio of C and H, so after oxidation, equal numbers of moles of CO_2 and H_2O are formed in both cases. However, palmitic acid and other fats have much less oxygen per C and H than glucose and other carbohydrates, so less mass of them is consumed per mole of CO_2 and H_2O formed. (The molar mass of palmitic acid is 256 g, and this is much less than the 480 g molar mass of the "equivalent carbohydrate" $C_{16}H_{32}O_{16}$.) Alternatively (and equivalently), we can say that fats are less oxidized, are more reduced, and have greater reducing power than carbohydrates.

These energies released per unit mass are the maximum energies available, and are really called *bomb calorimetry energies* because they are often measured in laboratory-controlled explosions. The amount available to the body is somewhat less and is defined as the *caloric value.* Table 6.2 shows the average caloric values and calorific equivalents of the types of basic body fuels.

The caloric value is also a bit lower than the bomb calorimetry value because of losses during digestion, which average 2% for carbohydrates, 5%

Table 6.2. Average caloric content of food. (Using data from [306])

food	net caloric value (kcal/g)	bomb calorimetry energy (kcal/g)	calorific equivalent (kcal/L O_2)	CO_2 production (kcal/L CO_2)	RER (L CO_2/L O_2)
carbohydrate	4.02	4.10	5.05	5.05	1.0
protein	4.20	5.65	4.46	5.57	0.80
ethanol	7.00	7.10	4.86	7.25	0.67
fat	8.98	9.45	4.74	6.67	0.71

RER is the respiratory exchange ratio.

for fats, and 8% for proteins plus an additional loss of 17% of protein energy in the urine. The small losses for ethanol are in the urine and exhaled air. Note that the bomb calorimetry energy of glucose is a bit under the average for carbohydrates. The caloric value for carbohydrates and proteins is $\simeq$4 kcal/g and that of fats is about $\simeq$9 kcal/g. (Each kcal is a food calorie, or Cal, which is called a calorie on food package labels.) Ethanol provides much caloric input, which explains why it is easy to gain weight by drinking alcohol, such as beer. If your caloric input is mainly from alcohol, you can "starve" to death in the sense that your caloric needs are being met, but not your micronutrient needs, such as vitamins and minerals. These caloric values can be compared to common heating and automotive fuels: gasoline, 11.4 kcal/g; coal, 8.0 kcal/g; and pine wood, 18.5 kcal/g. The energy released per L O_2 (calorific equivalent) is fairly constant for all body fuels, ranging from $\sim$4.5 to 5.5 kcal/L O_2. For a "mixed diet" the calorific equivalent can be estimated to be 4.83 kcal/L O_2 and the CO_2 production to be 5.89 kcal/L CO_2. The energy storage densities of different chemical fuels are compared in Table 6.3.

The *respiratory exchange ratio* (RER) (or *respiratory quotient* (RQ)) is another way to characterize metabolic processes. It is the number of moles of CO_2 produced/number of moles O_2 used, and is a measure of how much carbon dioxide needs to be released in respiration relative to how much oxygen

Table 6.3. Energy storage density for chemical fuels. (Using data from [308])

energy storage fuel	storage density (J/m^3)	storage density (J/kg)
ATP	1.4×10^8	1.0×10^5
H_2 gas, 10^3 atm.	4.9×10^9	1.2×10^8
nitroglycerine	1.0×10^{10}	6.3×10^6
glycine (amino acid)	1.0×10^{10}	6.5×10^6
wood	1.1×10^{10}	1.9×10^7
urea	1.4×10^{10}	1.1×10^7
methanol	1.8×10^{10}	2.2×10^7
vegetable protein	2.3×10^{10}	1.7×10^7
acetone	2.4×10^{10}	3.1×10^7
glucose	2.4×10^{10}	1.6×10^7
glycogen (starch)	2.5×10^{10}	1.8×10^7
animal protein	2.5×10^{10}	1.8×10^7
carbohydrate	2.6×10^{10}	1.7×10^7
gasoline	2.8×10^{10}	4.4×10^7
butane	3.0×10^{10}	4.9×10^7
fat	3.3×10^{10}	3.9×10^7
cholesterol (lipid)	4.2×10^{10}	3.9×10^7
H_2 solid (10^5 atm.)	7.2×10^{10}	1.2×10^8
diamond	1.2×10^{11}	3.3×10^7

Note that 10^{10} $J/m^3 = 2.39$ $kcal/cm^3$ and 10^7 J/kg = 2.39 kcal/g.

Table 6.4. Caloric value of 1 rich frosted Entenmann's^{TM} donut (in 2005)

18 g fat	×9 kcal/g	= 162 kcal
29 g carbohydrate	×4 kcal/g	= 116 kcal
2 g protein	×4 kcal/g	= 8 kcal
49 g total		= 286 kcal

the body needs to bring in by respiration. We see that RER = 6 L CO_2/6 L O_2 = 1.0 for glucose oxidation and 16 L/23 L = 0.7 for palmitic acid oxidation. It is typically 0.8 for protein oxidation.

We can put the caloric value in perspective by developing a standard unit: The Donut. We will use 1 Rich Frosted (i.e., chocolate frosted) Entenmann's^{TM} Donut as the standard donut. The package labeling says that a donut contains 18 g of fat, 29 g of carbohydrate, and 2 g of protein (in 2005), so we can estimate the caloric value of each donut, as in Table 6.4. We see that 49 g of the total 57 g mass of the donut has caloric content. Our calculation suggests there are 286 kcal (286 food calories) per donut. The package says there are 280 kcal per donut. These values are very consistent with each other because we have used average caloric values for the fat, carbohydrate, and protein. Also, we have rounded off the caloric values and they have provided values of the mass of each body fuel in grams rounded off to the nearest integer (as required by the US FDA (Food and Drug Administration)). (For more on rounding off see Problems 6.11 and 6.12.) Approximately 57% of the calories (162 kcal/286 kcal) come from fat. We will soon determine how much physical exertion is needed to remove the "fattening" consequences of eating a standard donut. (Problems 6.13 and 6.14 address other potential standard donuts.)

6.2.1 Metabolizable Energy and Energy Storage

We should be a bit more careful about defining how much of the energy from food is being metabolized and actually used, even though these relatively fine distinctions in body energetics may not be very significant to us here. The rate of *apparently digested energy* is the difference between the rate of intake of dietary energy and the rate of loss of energy in the feces. The rate of obtaining *metabolizable energy* (ME), $d(ME)/dt$, is the difference between the rate of intake of dietary energy and the sum of the rates of loss of energy in the feces, urine, and combustible gas. This is the actual rate at which energy is being made available to the body [298].

Table 6.5 shows the caloric content and components of several types of food. The caloric content per unit mass depends on the relative amounts of carbohydrates, proteins, and fats, and ethanol for alcoholic drinks, and the amounts of water and "ash." For comparison, our standard donut has 280 kcal, or 280 kcal/57 g = 4.91 kcal/g. Table 6.6 shows the enthalpy of combustion for different types of milk; this constitutes consumed and not necessarily

Table 6.5. Components and energy (kcal) of edible parts of common foods. (Using data from [311])

food, serving size	mass (g)	energy (kcal)	carb.[a] (g)	protein (g)	fat (g)	water (%)
grains and cakes						
bread, white, 1 slice	25	67	12	2	1	37
oatmeal, regular, 1 cup prepared	234	145	25	6	2	85
yellow cake, chocolate frosting, 1 piece	64	243	35	2	11	22
cheesecake, 1/6 of 17 oz cake	80	257	20	4	18	46
dairy						
milk, whole, 1 cup	244	150	11	8	8	88
butter, salted, 1/4 lb stick	113	813	Tr[b]	1	32	16
cheddar cheese, 1 oz	28	114	Tr	7	9	37
cottage cheese, 4%, 1 cup	225	233	6	28	10	79
ice cream, chocolate, 1/2 cup	66	143	19	3	7	56
eggs, raw, 1 large	50	75	1	6	5	75
meat and fish						
chicken, meat only, roasted 1/2 breast	86	142	0	27	3	65
beef, ground, 79% lean, broiled, 3 oz	85	231	0	21	16	56
salmon, broiled, 3 oz	85	184	0	23	9	62
fruits, vegetables, nuts, and oils						
apple, raw, unpealed, 1 whole	138	81	21	Tr	Tr	84
apricots, raw, without pits, 1 whole	35	17	4	Tr	Tr	86
apricots, dried, sulfured, 10 halves	35	83	22	1	Tr	31
orange, peeled, 1 whole	131	62	15	1	Tr	87
carrots, raw, $7\frac{1}{2}$ in long	72	31	7	1	Tr	88
potato, baked, with skin	202	220	51	5	Tr	71
French fries, medium portion	134	458	53	6	25	35
peanuts, dry roasted, 1 cup	146	854	31	35	73	2
walnuts, 1 cup chopped	120	785	16	18	78	4
canola oil, 1 cup	218	1,927	0	0	218	0
beverages						
cola, 12 fl oz	370	152	38	0	0	89
beer (regular), 12 fl oz	355	146	13	1	0	92
gin, vodka, whiskey, 86 proof, 1.5 fl oz	42	105	Tr	0	0	64
wine, red, 3.5 fl oz	103	74	2	Tr	0	89

[a] Carbohydrate.
[b] Trace.

metabolizable energy. The enthalpy available from fat, protein, and carbohydrates is quite different for each source. With a 42% fat/18% protein diet, about 89.3% of the consumed food is typically metabolizable – meaning that this is the fraction of the heat of combustion (caloric content) of food that is

Table 6.6. Food content of milk. (Using data from [298])

species	enthalpy of combustion (kcal/g)	distribution of enthalpy (%)			dry matter (%)
		fat	protein	carbohydrate	
human	0.69	54	7	39	12.4
cow	0.71	48	26	26	12.4
goat	0.69	50	22	27	12.0
horse	0.51	23	22	54	10.5
seal (northern fur)	5.09	88	11	0.1	61.0

metabolizable energy. About 5.8% of the consumed energy is lost in the feces, 4.5% in the urine, and 0.4% lost as methane. The average retention time of food in a human's digestive track is 46 h; the first appearance of the residues of food appears in about one third this mean time and the last in about four times this time.

Energy can be retained or secreted (as milk) by the body as the enthalpy of tissues R (which includes fat, protein, and CHO). (The enthalpy (or heat content) is the maximum thermal energy that is obtainable at constant pressure.) The rate at which enthalpy is retained $\mathrm{d}R/\mathrm{d}t$ is the difference between the rate of metabolizable energy input and heat production through the metabolism

$$\frac{\mathrm{d}R}{\mathrm{d}t} = \frac{\mathrm{d(ME)}}{\mathrm{d}t} - \frac{\mathrm{d}Q_{\mathrm{met}}}{\mathrm{d}t}. \tag{6.10}$$

(This ignores lactation.) We combine this with (6.3) to give

$$\frac{\mathrm{d}R}{\mathrm{d}t} = \frac{\mathrm{d(ME)}}{\mathrm{d}t} + \frac{\mathrm{d}Q_{\mathrm{loss}}}{\mathrm{d}t} - \frac{\mathrm{d}W}{\mathrm{d}t} - \frac{\mathrm{d}U}{\mathrm{d}t}. \tag{6.11}$$

When food is not eaten ($\mathrm{d(ME)}/\mathrm{d}t = 0$), $\mathrm{d}R/\mathrm{d}t$ is negative and its magnitude is the heat of catabolism of body tissues. As more food is eaten, $\mathrm{d}R/\mathrm{d}t$ increases, and less stored (or retained) energy is used and there is increased heat production; this increased heat production is the *specific dynamic or thermogenic effect of food.* When energy retention is zero over several days (so no net stored energy is used or formed and $\mathrm{d}R/\mathrm{d}t$ averages to zero), the input dietary energy ME is that required for *maintenance*. For greater values of ME, energy is stored in the body.

Table 6.7 shows the results of calorimetric experiments for people who are fasting (for whom stored energy in proteins and fats is turned into heat), eating who do no exercise (who have net storage of body fat), and eating who do exercise (for whom a small amount of stored energy is turned into heat, even though there is more food intake).

Figure 6.2 shows the retention of energy and the heat production (thermogenic effect) vs. food intake. A fraction of the metabolizable energy from food intake is retained in the body, K, and a fraction is lost as heat, $1 - K$.

Table 6.7. Calorimetric experiments each averaged over several runs (in kcal/day). (Using data from [298], from [295])

energy	eating, without exercise	eating, with exercise (bicycle)	fasting
intake energy	2,659	4,340	0
feces energy	107	176	0
urine energy	134	138	105
change in body protein	−16	−57	−463
change in body fat	176	−484	−1,892
heat produced	2,270	4,554	2,187
discrepancy	12	−13	−63

The slope of the retention curve k is

$$k = \frac{\mathrm{d}R/\mathrm{d}t}{\mathrm{d(ME)}/\mathrm{d}t} \tag{6.12}$$

and the slope of the heat production curve is $1 - k$. K is an averaged value of k. Because the curves in Fig. 6.2 are not linear, k and K are usually not equal. k, also called the *efficiency of the utilization of metabolizable energy*, varies with the amount of food consumed. Below maintenance, k is $\simeq$0.90 – where it represents the efficiency of using stored energy – and above it is $\simeq$0.75 – where it really represents the efficiency of forming stored energy. This means that below maintenance only about 10% of food goes to heat and above it about 25% goes into heat production – and so 75% goes into making us fatter (unless we are building up our muscles). These numbers are for average diets. For carbohydrates, $k = 0.94$ below maintenance and 0.78 above it. For fats, $k = 0.98$ below maintenance and 0.85 above it. For proteins, $k = 0.77$ below maintenance and 0.64 above it. The body is more

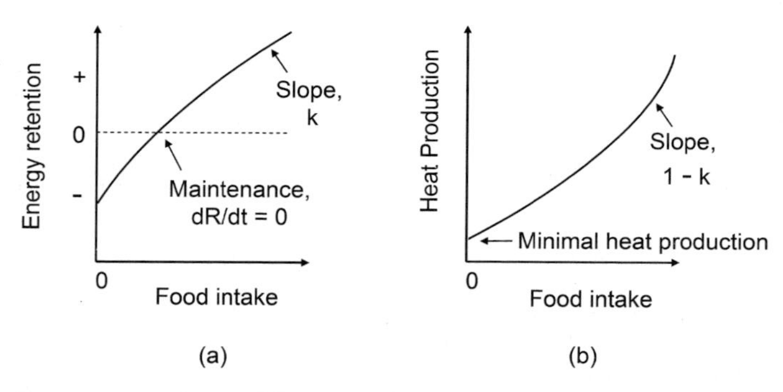

Fig. 6.2. Energy retention and heat production in the body vs. food intake. (Based on [298])

efficient in using stored energy than in depositing fat and proteins. Proteins are metabolized less efficiently than are carbohydrates and fat [298].

6.3 Energy Storage Molecules

6.3.1 How ATP is Produced and Used as an Energy Source

Catabolism. ATP, adenosine triphosphate (a-duh'-nuh-seen), is the basic unit of energy storage in the body and it enables the rapid release of energy. *Why does the body convert food fuel to ATP and not directly oxidize carbohydrates, fatty acids, and proteins?* The use of ATP is more controllable. Also, the unit of energy provided by ATP is small enough to be useful.

The structure of ATP is shown in Fig. 6.3. It consists of a five-carbon sugar, ribose, that is linked to the aromatic base, adenine – forming adenosine, and three phosphate groups. The two bonds linking the phosphate groups are unstable, high-energy bonds; the leftmost interphosphate bond in Fig. 6.3 splits in the hydrolysis of ATP

$$\mathrm{ATP} + \mathrm{H_2O} \rightarrow \mathrm{ADP} + \text{inorganic phosphate} + \text{energy}, \tag{6.13}$$

where ADP is adenosine diphosphate. This is the basic catabolic process for energy release, and was discussed as the driving force in the myosin power stroke in generating force in muscles in Chap. 5. The energy released, or more precisely the free energy, ranges from 7 to 14 kcal/mol of ATP, depending on conditions (see Problem 6.30). Under typical cellular conditions it can be $\simeq$12–14 kcal/mol [320, 322, 329].

Anabolism. After hydrolysis, ADP needs to be combined with a phosphate group to reform ATP for later use. On the average, each ATP molecule is recycled this way every minute (see Problem 6.36). *How does the body*

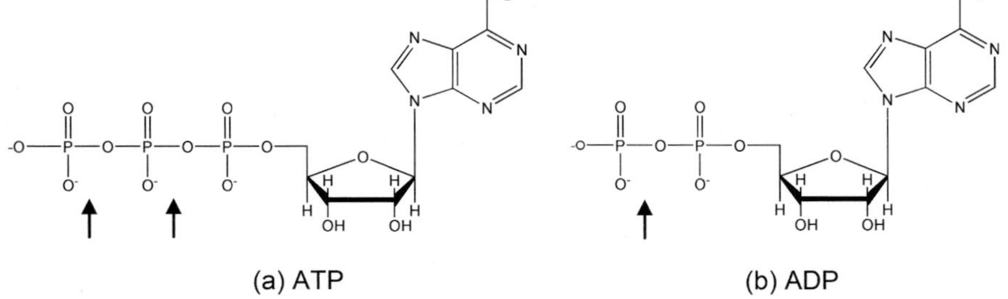

(a) ATP

(b) ADP

Fig. 6.3. The structures of (**a**) ATP and (**b**) ADP, showing that ATP has one more inorganic phosphate group P_i than does ADP. From left to right are the (**a**) three or (**b**) two phosphate groups, the five-carbon sugar, ribose, and adenine, which is the double-ringed structure. The two unstable, high energy bonds linking the phosphate groups in ATP and the one similar unstable bond in ADP are denoted by *arrows* *use food sources like glucose to do this?* This occurs by a series of chemical steps that can proceed to a limited extent without oxygen (*anaerobic glycolysis*) and to a greater extent with oxygen (*aerobic metabolism or respiration*) [301, 306, 321, 329, 340]. This utilizes glucose in the blood stream and glycogen, $(C_6H_{12}O_6)_n$ – a branched-chain polymerized sugar consisting of glucose molecules as monomers linked together with glycosidic (oxygen) bonds. Glycogen is stored in muscle cells, where it is used directly, and in the liver, where is it broken down into glucose by glycogenolysis, which is then delivered by the blood to the other cells. In aerobic metabolism, the products of the anaerobic steps are further metabolized (in the presence of oxygen) to complete the metabolism of stored glucose by a complex series of steps collectively called the *Krebs cycle* and the *electron transfer system (ETS)*; they are shown in Fig. 6.4 and described in a bit more detail later. The overall result is that in aerobic metabolism 1 mol of glucose can produce approximately 30–32 mol of ATP (depending on the details of the membrane shuttle mechanism). (In fact, 1 mol of carbohydrates from muscle glycogen can produce approximately 31–33 mol of ATP, but energy is used in forming glycogen from glucose; we

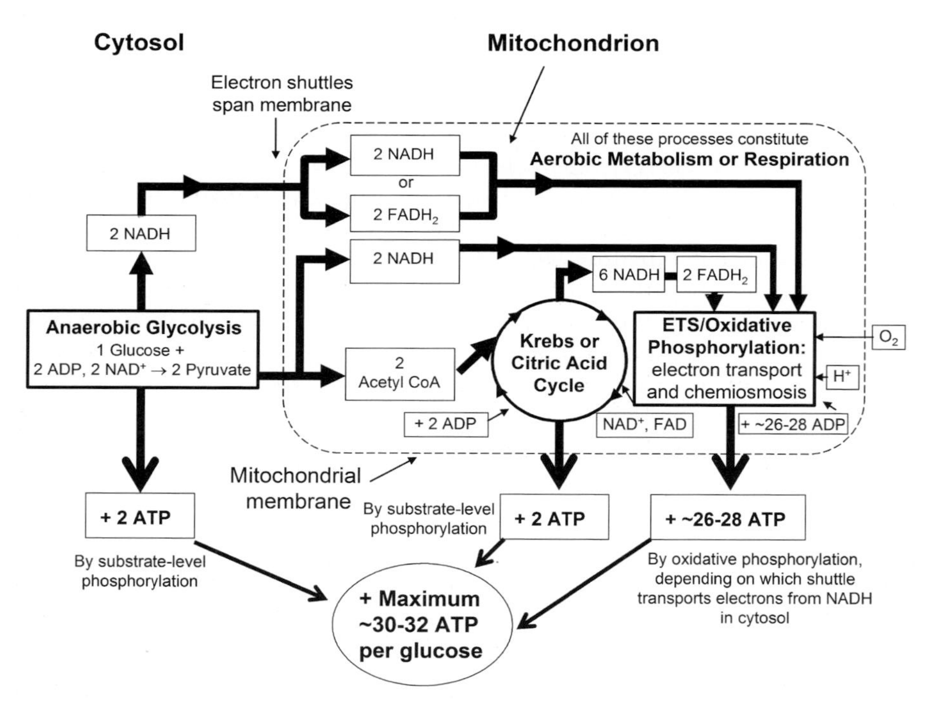

Fig. 6.4. Schematic of the processes involved in the resynthesis of ATP from ADP and inorganic phosphate by anaerobic glycolysis in the cytosol (the intracellular fluid) and aerobic metabolism (or respiration) in the mitochondrion. (Based on [301, 324, 329])

will consider only the metabolism of glucose.) Of these, 2 mol are produced by anaerobic processes and 28–30 additional moles are produced when there is sufficient oxygen. The bottom line is that the hydrolysis of these, say for now, 30 mol of ATP provide from glucose

$$30\,\text{mol ATP/mol glucose} \times 14.0\,\text{kcal/mol ATP} = 420\,\text{kcal/mol glucose.} \tag{6.14}$$

This should be compared to the energy from the combustion of glucose

$$180\,\text{g/mol glucose} \times 3.8\,\text{kcal/g glucose} = 686\,\text{kcal/mol glucose.} \tag{6.15}$$

This means that the efficiency of the body's usage of glucose to form available energy in the form of ATP is 420 kcal/686 kcal = 61%; for 32 ATP formed by glucose, this gives 448 kcal/686 kcal = 65%. This efficiency range of $\simeq$61–65% is for conditions in muscle cells. For other specific conditions the free energy of ATP can be a bit lower. (This is explored in Problems 6.30 and 6.31.) In any case, this efficiency is pretty good, although it is not perfect. (It is much better than the efficiency of typical heat engines, $\sim$10–20%.) The overall efficiency of using glucose to do mechanical work is much lower, because of other efficiency factors that we will discuss later.

6.3.2 How ATP is Actually Used by the Body

ATP is the ultimate source of energy for muscular motion, but it is not always the locally stored source. The body has a mechanism for using it directly, and then several levels of mechanisms for transferring energy from other molecules to the formation of ATP (from ADP), which is then used directly. There are four levels of steps, each of which can be used for successively longer times, although at successively lower levels of activity (Table 6.8) [306, 324, 329].

Step 1. Normally there is enough ATP in living skeletal muscles to supply energy for about 8 twitches. This may be enough for about 3 s or for about half of a 50 m dash. The energy comes from (Fig. 6.3)

$$\text{ATP} + \text{H}_2\text{O} \rightarrow \text{ADP} + \text{P}_\text{i} + \text{H} + 14\,\text{kcal/mol.} \tag{6.16}$$

Table 6.8. Estimated power and energy available from the body, for a 70 kg man with 30 kg of muscle, assuming 10 kcal/mol of ATP. (Using data from [306])

system	maximum power (moles of ATP/min)	maximum capacity (total moles of ATP)
phosphagen (ATP-PC) system	3.6	0.7
anaerobic glycolysis	1.6–2.5	1.2
aerobic metabolism from glycogen	1.0	90.0

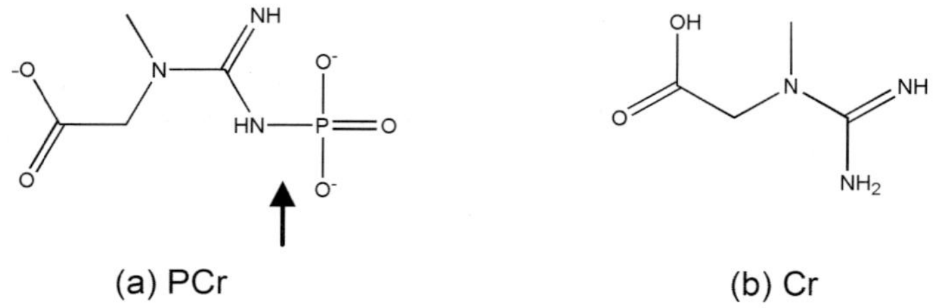

(a) PCr (b) Cr

Fig. 6.5. The structures of (**a**) phosphocreatine (PCr) and (**b**) creatine (Cr), showing that PCr has an inorganic phosphate group P_i. The unstable, high-energy bond linking the phosphate group in PCr is denoted by an *arrow*. The enzyme that catalyzes the breakdown of PCr to form ATP, which is the transfer of the phosphate group from PCr to ADP to form ATP (6.17), is called creatine kinase

Step 2. Now more ATP is needed. It is resynthesized from the local phosphocreatine (or creatine phosphate) (PCr) (Fig. 6.5) reservoir by

$$\mathrm{ADP} + \mathrm{PCr} \rightarrow \mathrm{ATP} + \mathrm{Cr}, \tag{6.17}$$

where Cr is creatine. This reaction is driven strongly to the right because it has an equilibrium constant >20. Muscle has enough PCr to provide ATP for about ∼100 twitches, or for the first 50 m (∼8–10 s) of a sprint.

Together, steps 1 and 2 (reactions (6.16) and (6.17)) constitute the *phosphagen system* or the ATP-PC system. It is the source of the highest peak power, with about 4 mol ATP used by a person every minute. It can be used for short times, 8–10 s, and at most for 100 m dashes or for limited jumping. PCr can be reformed from Cr and P_i by using ATP itself. For intense levels of exercise, this can occur only after the activity, and usually occurs by aerobic metabolism.

When more energy is needed than can be supplied by the phosphagen system alone, it is produced from glycogen that is stored in the muscle and from the metabolism of glucose and fatty acids in the blood stream. Two high-energy molecules are produced as intermediates during this formation of ATP (1) NADH (from NAD^+, nicotinamide adenine dinucleotide, a derivative of the vitamin niacin) and (2) $FADH_2$ (from FAD, flavin adenine dinucleotide, derived from riboflavin, a B vitamin). The modes of using these energy sources to form ATP are different for light and heavy activity.

Step 3. During "heavy" activity people need ATP fast and there is not enough oxygen for aerobic metabolism, so only anaerobic glycolysis (the Embden–Meyerhof pathway) occurs. One mole of glucose-1-phosphate (obtained from 1 mol of carbohydrates from glycogen) can convert only 3 mol of ADP to ATP. In this process 1 mol of ATP is used to make fructose-1, 6-diphosphate from fructose-6-phosphate (which is produced from glucose-1-phosphate), and 2 mol of ATP are formed in each of the two subsequent steps of glycolysis, leading to a net of 3 mol of ATP. Metabolism of blood glucose yields a net of only 2 mol of ATP because it takes 1 mol of ATP to convert it into glucose-6-phosphate (see Problem 6.33). This all occurs in the intracellular fluid in the muscle cell (the *cytosol*).

This degradation of the glycogen or glucose is only partial, and without sufficient oxygen the potential of producing the other 28–30 mol of ATP by the other steps in aerobic metabolism is totally lost. At the end of anaerobic glycolysis, energy is still stored in the 2 pyruvic acid molecules ($C_3H_4O_3$) (pyruvate ions in solution) formed from each glucose molecule, which are converted to 2 lactic acid molecules ($C_3H_6O_3$) (lactate ions in solution) if they cannot be used in the Krebs cycle in aerobic metabolism. In addition, 2 NADH molecules are formed per glucose molecule in anaerobic glycolysis, and the energy from these high-energy molecules is also wasted because they are not metabolized by the ETS in aerobic metabolism (see Fig. 6.4). (However, the NADH can remain until the oxygen debt is recovered.)

The advantage of this *anaerobic glycolysis or glycogen-lactic acid system* is that it provides a medium level of power, ∼2.5 mol ATP/min, which is ∼60% of that of the phosphagen system. It provides power for about 600 twitches, for an intermediate time ∼1.3–1.6 min – enough for a 400 m run. (The maximum anaerobic peak power is 2.1 hp for men and 1.7 hp for women. These values decrease dramatically after 25 years of age.) Anaerobic glycolysis has disadvantages. In addition to being inefficient, it produces lactic acid, which causes discomfort from acidosis, and causes fatigue. Because the pyruvate-lactate reaction is reversible, the lactic acid from anaerobic glycolysis can be converted to form pyruvate when oxygen is available during the recovery from heavy exercise (but primarily in the liver and not the muscle).

Step 4. During "light" exercise there is enough oxygen for aerobic metabolism to complete the oxidation of the carbohydrates. Aerobic metabolism consists of two complex processes in addition to anaerobic glycolysis: the Krebs cycle (which is also known as the tricarboxylic acid (TCA) cycle or the citric acid cycle) and the electron transfer system (ETS) (which is also known as the cytochrome pathway or oxidative phosphorylation) (Fig. 6.4). They both occur in the cell mitochondrion. Only the ETS directly requires oxygen, but without oxygen even the benefits of the Krebs cycle are lost due to back reactions. (By the way, Hans Adolf Krebs shared the Nobel Prize in Physiology or Medicine in 1953 for his discovery of the citric acid cycle.)

The 2 mol of NADH from anaerobic glycolysis are shuttled in through the mitochondrion membrane. Two moles of pyruvate from anaerobic glycolysis serve as the starting material for the Krebs cycle. Before entering this cycle, each pyruvate produces an acetyl CoA molecule, which enters the Krebs cycle, and 1 NADH, so 2 acetyl CoA and 2 NADH are produced per glucose molecule. For each pyruvate, the Krebs cycle produces 1 ATP, 3 NADH, and 1 $FADH_2$, along with 2 CO_2 molecules, which are waste products, and so for each glucose molecule the Krebs cycle produces 2 ATP, along with 6 NADH and 2 $FADH_2$ molecules.

This means that for each mole of glucose, a total of 10 mol of NADH and 2 mol of $FADH_2$ enter the ETS, where they are converted to ATP and

H_2O. (Alternatively, if the 2 mol of NADH from anaerobic glycolysis shuttled through the mitochondrion membrane are cycled through the membrane as 2 mol of $FADH_2$, a total of 8 mol of NADH and 4 mol of $FADH_2$ enter the ETS.) There are several series of alternative pathways in which these high-energy molecules are converted to ATP, which lead to effectively 2.5 ATP per NADH and effectively 1.5 ATP per $FADH_2$, or $2.5 \times 10 + 1.5 \times 2 = 28$ mol of ATP. There are also the 2 mol of ATP directly from anaerobic glycolysis and the 2 mol of ATP directly form the Krebs cycle, for a total of 32 mol of ATP. Overall, a total of 30–32 mol of ATP are produced from 1 mol of glucose (depending on the membrane shuttle mechanism). (This estimate of 30–32 mol of ATP per mole of glucose is cited in more current work [321, 329]; older estimates were of 36–38 mol of ATP produced per mole of glucose [320].)

ATP is also formed by the oxidation of fatty acids and some amino acids and proteins. (Fats and protein can be metabolized only in the presence of oxygen.) The role of protein metabolism is very minor during rest and does not contribute more than 5–10% of the total energy supply during normal exercise. Each fatty acid molecule is activated using 2 ATP molecules, and is then metabolized in successive passes through a metabolic cycle within the mitochondrion in which two carbon atoms are lost in each pass to produce a molecule of acetyl-CoA, NADH, and $FADH_2$ by β oxidation, until acetyl-CoA is left. For the 16-carbon chain palmitic acid, there are seven complete passes and so one mole of palmitic acid produces 8 mol of acetyl-CoA (including the one remaining after β oxidation), 7 mol of NADH, and 7 mol of $FADH_2$. Each mole of acetyl-CoA is then oxidized in the Krebs cycle to give 3 mol of NADH, 1 mol of $FADH_2$, and 1 mol of ATP. The NADH and $FADH_2$ enter the ETS system. For palmitic acid, this means that 31 mol of NADH and 15 mol of $FADH_2$ enter the ETS system. Overall, $2.5 \times 31 + 1.5 \times 15 = 100$ mol of ATP are formed in the ETS. Including the other 8 mol of ATP, 108 moles of ATP are formed per mole of palmitic acid. Including the 2 ATP moles used for activation, a net of 106 moles of ATP are formed [329].

A healthy man with proper training can provide ~50 mL oxygen/kg-min. This aerobic system delivers the lowest peak power, 1 mol ATP/min ~25% of the phosphagen system, but for a very long time. There is enough glycogen for ~10,000 twitches. It is useful for long-distance, endurance activities, including marathon running. The glycogen in the muscle is depleted after several hours (~1.5–4.0 h) of this level of activity. (Glucose in the blood can also be converted into pyruvic acid and then be used in the cell mitochondria in the presence of oxygen to form more ATP. This is an efficient, long-term process.)

Negligible oxygen is stored in the body. It must be brought in by the lungs and transferred to the blood in the arteries continuously. The oxygen is transferred to the cells, leaving oxygen-depleted blood in the veins. We can track the conservation of oxygen molecules in the body. The rate of body consumption of O_2, $\mathrm{d}V_{O_2}/\mathrm{d}t$ equals the product of the cardiac output (blood flow rate, Q_t, see Chap. 8) and the difference in the oxygen content (oxygen

Table 6.9. Estimated energy available from the body (per kg and also, in parentheses, total), for a 70 kg man with 30 kg of muscle, assuming 10 kcal/mol ATP. (Using data from [306])

system	muscular amounts (in mmol/kg muscle[a]) (total)	useful energy (in kcal/kg muscle) (total)
phosphagen (ATP-PC) system		
ATP	4–6 (120–180)	0.04–0.06 (1.2–1.8)
PC	15–17 (450–510)	0.15–0.17 (4.5–5.1)
Total: ATP + PC	19–23 (570–690)	0.19–0.23 (5.7–6.9)
anaerobic glycolysis		
ATP formation	33–38 (1,000–1,200)	0.33–0.38 (10.0–12.0)
aerobic metabolism		
from stored glycogen	13–15 g (400–450)	
ATP formation	2,800–3,200 (87,000–98,000)	28–32 (870–980)

[a]Unless otherwise noted.

partial pressure) in the arteries and veins, $p_\mathrm{a} - p_\mathrm{v}$:

$$\frac{\mathrm{d}V_{\mathrm{O}_2}}{\mathrm{d}t} = Q_\mathrm{t}(p_\mathrm{a} - p_\mathrm{v}). \tag{6.18}$$

During aerobic exercise, $\mathrm{d}V_{\mathrm{O}_2}/\mathrm{d}t$ increases linearly with Q_t. For a person with average fitness, the maximum oxygen use and blood flow rates are $(\mathrm{d}V_{\mathrm{O}_2}/\mathrm{d}t)_\mathrm{max} \approx 2.8\,\mathrm{L/min}$ (L of oxygen gas at an atmosphere) and $(Q_\mathrm{t})_\mathrm{max} \approx 19\,\mathrm{L/min}$, and for a highly fit person they are $\approx 4\,\mathrm{L/min}$ and $25\,\mathrm{L/min}$, respectively.

Table 6.9 gives the concentrations and total amounts of energy storage molecules and the energy available from them for a 70 kg man with 30 kg of muscle. Women have about the same concentrations of ATP and PC per kg muscle as do men, but have less overall muscle. The typical specific chemical energy resources stored in the body are listed in Table 6.10.

Table 6.11 lists the energy system that is primarily used in various sports activities. This is explored in more detail in Table 6.12, which gives the fraction of each system used in different sport activities. These fractions are given for shorter to longer distances in ice speed skating, swimming, and running in Tables 6.13–6.15. In each case, short distance sprints use the ATP-PC and anaerobic systems, while the longer distance events primarily use the aerobic system.

The metabolic steps involved in these systems are all regulated by a feedback and control mechanism (Chap. 13), and are followed by a recovery step. Table 6.16 gives the recovery time for returning resting energy reserves and O_2 and for reducing lactic acid after exercise. Anaerobic exercise is limited by the maximal lactic acid tolerance, which is about 2.0–2.3 g/kg muscle, and so this is 60–70 g for 30 kg of muscle. You can learn a bit about the recovery step

Table 6.10. Body stores of fuel and energy, for a 65 kg (143 lb) person with 12% body fat. (Using data from [340])

	amount (g)	energy (kcal)
carbohydrates		
liver glycogen	110	451
muscle glycogen	500	2,050
glucose in body fluids	15	62
carbohydrates total	625	2,563
fat		
subcutaneous and visceral	7,800	73,320
intramuscular	161	1,513
fat total	7,961	74,833

Table 6.11. Energy systems used in sports. (Using data from [314])

mostly phosphagen system
100-m dash
jumping
weight lifting
diving
football dashes
phosphagen and glycogen–lactic acid systems
200-m dash
basketball
baseball home run (running around the bases)
ice hockey dashes
mainly glycogen–lactic acid system
400-m dash
100-m swim
tennis
soccer
glycogen–lactic acid and aerobic systems
800-m dash
200-m and 400-m swim
1,500-m skating
boxing
2,000-m rowing
1,500-m/1-mile run
aerobic systems
10,000-m skating
cross-country skiing
marathon run (26.2 miles, 42.2 km)
jogging

Table 6.12. Percent emphasis of energy systems. (Using data from [306, 307])

sport or activity	ATP-PC and anaerobic glycolysis	anaerobic glycolysis and aerobic	aerobic
aerobic dance	5	15–20	75–80
baseball	80	15	5
basketball	60	20	20
diving	98	2	negligible
fencing	90	10	negligible
field hockey	50	20	30
football	90	10	negligible
golf	95	5	negligible
gymnastics	80	15	5
ice hockey			
forward, defense	60	20	20
goalie	90	5	5
lacrosse			
goalie, defense, attacker	50	20	30
midfielders, man-down	60	20	20
rowing	20	30	50
soccer			
goalie, wings, strikers	60	30	10
halfbacks or sweeper	60	20	20
stepping machine	5	25	70
tennis	70	20	10
field events, in track and field	95–98	2–5	negligible
volleyball	80	5	15
walking	negligible	5	95
wrestling	90	5	5

Table 6.13. Percent emphasis of energy systems for a range of distances in ice speed skating. (Using data from [306, 307])

activity	ATP-PC and anaerobic glycolysis	anaerobic glycolysis and aerobic	aerobic
ice speed skating			
500 m	80	10	10
1,000 m	35	55	10
1,500 m	20–30	30	40–50
5,000 m	10	25	65
10,000 m	5	15	80
in-line skating			
>10 km	5	25	70

Table 6.14. Percent emphasis of energy systems for a range of distances in swimming. (Using data from [306, 307])

swimming	ATP-PC and anaerobic glycolysis	anaerobic glycolysis and aerobic	aerobic
50 m	90	5	5
100 m	80	15	5
200 m	30	65	5
400 m	20	40	40
1,500 m	10	20	70

by running as fast as you can for as long as you can. You will then be huffing and puffing, breathing in air as fast as possible. This is part of recovery. (Consult with your physician before attempting this demonstration.)

6.4 Metabolic Rates

There are variations among people's metabolic rates (MR) due to their different weights, genes, etc. First, we will discuss the minimum metabolic rates for people and then address how activity increases the metabolic rate. The metabolic rate can also depend on food intake. For example, the MR is known to decrease under fasting conditions and this change counters some of the expected effects of dieting. Although widely called metabolic rates, these are more precisely catabolic rates.

6.4.1 Basal Metabolic Rate

The basal metabolic rate (BMR) is that of an inactive, awake body. The BMR for a 70 kg person is about 1,680 kcal/day ~70 kcal/h ~81 W. This means that

Table 6.15. Percent emphasis of energy systems for a range of distances in running. (Using data from [306, 307])

running	ATP-PC and anaerobic glycolysis	anaerobic glycolysis and aerobic	aerobic
100, 200 m	95–98	2–5	negligible
400 m	80	15	5
800 m	30	65	5
1,500 m (or mile)	20–30	20–30	40–60
3,000 m (or 2 miles)	10	20	70
5,000 m (or 3 miles)	10	20	70
10,000 m (or 6 miles)	5	15	80
marathon	negligible	5	95

Table 6.16. Recovery times after exhaustive exercise. (Using data from [306])

	minimum	maximum
restoration of phosphagen (ATP + PC)	2 min	5 min
muscle glycogen replenishment	5–10 h	24–46 h
liver glycogen replenishment	unknown	12–24 h
restoration of O_2 in plasma and myoglobin	10–15 s	1 min
duration of fast component of O_2 recovery	3 min	6 min
duration of slow component of O_2 recovery	30 min	1 h
reduction of lactic acid in blood and muscle	30–60 min[a]	1–2 h[a]

[a]Faster recovery with exercise and slower recovery with rest.

even at rest each of us gives off almost the same amount of heat as a 100 W incandescent light bulb. As we will see later, the heavier the person the higher the BMR. Approximately 85% of people of the same gender and weight have a BMR within 6–10% of the mean for their classification.

The exact numbers one finds for the BMR sometimes vary with the source. This may be due in part to the exacting conditions used to define the BMR and the lack of specification of the person's gender, age, weight, etc. To have your BMR measured you must:

(1) Have eaten no food for at least 12 h
(2) Have had a night of restful sleep and no strenuous activity thereafter
(3) Be resting completely in a reclining position for at least 30 min
(4) Be experiencing no excitement from psychic or physical factors
(5) Be in a room with a temperature from 20 to 27°C (68 to 80°F).

The BMR is greater than the metabolic rate during sleeping (when there is minimal digestion of food). Because of the stringent nature of BMR testing, "background" or "minimal" metabolic rates are sometimes tested under somewhat less exacting conditions and these measured rates are not cited as BMRs, but are termed differently although not always in a uniform way. The basal energy expenditure (BEE) is often defined to be equivalent to the BMR. The resting metabolic rate (RMR) is similar to the BMR, but is measured under somewhat less restrictive conditions – early in the morning after an overnight fast and 8 h of sleep – so it is easier to achieve and thus it is often used. The resting energy expenditure (REE) is usually measured as after 4 h of sitting and is a bit larger than the BMR. The BMR and metabolic rates during activities are often determined by measuring oxygen intake, because they are proportional to each other for aerobic activities.

The functioning of several organs contributes to this BMR in a resting person, as is seen from Table 6.17. Three-fourths of the metabolic activity takes place in organs with a total mass of 5 kg, which is ~8% of the total body mass. We will estimate these metabolic rates of the heart and lungs in Chaps. 8 and 9.

Table 6.17. Metabolism of a resting person. (Using data from [300])

system	percentage of BMR	met. rate (kcal/min)	organ mass (kg) for a 65 kg man
liver and spleen	27	0.33	–
brain	19	0.23	1.40
skeletal muscle	18	0.22	28.0
kidney	10	0.13	0.30
heart	7	0.08	0.32
remainder	19	0.23	–
		sum = 1.22	

The BMR is a function of mass, height, and gender, as we will see later in this chapter. It is also a function of body temperature. It changes by about 10% (some say 13%) per 1°C in body temperature. Consequently, the BMR increases by 30% (or 39%) when the body temperature increases from the normal 37°C (98.6°F) to 40°C (104.0°F) and decreases by 30% (or 39%) when the body temperature decreases from normal to 34°C (93.2°F).

The BMR of pregnant women is analyzed in Problem 6.50.

Scaling of the BMR

Before addressing how the BMR and MR depend on gender, weight, age, etc. we will temporarily expand our discussion of the BMR from that for humans to that for all mammals. The larger the animal the higher the metabolic (catabolic) rate, which is pretty obvious. Moreover, there is a way that the BMR scales with a physical attribute of size, specifically the body mass m_{b}.

Table 6.18 shows that the BMR of mammals increases sublinearly with m_{b}. Many think the variation for mammals is as the 3/4 power:

$$\mathrm{BMR} = cm_{\mathrm{b}}^{3/4}, \tag{6.19}$$

where $c \approx 90\,\mathrm{kcal/kg}^{3/4}$-day. This is known as *Kleiber's Law* [317, 318, 319], and is an example of an allometric relationship, as described in Table 1.13. It is valid from mice to elephants and some say to whales. (Regarding whales:

Table 6.18. BMR determined for several mammals. (See, for example [324])

species	mass	BMR (kcal/day)
mouse	20 g	3
reference woman (25 years)	55 kg	1,260
reference man (25 years)	65 kg	1,500
elephant	5,000 kg	70,000

Steinbeck's novel was valid only from mice to men. Also, there is great uncertainty about the BMR of whales. How do you measure it?) While this scaling relation was actually developed for comparing different species, it works fairly well intraspecies for humans, from children to adults. Others claim that the BMRs of mammals increase as mass to the 2/3 power. Those believing in Kleiber's Law point out that this latter scaling rule is off by a factor of two from mice to cows.

Both dependences give essentially the same prediction over the mass range of humans, so it matters little which we use for analysis. Adherents of both rules claim that the available BMR data support them. In any case, it is interesting to examine the physical reasons put forth to support the $m_{\mathrm{b}}^{2/3}$ and $m_{\mathrm{b}}^{3/4}$ scaling rules [304, 312]. There is both some sound and some highly questionable physical reasoning in both of these arguments.

Supporting Kleiber's $m^{3/4}$ Scaling Rule

The metabolic rate, and the BMR in particular, are supposed to scale as the maximum power output of muscles [323, 324]. Power usage is proportional to Fv, where F is the muscle force and v is the speed of muscle contraction. The muscle force $F = \sigma(\mathrm{PCA})$, where σ is the muscle force per unit area and PCA is the muscle cross-sectional area. Therefore,

$$\mathrm{BMR} \propto Fv = \sigma(\mathrm{PCA})v. \tag{6.20}$$

Studies show that for muscles of all species and size, v and σ do not vary substantially, so

$$\mathrm{BMR} \propto \mathrm{PCA}. \tag{6.21}$$

If the limbs of mammals, which of course contain many of these muscles, have a characteristic width d and length L, this area and the mass of the mammal scale m_{b} as

$$\mathrm{PCA} \propto d^2 \tag{6.22}$$

$$m_{\mathrm{b}} \propto d^2 L. \tag{6.23}$$

To relate the BMR to the mass, we need to relate L to d. Four arguments have been put forth suggesting that L scales with $d^{2/3}$. First, experimental data on primates show that L varies as d this way. Second, this scaling law also relates the height of a tree L with the width of the tree trunk d, and so this scaling law seems to be universal. Third, $L \propto d^{2/3}$ is also the scaling law for the condition for which a long vertical column of height L and width d buckles under its own weight when displaced from the vertical orientation.

A fourth argument is based on self-similarity, using the expression we derived earlier, (4.46). The end of the beam of length L and lateral dimension d in Fig. 4.40 deflects down by a distance $y(L) = -FL^3/3YI_{\mathrm{A}}$ when a force F

is applied at its end. Even though the area moment of inertia is different for different cross-sections – $I_A = \pi d^4/64$ for cylinders (from $I_A = \pi a^4/4$ for a circular cross-section of radius $a = d/2$ using (4.42)) and $d^4/12$ for rectangular solids with a square cross-section (from $I_A = wh^3/12$ for a rectangular cross-section of width w and height h – here with $d = w = h$, using (4.41)) – it makes no difference which we use because both have the same d^4 dependence. (We will formally use the circular cross-section.)

This downward angle of deflection θ is $\sim|y|/L$, giving

$$\theta = \frac{|y|}{L} = \frac{FL^2}{3YI_A}. \tag{6.24}$$

Self-similarity suggests that θ would be same for all species, and consequently equal to a constant. If F is due to gravity, then $F = m_b g = \rho V g$, where ρ is the beam mass density and V is its volume $= (\pi d^2/4)L$. If this force acted entirely at the end, then

$$\theta = \frac{(\rho((\pi d^2/4)L)g)L^2}{3Y(\pi d^4/64)} = \frac{16}{3}\frac{\rho g}{Y}\frac{L^3}{d^2}. \tag{6.25}$$

If θ were independent of L and d, we again get $L \propto d^{2/3}$.

Given that $m_b \propto d^2 L$ and $L \propto d^{2/3}$, we see that $m_b \propto d^2(d^{2/3}) = d^{8/3}$. Therefore $d \propto m_b^{3/8}$. With BMR $\propto$ PCA and PCA $\propto d^2$, we find that BMR $\propto d^2$. Using $d \propto m_b^{3/8}$, we see that

$$\mathrm{BMR} \propto (m_b^{3/8})^2 = m_b^{3/4}, \tag{6.26}$$

which is Kleiber's Law.

There are many gaps in this reasoning. For example, the power needed to operate skeletal muscles is actually a small fraction of the BMR (Table 6.17), so the scaling laws based on this power may not represent those of the whole BMR.

Supporting the $m^{2/3}$ Scaling Rule

The previous argument assumed that the BMR scaled as the metabolic power needs. This line of reasoning says that the BMR is limited by the rate of heat loss from the mammal. A mammal has a cross-sectional area that scales as L^2, where L is its dimension, while its volume and therefore mass m_b scale as L^3. Heat loss scales as the available surface area A, which clearly varies as $m_b^{2/3}$, and so

$$\mathrm{BMR} \propto m_b^{2/3}. \tag{6.27}$$

BMR Scaling in Humans

The BMR in humans depends on mass as well as other factors, such as gender, height, and age. The Harris–Benedict equations are commonly used to

Table 6.19. BMR (kcal/day) for different age groups. (Using data from [298], using [335])

age group (years)	BMR (males)	BMR (females)
Under 3	$59.5m_\mathrm{b} - 30$	$58.3m_\mathrm{b} - 31$
3–10	$22.7m_\mathrm{b} + 504$	$20.3m_\mathrm{b} + 486$
10–18	$17.7m_\mathrm{b} + 658$	$13.4m_\mathrm{b} + 693$
18–30	$15.1m_\mathrm{b} + 692$	$14.8m_\mathrm{b} + 487$
30–60	$11.5m_\mathrm{b} + 873$	$8.1m_\mathrm{b} + 846$
Over 60	$11.7m_\mathrm{b} + 588$	$9.1m_\mathrm{b} + 658$

m_b is the body mass in kg.

characterize measured human BMRs:

$$\text{For men:} \quad \mathrm{BMR} = 66.4730 + 13.7516m_\mathrm{b} + 5.0033H - 6.75505Y, \tag{6.28}$$

$$\text{For women:} \quad \mathrm{BMR} = 655.0955 + 9.5634m_\mathrm{b} + 1.8496H - 4.6756Y, \tag{6.29}$$

where the BMR is in kcal/day, m_b is the body mass in kg, H is the height in cm, and Y is the age in yr. This can also be expressed as

$$\text{For men:} \ \mathrm{BMR} = 71.2m_\mathrm{b}^{3/4}[1 + 0.004(30 - Y) + 0.010(S - 43.4)], \tag{6.30}$$

$$\text{For women:} \ \mathrm{BMR} = 65.8m_\mathrm{b}^{3/4}[1 + 0.004(30 - Y) + 0.010(S - 43.4)], \tag{6.31}$$

where S is the specific stature $= H(\text{in cm})/m_\mathrm{b}^{1/3}(m_\mathrm{b} \text{ in kg})$.

Another study re-examined the BMR in term of age groups and gender, as presented in Table 6.19. Table 6.20 gives the BMRs for adults of given

Table 6.20. BMR (kcal/day) for adult men and women of different ages, assuming Quételet's index (or BMI) Q is 22 for men and 21 for women and the relations in Table 6.19. (Using data from [298], from [305])

height (m) (also ft, in)	mass (kg) (also lb)	age (yr)		
		18–30	30–60	over 60
men				
1.5 (4′11″)	49.5 (109)	1,440	1,450	1,150
1.6 (5′3″)	56.5 (124)	1,540	1,530	1,250
1.7 (5′7″)	63.5 (140)	1,650	1,620	1,350
1.8 (5′11″)	71.5 (157)	1,770	1,710	1,450
1.9 (6′3″)	79.5 (175)	1,900	1,800	1,560
2.0 (6′7″)	88.0 (194)	2,030	1,900	1,670
women				
1.4 (4′7″)	41.0 (90)	1,100	1,190	1,030
1.5 (4′11″)	47.0 (104)	1,190	1,240	1,090
1.6 (5′3″)	54.0 (119)	1,290	1,300	1,160
1.7 (5′7″)	61.0 (134)	1,390	1,360	1,230
1.8 (5′11″)	68.0 (150)	1,500	1,420	1,310

heights as predicted by Table 6.19. The person's mass is chosen to make the Quételet's index (or BMI) $Q = m_b/H^2$ (m_b is the body mass (in kg) and H is the height (in m)) equal to 22 for men and 21 for women.

These relations show that younger people have higher BMRs than older people, which is true because they have more lean body mass. Similarly, tall thin people have higher BMRs and overweight people have lower BMRs. Pregnant women have higher BMRs than predicted above, as do people with fevers, those under stress, and those in hot and cold conditions. The BMR is lower than is predicted here for people who are fasting, starving, and malnourished.

The variations in the BMR and MR among humans are discussed below.

6.4.2 Metabolic Rates during Common Activities

Fortunately, we spend most of our time under conditions less restrictive than those needed to measure the BMR. Metabolic rates during activity are commonly expressed in several ways. (1) The total metabolic rate (MR) (including that due to the BMR) is expressed either as (a) a cumulative value (BMR + the additional MR due to activities), (b) a factor times the BMR – the factor being called the activity factor f, or (c) a factor times the RMR, with that factor called the MET, the metabolic equivalent. The RMR is more frequently used as the reference because it is much easier to measure than the BMR. (2) Only the net increase in metabolic rate over the BMR (or RMR) is given, such as MR−BMR.

Of course, each method should provide the same total MR. It is not surprising that the published values of MR expressed in these different ways are often somewhat inconsistent; this is true for several reasons. The exact metabolic rates may not be known very well, the details of the specific activity may not be well defined, the differences in rates among people of different genders, ages, and weights may not have been accounted for properly, or the exact differences between the reference metabolic rates may not be clear. For example, the differences in the BMR and RMR may not have been considered well. Also, sometimes the MET is defined relative to the RMR as defined earlier, right after sleep, and sometimes it is referenced to the MR for sitting. One more subtle reason is that the BMR varies as $m_b^{3/4}$, while the increase in metabolic rate due to activity is thought to depend linearly on m_b (and this may not always be exactly true). Consequently, scaling factors, such as the activity factor, cannot be rigorously correct. Nevertheless, examining the total MR is intriguing and we can learn much from it, even amidst these potential inconsistencies.

We see from Table 6.21 that the heavier the activity, the higher the metabolic rate and the greater the need for oxygen, and of course this is why we then need to breathe faster and harder. In fact, the air/oxygen consumption is proportional to the power exerted and the rate of heat

Table 6.21. Approximate total metabolic rates (MR) and oxygen consumption for different levels of activity for an average 70 kg person. (From [296] and [300])

activity	equivalent heat production		O_2 consumption (L/min)
	(kcal/h)	(W)	
very low level activity			
sleeping	71	83	0.24
sitting at rest	103	120	0.34
standing relaxed	108	125	0.36
light activity			
walking slowly, 5 km/h	228	265	0.76
moderate activity			
cycling, 15 km/h	344	400	1.13
moderate swim	400	465	1.32
heavy activity			
soccer	500	580	1.65
quite heavy activity			
climbing stairs, 116 steps/min	589	685	1.96
cycling, 21 km/h	602	700	2.00
basketball	688	800	2.28
extreme activity			
racing cyclist	1,400	1,600	4.62

being produced, if the power is low enough for only aerobic processes to be important (Tables 6.11–6.15). We can compare the MRs in this table with our BMRs.

The metabolic rate during exercise can be measured by using *direct calorimetry* (Fig. 6.6) or *indirect calorimetry* (Fig. 6.7). In direct (or "bomb") calorimetry the heat produced by a person exercising in a thermally insulated room is measured. The metabolic rate is determined assuming that ~40% of the energy liberated during metabolism is used to produce ATP, while the remaining 60% is converted to heat – which is being measured. The subject is in an enclosed, thermally insulated chamber. Heat produced by the subject's body eventually heats water flowing in tubing within the chamber walls and this water temperature is measured. Constructing such a chamber is relatively expensive and the time response of this method is slow, so metabolic rates during intense activity cannot be measured well.

In indirect calorimetry the rates of O_2 consumption, dV_{O_2}/dt, and CO_2 production, dV_{CO_2}/dt, are determined, so it is simpler to implement, as in Fig. 6.7. They are determined from the rates of inspiring and expiring volumes of air and from the fractions of O_2 and CO_2 in both the inspired and expired air (Problem 6.10). If the calorific equivalents of carbohydrates and

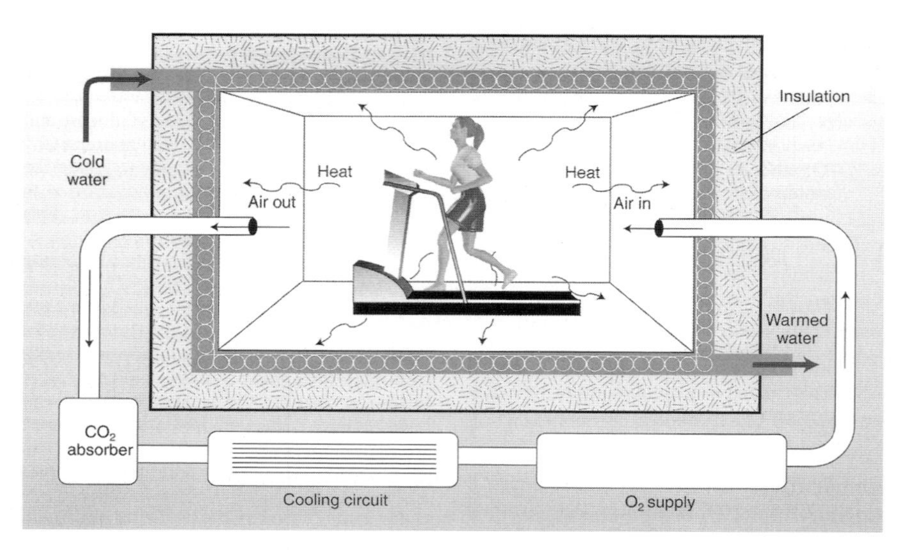

Fig. 6.6. In a calorimetric chamber for direct calorimetry, heat generated by the person is transferred to the air and chamber walls, and the heat produced – and therefore the metabolic rate – are determined by the measured temperature change in the air and the water flowing through the chamber. (From [340]. Used with permission)

fats were the same, the rate of heat energy produced would be $\mathrm{d}V_{\mathrm{O}_2}/\mathrm{d}t\times$ the calorific equivalent. Because they are not the same, the respiratory exchange ratio RER is measured (Table 6.2) to determine the relative amounts of carbohydrates and fats being metabolized. The RER can be expressed as

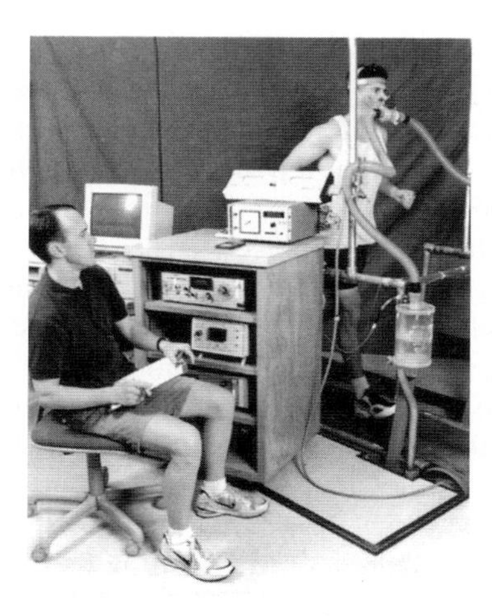

Fig. 6.7. Gas collection for indirectly measuring metabolic rates. (From [340]. Used with permission)

$(dV_{CO_2}/dt)/(dV_{O_2}/dt)$. An RER = 1.0 indicates that muscles are using only glucose or glycogen (carbohydrates). An RER = 0.71 indicates that only fats are being used. Values in between these limits indicate that both fats and carbohydrates are being used (Problem 6.10). This approach assumes that no proteins are being metabolized, which is not true for exercise lasting several hours, during which up to 10% of the metabolic energy can come from protein metabolism.

The aerobic capacities of normally active 18–22 year old college students are 44–50 mL/kg-min (of oxygen) for men and 38–42 mL/kg-min for women. For trained athletes this capacity increases to about 60 mL/kg-min and to 70–80 mL/kg-min and higher for elite male long distance runners. Poorly conditioned adults can have aerobic capacities below 20 mL/kg-min. The aerobic capacity decreases with age, by about 10% per decade, starting in the mid-teens for women and mid-20s for men. Among normally active men it decreases from about 47.7 mL/kg-min at 25 years of age, to 43.1 mL/kg-min at 35, 39.5 mL/kg-min at 45, 38.4 mL/kg-min at 52, 34.5 mL/kg-min at 63, and 25.5 mL/kg-min at 75 years of age.

All of these changes in body metabolic rates are linked to corresponding changes within cells, organs, and systems in the body. Table 6.22 lists the power output of various cells and tissues at various levels of activity.

As we discuss metabolic rates, we need to keep in mind that there are significant variations in the metabolism between people with the same body characteristics and activity levels. The usual variation in BMR is ±6–10%; this is presumably also true for the MR (for the same average activity levels). Table 6.23 shows a striking example in which the metabolic rates of two young women differ by 50%. Because most of this difference is due to metabolism during sleep at night, potential differences in activity levels do not seem very important.

Metabolic Rates during Locomotion

Figure 6.8 shows that the rate of heat production vs. walking speed is roughly linear, meaning that the energy cost per distance traveled is independent of speed. Above 3 m/s (7 mph) this curve becomes very superlinear. The energy cost of walking above that of standing is 2.13 J/kg-m at 1.2 m/s (2.6 mph), 2.49 J/kg-m at 1.8 m/s (4.0 mph), and 3.93 J/kg-m at 2.4 m/s (5.4 mph).

At high speeds, why do we prefer to run than walk? It is very hard to keep walking at these faster speeds because of the increased power needed relative to jogging and running (Fig. 3.25). In fact at 2.4 m/s (5.4 mph) the metabolic power needed to run is 14% less than that needed to walk. Phrased in terms of the energy needed per unit mass to move horizontally per meter, this is a decrease from 3.93 J/kg-m (0.94 cal/kg-m) to 3.40 J/kg-m (0.81 cal/kg-m). Figure 3.25 shows that the metabolic cost of locomotion per distance traveled is relatively independent of speed for a wide range of speeds when using the optimal gait, both for humans and horses. (Although it costs the same energy

Table 6.22. Estimated power output (metabolic rates) and power densities of biological cells and tissues (human, unless stated otherwise). (Using data presented in [308])

organelle, cell, or object	power output (picowatts)	volume (μm^3)	power density (W/m^3)
myosin muscle motor crossbridge	0.000001	5×10^{-7}	2×10^6
platelet (resting)	0.003–0.09	3	$0.1–3.0 \times 10^4$
red blood cell	0.008	94	8.5×10^1
E. coli bacterium (basal)	0.05	2	2.5×10^2
mitochondrian organelle	0.1–1.1	1	$0.1–1.1 \times 10^6$
platelet (activated)	0.7–7.0	3	$0.2–2.3 \times 10^6$
skin cell	1–3	1,000	$1.0–3.1 \times 10^3$
skeletal muscle (resting)	1–10	2,000	$0.5–4.9 \times 10^3$
typical tissue cell (basal)	30	8,000	3.8×10^3
intestine/stomach cell	46–52	8,000	$5.6–6.5 \times 10^3$
neuron cell (basal)	70–110	14,000	$5.0–7.9 \times 10^3$
heart muscle cell (typical)	87–290	8,000	$1.1–3.6 \times 10^4$
skeletal muscle cell (max., voluntary)	113	2,000	5.7×10^4
kidney cell	155–346	8,000	$1.9–4.3 \times 10^4$
neuron cell (maximum)	255–330	14,000	$1.8–2.4 \times 10^4$
typical tissue cell (maximum)	480	8,000	6.0×10^4
skeletal muscle cell (max., tetanic)	2,300	2,000	1.2×10^6
honeybee flight muscle cell	3,400	1,000	3.4×10^6
heart muscle cell (maximum)	3,500–5,000	8,000	$4.4–6.3 \times 10^5$
human brain	15–25 W	1.4×10^{-3} m^3	$1.1–1.8 \times 10^4$
human body (basal)	100 W	0.1 m^3	1.0×10^3
human body (maximum)	1,600 W	0.1 m^3	1.6×10^4
gasoline-powered automobile	200,000 W	10 m^3	2.0×10^4
The Sun	3.92×10^{26} W	1.41×10^{27} m^3	0.28

Note that 1 W = 0.86 kcal/h.

Table 6.23. Food intakes and heat production for two 23-year-old women of comparable physical size. (Using data from [298], from the result of Warwick as quoted by [309])

	"large eater"	"small eater"	ratio
body mass (kg)	54.2	52.7	1.03
lean body mass (kg)	45.1	43.6	1.03
average food intake (kcal/d)	2,370	1,550	1.54
heat production (kcal/d)	2,170	1,390	1.55
heat production at night (kcal/d)	1,720	1,100	1.56

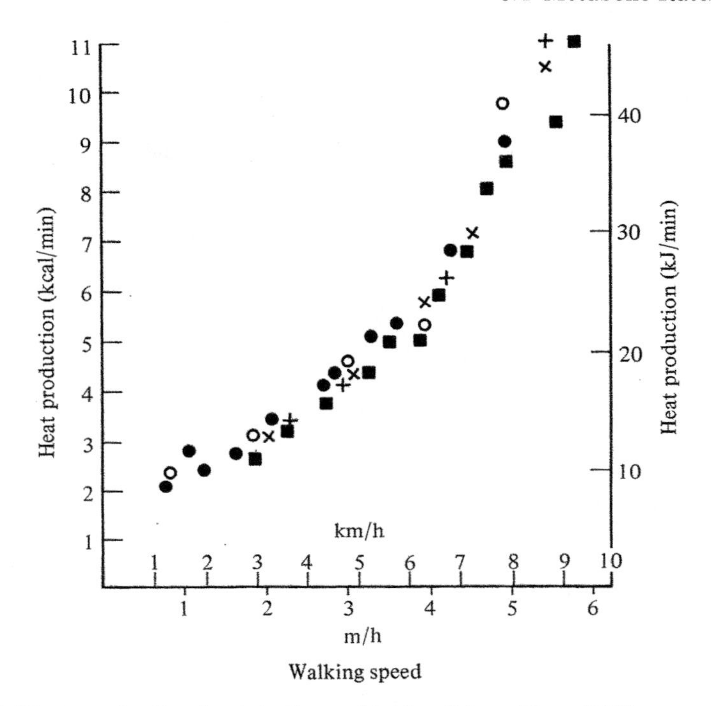

Fig. 6.8. Heat production (metabolic needs) of walking. (From [298], as summarized by [330])

to walk a mile and to run a mile, running provides better cardiovascular training.) The oxygen intake is about 5.8 mL/kg-min per mph running speed. Figure 6.9 shows metabolic needs vs. speed for several types of locomotion.

Concentric muscle contractions are said to do positive work, while eccentric contractions do negative work. (See Fig. 3.53.) The metabolic cost of positive work is greater than an equal amount of negative work. The average extra energy needed to move horizontally ~2.8 J/kg-m (relative to standing) is much less than that extra needed to move vertically ~30 J/kg-m. Obviously, walking uphill requires more power than on level terrain; walking downhill requires less power than on level terrain. The work and power required going uphill is about 2.7× the increase in potential energy and the concomitant rate of change of this energy; this means the efficiency of doing this type of work is about 1/2.7 = 36%. (As we will soon see, this says the average extra energy needed to move vertically is 26 J/kg-m, which is consistent with the value of ~30 J/kg-m that was just given.) When moving downhill, less power is needed than that needed on level terrain by ~50% of the rate of decrease of potential energy. All of the gained potential energy is not recovered because muscular work is needed to prevent us from falling downhill. Problem 6.58 explores the power needed in walking (or running) and gradients.

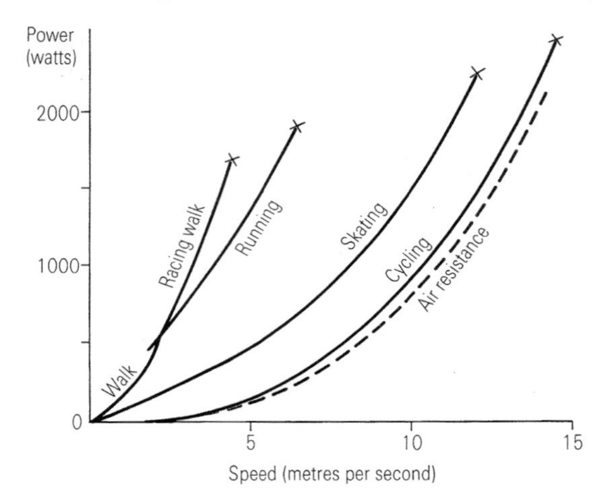

Fig. 6.9. Metabolic needs vs. speed for several types of locomotion, extrapolated to world record speeds for 5 km races (for running) or 10 km races (for the others). The *dashed line* shows the power needed to overcome air resistance only. (From [294], based on data from [332]. Used with permission)

Let us address the use of treadmills to assess the cardiovascular and cardiorespiratory capacity of healthy people and those with heart disease. The patient is wired with EKG (electrocardiogram) leads and EKG scans are continually taken as the patient walks/runs on a treadmill (see Chaps. 8 and 12 for more about EKGs). The treadmill speed and/or the uphill grade is increased after set periods of time (called stages), until either the complete cycle ends, the patient can no longer endure the workout, or an irregularity appears. In the Bruce Protocol, Table 6.24, the speed and grade increase with each stage. In another common protocol, the Naughton Protocol, the speed is fixed at 3.0 mph, and the grade increases from 2.5% in Stage I to 20.0% in Stage VIII, in steps of 2.5%, with each stage lasting 2 min. Such tiltable treadmills are also used to determine the efficiency of skeletal muscles in human locomotion. Oxygen consumption rates – and therefore metabolic rates – are compared for an ensemble of people moving at a given speed first on a level treadmill and then on a tilted treadmill. The efficiency is about 30%.

Table 6.24. The Bruce protocol for an exercise test on a treadmill. (Using data from [306])

stage	speed (mph)	grade (%)	cumulative time (min)
I	1.7	10	1–3
II	2.5	12	4–6
III	3.4	14	7–9
IV	4.2	16	10–12
V	5.0	18	13–15

As an example, we will determine the metabolic power and rate of oxygen consumption needed during Stage V of the Bruce protocol for a 75 kg male. The total rate is the sum of the BMR (or more precisely the MR for standing), the power needed to move horizontally, and the additional power needed to move uphill. The second term is $\sim$2.8 J/kg-m $\times$ (horizontal distance moved per unit time). This is 2.8 W $m_{\mathrm{b}}v$, where m_{b} is the body mass in kg and v is the speed of walking in m/s (or really jogging for this faster speed). The third term comes from how fast the potential energy $m_{\mathrm{b}}gh$ changes with time, which is $m_{\mathrm{b}}g(\mathrm{d}h/\mathrm{d}t)$, and the "inefficiency" factor of 2.7, and so it is 2.7 $m_{\mathrm{b}}g(\mathrm{d}h/\mathrm{d}t)$. A person moving a distance x along a ramp at angle θ elevates a distance $x\sin\theta$ (draw a diagram to prove this), so this third term becomes 2.7 $m_{\mathrm{b}}g\sin\theta(\mathrm{d}x/\mathrm{d}t)$. With $v = \mathrm{d}x/\mathrm{d}t$, this becomes 2.7 $m_{\mathrm{b}}gv\sin\theta$ = 26 W $m_{\mathrm{b}}v\sin\theta$. Combining the second and third terms gives $(2.8 + 26\sin\theta)m_{\mathrm{b}}v$ in W or $(2.4 + 22.8\sin\theta)m_{\mathrm{b}}v$ in kcal/h (using 1 W = 0.86 kcal/h). The speed is 5.0 mph = 2.23 m/s (using 1 mph = 0.447 m/s). A grade of 18% corresponds to an angle of $(0.18)90° = 16.2°$, and so $\sin\theta = 0.28$. For this 75 kg person, the second term is (2.4)(75)(2.23) = 401 kcal/h, and the third term is (22.8)(0.28)(75)(2.23) = 1,068 kcal/h. The burden of moving uphill rapidly is by far the largest contribution to the metabolic rate. The sum of these two terms is 1,469 kcal/h. Adding to this the MR for walking of about 100 kcal/min, the MR in Stage V is about 1,570 kcal/h, which is pretty high. (Of course, the patient is at this stage for no more than 3 min.) Assuming only aerobic metabolism, the rate of oxygen usage is this metabolic rate divided by the calorific equivalent. If only carbohydrates are being burned, the calorific equivalent is 5.05 kcal/L O_2 and we see that in Stage V $\mathrm{d}V_{O_2}/\mathrm{d}t$ = (1,570 kcal/h)/(5.05 kcal/L O_2) = 311 L O_2/h = 5.2 L O_2/min is needed.

Why is it hard to run into the wind? The answer is that it requires much more metabolic power (heat production, or equivalent oxygen consumption). Table 6.25 shows that for walking and running this power increases as the square of the wind speed. A fast walk into the wind can require as much power as a relatively slow run with no wind. Even still air affects metabolic needs; air resistance accounts for about 8% of the metabolic needs for middle distance runners and up to 15% for sprinters. (We examine this effect of drag in Chap. 7.)

MR Scaling with Metabolic Equivalents and Activity Factors

The actual heat production, or metabolic rate (MR), is higher than the BMR because of activity (Table 6.21). We can relate the MR and BMR by

$$\mathrm{MR} = f(\mathrm{BMR}), \tag{6.32}$$

where f is the activity factor.

The metabolic equivalent (MET) is another very closely related measure of activity. An average body consumes (3.5 mL/kg-min) $\times$ MET. Typical values of MET are given in Tables 6.26–6.30. It is often said that MET $\sim f/1.5$,

Table 6.25. The effect of wind on energy needs of walking and running. (Using data from [298], calculated using [333])

activity, speed (m/s) (or mph)	wind speed w (m/s) (or mph)	oxygen consumption (L/min)	heat production (W)	increase in energy/w^2 (W-s^2/m^2)
walking, 1.25 (2.80)	0.0 (0.0)	0.878	294	
	10.0 (22.4)	1.192	399	$\simeq$1.05
	14.1 (31.5)	1.505	504	$\simeq$1.05
walking, 2.08 (4.65)	0.0 (0.0)	1.649	552	
	10.0 (22.4)	2.221	744	$\simeq$1.91
	14.1 (31.5)	2.792	935	$\simeq$1.91
running, 3.75 (8.39)	0.0 (0.0)	2.836	950	
	10.0 (22.4)	3.243	1,086	$\simeq$1.36
	14.1 (31.5)	3.710	1,243	$\simeq$1.36
running, 4.47 (10.00)	0.0 (0.0)	3.010	1,008	
	10.0 (22.4)	3.710	1,243	$\simeq$2.52
	14.1 (31.5)	4.525	1,516	$\simeq$2.52

Note that the last column is the increase in heat production divided by the square of the wind speed for the particular walking or running condition.

because the activity factor f is referenced to the smaller BMR, but it is not clear that this ratio is as large as 1.5.

Daily-Averaged Activity Factor

The activity factor could refer to a given activity, as above, or to the daily average; we will call the daily average f_{av}. There is no universal agreement on the actual values of f_{av}. They depend on the type of work, amount of exercise, and access to automated conveniences. United Nations studies suggest for adult men $f_{\mathrm{av}} \sim 1.55$ for light occupational work, 1.78 for moderate occupation work, and 2.10 for heavy occupational work. Other studies suggest that the MR is much lower, with $f_{\mathrm{av}} \sim 1.3$ for low activity (sedentary), 1.5 for

Table 6.26. Typical MET values for self care. (Using data from [340])

self-care	MET
rest, sitting, standing, eating, conversation	1.0
dressing, undressing, washing hands and face, propelling wheelchair	2.0
walking 2.5 mph	3.0
showering	3.5
walking downstairs	4.5
walking 3.5 mph	5.5

Table 6.27. Typical MET values for housework. (Using data from [340])

housework	MET
handsewing	1.0
machine sewing, seeping floor	1.5
polishing furniture	2.0
peeling potatoes, scrubbing, hand washing clothes, kneading dough	2.5
scrubbing floors, cleaning windows, making beds	3.0
ironing, mopping, wringing wash by hand,	
hanging wash by hand	3.5
beating carpets	4.0

Table 6.28. Typical MET values for occupations. (Using data from [340])

occupational	MET
sitting at desk, writing, riding in car, watch repair	1.5
typing	2.0
welding, radio assembly, playing musical instrument	2.5
parts assembly	3.0
bricklaying, plastering	3.5
heavy assembly work, wheeling wheelbarrow (115 lb, 2.5 mph)	4.0
carpentry	5.5
mowing lawn with hand mower, chopping wood	6.5
shoveling	7.0
digging	7.5

Table 6.29. Typical MET values for exercise. (Using data from [340])

exercise (level conditions)	MET
walking 2 mph	2.5
cycling 5.5 mph	3.0
walking 2.5 mph, cycling 6 mph	3.5
walking 3 mph; calisthenics	4.5
cycling 9.7 mph; swimming, crawl, 1.0 ft/s	5.0
walking 3.5 mph	5.5
walking 4 mph	6.5
jogging 5 mph	7.5
running 7.5 mph; cycling 13 mph	9.0
swimming, crawl, 2.0 ft/s	10.0
running 8.5 mph	12.0
running 10 mph; swimming, crawl, 2.5 ft/s	15.0
running 12 mph; swimming, crawl, 3.0 ft/s	20.0
running 15 mph; swimming, crawl, 3.5 ft/s	30.0

Table 6.30. Typical MET values for recreation. (Using data from [340])

recreational	MET
painting, sitting	1.5
playing piano, driving car	2.0
canoeing 2.5 mph, horseback riding (walk)	2.5
volleyball (recreational), billiards	3.0
bowling, horseshoes	3.5
golf, cricket	4.0
archery, ballroom dancing, table tennis, baseball	4.5
tennis	6.0
horseback riding (trot), folk dancing	6.5
skiing, horseback riding (gallop)	8.0
squash	8.5
fencing, basketball, football	9.0
gymnastics, handball, paddleball	10.0

intermediate activity (some regular exercise), and 1.7 for a high level activity (regular activity or physically demanding job).

The highest of these values for heavy occupational work, $f_{\text{av}} \sim 2.10$, is still lower than that estimated for a Scottish coal miner in the 1950s before the occupation became more automated. The peak metabolic rate of the coal miner was $4.3 \times$ BMR during working hours and averaged a still very high $2.5 \times$ BMR during the day, as is calculated in Table 6.31. Most occupations are physically less demanding now.

Mechanical Work and Power

The first law of thermodynamics (6.1) shows that stored energy can be used to supply heat or work. For people, this is mechanical work and not any other type (such as thinking deeply about your homework). Mechanical work is the force you apply to an object $\times$ the distance you push or pull it.

In discussing metabolic rates we need to be careful about defining types of work and energy [341]. Figure 6.10 shows that in lifting an object, which we can call *external work*, work is also done about the joints of the body to lift the center of mass of the body itself, which is *internal work*. Also, concentric muscle contractions are said to do positive (mechanical) work, while eccentric contractions do negative work. Walking on level ground requires equal amounts of positive and negative work. Walking uphill requires relatively more positive work, while walking downhill requires relatively more negative work. The work efficiencies of positive and negative work are different.

Furthermore, the term *efficiency* can mean different things, so, in general, we need to be careful and see what is being defined [341]. *Metabolic or muscle efficiency* is the total mechanical work done by all muscles divided by the metabolic work done by the muscles. Another measure of efficiency is

Table 6.31. Analysis of the metabolic power needs of a Scottish coal miner in the early 1950s before mechanization. (Using data from [298], recalculated using the data of [310])

activity	time (min)	time (fraction of day, F)	MR/BMR $= f$	product $F \times f$
in bed	501	0.348	1.00	0.348
nonoccupational				
sitting	331	0.230	1.51	0.347
standing	19	0.013	1.71	0.022
walking	129	0.089	4.67	0.416
washing and dressing	43	0.030	3.14	0.094
gardening	17	0.012	4.76	0.057
cycling	21	0.015	6.28	0.094
occupational				
sitting	130	0.090	1.60	0.144
standing	18	0.012	1.71	0.021
walking	58	0.040	6.38	0.255
hewing	11	0.007	6.38	0.045
timbering	59	0.041	5.43	0.223
loading	104	0.072	6.00	0.432
Totals	1,440	1.000	–	2.497

MR is the metabolic rate.

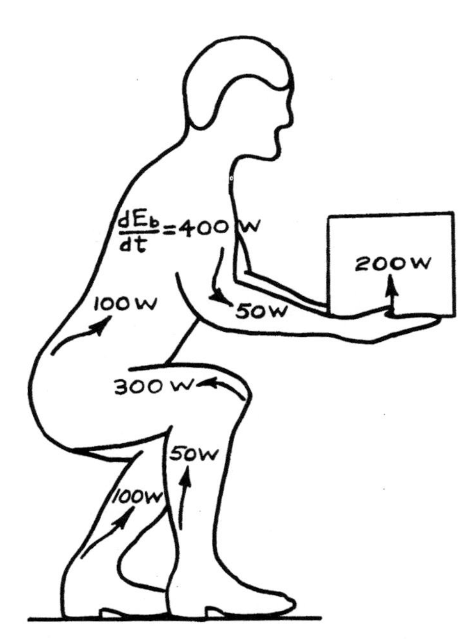

Fig. 6.10. "Internal" vs. "external" work in lifting an object. (From [341]. Reprinted with permission of Wiley)

Table 6.32. Mechanical power and efficiency of mechanical work. (Using data from [296])

task	mechanical power output		metabolic rate (W)	efficiency (%)
	(hp)	(W)		
cycling (typical)	0.15	112	505	19
tramming	0.12	90	525	17
shoveling sand	0.024	17.5	570	3

the *mechanical efficiency*, which is the total mechanical work done (internal + external) divided by the metabolic energy cost in excess of the resting metabolic cost. A more restrictive parameter is the *work efficiency*, which is the external mechanical work divided by the metabolic cost in excess of the zero-work metabolic cost. Inefficiency results from incomplete conversion of metabolic energy to mechanical energy and from neurological inefficiency in controlling this mechanical energy.

For our purposes we will simply define the efficiency, ϵ, as the rate of doing mechanical work divided by the metabolic rate. Tables 6.32 and 6.33 show that it rarely exceeds 20%, and it is usually much less than this. Cycling is relatively efficient, with ϵ approaching 22% for racing cyclists. Tramming is also efficient. (This involves pushing railroad cars loaded with coal down a railroad track.) Shoveling sand, a more common activity, is only 3% efficient. For comparison, steam engines have an efficiency of 17% and gasoline engines have $\epsilon = 38\%$.

Where does the balance of the energy go? The balance goes into the production of heat. This means that 80–100% of the final product of your metabolism is heat; this number is usually closer to 100%. In any case, the ~500 kcal mechanical work done by a physical laborer during a day translates into dietary needs of at least 2,500 kcal over the BMR, because the mechanical efficiency is usually <20%.

Table 6.33. Efficiencies of mechanical work. (Using data from [306])

exercise activity or task	efficiency (%)
cycling, level	24–34
cycling, uphill	19
ice skating	11
rowing	10–20
stepping machine	23
swimming, freestyle	2.9–7.4
walking, horizontal	20–35
walking, incline	21–43

Table 6.34. World record running speeds. (As of 2006)

distance (m)	average running speed (m/s)
100	10.22
200	10.35
400	9.26
1,500	7.28
10,000	6.32
42,200[a]	6.12
100,000	4.46

[a]Marathon, 26 miles, 385 yd

The rate of oxygen intake is limited by the rate the lungs can take up oxygen and the rate the heart can distribute this oxygen around the body for aerobic metabolism. One example of this is the decrease in human running speed with increasing distance (beyond ~200 m), which is due to the limitations on oxygen intake. This is clear from the world-record average running speeds at various distances shown in Table 6.34.

Table 6.35 shows that the maximum power of mechanical work a well-conditioned person can do decreases with the duration of the task. For short times we can do mechanical work at a higher rate because of anaerobic metabolic processes; longer-term work requires aerobic processes.

How to "Burn" off Food

Let us say you have just eaten a "standard" donut. You feel guilty and you want to burn off those extra calories. What can you do?

If you are sitting at rest you are naturally burning off ~103 kcal/h (70 kg man). If you decide to play basketball your metabolic rate increases to ~688 kcal/h, so you will be increasing your metabolic rate by ~585 kcal/h.

Table 6.35. Approximate maximum mechanical power from well-conditioned humans. (Using data from [296])

power		duration	total energy	
(hp)	(W)		(kJ)	(kcal)
2	1,500	6 s	9.0	2.2
1	750	1 min	45	10.8
0.35	260	35 min	546	130
0.2	150	5 h	2,700	645
0.1	75	8 h day of a laborer	2,160	516

You will burn off that standard 280 kcal donut in

$$\frac{280\,\mathrm{kcal}}{585\,\mathrm{kcal/h}} = 0.48\,\mathrm{h} = 29\,\mathrm{min}. \tag{6.33}$$

That donut will cost you a half an hour of real up-tempo basketball.

Let us say you want to "walk off" that donut. The metabolic rate during slow walking is $\sim$228 kcal/h, which exceeds that of sitting at rest by $\sim$125 kcal/h. To walk off that donut you would have to walk for

$$\frac{280\,\mathrm{kcal}}{125\,\mathrm{kcal/h}} = 2.24\,\mathrm{h} = 2\,\mathrm{h}\ 14\,\mathrm{min}. \tag{6.34}$$

Which is a little longer than most after-dinner strolls.

Why do people put on weight when they get older? One reason is the decrease in BMR with age (Tables 6.19 and 6.20). The activity level f_{av} often decreases with age. Also, sometimes people eat more (snacking) (see Problem 6.77). We now examine how caloric intake and activity combine to determine body weight.

6.4.3 Weight Gain and Loss

If your caloric intake exceeds your metabolic rate, you gain weight. If it is less, you lose weight. You can control your caloric intake by controlling what you put in your mouth. You have some control of your metabolic rate, by being more or less active in general and by exercising more or less. (In a daring example of his devotion to science and his readers, while writing this text the author showed how easy it is to gain 10 lb in 6 months merely by exercising less and by snacking more at night. He needed no dietary supplements to accomplish this feat! The author is considering the reverse experiment of weight loss through more exercise and less snacking.) As we will see later in this chapter, your metabolic rate is related to your body temperature and loss of body heat. (For example, if you are very cold, you shiver and this increases your MR.)

We will examine some very simple models of weight gain and loss here and in the problems. *How much weight in the form of body fat can you actually lose in a week? How is body weight controlled by salt intake? How does your body weight vary due to changes in caloric intake, age, and specific stature?* Our models are not only simple, they are overly simplistic – and yet it is a bit difficult to make them more realistic. For example, increases in body weight due to added muscle increase the BMR much more than those due to added fat. How can this be included in the model?

Is it reasonable to apply the relations characterizing how the average metabolic rate depends on body weight, age, and stature by using Kleiber's Law, Harris–Benedict equations, etc. to determine the steady-state weight and

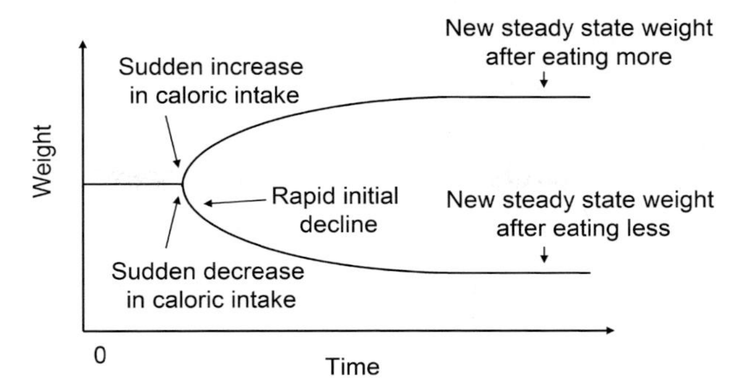

Fig. 6.11. Schematic of the time evolution of weight gain or loss, for a small sudden increase or decrease in metabolic intake (which is then maintained at this new level). Similar changes are expected for a small sudden decrease or increase in activity level, respectively

weight changes for a given individual? It is not clear that it is reasonable, because the body controls metabolic rates differently for different individuals – in response to stimuli such as dieting – making it easier for some, while harder for others, to lose weight. For now, let us assume that these relations also apply for a given individual, although it is not obvious that this is true. They can then be used to determine how the steady-state weight of a person changes with food intake, activity level, and age. *For a sudden change in metabolic intake or activity to a new level, how fast does the body reach its new steady state?* General changes are depicted in Fig. 6.11.

If your weight gain or loss involved only fat, the caloric value of fat of 9 kcal/g means that an increase or decrease in metabolic input by 4,090 kcal would result in a gain or loss of a pound (of fat). If your weight change involved only protein (4 kcal/g), a change in 1,820 kcal would change body mass by a pound. According to Wishnofsky's rule, when you increase or decrease your metabolic input by 3,500 kcal you gain or lose about a pound [342]. Consequently, Wishnofsky's rule suggests the gain or loss typically averages to about 3/4 fat and 1/4 muscle.

Because of the different muscle, fat, and fluid masses in men and women and in younger and older people, this long-used rule is now not thought to be really that universal. This average energy content of weight change is really significantly smaller for men (2,480 kcal/lb) than in women (3,380 kcal/lb). Also, it is about 10% larger for older people (2,610 kcal/lb, 70-year-old men; 3,510 kcal/lb, 70-yr-women) than younger people (2,360 kcal/lb, 35-yr-men; 3,260 kcal/lb, 35-yr-women) [331]. (All for these have been scaled to the same 90 kg, 170-cm tall person.)

Let us say you increase or decrease your food intake by 1,000 kcal/day. If it all went into increasing or decreasing your fat reserves, this would lead to a change of (1,000 kcal/day)/(9 kcal/g fat) or about 110 g of fat a day, which corresponds to 1/4 lb per day or roughly 2 lb per week. If you are on a diet, a decrease in caloric intake of 1,000 kcal/day is very large, so you can never expect to lose more than 2 lb a week. (Of course, exercise would also accelerate weight loss.)

Why do some diets claim faster weight loss, especially early on? Some diets may really involve the loss of body water, with very little loss of fat, carbohydrate, or protein. (Boxers who need to lose weight quickly to make a weight division often follow such diets.) Also, weight loss in a legitimate diet can be very fast early on. In the first days of dieting, burning 1,000 kcal involves burning glycogen in the small glycogen pool. Burning glycogen consumes 4.1 kcal/g glycogen, but because of the concomitant loss of water associated with the glycogen, only about 1 kcal is consumed per total g lost. Therefore, 1,000 kcal/(1 kcal/g) = 1,000 g is lost, which is about 2 lb. Later when fat is oxidized instead, this same deficit of 1,000 kcal results in a mass loss of only 110 g (1/4 lb). This suddenly slower rate of weight loss (in the approach to the new steady-state weight) can easily discourage dieters.

To paraphrase the advice given by Mr. Micawber to Charles Dickens' David Copperfield, "Daily caloric intake 2,000 kcal, daily energy expenditure 2,100 kcal, result happiness. Daily caloric intake 2,000 kcal, daily energy expenditure 1,900 kcal, result misery." (This is true at least for those way beyond their college years.)

During starvation, the body first utilizes all available carbohydrates – mostly glycogen in the liver and muscles – and this satisfies the body's energy needs for about half a day. Then fat and protein are depleted by the body (see Fig. 6.12); fat reserves are about 100× the initial carbohydrates reserves. Then with no fat remaining, protein is further depleted until death. Death occurs when about half of the proteins have been depleted because proteins are needed for cellular function.

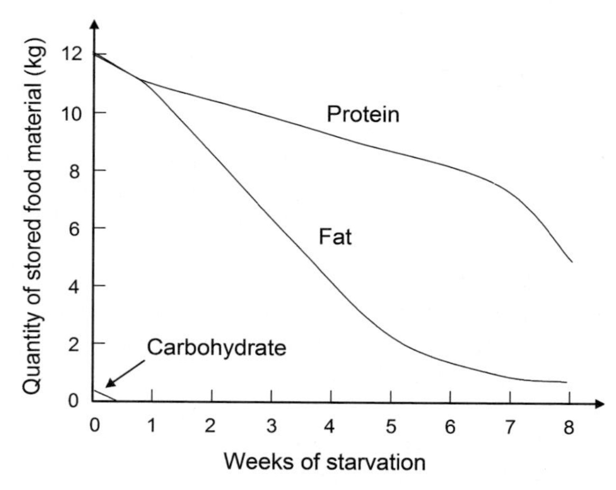

Fig. 6.12. Effect of starvation on energy supplies in the body. (Based on [314])

6.5 Loss of Body Heat

Why is the loss of body heat important? Virtually all of our metabolism goes into making heat, because very little energy usually goes into useful mechanical work. Consider the BMR of an $m_{\mathrm{b}} = 70\,\mathrm{kg}$ person, which is ∼1,680 kcal/day. This translates to 70 kcal/h. Using (6.5), the rate that body temperature would increase with time if all the heat stayed within the body is

$$\frac{\mathrm{d}T}{\mathrm{d}t} = \frac{1}{m_{\mathrm{b}}c}\left(\frac{\mathrm{d}Q}{\mathrm{d}t}\right) = \frac{\mathrm{BMR}}{m_{\mathrm{b}}c}, \tag{6.35}$$

where the rate of heat production $\mathrm{d}Q/\mathrm{d}t \approx \mathrm{BMR}$ and the average human specific heat is $c = 0.83\,\mathrm{kcal/kg\text{-}^\circ C}$. (To be consistent with the literature, we will usually express this in terms of degrees Kelvin in this section, which is 0.83 kcal/kg-K.) This heat capacity is $C = 0.83\,\mathrm{kcal/kg\text{-}^\circ C} \times 70\,\mathrm{kg} = 58\,\mathrm{kcal/^\circ C}$. The rate of temperature increase equals

$$\frac{\mathrm{d}T}{\mathrm{d}t} = \frac{70\,\mathrm{kcal/h}}{58\,\mathrm{kcal/^\circ C}} = 1.2^\circ\mathrm{C/h} = 2.1^\circ\mathrm{F/h}. \tag{6.36}$$

Our body temperature would go up very fast without effective modes of heat loss, even with some production of mechanical work ($0.8\,\mathrm{MR} < \mathrm{d}Q/\mathrm{d}t < \mathrm{MR}$).

The specific heat of the body is often cited as 0.829 kcal/kg-K. This is a bit higher than that calculated using the specific heat for each body component, which are (in kcal/kg-K) 0.507 for lipid, 0.299 for protein, 0.201 for mineral matter, 0.272 for carbohydrate, and 1.00 for water. These calculated values are 0.798 kcal/kg-K for a fat-free man (containing 72% water, 22% protein, and 6% minerals by weight), 0.765 kcal/kg-K for a thin man (containing 12% body fat), and 0.652 kcal/kg-K for an obese man (containing 50% body fat). The heat capacity of different body tissues are shown in Table 6.36 and that of other common materials in Table 6.37.

6.5.1 Modes of Heat Loss

There are four modes of heat loss. Their absolute magnitude and relative importance depend on clothing, environment, surroundings, etc. (Figs. 6.13 and 6.14). For general discussions of heat loss in people see [297, 298, 300, 306, 340].

1. Radiation loss, also known as black body radiation, is the thermal radiation emitted by an object in thermal equilibrium. At rest, about 54–60% of energy loss is typically through thermal radiation.
2. Convection and conduction of air from the body account for ∼25% of heat loss.
3. The evaporation of sweat accounts for ∼7% of heat loss.
4. The evaporation of water through breathing accounts for ∼14% of this loss.

Table 6.36. Thermophysical characteristics of body tissues and organs and other materials. (Using data from [308])

organ or tissue	thermal conductivity K (W/m-K)	specific heat c_v (MJ/m^3-K)	density (approximate) ρ (kg/m^3)
skin – very warm	2.80	3.77	1,000
skin – normal hand	0.960	3.77	1,000
skin – cold	0.335	3.77	1,000
subcutaneous pure fat	0.190	1.96	850
muscle – living	0.642	3.94	1,050
muscle – excised, fresh	0.545	3.64	1,050
bone – average	1.16	2.39	1,500
bone – compact	2.28	2.70	1,790
bone – trabecular	0.582	2.07	1,250
blood – water at 310 K	0.623	4.19	993
blood – plasma (Hct = 0%) at 310 K	0.599	4.05	1,025
blood – whole (Hct = 40%)	0.549	3.82	1,050
heart – excised, near fresh	0.586	3.94	1,060
liver – excised, near fresh	0.565	3.78	1,050
kidney – excised, near fresh	0.544	4.08	1,050
abdomen core	0.544	3.89	1,050
brain – excised, near fresh	0.528	3.86	1,050
brain – living	0.805	–	–
lung – excised, bovine	0.282	2.24	603
whole body (average)	–	4.12	1,156

Note that 1 MJ/m^3-K = 239 kcal/m^3-K and 1,000 kg/m^3 = 1 g/cm^3.

During very heavy activity, heat loss increases from the resting value of ~1.5 to ~15 kcal/min. The amount due to radiation remains about the same, 0.8 kcal/min, but it decreases to about 5% of the total. The loss to conduction and convection increases from ~0.3 to 2.2 kcal/min and this represents about 15% of the total loss. Losses due to the evaporation of sweat increase from ~0.3 to 12.0 mL/min and they constitute about 80% of the heat loss.

The principles of these modes of heat loss are totally determined by physics; the absolute magnitudes of the actual loss can be modified by biological processes.

In modeling heat loss, people are often modeled as upright cylinders, while four-legged animals are modeled as horizontal cylinders. In particular, a typical man is modeled as a cylinder that is 1.65 m high with a 0.234 m diameter. (Recall the comment in the introduction that to zero order a physicist might model a cow as a sphere and that our models here may get a bit more complex here; a cylinder is indeed more complex than a sphere. See Fig. 1.31 for a model of a man composed of many cylindrical components.)

Table 6.37. Thermophysical characteristics of materials. (Using data from [308])

material	thermal conductivity K (W/m-K)	specific heat c_{v} (MJ/m^3-K)	density ρ (kg/m^3)
air	0.009246	0.00119	1.18
cotton fabric at 310 K	0.0796	0.0267	160
rubber	0.156	2.41	1,200
ethanol at 310 K	0.163	1.96	789
teflon	0.399	2.20	2,180
concrete	0.934	1.93	2,310
glass, plate	1.09	1.94	2,520
ice at 249 K (−42°C)	2.21	1.76	913
sapphire (normal to c-axis) at 310 K	2–20	2.89	3,970
stainless steel	13.8	3.68	7,910
aluminum	204	2.45	2,710
silver	405	2.59	10,500
diamond, natural	2,000	1.82	3,510

Note that 1 MJ/m^3-K = 239 kcal/m^3-K and 1,000 kg/m^3 = 1 g/cm^3.

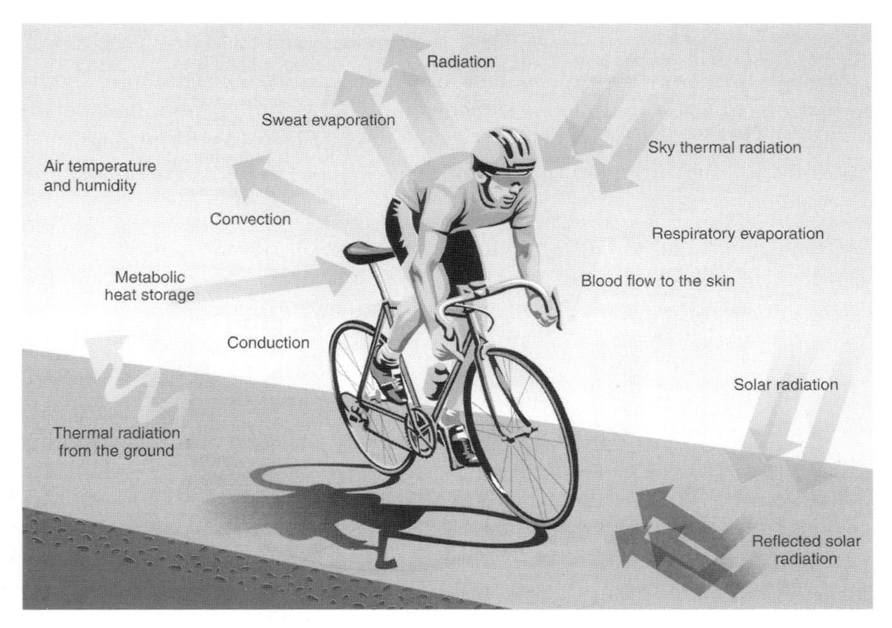

Fig. 6.13. Overall body modes of loss of heat and nonmetabolic gain of energy. (From [340], adapted from [313]. Used with permission)

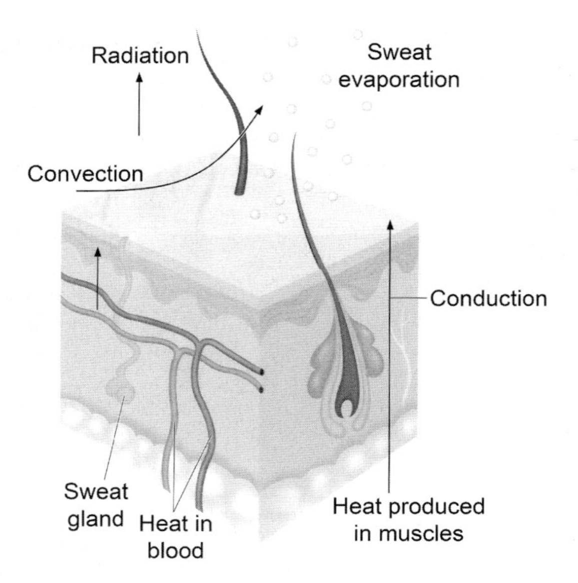

Fig. 6.14. More microscopic view of modes of heat loss by removal from the skin. When the skin temperature exceeds the environmental temperature, heat is removed by sweat evaporation, while when it is less than the environmental temperature there is net heat loss only by sweat evaporation. The skin is heated by arterial blood and by thermal conduction through subcutaneous (below the skin) tissue. (From [340]. Used with permission)

We will focus mostly on heat loss in one dimension. For example, the heat conduction equation (6.6) is one dimensional, and describes heat flow between two flat interfaces separated by a thickness Δx (Fig. 6.15a). Mathematically, heat flow radially through a thin cylindrical shell of thickness Δr on a cylinder

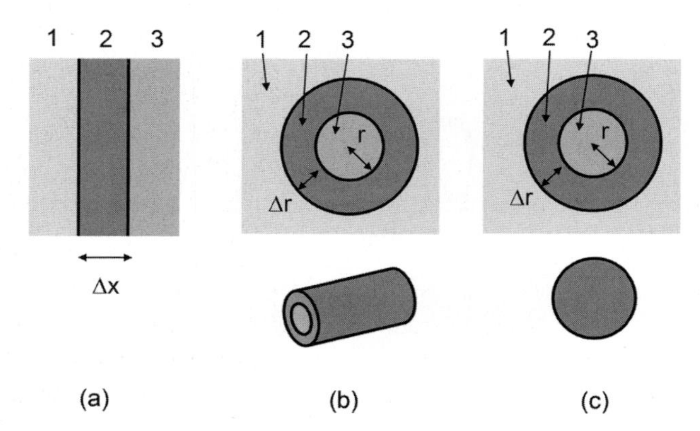

Fig. 6.15. Thermal flow between media 1 and 3, across medium 2, in different geometries: (**a**) rectangular slab, (**b**) cylindrical shell, (**c**) spherical shell. In common cases, Δr is much smaller than r

of radius r (Fig. 6.15b) is identical to that across flat interfaces if the Δx in (6.6) is replaced by $r \ln(1 + \Delta r/r)$. For conduction through a thin spherical shell of thickness Δr on a sphere of radius r (Fig. 6.15c), it needs to be replaced by $r\,\Delta r/(r + \Delta r)$. In both cases the replaced parameter reduces to Δr when $r \gg \Delta r$, i.e., when the new interface approximates a flat plane.

Heat Loss by Radiation

Bodies in thermal equilibrium emit a specific flow of energy per unit surface area and time depending on their temperature. They also receive a flow of thermal radiation from the outside world over their surface area that depends on the temperature of the surroundings.

All objects at a temperature T emit such thermal radiation spread over a range of wavelengths (or equivalently frequencies) according to the Planck black body distribution for intensity, which is

$$I(\lambda, T)\,\mathrm{d}\lambda = \frac{2hc^2\epsilon/\lambda^5}{\exp(hc/\lambda k_\mathrm{B}T) - 1}\,\mathrm{d}\lambda, \tag{6.37}$$

where ϵ is a parameter known as the emissivity, h is Planck's constant (6.626×10^{-34} J-s), k_B is Boltzmann's constant (1.381×10^{-23} J/K), and T is the temperature in degrees Kelvin (K) ($T(\mathrm{K}) = T(^\circ\mathrm{C}) + 273^\circ$). ($I(\lambda, T)$ is actually the intensity per unit wavelength range. $I(\lambda, T)\mathrm{d}\lambda$ is the intensity covering the wavelength range from λ to $\lambda + \mathrm{d}\lambda$, where $\mathrm{d}\lambda \ll \lambda$. Intensity is energy/area-time.)

Two features of this distribution are important to us. First, it peaks at a wavelength λ given by Wein's Law

$$\lambda_\mathrm{peak}(\text{in } \mu\text{m}) = \frac{2{,}898}{T\,(\text{in K})}. \tag{6.38}$$

This means that thermal radiation from a person or the inside wall of a house peaks at $10\,\mu$m (mid-infrared light) because $T \approx 300$ K, while that from the sun peaks near $0.6\,\mu$m (600 nm, visible light) because the temperature of the sun is $\sim$5,000 K. Second, the total energy flow per unit area and time (the energy flux) is determined by integrating (6.37) over all wavelengths to obtain the radiation flux, R, (which is energy/area-time)

$$R = \epsilon\sigma T^4, \tag{6.39}$$

where σ is the Stefan–Boltzmann constant $= 5.67 \times 10^{-8}$ W/m^2-K^4. This is known as Stefan's Law. At skin temperature (34°C), the thermal emission is 505 W/m^2 for $\epsilon = 1$.

Radiation incident on an object can be reflected by it, absorbed by it, or transmitted through it. The emissivity is the fraction of energy incident on the object that is absorbed. For a perfect black body it is 1. Most objects have a somewhat lower emissivity because of reflection and transmission, and an

Table 6.38. Infrared emissivities. (Using data from [298, 326, 327], and other sources)

fresh snow	0.89
ice	0.96
dry or wet sand	0.89
concrete	0.95
moist soil	0.97
grass surface	0.96
red brick	0.92
wood	0.90
white paint	0.93
aluminized paint	0.55
galvanized iron	0.28
aluminum foil	0.08
human skin	0.95–0.99

emissivity that actually varies with wavelength (and sometimes temperature). We are interested in the emissivity in the infrared (Table 6.38), because black body radiation from people and walls peaks near 10 μm, and that in the visible, because it is needed to understand how thermal radiation from the sun affects people and objects. In calculations the infrared emissivity of human clothing and skin is generally taken to be 0.95.

The energy loss per unit time due to thermal radiation from the body is

$$-\left(\frac{\mathrm{d}Q}{\mathrm{d}t}\right)_{\mathrm{loss}} = RA = \epsilon_{\mathrm{skin}}\sigma T_{\mathrm{skin}}^4 A_{\mathrm{skin}}, \tag{6.40}$$

where ϵ_{skin} is the skin (or clothing) emissivity, T_{skin} is the body skin temperature, and A_{skin} is the body surface area. (Remember that $\mathrm{d}Q/\mathrm{d}t < 0$ means that the body loses energy and becomes colder. The left-hand side of equations such as this are positive.) The correct temperature to use is that of the body skin ($34°\mathrm{C} = 307\,\mathrm{K}$), which is a bit lower than that of the core of the body ($37°\mathrm{C} = 310\,\mathrm{K}$). The average body surface area is $A_{\mathrm{skin}} \approx 1.85\,\mathrm{m}^2$, so the rate of body heat loss is

$$-\left(\frac{\mathrm{d}Q}{\mathrm{d}t}\right)_{\mathrm{loss}} = (1)(5.67 \times 10^{-8}\,\mathrm{W/m^2\text{-}K^4})(307\,\mathrm{K})^4(1.85\,\mathrm{m}^2) = 932\,\mathrm{W}, \tag{6.41}$$

assuming the emissivity ϵ_{skin} is 1. (Ignoring all sources of heat gain and other sources of heat loss, this radiative cooling alone would cool the body at a rate of 14°C/h, as is seen by using (6.35) and (6.36). Fortunately, this loss of heat is balanced by gains!)

The body also receives thermal radiation from its surroundings. The analysis is somewhat different for a person who is indoors, i.e., in an enclosed space, and one who is outdoors, where there are no enclosures and there can be direct heating by the sun during the day.

Receiving Thermal Radiation Indoors

Let us first assume that the body receives radiation from a source at room temperature of the same area $A_{\text{room}} = A_{\text{skin}}$. Then the rate of heat gain in a $24°\text{C} = 297$ K environment (T_{room}) is

$$\left(\frac{\mathrm{d}Q}{\mathrm{d}t}\right)_{\text{gain}} = \epsilon_{\text{room}}\sigma T^4_{\text{room}} A_{\text{room}} \tag{6.42}$$

$$= (1)(5.67\times 10^{-8}\,\text{W/m}^2\text{-K}^4)(297\,\text{K})^4(1.85\,\text{m}^2) = 816\,\text{W}, \tag{6.43}$$

assuming these temperatures and an emissivity of 1. (We will call the temperature in the environment T_{room}, whether or not the person is indoors or outdoors.) Then the net radiative heat loss by the body is given by

$$-\left(\frac{\mathrm{d}Q}{\mathrm{d}t}\right)_{\text{net}} = -\left(\frac{\mathrm{d}Q}{\mathrm{d}t}\right)_{\text{loss}} - \left(\frac{\mathrm{d}Q}{\mathrm{d}t}\right)_{\text{gain}} = 932\,\text{W} - 816\,\text{W} = 116\,\text{W}, \tag{6.44}$$

which we will now call $-(\mathrm{d}Q/\mathrm{d}t)_{\text{r}}$. This depends on the temperature of the environment. Because the body emissivity is a bit smaller than 1, this is a slight overestimate, but it is still consistent with the body's usual metabolic rate.

This commonly used analysis is actually flawed. The walls of the room have a much larger area, A_{room}, than that of the person, which would seem to increase the thermal radiation incident on the person relative to that emitted by the person. However, only part of the radiation from this room enclosure hits the person – most of it hits the other parts of the enclosure, and this decreases the thermal gain by the body. We have also ignored the likelihood that the emissivities of the walls of the room (or environment) and the person are likely different. These effects are included by replacing $\epsilon_{\text{room}}A_{\text{room}}$ in (6.42) by $(\epsilon_{\text{room}}A_{\text{room}})_{\text{effective}}$, by using Christiansen's equation

$$(\epsilon_{\text{room}}A_{\text{room}})_{\text{effective}} = \frac{\epsilon_{\text{skin}}A_{\text{skin}}}{1 + (A_{\text{skin}}/A_{\text{room}})(\epsilon_{\text{skin}}/\epsilon_{\text{room}} - 1)} \tag{6.45}$$

(which we will not derive here). This expression approaches $\epsilon_{\text{skin}}A_{\text{skin}}$ as the room area becomes very large relative to the body, because only a small amount of radiation from the room walls hits the body. The prediction of (6.44) is therefore reasonable. (Of course, this analysis needs to be modified when all the "walls" are not at the same temperature, such as when there are cold windows.) A poorly insulated house feels cold, even when there are no cold drafts leaking inside. With cold walls and windows, the air near our bodies may be warm, but the walls are cold and so thermal radiation from them is relatively low (Problem 6.84).

In general, the net radiative heat loss is

$$-\left(\frac{\mathrm{d}Q}{\mathrm{d}t}\right)_{\mathrm{r}} = \epsilon_{\mathrm{skin}}\sigma T_{\mathrm{skin}}^4 A_{\mathrm{skin}} - \epsilon_{\mathrm{room}}\sigma T_{\mathrm{room}}^4 A_{\mathrm{room}} \qquad (6.46)$$

and in the limit of a large room we use $(\epsilon_{\mathrm{room}} A_{\mathrm{room}})_{\mathrm{effective}}$ and this becomes

$$-\left(\frac{\mathrm{d}Q}{\mathrm{d}t}\right)_{\mathrm{r}} = \epsilon_{\mathrm{skin}}\sigma\left(T_{\mathrm{skin}}^4 - T_{\mathrm{room}}^4\right) A_{\mathrm{skin}}. \qquad (6.47)$$

This can be used directly (and exactly) or it can be approximated by "linearizing" it. This will also allow us to compare it with other modes of heat loss. With $\delta = T_{\mathrm{skin}} - T_{\mathrm{room}}$, then $T_{\mathrm{room}} = T_{\mathrm{skin}} - \delta$ and

$$\begin{aligned}
-\left(\frac{\mathrm{d}Q}{\mathrm{d}t}\right)_{\mathrm{r}} &= \epsilon_{\mathrm{skin}}\sigma\left(T_{\mathrm{skin}}^4 - T_{\mathrm{room}}^4\right) A_{\mathrm{skin}} \\
&= \epsilon_{\mathrm{skin}}\sigma(T_{\mathrm{skin}}^4 - (T_{\mathrm{skin}} - \delta)^4) A_{\mathrm{skin}} && (6.48) \\
&= \epsilon_{\mathrm{skin}}\sigma(4\delta T_{\mathrm{skin}}^3 - 6\delta^2 T_{\mathrm{skin}}^2 + 4\delta^3 T_{\mathrm{skin}} - \delta^4) A_{\mathrm{skin}} && (6.49) \\
&\simeq \epsilon_{\mathrm{skin}}\sigma(4\delta T_{\mathrm{skin}}^3) A_{\mathrm{skin}} = \epsilon_{\mathrm{skin}}\sigma(T_{\mathrm{skin}} - T_{\mathrm{room}})(4T_{\mathrm{skin}}^3) A_{\mathrm{skin}}, && (6.50)
\end{aligned}$$

where the last approximation follows because $|\delta| \ll T_{\mathrm{skin}}$. So we see that

$$\begin{aligned}
-\left(\frac{\mathrm{d}Q}{\mathrm{d}t}\right)_{\mathrm{r}} &\simeq (4\sigma T_{\mathrm{skin}}^3)\epsilon_{\mathrm{skin}} A_{\mathrm{skin}} (T_{\mathrm{skin}} - T_{\mathrm{room}}) && (6.51) \\
&= (4 \times 5.67 \times 10^{-8}\,\mathrm{W/m^2\text{-}K^4} \\
&\quad \times (307\,\mathrm{K})^3)\epsilon_{\mathrm{skin}} A_{\mathrm{skin}} (T_{\mathrm{skin}} - T_{\mathrm{room}}). && (6.52)
\end{aligned}$$

Also, we find

$$-\left(\frac{\mathrm{d}Q}{\mathrm{d}t}\right)_{\mathrm{r}} = h_{\mathrm{r}}' \epsilon_{\mathrm{skin}} A_{\mathrm{skin}} (T_{\mathrm{skin}} - T_{\mathrm{room}}) = h_{\mathrm{r}} A_{\mathrm{skin}} (T_{\mathrm{skin}} - T_{\mathrm{room}}), \qquad (6.53)$$

where h_{r}' is the *first power radiant heat transfer coefficient*

$$h_{\mathrm{r}}' = 4\sigma T_{\mathrm{skin}}^3 = 5.6\,\mathrm{kcal/m^2\text{-}h\text{-}{}^\circ C} \qquad (6.54)$$

for $T_{\mathrm{skin}} = 34°\mathrm{C}$. In (6.53) we have also combined $h_{\mathrm{r}}' \epsilon_{\mathrm{skin}}$ into one factor $h_{\mathrm{r}} = 4\sigma\epsilon_{\mathrm{skin}} T_{\mathrm{skin}}^3$ to cast it into a form similar to those we will use for heat transfer by convection, conductivity, and evaporation.

This linearization method is quite general and is applicable here because $|T_{\mathrm{skin}} - T_{\mathrm{room}}| \ll T_{\mathrm{skin}}$. However, because of the fortuitous factorization

$$T_{\mathrm{skin}}^4 - T_{\mathrm{room}}^4 = (T_{\mathrm{skin}} - T_{\mathrm{room}})(T_{\mathrm{skin}}^3 + T_{\mathrm{skin}}^2 T_{\mathrm{room}} + T_{\mathrm{skin}} T_{\mathrm{room}}^2 + T_{\mathrm{room}}^3), \qquad (6.55)$$

(6.53) becomes exact with $4T_{\mathrm{skin}}^3$ replaced by $(T_{\mathrm{skin}}^3 + T_{\mathrm{skin}}^2 T_{\mathrm{room}} + T_{\mathrm{skin}} T_{\mathrm{room}}^2 + T_{\mathrm{room}}^3)$, and so

$$h_{\mathrm{r}} = \sigma\epsilon_{\mathrm{skin}}(T_{\mathrm{skin}}^3 + T_{\mathrm{skin}}^2 T_{\mathrm{room}} + T_{\mathrm{skin}} T_{\mathrm{room}}^2 + T_{\mathrm{room}}^3) \simeq 4\sigma\epsilon_{\mathrm{skin}} T_{\mathrm{skin}}^3. \qquad (6.56)$$

Table 6.39. Reflectance in the visible. (Using data from [298])

from	reflectance
clipped grass	0.24
dry sand	0.40
dry tar macdam	0.12
dry concrete	0.24
fresh snow	0.85
black human skin	0.18 (0.10[a])
white human skin	0.35 (0.42[a])

[a]From [297] for very black and very white human skin.

Receiving Thermal Radiation Outdoors

There are generally no wall-type enclosures outdoors that emit significant thermal radiation, but there is thermal (infrared) radiation from the ground (asphalt, grass, snow cover, etc.), buildings, trees, etc. There is also heating by rays directly incident from the sun and scattered sun light; this depends on the latitude, time of year, cloud cover, etc. The heat gained by the body per unit area and time is

$$\frac{1}{A}\left(\frac{\mathrm{d}Q}{\mathrm{d}t}\right)_{\mathrm{gain}} = (1-\rho)(R_{\mathrm{S,D}} + R_{\mathrm{S,I}} + R_{\mathrm{S,R}}) + \alpha(R_{\mathrm{IR,U}} + R_{\mathrm{IR,D}}), \quad (6.57)$$

where the first term describes heating by the sun (mostly visible light) and the second term longer wavelength, infrared light. The solar light can be that directly from the sun ($R_{\mathrm{S,D}}$), indirect scattered or diffused light from the sky ($R_{\mathrm{S,I}}$), or that reflected from the terrain ($R_{\mathrm{S,R}}$). A fraction ρ of that light is reflected from the body (see Table 6.39). The contribution of each solar term is seen in Table 6.40. The infrared light can be coming "up" from the terrain ($R_{\mathrm{IR,U}}$) or "down" from the sky ($R_{\mathrm{IR,D}}$); this contribution is analogous to the thermal emission from the room walls in the previous section. The component of long wavelength, infrared radiation from the sky is seen in Table 6.41. A fraction α of this infrared light is absorbed by the body (and this includes reflection and angle averaging effects); this is typically $\sim 1/2$. Typically more of this radiant heat (solar plus infrared) is absorbed by the body than is emitted by the body during the day; the reverse is typically true at night.

Heat Loss by Convection and Conduction

Convection is the transport of heat through mass flow of the medium. Conduction is the transport of heat through a medium without mass flow (involving electrons, vibrations, local molecular motion, etc.). In gases and liquids both mechanisms can contribute. Convection is the larger of the two and it is not

Table 6.40. Solar radiation from clear skies on a fleeced sheep with its sides at right angles to the solar beam, in W/m^2. (Using data from [298], from [303])

solar elevation (angle, °)	$R_{S,D}$	$R_{S,I}$	$R_{S,R}$	R_S	$(1-\rho)R_S$
5	14	29	5	48	36
10	79	34	12	125	94
15	150	43	19	212	159
20	208	48	27	283	212
25	252	51	35	338	254
30	285	52	44	381	286
45	331	55	67	453	340
60	344	53	86	483	362
75	348	52	97	497	373
90	352	52	101	505	379

The direct, indirect reflected components and the total fluxes are, respectively, $R_{S,D}$, $R_{S,I}$, $R_{S,R}$, R_S, and $(1-\rho)R_S$. The atmospheric transmission coefficient was taken to be 0.7.

always simple to differentiate between the two. In solids only conduction contributes.

Convection

The convection heat flux consists of that from forced convection (air streams) and natural or free convection (which is due to movement near the body induced by thermal expansion caused by temperature gradients). In most general form, the convective heat transfer coefficient per unit area (in W/m^2-°C) is

$$h_{\text{c-forced convection}} = \frac{akw^n}{\upsilon^n L^{1-n}}, \tag{6.58}$$

Table 6.41. Incoming long wavelength radiation from clear skies and the equivalent radiation temperature. (Using data from [298], calculated using [339])

air T (°C)	incoming radiation (W/m^2)	radiant T (°C)
−30	67	−88
−20	108	−65
−10	155	−45
0	208	−28
10	266	−12
20	331	3
30	404	17

where w is the wind speed in m/s, L is the characteristic dimension of the object in m – which is the diameter of the cylinder in some models, v is the kinematic viscosity of air in m^2/s, and a, k, and n are parameters. This is used in

$$-\frac{1}{A}\left(\frac{\mathrm{d}Q}{\mathrm{d}t}\right)_{\mathrm{c}} = h_{\text{c-forced convection}}(T_{\text{skin}} - T_{\text{air}}), \tag{6.59}$$

which is derived [298] using scaling arguments involving the Reynolds number, a parameter that is discussed in Chap. 7. The natural convective heat loss per unit area A (in $\mathrm{W/m^2}$) from a man of height L (in m) has been predicted to be

$$-\frac{1}{A}\left(\frac{\mathrm{d}Q}{\mathrm{d}t}\right)_{\text{c-natural convection}} = \frac{1.69}{L^{0.25}}(T_{\text{skin}} - T_{\text{air}})^{1.25}, \tag{6.60}$$

which scales with temperature differently than (6.58) and (6.59) for forced convection. Usually, only the larger of forced or natural convection is considered in the analysis, and it is modeled in the form

$$-\frac{1}{A}\left(\frac{\mathrm{d}Q}{\mathrm{d}t}\right)_{\mathrm{c}} = h_{\mathrm{c}}(T_{\text{skin}} - T_{\text{air}}), \tag{6.61}$$

where h_{c} is the convective heat transfer coefficient per unit area and T_{air} is the air temperature very near the person, which is usually the same as T_{room}. With still air and normal clothing, $h_{\mathrm{c}} \sim 2.3\,\mathrm{kcal/m^2\text{-}h\text{-}^\circ C}$, which is $< h_{\mathrm{r}}$ (the radiant heat transfer coefficient). The coefficient h_{c} decreases with heavier clothing and increases with less and lighter clothing; when you are nude you get cold faster because of increases in h_{c}. It also increases with air speed, as is seen in Table 6.42. This leads to the *wind chill factor*. For a naked person

$$h_{\mathrm{c}} = 8.3 w^{0.5} \tag{6.62}$$

in $\mathrm{W/m^2\text{-}^\circ C}$, where w is the air speed in m/s. In another formulation [300],

$$h_{\mathrm{c}} = 10.45 - w + 10 w^{0.5} \tag{6.63}$$

in $\mathrm{kcal/h\text{-}m^2\text{-}^\circ C}$, for speeds between 2 m/s ($\simeq$4.5 mph) and 20 m/s ($\simeq$45 mph), where again v is in m/s. Both clothed and bare regions on a body need to be considered. The body "feels" the wind chill temperature, which is colder and is

$$T_{\mathrm{wc}} = 13.12 + 0.6215T - 11.37 w^{0.16} + 0.3965 T w^{0.16}, \tag{6.64}$$

where w is the wind speed in km/h and T the temperature in °C. (This is the relation as revised in 2001.) Also see Table 6.43. (In contrast, humidity makes you feel even warmer than expected from the ambient temperature, as is described by the heat index discussed later in this chapter.)

Table 6.42. The convective heat transfer coefficient (h_c) and air insulation $[1/(h_c + h_r)]$ of the skin–air interface for a naked man. (Using data from [298])

wind speed (m/s)	convective heat transfer coefficient (W/m^2-°C)	air insulation (m^2-°C/W)
0.1	2.6	0.123
0.2	3.7	0.109
0.4	5.2	0.093
0.6	6.4	0.084
0.8	7.4	0.077
1.0	8.3	0.072
2.0	11.7	0.058
3.0	14.4	0.50
4.0	16.6	0.045
5.0	18.6	0.041

The radiant heat transfer coefficient h_r is 5.5 W/m^2-°C.

With $T_{room} = T_{air}$, the total heat flux due to (net) radiation and convection comes from summing (6.53) and (6.61):

$$-\frac{1}{A}\left(\frac{dQ}{dt}\right)_{total} = -\frac{1}{A}\left(\frac{dQ}{dt}\right)_r - \frac{1}{A}\left(\frac{dQ}{dt}\right)_c = (h_r+h_c)(T_{skin}-T_{room}), \quad (6.65)$$

which gives

$$T_{skin} - T_{room} = -\frac{(1/A)(dQ/dt)_{total}}{h_r + h_c} \quad (6.66)$$

Convection within the Body We have been discussing the convection of heat from the outside of the body to air by the convection of air. There is also of convection *within* the body by the flow of blood. This does not change the

Table 6.43. Wind chill factor temperature, in °C, as a function of wind speed and temperature; the temperature is given in the first row for the 0 km/hr wind speed. (Using the scale revised in 2001 by the US National Weather Service)

wind speed (km/hr)	wind chill factor temperature (°C)					
0	10	0	−10	−20	−30	−40
20	7.4	−5.2	−17.9	−30.5	−43.1	−55.7
40	6.0	−7.4	−20.8	−34.1	−47.5	−60.9
60	5.1	−8.8	−22.6	−36.5	−50.3	−64.2
80	4.4	−9.8	−24.0	−38.2	−52.4	−66.6
100	3.9	−10.6	−25.1	−39.6	−54.1	−68.6

average temperature of the body, but the distribution of temperature within the body. Equation (6.5) says that heat flow $Q = mc(\Delta T)$ is needed into a body of mass m and specific heat c to heat the mass by ΔT and heat flow of this magnitude away from the body is needed to cool it. This means that if blood flows into a body part, such as the foot in Problem 6.95, at a rate of F_m (mass per unit time) with a temperature T_{blood} and leaves it (at the same rate, of course) at $T_{\text{blood}} + \Delta T_{\text{blood}}$, it will draw away heat from the part at a rate

$$-\left(\frac{\mathrm{d}Q}{\mathrm{d}t}\right)_{\text{blood flow}} = F_m c(\Delta T_{\text{blood}}). \tag{6.67}$$

Therefore heat is left in the body part by the blood when $\mathrm{d}Q/\mathrm{d}t > 0$ and the blood gets colder, $\Delta T_{\text{blood}} < 0$. Of course, this transfer of heat between the blood and local tissues occurs by thermal conduction.

This analysis has not addressed what $(\mathrm{d}Q/\mathrm{d}t)_{\text{blood}}$ and ΔT_{blood} actually are, but only how they are related. To determine them, we would have to consider the rates of thermal conduction between the blood in the arteries, capillaries, and veins and the surrounding tissue and the actual flow rates of blood in each of these vessels. If the flow is very fast, there may not be enough time for much heat flow to occur.

If the arteries flowing into a body part, such as a hand or foot, are physically displaced from the veins that return the blood to the heart, there is minimal transfer of heat from arteries to veins. If they are instead very near to each other, there can be substantial heat transfer from the warmer blood flowing in the arteries to the cooler blood flowing in the veins. (The venous blood is cooler because of heat loss to the local tissues which can be much colder than the core.) This is called *countercurrent heat exchange*, and is shown in Fig. 6.16 [325]. It lowers overall heat loss at the body part – particularly for extremities and surface regions – and therefore overall heat loss in the blood and the body, because it decrease the difference in temperature between the blood in the body part and the tissue (and locally, outside the body). As part of its temperature control system (Chap. 13), the body can decide whether to minimize heat loss, by using countercurrent cooling, or to maximize it, by diverting more blood flow nearer to the skin. Countercurrent heat exchange is modeled in Problem 6.102.

Figure 6.17 shows how this occurs in the forearm through two different sets of veins. If it is cold at the surface of the body, blood returns to the heart through veins that are very close to the arteries, so heat loss is minimized through countercurrent heat exchange. Because this minimizes the warming of the hands, this is one way you can get excessively cold hands. If it is warm, blood flow near the surface and heat transfer and cooling are maximized. Blood is directed to these "inner" and "outer" veins by vasodilating and vasoconstricting muscles surrounding the appropriate veins (to dilate and constrict them, respectively), as controlled by the temperature sensor and the temperature regulatory system (Chap. 12).

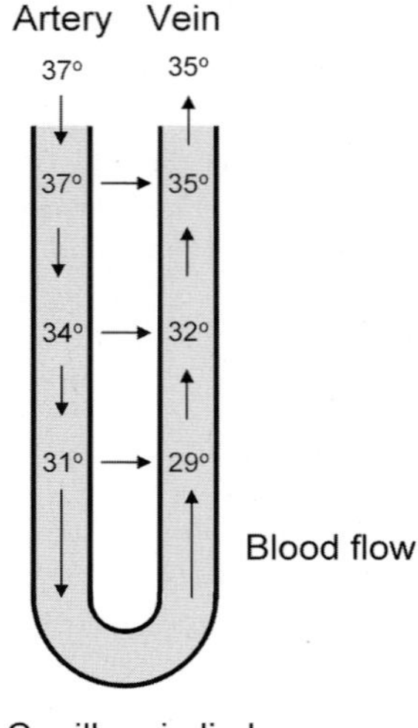

Fig. 6.16. Example of countercurrent heat flow from an artery to a nearby vein. Without this flow the temperature drop along the artery would be even greater under cold conditions because there would be more heat flow to the environment. All temperatures are in °C. Also see Fig. 6.17a. (Based on [325])

Counterflow heat exchange is also used to regulate the temperature of the testes. The testes are kept outside the core of the body because fertility is improved at a few degrees below core temperature. Before blood enters the testes region it often undergoes counterflow heat exchange so that cooler blood enters this region. This prevents excessive heating of the testes by warm blood, as well as the excessive loss of body heat from this "thermally" exposed region.

A counterflow multiplier mechanism is also used by the body in diffusion in the kidney to overcome large gradients in concentrations [325, 336].

Conduction

As shown in (6.6), the heat flow during thermal conduction is the product of the thermal conductivity K and the temperature gradient

$$-\frac{1}{A}\left(\frac{\mathrm{d}Q}{\mathrm{d}t}\right) = K\frac{\mathrm{d}T}{\mathrm{d}x} \sim K\frac{\Delta T}{\Delta x}. \tag{6.68}$$

For a well-defined distance $d = \Delta x$ between two regions of different but uniform temperature, say due to the thickness of clothing or an air boundary layer, we can define a heat transfer coefficient per unit area $h_{\mathrm{t}} = K/d$ and

$$-\frac{1}{A}\left(\frac{\mathrm{d}Q}{\mathrm{d}t}\right) = h_{\mathrm{t}}\Delta T, \tag{6.69}$$

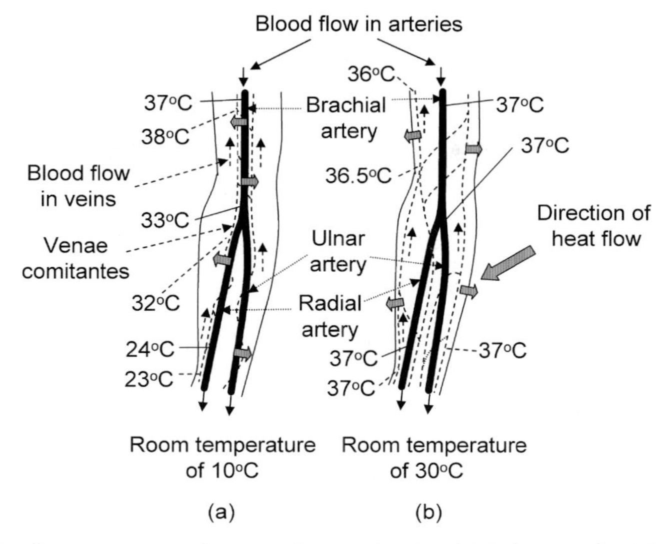

Fig. 6.17. Countercurrent heat exchange in the (right) arm for blood flowing through the brachial artery in (**a**). At the higher room temperatures (30°C) in (**b**) and during exercise the venous return flow occurs through superficial vessels to enable heat loss to the room. At the lower temperature (10°C) in (**a**) this return path is shifted to deep veins, which helps retain heat in two ways: Heat from the arteries is lost to the venous blood before it is lost to the room, which is countercurrent heat exchange (Fig. 6.16), and little heat is lost to the room from these veins because they are deep. Note that in the lower arm the temperature in the venae comitantes is 1°C lower than that in the artery. The venae comitantes are the veins accompanying the arteries in the limbs. Different types of *arrows* are used to denote blood flow in arteries (*solid, narrow*), blood flow in veins, (*dashed, narrow*), and heat flow (*wide, gray*). (Based on [299, 315])

as in (6.7). This looks similar to the form for heat loss due to (net) radiation (6.53) and forced convection (6.59).

Let us determine how much heat a person loses when he puts his hand and forearm in very cold water (4°C) for 2 min [293]. His hand and forearm can be modeled as a cylinder with a length of $L = 30$ cm and an average radius of 3 cm. Let us assume the body has a $\Delta r = 3$ mm thick layer of fat with thermal conductivity $K = 0.2$ W/m-°C that separates the cylindrical core (of radius $r = 2.7$ cm, at 37°C) from the cold water. The contact area between the arm and water is $A = 2\pi r L$. Using (6.68) modified for cylinders (by replacing Δx with $r \ln(1+\Delta r/r)$, as above), the rate heat leaves the arm is $(2\pi r L)(K\ \Delta T)/(r \ln(1+\Delta r/r)) \simeq (2\pi r L K\ \Delta T)/(\Delta r)$. The exact expression gives 118 W, while the approximation gives 112 W. In 2 min, the body loses (118 W)(120 s) = 14,200 J or 3.4 kcal.

Insulation I can be formally defined as $I = 1/h_{\mathrm{t}} = d/K$. It is sometimes expressed in terms of the clothing or Clo value, where 1 Clo is the insulation

Table 6.44. The insulation from animal coats, fabrics, body components, and other materials, in units of m^2-°C/W per cm of insulation. (Using data from [293, 298]; also see [302])

material	insulation per cm
cattle coat	0.07–0.10
sheep coat	0.13–0.24
husky dog coat	0.24
pig coat	0.04
wool	0.29
goose down	0.38
human clothing (average)	0.25
air	0.36/0.39
muscle, bone	0.01/0.024
body fat	0.01/0.05
water	0.01/0.017
ice	0.01/0.004
typical wood[a]	0.01/0.062

h_t is this 1/(insulation per cm × the thickness in cm). Thermal conductivity K is 1/(insulation per cm) or 0.01 × this value when expressed in units of W/m-°C.
[a]Balsa wood is higher; green lumber is smaller.

for a man in an average suit with no gloves or light gloves, wearing leather shoes and light socks [297, 302]. An insulation of 1 Clo = 0.155 m^2-°C/W (= 0.88 ft^2-°F/BTU). A nude person has 0 Clo insulation. A light working ensemble (undershorts, wool socks, cotton work shirt – shirt tail out and neck open – and work trousers) corresponds to 0.6 Clo. A typical light suit has 0.7–1.0 Clo insulation, while a heavy business suit has 1.2 Clo insulation. A polar weather suit made of heavy wool pile has 3–4 Clo insulation. Typical values of insulation are given in Table 6.44. The insulation provided by human clothing is about 0.25°C-m^2/W per cm of clothing thickness. The thermal conductivity of various body tissues is given in Table 6.36 and that of common materials in Table 6.37.

Heat Loss by Evaporation of Water: Sweating and Breathing

The amount of heat needed to evaporate 1 L of water is 540 kcal; this is the heat of vaporization. When we sweat there is loss of heat only if the water is allowed to evaporate. (Wiping off the sweat will not cool you.) This normally accounts for ∼7 kcal/h loss.

We normally breathe in air that is cooler than body temperature. It gets warmed up and is then exhaled: this is also a source of body cooling. Also, we breathe in relatively dry air, and exhale air saturated with water vapor. Because this water vapor is formed by the evaporation of liquid water, this is yet another source of cooling.

Respiratory heat loss due to the lungs can be described by

$$-\left(\frac{\mathrm{d}Q}{\mathrm{d}t}\right)_{\mathrm{l}} = \rho_{\mathrm{air}} c_{\mathrm{p,\,air}} (T_{\mathrm{exp}} - T_{\mathrm{insp}}) \frac{\mathrm{d}V_{\mathrm{air}}}{\mathrm{d}t} + \Delta H_{\mathrm{evap,\,water}} (\rho_{\mathrm{exp,\,water}} - \rho_{\mathrm{insp,\,water}}) \frac{\mathrm{d}V_{\mathrm{air}}}{\mathrm{d}t}, \qquad (6.70)$$

where ρ is the mass density (of air or of expired or inspired water vapor), $c_{\mathrm{p,\,air}}$ is the specific heat of air (2.4×10^{-4} kcal/g-°C), T is the temperature of expired or inspired air, $\mathrm{d}V/\mathrm{d}t$ is the rate of inhaling air, and $\Delta H_{\mathrm{evap,\,water}}$ is the heat of evaporation of water (0.54 kcal/g). The first term accounts for body cooling by the heating of inspired air before expiration (exhalation) and the second term accounts for body cooling by the evaporation of water in the lungs that increases the water content of inspired air before expiration.

When it is hot, why does it feel even hotter than the ambient temperature when it is humid? Humidity decreases the rate of heat loss from the body by decreasing the rate of heat flow from the body to the air, decreasing the evaporation of sweat (because the vapor pressure of water in the air decreases the difference between the vapor pressure above the sweat and in the air), and decreasing the rate of respiratory loss of heat due to the last term in (6.70). The *heat index* describes how hot it really feels at a given temperature and humidity [337, 338]. For example, when it is 90°F (32°C), it feels like it is 95°F (35°C) when the relative humidity is 50%, 109°F (43°C) when it is 75%, and 132°F (56°C) when it is 100%. (The relative humidity is the actual vapor pressure of water in air divided by the equilibrium vapor pressure of water (in contact with liquid) water, at the same temperature. It is usually expressed in per cent.) The likelihood of heat disorders, such as heat stroke and sunstroke, are characterized by warnings such as "caution" for heat index values of 80–90°F (27–32°C), "extreme caution" from 90 to 105°F (32 to 41°C), "danger" from 105 to 129°F (41 to 54°C), and "extreme danger" above 130°F (54°C). The heat index is explored more in Problems 6.108–6.111.

6.6 Body Temperature

The body is very good at controlling its temperature. Figure 6.18 shows that the nude body can maintain the core temperature between 97 and 100°F (36 and 38°C) for several hours with environmental temperatures ranging between 70 and 130°F (21 and 54°C) in dry air. However, we clearly have limits. Figure 6.19 shows that we can take only limited exposure to hot air, while Fig. 6.20 shows we can survive only limited exposure to cold water; in both cases the larger the temperature deviation from normal body temperature, the shorter the maximum possible exposure time until death. Figure 6.21 tracks how skin and inner body temperatures decrease during exposure to cold water, and how it recovers when the body is then placed in warm water. *How do the energetics of the body change to*

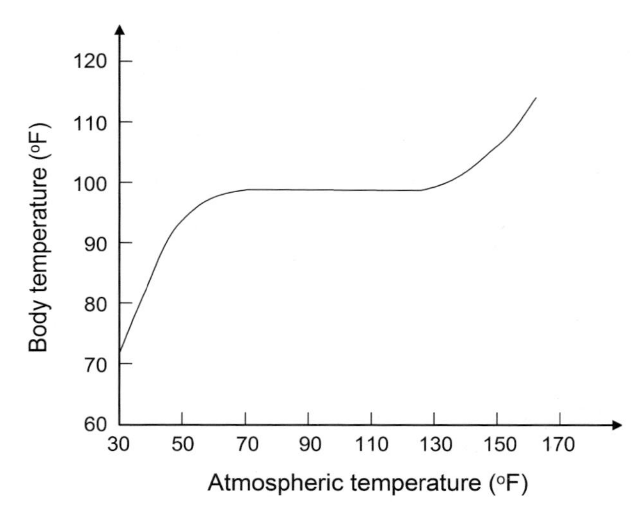

Fig. 6.18. Body control of core temperature for several hours with large variations in local environmental temperature, for dry air. Normal body temperature 98.6°F (37°C) is maintained for atmospheric temperatures from ~70 to 130°F (~21 to 54°C). The exact shape of this curve depends on the surroundings: humidity, air movement, thermal radiation to the body, and so on. (Based on [314])

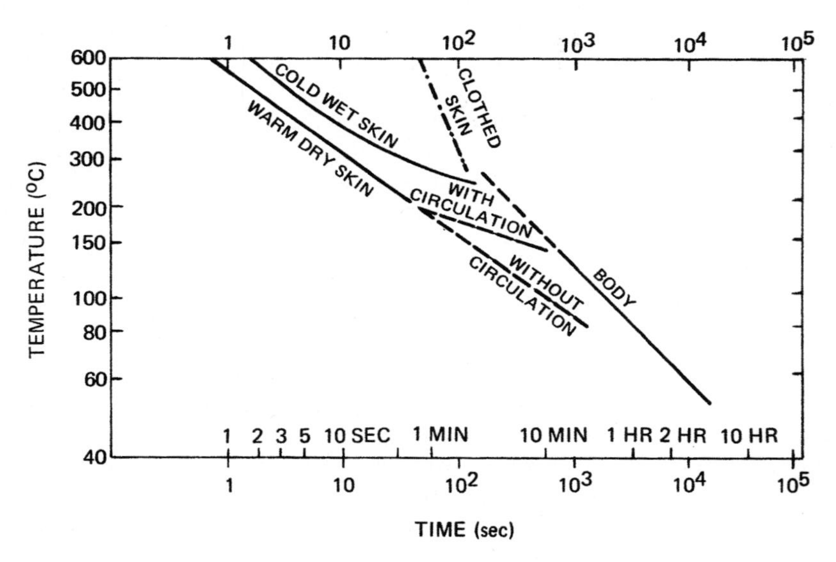

Fig. 6.19. Maximum tolerable elevated temperatures (of air) as a function of exposure times, for different body conditions. Tolerable conditions are below the lines for each skin condition and for the body, and intolerable conditions are above them. (From [297])

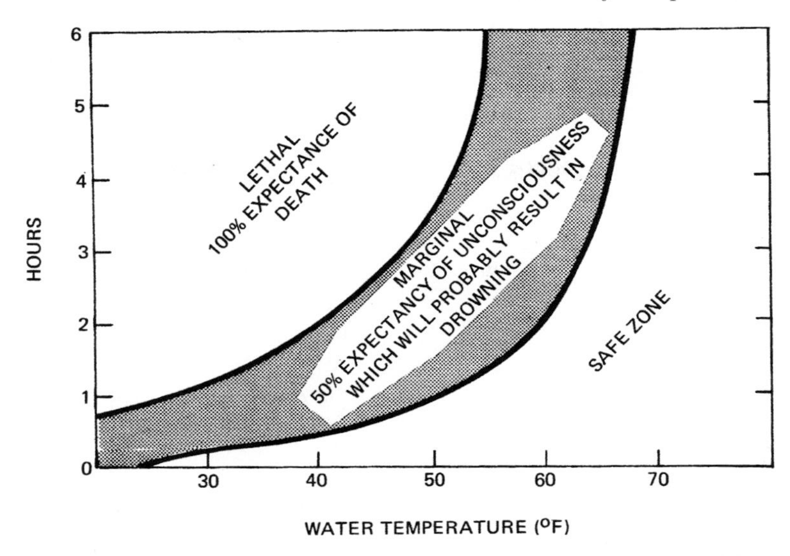

Fig. 6.20. Life expectancy of a person without an exposure suit in cold water for different times. (From [297])

maintain normal temperature and to resist changes that can be induced by extreme environmental temperatures or by exercise?

Each of the modes of body heat loss we have described is important for regulating the body temperature. You can control the convection route by changing your clothes. Some are controlled involuntarily by your body, such as the sweating and breathing modes. *What happens when the room temperature exceeds the body core and skin temperature?* The body is then heated by radiation. Convection needs to be re-evaluated. Exhaled air is no longer heated up. *What happens when you exercise and your metabolic rate increases?* The body has to respond in a way to increase the rate of losing body heat. All these modes of losing heat are important. A quantitative model of this regulation is presented in Chap. 13, as part of the discussion of feedback and control in the body.

Heat production varies with the local, environmental temperature, which we called T_{room} above. It is high at low temperatures, then decreases with increasing temperature until T_{c}, becomes constant, and then at T_{h} it increases again, as is shown in Fig. 6.22. The *critical temperature* T_{c} (or *lower critical temperature*) is therefore the environmental temperature at which heat production begins to rise when the environmental temperature falls. It is 26°C for a naked adult man and 35°C for a naked baby, assuming minimal air movement. Similarly, the *temperature of hyperthermal rise* T_{h} (or *upper critical temperature*) is the environmental temperature at which heat production begins to rise when the environmental temperature increases. Between T_{c} and T_{h} is the *zone of minimal metabolism* or the *thermoneutral zone.*

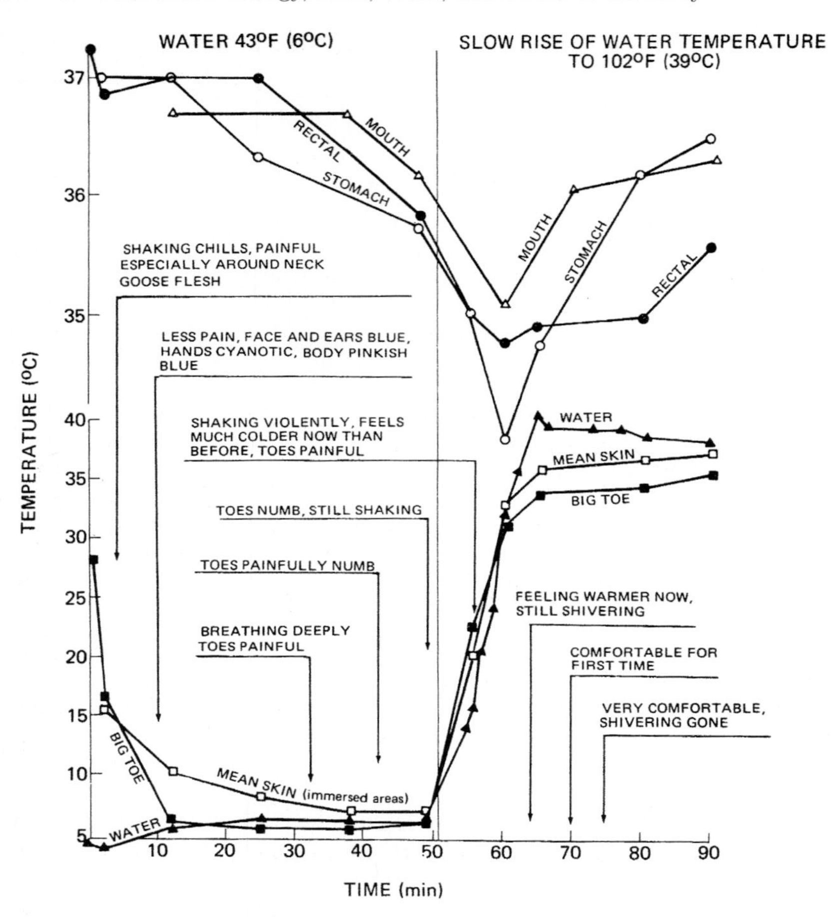

Fig. 6.21. Variation in skin and internal temperatures when a person is immersed in 6°C (43°F) water for 52 min and then warmed slowly using 39°C water. (From [297])

Below T_c, the heat loss increases due to nonevaporative pathways, such as radiation and convection. The increased metabolism is due to increases in muscle tension and shivering, which is technically the visible movement of subcutaneous muscle. This increases as the difference between body and environmental temperature increases, until the *summit metabolism* is reached for that environmental temperature. The summit metabolism is about $21m_b^{0.75}$, in W with m_b in kg, for man (and generally about $25m_b^{0.75}$ for most mammals). This is about 7 × the fasting rate of metabolism or 2–3 × the field rate of energy expenditure. (It is still much lower than the peak metabolic rates you can achieve during cycling, running, and so on.) At lower temperatures, the core body temperature cannot be maintained and there is body cooling and hypothermia. Above T_h the heat loss is dominated by evaporation.

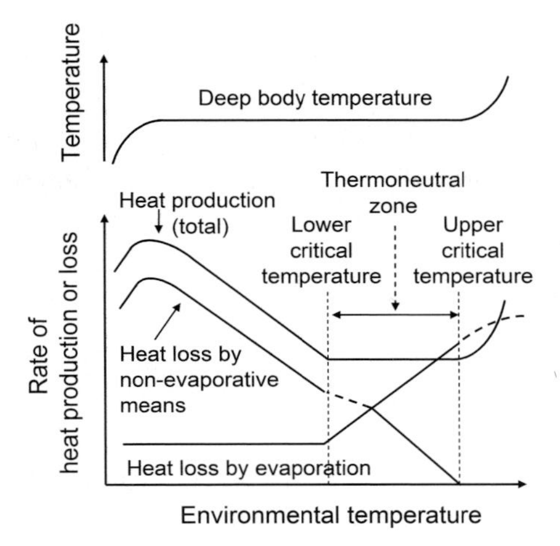

Fig. 6.22. Critical temperatures and relation between the environmental temperature and heat production in homeothermic animals (which are those having constant and relatively high body temperature, which means birds and mammals). (Based on [298, 328])

The critical temperature can be evaluated by equating heat loss and heat production by the metabolism. During exercise the work done and heat stored or liberated from tissues need to be considered also. The core body temperature can be related to the maximal aerobic capacity by

$$T_{\text{core}}(^{\circ}\text{C}) = 36.5 + 3.0 \times \frac{\text{oxygen consumption}}{\text{maximum possible oxygen consumption}}, \qquad (6.71)$$

because the oxygen consumption rates are proportional to the metabolic rates. Also, movement during exercise increases convective heat loss. A rise in body temperature of 1°C increases heat production by $\sim$13%. Heat production is also induced by pyrogens, etc., and by trauma, bone fractures, burns, etc.

The temperature T in (6.35) is the average body temperature. The temperature of the body skin $T_{\text{skin}}(34^{\circ}\text{C} = 307\,\text{K})$ is usually lower than that of the core (rectal) $T_{\text{core}}(37^{\circ}\text{C} = 310\,\text{K})$ – that you measure with a thermometer. Table 6.45 gives the normal temperature of different body organs and blood vessels. The temperature of inner organs can differ by $0.2 - 1.2^{\circ}$C and by 0.9°C in a given organ under normal conditions [308]. There are 1.4°C temperature variations in the brain, with the cortex being cooler than the basal regions. The average temperature of the body is the weighted average of the two temperatures, $T_{\text{av}} = (80 \pm 5\%)T_{\text{core}} + (20 \pm 5\%)T_{\text{skin}}$. The temperature throughout the body varies with the temperature in the environment. As shown in Fig. 6.23, these variations can be very large in a cold room. These internal distributions can be modeled using the thermal properties of the body (Table 6.36) and the thermal transport models presented here.

Table 6.45. Core temperatures within the human body. (Using data from [308])

body region	normal temperature (°C)
skin	32–35
scrotum	34.0
liver	36.4–36.8
oral cavity	36.5–36.6
superior vena cava	36.65
esophagus, lungs	36.75
heart (right ventricle)	36.75
aorta, inferior vena cava	36.75
pulmonary artery and vein	36.75
kidney	36.85
spinal cord	36.95
stomach, rectum (mean)	37.0
rectum (range)	36.2–37.8
brain, uterus	37.3

Let us interrelate the body temperature in the deep body or core, T_{core}, of the skin, T_{skin}, on the outside of our clothes, T_{clothes}, and in the air, T_{room}. This can viewed in terms of the model of Fig. 6.24 in which each region is a rectangular slab. More refined models would consider the body as a cylinder, as in Fig. 6.15b.

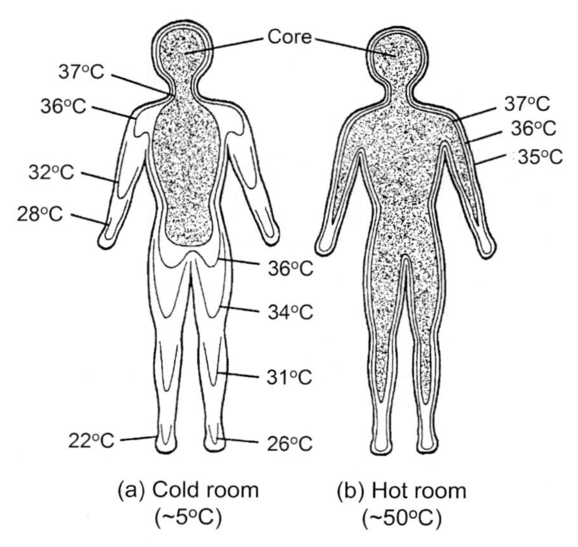

Fig. 6.23. Temperature in the body in cold and hot environments. (From [308]. Courtesy of Robert A. Freitas Jr., Nanomedicine, Vol. 1 (1999), http://www.nanomedicine.com)

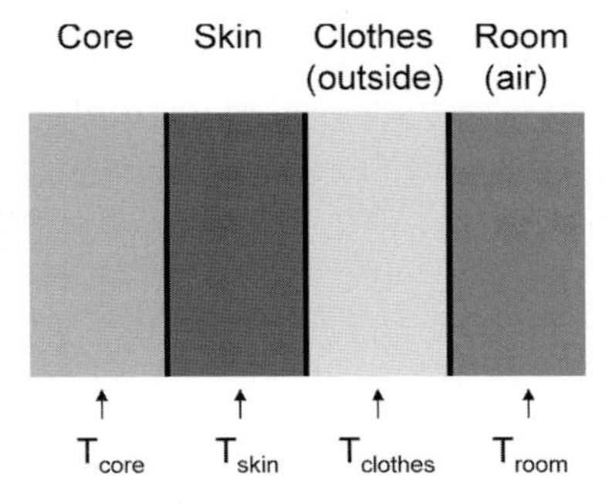

Fig. 6.24. Model of heat flow from the body core to air

Following (6.66), we can model the flow between two of these adjacent regions by

$$(T_1 - T_2) = -I_{12}\frac{1}{A}\left(\frac{\mathrm{d}Q}{\mathrm{d}t}\right), \tag{6.72}$$

where the heat flow, $\mathrm{d}Q/\mathrm{d}t$, across a cross-sectional area A between regions 1 and 2 is controlled by the insulation between them

$$I_{12} = \frac{1}{\sum h_{x,12}}, \tag{6.73}$$

where $h_{x,12}$ are the relevant heat transfer coefficients between the two regions, such as those due to radiation, convection, conduction, and evaporation.

The difference between the temperature in the body core and on the skin is determined by the flow of heat from the core to the skin and the tissue insulation I_t, and is

$$T_\mathrm{core} - T_\mathrm{skin} = -\frac{I_\mathrm{t}}{A}\left[\left(\frac{\mathrm{d}Q}{\mathrm{d}t}\right)_\mathrm{r} + \left(\frac{\mathrm{d}Q}{\mathrm{d}t}\right)_\mathrm{c} + \left(\frac{\mathrm{d}Q}{\mathrm{d}t}\right)_\mathrm{v}\right], \tag{6.74}$$

where $(\mathrm{d}Q/\mathrm{d}t)_\mathrm{r}$, $(\mathrm{d}Q/\mathrm{d}t)_\mathrm{c}$, and $(\mathrm{d}Q/\mathrm{d}t)_\mathrm{v}$ are, respectively, the flows of radiative heat loss, convective/conductive heat loss, and the loss of heat by vaporizing moisture on the skin surface. (The loss of heat from the lungs and respiratory passages is not included.) Because the heat flow is from the core to the skin for each, each term is negative and the skin is cooler than the core, as is expected. The insulation provided by human tissue has a maximum value of $0.10°\mathrm{C\text{-}m^2/W}$ in adults at cold temperatures (below the critical temperature) and a minimum value of $0.03°\mathrm{C\text{-}m^2/W}$ at high temperatures. In babies these high and low values are of $0.05°\mathrm{C\text{-}m^2/W}$ and $0.015°\mathrm{C\text{-}m^2/W}$, respectively.

The difference between the temperature on the skin and the outside of one's clothing is

$$T_\mathrm{skin} - T_\mathrm{clothes} = -\frac{I_\mathrm{c}}{A}\left[\left(\frac{\mathrm{d}Q}{\mathrm{d}t}\right)_\mathrm{r} + \left(\frac{\mathrm{d}Q}{\mathrm{d}t}\right)_\mathrm{c}\right], \tag{6.75}$$

where I_c is the insulation of the clothing (or fur for animals). The insulation provided per cm thickness of human clothing averages to 0.25°C-m^2/W-cm (Table 6.44).

The difference between the temperature outside of one's clothing and the air is

$$T_{\text{clothes}} - T_{\text{room}} = -\frac{I_a}{A}\left[\left(\frac{dQ}{dt}\right)_r + \left(\frac{dQ}{dt}\right)_c\right], \tag{6.76}$$

where I_a is the insulation of the air (Table 6.42).

6.7 Summary

The energy used by the body and body temperature are determined by the body metabolism, which includes the use of food, the energy needed to operate the body, and the production of heat, and by the loss of heat by the body. The energy value of food can be analyzed and compared to the production of ATP by the body and the use of ATP in different activities. The body metabolic rate is composed of many components, each of which can be modeled. This includes analyzing the rate that energy is needed for the operation of body functions for a person who is inactive, which combine to form the BMR, and the rate that energy is needed for the body to engage in physical activity. The body metabolism and food intake can be linked to models of the steady-state body weight and of weight gain and loss. The body can lose heat by radiation, convection, conduction, and the evaporation of water, each of which can be modeled and linked to models that determine the body temperature.

Problems

Heat Capacity

6.1. A person weighing 60 kg drinks 600 mL (0.6 kg) of water. Say the water is at a temperature that is either 25°C above or below the body temperature. How does the body temperature change? (Use the fact that the specific heat of the water is 1 kcal/kg-°C, so that it is supplying the body with an energy of 1 kcal/kg-°C × 0.6 kg × 25°C, and that the specific heat of the body is 0.83 kcal/kg-°C.) Would such a warm drink (+25°C) help with hypothermia? Would such a cold drink (−25°C) help with a fever?

6.2. What is the heat capacity of a typical 20 μm diameter human cell? Assume it is a sphere and has a specific heat equal to the body average.

6.3. In a typical 20 μm diameter human cell there is approximately 5×10^{-6} J, 2.0×10^{-5} J, and 8×10^{-6} J of energy available respectively from carbohydrates, protein, and fat. If all of this energy were converted to heat, what would be the temperature rise in the cell? Use the results of Problem 6.2 and ignore heat losses.

6.4. A British thermal unit (BTU) is the amount of heat needed to raise the temperature of 1 lb of water by 1°F. (This unit is therefore defined as being equal to the specific heat of water, which is 1.0 cal/g-°C.) The amount of heat needed to freeze water or melt ice (the latent heat of fusion (freezing)) is 80 cal/g = 335 J/g and that needed to vaporize water (the latent heat of vaporization) is 543 cal/g = 2,272 J/g.
(a) Show that 1 BTU = 1,055 J.
(b) Find the number of BTU required to melt a pound of ice.
(c) Find the number of BTU required to vaporize a pound of water.

Caloric Value

6.5. Consider the oxidation of the fat tributyrin: $C_3H_5O_3(OC_4H_7)_3 + 18.5\ O_2 \rightarrow 15CO_2 + 13H_2O + 1,941\,\text{kcal}$.
(a) Find the energy release/g, calorific equivalent, and respiratory exchange ratio (RER) (or respiratory quotient (RQ)).
(b) Are these values representative of a fat? Why or why not?

6.6. Consider the oxidation of the alcohol (ethanol): $C_2H_5OH + 3O_2 \rightarrow 2CO_2 + 3H_2O + 327\,\text{kcal}$.
(a) Find the energy release/g, calorific equivalent, and respiratory exchange ratio (RER) (or respiratory quotient (RQ)).
(b) Compare these values to those for carbohydrates, proteins, and fats.

6.7. Determine the energy content in one fluid ounce of 86 proof liquor from the caloric content of alcohol. (200 proof means that the volume fraction of alcohol is 100%.)

6.8. Show that the volume of alcohol in a (1 oz) shot of liquor, a glass of wine, and a bottle of beer are all about the same, and therefore they have the same energy content (use Table 6.5).

6.9. Are the energy densities per unit mass for carbohydrates, proteins, and fats listed in Table 6.3 consistent with the values in Table 6.2?

6.10. (a) Derive general expressions for the percentage of energy coming from carbohydrates and fats and the heat energy per L oxygen consumed for an arbitrary RER between 0.71 and 1.00.
(b) Determine the percentage of kcal coming from carbohydrates and fats and the heat energy per L oxygen consumed for each of these RERs: 0.71, 0.75, 0.80, 085, 0.90, 0.95, and 1.00.

6.11. Show how rounding off the carbohydrate, protein, and fat masses in the standard donut to the nearest integral value of grams can easily lead to the difference in the calculated and listed caloric content. (Show that this is true even ignoring the likely rounding off of the total caloric content to the nearest tens of kcal – so 280 kcal could mean any value from 275.1 kcal to 284.9 kcal.) Assume that carbohydrates and proteins have caloric content 4.0 kcal/g and fats have caloric content 9.0 kcal/g.

6.12. Repeat Problem 6.11, assuming that the mass contents are exact but now considering uncertainties in the caloric contents of the donut carbohydrates, proteins, and fats.

6.13. The Entenmann's™ Rich Frosted Donut designated in the text as our standard donut was sold as a stand-alone item for many years and as of 2006 was still available in the donut variety pack with the same food content. In 2006, the Rich Frosted Donut in stand-alone packaging had somewhat different food content: 26 g fat, 29 g carbohydrate, 2 g protein and a total mass of 64 g. Compare its caloric content to that of our standard donut.

6.14. The Entenmann's™ Glazed Donut has 13 g fat, 34 g carbohydrate, 2 g protein, and a total mass of 60 g (as of 2006).
(a) Compare its caloric content as listed on the package (260 kcal) to that of our standard donut.
(b) Find percentage of calories that come from fat.
(c) Find the percentage of donut mass that does not come from fat, carbohydrate, or protein.

6.15. For which foods in Table 6.5 do the caloric contents of carbohydrates, proteins, and fats not explain the total caloric value? Why?

6.16. We know that fats have about 9 kcal/g energy content and proteins and carbohydrates have about 4 kcal/g. Explain why the metabolizable energy per unit mass of the foods in Tables 6.5 and 6.6 are so different than these numbers? In particular why are the values for milk and carrots so low? What can we learn about oatmeal and flour? What do you learn about butter and nuts?

6.17. A certain fruit has a metabolizable energy of 0.4 kcal/g normally and 2.4 kcal/g after it has been dried. Assume the fruit consists of only water, sugar, and nonmetabolizable matter and find the fraction of each in the fruit and in the fruit after it is dried.

6.18. Determine the energy content per unit mass (kcal/g) for each nonalcoholic food in Table 6.5. Explain the variation in terms of relative amounts of carbohydrates/protein/fat, water, and "ash."

6.19. Determine the percentage of calories from fat for each nonalcoholic food in Table 6.5.

6.20. If nonfat milk is simply whole milk minus the fat, determine the caloric content, mass of carbohydrate, protein, and fat, and % water of 1 cup (245 g) of nonfat milk.

6.21. How much oil is there in French fries? More precisely, what fractions of the mass and caloric value of French fries are due to the oil and what fractions are due to the potato? Use the data given in Table 6.5. Ignore the skin of the potato and consider the possibility that water is lost in the frying process. Oil is all fat. How much water is lost in the frying process?

6.22. The yolk from a large raw egg has a mass of 17 g, of which 3 g is protein and 5 g is fat (and a trace amount is carbohydrate). It has a total of 59 kcal energy content and is 49% water. Determine the corresponding values for the white of the egg, using Table 6.5.

6.23. For the wine and beer listed in Table 6.5, calculate the number of kcal and mass (g) due to alcohol (ethanol) in each, and the fractional volume of alcohol in each, assuming ethanol has a mass density of $0.79\,\mathrm{g/cm^3}$.

6.24. You eat a pound of food. You know that 10% of its weight is water, 5% of its weight is from nondigestible solids, and 30% of its calories come from fat. How many calories have you ingested?

6.25. In a hypothetical proposed diet, a person is supposed to derive $40 \pm 5\%$, $30 \pm 5\%$, and $30 \pm 5\%$, of her calories from carbohydrates, proteins, and fat, respectively.
(a) Use Table 6.5 to design two different diet plans for a 2,000 kcal/day diet, using very different foods in each plan.
(b) Which diet has a daily lighter mass, and therefore is perhaps better for backpacking or space trips? Why is this so?

6.26. (a) You are at a salad bar and want to intake 600 kcal. If you assume the food is 20% water (by mass) and 5% ash, how much food should you take (in oz and g) if you assume that of the remaining amount 50% is carbohydrate, 20% is protein and 30% is fat?
(b) For this amount of food, what would the caloric value be if nonwater, nonash content were really all carbohydrate, all protein, or all fat?

6.27. The energy released per ton of detonated TNT (trinitrotoluene) is 4.18×10^9 J. The same food energy is found in how many tons of our standard donuts?

6.28. A body consumes 0.3 L-atm. of oxygen every minute.
(a) Determine the rate of heat production if only carbohydrates are being consumed.
(b) Determine the rate of heat production if only fats are being consumed.

6.29. From the discussion after (6.18), what is $p_{\mathrm{a}} - p_{\mathrm{v}}$ in the blood in the system (in mL of O_2/L of blood)?

Metabolic Rates

6.30. The Gibbs free energy ΔG of a chemical reaction is the energy available when it occurs at a constant temperature T and constant pressure (1 atm.). The free energy of ATP hydrolysis (6.16) is $\Delta G^0 = -7.3$ kcal/mol under the standard conditions of $T = 298$ K, 1 atm., pH $= 7.0$, and equal 1 M (molar, mol/L) concentrations of all reactants and products. For other conditions, it is $\Delta G = \Delta G^0 + RT \ln(K/K^0)$, where R is the gas constant (1.987 cal/mol-K), T is the temperature for the condition, K is the equilibrium for the condition, and K^0 is the equilibrium constant for standard conditions [329]. The equilibrium constant is the product of the concentrations of the products (ADP and P_i here) divided by that of the reactants (ATP here). (The concentrations of other products and reactants in (6.16), such as H_2O, are essentially the same for the different conditions and do not need to be included. Why?)
(a) Calculate the free energy at $T = 310$ K in the cell of a representative human erythrocyte (red blood cell) with concentrations of 2.25 mM (millimolar, millimol/L) for ATP, 0.25 mM for ADP, and 1.65 mM for P_i.
(b) In resting muscle, the concentration of free ADP is thought to be between 1 and 37 μM (micromolar). Calculate the free energy at $T = 310$ K if the ADP concentration is 25 μM (and other concentrations are the same as in (a)). How do these values in (a) and (b) compare to those given in the text? (Other than the difference in cell types, the major difference with part (a) is that the relevant concentration is that of free ADP and not total ADP, which includes free ADP and that bound to cellular proteins.)
(c) How does the use of such free energies under cellular conditions rather than those for standard conditions affect the calculation of the efficiency of converting glucose into ATP by the body relative to that of oxidizing glucose?

6.31. Find what the efficiency of body usage of glucose would be if in aerobic metabolism 1 mol of glucose produced 36–38 mol of ATP (the conversion factor often cited several years ago), instead of the currently accepted range 30–32 mol, as given in the chapter. Assume the energy released from ATP is 14.0 kcal/mol.

6.32. Assume that the oxidation of glucose extracts all the free energy possible. Find the maximum number of ATP molecules that could be formed if ATP hydrolysis were 100% efficient. Assume the energy released from ATP is 14.0 kcal/mol.

6.33. Say that in aerobic metabolism 1 mol of glucose can produce 32 mol of ATP, while under similar conditions 1 mol of carbohydrates from glycogen can produce 33 mol of ATP. Because glycogen is made by the body from glucose, is the body getting 1 additional mole of ATP for free by first converting glucose to glycogen and then using it or is something else really happening? Explain.

6.34. The standard free energy for the hydrolysis of ADP to AMP (adenosine monophosphate) is -7.8 kcal/mol. What is it for cellular conditions? Assume the conditions for an erythrocyte in Problem 6.30a and an AMP concentration of 0.2 mM [329].

6.35. Redo Problem 6.30 for palmitic acid.

6.36. (a) The molar mass of ATP is 507 g. If a 60 kg person has a metabolic rate of 2,000 kcal/day and ATP releases 14.0 kcal/mol during hydrolysis, what mass of ATP would be needed by that person per day if the ADP formed after ATP hydrolysis were not recycled back to ATP?
(b) How does this compare to the person's mass?
(c) How many cycles of ATP hydrolysis and re-creation does an average ATP molecule make per day and per minute if there are a total of 0.2 mol of ATP in the body?
(d) Repeat each part for ATP releasing 7.3 kcal/mol during hydrolysis.

6.37. (a) The molar mass of ATP is 507 g. How many kcal would be available to the body if all of the rest mass of one ATP molecule were converted to (useful) energy? Use $E = mc^2$.
(b) What mass of ATP (in g), would a person with a metabolic rate of 2,000 kcal/day use if all of the ATP were converted to energy?
(c) If a 1 GW nuclear power plant were to convert this ATP energy to electrical energy with 20% efficiency, how much ATP would it use (in g) per second and per year?
(d) In nuclear reactions, only of fraction of the rest mass is converted to energy. Repeat (a)–(c) if that fraction were 2.5×10^{-5}, which is the fraction of rest mass energy converted to kinetic energy in the alpha particle decay $^{241}\text{Am} \rightarrow \alpha + {}^{237}\text{Np}$.

6.38. Sketch on the same set of labeled axes the available rate of ATP usage (normalized to its maximum possible value) vs. time for a person beginning moderate physical activities at $t = 0$, for each of the following cases:
(a) Only the phosphagen system is available.
(b) There is not much oxygen, so only the phosphagen and anaerobic glycolysis systems are available.
(c) There is much oxygen available, so the phosphagen and aerobic metabolism systems are used.

6.39. The maximum anaerobic peak power for men is 2.1 hp and the maximum metabolic rate during anaerobic glycolysis is 1.6 mol of ATP/min. Are these rates consistent? Why? (Assume 10 kcal per mole ATP here).

6.40. Table 6.3 shows that ATP stores much less energy per unit volume and mass than other fuels. Why then is it so important?

6.41. Stearic acid is an 18-carbon chain fatty acid that undergoes 8 complete passes in the aerobic metabolism. How many moles of ATP are formed per mole of stearic acid?

6.42. Use the data in Table 6.22 to find what the maximum power output of a human (in W) would be if the entire human volume were composed of typical tissue cells operating at maximum power output. Does this exceed the listed maximum human body power output? Is this reasonable? Why?

6.43. (a) If the usual volume of a person were instead occupied only by tightly packed myosin muscle motors, what would the power consumed by the body be (in W) and how does this compare to the basal and maximum metabolic rates?
(b) What would be the rate of temperature rise of the body for such a volume, assuming the heat capacity is that of a normal human and there are no heat losses?

6.44. How many myosin cross bridges would be needed to power a 200 hp car? (Assume the data for the myosin molecules in Table 6.22.) How does this compare to the number of crossbridges in a typical human?

Basal Metabolic Rates

6.45. Use Kleiber's Law to compare the BMR per unit mass (with overall units kcal/kg-day) for a mouse (30 g), human (70 kg), and a blue whale (10^5 kg).

6.46. Compare the metabolic rate expected for the standard man in Chap. 1 using Kleiber's Law and the Harris–Benedict equations? Why are they different? Is this reasonable or not?

6.47. How many standard donuts must a 60 kg person eat daily to maintain that person's BMR, assuming Kleiber's Law.

6.48. How much food would you have to eat each day (in g) to balance a basal metabolic rate (BMR) of 1,700 kcal/day if you only ate (a) glucose or (b) (typical) fat?

6.49. Another formulation of the Harris–Benedict equations (6.28) and (6.29) utilizes rounded-off numbers:

$$\text{For men:} \quad \text{BMR} = 66 + 13.7 m_\text{b} + 5H - 6.9Y \tag{6.77}$$

$$\text{For women:} \quad \text{BMR} = 665 + 9.6 m_\text{b} + 1.7H - 4.7Y\,, \tag{6.78}$$

where the BMR is in kcal/day, m_b is the body mass in kg, H is the height in cm, and Y is the age in yr. How different are the two formulations for a 6-ft-tall, 25-yr-old male weighing 200 lb and a 5 ft 2 in., 50-yr-old female weighing 125 lb? Are these differences significant?

6.50. How much higher is the BMR for a pregnant women? Estimate this by considering the increase in her BMR from two factors. She must supply food to build the fetus along with the supporting structure in the womb and must supply the energy for the fetus' BMR. (She is literally eating for two.) Assume she is late in pregnancy and is gaining 4 lb a month of typical human tissue, assuming this does not include extra fluid retention outside of the womb. Assume the fetus has a BMR for a baby at birth that weighs 7 lb (and use the BMR charts for children under 3-yr-old).

6.51. You tell an 80-kg-man who has been sitting and inactive for some time that he is no different from an 80 W light bulb. How accurate is your statement in terms of:
(a) total power emitted
(b) power emitted per unit volume
(c) approximate cost of energy output (in terms of US cents/kW-h)
(d) types of energy that are output
(e) types of energy that are input
(f) spectrum of any light (electromagnetic radiation) that is emitted
(g) flow of gases to and from each object?

6.52. Use Fig. 6.12 to find the metabolic power (in kcal/day) a starving person has by metabolizing stored body energy reserves during the fourth week of starvation. How does this compare to a typical basal metabolic rate?

6.53. Does the basal metabolic rate per unit volume for a human make quantitative sense, given the basal metabolic rates for different cells and the fraction of the body that contains cells (Table 6.22)?

Metabolic Rates During Activities

6.54. In a high jump a 70-kg person elevates his center of mass by 51 cm during an extension phase (with constant acceleration) that takes 0.25 s.
(a) What is the person's kinetic energy upon takeoff?
(b) If the muscle efficiency is 20%, how much chemical energy is used to make one jump? (Express your answer in joules and in kcal (the usual food calories).)
(c) What is the average power generated by the person during the jump in watts and horsepower (1 horsepower (hp) = 746 W)?
(d) How frequently would someone have to jump (during a 12 h awake cycle) to increase his/her (daily averaged) metabolism rate to twice the basal value (which is 1,500 kcal/day)?

6.55. Use Fig. 3.25 to find the metabolic rates (in kcal/h) for running at speeds of 2 and 4 m/s, each for 50 and 90 kg people? (Have you included the BMR?)

6.56. Convert the data on aerobic capacities in the paragraph describing the capacities of people from mL/kg-min to kcal/h for 70 kg men and 50 kg women.

6.57. The metabolic rate during walking, MR in cal/min, has been determined by measuring the rate of oxygen consumption, as a function of walking speed, v in m/min :

$$\mathrm{MR} = 0.267v^2 + 2,160. \tag{6.79}$$

(a) Determine the walking speed at which the energy expended per unit distance is a minimum, in m/min, m/s, and mph (miles per hour) [316]. How does this compare to the minimum shown in Fig. 3.25?
(b) Determine the power consumption rate at this optimal speed in kcal/hr.

6.58. (a) Show that the rate of change of potential energy when walking or running at a speed v on a gradient at an angle θ (positive for uphill, negative for downhill) is $m_{\mathrm{b}}gv\sin\theta$.
(b) Consider the metabolic needs for a slow walk by a 50 kg female, assuming the energy cost of walking above that of standing is 2.13 J/kg-m at 1.2 m/s (2.6 mph). Assume the work and power required going uphill is about 2.7× the increase in potential energy and when moving downhill less power is needed than that needed on level terrain by about 50% of the rate of decrease of potential energy. Find the average (total) metabolic need in kcal/min for this person walking at this constant speed: (i) on a level surface, (ii) on a 10% grade uphill, (iii) on a 10% grade downhill, and (iv) when half is on the 10% uphill grade and half on the 10% downhill grade (so the final elevation is the same as the initial). (v) How does the answer in (iv) compare to that in (i)?
(c) If you wanted to maintain your metabolic rate whether you were walking on the level surface, uphill, or downhill in part (b), how fast would you walk up and downhill, assuming that on the level terrain you were moving with the speed v in part (b).

6.59. Consider the accelerating sprinter in Chap. 3. What is the metabolic rate of the sprinter (in kcal/min) if the efficiency of converting metabolic energy to mechanical energy is 20%? How much metabolic energy does the sprinter consume in a 100 m dash?

6.60. (a) Determine the metabolic rate and oxygen consumption rates for each stage of the Bruce Protocol for a 75 kg male.
(b) How many kcal are burned in each stage and what is the total energy burned if the patient finishes all stages?

6.61. (a) Determine the metabolic rate and oxygen consumption rates for each stage of the Naughton Protocol for a 75 kg male.
(b) How many kcal are burned in each stage, and what is the total energy burned if the patient finishes all stages?

6.62. How do the metabolic rates at the highest stages of the Bruce and Naughton Protocol compare?

6.63. (a) If the energy needed to supply the basal metabolism needs of a 70 kg male for one day were used to increase that person's gravitational potential energy with 100% efficiency, how high would that person be lifted?
(b) If a person can climb stairs with 20% mechanical efficiency, how high can the person climb using the daily metabolic energy described in part (a)? This corresponds to about how many floors?

6.64. In football, an offensive and defensive lineman, each 300 lb, face each other, crouched in a "stance," separated by 2 yards (which is a bit of an overestimate). When a "play" begins they rush at each other, each accelerating at $g/2$, and then they collide and spend the next 5–10 s pushing each other with, in this model, no motion. Consider one of the linemen, say the offensive lineman.
(a) How much kinetic energy does he develop before the collision (in J)?
(b) If he converts metabolic energy into this kinetic energy with 10% efficiency and plays 50 plays per game, how much metabolic energy does he use in a game (in kcal)?
(c) It is said that linemen use up to 10,000 kcal in a game. Does your answer in (b) support this statement? If not, why not?

6.65. As described in Chap. 7, a person tries to fly by flapping his arms. Say that at the beginning of each flap each arm starts with no kinetic energy and at the end of the flap each arm is moving at an average speed of 20 m/s.
(a) If this kinetic energy of the arms needs to be generated 3 times a second, and the body converts metabolic energy to this mechanical energy with 10% efficiency, how many kcal/min would a 70 kg body consume in this attempt to fly? (Use the anthropometric data in Chap. 1 to determine the mass of the arms.)
(b) Would it be difficult for a person to continue this fruitless and quite silly attempt to fly for a long time? Why?

6.66. A person lifts a 15-kg mass from the floor to over his head (a distance of 2 m). How many times does he have to do this to lose a pound of fat, assuming a muscle efficiency of 25%?

6.67. A 50 kg woman does 10 chin-ups in a minute (each raising her center of mass by 0.5 m). After 5 min how much mechanical work has she done (in J) and how much metabolic energy has she used (in kcal), assuming 25% muscle efficiency?

6.68. How much energy (gravitational potential energy) is required to lift a 70 kg person by 1 m (in kcal)? If your muscles can do this with 25% efficiency, how many times would you have to lift such a person to burn off the caloric content in one standard donut (280 kcal/donut)?

6.69. After a long run, you walk for 10 min at a leisurely pace. Is your metabolic rate the same as for walking in Table 6.21? Why?

6.70. You run for a half hour and expend 300 kcal/hr above the BMR during that time. After running, you huff and puff for another 5 min. Estimate the number of kcal you expend in excess of the BMR amount during this time. Is this significant?

6.71. Short-term peak metabolic rates are much greater than the average rates that we can maintain for long times. Let us estimate them by assuming that we operate at this peak rate when we throw a $m = 1$ kg ball a distance of $d = 60$ m. This is done by throwing it at a speed of v at takeoff angle of 45° and ignoring air resistance. Also ignore the difference between the takeoff and landing in the trajectory of the ball.
(a) Show that the peak mechanical power is $P_{\mathrm{mech}} = mv^2/t + mgv/\sqrt{2}$, where t is the time needed to launch the object by accelerating one's arms (at a constant rate) over a distance s. We will take $s = 1$ m.
(b) If the efficiency to do mechanical work is ϵ, show that the needed peak metabolic power is $P_{\mathrm{met}} = m(gd)^{3/2}/2\epsilon s + mg^{3/2}d^{1/2}/\sqrt{2}\,\epsilon$.
(c) Using the given values, show that the peak metabolic rate is about 14,000 W, and therefore about 200× the BMR. (The BMR is roughly the same – and is usually higher – than the maximum long-term rate of doing mechanical work. They would be equal if the daily-average activity factor were 2.)

Activity Factors

6.72. Estimate the average daily metabolic rate and activity factor of a 50 kg "homemaker." (Assume that f = MET.)

6.73. Calculate your activity factor f for a typical day. (Assume that f = MET.)

6.74. A student spends 6 h a day sleeping, 11 h sitting (attending classes (and awake), eating, playing video games, etc.), 0.5 h washing and dressing, 4.75 h walking, 1 h standing, 0.25 h playing basketball, and 0.5 h cycling. What are the student's MR and activity level? (Assume the BMR of a 55 kg, 20-yr-old female student and that f = MET.)

Weight Gain and Loss

6.75. Determine how many standard donuts a (70 kg) person should eat after playing basketball for 2 h to maintain the same energy reserve as the person
(a) had before the basketball playing or
(b) would have had after 2 h of sitting instead.

6.76. A 150 lb (68 kg) person consumes 2,500 kcal every day and maintains a steady weight. Now this person decides to eat an extra slice of bread every day (100 kcal) and eventually achieves a new steady-state weight.

(a) How much weight does that person gain (in steady state) if the person's entire metabolic rate scales, respectively, as (i) $m_b^{3/4}$ or (ii) m_b (m_b = body mass)?
(b) For both scaling relationships, how much weight will that person lose if he/she ate one less slice of bread each day?
(c) If the person wants to eat that extra slice of bread everyday, but does not want to gain the extra weight, how long would that person have to run each day to keep that extra weight off?

6.77. How much weight do people gain with age? Consider a 30-yr-old, 6 ft (1.83 m), 150 lb (68 kg) male with an activity factor $f = 1.5$ and assume that the person's BMR scales as in (6.30) – which has a strong $m_b^{3/4}$ dependence – and that the total metabolic rate MR equals this BMR times the activity factor. Find how much weight the person will gain (to his new steady-state weight):
(a) when he turns 50 (with the same food intake, activity level, etc.).
(b) if the activity level of this 30-year decreases from 1.5 to 1.4 (with the same food intake, etc.).
(c) if the 30-year old ($f = 1.5$) increases his food intake by 100 kcal/day (one big slice of bread a day).
(d) if each of (a)–(c) occurs (i.e., the 30-year-old becomes older (now 50), is less active (f now 1.4), and eats more (100 kcal/day more)).
(e) Do the separate changes in (a)–(c) sum to give that in (d)?
(f) If the person in (c) wants to eat that extra slice of bread everyday, but does not want to gain the extra weight, how long would that person have to run each day to keep that extra weight off?
You may assume that the specific stature remains constant, at its initial value, in parts (a)–(d).

6.78. If your daily metabolic rate exceeds your caloric intake by 1,000 kcal (which is a very large difference) for a week, how much weight will you lose? Assume that the change is due only to the use of body fat to make up this difference, with 9 kcal per gram of body fat, so there is no other concomitant loss of weight. (Are there other concomitant weight losses? Some weight loss supplements may also lead to more trivial loss of body water.)

6.79. If 5 lb (2.3 kg) of body fat were changed to 5 lb of body muscle, how would that decrease the level of other body fat because of the increased MR?

6.80. What is the mass in kg and weight in pounds of a hemisphere of human tissue with a diameter of 0.4 m (with the average density of such tissue being 0.9 g/cm^3)? (This is a model of the weight gain of a pregnant woman – aside from that due to water retention – or that due to a relatively small "beer belly.")

6.81. A change in body mass of 1 kg is caused by a change in metabolizable energy of about 3,500 kcal. This means that about 3,500 kcal of chemical energy is obtainable by oxidizing 1 kg of average body storage materials. Special relativity teaches that the total energy content of a mass m is $E = mc^2$. How much is this for 1 kg (in units of kcal)? How many times larger is this than the chemical energy?

6.82. The level of sodium ions in the body depends on salt intake and how the body controls the level. The body maintains a Na^+ concentration of about 10 mM (millimolar) for intracellular fluid (normally ~5 L) and about 145 mM for extracellular fluid (normally ~2.5 L, which is the blood plasma half of the 5 L of whole blood). Let us say that you eat a bag of potato chips that contains 2 g of sodium (as sodium ions) and let us assume that – for the moment – it all stays in the body, and, in particular, in these fluids. Now let us say that the body responds by trying to maintain the previous Na^+ concentrations, so you then drink (and retain) enough water to maintain these concentrations. How much weight will you gain, assuming all the new Na^+ (and therefore the added water) goes to either (a) the intracellular water, (b) the extracellular water, or (c) both, so as to keep the same ratio of intracellular and extracellular fluid volumes? (The recommended daily intake of Na is about 2.5 g.)

6.83. Repeat Problem 6.82 for eating a big bag of pretzels that contains 8 g of sodium.

Heat Loss

6.84. In (6.44) there was a net radiation deficit of 116 W for a wall (and air) temperature of 24°C. Find the deficit for a wall temperature of 0°C.

6.85. A standard person is in interstellar space. He emits thermal radiation based on his body surface temperature and receives thermal radiation from the 3 K black body background. How long would it take the person's temperature to decrease by 5 K? Assume no other sources of heat generation and loss, and ignore the fact that the person would not live very long for many other reasons. Also assume thermal transport in the body occurs on a much shorter time scale than the one you calculate here.

6.86. Consider a person sitting nude on a beach in Florida [300]. On a sunny day, visible radiation energy from the sun is absorbed by the person at a rate of 30 kcal/h or 34.9 W. The air temperature is a warm 30°C and the individual's skin temperature is 32°C. The effective body surface exposed to the sun is 0.9 m^2. (Assume this same area for sun absorption, radiative transfer, and convective loss. Is this a good assumption?)
(a) Find the net energy gain or loss from thermal radiation each hour. (Assume thermal radiative gain and loss according to (6.51) and an emissivity of 1.)

(b) If there is a 4 m/s breeze, find the energy lost by convection each h (use (6.61), with (6.63))
(c) If the individual's metabolic rate is 80 kcal/h (93.0 W) and breathing accounts for a loss of 10 kcal/h (11.6 W), how much additional heat must be lost by evaporation to keep the body core temperature constant?

6.87. How different are (6.62) and (6.63) over the range where the latter is valid?

6.88. A racing cyclist produces heat at a rate 1,300 kcal/h in excess of the normal rate. If all of this excess heat is lost by perspiration and evaporative cooling, how many liters of water must the cyclist drink every hour to maintain body fluids (at the level she would have had without cycling)?

6.89. What happens when room temperature exceeds the body temperature – for both the black body and breathing "cooling" routes?

6.90. In indirect calorimetry the rates of O_2 consumption, dV_{O_2}/dt, and CO_2 production, dV_{CO_2}/dt, are determined while the subject is engaged in a physical activity.
(a) Describe how they can be determined by measuring the rates of inspiring and expiring volumes of air, dV_{air}/dt, and the fractions of O_2 and CO_2 in both the inspired and expired air.
(b) A subject is inspiring and expiring air at a rate of 100 L/min, the inspired air is 21% O_2, and the expired air is 16% O_2. Show that dV_{O_2}/dt is 5 L/min.
(c) What is the metabolic rate of the subject in kcal/min?
(d) In what type of activity could the subject be engaging?

6.91. Compare the thermal conductivities of the materials in Table 6.37 to those of the parts of the body.

6.92. Use Table 6.37 to explain why is it bad to make clothing for cold weather out of aluminum foil, very good to make it out of cotton fabric, and great to make it out of layers of cloth separated by air?

6.93. (a) How thick does your clothing have to be on a cold day? Model a person by a cylinder 1.65 m high and 0.234 m in diameter. Assume that heat loss is 1,500 kcal/day and occurs at the circumference of the cylinder (but not top or bottom). Your skin temperature is 34°C and the outside temperature is 0°C. Assume the insulation values in Table 6.44 and that this is the only significant form of insulation. Use the ordinary heat flow equation for flat surfaces and planar regions (6.72).
(b) The radial heat flow in a cylinder of radius r can be modeled assuming flat surfaces if the thickness Δx is replaced by $r \ln(1 + \Delta r/r)$. How does this change the result in (a)?

6.94. The R-value is often used to characterize insulation (or thermal resistance) needed and used for insulating houses. It is formally the same as the

term insulation defined in this chapter. When expressed in SI units, the insulation value is called the RSI-value, so if the insulation is $3\,\mathrm{m}^2$-°C/W, the insulation value is RSI-3.
(a) In the US, the value of insulation in units of ft^2-°F-hour/BTU is used and called the R-value. Show that an R-value of 10 corresponds to an RSI-value of 1.761 and that an RSI-value of 1 corresponds to an R-value of 5.6745.
(b) Find the R-value and RSI-value of 1 Clo.
(c) R-19 insulation is typical of modest insulation needed for a house. Approximately how many polar weather suits made of heavy wool pile does this correspond to? (These suits would be in series of course. Assume they are fully open, so you have a single, and not double, layer of insulation from each.)

6.95. (a) How much energy (kcal) do you lose every hour by thermal conduction when you stand barefoot on ice? Assume the flow is between the plane of the ice, which is at a temperature 0°C, and a plane 5 mm deep into your body (across the skin), which is at 34°C [293, 325]. (Why is this temperature reasonable?) Assume the coefficient of thermal conductivity is 0.3 W/m-°C. (Why is this value reasonable?) Also assume that each foot contacts the ice over an area of $0.02\,\mathrm{m}^2$.
(b) You need to heat your body by this amount every hour to maintain your average body temperature. How much body fat (expressed in g and lb) would your body have to burn with 100% efficiency each hour to make up for this heat loss?
(c) How many donuts would you have to eat every hour to account for this heat loss (again assuming 100% efficiency in conversion to heat)?
(d) Assume that a burning a match provides 1 kJ of energy and determine how many matches would have to be burnt every hour to supply the body with heat to make up for this loss (assuming perfect transfer of energy from the match)?

6.96. Say you stand barefoot on ice [325]. Can blood flowing into your foot supply enough heat (by the convection of the flowing blood) to counter the loss of heat by thermal conduction from your foot to the ice, which we will assume here to be 20 W? Use (6.67). Assume that blood flows into the foot at a temperature of 36°C and leaves it at 32°C, so the foot is at 34°C, and the specific heat of blood is 1 cal/g°C. The mass flow rate of blood $F_m = \rho A u$, where $\rho = 1.06\,\mathrm{g/cm}^3$ is the blood mass density, A is the cross-sectional area of the artery, and u is the linear speed of blood flow. Assume a major artery modeled as cylindrical tube of radius 2 mm with a blood speed of 35 cm/s.

6.97. Assume your core body temperature varies linearly with distance from 38°C in your head to 31°C in your feet (so $\mathrm{d}T/\mathrm{d}x$ is constant). (You might say that you are "hot-headed" and have "cold feet," which is not a very good combination.) Assume you are 1.8 m tall, have a mass of 80 kg and a constant transverse body cross-sectional area estimated from the data in Chap. 1, and the average thermal conductivity coefficient in your body is

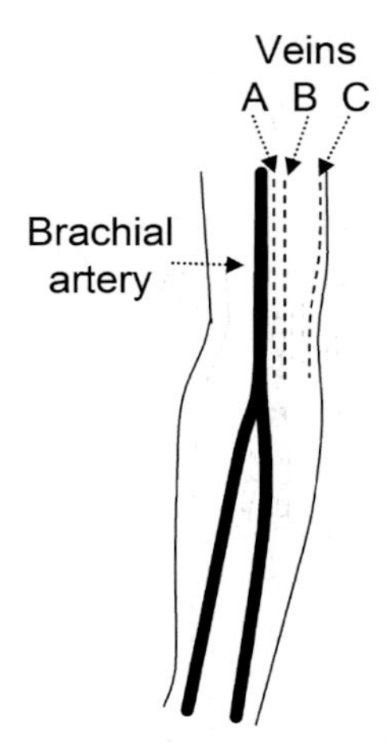

Fig. 6.25. Return blood flow in three hypothetical veins in the right arm, labeled A, B, and C)

0.3 W/m-°C. What is the heat flow in your body by thermal conduction and which way is the heat flowing [325]?

6.98. Model a person as a right circular cylinder [325]. In which case does the person lose heat faster: (i) the surface is 34°C everywhere or (ii) half of the surface is at 36°C and the other half is at 32°C, assuming loss either by
(a) radiation or
(b) conduction.

6.99. Use (6.64) to find a relation for the wind chill temperature in metric units, with temperature still in °C, but with wind speed in m/s.

6.100. Use (6.64) to find a relation for the wind chill temperature in English units, with temperature in °F and wind speed in mph.

6.101. Blood flow in which veins in Fig. 6.25 promotes
(a) relatively low heat loss to the outside world by direct thermal conduction
(b) high heat loss to the outside world by direct thermal conduction
(c) low heat loss by countercurrent flow?

6.102. (advanced problem) Determine the temperature profile in an artery ($T_{\mathrm{a}}(x)$) and a vein ($T_{\mathrm{v}}(x)$) that are close enough that they form a countercurrent heat exchange system [325], as in Figs. 6.16 and 6.17. Use the model in Fig. 6.26. The artery and vein are in such close proximity that there is rapid heat transfer between them. For simplicity, assume both vessels are identical and have cross-sectional area A. Blood flows in them with a speed v in the x

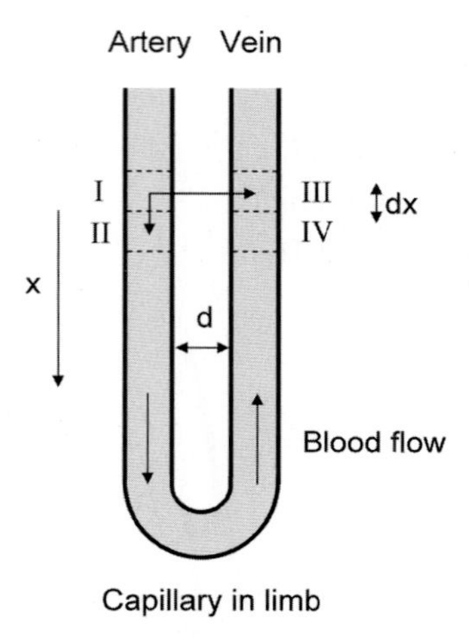

Fig. 6.26. Model of countercurrent heat exchange between an artery and vein. Heat from the artery flows either from region I to II by convection or from I to III by heat conduction. (Based on [325])

and $-x$ directions, respectively, and has a specific heat c and mass density ρ.
(a) Explain why conservation of energy implies that the rate of heat conduction from the warmer artery to the colder vein (from region I to III in Fig. 6.26) equals the loss of heat energy in the convection of blood flowing in the artery (from region I to II, because it gets colder along the x direction), and that this also equals the gain of heat energy in the convection of blood flowing in the vein (from region IV to III, because it gets warmer as it travels in the $-x$ direction).
(b) If the thermal conductivity is K, the distance between the vessels is d (approximately the combined thicknesses of the vessels), and the contact area is $w\,\mathrm{d}x$, where w is the width of the contact area along length $\mathrm{d}x$, show that the rate of heat conduction from the artery to the vein is

$$R_{\text{conduction}} = \frac{Ka}{d}(T_{\text{a}}(x) - T_{\text{v}}(x)) \tag{6.80}$$

(c) Show that the loss of heat flow in convection along the artery is

$$R_{\text{convection}} = -c\rho A v \frac{\mathrm{d}T_{\text{a}}(x)}{\mathrm{d}x} \tag{6.81}$$

(d) Combine (6.80) and (6.81) to obtain

$$\frac{\mathrm{d}T_{\text{a}}(x)}{\mathrm{d}x} = \frac{Kw}{c\rho A v d}(T_{\text{a}}(x) - T_{\text{v}}(x)) \tag{6.82}$$

and similarly for veins show that

$$\frac{\mathrm{d}T_{\mathrm{v}}(x)}{\mathrm{d}x} = \frac{Kw}{c\rho Avd}(T_{\mathrm{a}}(x) - T_{\mathrm{v}}(x)) \tag{6.83}$$

(e) Solve these two equations to show that the temperature difference between the artery and vein is constant along x and that the temperature in each decreases linearly with x, by

$$T_{\mathrm{a}}(x) = T_{\mathrm{a}}(0) - \frac{Kw}{c\rho Avd} \Delta T \, x \tag{6.84}$$

and

$$T_{\mathrm{v}}(x) = T_{\mathrm{v}}(0) - \frac{Kw}{c\rho Avd} \Delta T \, x, \tag{6.85}$$

where $\Delta T = T_{\mathrm{a}}(0) - T_{\mathrm{v}}(0)$.
(f) Sketch $T_{\mathrm{a}}(x)$ and $T_{\mathrm{v}}(x)$ vs. x along a limb of length L, alternately if $(Kw/c\rho Avd)L$ is $\gg$ or $\ll 1$. What do these two very different conditions indicate?

6.103. Someone has a daily metabolic rate of 2,200 kcal, which is all converted to heat. If 21% of heat loss is through evaporation of water (on the skin and in the lungs), how much water is evaporated each day (in L)?

6.104. Derive (6.70).

6.105. Equation (6.70) explicitly neglects any heating or cooling of the water vapor. Is this a reasonable approximation? Why?

6.106. Calculate the values of the two terms in (6.70) in kcal/day units assuming the person inspires air at 20°C (68°F) that is very dry (0% relative humidity) at a rate of 6 L/min and exhales air (at the same rate of course) at 37°C (98.6°F) that is totally saturated with water vapor (100% relative humidity, 47.1 mmHg partial pressure, 44.0 g/m^3 water vapor). How significant is each loss to the overall heat loss by the body?

6.107. Repeat Problem 6.106 for the following three cases.
(a) For a person inhaling air at 20°C that is totally saturated with water vapor (100% relative humidity, 17.5 mmHg partial pressure, 17.3 g/m^3 water vapor).
(b) For a person inhaling air at 40°C (104°F) that is very dry (0% relative humidity).
(c) For a person inhaling air at 40°C that is totally saturated with water vapor (100% relative humidity, 55.3 mmHg partial pressure, 51.1 g/m^3 water vapor).

(d) Compare the overall cooling rate for these three cases with each other and that in Problem 6.106.

6.108. Explain each of the factors leading to the heat index.

6.109. The heat index, HI, has been fit by

$$\begin{aligned}\mathrm{HI} = &-42.379 + 2.04901523T + 10.14333127R - 0.22475541TR \\ &- 6.83783 \times 10^{-3}T^2 - 5.481717 \times 10^{-2}R^2 + 1.22874 \times 10^{-3}T^2R \\ &+ 8.5282 \times 10^{-4}TR^2 - 1.99 \times 10^{-6}T^2R^2,\end{aligned} \tag{6.86}$$

with the HI and the ambient dry bulb temperature T in °F and the relative humidity R in % [337, 338]. (This assumes a wind speed of 5.65 mph (5 knots) and has an error of ±1.3°F. Exposure to full sunshine can increase the HI by up to 15°F.)
(a) Construct a table of the heat index in °F, with columns describing dry (bulk) temperatures ranging from 80–110°F (in steps of 2°F) and rows describing the relative humidity ranging from 40–100% (in steps of 5%).
(b) Mark the combinations of temperature and humidity in the table for each of the four health warning regions described in the text.

6.110. Repeat Problem 6.109 for 27–47°C (in steps of 2°C) and relative humidity ranging from 40–100% (in steps of 5%). (Convert the temperatures from °C to °F and then use (6.86).)

6.111. (a) Convert (6.86) from HI and T both being in °F to both being in °C.
(b) Repeat Problem 6.110, working directly in °C.

6.112. Consider two animals that we will model as cylinders [325]. Both have body temperatures of 35°C and live in an environment of 0°C. Animal A has a length of 1 m, diameter of 1 m, and a 1-cm thick layer of fat (outside the given diameter) for insulation. Animal B has a length of 6 m, diameter of 0.5 m, and a 4-cm thick layer of fat for insulation.
(a) Which animal has more thermal energy (heat)?
(b) Which animal loses heat at a faster rate? (Include losses around the diameter and at the ends, and use simple theory – i.e., assume planar flow and ignore the corrections needed for cylindrical flow.)
(c) Which animal loses a larger fraction of heat per unit time?
(d) Which animal do you expect to have the larger BMR per unit mass? Why?
(e) Are heat losses less for the more spherically shaped animals, such as animal A, or the more pencil-shaped animals, such as animal B? (Ignore the different thicknesses of the fat insulating layers.)

6.113. (a) Find the skin temperature assuming that 1 cm of fat, skin (with no thermal insulating properties assumed here), and 1 cm of clothing separate the body core at 37°C and the outside at 20°C. Take the thermal conductivity

K of fat to be 5×10^{-4} cal/cm-s-°C and that of the clothing to be equal to that of air 6×10^{-5} cal/cm-s-°C [325].
(b) Find the heat loss per cm^2 per day.
(c) Find the heat loss per day over the entire body using a surface area of $1.5\,m^2$. How does this compare to the BMR? Does this make sense?

6.114. Repeat Problem 6.113 assuming the person is wearing no clothing.

7

Fluid Pressure, Fluid Flow in the Body, and Motion in Fluids

In the following two chapters we will examine the flow of two fluids in the body: blood in the heart and circulatory system, and air in the lungs and respiratory system. Flow of fluids elsewhere in the body is also important, such as in the urinary system (urine, liquids through the kidneys, etc.). Such directed flow of material in the body occurs predominantly by fluid flow in systems of vessels. Directed transport also occurs on microtubules in the body by motor proteins, such as kinesin and dynein, as described in Chap. 5. Undirected motion occurs by diffusion. Over "longer" distances such direct transport is preferred to diffusion because it provides a directed motion and a motion that is faster than diffusion. Diffusion is important in the body only over very short distances, up to $\sim$100 μm.

In this chapter we will discuss the concept of pressure as it relates to fluids in the body. For example, the pressure of the vitreous humor in the eyeball serves several functions, including maintaining the shape of the eyeball. This pressure is similar to the stress we examined in Chap. 4, such as that in our long bones when we walk. They both describe a force per unit area. The pressure in the fluid is hydrostatic, i.e., the force per unit area is the same stress in all directions. In solids the stress is often anisotropic. We will review the basic physics of pressure and fluid flow, including the relationship of pressure and fluid flow, and diffusion [353]. We will also examine the flow of humans in fluids, i.e., swimming, along with the possibility of human flight.

7.1 Characteristic Pressures in the Body

7.1.1 Definition and Units

The pressure of a fluid column is given by (2.48), $P = \rho g h$, where ρ is the fluid density, g is the gravitational constant, and h is the height of the column.

For mercury ρ is 13.6 g/cm^3. For water $\rho = 1.00\,\mathrm{g/cm^3}$ at 4°C. The density of whole blood is a bit higher, 1.06 g/cm^3 at 37°C. The units of pressure are presented in Table 2.6.

So far we have been discussing *absolute pressure*, P_{abs}, which is the total force per unit area. In discussions concerning the body it is very common to cite the *gauge pressure*, P_{gauge}, which is the pressure relative to a standard, which is usually atmospheric pressure, and so $P_{\mathrm{gauge}} = P_{\mathrm{abs}} - 1\,\mathrm{atm}$. This is helpful because it is the difference in pressure that is the net force that acts on a unit area. In discussing blood pressure and the pressure of air in the lungs, it is assumed that the term pressure P refers to the gauge pressure relative to the local atmospheric pressure. During breathing in (which is called inspiration), the pressure in the lungs is lower than that outside the body and so the internal (gauge) pressure is <0. Table 7.1 gives typical pressures in the body.

Table 7.1. Typical (gauge) pressures in the body (in mmHg). (Using data from [345])

arterial blood pressure	
maximum (systolic)	100–140
minimum (diastolic)	60–90
capillary blood pressure	
arterial end	30
venous end	10
venous blood pressure	
typical	3–7
great veins	<1
middle ear pressure	
typical	<1
eardrum rupture threshold	120
eye pressure	
humors	20 (12–23)
glaucoma threshold range	∼21–30
cerebrospinal fluid pressure	
in brain – lying down	5–12
gastrointestinal	10–12
skeleton	
long leg bones, standing	∼7,600 (10 atm.)
urinary bladder pressure	
voiding pressure	15–30 (20–40 cmH_2O)
momentary, up to	120 (150 cmH_2O)
intrathoracic	
between lung and chest wall	−10

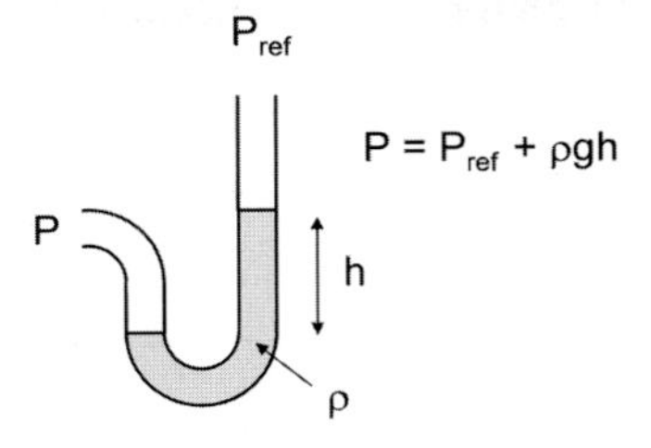

Fig. 7.1. Manometer

7.1.2 Measuring Pressure

One way of directly measuring pressure is with a manometer (Fig. 7.1). The measured pressure is that corresponding to the height of the fluid column plus the reference pressure, so

$$P = P_{\mathrm{ref}} + \rho g h. \tag{7.1}$$

The most common way to measure blood pressure is with a *sphygmomanometer* (sfig-muh-ma-nah'-mee-ter), which consists of a cuff, a squeeze bulb, and a meter that measures the pressure in the cuff (Fig. 7.2). The cuff is the balloon-like jacket placed about the upper arm above the elbow; this

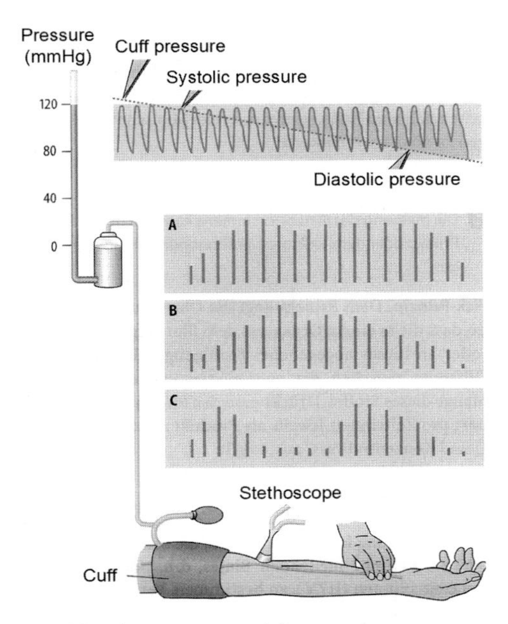

Fig. 7.2. Measuring blood pressure with a sphygmomanometer, listening to Korotkoff sounds (of varying levels during the turbulent flow shown in **A**–**C**). (Listening to sounds is called *auscultation*). (From [364])

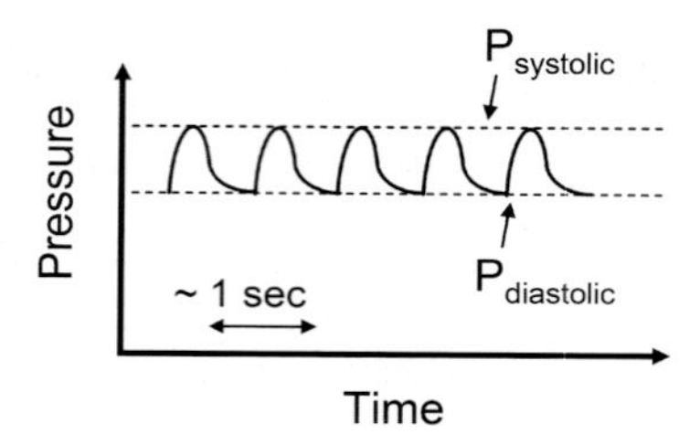

Fig. 7.3. Variation of blood pressure with time, for blood leaving the left heart for the systemic system, with the systolic and diastolic pressures shown

encircles the brachial artery. The cup of a stethoscope is placed on the lower arm, just below the elbow, to listen for the flow of blood. With no pressure in the cuff, there is normal blood flow and sounds are heard through the stethoscope. Gurgling sounds are heard after the cuff is pressurized with the squeeze bulb and then depressurized by releasing this pressure with a release valve in this bulb.

To understand when these sounds occur and their significance, we need to understand how the pressure in the main arteries varies with time. (This will be detailed in Chap. 8.) In every heart beat cycle (roughly 1/s), the blood pressure in the major arteries, such as the brachial artery, varies between the systolic pressure (∼120 mmHg) and the diastolic pressure (∼80 mmHg), as is depicted in Fig. 7.3. (The units of these cited gauge pressures are in mmHg – see (7.1) and Chap. 2.) When the pressure in the cuff exceeds the systolic pressure, there is no blood flow to the lower arm and consequently there are no sounds. When the pressure in the cuff is lowered with the release bulb to just below the systolic pressure, there is intermittent flow. During the part of the cycle when the arterial blood pressure is lower than the cuff pressure there is no flow; when it is greater, there is flow. This intermittent flow is turbulent and produces gurgling sounds. These sounds, the *Korotkoff* or K *sounds*, are heard by the stethoscope. As the cuff pressure is lowered further, the K sounds get louder and then lower, and are heard until the cuff pressure decreases to the diastolic pressure. Blood flow is not interrupted when the cuff pressure is less than the diastolic pressure and the K sounds cease because the blood flow is no longer turbulent. Therefore, the onset and end of the K sounds, respectively, denote the systolic and diastolic blood pressures. (This auscultatory method of Korotkoff was introduced by Russian army physician Korotkoff [362] who discovered a century ago that sound can be heard distally from a partially occluded limb [349].)

7.2 Basic Physics of Pressure and Flow of Fluids

In this section we overview the basics of fluids. Some of this will be a review for most. Some of the more advanced results are derived, while others are merely presented. These basics will be used in subsequent chapters.

Both gas and liquid fluids are important in the body. Gases will be treated by the ideal gas law

$$P = nRT, \tag{7.2}$$

where P is the pressure, n is the gas density, R is the gas constant ($= 8.31$ J/mol-K), and T is the temperature (in K). The gas density $n = N/V$, where N is the total number of molecules in a volume V. The gas constant $R = N_\mathrm{A} k_\mathrm{B}$, where N_A is Avogadro's number, 6.02×10^{23}, and k_B is Boltzmann's constant, 1.381×10^{-23} J/K.

One guiding principle is *Pascal's Principle*: the pressure applied to a confined fluid increases the pressure throughout by the same amount. Also quite important is *Archimedes' Principle*: the buoyant force on a body immersed in a fluid is equal to the weight of the fluid displaced by that object. Another important relation is the *Law of Laplace*, which relates the difference of pressures inside and outside a thin-walled object – of a given shape – to the tension in the walls of the object. We will also need to understand the properties of flowing fluids to be able to analyze the physics of the circulatory system.

7.2.1 Law of Laplace

The pressure inside blood vessel walls, P, exceeds that outside, P_ext, by $\Delta P = P - P_\mathrm{ext}$. *How large of a tension should the vessel walls be able to withstand to support this positive pressure difference in equilibrium?* The answer is provided by the *Law of Laplace* for hollow cylinders. It is derived here and then used in Chap. 8.

Consider a tube of radius R and length L. Figure 7.4a shows a section of this tube with angle $\theta \ll 1$. The outward force (upward in the diagram) on this

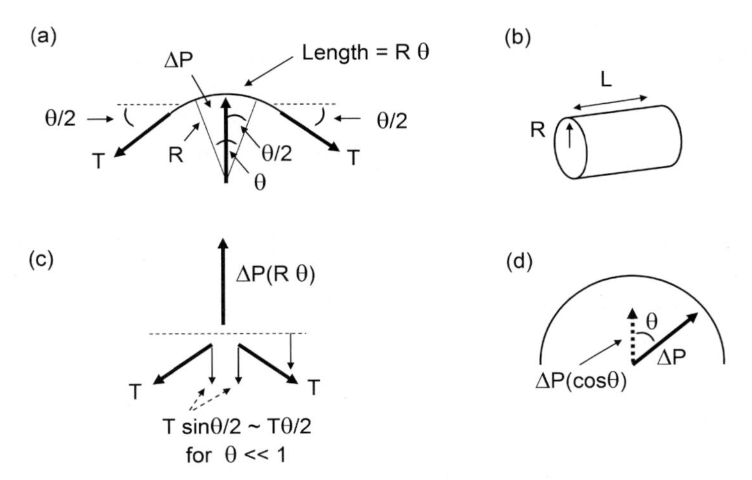

Fig. 7.4. Derivation of the Law of Laplace for the cylinder in (**a**), with the force diagram for a section of a cylinder in (**b**), leading to the force diagram in (**c**), and the resolution of pressures for analysis of a half cylinder in (**d**)

area is the pressure difference, ΔP, times the area, $(R\theta)L$. The circumferential tension T is the force per unit length (along the tube length). (Note that this use of the word "tension" has a different meaning than in earlier chapters, where it meant a force, often used to pull things apart.) This film tension has units of force/length or energy/area. It is equal to a circumferential stress $\sigma = T/w$, where w is the cylinder thickness (with $w \ll R$). These forces can be those within the blood vessel walls (Chap. 8). The horizontal components of the film tension to the left and right cancel. The vertical components are inward and each equal to $T \sin(\theta/2) \simeq T(\theta/2)$ for small angles. With both of these tension components multiplied by L, in static equilibrium force balance gives

$$\Delta P(R\theta)L = 2\left(T\frac{\theta}{2}\right)L. \tag{7.3}$$

This means

$$\Delta P = \frac{T}{R} \qquad \text{or} \qquad T = R(\Delta P). \tag{7.4}$$

This is a differential method. Alternatively we could integrate the forces over a half cylinder, as shown in Fig. 7.4d. The total downward force is the area of the walls, $2wL$, times the stress, σ, or $2wL\sigma$. The total upward force is the cross-sectional area, $2RL$, times the pressure difference, ΔP, or $2RL(\Delta P)$. In equilibrium

$$2RL(\Delta P) = 2wL\sigma \tag{7.5}$$

$$\Delta P = \frac{w\sigma}{R} = \frac{T}{R}, \tag{7.6}$$

which is the same as (7.4). (Figure 7.4d shows that the total upward force is really the integral of the upward force component, $\Delta P \cos\theta$, times the area element, $RL\mathrm{d}\theta$, integrated from $-90°$ to $90°$ or

$$(\Delta P)RL\int_{-90°}^{90°} \cos\theta \mathrm{d}\theta = (\Delta P)RL(\sin(90°) - \sin(-90°)) = 2(\Delta P)RL, \tag{7.7}$$

which turns out to be the same as $(\Delta P)2RL$.)

The Law of Laplace is also important in spheres, such as soap bubbles and the alveoli in the lungs. For a sphere of radius R and wall thickness w, we can balance the forces in the half sphere. The total downward force is the area of the walls, $2\pi Rw$, times the stress, σ, or $2\pi Rw\sigma$. The total upward force is the cross-sectional area, πR^2, times the pressure difference, ΔP, or $\pi R^2(\Delta P)$. In equilibrium

$$\pi R^2(\Delta P) = 2\pi Rw\sigma \tag{7.8}$$

$$\Delta P = \frac{2w\sigma}{R} = \frac{2T}{R}. \tag{7.9}$$

Table 7.2. Surface tension (γ) for several liquids. (Using data from [351, 358, 363])

liquid	T (°C)	γ (10^{-4} N/m)
water	0	7.56
	20	7.28
	60	6.62
	100	5.89
whole blood	20	5.5–6.1
blood plasma	20	5.0–5.6
lung surfactant	20	0.1
cerebrospinal fluid	20	6.0–6.3
saliva	20	1.5–2.1
benzene	20	2.89
mercury	20	46.4

This is the Law of Laplace for a sphere. We will use it in Chap. 9. (It is derived in more detail in Problem 7.12.)

For a spheroid with different radii of curvature, R_1 and R_2, (7.4) and (7.9) generalize to

$$\Delta P = \frac{T}{R_1} + \frac{T}{R_2}. \tag{7.10}$$

For a cylinder, $R_1 = R$ and $R_2 = \infty$ and this reduces to (7.4). For a sphere, $R_1 = R$ and $R_2 = R$ and it reduces to (7.9).

Our force balance arguments have made a direct connection between this tension, or really surface tension, and its units of force/length. Surface tension also has the same units as energy/area. This is reasonable because it is also the energy "cost" of making a unit area of a surface (or interface). Representative values of surface tension are given in Table 7.2.

7.2.2 Fluids in Motion

There are five attributes of the flow of fluids:

1. Flow can be *laminar/streamline/steady* or *turbulent/unsteady*. In laminar flow, a particle in the flow moves in a smooth manner along well-defined streamlines. In contrast, the motion is very random locally in turbulent flow. The Reynolds number Re is a dimensionless figure of merit that crudely divides the regimes of laminar and turbulent flow. It is the ratio between inertial force ($\rho u^2/2$; ρu^2 is used here) and viscous force ($\rho\eta u/d$) per unit volume on the fluid, where ρ is the fluid density, u is the average speed of flow, d is the tube diameter, and η is the fluid *coefficient of viscosity* or the *dynamic or absolute viscosity*, which is defined later. This

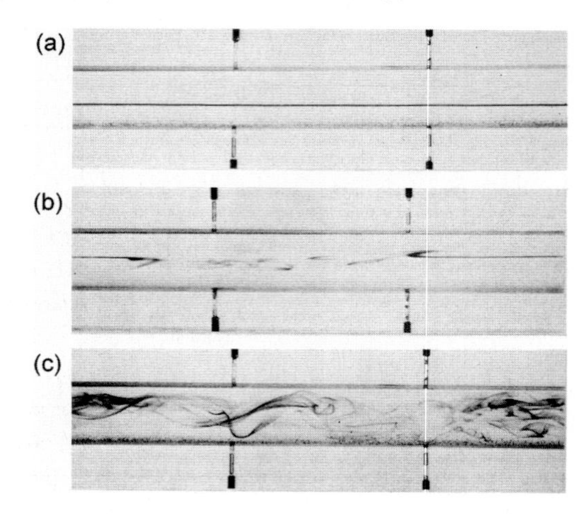

Fig. 7.5. Motion of a filament of dye in a straight pipe, showing (**a**) steady, laminar flow at low *Re*, (**b**) short bursts of turbulence for *Re* above the critical value, and (**c**) fully turbulent flow with random motion of the dye streak for higher *Re*. (From [346]. Used with permission of Oxford University Press)

gives

$$Re = \frac{\rho u^2}{\rho\eta u/d} = \frac{\rho u d}{\eta} = \frac{ud}{\upsilon}, \tag{7.11}$$

where $\upsilon = \eta/\rho$ is the coefficient of *kinematic viscosity*.
Although this dividing line is not hard and fast, generally, flow in a rigid tube with $Re < 2{,}000$ is laminar and that with $Re > 2{,}000$ is turbulent. This dividing region is often cited as being between 1,200–2,500, and in the higher range for smoother-walled tubes. Figure 7.5 shows flow in the laminar and turbulent regimes, and in the transition region between them.

2. Flow can be *compressible* or *incompressible*. Gases, such as air, are very compressible. Liquids are less compressible, and are often approximated as being incompressible.
3. Flow can be *viscous* or *nonviscous*. Fluids (other than superfluids) always have some viscosity, but in some cases it can be ignored totally, or first ignored and then considered as a perturbation.
4. Flow can be *rotational* or *irrotational*. In the cases we will consider there is no local rotation (such as vortices), so the flow will be irrotational.
5. Flow can be *steady* (constant in time) or *pulsatile* (with pulsing changes). Blood flow in the body is pulsatile, but is commonly treated as being in steady state in simple models. We will use both steady and pulsatile models in Chap. 8.

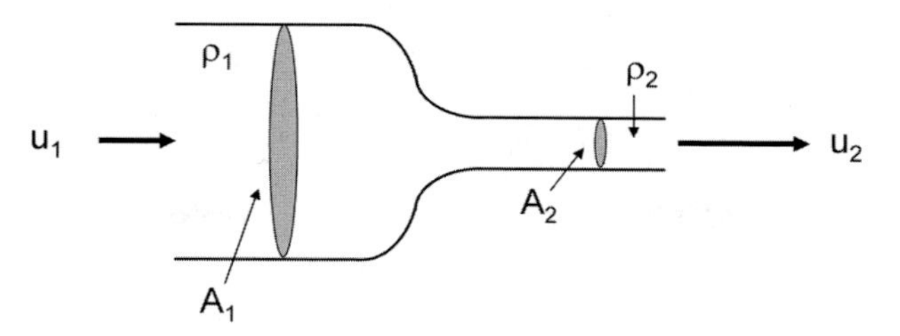

Fig. 7.6. Continuity of flow when the tube cross-sectional area changes

7.2.3 Equation of Continuity

The equation of continuity is a statement of the conservation of mass during flow. As seen in Fig. 7.6, when a fluid of a given mass density ρ moves with average speed u in a tube of cross-sectional area A, the product $\rho A u$ is constant (i.e., it is conserved). Because the speed is a longitudinal distance per unit time, Au is the volume flow per unit time (because $A \times$ distance $=$ volume). Consequently, $\rho A u$ is the mass per unit time. In steady state, the same mass flows into a volume and leaves it. For the regions marked 1 and 2 in Fig. 7.6, this means that

$$\rho_1 A_1 u_1 = \rho_2 A_2 u_2. \tag{7.12}$$

If the fluid is incompressible, the density in $\rho_1 A_1 u_1 = \rho_2 A_2 u_2$ does not change with pressure and is the same everywhere. With $\rho_1 = \rho_2$, we follow the volume or volumetric flow rate Q, which is now a constant. This means $Q_1 = A_1 u_1$ and $Q_2 = A_2 u_2$, and so the continuity equation becomes $Q = Q_1 = Q_2$ with

$$Q = A_1 u_1 = A_2 u_2. \tag{7.13}$$

7.2.4 Bernoulli's Equation

Bernoulli's Principle (or equation) relates the average flow speed u, pressure P, and height y of an incompressible, nonviscous fluid in laminar, irrotational flow (Fig. 7.7). At any two points

$$P_1 + \frac{1}{2}\rho u_1^2 + \rho g y_1 = P_2 + \frac{1}{2}\rho u_2^2 + \rho g y_2. \tag{7.14}$$

The densities $\rho_1 = \rho_2 = \rho$ for this incompressible fluid. (Bernoulli's equation actually applies to any two points along a streamline.)

There are three special cases of Bernoulli flow. (1) For static fluids ($u = 0$), and Bernoulli equation's reduces to $P_1 + \rho g y_1 = P_2 + \rho g y_2$. (2) It reduces to Torricelli's theorem when $P_1 = P_2$, namely $\rho u_1^2/2 + \rho g y_1 = \rho u_2^2/2 + \rho g y_2$.

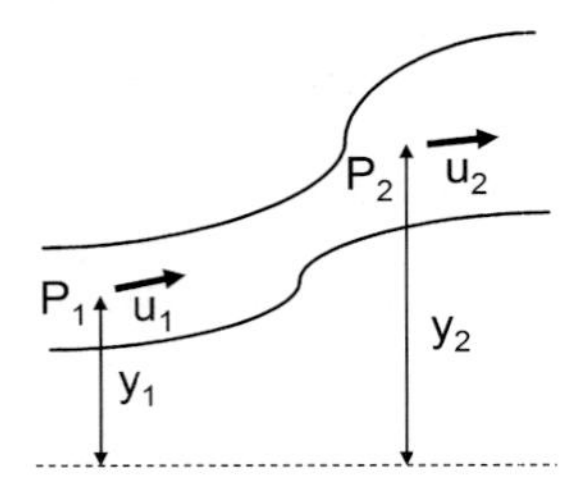

Fig. 7.7. For irrotational and nonviscous flow, the pressure, flow speed, and height are related by Bernoulli's equation along any streamline

(3) It reduces to the Venturi flow regime when $y_1 = y_2$ (Fig. 7.8), so

$$P_1 + \frac{1}{2}\rho u_1^2 = P_2 + \frac{1}{2}\rho u_2^2. \tag{7.15}$$

Because the continuity of flow in such a Venturi tube is $A_1 u_1 = A_2 u_2$

$$u_2 = \frac{A_1}{A_2} u_1. \tag{7.16}$$

Therefore we find

$$P_1 + \frac{1}{2}\rho u_1^2 = P_2 + \frac{1}{2}\rho \left(\frac{A_1}{A_2} u_1\right)^2 \tag{7.17}$$

and

$$P_2 - P_1 = \frac{1}{2}\rho u_1^2 \left(1 - \left(\frac{A_1}{A_2}\right)^2\right). \tag{7.18}$$

With $A_2 < A_1$, we see that $u_2 > u_1$ and $P_2 < P_1$. This shows that the flow becomes faster and the pressure becomes lower in clogged blood vessels.

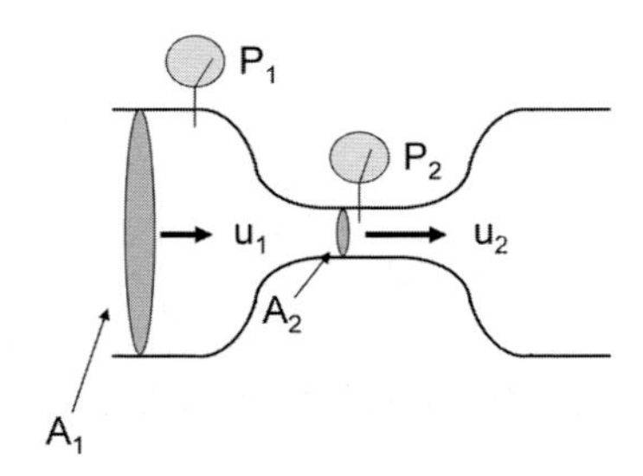

Fig. 7.8. Flow in a tube when the tube cross-sectional area changes. This is a Venturi tube, for which pressure and flow speed are related by Bernoulli's equation in the limit of constant height

7.2.5 Interactions among the Flow Parameters

Pressure P, volume V, and flow rate Q are all interrelated in flow through vessels, be it blood flow in the circulatory system or air flow in breathing. *Resistance* R_{flow} is the pressure difference ΔP needed to cause a given flow rate Q

$$R_{\text{flow}} = \frac{\Delta P}{Q}. \tag{7.19}$$

Compliance C_{flow} is the change in volume caused by a change in pressure in a vessel

$$C_{\text{flow}} = \frac{\Delta V}{\Delta P}. \tag{7.20}$$

Occasionally, the *inertance* L_{flow} is also defined. It is the change in pressure caused by a change in flow rate

$$L_{\text{flow}} = \frac{\Delta P}{\Delta Q}. \tag{7.21}$$

See Appendix D for an analog to electrical circuits.

7.2.6 Viscous Flow and Poiseuille's Law

Bernoulli's equation would predict that the pressure does not change during flow if the tube cross-section and height do not change. This is true for an ideal, nonviscous fluid. Viscosity is the friction during flow. It is always present and causes the pressure to drop during flow.

The coefficient of (dynamic or absolute) viscosity η is formally defined in (7.22), which gives the tangential or shear force F required to move a fluid layer of area A at a constant speed v, in the x direction, when that layer is a distance y from a stationary plate (Fig. 7.9) [350, 354]

$$F = \eta \frac{A}{y} v. \tag{7.22}$$

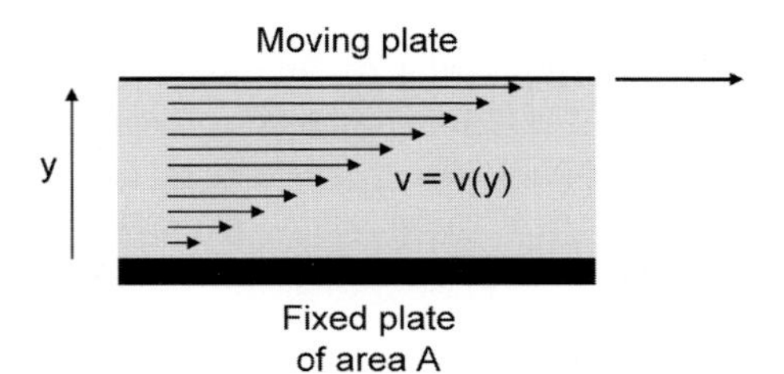

Fig. 7.9. Viscous fluid flow, with a linear gradient of fluid speed with position between a fixed and moving plate. This is shown for Newtonian flow

This equation is also written as

$$\tau = \eta \frac{dv}{dy}, \tag{7.23}$$

where $\tau = F/A$ is the shear stress, as in (4.5) and Figs. 4.10 and 4.11, and dv/dy is called the *shear rate*. (Check that the units of the shear rate are those that a rate should have, 1/s.) Fluids that are characterized by (7.22) and (7.23) are called "Newtonian fluids" and are said to undergo "Newtonian flow."

The SI units of η are (N/m^2)s, which is equal to kg/m-s and Pa-s; this is called a Poiseuille (PI), but this unit is not often used. More commonly used than this last unit is the poise (P) which is 10× smaller. It is a natural unit in the CGS units system with 1 poise $=$ 1 g/cm-s $=$ 0.1 (N/m^2)s $=$ 0.1 kg/m-s $=$ 0.1 Pa-s. Also common is the centipoise (cP), with 1 cP $=$ 0.01 poise $=$ 0.001 Pa-s, because the viscosity of water at 20°C is almost equal to 1 cP (and is actually 1.002 cP). We will usually use the units of Pa-s. Also, this viscosity coefficient is often called η by physicists (and is used as such here), whereas it is often called μ by biomedical engineers. It is also related to, but different from the viscosity damping constant for the dashpot c in (4.48).

Because of this drag, there must be a pressure difference (gradient) to maintain fluid flow in a tube. The relation between this pressure drop and the volumetric flow rate Q is given by Poiseuille's Law (or Hagen-Poiseuille's Law)

$$Q = \frac{\pi R^4}{8\eta L}(P_1 - P_2), \tag{7.24}$$

where R is the radius of the tube and L is its length (Fig. 7.10). This relation can be viewed as the flow rate for a given pressure drop. Alternatively, it can be viewed as the pressure drop when there is a flow Q in the tube

$$P_1 - P_2 = \Delta P = \frac{8\eta L}{\pi R^4} Q. \tag{7.25}$$

We will use this expression in Chap. 8 to determine the pressure drops in blood vessels during circulation. It is derived later as an advanced topic.

Equation (7.25) is formally analogous to Ohm's Law for resistors, $V = IR_{\text{elect}}$ (or in a manner more parallel to this equation, $V = R_{\text{elect}}I$), where V is the voltage or potential difference across the resistor and is the driving term (which is analogous to ΔP), R_{elect} is the electrical resistance (analogous to the resistance of flow $8\eta L/\pi R^4$ here, which we will call R_{flow}), and I is the electrical current, which is the flow resulting from the driving term (analogous to the volumetric flow Q here).

Consider a tube with cross-sectional area A. The net force on the fluid in it is $(\Delta P)A$. If this force moves the fluid a distance L, the work done on it is $FL = (\Delta P)AL$. If this volume AL is moved in a given time, the work needed

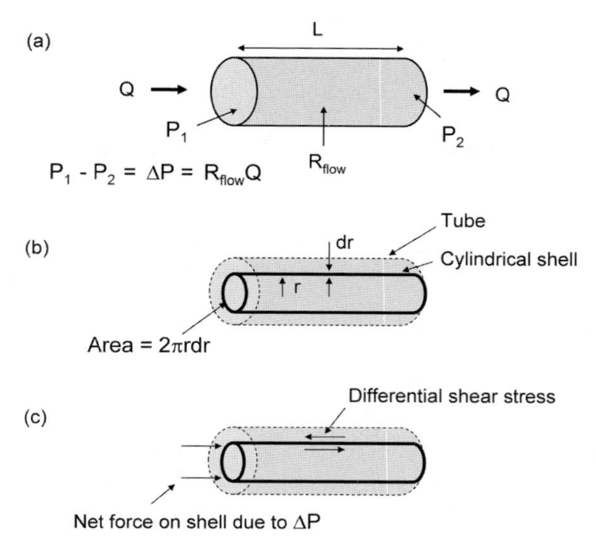

Fig. 7.10. Calculation of Poiseuille's Law for a tube in (**a**), using the cylindrical shell in (**b**), and balancing forces between the hydrostatic flow pressure force and the differential shear stress on the shell in (**c**)

to do this in this given time – the power – is

$$P_{\text{power, flow}} = (\Delta P)Q, \tag{7.26}$$

or $P_{\text{power, flow}} = Q^2 R_{\text{flow}} = (\Delta P)^2/R_{\text{flow}}$. These expressions are analogous to those for the power dissipated by an electrical resistor: $P_{\text{power, elect}} = VI = I^2 R_{\text{elect}} = V^2/R_{\text{elect}}$.

The coefficient of viscosity for water is 1.78×10^{-3} Pa-s at 0°C and it decreases with temperature, dropping to 1.00×10^{-3} Pa-s at 20°C and 0.65×10^{-3} Pa-s at 40°C. At 37°C, η is 1.5×10^{-3} Pa-s for blood plasma and 4.0×10^{-3} Pa-s for whole blood, which are both higher than that for water at the same temperature. (Blood is really thicker than water.) The coefficients of viscosity of common human body fluids and other materials are listed in Table 7.3. As is clear from the table, the viscosity of liquids decreases with increasing temperature T, because the kinetic energy of molecules increases with T and this can overcome intermolecular forces that slow down motion between the dense, adjacent layers. In contrast, viscosity increases with temperature for gases, as $T(\text{in K})^{1/2}$, because diffusion between adjacent layers increases with T.

Derivation of Poiseuille's Law (Advanced Topic)

Now consider flow in a tube of radius R (Fig. 7.10). The distance radially from the center line of the tube is r. Using (7.22) and (7.23), the shear force and

Table 7.3. Coefficient of viscosity η of common materials, in Pa-s (1 poise = 0.1 Pa-s). (Using data from [351, 358, 363])

material	T (°C)	η
water	0	1.78×10^{-3}
	20	1.00×10^{-3}
	37	0.69×10^{-3}
	50	0.55×10^{-3}
	100	0.28×10^{-3}
blood plasma	37	1.5×10^{-3}
whole blood[a]	37	$\sim 4.0 \times 10^{-3}$
low shear rate, Hct = 45%		$\sim 100 \times 10^{-3}$
low shear rate, Hct = 90%		$\sim 1{,}000 \times 10^{-3}$
high shear rate, Hct = 45%		$\sim 10 \times 10^{-3}$
low shear rate, Hct = 90%		$\sim 100 \times 10^{-3}$
cerebrospinal fluid	20	1.02×10^{-3}
interstitial fluid	37	$1.0–1.1 \times 10^{-3}$
human tears	37	$0.73–0.97 \times 10^{-3}$
synovial fluid[b]	20	>0.3
castor oil	20	1
motor oil, SAE 10	20	0.065
motor oil, SAE 50	20	0.54
machine oil, heavy	37	0.13
machine oil, light	37	0.035
ethylene glycol	37	0.011
mercury, liquid	37	1.465×10^{-3}
methanol	37	0.47×10^{-3}
ketchup	20	50
peanut butter	20	250
glass (anneal)	720–920 K	2.5×10^{12}
(blowing)	∼1,300 K	$\sim 1 \times 10^{6}$
(furnace)	1,500–1,700 K	$\sim 1 \times 10^{2}$
air	20	1.8×10^{-5}
	100	2.1×10^{-5}

Hct is the hematocrit, which is the volume fraction of red blood cells in blood.
[a] See Figs. 8.10 and 8.11.
[b] See Fig. 7.14.

stress are

$$F = \eta A \frac{\mathrm{d}v}{\mathrm{d}r} \tag{7.27}$$

$$\tau = \eta \frac{\mathrm{d}v}{\mathrm{d}r} \,. \tag{7.28}$$

Imagine a series of concentric cylinders within this tube of thickness $\mathrm{d}r$ and length L (centered about the center symmetry axis, Fig. 7.10b), with a pressure drop ΔP along L. The force pushing one of these cylindrical shells forward is this pressure drop, ΔP, times the area of the front (and back) cylinder face, $2\pi r \mathrm{d}r$, or $(\Delta P)2\pi r \mathrm{d}r$. The viscous drag force that the cylindrical shell feels from the other shells (i.e., from the liquid) is the difference between the shear stress felt on its inner and outer surfaces × its surface area, $2\pi r L$. Using (7.28), this difference is

$$\frac{\mathrm{d}(2\pi r L \tau)}{\mathrm{d}r}\mathrm{d}r = \frac{\mathrm{d}\left(2\pi r L \eta \frac{\mathrm{d}v}{\mathrm{d}r}\right)}{\mathrm{d}r}\mathrm{d}r = 2\pi L \eta \frac{\mathrm{d}\left(r \frac{\mathrm{d}v}{\mathrm{d}r}\right)}{\mathrm{d}r}\mathrm{d}r \tag{7.29}$$

assuming the viscosity does not depend on r.

In steady state, the force due to the pressure drop plus the drag force equals zero, so

$$(\Delta P)2\pi r \mathrm{d}r + 2\pi L \eta \frac{\mathrm{d}\left(r \frac{\mathrm{d}v}{\mathrm{d}r}\right)}{\mathrm{d}r}\mathrm{d}r = 0 \tag{7.30}$$

or

$$\frac{\mathrm{d}\left(r \frac{\mathrm{d}v}{\mathrm{d}r}\right)}{\mathrm{d}r} = -\frac{\Delta P}{\eta L} r. \tag{7.31}$$

Integrating gives

$$r\frac{\mathrm{d}v}{\mathrm{d}r} = -\frac{\Delta P}{2\eta L} r^2 + C \tag{7.32}$$

$$\frac{\mathrm{d}v}{\mathrm{d}r} = -\frac{\Delta P}{2\eta L} r + \frac{C}{r}. \tag{7.33}$$

The constant C must equal zero, because otherwise the second term would be infinite at the center.

Integrating again gives

$$v(r) = -\frac{\Delta P}{4\eta L} r^2 + D. \tag{7.34}$$

(see Appendix C). Because at the tube radius the velocity is zero ($v(R) = 0$), D is determined and this gives

$$v(r) = -\frac{\Delta P}{4\eta L} r^2 + \frac{\Delta P}{4\eta L} R^2, \tag{7.35}$$

so

$$v(r) = \frac{\Delta P}{4\eta L}\left(R^2 - r^2\right) = \frac{R^2 \Delta P}{4\eta L}\left(1 - \frac{r^2}{R^2}\right). \tag{7.36}$$

This speed is seen to be maximum in the center where $r = 0$. This maximum value of $R^2\Delta P/(4\eta L)$ decreases to 0 as r increases from 0 to R.

The flow rate in the tube Q equals uA when the speed is uniform across the area A. When it is not, as here, Q is obtained by integrating $v(r)$ across the cross-sectional area. This is done by multiplying $v(r)$ by the area element $2\pi r\mathrm{d}r$ (the circumference $\times$ the differential in r) and integrating r from 0 to R, which gives

$$Q = \int_0^R \frac{\Delta P}{4\eta L}\left(R^2 - r^2\right) 2\pi r\mathrm{d}r \tag{7.37}$$

$$Q = \int_0^R \frac{\pi\Delta P}{2\eta L}\left(rR^2 - r^3\right)\mathrm{d}r = \frac{\pi\Delta P}{2\eta L}\left(\frac{R^4}{2} - \frac{R^4}{4}\right) = \frac{\pi R^4\Delta P}{8\eta L}. \tag{7.38}$$

This is Poiseuille's Law $\Delta P = \left(8\eta L/\pi R^4\right) Q$ ((7.24) and (7.25)). Because Q is also equal to the area $\times$ the average speed, this average speed is

$$u = \frac{\pi R^4\Delta P/8\eta L}{\pi R^2} = \frac{R^2\Delta P}{8\eta L} \tag{7.39}$$

and

$$v(r) = 2u\left(1 - \frac{r^2}{R^2}\right). \tag{7.40}$$

This is depicted in the rightmost profile shown in Fig. 7.11.

Many fluids are non-Newtonian fluids (Fig. 7.12), which means they are not characterized by (7.22) and (7.23), but by other relations. We assumed earlier that a fluid could generate no shear stress at any shear or strain rate; this is a frictionless or nonviscous fluid, which is unrealistic except for superfluids. (Note that such shear or strain "rates" are really gradients with respect to the direction normal to flow, i.e., y, and not with respect to time t. However, this terminology is reasonable because they have the same units as strain rates and because of the scaling argument given in Problem 7.23.) In some real non-Newtonian fluids, the shear stress is $F/A = \eta(\mathrm{d}v/\mathrm{d}y)^n$, where n could be greater or less than 1, as in Fig. 7.12. This is sometimes phrased as $F/A = \eta'(\mathrm{d}v/\mathrm{d}y)$ where the effective viscosity $\eta' = \eta(\mathrm{d}v/\mathrm{d}y)^{n-1}$ depends on the strain rate; as such a Newtonian fluid

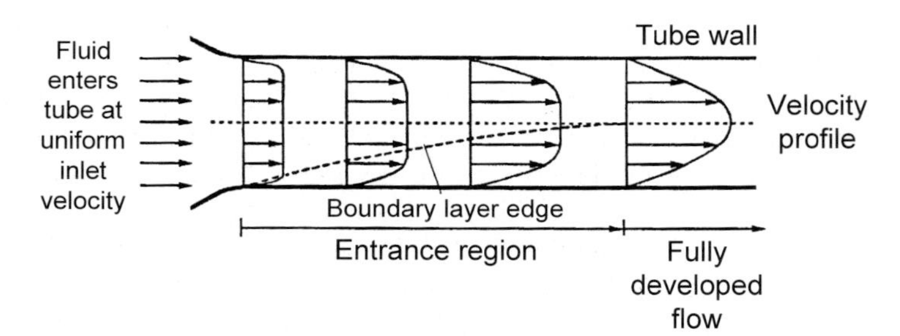

Fig. 7.11. Establishment of steady-state Newtonian flow into the parabolic velocity profile (in the fully developed flow). (From [351], based on [355]. Courtesy of Robert A. Freitas Jr., Nanomedicine, Vol. 1 (1999), http://www.nanomedicine.com)

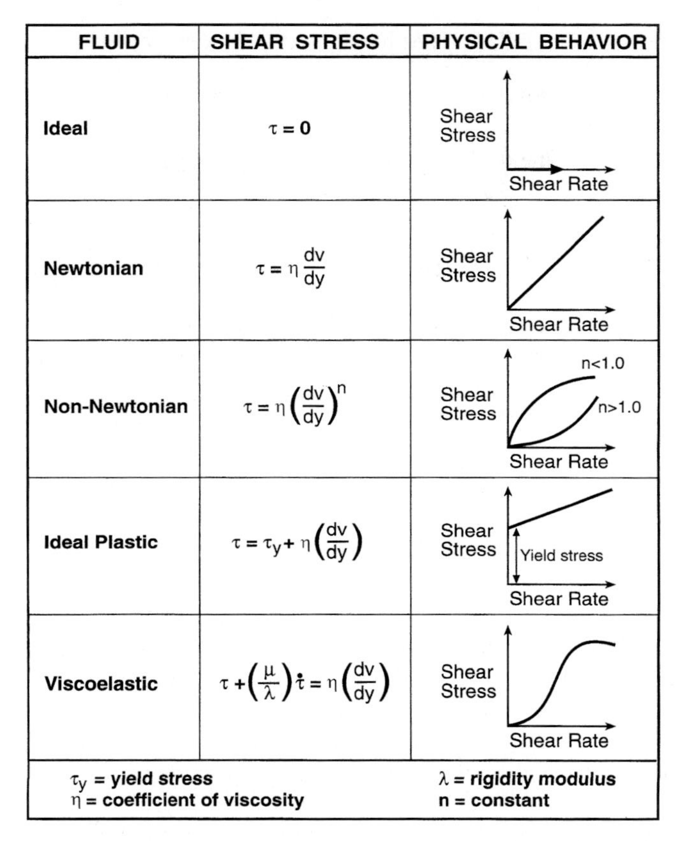

FLUID	SHEAR STRESS	PHYSICAL BEHAVIOR
Ideal	$\tau = 0$	Shear Stress / Shear Rate
Newtonian	$\tau = \eta \frac{dv}{dy}$	Shear Stress / Shear Rate
Non-Newtonian	$\tau = \eta \left(\frac{dv}{dy}\right)^n$	Shear Stress / Shear Rate; n<1.0; n>1.0
Ideal Plastic	$\tau = \tau_y + \eta \left(\frac{dv}{dy}\right)$	Shear Stress / Shear Rate; Yield stress
Viscoelastic	$\tau + \left(\frac{\mu}{\lambda}\right)\dot{\tau} = \eta \left(\frac{dv}{dy}\right)$	Shear Stress / Shear Rate

τ_y = yield stress
η = coefficient of viscosity
λ = rigidity modulus
n = constant

Fig. 7.12. Newtonian and non-Newtonian fluid flow. (From [357])

would have an effective viscosity that is independent of the strain rate. A dilatant or shear-thickening fluid has an effective viscosity that increases with increasing stress. A plastic or shear-thinning fluid has an effective viscosity that decreases with increasing stress. A Bingham plastic, such as toothpaste, has a finite yield stress even for $\mathrm{d}v/\mathrm{d}y = 0$, and above the yield stress it has a linear relationship with strain rate, $F/A = \alpha + \eta(\mathrm{d}v/\mathrm{d}y)$. The composition of blood makes it a non-Newtonian fluid; this is discussed in Chap. 8. Consequently, the flow pattern of blood is decidedly nonparabolic (Fig. 7.13).

Synovial fluid is one example of a non-Newtonian fluid. Figure 7.14 shows that its coefficient of friction is high at low shear rates and much lower at high shear rates. Figure 8.11 shows that whole blood is also a non-Newtonian fluid.

The dependence of flow on pressure drop within the laminar, intermediate, and turbulent regimes is shown in Fig. 7.15.

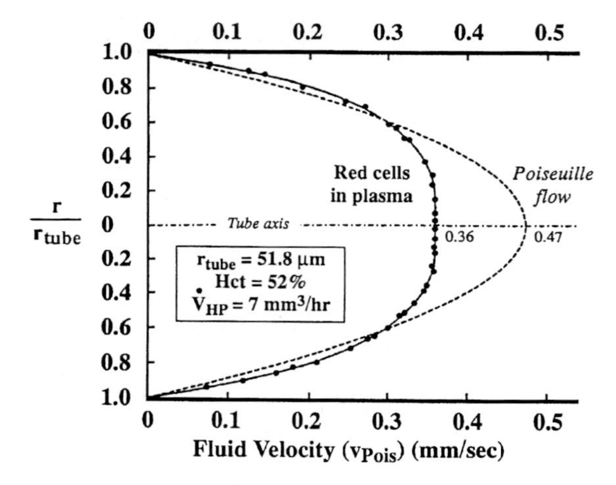

Fig. 7.13. Velocity flow profile of whole blood is blunted relative to the ideal parabolic flow of a Newtonian fluid. (From [351], based on [355]. Courtesy of Robert A. Freitas Jr., Nanomedicine, Vol. 1 (1999), http://www.nanomedicine.com)

Approach to Steady Flow

The results of Sect. 7.2.6 apply to steady, laminar flow. If a tube bifurcates – such as in branching arteries, the velocity profile we derived with its *boundary layer* at the tube circumference (where the flow velocity decreases to zero), will not represent the flow distribution immediately after the bifurcation. It will be valid only after a distance past the bifurcation called the entrance length, X [346]. Experimentally

$$X = 0.03d(Re) \tag{7.41}$$

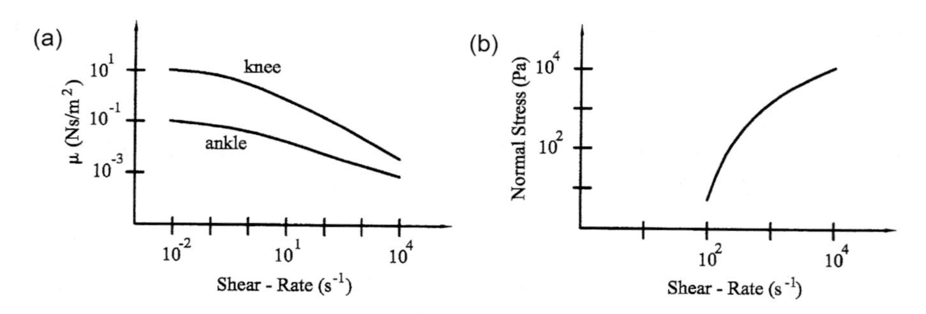

Fig. 7.14. (**a**) Synovial fluid is a non-Newtonian fluid, with a coefficient of friction that decreases with shear rate. (**b**) Another property of such a non-Newtonian fluid is that it can create a normal stress that depends on shear rate. (From [361])

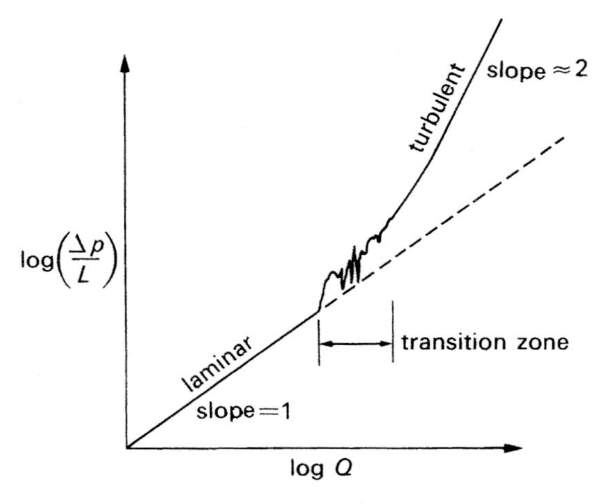

Fig. 7.15. Pressure drop per unit length vs. log of the flow rate for a long tube, showing a transition region between laminar and turbulent flow. (From [346]. Used with permission of Oxford University Press)

for a straight pipe, where d is the diameter ($d = 2R$). For the laminar flow regime with $Re < 10$, this is not valid and the entrance length is smaller; when $Re \ll 1$ and inertial forces can be ignored

$$X \sim d. \tag{7.42}$$

For $Re > 2{,}500$, the flow is likely turbulent and the entrance length (for steady state turbulent flow) is shorter than that for fast laminar flow

$$X = 0.693d(Re)^{1/4}. \tag{7.43}$$

The development of parabolic flow for a Newtonian fluid in the laminar flow region is illustrated in Fig. 7.11.

Flow in Curving Tubes such as Arteries

When you hold a hose with flowing water and try to change its direction you feel a resistance. This resistance is the force you need to apply to change the direction of the momentum of the water flow. This centripetal force becomes larger with faster flow rates (i.e., for larger hose areas and faster water flow speeds), as is well known to all firepersons. Curving arterial walls, such as the aorta, feel a pressure due to the difference in hydrostatic pressure inside and outside the vessel that arises from this force. This pressure is felt equally around the wall.

Consider a tube or artery of inner radius R that is turning with a radius $\mathcal{R}$. Figure 7.16 shows an arc of angle θ (in radians) $\ll 1$ of such a vessel. The average speed of flow is u and the mass density of blood is ρ. The magnitude

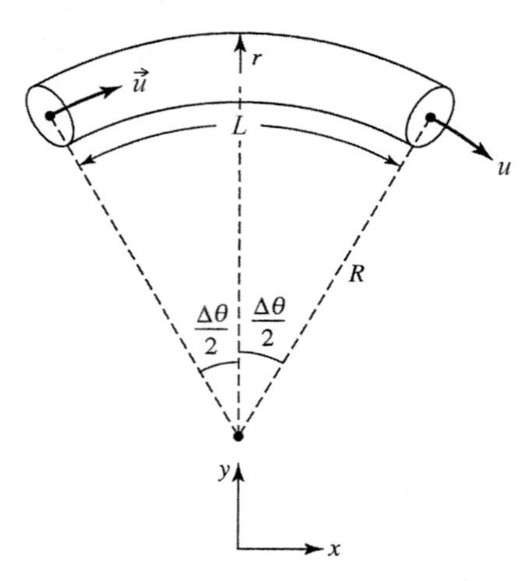

Fig. 7.16. Flow in a curved tube. (From [344])

of the momentum in the vessel per unit volume is $\rho u(AL)$, for a vessel with cross-sectional area $A = \pi R^2$ for a length L of blood flow. In traversing an angle θ, the momentum vector changes by $\sim(\rho u)(AL)\theta$. This occurs when the blood moves a distance $\mathcal{R}\theta$, given by the arc length, with a speed u, so this occurs in a time $\mathcal{R}\theta/u$. Consequently, the force needed to do this is the change of momentum per unit time, which is $(\rho u AL\theta)/(\mathcal{R}\theta/u) = \rho u^2 AL/\mathcal{R}$. Because the mass of this volume of blood is $m = \rho AL$, this looks like the centripetal force $mu^2/\mathcal{R}$. (It looks like it, because that is what it is.)

This force is distributed across the outer half of the inner arterial wall, which has a cross-sectional area πRL. Because the force is outward, there is a larger load on the outermost portions shown in Fig. 7.16 and a smaller load on the outer upper and lower regions. Therefore, the peak force per unit area is more accurate when you use a smaller effective area, say $\pi RL/2$. Consequently, the peak pressure is the force per unit area $\rho u^2 AL/\mathcal{R}$, with $A = \pi R^2$, divided by this area $\pi RL/2$, or

$$P_{\text{cent}} = \frac{\rho u^2 (\pi R^2) L/\mathcal{R}}{\pi RL/2} = 2\rho u^2 \frac{R}{\mathcal{R}}. \tag{7.44}$$

Flow of Objects in Fluids: Drag and Lift

The viscosity of a fluid also creates a drag force on objects that move in the fluid [343]. The reason for this is clear from (7.22); such objects are just like the plate in Fig. 7.9 in this functional definition of viscosity. Viscosity causes the boundary layer of the fluid near the ball (or plate) to move with it. If the object is moving at a speed u relative to the fluid, this drag force on the object is given by *Stokes Law*

$$F_{\text{drag,Stokes}} = 6\pi R\eta u, \tag{7.45}$$

where R is the hydrodynamic radius of the object, which is about half the typical lateral dimension D. This expression for Stokes friction is valid when the flow speed is slow enough that the streamlines about it are laminar. Here this means the Reynolds number $Re = \rho Du/\eta = 2\rho Ru/\eta$ is smaller than $\sim$100.

For Reynolds numbers much larger than 100, viscosity is no longer totally dominant and the main drag force is due to the formation of vortices that appear and trail the object, particularly as turbulent flow becomes important. This *hydrodynamic drag force* is

$$F_{\text{drag,hydrodyamnic}} = \frac{1}{2} C_{\text{D}} A \rho u^2, \tag{7.46}$$

where A is the frontal surface area and C_{D} is the drag coefficient. For $100 < Re < 2 \times 10^5$, $C_{\text{D}} \simeq 1.0$ for circular cylinders. For spheres, C_{D} decreases from 1.0 to $\simeq$0.5 as Re increases from 100 to 1,000 and it remains about 0.5 for $1,000 < Re < 2 \times 10^5$. For both cylinders and spheres, C_{D} becomes smaller at somewhat higher Re. These vortices or eddies are produced at the *Strouhal frequency*

$$f_{St} = \frac{(St)u}{D}, \tag{7.47}$$

where St is the Strouhal number. St depends on C_{D} and Re, and is typically between 0.12 and 0.23.

Problem 7.39 examines which drag regime dominates for human motion in fluids: walking and running in air and swimming in water.

Another source of drag that is present at all speeds is *skin friction*, which is due to the acceleration of the initially still fluid to the object speed u, because fluid in the boundary layer near the object sticks to it. This is different from Stokes drag, which is due to frictional losses in the fluid. This skin friction is

$$F_{\text{drag,skinfriction}} = \frac{1}{2} C_{\text{sf}} S \rho u^2, \tag{7.48}$$

where C_{sf} is the skin friction coefficient, which depends on the details of the flow, and S is the wetted surface area. When you swim at or near the surface, fluid builds up to a higher than ambient level in front of your head (as you push the water forward). The water is depressed to a level lower than ambient after your head, as it "ventilates." This *ventilation drag* force varies as u^4.

The power consumed by each of these drag forces is

$$P = F_{\text{drag}} u. \tag{7.49}$$

When a foil that is tilted up at an angle β moves in a fluid, an upward force is generated on it called *lift*, which is

$$F_{\text{lift}} = \frac{1}{2} C_{\text{lift}} S \rho u^2. \tag{7.50}$$

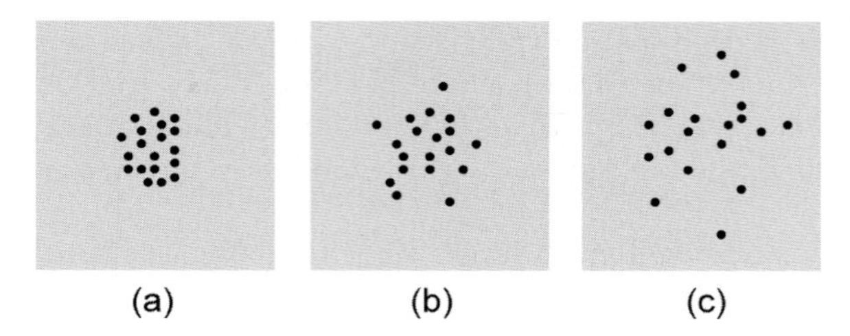

Fig. 7.17. Schematic of how the locations of particles vary at successively later times, from (**a**) to (**c**), as a result of diffusion

The lift coefficient, C_{lift}, varies linearly with this angle of attack. For small angles, it varies linearly from -0.4 to 1.2 for β varying from $-4°$ to $12°$ (for $Re = 1.7 \times 10^6$). Of course, $C_{\text{lift}} = 0$ for $\beta = 0°$. For β much larger than $12°$, the flow separates from the upper edge of the wing and there is stalling of the lift.

Chapter 3 discussed the lift force on spinning objects, such as thrown baseballs and such, which is commonly called the Magnus force.

7.3 Diffusion (Advanced Topic)

When the concentration of particles (or molecules) is not uniform, the random particle thermal motion leads to a net movement (or diffusion) of particles from regions of higher concentration to regions of lower concentration. The net effect is to make the concentration more uniform (Fig. 7.17). This diffusion flow rate increases with the nonuniformity or gradient of the concentration, which is *Fick's First Law of Diffusion*

$$J = -D_{\text{diff}} \frac{\partial n}{\partial x} \tag{7.51}$$

for flow in one-dimension, where J is the flux of particles (particle flow per unit area per unit time), D_{diff} is the diffusion coefficient, and $n(x, t)$ is the concentration of particles. (We must use partial derivatives here because everything depends on x and t.) The diffusion coefficient depends on the background medium, and is on the order of $\sim 10^{-1}\,\text{cm}^2/\text{s}$ in gas, $\sim 10^{-5}\,\text{cm}^2/\text{s}$ in liquid, and $\sim 10^{-9}\,\text{cm}^2/\text{s}$ in solid backgrounds.

During this flow the total number of particles must be conserved. Consider the cylindrical volume construct in Fig. 7.18, with its axis along the x-axis, and of length $\mathrm{d}x$ and cross-sectional area A. The total number of particles entering from the left in a unit time $\mathrm{d}t$ is $J(x)A(\mathrm{d}t)$ and the number leaving from the right in this same time is $J(x + \mathrm{d}x)A(\mathrm{d}t) \simeq (J(x) + (\partial J/\partial x)\mathrm{d}x)\, A(\mathrm{d}t)$. Therefore the net increase in the number of particles in the cylinder is the difference $-(\partial J/\partial x)(\mathrm{d}x)A(\mathrm{d}t)$. This must be accounted for by the change in

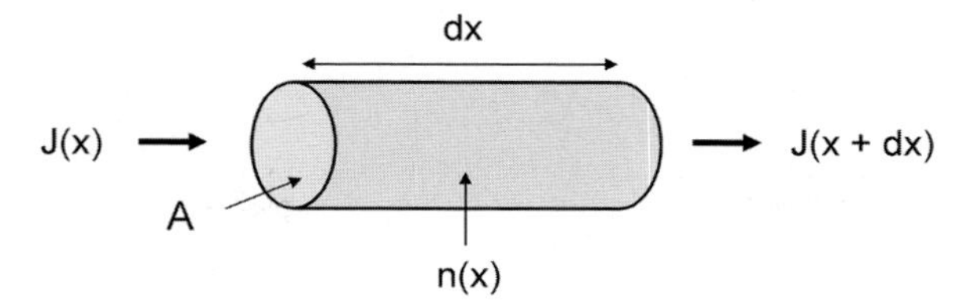

Fig. 7.18. Particles are conserved for any flow in and out of the cylinder through a change in concentration inside of it, for any flow process and for diffusion

density in this time in the volume, which is $[(\partial n/\partial t)\mathrm{d}t][A(\mathrm{d}x)]$. This gives the conservation of the number of particles

$$\frac{\partial n}{\partial t} = -\frac{\partial J}{\partial x}. \tag{7.52}$$

These two equations can be combined by differentiating (7.51) in space to get $\partial J/\partial x = -D_{\text{diff}}\, \partial^2 n/\partial x^2$ (assuming that D_{diff} does not depend on x) and replacing $\partial J/\partial x$ from (7.52). This gives the *Diffusion equation* (or *Fick's Second Law of Diffusion*)

$$D_{\text{diff}} \frac{\partial^2 n}{\partial x^2} = \frac{\partial n}{\partial t}. \tag{7.53}$$

Such diffusion leads to a slow gaussian-like, undirected spreading of the species over a distance $x \sim \sqrt{2D_{\text{diff}}t}$ in a time t. (A gaussian profile is of the general form $\exp\left(-x^2/a^2\right)$.) For a total number of particles N initially at $x = 0$ at $t = 0$, the concentration is approximately

$$n(x,t) \sim \frac{N}{\sqrt{2\pi D_{\text{diff}}t}} \exp\left(-x^2/2D_{\text{diff}}t\right). \tag{7.54}$$

As presented, this solution is not valid for small times. The exact solution is slightly more complicated in other ways as well, but it is essentially the same result when the initial spread of particles is very small (Fig. 7.19) (see Appendix C). If the initial distribution is gaussian, (7.54) becomes

$$n(x,t) = \frac{N}{\sqrt{2\pi\sigma^2(t)}} \exp\left(-x^2/2\sigma^2(t)\right), \tag{7.55}$$

where

$$\sigma^2(t) = \sigma^2(0) + 2D_{\text{diff}}t \tag{7.56}$$

and $\sigma(0)$ is the initial spread.

In three-dimensions, the spreading of particles by diffusion is described by

$$n(x,t) = \frac{N}{\left(2\pi\sigma^2(t)\right)^{3/2}} \exp\left(-r^2/2\sigma^2(t)\right), \tag{7.57}$$

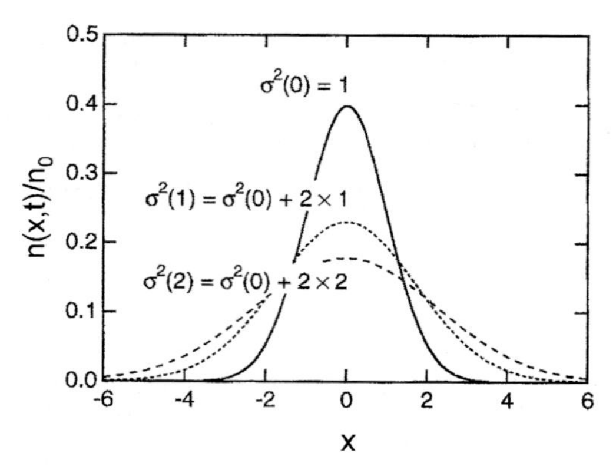

Fig. 7.19. Diffusion in one dimension, with gaussian spreading, with the initial distribution ($\sigma(0)$), one at time t during which σ^2 has tripled (in this particular example) ($\sigma(t)$), and one at time $2t$ ($\sigma(2t)$). (From [360])

with

$$\sigma^2(t) = \sigma^2(0) + 6D_{\text{diff}}t, \tag{7.58}$$

where $r^2 = x^2 + y^2 + z^2$.

Diffusion can be very important in the body over very small distances, on the order of 1–100 µm (~100 µm for oxygen diffusion), but is not very useful over much longer distances. The amount of material that can be transported from one place to another is limited by the lack of directionality of diffusion (Problem 7.26). It leads to an increase in disorder, whereas a functioning organism requires careful control and regulation within characteristic time frames. (The level or disorder is known as *entropy*, which is discussed in more detail in discussions of thermodynamics, statistical mechanics, and in several areas of biophysics.) Smelling object depends on the diffusion of molecules to your nose (Problem 7.30).

Diffusion is also important in flowing systems. This is illustrated in Fig. 7.20 for an artery.

7.4 Pressure and Flow in the Body

Table 7.1 gives characteristic pressures in the body. The blood pressure ranges from ~1–140 mmHg in different vessels and the speed of blood flow in these vessels ranges from ~0.05–50 cm/s. The overall volumetric flow rate is ~5 L/min. The relationship between pressure and flow in the circulatory system is detailed in Chap. 8. The characteristic pressure difference between the lungs and surrounding media is several mmHg and the volumetric flow rate of air into the lungs is ~6 L/min; this is discussed further in Chap. 9.

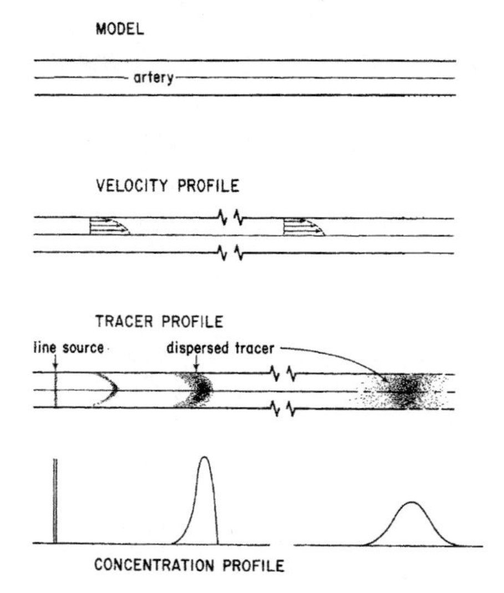

Fig. 7.20. Diffusion of an injected impulse, such as a dye, in an artery, with the shown line source initial distribution. The profile of the injection distorts as it adopts the velocity profile of the flow and it also diffuses. (From [353])

The flow rates in much of the human alimentary (digestive) system are quite slow (Table 7.4). Propulsive movements in this system are due to peristaltic action, with muscular contraction of the contractile ring around the gut sliding food forward, as diagrammed in Fig. 7.21. When there is a large amount of food in it, the gut stretches or distends and through sensors and feedback this stimulates a contractile ring 2–3 cm upstream. Mixing movements in the gut are caused by these peristaltic actions and by local constrictive contractions that occur every few cm in the gut and last for several seconds.

Table 7.4. Approximate flow rates and other properties of the human alimentary system, estimated for a 70 kg male. (Using data from [351])

component	length (cm)	external dimension or width (cm)	internal volume (cm^3)	luminal area (cm^2)	contents passage time	contents speed (cm/s)
mouth and pharynx	8	2–5	∼50	∼80	1–10 s	1–8
esophagus	25	1.3–2.5	∼100	∼200	5–20 s	3–5
stomach	12	8	230–1,000	∼600	2–6 h	∼0.001
small intestine	400	3–6	1,100	∼3,500	3–5 h	0.03
large intestine	∼150	5.0–7.5	300	∼2,000	10–20 h	0.004–0.008
rectum	16–20	2.5–3.8	40	∼100	∼1 h	0.006
total, average, or range	∼600	∼3.5	1,800–2,600	∼6,500	16–32 h	∼0.01

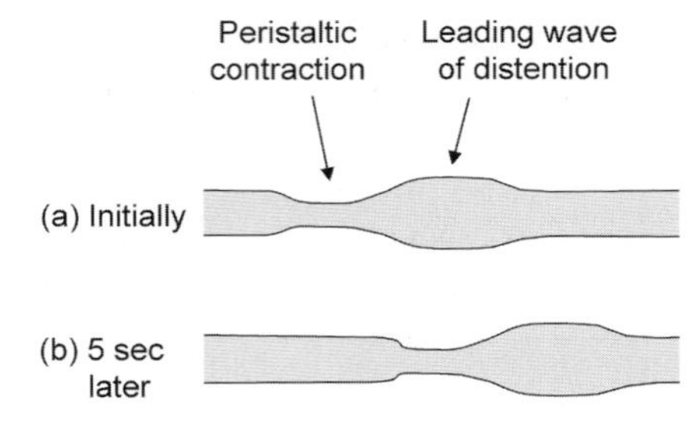

Fig. 7.21. Peristaltic action in the gut. (See Fig. 8.14 for peristaltic assistance in the return of venous blood to the heart). (Based on [356])

The relationship between volume and pressure is important in this digestive system. Pressure (tension) in the walls of the stomach increases during eating. The volume of the stomach of radius R increases as R^3. (This models the stomach as a sphere of volume V and ignores its finite radius with no food contents.) From (7.9), $\sigma = R(\Delta P)/2w = ((\Delta P)/2w)(3V/4\pi)^{1/3}$, so the tension in the stomach walls should increase, much slower, as R. Pressure in the stomach can also increase because of air swallowed during eating, which can lead to burping or belching. Bacterial action produces gas in the gut; at high enough pressure this causes flatulence.

As with the stomach, the pressure within the bladder increases slower than its volume, and this is seen in Fig. 7.22. The pressure rises to 5–10 $\mathrm{cmH_2O}$ when it is filled by 30–50 mL of urine. (The units of $\mathrm{cmH_2O}$ are commonly used in this area, with 1 $\mathrm{cmH_2O}$ = 0.738 mmHg.) Much additional urine can collect, 200–300 mL, with only a small rise in pressure. Above 300–400 mL the pressure increases rapidly. At $\sim$30 $\mathrm{cmH_2O}$ (3 kPa), there is an urge to urinate. Muscle contraction in the bladder (micturition reflexes) momentarily

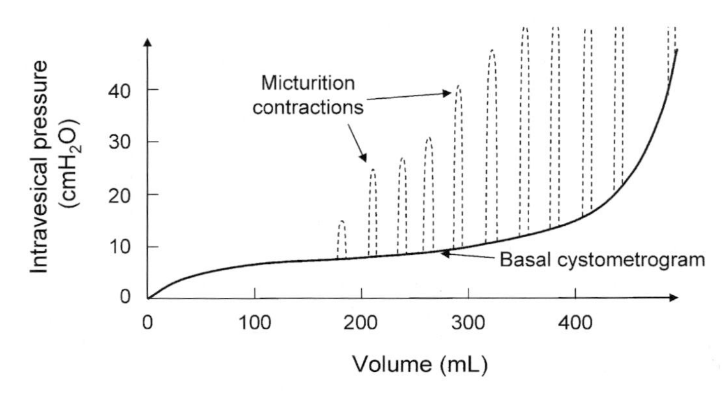

Fig. 7.22. Normal pressure–volume in the human urinary bladder (cystometrogram), also showing acute pressure waves (*dashed spikes*) caused by micturition reflexes. (Based on [356])

increases this pressure to 150 cmH_2O (15 kPa), with the normal voiding pressure being 20–40 cmH_2O (2–4 kPa). The wall tension increases with the volume of the bladder $V_{\mathrm{bladder}}^{1/3}$, as seen from the Law of Laplace assuming a constant wall thickness. Therefore, the sensors to signal the urge to urinate would seem to be in the wall, sensing wall stress, and not sensing the pressure inside the bladder, because the pressure is fairly constant.

7.5 Motion of Humans in Fluids

We have already encountered several examples of humans in fluids. One is the loss of heat by thermal conduction and convection to the surrounding air in Chap. 6. Drag is also important in walking, running, cycling, and so forth, as is clear from how wind increases the metabolic needs during walking and running (Table 6.25). Locomotion in water, i.e., swimming, and potential human flight are examples in which the effects of the fluid are paramount [384].

7.5.1 Swimming

We are fairly buoyant, but not all can float. To float we must have an average density less than that of water (1.0 g/cm^3). (Equivalently, we must have a specific gravity (= density/water density) <1.) Those with relatively more fat (with an endomorph body shape) can float, with face, chest, and toes above the surface, because fat ($\simeq$0.8 g/cm^3) has a density lower than water. Those who are relatively muscular or big-boned (a mesomorph) cannot float because the densities of muscle ($\simeq$1.0 g/cm^3) and bone $\simeq$1.5–2.0 g/cm^3 are, respectively, roughly equal to and larger that of water. People with an average density a bit higher than that of water may be able to float after taking in a deep breath because of the low density of air (0.0012 g/cm^3). Most men and women will float after taking in a deep breath, but most men will sink with just residual air in their lungs (after an normal exhalation, see Chap. 9). Very young and very old people are more likely to float because they have more fat, less muscle mass, and (for old people) lower long bone density. (Measuring body density and fat percentage is described in Problem 1.40.)

When floating (or almost floating) people push water parallel to the surface, in the "backwards" direction, they are propelled forward by the reaction force (Newton's Third Law). In other words, they swim. Because the arm and leg strokes are periodic, the forward propulsion is really periodic in theory, much like the periodic nature of blood rhythmically pumped by the heart. The net forward acceleration of the swimmer is due to the sum of this forward reaction response of the backward pushing of water and drag. (We are ignoring other lift forces [343, 347, 359].) For a person swimming with speed u

$$m_{\mathrm{b}}\frac{\mathrm{d}u}{\mathrm{d}t} = F_{\mathrm{forward\ propulsion}} - F_{\mathrm{drag}} \tag{7.59}$$

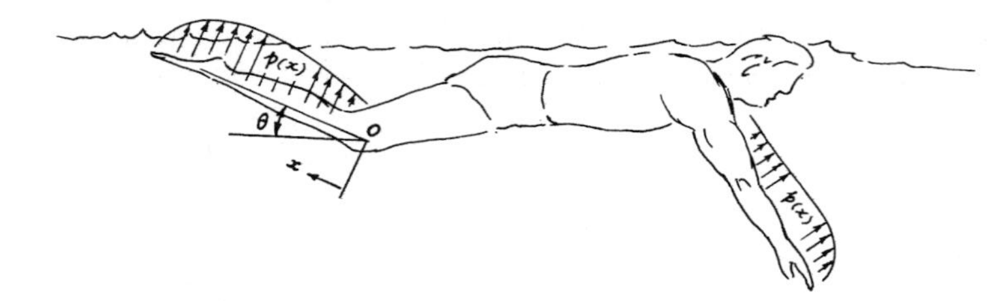

Fig. 7.23. Arm and leg motion during freestyle swimming (the crawl). Also see Problem 7.36. (From [353])

(More rigorously, $-F_{\text{drag}}$ is really $- \mid F_{\text{drag}} \mid$.) As is proved in Problem 7.39, the main source of drag is hydrodynamic, which scales as u^2, and not viscous Stokes-type drag, which scales as u; consequently, $F_{\text{drag,hydrodynamic}} = \frac{1}{2}C_D A \varrho_{\text{water}} u^2$ (7.46). Actually, there are three identifiable sources of drag that scale as u^2: that due to frontal resistance, eddy resistance (due to water not filling in the body's wake and forcing the body to drag along these eddies), and surface drag. The first two types are sometimes collectively called hydrodynamic drag. Assuming now that this propulsion is continuous (as opposed to cyclic), the left side of (7.59) is zero and the steady state speed is

$$u^2 = \frac{2F_{\text{forward propulsion}}}{C_D A \varrho_{\text{water}}}. \tag{7.60}$$

For freestyle swimming (which is technically called the "crawl," Fig. 7.23), the propulsion force during a stroke can be estimated as the momentum gained by the pushed water during the duration of the stroke T_{stroke}, divided by that stroke time. (Remember, $F = ma$ can be expressed in terms of the momentum $p = mv$, as $F = \mathrm{d}p/\mathrm{d}t$ or $\Delta p/\Delta t$.) The momentum of the water is the mass of water displaced, $\varrho_{\text{water}} V_{\text{water}}$, times the final water speed, v_{water} (relative to the swimmer), or $\varrho_{\text{water}} V_{\text{water}} v_{\text{water}}$. Therefore we find

$$F_{\text{forward propulsion, stroke}} \simeq \frac{\varrho_{\text{water}} V_{\text{water}} v_{\text{water}}}{T_{\text{stroke}}}. \tag{7.61}$$

Let us consider forward propulsion due to motion of the hands and arms only and ignore leg motion. Let us also assume that the swimmer's arm is straight during the stroke and rotates about the shoulder in a cylindrical sweeping motion with a radius of the arm length l_{arm} and a width equal to the hand width w_{hand} (which we will say is also roughly equal to the arm width, w_{arm}). Therefore, the volume of displaced water is $V_{\text{water}} = \pi l_{\text{arm}}^2 w_{\text{hand}}/2$. (The factor of two accounts for the half of the cylindrical volume that is in the water.) The speed of the end of the arm is roughly $l_{\text{arm}}/T_{\text{stroke}}$ and near the shoulder it approaches zero. Because the water is moved at the arm speed, the average speed of the water is $v_{\text{water}} \simeq l_{\text{arm}}/2T_{\text{stroke}}$. (Part of the water is pushed downward during an ideal circular motion of the arm and this does not contribute to this forward propulsion; we will ignore this because the stroke motion is not really circular.) Therefore, (7.61) becomes

$$F_{\text{forward propulsion,stroke}} \simeq \frac{\varrho_{\text{water}} \pi l_{\text{arm}}^3 w_{\text{hand}}}{4T_{\text{stroke}}^2}. \tag{7.62}$$

We can estimate the arm length as the sum of the lengths of the upper and lower arms and half the length of the hand (because it is cupped), and so using Table 1.6 we see that $l_{\text{arm}} = 0.386H$, where H is the body height. We estimate that $w_{\text{hand}} = 0.07H$. Excellent swimmers make about 60 strokes a minute, so $T_{\text{stroke}} = 1\,\text{s}$. Using $H = 1.8$ m, we find that $F_{\text{forward propulsion,stroke}} \simeq 27\,\text{N}$. This is what we would expect for an effective force from a muscle with a cross-section of $1.3\,\text{cm}^2$ going into this motion, which seems a bit low. The steady state speed u is obtained from (7.60), using $C_{\text{D}} = 1.0$ and the transverse area $A \sim 0.076\,\text{m}^2$ (from the shoulder width, $0.259H$, times the chest depth, $0.09H$, using $H = 1.8\,\text{m}$). The average speed during a stroke is then $u \simeq 0.8\,\text{m/s}$.

The next stroke, with the other arm, starts when the previous one has stopped. The arm of this previous stroke "recovers" to the forward position above the water line and so it does not provide propulsion in reverse. Therefore, the forward propulsion is really continuous and this average speed seems reasonable. This speed of 0.7 m/s is not that different from typical swimming speeds and is not that far from the speeds of world-class freestyle swimmers. (The average speed for world-record men's freestyle swimming (in 2006) is ~2 m/s, decreasing from 2.3 m/s for 50 m distances to 1.8 m/s for 400 m.) Drag may be less than estimated here – in particular C_{D} and A may be smaller – and more water is likely being pushed per stroke by good swimmers than we estimated here. Remember that we totally ignored propulsion by the kick of the feet and legs and any propulsion by the rest of the body. Also, our analysis has ignored the complication of the initial dive into the pool and of reversing directions at the ends of the pool, etc.

The stroke is not exactly as described here. Actually, the arm does not pull straight in any stroke (freestyle (crawl), butterfly, breaststroke, and backstroke); after starting straight, it bends midway through and then (except for the breaststroke) straightens again for the crawl. This suggests that good swimmers use their hands more like propellers than paddles and that this type of motion can make lift significant, which we have ignored here.

The allometric relation for the swimming speed u of aquatic animals is

$$u \simeq 0.5 m_{\text{b}}^{0.19}, \tag{7.63}$$

where u is in m/s and m_{b} is in kg. This suggests that a 70 kg aquatic animal, such as a common dolphin, would swim at about 1 m/s, which is not far off from typical human performance. (Bottle nose dolphins have the same mass, but swim several times faster.)

What happens if a swimmer stops stroking and just glides? With no forward propulsion, (7.46) and (7.59) combine to give

$$m_\mathrm{b}\frac{\mathrm{d}u}{\mathrm{d}t} = -\frac{1}{2}C_\mathrm{D}A\varrho_\mathrm{water}u^2. \tag{7.64}$$

Bringing the velocity terms to the left gives

$$\frac{\mathrm{d}u}{u^2} = -\frac{C_\mathrm{D}A\varrho_\mathrm{water}}{2m_\mathrm{b}}\mathrm{d}t \tag{7.65}$$

and integrating from the initial speed u_i at $t = 0$ to the speed at time t gives

$$-\frac{1}{u(t)} + \frac{1}{u_\mathrm{i}} = -\frac{C_\mathrm{D}A\varrho_\mathrm{water}}{2m_\mathrm{b}}t. \tag{7.66}$$

Therefore the swimmer's speed approaches zero as

$$u(t) = \frac{u_\mathrm{i}}{1 + \frac{C_\mathrm{D}A\varrho_\mathrm{water}u_\mathrm{i}}{2m_\mathrm{b}}t} \tag{7.67}$$

with a characteristic time of say $18m_\mathrm{b}/C_\mathrm{D}A\varrho_\mathrm{water}u_\mathrm{i}$, at which time $u = 0.1u_\mathrm{i}$ (see Appendix C).

Because $u = \mathrm{d}x/\mathrm{d}t$, we find

$$\mathrm{d}x = \frac{u_\mathrm{i}}{1 + \frac{C_\mathrm{D}A\varrho_\mathrm{water}u_\mathrm{i}}{2m_\mathrm{b}}t}\mathrm{d}t \tag{7.68}$$

Integrating from position $x = 0$ at $t = 0$ gives

$$x(t) = \frac{2m_\mathrm{b}}{C_\mathrm{D}A\varrho_\mathrm{water}}\ln\left(1 + \frac{C_\mathrm{D}A\varrho_\mathrm{water}u_\mathrm{i}}{2m_\mathrm{b}}t\right). \tag{7.69}$$

7.5.2 Human Flight

Why cannot we fly? (That is, why cannot we fly without the assistance of a jet or helicopter, or propulsion devices on our backs?) The answer is easy. We cannot generate enough vertical force to counter our weight to enable us to hover or fly. In principle, we could do this by pushing air down fast enough or by generating a vertical force by aerodynamic lift – which could be possible if we could propel ourselves forward fast enough.

What happens if we try to fly by pushing air down by flapping our arms up and down? The volume of air we could push down per arm flap is the arm area, which is length × width, times the distance pushed, which is approximately the arm length. This is roughly $0.7\,\mathrm{m} \times 0.1\,\mathrm{m} \times 0.7\,\mathrm{m}$ per arm or $\sim 0.1\,\mathrm{m}^3$ for both arms. The mass density of air is $10^{-3}\,\mathrm{g/cm}^3 = 1\,\mathrm{kg/m}^3$, so the mass displaced per flap is $\sim 0.1\,\mathrm{kg}$. If the ends of our arms attained a speed of

80 mph $\simeq$ 40 m/s (which is the speed of a fair major league fastball, and is clearly an overestimate), our average arm speed would be about 20 m/s. If we flapped our arms 3 times a second (which is also faster than expected), the change in momentum in the moved air per unit time would be (0.1 kg)(20 m/s)(3/s) = 6 N (assuming no air is moved when our arms return to their initial positions at the end of each flap). This is much less than the weight of a 70 kg person, which is 700 N. Let us say we wear lightweight wings that would increase the effective flapping area to 2 m^2 (1 m^2 per wing) and the volume of the air we would move increases to 2 m^3. We would then generate an upward force of 120 N from this downward draft, and so even with our wildly high estimates of wing speed and flapping rate, we could not even approach developing enough vertical force to counter gravity and fly (or at least hover). (Because water has a density that is 1,000× that of air, we can easily keep ourselves afloat by pushing water down, and this is also assisted by buoyancy.) Such hovering, by the reaction force to the down draft in air, is more difficult than flying because there is no upward lift. Perhaps we could flap and propel ourselves forward and develop some lift.

Could we at least "takeoff" after running fast with our artificial wings in place? Assume that a person accelerated to the world record speed of about 10 m/s and suddenly spread his or her 2 m^2 area wings. Using (7.50) under optimal conditions, we find $F_{\text{lift}} = \frac{1}{2}C_{\text{lift}}S\rho u^2 = (0.5)(1.2)(2\,\text{m}^2)(1\,\text{kg/m}^3)(100\,\text{m}^2/\text{s}^2)$, or 120 N of lift, which is still not enough. (Of course, even if the lift were enough and the person became airborne, forward deceleration due to drag would lead to a landing (or a crash).)

Clearly, any combination of wing flapping, for forward and some upward propulsion, and wing gliding for lift will also not lead to flight. The old saying, "If man (or woman) were meant to fly, he (or she) would have wings." is not true, because we could not fly even if we had wings. Of course people can hang glide with artificial wings; such gliding involves lift, drag, wind, and gravity.

Human-powered flight has indeed been demonstrated in the bicycle-powered aircraft built by the Paul MacCready team and cycled/flown by Bryan Allen, a champion bicyclist. In this aircraft the pedaling pilot propelled the propeller at the rear of the craft, which is connected to the cycle by a series of gears. Consequently thrust is created in this craft, which was optimized for lift, with minimal drag and weight. The "Gossamer Condor" flew for 7 min, 2.7 s in a closed course, and then on June 12, 1979 the "Gossamer Albatross" (with 30 m wingspan and 30 kg mass without the pilot) flew the first completely human-powered flight across the English Channel. It covered 35.6 km in 2 h 49 min, and thereby won the Kremer Prize established in 1959. This world-class cyclist provided 125 W of mechanical power, flew very close to the surface to take advantage of the "ground effect" – which is a temperature inversion near the surface – and was completely exhausted at the end of the flight.

7.6 Summary

The Law of Laplace, the equation of continuity and Bernoulli's equation for nonviscous flow, and Poiseuille's Law of viscous flow can be used to model the flows of fluids in the body, such as blood and air – which are described in Chaps. 8 and 9, and the movement of the body in fluids, such as swimming and flight, which is described in this chapter. The physics of pressure in fluids and diffusion are also used in these models.

Problems

Basic Fluidics and Pressure

7.1. Your blood pressure is measured with a sphygmomanometer, however with your upper arm pointed upward instead of downward. If your blood pressure is really 120 mmHg/80 mmHg, approximately what pressure would be measured?

7.2. The water level in a 4 m wide and 20 m long pool rises 0.75 mm when a person enters it and floats. What is the mass of that person?

7.3. You want to measure the volume of your whole arm by sticking it in an upright, long cylindrical tube with internal diameter of 15 cm, which is partially filled with water. The water level rises by 12.7 cm when a 50 kg female makes this measurement? What are the mass, weight, and volume of her arm? (See Chap. 1.)

7.4. Who is more buoyant and consequently floats higher: a large-boned, heavy muscled person with little body fat or a small-boned, lightly muscled person with more body fat?

7.5. Will retaining water affect a person's ability to float?

7.6. A 50 kg woman has a density of 1.01 g/cm^3 after normal exhalation. Does she float? Will she float after she inhales 2 L of air?

7.7. A 70 kg man with a density of 1.03 g/cm^3 ages. He gains 5 kg of fat. Will he float?

7.8. Will a person with an ectomorph shape float?

7.9. Why can all people float in the Dead Sea? (It has a specific gravity of 1.2–1.3. We have been assuming floating in water with no salt. Ocean water has a density of 1.027 g/cm^3.)

7.10. Three 50 kg women are airborne in a balloon filled with He. What is the minimum diameter of the balloon? (What assumptions are you making about the mass of the basket in which they are riding and the balloon itself?)

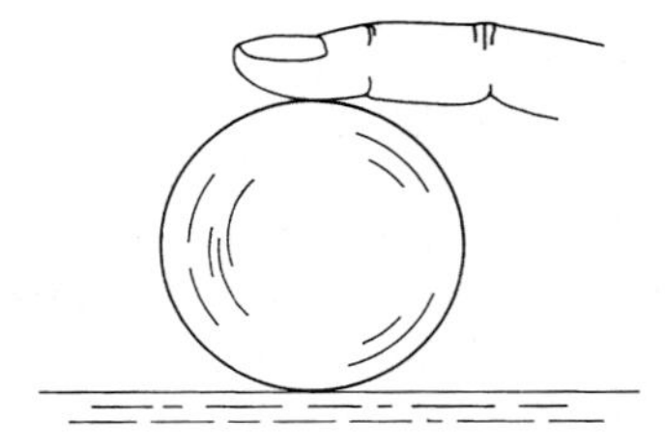

Fig. 7.24. Palpation of a blood vessel. (From [352].) For Problem 7.14

7.11. The gauge pressure inside a cylindrical tube is 100 mmHg and its radius is 1 mm, what is the tension in the tube wall at equilibrium (in SI units)?

7.12. (advanced problem) Derive the Law of Laplace for a sphere (7.9) by careful integration of the normal force on a hemisphere, in a manner analogous to the integration in (7.7) for a half-cylinder.

7.13. Over a large range of volumes, the pressure in the bladder is at a fairly constant value near 8 mmHg. If the thickness of the bladder is 5 mm, show that the wall tension is $\sigma = 600\,\mathrm{Pa/cm}\ V_{\mathrm{bladder}}^{1/3}$ where the bladder volume is in cm^3.

7.14. The internal pressure of an elastic vessel, such as an artery, vein, eyeball, aneurysm, or balloon, can be estimated by pushing down on it with your finger; this method is called *palpation* (Fig. 7.24):
(a) Show that the pressure felt by the finger is affected by the tension in the vessel wall.
(b) Show that the pressure you feel equals the pressure internal to the vessel when you push down on it so that the vessel wall is flat.

7.15. Assuming no viscosity and no changes in height, determine how the flow speed in a vessel changes if its diameter decreases by a factor of 4.

Viscous Flow

7.16. Compare the SI units of dynamic viscosity, η in (7.23), with those of the viscosity damping constant of the dashpot, c in (4.48).

7.17. One wants to use oil in car engines so the oil viscosity is a specific, optimized value – especially when the engine is started cold. Usually a heavyweight oil is used in very hot weather, such as SAE 50, and a lightweight oil in very cold weather, such as SAE 10. Using Table 7.3, estimate the viscosity needed at moderate temperature. Also estimate how the motor oil viscosity changes with temperature. (Nowadays, multiviscosity oils, such as SAE 10W/40 are used, which are suitable over a wide range of temperatures.)

7.18. Viscous flow with flow rate Q in a big tube of diameter D and length L, subdivides into N identical small tubes of length L with equal flow rates:
(a) What is the flow rate in each small tube?
(b) You are told that the pressure drops across the big tube and across the small tubes are the same (and both equal to ΔP). Find the diameter of the small tubes and determine if this is possible.
(c) If instead, the diameters and lengths of the small tubes are $\alpha\times$ and $\beta\times$ that of the big tube, what is the resistance across each small tube and across the whole small tube system in terms of the resistance across the big tube?

7.19. We are very sensitive to even small changes in core body temperature. Let us examine what happens when the viscosity of blood changes because of such temperature changes. It is known that the dynamic viscosity of whole blood decreases by 30% when temperature increases from 25°C to 37°C. What is the increase in systolic blood pressure, from its normal value of 120 mmHg, needed to pump blood throughout the body at the same rate if the core body temperature decreased to 25°C, with everything else being the same? (This temperature change will affect the body in many other ways even more dramatically; see Chap. 13.)

7.20. (a) How much force F (in N and lb) must be applied to a plunger to inject $1.0 \times 10^{-6}\,\mathrm{m}^3$ of the solution in 3.0 s with a hypodermic syringe? Apply Poiseuille's Law for the pressure drop across the needle, as in Fig. 7.25 [348]. The needle is injected into a vein with a (gauge) pressure of 14 mmHg (1,900 Pa). Assume the plunger has an area of $8.0 \times 10^{-5}\,\mathrm{m}^2$ and the syringe is filled with a solution with viscosity of 1.5×10^{-3} Pa-s. The needle has an internal radius of 4.0×10^{-4} m and a length of 0.025 m. Remember that you want to apply a (gauge) pressure in excess of the venous pressure to achieve the desired flow rate Q.
(b) Why are such injections performed intravenously and not intra-arterially?

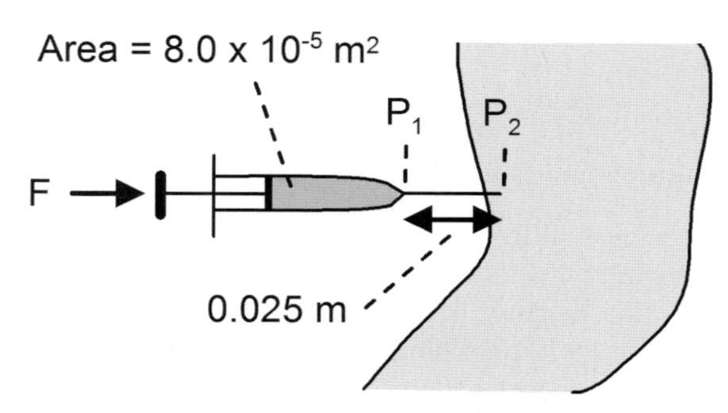

Fig. 7.25. Intravenous injection by a hypodermic syringe. (Based on [348].) For Problem 7.20

7.21. Calculate the Reynolds number for each component of the alimentary system. Assume the coefficient of viscosity is 1 N-m/s to the stomach and 10 N-m/s after the stomach. Is the flow streamline or turbulent?

7.22. (a) For Newtonian flow, calculate the shear stress on the wall of a tube of radius R, for an average fluid speed u and fluid viscosity η.
(b) Estimate this (in SI units) for a typical human artery.

7.23. (advanced problem) Show that the strain rate dv/dy used in flow is related to the time rate of change of strain $d\epsilon/dt$. (Hint: Express strain as the partial derivative of a deformation u, $\epsilon = \partial u/\partial y$ and speed as $v = \partial u/\partial t$. Then evaluate $\partial\epsilon/\partial t$, and switch the order of the y and t derivatives.)

Diffusion

7.24. Important molecules are formed in the middle of a 2 μm-diameter cell. How long does it take for them to diffuse throughout the cell? (Assume the cell contents are liquids and that the diffusion coefficient $D_{\text{diff}} = 10^{-5}\,\text{cm}^2/\text{s}$.) Is this fast enough to achieve normal metabolic activity rates?

7.25. In one-dimension, estimate the characteristic distances for diffusion in 1 s in a gas, liquid, and solid.

7.26. A 1 mm^3 volume of biological material must be transported 2 cm away to another 1 mm^3 region:
(a) If it flows in a vessel at a speed of 10 cm/s, how fast does it get there and what fraction of it arrives there?
(b) If it diffuses in a liquid with $D_{\text{diff}} = 10^{-5}\,\text{cm}^2/\text{s}$, approximately when will the maximum amount of it arrive and approximately what fraction of it will arrive?
(c) Which mode of transport is preferred and why?

7.27. Refer to Fig. 7.19. If σ and the abscissa are in cm and t is in s, what is D_{diff}?

7.28. Use substitution to confirm that (7.55) and (7.56) are the solution to the one-dimensional diffusion equation, (7.53).

7.29. (advanced problem) Use substitution to confirm that (7.57) and (7.58) are the solution to the three-dimensional diffusion equation

$$D_{\text{diff}} \frac{1}{r^2} \frac{\partial}{\partial r} \left(r^2 \frac{\partial n}{\partial r} \right) = \frac{\partial n}{\partial t}. \tag{7.70}$$

7.30. (a) You can detect 4×10^8 molecules of ethyl mercaptan (which causes the rotten fish smell) per cm^3, which corresponds to one molecule per 10^{11} molecules in air (because the air density is $5 \times 10^{19}/\text{cm}^3$). If 1 mm^3 of this

liquid is released 10 m away, how long will it take to notice this release? (Ethyl mercaptan, C_2H_5SH has 62.1 g/mole and is a liquid with a density of 1.01 g/cm^3. It has an odor resembling that of rotten eggs, and is added to natural gas and propane to give those normally odorless fuels a distinctive smell.)
(b) If your dog's nose is a thousand times more sensitive, when will she or he smell it?

Swimming, Flying, and Drag Forces

7.31. Repeat the analysis that determines the speed of a swimmer, but now assume that the effective force of 2 in diameter muscles is providing 405 N continuously. (Why is this force reasonable?) Does your answer make sense? Why?

7.32. Repeat the analysis of the speed of a swimmer, but now assume that Stokes friction is the only dominant drag force. Does your answer make sense? Why?

7.33. Go through all the steps in determining the position during gliding in swimming, from (7.68) to (7.69).

7.34. Using the parameters in the text, estimate the characteristic time needed for a world class freestyle swimmer who stops stroking and glides to slow down. Also estimate the distance she travels in that time.

7.35. Repeat the analysis of gliding, (7.64)–(7.69), assuming only Stokes drag.

7.36. In Fig. 7.23, the lower leg of a swimmer is hinged at the knee (at $x = 0$) and is acted on by forces that are normal to its axis with force per unit length of $p(x)$. Show that the work done by the leg, of length L, as is rotates by $\mathrm{d}\theta$ is [353]

$$\mathrm{d}W = \left(\int_0^L p(x) x \mathrm{d}x \right) \mathrm{d}\theta. \tag{7.71}$$

7.37. Could people fly on another planet using artificial wings? How would g and the mass density of the atmosphere ρ have to change? (Does a change in g imply the same or an oppositely signed change in ρ?) Ignore the impact of spacesuits, differences in temperature, changes in metabolism, muscle atrophy, and so on.

7.38. What is the Reynolds number of a piece of matter 1 μm in diameter, such as a cell in water or particulate in blood? Assume a density of 1 g/cm^3, a speed of 4 mm/s, and the viscosity of blood. What type of drag dominates?

7.39. (a) Is Stokes friction or hydrodynamic drag dominant for people walking and running in air?
(b) Which is dominant for people swimming in water?
(Make sure you calculate the Reynolds numbers in each case.)

7.40. (a) Estimate the hydrodynamic drag force on a very fast runner.
(b) How much power is lost to drag?
(c) How does this compare to the metabolic power needed for running?

7.41. Speed skaters often adopt a position with a nearly horizontal trunk and downhill skiers adopt the "egg" position with a hunched-down body and skis pointed backward when they are not maneuvering. Why?

7.42. A person without a parachute is dropped from a plane at an altitude of 1,000 m. Determine the "terminal" speed of the person by equating the forces of gravity and drag. (Which drag limit is appropriate? Is there enough time for a constant final speed to be attained?)

7.43. (a) A person with a parachute is dropped from a plane at an altitude of 1,000 m. Determine the final steady state speed of the person by equating the forces of gravity and drag. (Which drag limit is appropriate? Is there enough time for a constant final speed to be attained?) Assume the person has a mass of 70 kg and the parachute has a negligible mass and is 7 m across when it is open.
(b) What is the minimum height above ground that the parachute should be opened so the person lands with a speed no greater than 1.5 m/s? Assume it takes 2.5 s for the parachute to deploy fully.

8

Cardiovascular System

There are three components of the cardiovascular system. (a) Blood is the vehicle for transport. It transports fuel from the digested food to the cells, transports oxygen from the air in the lungs so it can combine with fuel to release energy, and it disposes of waste products – such as carbon dioxide from the fuel engine and other metabolic wastes. (b) The circulatory system is the distribution system, and consists of a series of branched blood vessels. (c) The heart is the four-chambered pump composed mostly of cardiac muscle that enables this circulatory flow. General descriptions of the cardiovascular system can be found in [368, 369, 372, 373, 376, 378, 384, 385, 388, 390, 395, 396, 402, 410, 417].

8.1 Overview of the Circulatory System and Cardiac Cycle

8.1.1 Circulation

Blood flow from the heart branches into two separate systems (Fig. 8.1). In the *pulmonary circulation* system, the right side of the heart pumps oxygen-poor ("blue") blood to the lungs to be oxygenated; oxygen-rich ("red") blood then returns to the left side of the heart. In the *systemic circulation* system, the left side of the heart pumps this oxygen-rich ("red") blood to the rest of the body where it is used; oxygen-poor ("blue") blood then returns to the right side of the heart. This occurs in a system of arteries that conducts the blood from the heart to the lungs and other organs and components, and a system of veins that returns the blood to the heart.

In the pulmonary system (Fig. 8.1), blood enters the *right atrium* (RA) of the heart (Fig. 8.2) through the *inferior and superior vena cava(e)* (vee'-na cae'vuh). The blood passes through the *right atrioventricular (or tricuspid) valve* to enter the *right ventricle* (RV). Blood is first pumped through the *pulmonary semilunar valve* to the pulmonary arteries, which branch out into a series of more minor arteries and arterioles, and then into capillaries in the

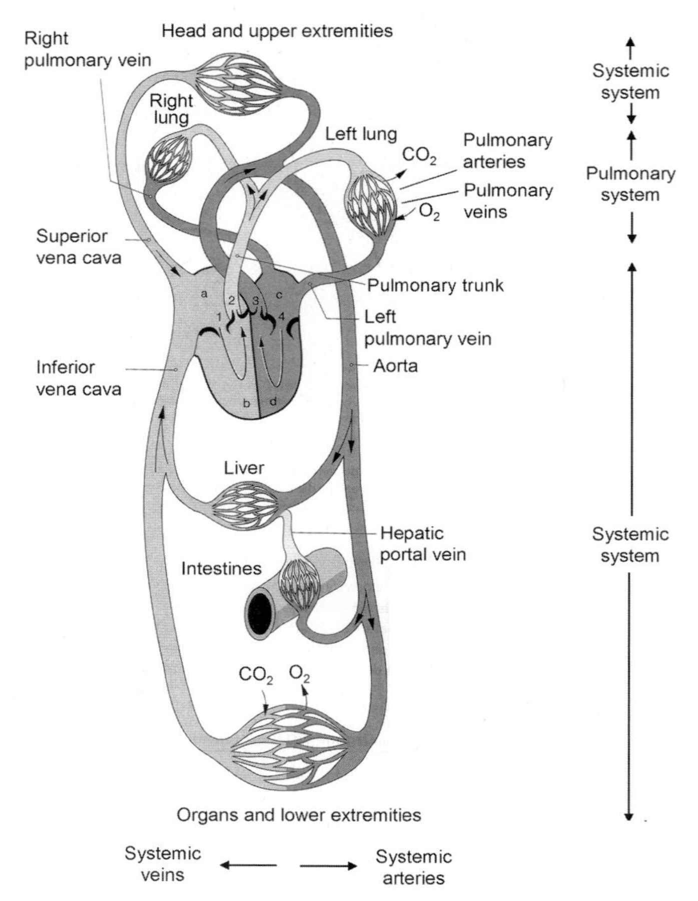

Fig. 8.1. Blood circulation system, and labeled within the heart: the (a) right atrium, (b) right ventricle, (c) left atrium, (d) left ventricle, (1) right atrioventricular (tricuspid) valve, (2) pulmonary semilunar valve, (3) aortic semilunar valve, (4) left atrioventricular (bicuspid, mitral) valve. (From [416])

lungs. These pulmonary capillaries combine into venules (veen'-yools), then into more major veins, and finally into the pulmonary veins.

In the systemic system (Fig. 8.1), blood enters the *left atrium* (LA) of the heart through the pulmonary veins. The blood passes through the *left atrioventricular (or bicuspid or mitral) valve* to enter the *left ventricle* (LV). Blood is pumped through the *aortic semilunar valve* to the *aorta*, which first branches out into a series of major and then minor arteries (with smaller diameters, the arterioles), and finally into a series of capillaries in the systems where gas exchange and diffusion occur. These systemic capillaries combine into venules, then more major veins, and finally into the superior (from above the heart) and inferior (from below the heart) vena cavae.

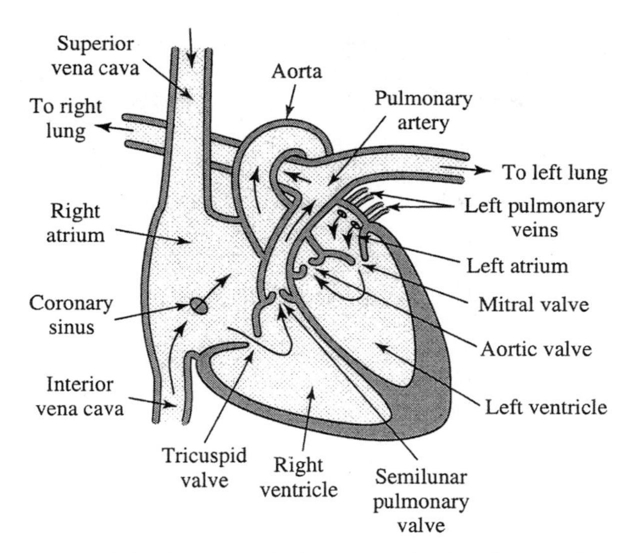

Fig. 8.2. Diagram of the heart, with its principle chambers, valves, and vessels. (From [367])

(A useful mnemonic for the flow of blood in the heart comes from knowing that the author once lived on Rahlves Drive in Castro Valley, California – a town approximately 20 miles south of Berkeley. The whole heart cycle starts with blood flowing into the right atrium (RA) and then getting oxygenated in the lungs, returning to the heart (H), and then continuing with the blood leaving the left ventricle (LV) and exiting (E) for the systems (S). Put together this spells RAHLVES. The most important concept here is that deoxygenated blood enters the heart through the right atrium (RA) and eventually oxygenated blood leaves through the left ventricle (LV) of the heart to be used by the body for metabolism. For some, it may be easier to remember that an American Daron Rahlves was the winner of the super-G downhill skiing competition in the 2001 World Championships.)

The systemic and pulmonary systems have similarities and differences. They have the same volumetric flow rate Q. (If they were not equal, blood would have to pile up somewhere.) In the systemic system the blood disposes of oxygen and receives carbon dioxide, while in the pulmonary system the blood disposes of carbon dioxide and receives oxygen. Table 8.1 shows that

Table 8.1. Normal resting values of blood pressure, with system volumes

	P (mmHg)	V (L)
systemic arteries	100	1.0
systemic veins	2	3.5
pulmonary arteries	15	0.1
pulmonary veins	5	0.4

the systemic system has higher pressures (in the arteries) and larger volumes than the pulmonary system (even with the same Q). This difference in pressure makes sense because the blood vessels need to be longer to get to more distant regions in the body in the systemic system. The left heart (LA + LV) is bigger (and is a larger pump) than the right heart (RA + RV) because of this need to generate higher pressure for systemic circulation. The heart walls consist mostly of the thick middle muscle layer, the *myocardium*, which is lined internally by a thin layer of tissue, the *endocardium*, and externally by a membrane, the *epicardium*. The two sides of the heart are separated by a wall called a *septum*. The difference in volume is due to the longer distance of travel and the much higher number of systems that receive blood in the systemic system. Table 8.1 also shows that arteries have higher pressure than the corresponding veins, whereas the veins have larger volumes. The total volume of blood is $\simeq$5 L.

For a person at rest, 12% of the blood is in the heart chambers, 2% in the aorta, 8% in the arteries, 1% in the arterioles, 5% in the capillaries, 50% in the systemic veins, and 18% in the pulmonary circulation.

Major arteries and veins are shown in Figs. 8.3 and 8.4. Tables 8.2 and 8.3 provide a very approximate quantification of the vessels in the circulatory system.

8.1.2 Cardiac Cycle

There is a highly controlled timing cycle in well-functioning hearts, the cardiac cycle, which lasts a time τ (Fig. 8.5). In the first stage of *diastole* (die-as'-toe-lee), the veins fill up both the right and left atria, while the right and left ventricles are relaxed. In the second stage, the cardiac muscle (myocardium) of the right and left atria contract and pump blood through the atrioventricular valves, into the right and left ventricles, respectively, at the same time $t = 0$. (This is actually a gross simplification of ventricular filling, because $\sim$75% of this blood flows into the ventricles from the atria before atrial contraction.) In the first step of *systole* (sis'-toe-lee, which has the same cadence as Sicily), both ventricles contract (isovolumetrically) at the same time Δ. In the second stage, they eject blood through the respective semilunar valves: the right ventricle into the pulmonary arteries and the left ventricle into the aorta. The *systolic* (sis-stah'-lic) *blood pressure* occurs in this second stage of systole, while the *diastolic* (die-uh-stah'-lic) *pressure* is that during diastole. (One way to measure the flow of blood ejected by the left ventricle is *ballistocardiography*, which is described in Problem 8.47. Another method is Doppler ultrasonography echocardiography, which is used more often clinically; it is described in Problem 10.23.)

The right and left hearts must work at exactly the same time to keep the flow rate Q the same in both systems. There is a timing mechanism in place to do this and to set the contraction times 0 and Δ for one beat, followed by τ and $\tau + \Delta$ for the next, 2τ and $2\tau + \Delta$ for the next,

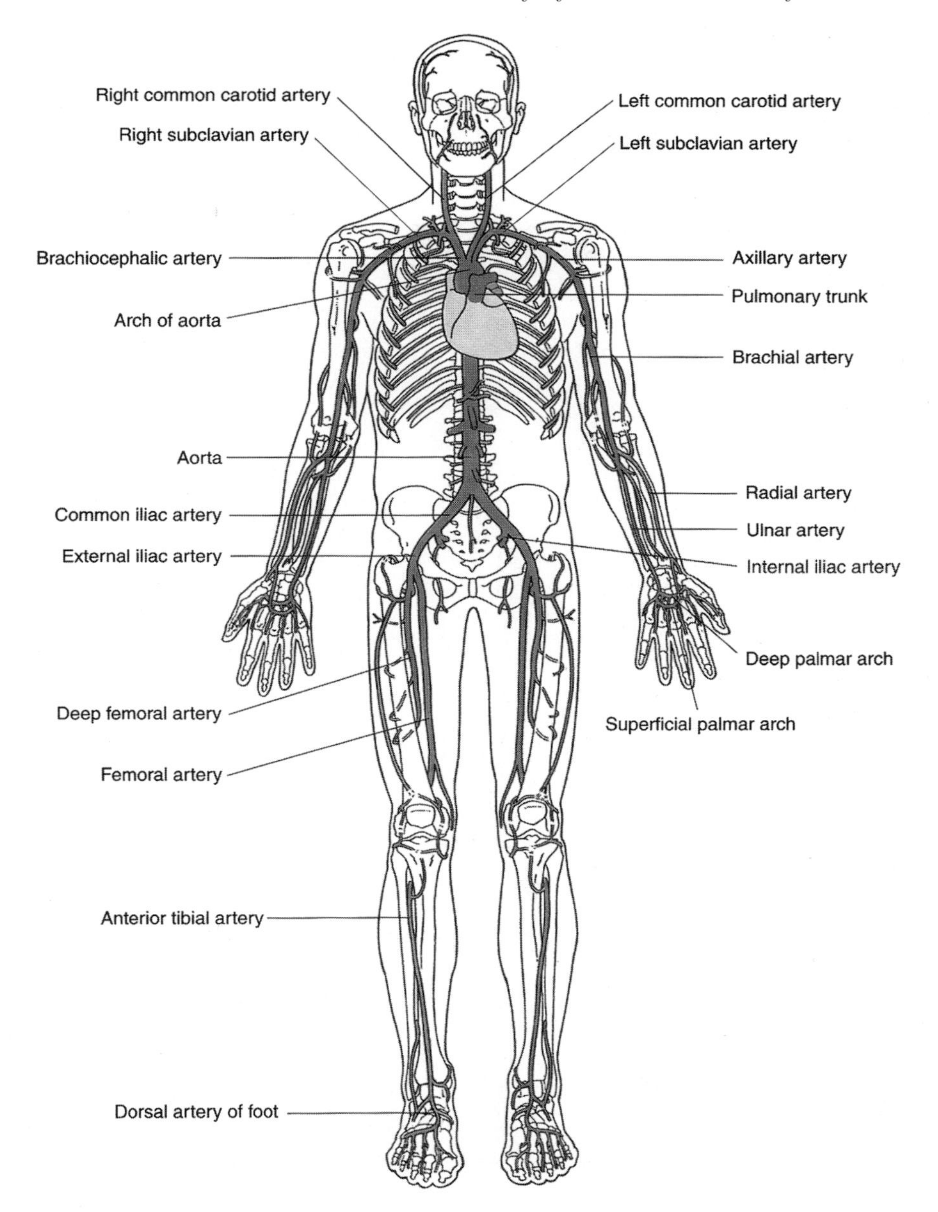

Fig. 8.3. Major arteries in the body. Arteries carry blood away from the heart in the systemic and pulmonary system. Many come in pairs, such as the right and left radial arteries. (From [408]. Used with permission)

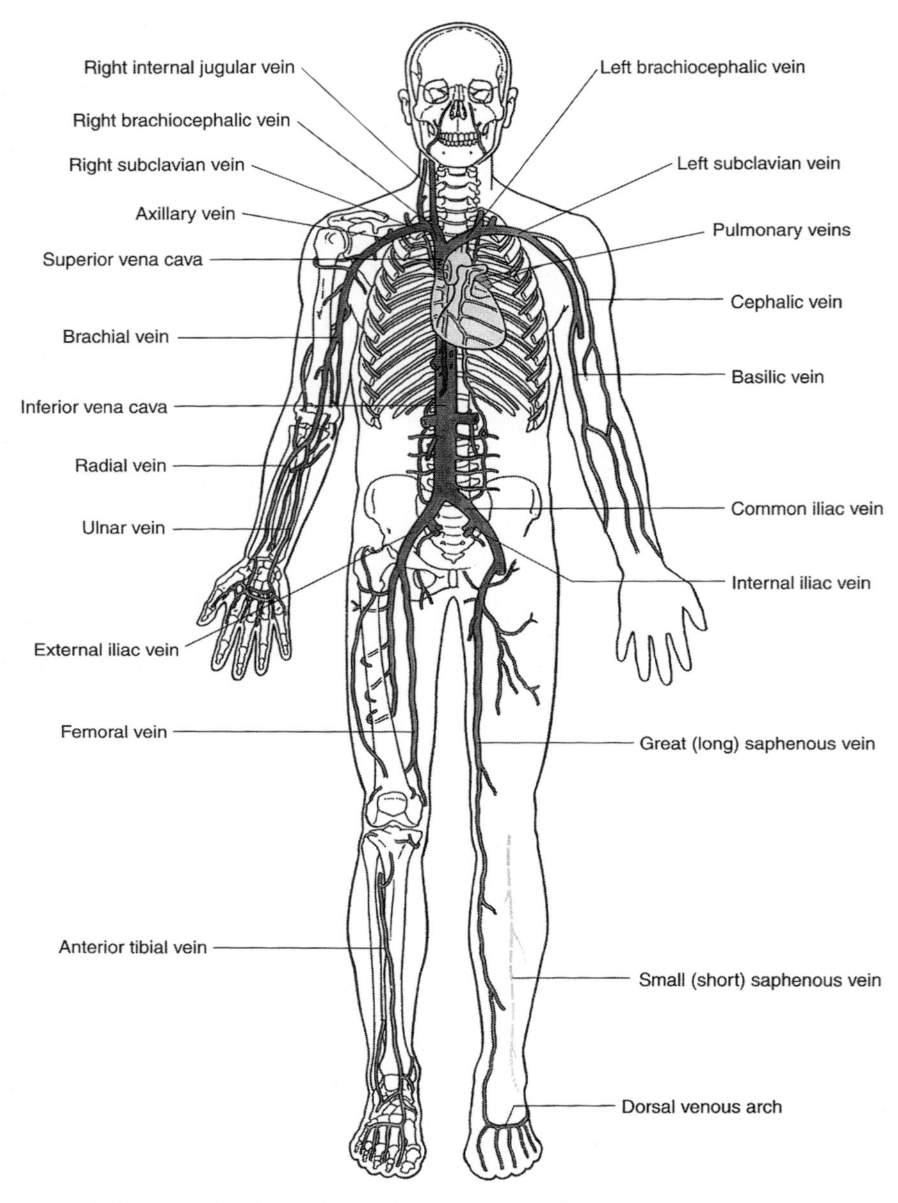

Fig. 8.4. Major veins in the body. Only the superficial veins are shown in the left limbs and only the deep veins are shown in the right limbs. Veins carry blood back to the heart in the systemic and pulmonary system. Many come in pairs, such as the right and left radial veins. (From [408]. Used with permission)

Table 8.2. Approximate quantification of individual vessels in the human circulatory system. (Using data from [382])

vessel	diameter (mm)	length (mm)	wall thickness (μm)	pressure (mmHg)
aorta	25.0	400	1,500	100
large arteries	6.5	200	1,000	100
main artery branches	2.4	100	800	95
terminal artery branches	1.2	10	125	90
arterioles	0.1	2	20	60
capillaries	0.008	1	1	30
venules	0.15	2	2	20
terminal venules	1.5	10	40	15
main venous branches	5.0	100	500	15
large veins	14.0	200	800	10
vena cava[a]	30.0	400	1,200	5
heart chambers	–	–	–	120

This is for a 30-yr-old male, with mass 70 kg and 5.4 L blood volume.
[a]There are really two vena cavae.

and so on. There is a heart pacemaker at the sinoatrial or sinus node (see the conducting system in Fig. 8.6), which sends an electrical signal to the atrial cardiac muscle of both atria for simultaneous atrial contraction. This electrical signal then travels to the atrioventricular or AV node,

Table 8.3. Approximate quantification of total vessel systems in the human circulatory system. (Using data from [382])

vessel	number	total length (mm)	total surface area (mm^2)	total blood volume (mm^3)
aorta	1	400	31,400	200,000
large arteries	40	8,000	163,000	260,000
main artery branches	500	50,000	377,000	220,000
terminal artery branches	11,000	110,000	415,000	120,000
arterioles	4,500,000	9,000,000	2,800,000	70,000
capillaries	19,000,000,000	19,000,000,000	298,000,000	375,000
venules	10,000,000	20,000,000	9,400,000	355,000
terminal venules	11,000	110,000	518,000	190,000
main venous branches	500	50,000	785,000	1,590,000
large veins	40	8,000	352,000	1,290,000
vena cava[a]	1[a]	400	37,700	280,000
heart chambers	–			450,000
Total		∼19,000 km	312,900,000	5,400,000

This is for a 30-yr-old male, with mass 70 kg and 5.4 L blood volume.
[a]There are really two vena cavae.

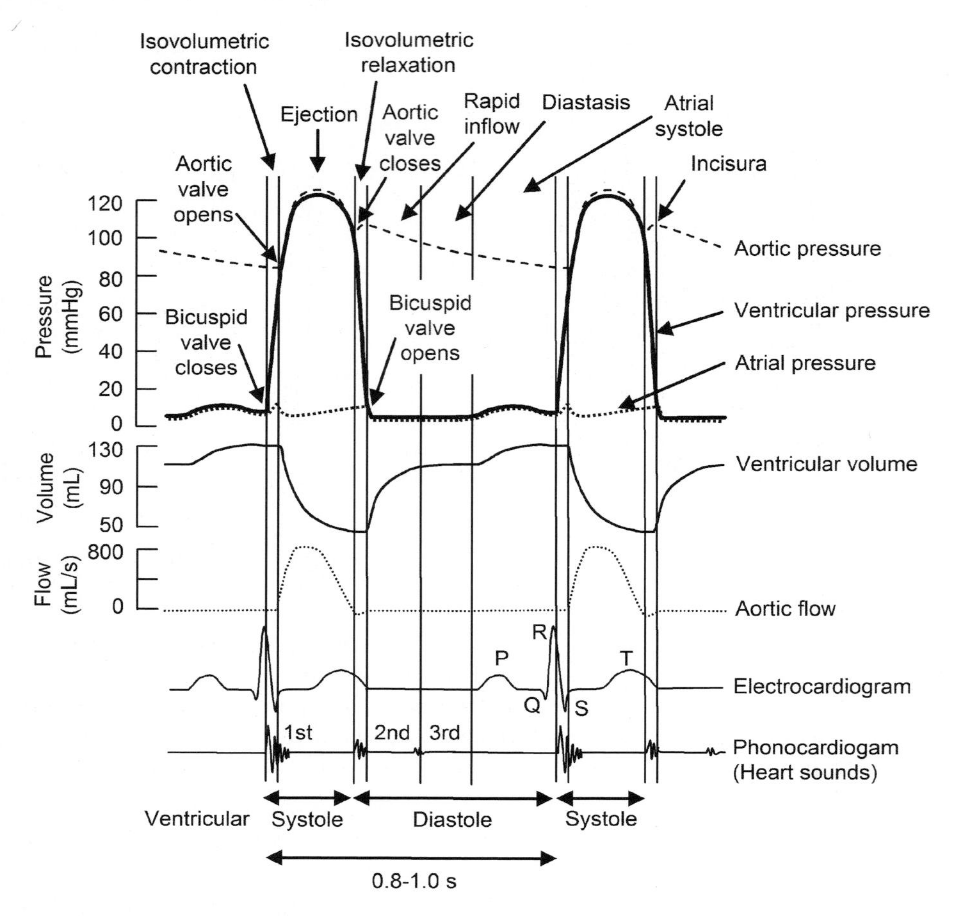

Fig. 8.5. The left ventricular and atrial pressures are plotted along with the left ventricular volume, aortic pressure and flow rate, the electrocardiogram and the phonocardiogram (which is the signal from heart sounds) in this Wiggers diagram. The opening and closing times of the aortic semilunar and bicuspid (mitral) valves are also shown. (Based on [390], [414], and [417])

is delayed there for a time Δ, and then the node sends a signal to the ventricular cardiac muscle of both ventricles for simultaneous ventricular contraction.

The electrocardiogram (EKG or ECG) is a measurement of these electrical signals, and their timing, as measured by probes on the body [379, 386, 401]. Figure 8.6 shows the EKG during one $\sim$1 s long heart beat (also see Fig. 12.28). The P wave is due to atrial depolarization (which is atrial contraction). The QRS complex is due to ventricular depolarization (contraction). The T wave is due to ventricular repolarization (relaxation). The atrial repolarization (relaxation) signal is masked by the larger QRS complex. Depolarization and

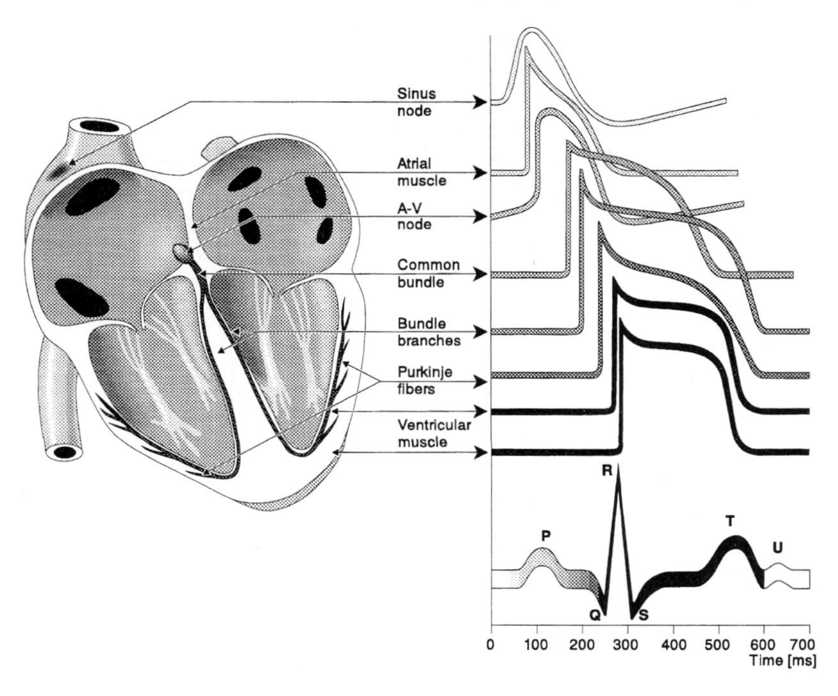

Fig. 8.6. Conducting system of heart with the sinoatrial/sinus and atriventricular/AV nodes, along with the electrical waveforms of the activity of each – including the EKG on the bottom. Also see Fig. 12.28. (From [404])

repolarization, and the electrical properties of the heart and the use of the EKG are discussed more in Chap. 12.

There are several cardiac arrhythmias (i.e., timing irregularities) of varying degrees of concern that affect this cardiac timing mechanism. In an *atrioventricular block*, there is injury to the atrioventricular (AV) fibers from the AV node to the ventricle or to the AV node itself (Fig. 8.6). In an incomplete AV block, the conduction time through the AV junction increases from the normal 0.16 s to 0.25–0.50 s and there are dropped ventricular beats; these sometimes lead to 2:1, 3:2, or 3:1 rhythms of atrial to ventricular beats. In a complete AV block, a person may faint until ventricular beats develop (with 40/min, compared to 100 beats/min in the atria). There can also be *premature contractions* of the atria or ventricles. In *paroxysmal tachycardia*, there are sudden increases in the heart rate, say from 95 to 150 beats/min in the atria or ventricles, which can cause serious ventricular (not atrial) damage. *Ventricular fibrillation* is the most serious arrhythmia and is fatal if not treated immediately. It can be caused by 60-cycle AC. There is uncoordinated muscle contraction of the ventricles, and so parts of them contract while other parts relax; this leads to little or no pumping of blood. Unconsciousness occurs in 4–5 s and the death of tissues begins in a few minutes. *Atrial fibrillation* involves similar uncoordinated muscle contraction, but it is less serious because most blood flows passively from the atria to ventricles. Blood flow decreases by only ∼20–30%. Resuscitation after *cardiac arrest* can occur in many cases by cardiac pulmonary resuscitation (CPR) [390].

Cardiac muscle is similar to the skeletal muscle described in Chap. 5 (Fig. 5.1b). In particular, the basic building block is the sarcomere with its sliding actin and myosin filaments. There are some differences, however. At the resting muscle length, the maximum tension for skeletal muscle is $\sim 20\,\mathrm{N/cm^2}$ or more, while it is only $\sim 7\,\mathrm{N/cm^2}$ for cardiac muscle. Also, the resting, passive tension is fairly large at the length of peak tension in cardiac muscle (as is depicted in Fig. 5.25a). Both of these differences can be attributed in part to the greater fraction of noncontractile tissue in heart muscle, which contains collagen and other fibrotic tissue. The first difference also arises from the nonparallel nature of cardiac muscle fibers. Another difference between skeletal and cardiac muscle is that it is usually not possible to tetanize cardiac muscle. The twitches merge only partially at very high stimulation frequency.

Echocardiography is the use of ultrasound to diagnose heart disorders and blood flows. It and related methods are described briefly in Chap. 10 (and in Problems 10.22 and 10.23).

8.1.3 Valves

There are four major valves in the heart (Fig. 8.7). The right atrioventricular valve controls flow between the right atrium and right ventricle. It has three flaps (or cusps) and is therefore also called the tricuspid valve. The pulmonary semilunar valve controls blood flow from the right ventricle to the left and right pulmonary arteries. The left atrioventricular valve controls flow from the left atrium to the left ventricle. It has two flaps and is therefore also called the

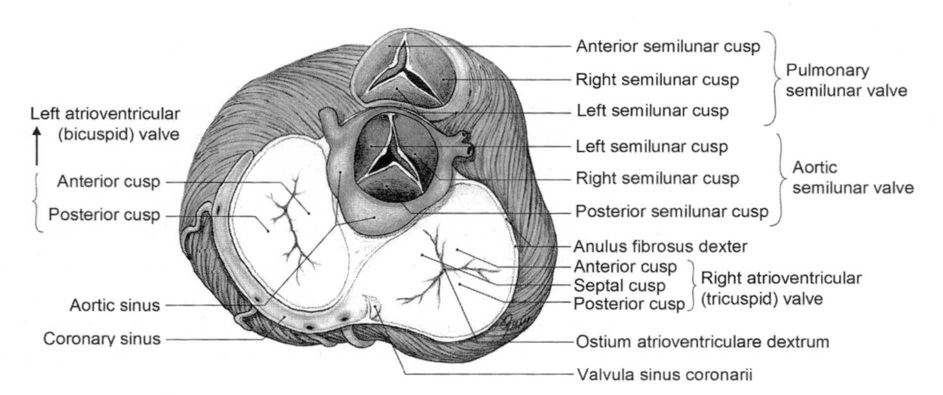

Fig. 8.7. Drawing of the four major heart valves, showing the cusps (flaps). (From [418])

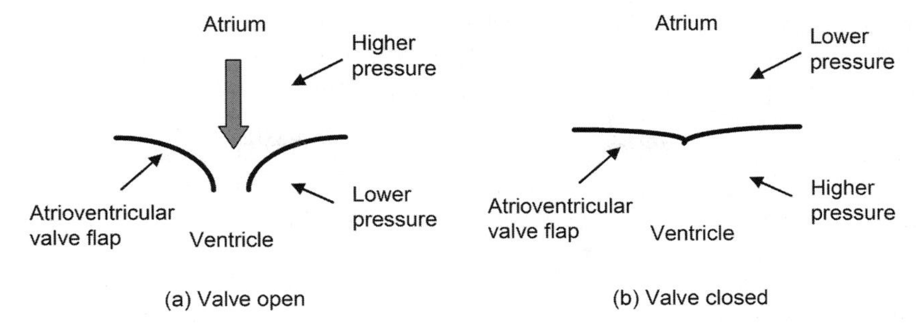

Fig. 8.8. Schematic of the unidirectional flow of an atrioventricular heart valve. (**a**) The pressure in the atrium exceeds that in the corresponding ventricle and the valve opens, with a jet of blood rushing in. Toward the end of diastole, the jet is broken. The deceleration of the blood creates a pressure, which tends to close the valve. (**b**) The valve is normally closed. (Based on [417])

bicuspid valve. Another name for this valve is the mitral valve, because it looks like a miter. The aortic semilunar valve controls flow from the left ventricle to the aorta.

These four valves share some common traits. They are one-way valves (Fig. 8.8) that allow blood flow in the described direction under some conditions, but never in the opposite direction (unless they are defective). We can imagine a flapped unidirectional valve that will not allow any back flow. With this type of valve we can see how the flaps will open, allowing this unidirectional flow, when the pressure in front of the valve exceeds that on the other side. In this way, the valve is closed until the pressure in the chamber increases due to contraction to a value greater than that after the valve. However, such a valve could not withstand very much back pressure. Backward opening of the atrioventricular valves is also prevented by the papillary muscles on the ventricular side that contract when the valve is closed, making the chordae tendineae that are attached to the flaps taut (Fig. 8.9). This prevents the flaps from bending backward, so there is no backward flow of blood.

Let us consider the cycle for the aortic semilunar valve. During ventricular relaxation the pressure in the left ventricle is ~0 mmHg. In the aorta the pressure is ~120 mmHg during systole and then decreases to ~80 mmHg during diastole, just before ventricular contraction. The valve is still closed. During ventricular contraction the pressure in the LV increases to 80 mmHg, continuing up to ~120 mmHg. Because the pressure in the aorta is ~80 mmHg, the aortic semilunar valve opens once the pressure in the LV exceeds 80 mmHg and then remains open. During this flow, the pressure in the LV and aorta become equal, ~120 mmHg, and then the valve closes as the flow cycle comes to an end.

The measurement of systolic and diastolic pressure by listening to Korotkoff sounds is described in Chap. 7. This method is the standard way

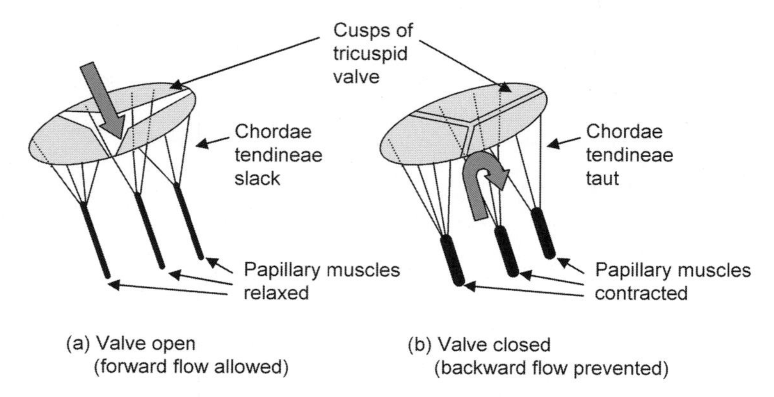

Fig. 8.9. The papillary muscles on the ventricular side of the atrioventricular valves contract when the values are closed, making the chordae tendineae taut, as in (**b**). The muscles and the chordae tendineae are relaxed when there is forward blood flow, as in (**a**). (Based on [417])

of determining blood pressure even though it routinely underestimates systolic pressure by 5–20 mmHg and overestimates diastolic pressure by 12–20 mmHg [377].

Cardiac valve openings and closings and the flow of blood in the heart create sounds that can be heard with a stethoscope, and are described in Chap. 10. (Also see Fig. 8.5.)

8.2 Physics of the Circulation System

We now examine the circulation system in more detail. First, we will examine how the blood pressure varies with distance along the arteries and veins, including within the capillaries. We then investigate the consequences of nonuniformities in arteries, such as clogged arteries and aneurysms – in the context of the strength of the artery walls. We next calculate the work done by the heart, to see how this contributes to the metabolic needs of the body. In the last section of this chapter we will develop a model of the entire circulatory system and the heart.

8.2.1 Properties of Blood

Blood is a non-Newtonian fluid, in part because of its complex, inhomogeneous composition. The blood solution consists of plasma, red blood cells (erythrocytes; 5 million/mm^3; 45% of total blood volume), white blood cells (leukocytes; 0.3%), and platelets (0.15%). (The red blood cell volume fraction is called the hematocrit.) The red blood cells are biconcave disks that are toroidal in shape with the center partially filled in, and have a diameter of

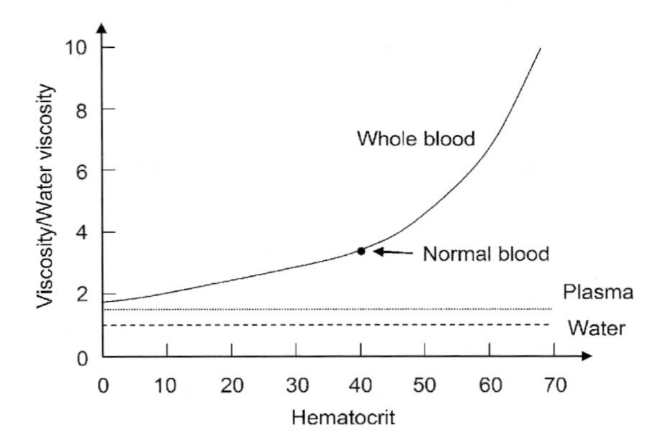

Fig. 8.10. Blood viscosity vs. hematocrit. (Based on [390])

7.5 μm and maximum thickness of 2 μm. Their diameter is about the same as the inner diameter of capillaries, but they can deform and flow in even smaller tubes. White blood cells are spherical, with a diameter of 7 μm, while the platelets are much smaller. The blood plasma is 90% water and behaves like a Newtonian fluid with a viscosity of 0.0012 Pa-s. The blood rheology is greatly altered by the red blood cells, and not much by the white blood cells or platelets because they comprise very small fractions of the blood volume. (Rheology is the study of the deformation and flow of materials, particularly unusual materials.) The blood viscosity increases with the hematocrit, as seen in Fig. 8.10.

The effective viscosity of blood decreases as the shear rate increases (Fig. 8.11). For very slow shear rates, this viscosity is more than 100× that of water, while at the high shear rates characteristic of flow in larger vessels it is about 4× that of water, with a value of 0.004–0.005 Pa-s.

The viscosity of some fluids changes even while the strain rate is constant. Blood is a thixotropic fluid, for which the shear stress decreases while the strain rate is constant. Still, for our purposes it will be adequate to treat blood as a Newtonian fluid, even though the velocity flow profile is not the ideal parabolic form for a Newtonian fluid (Fig. 7.13).

8.2.2 Blood Pressure and Flow in Vessels

Structure of Blood Vessels

Arteries contain inner layers that are 1–2 endothelial (lining) cells thick – along with elastic issue (composed of collagen and elastic proteins). This innermost region surrounding the opening – the lumen – is known as the tunica intima. Next in the wall comes a layer of circular, smooth muscle fibers interspersed with elastic tissue (the tunica media) and finally connective tissue (the tunica adventitia) (Fig. 8.12, also see Fig. 8.44). The walls of veins have a thickness

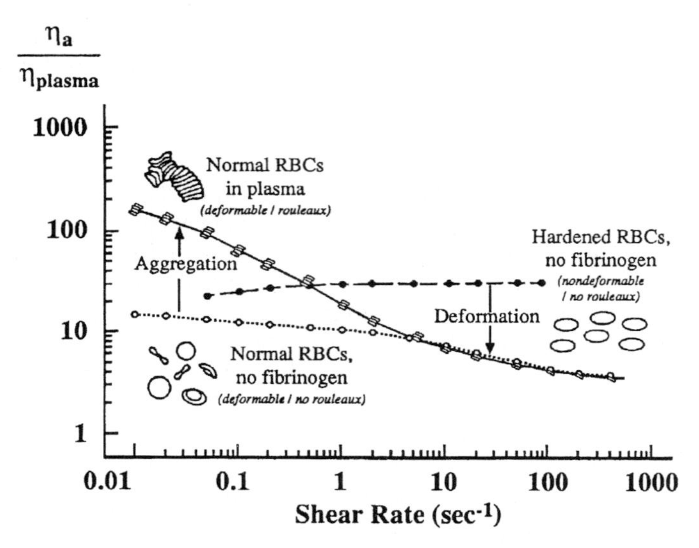

Fig. 8.11. Blood viscosity vs. shear rate for a hematocrit of 45%, at 310 K. (From [382]. Courtesy of Robert A. Freitas Jr., Nanomedicine, Vol. 1 (1999), http://www.nanomedicine.com, based on [375])

w that is typically $\sim d/10$, where d is the lumen diameter; they are thinner than the walls in arteries of corresponding diameters, for which $w \sim d/5$. The aorta and other large arteries contain much elastic tissue and stretch during systole and recoil during diastole. The walls of the arterioles contain

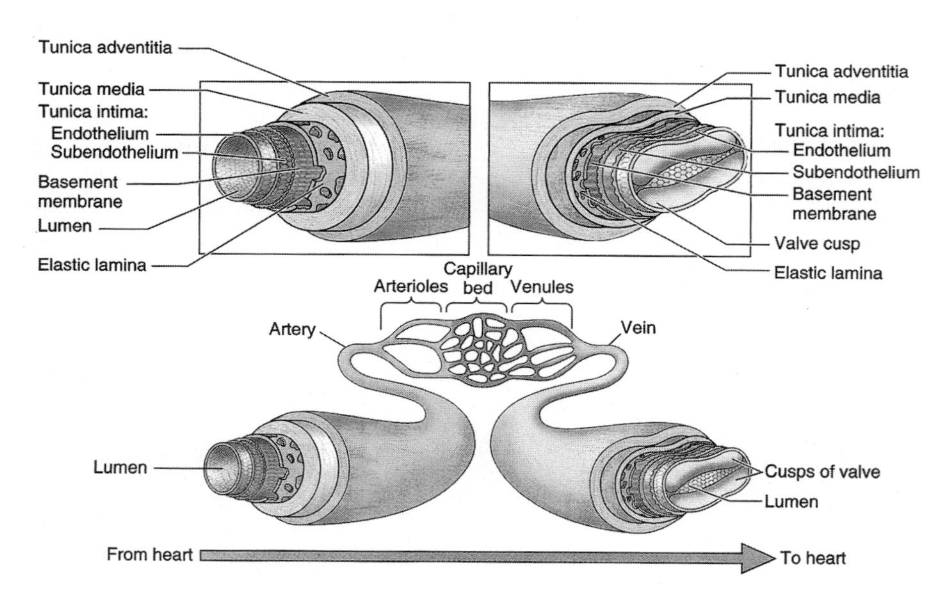

Fig. 8.12. Schematic of the walls of arteries and veins. (From [408]. Used with permission)

less elastic tissue and more smooth muscles, and stretch relatively little. The walls of the capillaries are composed of a single layer of endothelial cells. The diameters of the capillaries are so small that red blood cells can barely pass through them.

Approximately 70% of the walls of arteries and veins is composed of water, which is not elastic, except in how it withstands compression. The other 30% consists of the dry mass: elastin, collagen, and smooth muscle fibers, each having different materials properties. As discussed in Chap. 4, elastin is rubber-like and has a Young's modulus of $\sim 3 \times 10^5$ Pa; it can be stretched to twice its relaxed length. Collagen is much stiffer, with a Young's modulus of $\sim 1 \times 10^8$ Pa. Elastin has an ultimate tensile stress (UTS) less than 5% of that of collagen. Smooth muscle has a Young's modulus more like that of elastin, with $Y \sim 1 \times 10^5$ Pa when relaxed and $\sim 2 \times 10^6$ Pa when active. About half the dry mass in vessels is elastin and collagen, with more elastin than collagen in the aorta ($\sim 1.5\times$) and relatively less elastin in other arteries ($\sim 0.5\times$) and veins ($\sim 0.3\times$). Veins contain less elastin than arteries. The fraction of smooth muscle in the dry mass averages to $\sim 50\%$, and is $\sim 25\%$ in the aorta, and increases to $\sim 60\%$ more peripherally in the arteries and arterioles. The mechanical properties of these vessels (Chap. 4) also depend on the tissue to which they are attached. This is particularly significant for capillaries, because the vessel walls are essentially a single layer of endothelial cells.

Blood Pressure

Blood pressure is needed to push blood flow. Figure 8.13 is a schematic of the mean arterial and venous blood pressure at different positions in the

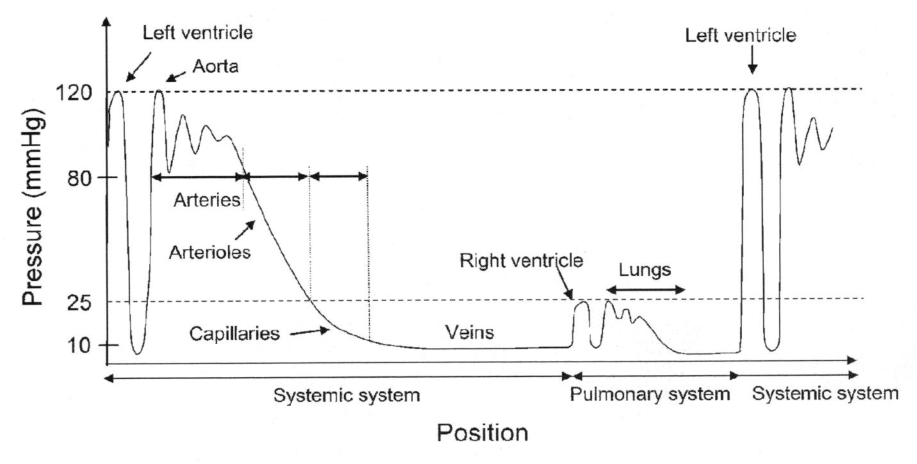

Fig. 8.13. Blood pressure along the circulatory system for a person lying horizontally. (Based on [371])

circulation cycle, for a person lying horizontally. The oscillations in blood pressure along the major arteries in systemic circulation reflect the oscillating pressure of this blood when it was leaving the aorta, at pressures between ~80 mmHg (P_{diastole}, at diastole) and ~120 mmHg (P_{systole}, at systole) (Fig. 8.13). Because systole lasts for about 1/3 of the cycle and diastole for about 2/3, the *mean blood pressure* is a weighted sum,

$$P_{\text{mean}} = \frac{P_{\text{systole}} + 2P_{\text{diastole}}}{3}, \tag{8.1}$$

or (1/3)120 mmHg + (2/3)80 mmHg ~ 94 mmHg in this example. This difference in pressure of 40 mmHg between systole and diastole is the *arterial pulse pressure* P_{pulse}. We will see that blood flows at a speed of ~20 cm/s in these systemic arteries, so with a heart rate of about 1 Hz = 1 cycle/s it is reasonable that there are quasiperiodic variations every 20 cm or so. Much of the pressure drop in the arterial system is in the arterioles (small arteries) and the capillaries. We will see that this can be attributed to viscous flow. There is very low pressure in the veins. It is too little pressure to pump the blood back to the heart – even with the large diameters of the veins and consequently low resistance to flow (7.24). There is a peristaltic pumping mechanism by muscles surrounding the large veins that assists the return of venous blood to the heart, with one-way valves to prevent backflow (Fig. 8.14). (Similarly, blood flow in the capillaries is usually not continuous, but is turned on and off every few seconds or minutes, due to sphincter muscles that can contract the feeding arterioles (*vasomotion*).) The pulmonary system mirrors this systemic circulation, except the pressures are all lower. Figures 8.13, 8.15, and 8.16 show the blood pressure and flow speed at different points in the arterial tree.

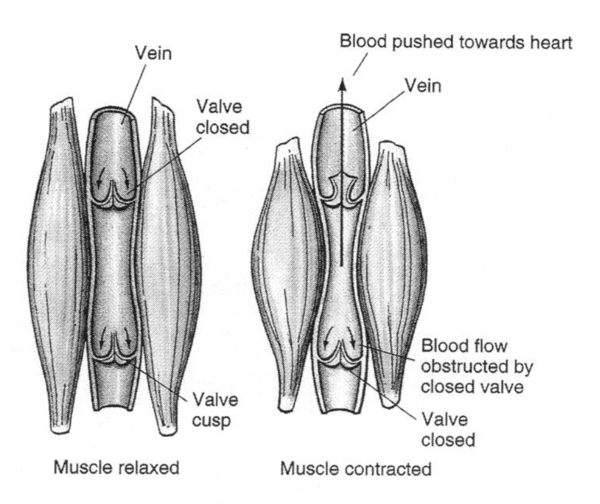

Fig. 8.14. Musculovenous pump of veins, with outward expansion of the bellies of contracting muscles pumping the blood back to the heart against gravity and distal valves closing to prevent backflow. (From [408]. Used with permission)

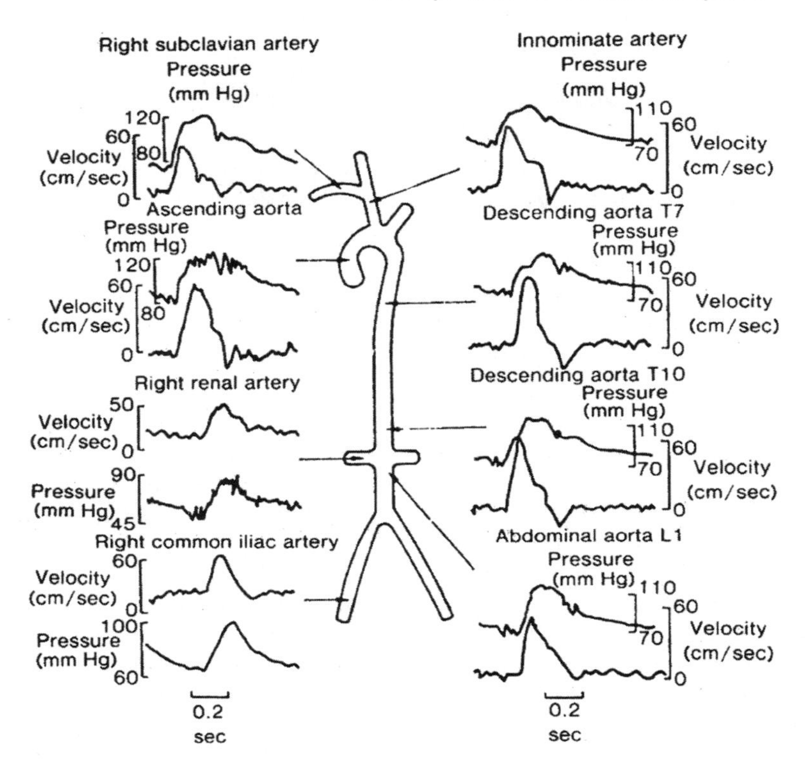

Fig. 8.15. Simultaneous pressure and flow velocity at different points in the human arterial tree for a person lying horizontally. All data were taken from one patient except for the right renal artery and the right common iliac artery. (From [391]. Adapted from [407])

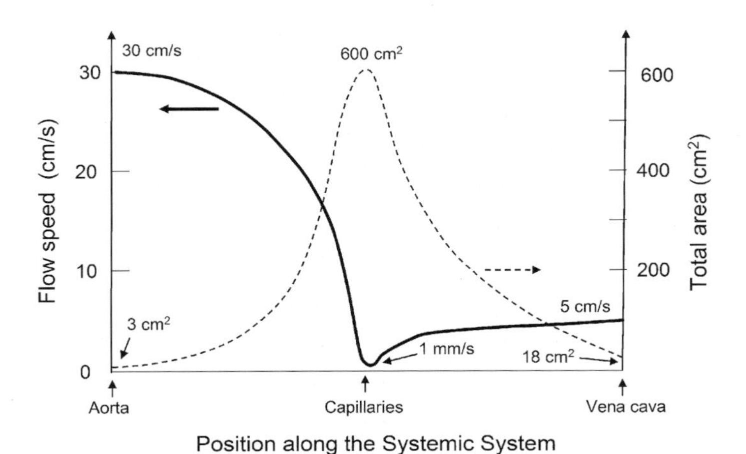

Fig. 8.16. Flow speed (*solid curve*) and total area (*dashed curve*) in the systemic circulation system. (Based on [371])

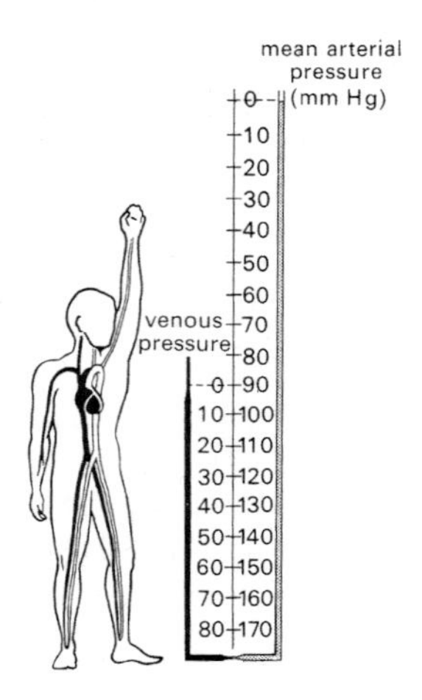

Fig. 8.17. Mean arterial and venous (gauge) pressures for a vertical person. (From [372], after [412]. Used with permission of Oxford University Press)

The pumping cycle sets up a pressure pulse wave in addition to the hydrostatic pressure variation. This pulse wave is independent of the speed of blood flow – and is faster than this blood flow speed: 4 m/s in the aorta, 8 m/s in the large arteries, and 16 m/s in the small arteries of young adults.

When you stand upright vertically (Fig. 8.17) there is an additional pressure $\rho g h$, where h is the height relative to the heart. This is approximately the height in the upper arm where blood pressure measurements are made. For $\rho = 1.06\,\mathrm{g/cm^3} = 1{,}060\,\mathrm{kg/m^3}$, $g = 9.8\,\mathrm{m/s^2}$ and $h = 1\,\mathrm{m}$, this pressure is $10{,}400\,\mathrm{N/m^2} = 10{,}400\,\mathrm{Pa} = 79\,\mathrm{mmHg}$ (with $1\,\mathrm{MPa} = 7{,}600\,\mathrm{mmHg}$). At any given height, the driving pressure difference from the arteries to the veins is unchanged. Also, this pressure change is not important when considering pressure changes between the inside and outside of a vessel because $\rho g h$ is added both inside and outside the vessel.

Still, this effect of gravity can be significant. The blood pressure at the aorta has to be high enough to pump the blood to the top of your brain. This distance is about $h = 40\,\mathrm{cm}$, so the pressure drop is about 30 mmHg (compared to the diastolic pressure of $\sim$80 mmHg). Problems 8.1 and 8.2 explore what happens to cranial blood circulation in humans in rapidly climbing jets and on more massive planets with higher g, and also in giraffes. One manifestation of this effect of gravity is potential fainting when you stand. When you stand up, the volume of blood in the leg veins increases and the pressure in the veins pumping blood back to the heart decreases. This can decrease the cardiac output and the flow of blood to the brain. This rarely happens because there is a reflex constriction of the veins in the legs (due to a contraction of the skeletal muscle surrounding the veins, Fig. 8.14) that limits the blood pool and an arteriolar constriction that increases flow resistance and lessens the decrease in arterial blood pressure. The effect of gravity on humans – who

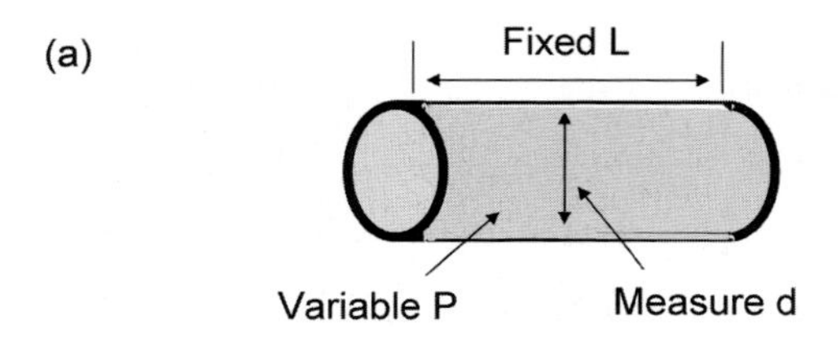

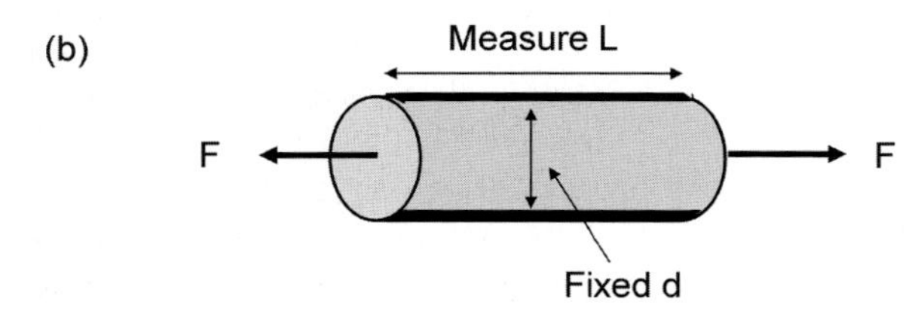

Fig. 8.18. Measuring the mechanical properties of blood vessels by fixing the vessel (**a**) length (which is called *inflation*) or (**b**) diameter (which is called *extension*). Some investigators do combined loading, with combinations of inflation, extension, and torsion

normally stand upright – also explains why standing on your head for long periods is not advisable (Problem 8.3). The veins in your head are not designed to pump blood back to the heart (as are those in the lower body). Also, your feet would stop getting blood. It also explains why varicose veins are worse when you stand upright, because blood then needs to be pumped up.

Body control of blood pressure is briefly described in Chap. 13.

Measuring Flow in Blood Vessels

The mechanical properties of blood vessels can be measured under two types of conditions (1) The length of a given vessel can be kept constant, while its diameter is measured as a function of the *distending pressure*. This leads to a tensile stress on the wall, directed around the circumference, which is called the circumferential or hoop stress. (2) The diameter of a vessel can be kept constant, while its length is measured as it is stretched longitudinally. Examples of both are shown in Fig. 8.18.

The flow of blood in arteries is affected by changes in the heart beat rate and stroke volume (which is the volume pumped per beat), and also by changes in the arteries themselves that control their diameters by chemical and neural mechanisms.

Modeling Flow in Blood Vessels

Figure 8.19 is a schematic describing the flow in a vessel of length L, with a volumetric flow rate Q_1 in and Q_2 out. In steady state $Q_1 = Q_2 = Q$. At the

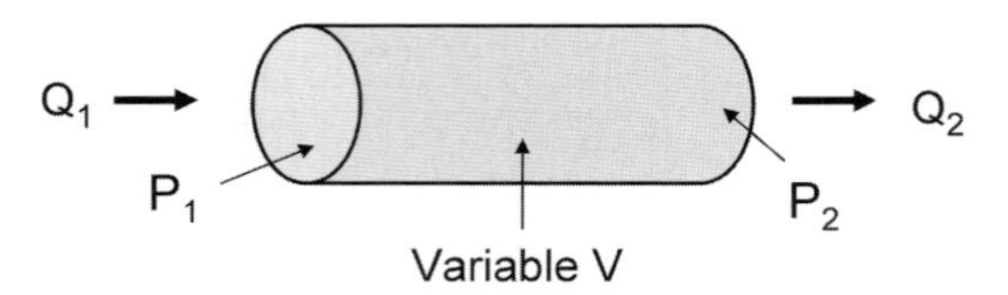

Fig. 8.19. Blood flow: general vessel

beginning of the vessel the pressure is P_1 and at the end it is P_2. The pressure outside the vessel is P_{ext}, which can be taken to be 0 when considering gauge pressure. The volume of the vessel is V. The relation between these various parameters depends on the properties of the vessels.

There are two physical attributes of blood vessels. (a) They have a *resistance* to flow, and so they need a pressure difference along the length of the vessel to drive the blood flow. (b) They have a *compliance* in response to a distending pressure. This is much like a balloon expanding when the pressure inside increases much above that outside.

One special case is a rigid vessel with constant volume V, which is called a *resistance vessel* (Fig. 8.20). Equation (7.25) applies to this vessel, so

$$P_1 - P_2 = R_{\text{flow}} Q \tag{8.2}$$

or

$$Q = \frac{1}{R_{\text{flow}}}(P_1 - P_2), \tag{8.3}$$

where the vascular resistance is $R_{\text{flow}} = 8\eta L/\pi r^4$ for a tube with radius r. The former equation has the same form as Ohm's Law $V_{\text{elect},1} - V_{\text{elect},2} = R_{\text{elect}} I$, which relates the drop in voltage, V_{elect}, when a current of charges I traverses a structure with electrical resistance R_{elect}.

A second special case is an elastic vessel that has no noticeable resistance, which is called a *compliance vessel*. There is no pressure drop, so $P_1 = P_2 = P$. One model (Fig. 8.21) of the properties of such a vessel is

$$V(P) = C_{\text{flow}}(P - P_{\text{ext}}) = C_{\text{flow}} P, \tag{8.4}$$

where C_{flow} is the compliance and P_{ext} is taken to be 0. Because the vessel usually has a volume with no pressure, called the dead volume V_{d}, a better

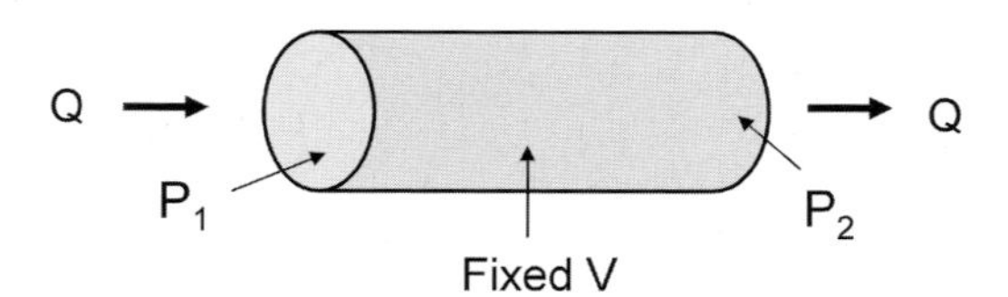

Fig. 8.20. Blood flow: ideal resistance vessel

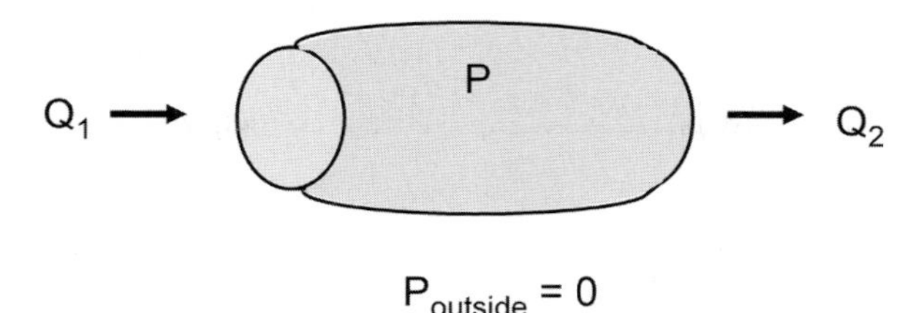

Fig. 8.21. Blood flow: ideal compliance vessel

relation is

$$V(P) = V_{\rm d} + C_{\rm flow}P. \tag{8.5}$$

Note that a property of the resistance vessel, namely the flow rate, is affected by the pressure drop along (and inside the vessel), while a property of the compliance vessel, its volume, is affected by the pressure difference between the inside and outside of the vessel. We can also describe compliance by changes in the radius r

$$r(P) = r_{\rm d} + \frac{C'_{\rm flow}}{2}P, \tag{8.6}$$

where $r_{\rm d}$ is the radius with no pressure difference. Therefore

$$\frac{{\rm d}r}{{\rm d}P} = \frac{C'_{\rm flow}}{2}. \tag{8.7}$$

Because $V = \pi r^2$ and $V_{\rm d} = \pi r_{\rm d}^2$, these two formulations can be interrelated.

Real vessels have some attributes of both types of vessels. Still, the aorta, large arteries and large veins are much like compliance vessels. We will see that the pressure drops along them are relatively small. Arterioles, capillaries, and venules act like resistance vessels. We will see that they, and in particular the arterioles, are the main sites of the pressure drop, and this is the reason why the heart needs to pump blood to such high pressures.

Pressure Drops in Arteries and Resistive Vessels

We will use Poiseuille's Law $\Delta P = \left(8\eta L/\pi r^4\right) Q$ (7.25) to estimate the pressure drop $\Delta P = P_1 - P_2$ across the aorta, large arteries, arterioles, and capillaries, and compare these results to the plot in Fig. 8.13 (also see Fig. 8.15). First we calculate the resistance $R_{\rm flow,0} = 8\eta L_0/\pi r_0^4$ for a standard radius $r_0 = 1$ cm and standard length $L_0 = 1$ cm and scale the results for each specific case. With the viscosity $\eta = 4.0 \times 10^{-3}$ Pa-s $= 4.0 \times 10^{-3}$ (N-s/m^2) $= 4.0 \times 10^{-2}$

poise for whole blood at 37°C

$$R_{\text{flow},0} = \frac{8 \times (4.0 \times 10^{-3}\,\text{N-s/m}^2)(1\,\text{cm})}{\pi(1\,\text{cm})^4} = \frac{1.02 \times 10^{-2}\,\text{N/m}^2}{\text{cm}^3/\text{s}} \tag{8.8}$$

$$= \frac{1.02 \times 10^{-8}\,\text{N/mm}^2}{\text{cm}^3/\text{s}} = \frac{7.7 \times 10^{-5}\,\text{mmHg}}{\text{cm}^3/\text{s}}, \tag{8.9}$$

where we have used $0.1\,\text{N/mm}^2 = 1\,\text{atm.} = 760\,\text{mmHg}$. The units in (8.9) are mmHg-s/cm^3. This is the resistance when the pressure difference is 1 mmHg and the flow rate is 1 mL/s, and is also known as a PRU, a peripheral resistance unit. This unit is commonly used in physiology.

The resistance of a vessel of an arbitrary length and radius is

$$R_{\text{flow}} = R_{\text{flow},0}\frac{L/L_0}{(r/r_0)^4} = R_{\text{flow},0}\frac{L(\text{in cm})}{r(\text{in cm})^4} \tag{8.10}$$

and so

$$\Delta P = R_{\text{flow}}Q = 7.7 \times 10^{-5}\,\text{mmHg}\,\frac{L(\text{in cm})}{r(\text{in cm})^4}\,Q\ (\text{in cm}^3/\text{s}). \tag{8.11}$$

The total flow rate, Q_t, from the aorta, enters the large arteries, and the whole flow from the large arteries enters the arterioles, and finally this whole flow enters the capillaries. In each level of flow we will model the arteries as n parallel vessels of roughly equivalent length and diameter carrying roughly the same flow (Fig. 8.22), where n increases for each successive level of flow. So in a given level of flow with n vessels, the flow rate in each vessel is $\sim Q_\text{t}/n$.

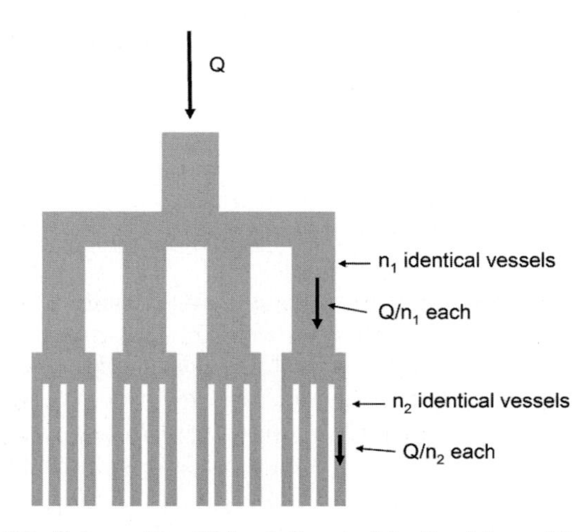

Fig. 8.22. Schematic of blood flow in idealized branching vessels

With $Q_t = 80\,\text{cm}^3/\text{s}$ and $Q = Q_t/n$, we get

$$\Delta P = 7.7 \times 10^{-5}\,\text{mmHg}\,\frac{L(\text{in cm})}{r(\text{in cm})^4}\,\frac{80}{n} \tag{8.12}$$

$$= \frac{0.0062\,\text{mmHg}}{n}\,\frac{L(\text{in cm})}{r(\text{in cm})^4}. \tag{8.13}$$

This is the pressure drop across any vessel in a given level of flow and, because they are in parallel, it is the pressure drop across the entire given level of arterial flow. We now determine this for the various levels of arteries.

Aorta. There is one aorta ($n = 1$) with $r \sim 1.25\,\text{cm}$ and $L \sim 10\,\text{cm}$, and so ΔP across the aorta is 0.025 mmHg, which is insignificant.

Largest arteries. There are about 200 large arteries with $r \sim 0.2\,\text{cm}$ and $L \sim 75\,\text{cm}$, and so ΔP is 1.4 mmHg, which is pretty insignificant.

Smallest arteries and arterioles. There are about 5×10^5 arterioles with $r \sim 30\,\mu\text{m}$ and $L \sim 0.6\,\text{cm} = 6\,\text{mm}$, and so ΔP is 91 mmHg, which is very significant.

Capillaries. There are about 10^{10} capillaries with $r \sim 3.5\,\mu\text{m}$ and $L \sim 0.2\,\text{cm} = 2\,\text{mm}$, and so ΔP is 8.2 mmHg, which is fairly significant.

We could have just plugged the parameters for each vessel directly into (7.25), without calculating $R_{\text{flow},0}$, but our scaling approach does give some new insight.

These estimates agree with what we would expect from Fig. 8.13. Also, arterioles and capillaries are seen to be well modeled as resistance vessels. The aorta and large arteries have very small pressure drops across them, and behave more like compliance vessels. Veins have larger diameters than the corresponding arteries, and consequently much lower resistances and pressure drops across them.

Along any vessel there is obviously a linear pressure drop with distance x along the vessel. This is seen from Poiseuille's Law, (7.25), $(\Delta P/L) = (8\eta/\pi r^4)\,Q$ or, recognizing that this change in pressure is negative,

$$\frac{\mathrm{d}P}{\mathrm{d}x} = -\frac{8\eta}{\pi r^4}Q. \tag{8.14}$$

This represents a "distributed" or "transmission-line" view of blood flow, in which flow is analyzed per unit length along the vessel, whereas in (8.2) and (8.11) flow was analyzed with the vessel as a "lumped" parameter (see Appendix D).

We can study the pressure drop in clogged arteries. There could be a larger pressure drop for the same Q or a smaller Q for the same pressure drop if r decreases, as occurs with clogged arteries, or with fewer vessels. This can stimulate an increase in blood pressure to maintain the flow rate or lead to a reduction in flow at a given inlet pressure, which is what actually happens in coronary artery disease.

How can we "optimize" the design of resistive vessels and how such vessels bifurcate and otherwise branch into smaller vessels (and what does it really mean to optimize the design)? See Problems 8.28–8.31.

Radial Profile of Blood Flow

Blood flow in vessels is not uniform. We have implicitly been assuming that the blood flow is parabolic (Fig. 7.11) because we are modeling blood as a Newtonian fluid. However, the parabolic profile of blood flow speed in a resistive vessel from (7.40) is not quite accurate. This assumes steady-state flow, which begins only a certain distance from a bifurcation ((7.41)–(7.43)). This approach to steady-state flow is depicted for a different initial condition in Fig. 7.11. Furthermore, because whole blood is not a Newtonian fluid, the steady-state profile is not parabolic, as is seen in Fig. 7.13.

Properties of a Compliance Vessel

We can show that (8.4) and (8.5) are reasonable models of an elastic compliance vessel and determine the compliance C_{flow} by examining a thin-walled cylindrical tube of inner radius r, thickness w, and length L, with a pressure difference P between the inside and outside of the vessel (Fig. 8.23). The Law of Laplace for cylinders (7.4) shows that the tension T in the walls of a cylinder in equilibrium is rP. We can conceptually slit the vessel along its length and see that this tension (force per unit length along L) corresponds to a force per unit area of T/w on the rectangular face with dimensions w and L. The length of the rectangular solid is $2\pi r$. Let us consider the stress–strain relation $\sigma = Y\epsilon$ in the context of this unfolded vessel. The stress is $\sigma = T/w$.

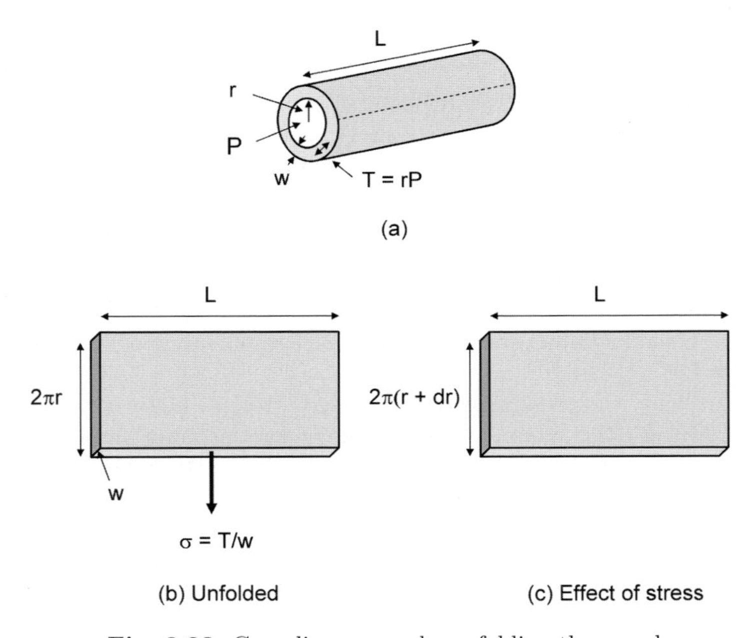

Fig. 8.23. Compliance vessels: unfolding the vessel

The strain is $\epsilon = \mathrm{d}(2\pi r)/2\pi r = \mathrm{d}r/r$ or really $\mathrm{d}r/r_0$, where r_0 is the radius with no pressure. The stress–strain relation is

$$\frac{T}{w} = Y\frac{\mathrm{d}r}{r_0} \tag{8.15}$$

or with $T = rP$

$$\frac{\mathrm{d}r}{r_0} = \frac{T}{wY} = \frac{rP}{wY} = \frac{r_0}{w}\frac{P}{Y}. \tag{8.16}$$

The internal volume of the vessel is $V = \pi r^2 L$. Therefore we see that $\mathrm{d}V = 2\pi r(\mathrm{d}r)L$ and $\mathrm{d}V/V = 2\mathrm{d}r/r = 2(r_0/w)(P/Y)$. For small changes in volume

$$V(P) = V_{\mathrm{d}}\left(1 + \frac{\mathrm{d}V}{V_{\mathrm{d}}}\right) = V_{\mathrm{d}}\left(1 + 2\frac{r_0}{w}\frac{P}{Y}\right) \tag{8.17}$$

$$= V_{\mathrm{d}} + 2V_{\mathrm{d}}\frac{r_0}{w}\frac{P}{Y} = V_{\mathrm{d}} + 2(\pi r_0^2 L)\frac{r_0}{wY}P, \tag{8.18}$$

with $V_{\mathrm{d}} = \pi r_0^2 L$. Using (8.5) the compliance is

$$C_{\mathrm{flow}} = \frac{2\pi r_0^3 L}{wY}. \tag{8.19}$$

How large is this expansion? The pressure in the aorta and large arteries is 120 mmHg = 0.0158 MPa during systole. The value of Y for such vessels is about 1 MPa (Table 4.2) and so $P/Y = 0.0158 \sim 1.6\%$. The thickness of arterial walls is typically 1/5 of the radius, so $r_0/w = 5$. This means that the fractional increase in radius of these vessels due to this internal pressure is 8% and the fractional increase in volume is 16% – both sizeable fractions. Also, this predicts that the radius changes by ~3% during each heart beat during the changes between systolic (120 mmHg) and diastolic (80 mmHg) pressure. Veins are also compliance vessels.

Distensibility

Such compliance changes are equally well described in terms of the distensibility D_{flow} of the tube. The cross-sectional area A of a tube increases by ΔA when the pressure difference between the inside and outside of the tube increases by ΔP. The distensibility is defined as the fractional change in area for a change in pressure:

$$D_{\mathrm{flow}} = \frac{\Delta A/A}{\Delta P}. \tag{8.20}$$

With $A = \pi r^2$ and $\Delta A = 2\pi r\,\Delta r$, we see that $\Delta A/A = 2\,\Delta r/r$ and using (8.16),

$$D_{\mathrm{flow}} = \frac{2\,\Delta r/r}{\Delta P} = \frac{2(r/w)(\Delta P/Y)}{\Delta P} = \frac{2r}{wY} = \frac{1}{Y(w/d)}, \tag{8.21}$$

where $d = 2r$ is the diameter, w is the wall thickness, and w/d ($\ll 1$).

A more exact analysis relates the Young's modulus for circumferential stretch, which we still call Y, to the external and internal diameters d_e and d_i, the change in external diameter Δd_e occurring with this change in pressure difference, and Poisson's ratio υ [372]. This gives

$$Y = \frac{\Delta P}{\Delta d_\mathrm{e}} \frac{2 d_\mathrm{e} d_\mathrm{i}^2}{d_\mathrm{e}^2 - d_\mathrm{i}^2} (1 - \upsilon^2). \tag{8.22}$$

For a thin-walled tube with wall thickness $w = (d_\mathrm{e} - d_\mathrm{i})/2 \ll d_\mathrm{i}$, and with $d_\mathrm{e} \sim d$ and $d_\mathrm{i} \sim d - 2w$, we find

$$Y = \frac{\Delta P}{\Delta d} \frac{d^2}{2w} (1 - \upsilon^2). \tag{8.23}$$

With $\Delta A/A = 2\,\Delta d/d$

$$D_\mathrm{flow} = \frac{\Delta A}{A} \frac{1}{\Delta P} = \frac{2\,\Delta d}{d} \frac{1}{\Delta P} = \frac{(1 - \upsilon^2)}{Y(w/d)}. \tag{8.24}$$

This reduces to (8.21) for small Poisson's ratios.

Flow with Resistance and Compliance

If a vessel is resistive and compliant [405], the change in pressure with distance is

$$\frac{\mathrm{d}P}{\mathrm{d}x} = \frac{\mathrm{d}P}{\mathrm{d}r}\frac{\mathrm{d}r}{\mathrm{d}x} = \frac{2}{C'_\mathrm{flow}} \frac{\mathrm{d}r}{\mathrm{d}x}, \tag{8.25}$$

using $\mathrm{d}P/\mathrm{d}r = 2/C'_\mathrm{flow}$ from (8.7). Setting this equal to $\mathrm{d}P/\mathrm{d}x$ from (8.14) and bringing the r terms to the left and the x terms to the right, gives

$$r^4 \mathrm{d}r = -\frac{4C'_\mathrm{flow}\eta}{\pi} Q\, \mathrm{d}x. \tag{8.26}$$

After integrating over a vessel length from $x = 0$ to $x = L$, we get

$$(r(x=0))^5 - (r(x=L))^5 = \frac{20 C'_\mathrm{flow}\eta}{\pi} QL \tag{8.27}$$

and after using (8.6)

$$\left(r_\mathrm{d} + \frac{C'_\mathrm{flow}}{2} P(x=0)\right)^5 - \left(r_\mathrm{d} + \frac{C'_\mathrm{flow}}{2} P(x=L)\right)^5 = \frac{20 C'_\mathrm{flow}\eta}{\pi} QL. \tag{8.28}$$

(See Appendix C.)

Both terms on the left side can be expanded to five terms. The first terms are both r_d^5, which cancel, and for relatively small compliance

($C'_{\text{flow}}P/2r_{\text{d}} \ll 1$) only the next two of the remaining four terms in each need to be retained, giving

$$Q = \frac{\pi r_{\text{d}}^4}{8\eta L}(P(0) - P(L))\left(1 + \frac{C'_{\text{flow}}}{r_{\text{d}}}(P(0) - P(L))\right). \tag{8.29}$$

This is Poiseuille's Law (7.24) with a correction for compliance. So, for a rigid wall vessel ($C'_{\text{flow}} = 0$) the flow rate Q is linear with the pressure drop, but when compliance is included, the variation with pressure drop is between linear and quadratic. This relation says that for a given pressure drop, the flow rate is increased due to the compliant nature of the vessel.

The electrical analog of blood flow is described in Appendix D.

The Strength of Blood Vessel Walls

The pressure inside blood vessel walls P exceeds that outside P_{ext}, by $\Delta P = P - P_{\text{ext}}$. *How large of a tension should the vessel walls be able to withstand to support this positive pressure differential?* Chapter 7 showed the answer is provided by the Law of Laplace for hollow cylinders (7.4). For a cylinder of radius of curvature R, this tension T is

$$\Delta P = \frac{T}{R}. \tag{8.30}$$

Table 8.4 shows that the tension capillaries need to withstand is very small because of their small radius. This circumferential stress, the tension (force

Table 8.4. Calculated tension in blood vessel walls. (Using data from [382] and [391])

vessel	diameter (mm)	wall thickness, w (mm)	internal pressure, ΔP (mmHg)	wall tension, T (dyne/cm)	T/w (kPa)
aorta	24.0	3.0	100	160,000	53
large artery	8.0	1.0	97	52,000	52
medium artery	4.0	0.8	90	24,000	30
small artery	2.0	0.5	75	10,000	20
arteriole	0.3	0.02	60	1,200	60
capillary	0.008	0.001	30	16	16
venule	0.02	0.002	20	27	13
small vein	3.0	0.2	18	3,600	18
medium vein	5.0	0.5	15	5,000	10
large vein	15.0	0.8	10	10,000	12
vena cava	30.0	1.5	10	20,000	13

The wall thickness w is $R/5$ for arteries and $R/10$ for veins (where R is the vessel radius) and 1 μm for capillaries. Also see Table 8.2.

per tube length) divided by the vessel wall thickness w, is surprisingly similar for these very different vessels. T/w can be compared to the UTS of such vessels.

Flow in Curving Arteries

In Chap. 7 we showed that the arterial walls feel a pressure due to the difference in hydrostatic pressure inside and outside the vessel. This pressure is felt equally around the wall. When blood flows in an artery that curves, a force equal to the centripetal force is felt on the arterial wall on the outer surface of the curve to change the direction (but not the magnitude) of the momentum vector of blood flow. *How large is this force? Is it comparable to the uniform hydrostatic pressure? Does it constitute a significant extra load on the arterial wall?*

For an artery of internal radius R that is turning with a radius $\mathcal{R}$ (Fig. 7.16), with blood of density ρ and average flow speed u, (7.44), this peak pressure is

$$P_{\text{cent}} = 2\rho u^2 \frac{R}{\mathcal{R}}. \tag{8.31}$$

This is largest for the fastest blood flow, which is in the aorta. Using $\rho = 1\,\text{g/cm}^3$, $R = 1.25\,\text{cm}$, $\mathcal{R} = 2\,\text{cm}$, and $u = 100\,\text{cm/s}$, we find $P_{\text{cent}} = 4.7\,\text{mmHg}$. This $\sim$5 mmHg is the extra pressure that must be supplied by the outer aorta wall to turn the blood around the aortic arch. This is small compared to the typical average aorta pressure of 100 mmHg, and does not likely promote pathological conditions such as aneurysms.

8.2.3 Capillaries and Osmotic Pressure

The purpose of systemic circulation is to supply blood to the capillary bed. We have seen that pressure is needed to bring the blood to the capillaries. There is diffusion and bulk flow between the blood in the capillaries and the interstitial fluid. Diffusion across the capillary wall transports oxygen, which is carried in red blood cells, and carbon dioxide, which is dissolved in the blood. (Not enough oxygen can be directly dissolved in the blood for our metabolic needs.) In the systemic capillaries there is net diffusion of oxygen out of the capillaries and carbon dioxide into them. In the pulmonary capillaries there is net diffusion of oxygen into the capillaries and carbon dioxide out of them. There is also bulk flow of fluid across the capillary walls due to the net pressure across the walls.

There are two forces driving this bulk transport: the force/area mechanical pressure P we have been discussing, which we will call *hydrostatic pressure* in this section, and a chemical driving force, called *osmotic pressure*, Π. The osmotic pressure characterizes the flow across a semipermeable membrane that occurs to equalize the concentrations of solutes on either side of the wall. This drives species that can permeate across the membrane, such as water, across it from the side of low concentration of solute to the side of high concentration. For low concentrations, the osmotic pressure is given by the van't Hoff equation

$$\Pi = n_{\mathrm{s}} RT, \tag{8.32}$$

where n_{s} is the density (or concentration) of the solute in solution in moles per unit volume. This looks deceptively similar to the ideal gas law (7.2). It is the difference in the sum of these on either side of the capillary walls, $P + \Pi$, that drives the net transport across these walls.

As seen in Fig. 8.24, the hydrostatic force in the capillary is always greater inside the vessel than outside, and it decreases from about 36 mmHg at the arteriole side to 15 mmHg at the venule end because of viscosity. If this were

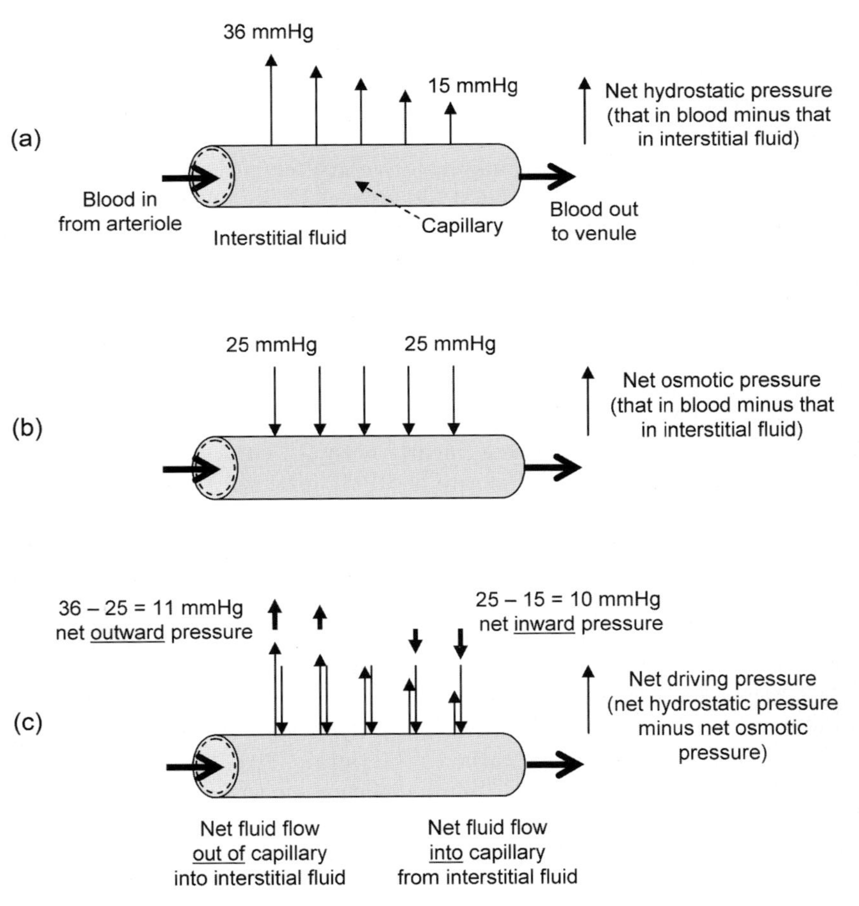

Fig. 8.24. Osmotic pressure in a capillary. (Based on [417])

the only driving force, there would be a large net flow of fluid from inside the capillary to the interstitial fluid! Small molecules, such as oxygen and carbon dioxide, are able to diffuse across the capillary wall, but larger molecules, such as proteins, cannot. Because there is a much higher density of proteins in the blood than in the interstitial fluid, there is a chemical driving force of fluid into the capillary to try to equalize these densities. This leads to a net osmotic pressure of about 25 mmHg into the capillary. In Fig. 8.24, the net pressure is 36 mmHg − 25 mmHg = 11 mmHg outward at the arteriole end and 15 mmHg − 25 mmHg = −10 mmHg inward at the venule end. Therefore, there is net flow out of the capillary in the arteriole end and net flow into the capillary in the venule end. There is a small imbalance in this and a small net bulk flow out of the capillary.

We have assumed that the flow in all blood vessels, including the capillaries, is laminar. This cannot be really true for capillaries because many capillaries have an inner diameter of 5 or 6 μm and the red blood cells have a diameter of about 7.5 μm. The red blood cells deform to pass through the capillary and the resulting flow is called *bolus flow.* The red blood cells form plugs and the blood plasma is trapped in the regions between these plugs and moves in streamlines. Nowhere else in the body is the multicomponent nature of blood more apparent.

One major function of this capillary blood flow is the transfer of oxygen to the cells, leaving oxygen-depleted blood in the veins. As in (6.18), the rate of body consumption of O_2, dV_{O_2}/dt equals the product of the cardiac output Q_t (see below) and the difference in the oxygen partial pressure in the arteries and veins, $p_a - p_v$

$$\frac{dV_{O_2}}{dt} = Q_t(p_a - p_v). \tag{8.33}$$

If the lungs are bringing in air fast enough, then $p_a - p_v$ is fixed, and during aerobic exercise dV_{O_2}/dt increases linearly with Q_t. For a person with average fitness, the maximum blood flow rate is ≈19 L/min, for a highly fit person it is ≈25 L/min, and for an elite athlete it can be 35 L/min.

Oxygen combines with hemoglobin in the red blood cells in the lungs where the partial pressure of oxygen is high, about 100 mmHg. It is transported in the arteries to the tissues where it is released because the partial pressure of oxygen is low – and it is then used. The blood in the veins is then depleted in oxygen. Figure 8.25a shows the hemoglobin–oxygen dissociation curve. Clearly, hemoglobin is over 90% saturated with O_2 for partial pressures above 60 mmHg O_2. Increased CO_2 levels, increased temperature, and decreased pH all shift this curve to the right (Fig. 8.25b), which improves body performance. In Fig. 8.25a the dissociation curve for the lung is seen to be to the left of that in the tissues because the pH is higher and the CO_2 level is lower in the lung, increasing oxygen binding in the lungs relative to that in the tissues. During exercise, the muscle tissue pH falls and the local partial pressure of CO_2 and the local temperature increase. All of these changes move the curve to the right and this leads to more oxygen release (Fig. 8.25b).

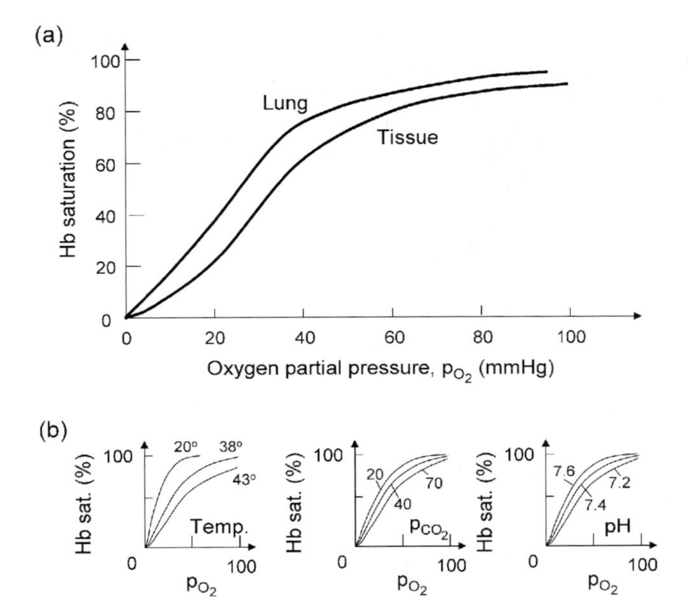

Fig. 8.25. (**a**) Hemoglobin–oxygen equilibrium in the lungs and tissue. During exercise oxygen intake is improved by the lung curve moving to the left and the tissue curve (exercising muscle) moving to the right due to increasing temperature, CO_2 partial pressure, and 2,3-diphosphoglycerate (DPG) (an end-product of red blood cell metabolism), and decreasing pH. (**b**) The hemoglobin–oxygen equilibrium shifts to the right with increasing temperature, increasing CO_2 partial pressure, and increasing DPG (not shown), and decreasing pH. (Based on [411] and [419])

During exercise $p_{\mathrm{a}} - p_{\mathrm{v}}$ increases from the resting value of about 50 mL of oxygen per L of blood to 150 mL/L in normal people at $(\mathrm{d}V_{\mathrm{O}_2}/\mathrm{d}t)_{\mathrm{max}}$ (and to 160–170 mL/L in very fit people), in part because blood flow is being diverted from the organs to the muscles, where oxygen extraction is higher because of the exercise (see Fig. 8.25).

8.2.4 Blood Flow Rates and Speeds

The heart pumps about 80 mL (= 80 cm^3) of blood per contraction; this quantity is called the *stroke volume* V_{stroke}. The pump rate is the *heart beat rate* F of about 60/min or 1/s = 1 Hz. The *cardiac output* or total volumetric flow rate Q_{t} is the product of these two

$$Q_{\mathrm{t}} = F V_{\mathrm{stroke}} \tag{8.34}$$

or about 80 cm^3/s = 4.8 L/min. The total volume of blood is about 4.5–5.0 L, so all the blood is pumped throughout the body every minute. The flow rate in the arteries, arterioles, capillaries, venules, and veins are all the same because of the continuity of flow. (Q_t is actually a little less in the venules and veins because of the net fluid loss in the capillaries.)

The total flow in each of these vessel systems, Q_t, equals the total cross-sectional area A times the blood speed u, $Q_t = Au$. The parameters A and u are plotted in Fig. 8.16, which shows this inverse relationship for a flow rate of $90\,\text{cm}^3/\text{s}$. The cross-sectional area of the aorta is $3\,\text{cm}^2$, so in the aorta $u = (90\,\text{cm}^3/\text{s})/3\,\text{cm}^2 = 30\,\text{cm/s}$. In the capillaries the flow speed is much slower, $(90\,\text{cm}^3/\text{s})/4{,}000\,\text{cm}^2 = 0.02\,\text{cm/s} = 0.2\,\text{mm/s}$. The net cross-sectional area in the capillaries is larger ($\sim$4,000 cm^2) even though they are very small ($\sim$3.5 μm in radius) because there are so many of them ($\sim 10^{10}$). In the vena cava the flow speed is relatively fast $(90\,\text{cm}^3/\text{s})/18\,\text{cm}^2 = 5\,\text{cm/s}$. In the arterial and venous systems, the smaller the vessel radius, the larger the total cross-section of all vessels in that order and the slower the blood speed. These are actually average blood flows during each cycle.

The maximum Reynolds number ($Re = \rho u d/\eta$, (7.11)) over a cardiac cycle ranges from $\sim$6,000 in the heart and aorta to $<10^{-3}$ in the capillaries. The nominal lower threshold for turbulent flow is $Re \sim 2,000$, so it is possible that flow in the aorta is turbulent.

The overall flow in the systemic arterial system can be described by relating the total cardiac output Q_t to the systemic arterial pressure P_{sa}, by

$$P_{sa} = (\text{TPVR})Q_t, \tag{8.35}$$

where TPVR is the total peripheral vascular resistance – which is due to the combined effect of all the organ beds of systemic circulation (mostly arterioles and capillaries). (P_{sa} should really be replaced by the pressure drop in the system. See Problem 8.20.) Normal values for the systemic system range from 700 to 1,600 dyne-s/cm^5, and analogous normal values for the pulmonary system range from 20 to 130 dyne-s/cm^5. Equation (8.2) applies to an individual vessel, while this describes the entire systemic system. The body regulates P_{sa} by controlling the cardiac output and this peripheral resistance. When we lie down, a large volume of blood is transiently stored in the lower extremities and abdomen, and so when we stand there is initially less flow of blood to the heart and a drop in blood pressure, which can make you faint. Even though Q_t decreases, P_{sa} drops only mildly because there is a prompt reflex that increases the TPVR (vasoconstriction). In contrast, when blood pressure rises suddenly, feedback tends to decrease the overall vascular resistance, to restore a lower blood pressure. These are two examples of body feedback and control, as described in Chap. 13.

We can also evaluate the overall compliances of the vascular systems, such as those of the systemic arterial and venous systems. The compliance is the reciprocal of the slope of a pressure–volume curve in Fig. 8.26. The smooth muscles surrounding a large vessel can change the volume of the vessel at a given pressure, either decreasing it (by stimulating the muscles) or increasing it (by inhibiting the muscles).

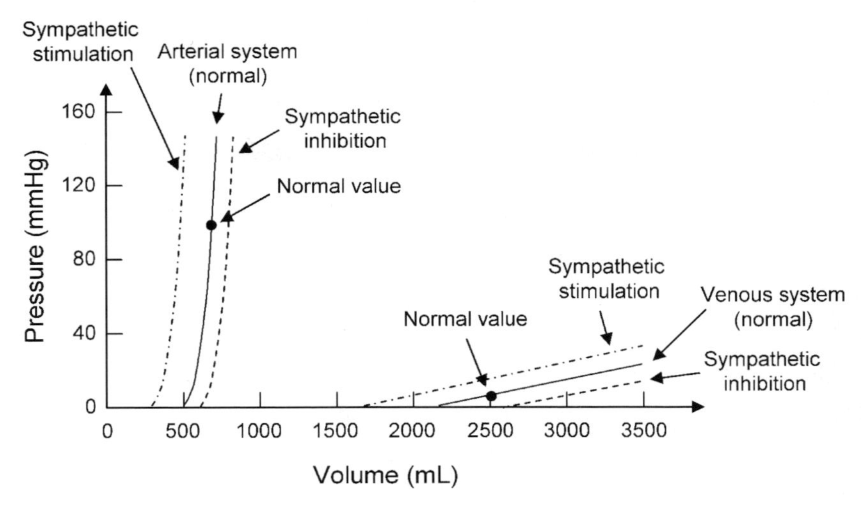

Fig. 8.26. Volume–pressure curves for the systemic arterial and venous systems, for normal conditions and for sympathetic stimulation and inhibition. (Based on [390])

During even moderate exercise the blood flow rate increases substantially, as seen in Fig. 8.27, and the absolute and relative distribution of blood to different parts of the body also changes radically. Figure 8.28 shows an example in which the flow rate increases from 5 to 12.5 L/min during exercise. There are extremely large increases of blood flowing to the skeletal muscle to supply oxygen for aerobic metabolism (up 1,066%), to the heart so it can pump faster (up 367%), and to the skin (up 370%) to assist cooling (which increases

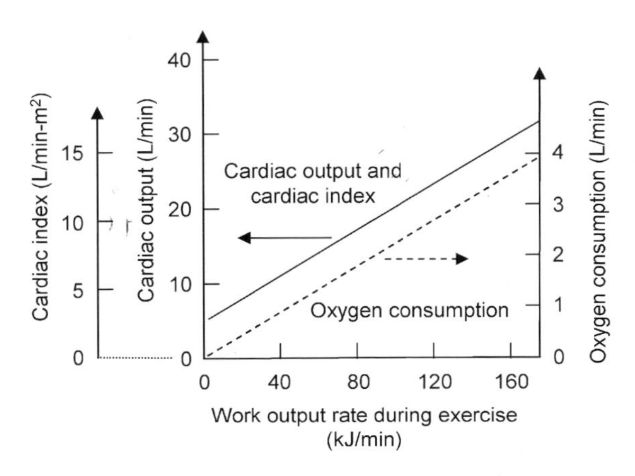

Fig. 8.27. Variation in cardiac output (and cardiac index) and oxygen consumption needed during varying levels of exercise with work output. (The cardiac index is the cardiac output divided by the person's surface area.) (Based on [389] and [390])

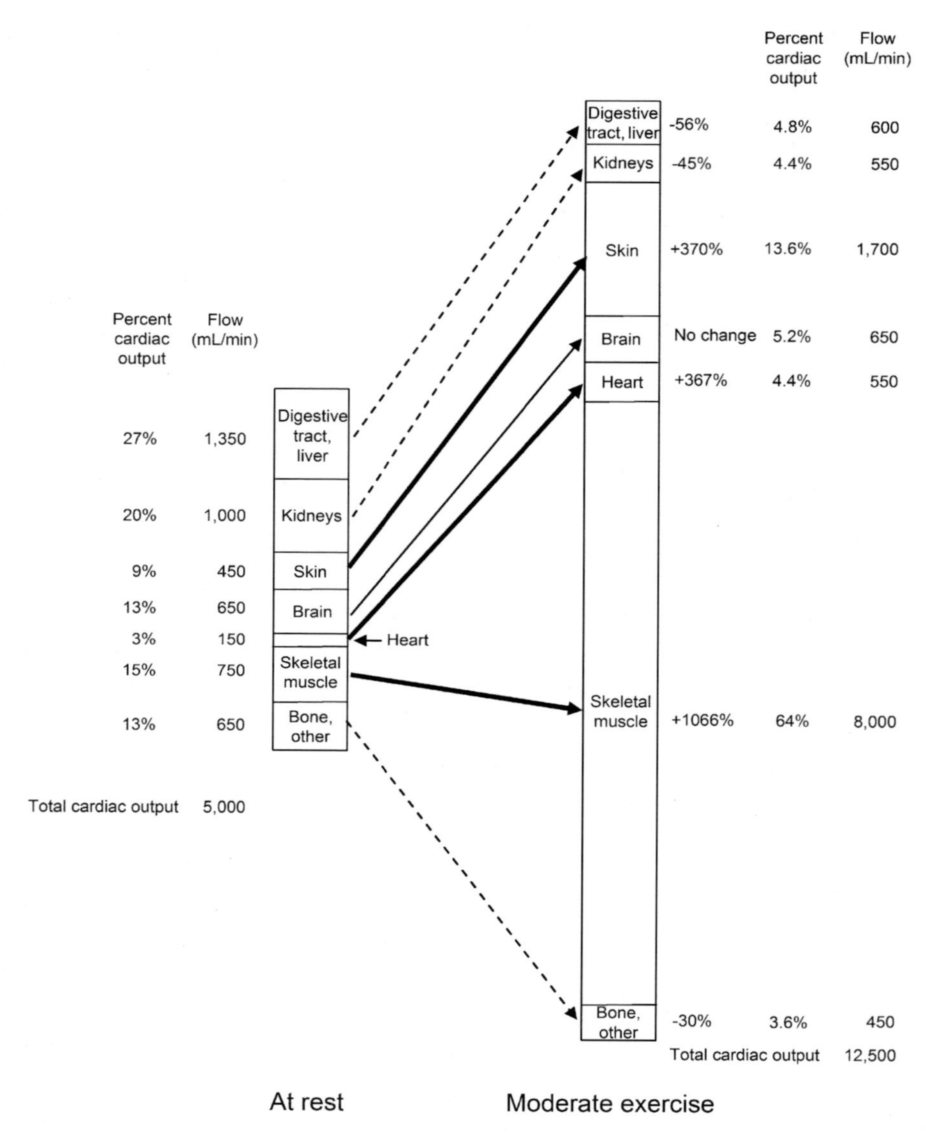

Fig. 8.28. Blood flow to different organs at rest and during moderate exercise, showing no change to the brain, increases to the skin, heart, and skeletal muscles (*thick arrows*), and less blood flow elsewhere (*dashed arrows*). (Based on [417])

the skin temperature, thus accelerating radiative and convection conduction from the body). Blood flow to the brain is unchanged. In contrast, blood flow to the digestive track, liver, and kidneys decrease by a factor of $\sim$2. These changes are also seen in Tables 8.5 and 8.6. Figure 8.29 shows that the blood

Table 8.5. Total cardiac flow (mL/min) for organs during exercise, including percentage of total flow. (Using data from [381])

organ	rest	light exercise	heavy exercise	maximal exercise
brain	750 (13%)	750 (8%)	750 (4%)	750 (3%)
heart	250 (4%)	350 (3.5%)	750 (4%)	1,000 (4%)
muscle	1,200 (21%)	4,500 (47%)	12,500 (72%)	22,000 (88%)
skin	500 (8.5%)	1,500 (16%)	1,900 (11%)	600 (2.5%)
kidney	1,100 (19%)	900 (9.5%)	600 (3.5%)	250 (1%)
abdomen	1,400 (24%)	1,100 (11.5%)	600 (3.5%)	300 (1.2%)
other	600 (10.5%)	400 (4%)	400 (2%)	100 (0.4%)
Total	5,800 (100%)	9,500 (100%)	17,500 (100%)	25,000 (100%)

flow to the calf during rhythmic exercise is higher than normal and it varies with time.

This increase in cardiac output occurs because of increases in both the heart rate and the stroke volume; the blood speed also increases because $Q_{\mathrm{t}} = Au$. The faster the heart rate, the shorter is diastole, while the duration of systole does not change. For short term (5–10 min) submaximal exercise, the cardiac output increases from 5 L/min to a new steady-state value in about 2 min. For a steady-state cardiac output of 18 L/min, the stroke volume increases from about 70 to 120 mL/beat and the heart rate F from about

Table 8.6. Approximate blood flow (perfusion) for tissues and organs, per gram. (Using data from [382])

tissue type	location or organ	specific blood flow rate (mm^3/s-g)
adipose tissue	abdomen, ~20 mm thick	0.51
	abdomen, >40 mm thick	0.31
	thigh, ~20 mm thick	0.33
bone	humerus, marrow flow only	0.055
connective tissue	typical basal (max)	0.50 (2.5)
muscle	typical basal (max)	0.50 (10)
organ	brain, basal (max)	9.1 (18.3)
	gastrointestinal track, basal (max)	6.7 (26.7)
	heart, basal (max)	13.7 (64.0)
	kidney, basal (max)	68 (100)
	liver, basal (max)	12 (54)
	lung, basal (max)	90 (490)
skin	typical resting flow (max)	1.7 (25.0)

In some cases the basal rate is given, along with the maximum rate in parentheses.

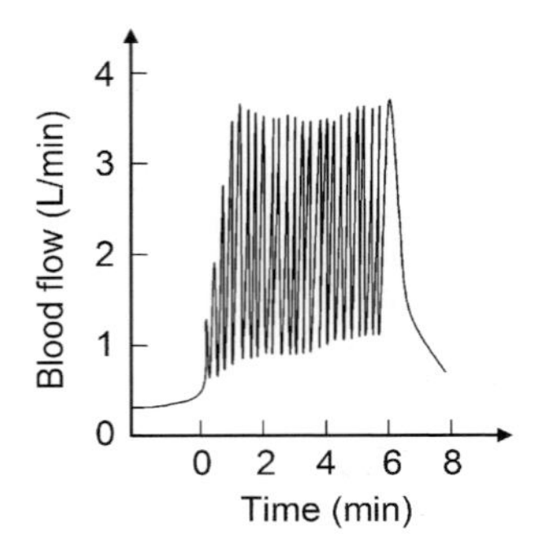

Fig. 8.29. Blood flow to the calf during rhythmic contraction exercises, showing less blood flow during contractions than between them. (Based on [366] and [390])

70 to 150 beats/min. The cardiac output returns to the resting value in 5–10 min after exercise.

For a longer submaximal workout (30–60 min), the new steady-state cardiac output is maintained, but the stroke volume slowly decreases and the heart rate gradually increases with time, particularly in warmer environments. This *cardiovascular drift* is caused by a decrease in the venous return of blood to the heart, which decreases the stroke volume and so the heart rate must increase to maintain the same cardiac output. This decrease in venous return is caused by two factors. (1) During such exercise more blood flows under the skin to help lower the increase in body core temperature caused by the increased metabolic activity (Chap. 6), and this lowers the steady-state flow of blood back. (2) During exercise water flows from the blood to the surrounding cells and tissues because of increased arterial pressure and the compression of venules due to muscle action. This produces a steady-state decrease in blood plasma and blood volume, and a steady-state decrease in blood returning to the heart and the stroke volume.

Stroke volume and cardiac output are determined by the preload and afterload conditions (see later), contractility (ability to contract), and heart rate. The cardiac output is not directly regulated, but there is a feedback and control system that regulates arterial pressure that affects the heart rate and contractility, as well as afterload and other factors that control the preload [415].

The Frank–Starling mechanism (or Starling's Law of the heart) states that the larger the end-of-diastole volume or pressure (the *preload*), the larger the stroke volume, as is seen in Fig. 8.30. Furthermore, the larger the aortic pressure, the less blood can be ejected by the left ventricle (the *afterload*), as is seen in Fig. 8.31. An increase in the heart rate also increases cardiac output, however, the increase is sublinear because the stroke volume decreases (Fig. 8.32) due to the above preload and postload factors. With greater cardiac output there is less blood in the veins to return to the heart for diastole (lower preload) and the arterial pressure is higher so the heart

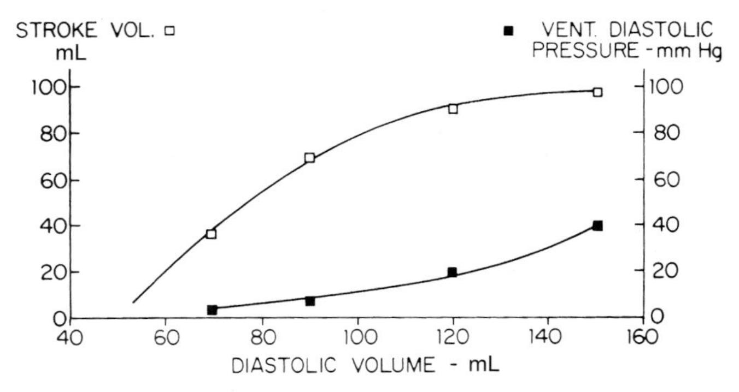

Fig. 8.30. The stroke volume increases with diastolic volume and pressure, as seen with data for four heart beats. This dependence, along with the explanation of it, is known as the Frank–Starling mechanism. (Reprinted from [415]. Used with permission of Elsevier)

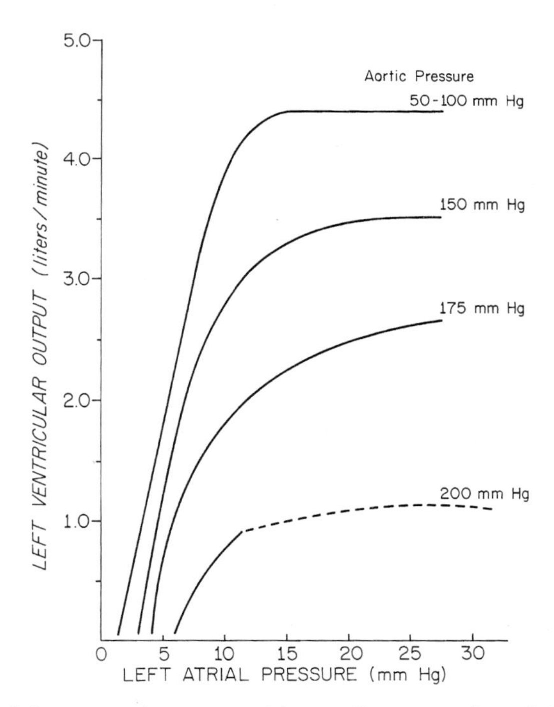

Fig. 8.31. The left ventricular output (the cardiac output) vs. left atrial pressure for different aortic pressures. It increases with this atrial pressure and then levels off, and decreases with increasing aortic pressure. (Reprinted from [415]. Used with permission of Elsevier; adapted from [413])

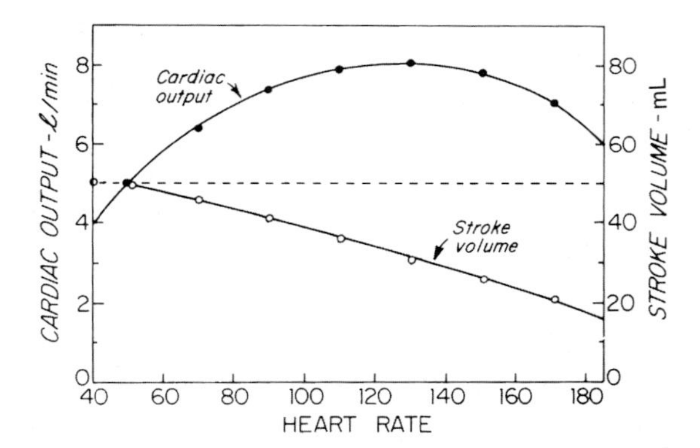

Fig. 8.32. Typical dependence of cardiac output and stroke volume on heart rate. (Reprinted from [415]. Used with permission of Elsevier)

can eject less blood (higher afterload), assuming the peripheral resistance is constant.

For the most part, these are changes in the systemic system. Similar increases in cardiac output have to occur in the pulmonary system. (Why?) This occurs by an increase in the number of open capillaries in the lung, by up to a factor of three, and by a distending of all the pulmonary capillaries, which increases the flow in each capillary by up to a factor of two, with very little change in the pulmonary arterial pressure (Fig. 8.33).

The maximum heart rate $F_{\max}$ (in beats/min) depends on age Y (in years) as

$$F_{\max} = 220 - Y. \tag{8.36}$$

The standard error in this relation is ± 10 beats/min, which means 67% of people have a maximum rate ± 10 beats/min within the value predicted by

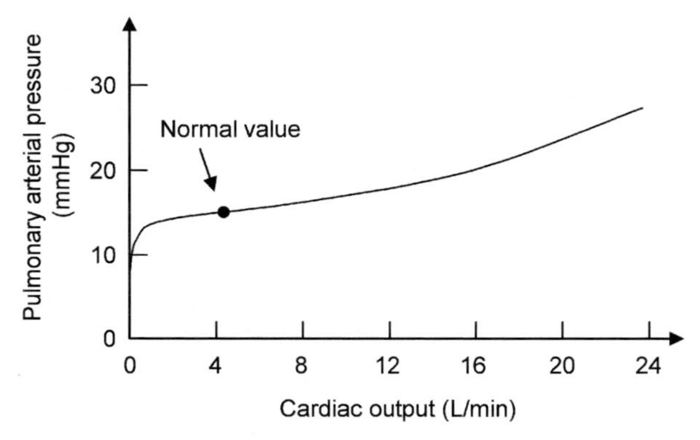

Fig. 8.33. The pulmonary arterial pressure vs. cardiac pressure, showing it does not change much during exercise. (Based on [390])

this relation and 95% of all people have a rate that is ±20 beats/min within the predicted value. Because F does not depend on the level of fitness, athletes increase their maximum cardiac output by increasing their stroke volumes. Stroke volumes in untrained athletes are 50–70 mL at rest and reach 80–110 mL during heavy activity. For trained and highly trained athletes these stroke volumes increase to 70–90 mL and 90–110 mL at rest and 110–150 mL and 150–220 mL during heavy activity.

In steady state, the cardiac output Q_t must equal FV_{stroke} (8.34), as well as $P_{sa}/TPVR$ (8.35). During heavy exercise, Q_t increases from 5 to 20 L/min, so not only must F and V_{stroke} increase, as we have described, but $P_{sa}/TPVR$ must increase accordingly. Diastolic pressure changes little during exercise, remaining within ±10 mmHg of the resting value. The systolic pressure increases to about 200 mmHg for men and 180 mmHg for women. Using (8.1), the mean arterial pressure then increases only to 140–150 mmHg, which cannot account for most of the increase in blood flow rate. During heavy exercise the systemic vascular resistance TPVR decreases to about 40% of its resting value because of the widening of muscular vascular beds that are normally constricted at low levels of activity.

Figure 8.34 show that blood pressure typically increases with age. Hypertension begins with systolic pressure ≥140 mmHg or diastolic pressure ≥90 mmHg. In essential hypertension this blood pressure is heightened for no obvious reason. The average blood pressure is the product of the total peripheral vascular resistance and the cardiac output, $P_{sa} = (TPVR)Q_t$ (8.35). In people under 40 years of age, hypertension is driven by increased cardiac output, with normal TPVR. In older people, the cardiac output is normal or reduced, but the TPVR is high.

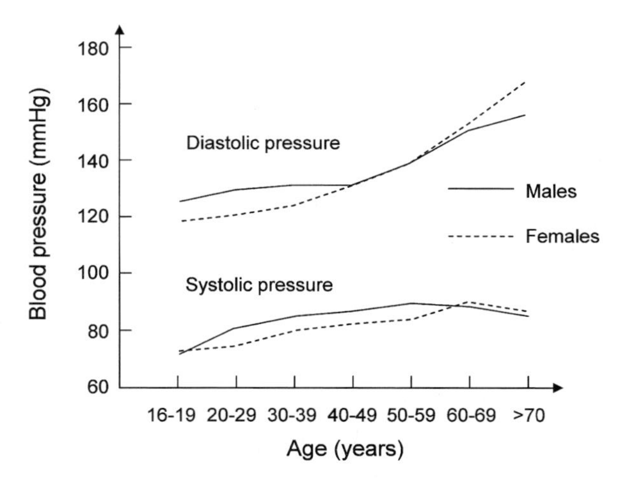

Fig. 8.34. Systolic and diastolic blood pressure is shown for males and females, averaged over age groups. The trend is to increased blood pressure with age. (Based on [400] and [403])

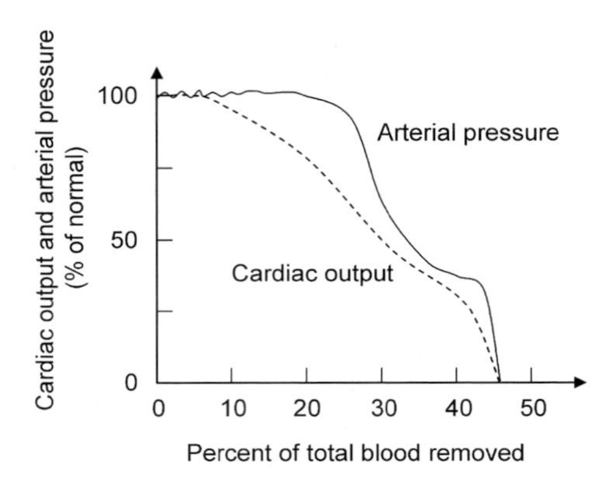

Fig. 8.35. Cardiac output and arterial pressure with decreased blood volume from hemorrhaging. (Based on [390])

Circulatory shock occurs when there is too little blood flowing generally in the body, and this results in tissue damaged from the inadequate delivery of oxygen and nutrients to the cells and the inadequate removal of waste products. Such shock can occur from inadequate pumping of blood by the heart or by inadequate venous return of blood to the heart, such as due to diminished blood volume (*hypovolemia*), decreased capillary vasomotion, or obstructed circulation. *Hemorrhage* is often the cause of the diminished blood volume. Figure 8.35 shows that cardiac output and arterial pressure can withstand a $\sim$10% blood loss – if this were not so you would not be able to donate blood – but decrease for larger losses and approach zero with 35–45% blood loss. The localized loss of blood flow to the brain, *strokes*, is discussed later. This occurs due to clogged arteries and hemorrhaging. The localized loss of blood flow to the heart results in a *myocardial infarction* (heart attack).

8.2.5 Consequences of Clogged Arteries

Atherosclerosis (a-thear-oh'-scler-oh-sis) occurs when a deposit or atheroma (a-thear-oh'-ma) (or plaque) forms on an arterial wall (Fig. 8.36). The smaller

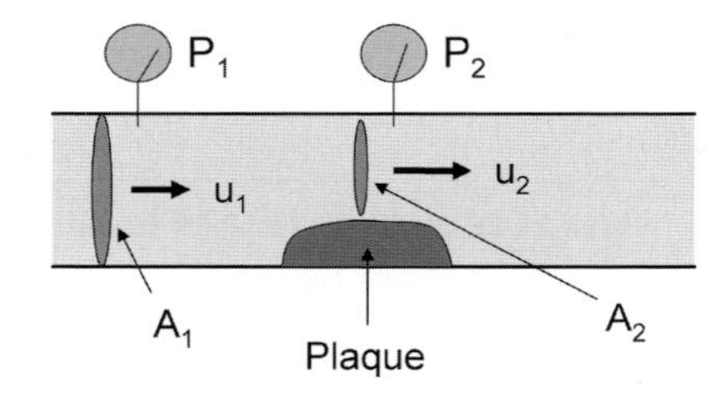

Fig. 8.36. Sketch of flow in a clogged artery

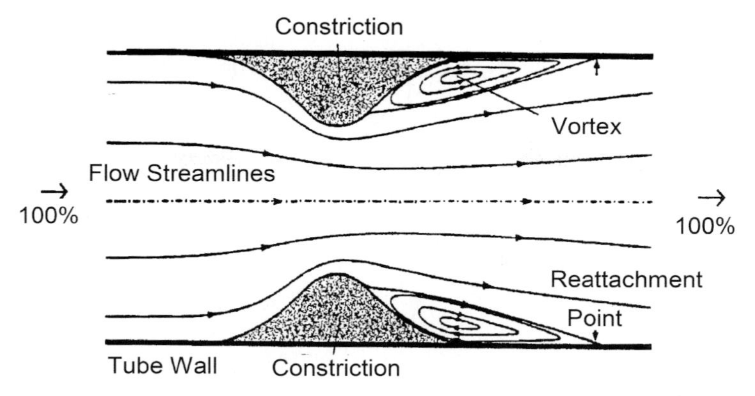

Fig. 8.37. The flow is partially turbulent in clogged arteries. (From [382]. Adapted from [398]. Courtesy of Robert A. Freitas Jr., Nanomedicine, Vol. 1 (1999), http://www.nanomedicine.com)

cross-sectional area at this site, because $A_2/A_1 < 1$, leads to a faster flow speed due to continuity of flow, with

$$u_2 = \frac{A_1}{A_2} u_1 \tag{8.37}$$

from (7.16). For $A_2/A_1 = 1/3$, we find that $u_2 = 3u_1$. Also from Bernoulli's equation (7.18) we find that

$$P_2 - P_1 = \frac{1}{2} \rho u_1^2 \left(1 - \left(\frac{A_1}{A_2} \right)^2 \right) \tag{8.38}$$

and so $P_2 < P_1$ and for $A_2/A_1 = 1/3$ we see that $P_2 - P_1 = -4\rho u_1^2$. This pressure drop increases with blood speed and so it is expected to increase with increased physical activity. This flow is not necessary laminar in the occluded region, as is seen in Fig. 8.37.

In 1954 Arturo Toscanini was conducting the NBC Symphony Orchestra. (This orchestra was pretty prestigious then, performing on radio and the then-new television, but it does not exist now.) He was vigorously waving his arms, as conductors often do, and he fainted. Why? Equation (8.38) contains the answer and Fig. 8.38 illustrates why. He suffered a *transient ischemic attack*, or TIA [387, 399]. *Ischemia* (iss-kee'-mee-uh) is the local decrease in blood flow. TIA is the temporary loss of blood to the brain by the "subclavian steal syndrome." It results in temporary dizziness, double vision, headache, and weakness in the limbs. By its nature it is only temporary, but it indicates a more severe problem.

The left and right carotid arteries are two major arteries supplying blood to the anterior brain. The left and right vertebral arteries supply blood to the posterior part of the brain. They branch off from the subclavian arteries that

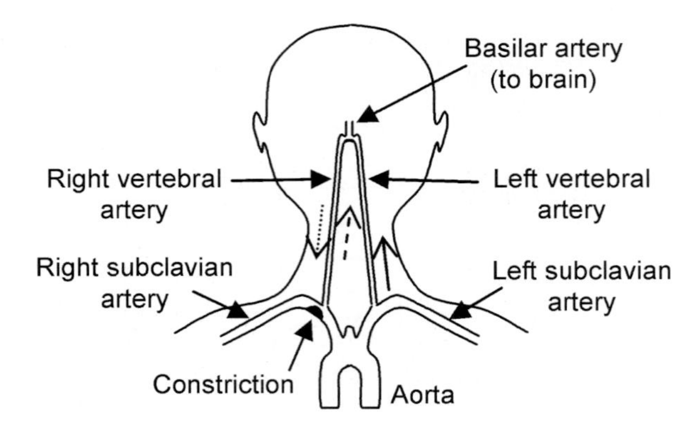

Fig. 8.38. Anterior view of the blood flow to the brain with the subclavian steal syndrome, resulting in a transient ischemic attack. Blood flow in the left vertebral artery is shown by the *unbroken line arrow*. Without the constriction, the blood flow in the right vertebral artery is normal, as shown by the *dashed arrow*, so there is normal blood flow to the basilar artery. With the constriction, there can be blood flow from the left vertebral artery to the right vertebral artery (*dotted arrow*), and there is no blood flow into the basilar artery. (Based on [387])

also supply blood to the arms. The internal carotid and vertebral systems join with each other at the base of the brain, forming the circle of Willis (Fig. 8.39, also see Fig. 8.43). Posteriorly, the flow in the left and right vertebral arteries merge to form a single basilar artery to the brain.

Say there is a constriction in the right subclavian artery near where the vertebral artery branches off (Fig. 8.38). The pressure before the constriction

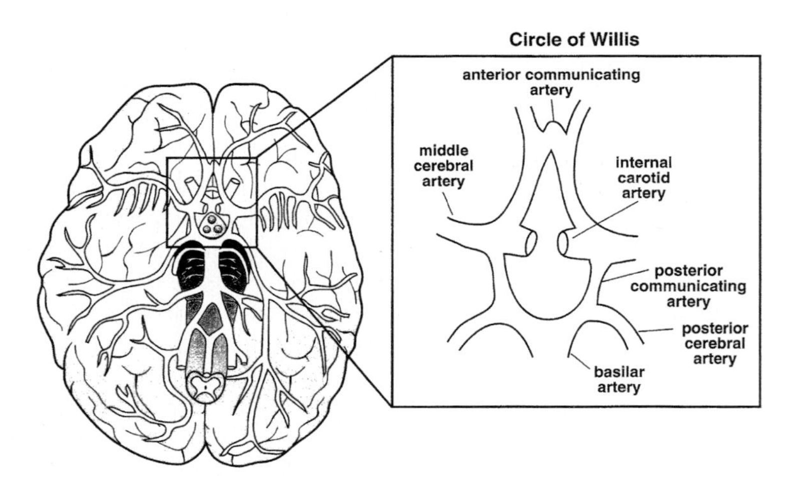

Fig. 8.39. Circle of Willis in the brain. (From [391])

and in the same region in the normal right side is P_1. Because of the constriction, the pressure on the left side is $P_2 < P_1$. This difference increases with faster blood flow. With vigorous motion of the arms, there is a need for greater blood flow in the subclavian arteries to supply more oxygen to the skeletal muscle in the shoulder and arms, and the blood speed in these arteries u_1 increases. If u_1 increases enough, P_2 becomes so much smaller than P_1 that flow in the left vertebral artery is diverted to the right vertebral artery – and does not flow to the basilar artery to the brain. When this happened to Toscanini, he fainted. He stopped waving his hands, of course. Blood flow to his arms then slowed down. The difference in P_1 and P_2 decreased to its usual smaller value (even with this constriction). Blood from both vertebral arteries then flowed to his basilar artery. His brain started receiving a normal flow of blood again, and he regained consciousness. All was fine – but this constriction had to be removed. (Toscanini never conducted again. The underlying reason for his fainting spell was not known in 1954; TIA was first explained in 1961.)

More examples of the effects of obstructions in arteries will be examined later in this chapter.

8.2.6 Work Done by the Heart and the Metabolic Needs of the Heart

How much work is done by the heart? Consider the left ventricle, which is a pump during systole, as diagrammed in Fig. 8.40.

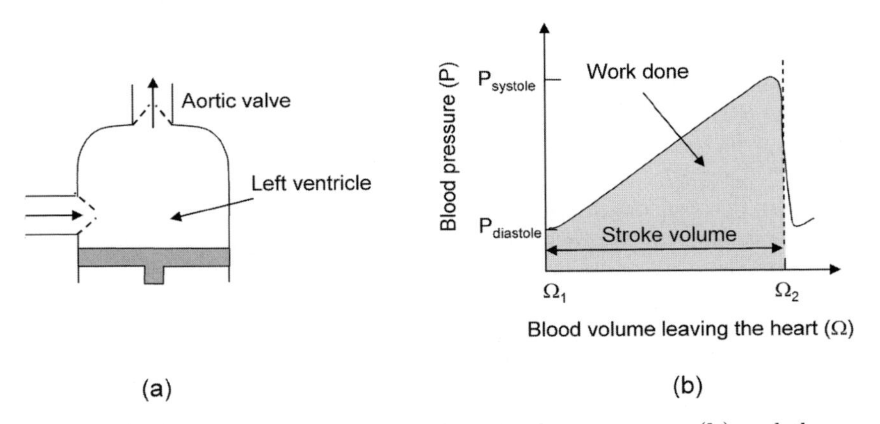

Fig. 8.40. (**a**) Schematic of the heart left ventricle as a pump, (**b**) and the pressure and volume of the left ventricle during systolic contraction during systole (from t_1 to t_2), showing the work done by the heart (*shaded area*). (Based on [367])

Work is done by this pump, with a force pushing the piston from the initial to final positions L_i to L_f to give

$$W = \int_{L_\mathrm{i}}^{L_\mathrm{f}} F\,\mathrm{d}L = \int_{L_\mathrm{i}}^{L_\mathrm{f}} (F/A)(A\,\mathrm{d}L) = \int_{V_\mathrm{i}}^{V_\mathrm{f}} P\,\mathrm{d}V, \tag{8.39}$$

where the force/area is the pressure, P, and the distance times the area is the ventricular volume, V. During systole the ventricular pressure increases from a very low value (that in the left atrium), to $P_\mathrm{diastole} \sim 80\,\mathrm{mmHg}$ (at time t_1), and then up to $P_\mathrm{systole} \sim 120\,\mathrm{mmHg}$, and it stays at this value until the end of systole (at time t_2). The aortic valve first opens when this pressure rises above P_diastole (at t_1), and blood is pumped out until systole is over (at t_2). At t_1, the volume of pumped blood is $\Omega_\mathrm{i} = 0$ and at t_2 it is $\Omega_\mathrm{f} = V_\mathrm{i} - V_\mathrm{f}$. Figure 8.40 shows this evolution of the ventricular pressure and pumped volume from t_1 to t_2.

The area under this curve is the work done and so $W = P_\mathrm{av} V_\mathrm{stroke}$, where P_av is average pressure during this cycle (averaged over the volume displaced and which does not necessarily scale linearly with time), and $V_\mathrm{stroke} = \Omega_\mathrm{f}$, the stroke volume. Clearly, $P_\mathrm{diastole} < P_\mathrm{av} < P_\mathrm{systole}$. For a linear variation, we see that $P_\mathrm{av} = (P_\mathrm{diastole} + P_\mathrm{systole})/2 \sim 100\,\mathrm{mmHg} = 1.3 \times 10^4\,\mathrm{N/m^2}$. (This averaging is different from that in (8.1) because of the simplicity of this model. This leads to an $\sim$20% uncertainty.) We take $V_\mathrm{stroke} = 80\,\mathrm{cm^3} = 8 \times 10^{-5}\,\mathrm{m^3}$, so $W = (1.3 \times 10^4\,\mathrm{N/m^2})(8 \times 10^{-5}\,\mathrm{m^3}) = 1.04\,\mathrm{J}$ per cycle. With a heart rate of $60/\mathrm{min} = 1/\mathrm{s}$, the rate the left ventricle does work is $P_\mathrm{power,mech,av} = (1.04\,\mathrm{J}$ per cycle$)(1\,\mathrm{cycle/s}) = 1.04\,\mathrm{W}$.

The efficiency, ϵ, of converting metabolic energy into this mechanical work is approximately 20% (and sometimes this range is given as 12–30%), and so the metabolic power needed to run the left ventricle is $P_\mathrm{power,metab,av} = P_\mathrm{power,mech,av}/\epsilon = 5$ W.

The heart pumps for about 1/3 of the cardiac cycle and rests for the other 2/3 of the time. Therefore the peak powers are higher than these average values by a factor of 3, with $P_\mathrm{power,mech,peak} = 1.5\,\mathrm{W}$ and $P_\mathrm{power,metab,peak} = 15\,\mathrm{W}$.

The energy consumed to run the left ventricle is $(86{,}400\,\mathrm{s/day})(5\,\mathrm{W}) = 4.32 \times 10^5\,\mathrm{J/day} = 104$ kcal/day. So far we examined the work done by only the left ventricle. The right ventricle pumps the same volume per cardiac cycle (to maintain the steady-state flow throughout), but at a pressure 1/5 times that of left ventricle, so the work and all of these powers are smaller by a factor of five. This increases the required metabolic power by 20%. Similarly, the pressures for the two atria are also relatively very small. Overall, with 20% muscle efficiency we expect to need $\sim$125 kcal/day to run the heart; with 10% muscle efficiency it would be $\sim$250 kcal/day.

The experimental value for the BMR contribution for the entire heart is $\sim$117 kcal/day (Table 6.17), which is close to our estimate. The biggest uncertainty here is the efficiency of the cardiac muscle.

What happens with strenuous exercise? The blood pressure can increase by 50% and the blood flow rate can increase by a factor of 5. Therefore, the mechanical power exerted by the heart and the associated metabolic requirements can increase by a factor of 7.5.

8.3 Strokes and Aneurysms

Cerebral blood flows from the internal carotid and vertebral arteries and through the circle of Willis at the base of the brain (Fig. 8.39), and then permeates the brain through a complex series of capillaries.

Any severe restriction of blood to the brain is called a *stroke* [391, 393]. *Ischemia* means there is a lack of blood flow. The nearby tissue becomes deficient in oxygen and metabolites, and has excessive metabolic waste products. (We saw an example of ischemia earlier this chapter with the Toscanini TIA. Because TIA is transient, there is an oxygen deficiency due to the stroke but the patient can still recover with little or no brain damage.) *Hypoxia* means a lack of oxygen, and it can result from ischemia or other causes, such as high altitude (see Chap. 9) or CO poisoning. *Infarction* means that the stroke causes permanent brain damage. (The terms *ischemia* and *infarction* are actually more general and also apply to tissues outside the brain, with a myocardial infarction in the heart as one example.) The transition from the reversible event to the irreversible infarct with the formation of necrotic tissue occurs when the stroke is particularly long or of particularly large magnitude. (Necrotic tissue is dead tissue that did not die in a manner programmed by the body, which is in contrast to apoptotic tissue which the body kills as part of the life cycle.) The occurrence of a transient ischemic attack, or ministroke, sometimes means a more damaging stroke is imminent.

Hemorrhagic strokes are due to a ruptured vasculature (blood vessels) within the brain, attributed to an aneurysm or weakened blood vessel (Fig. 8.41). An *aneurysm* is an enlarged blood vessel. In addition to the loss

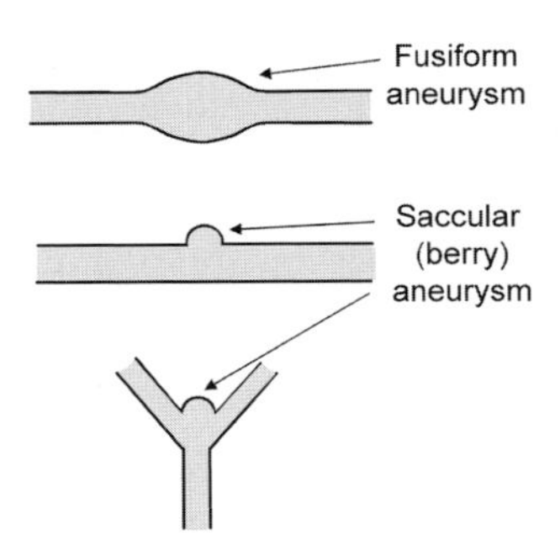

Fig. 8.41. Fusiform and saccular/berry aneurysms, the latter in a vessel and for bifurcated flow

of blood to the targeted regions, during hemorrhages blood fills the surrounding spaces and compresses the surrounding tissues. This accumulating blood compresses other blood vessels, decreasing their diameters and reducing the flow of blood to other parts of the brain, and increases the intracranial pressure, which leads to neurological complications. About 20% of all strokes are hemorrhagic; they occur mostly in the young and middle-aged, due to vascular *lesions* such as arteriovenous malformations and aneurysms. (Lesions are entities of diseased or abnormal tissue.) In the elderly, blood vessels are brittle and less distensible due to atherosclerotic deposits, and this can lead to possible spontaneous rupture of these vessels in the brain and hemorrhagic stroke. In *atherosclerosis* ("hardening of the arteries") lipid or fatty deposits in the blood accumulate on the inner vessel wall and eventually form hard arterial plaques.

During an *ischemic stroke* there is cessation of blood flow in arteries transporting blood to the brain due to a luminal obstruction or clogging. (As alluded to earlier, the *lumen* is the opening of a blood vessel.) About 80% of all strokes are ischemic. An *embolus* is a gaseous (air bubble), particulate matter, or blood clot that travels within a blood vessel and causes the obstruction of blood flow. For example, artherosclerotic lesions (in the brain and elsewhere in the body) cause an irregular inner vessel surface and blood platelet aggregation due to turbulence, that can produce emboli that are platelet aggregates. Such emboli can be formed outside the brain, in the heart, lungs, and systemic circulation, and travel to the brain until they reach vessels too small for further travel; this prevents blood flow to more distal (downstream) regions in the brain. A *thrombus* is blood coagulation that can produce a local fibrin clot; this can also cause an ischemic stroke. (If the thrombus forms and moves elsewhere, such as to a smaller diameter vessel or a partially occluded vessel, it is an embolus.)

The majority of cerebral aneurysms are *saccular (or berry) aneurysms* that most often occur where large cerebral arteries bifurcate (Figs. 8.42 and 8.43). In *fusiform aneurysms* there is uniform ballooning of the circumference of the vessel walls, instead of in localized regions of the vessels as in saccular aneurysms; this leads to ellipsoidal or football shaped aneurysms. (Fusiform means tapering at each end. Here it indicates a cylindrically symmetric aneurysm that tapers to the normal vessel at either end, while for the fusiform muscles in Chap. 5 it indicates a cylindrically symmetric muscle that tapers to tendons on each end.) Fusiform aneurysms are less common in the brain than are saccular aneurysms, but are common elsewhere in the body. The abdominal aortic aneurysm, which develops along the aorta in the abdominal or gut region, is the most common aneurysm found in the body and is a fusiform aneurysm (Fig. 8.44).

Healthy arteries contain the structural proteins elastin and collagen. Collagen has the larger Young's modulus (Table 4.2) and is expected to dominate the elastic properties of arteries. Still, the elastin contributes to the distensibility of the artery. The resistance to stretching at low pressures seems to be due to the elastin fibers, at normal physiological pressures it is due to elastin

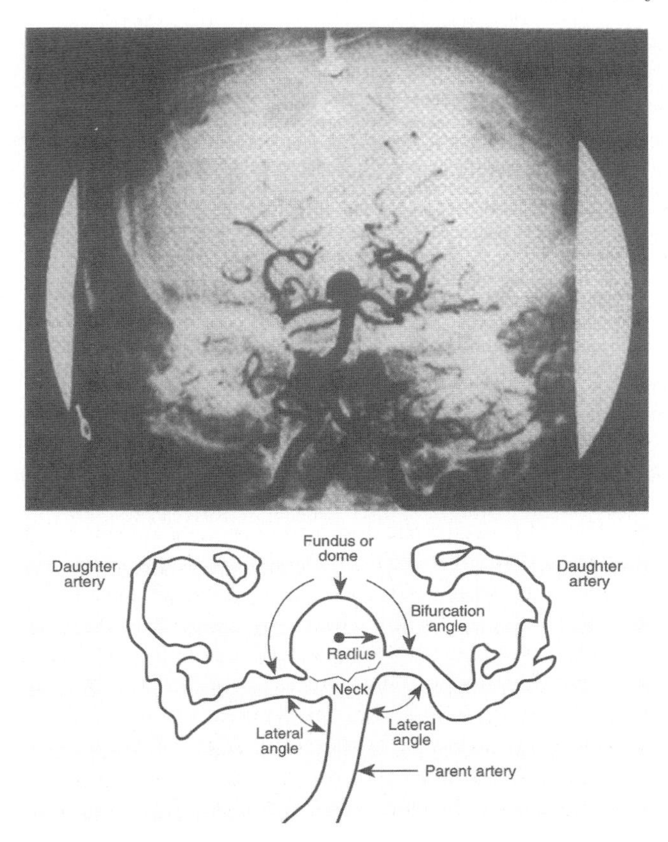

Fig. 8.42. Saccular (berry) aneurysm at an apex of a branching vessel, showing an angiographic projectional image (*top*) and a model (*bottom*). (From [391])

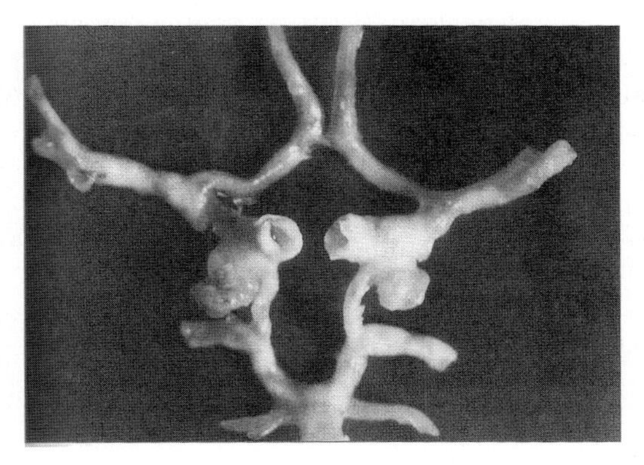

Fig. 8.43. Photograph of an inferior view of an excised human circle of Willis. Bilateral (i.e., on both sides) saccular aneurysms are seen near the junction between the internal carotid artery and the circle; the larger lesion (the one on the *left side* of the photograph) had ruptured. (From [395])

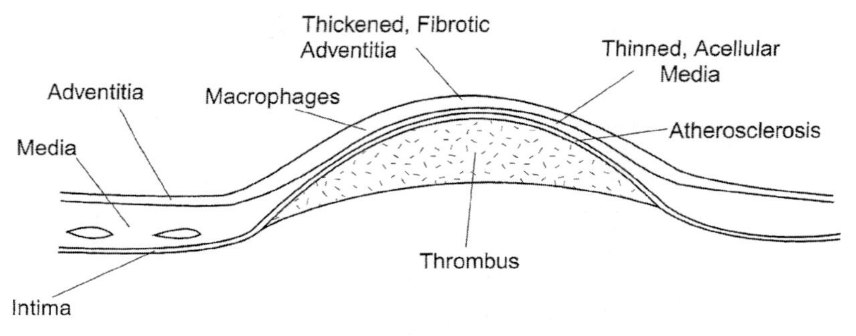

Fig. 8.44. Schematic of an abdominal aortic aneurysm, showing an attenuated media and an intraluminal thrombus. (From [395])

and collagen fibers, and at even higher pressures it is due to collagen. When the artery wall balloons or sacculates (i.e., it balloons in one circumferential part of the wall) as the aneurysm develops, the elastin becomes less effective in maintaining structural integrity of the artery and the collagen takes on most of the load. (The artery becomes less distensible and this translates to greater stress for the same strain, thereby accelerating structural fatigue.) This process accelerates the load on the arterial wall, and leads to rupture. A possible scenario for the formation and rupture of a saccular aneurysm is shown in Fig. 8.45. Figure 8.46 shows the equilibrium circumferential tension for a vessel assuming the Law of Laplace, for a normal artery, and one with an aneurysm. Figure 7.11 shows that the maximum blood flow velocity is in the center, where the wall shear stress is minimum. The minimum blood flow velocity is near the wall, where the wall shear stress is maximum.

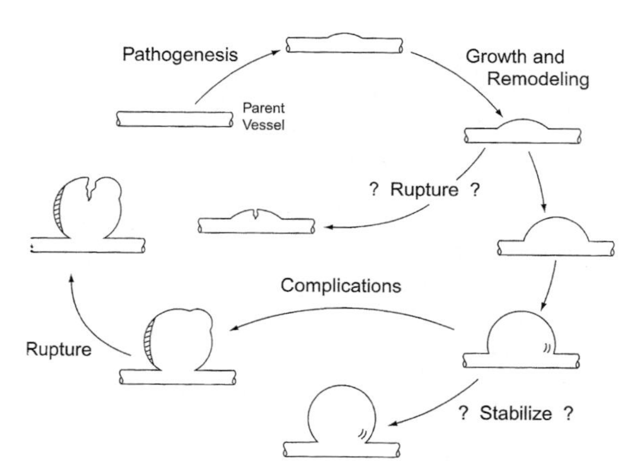

Fig. 8.45. A possible natural history for the development of a saccular aneurysm. A local weakening of the vessel wall, leading to a mild dilatation, can be caused by an initial "insult" from one of several causes. (From [395])

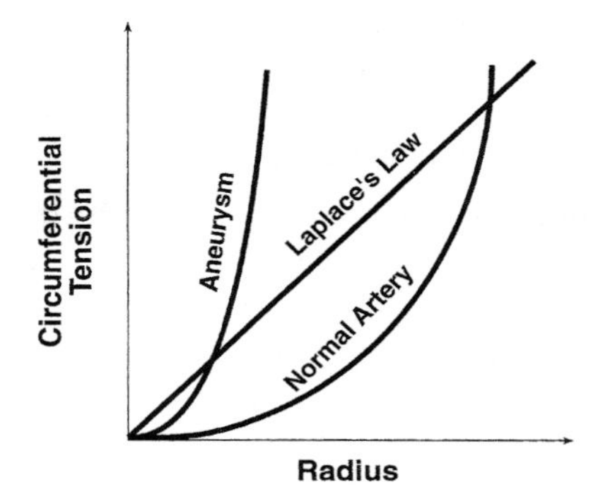

Fig. 8.46. Circumferential tension needed for equilibrium as a function of vessel radius, assuming the Law of Laplace, a normal artery, and an artery with an aneurysm. (From [391, 392])

There are (at least) four physical reasons why the larger radius of a fusiform aneurysm can lead to rupture. All are related to the stress relation $T/w = r(\Delta P)/w$, where w = wall thickness and r is local vessel radius. (1) r is larger so more tension is needed to withstand even an unchanged pressure difference (because $T = r(\Delta P)$). (2) This increase in vessel radius can be accompanied by thinning walls (if the volume of vessel wall per unit length, $\sim 2\pi rw$, is relatively unchanged by the aneurysm), so the stress T/w increases even more. (3) With this wall thinning there may be damage that lowers the UTS locally, and the UTS needs to be $\gg T/w$ to avoid rupture. (4) A larger r, and the concomitant larger cross-sectional area A, leads to a slower blood speed u, through volumetric continuity (7.16). This in turn leads to a larger pressure P, through Bernoulli's equation (7.18). This increases the tension that the vessel must withstand. The magnitudes of several of these effects are evaluated in Problem 8.32.

8.3.1 Arterial Bifurcations and Saccular Aneurysms

Two of the reasons for the formation of saccular aneurysms are the forces on the arterial walls caused by the change of momentum (like the fire hose effect in curving arteries) and shear stress. Figure 8.47 depicts a "parent" artery with cross-sectional area A_1 in which blood flows at an average speed u_1, which divides into two "daughter" vessels at an angle θ to the parent (and 2θ to each other). (The half-angle is shown as θ_2 in the figure.) Each daughter vessel has cross-sectional area A_2 and blood speed u_2. The angle 2θ usually ranges from 30 to 120° (also see Problems 8.28–8.31). The apex of the bifurcation (Fig. 8.47) is the site of maximum stress due to the impact, deflection, and separation of the flow, and possible turbulence and vortex formation (Fig. 8.48). Conservation of flow rate Q (volume flow/time)

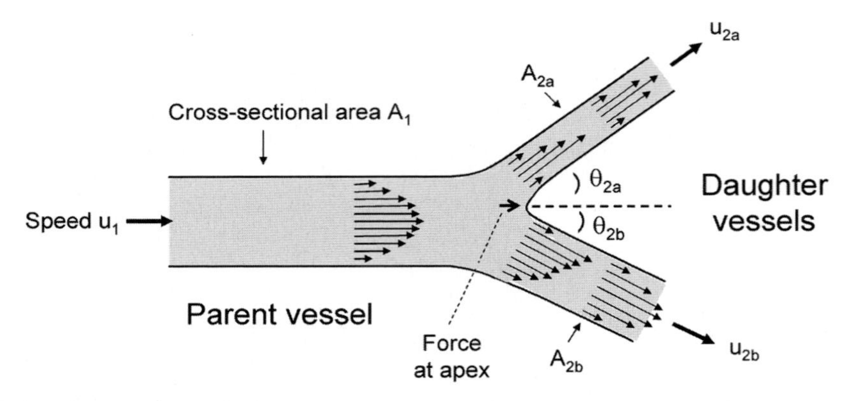

Fig. 8.47. Diagram of asymmetric bifurcation in vessel, with bifurcation angle $\theta_{2a} + \theta_{2b}$. For the symmetric bifurcation described in the text the "a" and "b" labels can be ignored, and the half-angle is called θ. (See Problems 8.28–8.31)

means

$$Q = u_1 A_1 = 2u_2 A_2. \tag{8.40}$$

In these vessels, usually $2A_2 > A_1$, so $u_2 < u_1$.

How large is the force on the arterial wall at the apex? The linear momentum per unit volume of blood in the parent artery (which we will say is in the x direction) is ρu_1, where ρ is the blood mass density. The momentum per unit volume carried in each daughter artery is ρu_2, of which $\rho u_2 \cos\theta$ is along the x direction. This change in momentum causes a force on the arterial wall. The force this flow exerts on an imaginary screen across the vessel in the parent artery is the change of this momentum per unit time, which equals this linear momentum per unit volume × the flow rate

$$F_{z,\text{parent}} = \rho u_1 Q. \tag{8.41}$$

Because the flow rate in each vessel is $Q/2$ and there are two of them, the force of the flow in the daughter arteries is

$$F_{z,\text{daughters}} = 2\,\frac{\rho u_2 \cos\theta\, Q}{2} = \rho u_2 \cos\theta\, Q. \tag{8.42}$$

The difference of these forces is

$$\begin{aligned} F_{\text{arterial wall}} &= F_{z,\text{daughters}} - F_{z,\text{parent}} \\ &= \rho u_2 \cos\theta\, Q - \rho u_1 Q = \rho Q(u_2 \cos\theta - u_1) \end{aligned} \tag{8.43}$$

or with (8.40)

$$F_{\text{arterial wall}} = \rho Q u_1 \left(\frac{A_1}{2A_2}\cos\theta - 1\right) = \rho A_1 u_1^2 \left(\frac{A_1}{2A_2}\cos\theta - 1\right). \tag{8.44}$$

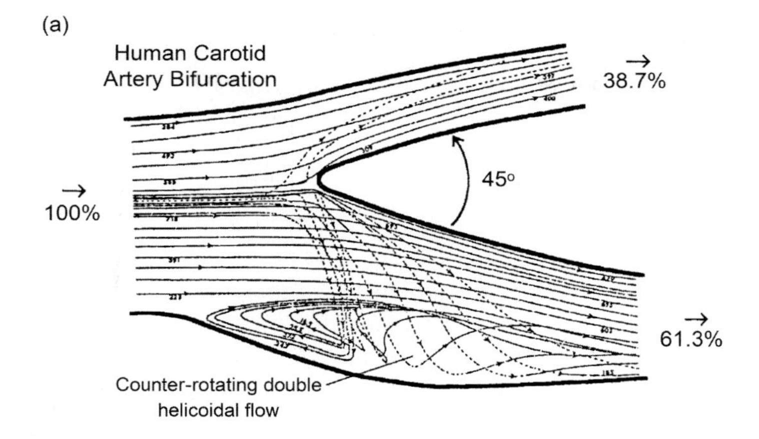

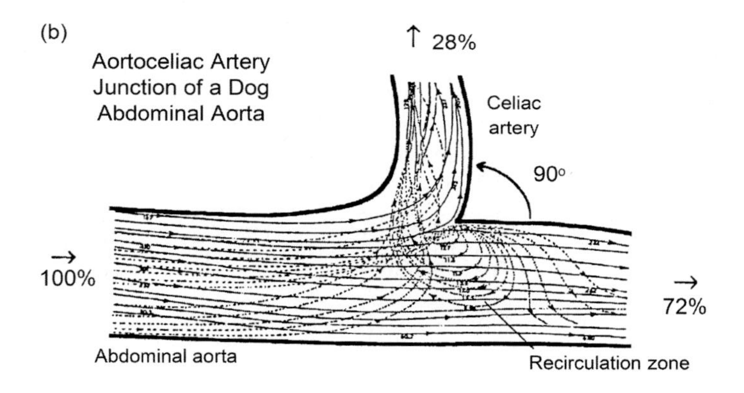

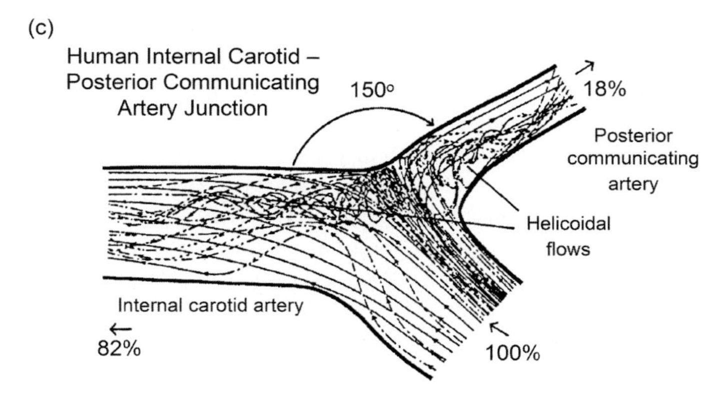

Fig. 8.48. Disturbed flow streamlines for progressively large angle bifurcations at bifurcations, and approach to steady flow afterward. (From [382], based on (**a**) [409], (**b**) [397], and (**c**) [398]. Courtesy of Robert A. Freitas Jr., Nanomedicine, Vol. 1 (1999), http://www.nanomedicine.com)

This has a maximum negative value of $-\rho Q u_1 = -\rho A_1 u_2^2$ when $\theta = 90°$ (which is a bit larger than the typical maximum angles). Because pressure is force/area, if this force is exerted on the vessel wall of cross-sectional area A_1, as in Fig. 8.42, there is a pressure on the arterial wall of

$$P_{\text{arterial wall}} = \frac{\rho Q u_1}{A_1} = \frac{\rho A_1 u_1^2}{A_1} = \rho u_1^2 \tag{8.45}$$

(which is also the kinetic energy per unit volume). This can lead to the formation of an aneurysm there. This mechanism of momentum change is the same as that causing a force on an arterial wall during flow in a curving artery. The saccular aneurysm gets larger and larger until it ruptures.

The shear stress near the bifurcation apex due to viscosity is another factor in the formation of a saccular aneurysm. Figure 8.47 shows that in steady flow the velocity is small near the walls and the velocity gradient and consequently the shear stress is large near the walls. After bifurcation and momentum transfer to the apex, the blood velocity, velocity gradient, and shear are larger near the vessel walls after the apex. The shear stress is large, until the flow pattern rearranges to give (7.40) (Fig. 7.11). This is also seen in Fig. 8.48 for bifurcations at increasingly large angles. Equation (7.41) can be used to estimate the distance from the apex for steady-state flow. Note that the Law of Laplace for spheres applies to saccular aneurysms.

8.3.2 Stenosis and Ischemic Strokes

Strokes can also occur by *stenosis* (narrowing or closure of lumens) or *occlusions* (closures or obstructions) (Fig. 8.49). They are most commonly due to artherosclerotic lesions. Such lesions are irregularly distributed masses of calcified fatty deposits that narrow the arterial lumen. If the normal inner diameter of the vessel is d_{norm} and the minimum diameter due to the stenotic lesion is d_{sten}, then the % stenosis is defined as: $((d_{\text{norm}} - d_{\text{sten}})/d_{\text{norm}}) \times 100\%$. They are characterized as being mild (1–39%), moderate (40–59%), severe (60–79%), critical (80–99%), and occluded (100%). Poiseuille's Law (7.24) shows that the flow decreases with decreasing lumen diameter for a given initial pressure. Flow decreases dramatically above the onset of critical stenosis, and the pressure drop across the stenosis increases, resulting in a need for a greater blood pressure to maintain the same flow rate.

This calcified lesion stiffens the vessel and abruptly changes the flow pattern as the blood flows from an elastic, distensible region of a vessel to this rigid and narrower region and back to a distensible vessel. Overall blood flow is slower because of the stenosis (unless the pressure increases), which can lead to clotting. The resulting clot (or thrombus) does not adhere well to the vessel wall and can move to elsewhere in the flow stream and this embolus can lead to a stroke. Within the stenosis itself, the blood flow is faster than just before it, from Bernoulli's Principle. This increased blood flow has

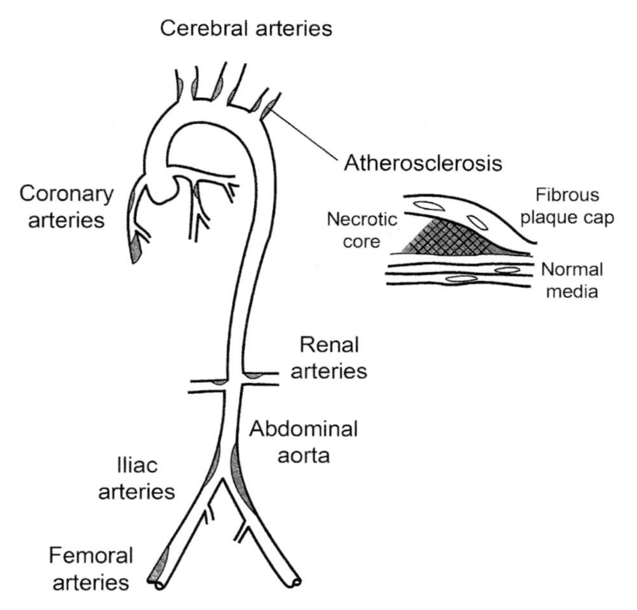

Fig. 8.49. Clogging of arteries by plaque. A schematic of the atherosclerosis is shown, along with some of the preferred sites of atherosclerosis in the vasculature (blood vessels). (From [395])

several negative consequences. As this increased kinetic energy (per unit volume) in the stenotic region decreases in the poststenotic normal vessel (due to Bernoulli's Principle), there can be structural fatigue in this latter region leading to distention and this possible dilatation can lead to a fusiform aneurysm. The increased stenotic flow speed and irregular geometry can increase the Reynolds number, resulting in turbulent flow and the eventual disengaging of arterial plaque, which then becomes a particulate embolus. The increased speed also leads to higher shear stresses on the lesion.

8.3.3 Equation of Motion of Arteries and Aneurysms during Pulsatile Flow (Advanced Topic)

For steady flow, the tension in the arterial wall balances the pressure difference inside and outside the artery, as described by the Law of Laplace. Because the pressure inside the artery really varies within every cycle (during the arterial pulse), this balance needs to be evaluated to account for these cyclic pressure variations. Let us model a saccular aneurysm as a sphere of radius R, with a thin wall of thickness w. (We could similarly model a fusiform aneurysm as a cylinder of radius R.)

If the change in radius due to this time-varying pressure is r, then the circumference of the sphere changes from $2\pi R$ to $2\pi(R+r)$. The circumferential strain ϵ is $(2\pi(R+r) - 2\pi R)/2\pi R = r/R$, and the circumferential stress σ is related to this strain by

$$\sigma = Y\epsilon = Y\frac{r(t)}{R}. \tag{8.46}$$

The response of the arterial wall has a component that balances the average pressure P_{av} – which is a time-weighted average of the systolic and diastolic pressures for larger arteries – and a time varying component – due to the pulse $P_{\mathrm{pulse}}(t) = P_{\mathrm{p}} \cos \omega_{\mathrm{p}} t$, where P_{p} is a time-weighted difference of the systolic and diastolic pressures for larger arteries and ω_{p} is the pulse (radial) frequency. This time-varying pressure is $P(t) = P_{\mathrm{ss}} + P_{\mathrm{p}} \cos \omega_{\mathrm{p}} t$.

The inertial response force ma, per unit area on the sphere is the mass per unit area of the spherical shell, ρw, where ρ is the mass density, times the radial acceleration, $\mathrm{d}^2r/\mathrm{d}t^2$, or $\rho w\, \mathrm{d}^2r/\mathrm{d}t^2$. The hydrostatic pressure term $P(t)$ tends to increase the radius, while the stress due to the wall resists this change, so the inertial response is

$$\rho w \frac{\mathrm{d}^2 r_{\mathrm{p}}(t)}{\mathrm{d}t^2} = P_{\mathrm{av}} + P_{\mathrm{p}} \cos \omega_{\mathrm{p}} t - Y\frac{r(t)}{R} \tag{8.47}$$

because $\mathrm{d}^2 r_{\mathrm{av}}/\mathrm{d}t^2 = 0$.

The perturbation $r(t)$ has a component in response to the average pressure P_{av} and a time-varying part in response to the pulse pressure $P_{\mathrm{p}} \cos \omega_{\mathrm{p}} t$. Substituting $r(t) = r_{\mathrm{av}} + r_{\mathrm{p}}(t)$ into (8.47), we get

$$\rho w \frac{\mathrm{d}^2 r_{\mathrm{p}}(t)}{\mathrm{d}t^2} = P_{\mathrm{av}} + P_{\mathrm{p}} \cos \omega_{\mathrm{p}} t - Y\frac{r_{\mathrm{av}} + r_{\mathrm{p}}(t)}{R}. \tag{8.48}$$

Equating the time-varying terms to give one equation and the average terms to give another leads to

$$\rho w \frac{\mathrm{d}^2 r_{\mathrm{p}}(t)}{\mathrm{d}t^2} = P_{\mathrm{p}} \cos \omega_{\mathrm{p}} t - Y\frac{r_{\mathrm{p}}(t)}{R} \tag{8.49}$$

and

$$0 = P_{\mathrm{av}} - Y\frac{r_{\mathrm{av}}}{R}. \tag{8.50}$$

Equation (8.50) is related to the Law of Laplace for a sphere (7.9) ($\Delta P = 2T/R$). Equation (8.49) can be rewritten as

$$\frac{\mathrm{d}^2 r_{\mathrm{p}}(t)}{\mathrm{d}t^2} + \omega_0^2 r_{\mathrm{p}}(t) = \frac{P_{\mathrm{p}}}{\rho w} \cos \omega_{\mathrm{p}} t, \tag{8.51}$$

where $\omega_0^2 = Y/R\rho w$. This looks like the equation of motion for a simple harmonic oscillator of frequency ω_0 plus an extra term (the last one), which is due to the pulse driving force at frequency ω_{p}. The steady-state solution to this equation is

$$r_{\mathrm{p}}(t) = \frac{P_{\mathrm{p}}/\rho w}{\omega_0^2 - \omega_{\mathrm{p}}^2} \cos(\omega_{\mathrm{p}} t), \tag{8.52}$$

which can be verified by substitution (see Appendix C). If r_{p} is large, it could lead to rupture. In principle, r_{p} can become larger as the aneurysm develops because Y, R, ρ, or w change (in $\omega_0^2 = Y/R\rho w$). Without the driving term ($P_{\mathrm{p}} = 0$), the solution is the usual harmonic solution: $r_{\mathrm{p}}(t) = A\cos(\omega_0 t + \phi)$.

If ω_0 were to approach ω_{p}, (8.52) indicates that the change in radius would become very large because of this resonance. If this were to occur (and it does in some examples of driven oscillators but it really does not for aneurysms), the viscous (or damping) properties of the vessel wall would have to be included through a term $-\gamma\, \mathrm{d}r_{\mathrm{p}}/\mathrm{d}t$, leading to the new equation of motion

$$\frac{\mathrm{d}^2 r_{\mathrm{p}}(t)}{\mathrm{d}t^2} + \gamma \frac{\mathrm{d}r_{\mathrm{p}}(t)}{\mathrm{d}t} + \omega_0^2 r_{\mathrm{p}}(t) = \frac{P_{\mathrm{p}}}{\rho w} \cos \omega_{\mathrm{p}} t \tag{8.53}$$

with steady-state solution

$$r_{\mathrm{p}}(t) = \frac{\left(\omega_0^2 - \omega_{\mathrm{p}}^2\right) P_{\mathrm{p}}/\rho w}{\left(\omega_0^2 - \omega_{\mathrm{p}}^2\right)^2 + (\gamma\omega_{\mathrm{p}})^2} \cos(\omega_{\mathrm{p}} t). \tag{8.54}$$

(See Appendix C.) This now includes the viscoelastic properties of the arterial wall, which dampens the resonance a bit.

Far above the $\omega_0 \simeq \omega_{\mathrm{p}}$ resonance (with $\omega_0 \gg \omega_{\mathrm{p}}$), (8.52) and (8.54) give

$$r_{\mathrm{p}}(t) \simeq \frac{P_{\mathrm{p}}}{\rho w \omega_0^2} \cos(\omega_{\mathrm{p}} t). \tag{8.55}$$

Without the driving term ($P_{\mathrm{p}} = 0$), the solution is that of a damped harmonic oscillator: $r_{\mathrm{p}}(t) = A\exp(-\gamma t/2)\cos(\omega_0 t + \phi)$, for $\omega_0 \gg \gamma$. This harmonic oscillation damps in a time $\sim 1/\gamma$, which corresponds to about $\omega_0/(2\pi\gamma)$ cycles; this last number is often called the quality factor Q of the system, as is discussed in the Chap. 10 discussion of acoustic resonances and in Appendices C and D.

8.4 Modeling the Circulatory System and the Heart

The branching in the circulatory system is very complex. Still, there is an orderly transition from larger to smaller arteries and then from smaller to larger veins, and so we can imagine an overall model of the circulation with all arteries or veins of a given diameter combining to form a subsystem. We have seen that larger vessels can be modeled quite well as ideal compliance vessels, while smaller vessels can be modeled as resistance vessels. We will now use these models to develop a comprehensive model of circulation, which can handle steady-state flow and changes that depend on time, like the arterial pulse. It can be used to understand the control of circulation. For a complete circulation model, we need to include the action of the heart, which is really two separate pumps: the right heart for the pulmonary system and the left heart for the systemic system. We first develop static and dynamic models of the left and right hearts.

8.4.1 Model of the Heart

Let us consider the left ventricle, which is the major pump in the left heart; the treatment of the right heart is analogous. During systole, the mitral valve is closed and the aortic valve is open. The pressure that develops is essentially that in the systemic arteries (sa) P_{sa} because of the very small pressure drop. During diastole the aortic valve is closed and the mitral valve is open. The left ventricle receives blood from the left atrium at a pressure that is pretty low, and is essentially equal to that in the pulmonary veins (pv), $P_{\mathrm{pv}} \sim 5\,\mathrm{mmHg}$, that feeds the left atrium.

Static Model of the Ventricles

We will model the left (or right) ventricle as a hemispherical shell, with an inner radius r_{i} and outer radius r_{o}, with a very thick wall of thickness $r_{\mathrm{o}} - r_{\mathrm{i}}$. Assume the open side is facing upward, as in Fig. 8.50. The (gauge) pressure inside P_{i} pushes the ventricle down and reaches a maximum during systole of 120 mmHg. The pressure outside (acting on the round bottom surface) P_{o} is from the pericardium and pushes the ventricle up and is approximately the pleural pressure, which can be negative (i.e., less than an atmosphere). The circumferential wall stress σ acts vertically and pushes the ventricle up. These arguments are the same as those for the Law of Laplace for a sphere (7.9), except we are now assuming the wall has finite thickness.

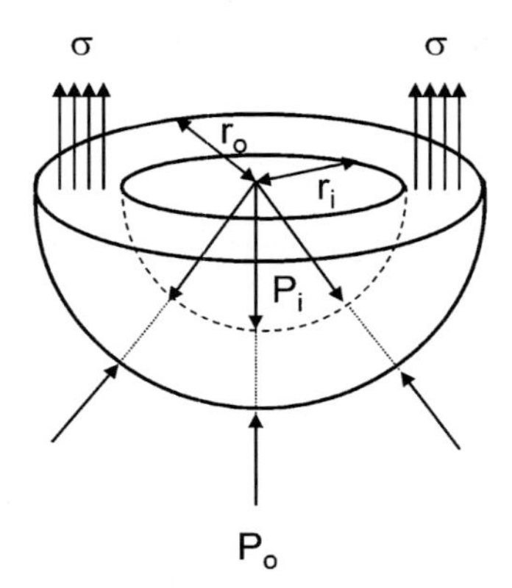

Fig. 8.50. Model of the left ventricle. (Based on [405])

Force balance in the vertical direction gives

$$(\pi r_{\mathrm{i}}^2)P_{\mathrm{i}} = (\pi r_{\mathrm{o}}^2)P_{\mathrm{o}} + \left(\pi(r_{\mathrm{o}}^2 - r_{\mathrm{i}}^2)\right)\sigma \tag{8.56}$$

and so the circumferential wall stress is

$$\sigma = \frac{P_{\mathrm{i}}r_{\mathrm{i}}^2 - P_{\mathrm{o}}r_{\mathrm{o}}^2}{r_{\mathrm{o}}^2 - r_{\mathrm{i}}^2}. \tag{8.57}$$

The first two terms in (8.56) are not just the surface areas of the respective areas, $2\pi r_{\mathrm{i,o}}^2$, times the hydrostatic pressure, $P_{\mathrm{i,o}}$, but half of that, because only part of the force due to the pressure is in the vertical direction (as in the Law or Laplace for spheres and Problem 7.12). Because $P_{\mathrm{i}} \gg P_{\mathrm{o}}$

$$\sigma \simeq \frac{P_{\mathrm{i}}}{(r_{\mathrm{o}}/r_{\mathrm{i}})^2 - 1}. \tag{8.58}$$

Because the material in the heart wall is incompressible, the volume of the heart wall V_{wall} does not change with pressure. From the difference of the volumes of the outer and inner hemispherical shells, we know that $V_{\mathrm{wall}} = 2\pi(r_{\mathrm{o}}^3 - r_{\mathrm{i}}^3)/3$ and so $(r_{\mathrm{o}}/r_{\mathrm{i}})^2 = (1 + V_{\mathrm{wall}}/V_{\mathrm{i}})^{2/3}$, where the inner volume of the left ventricle is $V_{\mathrm{i}} = 2\pi r_{\mathrm{i}}^3/3$. Therefore we see

$$\sigma \simeq \frac{P_{\mathrm{i}}}{(1 + V_{\mathrm{wall}}/V_{\mathrm{i}})^{2/3} - 1}. \tag{8.59}$$

Using the expansion $(1+x)^n \simeq 1 + nx$ for $|\, x \,| \ll 1$ and the fact that the internal volume of the left ventricle is much larger than the volume of the heart wall, we find

$$\sigma \simeq \frac{3V_{\mathrm{i}}}{2V_{\mathrm{wall}}} P_{\mathrm{i}}. \tag{8.60}$$

This shows how excessive systolic pressure or the enlargement of the left ventricle (and of the heart) will lead to excessive cardiac wall stress (for a constant wall thickness).

Dynamic Model of the Ventricles

We will now model a ventricle as a compliance vessel with a compliance that changes with time [394], so we use (8.5)

$$V(t) = V_{\mathrm{d}} + C(t)P(t). \tag{8.61}$$

(We will simply call the flow resistance R_{flow} and compliance C_{flow}, respectively R and C in this section.) Figure 8.51 shows how the compliance of the ventricle changes during a cardiac cycle. During systole the compliance becomes low, which causes a high pressure to develop because the volume

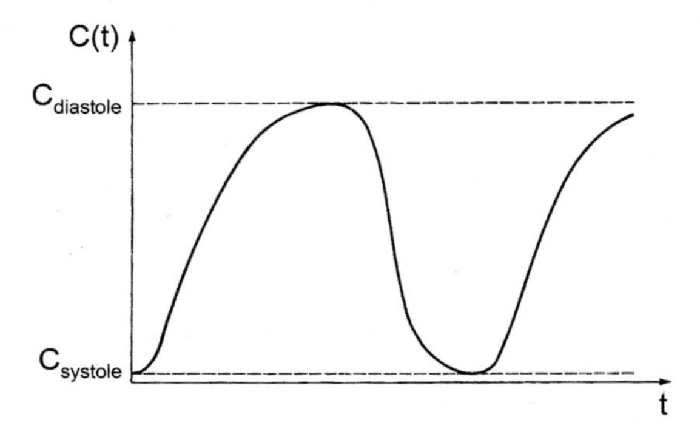

Fig. 8.51. Changes in ventricle compliance during the cardiac cycle, from a small value while it is contracting during systole to a large value when it is relaxing during diastole. (From [394])

remains pretty constant. In diastole the ventricle relaxes, which is associated with a large compliance that induces a low pressure.

The changes of volume and pressure in the ventricle during a cardiac cycle are modeled in Fig. 8.52. Stage A is when the inflow valve (which is the mitral valve for the LV) closes, which marks the end of diastole (ED) and the beginning of systole. The ventricle volume is a maximum

$$V_{\mathrm{ED}} = V_{\mathrm{d}} + C_{\mathrm{diastole}} P_{\mathrm{pv}}. \tag{8.62}$$

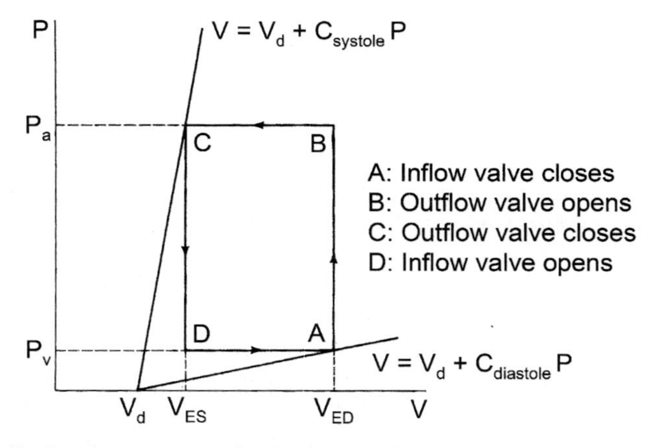

Fig. 8.52. Cycle of pressure and volume in either ventricle is given by the ABCD rectangle in this simplified model. The slanting lines radiating from the dead volume are the ventricle pressure–volume relationships at the end of systole and diastole. (Similar slanting lines with different slopes (not shown) characterize the ventricle at other times.) The venous pressure is the same as the inflow atrial and venous pressures during diastole. The arterial pressure is the outflow ventricle pressure during systole. (From [394])

During systole, the compliance decreases isovolumetrically, so the pressure increases. At Stage B, the outflow valve (which is the aortic valve for the LV) opens; the pressure remains constant as blood leaves the ventricle and concomitantly the volume of the ventricle decreases. Stage C is when the outflow valve closes, which marks the end of systole (ES) and the beginning of diastole. The ventricle volume is a minimum:

$$V_{\mathrm{ES}} = V_{\mathrm{d}} + C_{\mathrm{systole}} P_{\mathrm{sa}}. \tag{8.63}$$

During diastole, the compliance increases isovolumetrically, so the pressure decreases. At Stage D, the inflow valve opens; the pressure then remains constant as blood enters the ventricle and concomitantly the volume of the ventricle increases – until the end of diastole is reached, Stage A again.

The stroke volume is

$$V_{\mathrm{stroke}} = V_{\mathrm{ED}} - V_{\mathrm{ES}} = C_{\mathrm{diastole}} P_{\mathrm{pv}} - C_{\mathrm{systole}} P_{\mathrm{sa}}. \tag{8.64}$$

Because $C_{\mathrm{systole}} \sim 0$, we can take

$$V_{\mathrm{stroke}} = C_{\mathrm{diastole}} P_{\mathrm{pv}}. \tag{8.65}$$

With a heart rate F, the volumetric flow is

$$Q = F V_{\mathrm{stroke}} = F C_{\mathrm{diastole}} P_{\mathrm{pv}}. \tag{8.66}$$

Calling $K = F C_{\mathrm{diastole}}$ the pump coefficient, we can model the left ventricle by

$$Q_{\mathrm{L}} = K_{\mathrm{L}} P_{\mathrm{pv}}. \tag{8.67}$$

Analogously for the right heart, the right ventricle, which is fed by the systemic veins (sv), is modeled by

$$Q_{\mathrm{R}} = K_{\mathrm{R}} P_{\mathrm{sv}}. \tag{8.68}$$

8.4.2 Model of the Overall Flow in the Circulatory System

We will model the eight subsystems shown in Fig. 8.53, two hearts (L and R), large arteries in the systemic and pulmonary systems (sa and pa), large veins in the systemic and pulmonary systems (sv and pv), and the small vessels (arterioles/capillaries/venules) in the systemic and pulmonary systems [394]. Blood flows from the left ventricle successively through the systemic large arteries, small vessels, and large veins, and to the right ventricle. Then blood goes from the right ventricle successively through the pulmonary large arteries, small vessels, and large veins, and to the left ventricle.

The flows through the ventricles are determined by the pressures in the veins feeding them (through the atria)

$$Q_{\mathrm{L}} = K_{\mathrm{L}} P_{\mathrm{pv}} \qquad \text{and} \qquad Q_{\mathrm{R}} = K_{\mathrm{R}} P_{\mathrm{sv}}. \tag{8.69}$$

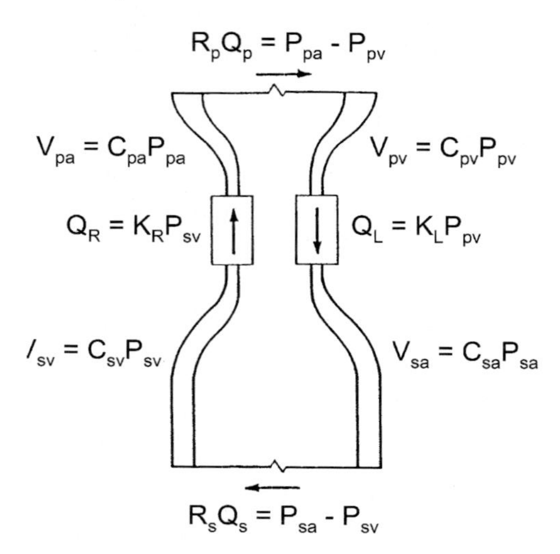

Fig. 8.53. The eight subsystems in systemic and pulmonary flow, with their model equations. (From [394])

The volumes in the large arteries are determined by their compliances and the pressures within them

$$V_{\rm sa} = C_{\rm sa} P_{\rm sa} \qquad \text{and} \qquad V_{\rm pa} = C_{\rm pa} P_{\rm pa}, \tag{8.70}$$

as are the volumes of the veins

$$V_{\rm sv} = C_{\rm sv} P_{\rm sv} \qquad \text{and} \qquad V_{\rm pv} = C_{\rm pv} P_{\rm pv}. \tag{8.71}$$

(A more refined model would include the dead volume in these large vessels.) The pressure drop across the small vessels is the difference in pressure between the large arteries and large veins. For the respective systemic and pulmonary systems, the pressure drop is determined by their resistances ($R_{\rm s}$ and $R_{\rm p}$) and the flows through them ($Q_{\rm s}$ and $Q_{\rm p}$)

$$P_{\rm sa} - P_{\rm sv} = R_{\rm s} Q_{\rm s} \qquad \text{and} \qquad P_{\rm pa} - P_{\rm pv} = R_{\rm p} Q_{\rm p}. \tag{8.72}$$

We see a symmetry here. There are three types of variables, the flow Q, volume V, and pressure P. Each subsystem depends on relations between two of them: Q and P for the ventricles, V and P for the large vessels, and P and Q for the small vessels.

There are 12 unknowns: four flows: $Q_{\rm R}$, $Q_{\rm L}$, $Q_{\rm s}$, $Q_{\rm p}$; four pressures: $P_{\rm sa}$, $P_{\rm sv}$, $P_{\rm pa}$, $P_{\rm pv}$; and four volumes: $V_{\rm sa}$, $V_{\rm sv}$, $V_{\rm pa}$, $V_{\rm pv}$. We need 12 equations to solve for these 12 unknowns, but have only eight here and need four more. However, for steady-state flow, the flow in each region is the same, so $Q_{\rm R} = Q_{\rm L} = Q_{\rm s} = Q_{\rm p}$. These are really the three independent equations

$$Q_{\rm R} = Q_{\rm L}, \qquad Q_{\rm R} = Q_{\rm s}, \qquad \text{and} \qquad Q_{\rm R} = Q_{\rm p}, \tag{8.73}$$

with each flow rate equal to Q. The fourth equation describes the constant total volume and is

$$V_0 = V_{\mathrm{sa}} + V_{\mathrm{sv}} + V_{\mathrm{pa}} + V_{\mathrm{pv}}. \tag{8.74}$$

We can now solve these equations. From the pump (8.67) and (8.68), the venous pressures are

$$P_{\mathrm{sv}} = \frac{Q}{K_{\mathrm{R}}} \quad \text{and} \quad P_{\mathrm{pv}} = \frac{Q}{K_{\mathrm{L}}}. \tag{8.75}$$

Inserting these venous pressures into the resistance (8.72) gives

$$P_{\mathrm{sa}} = \frac{Q}{K_{\mathrm{R}}} + R_{\mathrm{s}}Q \quad \text{and} \quad P_{\mathrm{pa}} = \frac{Q}{K_{\mathrm{L}}} + R_{\mathrm{p}}Q. \tag{8.76}$$

Inserting these venous and arterial pressures into the compliance (8.70) and (8.71) gives

$$V_{\mathrm{sa}} = QC_{\mathrm{sa}}\left(\frac{1}{K_{\mathrm{R}}} + R_{\mathrm{s}}\right) \quad \text{and} \quad V_{\mathrm{pa}} = QC_{\mathrm{pa}}\left(\frac{1}{K_{\mathrm{L}}} + R_{\mathrm{p}}\right) \tag{8.77}$$

$$V_{\mathrm{sv}} = QC_{\mathrm{sv}}\left(\frac{1}{K_{\mathrm{R}}}\right) \quad \text{and} \quad V_{\mathrm{pv}} = QC_{\mathrm{pv}}\left(\frac{1}{K_{\mathrm{L}}}\right). \tag{8.78}$$

These can be expressed as

$$V_{\mathrm{i}} = T_{\mathrm{i}}Q \quad \text{for i} = \text{sv, pv, sa, pa}, \tag{8.79}$$

with

$$T_{\mathrm{sa}} = C_{\mathrm{sa}}\left(\frac{1}{K_{\mathrm{R}}} + R_{\mathrm{s}}\right) \quad \text{and} \quad T_{\mathrm{pa}} = C_{\mathrm{pa}}\left(\frac{1}{K_{\mathrm{L}}} + R_{\mathrm{p}}\right) \tag{8.80}$$

$$T_{\mathrm{sv}} = C_{\mathrm{sv}}\left(\frac{1}{K_{\mathrm{R}}}\right) \quad \text{and} \quad T_{\mathrm{pv}} = C_{\mathrm{pv}}\left(\frac{1}{K_{\mathrm{L}}}\right). \tag{8.81}$$

Inserting these equations in (8.74) for the total blood volume gives

$$V_0 = V_{\mathrm{sa}} + V_{\mathrm{sv}} + V_{\mathrm{pa}} + V_{\mathrm{pv}} \tag{8.82}$$

$$= Q(T_{\mathrm{sa}} + T_{\mathrm{sv}} + T_{\mathrm{pa}} + T_{\mathrm{pv}}) \tag{8.83}$$

or

$$Q = \frac{V_0}{T_{\mathrm{sa}} + T_{\mathrm{sv}} + T_{\mathrm{pa}} + T_{\mathrm{pv}}}. \tag{8.84}$$

All the volumes are obtained from this and (8.79)

$$V_{\mathrm{i}} = \frac{T_{\mathrm{i}}V_0}{T_{\mathrm{sa}} + T_{\mathrm{sv}} + T_{\mathrm{pa}} + T_{\mathrm{pv}}} \tag{8.85}$$

and all of the pressures from $P_{\mathrm{i}} = V_{\mathrm{i}}/C_{\mathrm{i}}$ (from (8.70) and (8.71))

$$P_{\mathrm{i}} = \frac{1}{C_{\mathrm{i}}} \frac{T_{\mathrm{i}}V_0}{T_{\mathrm{sa}} + T_{\mathrm{sv}} + T_{\mathrm{pa}} + T_{\mathrm{pv}}}. \tag{8.86}$$

With known values of the flow rate and the total volume, the model parameters can be determined – see Table 8.7.

Table 8.7. Normal resting parameters of the circulation model. (Using data from [394])

	systemic system	pulmonary system
resistance, mmHg/(L/min)	$R_s = 17.5$	$R_p = 1.79$
compliance, L/mmHg		
arterial	$C_{sa} = 0.01$	$C_{pa} = 0.00667$
venous	$C_{sv} = 1.75$	$C_{pv} = 0.08$
heart	$K_L = 1.12$	$K_R = 2.8$
total volume: $V_0 = 5.0$ L		

8.4.3 The Arterial Pulse

This model is capable of analyzing much more than just this idealized average flow. We can use it to understand the arterial pulse, which is the periodic deviation of the systemic arterial pressure from the diastolic value due to systole (Fig. 8.54). This pulse pressure is

$$P_{\text{pulse}} = P_{\text{systole}} - P_{\text{diastole}}, \tag{8.87}$$

which is ~40 mmHg (= 120 mmHg − 80 mmHg). These steady-state assumptions we just made are no longer appropriate here for this *pulsatile* flow.

Consider the systemic artery system, which has volume V_{sa} and pressure P_{sa}. As a compliance vessel its volume can change, and it will change when the flow into it does not equal to the flow leaving it (Fig. 8.55). The flow into it is that from the left heart Q_L and the flow out of it goes into the (noncompliant) small vessel system, which has flow Q_s. Conservation of volume gives

$$\frac{\mathrm{d}V_{sa}(t)}{\mathrm{d}t} = Q_L(t) - Q_s(t). \tag{8.88}$$

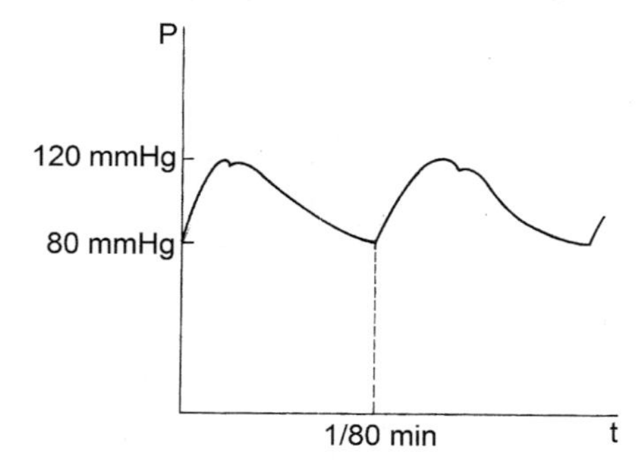

Fig. 8.54. The systemic arterial pulse. Also see the aortic pressure in Fig. 8.5. (From [394])

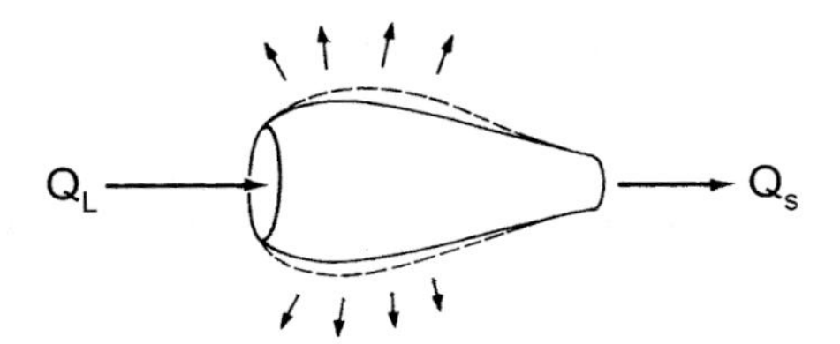

Fig. 8.55. During pulsatile flow, the change of arterial volume is equal to the volumetric inflow minus the outflow, as required by the conservation of volume (mass). (From [394])

For this compliance vessel we know that

$$V_{\mathrm{sa}}(t) = V_{\mathrm{sa,d}} + C_{\mathrm{sa}} P_{\mathrm{sa}}(t), \tag{8.89}$$

where we have now included the dead volume. Differentiating this equation with respect to time and combining it with the previous equation gives

$$\frac{\mathrm{d}V_{\mathrm{sa}}(t)}{\mathrm{d}t} = C_{\mathrm{sa}} \frac{\mathrm{d}P_{\mathrm{sa}}(t)}{\mathrm{d}t} = Q_{\mathrm{L}}(t) - Q_{\mathrm{s}}(t). \tag{8.90}$$

Using (8.72) gives

$$R_{\mathrm{s}} Q_{\mathrm{s}} = P_{\mathrm{sa}} - P_{\mathrm{sv}} \sim P_{\mathrm{sa}} \tag{8.91}$$

since $P_{\mathrm{sv}} \ll P_{\mathrm{sa}}$. Therefore we arrive at

$$C_{\mathrm{sa}} \frac{\mathrm{d}P_{\mathrm{sa}}(t)}{\mathrm{d}t} = Q_{\mathrm{L}}(t) - \frac{P_{\mathrm{sa}}}{R_{\mathrm{s}}}. \tag{8.92}$$

This determines the time dependence of the systemic artery pressure, and consequently the arterial pulse, if the flow rate out of the left ventricle is known.

We will assume that systole occurs very fast (for a very small fraction of the cardiac cycle of duration T, with $T = 1/F$) with a very large $Q_{\mathrm{L}}(t)$ for $t \sim 0$, and that it is zero for the rest of the cardiac cycle. This idealized model is amenable to simple analysis. Although it is not very accurate, some of the features it predicts are accurate.

When $Q_{\mathrm{L}}(t) = 0$, (8.92) is

$$C_{\mathrm{sa}} \frac{\mathrm{d}P_{\mathrm{sa}}(t)}{\mathrm{d}t} = -\frac{P_{\mathrm{sa}}}{R_{\mathrm{s}}} \tag{8.93}$$

or

$$\frac{\mathrm{d}P_{\mathrm{sa}}(t)}{\mathrm{d}t} = -\frac{P_{\mathrm{sa}}}{R_{\mathrm{s}} C_{\mathrm{sa}}} = -\frac{P_{\mathrm{sa}}}{\tau}, \tag{8.94}$$

where $\tau = R_{\mathrm{s}} C_{\mathrm{sa}}$. Therefore

$$P_{\mathrm{sa}}(t) = P_{\mathrm{sa}}(0) \exp(-t/\tau), \tag{8.95}$$

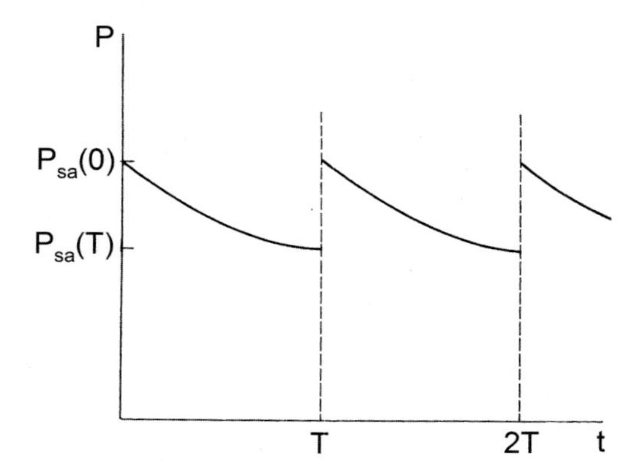

Fig. 8.56. Systemic arterial pulse from idealized model. (From [394])

as is depicted in Fig. 8.56 (see Appendix C). In this model, at the start of the cardiac cycle $P_{\rm sa}(0) = P_{\rm systole}$ and at the end of the cardiac cycle $P_{\rm sa}(T) = P_{\rm diastole} = P_{\rm systole}\exp(-T/\tau)$. The pressure of the arterial pulse is $P_{\rm pulse} = P_{\rm systole} - P_{\rm diastole}$.

We can determine these pressures in terms of the stroke volume using (8.62)–(8.64) and (8.70)

$$V_{\rm sa}(0) = V_{\rm sa,d} + C_{\rm sa}P_{\rm sa}(0) \tag{8.96}$$

$$V_{\rm sa}(T) = V_{\rm sa,d} + C_{\rm sa}P_{\rm sa}(T). \tag{8.97}$$

The difference is the stroke volume

$$\begin{aligned} V_{\rm stroke} &= V_{\rm sa}(0) - V_{\rm sa}(T) = C_{\rm sa}(P_{\rm sa}(0) - P_{\rm sa}(T)) \\ &= C_{\rm sa}P_{\rm sa}(0)[1 - \exp(-T/\tau)]. \end{aligned} \tag{8.98}$$

Consequently, we find

$$P_{\rm systole} = P_{\rm sa}(0) = \frac{V_{\rm stroke}}{C_{\rm sa}[1 - \exp(-T/\tau)]} \tag{8.99}$$

$$P_{\rm diastole} = P_{\rm sa}(T) = \frac{V_{\rm stroke}\exp(-T/\tau)}{C_{\rm sa}[1 - \exp(-T/\tau)]}. \tag{8.100}$$

With (8.87) we see

$$P_{\rm pulse} = P_{\rm systole} - P_{\rm diastole} = \frac{V_{\rm stroke}}{C_{\rm sa}} \tag{8.101}$$

and so $P_{\rm systole}$, $P_{\rm diastole}$, $V_{\rm stroke}$, the heart rate $1/T$, $C_{\rm sa}$, and $R_{\rm s}$ $(= \tau/C_{\rm sa})$ are all interrelated.

Let us consider a numerical example. For $V_{\rm stroke} = 70\,{\rm cm}^3 = 0.070\,{\rm L}$ and $P_{\rm pulse} = 40\,{\rm mmHg}$, (8.98) gives $C_{\rm sa} = V_{\rm stroke}/P_{\rm pulse} = 0.00175\,{\rm L/mmHg}$.

Using R_s from Table 8.7 of 17.5 mmHg/(L/min), we know that $\tau = R_\mathrm{s}C_\mathrm{sa} = 0.0306\,\mathrm{min}$. With $T = 1/80\,\mathrm{min} = 0.0125\,\mathrm{min}$, (8.99) and (8.100) give the last remaining parameter, which could be either P_systole or P_diastole, because they are related by $P_\mathrm{pulse} = P_\mathrm{systole} - P_\mathrm{diastole}$. From (8.101), we get $P_\mathrm{systole} = 120\,\mathrm{mmHg}$. As expected from self-consistency, $P_\mathrm{diastole} = P_\mathrm{systole}\exp(-T/\tau) = 120\,\mathrm{mmHg} \times \exp(-0.0125\,\mathrm{min}/(-0.0306\,\mathrm{min})) = 80\,\mathrm{mmHg}$.

A better model for the arterial pulse is described in Problem 8.49, in which the flow of blood from the left ventricle to the systemic arteries occurs with a finite (nonzero) duration in the cardiac cycle. It gives predictions that agree with Fig. 8.54 better than do those in Fig. 8.56. Even this improved model does not explain the small increase in pressure just after systole that is seen in Figs. 8.5 and 8.54, which is known as the "incisura" or "dicrotic notch." When the semilunar valve closes, some backward flowing blood bounces off the elastic aorta walls, setting up a pressure wave in the aorta; this pressure blip can be included with even more refinement in the model.

8.4.4 Windkessel Model

This simplified model of the arterial pulse is also known as the Windkessel Model [370], which was the first real model of blood flow. In this model blood flows from the left ventricle at a rate $Q_\mathrm{in}(t)$ into an elastic chamber of compliance C (i.e., the larger arteries) of volume $V(t)$ and leaves it at a rate $Q_\mathrm{out}(t)$ to enter a resistive element of resistance R_p (peripheral resistance of the arterioles and capillaries). It assumes that all pressure changes in the arteries occur at the same time. Because conservation of volume for an incompressible fluid gives

$$Q_\mathrm{in}(t) = \frac{\mathrm{d}V(t)}{\mathrm{d}t} + Q_\mathrm{out}(t), \tag{8.102}$$

with $C = \mathrm{d}V(t)/\mathrm{d}P(t)$ (8.61) and $Q_\mathrm{out}(t) = P(t)/R_\mathrm{p}$ (8.72), (8.102) becomes

$$Q_\mathrm{in}(t) = C\frac{\mathrm{d}P(t)}{\mathrm{d}t} + \frac{P(t)}{R_\mathrm{p}}. \tag{8.103}$$

The formal solution to this is

$$P(t) = \exp(-(t/\tau))\left(P(0) + \frac{1}{C}\int_0^t \exp(t'/\tau)Q_\mathrm{in}(t')\mathrm{d}t'\right), \tag{8.104}$$

where $\tau = R_\mathrm{p}C$. This can be shown to be the solution by substitution and by the method shown in Appendix C. Still it is simple and instructive to examine what happens after ventricular ejection, starting at $t = 0$ when $Q_\mathrm{in}(t) = 0$ and $P = P(0)$, until the next cycle starts at $t = T$. Then

$$\frac{\mathrm{d}P(t)}{\mathrm{d}t} + \frac{P(t)}{R_\mathrm{p}C} = 0 \qquad \text{for } 0 < t < T. \tag{8.105}$$

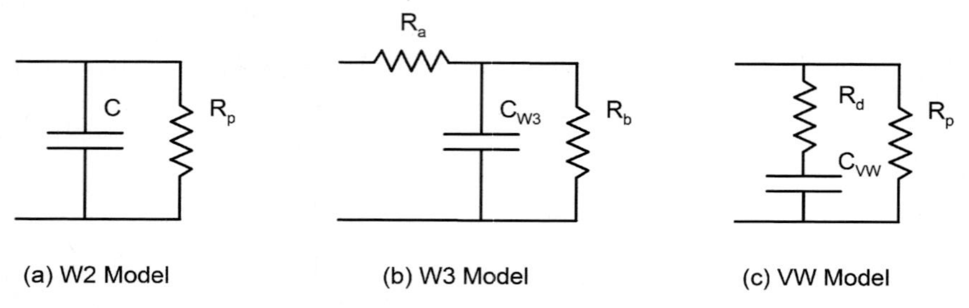

Fig. 8.57. Electrical analogs of the (**a**) classic or elastic Windkessel Model (W2), (**b**) three-element (or improved) Windkessel Model (W3), and (**c**) viscoelastic Windkessel Model (VW). The electrical resistances represent flow resistances and the electrical capacitances represent flow compliances. (See Appendix D)

The solution is

$$P(t) = P(0)\exp(-(t/\tau)) \qquad \text{for } 0 < t < T, \tag{8.106}$$

which is the same as (8.95) (see Appendix C).

This two-element classic Windkessel Model is formally equivalent to the electrical circuit, where the electrical resistance R_{elect} maps into the flow resistance in the capillaries and the electrical capacitance C_{elect} maps into the flow compliance (Table D.1, Appendix D). (This is how the model is expressed in Fig. 8.57a.) The voltage V_{elect} in the electrical model corresponds to the pressure P in the flow model and the current I corresponds to the volumetric flow rate Q. Ohm's Law relates V_{elect} and I across the resistor by $V_{\text{elect}} = IR_{\text{elect}}$. The voltage across the capacitor is $V_{\text{elect}} = q/C_{\text{elect}}$, where the charge on the capacitor is q. The voltages across both elements sum to zero (Kirchhoff's 2nd Law, Chap. 12), so $IR_{\text{elect}} + q/C_{\text{elect}} = 0$. Because $I = \mathrm{d}q/\mathrm{d}t$, we have $(\mathrm{d}q/\mathrm{d}t)R_{\text{elect}} + q/C_{\text{elect}} = 0$ or with $V_{\text{elect}} = q/C_{\text{elect}}$

$$\frac{\mathrm{d}V_{\text{elect}}}{\mathrm{d}t} + \frac{V_{\text{elect}}}{R_{\text{elect}}C_{\text{elect}}} = 0. \tag{8.107}$$

The two-element Windkessel Model can be refined by adding more elements to the electrical analog. The three-element Windkessel Model shown in Fig. 8.57b predicts a more realistic arterial pulse. The vascular resistance R_{a} of the aorta is added in series, and the value of the resistance in parallel is now R_{b}, which equals $R_{\text{p}} - R_{\text{a}}$. The viscoelastic Windkessel Model (Fig. 8.57c) represents a different type of improvement of the two-element Windkessel Model. The compliant arterial systems are represented by a capacitance (in the electrical analog of flow compliance) that is in series with a resistor R_{d}, which represents the viscous wall motion to more fully represent the viscoelasticity of the aorta.

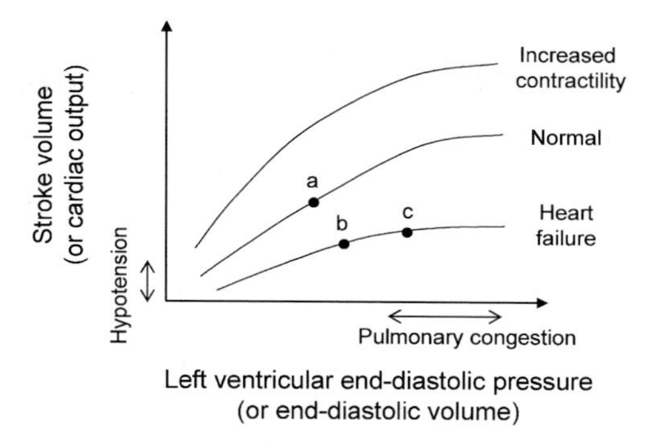

Fig. 8.58. Left ventricular performance (Frank–Starling) curves relate cardiac performance to preconditions (preloading), such as the pressure or volume of the left ventricle after diastole. (Based on [380])

8.4.5 Modeling the Malfunctioning Heart

There are many ways a heart can malfunction [402]. These conditions can be characterized quantitatively and, in principle, can be included in our models of the heart and circulation. Cardiac performance can be characterized by several parameters, such as the stroke volume V_{stroke} and cardiac output $Q_{\mathrm{t}} = FV_{\mathrm{stroke}}$, described earlier.

Figure 8.58 shows cardiac performance as a function of the conditions at the end of diastole in the left ventricle. The middle curve describes the operation of that person's heart, and point a represents that normal person at rest. After heart failure, such as after a myocardial infarction, the curve shifts down due to lessened left ventricle contractility. Point b represents the person described by point a after heart failure. Increased circulatory volume in this person is represented by point c. The stroke volume is increased due to increased contractility in the uppermost curve, perhaps caused by the infusion of the drug norepinephrine.

The solid line pressure–volume loops in Fig. 8.59 represent normal heart function, and are more accurate representations than the simplified rectangular one depicted in our model in Fig. 8.52. In systolic dysfunction, the systolic curve shifts from 1 to 2 in (a) due to decreased cardiac contractility. As seen in the dashed loop, the volume at the end of systole increases. In diastolic dysfunction, the diastolic curve shifts from 1 to 2 in (b) due to increased stiffness (decreased compliance). As seen by the dashed loop, the ventricular pressure is higher than normal at any diastolic volume.

Several series of problems occur when there is leakage between heart chambers. In valvular *regurgitation* there is backflow through one of the heart valves, such as mitral regurgitation and aortic regurgitation in the left heart valves. The regurgitant fraction for mitral regurgitation is the volume of blood flowing back through the mitral valve divided by the left ventricle stroke

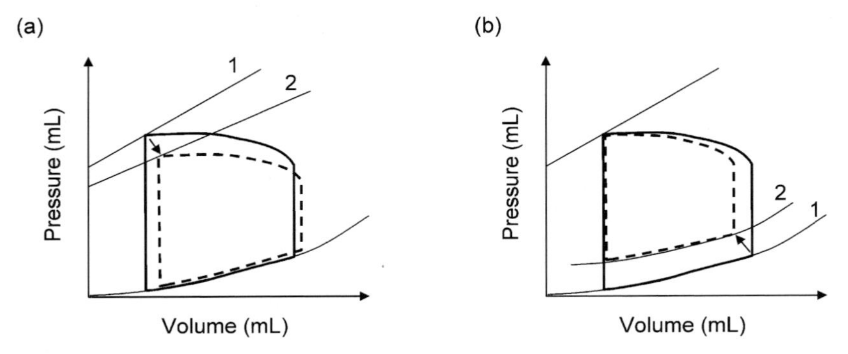

Fig. 8.59. Pressure–volume curves for the left ventricle for normal performance (*solid loop*), systolic dysfunction (*dashed loop* in (**a**)), and diastolic dysfunction (*dashed loop* in (**b**)). (Based on [380])

volume. There can also be openings in the septum between the atria or the ventricles, called a septal defect [374]. For example, a ventricular septal defect is a congenital condition in which there is an opening between the left and right ventricles (Fig. 8.60). During systole blood flows from the left to right ventricle because of the higher pressure in the former. This leads to increased blood return to the left side of the heart which causes the left atrium and ventricle to enlarge. The ejection fraction (EF) is the stroke volume divided by the ventricular volume at the end of diastole (when the ventricle is full). It is normally 55–75%, but can be less with a septal defect.

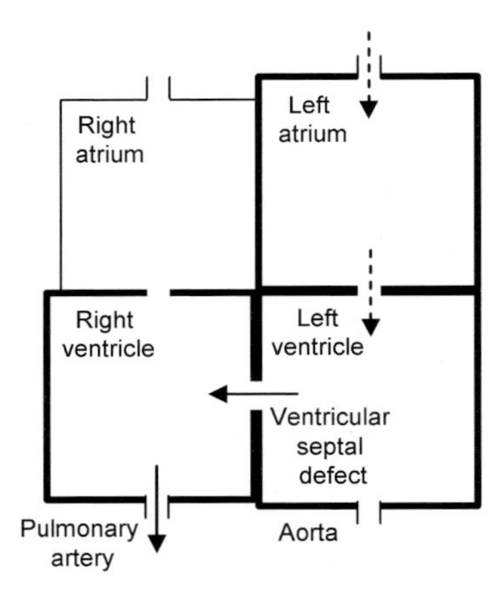

Fig. 8.60. With a ventricular septal valve defect blood flows from the left to right ventricle. (Based on [374])

8.5 Summary

The flow of blood in the vessels of the circulation system can be modeled using the resistive and compliant nature of the flow. This can be combined with a model of the heart to understand the flow in the entire circulation system. These models can also be used to understand the physical consequences of a malfunctioning heart and problems in the circulation system, such as clogged and weakened arteries, that can lead to strokes and aneurysms. The energy needed to operate the heart can also be modeled.

Problems

Blood Pressure

8.1. (a) The brain in a human is 55 cm above the heart. If the average blood pressure in the major arteries near the heart is 100 mmHg, what is the blood pressure in major arteries in the brain (in both mmHg and cmH_2O) when a person is either lying down or standing up.
(b) Repeat part (a) for an erect human on the moon ($g = g_{\text{Earth}}/6$) and on Jupiter ($g = 2.34 g_{\text{Earth}}$).
(c) A pilot coming out of a dive experiences an upward centripetal acceleration a of magnitude v^2/r, where v is the speed of the jet and r is the radius of curvature of the trajectory, that adds to gravity (effectively increasing g to $g + a$). What is the arterial pressure in the pilot's brain for $v = 200$ m/s and $r = 2$ km? What could happen to the pilot during this recovery from the dive? Would you expect dizziness because of a lack of blood to the head? (See Fig. 8.61.)
(d) What must the pressure in the aorta in a giraffe be (on Earth) for its brain to receive blood? (How can you estimate the elevation of its brain above its aorta?)

8.2. In Problem 8.1d we saw that the pressure of the blood leaving the heart of a giraffe and entering its systemic system must be much larger than that for humans because of its long neck.

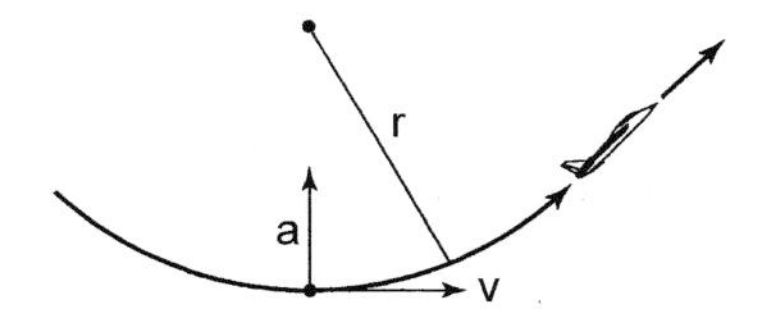

Fig. 8.61. Trajectory of a pilot coming out of a dive, with the acceleration and velocity vectors shown for the low point of the dive. For Problem 8.1. (From [367])

(a) Would a blood pressure of 280 mmHg/180 mmHg account for pumping the blood up this long neck in large arteries and then for the pressure drop that occurs in the very small arteries in the brain?
(b) The giraffe has this relatively high blood pressure because of this long neck, which is usually nearly vertical. However, we could expect that there would be a rush of blood to the brain because of this high pressure when the giraffe lowers its head by almost 7 m when it bends down to drink water, and that this could lead to rupture of the arteries in the brain. Why?
(c) This artery rupturing does not occur because of compensating effects. Explain this by considering the following (i) The elastic walls of the long giraffe carotid artery help force blood upward (which is a peristaltic action), and this also means that this artery can swell to absorb excess blood when the head is lowered because it is very compliant. (ii) The giraffe jugular vein contains a series of one-way valves that prevent back-flow of the blood when the giraffe's head is down.

8.3. When you stand on your head, why does your head become red and why do your legs become pale?

8.4. Why is blood pressure measured using major arteries in the upper arm, rather than those in the lower arm or leg?

8.5. You are told that your blood pressure is 880 mmHg/840 mmHg. You are quite understandably concerned because these values are astronomically high, but you are told not to worry because your blood pressure is normal. Should you be concerned?

8.6. (a) An intravenous infusion is made under gravity. If the fluid to be delivered has a density of 1.0 g/cm^3, at what height above the vein, h, should the top surface of the fluid in the bottle be positioned so the fluid just barely enters the vein? The gauge pressure in the vein is 18 mmHg. (Assume the needle entering the vein has a "large" inside diameter.) (See Fig. 8.62.)
(b) If this needle has a "very small" inside diameter, should the bottle be placed higher, lower, or at the same height? Why?
(c) Why are such infusions performed intravenously and not intra-arterially?

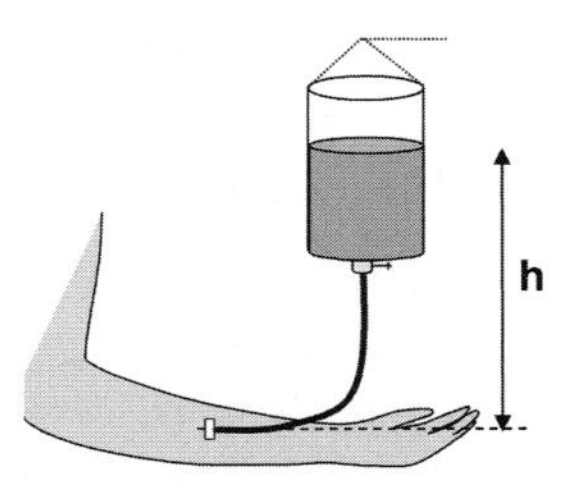

Fig. 8.62. Intravenous infusion under gravity. (Based on [387].) For Problem 8.6

8.7. You are lying down and are injured in such a way that blood from a major artery squirts upward. How high can it spurt?

8.8. Twirl one arm around as fast as you can many times until you see your fingers on that arm turn red. (Continue even if you do not see them getting redder.) [365]
(a) Estimate the centrifugal acceleration at the end of your finger tips, $v_{\text{radial}}^2/r = v_{\text{radial}}^2/l_{\text{arm}}$, where l_{arm} is your arm length (to your fingertips) and v_{radial} is the radial speed of your finger tips. (Why is $v_{\text{radial}} = 2\pi l_{\text{arm}}/T$, where T is the period for a complete cycle of this motion?)
(b) Express this acceleration in units of g.
(c) Calculate the effective pressure pushing your blood to your fingers by this motion. This is the apparent outward force divided by the cross-sectional area of your fingers, A_{fingers}, or $(m_{\text{fingers}}v_{\text{radial}}^2/l_{\text{arm}})/A_{\text{fingers}}$. Because $m_{\text{fingers}} = \rho_{\text{fingers}}l_{\text{fingers}}A_{\text{fingers}}$, this pressure is $\rho_{\text{fingers}}v_{\text{radial}}^2(l_{\text{fingers}}/l_{\text{arm}})$.
(d) Compare this to the systolic pressure 120 mmHg and explain why your fingers (could have or should have) turned red.

Blood Pressure Drop During Flow

8.9. Calculate the pressure drop (in mmHg) across the following arterial systems using Poiseuille's Law with $\eta_{\text{blood}} = 4 \times 10^{-3}$ Pa-s, for a total flow of 80 cm^3/s across each system:
(a) aorta (internal radius $r = 1.25$ cm, length $L = 10$ cm, all of the flow in this one aorta)
(b) large arteries ($r = 0.2$ cm, $L = 75$ cm, $n = 200$ of them, each with equal flow and the same dimensions)
(c) arterioles ($r = 30\,\mu$m, $L = 0.6$ cm, $n = 5 \times 10^5$)
(d) capillaries ($r = 3.5\,\mu$m, $L = 2$ mm, $n = 10^{10}$).

8.10. In estimating pressure drops across the different arterial branches we assumed specific numbers of arteries of given diameters and lengths. There is really a wide range of arterial diameters and lengths. How does this affect the pressure drops in the systemic arterial system?

8.11. In estimating pressure drops across the arterioles we assumed a specific number of arterioles with the same diameter and length.
(a) Let us say that all arterioles have the same radius, but their lengths (instead of all being L) range between $0.8L$ and $1.2L$ (with equal probability throughout). How does this change the overall resistance of the arteriole system?
(b) Let us say that all arterioles have the same length, but their radii (instead of all being r) range between $0.8r$ and $1.2r$ (with equal probability throughout). How does this change the overall resistance of the arteriole system?
(c) The pressure drop across each arteriole in the system must be the same because each is fed by the large arteries, whose pressure is set by the left

ventricle, and by the arterial side of the capillaries, whose pressure is also set. If the overall pressure drop across the arterioles is unchanged (by the changes in (a) or (b)), how is the overall flow rate in the arteriole system changed and what is the flow in each arteriole, for the situations alternately described in (a) and (b).
(d) If you wanted the flow rate to stay the same in each arteriole in (a) and (b) for the given distributions of lengths and radii, how would you have to change the distributions of radii and lengths in each, respectively?

8.12. Find the pressure drop across the arterioles in Problem 8.9c, if – with the same total flow in both cases – and either
(a) all the arterioles become clogged in such a way that their radii decrease to 28 μm or
(b) the number of the arterioles decreases to 4×10^5.
(c) By how much would the pressure in the main arteries need to change if the body responded to either change by maintaining the same flow rate?

8.13. Assume that the diameter of each blood vessel in a person is doubled and the total volumetric flow rate is not changed.
(a) What is the new total volume of blood? (Assume the base line parameters in the chapter.)
(b) What is the new circulation time for blood (total blood volume/total volumetric flow rate)?
(c) How do the resistances of the arterioles and capillaries change?
(d) How does the pressure drops across the arterioles and capillaries change?
(e) How does the work done by the heart change?

8.14. Repeat Problem 8.13 if instead the length of each blood vessel is doubled.

8.15. The length of a blood vessel is doubled and its diameter is doubled.
(a) How does the flow resistance change?
(b) If the flow through it is unchanged, how does the pressure drop change?
(c) How does the flow through it change if instead the pressure drop is unchanged?

8.16. Your internal body temperature increases from 37 to 40°C. Assuming that the only thing that changes is the viscosity of blood, how must the blood pressure change to ensure the flow rate remains unchanged?

8.17. Use (8.29) to relate $P(L)$ to $P(0)$ and other flow terms.

8.18. (a) Estimate how much the flow is changed in small arteries by including the influence of compliance by using (8.29) and assuming the same pressure drop.
(b) Repeat this estimate for how much the pressure drop changes with this analysis assuming the same flow rate.

8.19. Express the flow resistance units of N-s/m^5 in terms of (N/m^2)/(cm^3/s), dyne-s/cm^5, and PRU (with 1 PRU = 1 mmHg-s/mL).

8.20. (a) Calculate the total peripheral vascular resistance in the systemic and pulmonary systems for someone with a steady-state blood flow rate of 5 L/min, and with 120 mmHg/80 mmHg blood pressure in the systemic system and 25 mmHg/8 mmHg pressure in the pulmonary system. Express your answer in the units of dyne-s/cm^5, which are CGS units and those that are often used by cardiologists.
(b) The expression given in the text for this vascular flow resistance should be corrected because it uses the average pressure at the beginning of the system instead of the pressure drop across the system. For the systemic system the final pressure is that at the right atrium (2 mmHg) and for the pulmonary system it is that at the left atrium (5 mmHg). How does using the actual pressure drop affect your calculations in part (a) (both qualitatively and quantitatively)?

8.21. (a) The pulmonary vascular resistance changes with lung volume. Figure 8.63 shows that it increases much with larger lung volumes, in part because the larger alveoli stretch the pulmonary capillaries. It also increases at very small lung volumes because these capillaries surrounding the alveoli become narrow. Calculate the range of pulmonary blood flow rates (in L/min), assuming this range of resistances and assuming that the pulmonary pressures are the same as in Problem 8.20.
(b) The pulmonary pressures actually change with lung volume. If they changed in a manner to keep the average flow the same as it is for a 110 mL lung volume, determine this change. (For the purpose of this calculation,

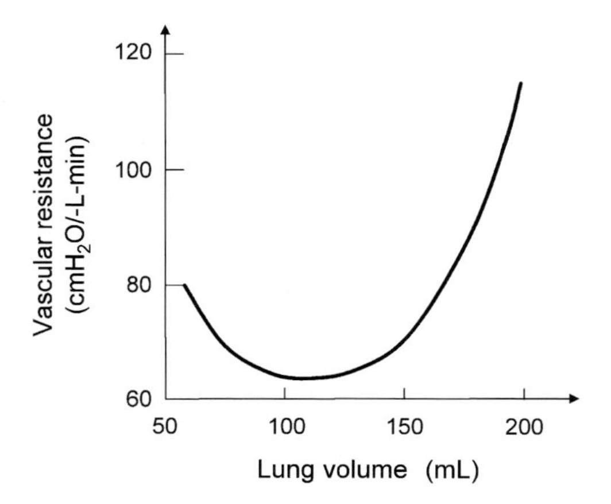

Fig. 8.63. Effect of lung volume on pulmonary vascular (blood flow) resistance (from an animal lobe preparation). (Based on [419].) For Problem 8.21

assume that the ratio of systolic and diastolic pressures remains a constant and that the left atrium pressure remains the same.)

8.22. (a) Determine the overall compliances of the systemic arterial and venous systems by using Fig. 8.26.
(b) Is the ratio of these two compliances reasonable, given our model for compliance and the data for the vessels in both groups? (Consider only the large vessels in both groups.)

8.23. Does Fig. 8.26 suggest that sympathetic stimulation and inhibition change the vessels' compliances, dead volumes, or both?

Flow and Pressure

8.24. What is the average time blood spends in a capillary?

8.25. The *cardiac index* is the cardiac output divided by the person's surface area. It normally ranges from 2.6 to 4.2 (L/min)/m^2. Use this to determine the cardiac output of a standard human. How does this value compare to the normal cardiac output we have assumed?

8.26. An artery with radius r_1 and blood speed u_1 divides into n arteries of equal radius. Find the radius r_2 and blood speed u_2 in these daughter arteries assuming that the pressure drop per unit distance $\mathrm{d}P/\mathrm{d}x$ is the same in the initial artery and each daughter artery.

8.27. Four veins with radius r_1 and flow speed u_1 combine to form one vein with radius $r_2 = 4r_1$. Find the flow speed in the larger vein.

8.28. The design of blood vessels is sometimes optimized by minimizing a "cost function," F, which is the sum of the rate work is done on the blood in the vessel and the rate that energy is used by metabolism through the blood in the vessel [385]. The first term is $Q(\Delta P)$, for flow rate Q and pressure drop ΔP, and the second term is assumed to be proportional to the volume of the vessel of radius r and length L, $K\pi r^2 L$, where K is a constant. Consequently, the cost function can be written as

$$F = Q(\Delta P) + K\pi r^2 L. \tag{8.108}$$

(a) For a resistive vessel, show that this becomes

$$F = \frac{8\eta L}{\pi r^4} Q^2 + K\pi r^2 L. \tag{8.109}$$

(b) For a fixed vessel length and flow rate, show the optimal radius is

$$r_{\mathrm{opt}} = \left(\frac{16\eta}{\pi^2 K}\right)^{1/6} Q^{1/3} \tag{8.110}$$

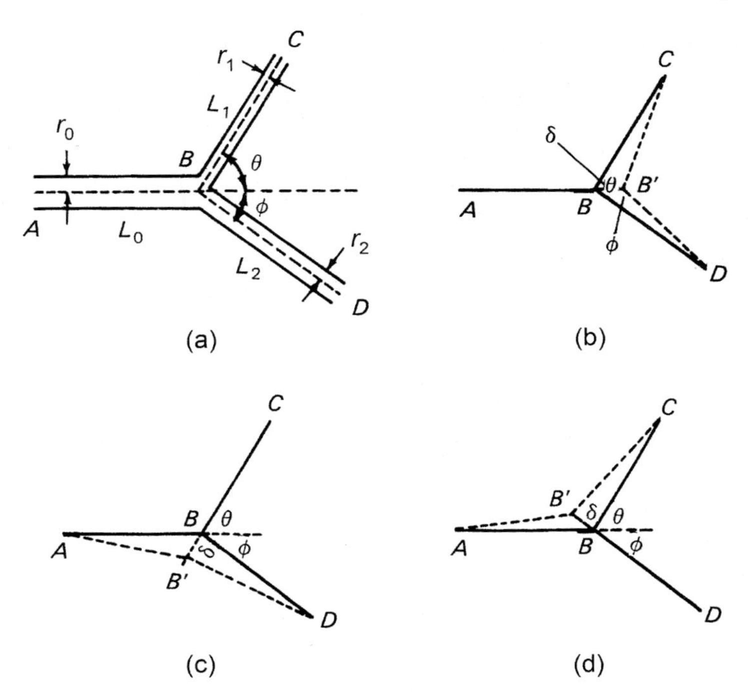

Fig. 8.64. (**a**) A planar, bifurcating vessel. Determining variations in the length of each vessel for small planar displacements of B to B' in the (**b**) AB, (**c**) AC, and (**d**) DB directions. (Based on [385].) For Problem 8.29

and the minimum cost function is

$$F_{\min} = \frac{3\pi}{2} K L r_{\text{opt}}^2. \tag{8.111}$$

8.29. (advanced problem) We will use the cost function defined in Problem 8.28 to optimize the flow in a vessel of radius r_0 and length L_0 with flow rate Q_0, that bifurcates into a vessel of radius r_1 and length L_1 with flow rate Q_1 at an angle θ to the first vessel and one with radius r_2 and length L_2 with flow rate Q_2 at an angle ϕ to the first vessel, as seen in Fig. 8.64a [385]. Having straight, coplanar vessels minimizes the vessel lengths.
(a) Show that the total cost function is

$$F_{\min} = \frac{3\pi K}{2}(r_0^2 L_0 + r_1^2 L_1 + r_2^2 L_2). \tag{8.112}$$

(b) We can optimize the lengths and angles of the vessels by considering how the displacements of point B in Figs. 8.64b–d change the cost function. Show that any such movement of point B causes length changes δL_0, δL_1, and δL_2

that lead to a change of the cost function of

$$\delta F_{\min} = \frac{3\pi K}{2}(r_0^2(\delta L_0) + r_1^2(\delta L_1) + r_2^2(\delta L_2)). \tag{8.113}$$

This is optimized by setting $\delta F_{\min,\text{opt}} = 0$.
(c) Show that moving point B along AB to B' by a distance δ as shown in Fig. 8.64b gives $\delta L_0 = \delta$, $\delta L_1 = -\delta\cos\theta$, and $\delta L_2 = -\delta\cos\phi$, and $\delta F_{\min,\text{opt}} = (3\pi K\delta/2)(r_0^2 - r_1^2\cos\theta - r_2^2\cos\phi) = 0$, and so it is optimized by

$$r_0^2 = r_1^2\cos\theta + r_2^2\cos\phi. \tag{8.114}$$

(d) Show that moving point B along CB to B' by a distance δ as shown in Fig. 8.64c gives $\delta L_0 = -\delta\cos\theta$, $\delta L_1 = \delta$, and $\delta L_2 = \delta\cos(\theta+\phi)$, and $\delta F_{\min,\text{opt}} = (3\pi K\delta/2)(-r_0^2\cos\theta + r_1^2 + r_2^2\cos(\theta+\phi)) = 0$, and so it is optimized by

$$-r_0^2\cos\theta + r_1^2 + r_2^2\cos(\theta+\phi) = 0. \tag{8.115}$$

(e) Show that moving point B along DB to B' by a distance δ as shown in Fig. 8.64d gives the optimization condition

$$-r_0^2\cos\phi + r_1^2\cos(\theta+\phi) + r_2^2 = 0. \tag{8.116}$$

(Note the symmetry in the last two equations.)
(f) Show that (8.114)–(8.116) can be solved to give

$$\cos\theta = \frac{r_0^4 + r_1^4 - r_2^4}{2r_0^2 r_1^2}, \tag{8.117}$$

$$\cos\phi = \frac{r_0^4 - r_1^4 + r_2^4}{2r_0^2 r_2^2}, \tag{8.118}$$

and

$$\cos(\theta+\phi) = \frac{r_0^4 - r_1^4 - r_2^4}{2r_1^2 r_2^2}. \tag{8.119}$$

(g) Use continuity of flow and (8.110) to show that

$$r_0^3 = r_1^3 + r_2^3. \tag{8.120}$$

(h) Show that (8.117) then becomes

$$\cos\theta = \frac{r_0^4 + r_1^4 - (r_0^3 - r_1^3)^{4/3}}{2r_0^2 r_1^2} \tag{8.121}$$

and find the analogous relations for (8.118) and (8.119).

8.30. Use Problem 8.29 to show that for optimized bifurcating vessels if [385]
(a) $r_1 = r_2$, then $\theta = \phi$,
(b) $r_2 > r_1$, then $\theta > \phi$,
(c) r_2 is much greater than r_1, then r_2 approaches r_0 and ϕ approaches $\pi/2$,
(d) $r_1 = r_2$, then $r_1/r_0 = 2^{-1/3} = 0.794 = \cos\theta$ and so $\theta = 37.5°$.
These results generally agree with observations.

8.31. Use Problem 8.30d to show that it would take ~30 generations of symmetric bifurcations starting with a vessel with the aorta radius of 1.5 cm to arrive at a vessel with the capillary radius of 5×10^{-4} cm [385]. (Note, however, that such arterial divisions are usually not simple symmetric bifurcations.)

8.32. There is a fusiform aneurysm in an aorta where the internal radius increases from $r_1 (= 1.25\,\text{cm})$ in the normal section to $r_2 = 1.3r_1$ in the diseased section, while staying at the same vertical height. The speed of blood flow is $v_1 = 0.4\,\text{m/s}$ in the normal section and the (gauge) pressure P_1 is 100 mmHg. The blood density is 1,060 kg/m^3.
(a) Find the speed of blood flow v_2 in the aneurysm.
(b) Find the pressure P_2 in the aneurysm.
(c) Use the Law of Laplace to find the tensions required in the normal part of the aorta and in the aneurysm to maintain the pressure difference (from inside to outside the vessel). Compare these values.
(d) Describe how this increase in the tension needed in the aneurysm wall and the decreased strength of the wall (due to the thinner aorta wall in the aneurysm) can lead to an unstable situation.

8.33. The normal inner radius of a large artery is 2 mm. It is 75 cm long, and the flow through it is 1/200 of the total blood flow. How would the pressure drop across it change if the flow through it were unchanged and there were severe stenosis in the artery
(a) across its entire length or
(b) across 5 cm of it?
(c) In each case, if the pressure at the beginning of the artery were 75 mmHg, would the pressure drop be severe enough to affect flow in the arterioles and capillaries?
(d) In each case, what added pressure would be needed at the beginning of the artery to maintain an unchanged flow in these arterioles and capillaries?

8.34. Arteriosclerotic plaque narrows down a section of an artery to 20% of its normal cross-sectional area. What is the pressure in that section if immediately before it the pressure is 100 mmHg and the flow speed is 0.12 m/s?

8.35. The osmotic pressure of blood is 25 mmHg higher than that of interstitial fluid because it has a higher density of proteins. What is the difference in their densities of proteins that accounts for this?

8.36. (more advanced) The blood hematocrit is usually higher nearer the center of a blood vessel than at the blood vessel wall and has a distribution that we will take as $h(r) = H[1 - (r/R)^2]$ from $r = 0$ to R [406]. (The reason why flowing suspended particles, such as red blood cells, have higher concentrations near the center, called the Fahraeus–Lindquist effect, is not obvious.)
(a) The volume flow of a cylindrical shell in the vessel is $2\pi r v(r)\mathrm{d}r$, where $v(r) = 2u\left(1 - r^2/R^2\right)$ from (7.40), so this flow weighted for the hematocrit is $2\pi r h(r) v(r)\mathrm{d}r$. Therefore the average hematocrit in the transported blood is $h_{\mathrm{av}} = \int_0^R 2\pi r h(r) v(r)\mathrm{d}r / \int_0^R 2\pi r v(r)\mathrm{d}r$. Show that $H = 3h_{\mathrm{av}}/2$.
(b) Now find the average value of the hematocrit at any given time in the blood vessel by calculating $\int_0^R 2\pi r h(r)\mathrm{d}r / \int_0^R 2\pi r\,\mathrm{d}r$. Show that for the parabolic distribution of hematocrit this volume-averaged hematocrit is $3h_{\mathrm{av}}/4$.
(c) The result in (b) states that the average hematocrit of the blood in the vessel at any given time is less than that in the blood that is being transported. Is this a contradiction?

8.37. (more advanced) Repeat parts (a) and (b) in Problem 8.36 assuming $h(r) = H\exp(-r/R)$ and show the volume-averaged hematocrit is $0.88h_{\mathrm{av}}$ [406].

8.38. (more advanced) The analysis in Problems 8.36 and 8.37 assumed that the parabolic $v(r)$ we derived earlier assuming a constant viscosity is still valid when the hematocrit – and consequently the viscosity – decreases with radius. This should not be true. Qualitatively, how would you expect the spatially varying hematocrit and viscosity to affect the parabolic flow rate?

The Heart and Circulation

8.39. Would you expect cardiac muscle to be most similar to Type I, IIA, or IIB skeletal muscle? Why?

8.40. Compare the total mechanical and metabolic powers needed by the left heart and the right heart to pump blood.

8.41. Determine all the pressures, volumes, and flow rates in the overall body circulation model using the data provided in the Table 8.7. Do your answers agree with your expectations, such as the values in Table 8.1?

8.42. The volumetric flow rate out of a ventricle has been described in terms of the heart rate F and stroke volume V_{stroke} as $Q = FV_{\mathrm{stroke}}$, while flow rates have also been expressed in terms of the vessel cross section A and flow speed u as $Q = Au$. Explain why these two characterizations are either consistent or inconsistent.

8.43. If the cardiac output is 5 L/min and heart rate is 1 Hz, determine the volume of the left ventricle at its peak if the ejection fraction is 65%.

8.44. There is a hole in the septum that separates the left ventricle and right ventricle (Fig. 8.60).
(a) Explain why you would expect the pressure in the left ventricle to decrease and that in the right ventricle to increase.
(b) Explain why you would expect the stroke volume from the left ventricle to decrease because of this.
(c) Explain why the oxygenation of the blood in the left ventricle would decrease and that in the right ventricle would increase.
(d) If during systole the (gauge) pressure, stroke volume, and oxygenation levels (relative to that in the vena cavae) in the left ventricle each decreases by 10% as a result of this, explain quantitatively how the body could try to compensate for this?

8.45. (a) If the inner volume of the left ventricle is $100\,\mathrm{cm}^3$ and the wall volume is $30\,\mathrm{cm}^3$, find the inner radius, outer radius, and wall thickness for the ventricle modeled as a hemispherical shell.
(b) Find the wall stress during systole.

8.46. The cardiac output of a woman remains at 5 L as she ages from 25 to 65 years of age, while her blood pressure increases in the average way.
(a) How does her total vascular resistance change?
(b) What fractional changes in vessel radius do this correspond to? (Assume conditions for arterioles.)

8.47. When blood is pumped out of the left ventricle, it travels "upward" a distance of about 10 cm in the aorta during the ~0.2 s duration of the peak of systole, stretching the walls of this very compliant vessel. There are no external forces on the body during this time, so the center of mass of the body does not move. Consequently, when the stroke volume of blood (of mass $m_{\mathrm{blood}} \simeq 70\,\mathrm{g}$) is ejected upward, the rest of the body (of mass $m_{\mathrm{rest}} \simeq 70\,\mathrm{kg}$) moves "downward" (ignoring gravity and frictional forces). This is the basis of the diagnostic method called *ballistocardiography*, in which a person rests horizontally on a light, very low friction horizontal suspension [367]. (Such devices have been used to develop methods that assess heart function, but are not in clinical use.) Assume the person is lying along the x direction on this "couch" – with his head pointing in the positive direction – and the center of mass of ejected blood is at x_{blood}, that of the rest of the body is at x_{rest}, and that of the entire body is x_{body}.
(a) Show that $x_{\mathrm{rest}} = (x_{\mathrm{blood}} m_{\mathrm{blood}} + m_{\mathrm{rest}} x_{\mathrm{rest}})/(m_{\mathrm{blood}} + m_{\mathrm{rest}})$ always.
(b) Now let us call the positions in (a) those before systole. At the end of the main part of systole, the blood and rest of the body have moved by

Δx_{blood} and Δx_{rest}, respectively. Show that the body has moved by $\Delta x_{\text{rest}} = -(m_{\text{blood}}/m_{\text{rest}})\Delta x_{\text{blood}}$ and that this is -0.1 mm.
(c) Because the blood moves with a constant velocity in this motion in the aorta, show that the velocity of the body during systole is -0.5 mm/s in the x direction.
(d) What is the average of Δx_{rest} during a full cardiac cycle? Why?

8.48. Someone wants to donate two pints of blood instead of the usual (and allowed maximum of) one. What consequences could this have?

8.49. (advanced problem) (a) Solve (8.92) assuming that the flow $Q_{\text{L}}(t)$ is a constant a from $t = 0$ to αT, and 0 from $t = \alpha T$ to T, where $0 \leq \alpha \leq 1$. (This repeats for every heart beat.) Note that the pressure at the beginning and end of each cardiac cycle is P_{diastole} and it becomes P_{systole} at $t = \alpha T$. (Hint: The analysis is similar to that for exciting an isometric muscle in Chap. 5 (see (5.11)–(5.13)) and temperature regulation in Chap. 13 (see (13.18)); also see Appendix C.)
(b) Show that $a = V_{\text{stroke}}/\alpha T C_{\text{sa}}$.
(c) Show the solutions from (a) lead to the relations

$$P_{\text{diastole}} = P_{\text{systole}} \exp(-(1-\alpha)T/\tau) \qquad \text{and} \tag{8.122}$$

$$P_{\text{systole}} = V_{\text{stroke}}\tau/\alpha T C_{\text{sa}} + (P_{\text{diastole}} - V_{\text{stroke}}\tau/\alpha T C_{\text{sa}}) \exp(-\alpha T/\tau). \tag{8.123}$$

(d) Sketch $P_{\text{sa}}(t)$ for several heart beats for $\alpha = 1/3$. Compare this sketch with those from the simple model in Fig. 8.56 and the real pulse in Fig. 8.54. Is this model better? Why?
(e) Show that when $\alpha = 0$ the solutions in (a) and (c) give the results presented in the text for the simpler model.

8.50. The solution to the classic Windkessel Model for steady-state flow that is suddenly turned off is exponential decay of the flow, as we saw in the simple model of the arterial pulse. In the electrical analog in Fig. 8.57 this corresponds to tracking the current when a constant voltage is initially applied and is suddenly turned off. Analyze this electrical problem analog and show that it has the same solution as the flow problem.

8.51. (advanced problem) Solve the electrical analog in Problem 8.50 for the three-element Windkessel Model.

8.52. (advanced problem) Solve the electrical analog in Problem 8.50 for the viscoelastic Windkessel Model.

Scaling

8.53. Calculate the heart mass and heart beat rate (in beats per minute) for a man (70 kg), woman (50 kg), and an infant (5 kg) using the allometric relation parameters in Table 1.13.

8.54. The heart rate of mammals F is known to decrease with body mass as $m_{\mathrm{b}}^{-1/3}$. This seems to be true interspecies and also within a species. The human heart rate is known to decrease from infancy, through childhood and to maturity in a manner described better by body mass than age. Derive this relation using the dimensional analysis methods presented in Chap. 1. Assume that the stroke volume scales as body mass. Assume that a primary function of circulation is to bring warm blood from the core to the body surface to keep it warm. This means that the total blood flow rate scales as the rate of heat loss from the body.

9

Lungs and Breathing

Our lungs serve several important functions. They interact with blood by exchanging carbon dioxide for oxygen (Chap. 8) and they maintain the blood pH. The lungs are involved in heat exchange and fluid balance in the body, because relatively dry and usually cooler air is inhaled and air at the body temperature saturated with water vapor is exhaled (Chap. 6). They are also a key element in voice production (Chap. 10). We will highlight the mechanics of breathing [423, 424, 428, 429, 430, 432, 434, 435, 439, 443, 444].

We typically breathe in 6 L/min of air. (This compares to the ∼5–6 L of blood pumped per min in the pulmonary circulation through the lungs.) Because air is ∼20% oxygen, we inhale (inspire) 1.2 L oxygen/min. The breathing rate is typically 12/min for men, 20/min for women, and 60/min for infants. The air we inhale has 80% N_2/20% O_2 (or more precisely 78.084% N_2/20.947% O_2/0.934% Ar/0.035% CO_2), and the air we exhale (expire) has 80% N_2/16% O_2/4% CO_2. (If the air we exhaled had little or no oxygen, we could not use it for mouth-to-mouth resuscitation.) We breathe in roughly 10 kg air/day, with ∼2 kg O_2/day. The lungs absorb about 0.5 kg O_2/day (400 L). We exhale air with ∼0.5 kg water vapor/day.

Because we inspire and expire air at the same rate (if not where would the difference go?), it is clear the body uses ∼0.3 L oxygen/min during usual sedentary activity, delivered by the ∼5–6 L of blood pumped per min. We have called this rate of oxygen consumption in the body dV_{O_2}/dt in Chaps. 6 and 8. During aerobic exercise dV_{O_2}/dt increases linearly with cardiac output Q_t (see Fig. 8.27, and (6.18) and (8.33)). The maximum rate of oxygen usage is ≈2.8 L/min for a person of average fitness and ≈4 L/min for a highly fit person. This assumes the lungs bring in air at a rate fast enough to maintain the needed oxygenation of arterial blood.

Gauge pressures, relative to atmosphere, are usually used in discussing breathing. Two roughly equal types of units are commonly used, mmHg and cmH_2O, with 1 mmHg = 1.36 cmH_2O.

9.1 Structure of the Lungs

Air is inhaled through the nose or mouth and then through the pharynx, larynx, and the trachea (windpipe) (Fig. 9.1). The trachea divides into the right and left bronchus (Fig. 9.2), each of which continues to bifurcate into smaller and smaller bronchi and bronchioles over 23 levels of bifurcation ($2^{24} = 1.6 \times 10^8$) (Table 9.1, Figs. 9.2 and 9.3) until they form alveoli (which is the plural of alveolus) (al-vee-oh'-lie (lus)), which are the actual operating units of the lungs. The average diameter of the airways decreases with generation z, as $d(z) = 2^{-z/3}d(0)$ until generation 16. This relation is the optimal design of a branched system of tubes in hydrodynamics. There are about 3×10^8 alveoli, each ~0.2–0.3 mm in diameter, with walls that are

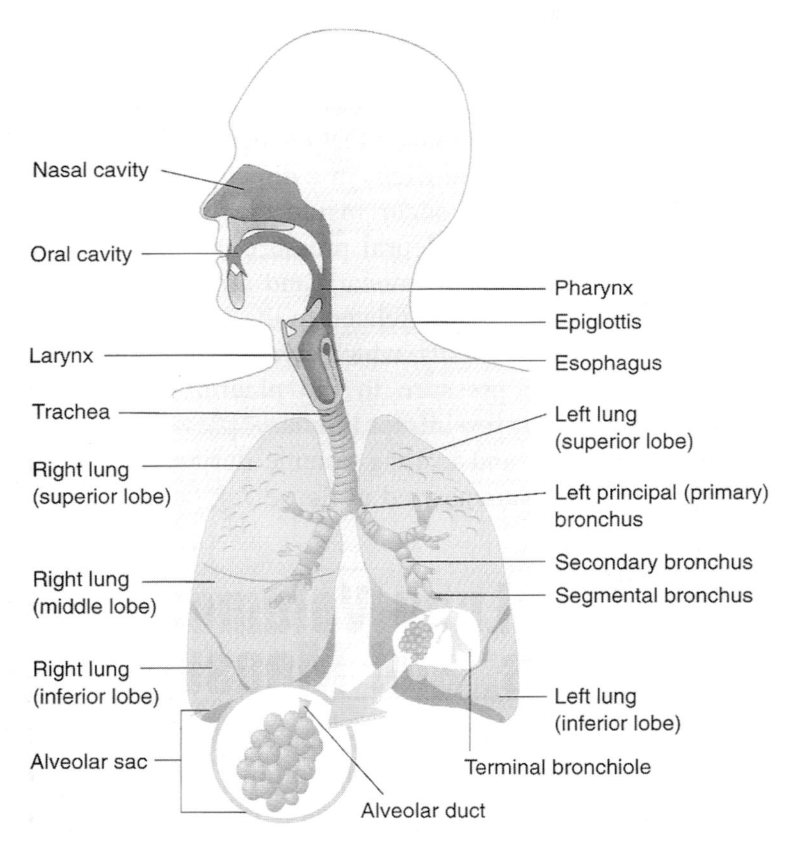

Fig. 9.1. Diagram of parts of the respiratory system. (These components are also important in voice production (Chap. 10). The vocal cords (or vocal folds) used in speaking are in the larynx.) (From [425]. Used with permission)

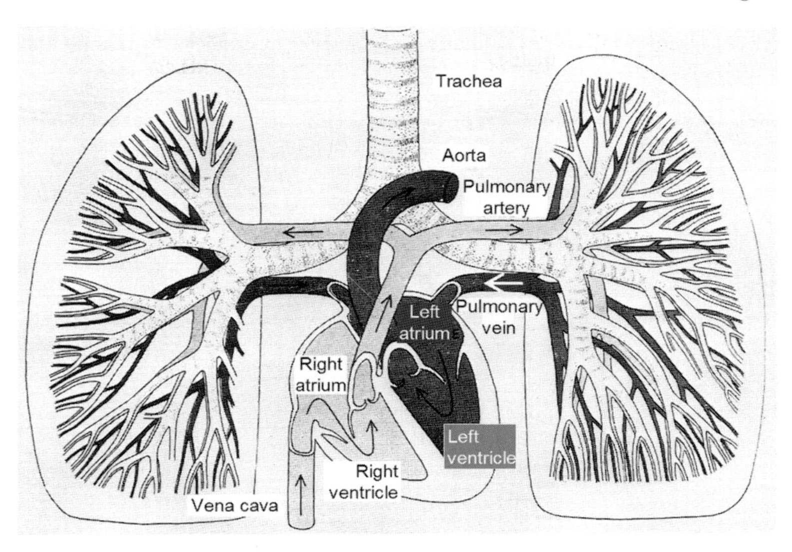

Fig. 9.2. The relationship between the lung and heart is shown. The first few generations of the branching of the air vessels in the lungs, pulmonary arteries, and the pulmonary veins are shown. These three systems can be called the three "trees" of the lung. Note that the pulmonary arteries are close to the bronchi, while the pulmonary veins stand alone. (From [427])

$\sim$0.4 μm thick. They are in contact with blood in the pulmonary capillaries (Fig. 9.4), which themselves form after subdividing in 17 branches (Table 9.2, Figs. 9.2 and 9.5). Oxygen diffuses from the alveoli to the red blood cells, while carbon dioxide diffuses from the blood into the air in the alveoli. The total surface area of the alveoli is $\sim 80\,\mathrm{m}^2$ (ranging from 50–100 m^2). The total external surface area of the lungs is only $\sim 0.1\,\mathrm{m}^2$, so subdividing into alveoli results in a tremendous increase in the surface area in contact with the blood, by a factor of almost 1,000. This is also the factor by which the oxygen intake increases. Without this, we would never even come close to meeting our metabolic needs for oxygen. Our chests expand when we breathe because incoming air filling the alveoli makes each one bigger, just as with ordinary bubbles.

The circulatory system is the conduit for the transfer of O_2 and CO_2 between the alveoli and tissues, and so we should track the partial pressure in each system. Within the alveoli the partial pressure of O_2 is $\simeq$105 mmHg, which is smaller than that in the atmosphere (159 mmHg = 21% of 760 mmHg) because of the dead volume in the respiratory system. The partial pressure of O_2 blood in the pulmonary capillaries increases from 40 to $\simeq$100 mmHg after O_2 is transferred from the alveoli, and this is the partial pressure in the pulmonary veins and systemic arteries. The partial pressure of O_2 in tissue is 40 mmHg, so that after transfer of O_2 from the capillaries to surrounding

Table 9.1. Approximate quantification of the bronchial system. (Using data from [426, 440]. Also see [441, 442])

pulmonary branch	generation z	branch diameter (mm)	branch length (mm)	total cross-sectional area (cm^2)	volume (cm^3)	air speed (cm/s)
trachea	0	18.0	120.0	2.5	31	393
main bronchus	1	12.2	47.6	2.3	11	427
lobar bronchus	2	8.3	19.0	2.1	4.0	462
	3	5.6	7.6	2.0	1.5	507
segmental bronchus	4	4.5	12.7	2.5	3.5	392
	5	3.5	10.7	3.1	3.3	325
bronchi	6	2.8	9.0	4.0	3.5	254
w/cartilage in wall	7	2.3	7.6	5.1	3.8	188
	8	1.86	6.4	7.0	4.4	144
	9	1.54	5.4	9.6	5.2	105
	10	1.30	4.6	13	6.2	73.6
terminal bronchus	11	1.09	3.9	20	7.6	52.3
	12	0.95	3.3	29	9.8	34.4
bronchioles	13	0.82	2.7	44	12	23.1
w/muscle in wall	14	0.74	2.3	69	16	14.1
	15	0.66	2.0	113	22	8.92
terminal bronchiole	16	0.60	1.65	180	30	5.40
respiratory bronchiole	17	0.54	1.41	300	42	3.33
respiratory bronchiole	18	0.50	1.17	534	61	1.94
respiratory bronchiole	19	0.47	0.99	944	93	1.10
alveolar duct	20	0.45	0.83	1,600	139	0.60
alveolar duct	21	0.43	0.70	3,200	224	0.32
alveolar duct	22	0.41	0.59	5,900	350	0.18
alveolar sac	23	0.41	0.50	12,000	591	0.09
alveoli, 21 per duct		0.28	0.23		3,200	

The air speed is assumed to be 1 L/s. The data include that for both lungs. The number in each generation is 2^z (for generations $z = 0$–23), and 300×10^6 for the alveoli.

tissues, the partial pressure in the systemic veins and pulmonary arteries is also $\simeq$40 mmHg – and then it is again increased to 100 mmHg in the lungs.

Similarly, within the alveoli the partial pressure of CO_2 is $\simeq$40 mmHg; this is much larger than that in the atmosphere ($\sim$0.25 mmHg), again because of the dead volume. The partial pressure of CO_2 blood in the pulmonary capillaries decreases from 46 to $\simeq$40 mmHg after CO_2 is transferred to the alveoli, and this is the partial pressure in the pulmonary veins and systemic arteries. The partial pressure of CO_2 in tissue is 46 mmHg, so that after transfer of CO_2 into the capillaries from the tissues, the partial pressure in the systemic veins and pulmonary arteries is also $\simeq$46 mmHg – and then it is again decreased to 40 mmHg in the lungs.

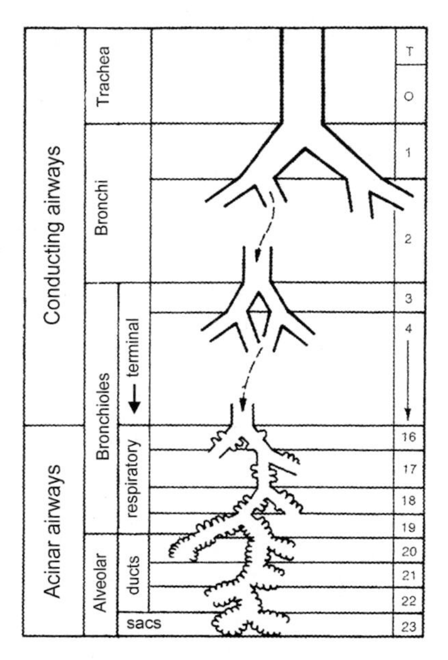

Fig. 9.3. Bifurcations of lung airways, showing generation number z. (From [436])

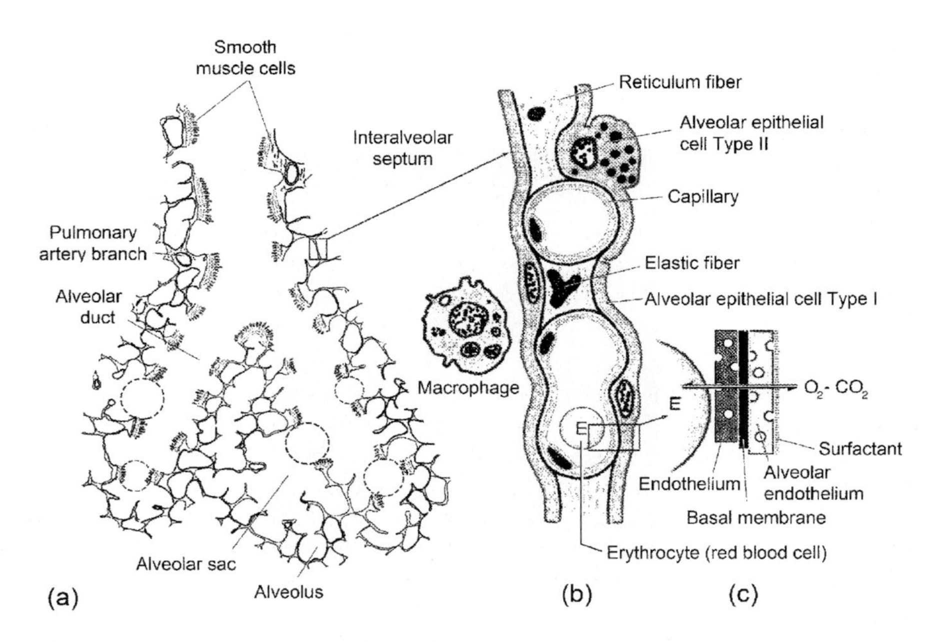

Fig. 9.4. The details of the alveolar bifurcation are shown in (**a**). These alveoli are sacs imbedded in capillary beds. The details of the interaction between the alveoli and capillaries are depicted in (**b**) and (**c**). (From [436])

Table 9.2. Branching structure of the pulmonary arterial network. (Using data from [426, 438])

pulmonary branching order	number of branches of each order	vessel length (mm)	vessel diameter (mm)
1	1	90.5	30.0
2	3	32.0	14.83
3	8	10.9	8.06
4	20	20.7	5.82
5	66	17.9	3.65
6	203	10.5	2.09
7	675	6.6	1.33
8	2,290	4.69	0.85
9	5,861	3.16	0.525
10	17,560	2.10	0.351
11	52,550	1.38	0.224
12	157,400	0.91	0.138
13	471,300	0.65	0.086
14	1,411,000	0.44	0.054
15	4,226,000	0.29	0.034
16	12,660,000	0.20	0.021
17	300,000,000	0.13	0.013

Fig. 9.5. A silicone elastomer cast of the venous tree of the lung of a cat. The venous pressure was $-7\,\mathrm{cmH_2O}$ ($= -5\,\mathrm{mmHg}$), the airway pressure was $10\,\mathrm{cmH_2O}$ ($= 7\,\mathrm{mmHg}$), and the pleural pressure was $0\,\mathrm{cmH_2O}$. (From [427])

9.2 The Physics of the Alveoli

The alveoli are similar to interconnected bubbles. Inside them the pressure is P_{in} and outside the pressure is P_{out}, with $\Delta P = P_{\text{in}} - P_{\text{out}}$, and they have a radius R. The Law of Laplace for a sphere (7.9) is

$$\Delta P = \frac{2T}{R}, \tag{9.1}$$

where T is the tension in the sphere walls. The main source of this tension in the alveoli is not within the walls but on the surfaces. This contribution is called the surface tension γ, which has the same units as T – of force/length or energy/area. In typical bubbles, such as soap bubbles, both surfaces contribute the same surface tension and so T is replaced by 2γ. Therefore we find

$$\Delta P = P_{\text{in}} - P_{\text{out}} = \frac{4\gamma}{R}. \tag{9.2}$$

For the water/air interface $\gamma \simeq 7.2\times10^{-4}$ N/m (Table 7.2). In alveoli, however, only the surface tension of the inner surface is really important because it is a fluid/air interface with larger surface tension than the fluid/fluid interface of the outer surface, and so

$$\Delta P_{\text{alveoli}} = P_{\text{in}} - P_{\text{out}} = \frac{2\gamma}{R}. \tag{9.3}$$

There is an apparent instability that seemingly leads to an unreasonable situation in interconnected bubbles or alveoli. Consider two bubbles that are initially not interconnected, as in Fig. 9.6, because there is a plug between them. Bubble #1 has an internal pressure P_1 and radius R_1, and Bubble #2 has an internal pressure P_2 and radius R_2. (Because the difference between the pressure inside and outside the bubble is what is significant, the external

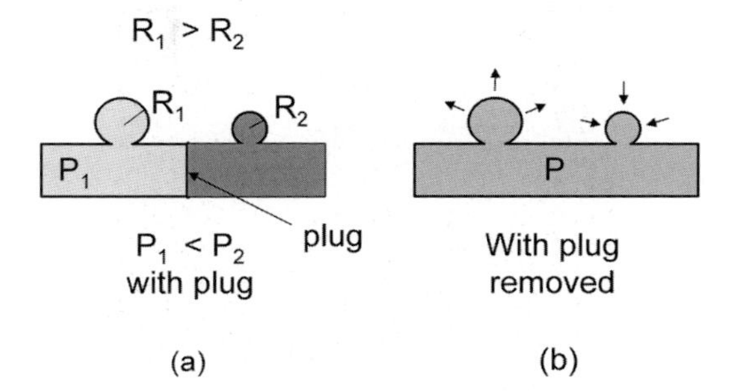

Fig. 9.6. Instability of bubbles, according to the Law of Laplace. This assumes that the surface tension does not change with bubble (or alveolus) radius. The external pressure is 0

pressure is equal to zero.) In equilibrium, the internal pressure $P_{\text{internal}} = 4\gamma/R$ for each bubble. (Whether this factor is 4 or 2 is not significant here.) Say Bubble #2 is the smaller bubble. Because $R_2 < R_1$, in equilibrium $P_2 > P_1$; the smaller bubble has the higher internal pressure. If the plug is opened, air will flow from higher pressure to lower pressure, and therefore from the smaller bubble to the larger bubble. The loss of air in Bubble #2 makes it smaller. With this smaller radius, the equilibrium internal pressure increases. Because this pressure is still higher than in Bubble #1, air continues to flow from the smaller bubble to the larger bubble, until it collapses.

This implies that the largest of the hundreds of millions of alveoli would get ever larger at the expense of all of the smaller ones and the system of alveoli we have described for the lungs could not be stable. What is wrong? There is no error in our reasoning; however, we have made one assumption that is not accurate for alveoli. We implicitly assumed that the surface tension is not a function of radius R. There is a surfactant on the surfaces of the alveoli of healthy people, containing dipalmitoyl phosphatidycholine or DPPC, that causes $\gamma(R)$ to decrease for decreasing R. With $\Delta P = 2\gamma(R)/R$, as R of the smaller bubble or alveolus becomes smaller in Fig. 9.6, eventually $\gamma(R)$ decreases with smaller R faster than R does itself, as in Fig. 9.7, so ΔP begins to decrease with smaller R. Such a system of interconnected alveoli is stable.

We can see how such a dependence of $\gamma(R)$ can occur with the following model. The surface of an alveolus can be covered either with a lipoprotein or by water; the surface tension of the lipoprotein is much lower ($\gamma_{\text{lung}} = 1 \times 10^{-3}$ N/m) than that of water ($\gamma_{\text{water}} = 7.2 \times 10^{-2}$ N/m (= 72 dynes/cm)). Assuming the alveolus is spherical, for one particular radius R_0 there is exactly one monolayer of lipoprotein on the whole surface and at that radius the surface tension is γ_{lung} over the $4\pi R_0^2$ surface area. If this alveolus becomes smaller, so $R < R_0$, it has several monolayers of lipoprotein on its surface and

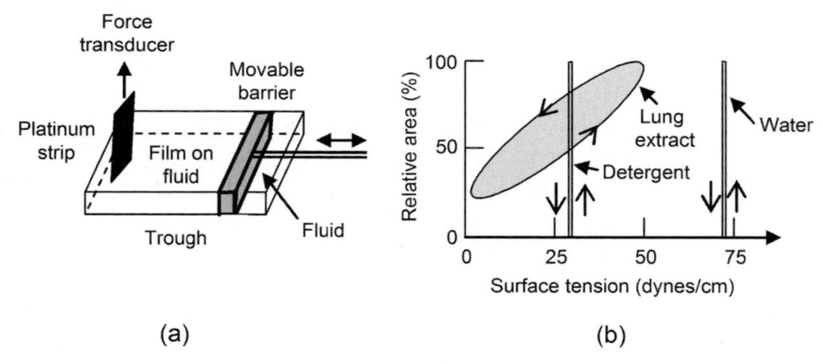

Fig. 9.7. Surface tension on alveoli walls (lung extract) in (**b**), as measured by the surface balance in (**a**) which measures surface tension vs. area. Similar measurements for detergent and water are also shown. (Based on [430, 443])

its surface tension is still γ_{lung}, and so

$$\gamma(R) = \gamma_{\text{lung}} \quad \text{for } R \leq R_0. \tag{9.4}$$

If this same alveolus instead becomes larger, so $R > R_0$, it has a monolayer of lipoprotein over only a portion of its surface (of surface area $4\pi R_0^2$ because the layer cannot become smaller than a monolayer) and water over the rest of the surface (of area $4\pi R^2 - 4\pi R_0^2$). So the average surface tension is

$$\gamma(R) = \frac{4\pi R_0^2 \gamma_{\text{lung}} + (4\pi R^2 - 4\pi R_0^2)\gamma_{\text{water}}}{4\pi R^2} \quad \text{for } R > R_0 \tag{9.5}$$

or

$$\gamma(R) = \gamma_{\text{water}} - \frac{R_0^2}{R^2}(\gamma_{\text{water}} - \gamma_{\text{lung}}) \quad \text{for } R > R_0. \tag{9.6}$$

This approaches the much smaller γ_{water} for $R \gg R_0$ at a rate that is faster than $1/R$, so the alveoli will be stable.

Because this lipoprotein is only on one of the surfaces, the stability condition is $\Delta P = 2\gamma/R$. For $R > R_0$, there is a stable equilibrium when $\mathrm{d}(\Delta P)/\mathrm{d}R = \mathrm{d}[2\gamma_{\text{water}}/R - 2(R_0^2/R^3)(\gamma_{\text{water}} - \gamma_{\text{lung}})]/\mathrm{d}R = 0$ or

$$\frac{\mathrm{d}(\Delta P)}{\mathrm{d}R} = -\frac{2\gamma_{\text{water}}}{R^2} + 6\frac{R_0^2}{R^4}(\gamma_{\text{water}} - \gamma_{\text{lung}}) = 0 \tag{9.7}$$

or

$$R_{\text{eq}} = \sqrt{3\frac{\gamma_{\text{water}} - \gamma_{\text{lung}}}{\gamma_{\text{water}}}}\, R_0. \tag{9.8}$$

Because $\gamma_{\text{water}} \gg \gamma_{\text{lung}}$, the equilibrium radius $R_{\text{eq}} \simeq \sqrt{3}R_0$.

Figure 9.7 shows that this surface tension of the surfactant in the lung decreases from 5×10^{-2} N/m (50 dynes/cm) to zero as the area of the film gets smaller. Alveoli are typically stable at approximately 1/4 of their maximum size.

One function of the surfactant is to provide alveolus stability. Another function is to lower the amount of force needed to be supplied by the diaphragm to inflate the alveoli. With $\gamma_{\text{water}} = 7.2 \times 10^{-2}$ N/m and $R = 0.05$ mm of the alveoli when they are collapsed (and need to be inflated), (9.3) gives $\Delta P_{\text{alveoli}} = 2.9 \times 10^3$ N/m^2 = 22 mmHg. The area of an adult diaphragm muscle is about 500 cm^2, so the force it needs to exert to expand the alveoli for breathing is ∼150 N – which corresponds to a weight of 15 kg. With the lower surface tension of the lung surfactant, this force is over an order of magnitude smaller and breathing is easier, especially for infants. This explains why people with insufficient surfactant – with hyaline membrane disease – have difficulty breathing.

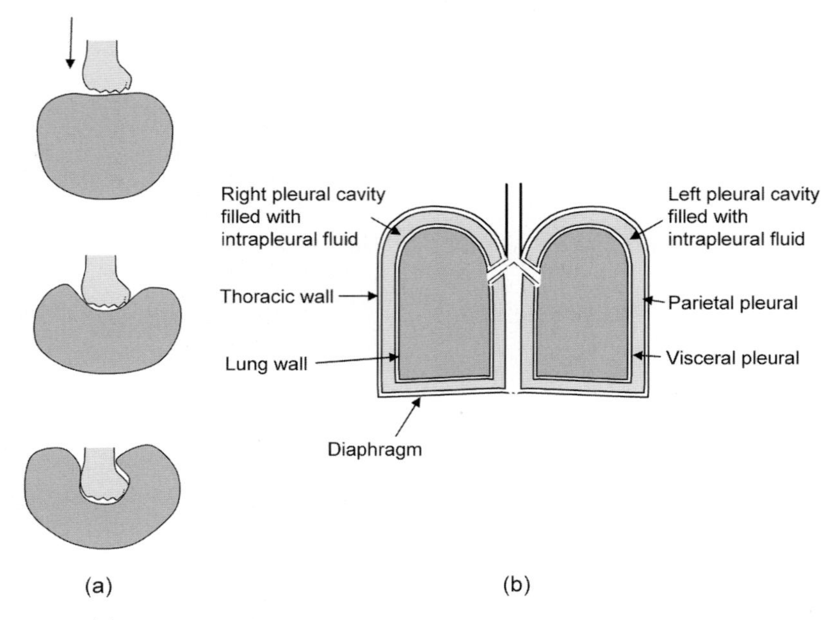

Fig. 9.8. (**a**) Pushing a fist into a balloon is analogous to the lungs in the pleural cavity. (**b**) Schematic of the lungs in the pleural cavities. (Based on [439])

9.3 Physics of Breathing

Each lung is surrounded by a sac membrane within the thoracic cavity. We can picture the pleural sac as a balloon, as in Fig. 9.8, filled with intrapleural fluid. The inside wall of this sac, the visceral pleura (membrane), attaches to the outer lung wall. The outside wall of this sac, the parietal pleura (membrane), attaches to the thoracic wall. It is the springiness of the lung that pulls the two pleural membranes apart, and this causes a slight decrease of pressure of the pleural sac relative to atmospheric pressure of -4 mmHg to -6 mmHg. This pressure difference is what keeps the lungs expanded, and keeps them from collapsing. The mechanical "driving force" in controlling lung volume is the *transpulmonary pressure*, which is the difference in pressure in the alveoli in the lungs and that around the lung in the pleural sac, which is called the *intrapleural (or pleural) pressure*. (The alveolar and pleural pressures are gauge pressures, referenced to atmospheric pressure.)

The lungs are expanded and contracted by the motion of structures surrounding them by way of inspiratory and expiratory muscles. This occurs in two ways (Fig. 9.9), of which only the first is used during quiet breathing (1) The diaphragm moves downward to lengthen the chest cavity (by pulling the bottom of the lungs downward) during inspiration. During quiet breathing, the lungs contract by the natural elastic recoil of the lungs and chest wall, with the diaphragm relaxed, while in heaving breathing this contraction is

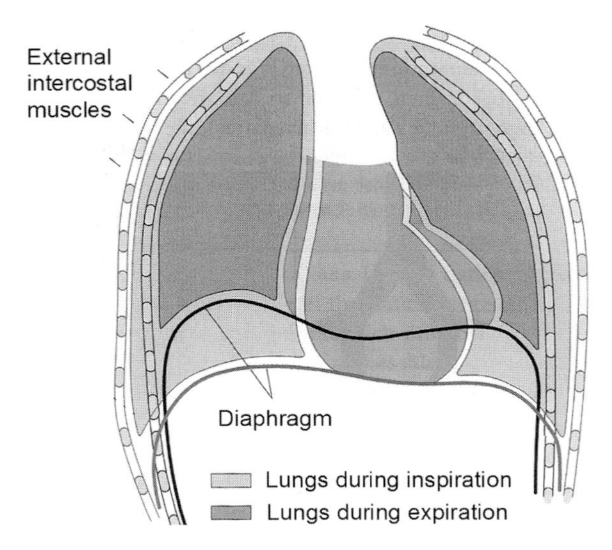

Fig. 9.9. Expansion and contraction of the thoracic cage during expiration and inspiration, showing the ribs, lungs and heart, the external intercostal muscles (that contract during inspiration to elevate the rib cage and widen it laterally so the cage increases in all three dimensions), and the diaphragm (that contracts to increase the vertical dimension of the cage during inspiration). (From [437])

accelerated by the contraction of the abdominal muscles that push the abdominal contents and then the diaphragm upward to shorten the chest cavity. (2) The ribs are elevated by the neck muscles to increase the anteroposterior (front-to-back) diameter of the chest cavity and are depressed (lowered) by the abdominal recti to decrease it. This causes chest cavity expansion and contraction, respectively, because the ribs slant outward and have larger transverse cross-sectional areas in the lower sections; this can increase the anterior–posterior chest thickness by about 20% during inspiration.

How does this help bring air into the lungs? Before inspiration, there is atmospheric pressure in the lungs. The attractive force of the visceral pleura for the parietal pleura and the outward force of the outer lung wall due to the lower-than-atmospheric pressure in the pleural sac (~ -4 mmHg) cause each lung to expand. In equilibrium their sum is balanced by the tendency of the lungs to contract due to their springiness. This preinspiration force balance is shown in Fig. 9.10. They are no longer in balance during inspiration.

The steps in inspiration (inhaling) are shown in Fig. 9.11. The inspiratory muscles (diaphragm and external intercostals) increase the dimensions of the rib cage (the thoracic cavity). This causes the visceral and parietal pleurae to separate. The lung volume then increases because (1) the attraction of the visceral and parietal pleurae increases as they are separated further and (2) this separation causes $P_{\text{lung}} - P_{\text{pleura}}$ to decrease even more, from ~ -4 to

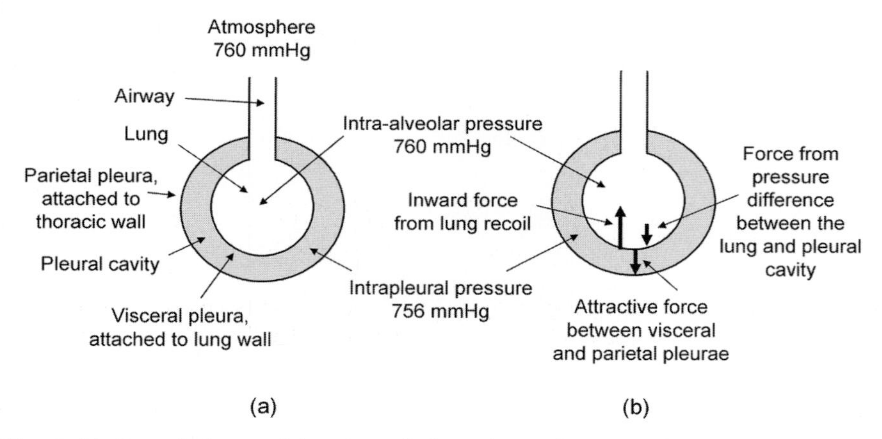

Fig. 9.10. Force balance of the visceral pleura/outer lung wall during preinspiration. Note that the forces are really normal to the wall everywhere, not just at the bottom as depicted. (Based on [439])

~ -6 mmHg (i.e., from $\sim$756 to 754 mmHg absolute pressure). Because both of these forces in the direction of lung expansion increase, they now overcome the springiness of the lungs that favors lung contraction – and the lung expands. The pressure in the lungs and alveoli decreases from $\sim$0 to ~ -1 mmHg (i.e., from $\sim$760 to 759 mmHg absolute pressure), and then air flows from the

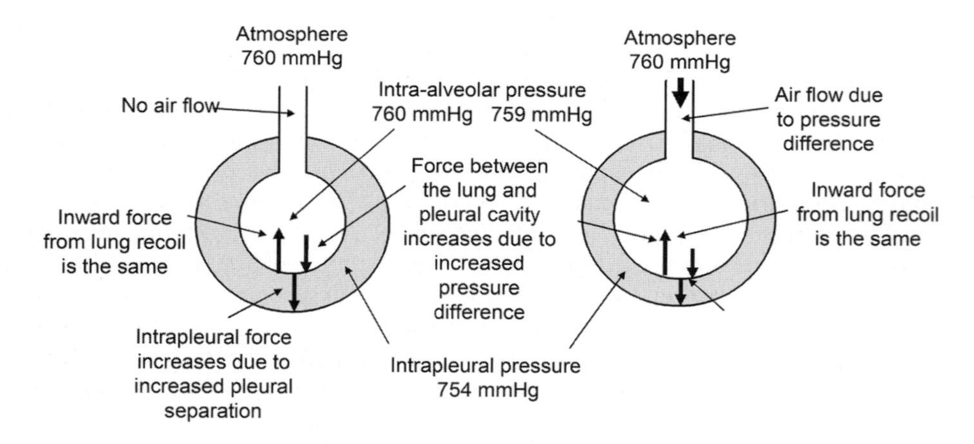

Fig. 9.11. (**a**) Force imbalance at the onset of inspiratory muscle contraction (and expansion of the thoracic wall/parietal pleura) leads to a (**b**) subatmospheric pressure in the lungs and flow of air into the lungs. (Compare this to the preinspiration force balance in Fig. 9.10). (Based on [439])

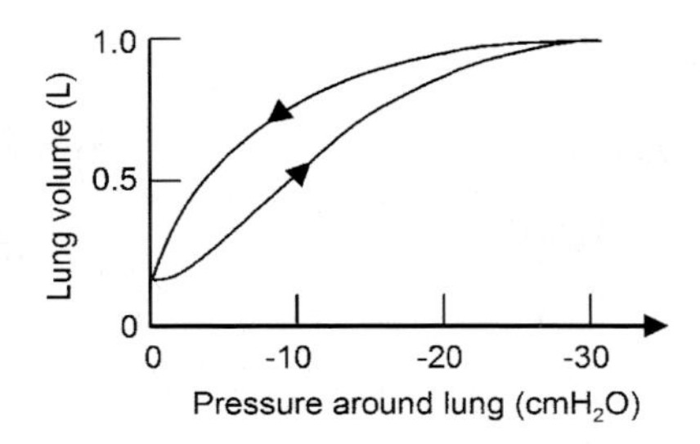

Fig. 9.12. The measurement of the pressure–volume curve of an excised lung, which shows hysteresis in inflation and deflation. (Based on [443])

mouth and nose into the lungs. During normal breathing exhaling is automatic, requiring no contraction by muscles. Muscle contraction is necessary during heavy exercise to inhale more fresh air and to actively exhale stale air.

9.4 Volume of the Lungs

The volume of the lungs depends on the transpulmonary pressure, as is seen in Fig. 9.12 for an excised lung. The inflation and deflation curves are not the same; as in Chap. 4, this is called hysteresis.

The volume of the lungs during different stages of normal and deep breathing is a good diagnostic of lung functionality. It is easily measured using a *spirometer* (Fig. 9.13). Figure 9.13 shows one such measurement during different types of breathing. In this example, during normal breathing the lung volume is seen to oscillate between 3.2 L after normal inspiration and 2.2 L after normal expiration. The difference is the tidal volume (TV) (∼1 L), which is the usual lung volume used during breathing when at rest. The volume after normal expiration is the functional residual capacity (FRC). After a deep inspiration the lung volume is the total lung capacity (TLC), ∼6 L. This exceeds the volume after normal inspiration by the inspiratory reserve volume (IRV, which is also one of the author's nicknames). After a deep expiration, the remaining volume is the residual volume (RV), ∼1 L. The difference in lung volumes after deep inspiration and deep expiration is the vital capacity (VC) ∼5 L, which also equals the total lung capacity minus the residual volume. After deep expiration the lung volume is smaller than that after normal expiration by the expiratory reserve volume (ERV), which also equals the functional residual capacity minus the residual volume.

The vital capacity is an important measure of how well the lungs are functioning. (More importantly, you need a robust vital capacity to inflate balloons.) The functional reserve capacity is the volume of stale air that normally mixes with new air (the tidal volume). There is also dead space. Some is anatomic (0.15 L), due to the trachea and bronchii, and some is physiological alveoli dead space, where the alveoli have no access to blood.

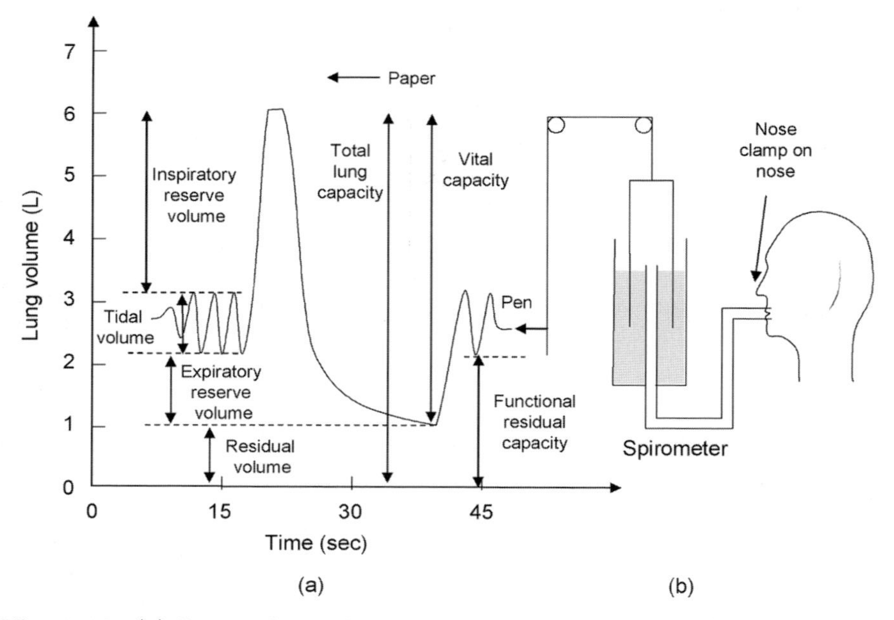

Fig. 9.13. (**a**) Lung volume changes during breathing cycles, (**b**) along with a schematic of a spirometer. (Based on [422, 443])

We know that we inhale air that is 80% N_2/20% O_2/0% CO_2, it mixes with stale air, and after diffusion across the alveoli walls we exhale air that is 80% N_2/16% O_2/4% CO_2. *What is the composition of this stale air?* Let us call its composition 80% $N_2/x\%$ $O_2/y\%$ CO_2. After inspiration the lungs have a tidal volume of 80% N_2/20% O_2/0% CO_2 air that has mixed with a functional reserve capacity of 80% $N_2/x\%$ $O_2/y\%$ CO_2 air. If the tidal volume is a fraction α of this volume, the functional reserve capacity volume fraction is $1-\alpha$. After inhalation the composition of air in the lungs is 80% $N_2/[20\alpha+(1-\alpha)x]\%$ $O_2/(1-\alpha)y\%$ CO_2. Say a fraction β of all of the inhaled air is absorbed by the lungs. This means that a fraction β of the $20\alpha\%$ O_2 in the lungs is absorbed, which is $20\alpha\beta\%$ O_2. There is an equal $20\alpha\beta\%$ increase in CO_2 that diffuses into the lungs. So, after the oxygen and carbon dioxide transfer, the air in the lungs has a composition of 80% $N_2/[20\alpha-20\alpha\beta+(1-\alpha)x]\%$ $O_2/[20\alpha\beta+(1-\alpha)y]\%$ CO_2, which is exhaled. Therefore, the oxygen and carbon dioxide fractions are, respectively,

$$16\% = [20\alpha - 20\alpha\beta + (1-\alpha)x]\% \tag{9.9}$$

$$4\% = [20\alpha\beta + (1-\alpha)y]\%. \tag{9.10}$$

The tidal volume fraction is $\alpha = 1/3$ and the fraction of inhaled oxygen that is absorbed by the lungs is $\beta = 1/4$. This gives $x = 16.5$ and $y = 3.5$, so the stale air in the lungs has a composition 80% N_2/16.5% O_2/3.5% CO_2.

(This air is clearly oxygenated enough to be useful during mouth-to-mouth resuscitation.)

9.5 Breathing Under Usual and Unusual Conditions

9.5.1 Flow of Air During Breathing

During inspiration, air flows because the pressure is lower in the lungs and alveoli by a positive amount ΔP than in the atmosphere. The amount of air that flows is determined by the resistance and compliance of the respiratory system.

The compliance of the lung is $\sim$0.2 L/cmH_2O and it decreases for a normal person with higher expanding pressures, as is seen by the decreasing slope in Fig. 9.12. In trying to evaluate lung performance, the *specific compliance* of the lung is perhaps more meaningful, for which the lung compliance is normalized by a characteristic of the person's size, such as a characteristic lung volume (FRC, VC, TLC, etc.), the lung dry weight, or the body weight. The elasticity of the chest, as well as that of the lung, contributes to the lung compliance.

Airway resistance is dominant in the generation of the intermediate sized bronchii, as seen in Fig. 9.14a and Problem 9.21. Poiseuille's Law (7.24) can be used to calculate the airway resistance in the lungs (Problems 9.21–9.23) and other passages (Problems 9.16 and 9.18). The total airway resistance is typically $\sim$2 cmH_2O/(L/s) during normal breathing and it decreases with increasing lung size (Fig. 9.14b). (It is measured as in Fig. 9.17 later).

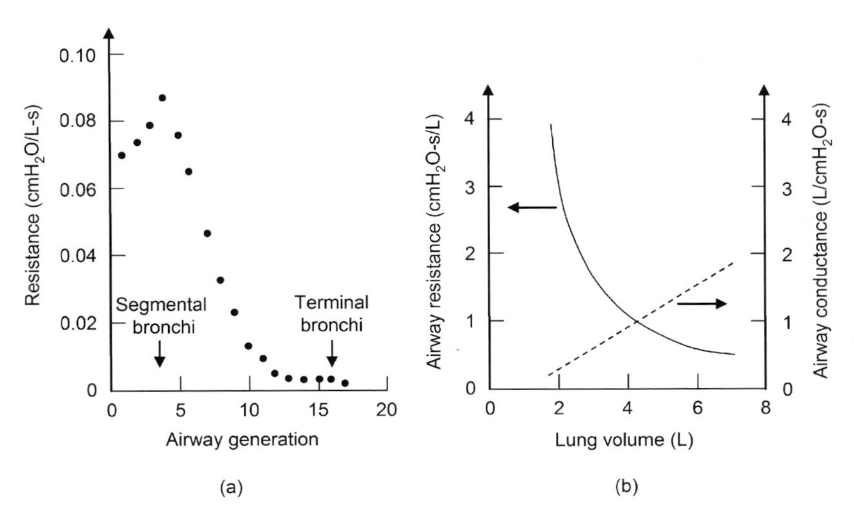

Fig. 9.14. (**a**) Airway resistance for each bronchus generation, and (**b**) total airway resistance and conductance vs. lung volume. (Based on [443], from (a) [433], (b) [421])

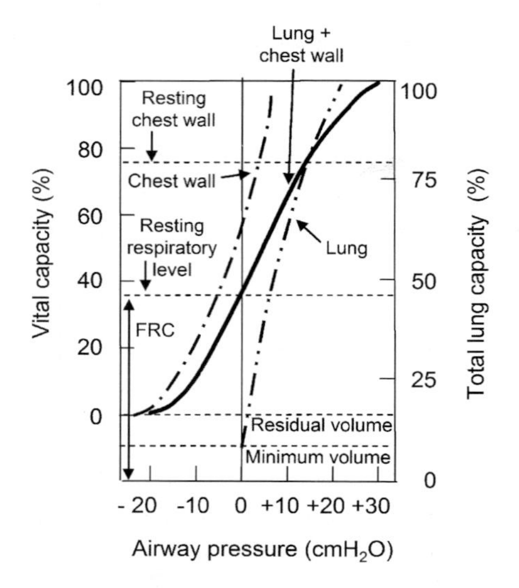

Fig. 9.15. Pressure–volume curves of the lungs, chest wall (no lungs), and chest wall with lungs, with relaxed respiratory muscles. The lung data are the same as those in Fig. 9.12, except no hysteresis is shown and the airway pressure is of the opposite sign to the pressure around the lung. FRC is the functional residual capacity. The measurement is made with a spirometer-like apparatus, similar to that in Fig. 9.13b. (Based on [443])

This airway resistance is about 80% of the total *pulmonary resistance*. The other 20% is due to viscous forces in chest and lung wall movement, and this is called *tissue resistance*.

The elastic properties of the thoracic cage (with the chest well) are important in breathing, as are those of the lungs. Figure 9.15 shows the volume of the lungs, chest walls alone, and the lungs in the chest wall as a function of pressure. This is measured after inspiration or expiration with a spirometer and subsequent relaxing of respiratory muscles. At every volume the pressure (the relaxation pressure) of the lung/chest wall combination is the sum of those for the lungs and chest walls separately.

Is the pressure difference between the alveoli and atmosphere large enough to drive the right amount of air into our lungs each breath? (Under normal conditions, it had better be.)

During each breath, this pressure difference starts at zero, increases to a maximum $\Delta P_{\max}$, and then decreases to zero again at the end of the breath. Let us determine the average pressure difference in this sequence. We will model inhalation as a half cycle of a sine wave with: $\Delta P(t) = \Delta P_{\max}\sin(2\pi ft) = \Delta P_{\max}\sin(\pi t/T_{\text{half period}})$, which lasts a half-cycle time $T_{\text{half period}} = 1/(2f)$. (The parameter f is the same as the breathing or respiratory rate only if the inhalation and exhalation times are the same.) For

$f = 0.25$ Hz, $T_{\text{half period}} = 2$ s. The average pressure difference during this inhalation is

$$\Delta P_{\text{av}} = \frac{1}{T_{\text{half period}}} \int_0^{T_{\text{half period}}} \Delta P_{\text{max}} \sin(\pi t/T_{\text{half period}}) \mathrm{d}t \tag{9.11}$$

$$\Delta P_{\text{av}} = \frac{\Delta P_{\text{max}}}{T_{\text{half period}}} \frac{T_{\text{half period}}}{\pi} (\cos(\pi T_{\text{half period}}/T_{\text{half period}}) - \cos(0))$$
$$= \frac{2}{\pi} \Delta P_{\text{max}}, \tag{9.12}$$

so for $\Delta P_{\text{max}} = 1.1$ mmHg, we see that $\Delta P_{\text{av}} = 0.7$ mmHg.

Let us say that the inflow of air per breath is V_{in}. Then the average flow rate is $Q_{\text{av}} = V_{\text{in}}/T_{\text{half period}}$. For $V_{\text{in}} = 0.5$ L, this is 0.25 L/s. If the flow rate is proportional to the pressure drop, Q and ΔP have the same dependence on time and so $Q_{\text{max}} = (\pi/2)Q_{\text{av}}$, which is $(\pi/2)0.25$ L/s $= 0.4$ L/s, and $Q(t) = Q_{\text{max}} \sin(2\pi f t) = Q_{\text{max}} \sin(\pi t/T_{\text{half period}})$. Moreover, $Q = \Delta P/R_{\text{flow}}$, where R_{flow} is the total resistance to flow in the nasal passages, trachea, and so on. The resistance to flow is $R_{\text{flow}} = \Delta P/Q$, and so using average values we see that $R_{\text{flow}} = 0.7\,\text{mmHg}/(0.25\,\text{L/s}) = 3.7 \times 10^5$ Pa-s/m^3.

Is the flow laminar or turbulent? If the trachea has a radius of 9 mm, the air flows at a maximum speed $u_{\text{max}} = Q_{\text{max}}/A = (400\,\text{cm}^3/\text{s})/(\pi(0.9\,\text{cm})^2) = 160$ cm/s. The Reynolds number (7.11) is $Re = \rho u d/\eta$. Using the mass density of air at body temperature $\rho = 1.16 \times 10^{-3}\,\text{g/cm}^3$ and the air viscosity 2×10^{-5} Pa-s $= 2 \times 10^{-5}(\text{N/m}^2)$s, we find that $Re = (1.16 \times 10^{-3}\,\text{g/cm}^3)(160\,\text{cm/s})(1.8\,\text{cm})/(2 \times 10^{-4}(\text{dyne/cm}^2)\text{s}) = 1{,}600$, so the flow would generally be expected to be laminar in the trachea, as well as in the nasal passages and pharynx. However, some turbulence is expected because the walls of these passages are not smooth.

9.5.2 Mechanical Model of Breathing and Model Parameters

We have just examined only the resistance to flow during breathing. Figure 9.16b shows a more complete mechanical model of the lungs and breathing. It is a compliance vessel described by $V(t) = V_{\text{d}} + C_{\text{flow}} P(t)$, attributed to the springiness of the lungs, in series with an inertial element. They are in parallel with a resistive element, attributed to the airway resistance we just examined. These model elements are driven by a pressure determined by the inspiratory muscles.

9.5.3 Inspiration/Expiration Cycle

Figure 9.17 shows the lung pressure, rate of flow of air into the lungs, and lung volume vs. time during a cycle of inspiration and expiration. If the effect of airway resistance were neglected, the alveolar pressure would be zero and the intrapleural pressure would follow the broken curve, which is determined by the elastic recoil of the lung.

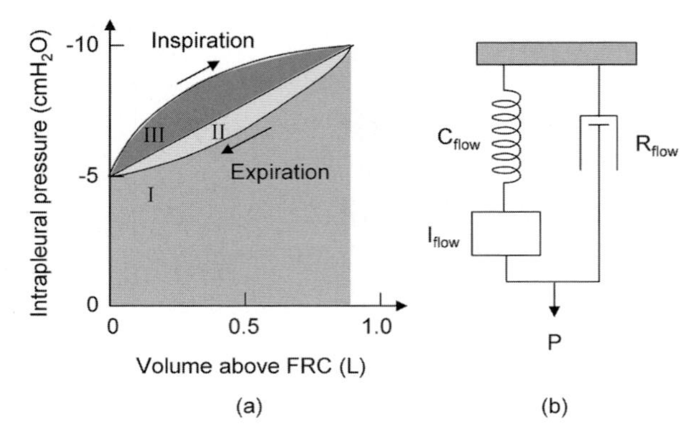

Fig. 9.16. (**a**) Work done during inspiration (areas I + II + III) and work recovered during expiration (area I). With no viscous, resistive forces, the work in inflating the lung would be areas I + II, and this is associated with the lung compliance. The extra work done overcoming respiratory flow resistance is area III. (**b**) Mechanical model of breathing has the lumped compliance (elastance) C_{flow}, resistance R_{flow}, and inertance I_{flow}, and P represents the inspiratory muscles. The inertance can be neglected except for large flows. (Based on [430, 444])

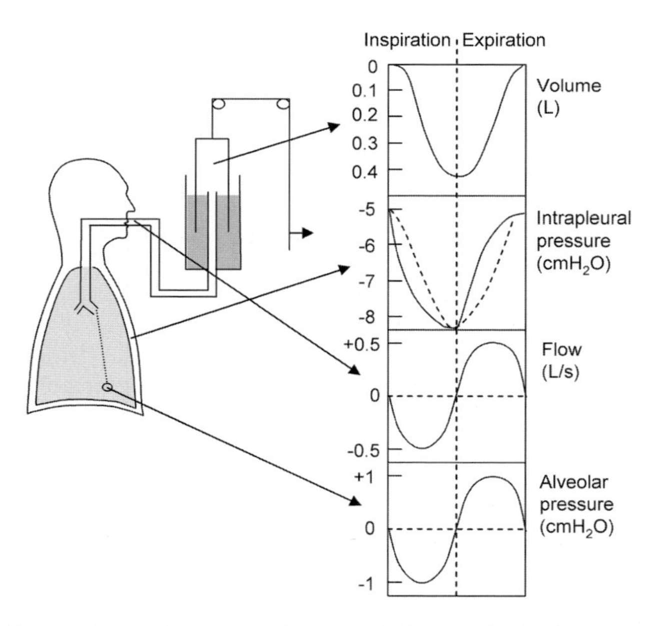

Fig. 9.17. Lung volume, intrapleural pressure, flow, and alveolar pressure vs. time during a breathing cycle. If the effect of airway resistance were neglected, the alveolar pressure would be zero and the intrapleural pressure would follow the broken curve. (Based on [443])

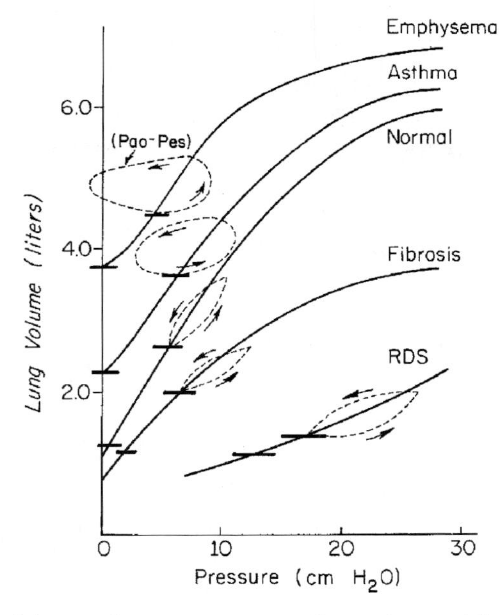

Fig. 9.18. Typical lung volume vs. pressure for patients with various respiratory conditions. The monotonically increasing curves are for static conditions, with the lower horizontal bar being the residual volume and the upper horizontal bar the functional residual capacity. Representative dynamic breathing loops (enclosed *dashed curves*) for tidal volume breathing are shown for each condition. (Reprinted from [430], with permission of Elsevier)

It is also instructive to plot these variables as functions of each other. Lung volume is plotted vs. lung pressure for all times during a breathing cycle in the dashed cycle trajectories in Fig. 9.18. Time is an implicit variable along the trajectories. Such plots are useful because the model of the lungs includes a compliance vessel in which volume and pressure are interrelated. The differences in various modes of breathing are easily seen in such plots.

9.5.4 Breathing with a Diseased Lung

The static and dynamic pressure–volume curves in Fig. 9.18 indicate how different diseases affect lung compliance, lung volume, and airway resistance. *Obstructive* disorders are due to airway obstructions, and include chronic bronchitis (excessive mucus production in the bronchial tree), emphysema (enlargement of air spaces after the terminal bronchiole, with the destruction of respiratory system walls, such as those of the alveoli), and asthma (widespread narrowing of airways, sometimes spontaneously). Such obstructions (Fig. 9.19) can be due to excessive secretions (due to chronic bronchitis), thickening of airway walls (edema or muscle hypertrophy), and outside

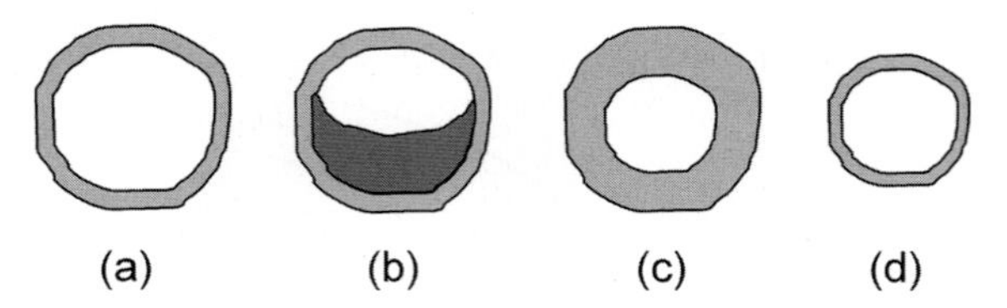

Fig. 9.19. (**a**) Normal airway and obstructed airways due (**b**) blocking, (**c**) airway wall thickening, and (**d**) outside abnormality. Also see Fig. 9.22. (Based on [444])

abnormalities (edema, enlarged lymph nodes, or destruction of lung alveoli tissue as in emphysema). (Edema is an excessive accumulation of fluid in tissue spaces or a body cavity.) These obstructions and the loss of small airways due to the destruction of lung tissue all increase airway resistance. The breakdown of elastic alveoli walls also reduces the springiness, and therefore also the compliance. *Restrictive* disorders are those in which the expansion of the lung is restricted. The decreased compliance (slope) in *pulmonary fibrosis* and idiopathic respiratory distress syndrome (RDS) is clear in Fig. 9.18. They are characterized by a lower vital capacity, but airway resistance (per lung volume) is not increased. There are also *vascular disorders*, such as pulmonary edema, which is the abnormal accumulation of fluid in the lungs.

Poor breathing due to a diseased lung is manifest in different ways. As seen in Fig. 9.20, the inspired volume is very low if the compliance is less than normal and/or the *airway resistance* is greater than normal. A spirometer with a low resistance can also be used to test for lung malfunctions in a manner that is slightly different from that used in Fig. 9.13. After a very deep inhalation, a person forces air out as fast as possible. The total volume exhaled is the forced vital capacity (FVC), which can be a bit less than the vital capacity

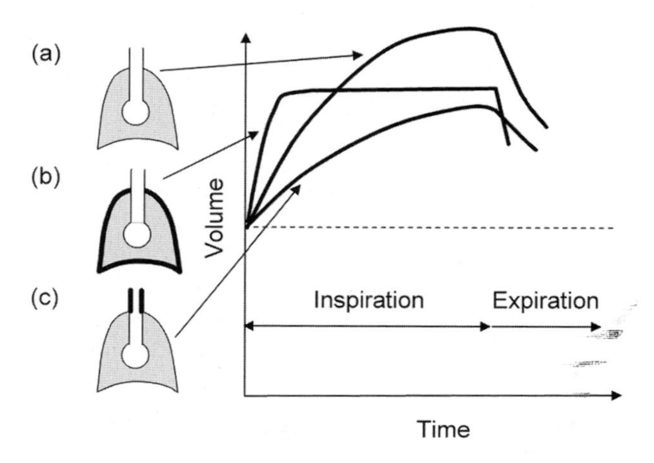

Fig. 9.20. Lung volume during inspiration for (**a**) normal conditions, (**b**) decreased compliance, and (**c**) increased airway resistance. (Based on [443])

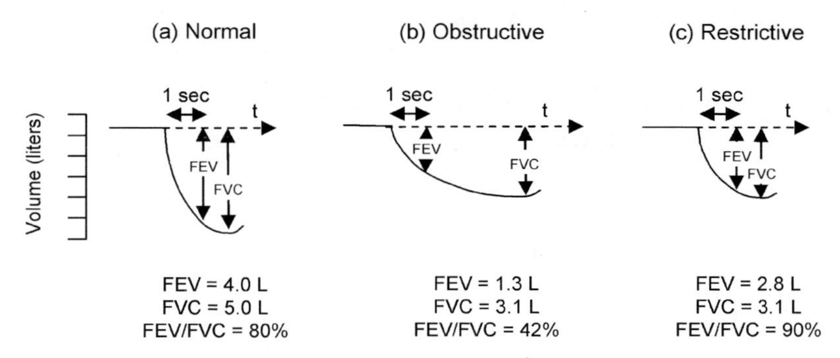

Fig. 9.21. Forced expiration for (**a**) normal, (**b**) obstructive, and (**c**) restrictive patterns. (Based on [444])

measured with slower expiration. Also of note is the volume exhaled in the first 1 s, which is the forced expiratory volume (FEV) (or FEV_1, which denotes specifically that this volume was expired in 1 s), and the ratio FEV/FVC. As seen in Fig. 9.21, for a normal lung FVC = 5.0 L and FEV/FVC = 80%. The example of an *obstructive* pattern has a lower FV, 3.1 L, and smaller FEV/FVC, 42%. The example of a *restrictive* pattern also has a lower FV, 3.1 L, but a high FEV/FVC, 90%. The flow rates for the obstructive pattern are also abnormally low (Problem 9.27).

Figure 9.18 shows cycles that typify the breathing cycles of a normal person and of people with lung disorders. The divisions between the alveoli break down in people with *emphysema*. Consequently, the lungs become less springy and more compliant, and the airway resistance contribution dominates breathing. In pulmonary fibrosis, the compliance is reduced by an increase in fibrous tissue. This condition increases in pulmonary emphysema and in normal aging, due to a change in elastic tissue in the lungs. The volume/pressure locus is also shown for idiopathic respiratory distress syndrome (RDS). It is seen to move to higher pressure due to a lack of alveoli surfactant, as occurs in some premature babies. In such infants the minimal surface tension is only 2×10^{-4} N/m, compared to $\ll 0.5 \times 10^{-4}$ N/m for normal lungs. With less surfactant, the alveolus surface tension decreases and, at the same pressure difference the alveolus is smaller. Figure 9.22 shows that the airways tend

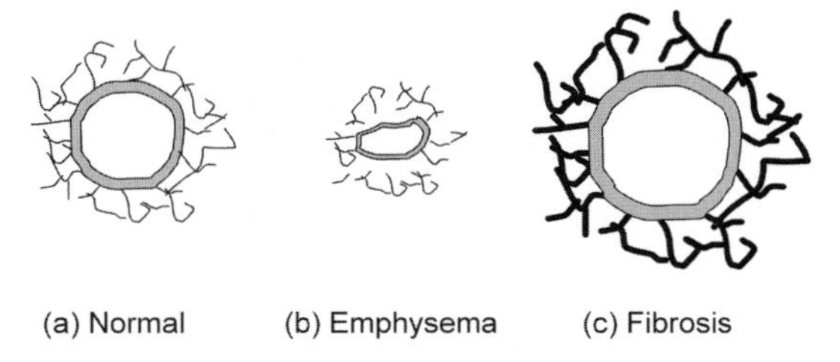

Fig. 9.22. Airways dimensions for different patients. (Based on [444])

to collapse in emphysema because of the loss of radial traction with exterior structures, while in interstitial fibrosis the airways can become large in diameter, due to excessive radial traction, making the airways large in volume relative to the lungs.

9.5.5 Breathing at Higher Elevations

The partial pressure of oxygen is 161 mmHg (21.2 kPa) at sea level (300 K). Hypoxia (which is the lack of oxygen reaching living tissues) occurs below a critical partial pressure of 57 mmHg (7.5 kPa), for a normal, relaxed breathing rate. (Another way of saying this is: Our bodies have been designed so that our rate of breathing air, rate of transferring oxygen to the blood, the capacity of the blood to hold oxygen, the rates of blood flow to tissues, and so on leads to a metabolism that functions well when the partial pressure of oxygen is above this critical value.) *At what elevation above sea level does hypoxia occur?*

The pressure of a fluid column of constant fluid density ρ and constant gravitational constant g is given by (2.48), $P = \rho g h$, where h is the height of the column. In Chap. 8 we considered a column of blood; now let us consider a column of air at a height z and above, where $z = 0$ at sea level. The change in pressure for a change in height is

$$\mathrm{d}P = -\rho g \mathrm{d}z. \tag{9.13}$$

Because we will be considering heights above sea level that are much smaller than the radius of the earth, we have ignored the dependence of the gravitational acceleration g on z. (It is considered in Problem 9.35.)

The ideal gas law (7.2) is $P = nRT$, where n is the density, R is the gas constant, and T is the temperature, or $P = \rho RT/m$

$$P = \frac{RT}{m}\,\rho, \tag{9.14}$$

where $\rho = mn$ is the mass density, with m the molecular mass. We will assume that the atmospheric temperature (300 K) does not vary with height (see Problem 9.40). Dividing (9.13) and (9.14) gives

$$\frac{\mathrm{d}P}{P} = -\frac{mg}{RT}\,\mathrm{d}z. \tag{9.15}$$

Integrating from sea level, $z = 0$ with pressure $P_{\text{sea level}}$, to a height h, with pressure $P(h)$, gives

$$P(h) = P_{\text{sea level}} \exp\left(-\frac{mg}{RT}\,h\right). \tag{9.16}$$

Using the partial pressure of oxygen at sea level and $m = 32$ g/mol for oxygen, this shows that hypoxia occurs at a height of 8.25 km (= 27,100 ft = 5.1 miles).

9.6 Work Needed to Breathe

During inspiration the thorax and abdomen do positive work to expand the lungs [420, 430, 431, 443]. The body does no work during normal expiration, but during forced breathing work is also done to contract the lungs during expiration. Because $V = V_{\rm d} + C_{\rm flow}P$, for the breathing cycle with volume changing between the functional residual capacity volume, $V_{\rm FRC}$, and that plus the tidal volume, $V_{\rm FRC} + V_{\rm t}$, we can write $V(t) = V_{\rm FRC} + C_{\rm flow}P(t)$. We see that $\Delta V(t) = V(t) - V_{\rm FRC} = C_{\rm flow}P(t)$ or $P(t) = \Delta V(t)/C_{\rm flow}$. The work done during inhaling a tidal volume $V_{\rm t}$ is

$$W = \int_0^{V_{\rm t}} P\mathrm{d}(\Delta V) = \int_0^{V_{\rm t}} \frac{\Delta V}{C_{\rm flow}}\mathrm{d}(\Delta V) = \frac{V_{\rm t}^2}{2C_{\rm flow}}. \tag{9.17}$$

In Fig. 9.16a, the work done overcoming these elastic (compliance) effects is the area defined by regions I + II, and this is what we have derived here (and will use later). The work is really larger, the area represented by regions I + II + III, because of viscous (resistive) effects. (These viscous/resistive effects also lead to the hysteresis here and in Fig. 9.12, just as in Chap. 4.)

For a breathing rate of f, the rate of doing work for inspiration is

$$\frac{\mathrm{d}W}{\mathrm{d}t} = f\frac{V_{\rm t}^2}{2C_{\rm flow}}. \tag{9.18}$$

With a breathing rate of 20/min, tidal volume of 500 cm^3, and lung compliance of $0.1\,\mathrm{L/cmH_2O} = 0.1\ \mathrm{cm}^3/(\mathrm{dyne/cm}^2)$, this is

$$\frac{\mathrm{d}W}{\mathrm{d}t} = \frac{(20/\mathrm{min})(500\ \mathrm{cm}^3)^2}{2 \times 0.1\ \mathrm{cm}^3/(\mathrm{dyne/cm}^2)} = 3.6 \times 10^3\ \mathrm{J/day} = 0.86\ \mathrm{kcal/day}. \tag{9.19}$$

The respiratory muscle efficiency is ϵ = 5–10%, and so the metabolic need is

$$\frac{\mathrm{d}E}{\mathrm{d}t} = \frac{\mathrm{d}W/\mathrm{d}t}{\epsilon} = \frac{0.86\ \mathrm{kcal/day}}{0.05} = 17\ \mathrm{kcal/day} \tag{9.20}$$

assuming 5% efficiency; this calculation is very sensitive to the values chosen for $C_{\rm flow}$ and ϵ. This result is about 1% of the BMR; however this value is really about 2% of the BMR, so maybe ϵ is closer to 3% or other effects need to be considered, such as dissipation due to resistance of the flow and the viscous nature (of the overall viscoelasticity) of the lungs and chest wall. Airflow resistance in the nose seems to be responsible for about half of the work needed to breath.

During normal breathing there is no mechanical work done in expiration. During heavy workouts and strenuous exercise the metabolic needs increase because there is (1) also work done during expiration, (2) a faster breathing rate, and (3) a larger tidal volume. The work done to breathe can use 25%

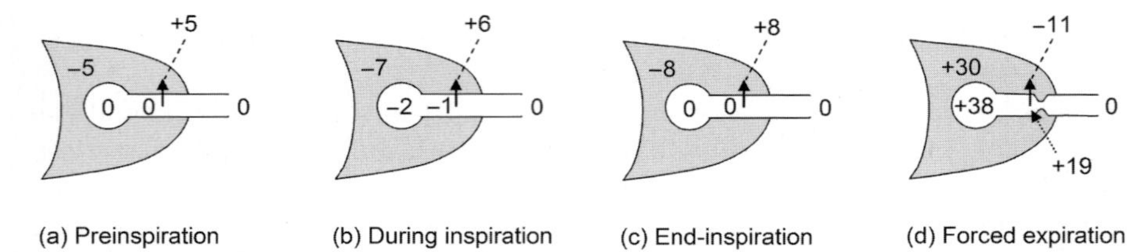

Fig. 9.23. Intrapleural, alveoli, airway, and atmospheric pressures in cmH_2O during inspiration and forced expiration. (Based on [443])

of the total body energy consumption. Such a large increase is suggested by Fig. 8.27. Rapid shallow breathing (as recommended in Lamaze training for childbirth) requires more energy than does normal breathing, to overcome the resistive nature of flow in the air passages. Similarly, slow, deep breathing requires more energy to overcome the elastic nature of the lung and chest. Problem 9.12 compares other relative advantages and disadvantages of these last two modes of breathing.

Another interesting thing can occur during forced expiration. In normal breathing the pressure in the airways always exceeds the intrapleural pressure during inspiration and expiration (as is seen in Fig. 9.23 during normal inspiration). However, during forced expiration the intrapleural and alveoli pressure both increase by the same amount (38 cmH_2O) and so they are both positive relative to the atmosphere (+30 and +38 cmH_2O in the figure). There is now a large pressure drop from the alveoli to the lips and nostrils. At some point along this route the pressure in the airway will be lower than 30 cmH_2O and this will compress the airways (shown where the pressure is +19 cmH_2O in this figure) and increase the airway resistance.

9.7 Summary

The macroscopic basis of lung function and breathing can be understood by analyzing the volume, pressure, and air flow during breathing, by using models of fluid flow and lung expansion. The physical nature of the individual operating units in the lungs, i.e., the alveoli, can also be understood this way. These models can also be used to understand the physical consequences of a diseased lung. The energy needed to operate the lungs can also be modeled.

Problems

Lungs

9.1. Calculate the effective lung volumes and breathing rates for a man (70 kg), woman (50 kg), and an infant (5 kg) using the allometric relation parameters in Table 1.13. How do the breathing rates compare with those given above?

9.2. If there are 3×10^8 alveoli in a lung with a functional residual capacity (FRC) of 2.5 L, calculate the average volume and radius of an alveolus.

9.3. Use Table 9.1 to show that the air travels a total distance of 273 mm from the trachea to the alveoli.

9.4. What is the total volume of the lungs described in Table 9.1? Where is most of the volume?

9.5. Is continuity of flow obeyed by the data for the lungs in Table 9.1? Check this using the data for bronchial generations 0, 1, 2, 3, 4, 5, 10, 16, 20, and 23.

9.6. Calculate the Reynolds number for the bronchial generations listed in Problem 9.5. Is the flow laminar or turbulent in the respiratory system?

9.7. Calculate the pressure drop across pulmonary arterial orders 1, 4, 10, 13, 16, and 17, assuming a total blood flow of 5 L/min.

9.8. The CO_2 level in the atmosphere was ~280 ppm (parts per million) in preindustrial times and is ~380 ppm now. Express these levels in terms of mmHg. Would this change be expected to affect the exchange of CO_2 in the lungs in any significant manner?

Alveoli and Surface Tension

9.9. Derive (9.8) from (9.7).

9.10. Estimate the force the adult diaphragm would need to exert if there were no lung surfactant.

Breathing

9.11. Use Fig. 9.13b to explain how a spirometer works. How much should the water in the spirometer rise and fall during breathing cycles? (Assume reasonable dimensions for the instrument.)

9.12. During breathing, the pulmonary ventilation, V_p (in L/min) (the rate at which air enters the trachea), equals the respiratory rate, R (in units of per min), times the tidal volume, V_t (in L). Because of the anatomical dead space volume V_d, only $V_t - V_d$ enters the alveoli (and is thus of use). Therefore, a more meaningful ventilation rate is the alveolar ventilation $V_a = R(V_t - V_d)$:

Table 9.3. Examples of breathing cycles

	activity	R (per min)	V_t (L)
(i)	at rest, quiet breathing	12	0.5
(ii)	at rest, with rapid, shallow breathing	24	0.25
(iii)	at rest, with very rapid, very shallow breathing	40	0.15
(iv)	at rest, with slow, deep breathing	6	1.0
(v)	exercising, with rapid, shallow breathing	24	0.5
(vi)	exercising, with slow, deep breathing	12	1.0

(a) Find V_p and V_a for the conditions in Table 9.3, assuming $V_d = 0.15$ L.
(b) Compare the pulmonary ventilation for the four breathing patterns in this table for the person at rest. (Patterns (ii)–(iv) require more metabolic power than does (i), because of increased work due to resistance to flow and resistance in the tissues for (ii) and (iii), and increased work due to compliance (elastic) forces of the lung and chest in (iv)) Which of the four are clearly inadvisable because of poor alveolar ventilation?
(c) During exercise, both the respiratory rate and tidal volume increase. Based on the results in part (a) for (i)–(iv) and for (v)–(vi), do you gain more by breathing faster or deeper for a given pulmonary ventilation?

9.13. What are the maximum and average air flows for each breathing cycle in Problem 9.12, assuming the inhalation and exhalation periods are the same?

9.14. (a) When you take in a deep breath of say 1 L, how much does your mass (in kg) and weight (in N and lb) increase?
(b) Does your average density increase, decrease, or stay the same? If there is a change, estimate it.

9.15. (a) What does Fig. 8.27 say about the amount of oxygen that can be consumed per amount of cardiac output?
(b) What does it say about how much oxygen is needed to do work? Is this consistent with what is presented in the text?
(c) How is work output defined in this figure?

9.16. Estimate the resistance of the trachea using Poiseuille's Law, assuming it has a radius of 9 mm and a length of 110 mm. How does this compare to the total resistance?

9.17. Estimate the resistance of the vocal tract using Poiseuille's Law, assuming it can be modeled as three tubes in series with respective lengths 6, 3, and 6 cm and cross-sectional areas 5, 1, and 5 cm^2. (Also sketch this model.)

9.18. Estimate the resistance of the nasal passage using Poiseuille's Law, assuming it has a radius of 4 mm and a length of 3 cm. How does this compare to the total resistance and is it a limiting factor in the resistance to flow?

9.19. If you model the breathing airway as a series of sequential passages, the nasal or mouth passage, the pharynx, larynx, and then trachea, each with a resistance to air flow, what is the total resistance to air flow in terms of these individual resistances?

9.20. In both inspiration and expiration, a pressure difference of 0.4 cmH_2O causes a flow of 0.15 L/s in the nose. Determine the flow resistance in it.

9.21. Consider the lung bifurcation generations 1–19 in Table 9.1:
(a) In which generation is the flow resistance largest? What is its value?
(b) Do your results agree with those in Fig. 9.14a?
(c) In which generation is the pressure drop greatest, and generally in what range of bifurcations is most of the pressure drop?

9.22. The total airway resistance is the sum of those in each generation. Do the resistances in Fig. 9.14a add to give you a total resistance consistent with that in Fig. 9.14b?

9.23. Calculate the resistance for generation 4 using Poiseuille's Law and compare it to the values given in the chapter.

9.24. (a) Use Fig. 9.15 to determine the compliance of the lungs ($C_{\text{flow,lung}}$) and chest walls ($C_{\text{flow,chest wall}}$) at 0, 20, 40, 60, and 80% of vital capacity.
(b) Determine the compliance of the combined lung/chest wall system ($C_{\text{flow,lung/chest wall}}$) at these volumes, and compare these values with those from part (a) by using $1/C_{\text{flow,lung/chest wall}} = 1/C_{\text{flow,lung}} + 1/C_{\text{flow,lung/chest wall}}$.

9.25. (a) Use Fig. 9.18 to determine the compliance of the lungs for each condition (within the lowest 5 cmH_2O pressure range shown for each).
(b) How does the compliance vary for each over the pressure range shown?

9.26. Compare the specific lung compliances of a 65 kg man and 20 g mouse, with respective compliances of 0.2 L/cm-H_2O and 0.0001 L/cm-H_2O.

9.27. Show that the forced expiratory flow (FEF) rates for the normal, obstructive, and restrictive flows in Fig. 9.21 are 3.5, 1.4, and 3.7 L/s, respectively. Do this by determining the slopes of the three curves in this figure. (Use a straight-line fit between points that have decreased by 25% and 75% on the way to the FVC.)

9.28. Determine the air flow resistance from the flow rate and alveoli pressure in Fig. 9.17.

9.29. Consider only the compliance in the work of breathing and assume that the compliance C_{flow} for normal lungs is 0.1 $cm^5/dyne$:
(a) In fibrosis of the lungs the compliance of the lungs decreases. For a given tidal volume, how does the rate of work of breathing change if the compliance decreases by x%?
(b) Compare the rate of work done in breathing (J/day) and the associated rate of metabolism (kcal/day) (if the muscles associated with breathing are 5% efficient) for cases (i) and (iv) in Problem 9.12 for normal lungs.

9.30. (advanced problem) Write down the equation of motion for the mechanical model in Figure 9.16b and solve it for inspiration.

9.31. (advanced problem) Show that the solution in Problem 9.30 qualitatively agrees with the trends seen in Fig. 9.20: with decreased compliance, the time constant decreases and the volume breathed during a cycle decreases, while with increased airway resistance, the time constant increases and the volume breathed during a cycle decreases.

9.32. Estimate the rate of energy consumed by the lungs during exercise with a breathing rate of 40/min and tidal volume of 1,000 cm^3.

9.33. Use a blood circulation rate of 5 L/min and the known change in the partial pressures of O_2 and CO_2 in the systemic capillaries to find the number of liters of O_2 consumed and CO_2 exhaled each day. How do your results change if you instead use the change in the partial pressures of O_2 and CO_2 in the pulmonary capillaries? Explain why.

9.34. If your chest wall and parietal pleura of a lung are punctured, the intrapleural pressure will increase to atmospheric pressure and that lung will collapse. Explain why. Also draw a diagram explaining this.

Breathing at High Elevation

9.35. (a) Show that the gravitational acceleration constant g varies with height z above sea level as $g(z) = g\left(R_{\text{Earth}}/(R_{\text{Earth}} + z)\right)^2$, where the radius of the earth is $R_{\text{Earth}} = 6,378\,\text{km}$.
(b) Show that this variation does not affect the analysis of oxygen deprivation at high elevations, described in the text.

9.36. What is the atmospheric pressure in the "mile-high" city of Denver? What is the partial pressure of oxygen there?

9.37. Commercial jets typically cruise at an altitude of ~10,700 m (~35,000 ft). What are the total pressure and partial pressure of oxygen at that height? Why are jets pressurized? Why are oxygen masks made available just in case the cabin is depressurized?

9.38. Why do some athletes train at high elevations?

9.39. Apply (9.16) to the variation of the partial pressure of nitrogen, using $m = 28\,\text{g/mol}$. Let us say here that the ratio of oxygen to nitrogen is 20.9%/78.1% = 0.268 at sea level. What is this ratio at the critical height for hypoxia?

9.40. The temperature of the troposphere (the atmosphere up to roughly 11 km) decreases with height, by a bit less than 1 K per 100 m of elevation. In

Table 9.4. Total and partial pressures at different elevations. (Using data from [445])

altitude	sea level	at 2,500 m
atmospheric pressure (total)	760	560
atmospheric pressure (O_2)	159	117
in alveoli (O_2)	105	77
in arterial blood (O_2)	100	72
in venous blood (O_2)	40	40

the standard atmosphere $T(z) = T_{\text{sea level}} + \alpha z$, with $T_{\text{sea level}} = 288.19\,\text{K}$ and $\alpha = -0.00649\,\text{K/km}$. (For a dry atmosphere, $\alpha = -0.0098\,\text{K/km}$.):
(a) Use this temperature variation in (9.15) to show that

$$P(h) = P_{\text{sea level}} \left(T_{\text{sea level}}/(T_{\text{sea level}} + \alpha h)\right)^{gm/R\alpha} . \tag{9.21}$$

(b) Show that hypoxia occurs at a lower elevation, 7.21 km.

9.41. Table 9.4 compares the partial pressure of oxygen (in mmHg) in the air and in the body at sea level and at an elevation of 2,500 m:
(a) Justify the values given for total pressure and O_2 partial pressure at 2,500 m.
(b) Justify the O_2 partial pressure in the alveoli at 2,500 m by using the pressure at sea level.
(c) At sea level your blood flows at a rate of 5 L/min. How fast would it have to flow at an elevation of 2,500 m to provide the same flow of oxygen to the tissues? (Assume no change in the red blood cell and hemoglobin concentrations in the blood. These increase as part of adapting to higher elevations.)
(d) How much faster would you have to breathe at this elevation (in liters of air per min) to maintain the same rate of oxygen delivery? How could this translate into changes in the breathing rate and tidal volume?

10

Sound, Speech, and Hearing

In this chapter we will ask: *How do we speak? How do we hear?* To answer these questions we will have to probe into the physics of sound waves, which is called acoustics, because speaking is creating sound and hearing is detecting sound. We will first review the basic physics of sound waves and will then examine the physics of speaking and hearing by humans in more detail.

Sound is also important in medical diagnostics. Physicians use stethoscopes to listen to sounds in the body (*auscultation*; see the end of this chapter) and they tap on parts of the body, as one would a drum, as a percussive diagnostic. Another important medical diagnostic using sound is *ultrasonic mapping* or *ultrasonography.* It is a nondestructive imaging method that makes use of sound at frequencies (~1–10 MHz), way above our hearing range (20 Hz–20 kHz), that provides images with the very useful spatial resolution of ~1 mm [489, 506]. Waves are sent to an object and reflected, with the delay time between the transmission of the probe beam and the arrival of the reflected acoustic pulses at the detector giving the relative location of the object. For example, in analyzing the heart the use of a scanned single beam gives valuable, yet limited information, such as the wall thickness and chamber diameters (*M-mode echocardiography*), while the use of multiple beams transmitted through a wide arc provides two-dimensional images of the heart (*2-D echocardiography*). The shifting of the acoustic frequency when the ultrasound reflects from a moving target (*the Doppler effect*) is the basis for measuring blood flow direction, turbulence, and speed (*Doppler ultrasonography*). (See Problems 10.22 and 10.23, and Fig. 10.56.)

10.1 The Physics of Sound Waves

Sound is a compressional wave in a gas, liquid, or solid. A wave is a periodic disturbance that travels in space, say in the z direction. It is periodic in space, which means that at any given time t, the disturbance is periodic with z, as in Fig. 10.1. It is periodic in time, which means that at any given position z,

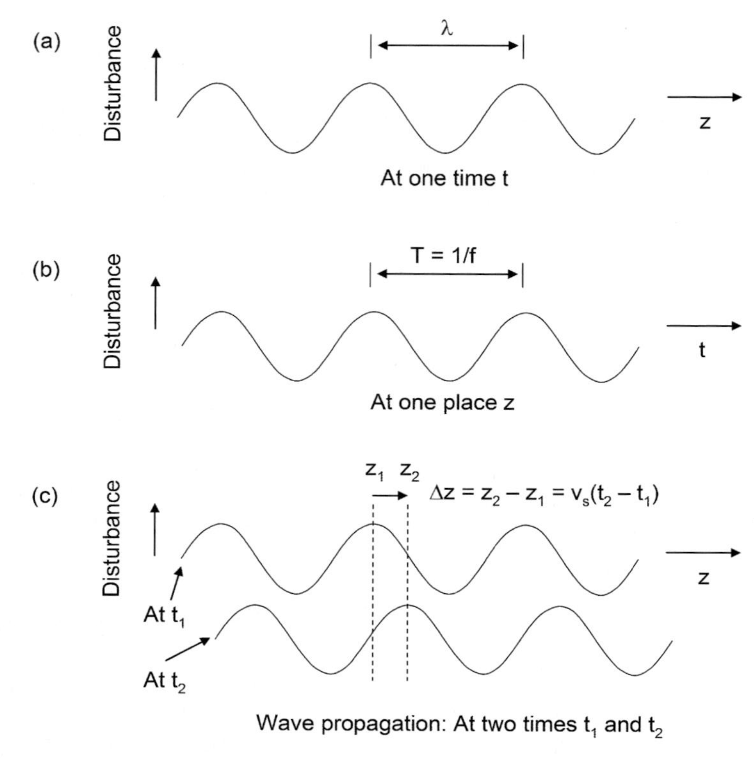

Fig. 10.1. Waves at (**a**) one time, (**b**) one place, and (**c**) two different times, showing wave propagation

the disturbance is periodic with time t. The disturbance travels with a speed v, the speed of sound, so from time t_1 to time t_2 the disturbance travels a distance $\delta z = z_2 - z_1 = v(t_2 - t_1)$. The quantity $z - vt$ does not change for the disturbance as it "travels" with the wave.

So far we have described the propagation of a disturbance by a generic type of wave, but have not specified what is being disturbed. In sound waves, these disturbances are local changes in pressure, $\delta P = P(z,t) - P_{\text{ambient}}$, mass density, $\delta\rho$ (or molecular density, δn), and displacement, δz, from their ambient values. Sound waves are *compressional* in that these changes occur in the same direction as the wave propagates, here in the z direction. If you pluck a string, the wave propagates along the string, but the actual disturbance of the string is perpendicular to it, making it a *transverse* wave. Figure 10.2 shows the longitudinal motion of the molecules during a sound wave. These pressure and density variations are in phase with each other, meaning that they both increase (compression) or decrease (rarefaction) from the ambient values together. (The equations of state of materials, such as $P = nRT$ for ideal gases, usually show that density increases with pressure.) In contrast,

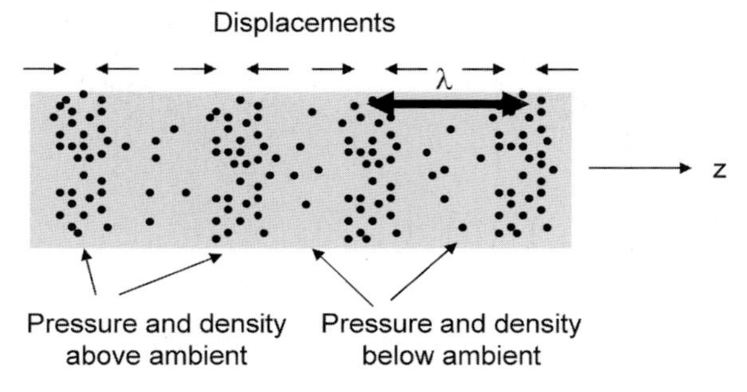

Fig. 10.2. Sound waves are compressional waves

the displacement of molecules is out of phase with density and pressure; the displacement spatially varies as a cosine wave if the density and pressure vary as sine waves. Where the density and pressure are at a maximum, the displacement is zero, but the displacement is positive just to the left and negative just to the right – which maximizes the density and pressure. Where the density and pressure are at a minimum, the displacement is also zero, but the displacement is negative just to the left and positive just to the right. This is seen in Fig. 10.2.

We have chosen to examine the simplest waves, for which the periodic disturbance varies as a sine function (or equivalently a cosine function). While sound waves from speaking are complex sums of such waves at different frequencies, to a very good approximation we can examine the physics of each frequency by itself and sum the effects. (This is called linearity or linear superposition.)

10.1.1 The Speed and Properties of Sound Waves

Sound waves move at a speed v_{s} that is determined by the properties of the medium. In general the sound speed is

$$v_{\mathrm{s}} = \sqrt{\frac{C}{\rho}}, \tag{10.1}$$

where C is a constant describing the stiffness of the material (when there is no heat flow, which are "adiabatic" conditions) and ρ is the mass density. In solids, this stiffness constant can depend on the direction the sound wave propagates. It equals Young's modulus Y for the propagation of compressional waves down a rod that is much longer than it is wide. For steel, $v_{\mathrm{s}} = 5{,}960\,\mathrm{m/s}$. In fluids (liquids and gases), C is the (adiabatic) bulk modulus B, which describes how much pressure is needed to achieve a given fractional decrease in volume. In gases, $B = \gamma P$, where γ is the ratio of the specific heats at constant pressure (c_p) and volume (c_v). (The ratio c_p/c_v ranges from 1, for very large molecules, to 5/3, for an ideal monatomic gas, and is 1.4 for air, which is composed of diatomic gases.) Consequently, the speed of sound in gases is

$$v_{\mathrm{s}} = \sqrt{\frac{\gamma P}{\rho}} = \sqrt{\frac{\gamma RT}{m}}, \tag{10.2}$$

where R is the constant in the ideal gas law (7.2) ($R = 8.31\,\mathrm{J/mol\text{-}K}$), and m is the molecular mass. The speed of sound in air is 343 m/s (at 20°C), which is 15× slower than that in steel, while in water it is 1,482 m/s (see Problems 10.4 and 10.5).

Waves are periodic in space (at one t), which means they repeat with a spatial periodicity called the wavelength λ (Fig. 10.1). They are also periodic in time (at one z) with a temporal periodicity called the period T, which corresponds to a frequency f with $f = 1/T$ (Fig. 10.1). Mathematically, a disturbance moving to the right (larger z) can be expressed as $g(z - v_{\mathrm{s}}t)$ – for any wave of functional form g, because $z - v_{\mathrm{s}}t$ remains constant as the wave moves to the right at speed v_{s}. One example of this is $\cos(z - v_{\mathrm{s}}t)$. Similarly, a disturbance moving to the left (smaller z) can be expressed as $g(z + v_{\mathrm{s}}t)$, because $z + v_{\mathrm{s}}t$ remains constant as the wave moves to the left.

Frequency has units of cycles per second (cps) = Hertz (Hz). (Sometimes we will use the radial frequency ω, which has units of rad/s or just 1/s, and which is related to the frequency by $\omega = 2\pi f$.) The wavelength, frequency, and speed of a wave are interrelated by

$$v_{\mathrm{s}} = \lambda f. \tag{10.3}$$

So, low-frequency waves have long wavelengths, while high frequency waves have short wavelengths. Using the speed of sound in air, $f = 1{,}000$ Hz sound waves in air have a wavelength of 0.34 m $\simeq$1 ft. For some types of waves, v_{s} depends on f, and so $v_{\mathrm{s}} = v_{\mathrm{s}}(f)$, but it does not for sound in air. When sound waves travel from one medium to another, the frequency stays the same, but the wavelength changes with the change in sound speed (10.3). Note that for acoustic waves it is the disturbance that is propagating; the actual molecules move very little (and do not physically travel with the disturbance).

10.1.2 Intensity of Sound Waves

The intensity I of a sound wave is the energy carried by the wave per unit area and per unit time (in units of $\mathrm{J/m^2\text{-}s}$ or $\mathrm{W/m^2}$). At a distance R from an isotropic source of average acoustic power P_{power}, the intensity is

$$I = \frac{P_{\mathrm{power}}}{4\pi R^2}. \tag{10.4}$$

Table 10.1. Mass density, sound speed, and acoustic impedance. (Using data from [467, 489])

material	ρ (kg/m^3)	v_s (m/s)	Z (= ρv_s) (kg/m^2-s)
air (20°C)	1.20	343	413
water	1.00×10^3	1,480	1.48×10^6
fat	0.92×10^3	1,450	1.33×10^6
muscle	1.04×10^3	1,580	1.64×10^6
bone	2.23×10^3	3,500	7.80×10^6
blood	1.03×10^3	1,570	1.61×10^6
soft tissue (avg.)[a]	1.06×10^3	1,540	1.63×10^6
lung	286	630	1.80×10^5

[a]The soft tissue value is representative of the skin, kidney, liver, and the brain.

The acoustic intensity is also equal to the kinetic energy of the wave per unit volume, $\rho u_{\text{max}}^2/2$, times the wave speed, v_s, or

$$I = \frac{1}{2}\rho u_{\text{max}}^2 v_s, \tag{10.5}$$

where $u_{\text{max}} = (\delta z_{\text{max}})\omega$ is the maximum speed of the molecules for a maximum displacement δz_{max} during the disturbance. Therefore, we see that

$$I = \frac{1}{2}\rho v_s[(\delta z_{\text{max}})\omega]^2. \tag{10.6}$$

The *acoustic impedance* of a medium Z is given by the product of the mass density and sound speed for that medium, so

$$Z = \rho v_s. \tag{10.7}$$

Table 10.1 lists the mass density, sound speed, and acoustic impedance for air, water, fat, muscle, and several other body materials. The acoustic impedance for water, fat, and muscle are ~3,500× that for air. We will see later that this mismatch is responsible for the reflection of sound between air and these other media.

For simplicity, we will relabel the gauge pressure δP $(= P - P_{\text{atmosphere}})$ as P for the rest of this chapter and call it the sound pressure. Moreover, P will actually denote the maximum pressure in the cycle, P_{max}. The magnitude of the maximum pressure variation in the sound wave is related to this maximum (out of phase) displacement by

$$|P| = (\rho v_s)\omega|\delta z_{\text{max}}| = Z\omega|\delta z_{\text{max}}|. \tag{10.8}$$

We can understand this because pressure is the force per unit area and force is the change of linear momentum with time, and so pressure is the change in linear momentum per unit area per unit time. Using (10.8), the acoustic intensity is

$$I = \frac{1}{2}Z(\delta z_{\mathrm{max}})^2\omega^2 = \frac{P^2}{2Z}. \tag{10.9}$$

We can present the sound intensity I in units of W/m^2 or any other equivalent units. It is also very common to characterize the sound intensity in a more physiologically-based manner, in which I is referenced to $I_{\mathrm{ref}} = 10^{-12}$ W/m^2; I_{ref} is a sound intensity that is barely audible at 3,000 Hz. Because sound intensities in our everyday experience can be many orders of magnitude larger than this reference intensity, we usually use a logarithmic scale – referenced to base 10 – to characterize I. In units of bels, named after Alexander Graham Bell, $I(\text{in bels}) = \log_{10}(I/I_{\mathrm{ref}})$. It is, in fact, standard to use a finer scale in tenths of bels, called decibels or dB, with

$$I(\text{in dB}) = 10\log_{10}\frac{I}{I_{\mathrm{ref}}}. \tag{10.10}$$

For example, for $I = 10^{-8}$ W/m^2 $= 10^4 I_{\mathrm{ref}}$ we see that $I(\text{in dB}) = 10\log_{10}(10^4) = 10 \times 4 = 40$, or 40 dB.

The dB scale is also used to denote the relative magnitude of intensities, such as that of I_2 relative to I_1. With

$$I_1(\text{in dB}) = 10\log_{10}\frac{I_1}{I_{\mathrm{ref}}} \quad \text{and} \quad I_2(\text{in dB}) = 10\log_{10}\frac{I_2}{I_{\mathrm{ref}}} \tag{10.11}$$

we see that

$$I_2(\text{in dB}) - I_1(\text{in dB}) = 10\log_{10}\frac{I_2}{I_{\mathrm{ref}}} - 10\log_{10}\frac{I_1}{I_{\mathrm{ref}}} = 10\log_{10}\frac{I_2}{I_1}. \tag{10.12}$$

So, a 20 dB increase in sound intensity indicates a factor of $10^2 = 100$ increase in I. Because the dB scale is also used to indicate relative magnitudes, this absolute acoustic unit in (10.10) is often denoted as dB SPL (sound pressure level). Table 10.2 shows the acoustic intensities of common sounds. Normal background noise is 50–60 dB SPL. Normal conversation is 60–70 dB SPL. Speech is around 70–80 dB SPL at 1 m. The threshold of pain is about 120 dB SPL. Windows break at about 163 dB SPL. Shock waves and sonic booms cause levels of about 200 dB SPL at a distance of 330 m.

Apollo astronauts were exposed to very loud sounds during liftoff, over 85 dB SPL for about 80 s after liftoff – with maximum levels below 100 Hz [468, 480]. About 60 s after launch the sound in the crew area was about 123 dB over a wide frequency range, but astronauts heard much less intense sounds (because they wore helmets and suits), from 119 dB at 63 Hz to 89 dB at 2,000 Hz. On the gantry, 10 m above ground, the maximum levels were 158–168 dB SPL from 2–2,000 Hz and then down to 152 dB at 8,000 Hz [480, 482]. The levels were 12–31 dB lower 400 m from the gantry.

Table 10.2. Typical sound intensities

	intensity (W/m^2)	intensity level (dB SPL)
sound barely perceptible, human with good ears	10^{-12}	0
human breathing at 3 m	10^{-11}	10
whisper at 1 m, rustling of leaves, ticking watch	10^{-10}	20
quiet residential community at night, refrigerator hum	10^{-8}	40
quiet restaurant, rainfall	10^{-7}	50
normal conversation at 1 m, office, restaurant	10^{-6}	60
busy traffic	10^{-5}	70
loud music, heavy traffic, vacuum cleaner at 1 m	10^{-4}	80
loud factory	10^{-3}	90
fast train, pneumatic hammer at 2 m, disco, blow dryer	10^{-2}	100
accelerating motorcycle at 5 m, chainsaw at 1 m	10^{-1}	110
rock concert, jet aircraft taking off at 100 m	$1 = 10^0$	120
jackhammer	10^1	130
shotgun blast, firecracker	10^2	140
jet engine at 30 m	10^3	150
rocket engine at 30 m	10^6	180

Equation (10.9) relates the sound intensity and sound pressure by $I = P^2/2Z$. For the reference intensity $I_{\text{ref}} = P_{\text{ref}}^2/2Z$, and with $I_{\text{ref}} = 10^{-12}\,\text{W/m}^2$ and $Z = 413\,\text{kg/m}^2\text{-s}$ for air, we see that the reference pressure $P_{\text{ref}} = 2.9 \times 10^{-5}\,\text{N/m}^2$ (Pa). Dividing these two relations for intensity and pressure gives

$$\frac{I}{I_{\text{ref}}} = \frac{P^2}{P_{\text{ref}}^2} \tag{10.13}$$

and using (10.10) gives

$$I(\text{in dB SPL}) = 20 \log_{10} \frac{P}{P_{\text{ref}}}. \tag{10.14}$$

Sometimes a distinction is made between the expression for the intensity, (10.10), which gives the intensity level (IL) in units of dB IL, and that from (10.14) for pressure, which gives the sound pressure level in dB SPL, but we will not make such a distinction here and the units will be called dB SPL. Generalizing (10.12) gives

$$I_2(\text{in dB SPL}) - I_1(\text{in dB SPL}) = 10 \log_{10} \frac{I_2}{I_1} = 20 \log_{10} \frac{P_2}{P_1}. \tag{10.15}$$

When the sound intensity increases by 60 dB, I increases by 10^6 and P increases by 10^3.

There is an analog to (10.9) in electronics with power $P_{\text{elect}} = V_{\text{elect}}^2/2R_{\text{elect}}$, where V_{elect} is the voltage drop and R_{elect} is the resistance. (The factor of 2 in the denominator is present for sinusoidal voltages and absent for DC voltages.) As in acoustics, the dB scale is commonly used in electronics analogous to (10.10), (10.12), and (10.15).

Absorption of Sound

Within a given medium the sound wave can be attenuated by absorption and scattering. In scattering, part of the propagating beam is redirected into many directions, without being absorbed. The amplitude of the acoustic wave ($A =$ magnitude of the pressure δP_{max} or displacement δz_{max}) decreases exponentially with the distance the wave propagates, z. This can be written as

$$A(z) = A(z = 0) \exp(-\gamma_{\mathrm{sound}} F z), \tag{10.16}$$

where γ_{sound} is the absorption coefficient and F is the frequency dependence. For pure liquids, $F = f^2$ in Hz^2 and for soft tissues, $F \sim f$ in Hz. Typical values of γ_{sound} are given in Table 10.3 for body tissues and in Table 10.4 for other materials. Using (10.13), the acoustic power or intensity varies as

$$I(z) = I(z = 0) \exp(-2\gamma_{\mathrm{sound}} F z). \tag{10.17}$$

This variation also applies to the absorption of light in media and is called Beer's Law. In Beer's Law jargon the intensity absorption coefficient $2\gamma_{\mathrm{sound}}F$ in (10.17) is called α_{sound}, and so, for instance, $\alpha_{\mathrm{sound}} = 2\gamma_{\mathrm{sound}}f$ for tissue.

Table 10.3. Amplitude absorption coefficient γ_{sound} for tissues. (Using data from [467])

tissue	γ_{sound} (s/m)
aqueous humor	1.1×10^{-6}
vitreous humor	1.2×10^{-6}
blood	2.1×10^{-6}
brain (infant)	3.4×10^{-6}
abdomen	5.9×10^{-6}
fat	7.0×10^{-6}
soft tissue (average)	8.3×10^{-6}
liver	1.0×10^{-5}
nerves	1.0×10^{-5}
brain (adult)	1.1×10^{-5}
kidney	1.2×10^{-5}
muscle	2.3×10^{-5}
crystalline eye lens	2.6×10^{-5}
bone	1.6×10^{-4}
lung	4.7×10^{-4}

It is multiplied by the frequency f (in Hz) to obtain the amplitude absorption coefficient per unit length.

Table 10.4. Amplitude absorption coefficient γ_{sound} for fluids. (Using data from [467])

fluids	γ_{sound} ($\mathrm{s^2/m}$)
water	2.5×10^{-14}
castor oil	1.2×10^{-11}
air (STP)	1.4×10^{-10}

It is multiplied by f^2, where f is the frequency (in Hz), to obtain the amplitude absorption coefficient per unit length. STP is standard temperature and pressure.

Beer's Law for light is usually expressed as:

$$I(z) = I(z = 0)\exp(-\alpha_{\mathrm{light}} z). \tag{10.18}$$

In analyzing the measured fraction of the intensity of sound or light transmitted through a medium, all of the loss mechanisms must be considered: the reflection from interfaces (see later), absorption, and scattering. Because the wavelengths of sound waves are of the same order of magnitude as everyday objects, such as windows, pipes, heads, and so on, the diffraction and interference of sound waves is very common.

Components of Impedance (Advanced Topic)

The concept of impedance is more generally used to assess the characteristics of a medium that "opposes" energy flow in a system. It is also used to characterize electrical components and circuits, where the electrical resistance is the real part of the impedance and the capacitance and inductance of a system correspond to the imaginary part of the electrical impedance. Analogous concepts are used in optics to characterize the transmission of light in media. In acoustics (and other areas), the reciprocal of the *impedance* Z, is called the *admittance* $Y = 1/Z$, which describes the ease of energy flow. *Immittance* is refers to either the impedance or admittance, and is used as a general term to describe how well energy flows in a medium.

The impedance is generally a complex parameter. The real part of Z is the *resistance* R. Out of phase to the resistance by 90° (or $\pi/2$ rad) are the *mass (positive) reactance* (X_{m}), which is proportional to frequency f, and the *stiffness (or negative) reactance* X_{s}, which is inversely proportional to f; X_{m} and X_{s} are 180° (or π rad) out of phase to each other. Overall, $Z = R + i(X_{\mathrm{m}} - X_{\mathrm{s}})$. The net reactance is $X_{\mathrm{net}} = \mid X_{\mathrm{m}} - X_{\mathrm{s}} \mid$. The magnitude of the impedance is related to these component parts by

$$\mid Z \mid = \sqrt{R^2 + X_{\mathrm{net}}^2}\,. \tag{10.19}$$

Similarly, the admittance $Y = G + i(B_{\mathrm{m}} - B_{\mathrm{s}})$. The real part G is the *conductance*, and the imaginary parts B_{m} and B_{s} are the *mass susceptance* and *stiffness (compliant) susceptance*, respectively. (When the impedance has only a resistive term, the conductance is $G = 1/R$.) The net susceptance is $B_{\mathrm{net}} = \mid B_{\mathrm{m}} - B_{\mathrm{s}} \mid$. The components of the admittance are related to the impedance terms by $G = R/(R^2 + X_{\mathrm{net}}^2)$, $B_{\mathrm{m}} = -X_{\mathrm{m}}/(R^2 + X_{\mathrm{net}}^2)$, and $B_{\mathrm{s}} = -X_{\mathrm{s}}/(R^2 + X_{\mathrm{net}}^2)$. The magnitude of the admittance is related to these component parts by

$$\mid Y \mid = \sqrt{G^2 + B_{\mathrm{net}}^2} \, . \tag{10.20}$$

Problems 10.26–10.29 address these and other relations involving impedances.

This discussion can apply to the impedance in acoustics, electronics, or flow (Appendix D). For example, this terminology can be applied to the analog in mechanics of a mass attached to a spring that slides on a rough surface. The friction due to sliding on the rough surface is the resistance. The inertia due to the mass is the mass reactance and the resistance to movement due to the spring is the stiffness reactance. When used in audiology and other aspects of sound, these parameters are also characterized by the term "acoustic," such as acoustic impedance, acoustic admittance, acoustic resistance, mass (positive) acoustic reactance, and so on. To avoid confusion we could use subscripts, such as R_{acoust} for acoustic resistance, and so on.

The units of impedance Z is ohms (Ω), just as in electronics. The units of admittance Y is the 1/ohm = 1 mho. R, X_{m}, and X_{s} all have units of ohms, and G, B_{m}, and B_{s} all have units of mhos. In audiology the admittance values are small and the more useful unit is the millimho or mmho.

The Impedance in a Harmonically Driven System (Advanced Topic)

We now examine the concept of impedance and derive (10.19) by considering the motion of a body of mass m attached to a spring of force constant k. It is subject to a viscous, frictional force, characterized by η, and a driving term, $F_{\mathrm{A}} \exp(\mathrm{i}\omega t)$. This motion is described by

$$m\frac{\mathrm{d}^2 x}{\mathrm{d}t^2} + \eta\frac{\mathrm{d}x}{\mathrm{d}t} + kx = F_{\mathrm{A}} \exp(i\omega t). \tag{10.21}$$

The driving term has a real part $F_{\mathrm{A}} \cos(\omega t)$, because $\exp(i\omega t) = \cos(\omega t) + i\sin(\omega t)$. This model is similar to the Voigt model in Chap. 4 with $F = \eta\, \mathrm{d}x/\mathrm{d}t + kx$ (from (4.57)). (It can be represented by the model presented later in Fig. 10.13, with the force driving term acting laterally on the mass.)

Substituting a potential solution $x = x_0 \exp(i\omega t)$ in this equation, gives the ("particular" or steady state) solution

$$x = \frac{F_{\mathrm{A}}}{-m\omega^2 + \mathrm{i}\eta\omega + k} \exp(i\omega t). \tag{10.22}$$

(See Appendix C.) Differentiating this, the speed of the body is

$$\frac{\mathrm{d}x}{\mathrm{d}t} = \frac{\mathrm{i}\omega F_{\mathrm{A}}}{-m\omega^2 + \mathrm{i}\eta\omega + k} \exp(\mathrm{i}\omega t) \quad (10.23)$$

$$= \frac{F_{\mathrm{A}}}{\eta + \mathrm{i}m\omega + k/\mathrm{i}\omega} \exp(\mathrm{i}\omega t). \quad (10.24)$$

The ratio of the driving force, $F_{\mathrm{A}} \exp(i\omega t)$, to the speed, $\mathrm{d}x/\mathrm{d}t$, is the mechanical impedance Z

$$\frac{F_{\mathrm{A}} \exp(\mathrm{i}\omega t)}{\mathrm{d}x/\mathrm{d}t} = \eta + \mathrm{i}m\omega + k/\mathrm{i}\omega = Z. \quad (10.25)$$

This impedance is sometimes expressed as

$$Z = R + \mathrm{i}\omega M + S/\mathrm{i}\omega = R + \mathrm{i}(\omega M - S/\omega). \quad (10.26)$$

Here, $R = \eta$ provides the resistance, $M = m$ is the inertial term (or inertance), and $S = k$ is the stiffness. In terms of the *mass (positive) reactance* (X_{m}), which is proportional to frequency f, and the *stiffness (or negative) reactance* X_{s}, which is inversely proportional to f, we can say

$$Z = R + \mathrm{i}X_{\mathrm{m}} + X_{\mathrm{s}}/\mathrm{i} = R + \mathrm{i}(X_{\mathrm{m}} - X_{\mathrm{s}}), \quad (10.27)$$

with $X_{\mathrm{m}} = \omega M = 2\pi f M$ and $X_{\mathrm{s}} = S/\omega = S/2\pi f$. Equation (10.19) follows from the magnitude of a complex number.

10.1.3 What Happens when Sound Travels from One Medium to Another?

Sound transmission from one medium to another is very important in hearing, because sound is transmitted from the air in the auricle and ear canal into the middle ear and then into the inner ear. This can be understood by seeing what happens to a sound wave incident on the planar interface between two different semi-infinite media, such as media 1 and 2 in Fig. 10.3.

This sound wave travels in medium 1 with intensity I_{i} and pressure P_{i} incident on the interface. The part that is reflected back into the same medium has intensity I_{r} and pressure P_{r}, and the part that is transmitted into medium 2 has intensity I_{t} and pressure P_{t}. These pressures are related by "matching the boundary conditions" at the interface, as we will see very soon. As stated earlier, the frequency is the same in both media, but the wavelength changes with the change in sound speed by (10.3), so $\lambda_1 = v_1/f$ and $\lambda_2 = v_2/f$. These frequencies need to be the same because the sinusoidal oscillations of pressure and matter movement on both sides of the interface must always match. The magnitudes of the pressures must match so there is no net force

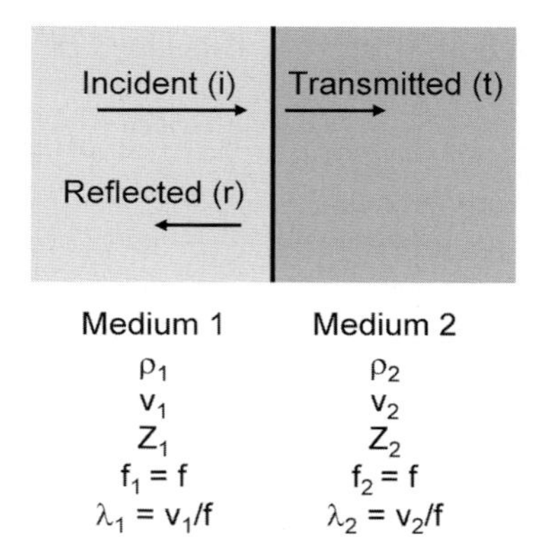

Fig. 10.3. Schematic of acoustic wave transmission and reflection

on the interface. This gives

$$P_{\mathrm{i}} + P_{\mathrm{r}} = P_{\mathrm{t}}. \tag{10.28}$$

The motion at the interface caused by all three waves must match. The speed of this lateral displacement is $|\omega(\delta z)|$ and for a wave initially moving to the right it is positive for the incident and transmitted waves and negative for the reflected wave. Using (10.8), $\delta z \sim P/(Z\omega)$, we know that $|\omega(\delta z)| \sim P/Z$, and so matching displacement gives

$$\frac{P_{\mathrm{i}}}{Z_1} - \frac{P_{\mathrm{r}}}{Z_1} = \frac{P_{\mathrm{t}}}{Z_2}. \tag{10.29}$$

Solving these two equations gives

$$\frac{P_{\mathrm{r}}}{P_{\mathrm{i}}} = \frac{Z_2 - Z_1}{Z_1 + Z_2} \tag{10.30}$$

$$\frac{P_{\mathrm{t}}}{P_{\mathrm{i}}} = \frac{2Z_2}{Z_1 + Z_2}. \tag{10.31}$$

Using (10.9) and (10.30), the fraction of intensity that is reflected is

$$R_{\mathrm{refl}} = \frac{I_{\mathrm{r}}}{I_{\mathrm{i}}} = \frac{P_{\mathrm{r}}^2/2Z_1}{P_{\mathrm{i}}^2/2Z_1} = \frac{P_{\mathrm{r}}^2}{P_{\mathrm{i}}^2} \tag{10.32}$$

$$= \left(\frac{Z_2 - Z_1}{Z_1 + Z_2}\right)^2 = \left(\frac{1 - Z_2/Z_1}{1 + Z_2/Z_1}\right)^2. \tag{10.33}$$

Using (10.31), the fraction of intensity that is transmitted is

$$T_{\mathrm{trans}} = \frac{I_{\mathrm{t}}}{I_{\mathrm{i}}} = \frac{P_{\mathrm{t}}^2/2Z_2}{P_{\mathrm{i}}^2/2Z_1} = \frac{Z_1}{Z_2}\frac{P_{\mathrm{t}}^2}{P_{\mathrm{i}}^2} \tag{10.34}$$

$$= \frac{Z_1}{Z_2}\left(\frac{2Z_2}{Z_1 + Z_2}\right)^2 = \frac{4Z_2/Z_1}{(1 + Z_2/Z_1)^2}. \tag{10.35}$$

Table 10.5. Representative fractions of reflected and transmitted acoustic energy at tissue interfaces. (Using data from [467, 489, 508])

tissue interface	reflected fraction (in %)	transmitted fraction (in %)
water/soft tissue	0.23	99.77
fat/muscle	1.08	98.92
bone/muscle	41.23	58.77
soft tissue/bone	43.50	56.50
bone/fat	48.91	51.09
soft tissue/lung	63.64	36.36
air/muscle	98.01	1.99
air/water	99.89	0.11
air/soft tissue	99.90	0.10

We see that $I_{\mathrm{i}} = I_{\mathrm{r}} + I_{\mathrm{t}}$ and $R_{\mathrm{refl}} + T_{\mathrm{trans}} = 1$, which means that sound energy (and intensity) is conserved. Also, the reflected fraction is large (and approximately equal to 1) when Z_2/Z_1 is either much larger or smaller than 1, i.e., when there is a very large acoustic impedance mismatch. We will see this means the large mismatch of acoustic impedances between air and any solid or liquid medium (Table 10.1) has important consequences in the design of the ear. The term "to match impedance" means to minimize R_{refl}, and this is accomplished by making Z_1 and Z_2 approximately equal (10.33). Examples of this mismatch in the body are given in Table 10.5. Most sound incident on the body is reflected. Little sound is reflected between soft tissues (brain, skin, kidney, liver, and so on) and between soft tissues and blood. There is significant reflection at interfaces of soft tissues with the lungs and with bones.

10.1.4 Resonant Cavities

Many properties of sound waves depend on the medium the wave propagates in and on the characteristics of the enclosure. This is important in the production of sound – as in human speaking, musical instruments, megaphones, and sounds produced in echo chambers – and in the collection of sound waves – as in human hearing in the outer ear. One way to analyze these properties is to consider the properties of sound waves in resonant cavities; more generally, these properties of *resonant cavities* can be applied to any waves in any resonant structure.

What is a resonant structure? Let us consider a wave on a string of length L. The string can be plucked so a localized wave can propagate along the length of the string. Alternatively, one can pluck the string so a (transverse wave) oscillates everywhere periodically. If the string is rigidly fixed at both ends, the lowest order periodic motion (or mode – for which the transverse displacement at any point along the string varies as $\cos \omega t$) is one with a

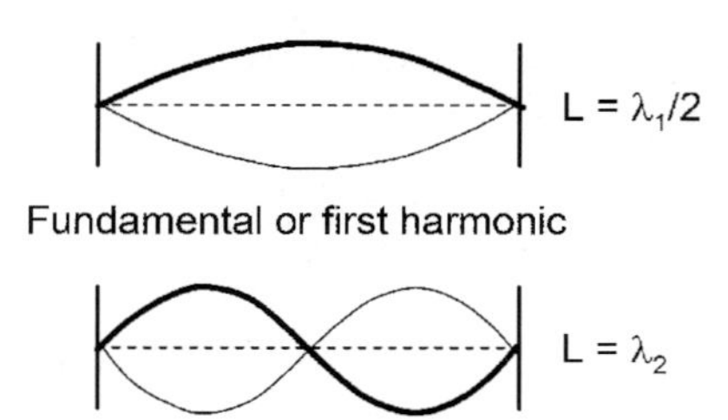

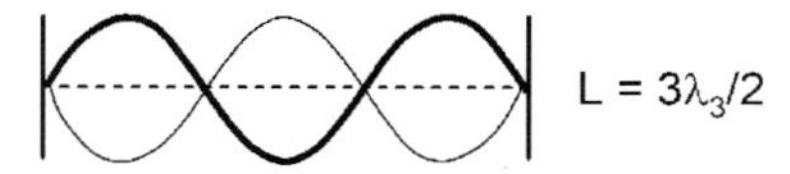

Fig. 10.4. Lowest three modes on a string of length L, each shown at times with maximum and zero excursions. (Based on [471])

half wave between the two ends (Fig. 10.4). Here $\lambda/2 = L$, so the wavelength $\lambda = 2L$. The "boundary conditions" for this motion are zero-transverse motion at each end; these places with no motion are called *nodes*. Higher order modes with higher integral numbers of half wavelengths are possible. In general, for n half wavelengths, the wavelength is $n(\lambda_n/2) = L$, and so the resonant wavelengths are

$$\lambda_n = \frac{2L}{n} \qquad \text{with } n = 1, 2, 3, \ldots . \tag{10.36}$$

Since $v_\mathrm{s} = \lambda f$, the resonant or mode frequencies are

$$f_n = \frac{v_\mathrm{s}}{\lambda_n} = n\frac{v_\mathrm{s}}{2L}, \tag{10.37}$$

which is $v_\mathrm{s}/2L$, $2(v_\mathrm{s}/2L)$, $3(v_\mathrm{s}/2L)$, ... for $n = 1, 2, 3, \ldots$; these are also called the *fundamental frequency*, the *first harmonic*, the *second harmonic*, etc. The mode frequencies are equally spaced by $v_\mathrm{s}/2L$.

If one of the two ends of the massless string is not fixed, but is free to move (only transverse to string axis), this different boundary condition leads to a different set of resonant wavelengths and frequencies. At the fixed end, the transverse displacement is still zero, which is a node. At the free end, the transverse displacement now has maximum magnitude (and zero slope) – and this is called an *antinode*, as in Fig. 10.5. The lowest order mode now has a quarter wavelength between the two ends separated by the string length, and so $\lambda/4 = L$ and the wavelength $\lambda = 4L$. The next lowest order mode has an extra half wavelength cycle with $3\lambda/4 = L$, so the wavelength $\lambda = 4L/3$, etc. Now the resonant condition is $(m/2 + 1/4)\lambda_m = L$, so the resonant wavelengths are

$$\lambda_m = \frac{2L}{m + 1/2} \qquad \text{with } m = 0, 1, 2, 3, \ldots . \tag{10.38}$$

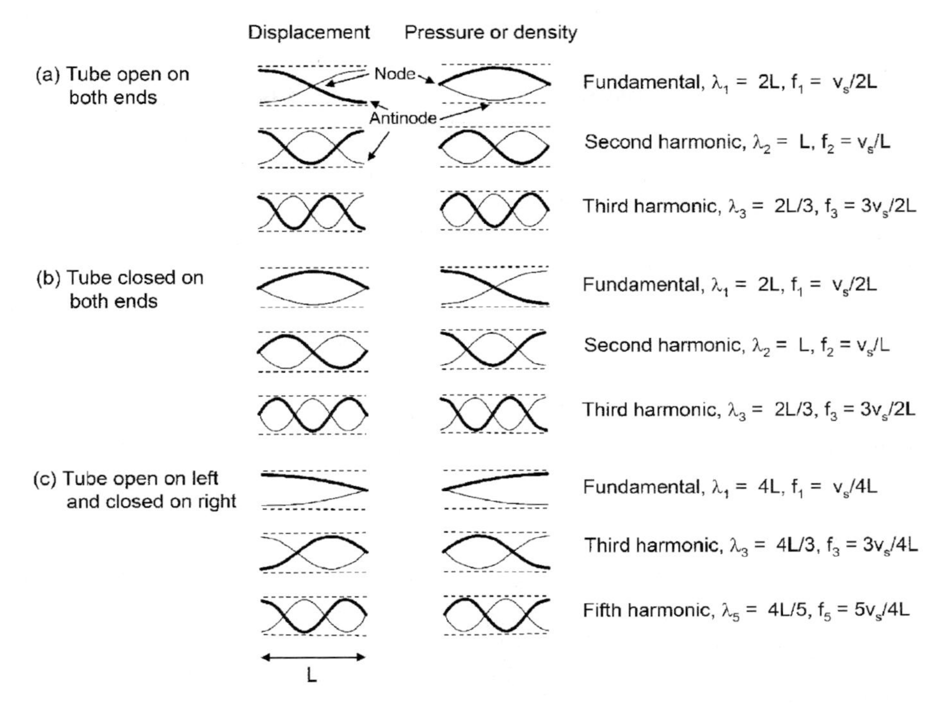

Fig. 10.5. Wave modes in a tube, for tubes (**a**) open on both ends, (**b**) closed on both ends, and (**c**) open on the left side and closed on the right side. The mode displacements of air are shown on the left for the first overtone or fundamental mode and for the next two overtones, and the corresponding changes in pressure and density for these modes are shown on the right. (Based on [471])

The mode frequencies are

$$f_m = \frac{v_\mathrm{s}}{\lambda_m} = \left(m + \frac{1}{2}\right)\frac{v_\mathrm{s}}{2L}, \tag{10.39}$$

which is $v_\mathrm{s}/4L$, $3v_\mathrm{s}/4L$, $5v_\mathrm{s}/4L$, $\ldots$ for $m = 0, 1, 2, \ldots$. Although the mode frequencies are different from the previous case, they are still equally spaced by $v_\mathrm{s}/2L$. The lowest frequency is now $v_\mathrm{s}/4L$ and only odd harmonics are present. Figure 10.6 shows the mode frequencies for both boundary conditions.

Hollow tubes are good first-order models for the tubes in the body where longitudinal sound waves propagate (similar to pipes in a pipe organ) (Fig. 10.6). If the tube of length L is open at both ends, the pressure is fixed at the ambient pressure at each end and $P = 0$ at the ends. This is formally equivalent to the string with fixed ends, so the resonant wavelengths and frequencies are given by (10.36) and (10.37). The density oscillation is also zero

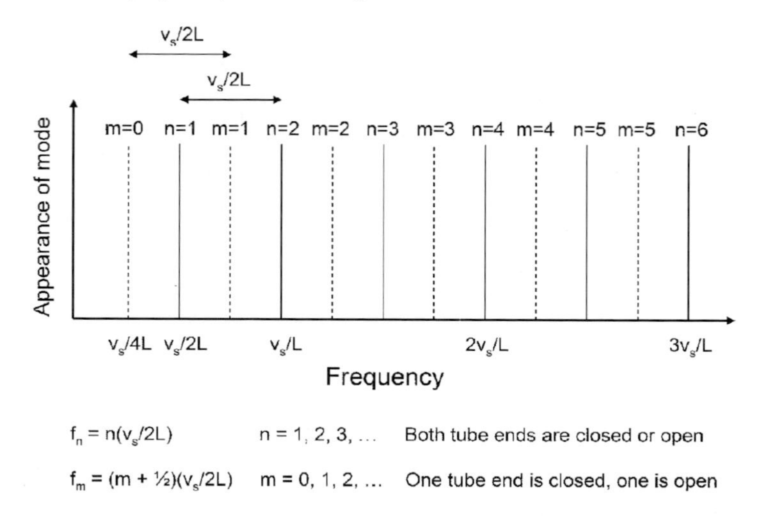

Fig. 10.6. Mode frequencies for a tube closed on both ends, open on both ends, or closed on one end and open on the other

at each end (nodes) and the molecular displacement is of maximum magnitude at each end (antinodes). If the tube is open at one end and closed at the other, P is still zero at the open end (a node). At the closed end, the lateral displacement is fixed at zero (a node), while the pressure and density changes have maximum magnitude (an antinode). This is formally equivalent to the string fixed at one end and free at the order, so the resonant wavelengths and frequencies are given by (10.38) and (10.39). (The resonant wavelengths and frequencies for a tube closed at both ends are the same as for one that is open at both ends, but the identifications of nodes and antinodes are interchanged.) This is the reason why the tube length and boundary conditions are important in determining these resonances, and why these resonances can be changed by varying the tube conditions (length, varying cross-sectional area and shape) – as can happen in the region between the vocal cords and the mouth opening. For details about acoustic resonances see [491, 492].

The mathematics of resonant excitation of an acoustic medium are the same as those used in Chap. 8 to analyze (8.52) and (8.54), and are due to acoustic losses. This damping leads to the finite widths of the resonances for acoustic energy, which are $\sim\gamma/2\pi$, in Hz, between the points with half the maximum response. Figure 10.7 compares the unrealistically narrow resonances to the more realistic, broader, damped resonances. The number of cycles that an excitation lasts during its decay (to $1/e$ of the initial energy) is $Q/2\pi$, where the quality factor $Q = \omega/\gamma = 2\pi f/\gamma$, where ω and f are the resonant frequencies. (See Appendix D for more details.)

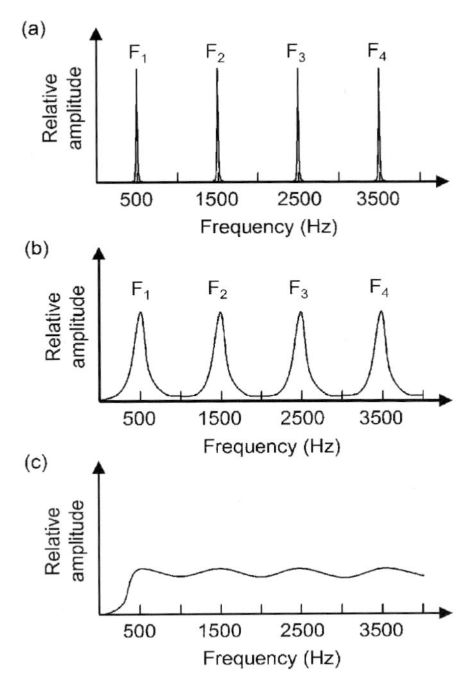

Fig. 10.7. (**a**) Loss-free (infinite Q), (**b**) moderate loss (moderate Q), and (**c**) very lossy (very low Q) transmission resonances for a tube. (Based on [504])

10.2 Speech Production

10.2.1 Types of Sounds

Human speech is made by air from the lungs as it passes through the *trachea* (windpipe), *larynx* (which houses the *vocal folds or cords*), and *pharynx* (throat), through the mouth and nasal cavities, and then out of the mouth and nose (Fig. 9.1). For general sources about speech see [450, 451, 460, 461, 466, 472, 474, 501, 504, 510].

The vocal folds (or vocal cords) are folds of ligament extending on either side of the larynx, with the space in between them called the *glottis*. The structure above the larynx is called the *vocal tract*. In adult females, the average length of the pharynx is 6.3 cm and that of the oral cavity is 7.8 cm, for a total vocal tract length of 14.1 cm. In adult males, the average length of the pharynx is 8.9 cm and that of the oral cavity is 8.1 cm, for a total vocal tract length of 16.9 cm.

The shape of the vocal tract can be varied by moving the soft palate, tongue, lips, and jaw, and these are the *articulators*. Adjusting the vocal tract to produce speech sounds is called *articulation*. The basic elements of speech are classified as the (1) phonemes (the basic sounds), (2) phonetic features (how the sounds are made), and (3) the acoustic signal (the acoustic nature of the sounds). We will first discuss the phonemes and the acoustic signal, and then describe how we make sound.

Phonemes are the shortest segments of speech, which in General American English are the 14 vowel sounds and 24 consonant sounds. Each phoneme is produced by distinctive movements of the vocal tract, which are the *phonetic features of speech.* Vowels are produced with the vocal tract relatively open, with different shapes of the opening. Consonants are produced by a constriction or closing of the vocal tract. The production of phonemes is characterized by three phonetic features (a) voicing, (b) place of articulation, and (c) manner of articulation.

Sounds produced as air rushes though the vibrating vocal folds in the larynx (Fig. 9.1) are called *voiced sounds.* All vowels are voiced sounds. (Touch the middle of your throat as you speak them.) Many consonants are also voiced, such as the "d," "m," "w," and "v" sounds. (We will not use the formal notation for phonemes [450, 461, 504].) Sound produced without the vocal folds vibrating and only involving air flow through constrictions or past edges produced by the tongue, teeth, lips, and palate are called *unvoiced sounds*, such as for the "t" and "f" sounds.

The obstructions needed to produce consonants are formed in different *places of articulation*, including (in order from the front to the back of the mouth) *bilabial* (both lips, or labial) for the "p," "b," "m," and "w" sounds, *labiodental* (the bottom lip and upper front teeth) for the "f" and "v" sounds, *dental* or *interdental* (teeth) for both "th" sounds (as in "thin" and "them"), the alveolar ridge or *alveolar* (upper gums near the teeth) for the "d," "t," "s," "z," "n," "r," and "l" sounds, *palatal* or *alveopalatal* (hard palate, which is behind the upper gums) for the "sh," "zh" (as in "vision"), "ch," "j," and "y" sounds, *velar* (soft palate, which is behind the hard palate) for the "k," "g," "ng," and "w" sounds, and *glottal* for the "h" sound.

The mechanical means by which consonants are formed, including the way air is pushed through the opening, is the *manner of articulation. Plosive* or *stop* sounds such as "d," "b," "p," "g," "t," and "k" have a staccato nature because they are formed by blocking air flow and then letting a slight rush of air. *Fricative* sounds such as the voiceless "f," "th" (as in "thin"), "s," and "sh," and the voiced "v," "th" (as in "them"), "z," and "zh" (each set in order from the front to the back of the mouth) have a hissing nature because air flow is constricted at the place of articulation, making the air flow turbulent. The "m" sound is *nasal* because the soft palate is lowered to couple the nasal cavities to the pharynx and air is suddenly released to flow through the nose. Some unvoiced sounds have a combination of initially plosive and then fricative character (and are *affricative*), such the "ch" and "j" sounds and other gutterals. The *approximants* are produced by moving one articulator to another without creating a closed constriction, such as for the "w," "y,"

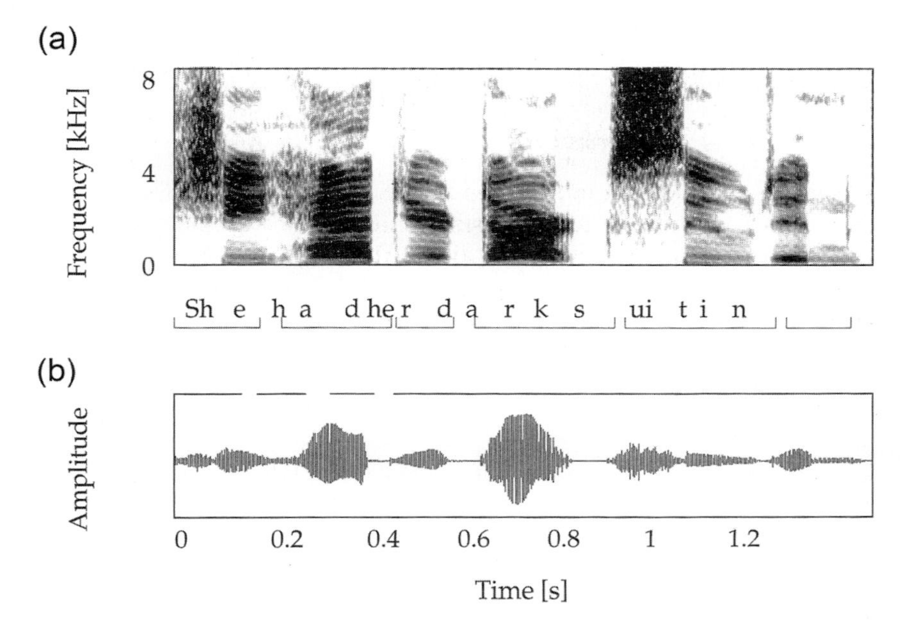

Fig. 10.8. Sound spectrogram of a female speaker saying "She had her dark suit in..." in (**a**), with the corresponding time-domain waveform signal in (**b**). (From [449])

"r," and "l" sounds. The "w" sound is also said to be formed in a *semivowel* manner.

The *acoustic signal* is the set of acoustic frequencies and intensities as a function of time. For example, Fig. 10.8a shows a sound spectrogram – a plot of the frequency components of sound vs. time – for the expression "She had her dark suit in..." Distinctively different signals are seen for the different vowels and consonants. This spectrogram clearly has more useful information than the time-domain waveform for the same expression in Fig. 10.8b (even though they have the same information content).

These different contributions of vowels and consonants are more clearly seen in Fig. 10.9 [449]. In part (a) the "sh" and "ch" consonant spectra are seen to be different and have different durations; also there is an interval of silence before the "ch." They are both very different from the "s" spectrum. Each vowel has characteristic bands. For example, the long a in "say" in (a) has three frequency bands, near 500, 1,700, and 2,500 Hz. These can be, respectively, labeled as F_1, F_2, and F_3. Such multiple bands are characteristic of vowels and are called *formants*, the first-formant (F_1), the second-formant (F_2), and so on. These formant frequencies differ for the different vowels, as is seen for this long a, the "ah" in "father" and "pot" in part (a) and /ba/ in part (c), and the long i in "mite" in part (b). As we will see later, these formants define the vowels and the characteristic feature in the perception of

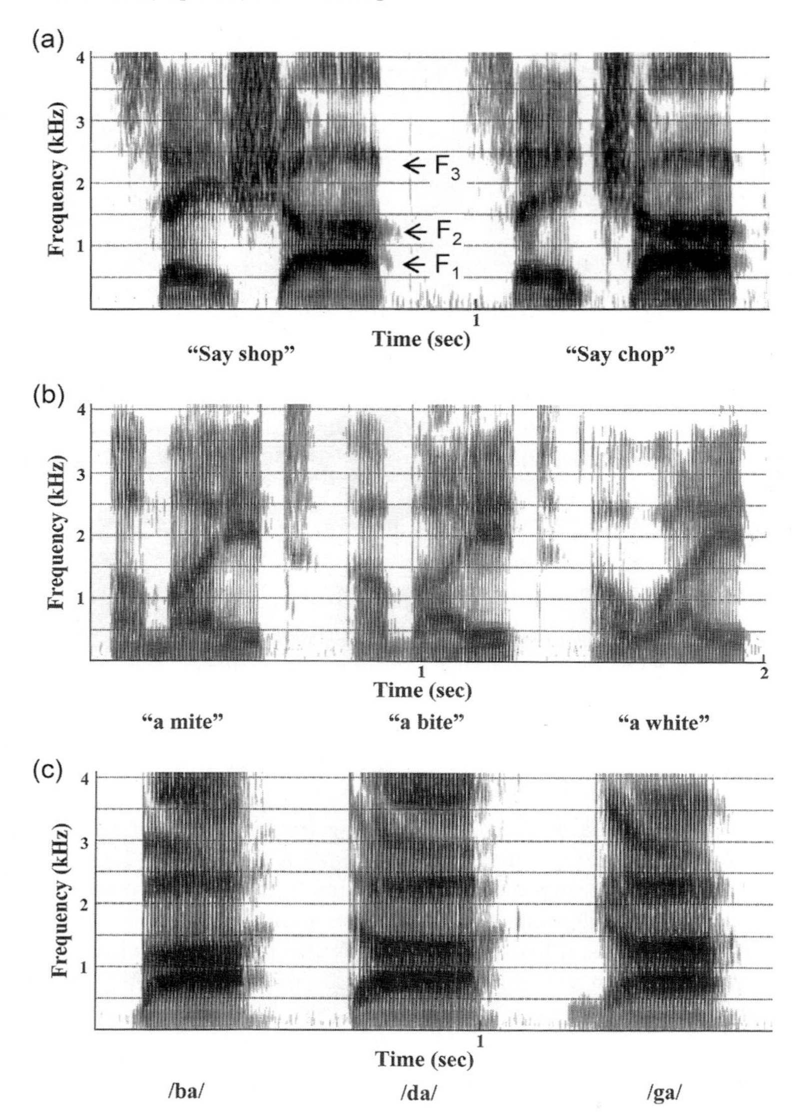

Fig. 10.9. Sound spectrograms of (**a**) "say shop" and "say chop," (**b**) "a mite," "a bite," and "a white," and (**c**) /ba/, /da/, and /ga/ as in "father" and "pot." The F_1, F_2, and F_3 formant frequencies for the vowel in 'shop" are *denoted* in part (a). (From [462])

vowels. For all three syllables in part (c), the lowest part of the first spectrum of the consonant increases into the F_1 of the vowel. However, this variation is different for the second and third formants and this helps define these different sounds. For /ba/, F_2 and F_3 rise into the vowel; for /da/,

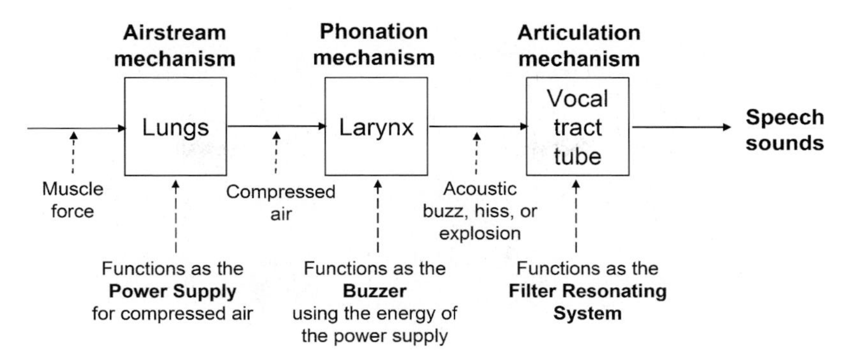

Fig. 10.10. The three neuromuscular systems in voice production. (Based on [460])

F_2 and F_3 fall into the vowel; and for /ga/, F_2 falls, while F_3 rises into the vowel.

10.2.2 Systems in Speech Production

There are three sequential neuromuscular systems involved in speech production (Fig. 10.10). (a) The lungs are the *airstream mechanism* in which muscle force is used to produce a stream of compressed air. (b) The larynx is the *phonation mechanism*, which takes the compressed air and turns it into an acoustic buzz, hiss, or explosion. (c) The vocal tube track is the *articulation mechanism*, which takes the larynx sounds and turns them into speech sounds. These three systems, respectively, function as a power supply of compressed air, a buzzer, and a filter resonating system. We have described the properties of the lungs in Chap. 9. We will examine the phonation and articulation mechanisms now.

The Acoustic Buzzer

The separation of the vocal folds (vocal cords, glottis opening) varies with our state of speaking (Fig. 10.11). The open glottis is V-shaped because the vocal folds are held together in the front of the larynx and move apart in the back. When we are *voiceless*, the vocal folds are totally open for normal breathing. During *whispering* the folds are closer together. Air moves through the glottal constriction (picks up speed) and rotates in turbulent eddies to give it its distinctive sound. As detailed later, our *voice* during the normal speaking of voiced sounds is created by the folds periodically opening and closing, to give periodic bursts of air. No air flows when there is a *glottal stop*.

During speaking, air rushing through vocal folds in the larynx from the trachea to the pharynx causes the vocal folds to vibrate, which in turn leads to a modulation of the air flowing into the trachea (Fig. 10.12). Initially the folds are apart and the pressure in the trachea and pharynx are equal, and both are

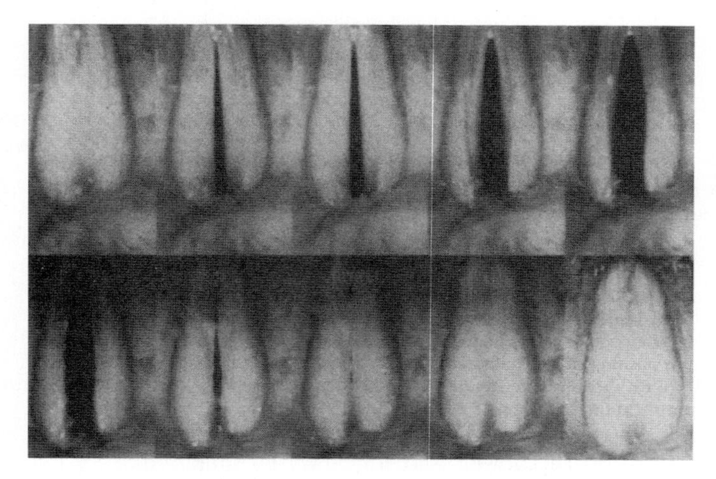

Fig. 10.11. A series of video frames of vocal-fold movement during a normal glottal cycle. Note that the opening is asymmetric, with the glottis more widely open at the bottom than at the top. During whispering, the glottis is even more open than during normal speaking (rightmost in top row), and it is even more open during forced inhalation. (From [504], photo by Debra K. Klein. The University of Iowa Hospitals and Clinics. Used with permission)

equal to atmospheric pressure. Because of the separation, the pressure is the same before and after the folds, even with air flowing from the lungs. Muscle contractions cause the folds to move to the midline of the tube. Because of this constricted air flow, the pressure before the glottis (subglottis pressure) then exceeds the oral pressure. This pressure difference forces the folds to separate rapidly, leading to a rapid burst of air. This produces an overpressure above the glottis and an acoustic shockwave that moves up the vocal tract. The folds then rebound back to their initial positions because of their elastic recoil properties and the pressure gradient. This pressure gradient arises from the combination of Bernoulli's equation in the Venturi limit and the equation describing continuity of flow, as in (7.18):

$$P_{\mathrm{glottis}} - P_{\mathrm{subglottis}} = \frac{1}{2}\rho v_{\mathrm{subglottis}}^2 \left[1 - \left(\frac{A_{\mathrm{subglottis}}}{A_{\mathrm{glottis}}}\right)^2\right] < 0. \qquad (10.40)$$

Because the open flow area in the glottis (between the folds) is much smaller than that below it, the flow speed in this air burst is faster and the pressure in the glottis is smaller than that beneath it. The folds return to their initial positions – with the air flow continuing. The cycle repeats and the modulated flow of air results in a buzzing sound. Figure 10.11 shows photographs of the glottis during different stages of this vibration. The motion of the glottis can be modeled with springs and other mechanical components, as shown in Fig. 10.13.

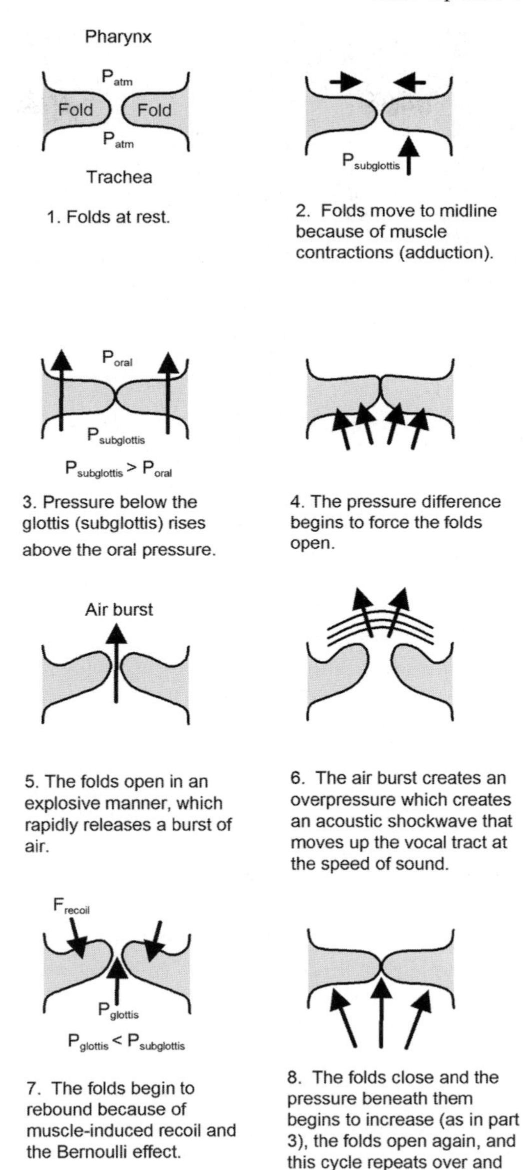

Fig. 10.12. The sequential stages of glottal vibration. (Based on [460])

The frequency of this buzzing sound depends on the mass of the vocal folds and their tension. The fundamental frequency f_{fund} (or tone or pitch) of this buzzing is typically ~125 Hz in men and ~250 Hz in women. The lowest fundamental frequency (bass singer) is ~64 Hz (which is a low C), while the

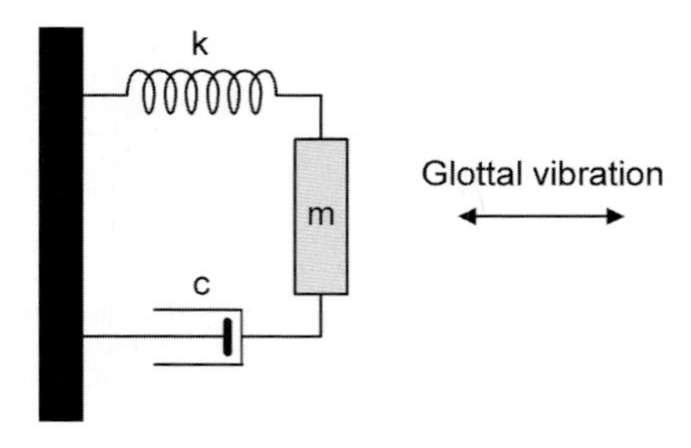

Fig. 10.13. A one-mass mechanical model of glottal vibration, with pressure against the tissue wall. (Based on [504])

highest fundamental frequency (soprano) is ~2,048 Hz (five octaves above low C). (An *octave* extends from any given frequency to one twice that.) The acoustic buzz contains this fundamental frequency along with many harmonics (integral multiples of the fundamental).

Each vocal fold can be modeled as a two-dimensional ribbon-like object, with vibrations along the length of the fold and in a perpendicular direction across the height of the fold. To first-order, each of these vibrations can be treated with our one-dimensional models of vibration.

Using (10.37), the fundamental frequency of a string rigidly fixed at both ends, of mass density ρ, length L, and radius r under tensile force F, is $v_s/2L$, where v_s is speed of waves along the string. For such a string, we know that $v_s = \sqrt{F/\mu}$, where μ is the mass density per unit length. Using $\mu = \pi r^2 \rho$ and $v_s = \sqrt{F/(\pi r^2 \rho)} = \sqrt{[F/(\pi r^2)]/\rho} = \sqrt{\sigma/\rho}$ – where the stress $\sigma = F/(\pi r^2)$ – the fundamental frequency is

$$f_{\text{fund}} = \frac{1}{2L}\sqrt{\frac{\sigma}{\rho}}. \tag{10.41}$$

Most of the length of the vocal fold consists of membranous material. This length is about 16 mm for adult males and 10 mm for adult females. Figure 10.14 shows that the variation of the fundamental frequency of vocal-fold oscillation with this membranous length has approximately the $1/L$ dependence predicted by (10.41). (Also see Problem 10.37.) Note this length is longer in men than women and grows with age during childhood, so the fundamental frequency is lower in men than women and decreases into adulthood. The typical stress–strain curves for human vocal-fold tissues given in Fig. 10.15 provide elastic constant information for this model (Problem 10.38). Muscular control of the folds must also be included in such models, because the oscillation is not really free.

What does the airstream leaving the glottis look like? Say that air flows into the lungs of a male during inspiration at a rate of 100 cm^3/s. This is also the average flow of air leaving the lungs and passing through the glottis. If the glottis vibrates at 125 Hz, a volume of (100 cm^3/s)/125 Hz = 0.8 cm^3 leaves in each air puff in a cycle. These air puffs last for 3 ms (within the 8 ms cycle

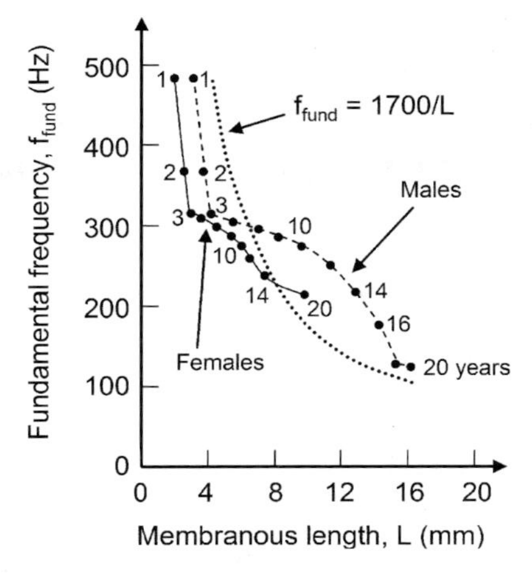

Fig. 10.14. Vocal-fold frequency as a function of length, with age as an implicit variable. (Based on [504])

for 125 Hz), so the peak volumetric flow rate of air is $(0.8\,\text{cm}^3/3) \times 10^{-3}\,\text{s} = 270\,\text{cm}^3/\text{s}$. Because the glottis opening is typically $\sim$0.3 cm wide and it is $\sim$2 cm high, this air flows through an area of $\sim$0.6 cm^2. Using (7.13), the maximum speed of air leaving the glottis is $(270\,\text{cm}^3/\text{s})/(0.6\,\text{cm}^2) = 450\,\text{cm/s} = 4.5\,\text{m/s}$.

Voice-Filtering Theory (Advanced Topic)

The glottal wave (i.e., the sequence of vocal-fold pulses) from the larynx is schematically shown in the time domain in Fig. 10.16. This could represent the acoustic intensity or pressure. Let us say that it is the intensity $I(t)$. This

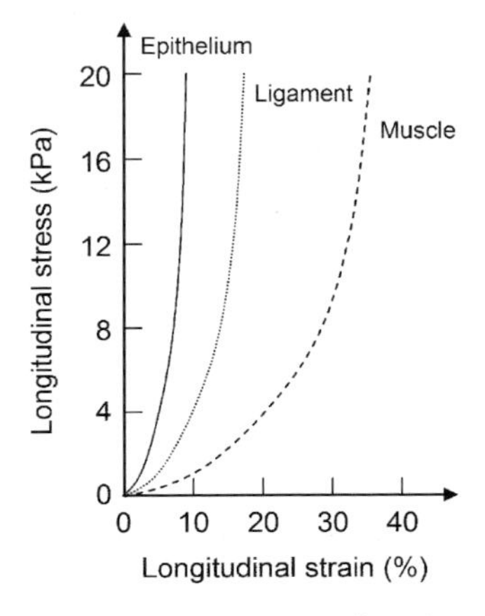

Fig. 10.15. Hypothetical stress–strain curves for tissues in human vocal folds. (Based on [504])

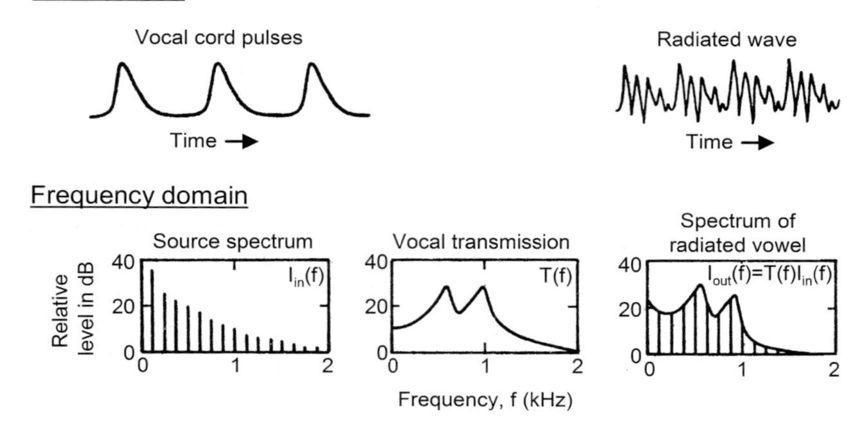

Fig. 10.16. Modification of the glottal wave by the vocal tract in the voice-filtering theory of speaking. The signals plotted vs. time (in the time domain) are in the upper row, while those plotted vs. frequency (in the frequency domain) are in the lower row. The source spectrum from the larynx (shown as vocal-fold pulses in the time domain) is $I_{\text{in}}(f)$, where f is frequency. The formation of the first two formants is seen in the vocal transmission ($T(f)$) and the radiated vowel ($I_{\text{out}}(f)$). The radiated vowel is shown as a radiated wave in the time domain. It is precisely our inability to conjure up a time-domain analog of the frequency-domain vocal transmission box that makes this type of frequency analysis so important. (Adapted from [464]. Used with permission)

description is in the time domain. Alternatively, we could characterize this same intensity wave using a different approach, in the frequency domain by looking at the frequencies of the sine (or cosine) waves that can be added to give the same pattern in time. The frequency components of this cyclic-glottal pattern in time can be determined by a method called Fourier analysis, to determine the strength of the fundamental I_1 (at f_{fund}) and each harmonic I_n (at nf_{fund} for $n = 2, 3, 4, \ldots$)

$$I(t) = \sum_{n=1 \text{ to } \infty} I_n \sin(2\pi n f_{\text{fund}} t) \tag{10.42}$$

$$= I_1 \sin(2\pi f_{\text{fund}} t) + I_2 \sin(4\pi f_{\text{fund}} t) + I_3 \sin(6\pi f_{\text{fund}} t) + \cdots . \tag{10.43}$$

(More generally, all functions of time can be *Fourier analyzed* in this manner.) This frequency spectrum associated with the glottal wave is shown in the frequency domain in Fig. 10.16, with the intensity components I_n expressed in dB SPL. (The differences in the intensity components would be more pronounced in a linear plot.) For comparison, Fig. 10.17 shows that a sine wave (or cosine wave) has a single frequency component – which is precisely the frequency of the wave, a square wave – with the same period as the sine wave – is composed of the same frequency component plus odd harmonics of it, and

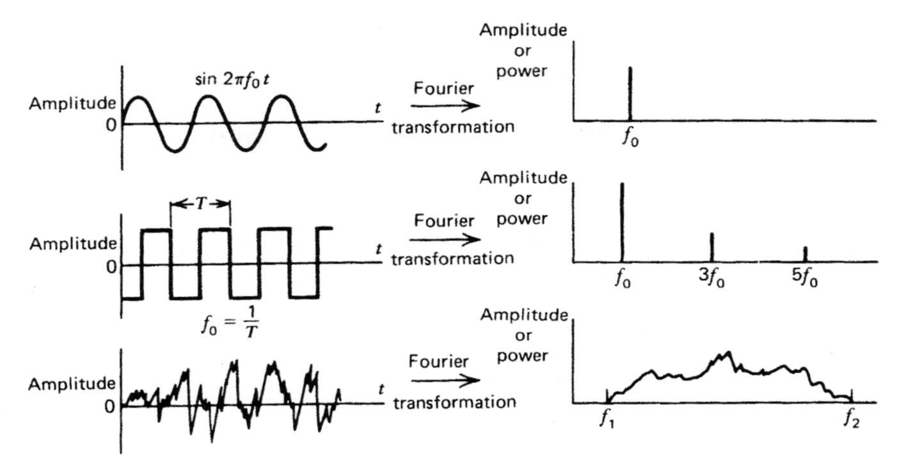

Fig. 10.17. Three waveforms on the left with their variations with time (the time domain), a sine (or cosine) wave, a square wave, and a more random pattern, with their respective frequency components (as Fourier analyzed and in the frequency domain). (From [507]. Reprinted with permission of Wiley)

a more random, aperiodic waveform in time is composed of a wide range of frequency components.

Within the voice-filtering theory of articulation, this glottal wave (Fig. 10.16) is the input to the pharynx, and the pharynx, oral, and nasal cavities act as resonant cavities that transmit some frequencies (that eventually leave our mouths and noses) better than other frequencies. We examine the frequency components of the glottal wave because it is easier to understand how this sound wave gets modified in the vocal tract by tracking each frequency component (with the transmission factor in Fig. 10.16) than by tracking the response as a function of time. The reason for this should be clear from our discussion of the simple resonant cavities formed by open and close-ended tubes.

In the idealized cavity of a tube, say a tube that is open on both sides, part of the sound entering one end propagates through and leaves the other end. Only those frequency components of the entering sound wave that coincide with the various resonant frequencies of the tube (the fundamental and its harmonics) will propagate through the tube and leave; more precisely, they lead to large pressures in the tube, which are then emitted by the tube. The other frequencies will be dissipated. This modification in the frequency components of the sound wave leads to a corresponding change in the sound wave in the time domain. The frequency components at each frequency f are multiplied by the transmission factors at that frequency $T(f)$

$$I_{\text{out}}(f) = T(f) I_{\text{in}}(f). \tag{10.44}$$

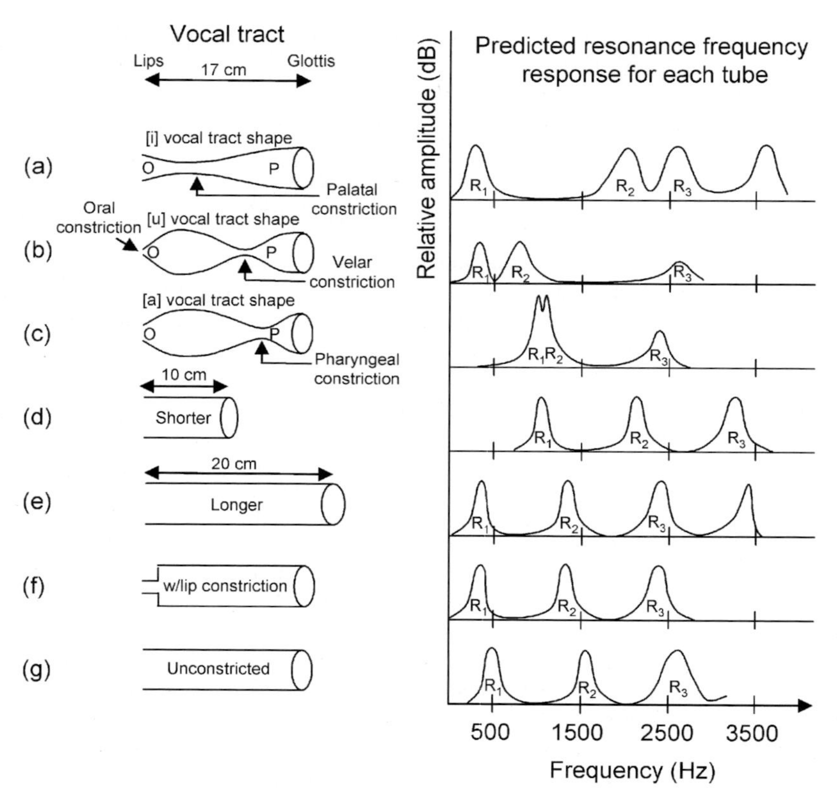

Fig. 10.18. Resonance frequencies for vocal tracts with different lengths and constrictions, relative to the unconstricted vocal track in (g). The vowels are (**a**) [i] as in "sit," (**b**) [u] as in "put," and (**c**) [a] (or "ah") as in "father" and "pot." (Based on [460])

In reality, transmission does not occur only at precisely the resonant frequencies but in a frequency region centered about the resonant frequencies. The width of range is sometimes expressed as the frequency f divided by the Q of the cavity. These widths are relatively small in a high Q cavity. This is shown schematically in Figure 10.7 for a generic tube. Moderate Q resonances are shown in Fig. 10.18 for several types of open-ended tubes.

The vocal tract is not a tube with these equally separated resonances, so its transmission curve is different from that of a constant diameter tube (Fig. 10.18g). The Q is quite low and so the resonances are quite broad, because of (1) the irregular shape of the vocal cavity, (2) the soft irregular nature of the walls, and (3) acoustic losses due to (a) transmission through the mouth and back through the glottis to the lungs, (b) the viscoelastic nature of the vibrating soft tissues, and (c) the viscosity of air. Because we are capable of changing the shape of the vocal tract, we routinely change this transmission curve during speech. For a given transmission curve $I_{\text{out},n} = T(nf_{\text{fund}})I_{\text{in},n}$ with

$$I_{\text{in}}(t) = \sum_{n=1 \text{ to } \infty} I_{\text{in},n} \sin(2\pi n f_{\text{fund}} t) \quad (10.45)$$

$$I_{\text{out}}(t) = \sum_{n=1 \text{ to } \infty} T(nf_{\text{fund}}) I_{\text{in},n} \sin(2\pi n f_{\text{fund}} t). \quad (10.46)$$

Figure 10.16 shows schematically the corresponding representations of the buzz in the time to frequency domains, the multiplication of the frequency components by the transmission factor, and the output sound signals in the frequency and time domains.

The output sound wave has the same frequency components as the input wave, including the same fundamental frequency. However, the relative strengths of each frequency component are very different. In the input wave, $I_{\text{in},n}$ decreases monotonically with n (the higher the frequency the smaller the intensity). This is usually no longer true in the output wave. In fact, $I_{\text{out},n}$ peaks at several harmonics, which are called formants. The lowest frequency peak is the first formant, the second is the second formant, and so on, as is seen in Fig. 10.16.

What makes an "a" sound as in "say" different from an "e" as in "see"? They have different formants, as is illustrated for several vowels in Fig. 10.9. The actual frequencies of these formants depend on the type of sound, and they profoundly change the time variations of the sound intensity.

Figure 10.19 plots the frequency of the second formant F_2 vs. that of the first formant F_1 for several vowel sounds. For each sound there is an oval showing the range of these formant frequencies for different people. For example, for the "ah" sound in "father" and "pot," the range in F_1 is 600–1,300 Hz, spanning the range from adult males to children, while for F_2 this range is 800–1,700 Hz. Even with these variations, the loci of frequencies for each sound are very different from each other (i.e., they do not overlap) – which makes it possible for us to differentiate one vowel from another spoken by anybody, even with these natural variations.

Figure 10.20 shows the acoustic spectra for soprano, alto, tenor, and bass human voices saying the vowel sound "ah" sound in "father" and "pot." In each, the fundamental and harmonics are seen to have different intensities because of the vowel formation. Evidence of the clustering of the peaks near formant frequencies is seen.

It is essential to differentiate between the different roles of the vocal-fold oscillation frequency and the vocal tract formant resonances. Figure 10.21 shows the acoustic signal in time and its frequency spectrum for the same uttered vowel with two different vocal fold fundamental frequencies. With a finer range of frequency components the short-time behavior in the time domain is smoother, but the envelope of both are similar. In contrast, Fig. 10.22 shows them for two different vowels with the same vocal fold fundamental

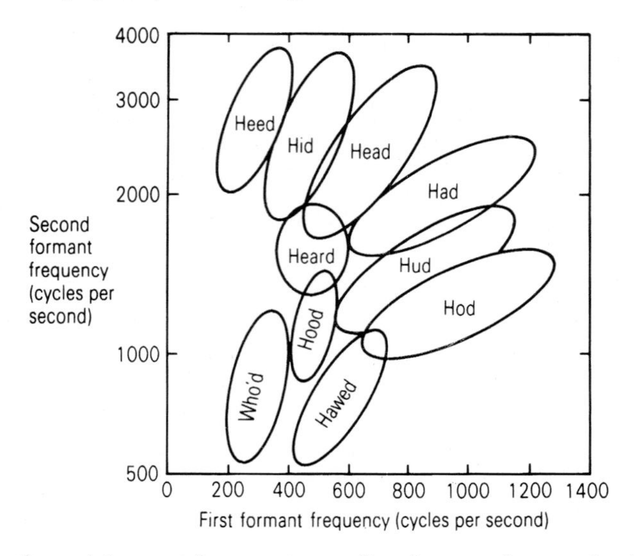

Fig. 10.19. Second-formant frequencies vs. first-formant frequencies for 10 American English vowels, as spoken by a wide range of 76 people: men, women, and children. The enclosures include 90% of the sounds spoken. (From [447] as adapted from [493], and used with permission)

frequency. The time domain profiles have the same short-term structure, but different envelopes. (The peaks of the envelopes are the formant frequencies.)

Figure 10.23 shows the spectral pattern of a man and woman saying the same message. The oscillation pattern for the man is more rapid than that for the woman, because the pattern consists of the harmonics of the fundamental buzzing frequency, which is lower in the man than the woman. However, the spectral envelope is the same for both, and this envelope carries the most important information for understanding speech.

Consider a vowel spoken by someone with a fundamental frequency of 128 Hz. If the vowel has a first formant at 512 Hz, then it is at the third harmonic of the fundamental at 128 Hz. In the time domain, this 512 Hz component is seen in the time domain signal as a fine ripple every 2.0 ms (1/(512 Hz)) superimposed on the slower repeated cycle every 7.8 ms (= 1/(128 Hz)).

In the simplest model of vowel formation in the vocal tract, the region from the vocal folds to the mouth is a cylinder of constant diameter and length L that is open on one end (mouth) and closed on the other (vocal folds). (This ignores the 90° bend in the vocal tract from the pharynx to the mouth.) Using (10.39), the resonant frequencies would be: $f_m = (m + \frac{1}{2})(v_s/2L)$ which is $v_s/4L$, $3v_s/4L$, $5v_s/4L, \ldots$ for $m = 0, 1, 2, \ldots$. The formant frequencies of sound transmitted through this cavity would have ratios 1:3:5 and so on. Figure 10.24 shows that the vocal tract has very different shapes for the three depicted vowels, and consequently these vowels have very different formant

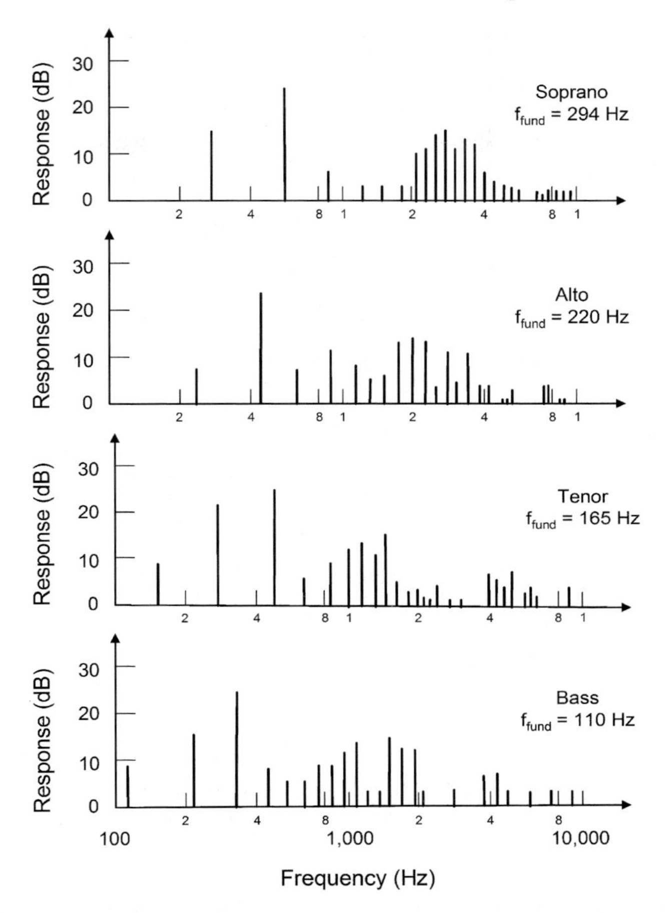

Fig. 10.20. The acoustic spectra for four voices, with the given fundamental frequencies, producing the vowel "ah" as in "father" and "pot." (Based on [492])

frequencies. It also shows that this one-tube model gives $F_1 = 500\,\text{Hz}$ and $F_2 = 1,500\,\text{Hz}$ for these vowels. Calling $f_r = v_s/4L$, these correspond to f_r and $3f_r$. However, the real formant frequencies are clearly shifted from these values.

Two-Tube Model (Advanced Topic)

How can we improve the one-tube model? A better model for the "ah" as in "father" and "pot," is the two-tube approximation in which the tract is modeled as a smaller diameter pharynx tube, which is closed on one end, connected to an equally long larger diameter mouth tube (Fig. 10.25a). This makes it look a bit more like the actual vocal tract in Fig. 10.24c.

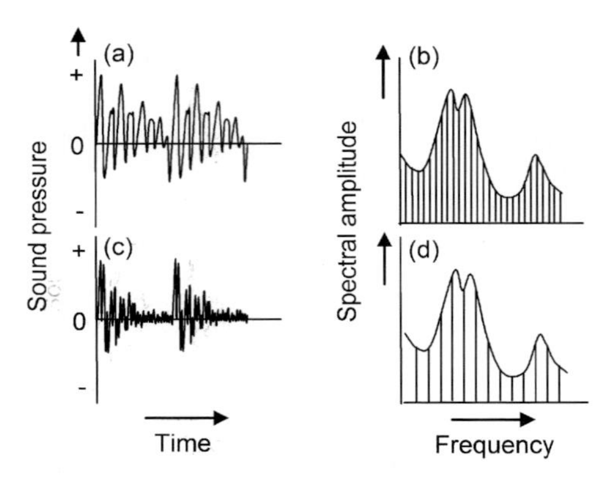

Fig. 10.21. Wave shapes (**a**,**c**) and corresponding spectra (**b**,**d**) for the same vowel "aw" as in "bought" and "awe," but with two different vocal fold frequencies of (a,b) 90 Hz and (c,d) 150 Hz. (Based on [461])

Before determining the mode frequencies for this two-tube model, we need to analyze how the pressure changes when an acoustic wave travels from a tube of one diameter with cross-sectional area A_1 to one of another diameter and cross-sectional area A_2, as in Fig. 10.25a,b. This can be treated as a generalization of the earlier analysis of the transmitted and reflected acoustic beams at an interface between two semi-infinite media with different acoustic impedances. Continuity of pressure, $P_\mathrm{i} + P_\mathrm{r} = P_\mathrm{t}$, at the interface between the two tubes – as in (10.28) – is still needed; however, (10.29) needs to be modified. The motion of the media is still described by P/Z, but this quantity must be multiplied by the area A, so continuity of motion of the media requires

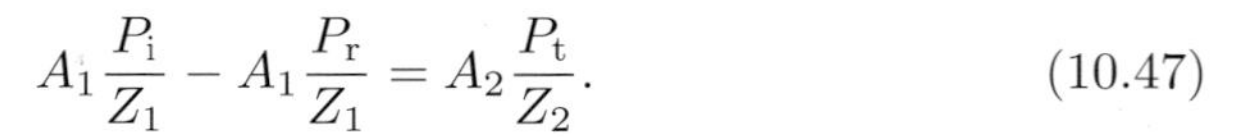

$$A_1 \frac{P_\mathrm{i}}{Z_1} - A_1 \frac{P_\mathrm{r}}{Z_1} = A_2 \frac{P_\mathrm{t}}{Z_2}. \tag{10.47}$$

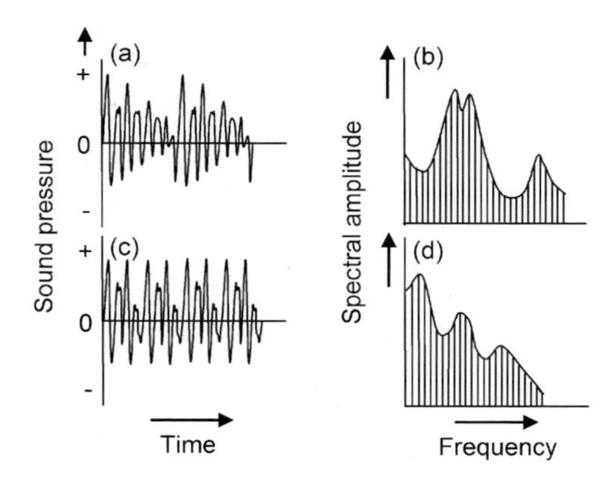

Fig. 10.22. Wave shapes (**a**,**c**) and corresponding spectra (**b**,**d**) for different vowels (a,b) "aw" as in "bought" and "awe" and (c,d) "uh" as in "but" and "about," but with the same vocal fold fundamental frequency of 90 Hz. (Based on [461])

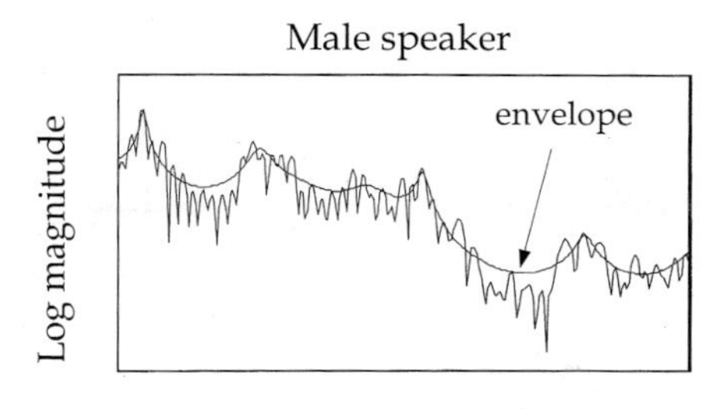

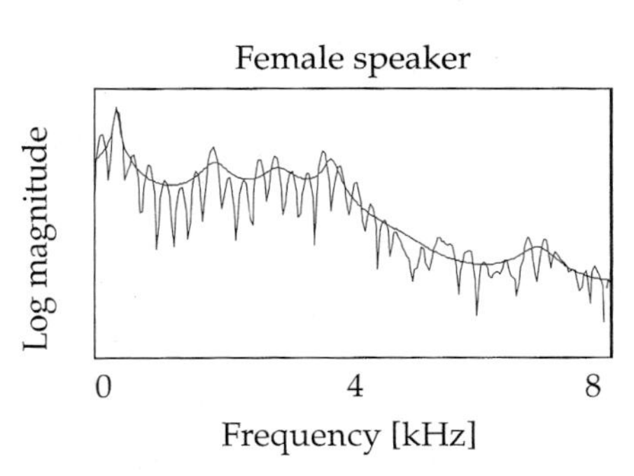

Fig. 10.23. The short-time spectra of the same linguistic pattern by a male and female speaker, with the spectral envelope shown. (From [449])

This gives

$$\frac{P_\mathrm{r}}{P_\mathrm{i}} = \frac{Z_2/A_2 - Z_1/A_1}{Z_1/A_1 + Z_2/A_2} \tag{10.48}$$

and

$$\frac{P_\mathrm{t}}{P_\mathrm{i}} = \frac{2Z_2/A_2}{Z_1/A_1 + Z_2/A_2}. \tag{10.49}$$

Here the impedance is the same in both tubes, so

$$\frac{P_\mathrm{r}}{P_\mathrm{i}} = \frac{A_1 - A_2}{A_1 + A_2} \tag{10.50}$$

and

$$\frac{P_\mathrm{t}}{P_\mathrm{i}} = \frac{2A_1}{A_1 + A_2}. \tag{10.51}$$

This means the pressure in the transmitted wave is smaller than that in the incident wave when the second tube is wider (expansion, $A_2 > A_1$) (Fig. 10.25a) and it is greater than it when the second tube is narrower (contraction, $A_2 < A_1$) (Fig. 10.25b).

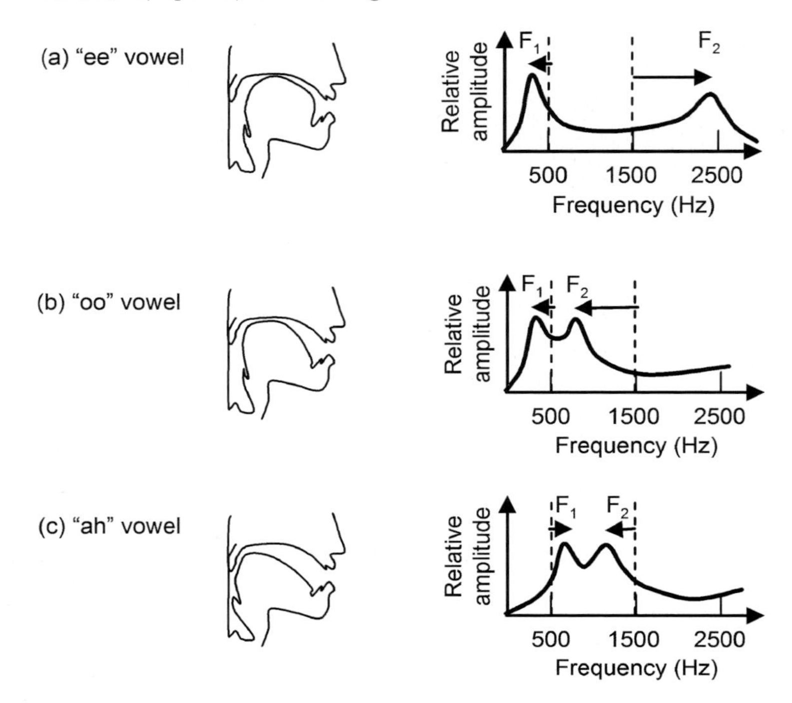

Fig. 10.24. Vocal tract shapes and spectra for three vowels: (**a**) "ee" as in "see," (**b**) "oo" as in "too," and (**c**) "ah" as in "father" and "pot." The difference of the first two formants are shown here relative to those for a tube with constant cross-section. These are similar to Fig. 10.18a–c, except for the more realistic vocal tract depicted here and the mouth is on the right of the larynx here. (Based on [504])

The upward shift in F_1 and downward shift in F_2 for "ah" as in "father" and "pot" relative to that of a cylinder of uniform diameter and length L, can be viewed from two perspectives. (1) The lowest two modes for a cylinder of length L change because the pressure in the wider mouth is lower than that in the narrower pharynx. This perturbs the mode structure from that of a uniform cylinder. For the lowest mode, the pressure in the mouth region is relatively less than indicated by this mode, so the opening effectively occurs (earlier in the air stream and) in the mouth, making L effectively shorter and the F_1 higher than f_r. For the next lowest mode, the node (zero gauge pressure) in the pharynx is effectively shifted to the right (toward the mouth) because the pressure should be larger in the pharynx. The length of this half wave is longer, making L effectively longer and F_2 lower than $3f_r$. This viewpoint is depicted in Fig. 10.25c,e. (2) Alternatively, we can treat the pharynx and mouth as two separate cylinders of length $L/2$ that couple weakly and therefore perturb each other. The pharynx is closed at the vocal fold end and almost open on the mouth side. Using (10.39), the lowest mode is at

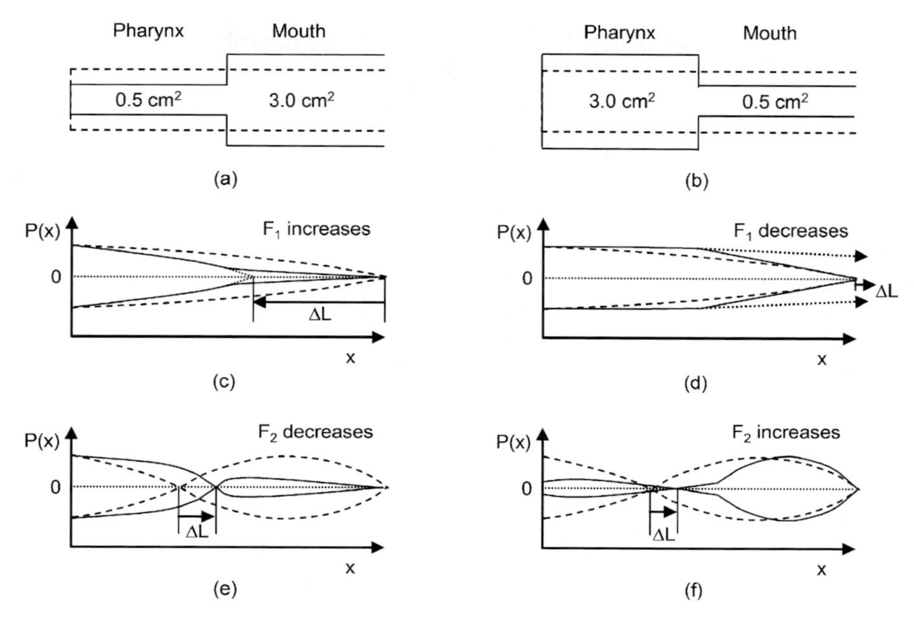

Fig. 10.25. Two-tube models of (left side; **a**,**c**,**e**) "ah" as in "father" and "pot," and (right side; **b**,**d**,**f**) "ee" as in "see," each showing the two-tube model with noted cross-sectional area in the equal length tubes in (a,b), modification of the mode for the one-tube model (*dashed lines*) to the mode in the two-tube model (*solid lines*, with changes in effective length denoted by the *dotted lines*) for the first (c,d) and second (e,f) formants. (Based on [504])

$v_s/(4(L/2)) = v_s/2L = 2f_r$, but since the mouth end is not totally open and there is pressure leakage to the mouth, the effective length is a bit longer, so this mode is at a somewhat lower frequency than $2f_r$, such as the F_1 formant in this case. Furthermore, the mouth is open on the tooth end and almost closed on the pharynx end (because the pharynx is narrower). Again the lowest mode would be at $2f_r$ if the pharynx side were totally closed. Because it is not and the pressure is higher on the pharynx side, this left node is pushed into the mouth cavity and L is effectively shorter and the lowest mode is a bit higher than $2f_r$, such as the F_2 formant for this vowel. A different two-tube model seems to work for other vowels (Fig. 10.25b,d,f, Fig. 10.24a, Problem 10.45 for the "ee" sound as in "see"). For more on voice-filtering models, see [460, 464, 504].

10.2.3 Parameters of the Human Voice

Each physiological feature that helps create the voice modulates a particular feature of the vocal sound waves, and each of these acoustic features is linked to a specific type of perception. The tension in the vocal folds affects the fundamental frequency of the vocal folds – and of their harmonics, and this frequency is perceived as the *pitch* of the voice. The medial compression of the vocal folds and the pressure from the lungs affect the intensity of phonation, which leads to perceived *loudness* of the voice. The vocal-fold adduction or abduction creates an acoustic spectral difference, and this is perceived as the *quality* or timbre of the voice.

Vocal frequency. The greater the tension in the vocal folds, the greater the elastic recoil force and the thinner the folds; both effects contribute to higher oscillation frequencies. This glottal buzz rate is the fundamental frequency of the human voice. The *modal* vocal-fold frequencies, f, are those glottal buzz rates that we are most comfortable with and use most often, as in normal conversation. They typically can be varied by a factor of 2.5, with $f_{\text{highest}} = 2.5 f_{\text{lowest}}$, by changing the vocal-fold tension; this means that people typically have a modal frequency range of 2.5 octaves. (The scale "do-re-mi-fa-so-la-ti-do" corresponds to one octave.) Even though this acoustic buzz frequency can be changed, with a concomitant change in perceived pitch, the formant frequencies do not change. (That they do not change by this factor of 2.5 should be clear from Fig. 10.19.) The modal frequency is in the lower 1/4–1/3 of the speaker's vocal range, so people typically phonate near the lower end of possible frequencies.

The average modal frequency for men is 128 Hz and that for women is 260 Hz, but there is a wide distribution about these means for men and women. Female voices are higher in pitch than male voices, because women on the average have smaller larynxes and shorter, thinner, and less massive vocal folds. The average speaking model frequencies for different voice classifications are: bass: 98 Hz, baritone: 123 Hz, tenor: 165 Hz, contralto: 175 Hz, mezzo-soprano: 196 Hz, and soprano: 247 Hz. Figure 10.20 plots the acoustic spectra for human voices in several of these categories.

Vocal intensity. The greater the medial compression of the vocal folds, the greater the subglottal pressure the closed folds can sustain before they open, and therefore the greater the loudness. Each doubling of the (gauge) pressure from the lungs increases the voice intensity by 8–12 dB, so the vocal intensity I varies roughly as the cube of the lung pressure P_{lung}, $I \propto P_{\text{lung}}^3$. Soft speech corresponds to 2–3 cmH_2O, loud speech to 15–20 cmH_2O, and shouting to 40–60 cmH_2O. At a distance of 1 m, conventional speech corresponds to a sound intensity of about 60 dB SPL and shouting to 108 dB SPL. (Also see Fig. 10.30 later.)

Phonation type. The vocal folds are totally open during forced inhalation and are still very open during normal breathing. They are closer together, but are still far apart during whispering and are much closer together during normal speaking – and form an obstruction to the airway. The degree that the vocal folds are open (abduction) or closed (adduction) affects the spectral distribution of the voice and therefore its quality, such as it being perceived as "noisy," "breathy" (noisy and buzzy), or "creaky" (irregularly spaced low-frequency air bursts).

10.2.4 The Energetics of Speaking

What fraction of the total energy in the air flow from the lungs during speaking is converted into the oscillatory acoustic waves that are radiated from the mouth? Say the entire vital capacity (about $5\,\mathrm{L} = 0.005\,\mathrm{m}^3$) were expelled in 1 s with a lung pressure of 2 kPa (about $20\,\mathrm{cmH_2O}$). This "aerodynamic" power would be $0.005\,\mathrm{m}^3 \times 2\,\mathrm{kPa} = 10\,\mathrm{W}$ during that 1 s. However, this volume could be expelled only with the glottis fully open, so with the glottis mostly closed during speaking the flow rate would be much lower than $0.005\,\mathrm{m}^3/\mathrm{s}$, and perhaps 0.0001–$0.0005\,\mathrm{m}^3/\mathrm{s}$. With the same pressure differences, the "aerodynamic" power would not exceed 1 W. Therefore, continuous speaking would add only about 1% to our metabolic rate (ignoring the efficiency of forming this air flow). Conversational speech is about 60 dB SPL at 1 m, which corresponds to $10^{-6}\,\mathrm{W/m}^2$, and so using (10.4) the power is about $\sim 10^{-5}\,\mathrm{W}$. Defining the glottal efficiency as the ratio of this acoustic power to the aerodynamic power, the efficiency is $\sim 10^{-5} = 0.001\%$. Over a range of conditions, this efficiency is often given as ranging from 0.0001–1%.

10.3 Hearing

The ear consists of the outer, middle, and inner ears (Figs. 10.26 and 10.27). The *outer ear* is the external *auricle* (or *pinna*), which "funnels" sound into the *auditory canal* or *ear canal tube* (or *external canal* or *meatus* – mee-ay'-tus – or external auditory meatus). The *middle ear*, also known as

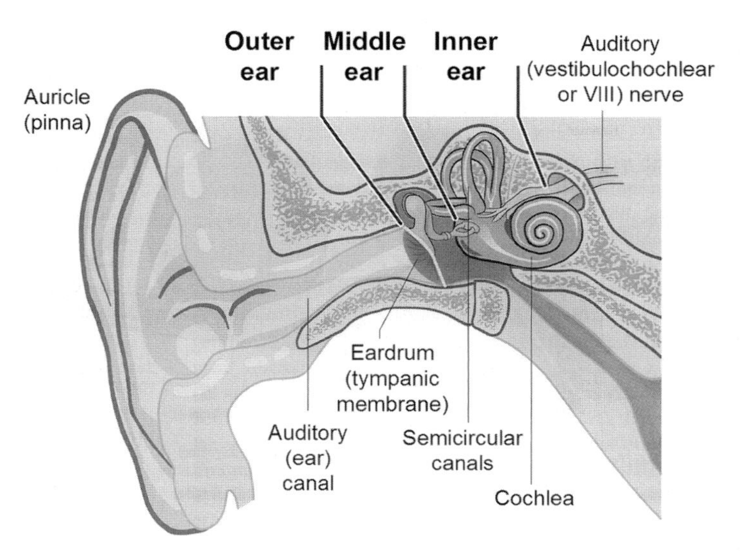

Fig. 10.26. Diagram of the outer, middle, and inner ear. (From [497])

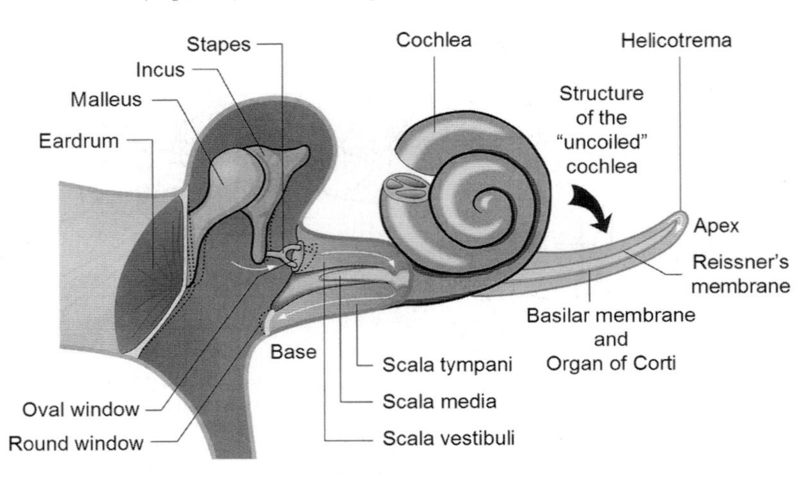

Fig. 10.27. Middle ear in detail and visualization of partial and total uncoiling of the cochlea. The cut in the partially uncoiled cochlea shows the cross-section of the tubes, which is seen in more detail in Fig. 10.32. (From [497])

the tympanic cavity, begins with the *tympanic membrane* (*eardrum*), and continues with the three connected small bones or ossicles, the ossicular chain: the *hammer* (*malleus*), *anvil* (*incus*), and *stirrup* (*stapes*). On one side the tympanic membrane interfaces with the ear canal tube and on the other side it is in contact with the hammer. The *inner ear* consists of the *cochlea* and the *auditory nerve* that travels from the cochlea to the brain. The stirrup is in contact with the oval window of the cochlea. The cochlea begins medially with the vestibule, continues with the anterior snail-shaped region and then in the posterior, three semicircular canals. Sound is brought into the ear by the outer ear and the actual processing of the sound begins in the inner ear. The outer and middle ear are collectively called the *conductive system*, which conducts the sound to the inner ear. We will see that the middle ear "improves" the transmission of sound from the outer ear to the inner ear. The cochlea and the eighth cranial nerve – which includes the auditory nerve as well as the vestibular nerve – are called the *sensorineural system*, which senses the sound and then creates an electrical signal and sends it to the brain. For general sources about hearing see [456, 460, 463, 465, 466, 469, 470, 472, 473, 474, 478, 479, 496, 509, 510].

There are four steps in the hearing process within the ear. (1) The sound wave enters the outer ear. (2) The movement of the tympanic membrane is transferred by conduction through the ossicles to the oval window of the cochlea. (3) The movement of the oval window generates a compressional (sound) wave in the fluid of the cochlea. (4) This wave moves the basilar membrane upon which the primary auditory receptors (the hair cells) are located, and the electrical signals generated by the hair cells are sent to the brain. This conversion of sound into electrical signals is called

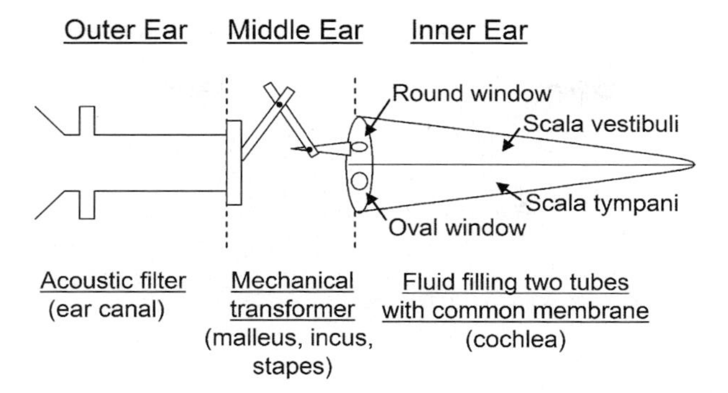

Fig. 10.28. Mechanical analogs of the outer, middle, and inner ear. (Based on [473])

mechanoelectrical transduction. Mechanical model analogs of the ear are depicted in Fig. 10.28.

10.3.1 Auditory Sensitivity

The normal-human auditory system is sensitive from about 20 Hz to 20 kHz, with maximum sensitivity in a broad region near 3,000 Hz, as is seen in the audiogram of pure tone thresholds in Fig. 10.29. Between 100 and 10,000 Hz

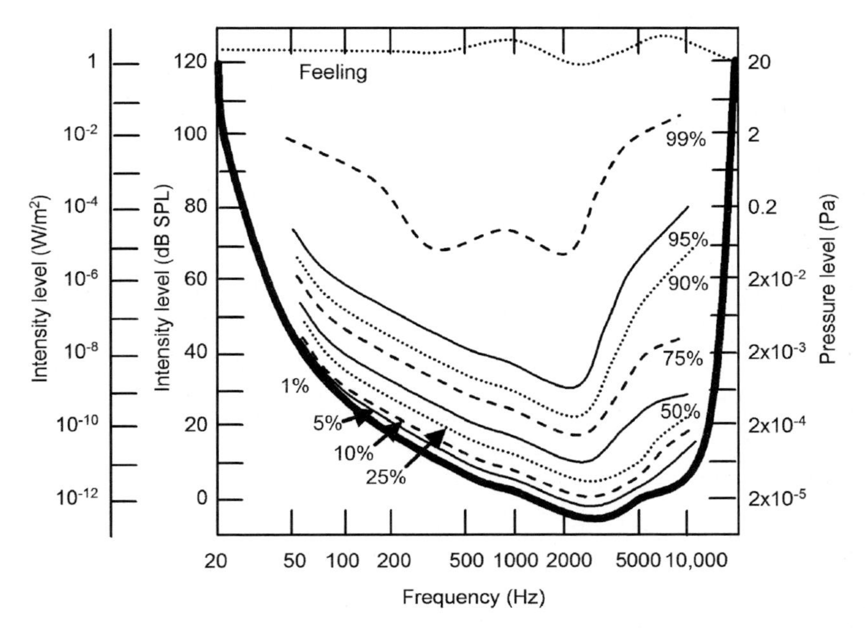

Fig. 10.29. Absolute auditory threshold for typical US residents. The curves show the percentage of people who could hear sounds below the level of the curve; the top curve shows the threshold for "feeling" in the ear. (Based on [461])

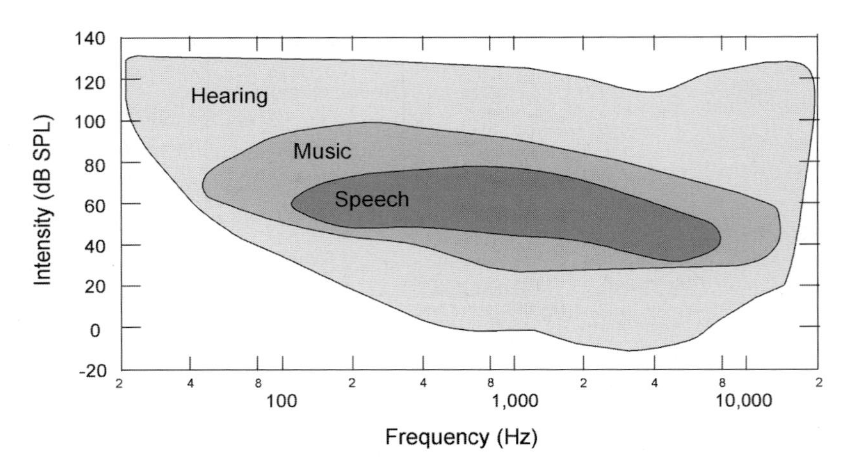

Fig. 10.30. The boundaries for normal hearing, in terms of limits of acoustic intensity and frequency, compared to the frequency and volume ranges of speech and orchestral music. (Based on [454, 491])

the normal threshold varies from about 0 to 20 dB SPL. The threshold for uncomfortable loudness is about 100 dB SPL. From 120 to 140 dB SPL, the threshold for "feeling," "prickling," and "tickling" in the ear is reached.

Related to this is the issue of how sensitive our hearing needs to be. Our ears must be sensitive to hear our voices, and of course our voices need to be loud enough to be heard. Figure 10.30 plots the ranges of normal hearing sensitivity vs. frequency, along with the ranges of intensity and frequency of speech and music, more specifically orchestral music.

Outer Ear

In spite of its funnel-like appearance, the auricle or pinna provides negligible sound collection and amplification capability. It does provide hints as to sound source localization. (Cats orient their pinnae toward a sound source to help determine its direction.) This localization depends on the asymmetrical and irregular shape of the pinna. One important feature of the pinna is that it attenuates the high-frequency components of sounds that come from the rear relative to those from the front.

The external canal in the outer ear is about 9 mm high, 6.5 mm wide, and about 2.5–3.5 cm long. It is open on one side (the outside) and terminated on the other side by the eardrum. Even though it is curved, we will model it as a straight tube, a tube open on one side and closed on the other (Fig. 10.28). With $L = 2.5\,\mathrm{cm}$, the fundamental mode has wavelength $=$ $4L = 10\,\mathrm{cm}$. The fundamental frequency $f = v_\mathrm{s}/4L$ ((10.39) with $m = 0$), and with $v_\mathrm{s} = 343\,\mathrm{m/s}$ for air, we see that $f = (330\,\mathrm{m/s})/10\,\mathrm{cm}= 3,430\,\mathrm{Hz}$. This is very close to the peak auditory sensitivity of humans, which is near

3,000 Hz (Fig. 10.29). This makes sense because sounds near 3,400 Hz are more efficiently transported from the outer ear to the inner ear. This canal has a low Q and a rather broad resonance for several reasons, including that it is not perfectly closed.

A model of the admittance of the ear canal is described later.

Middle Ear

Why is there a middle ear? Perhaps, it is better to ask a related question that is easier to answer: *Why does hearing become poor if the middle ear is damaged?* Without the tympanic membrane and conduction through the ossicles, sound from the external canal would be directly incident on the oval window of the cochlea and hearing sensitivity would decrease by almost 30 dB, because the oval window has a much smaller area than the tympanic membrane. This is enough of a loss to make a medium loud voice barely perceptible.

The eardrum is tilted at an angle of about 55° to the ear canal. It has an average thickness of about 0.074 mm, and is a bit taller (≈9–10 mm) than it is wide (≈8–9 mm). The malleus is ≈8–9 mm long and has a mass of 25 mg. The handle (*manubrium*) of this hammer is embedded in the eardrum by fibrous and mucous membrane layers, and is nearly upward – at 1 o'clock in the right ear and 11 o'clock in the left ear (Fig. 10.27). The incus is ≈7 mm long and has a mass of 30 mg, and looks more like a tooth with a body and two roots than an anvil. The head of the malleus connects to the body of the incus through the fairly rigid *incudomallear articulation*, which is a double-saddle joint. The longer root (or process) of the incus ends in a rounded module that forms the ball in the connection to the head of the stapes (the *incudostapedial* joint). The stapes looks like a stirrup, is ≈3.5 mm long, has a mass of 3–4 mg, and has a footplate of ∼3.2 mm^2 area that attaches to the oval window by the annular ligament.

In direct conduction most of the incident sound would be reflected, and very little would be transmitted for analysis by the cochlea. Acoustically, the oval window and the cochlear fluid can be approximated as being the same as water. (This is an oversimplified model. Nonetheless, cochlear fluid gives the dominant contribution to the impedance of the cochlea above 1 kHz [486].) With $Z_1 = Z_{\text{air}} = 413\,\text{kg/m}^2\text{-s}$ and $Z_2 = Z_{\text{water}} = 1.48 \times 10^6\,\text{kg/m}^2\text{-s}$ (Table 10.1), $Z_2/Z_1 = 3{,}580$ and the fraction of sound intensity transmitted into the cochlea is ((10.34) and (10.35))

$$T = \frac{4Z_2/Z_1}{(1 + Z_2/Z_1)^2} = \frac{4 \times 3{,}580}{(1 + 3{,}580)^2} = 1.1 \times 10^{-3} = 0.11\%. \tag{10.52}$$

This is a very large loss. Expressed in decibels it is a $|10 \log_{10}(I_{\text{coupled}}/I_{\text{in}})| = |10 \log_{10} 0.012| = 10 \times 2.95 = 29.5\,\text{dB}$, or about a 30 dB loss – which is very big indeed.

How does the middle ear improve this? Using jargon common in electronics, it is often claimed that the middle ear reduces the impedance mismatch by "impedance matching." This is not true. Even with the middle ear, most of the sound incident from the air on the higher impedance eardrum/ossicles/cochlear window/cochlear fluid is lost by reflection. The middle ear does improve hearing by increasing the pressure incident on the cochlear oval window via "funneling" the sound incident on the larger eardrum to the smaller cochlear window. There are several physical arguments for this (which we will present soon).

The importance of the middle ear in improving hearing is seen by the large hearing loss (~30 dB) that results when the hammer, anvil, and stirrup are not in contact with each other in series. The *acoustic reflex* is a mechanism that intentionally (and temporarily) reduces this transmission through the ossicular system to limit the pressure on the oval window to protect the cochlea from very loud sounds. The stapedius and tensor tympani muscles respond in 40–80 ms as a reflex action to such loud sounds. The stapedius muscle pulls the stapes outward and the tensor tympani muscle pulls the handle of the malleus inward [476]. This mechanism also reduces low frequency sounds that often constitute background noise in a loud environment, so that information-bearing high-frequency sounds can be detected. The importance of the middle ear is also seen by one mode of permanent hearing loss, of ~30 dB, that occurs when these bones are permanently not in contact to the adjacent ossicle(s).

One mechanism of middle ear conduction is the transmission of the force on the large area eardrum (of total area ~85 mm^2, of which only 2/3 of the area vibrates, so its effective area is $A_\mathrm{e} \sim 55\,\mathrm{mm}^2$) to the stapes, which has a smaller area ($A_\mathrm{s} \sim 3.2\,\mathrm{mm}^2$) in contact with the oval window. The force on the eardrum is $F_\mathrm{e} = P_\mathrm{e}A_\mathrm{e}$, where P_e is the sound pressure on it (Fig. 10.28). If this force is directly transmitted to the stapes then

$$F_\mathrm{e} = P_\mathrm{e}A_\mathrm{e} = F_\mathrm{s} = P_\mathrm{s}A_\mathrm{s} \tag{10.53}$$

or

$$P_\mathrm{s} = \frac{A_\mathrm{e}}{A_\mathrm{s}}\,P_\mathrm{e} = \frac{55\ \mathrm{mm}^2}{3.2\ \mathrm{mm}^2}\,P_\mathrm{e} = 17P_\mathrm{e}. \tag{10.54}$$

With the area of the stapes on the oval window being approximately the area of the oval window, the eardrum is seen to be a much bigger sound-collection device than the smaller oval window.

This force-funneling mechanism could occur with only one ossicle connecting the eardrum and oval window. A second mechanism relies on the geometry afforded by having more than one bone in the ossicular chain. This geometry is such that the chain acts as a lever, in which torques – and not forces – are actually transmitted to the oval window. With the first lever arm being L_1 and the second L_2, torque balance gives $F_\mathrm{e}L_1 = F_\mathrm{s}L_2$ and $P_\mathrm{e}A_\mathrm{e}L_1 = P_\mathrm{s}A_\mathrm{s}L_2$ or

$$P_\mathrm{s} = \frac{A_\mathrm{e}}{A_\mathrm{s}}\,\frac{L_1}{L_2}\,P_\mathrm{e} = (17)(1.3)P_\mathrm{e} = 22P_\mathrm{e}, \tag{10.55}$$

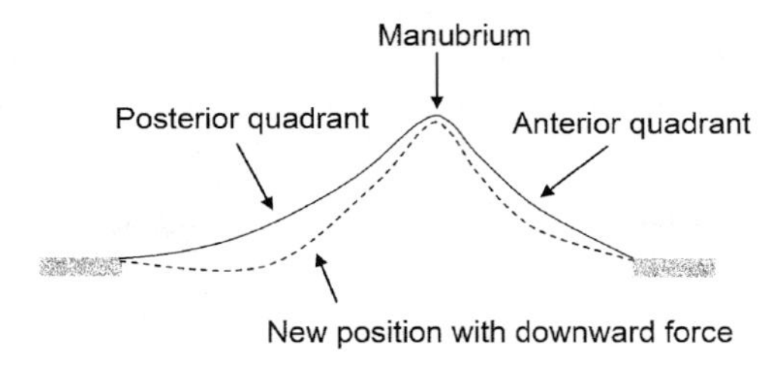

Fig. 10.31. Curved membrane buckling principle for the eardrum, with the position of the manubrium of the malleus shown. The membrane is shown without a force (*unbroken lines*) and with a (downward) force (*dashed lines*). (Based on [470, 505])

using $L_1/L_2 = 1.3$. Improved hearing due to this lever mechanism seems to be one reason why nature has developed an ossicular chain of bones.

These two mechanisms in the middle ear improve conduction of sound by $20 \log_{10}(P_\mathrm{s}/P_\mathrm{e}) = 20 \log_{10} 22 = 27\,\mathrm{dB}$. This is roughly equal to the 30 dB loss that is incurred by reflection losses from air to the cochlea; this near-equality may just be a coincidence of nature.

Other reasons for the improvement of hearing by the middle ear are sometimes cited, such as the curvature of the tympanic membrane, which leads to the *curved membrane buckling* mechanism (Fig. 10.31). The manubrium of the malleus is attached to the center of the curved eardrum, and as a consequence of this placement it moves back and forth less than (and about half as much as) the average back and forth displacement of the eardrum. The argument continues that the force on the manubrium is double that of the membrane because the same torque is transmitted from the eardrum to the malleus and the manubrium is moving half as much as the membrane. Some think the curved membrane buckling effect doubles the overall middle ear mechanical advantage from 22 to about 44; if it does, this leads to a total increase of 33 dB due to the middle ear.

We will now analyze the vibration of an eardrum under tension, which is a vibration that is normal to a 2D surface, by modeling it as the vibration of a string under tension, which is a vibration normal to a 1D object. Using (10.41), the fundamental frequency of a string under tensile stress σ that is rigidly fixed at both ends is $(1/2L)\sqrt{\sigma/\rho}$, for mass density ρ and length L. For a circular membrane of radius a, we substitute $2a$ for L, to get $(1/4a)\sqrt{\sigma/\rho}$. The fundamental frequency of transverse vibration of a circular membrane of radius a, thickness d, and mass density ρ under tension T is actually

$$f_\mathrm{fund} = \frac{2.405}{2\pi a}\sqrt{\frac{\sigma}{\rho}}, \tag{10.56}$$

where the stress $\sigma = T/d$. (The derivation of this exact result is beyond our scope.) The agreement with the simple string model is surprisingly good. Consider an eardrum with a diameter of 8 mm, a thickness of 20 μm, and mass density $\rho = 1,000\,\text{kg/m}^3$. For it to have a fundamental resonance at 1,000 Hz, the stress must be $1.1 \times 10^5\,\text{N/m}^2$ and the tension must be 2.2 N/m. One would expect the eardrum to have a resonant frequency roughly near the 3,000 Hz peak auditory sensitivity and the resonance of the ear canal.

Inner Ear

The vibration transmitted to the stapes is transmitted to the fluid in the cochlea by the stapes being in contact with the oval window. The vibration in the cochlear fluid induces motion which produces an electrical signal that causes the release of a chemical transmitter (Chap. 12). This transmitter excites nerve cells that are connected to the brain. Models of the cochlea are described in [448, 456, 459, 499]. (Much ground-breaking work on the cochlea was performed by Georg von Békésy, who was awarded the Nobel Prize in Physiology or Medicine in 1961 for his discoveries of the physical mechanism of stimulation within the cochlea.)

Cochlear Tube Vibrations

The cochlea can be viewed as a tube rolled into 2 3/4 turns. Uncoiled, as shown in Fig. 10.27, it would be a cylinder ~2 mm in diameter and 35 mm long. It has an "upper half" tube, the *scala vestibuli*, and a "lower half" tube, the *scala tympani*, that contain a fluid called *perilymph* and are separated by a cochlear partition consisting of the *scala media* and the *organ of Corti*, which contain a fluid called *endolymph* (Fig. 10.32). The scala vestibuli and scala tympani flow into each other at the *helicotrema*, which is at the end of the tube at the apex. The organ of Corti contains the hair cells and abuts the scala tympani. Sound is transmitted by the stapes into the scala tympani through the oval

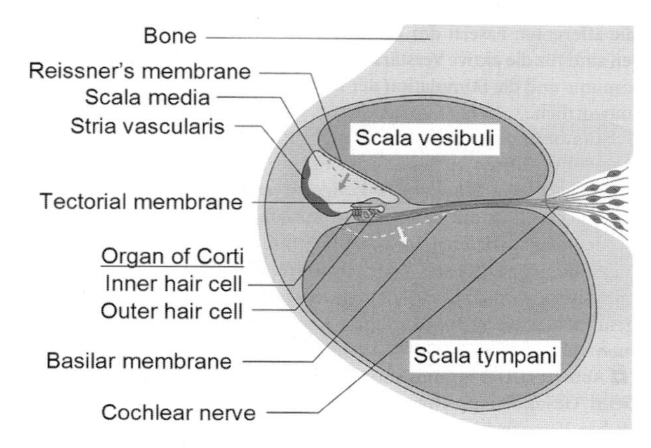

Fig. 10.32. Cochlea cross-section. (The cochlear nerve combines with nerves from the semicircular canals to form the auditory nerve.) (From [497])

Fig. 10.33. Scanning electron micrograph of a single outer hair cell bundle from a mammalian cochlea. (From [477]. Used with permission)

window and the coupling between the scala tympani and scala vestibuli sets the *basilar membrane* (between the scala tympani and the hair cells) moving transverse to the membrane, thereby exciting the hair cells (Fig. 10.33). This motion is like the traveling wave set in motion by the snapping of a rope.

There are 3,500 flask-shaped *inner hair cells* and 12,000 tube-like *outer hair cells*, which physically differ in how the hair in the cells protrude [465]. The inner hair cells relay information about the sounds transmitted to the cochlea to the brain. The outer hair cells increase the amplitude and frequency selectivity of the vibration of the basilar membrane for low-level sounds. About 95% of the auditory nerve cells receive their signals from the (relatively fewer) inner hair cells because each inner hair cell connects to about 8–30 auditory nerve fibers, while several outer hair cells connect to the same nerve fiber (Fig. 10.34). When the hair bends in one direction it depolarizes and in the opposing direction it hyperpolarizes, and this alternation causes the hair cell to release the neurotransmitter. (See Chap. 12 for a description of polarization.) Motions as small as 10^{-10} m can generate a response in a hair cell.

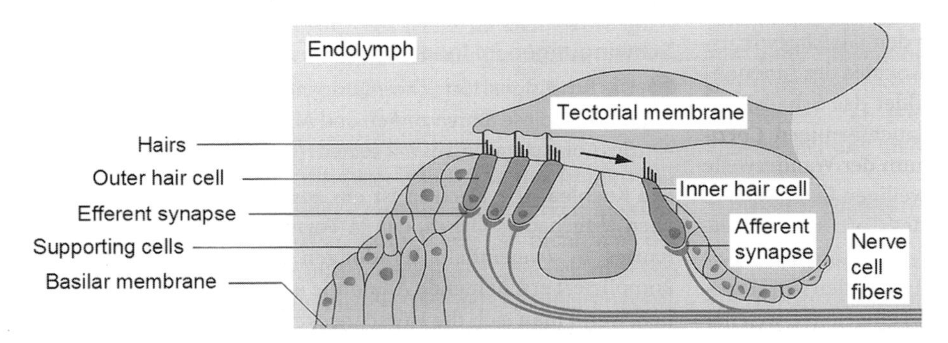

Fig. 10.34. Inner and outer hair cells in Organ of Corti. (From [497])

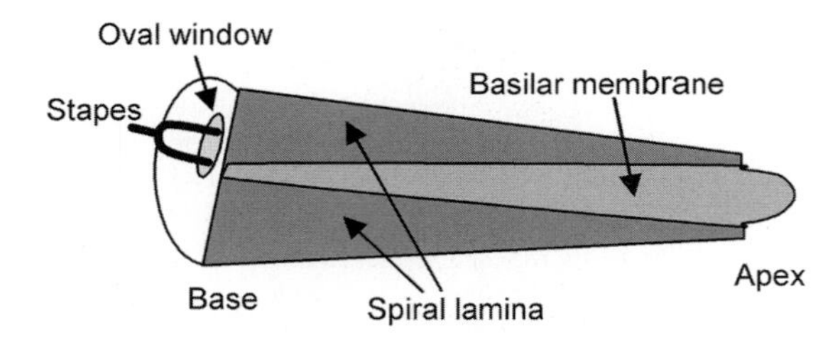

Fig. 10.35. The uncoiled cochlea gets narrower from the base at the stapes to the apex, as the basilar membrane gets wider. (Based on [472, 498])

The basilar membrane becomes 3–4× wider and a hundred-fold less stiff along the "uncoiled" cochlea from its beginning near the stapes (the base) to its end (the apex) (Fig. 10.35). Both of these characteristics help resolve the frequency components of the sound wave in such as way that the higher frequency components have larger amplitude vibrations nearer the stapes and the base, while the lower frequency components have larger amplitude vibrations nearer the apex (Figs. 10.36 and 10.37). This spatial separation of the

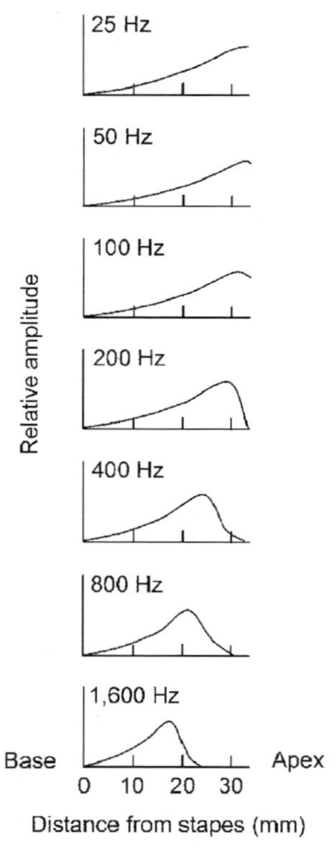

Fig. 10.36. Envelope of basilar membrane vibration vs. frequency. (Based on [453, 472])

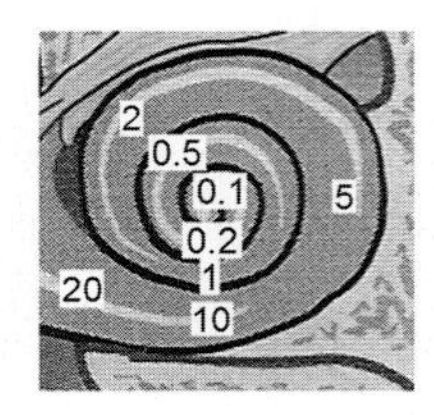

Fig. 10.37. Tonotopic map of cochlear sensitivity to audio frequencies (shown in kHz). (From [497], and based on [457, 465, 472])

frequency components means that the hair-cell-excited neurons progressively farther down the cochlea are selectively excited by progressively lower frequency components, and these frequency analyzed signals are sent as such to the brain. This frequency analysis is improved further by the changing characteristics of the hair cells along the cochlea, because the resonant frequencies of the hair cells are tuned to match the changing optimized local frequency. This is called a *place code* for frequency. There is also a *time code*: the higher the acoustic frequency, the higher the neural firing rate. (Because the maximum firing rate of these neurons, 500 Hz, is too slow for a single neuron to fire at most moderate to high acoustic frequencies, several neurons cooperatively fire. For instance, for a 3,000 Hz wave, this *volley principle* suggests that six neurons could fire successively, with say neuron #1 firing at cycles $1, 7, 13, \ldots$, neuron #2 firing at cycles $2, 8, 14, \ldots$ and so on.)

The higher the acoustic intensity, the faster, in general, the rate auditory neurons fire. However the dynamic range of response is only about 40 dB, and most fibers reach saturation above 40–50 dB SPL and do not fire at faster rates at higher intensities. Higher intensities are recognized as such (possibly) because some neurons do not begin to fire until the mid-intensity range or some neurons are located in places where they do not fire much until the intensity is high.

We could try to calculate the resonance modes of the basilar membrane by analyzing the waves in a structure that looks like an uncoiled cochlea. One simplified model is a tapered circular-sector membrane of uniform stiffness, as is shown in Fig. 10.38. Obtaining the analytical solution for the vibrations of this structure is beyond the level of this text; it is plotted in this figure. (Also see [491, 492].) The waves in this model semiquantitatively agree with the data in Fig. 10.36, the lower the frequency of the wave, the closer its maximum is to the apex. Inclusion of the stiffness variation in the model, requires a numerical solution, and this is expected to accentuate the spatial separation even more.

The outer hair cells respond to vibration much as the inner hair cells, but contribute little to the signals sent to the brain. These cells become shorter when they are depolarized and longer when they are hyperpolarized, by as much as 4% when they are isolated. This *cochlear amplifier* mechanism may help magnify the basilar membrane motion and improve the frequency selectivity of the auditory detection process, as is illustrated in Fig. 10.39.

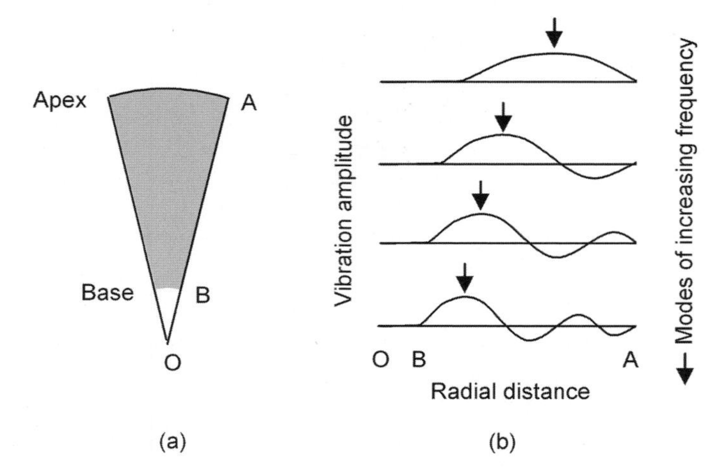

Fig. 10.38. Mechanical vibration model of the uncoiled cochlea in Fig. 10.35 modeled as a tapered membrane with uniform stiffness in (**a**), with predictions of the first few modes in (**b**). The principal maxima (at the positions of the arrows) of the modes move to the base with increasing frequency, which agrees with the trend in Fig. 10.36. For an untapered membrane, the modes would look the same except each antinode (maximum and minimum) in the mode would have the same magnitude. (Based on [466])

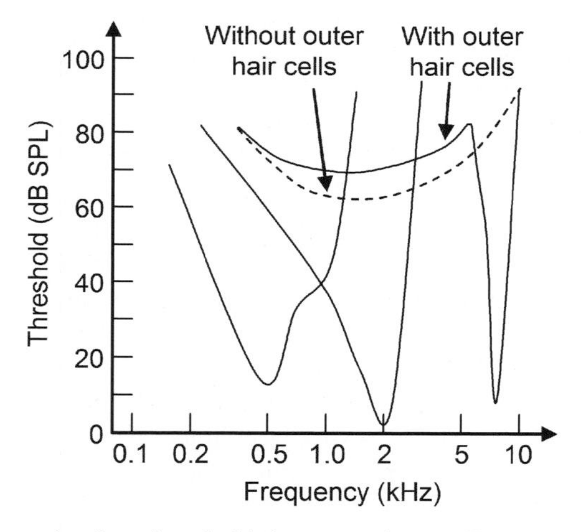

Fig. 10.39. The excitation threshold for nerve hair cells or auditory nerve fibers with three different resonant or characteristic frequencies. Without outer hair cells, the sound threshold is much higher and the frequency selectivity is much poorer, as is illustrated for the fiber with the highest of the three resonant frequencies. (Based on [465])

Hair Cell Vibrations

We will first try to model the transverse vibrations of hair fibers of length L (in hair cells) as the acoustic longitudinal vibrations of a tube that is closed at one end (into the body of the cell) and open on the other (to sense the vibrating fluid). (This is also analogous to the transverse vibration of a string that is fixed on one end and free on the other.) Using (10.38) and (10.39), the resonant wavelengths are $4L$, $4L/3$, $4L/5$, ... and the frequencies are $v_s/4L$, $3v_s/4L$, $5v_s/4L$, ... With $v_s = \sqrt{Y/\rho}$ the lowest frequency is $v_s/4L = (\sqrt{Y/\rho})/4L$. However, this is really the resonant frequency for longitudinal vibrations across a wide bar of length L, with one face fixed and the other free.

Transverse oscillations of a bar are quite different [491, 492]. For a cylinder of length L and radius a, the lowest resonant transverse frequency is

$$f_{\text{fund}} = \frac{\pi a \beta^2}{4L^2}\sqrt{\frac{Y}{\rho}}, \tag{10.57}$$

where $\beta = 0.597$. (The derivation of this is beyond our scope.) This assumes the bar is attached to a base with no compliance. As the base becomes more compliant, the value of β decreases, down to $\sim$0.35 for equally compliant bases and rods, down to $\sim$0.2 for bases that are 10-fold more compliant than the rod, and so on. Assuming hair is composed of keratin with $Y = 2 \times 10^9$ Pa and density $\rho = 1{,}000\,\text{kg/m}^3$, the lowest resonant frequency of a hair of length 2 mm and diameter 20 µm that is firmly mounted in an incompliant base is 1.0 kHz. (The cell is likely a bit compliant, so its resonant frequency is likely a bit lower.)

Admittedly, this model is perhaps too simplified. Several features should be added to it, including the viscosity of the cochlear fluid and physical connections between the hair cells. Moreover, it is not clear if there is a connection between hair cell vibration frequencies and auditory frequency selection in mammals.

The Immittance of the Human Ear (Advanced Topic)

We have analyzed the mechanical behavior of several components of the outer, middle, and inner ears. *How do these and other components combine to determine the acoustic immittance (impedance, Z, or admittance, $Y = 1/Z$) of the ear?* The stiffness (the stiffness (or negative) reactance, X_s or, equivalently, the stiffness (compliant) susceptance, B_s) comes from the air in the middle and outer ear spaces, the eardrum, and the tendons and ligaments in the ossicular chain. The inertial mass terms (the mass (positive) reactance, X_m, or the mass susceptance, B_m) come from the ossicles, the pars flaccida part of the eardrum, and the fluid in the scala vestibuli and scala tympani in the cochlea – the perilymph. The resistive, frictional component (the resistance, R, or the conductance, G) is produced by the perilymph, the mucous membrane linings of the middle ear, the narrow passages between the middle ear and mastoid air cavities, the eardrum, and the tendons and ligaments in the middle ear, and is mainly due to the absorption of the energy of the traveling wave by the cochlea.

It is easier to work with admittances here than with impedances because they are additive. (Why?) The admittance is measured with a probe that sends a 85 dB SPL signal, usually at 226 or 220 Hz, into the outer ear, and monitors the reflected signal with a microphone [470, 488]. This is all performed as a function of air pressure in the ear canal, with this (gauge) static (and not sound) pressure varied in time from about +200 daPa to −300 daPa. (The unit dekapascals (daPa) is common in audiology, with 1 daPa = 10 Pa = 0.075 mmHg = 1.02 mmH$_2$O.) Middle ear pathologies and middle ear muscle contraction (the acoustic reflex) can be analyzed with these *tympanograms*. The middle ear admittance, $Y_{\text{middle ear}}$, is determined by measuring the total admittance of the outer and middle ears at ambient pressure (the eardrum is relatively acoustically transparent then) to get $Y_{\text{total}} = Y_{\text{outer ear}} + Y_{\text{middle ear}}$, and subtracting from that the admittance of the outer ear only. $Y_{\text{outer ear}}$ is measured by pressurizing the outer ear, because the pressurized eardrum is acoustically opaque to the middle ear. From Fig. 10.40, $Y_{\text{outer ear}} = 1.0$ mmho (its value at 200 daPa) and $Y_{\text{total}} = 1.75$ mmho at 0 daPa, and we find $Y_{\text{middle ear}} = 0.75$ mmho. (At −300 daPa the admittance is actually lower than at 200 daPa, and it may provide a better value of $Y_{\text{outer ear}}$ than at 200 daPa.)

The ear canal volume expressed in cm^3 is the same number as the outer ear admittance expressed in mmho, when it is measured at 226 Hz. This

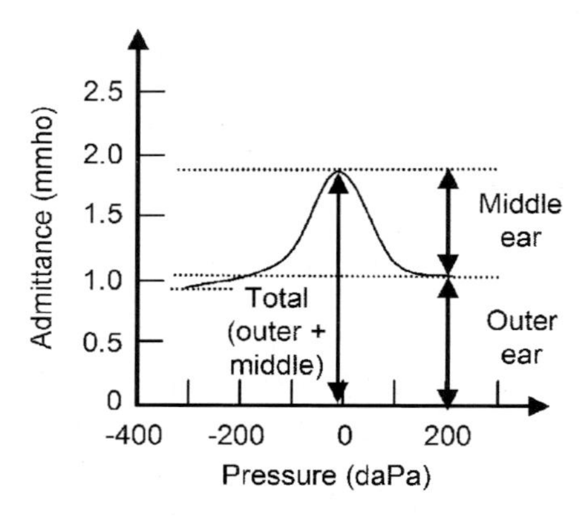

Fig. 10.40. Typical admittance tympanogram at 226 Hz, showing the different contributions to the acoustic admittance. Note that the outer ear admittance determined at −300 and −400 daPa is lower than that determined at +200 daPa, so the middle ear admittance determined at −300 and −400 daPa would be larger than that at +200 daPa. (Based on [470])

means that $Y_{\text{outer ear}} = 1.0\,\text{mmho}$, which indicates that the ear canal volume is $1.0\,\text{cm}^3$ ($= 1.0\,\text{mL}$). Is this a coincidence or something that is deep? We will now see that this is in part a coincidence of the natural units and in part caused by choosing $f = 226\,\text{Hz}$.

A Model of Air Vibrating in the Ear Canal (Advanced Topic)

How can we determine the admittance of the outer ear Y_{outerear}? Let us model the ear canal as a closed tube of volume $V = AL$, where A is its (constant) cross-section and L is its length [475, 491, 492]. Initially, the internal and external pressures are the same, P. If this tube is squeezed, the pressure inside increases by an amount ΔP and this will lead to a restoring force $(\Delta P)A$, like the restoring force of any spring-like material.

The ideal gas law (7.2) is $P = nRT$, where P is the pressure, n is the density ($= N/V$, where N is the total number of molecules in the volume V), R is the gas constant, and T is the temperature. If the volume were distorted to $V + \Delta V$, with the number of molecules kept constant and the temperature also constant (which is called an *isothermal* condition), the pressure would change to $P + \Delta P$ with PV and $(P + \Delta P)(V + \Delta V)$ both equal to NRT and to each other. So

$$PV = (P + \Delta P)(V + \Delta V) \simeq PV + (\Delta P)V + P(\Delta V), \tag{10.58}$$

where the term $(\Delta P)(\Delta V)$ is small and has been neglected. This gives

$$\Delta P = -\frac{P(\Delta V)}{V} = -\frac{P(\Delta L)}{L} = -\frac{nRT(\Delta L)}{L} \tag{10.59}$$

using the change in volume, $\Delta V = A(\Delta L)$, assuming area A is constant in the distortion and $P = nRT$.

It is conventional to normalize all terms by A^2, including the driving force $(\Delta P)A$, so

$$\frac{(\Delta P)A}{A^2} = \frac{\Delta P}{A} = -\frac{nRT}{V}(\Delta L) = -\frac{nRT}{V}x = -\frac{P}{V}x, \tag{10.60}$$

where now ΔL is called x. The coefficient $nRT/V = P/V$ is the spring constant k in (10.21), which was also called S in the earlier analysis.

We have assumed that the temperature does not change during the squeezing. If we instead assumed the more realistic condition that heat flow across the tube walls can be neglected during the squeezing and response (which is called an *adiabatic* condition), then

$$k = \frac{\gamma nRT}{V} = \frac{\gamma P}{V}, \tag{10.61}$$

where γ is the parameter in (10.2). (This is shown in Problem 10.59.) With $k = S$ and $X_\mathrm{s} = S/\omega = S/2\pi f$

$$X_\mathrm{s} = \frac{\gamma P}{\omega V} = \frac{\gamma P}{2\pi f V}. \tag{10.62}$$

When this is the main term in the magnitude of the impedance, then $\mid Z \mid = X_\mathrm{s}$ and the magnitude of the admittance is then $\mid Y \mid = B_\mathrm{s} = 1/X_\mathrm{s}$, giving

$$\mid Y \mid = B_\mathrm{s} = \frac{2\pi f}{\gamma P} V. \tag{10.63}$$

An ambient pressure P of 1 atm. = 0.1 N/mm^2 = 1×10^6 g/cm-s^2. For air $\gamma = 1.4$ and so at 226 Hz we find that $2\pi f/\gamma P = (2\pi)(226\,\mathrm{Hz})/(1.4\,(1 \times 10^6\,\mathrm{g/cm\text{-}s^2})) = 1.0 \times 10^{-3}$ cm-s/g. Because 1 mho = 1.0 cm^4-s/g and 1 mmho = 1.0×10^{-3} cm^4-s/g, we see that a volume V = 1.0 cm^3 has an admittance of 1.0 mmho. Using the scaling arguments developed in Chap. 8 (8.11), the admittance of the outer ear expressed in mmho gives the volume of the ear canal expressed in cm^3.

Hearing Loss

Sound levels above 85 dB SPL are considered harmful, those above the pain threshold of 120 dB SPL are unsafe, and those above 150 dB SPL cause physical damage to the human body. To avoid hearing damage, safety organizations recommend exposure to no more than 85–90 dB SPL for 8 h a day, 100 dB SPL for 2 h a day, or 110 dB SPL for 30 min a day. Eardrums rupture at 190 dB SPL to 198 dB SPL and sound levels of around 200 dB SPL can cause death to humans. Such levels are generated near bomb explosions, such as 3 m away from the detonation of 23 kg of TNT. Other potential causes of hearing loss are suggested in the cartoon in Fig. 10.41.

The main way hearing loss is evaluated is by determining the threshold for hearing pure tones from about 125 Hz to 8,000 Hz. The hearing level is usually expressed as the threshold value of hearing expressed in dB HL vs. audio frequency, with the clinical plot showing increasing threshold intensities toward lower ordinate values (Fig. 10.42b), rather than toward higher ordinate values, as in Figs. 10.29 or 10.42a. HL stands for "hearing level." At each frequency, the threshold expressed in dB HL is that in dB SPL minus the reference normal value in dB SPL. This accounts for the variation in threshold intensity with frequency that is seen in Figs. 10.29 and 10.42a. These reference values are about 7.5 dB SPL from 750 to 1,500 Hz. At lower frequencies they increase to $\simeq$13 dB at 500 Hz, 26 dB SPL at 250 Hz, and 47 dB SPL at 125 Hz. At higher frequencies they increase to $\simeq$10.5 dB SPL from 2,000 to 4,000 Hz and 13.5 dB SPL from 6,000 to 8,000 Hz.

Normal hearing is therefore defined as 0 dB HL at all "audible" frequencies (top, flat curve in Fig. 10.42b), but people with an average frequency

Fig. 10.41. Potential sources of hearing loss. (For sure, the current reader of this book does not share the complaint of the "dude" on the right. Still, we should interpret the interaction between the dude on the right and his mother from a physics perspective. His mother wants him to decrease the entropy (or disorder) of his room. He has (unstated) concerns about this because he will need to use energy to decrease the room entropy and he knows that the overall entropy of the universe would increase as a result of any effort on his part to clean up his room.) (From Beattie; Daytona Beach News-Journal, Copley News Service, NJ-CENTER.com, copyright 1999. Reprinted with permission)

response $\leq$15 dB HL are still said to have "normal hearing." Hearing loss is defined as slight (average value of 16–25 dB HL), mild (26–40 dB HL), moderate (41–55 dB HL), moderately severe (56–70 dB HL), severe (71–90 dB HL), and profound (>90 dB HL). An example of a real audiogram for a person with normal hearing is shown in Fig. 10.43. Figure 10.44 shows how hearing sensitivity decreases with age, which is called *presbycusis*. One such example is plotted in dB HL in Fig. 10.45.

Modes of hearing loss can be caused by problems in the middle ear or the inner ear. The regular route of receiving sound from the outer ear to the inner ear is called *air-conduction*. Sound can also be conducted to the inner ear by *bone-conduction*, in which sound is transmitted by vibrations of the bones of the skull. Both induce the same cochlear activity. In clinical audiology, air-conduction is tested by using earphones or loudspeakers, while bone-conduction is tested by a vibrator placed on the front of the skull. The modes of skull vibration change with vibrator frequency, as is seen in Fig. 10.46. At 200 Hz the skull vibrates in unison with the same phase in the forward and backward directions. When the frequency is increased to 800 Hz, the front and the back of the skull are out of phase with each other, and at 1,600 Hz there are

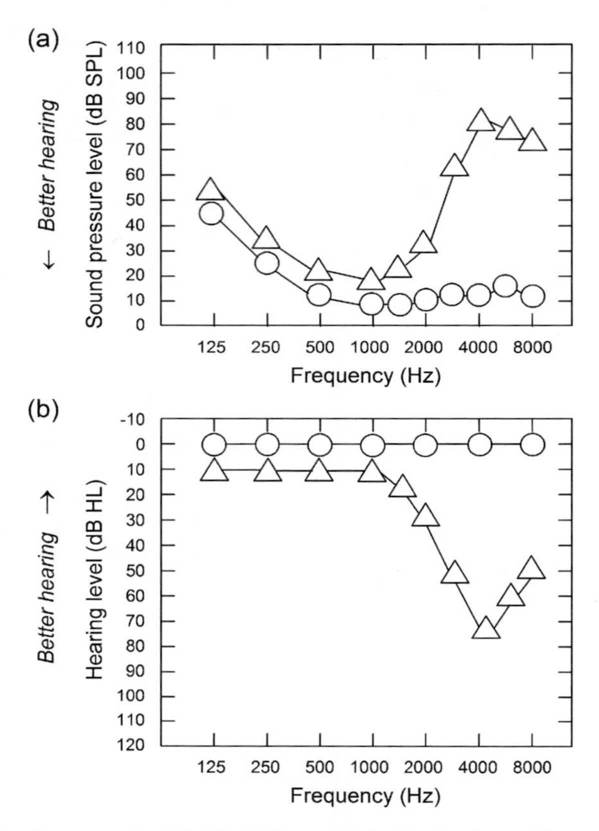

Fig. 10.42. Audiograms in (**a**) dB SPL and (**b**) dB HL units for a normal person (*circles*) and one with high-frequency hearing loss (*triangles*). (Based on [469, 470])

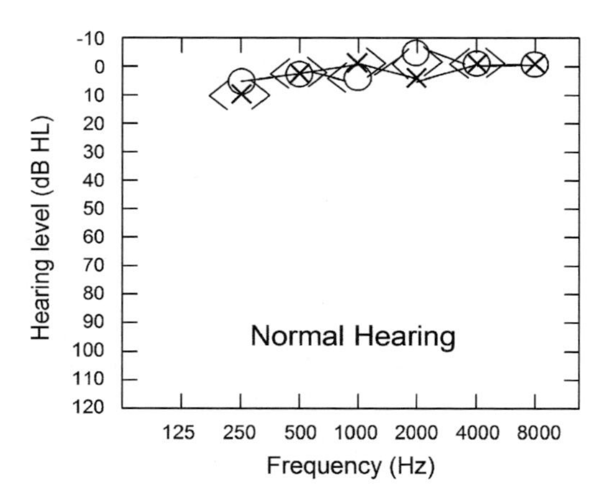

Fig. 10.43. Example of a real audiogram for a person with normal hearing. Audiogram key: right air-conduction (*open circles*), left air-conduction (*cross symbols*), right bone-conduction (*left angular brackets*), and left bone-conduction (*right angular brackets*), all unmasked. (Based on [470])

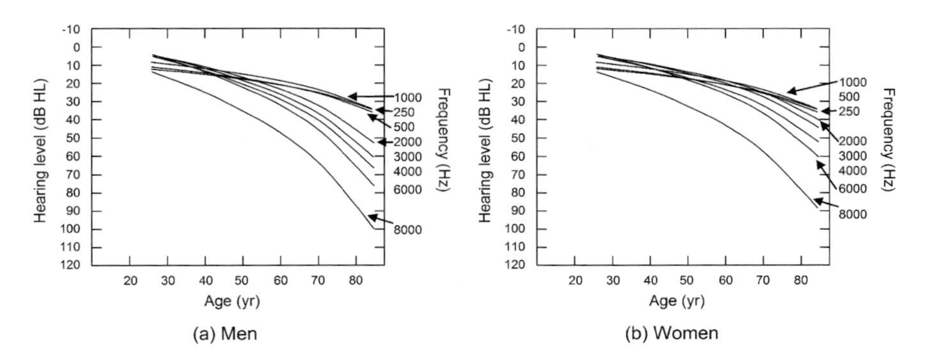

Fig. 10.44. Hearing levels vs. age for (**a**) males and (**b**) females. (Based on [470, 483, 500]. Also see [495] for more data)

forward–backward and right–left vibrations. Above ~1,000 Hz, skull vibrations lead to important hearing response by excitations of the outer, middle, and inner ear; at lower frequencies the middle and outer ear mechanisms are most important.

If someone has poor hearing due to a problem in the middle ear (or excessive wax in the ear canal), sound can still propagate through the bones in the head to the inner ear. Thresholds for conduction to the ear by air and by direct transmission to bone can be tested to distinguish between conductive and sensorineural system loss. If the threshold is the same for air-conduction and bone-conduction tests, thresholds above 0 dB HL can be attributed to a

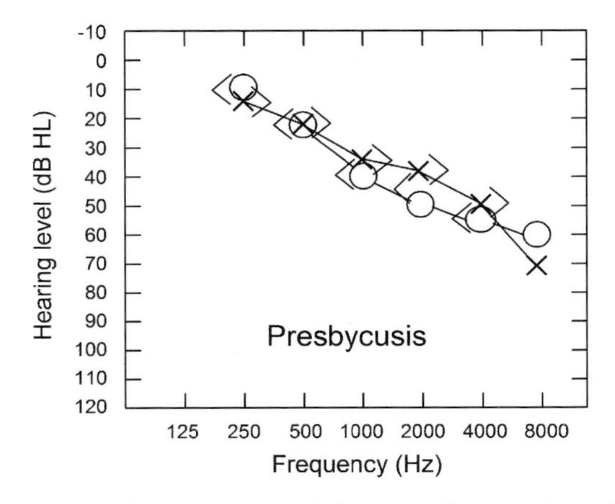

Fig. 10.45. Audiogram of a patient with bilateral hearing loss (i.e., loss in both ears) from (essentially symmetrical bilateral sloping) sensorineural hearing loss associated with presbycusis. Audiogram key: right air-conduction (*open circles*), left air-conduction (*cross symbols*), right bone-conduction (*left angular brackets*), and left bone-conduction (*right angular brackets*), all unmasked. (Based on [470])

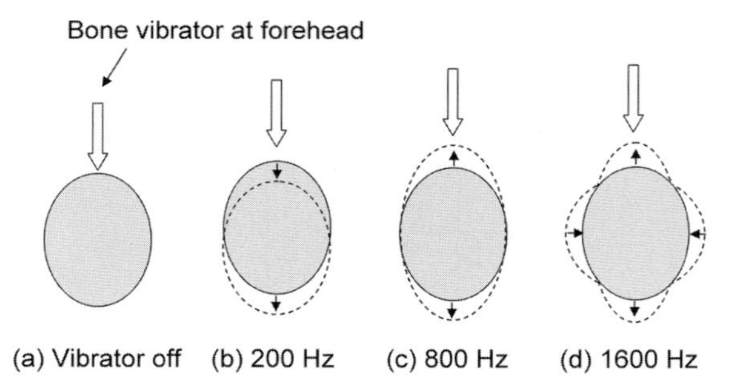

Fig. 10.46. Skull vibration patterns induced by (**a**) a bone vibrator positioned on the forehead, vibrating at different applied frequencies (**b**) 200 Hz, (**c**) 800 Hz, (**d**) 1,600 Hz. In (b)–(d) the vibrating skull (*dashed lines*) is shown relative to the still skull (*unbroken lines*) for a particular phase in the vibration. The small arrows depict the displacement, a displacement that is greatly exaggerated to illustrate it. In the opposing phase, the motion is reversed, with all arrows reversed (just as for a sine wave). (Based on [452, 470])

problem in the sensorineural part of the ear. If the thresholds are different, the difference can be attributed to a problem in the conductive system (middle ear). There can also be problems in both systems. Figure 10.47a shows the hearing response for someone with normal hearing in the left ear, and

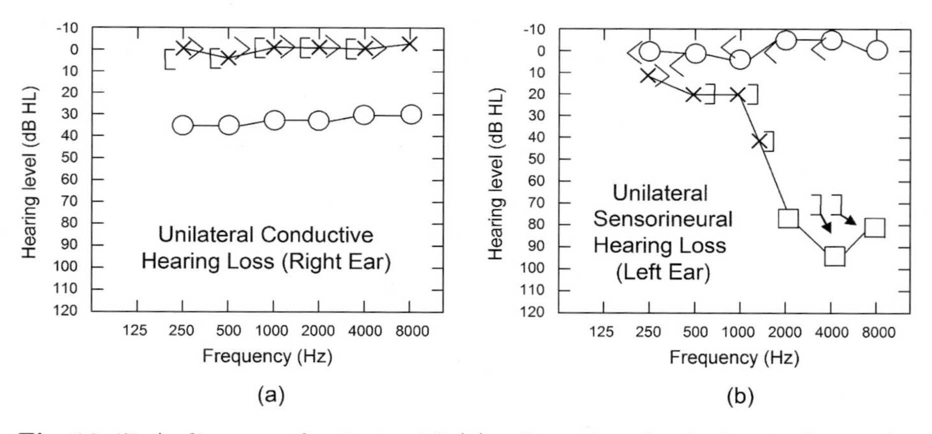

Fig. 10.47. Audiograms of patients with (**a**) unilateral conductive hearing loss in the right ear, with a normal left ear, and (**b**) unilateral sensorineural hearing loss in the left ear, with a normal right ear. Audiogram key: right air-conduction (*open circles*), left air-conduction (*cross symbols*), right bone-conduction (*left angular brackets*), and left bone-conduction (*right angular brackets*), all unmasked; left air-conduction (*open squares*), right bone-conduction (*left square brackets*), left bone-conduction (*right square brackets*), no response (*arrows*), all masked. (In masking the nontest ear is presented a constant noise signal to prevent it from detecting the signal in the test ear, so that only the test ear can respond.) (Based on [470])

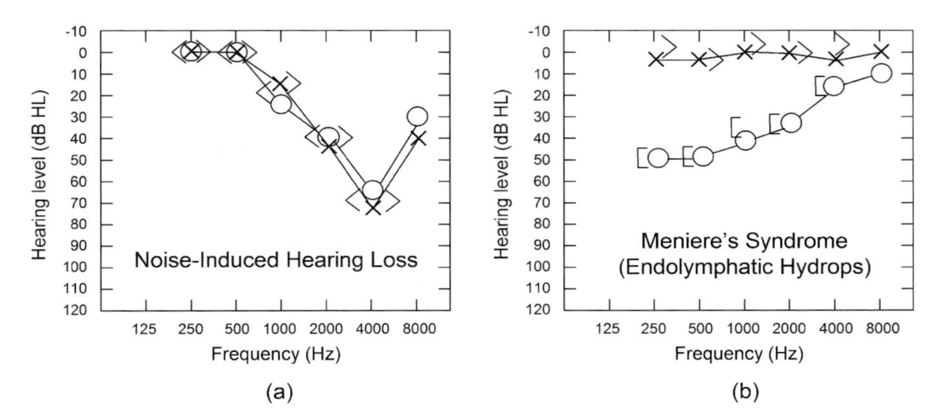

Fig. 10.48. Audiograms of patients with (**a**) bilateral sensorineural hearing loss associated with noise exposure and (**b**) Meniere's disease exhibiting low-frequency sensorineural hearing loss in the right ear. Audiogram key: right air-conduction (*open circles*), left air-conduction (*cross symbols*), right bone-conduction (*left angular brackets*), and left bone-conduction (*right angular brackets*), all unmasked; right bone conduction (*left square brackets*), masked. (Based on [470])

(unilateral, meaning in one ear) hearing loss in the right ear attributed to a problem in the right-conductive system. The sensorial system is normal, as is seen by the bone-conduction tests. Figure 10.47b shows another audiogram that indicates significant hearing loss only in one ear, the left ear, and this time it is due to hearing loss in the left sensorineural system. Here, hearing loss is seen in the left ear even with bone conduction. Note the sensorial hearing loss has a more striking dependence on frequency than the conductive hearing loss. The audiogram of a patient with hearing loss from high noise exposure (such as from too much loud rock – or classical – music) is shown in Fig. 10.48a. The loss is bilateral (in both ears) and is due to sensorineural malfunction. In contrast to these other examples, people with Meniere's disease (an inner ear disease) show sensorineural hearing loss at low frequencies (Fig. 10.48b).

Tinnitus, which is also known as "ringing in the ears," is not uncommon. Those with severe cases can perceive a tone in one or both ears of 90 dB or louder, which is louder than the 60–70 dB of normal conversation.

10.3.2 Connections to Hearing Perception

In *psychophysics* (or *psychoacoustics*), there are four main perception attributes of basic sounds: loudness, pitch, timbre, and localization [472, 510]. Each attribute is nonlinearly related to the physical characteristics of sound, such as acoustic intensity, frequency, and distribution of frequencies:

1. *Loudness* is related to the sound intensity in a nonlinear manner. As characterized by Stevens' Law((1.6), Table 1.15) the perception of loudness is very sublinear with stimulus. The lowest curve in Fig. 10.29 is the audibility curve, which denotes the threshold of hearing; this is the lowest intensity "equal loudness curve." This and the other "louder" "equal loudness curves" provide the dB SPL levels of acoustic intensity at each frequency (for pure tones, i.e., for single frequencies) needed to achieve the same level of perceived loudness. The degree of loudness of a sound is expressed as the dB SPL level at 1,000 Hz that has the same loudness. If this acoustic intensity at 1,000 Hz with the same loudness is called I, then the loudness in *phons* is

$$L_{\mathrm{p}}(\text{in phons}) = 10 \log \frac{I}{I_{\mathrm{ref,p}}}, \tag{10.64}$$

where $I_{\mathrm{ref,p}} = 1 \times 10^{-12}\,\mathrm{W/m^2}$. (This equation is clearly very similar to (10.10).) As seen in Fig. 10.49, above ~80 phons, approximately the same acoustic intensity is needed in low, mid, and high frequencies to perceive the same loudness in each region. Below ~80 phons, higher intensity is needed at low and high frequencies to achieve the same perceived loudness as in mid frequencies. This explains why one needs to increase the bass and treble in stereo tuners (such as by pushing the "loudness" button) at

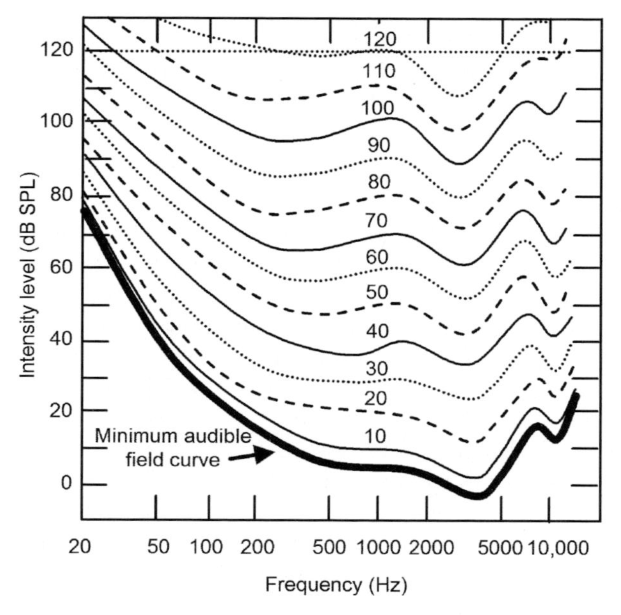

Fig. 10.49. Loudness and phon curves vs. frequency with phon (loudness level) labels. (Based on [470, 481, 494])

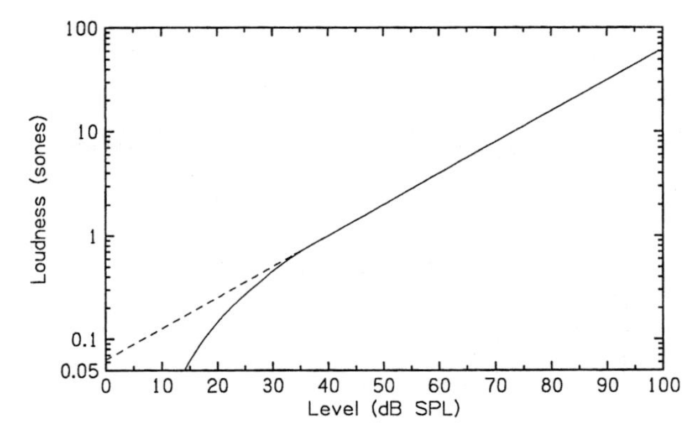

Fig. 10.50. Perceived loudness (in sones) vs. loudness level of stimulus (in phons) for a 1,000-Hz tone from [502, 510], shown as the *solid line*, along with the predictions of the approximate 0.3 power law shown as the *dashed line*, which is the basis for the national and international sone scale. (From [478])

lower intensity to perceive the same, except quieter, music. Figures 10.44 and 10.45 show that older people typically have less sensitivity at high frequencies (presbycusis). Several disorders, such as Meniere's disease, also deteriorate hearing (Fig. 10.48b).
The phon scale gives no indication of relative loudness. This is provided by the unit called *sones*, which is related to phons in Fig. 10.50. A loudness sensation of 1 sone is provided by 40 dB of 1,000 Hz pure tone. A sound of 10 sones is perceived as being twice as loud as one of 5 sones. A nearly exact formulation of loudness in sones L_s is

$$L_s(\text{in sones}) = \left(\frac{I}{I_{\text{ref,s}}}\right)^{0.3}, \tag{10.65}$$

where $I_{\text{ref,s}} = 1 \times 10^{-8}\,\text{W/m}^2$ and again I is the acoustic intensity at 1,000 Hz with the same loudness [478]. (This is a little different from the near square-root scaling of perceived loudness with the acoustic intensity indicated in Table 1.15.) Another measure of loudness is the *just noticeable difference* (JND) or *difference limen* (DL), which denotes how much more intense a sound needs must be to be perceived as being louder. For a 1,000 Hz tone, the DL for a 5 dB SPL sound is 5 dB SPL, but for 100 dB SPL it is much smaller fraction, 6 dB SPL.

2. *Pitch* denotes if a tone sounds high or low. A pressure wave that is purely sinusoidal is a pure tone; it is perceived as being low for low frequencies and high for high frequencies. The perception of pitch is expressed in units of *mels*, with the pitch of 1,000 Hz at 40 dB SPL defined to have a pitch of 1,000 mels. There is a very nonlinear relationship between mels and frequency for a pure tone (Fig. 10.51). A tone of 2,000 mels (corresponding to about 3,500–4,000 Hz) sounds like it is twice the pitch of a

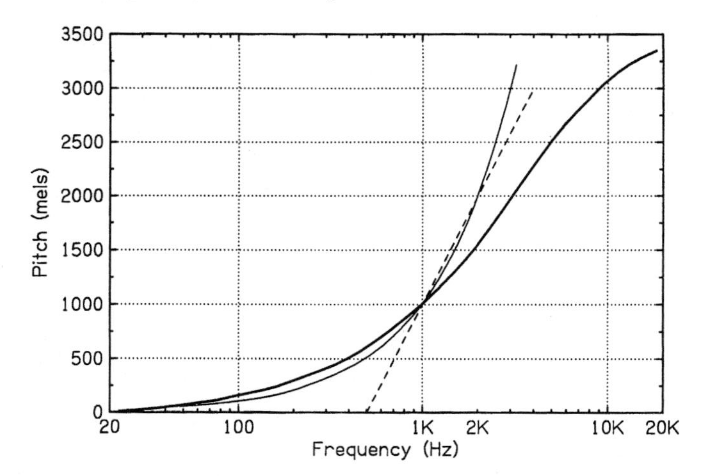

Fig. 10.51. Perceived pitch (in mels) vs. frequency [503] is shown as the *heavy line* (extending from 20 Hz to almost 20 kHz), along with the *thin line* plotting pitch equal to frequency and the *dashed line* showing pitch proportional to the octave number. (From [478])

tone of 1,000 mels, which itself sounds twice as high as a tone of 500 mels (corresponding to 400 Hz).

There is a minimum detectable frequency change for pure tones, of about 2–3 Hz for tones below 1,000 Hz and about 1/3% fractional change for higher frequency tones. At a constant loudness of 40 phons, one can perceive 1,400 different frequencies. Also, at a constant frequency of 1,000 Hz one can perceive about 280 different levels of loudness. Consequently, people can perceive almost $1{,}400 \times 280 = 400{,}000$ different variations of pitch and loudness.

The frequency increases by a factor of two progressing from one musical note to the same note in the next octave, such as in going from A, B, C, D, E, F, G, and back to A; this continual increase in frequency with successive notes is recognized as an increase in *tone height*. In going from one octave to another, such as A_5 on a piano with a fundamental frequency at 440 Hz, to A_6 at 880 Hz, to A_7 at 1,760 Hz, and so on, there is a repetition of notes that is perceptually similar, i.e., they exhibit the same properties of *tone chroma* (here specifically here that of the note A), but with different tone heights. These notes can be graphically represented as a vertical spiral with successive notes marked: each complete octave is a loop in the spiral, and all notes with the same tone chroma (all A's, all B's, etc.) are immediately above one another in successive loops.

Most musical notes have a range of frequencies, and consist of a fundamental frequency and its harmonics. Such complex tones are perceived to be at the fundamental frequency, so a 400 Hz tone and its harmonics at 800, 1,200, 1,600 Hz, and so on is perceived to have a pitch of 400 Hz.

Moreover, other complex series of frequencies are perceived to be at the minimum common difference frequency, so a complex tone with 800, 1,200, 1,600, and 2,000 Hz is perceived to have a pitch of 400 Hz, as is a complex tone with frequencies 2,000 and 2,400 Hz. These concepts are known as *periodicity pitch* or the *effect of the missing fundamental.*
Tones heard at the same time can also *mask* one other. Tones tend to mask nearby frequencies the best. Also, low-frequency tones mask high frequency tones very well, but high frequency tones do not mask low frequency tones well.

3. Complex tones containing many frequencies can be perceived as having the same loudness and pitch, but still be perceived as being different. This is characterized as being a difference in quality or *timbre*. One reason for this is that different complex tones with the same fundamental frequency consist of harmonics with different relative intensities. These different distributions are seen in Fig. 10.52 for a 196 Hz tone from a guitar, bassoon, and alto saxophone; this explains why these different instruments sound different when the same note (fundamental frequency) is played. In fact, when musical tones of the same frequency can be played in different ways on the same instrument they sound differently, such as the 440 Hz tone

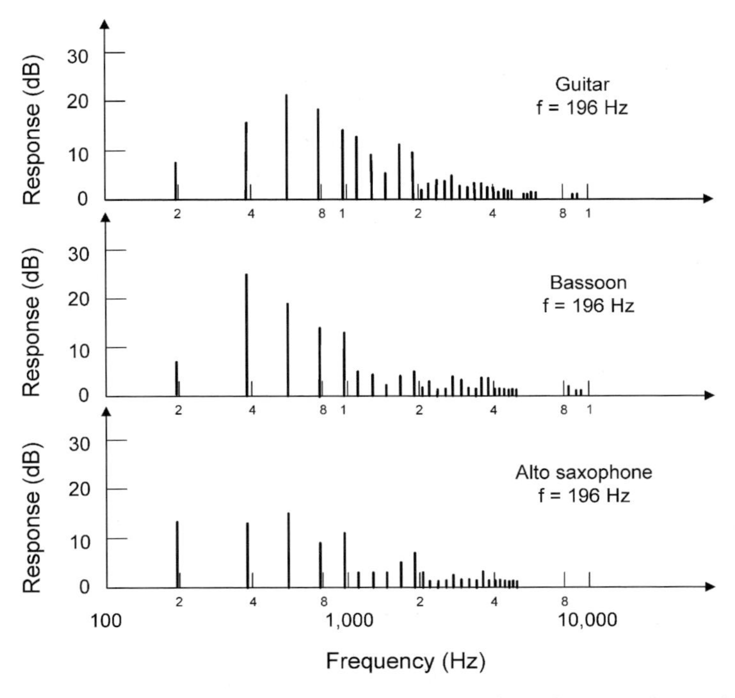

Fig. 10.52. Fourier spectra for a guitar, bassoon, and an alto saxophone playing at a fundamental frequency of 196 Hz. (Based on [492])

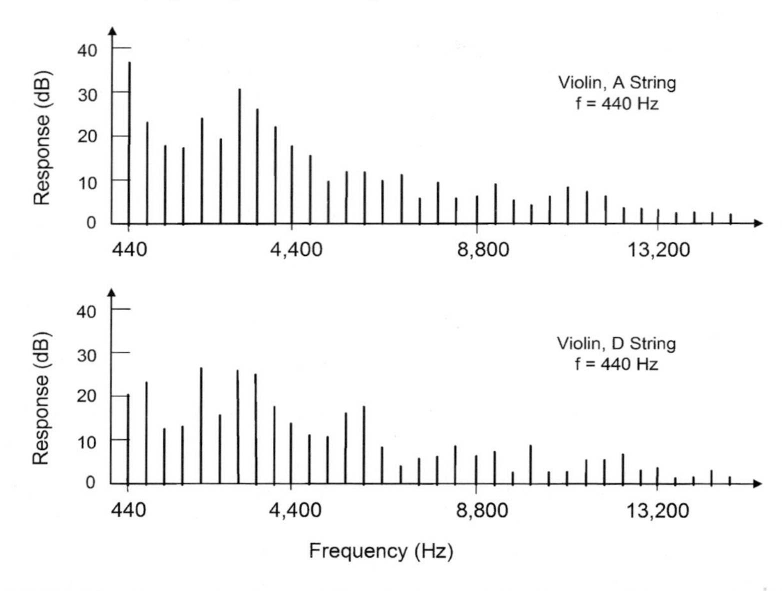

Fig. 10.53. Fourier spectra for a violin playing at a fundamental frequency of 440 Hz using the A and D strings. (Based on [492])

played on the A and D strings of a violin, as is seen in Fig. 10.53. Moreover, perception is not only based on these distinctive harmonic structures, but also on the distinctively different buildup and decay times for the various harmonics played on a given instrument.

4. Sound is perceived as coming from a given spatial location, which is called *localization.* Binaural hearing helps people localize the origin of a sound within about 10°. The brain localizes the source on the basis of information from both ears, using the difference in arrival times – with a longer propagation time to the farther ear – and intensity – with a weaker acoustic intensity arriving at the farther ear. This is examined further in Problem 10.79.

   Perception of more complex acoustic stimuli, as in speech, is much more involved. The basic elements relate to the signal processing of the frequency components.

10.4 Other Vibrations in the Body

10.4.1 Cardiac and Other Sources of Sounds

The body can be the source of many sounds, other than from speaking, such as from chewing, stomach growls, walking and running, hand clapping, hiccups, heart murmurs, body contact with anything, flapping eyelids, using towels,

Table 10.6. Estimated acoustic parameters from sources in the body. (From [467])

source	maximum pressure change (atm.)	typical frequency (Hz)
shouting	0.05	1,000
talking	0.005	1,000
whispering	0.0005	1,000
running	2.0	4
walking	0.4	1
clapping hands, vigorously	0.2	2
chewing crunchy food	0.0001	1,000
respiratory airflow turbulence	0.00004	1,000
arterial pulse	0.02	1

clicking knees and finger joints, vomiting, coughing, flatus, nose blowing, humming, whistling, sneezing, normal breathing, wheezing, and laughing. Some of these are listed in Table 10.6. The loudest of these acoustic vibrations can be sensed by the body. Some are heard and some are felt by the mechanical receptors in the body (Chap. 2). Heart sounds are described in [484] and lung sounds in [485]. Vibrations can also cause pain to be sensed in the body, as illustrated in Fig. 10.54.

Auscultation is the listening to sounds made by internal organs for medical diagnosis. Modern stethoscopes have two chest pieces for such diagnostics: a "bell" chest piece applied lightly to the skin to pick up low frequency sounds and a "diaphragm" chest piece pressed firmly to the skin to minimize low frequency sound and therefore accentuate hearing high-frequency sounds.

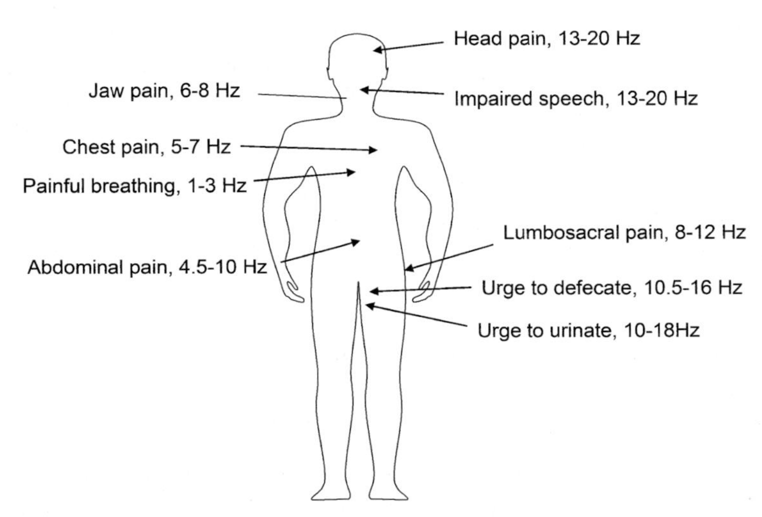

Fig. 10.54. Pain symptoms from vibrations from 1 to 20 Hz. (Based on [455, 487])

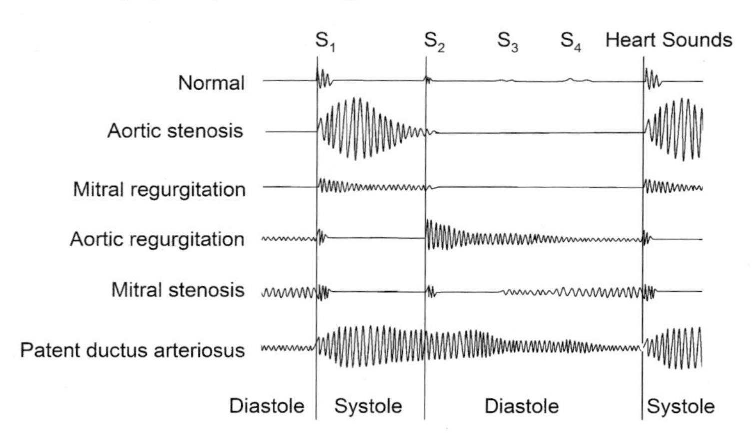

Fig. 10.55. Idealized phonocardiograms (sound traces) for normal and abnormal hearts, with left heart valve stenosis (partial blockage) or regurgitation (backflow) or with *patent ductus arteriosus*. (Patent ductus arteriosus is the condition when the ductus arteriosus (arterial shunt of pulmonary arterial blood flow from pulmonary artery to the aorta in a fetus) does not close after birth.) The times of the first (S_1), second (S_2), third (S_3), and fourth or atrial (S_4) heart sounds are shown. Also see the phonocardiogram in Fig. 8.5. (Based on [476])

Normally-functioning hearts make sounds that are easily heard with a stethoscope (Fig. 10.55) [484]. These are due to the opening and closing of heart valves and the flow of blood. The first heart sound (S_1) is a high-frequency sound heard in early systole due to the closing of the mitral and tricuspid atrioventricular valves, which occurs when the ventricular pressure exceeds the atrial pressure. The mitral valve closes about 10 ms before the tricuspid valve due to the earlier electrical excitation and contraction of the left ventricle. These different closing times are audible only when there is a disorder, such a right bundle branch block, which delays the closure of the tricuspid valve. The intensity of the S_1 sound can be accentuated or diminished by several heart disorders.

The second sound (S_2) is heard at the end of ventricular ejection and systole when the pulmonary and aortic valves close. The aortic component (A_2) precedes the pulmonary component (P_2) because the pressure gradient between the aorta and left ventricle is larger than that between the pulmonary artery and right ventricle. These two components are normally fused into one during expiration but are normally audibly split during inspiration due to the decrease in intrathoracic pressure that further delays P_2 and advances A_2. The decrease in intrathoracic pressure causes a larger pressure drop from inside to outside the pulmonary arteries and veins, and these compliance vessels increase in volume at the same internal pressure. This delay in P_2 is caused by the longer time needed to fill up the pulmonary artery. The advancing of A_2 is due to the smaller blood flow from the even-more blood-laden pulmonary veins to the left atrium and then the left ventricle, and the resulting smaller stroke volume and the smaller time needed for the left ventricle to become empty. The time of systole is approximately that between S_1 to S_2 and that of diastole from S_2 to S_1 in the next cycle. These main heart sounds, S_1 and S_2, are high-frequency sounds best heard with the stethoscope diaphragm.

The third heart sound (S_3) is a dull, low-pitch sound best heard with the bell of the stethoscope, heard after opening of the atrioventricular valves during ventricular filling; it is normal in children and young adults and abnormal in middle-aged and older adults. The fourth heart sound (S_4), which usually indicates cardiac disease, is generated by an atrium contracting against a stiffened ventricle late in diastole. *Heart murmurs* are generated by the presence of turbulent blood flow rather than laminar flow, and may indicate flow across a partial obstruction, such as a stenosis (narrowing) of the aorta, increased flow rate, blood ejection into a dilated chamber, regurgitant flow (backflow) in a defective heart valve, or abnormal shunting of blood from higher-pressure to lower-pressure chambers through a membrane divider (septum) [484]. (Remember from Chap. 8 that flow in the aorta is sometimes turbulent and can be heard.) As seen in Fig. 8.5, the loudest heart sounds have frequencies mostly below 30 Hz and are therefore inaudible. The pulsatile turbulent flow heard during the measurement of blood pressure was described in Chap. 8.

10.5 Summary

The physics of sound, i.e., acoustics, describes the properties of sound waves – including how sound intensity is expressed in the units of dB – and how sound waves propagate between different media and in cavities. Acoustic models can be used to understand how sound is produced in speech, which is partly explained by the voice filtering theory, and how to characterize the human voice. They can also be used to understand sound propagation in the outer and middle ears, how nerve signals are generated in the inner ear, and modes of hearing loss. Vibrations are important elsewhere in the body and in medical diagnostics, such as in the interpretation of heart sounds and the use of ultrasound for imaging.

Problems

Sound Waves

10.1. Find the distance molecules in air move for 20 dB SPL and 120 dB SPL sound levels, both at 500 and 5,000 Hz.

10.2. A musician with perfect pitch can identify a 1 kHz pure tone in 4 ms. How many periods of the sound wave is this?

10.3. Use (10.6) and the mass density of air of $1.3 \times 10^{-3}\,\mathrm{g/cm^3}$ to calculate the characteristic displacement of air for 0 dB SPL and 120 dB SPL for 4 kHz.

10.4. (a) Show that the speed of sound in air $v_{\mathrm{air}} \propto \sqrt{T}$ and the mass density varies as $\rho_{\mathrm{air}} \propto 1/T$, and so the acoustic impedance of air varies as $Z_{\mathrm{air}} \propto 1/\sqrt{T}$, where T is in K.
(b) Find each of these parameters at 0, 20, and 25°C.
(c) Does this variation of Z_{air} with T significantly affect the reflection and transmission of sound from air to the body components? Why or why not?

10.5. Find the speed of sound in water using the bulk modulus of water of 2.26 GPa. Is this the expected result?

Sound Intensity

10.6. The scale for acoustic intensity in Fig. 10.30 ranges from −20 to 140 dB SPL. What pressure range does this correspond to in dyne/cm^2?

10.7. Sound with intensity 60 dB SPL in air is incident on water. How much of it is transmitted into the water (in dB SPL and W/m^2)?

10.8. Hammering on a steel plate produces sounds that two feet away reach a maximum of 115 dB SPL. If this acoustic intensity is isotropic, what is the total power of this acoustic wave in W?

10.9. What is the change in dB if the intensity of a sound wave is:
(a) halved
(b) doubled
(c) tripled
(d) quadrupled?

10.10. Why there is a 6 dB SPL decrease in sound level for each doubling of distance from a small isotropic source?

10.11. The acoustic intensity is 60 dB SPL at a given distance from an isotropic source. What would the intensity be if this same level source were to radiate into only 1/10 of all space?

10.12. Which sound is more intense 20 m from its isotropic source: a 10 Hz sound that is 80 dB SPL a distance 4 m from its source or a 4,000 Hz sound that is 60 dB SPL a distance 3 m from its source?

Sound Intensity for Threshold Hearing Sensitivity

10.13. (a) At what distance from an isotropic 10 μW acoustic source is the sound at the audibility threshold for a human?
(b) The hearing threshold for dogs is $1 \times 10^{-15}\,\mathrm{W/m^2}$. At what distance can a dog hear this source?

10.14. (a) Express music's *triple forte* (very loud, $1 \times 10^{-2}\,\mathrm{W/m^2}$) and *triple piano* (very soft, $1 \times 10^{-8}\,\mathrm{W/m^2}$) sound levels in dB SPL.
(b) At an outdoor concert with no sound amplification, the audience sitting 4 m from the orchestra hears triple forte and triple piano sounds. What are the respective acoustic intensities (in dB SPL) for those in the audience sitting 60 m away? Treat the orchestra as a point source.

10.15. (a) Show that 1 W from an isotropic acoustic power radiator produces an intensity of 115 dB SPL a distance 0.5 m from the source.
(b) Find this acoustic intensity 1 m from the source.
(c) How far can you be from the source and still barely hear it (for a 1,000 Hz source)?

10.16. Determine P_{ref} and the pressure (changes) at 100 dB SPL in mmHg.

Sound Transmission, Reflection, and Ultrasound

10.17. Table 10.3 gives the absorption coefficient α for bone as 1.6×10^{-4} s/m. Other sources give it as 14 dB/cm at 1 MHz frequency. Are these two values consistent? Why?

10.18. Can you talk through a person? (In other words, are people good acoustic shielding?) Estimate the dB loss for 3,000 Hz sound transmitted through your chest to help answer this question.

10.19. For each tissue in Table 10.3, determine the thickness of tissue needed to decrease the intensity of a 5 MHz ultrasound wave by half. (Assume losses are due only to absorption, and not due to reflection at interfaces.)

10.20. Often in ultrasound measurement the ultrasound transducer and detector are on the same probe, so reflected sound waves are detected. Determine the fraction of initial sound intensity that is detected for the following cases by tracking the beam that is transmitted through material X in the body, reflected at normal incidence from the interface of X with Y, and then again transmitted through material X. (Assume that all the sound from the transducer enters X and all leaving X after reflection from Y enters the detector. Consider attenuation in the medium, as well as reflection at the X–Y interface.):
(a) X is 1 cm of muscle and Y is bone, for 1 MHz sound
(b) X is 1 cm of muscle and Y is bone, for 10 MHz sound
(c) X is 5 cm of fat and Y is muscle, for 1 MHz sound
(d) X is 2 cm of blood and Y is muscle, for 5 MHz sound
(e) X is 1 cm of bone and Y is muscle, for 1 MHz sound.

10.21. Calculate the relative delay times between sound reflecting at the beginning of the X medium and that reflecting at the X–Y interface, for each case in Problem 10.20.

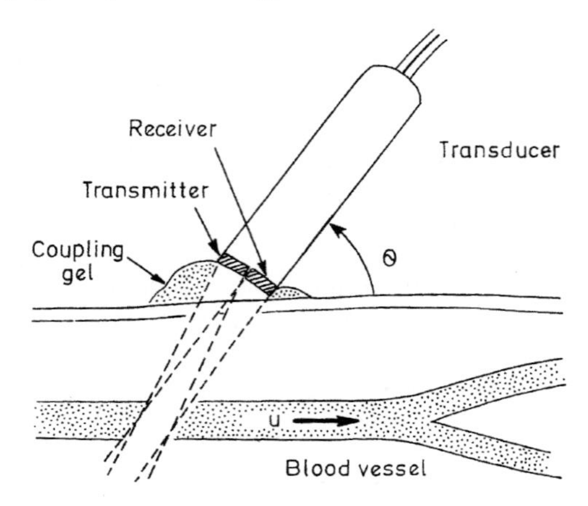

Fig. 10.56. Doppler examination of blood flow in a vessel at a speed u, with scanning shown. (Reprinted from [489]. Used with permission of Elsevier)

10.22. Estimate the relative delay times for 1 MHz sound reflecting normally from the first outer surface of the aorta, the first inner surface of the aorta, the second inner surface of the aorta, and second outer surface of the aorta.

10.23. When an ambulance siren blaring at a frequency f approaches you at speed v, you hear the frequency upshifted to $f' = f(1 + v/v_s)$, while when it is distancing itself from you, you hear the frequency downshifted to $f' = f(1 - v/v_s)$. This is the Doppler effect.
(a) Show that this is consistent with $f' = f(1 - (v/v_s)\cos\theta)$, where θ is the angle between the velocity vector of the moving object and the position vector from you to it.
(b) Show that Doppler ultrasonography echocardiography (Fig. 10.56) can be used to determine the blood flow speed v to be

$$v = \frac{(\delta f)v_s}{2f\cos\theta}, \tag{10.66}$$

where δf is the measured Doppler shift $(f' - f)$ and v_s is the speed of sound in body tissue.
(c) Calculate the maximum Doppler shift for blood flowing in the aorta, using 1 MHz ultrasound.

10.24. A generous dab of gel is put on the ultrasound probe head before it is placed on the skin (Fig. 10.56):
(a) Why?
(b) What must be the desired acoustic properties of this gel?

10.25. Express the unit of acoustic admittance, the mmho, in SI units.

10.26. (advanced problem) With $Z = R + \mathrm{i}X_m + X_s/\mathrm{i}$, derive (10.19).

10.27. (advanced problem) With $Z = R + \mathrm{i}(X_\mathrm{m} - X_\mathrm{s})$ and $Y = G + \mathrm{i}(B_\mathrm{m} - B_\mathrm{s})$, derive:
(a) Equation (10.20)
(b) G, B_m, and B_s in terms of R, X_m, and X_s
(c) R, X_m, and X_s in terms of G, B_m, and B_s.

10.28. (advanced problem) With $Z = R + \mathrm{i}\omega M + S/\mathrm{i}\omega$, where ω is the radial frequency in rad/s and $\omega = 2\pi f$ where f is the frequency in Hz or cycles per second, show that

$$| Z | = \sqrt{R^2 + (2\pi f M + S/2\pi f)^2}. \tag{10.67}$$

In electronics problems, R is the resistance, M corresponds to the inductance L, and S corresponds to the reciprocal of the capacitance C ($S = 1/C$).

10.29. (advanced problem) Use (10.24) to show that the resistive term of the speed varies as a cosine wave and the inertial and stiffness terms both vary as sine waves, but with opposite signs.

10.30. The speeds of sound in the brain and skull bone are 1,550 and 4,090 m/s, respectively. What fraction of sound is lost in reflection from air to the skull bone and then from the skull bone to the brain? Assume the densities of the brain and skull bone are 1 g/cm^3.

10.31. We are usually concerned with light entering the eye, but what happens when sound enters the eye? Calculate the reflection coefficient at each interface between the air/cornea/aqueous humor/eye lens/vitreous humor. The speeds of sound in the aqueous humor, eye lens (crystalline lens), and vitreous humor are 1,510, 1,630, and 1,540 m/s, respectively. Assume that the density of each medium in the eye is 1 g/cm^3 and that the cornea and eye lens have the same properties.

10.32. The speed of sound in collagen is 3,640 m/s along the fiber axis and 2,940 m/s across this axis. What is the reflection coefficient between blood and collagen, for sound traveling in both directions in the collagen?

Speaking

10.33. The oral cavity of a child is 8 cm long, as measured from her lips to vocal folds. What is the fundamental oscillation frequency of this cavity? (Treat the oral cavity as a cylinder open at one end and closed on the other.) Does this make sense in light of the differences of the voices of children and adults?

10.34. Plot on the same set of axes all of the resonant frequencies below 5 kHz for an 18-cm long cylinder that is open on both ends and a 16-cm long cylinder that is open on one end and closed on the other.

10.35. Smokers may have vocal folds that are slightly swollen and inflamed, and therefore, perhaps have folds that are more massive than those of non-smokers. How would this affect the vibration frequency of their vocal folds? Would the voices of smokers be relatively deeper or higher pitched?

10.36. Explain why people speak in a high pitch after taking a breath of helium. The speed of sound in helium is about 970 m/s. (Show this, given $\gamma = c_p/c_v$ is 5/3 for helium.)

10.37. In the text, the inverse relationship between the vocal-fold frequency and vocal-fold length was explained by considering the resonant frequencies of an oscillating string. Use (4.4) to show that this relationship is also expected if the vocal folds are modeled as a free, spring like object oscillating length-wise.

10.38. Estimate the vibration frequency of the vocal folds, by assuming they are a spring-like object oscillating length-wise that is 1 cm long, 0.3 cm wide, and 0.3 cm thick, with a Young's modulus of 100 kPa, and a mass density of 1 g/cm^3 [446]. (See (4.4).)

10.39. We can describe the production of the "m" sound as being voiced bilabial nasal. In what way does the production of the "n" sound differ?

10.40. Compared to men, do women use higher or lower harmonics of their fundamental buzzing frequency to produce the same vowel formant? Why?

10.41. Use the vowel formant plot (Fig. 10.19) to sketch the transmission curve of the vocal tract for three vowels.

10.42. Describe the frequency spectra of the vowels and consonants in Fig. 10.9. Point out their similarities and differences.

10.43. Plot the first- and second-formant frequencies for each vowel in Fig. 10.9 on the same set of axes.

10.44. Plot the first- and second-formant frequencies for the vowel in Fig. 10.20 on the same set of axes.

10.45. (advanced problem) Explain why the two-tube model in Fig. 10.25b,d,f, explains the mode shifts for the "ee" sound as in "see," which are shifted from the predictions of the one-tube model in Fig. 10.24a.

Hearing Mechanism

10.46. What is the resonant frequency of the 1.3-cm long ear canal of a baby? How does it compare to that of an adult?

10.47. The amplitude of motion of the eardrum is 0.03 μm when measured at 100 dB SPL for 3,000 Hz. Assuming the amplitude is linear with the total force on the eardrum (which is a good assumption for pressures below 130 dB SPL), find the amplitude of eardrum motion at 0 dB SPL in m.

10.48. Model the motion of the eardrum as a flat membrane that is fixed at its ends, with a displacement that increases linearly from the edge to the center of the eardrum. (The motion is really more closely sinusoidal than this triangular mode shape.) If this maximum displacement at the center is 0.03 μm, as in Problem 10.47, find the full change of angle the eardrum makes during its motion (in radians and degrees).

10.49. Compare the potential energy of motion of the eardrum at 3,000 Hz to that at 1 Hz for the same amplitude of motion. (Assume the motion can be modeled as a simple harmonic oscillator.)

10.50. Is modeling the eardrum as a freely vibrating object reasonable, given that the stapes touches it? Why?

10.51. Do people have good auditory sensitivity at the fundamental frequencies of their voices? Is this important? Why?

10.52. What is the gain in dB due to the middle ear if only the force enhancement in the middle ear (and not the entire torque enhancement) is considered?

10.53. Sketch the hair cell responses in Fig. 10.39 on a linear–linear plot of acoustic intensity vs. frequency.

10.54. How are the three sharp hair cell responses in Fig. 10.39 related to the responses for a oscillator as shown in Figs. 10.7 and D.3?

10.55. Design hair cells composed of keratin with resonant frequencies ranging from 20 Hz to 20 kHz with:
(a) A fixed length of 2 mm
(b) A fixed diameter of 20 μm or
(c) Dimensions of 2 mm length and 20 μm diameter for the hair resonant at 1 kHz, with variations in these two parameters that cause equal changes in frequency for different hair cells.

10.56. Calculate the fundamental frequency for lateral vibrations of a rod of solid bone of length 10 mm and diameter 1 mm ($Y = 1 \times 10^{11}$ Pa and density $\rho = 3{,}000\,\mathrm{kg/m^3}$) fixed at one end to an incompliant base.

10.57. (a) Show that 1 mmH_2O = 0.98 daPa.
(b) Express the tympanogram range of +200 daPa to −300 daPa in mmH_2O.

10.58. Show that the impedance relationship equivalent to the admittance relationship for tympanograms: $Y_{\text{middle ear}} = Y_{\text{total}} - Y_{\text{outer ear}}$, is $Z_{\text{middle ear}} = Z_{\text{outer ear}} Z_{\text{total}} / (Z_{\text{outer ear}} - Z_{\text{total}})$.

10.59. (advanced problem) For adiabatic conditions (no heat flow) PV^{γ} is a constant. Show that (10.59) becomes $\Delta P = -\gamma P (\Delta V)/V$ and that this leads to (10.61).

Hearing Levels and Perception

10.60. A student in a class wants to set the tone of his cell phone ringer so that he could hear it, but his instructor, who is a bit older than he is, cannot. Would a suitable fundamental frequency be 250, 1,000, 17,000, or 30,000 Hz? Why?

10.61. The acoustic power incident on the eardrum at threshold (0 dB SPL) is equivalent to how many optical photons [490]? (Hint: Use the known threshold intensity and the dimensions of the eardrum.)

10.62. Compare the intensity of the crack of a bat hitting a baseball as heard by the catcher and by a fan in the bleachers.

10.63. A noisy elevated train (in the open air) in Brooklyn in New York City causes acoustic discomfort to those 5 ft away from it. How many city blocks away can it be heard? (Treat the train as a point source, even though this is clearly an approximation. There are 20 canonical city blocks in a mile. Also, remember that background noises in the city could mask the sound of the train, so the threshold intensity for hearing is much above the threshold of hearing pure tones in a quiet room.)

10.64. The explosion of 23 kg of TNT creates a sound level of 200 dB SPL a distance 3 m away from the detonation. Assume the total acoustic energy produced in such explosions is proportional to the mass of the TNT:
(a) How far away must you be to avoid the threshold of pain from a blast from 100 kg of TNT?
(b) What fraction of the energy released from the detonation is in the form of acoustic energy over the hemispherical release region? Assume the blast is 50 ms long and a ton of TNT releases 4.18×10^9 J.
(c) It has been reported that 1 ton of TNT produces 120 dB SPL at 15 km. Is this consistent with the data given in this problem?

10.65. A static pressure of 8×10^3 Pa across the eardrum can cause it to rupture [455]. How does this compare to the sound pressure from a 160-dB SPL sound that can also cause the eardrum to rupture?

10.66. According to Fig. 10.30, what range of frequencies is needed to hear speech and over what overall volume range in dB SPL is this required?

10.67. According to Fig. 10.30, what range of frequencies is needed to hear orchestral music and over what overall volume range in dB SPL is this required?

10.68. Resketch the normal hearing range in Fig. 10.30 along with the music range for rock and roll music (which had not been developed in 1934 when this curve was made).

10.69. Show that (10.65) is referenced to a loudness corresponding to 40 phons.

10.70. Show that the loudness in phons, L_p, and sones, L_s, are related by: $L_p = 33.3 \log L_s + 40$, by using (10.64) and (10.65).

10.71. Which is loudest and which is the most quiet for these three sounds: one at 7,000 Hz with loudness of 60 phons, one at 4,000 Hz with a loudness of 8 sones, or one at 1,000 Hz with intensity 50 dB SPL? (As part of this problem, express each of the three sets of data into phon and sone loudness units.)

10.72. Which 1,000 Hz sound is louder 50 ft from its isotropic source: one with an intensity 80 dB SPL a distance 5 ft from the source or one that is 65 dB SPL a distance 25 ft from the source? (In both cases, calculate the intensity in dB SPL at 50 ft.)

10.73. Approximately 80 dB SPL is needed to achieve a loudness of 80 phons at 100 and 1,000 Hz. To achieve 40 phons, 40 dB SPL is needed at 1,000 Hz. How much more acoustic intensity is needed to attain this loudness at 100 Hz? Express your answer in dB and by the factor increase in intensity needed at 100 Hz relative to 1,000 Hz.

10.74. You have a radio with inexpensive speakers that produce sound only in the 250–5,000 Hz frequency range. Will you be able to hear notes with 100 Hz fundamental frequencies, and if so, why?

10.75. Does the relative loudness scale given by sones agree with the power law dependence described in Table 1.14?

10.76. You want to design an organ with a 55 Hz tone:
(a) Show that you need to use a 3-m long length of pipe (open at both ends) to produce this resonance frequency.
(b) Say this 3-m long pipe is too long to use in the organ, but you still want to perceive a 55-Hz note tone. Explain why you can play notes from pipes that are 1.0 and 1.5 m long at the same time and hear a 55-Hz tone.

10.77. The musical scale of "Just Intonation" consists of tones that sound pleasing when sounded together or immediately after one another. Such pleasing combinations occur when the notes are harmonics of each other or have frequencies that are related by fractions with relative small integral numerators and denominators [492]. The frequencies of the notes in one octave in the Tonic C in the major scale of Just Intonation are f for a C tone, $9f/8$ for D, $5f/4$ for E, $4f/3$ for F, $3f/2$ for G, $5f/3$ for A, $15f/8$ for B, and $2f$ for C:
(a) Find the frequencies in this scale if the A tone has a frequency of 440 Hz.
(b) Calculate the lengths of pipes with fundamental frequencies at the frequencies of each of these notes (with the tubes open on both ends).

10.78. In the text, the difference limen for a 1,000 Hz tone was cited to be 5 dB SPL for a 5 dB SPL sound and a much smaller fraction, 6 dB SPL, for the much louder 100 dB SPL sound. Determine and then compare the absolute increase in acoustic intensity (in W/m^2) in both cases.

10.79. A person, 2 m from your left ear, speaks to you, and his voice reaches your left ear with an intensity 60 dB SPL:
(a) If your right ear receives his voice delayed by 0.05 ms, where is this speaker located relative to you? (Say a person directly in front of you is at 0° and immediately to your right is at 90°. Assume your ears are separated by 20 cm.)
(b) If the intensity of the voice decreases inversely as the square of the propagation distance, how much lower (in dB) is the sound arriving in the right ear?

11

Light, Eyes, and Vision

Our eyes image a source onto light-sensitive retinas. Cells in the retina convert the imaged light into electrical signals. This information is carried to the brain by neurons via the optic nerve. The visual cortex in the brain processes this information, and we somehow perceive a visual image. Each of these four steps in the vision process is important and contains interesting physics. The physics of the first step is the optics of imaging the source onto the retina and the second is the quantum physics of the absorption of light by the retina. The physics of the conduction of electrical signals in nerves is step three and we will discuss this in Chap. 12. Physical processes and processing are also important in the fourth step, within the brain [530, 531, 532, 560, 562]. (By the way, David H. Hubel and Torsten N. Wiesel shared the Nobel Prize in Physiology or Medicine in 1981 for their discoveries concerning information processing in the visual system.) In this chapter, we will focus on the physics of imaging and briefly consider the absorption of light by retinal cells. Feedback and control (Chap. 13) are also important in vision. For example, the body controls the focal length of the eye lens (*accommodation*) to enable the imaging of both near and far objects and changes the diameter of the pupil to adjust the amount of light entering the eye.

For general references on vision see [513, 516, 520, 523, 527, 537, 542, 549, 550, 552, 560]. For general references on optics and the optics of vision see [511, 513, 516, 518, 529, 534, 535, 543, 544, 553, 555].

11.1 Structure of the Eye

Figure 11.1 shows the structure of the human eye. Light enters the cornea and passes through the aqueous humor in the anterior chamber, the pupil in the iris, the lens (commonly called the *crystalline lens* although it is not a crystal), and the vitreous humor, and images on the retina, which contains *rod* and *cone* photoreceptor cells. The three pairs of muscles that control eye

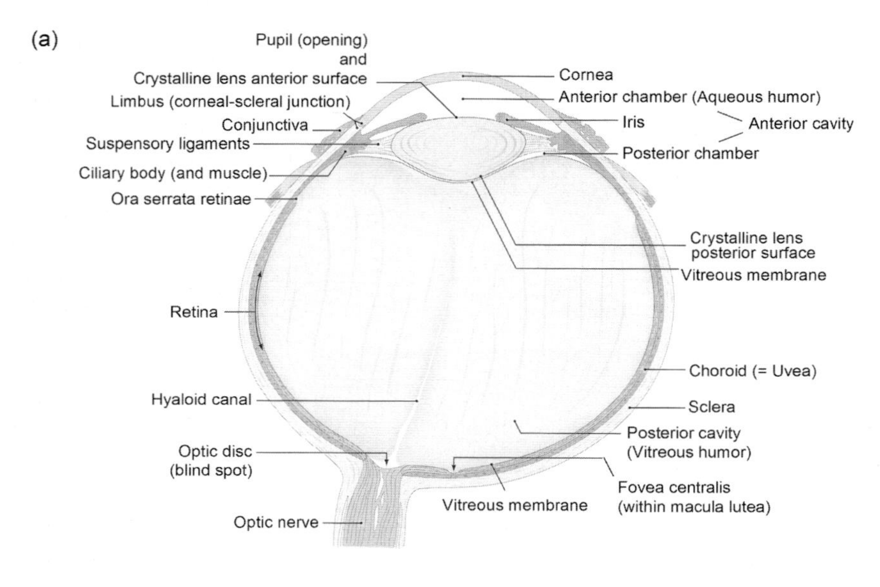

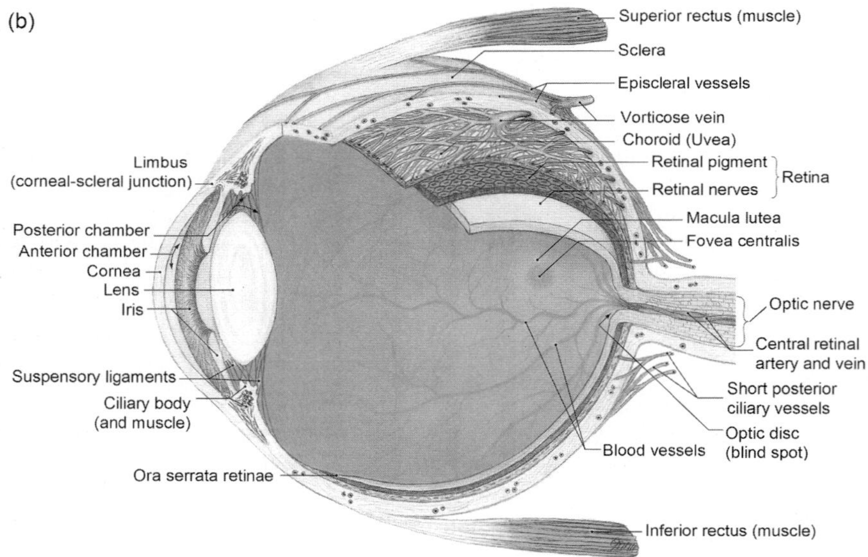

Fig. 11.1. Structure of the eye: (**a**) sagittal section, (**b**) three-dimensional perspective. (From [558])

motion were described in Fig. 1.14 and Tables 1.3 and 1.4. The eye has an average radius of 12 mm.

The adult cornea is 0.52 mm thick in the center and 0.65 mm thick in the periphery, and is about 12.6 mm in diameter horizontally and 11.7 mm vertically. It is composed of several layers, from anterior to posterior: the outer epithelium, the basement membrane, anterior limiting lamina (or the Bowman's layer), corneal stroma, posterior limiting lamina (or the Descemet's membrane), and the epithelium. The corneal stroma constitutes 90% of the corneal thickness and is composed of 50 layers, each with similarly oriented collagen fibers, with the fibers always parallel to the cornea surface. The cornea is transparent because it is uniform in structure, avascular (i.e., it has no blood vessels) except in the extreme periphery, and relatively dehydrated. It is covered by a 7–10 μm thick layer of tears, which, among other things, smoothes over optical irregularities on the anterior surface of the cornea and supplies the cornea with oxygen. The average radius of curvature of the anterior surface of the cornea is about 7.8 mm in the central region, with a variation among people of about ±0.4 mm, and is flatter in the periphery.

The aqueous humor fills the anterior chamber (with a volume of $0.3\,\mathrm{cm}^3$) bounded by the cornea, iris, and the anterior surface of the crystalline lens, and the posterior chamber ($0.2\,\mathrm{cm}^3$) on the periphery of the lens. It has many fewer proteins (0.1 g/L) than blood plasma (60–70 g/L). The pupil in the iris is usually slightly nasal and inferior to the center of the iris, and can vary roughly from 1.5 to 10 mm in diameter. The diameter of the pupil is controlled by an opposing pair of smooth muscles: the sphincter pupillae (which is a ring of muscles that encircle the pupil) contracts it and the dilator pupillae (which has the form of a thin disc) widens it.

The crystalline lens is suspended from the ciliary body by zonular fibers and rests on the posterior surface of the iris. It is composed of about 66% water and 33% protein. This crystalline lens is about 4 mm thick and 9 mm in diameter. It continues to grow during life, with new layers growing on older layers, forming a layered structure like an onion (Fig. 11.1). At 30 years of age, the lens has a mass of 170 mg, which increases by about 1.2 mg per year; similarly the lens width is about 4 mm and increases by about 0.02 mm per year. The crystalline lens is avascular and almost completely transparent. Still, it is slightly birefractive (i.e., it has slightly different refractive indices for different polarizations of light), becomes more yellow with age, and can become opaque (and this forms a cataract). Aphakia describes the condition when the crystalline lens is absent. The vitreous humor is about 99% water, with the remaining 1% composed of collagen (0.5 g/L proteins) and hyaluronic acid; the latter gives it its gelatinous, viscous physical characteristics. This humor accounts for about $5\,\mathrm{cm}^3$ of the 7–$8\,\mathrm{cm}^3$ volume of the eye.

As we will see below, the formation of an image on the retina is determined by the indices of refraction of each eye component that the light passes through and by the shapes of the surfaces of these elements. The cornea and crystalline lens are the actual focusing elements in the eye. The cornea performs about two-thirds of the focusing and the crystalline lens the remaining one-third. The shape and consequently the focal length of the crystalline lens are adjustable and do the fine-tuning of imaging for accommodation. The measured refractive index of the tears and the vitreous humor is about 1.336

and that of the aqueous humor is a bit higher, 1.3374. The refractive indices of the cornea, about 1.3771, and the crystalline lens are higher. At the center of the crystalline lens (which is called the nuclear region), the index is about 1.40–1.41 and it decreases to 1.385–1.388 in the direction towards the "poles" and to 1.375 in the direction toward the "equator"; it is 1.360 in the capsule, which is the elastic membrane that encloses the crystalline lens. There is still some uncertainty in these values; optical models of the eye use values close to these cited numbers.

Only about 50% of visible light (400–700 nm) incident on the eye actually reaches the retina as direct light. Then light must pass through the (transparent) ganglion and other retinal neurons before reaching and forming an image on the backward-facing photoreceptors on the retina (Fig. 11.2). The *fovea* or fovea centralis is the central region of the retina, and the region of sharpest vision because it has the highest density of cone cells on the retina (Fig. 11.3). The optic nerve leaves the eyeball at a *blind spot* (*optic disk*), a region with no rods or cones (Fig. 11.3); it is 13–18° away from the fovea in the "nasal" direction.

We are usually not aware of the blind spot when we use both eyes because the part of the image that forms on the blind spot in one eye is located in

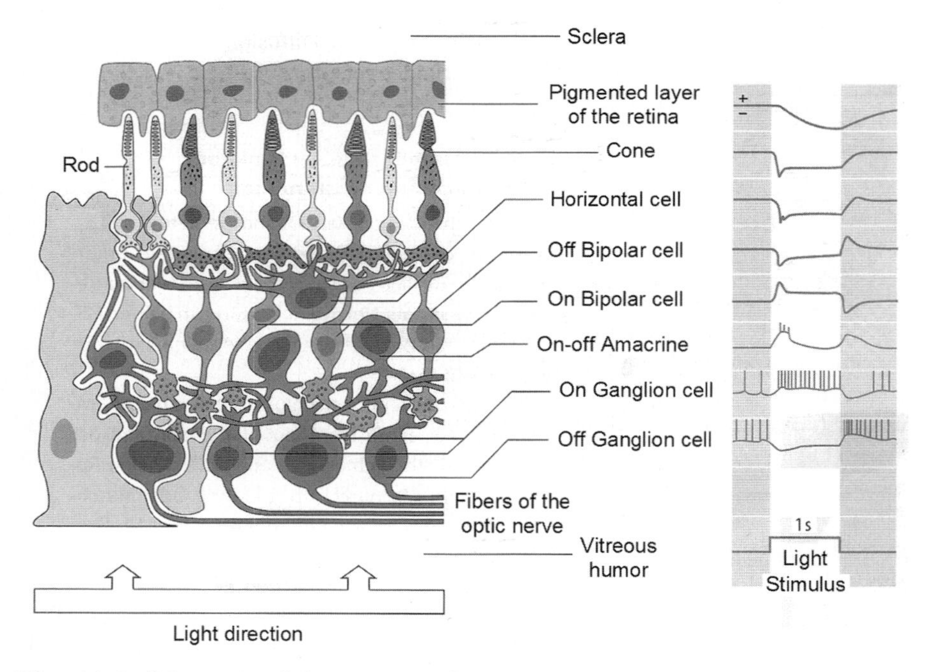

Fig. 11.2. Schematic of the retina in the eye, with the arrangement of rods and cones and other neurons, along with electrical excitation by the shown light stimulus. (From [551])

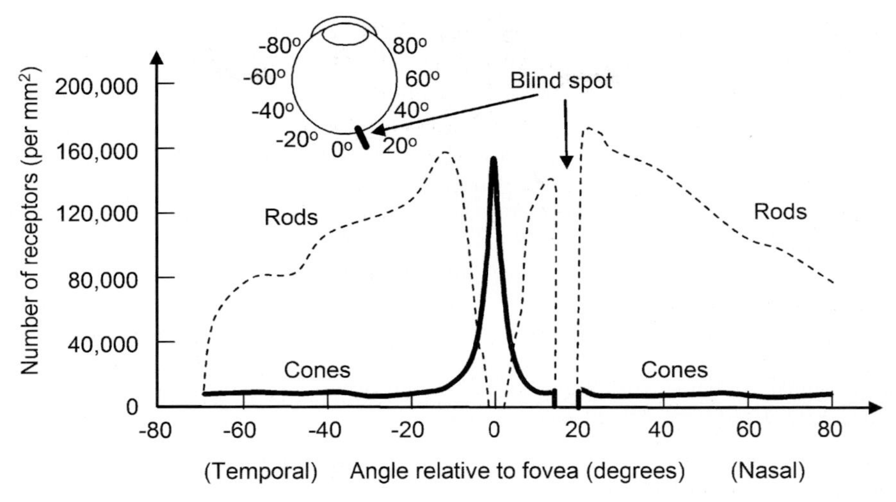

Fig. 11.3. Distribution of rods and cones on the retina, and the location of the blind spot. (Based on [523] and [560])

a functional region in the other eye and the brain fuses the images of the two eyes. It is easy to prove the existence of the blind spot. Close your right eye and use your left eye to look at the dot in Fig. 11.4. When you move the book about 10 cm from your left eye, you will find one position where the x disappears because of the blind spot in your right eye.

Fig. 11.4. Fixate on the x using your left eye, with your right eye closed. Keep the book about 10 cm from your left eye, and then move it back and forth until you do not see the central spot. This spot is then on the blind spot. The spots above and below it are still visible, but fuzzy because of the lower visual acuity outside the fovea. (Based on [560])

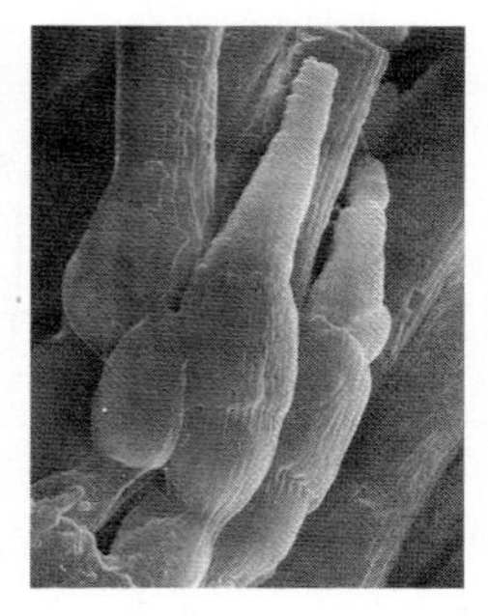

Fig. 11.5. Scanning electron micrograph of rod and cone outer segments, with the cone seen (with its tapered end) in the center and the end and beginning of two rods (which are longer than the cones) seen beneath and to the left of it. (Reprinted from [536]. Used with permission of Elsevier)

There are about 120 million rod cells per retina (Fig. 11.5). They have high sensitivity, low spatial acuity, and are relatively more numerous in the periphery of the retina. The sensitivity of rods peaks near 500 nm (Fig. 11.6). Vision using only rods results in various shades of gray. Night vision and peripheral vision are mostly due to rods. Rods are about 2 μm in diameter. Far from the fovea the rods become more widely spaced and many (in some cases several hundred) rods are connected to the same nerve fiber. Both factors decrease visual acuity in the outer portions of the retina.

There are about 6.5 million cone cells per retina. They have low sensitivity – about 1,000× lower than rods, high spatial acuity, and are concentrated in the fovea. There are three types of cone cells, with spectral sensitivities peaking near 445 nm (blue or S cones – S for short wavelength peak sensitivity), 535 nm (green or M cones – M for middle wavelengths), and 570 nm

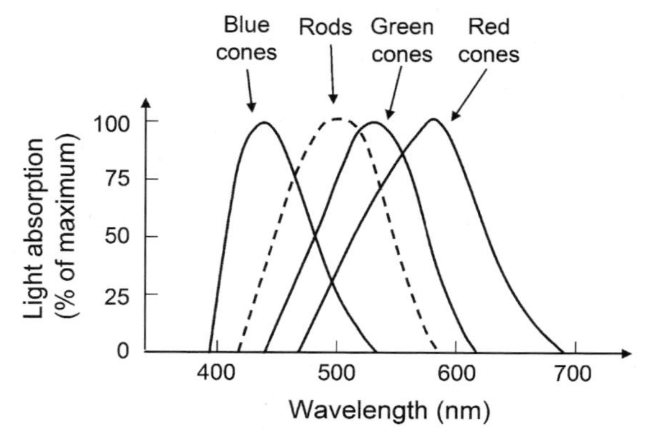

Fig. 11.6. Relative spectral sensitivity of rods and cones. The absolute sensitivity of rods is 1,000× larger those of the cones. (Based on [515], [526], and [540])

(red or L cones – L for long wavelengths) (Fig. 11.6). The overall spectral sensitivity due to the rods and cones of humans closely matches the spectrum of solar light reaching land. Sharp vision and color vision are due to cones, and consequently damage to the fovea leads to visual images that are fuzzy. Cones are about 1.0–1.5 μm in diameter and are about 2.0–2.5 μm apart in the fovea. There are only about 1 million nerve fibers in the eye, so there are some cones (as well as rods) connected to the same nerve cells. We will not delve into the cellular structure of the rods and cones, but will focus on two physical aspects of these sensors: the absorption of light and acuity of vision.

The absorption of light by the rods and cones is a fundamental quantum-mechanical process in which one photon (or quantum) of light is absorbed by the pigment rhodopsin. Quantum mechanics is the physics of small-scale objects, and has features that are distinct from the physics of larger-scale objects, which is the classical physics we have been using throughout this book. One feature of quantum physics is the quantization of energy levels in molecules, which means that a molecule can have only distinct energies. Consequently, a molecule can absorb light only at those specific energies (or frequencies) corresponding to the differences of its energy levels. Moreover, in quantum mechanics, light acts like light packets, called photons. The energy of a photon is

$$E = h\nu = \frac{hc}{\lambda}, \tag{11.1}$$

where h is Planck's constant (6.626×10^{-34} J-s, as in (6.37)), ν (or f) is the frequency of the light, c is the speed of light (3.0×10^{8} m/s), and λ is the wavelength of light. The last two parts of this equation reflect the relationship between frequency, wavelength, and propagation speed for these electromagnetic waves,

$$c = \lambda\nu, \tag{11.2}$$

as in (10.3). Absorption occurs when the photons have energy in ranges that can be absorbed by the photosensitive molecules in these cells.

Rhodopsin consists of a chromophore (i.e., the part of the molecule responsible for its color) covalently attached to the protein opsin. The chromophore is retinal, which is a derivative of vitamin A, and the absorption of a single photon of light isomerizes it (i.e., changes its molecular conformation) from 11-*cis* retinal to all-*trans* retinal. This isomerization triggers a change in the conformation of rhodopsin that starts a sequence of sensory transduction processes (Fig. 11.7). Proteins themselves have absorption bands in the ultraviolet, and cannot absorb in the visible. The absorption of free 11-*cis* retinal is in the near ultraviolet, 360–380 nm; however, the binding of the retinal to the protein red shifts the absorption by about 200 nm to the visible. Differences in the opsin proteins in the rods and the three cones cause the different wavelength responses for these four types of photoreceptor cells. (By the way,

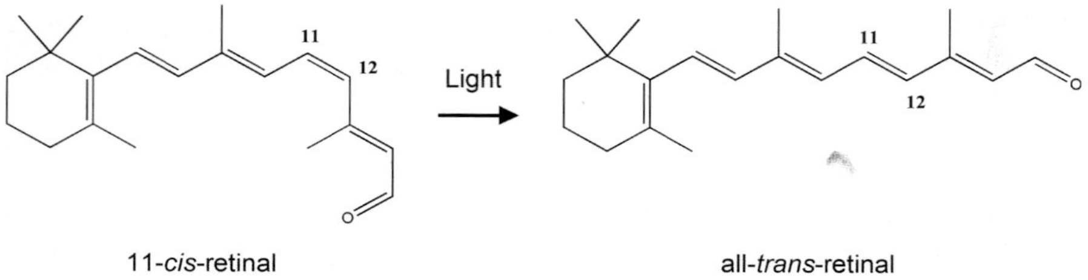

11 12 Light 11 12

11-*cis*-retinal all-*trans*-retinal

Fig. 11.7. The chromophore 11-*cis* retinal is photoisomerized by light to all-*trans* retinal (11-*trans* retinal)

Ragnar Granit, Haldan Keffer Hartline, and George Wald were awarded the Nobel Prize in Physiology or Medicine in 1967 for their discoveries concerning the primary physiological and chemical visual processes in the eye.)

The pressure in the eyeball maintains its shape. It is normally about 15 mmHg (ranging from 10 to 20 mmHg), and is determined by the rates of formation of the aqueous humor (about 1% of the total volume is produced per minute) and drainage of the aqueous humor through the canal of Schlemm. If the exit of the aqueous humor is impaired, the eyeball pressure increases, leading to glaucoma and possible blindness (as is addressed below). Intraocular pressure (IOP) is measured by the amount of force needed to flatten to a given area (or the area flattened by a given force) by using a tonometer. (This is explored in Problem 11.47.)

11.2 Focusing and Imaging with Lenses

11.2.1 Image Formation

Figure 11.8 shows how an object or source is imaged by a *convex (converging or positive) lens*. By convention in optics the object is placed a positive distance d_1 to the left of the lens and optical rays propagate from the left to the right. The object has a size (or height) y_1. For a convex lens the *focal length* f is positive, hence the name positive lens. The central axis (the z-axis) is known

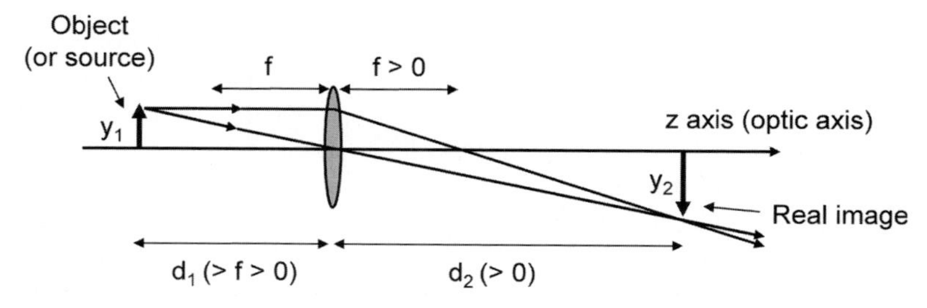

Fig. 11.8. Imaging by a thin, positive lens

as the *optic axis*. All rays passing through the lens form an image a positive distance d_2 to the right of the lens, where d_2 is given by the lens equation

$$\frac{1}{d_1} + \frac{1}{d_2} = \frac{1}{f}. \tag{11.3}$$

A *real image* forms at d_2 when $d_1 > f$, which means that you will see the image at d_2 if you place a screen, such as a piece of paper, there. The image is inverted and its size y_2 is magnified by $M = d_2/d_1$ (the *transverse magnification*). This can be seen from the triangles in Fig. 11.8 that give $y_1/d_1 = y_2/d_2$, so the magnification is

$$M = \frac{d_2}{d_1} = \frac{y_2}{y_1}. \tag{11.4}$$

(Actually, it is magnified when $d_2/d_1 > 1$ and reduced in size (minified) when $d_2/d_1 < 1$.) When $d_1 = \infty$, parallel rays are incident on the lens and an image forms at $d_2 = f$ (Fig. 11.9a). When the object is at the focus and so $d_1 = f$, the image is at ∞ (Fig. 11.9b). When the object is closer to the lens than the

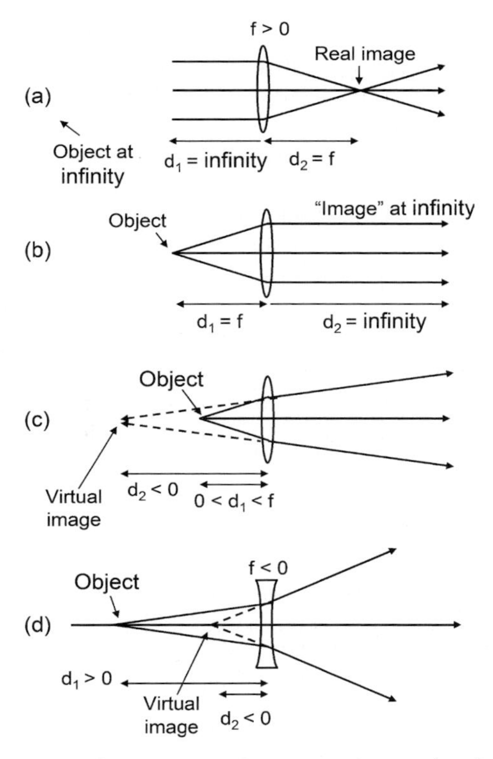

Fig. 11.9. Special cases of imaging with positive lenses (**a**–**c**), and imaging with negative lenses (**d**)

focal point and so $d_1 < f$, then $d_2 < 0$ and the image is to the left of the lens (Fig. 11.9c). This is a *virtual image.* Placing a screen there will give no image. However, if the light rays to the right of the lens are traced backward to the left of the lens, they will seem to emanate from this virtual image.

This same lens equation (11.3) can be used to determine the location of the image for a *concave (diverging or negative) lens*, which has a negative focal length f. Concave lenses produce virtual images (Fig. 11.9d).

The cornea and crystalline lens in the eye are positive lenses, because they need to form a real image on the retina. Corrective lenses (eyeglass lenses and contact lenses) can have positive or negative focal lengths, depending on the necessary correction. We will explore this later in this chapter. Focal lengths are expressed as distances, in cm or m. We will see that in discussing the eye and corrective lenses it is very common to discuss $1/f$ and use units of *diopters* (D), with $1\,\mathrm{D} = 1/\mathrm{m}$.

In a very simple model of the eye imaging system, the eye is treated as a thin lens with a 17 mm focal length in air (*Standard eye* model). For an image at $d_1 \gg f$, (11.3) shows that $d_2 \simeq f$. We will see that two points are resolvable by at best $\Delta y_2 = 2\,\mu\mathrm{m}$ on the fovea. So for a source that is 10 m away, two points are resolvable when separated by at least $\Delta y_1 = (d_1/d_2)\Delta y_2 = (10\,\mathrm{m}/17\,\mathrm{mm})(2\,\mu\mathrm{m}) = 1.2\,\mathrm{mm}$; this corresponds to an angle of $1.2\,\mathrm{mm}/10\,\mathrm{m} = 0.12\,\mathrm{mrad} = 25\,\mathrm{s}$ of arc. The $\sim 300\,\mu\mathrm{m}$ foveal diameter corresponds to a lateral separation of $\sim(10\,\mathrm{m}/17\,\mathrm{mm})(300\,\mu\mathrm{m}) = 18\,\mathrm{cm}$ at 10 m or $\sim 18\,\mathrm{cm}/10\,\mathrm{m} = 18$ mrad $\sim 1°$ of arc.

We have assumed *geometric optics*, which ignores the wave-like features of light due to optical diffraction; this is a good approximation for very short wavelengths and for much of the imaging in the eye. Our analysis also assumes only *paraxial rays*, i.e., all rays are near the optic axis and make small angles to it. (Rays that are farther away from the optic axis – nearer where the maximum amount of light is transmitted – are called *zonal* rays, and those at the margin of the lens are *marginal* rays. We will evaluate below the importance of diffraction and of these zonal and marginal rays.)

11.2.2 Scientific Basis for Imaging

We will trace rays by following how they propagate in straight lines in uniform media and how they refract at interfaces by using Snell's Law. Snell's Law of refraction shows that light from medium 1 with index of refraction n_1 impinging at an angle θ_1 (relative to the normal) on a flat interface with medium 2 with refractive index n_2, is refracted to an angle θ_2 given by (Fig. 11.10)

$$n_1 \sin\theta_1 = n_2 \sin\theta_2 \tag{11.5}$$

and for small angles (θ_1, $\theta_2 \ll 1$)

$$n_1\theta_1 = n_2\theta_2. \tag{11.6}$$

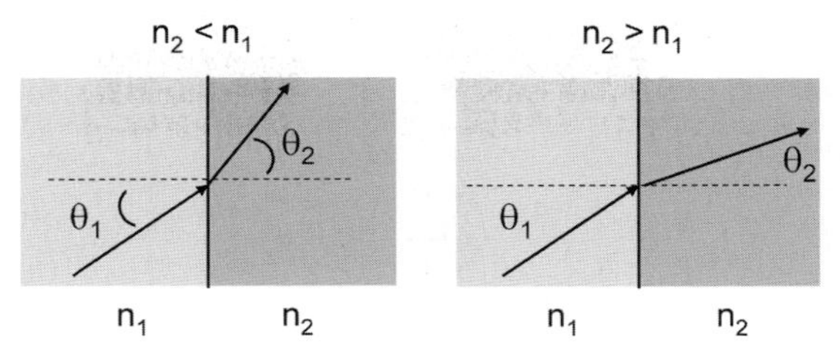

Fig. 11.10. Snell's law

Some important indices of refraction are 1.0 for air, 1.33 for water, 1.5–1.6 for different types of glass, and 1.44–1.50 for plastics. Refractive indices actually vary some with wavelength and temperature, but we will ignore those variations at present.

Imaging can occur when the interfaces are curved. Let us consider the refraction of paraxial rays at the interface in Fig. 11.11 from medium 1 to medium 2, which has a spherical radius of curvature R_{12}. (This region can be formed by slicing off a section from a sphere with radius R_{12}, composed of material 2.) As shown here, this radius is defined to be positive (see the Fig. 11.11 inset). Equation (11.5) still applies, but the angle of incidence for a light ray parallel to the optic axis varies with the distance y the ray is displaced from this axis. We see that $\theta_1 = y/R_{12}$. For $y \ll R_{12}$ (so $\theta_1 \ll 1$), (11.6) gives

$$\theta_2 = \frac{n_1}{n_2}\,\theta_1 = \frac{n_1}{n_2}\frac{y}{R_{12}}. \tag{11.7}$$

This refracted ray makes an angle $\theta_1 - \theta_2 = (1 - n_1/n_2)(y/R_{12}) = [(n_2 - n_1)/n_2](y/R_{12})$ with the horizontal (Fig. 11.11). Geometry shows that it hits the optic axis a distance $F(y) = y/(\theta_1 - \theta_2)$ after the interface. This distance

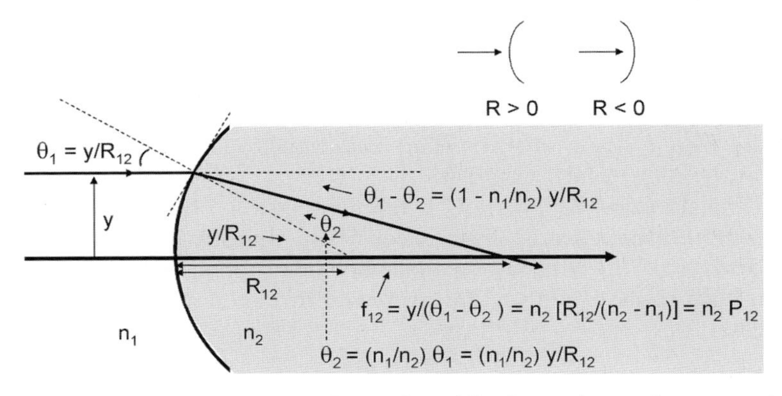

Fig. 11.11. Refraction at a curved interface. The inset shows the convention for the radius of curvature. In this figure, $R_{12} > 0$ and $n_2 > n_1$, and $\theta_1 \ll 1$ so $R_{12} \gg y$

is independent of y, and so all parallel rays impinging on the curved surface hit the optic axis at this same distance, which is called the focal length f (or f_{12} for this interface)

$$f_{12} = \frac{y}{[(n_2 - n_1)/n_2](y/R_{12})} = n_2 \frac{R_{12}}{n_2 - n_1} = \frac{n_2}{P_{12}}. \tag{11.8}$$

The last expression has been written in terms of the *refractive power* (or sometimes called the *convergence*) of the interface

$$P_{12} = \frac{n_2 - n_1}{R_{12}}. \tag{11.9}$$

The focal length is defined in terms of the refractive power of the interface and the refractive index of the medium the ray enters. The ratio of the indices of refraction of the two media is important and not their individual values. (For $n_2 = 1$, we see that $f_{12} = 1/P_{12}$.) The units of the refractive power are diopters ($1\,\mathrm{D} = 1/\mathrm{m}$).

We see that this derivation giving the focal length in (11.8) is similar to that for (11.3), except that the object and image are in regions with different refractive indices here. If n_1 and n_2 were interchanged, the sign of the focal length would change and its magnitude would change to $|\, n_1/P_{12} \,|$. The analog of (11.3) at this refractive interface is

$$\frac{n_1}{d_1} - \frac{n_2}{d_2} = -\frac{n_2 - n_1}{R_{12}}. \tag{11.10}$$

For $P_{12} > 0$, when $d_1 = \infty$, a real image occurs at $d_2 = f_{12} > 0$ (for rays traveling from left to right, as is the convention). For rays traveling from right to left and $d_2 = \infty$, a virtual image occurs at $d_1 = -f_{21} < 0$. So in general we see that

$$P_{12} = \frac{n_2 - n_1}{R_{12}} = \frac{n_2}{f_{12}} = \frac{n_1}{f_{21}}. \tag{11.11}$$

(For $P_{12} > 0$, this gives $f_{12} > 0$ and $f_{21} > 0$. Another popular convention is to define f_{21} as the negative of our definition, so the last term in (11.11) would be $-n_1/f_{21}$, and rays traveling from right to left would form a virtual image at $d_1 = f_{21}$, which would still be < 0.)

In clinical optics, $L_1 = n_1/d_1$ is often called the object *vergence* (which signifies ray convergence or divergence). (Sometimes it is also known as the *object proximity.*) $L_2 = n_2/d_2$ is called the image vergence (or *image proximity*). In both cases d_1 and d_2 are expressed in meters. Then (11.10) can be written as

$$L_2 = L_1 + P_{12}, \tag{11.12}$$

where all quantities are in diopters. This means that the propagation of light from an object to the image increases the vergence by an amount equal to the

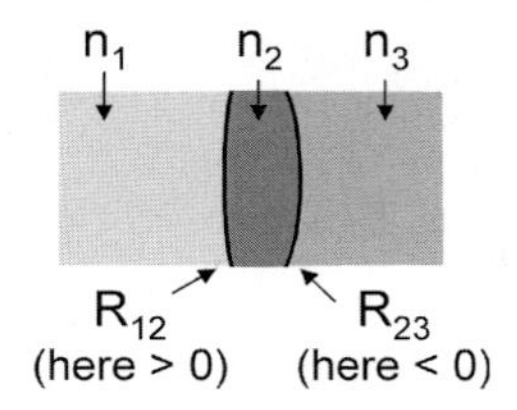

Fig. 11.12. Refraction at two spherical interfaces. As drawn here, the radii of curvature $R_{12} > 0$ and $R_{23} < 0$

power of that interface (or more generally, that of the optical system). The magnification, M, is modified from (11.4) to give

$$M = \frac{d_2/n_2}{d_1/n_1} = \frac{y_2}{y_1} = \frac{L_1}{L_2} = \frac{L_1}{L_1 + P_{12}}, \tag{11.13}$$

using (11.12).

What happens when there are two refracting interfaces in succession (Fig. 11.12)? If they are separated by a distance D that is "very small," the same reasoning gives an overall focal length f for this lens

$$f = \frac{n_3}{P_{12} + P_{23}}, \tag{11.14}$$

where P_{12} is given by (11.9) and

$$P_{23} = \frac{n_3 - n_2}{R_{23}}. \tag{11.15}$$

This is called the *thin lens approximation.* Equation (11.14) can be expressed as $f = n_3/P_{\text{total}}$, where

$$P_{\text{total}} = P_{12} + P_{23}, \tag{11.16}$$

so the refractive powers add in this approximation.

For a thin lens of refractive index $n_2 = n$ in air or vacuum (with refractive index $n_1 = n_3 = 1$, so $f_{13} = f_{31}$), (11.14) reduces to

$$\frac{1}{f} = (n-1)\left(\frac{1}{R_{12}} - \frac{1}{R_{23}}\right), \tag{11.17}$$

which is known as the *Lensmaker's equation.* This focal length can also be expressed as

$$f = \frac{1}{P_{12} + P_{23}}, \tag{11.18}$$

with $P_{12} = (n-1)/R_{12}$ and $P_{23} = -(n-1)/R_{23}$.

Lenses can have a range of shapes even for the same focal length (Fig. 11.13). For positive focal lengths, they can be either biconvex ($R_{12} > 0$ and $R_{23} < 0$),

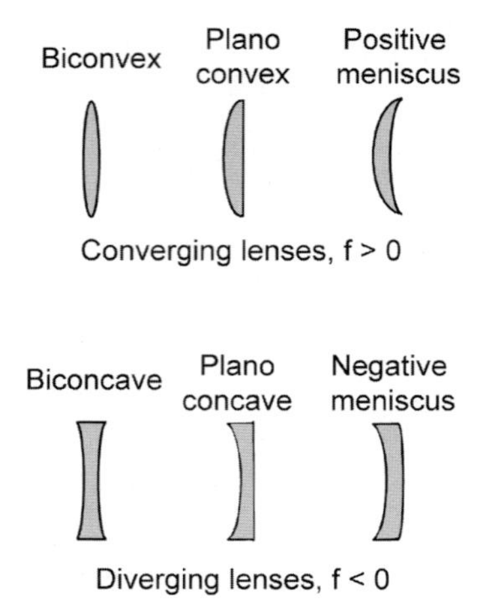

Fig. 11.13. Types of positive and negative lenses

planoconvex (one side flat), or positive meniscus ($R_{12}, R_{23} > 0$). For negative focal lengths, they can be either biconcave ($R_{12} < 0$ and $R_{23} > 0$), planoconcave (one side flat), or negative meniscus ($R_{12}, R_{23} < 0$). The cornea is a positive meniscus lens and the crystalline lens is an asymmetric biconvex lens.

The length of the eyeball is approximately 24 mm, so the distances from the cornea/crystalline lens to the retina and the focal length of optical rays focused by the cornea/crystalline lens propagating in the vitreous humor and imaging on the retina, are also about 24 mm. The refractive index of the vitreous humor is approximately 1.33. Equation (11.14) shows that a system with the same refractive power has a focal length that is proportional to this refractive index, so in air the focal length would be smaller by a factor of 1.33, or 24 mm/1.33 = 17 mm. That is why a model eyeball can be treated as if the cornea/crystalline lens system had an effective focal length of 17 mm, with the effective lens separated by 17 mm of air from the retina; we will call this the *Standard eye* model. (This differs from the eye models in Table 11.1.)

Moving Lenses

How does the imaging *effectivity* (i.e., effectiveness) of a lens change when you move it and why is this important to us? The corrective prescriptions for eyeglasses and contact lenses are different because eyeglasses are placed about 1.5 cm anterior to the cornea, while contact lenses sit right on the cornea. Let us consider parallel rays hitting a converging lens with power P in a medium with refractive index n. They form an image a distance n/P after the lens. If we move the same lens a distance D to the left and want the rays to focus in the same place, we now need them to focus a distance $n/P + D$ after the lens and so the lens will need to have a power P' such that $n/P' = n/P + D$.

Therefore we see that

$$P' = \frac{P}{1 + \frac{D}{n}P}. \tag{11.19}$$

This is known as the *effectivity formula.* (This equation is sometimes displayed with a negative sign in the denominator because the lens is being moved a distance D to the right.)

11.2.3 Combinations of Lenses or Refractive Surfaces

The eye itself is a combination of four distinct curved imaging interfaces. When imaging on the retina is not perfect, corrective lenses (with two additional interfaces) are chosen so the combined effect of these supplemental lenses and the eye imaging system produces a more perfect image on the retina. To first order we can apply (11.18), including all of the curved interfaces (for the eye or eye + corrective lens), with the ray finally entering a medium j with index of refraction n_j. We see that the focal length is

$$f = \frac{n_j}{\sum_{i=1 \text{ to } j-1} P_{i,i+1}}. \tag{11.20}$$

Merely using (11.20) to determine the imaging properties of these complex systems of curved surfaces or lenses is not sufficient, because its thin lens approximation ignores the propagation of rays from one refracting surface to the next. Because the refractive interfaces are displaced from one another, this usually needs to be corrected.

Two Thin Lenses

Before we explain how to include the separation of the curved interfaces in a very general way, let us consider the imaging by two separated thin lenses. This approach is rigorous but it is cumbersome when extended to more than two lenses. It can be applied, for example, to the two refractive surfaces of the cornea or the crystalline lens. We will use it to learn how to correct vision by considering the combined effect of eyeglasses or contact lenses and the eye, with both modeled as simple thin lenses.

Let us consider two thin lenses with focal lengths f_1 and f_2 that are separated by a distance D (Fig. 11.14). An object is placed at a distance d_{a} to the left of the first lens and the final image is formed at a distance d_{b} to the right of the second lens. In the *forward propagation approach,* we consider propagation from the left to the right in Fig. 11.14a. The first lens forms an image at a distance d_{i} to the right of this lens, where

$$\frac{1}{d_{\mathrm{a}}} + \frac{1}{d_{\mathrm{i}}} = \frac{1}{f_1}. \tag{11.21}$$

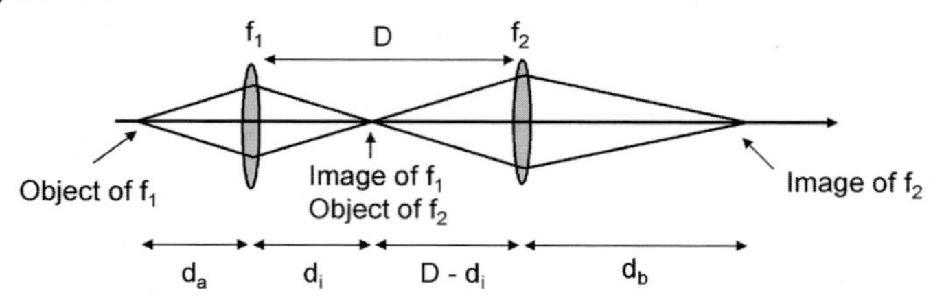

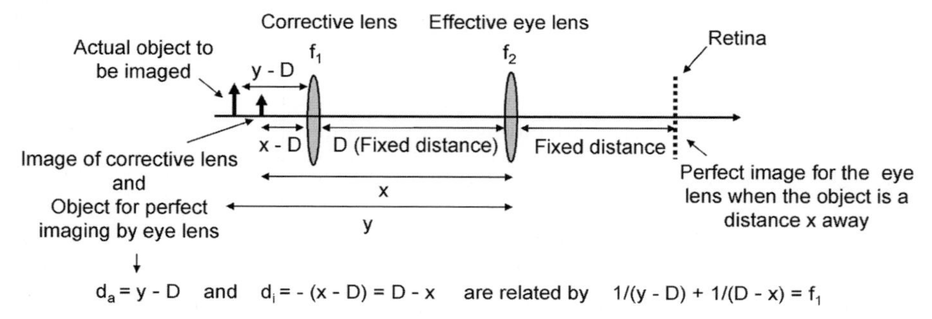

Fig. 11.14. Imaging by two lenses using (**a**) forward and (**b**) backward propagation

This intermediate image is at a distance $D - d_\mathrm{i}$ to the left of the second lens, and serves as the object for this second lens. Therefore, the second lens forms an image at a distance d_b to its right, where

$$\frac{1}{D - d_\mathrm{i}} + \frac{1}{d_\mathrm{b}} = \frac{1}{f_2}. \tag{11.22}$$

This distance d_b is determined by inserting d_i from (11.21) into (11.22)

$$D = \frac{d_\mathrm{a} f_1}{d_\mathrm{a} - f_1} + \frac{d_\mathrm{b} f_2}{d_\mathrm{b} - f_2}. \tag{11.23}$$

This interrelates all five parameters (D, d_a, d_b, f_1, f_2) and can be used to determine the fifth parameter, such as d_b, when the other four are known.

Sometimes it is simpler to work backward from the final image to the initial source, especially when the location of the final image is fixed (Fig. 11.14b). For example, consider using this *backward propagation approach* when corrective lenses (lens 1) are needed to correct the images of the effective eye lens (lens 2) on the retina. Let us say we know that the eye lens images perfectly on the retina when an object is at a distance z_1 away (to the left of the effective eye lens), but we would like to have clear images of objects at a distance z_2 away. The eye by itself cannot accomplish this, as we will soon see, due to insufficient accommodation or to myopia or hyperopia. The distance from the corrective lens to the eye lens is fixed, $D \sim 1.5\,\mathrm{cm}$ for eyeglass lenses and $D \sim 0\,\mathrm{cm}$ for contact lenses. Therefore this distance z_1 to the left of lens 2 is at a known distance $z_1 - D$ to the left of lens 1. The focal length of corrective lens 1, f_1, is chosen so that it takes an object $d_\mathrm{a} = z_2 - D$ to the left of it and forms an image $d_\mathrm{i} = D - z_1$ to the right of it. (The eye can then image this intermediate object quite well by itself.) Using (11.3), this condition becomes

$$\frac{1}{z_2 - D} + \frac{1}{D - z_1} = \frac{1}{f_1}, \tag{11.24}$$

and so the focal length of the necessary corrective lens is $f_\mathrm{corrective} = f_1$. One or both of the terms on the left-hand side of this equation can be negative. We will use this approach below to prescribe corrective lenses.

Complex Optical Systems (Advanced Topic)

We will now see that any system with several refracting surfaces, such as the eye, can be reduced to a simple effective optical system that is similar to a thin lens (Fig. 11.15). The effective thin lens system is analyzed by tracing optical rays with the help of six reference points on the optic axis called *cardinal points*. There are two principal points (P, P′), two focal points (F, F′), and two nodal points (N, N′).

Rays traveling parallel to the optic axis (from left to right) in the first medium (with refractive index n where the object is located) refract in the optical system and travel as straight rays that converge at the second (or

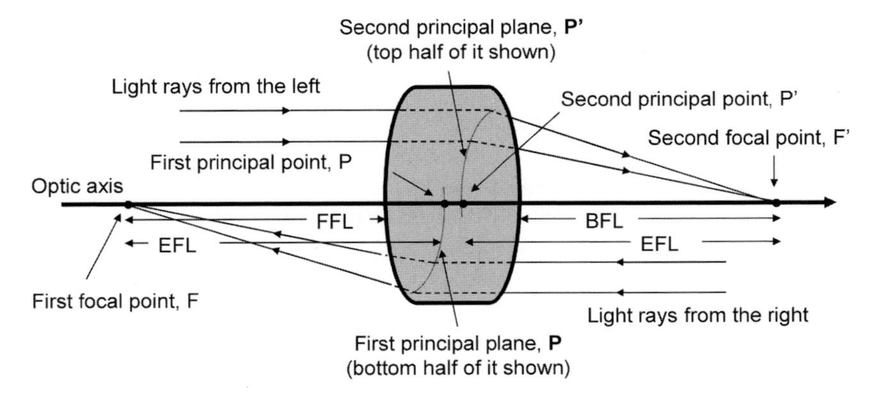

Fig. 11.15. Optical planes and imaging in a thick lens (or any multielement optical system) for rays that are parallel to the axis. The second (or first) focal length is the distance from the second (or first) principal point to the second (or first) focal point. If the source and image media refractive indices, n and n' are the same, both are called the effective focal length (EFL). The back (or front) focal length (BFL or FFL) is measured from the last (first) vertex of the last (first) optical element

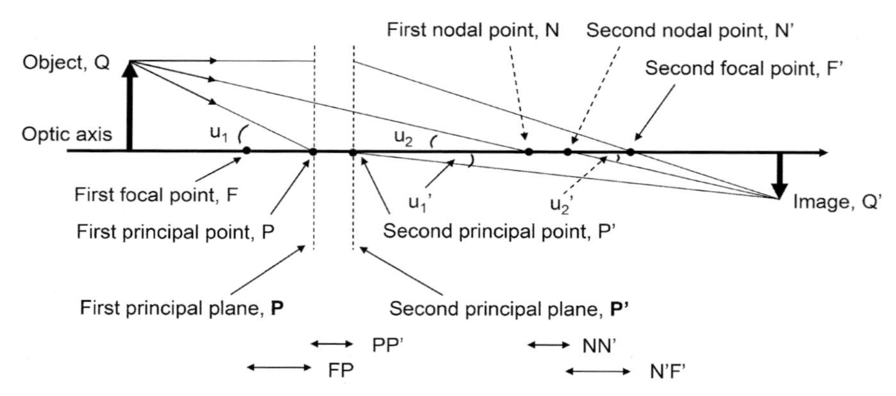

Fig. 11.16. Image formation using cardinal points. The location and size of the image (in a medium with refractive index n') can be determined by tracing rays from the object (in a medium with refractive index n): the object ray parallel to the optic axis and either the ray going through principal point P at angle u_1 and emerging from P′ at angle $u_1' = (n/n')u_1$ or that going through nodal point N at angle u_2 and emerging from N′ at the same angle $u_2' = u_2$. This example is similar to that of Schematic eye 2 depicted in the top half of Fig. 11.20, with $n' > n$

image) *principal focus* at point F′ (after the last interface in the last medium with refractive index n') (Fig. 11.15). When these rays are backtracked, they each intersect the initial parallel rays at a surface called the second (or image) *principal plane* **P**′. Similarly, rays traveling from right to left in this last medium intersect in the first medium at the first (or object) *principal focus* at point F, and when these rays are backtracked, they each intersect the initial parallel rays at a surface called the first (or object) principal plane **P**. (These "planes" are really curved surfaces that are approximately planar near the optic axis.) The first and second (or object and image) *principal points* P and P′ are the intersections of these planes with the optic axis. There are also *nodal points* N and N′ on the optic axis, which help in analyzing image formation (Fig. 11.16).

We will adopt a sign convention that is convenient for biconvex, positive lenses, so that all focal lengths will be defined to be positive for such lenses. (This is consistent with (11.11), but is not universal notation.) The first principal focus F is a distance f to the left of P when $f > 0$ and the second principal focus at F′ is a distance f' to the right of P′ when $f' > 0$. We will call distances along the optic axis, say between P and another point A – which is PA, positive when A is to the right of P and negative when it is to the left of it; therefore, AP $= -$PA. There is symmetry in the positions of the six cardinal points, with the separations PP′ $=$ NN′ and FP $=$ N′F′, as is depicted in Fig. 11.16.

Rays traveling from left to right emanate from the same object point Q and intersect to form an image at Q′. Any ray from Q hits plane **P** and emerges from plane **P**′ as if it were refracted by a thin lens using (11.3), with a "no man's land" between the two principal planes. So, with d_1 = QP (which is the distance along the optic axis) and $d_2 = \mathrm{P'Q'}$, the distance from object to image is really

$$\mathrm{QQ'} = \mathrm{QP} + \mathrm{PP'} + \mathrm{P'Q'} = d_1 + d_2 + \mathrm{PP'}. \tag{11.25}$$

Rays that hit point P, at an angle u, act as if they "emerge" from P′ at an angle u' (Fig. 11.16), where

$$n'u' = nu. \tag{11.26}$$

This can be viewed as Snell's Law in this paraxial limit. Nodal points are also defined in a way that rays on course to hit the first nodal point N, act as if they "emerge" from the second nodal point N′ with no change in angle. Again, these two sets of rays emanating from the same point Q intersect to form an image at Q′, as is seen in Fig. 11.16. The principal and nodal points coincide when the media to the left and right have the same refractive index; this is not the case for imaging by the eye. (For a thin lens in air, P, P′, N, and N′ merge into one point (inside the lens for a biconvex lens), and $f' = f$.)

Two Arbitrary Optical Refractive Systems (Advanced Topic)

Let us now consider two arbitrary optical systems [535] (Fig. 11.17). The first has principal points $\mathrm{P_a}$ and $\mathrm{P'_a}$, object and image refractive indices n_a and n'_a, and refracting power P_a. It is followed by the second with principal points $\mathrm{P_b}$

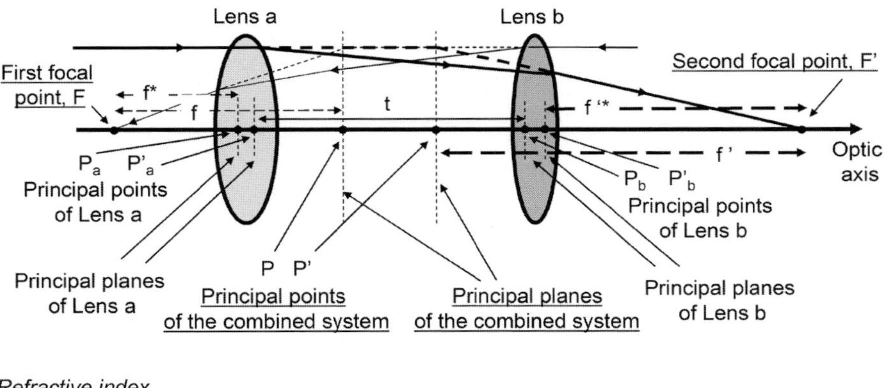

Fig. 11.17. Analyzing the combination of two optical systems, represented here by two "lenses." The notation for the planes and points of the combined systems is *underlined*. The *solid line* ray refracts at the principal planes of the individual systems, while the *dashed* ray refracts at the principal planes of the overall system

and P'_b, with object and image refractive indices n_b and n'_b, and refracting power P_b. The reduced distance between the second principal point of the first system and the first principal point of the second system is defined as

$$\delta = \frac{t}{n'_\mathrm{a}} = \frac{t}{n_\mathrm{b}} = \frac{t}{n_2}, \tag{11.27}$$

with $t = \mathrm{P}'_\mathrm{a}\mathrm{P}_\mathrm{b}$. We will now call $n_\mathrm{a} = n_1$, $n'_\mathrm{a} = n_\mathrm{b} = n_2$, and $n'_\mathrm{b} = n_3$.

Optical analysis that is beyond our scope gives general expressions for combining these two optical systems. The effective refractive power of the entire system can be shown to be

$$P_\mathrm{eff} = P_\mathrm{a} + P_\mathrm{b} - \delta P_\mathrm{a} P_\mathrm{b} \tag{11.28}$$

or

$$P_\mathrm{eff} = P_\mathrm{a} + P_\mathrm{b} - \frac{t}{n_2} P_\mathrm{a} P_\mathrm{b} = \frac{n_3}{f'} = \frac{n_1}{f}. \tag{11.29}$$

(This is consistent with (11.11). Using some other sign conventions for focal lengths, this last term is expressed as $-n_1/f$.)

Where are the principal points of the combined system? The position of the second principal point of the overall system P' relative to the second principal point of the second system is

$$\frac{\mathrm{P}'_\mathrm{b}\mathrm{P}'}{n_3} = -\frac{t}{n_2} \frac{P_\mathrm{a}}{P_\mathrm{eff}}, \tag{11.30}$$

which means that P' is to the left of P'_b in Fig. 11.17 when $P_\mathrm{a}/P_\mathrm{eff} > 0$ (such as for a biconvex lens).

Similarly, the position of the first principal point of the overall system P relative to the first principal point of the first system is

$$\frac{\mathrm{P}_\mathrm{a}\mathrm{P}}{n_1} = +\frac{t}{n_2} \frac{P_\mathrm{b}}{P_\mathrm{eff}}, \tag{11.31}$$

which means that P is to the right of P_a in Fig. 11.17 when $P_\mathrm{b}/P_\mathrm{eff} > 0$ (such as for a biconvex lens).

What are the new focal lengths? The second principal focus of the combined system F' is a distance

$$\mathrm{P}'\mathrm{F}' = f' \tag{11.32}$$

to the right of new second principal point P'. The new first principal focus F is a distance

$$\mathrm{FP} = f \tag{11.33}$$

to the left of new first principal point P. The focal lengths are given by (11.29). If $n = n'$, these focal lengths are the same (and are the effective focal length (EFL), as seen in Fig. 11.15).

Where are these foci relative to the principal points of the original two systems? Using (11.29) and combining (11.30)–(11.33) gives the position of the second focal point relative to the second principal plane of the second system $\mathrm{P}'_{\mathrm{b}}\mathrm{F}' = \mathrm{P}'_{\mathrm{b}}\mathrm{P}' + \mathrm{P}'\mathrm{F}'$:

$$\mathrm{P}'_{\mathrm{b}}\mathrm{F}' = f'\left(1 - \frac{t}{n_2}P_{\mathrm{a}}\right) = f'^{*} \tag{11.34}$$

and the position of the first focal point relative to the first principal plane of the first system $\mathrm{FP}_{\mathrm{a}} = \mathrm{FP} + \mathrm{PP}_{\mathrm{a}}$:

$$\mathrm{FP}_{\mathrm{a}} = f\left(1 - \frac{t}{n_2}P_{\mathrm{b}}\right) = f^{*}. \tag{11.35}$$

The parameters f^* and f'^* are sometimes called the front and back (vertex) focal lengths (FFL and BFL in Fig. 11.15) because for a thick lens these are the distances from the front and back surfaces. In analogy with (11.29), sometimes the front and back vertex powers are defined as

$$P^* = \frac{n_1}{f^*} \tag{11.36}$$

and

$$P'^* = \frac{n_3}{f'^*}. \tag{11.37}$$

Combining these with (11.29), and using (11.34) and (11.35) shows that

$$P^* = \frac{P_{\mathrm{b}}}{1 - \delta P_{\mathrm{b}}} + P_{\mathrm{a}} \tag{11.38}$$

and

$$P'^* = \frac{P_{\mathrm{a}}}{1 - \delta P_{\mathrm{a}}} + P_{\mathrm{b}}, \tag{11.39}$$

with $\delta = t/n_2$. This is an equivalent way of determining the location of the focal points of the combined optical system.

Equations (11.38) and (11.39) look very similar to the effectivity formula, (11.19), and for good reason: they are very closely related. Let us consider the effectivity formula for a thin lens "a" with refractive power P_{a} that is moved a distance $t = n_2\delta$ to the right, so it would be coincident with thin lens "b" with refractive power P_{b}. If its effectivity were not to change, its power must change from P_{a} to $P'_{\mathrm{a}} = P_{\mathrm{a}}/[1 - (t/n)P_{\mathrm{a}}]$, as given by the effectivity formula with $D = -t$. (Remember that D corresponds to a leftward displacement.) The new refractive power at the second lens is then $P'_{\mathrm{a}} + P_{\mathrm{b}}$. This gives (11.39). (Of course, now everything is referred to the principal planes of the optical systems rather than the location of the thin lenses.) Equation (11.38) is similarly obtained.

Because $P'P'_b$, $F'P'$, P_aP, and FP are known, the nodal points F and F′ can be found using $PP' = NN'$ and $FP = N'F'$, with F′ following F.

This approach can be used to evaluate complex optical systems. For example, for the eye it can be applied first to the cornea as a lens (see below), then to the crystalline lens, and then to combine these two optical elements to describe the paraxial optics of the entire eye. If the subject wears corrective lenses, the combined effect of these lenses and the eye can then be evaluated.

Thick Lenses. The focal length of a thick lens of thickness t follows from (11.29) and substituting $P_a = (n_2 - n_1)/R_{12}$ and $P_b = (n_3 - n_2)/R_{23}$ from (11.9)

$$\frac{n_3}{f_{\text{eff}}} = \frac{n_2 - n_1}{R_{12}} + \frac{n_3 - n_2}{R_{23}} - \frac{(n_2 - n_1)(n_3 - n_2)t}{n_2 R_{12} R_{23}} \tag{11.40}$$

and the back focal length is

$$f_{\text{bfl}} = f_{\text{eff}} - \frac{f_{\text{eff}}(n_2 - n_1)t}{n_2 R_{12}}. \tag{11.41}$$

The front focal length can similarly be determined.

For a single lens of refractive index n in air, the distance rays propagate between refractive surfaces affects the focal length. With the refractive surfaces separated by a distance t, the Lensmaker's equation, (11.17), for thin lenses can now be generalized to the *thick lens equation*

$$\frac{1}{f_{\text{eff}}} = (n - 1)\left(\frac{1}{R_{12}} - \frac{1}{R_{23}} + \frac{(n-1)t}{nR_{12}R_{23}}\right). \tag{11.42}$$

This effective focal length is the distance from the second principal point. The distance from the second surface to the focus is the (back) focal length f_{bfl}

$$f_{\text{bfl}} = f_{\text{eff}} - \frac{f_{\text{eff}}(n - 1)t}{nR_{12}}, \tag{11.43}$$

so for incident parallel rays the image forms a distance f_{bfl} past the second surface of the lens.

11.3 Imaging and Detection by the Eye

11.3.1 Transmission of Light in the Eye

Figure 11.18 shows the percentage of incident light that reaches the aqueous humor, the crystalline lens, the vitreous humor, and the retina. Perhaps it is surprising that only about 50% of visible light (400–700 nm) incident on the eye actually reaches the retina to form an image. These losses are due to reflection of light at the interfaces between the different ocular media, and the

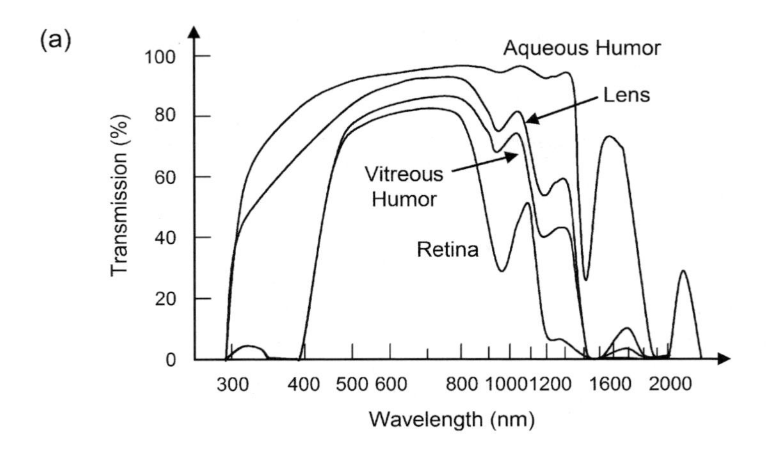

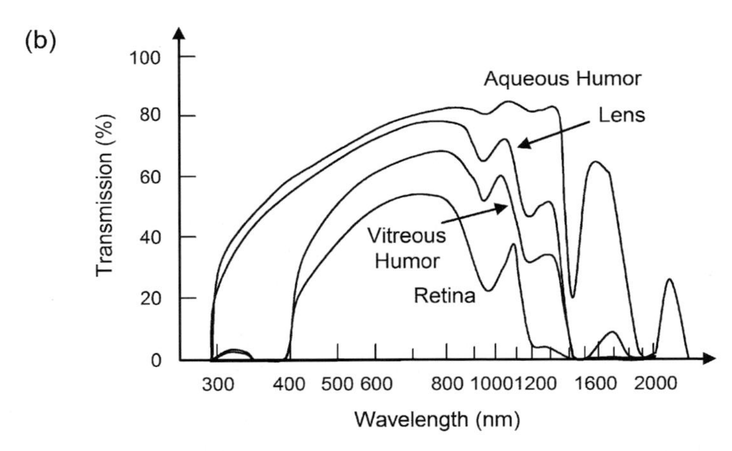

Fig. 11.18. Transmission of near-ultraviolet, visible, and near-infrared light through the eye up to the labeled part of the eye. In (**a**) only transmission losses due to absorption are included, while in (**b**) all transmission losses are accounted for, including those due to absorption and scattering. (Based on [514] and [527])

absorption and scattering in these ocular media. Much of this loss is due to scattering in the eye. This scattered light does not contribute to the desired image even if it hits the retina.

The expression for the fraction of light that reflects from a planar interface between two semi-infinite media with refractive indices n_1 and n_2 is actually the same as that for the reflection of acoustic waves between the interface of media. In Fig. 10.3 and (10.33) the acoustic impedances Z_1 and Z_2 are replaced by the respective refractive indices n_1 and n_2 to give

$$R = \left(\frac{n_2 - n_1}{n_2 + n_1} \right)^2 . \tag{11.44}$$

The reflected fraction at the air/cornea interface $[(1.337-1)^2/(1.337+1)^2] \sim 2\%$ is much larger than that at any other interface in the eye because of the very large difference in refractive indices between air and the cornea. The actual reflected fraction is really a bit larger because the reflected fraction increases for other angles of incidence (and all light rays do not enter the eye at normal incidence) and the ocular interfaces are curved. Still, reflection accounts for little of the transmission losses.

When a penlight is shined into an eye, four images, called *Purkinje images*, are formed from the reflection off the anterior and posterior surfaces of the cornea and crystalline lens. Purkinje image I from the anterior cornea surface is the strongest, as was just shown; it is a virtual, erect image 3.85 mm from the corneal apex. Purkinje image II from the posterior cornea surface is the next strongest; it is a virtual, erect image 3.77 mm from the corneal apex. Purkinje images III and IV from the anterior and posterior crystalline lens surfaces are the weakest; image III is a virtual, erect image 10.50 mm from the corneal apex and image IV is a real, inverted image 3.96 mm from the corneal apex.

Figure 11.18 shows that most of the light from 300 to 400 nm is absorbed by the crystalline lens, mostly by the yellow macular pigment xanthophyll. There are no important sources of absorption in the visible (which is one reason why this light is "visible"). Most of the transmission losses in the visible are due to scattering within the eye components due to inhomogeneities in the index of refraction, such as those caused by cells and submicroscopic particles. The efficiency of scattering attributed by changes in the refractive index over distances $\ll\lambda$, is called Rayleigh scattering. Scattering by a concentration N of nonabsorbing particles of diameter D and refractive index n_{p} in an ambient medium of refractive index n_{a} (such as that of the crystalline lens, cornea, or the humors) decreases the transmission of light intensity (I in $\mathrm{W/m^2}$) according to Beer's Law (10.18)

$$I(z) = I(z=0)\exp(-\alpha_{\text{light scattering}} z), \tag{11.45}$$

where I is the incident intensity (say in $\mathrm{W/m^2}$). For a light of wavelength λ (as measured in vacuum or air)

$$\alpha_{\text{light scattering}} = \frac{8\pi^4}{3}\frac{ND^4}{\lambda^4}\left(\frac{{n_{\mathrm{p}}}^2 - n_{\mathrm{a}}^2}{{n_{\mathrm{p}}}^2 + 2n_{\mathrm{a}}^2}\right)^2. \tag{11.46}$$

(The derivation of this is beyond the current scope.) The important point here is that the efficiency of scattering varies as $1/\lambda^4$ or, using (11.2), as ν^4. So blue light scatters more efficiently than red light, and this explains the greater loss in transmission at shorter visible wavelengths in Fig. 11.18b. (This is also one of the reasons why the sky is blue; blue light scatters more efficiently.)

Table 11.1. Dimensions for Schematic (Schem.) and Reduced (Red.) eyes. (Using data from [513], [525], and [535])

	Schem. exact eye	Schem. eye 1	Schem. eye 2	Schem. eye 2′	Schem. eye 3	Red. eye
radii of surfaces						
anterior cornea	7.70	7.80	7.80	7.80	7.80	5.55
posterior cornea	6.80	6.50	–	–	–	–
anterior lens	10.00	10.20	10.00	5.00	11.00	–
first internal lens	7.911	–	–	–	–	–
second internal lens	−5.76	–	–	–	–	–
posterior lens	−6.00	−6.00	−6.00	−5.00	−6.476	–
distance from anterior cornea						
posterior cornea	0.50	0.55	–	–	–	–
anterior lens	3.60	3.60	3.60	3.20	3.60	–
first internal lens	4.146					
second internal lens	6.565					
posterior lens	7.20	7.60	7.20	7.20	7.30	–
retina	23.9	24.20	23.89	23.89	24.09	–
first principal point P	1.348	1.59	1.55	1.78	1.51	0
second principal point P′	1.602	1.91	1.85	2.13	1.82	0
first nodal point N	7.078	7.20	7.06	6.56	7.11	5.55
second nodal point N′	7.332	7.51	7.36	6.91	7.42	5.55
first focal point F	−15.707	−15.09	−14.98	−12.56	−15.16	−16.67
second focal point F′	24.387	24.20	23.89	21.25	24.09	22.22[a]
refractive indices						
cornea	1.376	1.3771	–	–	–	4/3
aqueous humor	1.336	1.3374	1.3333	1.3333	1.336	4/3
crystalline lens – anterior	1.386	1.4200	1.4160	1.4160	1.422	4/3
crystalline lens – nucleus	1.406					
crystalline lens – posterior	1.386					
vitreous humor	1.336	1.3360	1.3333	1.3333	1.336	4/3

All eyes are accommodated for distant vision (unaccommodated), except Schematic eye 2′, which accounts for accommodation in Schematic eye 2 for near vision (accommodated). Distances are in mm.
[a]For the Reduced eye, the second focal point is 1.67 mm + 22.22 mm = 23.9 mm after the real anterior surface of the cornea.

11.3.2 The Eye as a Compound Lens

Parameters for several progressively simpler *schematic* models for the imaging of paraxial rays in the eye are given in Table 11.1 [513, 516, 518, 519, 525, 535]. In each case the eye is assumed to be fully relaxed (or unaccommodated), with the exception of Schematic eye 2′, which accounts for accommodating Schematic eye 2 from imaging distant objects to imaging near objects. The most complete model is the Gullstrand *exact eye* (Schematic exact eye), which has six refractive surfaces: at the anterior and posterior surfaces of both the

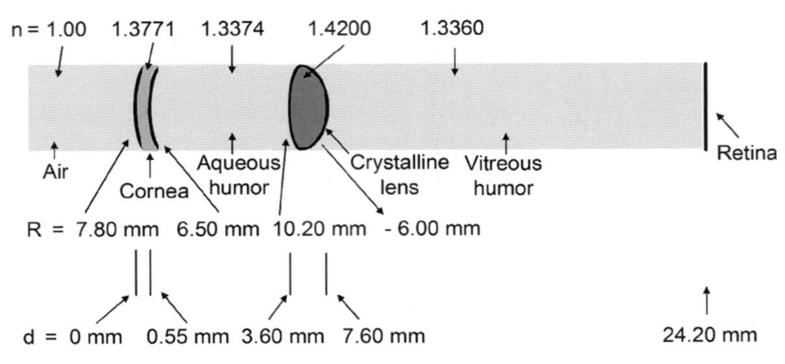

Fig. 11.19. Schematic eye 1

cornea and crystalline lens and two within the crystalline lens, so the variation of the refractive index within the lens is included (in a mathematical, but not a totally accurate physical, manner). (By the way, Allvar Gullstrand was the recipient of the 1911 Nobel Prize for Physiology or Medicine for his work in this area, the dioptrics of the eye. Dioptrics is the branch of geometrical optics dealing with the formation of images by refraction, especially by lenses.)

We will call the next simplest model Schematic eye 1 (the classic Emsley model), for which the refractive index within the crystalline lens is uniform, so there are only four refractive surfaces: at the anterior and posterior surfaces of both the cornea and crystalline lens. Figure 11.19 is a diagram of this model, showing the radii or curvature and locations of the four interfaces and the refractive indices of the five media. The given refractive index of the crystalline lens, 1.4200, is an averaged value; it ranges from about 1.406 near the center to about 1.386 far away from the center.

The cornea and aqueous humor are treated as one region in the *simplified schematic eyes* (Schematic eyes 2, 2′, and 3) so it has three refractive surfaces: at the cornea and at the two surfaces of the crystalline lens. Schematic eyes 2 and 2′ are the Gullstrand–Emsley models of relaxed and accommodated eyes and Schematic eye 3 is the revised relaxed eye model by Bennett and Rabbetts. The top part of Fig. 11.20 shows the location of the cardinal points for Schematic eye 2.

The simplest model is the *Reduced eye* (Fig. 11.21), which has only one refractive interface: at the "cornea" – which is actually 1.67 mm after the real cornea. This is seen along with the cardinal points for this Reduced eye in the bottom part of Fig. 11.20. For this eye all distances are relative to the single refractive interface (1.67 mm after the real anterior surface of the cornea), so the second focal point is 1.67 mm + 22.22 mm = 23.9 mm after this corneal surface.

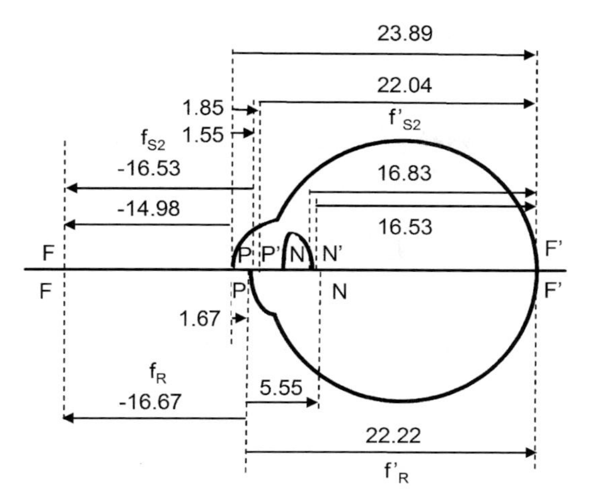

Fig. 11.20. Comparison of the cardinal points of the three-surface, relaxed Gullstrand–Emsley Schematic eye 2 in the *top* half and the Reduced eye in the *bottom* half. Distances are in mm. (Based on [513])

Thin Lens Approximation of the Schematic Eye

Consider Schematic eye 1 with four interfaces (Table 11.1). The refractive power of each interface is

$$P_{i,i+1} = \frac{n_{i+1} - n_i}{R_{i,i+1}}. \tag{11.47}$$

We will initially ignore the distance the rays propagate between the refractive interfaces. For the air/anterior cornea interface:

$$P_{12} = \frac{1.3771 - 1.0}{0.0078\,\mathrm{m}} = 48.35\,\mathrm{D}. \tag{11.48}$$

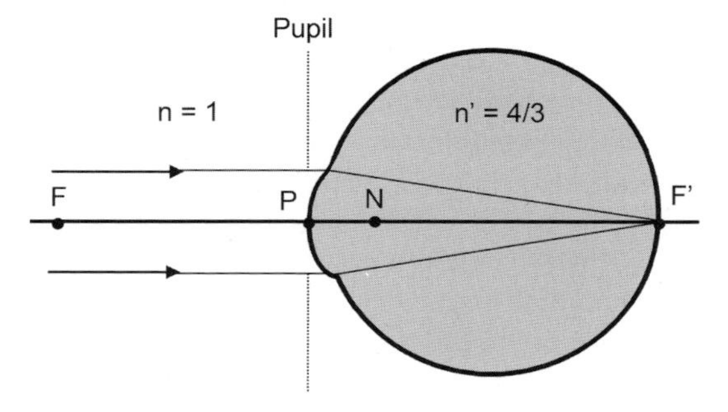

Fig. 11.21. Reduced eye. (Based on [513])

For the posterior cornea/aqueous humor interface:

$$P_{23} = \frac{1.3374 - 1.3771}{0.0065\,\mathrm{m}} = -6.11\,\mathrm{D}. \tag{11.49}$$

For the aqueous humor/anterior crystalline lens interface:

$$P_{34} = \frac{1.4200 - 1.3374}{0.0102\,\mathrm{m}} = 8.10\,\mathrm{D}. \tag{11.50}$$

For the posterior crystalline lens/vitreous humor interface:

$$P_{45} = \frac{1.3360 - 1.4200}{-0.0060\,\mathrm{m}} = 14.00\,\mathrm{D}. \tag{11.51}$$

The refractive power of the cornea, ignoring its thickness, is (11.16)

$$P_{\mathrm{cornea}} = P_{12} + P_{23} = 48.35\,\mathrm{D} - 6.11\,\mathrm{D} = 42.24\,\mathrm{D}. \tag{11.52}$$

The refractive power of the crystalline lens, again ignoring its thickness, is

$$P_{\mathrm{lens}} = P_{34} + P_{45} = 8.10\,\mathrm{D} + 14.00\,\mathrm{D} = 22.10\,\mathrm{D}. \tag{11.53}$$

This shows that two-thirds of the refractive power is due to the cornea and one-third is due to the lens.

The total refractive power of the eye is

$$P_{\mathrm{eye}} = P_{\mathrm{cornea}} + P_{\mathrm{lens}} = 42.24\,\mathrm{D} + 22.10\,\mathrm{D} = 64.34\,\mathrm{D}. \tag{11.54}$$

The focal length is (11.20)

$$f = \frac{n_j}{\sum_{i=1 \text{ to } j-1} P_{i,i+1}} = \frac{1.336}{64.34\,\mathrm{D}} = 0.0208\,\mathrm{m} = 20.8\,\mathrm{mm}. \tag{11.55}$$

The image from this compound lens falls on the retina, which is 24.20 mm from the anterior surface of the cornea and so this calculated focal length is not exactly correct. The compound lens is 7.6 mm long (anterior surface of the cornea to the posterior surface of the crystalline lens), so we would expect that the focal length is really measured for this type of compound lens from somewhere between the cornea and crystalline lens. We will estimate that it is from the middle (at the position 3.8 mm), so we would expect the image to fall 3.8 mm + 20.8 mm = 24.6 mm from the anterior surface of the cornea, compared to 24.20 mm. This agreement is surprisingly good. (Because the refractive power of the cornea and crystalline lens are not equal, the "starting point" is not exactly in the center, but this is a reasonable first guess.) We have ignored the propagation of light between the curved, refractive interfaces, which is not insignificant here, and will address it in the next section.

What happens when you swim in water? The refractive power of the first (air/anterior cornea) interface changes to

$$P_{12} = \frac{1.3771 - 1.331}{0.0078\,\mathrm{m}} = 5.91\,\mathrm{D}, \tag{11.56}$$

which is a loss of 42.44 D of refractive power. The refractive power of the cornea is $5.91\,\mathrm{D} + (-6.11\,\mathrm{D}) = -0.2\,\mathrm{D}$, which means the cornea has essentially no refractive power under water. The total refractive power of the eye is only $-0.2\,\mathrm{D} + 22.10\,\mathrm{D} = 21.90\,\mathrm{D}$, and the eye sees very blurred images because the focused image would be beyond the retina. (Why can we see much better in water when wearing ordinary goggles?) The images in water are made even blurrier when the water is not perfectly still, because the movement of water causes local variations in the index of refraction.

More Exact Analysis of the Schematic Eye (Advanced Topic)

How important is the finite separation of the refractive elements in the eye, as represented by Schematic eye 1 (Fig. 11.19)?

Using (11.29), (11.48), and (11.49), the effective power of the cornea is

$$P_{\mathrm{cornea}} = P_{12} + P_{23} - \frac{t_{\mathrm{cornea}}}{n_{\mathrm{cornea}}} P_{12}P_{23} \tag{11.57}$$

$$= 48.35\,\mathrm{D} - 6.11\,\mathrm{D} - \left(\frac{0.00055\,\mathrm{m}}{1.3771}\right)(48.35\,\mathrm{D})(-6.11\,\mathrm{D}) = 42.36\,\mathrm{D} \tag{11.58}$$

for a cornea thickness $t_{\mathrm{cornea}} = 0.55\,\mathrm{mm} = 0.00055\,\mathrm{m}$ and refractive index $n_{\mathrm{cornea}} = 1.3771$. (Remember, $1\,\mathrm{D} = 1/\mathrm{m}$.) This $+0.12\,\mathrm{D}$ correction to $P_{12} + P_{23} = 42.24\,\mathrm{D}$ from (11.52) is very small because the cornea is very thin. Using (11.30) and (11.31), the principal points of the cornea $\mathrm{P}_{\mathrm{cornea}}$ and $\mathrm{P}'_{\mathrm{cornea}}$ are $-0.058\,\mathrm{mm}$ and $-0.060\,\mathrm{mm}$ to the right of the anterior surface of the cornea, meaning they are 0.058 mm and 0.060 mm to the left of this surface.

Using (11.29), (11.50), and (11.51), the effective power of the crystalline lens is

$$P_{\mathrm{lens}} = P_{34} + P_{45} - \frac{t_{\mathrm{lens}}}{n_{\mathrm{lens}}} P_{34}P_{45} \tag{11.59}$$

$$= 8.10\,\mathrm{D} + 14.00\,\mathrm{D} - \left(\frac{0.0040\,\mathrm{m}}{1.420}\right)(8.10\,\mathrm{D})(14.00\,\mathrm{D}) = 21.78\,\mathrm{D} \tag{11.60}$$

for a crystalline lens thickness $t_{\mathrm{lens}} = 4.0\,\mathrm{mm} = 0.0040\,\mathrm{m}$ and refractive index $n_{\mathrm{lens}} = 1.420$. This includes a $-0.32\,\mathrm{D}$ correction to $P_{34} + P_{45} = 22.10\,\mathrm{D}$ from (11.53). Using (11.30) and (11.31), the principal points of the crystalline lens $\mathrm{P}_{\mathrm{lens}}$ and $\mathrm{P}'_{\mathrm{lens}}$ are 2.42 mm and 2.60 mm to the right of the anterior surface of the lens (meaning they are inside the lens) and they are 6.02 mm and 6.20 mm to the right of the anterior surface of the cornea.

Combining these two lenses using (11.29), the effective power of this Schematic eye 1 is

$$P_{\mathrm{eye}} = P_{\mathrm{cornea}} + P_{\mathrm{lens}} - \frac{t_{\mathrm{aqueous\ humor}}}{n_{\mathrm{aqueous\ humor}}} P_{\mathrm{cornea}} P_{\mathrm{lens}} \tag{11.61}$$

$$= 42.36\,\mathrm{D} + 21.78\,\mathrm{D} - \left(\frac{0.00608\,\mathrm{m}}{1.3374}\right)(42.36\,\mathrm{D})(21.78\,\mathrm{D}) = 59.95\,\mathrm{D}. \tag{11.62}$$

Referencing the six cardinal points to the anterior surface of the cornea (a positive number means the point is to the right of it and a negative number means it is to the left of it), the principal points P and P′ are at 1.59 mm and 1.91 mm. The first and second principal foci F and F′ are at −15.09 mm and 24.20 mm, which means they are respectively $f = 16.68$ mm to the left of P and $f' = 22.29$ mm to the right of P′. Using PP′ = NN′ and FP = N′F′, the nodal points N and N′ are at 7.20 mm and 7.52 mm, respectively. (Except for round-off error, these are the same as the values listed in Table 11.1.)

11.3.3 Accommodation

The controllability in the focal length of the eye crystalline lens, and the corresponding changes in overall eye focal length, is called *accommodation.* The eye crystalline lens is suspended by ligaments and the tension in these ligaments controls the curvature of the crystalline lens surfaces and consequently its focal length. These ligaments are attached to the ciliary muscles and the state of these muscles determines the tension in these ligaments (Fig. 11.22). When the ciliary muscles are relaxed (Fig. 11.22b), the tension in the suspensory ligaments is at a maximum, the crystalline lens flattens and the focal length is at a maximum. The lens equation shows that this flatter lens will produce clear images of distant objects. When the ciliary muscles are contracted, the ligaments are relaxed (Fig. 11.22a) and the crystalline lens assumes its more normal spherical shape. The larger curvature of the crystalline lens surfaces produces a shorter focal length, which produces clear images of nearby objects. Schematic eye model 2′ for near vision (accommodated) in Table 11.1 differs from Schematic eye models 2 for distant vision (unaccommodated) by a large change in the curvature of the anterior surface of the lens and a smaller change in the curvature of the posterior surface. In some people the change in the posterior surface may be very small and there may be a change in the effective refractive index of the crystalline lens in accommodation. (Analysis of accommodation is addressed further in Problems 4.5, 11.13, 11.15, and 11.43.)

Such tunability in eye focal length should allow people to see objects clearly both far and near. The maximum distance from the eye where objects form sharp images is called the *far point* (FP) – and we would like our FP to be ∞. The nearest distance where objects are clear is the *near point* (NP) – and we would like our NP to be 25 cm (= 10 in) for convenient reading and such.

How much accommodation do we need? First we need to find what the focal length of the eye f_{eye} – due to the combined focusing of the cornea

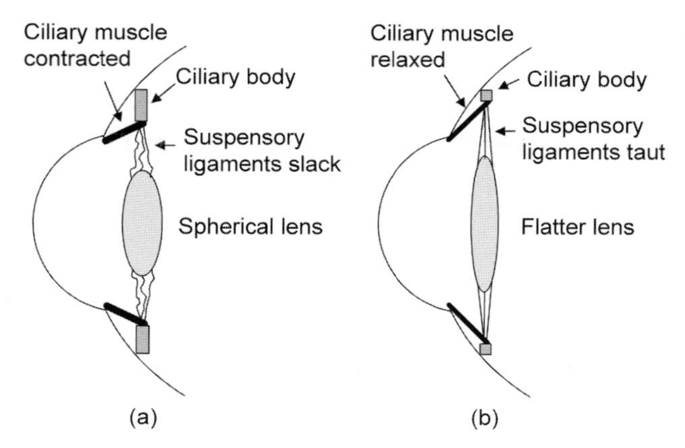

Fig. 11.22. Schematic of the physiological mechanism for accommodation by changing the crystalline lens focal length. (**a**) For imaging nearby objects, the ciliary muscles contract, which reduces the tension on the suspensory ligaments. This allows the lens to be rounder (its natural shape), which decreases the eye focal length and enables the eye to focus on nearer objects. (**b**) For imaging distant objects, the ciliary muscles relax, which increases the tension on the suspensory ligaments. This makes the lens flatter, which increases the eye focal length and enables the eye to focus on more distant objects. The ciliary muscle and suspensory ligaments encircle the eye, and are not as depicted here

and crystalline lens – needs to be to achieve far and near point vision (Fig. 11.22). In both cases the eye lensing system must image the object on the retina. In the Standard eye model, the effective eye lens acts as if it were in air a distance $L_{\mathrm{eye}} = 17\,\mathrm{mm} = 0.017\,\mathrm{m}$ before the retina. At the far point, FP $= d_1 = \infty$ and $d_2 = L_{\mathrm{eye}} = 0.017\,\mathrm{m}$, so (11.3) shows that $f_{\mathrm{eye}} = d_2 = 0.017\,\mathrm{m}$ or $1/f_{\mathrm{eye,FP}} = 1/0.017\,\mathrm{m} = 58.8/\mathrm{m} = 58.8\,\mathrm{D}$. At the near point, NP $= d_1 = 25\,\mathrm{cm} = 0.025\,\mathrm{m}$ and $d_2 = L_{\mathrm{eye}} = 0.017\,\mathrm{m}$, so $f_{\mathrm{eye}} = 15.9\,\mathrm{mm} = 0.0159\,\mathrm{m}$ or $1/f_{\mathrm{eye,NP}} = 1/0.0159\,\mathrm{m} = 62.8/\mathrm{m} = 62.8\,\mathrm{D}$. The necessary accommodation is

$$\frac{1}{f_{\mathrm{eye,NP}}}\,(= 62.8\,\mathrm{D}) - \frac{1}{f_{\mathrm{eye,FP}}}\,(= 58.8\,\mathrm{D}) = 4.0\,\mathrm{D}. \qquad (11.63)$$

Table 11.2 shows that people 45 years old and older tend to have less than this amount of accommodation, and therefore need to wear corrective lens to be able to see both near and far. If vision at the far point at infinity is clear (as it often is), then 55 year olds need to wear convex corrective lenses with ($4.0\,\mathrm{D} - 1.3\,\mathrm{D} =$) $2.7\,\mathrm{D}$ correction to read a book held 25 cm from their eyes. This lack of appropriate accommodation is called *presbyopia* ("old eyes" or "old age vision"). Most of the loss of accommodation with age can be explained by the increase in the Young's moduli of the (asymmetric) crystalline lens, a flattening of the lens, and a decrease in the Young's modulus of the lens capsule with increasing age [521]. Typical amounts of accommodation cannot

Table 11.2. Mean accommodation vs. age. (Using data from [559])

age (years)	mean accommodation (diopters, D)
8	13.8
25	9.9
35	7.3
40	5.8
45	3.6
50	1.9
55	1.3

correct the loss of refractive power for seeing in water, even for very young people.

11.3.4 Field of View and Binocular Vision

Our *field of view* of vision is quite amazingly large and our vision is quite good over this range, given the imaging distortions described later. For each eye the field of vision is approximately an ellipse that is about 150° high and 210° wide. The binocular field of vision, seen by both eyes together, is approximately a circle that is 130° in diameter. It is difficult to achieve this large field of view in optical systems unless the imaging surface is curved, and not flat as in film, film plates, or charge-coupled device (CCD) arrays. This is related to the *curvature of field* aberration. A large field of view with minimal distortion is achieved in our eyes because of the spherical surface of the retina. Of course, we form sharp images only in the region we fixate on because that region images on the fovea. In humans, the ocular muscles can rotate the field of view by 90° in each eye.

In most people, the brain *fuses* the different retinal images from both eyes to give a single perceptual image. This *binocular* vision provides depth perception beyond that from the monocular vision clues from the sizes of objects, the occlusion of a farther object by a nearer one, perspective (such as parallel lines appearing to converge far away), and motion parallax (that nearby objects tend to move more than objects that are farther away as the viewer moves). Such fusion does not occur in people with *diplopia*, who instead perceive two separate images. In people with *strabismus* the motion of the two eyes is not coordinated well and they cannot fixate in a binocular manner on objects. (Some of their ocular muscles are surgically shortened to help trick the brain to maintain binocular vision.) From now on we will address only each individual eye.

11.3.5 Adjustments of Light Levels

So far we have focused on forming a sharp image on the retinal detector. For proper vision it is equally important that this image is neither too dim nor too bright – but exactly what does too dim or too bright mean? Conventional photographic emulsion film has a dynamic range over which it can detect signals. If the incident light is too weak, no image is observable, and if it is too strong, everything is bleached (all is white). It is similar in the charge-coupled detector (CCD) elements in digital cameras. The retinal rods and cones also have a dynamic range over which they can lead to high-quality perceived images. In a still camera, the light level incident on the detector needs to be adjusted by changing the aperture diameter, altering the length of exposure time, and possibly by using a flash. In a motion camera, it needs to be adjusted by changing the aperture diameter and possibly by using external lighting. Humans adjust the level of light needed for vision in several conscious ways (changing ambient light levels) and subconscious ways (adjusting apertures and the bleaching and recovering of detector elements).

While we do not have flashes, we frequently consciously adjust light levels, increasing them by turning on more room lights or by using flashlights and decreasing them outside by putting on sunglasses. Our brains try to do the rest subconsciously.

Decreasing pupil size decreases light transmission to the retina. A change from the maximum to minimum pupil diameter, from roughly 8 to 2 mm, corresponds to a 16-fold change in area and a 16-fold decrease in the amount of light incident on the retina. In one typical example, the pupil diameter decreases from 7 to 4 to 3 to 2 mm as the luminance increases from 10^{-3} to 10 to 10^2 to $10^4\,\mathrm{cd/m^2}$. (The units of luminance are described later in this chapter; see Fig. 11.50 and Table 11.4.) Even with the change in pupil size, the amount of light entering the eye and imaging on the retina is much larger at high light levels. In the cited range the luminance × the pupil area increases, from $(\pi/4)(7\,\mathrm{mm})^2(10^{-3}\,\mathrm{cd/m^2}) = 4\times10^{-8}\,\mathrm{cd}$ to $(\pi/4)(2\,\mathrm{mm})^2(10^4\,\mathrm{cd/m^2}) = 3\times10^{-2}\,\mathrm{cd}$; this is a range of a million. Either the dynamic range of light levels over which our photoreceptors are sensitive is extraordinarily large or the sensitivity of these retinal detector elements, the rods and cones, must also decrease with increasing light intensity.

Rods are primarily important in dark-adapted (scotopic) vision, and cones for light-adapted (photopic) vision. The most important factor in our ability to see under a very wide range of light intensities is the combination of the bleaching of rhodopsin and its subsequent recovery by the processes in Fig. 11.23, and changes in retinal networking; changes in pupil diameter have a much smaller effect. After bright light exposure, all-*trans* retinal is released from the photopigment and is converted to all-*trans* retinol, which is transported to an adjacent cell layer. The all-*trans* retinol is converted to 11-*cis* retinal, which returns to the photoreceptor and combines with opsin to reform rhodopsin. For bright normal lighting, much of the rhodopsin in rods is "bleached" and that in cones is relatively "unbleached." The recovery time for rod rhodopsin is about 20–30 min, and because vision is most sensitive when the supply of rhodopsin in rods is maximum, the adaptation of eyes to night vision takes up to 30 min (see Fig. 11.51).

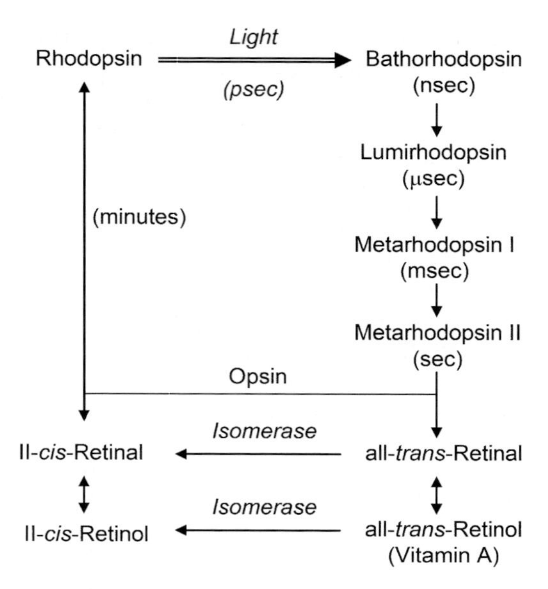

Fig. 11.23. After retinal exposure to light, rhodopsin in the rods reforms slowly by these chemical processes, with the time scales shown. (Based on [526])

The variation of the fraction of available photopigment p with time is described by

$$\frac{\mathrm{d}p(t)}{\mathrm{d}t} = -\frac{I(t)p(t)}{Q} + \frac{1 - p(t)}{\tau}, \tag{11.64}$$

where $I(t)$ is the level of illumination on the retina, Q is the energy in a flash required to reduce the fraction of unbleached photopigment to 1/e, and τ is the regeneration time constant. There are separate relations for rods and cones, with different Q and τ; τ is approximately 2 min for cones and 6 min for rods. For steady-state illumination at level I_{ss} (with the left-hand side of (11.64) set equal to zero)

$$p(I_{\mathrm{ss}}) = \frac{I_0}{I_0 + I_{\mathrm{ss}}}, \tag{11.65}$$

where $I_0 = Q/\tau$. When the light is turned off at $t = 0$ and so $I(t) = 0$, p recovers from its initial value, say p_{i}, to its unbleached value p_{max} as

$$p(t) = p_{\mathrm{i}} + (p_{\mathrm{max}} - p_{\mathrm{i}})(1 - \exp(-t/\tau)). \tag{11.66}$$

In addition to this regeneration of rhodopsin during recovery that reverses bleaching, there are reversible changes in retinal networking that control the summation of retinal signals. This is discussed a bit further in Sect. 11.5.

We could express the current signal strength of the image going to the brain, S (at time t_{now}), as a function f of the light intensity I that is currently incident on the eye, the current pupil diameter d, and the intensity entering the eye in the previous 30 or so min: $S(t_{\text{now}}) = f[I(t_{\text{now}}), d(t_{\text{now}}), I(t_{\text{now}} - 30 \text{ min} < t < t_{\text{now}})]$. Clearly, this is a fairly complex function. It is used by the brain for conscious and subconscious feedback and control of vision, such as to control the pupil diameter $d(t_{\text{now}})$ (Chap. 13).

11.3.6 Limitations to Visual Acuity

There are several fundamental factors that limit *visual acuity* even in those with the best of vision. Under the best of circumstances the imaging process is limited by optical diffraction. There are additional limitations in the image formation process, called *aberrations* (just as in cameras, microscopes, and related imaging systems). Diffraction is the limiting factor for pupil diameters smaller than $\simeq$3 mm, whereas aberrations limit acuity for larger diameters. Moreover, the transfer of the image to the brain is limited in principle by the discrete and finite nature of the rod and cone photodetectors (*pixel size*) and by whether their outputs are interconnected to the same neurons.

Diffraction and Pixel Size

When water waves, say with a linear wave front and of wavelength λ impinge on a slit with width D, the transmitted wave spreads out in angle; this is diffraction (Fig. 11.24). The smaller the slit, the wider the diffraction angle. At a screen placed a distance z downstream there is a sinusoidal-like pattern of

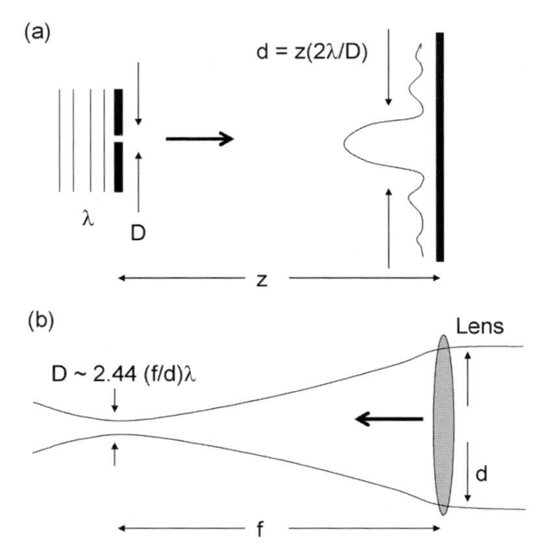

Fig. 11.24. Diffraction of water waves in (**a**) applied to light in (**b**)

waves, whose magnitude gets progressively smaller away from the center of the pattern. The width of the central region is $d = z(2\lambda/D)$, where $2\lambda/D$ is called the diffraction angle. We can run the experiment in reverse and somehow take these water waves with a linear wave front and send them through a "water lens" that focuses them to an angle $2\lambda/D$. After traveling a distance z the wave will converge to a minimum width D given by this same equation, with $D = z(2\lambda/d) = 2(z/d)\lambda$. This distance z is then the focal length f of the lens, so $D = 2(f/d)\lambda$.

This wave picture also describes the propagation of light. A lens of diameter d and focal length f can focus a light beam of wavelength λ to a diameter no smaller than a diameter D, given by

$$D \simeq 2.44\,\frac{f}{d}\,\lambda. \tag{11.67}$$

This is known as the diffraction-limited spot size, because the diffraction of light waves is the fundamental process that limits how small this focus can be even with a "perfect" lens. (d actually refers to the smaller of the lens diameter and the diameter of the aperture (pupil) that may be before or after it. The numerical factor in this equation is instead 1.22 when d is replaced by the appropriate radius.) The quantity f/d is also known as the *f-number* or $f\#$ of the lens or imaging system. Equation (11.67) can also be expressed as $D \simeq f(2.44\lambda/d)$, where $2.44\lambda/d$ is the (full) diffraction angle. This is closely related to the Rayleigh criterion for distinguishing between the diffraction-limited images of nearby points or angular regions; this is discussed further in Problem 11.27. Our earlier tracing of rays was within the *geometric optics* limit, which becomes more and more valid as λ decreases and approaches 0, which causes D to decrease toward 0.

For an eye crystalline lens of diameter of $\approx$8 mm and $f \approx 17$ mm (Standard eye model), this limit is $\approx$2.5 μm for 500 nm light. Because the crystalline lens is not perfect, this image diameter is measured to be larger, $\sim$11 μm. However, because this image is not uniform in brightness – it is brightest in the center and shades off nearer the edges – the resolution is a bit better than 11 μm. In the fovea the average diameter of cones is $\approx$1.5 μm, so images separated by $\sim$2 μm on the fovea can sometimes be barely resolved. We saw earlier that this corresponds to the normal visual acuity in discriminating between point light sources separated by 25 s of arc (corresponding to barely distinguishing two spots that are 1.5–2.0 mm apart and 10 m away). (See the discussion of measuring visual acuity below.)

Because the fovea is $\sim$300 μm in diameter, this high acuity occurs only for 1° of the visual field, corresponding to a width of 20 cm a distance 10 m away. Just outside the fovea, the visual acuity of the retina is worse by a factor of 5–10, and becomes progressively even worse toward the periphery (Fig. 11.25). This occurs because many rods and cones are connected to the same optic nerve fiber outside the fovea [526].

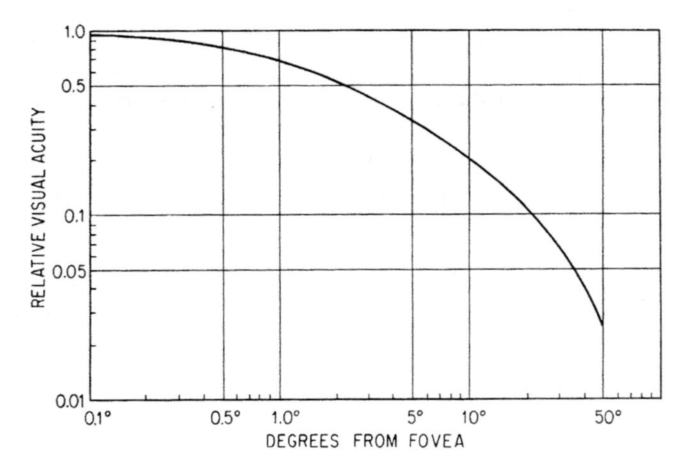

Fig. 11.25. The variation of visual acuity with retinal position relative to the fovea. (From [555]. Reprinted with permission of McGraw-Hill)

Aberrations in Image Formation

The imaging elements in the eye are quite good, but they have imperfections. Two of the more important sources are spherical and chromatic aberration.

In the paraxial limit all rays from the same point image at the same place. Rays making larger angles to the optic axis, image at different places. Such *monochromatic aberrations* depend on the details of the imaging surface and are present even with one color of light. *Spherical aberration* (SA) is due to the inherently imperfect imaging by a spherical surface. Only paraxial rays near the optic axis image according to (11.8) at the nominal focal length f. Horizontal light rays farther from the optic axis (a larger distance y – or half aperture height – away, corresponding to zonal and marginal rays) are refracted by a spherical convex lens more than is predicted by (11.7) and therefore cross the optic axis before the rays nearer the axis cross it (Fig. 11.26). With this definition, SA is positive for spherical convex lenses and negative for spherical

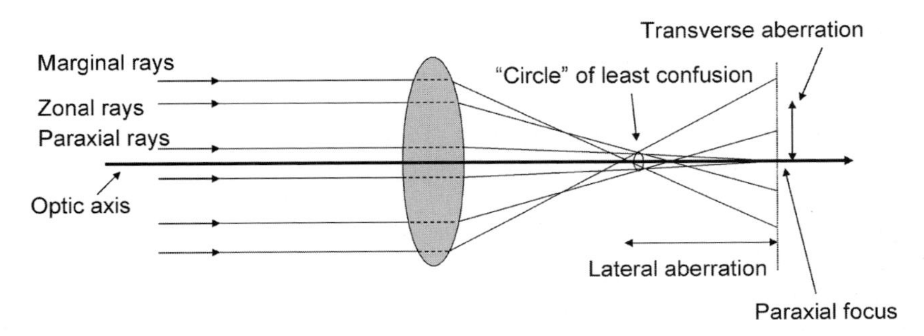

Fig. 11.26. Ray tracing and imaging with spherical aberration present

concave lenses. The distance along the optic axis between the places where these zonal and marginal rays and the paraxial rays hit the optic axis is the *longitudinal* (or axial) spherical aberration L·SA. The height above or below the optic axis a ray hits a screen placed at this focal length is the *transverse* (or lateral) spherical aberration T·SA. (Longitudinal and transverse spherical aberration become larger rapidly with semiaperture size y, as $\propto y^2$ and $\propto y^3$, respectively.) The image is sharpest somewhere between f and f − L·SA, where the image is called the *circle of least confusion.*

Spherical aberration can be limited by using compound lenses (multiple lenses in which the net effect of the aberration is negated), lenses with aspherical surfaces, or simple lenses preceded by apertures that transmit light only near the optic axis. (However, smaller apertures also limit the amount of transmitted light.) Notably, spherical and other aberrations can be greatly affected by the lens shape. From (11.17) it is clear that you can make a lens with a certain focal length, by using infinite sets of pairs of R_{12} and R_{23}. Figure 11.27 shows that for parallel rays incident on a positive lens, spherical aberration is minimized for a convex anterior surface and nearly planar posterior surface (so the rays make nearly equal angles to both surfaces). The eye crystalline lens is biconvex ($R_{12} > 0$ and $R_{23} < 0$ as in Fig. 11.11), but the posterior surface is much flatter than the anterior surface, and this

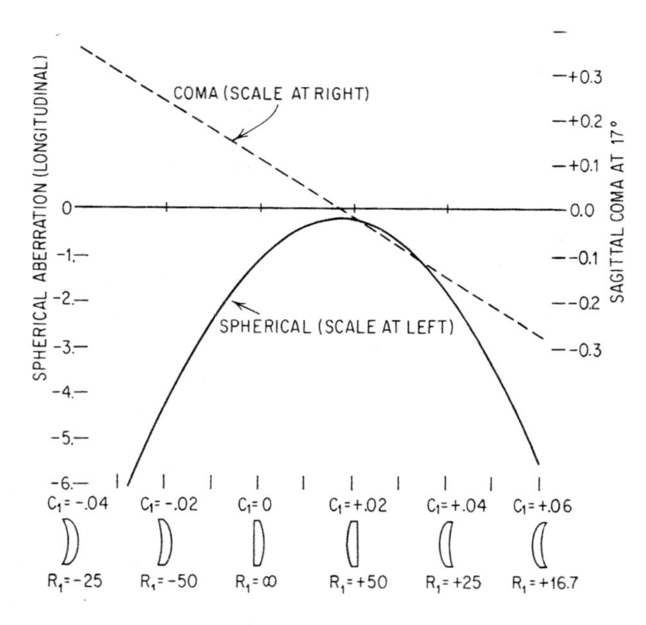

Fig. 11.27. Spherical aberration and coma as a function of lens shape for a 100-mm focal length lens (with a stop at the $f/10$ lens, covering a $\pm 17°$ field). The radius of curvature of the left interface in mm is labeled as R_1 – which is given in the text as R_{12}, and its reciprocal is shown as $C_1 = 1/R_1$. (From [555]. Reprinted with permission of McGraw-Hill)

decreases the usual positive spherical aberration. Man-made lenses usually have a uniform index of refraction. Nature has decreased spherical aberration in the human crystalline lens even further by decreasing the index of refraction from its center in all three directions and decreasing the magnitude of the radius of curvature towards the margin of the lens. Both variations reduce the surface refracting power at the lens margins to decrease SA. In fact, they overcorrect for the SA to compensate for the undercorrected SA at the anterior surface of the cornea. Spherical aberration in the eye decreases with accommodation for more distant vision.

There are several other monochromatic aberrations. Off-axis object points image to a comet-shaped image because the peripheral rays focus closer to the optic axis than the central rays. This is called *coma.* Coma is also present for objects points on the optic axis when the optical elements are not centered and are, in fact, tilted relative to each other, and this occurs in the eye. Points in the object plane that are far from the axis are farther from the lens than those near the optic axis, so they image at shorter distances after the lens. The surface of the image is not planar but curved (Petzval surface), even for rays passing through the center of the lens for this *curvature of field* aberration. A square grid images to the distorted grid shapes with *barrel distortion* or *pincushion distortion*, generally for negative and positive spherical lenses, respectively, because the lateral magnification of the lens varies from its center to periphery.

Moreover, simple, single component lenses have different focal lengths at different wavelengths, because the index of refraction of the lens material differs at different wavelengths. This is known as *dispersion.* Therefore the image can be sharp at one wavelength but a bit blurred at another because of this *chromatic aberration* (CA) (Fig. 11.28). This is common in glass lenses because blue light has a larger refractive index than red light in glass, and consequently glass lenses have a shorter focal length in the blue (Problem 11.32). The distance between where rays at two wavelengths (such as in the red and blue) hit the optic axis is the *longitudinal* (or axial) chromatic aberration L·CA [512]. The vertical distance between the images of the two colors on a screen placed at the focal length is the *transverse* (or lateral) chromatic aberration

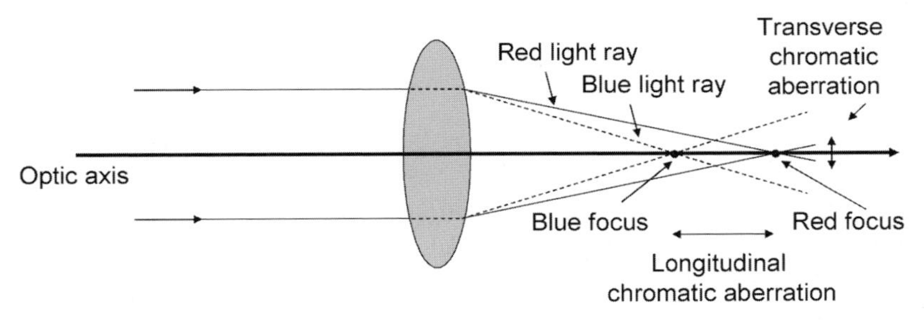

Fig. 11.28. Ray tracing and imaging with chromatic aberration present

Table 11.3. Refractive indices (n) in the ultraviolet (380 nm), visible, and near infrared (780 nm), and, in the last row, the dispersion for ocular media. (Using data from [513])

wavelength (nm) (spectral line)	humors	crystalline lens
380	1.3450	1.4334
480.0 (F′)	1.3378	1.4221
587.6 (d)	1.3333	1.4160
643.8 (C′)	1.3315	1.4138
780	1.3285	1.4101
dispersion constant	52.9	50.1

T·CA or lateral color. In camera and microscope lenses this is corrected by using compound lenses composed of lenses made from different types of glass – with different dispersion. The human aqueous and vitreous humors and the crystalline lens have small, yet significant, dispersion, as seen in Table 11.3. In the crystalline lens this chromatic aberration contributes $\sim$2 D change in optical power across the visible, as described by

$$D(\lambda) = p - \frac{q}{\lambda - c} \tag{11.68}$$

for the defocus in D, with wavelength λ (in μm), $p = 1.7312$, $q = 0.63346$, and $c = 0.21410$, relative to that at 578 nm [557]. The overall change in the refractive power of the schematic eye is plotted in Fig. 11.29. The dispersion

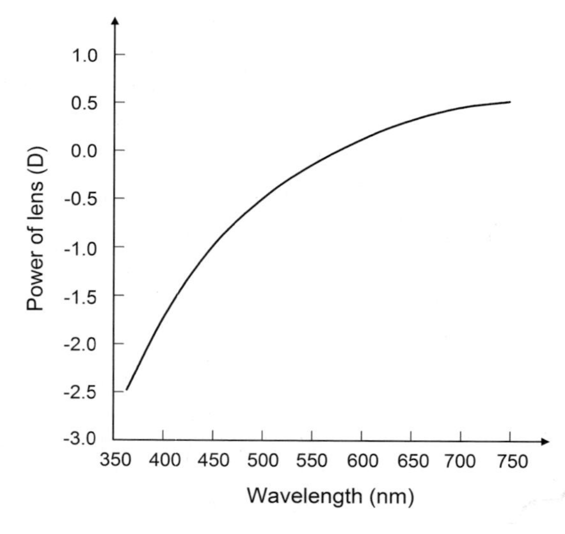

Fig. 11.29. Chromatic aberration of the human eye as quantified by the optical power that must be added to the human eye to bring the eye in the same focus as 587 nm, depicted as the *curve* given by (11.68). (Based on [560])

constant, defined as $[n(\lambda = 587.6\,\text{nm}) - 1]/[n(\lambda = 480.0\,\text{nm}) - n(\lambda = 643.8\,\text{nm})]$, and related matters are discussed in Problems 11.32–11.35.

Quantitative Evaluation of Image Acuity

Because of optical diffraction and these other imaging imperfections, a point really images to a blur with diameter w_B. Optical diffraction contributes $\sim 2.44(f/d)\lambda$. Spherical aberration contributes $\sim 2\text{T} \cdot \text{SA}$ (suitably averaged from the axis to y above it) to w_B and chromatic aberration contributes $\sim 2\text{T} \cdot \text{CA}$ (suitably averaged over the wavelength spread) to it. A more quantitative evaluation of image distortion is often needed than this poorly defined blur diameter. This is provided by measuring the image from a very thin line source, to obtain the *line spread function* (Fig. 11.30), or a point source, to obtain the *point spread function*. These functions provide the relative image intensity vs. distance or visual angle in one or two dimensions, respectively, and have the same full width $\sim w_\text{B}$. Alternatively, we can image a source with an intensity that periodically varies laterally (in the y direction) with spatial frequency k, such as by $(1 + \sin ky)/2$ or the square wave variation shown in Fig. 11.31b, and see how much of the modulation remains in the image. For $k \ll 1/w_\text{B}$ there is much modulation (11.31d), while for $k \gg 1/w_\text{B}$ there is very little modulation (Fig. 11.31e). Figures 11.30 and 11.31 track these changes as a function of position. They could also be tracked in terms of the spatial frequency k. This modulation variation vs. k is the *optical transfer function* or, for symmetric systems – as we will assume here, the *modulation transfer function*; they have widths $\sim 1/w_\text{B}$. Figure 11.32 shows this function for the eye with the line spread function of Fig. 11.30. This inverse relationship between the spread in real space (spread functions) and in spatial frequency

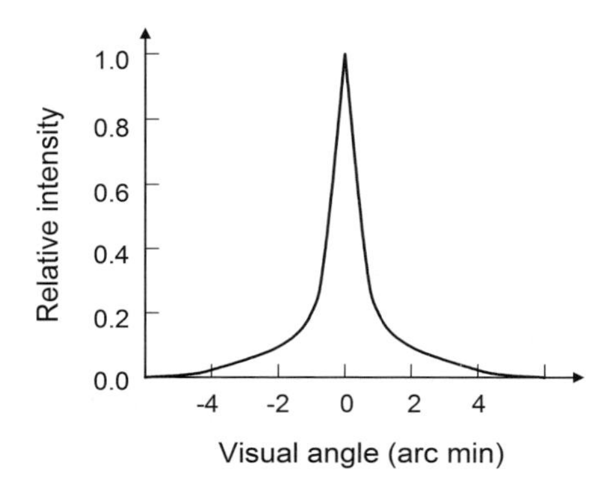

Fig. 11.30. The human line spread function for an eye with 3.0 mm pupil diameter, using a model in [561] (see Problem 11.30) that is based on experiment. (Based on [560] and [561])

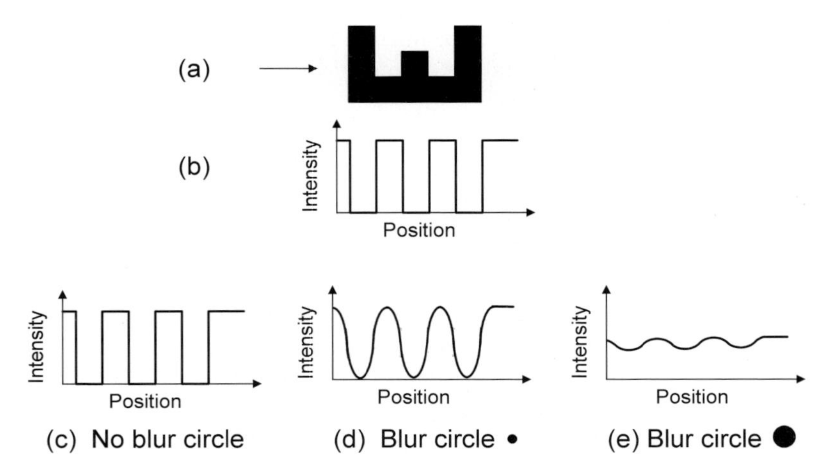

Fig. 11.31. (**a**) The horizontal letter E on it side, and (**b**) the light level in the direction of the *arrow* in (**a**) showing a periodic square wave object. This leads to potential images on the retina for (**c**) ideal imaging, (**d**) imaging with a small blur spot, and (**e**) imaging with a large blur spot

space (transfer functions) is similar to that in the Chap. 10 analysis of the voice intensity in time vs. frequency. In both cases the inverse functions (time vs. frequency there or space vs. spatial frequency here) are related by Fourier analysis as in (10.42) and (10.43) (where t would be replaced by x, $2\pi f$ by k, and the summation by an integral).

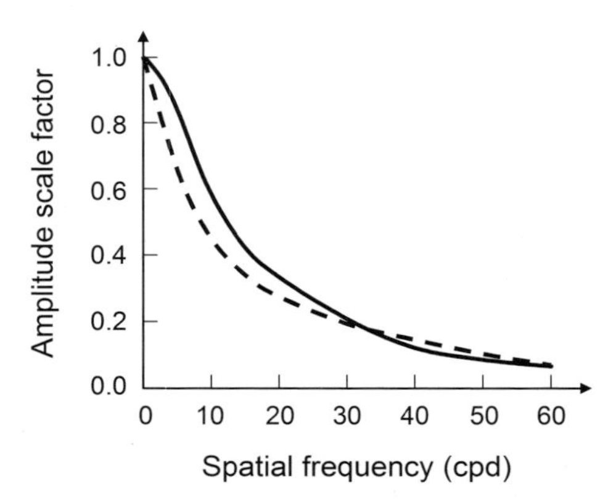

Fig. 11.32. Modulation transfer function of the human eye, as measured in [563] (*dashed curve* drawn through data points (which are not shown), as drawn in [560]) and compared to the predictions using the human line spread model shown in Fig. 11.30 (*solid curve*). The abscissa axis is in units of cycles per degree (cpd). (Based on [560])

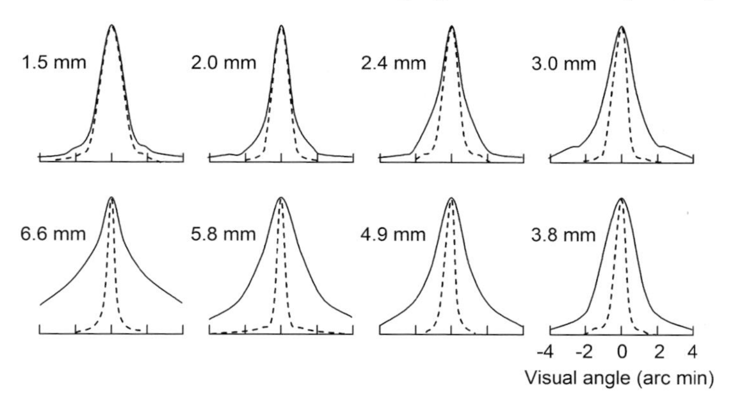

Fig. 11.33. Line spread function for the human eye for different pupil diameters. The *solid curves* are the total functions. The *dashed curves* are the line spread functions for those apertures assuming diffraction-limited conditions. (Based on [517] and [560])

Effects of Changing Pupil Size, Including Depth of Field

The human pupil can change from $\simeq 8$ to $\simeq 2$ mm in diameter. Decreasing the pupil size affects vision in several ways.

1. Decreasing pupil size decreases light transmission to the retina, as described earlier.
2. The size of the diffraction-limited spot varies inversely with aperture diameter (11.67). With the pupil closed all the way to a 2 mm diameter, this diffraction-limited spot diameter is $\approx 10\,\mu$m. Figure 11.33 shows that the contribution to the line spread function by diffraction increases with decreasing pupil size as expected, and that it becomes the dominant contribution to the blur at small pupil sizes.
3. Decreasing pupil size lessens the effect of lens aberrations. Transverse spherical aberration varies as the square or cube of the distance of rays from the optic axis. Figure 11.33 shows that contributions to the line spread function other than diffraction, i.e., due to these and related aberrations, increase remarkably with pupil diameter and become the dominant contribution to the blur at large pupil sizes.
4. The range in object (or source) distances over which good images are formed is called the *depth of field*. This depth increases by decreasing the size of the aperture (pupil) before the crystalline lens. (Conjugate to this is the range of image distances over which the images over the depth of field form, and this is called the *depth of focus*.) Figure 11.34 shows that the blur on the retina due to the different image locations with different source distances decreases with pupil diameter – and this is one manifestation of the depth of focus. Closely related to this is the change in image distance with source distance, as plotted in Fig. 11.35; the depth of focus is related

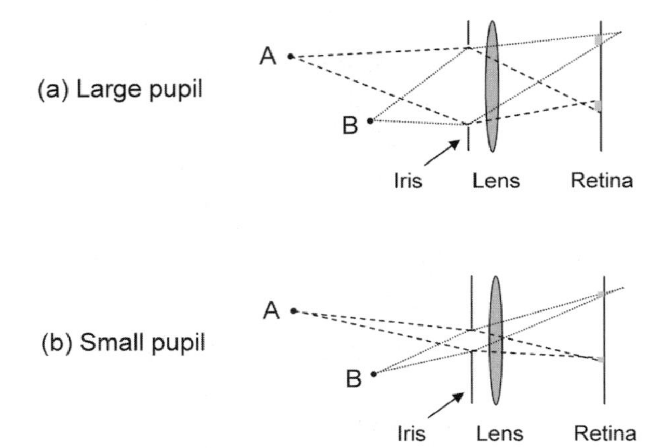

Fig. 11.34. Schematic showing the image blur for two object distances, that alternately image (A) before or (B) after the retina, for a relatively (**a**) large pupil and (**b**) small pupil. This provides an assessment of the depth of field vs. pupil diameter because the blur increases with pupil diameter

to the slope of the curves, so it is large for distant sources and small for nearby objects (and this is not ideal). This can also be easily understood using simple image formation using (11.3).

The pupil loses its ability to dilate with age. The maximum pupil diameter decreases from, ∼10 mm at 1 yr, to 6 mm at 20 yr, to 3 mm at 50 yr, and to 1 mm at 80 yr.

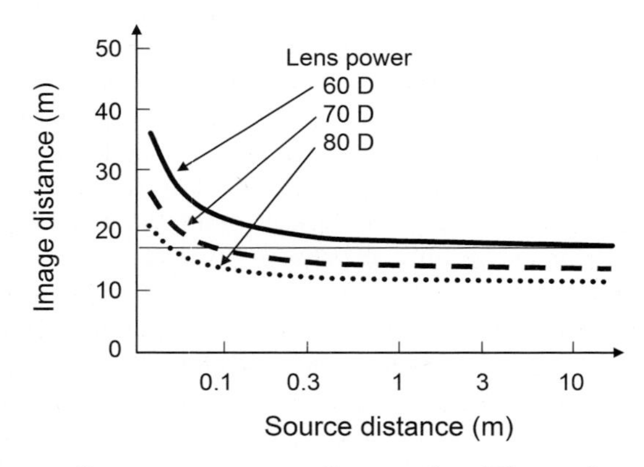

Fig. 11.35. Image distance vs. source distance for different lens, which provides information on the depth of field of the human eye. An extra *horizontal line* is shown for a distance of ≃17 mm, which is the focal length of the Standard eye in air. (Based on [560])

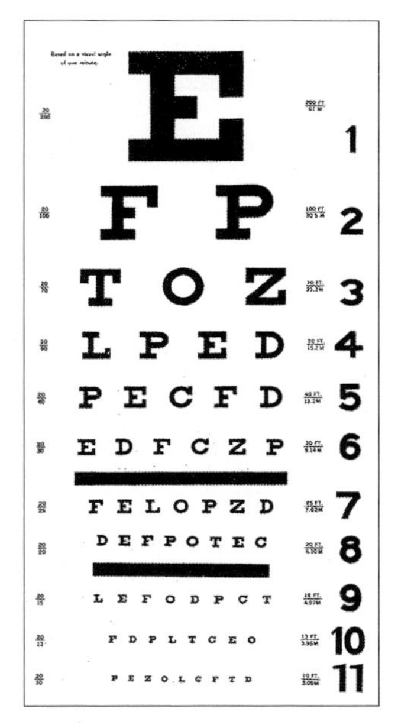

Fig. 11.36. The Snellen eye chart used to test visual acuity. The label of the line with the smallest letters you can read at 20 ft is your vision for that eye, with lines 1–11, respectively, corresponding to 20/200, 20/100, 20/70, 20/50, 20/40, 20/30. 20/25, 20/20, 20/15, 20/13, and 20/10. (From [545])

11.3.7 Imperfect Human Vision

Visual acuity (VA) in human vision is frequently tested by reading black letters (that have 4% reflectance) on a white background (84% reflectance) from a Snellen chart placed 20 ft (or 6 m) away under good light illumination (480 lux is a standard, and some say 800–1,000 lux) (Fig. 11.36). (These units will be explained later in this chapter.) For each eye, the vision is said to be $20/z$ if the smallest letters you can read at 20 ft are what people with "perfect" 20/20 vision (or 6/6 vision in the metric system) in that eye can read z feet away. This is a test designed to assess near-sightedness (or more precisely, lack of far-sightedness); the optical object infinity is considered to be 20 ft. Visual acuity is defined as the reciprocal of the angular size (in minutes of arc at 20 ft) of one of the elements of the letters on the chart. The letter E has three horizontal solid bars and two horizontal open bars (Figs. 11.31a and 11.36). In the 20/20 line, each of these is an element that subtends 1 min of arc and the whole letter subtends 5 min of arc at the eye, at a distance of 20 ft. Each bar in the E is about 1.7 mm wide and the E is 8.8 mm high in the 20/20 line. Someone with 20/40 vision would be able to resolve elements of no less than 2 min of arc and letters of 10 min of arc from the eye. Clearly, this is closely related to the ability to view square wave patterns, as in Fig. 11.31. Normal

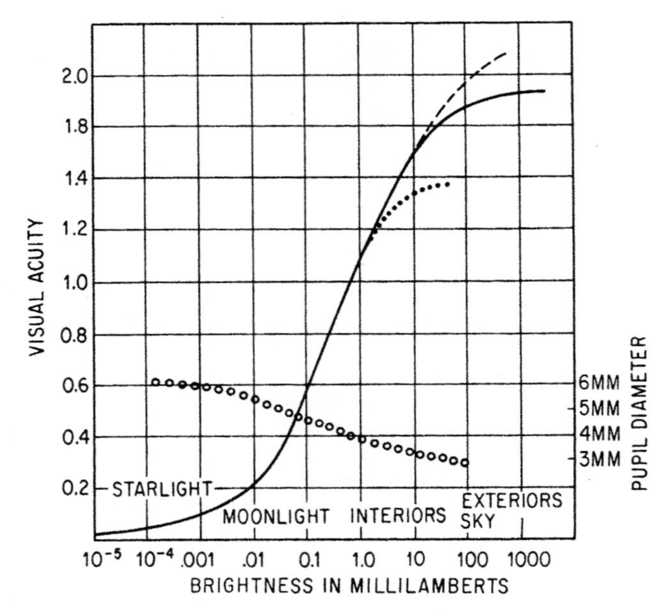

Fig. 11.37. Visual acuity (in 1/min) and pupil diameter (*circles*) as a function of object brightness. The *dashed* and *dotted* acuity lines are, respectively, for increased and decreased surround brightness of 1 milliLambert. (See Table 11.4: 1 milliLambert = 3.183 nit = 3.183 cd/m^2). (From [555]. Reprinted with permission of McGraw-Hill)

VA is 1.0 under normal indoor lighting conditions and decreases with poorer lighting (Fig. 11.37). Someone with 20/40 vision has a VA of 0.5. Under ideal conditions, some people have a VA of 2 or 3. Figure 11.25 illustrates that visual acuity is worse in the periphery than in the fovea. (Can you use the results from the test in this figure to estimate the dimensions of the fovea?) The analogous test for far-sightedness (i.e., lack of near-sightedness) is the Jaeger test, in which one reads lines of various font sizes at normal reading distance.

Emmetropes have perfect imaging (VA $\simeq$ 1), meaning their eyes need a correction less than a stated small amount, such as 0.5 or 1.0 D. They have a FP at infinity and a NP at 25 cm, with at least 4 D of accommodation (Fig. 11.38a). All other people are *ametropes*, and need corrective lenses to see clearly.

People with *myopia* are "near-sighted," meaning that their vision 25 cm away (their NP) is fine, but it needs correction for objects near infinity (VA < 1); the smallest Snellen eye chart line they can read is above the 20/20 line. They have good vision for objects from 25 cm to their far point, which is $<\infty$. Distant objects ($d_1 = \infty$) form images in the eyeball "before" the retina (Fig. 11.38) and negative lenses are needed to project the image on the retina. If the person has 4 D or more of accommodation, the image at 25 cm

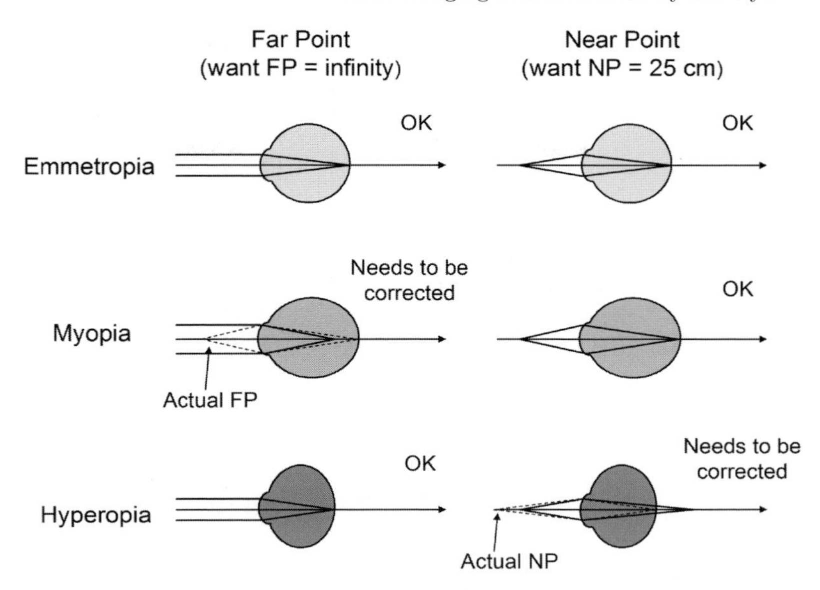

Fig. 11.38. Imaging in emmetropic, myopic, and hyperopic eyes. For the latter two both possible reasons for the conditions are illustrated: a long eyeball and short focal length for myopia and a short eyeball and long focal length for hyperopia

will also be fine with these lenses. Myopia occurs because the eyeball is too long (axial myopia, which is the most common reason) or the cornea is too curved and there is too much refractive power (refractive myopia). One potential contributing factor for myopia is the mechanical stress (Chap. 4) on the posterior sclera due to the muscles that control the eye motion – particularly the oblique muscles – and high vitreous pressure [524]. With possibly lessened tension in the sclera (perhaps due to weakening by the muscles) or higher pressure, the local radius of curvature of the posterior eyeball decreases in equilibrium according to the Law of Laplace (7.9), and this bulging leads to myopia.

People with *hyperopia* are "far-sighted," meaning that their vision at infinity (their FP) is fine, but it needs correction with objects that are near. They have good vision for objects from their near point, which is >25 cm, to their far point at ∞. Near objects ($d_1 = 25\,\mathrm{cm}$) form images "beyond" the retina (Fig. 11.38) and positive lenses are needed to project the image on the retina. If the person has 4 D or more of accommodation, the image at ∞ will also be fine with these lenses. Hyperopia occurs because the eyeball is too short (axial hyperopia, which is the most common reason) or the cornea is too flat and there is too little refractive power (refractive hyperopia).

With the optic axis (horizontal line in each part of Fig. 11.38) called the z-axis, the lateral extents of the object, lens, and image are described by the x and y axes. In everything we have said so far the x and y axes are perfectly interchangeable, and they still would be interchangeable after being rotated by any arbitrary angle about the z-axis. This is no longer true for people with *astigmatism*. Astigmatism is caused by ocular optics that lack rotational symmetry – often because their corneas have unequal curvatures in the x and y axes (a *toric* cornea – a toric lens is also possible) or because an optical element is tilted or displaced from the optic (symmetry) axis. There is a plane (or *meridian*) containing the optic axis in which there is a maximum optical power and a plane normal to that with minimum optical power, for *regular astigmatism*. There is a tendency, but not certainty, for these meridians to be rotated near the transverse and sagittal planes. The power of the vertical (sagittal) meridian is greater for *with-the-rule astigmatism* and that along the transverse meridian is greater for *against-the-rule astigmatism*. In *oblique astigmatism* the highest power meridian is neither vertical nor horizontal. Rarer is *irregular astigmatism*, for which the principal meridians are not normal to each other; this condition is caused by trauma, surgery, or disease. Only regular astigmatism can be corrected by spherocylindrical lenses (see below).

Keratoconus is a condition in which the cornea thins near the center to form a cone and can be locally wavy because of corneal scarring (Fig. 11.39).

Visual efficiency (VE) is sometimes calculated for legal and industrial purposes by using the results of visual acuity, visual field, and ocular motility (diplopia field, binocular field) measurements for corrected eyes. Each of these three measurements is described by a factor ranging from 0 to 1, with

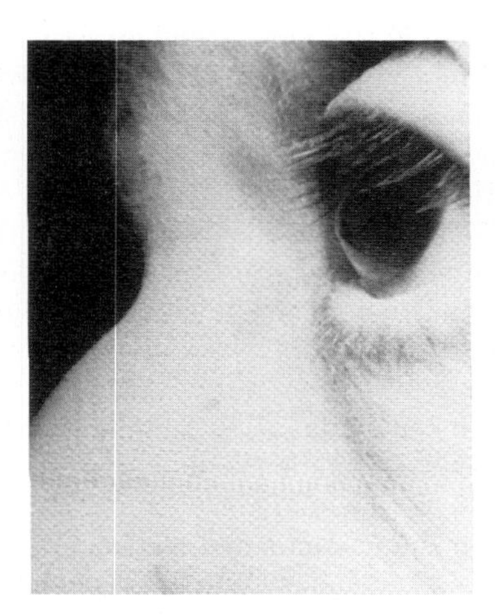

Fig. 11.39. Eye of one with severe keratoconus (pre-1973), showing a very conical cornea that protrudes out much (in a pointed, nonspherical manner) and that exhibits some drooping. The conical anterior surface of the cornea forms a very poor image. Nowadays, patients usually have a corneal transplant before keratoconus reaches this stage. (From [528]. Used with the permission of Jack Hartstein, M. D.; also see [556])

1 representing no loss of vision. They are multiplied together to give VE for an eye. VE for two eyes is a weighted sum of that for each eye, weighting the better eye (higher VE) by 3 and the poorer eye by 1. The visual loss is $1-\mathrm{VE}$. The VA (visual acuity) factor is an average from the Snellen (far) and Jaeger (near) vision tests. For example, the factors from the Snellen test are 1.0 for 20/20 vision, 0.75 (for 25% loss) for 20/50, 0.50 for 20/100, and 0.10 for 20/400. The visual field factor is the fraction of remaining visual field. The ocular motility factor is the fraction of remaining coordinated ocular movement (which is the ability to see one object as one and not two).

Absence of the photopigment in the three cone cells causes various types of color blindness, which affects about 10% of men and 1% of women. *Dichromats* are missing one cone photopigment. *Protanopes*, *deuteranopes*, and (the rarer) *tritanopes* are absent the long-, medium-, and short-wavelength photopigments, respectively. *Monochromats* lack two or all three of these cone photopigments and have no color vision. Color blindness is tested using pseudoisochromatic plates (for example, see [552]).

11.3.8 Correction of Vision by Eyeglasses, Contact Lenses, and Other Means

We will now analyze how eyeglasses, soft contact lenses, and hard/gas-permeable contact lenses can be used to correct vision [528, 538, 539, 541, 543, 556]. The aim is to achieve perfect vision, using a compound lens consisting of the eye (Standard eye model) and the corrective lens, with the far point at $\mathrm{FP} = \infty$ ($f \sim 17.0\,\mathrm{mm}$) and the near point at $\mathrm{NP} = 25\,\mathrm{cm}$ ($f \sim 15.9\,\mathrm{mm}$).

In eyeglasses (or spectacles), the distance between the back (ocular) surface of the lens and the cornea is 12–15 mm; if they were closer the lenses would be brushed by eyelashes during blinking. The topography of the lens surfaces is chosen so the eye encounters the same prescription for any direction the eye gazes through the spectacles. The front surface is usually spherical and the back surface toric. Glass lenses for eyeglasses are usually made of ophthalmic crown glass, which has a refractive index 1.523. Higher index material is sometimes used to reduce the thickness and weight of high power eyeglasses, such as with high index glass with indices >1.70. Because impact resistant eyeglasses require thicker glass, lighter plastic lenses are commonly used for safety glasses. CR39 plastic, with a refractive index of 1.498, is often used; polycarbonate, with a high index 1.586, is highly impact resistant, although it is soft and has large chromatic aberration.

Soft contact lenses adjust to the shape of the cornea, while hard/gas permeable contacts do not. These soft lenses are composed of hydrogel, a gel-like material that is soft and flexible when it is hydrated, and this material often includes 2-hydroxyethyl methacrylate (HEMA). The inner surface of a soft contact lens molds to the outer surface of the cornea, so the corneal topography and the controllable lens thickness (from the lens center) determine the radius of curvature of the outer lens surface, and therefore the imaging correction. The eye tolerates a soft lens, ≃13.0–14.5 mm in diameter, which is larger than the cornea (and which extends to the sclera); this larger size optimizes lens centering and stability. The center thickness of soft contact lenses start from ≃0.03 mm. Most of the oxygen needed by the cornea is supplied by oxygen diffusion through these very thin lenses. Because of this large size, there is very little motion of oxygen-rich tears to the cornea during blinking, and this motion supplies only ∼4–6% of the total oxygen to the cornea.

Rigid or hard lenses are much less pleasant to wear, but are necessary for irregular corneas – as for those with keratoconus (Fig. 11.39) or high levels of astigmatism. Rigid or hard lenses have traditionally been composed of poly(methyl methacrylate) (PMMA). Unfortunately, the oxygen needed by the surface of the cornea cannot penetrate the lens and can be supplied only by tears that slip to the cornea as the lens moves on the cornea during blinking. Because this level of oxygen is lower than needed, the corneas become irritated and such rigid lenses cannot be worn for long continuous stretches of time.

Gas permeable contacts are rigid lenses composed of materials such as silicone acrylate – a combination of silicone and PMMA – or fluorocarbonate silcone acrylate, that let oxygen diffuse to the cornea. This increases the amount of oxygen reaching the cornea to about 10× that by tears alone. Because such lenses can be worn only for most of the waking hours, ∼10–12 h, without excessive discomfort, gas permeable contact lenses are replacing PMMA hard contacts in most cases. These gas permeable lenses are 8–10 mm in diameter and only 0.15 mm thick (with thickness starting at 0.10 mm), and (should) remain over the cornea, over most of the iris. The radius of curvature of the posterior surfaces of a rigid contact lens is chosen to best fit the anterior surface of the cornea. The anterior surface is chosen to produce a perfect image on the retina. (If the topography of the posterior surface of the hard contact lens is different than that of the cornea anterior surface, in some cases it will shape this surface and change the refractive power in this way too. However, when there is a poor fit – especially when fitting to the nonspherical corneas in keratoconus as seen in Fig. 11.39, dust particles can be swept by tears to the region between the contact lens and cornea and this is very painful.) Figure 11.40 shows a three-point fit, which is ideal for keratoconus and similar eye disorders.

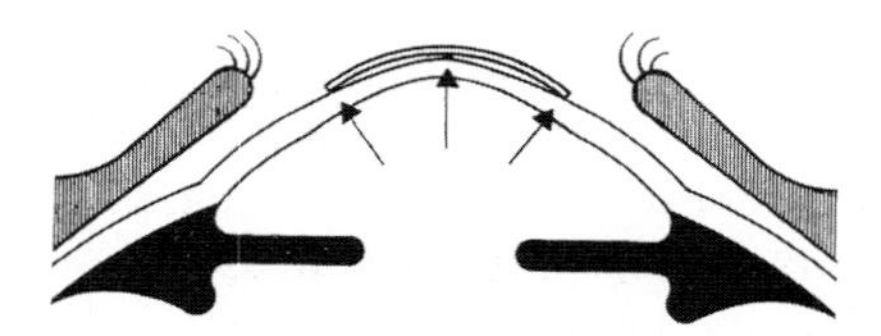

Fig. 11.40. Three point fit of contact lens – apical (apex) touch to the cone plus peripheral touch – for an eye with keratoconus. (Reprinted from [556]. With permission of Elsevier)

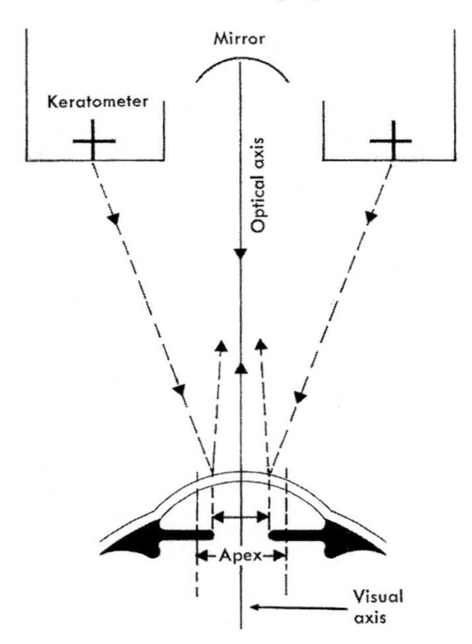

Fig. 11.41. Optical paths in a keratometer. (Reprinted from [556]. Used with permission of Elsevier)

In *keratometry* (which means "cornea measurement") the curvature of the anterior surface of the cornea is determined by looking at the reflection giving Purkinje image I (Fig. 11.41) (Problem 11.63). This curvature is used to determine the posterior surface (base curve) of the contact lens, and consequently to "fit" it. It is often converted to a *K reading or number* (in D); this is the refractive power of a curved interface with this measured radius, between air and a material with the refractive index of tears (and not the refractive index of the cornea) (11.9).

Why do not contact lenses fall off your eye? When you look down with open eyes, your contact lenses do not fall off because the force due to the surface tension between the contact lens and tears exceeds the downward force of gravity on the lens (Problem 11.66). *Why do contact lenses remain centered on the pupil when your head is upright and it does not slide off to the sclera (the white portion)?* This is particularly important for small lenses that cover most of the iris but do not extend to the sclera. Figure 11.42 shows some of the relevant forces. Gravity pushes the lenses down. Blinking of the eye lid pushes the lens downward and then upward in the blinking cycle. Viscosity of the tear volume between the cornea and contact lens resists downward motion. Slipping of the lens to a region of different curvature (i.e., the sclera, for the smaller gas-permeable lenses) is resisted by friction and changes in surface tension. Surface tension depends on tear viscosity, the topography of the cornea, and how well it is matched by the contact lens fit.

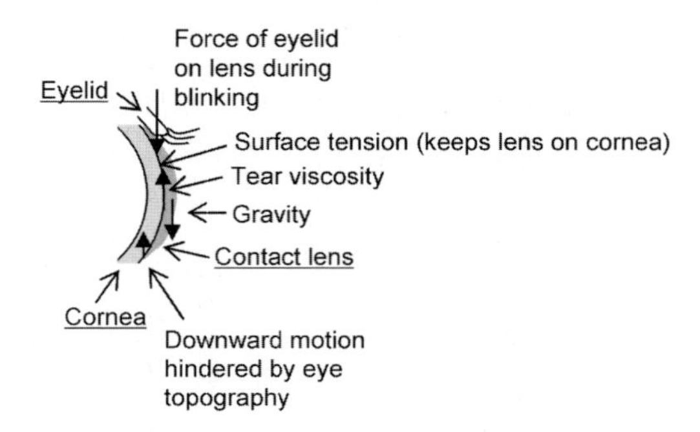

Fig. 11.42. Forces affecting contact lens position. (Based on [543])

Permanent changes in the optical power of the eye can be made by changing the curvature of the anterior surface of the cornea by (1) making four or eights radial slit incisions in it with a scalpel to flatten it (which is called *radial keratotomy* or *RK*), which will decrease its optical power and can help correct myopia only or (2) sculpting its surface with an excimer laser by either (a) *photorefractive keratectomy* (*PRK*) or (b) *laser assisted in situ keratomileusis* (*LASIK*), which can either decrease (for myopia) or increase (for hyperopia) its curvature. In both PRK and LASIK, the middle layer of the cornea, the stroma, is exposed before it is sculpted by the laser via vaporization. In PRK, the epithelium (the top layer of the cornea) is scraped away by laser vaporization before the laser sculpts the stroma, while in LASIK a flap of the top 20% of the cornea is cut using a knife called a microkeratome and this flap is then folded to one side. The exposed cornea is sculpted with the laser to change the radius of curvature of the anterior surface of the cornea and the flap is flapped back to its original position after surgery (Fig. 11.43). Such

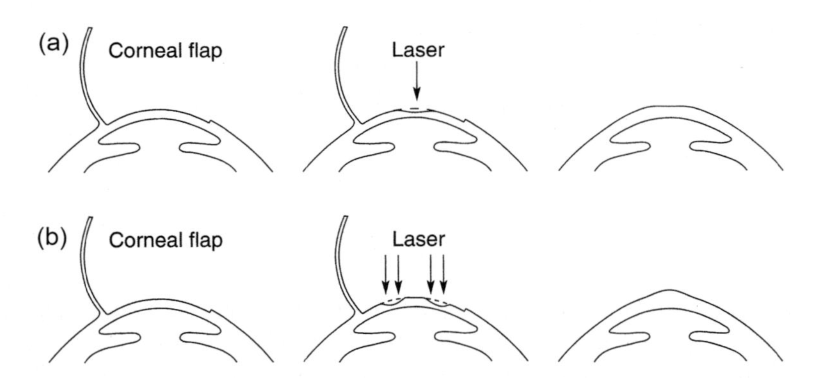

Fig. 11.43. In LASIK surgery the central cornea is made (**a**) more flat to correct for myopia and (**b**) more curved to correct for hyperopia. (From [553]. Reprinted with permission of McGraw-Hill)

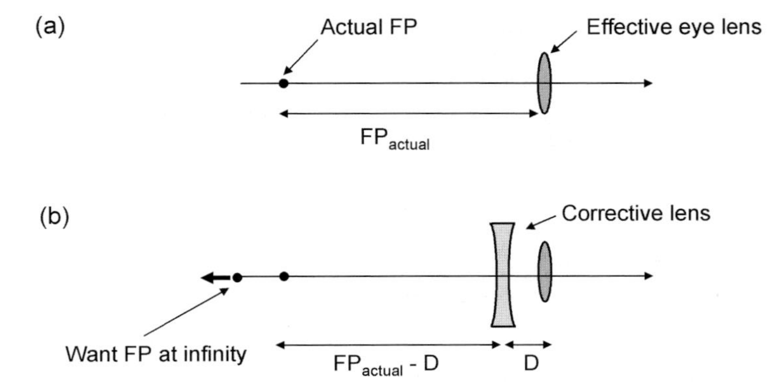

Fig. 11.44. (**a**) The problem in myopia and (**b**) the goal of choosing corrective lenses to correct it. The effective eye lens is that from the Standard eye model

cornea surgery cannot correct for the decreasing accommodation with age, so some sort of corrective lenses are eventually necessary.

Myopia

A person with myopia has an adequate near point (at 25 cm), but a far point that is too near. Say the FP = 530 cm from the eye, when the eye has accommodated as well as it can for infinite distances. The eye can image an object that is 530 cm away, but needs a corrective lens so with this accommodation it can see a farther distance, ∞ away (Fig. 11.44). The corrective lens is D to the left of the eye, with $D \sim 1.5$ cm for eyeglass lenses and $D \sim 0$ cm for contact lenses. We want this lens to take an object $D = \infty$ to the left of it and image it a distance $D - 530$ cm to the right of it. Using this backward propagation approach and (11.24), with f_1 replaced by $f_{\text{corrective}}$, this condition is

$$\frac{1}{\infty} + \frac{1}{D - \text{FP}} = \frac{1}{f_{\text{corrective}}} = P_{\text{corrective}}. \tag{11.69}$$

So we find

$$f_{\text{corrective}} = D - \text{FP} \tag{11.70}$$

and $f_{\text{corrective}} = D - 530$ cm here. For eyeglasses, we see that $f_{\text{corrective}} = -528.5\,\text{cm} = -5.285\,\text{m}$ and the corrective power is $P_{\text{corrective}} = 1/f_{\text{corrective}} = -0.1892\,\text{D} \simeq -0.19\,\text{D}$. For contact lenses, we find that $f_{\text{corrective}} = -530\,\text{cm} = -5.3\,\text{m}$ and $P_{\text{corrective}} = 1/f_{\text{corrective}} = -0.1887\,\text{D} \simeq -0.19\,\text{D}$. These are negative lenses, with similar prescriptions for the eyeglass and contact lenses for this particular example.

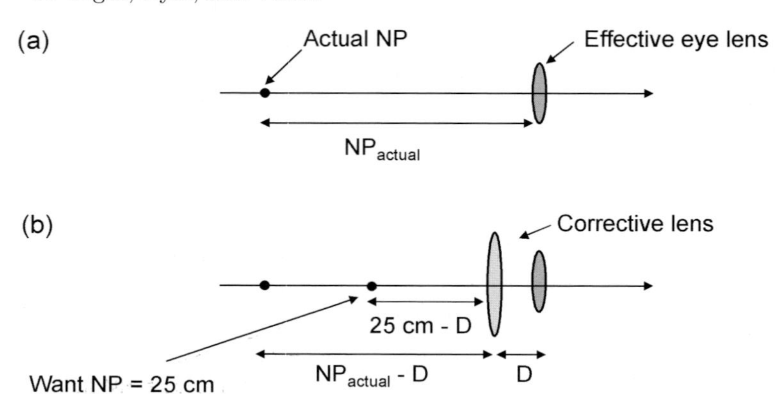

Fig. 11.45. (**a**) The problem in hyperopia and (**b**) the goal of choosing corrective lenses to correct it. The effective eye lens is that from the Standard eye model

Hyperopia

A person with hyperopia has an adequate far point (at ∞), but a near point that is too far away. Say NP $= 190\,\text{cm}$ from the eye, when the eye has accommodated as well as it can for near distances. The eye can image at $x = 190\,\text{cm}$ away, but needs a corrective lens so with this accommodation it can see a nearer distance, 25 cm away (Fig. 11.45). We want the corrective lens to take an object $25\,\text{cm} - D$ to the left of it and image it a distance $D - 190\,\text{cm}$ to the right of it. Using (11.3), this condition is

$$\frac{1}{25\,\text{cm} - D} + \frac{1}{D - \text{NP}} = \frac{1}{f_{\text{corrective}}}. \tag{11.71}$$

For eyeglasses, we see that $f_{\text{corrective}} = +26.8\,\text{cm} = 0.268\,\text{m}$ and $1/f_{\text{corrective}} = +3.72\,\text{D}$. For contact lenses, we find that $f_{\text{corrective}} = +28.8\,\text{cm} = 0.288\,\text{m}$ and $1/f_{\text{corrective}} = +3.47\,\text{D}$. These are positive lenses, with quite different prescriptions for the eyeglass and contact lenses for this particular example.

Differences in Prescriptions for Contact Lenses and Eyeglasses

Correcting the same eye by contact lenses and eyeglasses requires different prescriptions because the eyeglasses are usually worn about 1.5 cm anterior to the contact lenses. This can be included in correcting vision by using D in (11.70) and (11.71). Also, using the effectivity formula, (11.19), with $D = 1.5\,\text{cm}$ and $n = 1$, we see that

$$P_{\text{eyeglasses}} = \frac{P_{\text{contact lens}}}{1 + (1.5\,\text{cm})P_{\text{contact lens}}}. \tag{11.72}$$

A contact lens prescription of $-5.00\,\text{D}$ for a myopic eye becomes a stronger prescription of $-5.00\,\text{D}/[1 + (0.015\,\text{m})(-5.00\,\text{D})] = -5.41$ for eyeglasses.

A contact lens prescription of +5.00 D for a hyperopic eye becomes a weaker prescription of +5.00 D/[1 + (0.015 m)(+5.00 D)] = +4.65 D for eyeglasses.

Examples of eyeglass and contact lens prescriptions are provided below.

Analysis Using Vergences

Let us examine corrections for the Reduced eye in Table 11.1 by using the notation of vergences [553]. (We will ignore the 1.67 mm distance between the cornea and the refractive surface in this model.) From (11.9), the refractive power is $P_{12} = (1.333 - 1.00)/0.00555\,\text{m} = 60.00\,\text{D}$. For an object at infinity ($d_1 = \infty$), the object vergence $L_1 = n_1/d_1 = 0$, so the image vergence, from (11.12), is $L_2 = L_1 + P_{12} = 0\text{ D} + 60.00\,\text{D} = 60.00\,\text{D}$. Because $L_2 = n_2/d_2$ and $n_2 = 1.333$, the image will be at $d_2 = n_2/L_2 = 1.333/60.00\,\text{D} = 22.22\,\text{mm}$ (using $1/1\,\text{D} = 1\,\text{m}$). For an emmetrope, the retina is at 22.22 mm.

For an eye with myopia with the same refractive power but with the retina 23.22 mm away, an object at infinity will still image 22.22 mm after the interface, or 1.00 mm anterior to the retina. Now we want $L_2 = n_2/d_2 = 1.333/23.22\,\text{mm} = 57.41\,\text{D}$, so the image will fall on the retina and vision will be corrected if you place a contact lens with refractive power $57.41\,\text{D} - 60.00\,\text{D} = -2.59\,\text{D}$ at the interface.

For an eye with hyperopia with the same refractive power but with the retina 21.22 mm away, an object at infinity will again image 22.22 mm after the interface, which is now 1.00 mm posterior to the retina. Now $L_2 = n_2/d_2 = 1.333/21.22\,\text{mm} = 62.82\,\text{D}$, so the image will fall on the retina and vision will be corrected if you place a contact lens with refractive power $62.82\,\text{D} - 60.00\,\text{D} = 2.82\,\text{D}$ at the interface.

Astigmatism, Keratoconus, Presbyopia, and Prescriptions

For people with astigmatism, corrective lenses composed of spherical lenses can provide sharp imaging in one direction in the xy plane but not in any other direction. Vision can be corrected with lenses having unequal radii of curvature in the x and y axes, as defined by a suitable rotation about the z-axis (optic axis). Cylindrical lenses are curved in one direction and flat in the other (Fig. 11.46). The corrective lenses for those with astigmatism have a certain

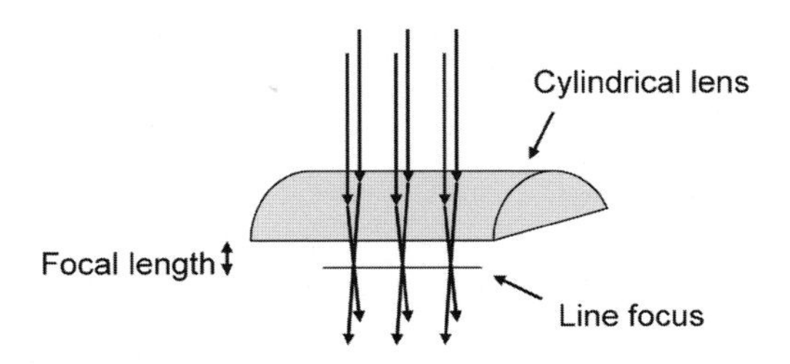

Fig. 11.46. A cylindrical lens

base level of curvature (like that of a spherical lens) plus added curvature in one direction (like that of an added cylindrical lens, rotated the correct way), and as such are called spherocylindrical lenses. Because of the need for this specific correction axis, correction for astigmatism has been possible with eyeglasses or hard/gas permeable contact lenses, but not with soft contact lenses. However, astigmatism can be corrected with new toric soft contact lenses that are oriented on the eye in a stabilized manner. One way to achieve this is by the action of the lids during blinking on the superior-to-inferior position thickness gradient.

A possible eyeglasses prescription for a nearsighted person with astigmatism is

$$\begin{aligned} &\text{O.D.} \quad -3.00 + 2.25 \times 175 + 1.25 \\ &\text{O.S.} \quad -2.50 + 1.75 \times 160 + 1.25, \end{aligned}$$

where O.D. (oculus dexter) stands for the patient's right eye and O.S. (oculus sinister) for the left eye. The first number for each is the refractive power of the spherical lens corrections, -3.00 D for the right eye and -2.50 D for the left. The second term for each is the refractive power of the cylindrical lens correction (and so the lens is a spherocylindrical lens when this term is not zero) and orientation for astigmatism for each eye. For the right eye, 2.25 D is added with the cylindrical axis at an angle 175° from the horizontal in the manner shown in Fig. 11.47. As seen in this figure, along one axis (the cylindrical axis – which has no added cylindrical refractive power), the refractive power is that of the spherical lens, -3.00 D. Along the perpendicular axis, the refractive power is -3.00 D $+2.25$ D $= -0.75$ D. For the left eye, 1.75 D is added with the cylindrical axis at an angle 160° from the horizontal. As presented, this is a prescription for bifocals. The added 1.25 D for each lens is in the lower, bifocal region, to compensate for insufficient accommodation for reading, and is needed because of presbyopia. When corrections for astigmatism and/or presbyopia are not needed, they are not designated on the prescription.

One major stimulus to the development of (initially hard) contact lenses was the need to correct the blurs in those with keratoconus (Fig. 11.39). For people with keratoconus, eyeglasses and soft contact lenses can correct for the average optical deviation of the cornea, but cannot produce sharp images on the retina. Blurred or double images would still form on each retina with either

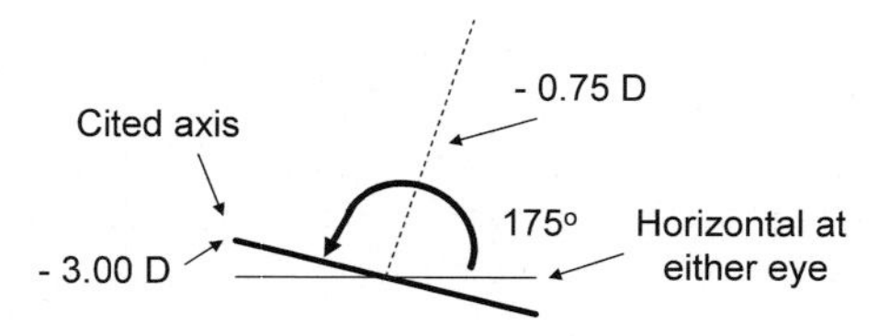

Fig. 11.47. Angles used in prescription for corrective lenses for astigmatism

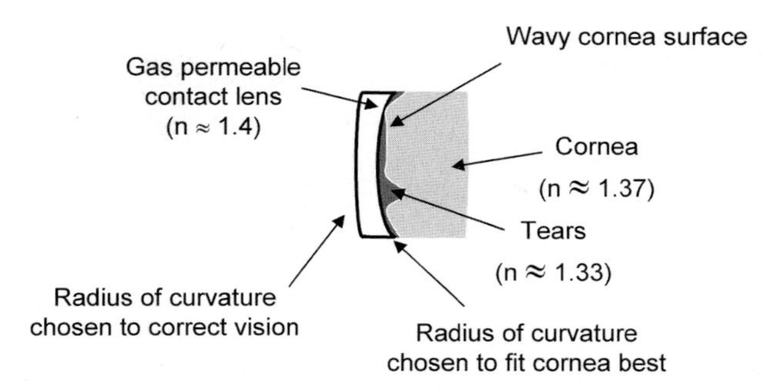

Fig. 11.48. Gas permeable (semirigid) contact lenses for correcting conditions with nonspherical corneas, such as keratoconus, with tears performing refractive index matching

correction, so sharp vision is possible only with hard/gas permeable contact lenses. Again, the inner radius of curvature is chosen to optimize the fit to the cornea, which is often not perfect (Fig. 11.40). Tears fill the gaps between the wavy cornea and the rigid inner lens surface (Fig. 11.48). Figure 11.48 shows how a gas permeable (semirigid) contact lens made of fluorosilane acrylate or silicon acrylate with spherical anterior and posterior surfaces can correct for the unequal radii of curvature in the cornea in astigmatism and myopia or hyperopia, and the waviness in the cornea in keratoconus. The tears (refractive index $n \approx 1.33$) very nearly index match the gap between the contact lens ($n \approx 1.4$) and the cornea ($n \approx 1.37$). Equation (11.15) shows that the refractive powers of the posterior surface of the contact lens/tears and tears/anterior surface of the cornea are very small because of the nearly matching indices of refraction, so the effects of the imperfections in the cornea are minimized.

Prescriptions for contact lenses provide optical information similar to those for eyeglasses – after correction for their different placement relative to the cornea, plus information needed to fit the lens well. One example is

$$\begin{array}{ll} \text{O.D.} & 8.6/14.0/-1.00 \\ \text{O.S.} & 8.6/14.0/-1.50 \end{array}$$

which indicates base curves of 8.6 mm and lens diameters of 14.0 mm for both eyes, and optical powers of -1.00 D for the right eye and -1.50 D for the left eye. Given the large diameter of this lens, it is clearly a soft contact lens. Astigmatism could be corrected with a rigid contact lens.

We have already described how people with *presbyopia* do not have sufficient accommodation (see Table 11.2). For those with good far vision, corrective lenses correct the near point for reading ("reading glasses"). Bifocals are commonly used by those who also need correction at infinity, where the main lenses are useful for far vision and the small "bifocal" parts are for reading (as in the above eyeglasses prescription). In a trifocal, there is also correction for intermediate distances. Eyeglasses are also available with a more continuous change in optical power from the upper region (for long-range vision) to the lower region (for reading), using *progressive lenses*, which do not have the bifocal discontinuity. There are several ways a contact lens wearer can adapt to insufficient accommodation. One is to use reading glasses over the contact lens. Rigid contact lenses shaped to be bifocals, called *alternating vision lenses*, can be put into place by the lower eyelid when the wearer looks down; however, these are difficult to fit well. Corrections for near and far vision are built everywhere with *simultaneous vision* rigid or soft lenses, so the wearer sees sharp and blurred images of near and far images. In *monovision*, one contact lens is corrected for distance and the other is corrected for reading, so that one eye always sees a sharp image and the other a blurred image. Many people adapt to this well.

Magnification by Corrective Lens

A magnifying glass is a lens with positive focal length that you try to place in an optimal position between the source and your eye. Similarly, the positive focal length eyeglass lenses used to correct hyperopia produce a larger, magnified image on the retina and the negative focal length lenses used to correct form myopia produce a smaller, minified image on the retina. Contact lenses neither magnify nor minify the image because they are placed on your cornea.

11.4 Types of Vision Impairment

One type of vision impairment is due to imperfect imaging on the retina. We have already discussed its origin and how to correct it. Imperfect transmission of the optical components of the eye, retinal damage, and optic nerve damage are three other sources of vision problems.

The optical properties of each component of the healthy eye are fairly uniform, so a high-quality image is transmitted with relatively little scattering. Scarring of the cornea due to corneal injury or disease and clouding of the crystalline lens due to cataracts lessen image quality and the amount of light imaged on the retina. Scar tissue on the cornea makes any transmitted light appear to have a halo. The only solution to this problem is to remove the damaged area, usually a disk with 6–8 mm diameter, and to transplant a section from another cornea. The treatment for cataracts is removal of the cloudy crystalline lens, with replacement by a plastic lens.

There are four major modes of retinal damage. (1) Almost a half of diabetics suffer from diabetic retinopathy, in which retinal capillaries swell and leak a bit. This sometimes progresses to *neovascularization*, in which abnormal blood vessels form that do not supply the retina with enough oxygen, and are fragile and bleed into the vitreous humor, thereby hindering the transmission of light to the retina. Neovascularization can also scar and detach the retina. Argon ion lasers are sometimes used to photocoagulate these thin fragile vessels periodically. (2) The macula is the central region of the retina, $\sim$5 mm in diameter including the fovea, where the central region of the object you are viewing is imaged. *Macular degeneration* is most prevalent in older people, with the thinning of the cone cells and sometimes the formation of small new blood vessels; this is similar to neovascularization and can be treated as such. (3) In a *detached retina*, the retina physically detaches from the underlying pigment epithelium, and sometimes tears. Because it has moved from the optimized imaging surface, the image formed on the detached part of the retina is poor. Furthermore, the regeneration of the visual pigments in the detached part of the retina cannot occur because it is not in contact with the pigment epithelium – and this can lead to blindness. The detached retina can be reattached by applying a cooling or heating probe to the outside of the eyeball in the region of the detachment. (4) *Retinitis pigmentosa* is a hereditary disease that first attacks rods (hurting night vision) and later cones, thereby leading to blindness.

Optic nerve damage can result from the increase in the aqueous humor pressure in glaucoma. The overall increase in intraocular pressure in a person with glaucoma compresses the blood vessels that nourish the retina and also increases the pressure on the head of the optic nerve. Both effects cause the optic nerve fibers to degenerate and can result in blindness. The risk of glaucoma increases exponentially with intraocular pressure (IOP), with it being $40\times$ greater for a person with an IOP of 40 mmHg than one with a relatively normal value of 15 mmHg.

High intensities of light entering the eye can affect various ocular elements causing temporary or permanent impairment of vision [554]. *Photokeratitis* is a painful but usually transient 1–2 day irritation of the cornea, due to photochemical changes caused by ultraviolet light ($\lambda = 180 - 400$ nm). For obvious reasons, it is also known as *welder flash* or *snow blindness*. (Remembering the discussion about the seemingly formidable nature of medical terminology in Chap. 1, photokeratitis merely means light (photo) + cornea (kerat) + inflammation (itis), or cornea inflammation caused by light. Similarly, keratoconus merely means conical cornea.) Cataracts can also be caused by photochemical injury of the crystalline lens by chronic exposure to ultraviolet light, from 295 to 325 nm and perhaps to 400 nm. The retina can be damaged by photochemical injuries by blue light, 400–500 nm, *photoretinitis*, as can occur by looking briefly at the sun (*solar retinitis*). The retina can also be damaged by thermal injuries from high intensity – and often pulsed – lasers, from 400 to 1,400 nm. The high coherence of lasers leads to very small foci on the retina, near the diffraction limit given by (11.67). This high intensity leads to damage by highly localized absorption and subsequent localized heating. Consequences are particularly severe when the person looks at the laser – which can occur automatically when the eye tracks to the bright source – because then the laser focuses on the fovea, which can become permanently damaged. If this occurs, the person has no sharp vision in the affected eye, but only fuzzy peripheral vision. Near infrared light, $\sim$800–3,000 nm, can thermally damage the crystalline lens. Also, the corneas of glassblowers and steelworkers were once prone to thermal injuries (burns) due to long-term exposure of high intensities of longer wavelength electromagnetic radiation, ~1,300 nm–1 mm.

11.5 Connections to Visual Perception

Consider the image formed on the retina as the input data for vision. The first factor that affects how this image is processed is the spatial variation of rods and cone detectors across the retina (Fig. 11.3) and whether these cells are connected individually or in an interconnected manner to ganglion cells. Much information processing occurs in the brain [530, 531, 532, 560, 562]. For example, an upright object forms an inverted image on the retina, but is perceived to be upright by the brain. If you wear prismatic eyeglasses that invert the object, you will first perceive an inverted image, but then after a week or so you will again perceive the object as being upright (until you take the glasses off).

The perception of brightness is another example of how the image on the retina is processed: brightness increases only sublinearly with the intensity of the light source. This is an important example of the more general Steven's Law, (1.6) $P = K(S - S_0)^n$ that describes how the perceived strength P varies with the intensity of a stimulus S for a given sensation above a threshold S_0. Experiments show that $n = 0.33 - 0.5$ for brightness (Table 1.15).

Another factor in brightness perception is color. Human sensitivity to different wavelengths is described by the standard luminosity curve $L(\lambda)$ (Fig. 11.49), with $L(\lambda = 555\text{ nm}) = 1$ describing the highest sensitivity and

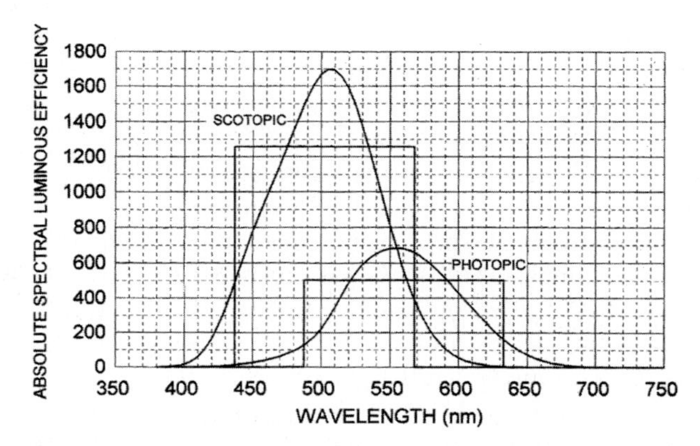

Fig. 11.49. Absolute spectral luminous efficiency of the eye for photopic vision (for normal levels of illumination, also known as the "luminosity curve") and scotopic vision (for dark conditions). The higher the number the more the eye is sensitive to a given light power or intensity at this wavelength. In some presentations each curve is given with its peak normalized to 1. The rectangles associated with each curve are approximations to the real curves. (From [548]. Reprinted with permission of McGraw-Hill)

$L < 1$ for all other wavelengths. Light with a power of 1 W (W being a *radiometric* unit, see next paragraph) at the peak psychological sensation, 555 nm, corresponds to 683 lumens (lumens being a *photometric* unit). A power of 1 W produces a response of $683L(\lambda)$ lumens, which is <683 W away from the peak response wavelength 555 nm. This response curve is for normal levels of illumination (*photopic* or diurnal vision) and is representative for vision dominated by the cones. Figure 11.49 shows the absolute (non-normalized) photopic luminosity curve. Under dark (*scotopic* or twilight) conditions, there is a blue shift in this curve to about 507 nm, at which 1 W gives a sensation of 1,700 lumens (Fig. 11.49). (This peak response is much larger than that at the peak of the photopic curve, but, of course, light levels are much lower for scotopic conditions.) This blue shift, the *Purkinje shift*, reflects the spectral sensitivity of rods.

There are two different sets of units used in studies of vision to denote the strength of the light beams (Fig. 11.50) [516, 546, 548, 549, 552]. Optical studies involving physical systems use MKS-SI units, which are also called *radiometric* units. In contrast, *photometric* units are scaled to account for the level of human response under normal, photopic conditions. Each set describes the same physical parameters, such as power from a source (either total or that per unit solid angle) or power incident onto a surface (either that incident

	Definition	**Common photometry units**	**Radiometric equivalent (units)**
Luminous power 	Total light power produced by a source	lumens	Radiant power (joules/second = watts)
Luminous intensity 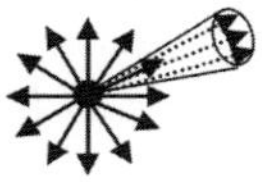	Light power produced in a solid angle by a point source	lumens/steradian candelas (1 lumen/steradian = 1 candela)	Radiant intensity (watts/steradian)
Luminance 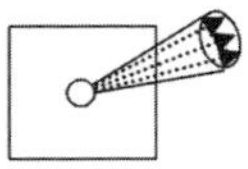	Luminous intensity per unit projected area of an extended source	candelas/square meter foot-lamberts	Radiance (watts/steradian-square meter)
Illuminance 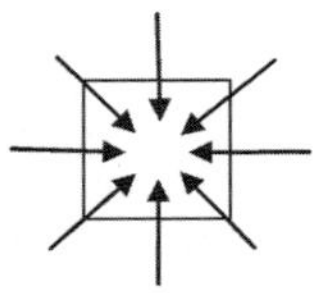	Luminous power falling on a surface	lumens/square meter lumens/square foot	Irradiance (watts/square meter)

Fig. 11.50. Terms and units in photometry (in common units column) and radiometry. (Based on [552])

per unit solid angle per unit area or per unit area). (All of the different terms can make understanding this a bit confusing.)

The units for the following radiometric parameters are:

Energy or radiant flux, or radiant power: for emitted energy/time, in W
Energy or radiant intensity: for emitted power/unit solid angle, in W/sr
Energy or radiant emittance: for emitted power/unit area, in W/m^2
Radiance: for incident power/unit solid angle-unit area, in W/m^2-sr
Irradiance: for incident power/unit area, in W/m^2

where sr stands for steradian. (All of angular space is covered by 4π steradians.) When working with lasers, variations with solid angle are usually not important and the relevant parameters are simply *power* for energy/time (W) and *intensity* for power/unit area (W/m^2).

Photometric units are normalized by the standard (photopic) luminosity curves, with L being 1 at 555 nm and smaller at other wavelengths. Photometric terms often involve the adjective *luminance*, which corresponds to the term *radiance* sometimes used in radiometric units. They are based on the *luminous intensity* unit, the candela (cd), which is light emitted at 555 nm with an energy intensity of 1/683 W/sr. The *luminous flux* is emitted light in units of the lumen (lm), which is 1 cd-sr (and so it corresponds to 1/683 W for 555-nm light). Incident light has units of nit = cd/m^2 = lm/m^2-sr for luminance and lux (lx) = lm/m^2 for illuminance.

The units for the following photometric parameters are:

Luminous flux or power: for emitted energy/time, in lumen (lm)
Luminous intensity: for emitted power/unit solid angle, in candela (cd) = lm/sr
Luminance: for incident power/unit solid angle–unit area, in nit = cd/m^2 = lm/m^2-sr
Illuminance, illumination (brightness) or luminous flux density: for incident power/unit area, in lux (lx) = lm/m^2

For a more complete set of units see Table 11.4.

The luminous flux density of the sun at noon at the equator is about 10^5 lux, while that from a full moon is about 0.2 lux. This luminous flux density in daylight is 10^4 lux, on an overcast day 10^3 lux, on a very dark day 10^2 lux, during starlight 10^{-3} lux, and on a moonless overcast night 10^{-4} lux. The crossover from photopic to scotopic vision occurs near 1 lux. Artificial illumination is usually about 100–2,000 lux. Light bulbs are rated by the input electrical power in W and by their light output in lumens. A 100 W incandescent light bulb produces about 1,600 lumens. This is about 2.4 W of visible light power, which shows that such light bulbs are only about 1–3% efficient. The illuminance a meter away from this light bulb is

Table 11.4. Equivalent photometric units for luminance and illuminance

luminance
$1\,\text{nit} = 1\,\text{cd/m}^2$
$1\,\text{stilb (sb)} = 1\,\text{cd/cm}^2 = 10^4\,\text{nit}$
$1\,\text{apostilb (asb)} = 1/\pi\,\text{cd/m}^2 = 1/\pi\,\text{nit} = 0.3183\,\text{nit}$
$1\,\text{lambert (L)} = 1/\pi\,\text{cd/cm}^2 = 10^4/\pi\,\text{nit} = 3{,}183\,\text{nit} = 1{,}000\,\text{mL}$ (milliLambert)
$1\,\text{footlambert (fL)} = 1/\pi\,\text{cd/ft}^2 = 10.764/\pi\,\text{nit} = 3.426\,\text{nit}$
$1\,\text{candela/ft}^2\,(\text{cd/ft}^2) = 10.764\,\text{nit}$
1 equivalent phot = 1 lambert (L)
1 equivalent lux = 1 blondel = 1 apostilb (asb)
1 equivalent footcandle = 1 footlambert
illuminance
$1\,\text{lux (lx)} = 1\,\text{lm/m}^2$
$1\,\text{phot (ph)} = 1\,\text{lm/cm}^2 = 10^4\,\text{lx}$
$1\,\text{milliphot (mph)} = 10^{-3}\,\text{lm/cm}^2 = 10\,\text{lx}$
$1\,\text{footcandle (fcd, fc, or ft-cd)} = 1\,\text{lm/ft}^2 = 10.764\,\text{lx}$

1,600 lumens divided by the total surface area of a sphere with 1 m radius, so it is $1{,}000\,\text{lumens}/(4\pi \times 1\,\text{m}^2) = 127\,\text{lux}$. For simple orientation and visual tasks 30–100 lux are needed, for common visual tasks 300–1,000 lux are needed, and for special visual tasks 3,000–10,000 lux are needed [549]. Table 11.5 shows recommended lighting levels for specific tasks.

A well-lit room has 200–500 lux (or 20–50 footcandles), and we usually need about 50 lux for reading. A dark-adapted human eye can see about 10^{-9} lux. If the pupil has a radius r of 2 mm, then the luminous flux through the pupil is $\pi(2 \times 10^{-3}\,\text{m})^2(10^{-9}\,\text{lm/m}^2) = 10^{-14}\,\text{lm}$. This corresponds to about 1.5×10^{-17} W of light. Using (11.1), a visible photon with 600 nm wavelength has an energy of $\approx 3 \times 10^{-19}$ J. This means that a dark-adapted eye can see $\sim$30 photons/s.

Table 11.5. Recommended lighting levels. (Using data from [543] and [549])

task or location	level (lux)
dance hall, residence entrance	50
auditorium, church/synagogue	150
cafeteria, chemical laboratory, eye examining room	500
court room, reading, classroom	700
close work laboratory	1,000
office, kitchen	1,500
autopsy table	10,000
operating table	25,000

The illuminance levels are given in lux; values in footcandles are $\approx 0.1\times$ these values, so 1,000 lux $\sim$ 100 footcandles.

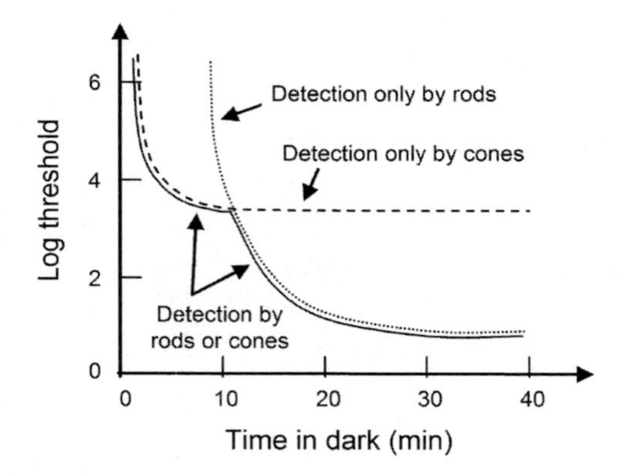

Fig. 11.51. Representative dark adaption curve, showing contribution from rods only (*dotted line*), cones only (*dashed line*), and from rods and cones (*solid line*). Typically, the subject is pre-exposed for several minutes to wide-angle white light at 1,500 mL. Typical vertical scales for threshold flux are in the pL–μL range, for various ranges of narrow-angle illumination. (Based on [516] and [552])

The eye has excellent vision over a billion-fold change in intensity. Adaptation to the dark can take up to 30 min (Fig. 11.51). In the first 5–10 min sensitivity improves by about a factor of 50 during adaptation in the cone-dominated photopic regime. In the next 20 or so minutes, sensitivity improves by about a factor of 1,000 during adaptation in the rod-dominated scotopic regime. These are determined by the regeneration of rhodopsin and changes in retinal networking that control the summation of retinal signals. The eye adjusts to higher fluxes of light within 3–5 min, with a decrease in the gain achieved in retinal networking at lower light levels.

The color content of light entering our eyes depends on the illumination and reflectance of objects (Figs. 11.52 and 11.53) and the absorption by each cone. Figure 11.54 shows the degree of stimulation of each cone for different monochromatic lights. The visual perception of color is a fairly complex function of its spectral components. Different mixtures of three different colors of lights can produce the same visual sensation of color and brightness (*metameric matches*) if they produce the same excitations of the three cones.

The perception of color by using different types of light was once described by the *R,G,B system*, which uses varying mixtures of real primaries: red (645 nm), green (526 nm), and blue (444 nm) to match colors. Because the amounts of each (called the *tristimulus values*) needed for matching (*color matching functions*) are negative in some cases, this system has been supplanted by the *CIE Color Specification System*, with "imaginary" primaries X, Y, Z (Fig. 11.55) [522, 533, 547, 564]. (CIE stands for Commission Internationale de L'Éclairage.) The amount of each primary needed to make a monochromatic color (spectral or monochromatic hue) is given by the x (for

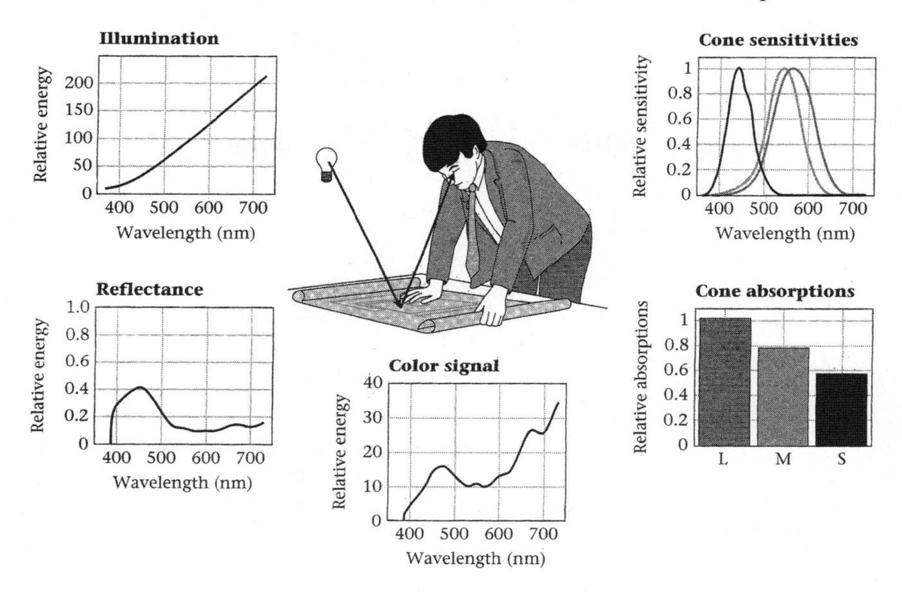

Fig. 11.52. The relative excitation of the cones – "Cone absorptions" – depends on the spectral distribution of light entering the eye – "Color signal" – and "Cone sensitivities," while the "Color signal" is the product of spectral power distribution of the light source – "Illumination" – and the "Reflectance" of the surface. (From [560]. Used with permission)

the amount of X) and y (for Y) on the arc perimeter of the *chromaticity diagram* in Fig. 11.56, with z (for Z) given by $1 - x - y$. (Also, see Fig. 11.57.) Non-monochromatic colors are shown inside the diagram. (The X,Y,Z "primaries" are outside of this curve because they are not real. While this system is mathematically consistent and correct, it is physically unappealing.) Figure 11.58 shows an example of using the chromaticity curve. Mixing monochromatic colors, 490 and 557 nm here, forms a mixture that has a color with a *dominant perceived wavelength* (or hue), which is 550 nm here, as described in the figure.

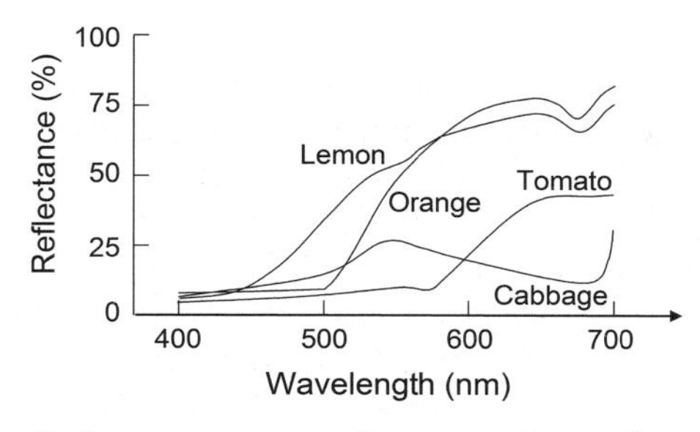

Fig. 11.53. Reflectance curves of common foods. (Based on [523])

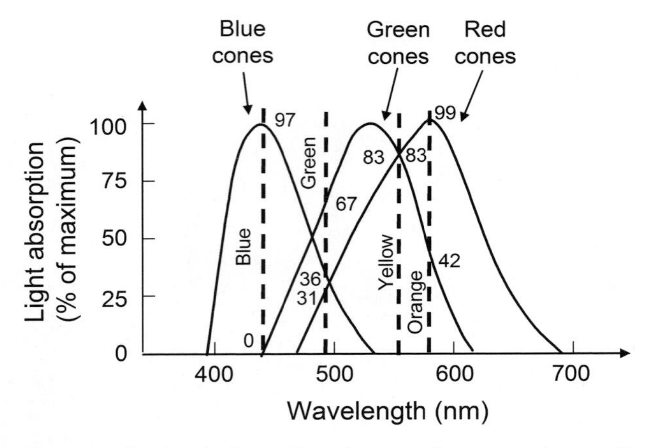

Fig. 11.54. Degree of stimulation of each cone for monochromatic blue, green, yellow, and orange light. (Based on [526])

Another factor in visual perception is response time. The *critical flicker fusion frequency* (CFFF or CFF) is the fastest rate of light pulses a person perceives with no fluctuation in light intensity (and so there is no flickering). It never exceeds 60 Hz even at very high light intensities, and is usually much

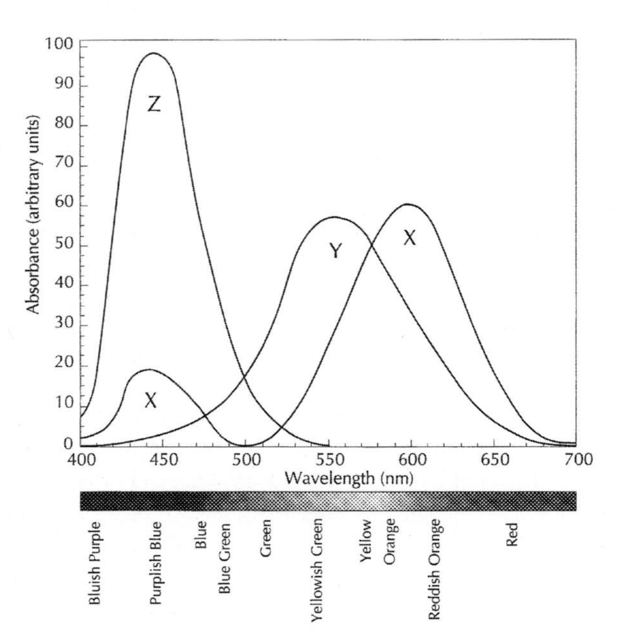

Fig. 11.55. Color matching functions for the imaginary primaries X, Y, and Z. They were arbitrarily chosen for the 1931 CIE scale, and are similar to the spectral responses of the cones – but with obvious differences, such as the short-wavelength second lobe of the X curve. (From [522])

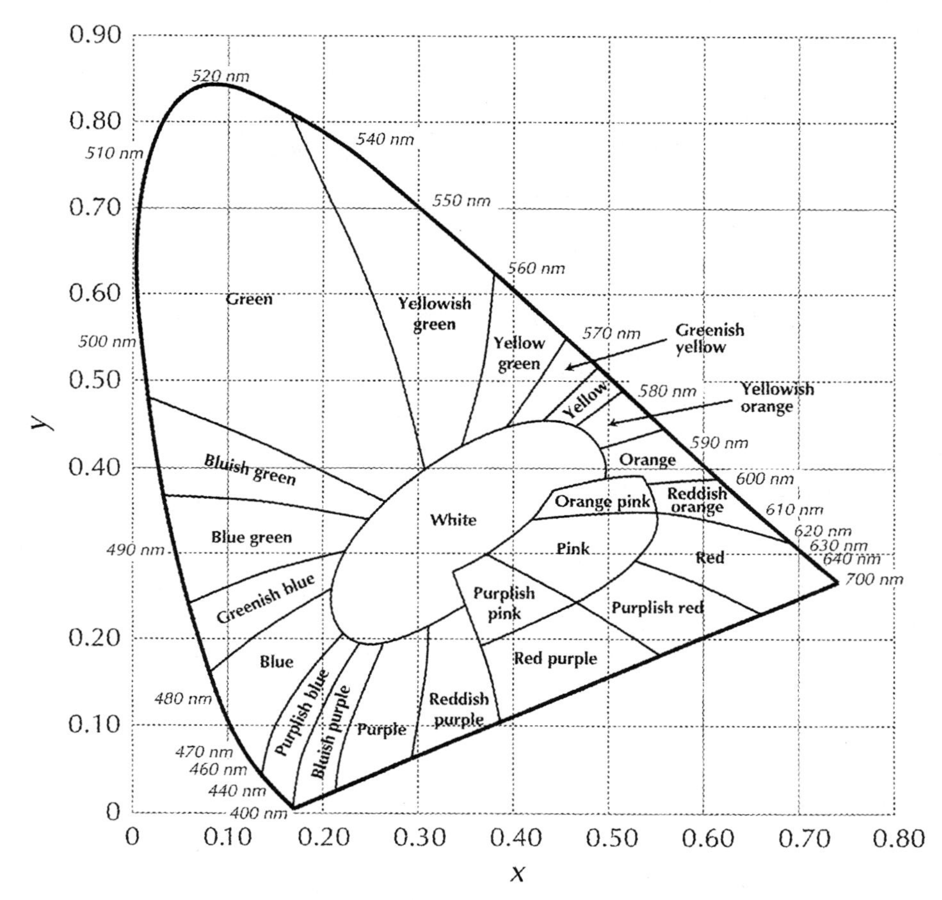

Fig. 11.56. CIE chromaticity curve (outer perimeter), with color labels and regions added inside. This is the version from 1931, which is still widely in use. (From [522])

lower <20–40 Hz. The CFF in the fovea is about 50 Hz, as limited by the response recovery of cones. It is about 20 Hz about 20° outside the fovea. At low light levels it decreases to about 10 Hz. Every other horizontal line in a television is refreshed every 1/60 s (one cycle in 60 Hz), so a new picture appears every 1/30 s. If our response were faster under normal conditions of watching television, the refresh rate would have to be faster to avoid perceived flickering. People with particularly fast visual responses sometimes see – and are bothered by – the flickering from fluorescent lights.

11.6 Vision in Other Animals

One can appreciate better the optical design features of the human eye by comparing it to that of other animals. Primitive eyes, as in planaria,

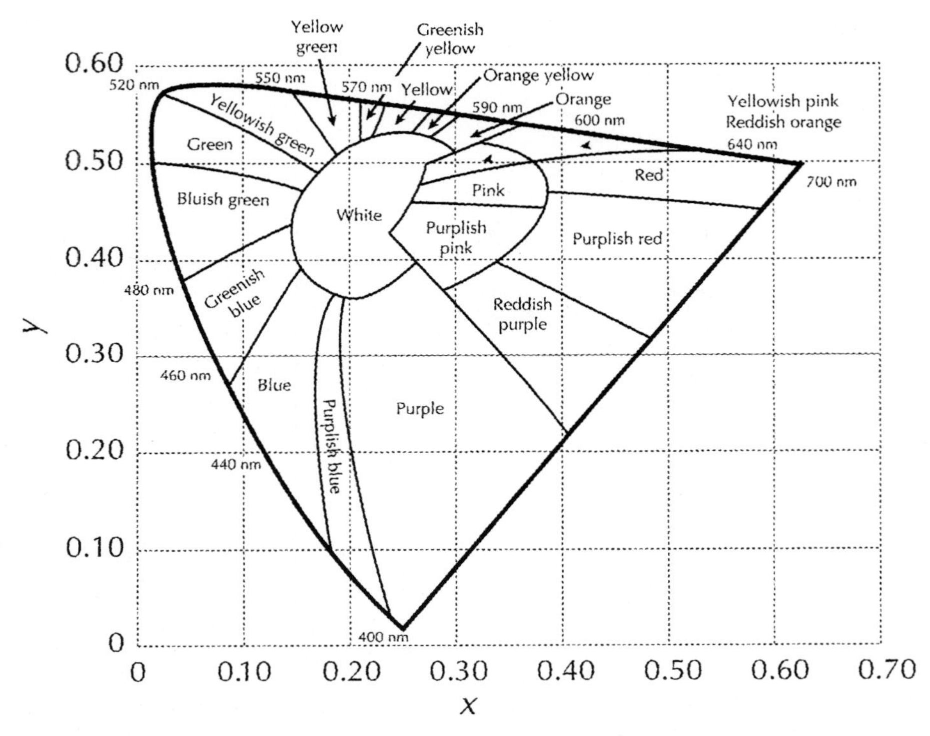

Fig. 11.57. CIE chromaticity curve from 1975. It is less biased to the green shades than the 1931 version is, so the change in color hue between any two points in this diagram is now roughly equal to the distance between the two points in the diagram. Nevertheless, the 1931 CIE is still predominantly used. (From [522])

consist of a series of photoreceptor cells in pits of pigmented cells known as eye spots. In more complex eyes, light is collected and imaged onto an array of photoreceptor cells. In camera or simple eyes, there is one imaging element, either a pinhole (as in a pinhole camera discussed in Problem 11.71) – such as found in the nautilus – or a combination of the cornea and a crystalline lens – as is found in all vertebrates. In compound eyes, there are a series of lenses that image those local regions onto photosensitive detection cells. Vertebrates have only camera eyes, while invertebrates can have camera and/or compound eyes. Different features are optimized in different animals.

1. *Night vs. day:* The eyes of diurnal (active during the day) animals such as lizards and chipmunks are optimized for visual acuity. They have cone-dominated retinas (because cones are smaller than rods) with very little interconnection of neurons leaving the eye (i.e., few convergences), and large eyes. The eyes of nocturnal (active during the night) animals have

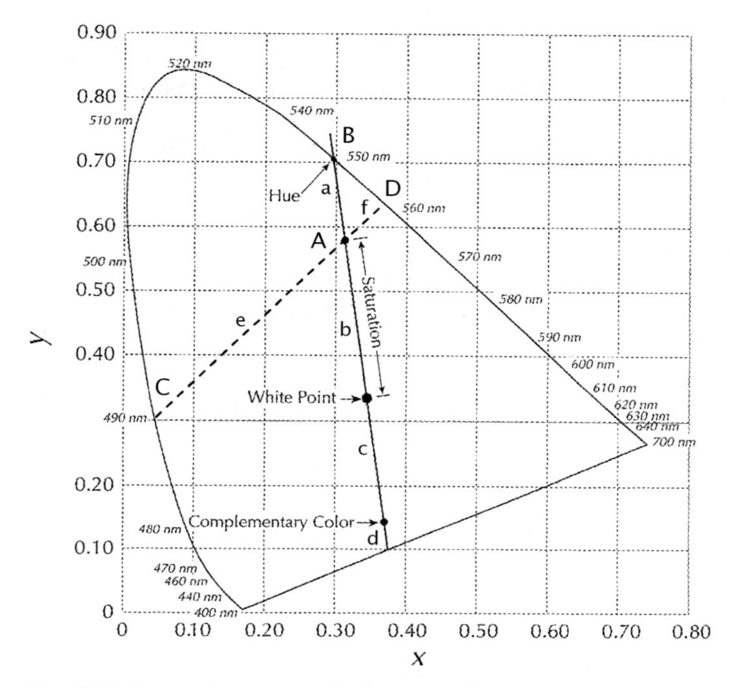

Fig. 11.58. CIE chromaticity curve (1931 version) illustrates several things about a particular color in the diagram, represented by point A. The line through A and the point representing white (White Point) intersects the CIE boundary at point B, showing that its hue (apparent color) is 550 nm. The saturation of this color (which denotes how close it is to being the pure monochromatic hue) is the ratio of the distance from the white point to the color (at A) to that to the hue (at B). This is $b/(a+b)$, where a, b, c, and d are the four line segment distances shown. The color complementary to A is the color on this line on the other side of the white point, which has the same relative distances to the CIE boundary as A, so $c/(c+d) = b/(a+b)$. This figure also shows how colors mix. When you mix 490 nm light (point C) with 557 nm light (point D), you end up with light perceived to be a color that is a point on the *CD dashed line*, and the location of the point on the line depends on how much light of each is used. If the relative intensities are $f/(e+f)$ and $e/(e+f)$, for the line segment distances e and f as shown, the color is A, which is seen to have a hue of 550 nm (point B). (In this example, when ~20% of 490 nm is mixed with ~80% of 557 nm, you see 550 nm). (From [522])

large corneas and crystalline lenses (to gather much light) and only rods (because they are more sensitive than cones). In some animals, such as cats, there is a mirror-like layer behind the retina (the *tapetum*) that reflects transmitted light back through the retinal cells, to increase retinal sensitivity; this produces *eye-shine* (bright eyes in the dark). They have pupils that close to slits, which can protect their retinas from the bright daylight sun much better than do round pupils (which cannot close as much). (Some truly nocturnal animals, such as owls, function well with round pupils because they do not come out during the day.)

2. *Air vs. water:* Humans cannot see well under water (aside from turbulence and waves in the water), because we have only about 1/3 of our imaging power under water, and that is mostly the power due to the crystalline lens. The reason for this is the different refractive indices of air and water, as we saw earlier. Vision can be much better if we wear corrective goggles. Fish and other animals living in the water have more spherical crystalline lenses than humans to provide the extra optical power needed to image in water. In species that live in water the crystalline lens is the main source of refractive power. In fish the crystalline lens is indeed spherical. Spherical aberration is minimized by a decrease in refractive index from the crystalline lens center to the periphery (i.e., away from the optic axis). Some animals can see well both in air and water. The "four-eyed" fish Anableps swims along the surface with the upper half of their eyes in air and the lower half in water. This fish has an elliptically shaped crystalline lens (when viewed from the side). Light transmitted through the upper (air) pupil impinges on the flattened surfaces of the crystalline lens (and so has less refractive power), while that transmitted through the lower (water) pupil impinges on the more curved surfaces of the crystalline lens (and so has more refractive optical power).
3. Accommodation occurs in fish, as well as in snakes and adult amphibians, by translating the crystalline lens forward and backward. In many birds and reptiles the ciliary muscles compress the crystalline lens at its equator, moving the anterior surface forward and increasing its curvature. The front of the crystalline lens can actually touch the iris, which is made rigid by sphincter muscles, and this can also increase the curvature of the crystalline lens. In this way diving birds can increase the refractive power of their eyes by 70–80 D, so they can see well in both air and water.
4. *Eye placement:* In some animals the placement of the two eyes is frontal (including cats and humans), while in others it is lateral – i.e., on the sides of their heads (such as in rabbits, birds, lizards, rodents, etc.). Frontal eyes have overlapping views, which provide good depth perception, while lateral eyes give a better panoramic view – sometimes with independent motion of each eye.

11.7 Summary

Geometric optics, using Snell's Law, can explain much of how the cornea and crystalline lens image light on the retina. This includes developing models of multiple lens systems and optical models of the eye. Image acuity can be lessened by lens aberrations and the effects of diffraction. Imperfect vision can be corrected by using corrective lenses designed by optics models. Models of image formation and the properties of the eye can be linked to vision and visual perception, including the physics of the perception of color. Molecular processes in the rod and cone cells in the retina, which detect the light and send electrical signals to the brain, can be characterized and modeled to further the understanding of vision.

Problems

The Eye and Basic Imaging

11.1. Let us explore our blind spot a bit further.
(a) Why does the experiment demonstrating the blind spot in your left eye (using Fig. 11.4) prove it is "nasal?"
(b) Assume the eye acts as a lens in air with a 17 mm focal length (Standard eye model) and determine how far your blind spot is from your fovea (in mm and degrees)? Do you confirm the blind spot is 13–18° away from the fovea?
(c) Repeat the experiment described in Fig. 11.4 using your right eye, with your left eye closed. How do you explain your findings? How can you modify the experiment so you can determine the blind spot in that eye? Does this prove that the blind spot in your right eye is also nasal and located 13–18° away from the fovea. Why?

11.2. (a) Derive Newton's relation for a thin lens in air that forms a real image, $xx' = f^2$, where $x = d_1 - f$ (the distance the source is in front of the focal point to the left of the lens) and $x' = d_2 - f$ (the distance the image is after the focal point). (Hint: Draw a diagram from a source with a ray that is parallel to the optic axis before the lens and one that is parallel to it after the lens, and consider similar triangles.)
(b) If a source is 20 cm in front of (to the left of) the primary focal point and the image is 5 cm to the right of the secondary focal point, find the focal length and power of the lens and also confirm that (11.3) is satisfied.

11.3. Snell's Law, $n_1 \sin\theta_1 = n_2 \sin\theta_2$ (11.5), predicts that rays will undergo total internal reflection (and not be transmitted) when $n_1 > n_2$ for incident angles $\theta_1 > \arcsin(n_2/n_1)$ (at the local planar interface). Show that light rays inside the eye that hit the cornea at angles exceeding 49° cannot leave the eye (assuming the medium in the eye has refractive index 1.33) (Fig. 11.59a). (This total internal reflection can affect the visual inspection of eyes requiring these large angles, except if a *goniolens* is used (Fig. 11.59b). These large-angle rays are transmitted through the eye into a saline/contact lens assembly in this goniolens and then they are reflected to more shallow angles for inspection.)

11.4. A thin lens in air is composed of material with an index of refraction of 1.062. For each posterior radius of curvature R_{23} given in (a)–(c), find the anterior radius of curvature R_{12} needed so this lens has a focal length of 50 mm.

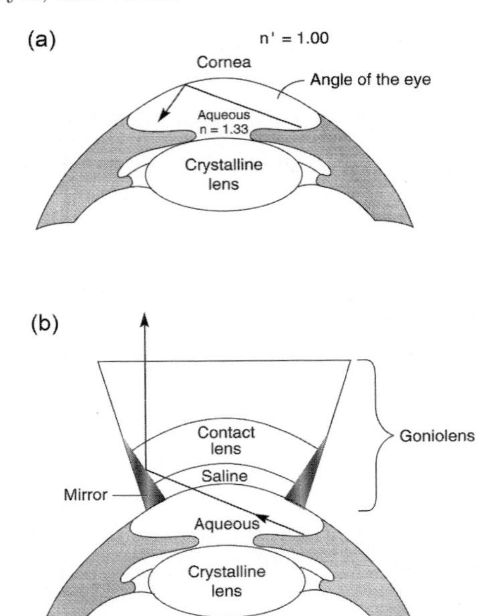

Fig. 11.59. (**a**) Total internal reflection in the eye and a (**b**) diagram describing how a goniolens works to inspect the eye, overcoming such total internal reflection. (From [553]. Reprinted with permission of McGraw-Hill.) For Problem 11.3

Also, for each case sketch the shape of the lens and describe it as being either biconvex, planoconvex, or positive meniscus.
(a) $-0.0060\,\mathrm{m}$
(b) ∞
(c) $+0.0060\,\mathrm{m}$.
(The parameters in this problem are equivalent to those for the eye crystalline lens, with the ratio of refractive indices of the lens to the humors being $1.42/1.337 = 1.062$.)

11.5. Show the equations for a thick lens in air, (11.42) and (11.43), reduce to the Lensmaker equation for a thin lens, (11.17), under appropriate conditions.

11.6. A glass lens in air has $n = 1.5$, $R_{12} = 200\,\mathrm{cm}$, and $R_{23} = -200\,\mathrm{cm}$.
(a) Assume it is a thin lens and find its focal length.
(b) Now assume that the lens has a thickness $t = 5\,\mathrm{mm}$. Find its effective and back focal lengths.
(c) Will the rays cross the axis first, i.e., focus first, for the lens described in (a) or (b)? (Compare the focal length for the lens in (a) to the distances in the lens in (b) that incident parallel rays focus beyond the (i) back surface, (ii) middle plane, and (iii) front surface of the lens.)

11.7. (a) Estimate the loss of light at each of the four ocular interfaces. Assume that each interface is flat and that light hits each at normal incidence.

(Use the optical data presented in Table 11.1 for Schematic eye 1.)
(b) Does this change significantly with wavelength?

11.8. Show that the thick lens equations for arbitrary media, (11.40) and (11.41), reduce to results for a thick lens in air, (11.42) and (11.43), when $n_1 = n_3 = 1$ and $n_2 = n$.

11.9. Use the parameters in the text for Schematic eye 1 to determine how far past the anterior surface of an optical element incident parallel rays converge to the optic axis to form a focus when the element is assumed to have a finite thickness and compare this to that assuming the thin lens approximation. Do this for the:
(a) Cornea.
(b) Crystalline lens.
(c) Does this suggest that the correction for a thick lens is significant?

11.10. Model the tears on the cornea as a 7 μm thick layer with refractive index 1.33, with anterior and posterior radii of curvature the same as that of the anterior surface of the cornea for Schematic eye 1.
(a) What is the optical power of this tear layer?
(b) What does it correspond to in D?
(c) Do tears affect imaging significantly?

11.11. Thomas Young, famous for seminal experiments and interpretations in optics almost 200 years ago, showed that ocular accommodation was due to the crystalline lens and not the cornea because he could not focus on nearby objects when he immersed his eye in water. We know that his conclusion was right, but was his reasoning faulty? Use (11.48) and (11.56), and what you (should) know about accommodation.

11.12. Equation 11.63 shows that we need 4 D of accommodation for our eyes to image at all distances ≥25 cm. Can the similar expression, $1/f_{\mathrm{eye,NP}} - 1/f_{\mathrm{eye,FP}}$, which uses the focal lengths of a person's own near and far points, be used to determine that person's own accommodation? (Hint: Consider a very young person. Also, consider a person with more than 4 D accommodation who still needs corrective lenses and perhaps also bifocals.)

11.13. Estimate the change in the radii of curvature of the crystalline lens surfaces needed to achieve 4 D of accommodation (so a person viewing a source at her far point can then view a source at her near point). Use (11.50) and (11.51) and alternately assume (a) that the curvatures of the anterior and posterior surfaces change by the same percentage or (b) that only the anterior surface changes. (Because accommodation is usually cited for an effective imaging element in air, the amount of accommodation required within the eye is larger by the refractive index of the humors, so $\simeq 4\,\mathrm{D} \times 1.34 = 5.5\,\mathrm{D}$ is needed.)

11.14. In a model of the crystalline lens, the lens has a refractive index of 1.42 and the humors before and after it have a refractive index of 1.337, and the magnitudes of the radii of curvature of the two lens surfaces are the same (but they have different signs). Find the radii of curvatures of the anterior and posterior surfaces if the total refractive power of the crystalline lens is the same as in (11.53) (ignoring the thickness of the lens).

11.15. A crystalline lens is modeled here as in Fig. 11.19, except it is biconvex with radii of curvature that always have equal magnitudes. This magnitude is 7.54 mm when a person views a source at her far point. (The crystalline lens still has an overall refractive power of 22.10 D, ignoring the thickness of the lens.) Find the radius of curvature magnitude when the lens adapts to give 4 D of accommodation (to then allow the person to see her near point), which is equivalent to 5.5 D in the eye (as is explained in Problem 11.13).

11.16. The argument is made that extremely little ultraviolet light with wavelengths from 300 to 400 nm reaches the retina, so it does not cause retinal damage and consequently this light does not damage eyes. Discuss the merits of this argument. Also, where is this light absorbed in the eye?

11.17. The light from a red laser pointer has a wavelength near 650 nm. Where in the eye is most of this wavelength absorbed?

11.18. How well is the transmission spectrum of the eye matched to the spectral responses of rods and cones?

11.19. How well are the spectral responses of the rods and cones matched to the spectrum of solar light? How well are they matched to the spectrum of solar light that actually reaches sea level?

11.20. A nanowatt of light, equally spread across the wavelengths 400–700 nm, is incident on your eye. Approximately what power impinges on your retina?

11.21. Which can cause more retinal damage, light that is highly spatially coherent that can focus according to (11.67) or light that is less coherent that cannot be focused as well. Why?

11.22. A laser enters your eye. Is more damage done if you turn your eye to look at the laser source (which is a reflex) or if you do not stare into the beam? Why?

Visual Acuity and Aberrations

11.23. In a Snellen chart a letter is 8.8 mm high in the 20/20 line. How high is it in the 20/200 line?

11.24. In the text it was said that a 8.8 mm letter in the 20/20 line of the Snellen chart subtended a 5 min arc at the eye of a person standing 20 ft from the chart. This 20/20 (i.e., 20 ft/20 ft) line is also known as the 6/6 (i.e., 6 m/6 m) line. Does this 5 min arc angle more nearly refer to a person standing 20 ft or 6 m from the chart?

11.25. (a) The letters in the 20/20 line of the Snellen chart are 8.8 mm high and wide. Show that this letter is about 25 μm high and wide on the retina.
(b) Explain why this is considered excellent vision.

11.26. If you use an illuminance much greater than 480 lux in conducting a Snellen eye test, your pupil diameter begins to decrease. Discuss how this could affect the visual acuity you are trying to measure in this test, given the effect of pupil diameter on diffraction and lens aberrations. (Most people think you see better in bright light.)

11.27. (a) Show that the 8.8 mm high letters in the 20/20 (or 6/6) line of the Snellen chart subtend an angle of 1.47 mrad when you are 6 m away (where 1 mrad = 1 milliradian = 1×10^{-3} rad).
(b) The Rayleigh criterion for barely resolving objects separated by an angle θ is that θ needs to be at least $\theta_{\mathrm{R}} = 1.22\lambda/d$, where d is the diameter of the lens (or that of the aperture limiting the lens). Show how this criterion is related to the Airy diameter (11.67).
(c) This Rayleigh criterion is a strict limit imposed by diffraction. Show that for 500 nm light, this angle is 0.122 mrad for an aperture diameter of 0.5 cm. How many times the Rayleigh criterion limit is the arc angle needed for 20/20 vision?
(d) The resolution limit for most people is 0.5 mrad, and for the most acute vision under optimum conditions it is 0.2 mrad. How many times the Rayleigh criterion limit are these?

11.28. How small would the pupil diameter need to be to affect the ultimate diffraction limit of the eye lensing system? Given the sizes of cones, would this change make a practical difference?

11.29. (a) Assume for the moment that the rods and cones are 0.1 μm in diameter, that they are tightly packed (with no space between them), and that each is individually connected to the brain by a single neuron. Ignoring optical aberrations, would you see images sharper in the blue or the red? Why?
(b) Ignore the assumptions in part (a), returning to the normal conditions, and address the same questions again.

11.30. (a) Reference [561] (as cited in [560]) used experimental data measured using a human eye with 3.0 mm pupil diameter to obtain an analytic fit to the line spread function: $I(i) = 0.47\exp(-3.3i^2) + 0.53\exp(-0.93|i|)$. The

distance on the retina from the fovea is i, in arc min. Show that Fig. 11.30 accurately plots this function.
(b) Plot on this same set of axes this same function after each point has been broadened by 1 arc min, and compare them. (One relatively crude, way to do this is divide the x-axis into bins that are 0.1 arc min wide, so you have a set of rectangles that have heights $I(i)$ and widths 0.1 arc min centered at $i = 0.0, \pm 0.1, \pm 0.2, \pm 0.3, \ldots$ arc min. Replace these by rectangles with widths of 1.0 arc min. Sum the contributions at each i and then normalize all points by the value at $i = 0$. This can be made less crude by using a smaller width or a broadening function that is smoother than a rectangle.)

11.31. In Schematic eye 1 and other models of the eye the radius of curvature of the anterior surface of the (biconvex) crystalline lens is larger in magnitude than that of the posterior surface. Use Fig. 11.27 to explain qualitatively why this asymmetry helps lessen spherical aberration in the eye.

11.32. Dispersion in glass is usually presented by the refractive index at the center of the yellow line doublet (D) from a sodium lamp at 589.3 nm, $n(D)$, and the dispersion constant (or Abbe number or V-number or constringence) $V = (n(D) - 1)/\Delta n$, where $\Delta n = n(F) - n(C)$ is the difference in the indices in the blue (F, 486.1 nm) and red (C, 656.3 nm) lines from a hydrogen lamp. The refractive index is assumed to vary linearly between the red and the blue. For borosilicate crown glass BSC-2: $n(D) = 1.517$ and $V = 64.5$ and for dense flint glass DF-2: $n(D) = 1.617$ and $V = 36.6$. (Chromatic aberration can be minimized in a doublet lens (and achromat), with the two component lens made of two glasses, such as these, with very different dispersion.)
(a) If the focal length of a simple single-component lens composed of either type of glass is designed to be 17 mm in the yellow, what is its focal length in the red and the blue?
(b) If the screen for the image is 17 mm away from the lens (as for the retina), what is the size of the image size (blur) for each range of color? (Ignore the limitations of diffraction. Assume the source is at infinity.)

11.33. The dispersion constants for ocular media in Table 11.3 use a revised definition of dispersion that differs slightly from the earlier definition used in Problem 11.32. Now $V = (n(d) - 1)/\Delta n$, where $\Delta n = n(F') - n(C')$ is the difference in the indices in the blue (F', 480.0 nm) and red (C, 643.8 nm) lines from a cadmium lamp and $n(d)$ is that at the center of the yellow line doublet (d) from a sodium lamp at 587.6 nm. Find the refractive powers at each of the four interfaces in the eye, and the sum of these refractive powers, at each of these three wavelengths. (Assume the refractive index of the cornea is at 587.6 nm and that dispersion constant of the cornea is 51.5.)

11.34. Use (11.68) (describing chromatic aberration in the eye in diopters) to estimate the change in index of refraction for the eye crystalline lens in the red (630 nm) and blue (470 nm) relative to the yellow (578 nm). Compare this to that of glass in Problem 11.33

11.35. Spectral dispersion in water is characterized by $n(D) = 1.333$, $V = 55$ (see Problem 11.32 for definitions). Model the air/anterior cornea interface instead as an air/water interface. Evaluate the chromatic aberration by determining the change in focal length across the visible (in D). How does this compare to the $\sim$2 D of chromatic aberration in the eye?

Schematic Eyes

11.36. Apply simple theory (and ignore the distance between refractive surfaces) to the schematic eyes to find the total refractive power for:
(a) Schematic eye 2
(b) Schematic eye 2′ (accommodated version of part (a)).
(c) Schematic eye 3.
(d) How much refractive power is added in the accommodated version of the Schematic eye 2? Is this enough so these two models can accommodate for near and far vision?

11.37. (advanced problem) Repeat Problem 11.36 using the more exact treatment of the schematic models (i.e., account for the distances between refractive surfaces).

11.38. Find the refractive power of the cornea, crystalline lens, and the entire eye for the Schematic exact eye, ignoring the distances between refractive surfaces. How do these compare to the values with more exact treatment, 43.05 D, 19.11 D, and 58.64 D, respectively?

11.39. (advanced problem) Show that the refractive powers of the cornea and crystalline lens combine to give the overall refractive power of the eye for the Schematic exact eye. The first and second principal points of the cornea are at -0.0496 mm and -0.0506 mm, respectively, and those of the crystalline lens are at 5.678 mm and 5.808 mm, respectively. Also use the data provided in Problem 11.38 and Table 11.1. As part of this problem, show that the positions of the principal points of the overall system agree with those given in Table 11.1.

11.40. Show that PP′ = NN′ and FP = N′F′ for each model in Table 11.1.

11.41. Model an effective eye lens as a cornea thin lens with focal length 25 mm followed by a "crystalline lens" thin lens with focal length 50 mm, with both reduced lenses in air:
(a) What is their combined effective focal length if their separation is zero?
(b) What is their combined effective focal length if their separation is 9 mm (as in the eye)?
(c) What is this difference in diopters and is it significant?

11.42. Another published set of refractive indices for the eye components has $n_{\text{cornea}} = 1.376$, $n_{\text{aqueous}} = 1.336$, $n_{\text{lens}} = 1.40$, $n_{\text{vitreous}} = 1.337$, and the four

Table 11.6. More data for simplified schematic eyes for unaccommodated (for distant vision) and accommodated (for near vision) conditions. (Using data from [535]). For Problem 11.43

	subject H unaccomm.	subject H accomm.	subject M unaccomm.	subject M accomm.
radii of surfaces (mm)				
cornea	7.74	7.74	7.6	7.6
anterior lens	11.62	6.90	12.0	5.5
posterior lens	−5.18	−5.05	−5.74	−4.87
distance from anterior cornea (mm)				
anterior lens	3.68	3.34	3.33	3.06
posterior lens	7.34	7.58	7.17	7.26
first principal point P	1.61	1.96	1.42	1.69
second principal point P′	1.88	2.33	1.70	2.01
refractive indices				
humors	1.336	1.336	1.336	1.336
crystalline lens	1.413	1.432	1.413	1.415
power of the complete eye (D)	60.44	69.62	60.04	68.95

radii of curvature as (from anterior to posterior) 7.8 mm, 6.4 mm, 10.1 mm, and −6.1 nm. How do the refractive power and focal length differ from those calculated earlier, using (11.54)–(11.55).

11.43. Let us explore accommodation using the simplified schematic eye (such as Schematic eye 2) for two people, subjects H and M, both having about 9 D of accommodation, by using the data in Table 11.6 [535].
(a) Estimate the ages of the subjects.
(b) Qualitatively, how do the subjects differ in how their eyes accommodate?
(c) Find the refractive power of each interface in each case.
(d) Use simple theory (and ignore the distance between the refractive surfaces) to find the refractive power of the eye in each case and the amount of accommodation in both subjects.
(e) (advanced problem) Accounting for the distance between the refractive surfaces, confirm the total refractive powers and the principal point locations of each eye, as listed in Table 11.6.

More Vision Models

11.44. Use simple theory (and ignore the distance between the refractive surfaces) to determine how the total refractive power of the eye would change if the refractive index of only one of the following optical elements were increased by 1%?
(a) cornea
(b) aqueous humor

(c) Crystalline lens.
(d) Vitreous humor.
Start with the values for Schematic eye 1.

11.45. People from Planet X have eyes that are constructed much like ours. They have the same components and the refractive indices of each medium are the same as ours. One difference is that their eyeballs are twice as long as ours. Assume geometric optics theory and ignore the distances between eye components to explore in what other ways their eyes are different. Assume that the radii of curvature of their four refractive interfaces (anterior and posterior surfaces of the cornea and crystalline lens) are in the same proportion as ours.

11.46. Ophthalmologists and optometrists treat 20 ft (or 6 m) as the "optical infinity." For the human eye, what is the difference in refractive power (in D) needed to see at this optical infinity vs. the real infinity?

11.47. In a Goldmann tonometer, the force needed to make a circular region of the cornea (with diameter 3.06 mm) flat is measured. The resulting pressure needed to make it flat is equal to the intraocular pressure, and therefore this is a good test for glaucoma. Show that for an intraocular pressure of 15 mmHg, this applied force corresponds to the gravitational force of 1.5 g applied over this area.

Correcting Vision

11.48. A person with myopia has a far point 60 cm from her eye. What corrective eyeglasses and contact lenses should she wear? (Give your answer in D.)

11.49. A person with hyperopia has a near point 500 cm from his eye. What corrective eyeglasses and contact lenses should he wear? (Give your answer in D.)

11.50. Aside from corrections for aberrations, what are the potential advantages of using eyeglasses with positive or negative meniscus lenses, rather than biconvex/biconcave or planoconvex/planoconcave lenses?

11.51. Must contact lenses always be meniscus lenses?

11.52. Prove that you can make positive or negative focal length contact lenses even though R_{23} (in Fig 11.12) is always >0.

11.53. For each case depicted in Fig. 11.60, describe the status of the patient's vision and how it could be corrected, if necessary.

11.54. Find the prescription in diopters (D) to correct the eyesight for:
(a) A myopic person with a far point of 2 m (using contact lenses).

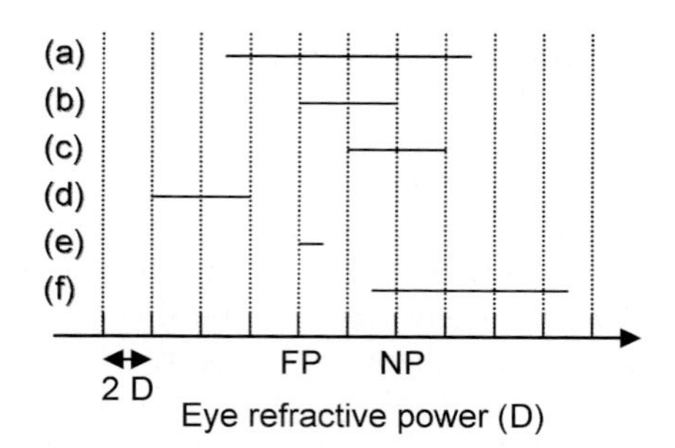

Fig. 11.60. Ranges of eye refractive powers for patients (a)–(f). The designated NP and FP are the refractive powers that are required to have good vision at the desired near and far points. For Problem 11.53

(b) A hyperopic person with a near point of 1 m who wants to read material 25 cm away and who has very good crystalline lens accommodation (using eyeglasses).
(c) A person with perfect vision for far points who, because of poor accommodation (presbyopia), has a near point of 1 m, and who wants to read material 25 cm away (using eyeglasses).
(d) How do the people in parts (b) and (c) differ? (Could they both use their eyeglasses while attending a baseball game and working at a computer terminal?)
(e) What is the accommodation of the person in part (c) (assuming a standard 17 mm long eyeball for the Standard eye model, with air replacing the humors)?
(f) If the near point for the person in part (a) is 15 cm without contact lenses, what is it when the prescribed contact lenses are worn?

11.55. An ophthalmologist estimates the prescription for a patient based on the smallest lines she can read in the Snellen chart: for 20/20: 0 D to −0.25 D, for 20/30: −0.50 D, for 20/40: −0.75 D, for 20/50: −1.00 D to −1.25 D, for 20/100: −1.75 D to −2.00 D, and for 20/200: −2.00 D to −2.50 D. Find the far point for patients who are able to read each given Snellen chart line (and no better without eyeglasses), using average prescriptions when ranges are given. (Note that the optical infinity we are using in our analysis is not the 20 ft value assumed in the Snellen chart. See Problem 11.46.)

11.56. A myopic person has an eyeglass prescription for −3.00 D. What is the appropriate prescription for contact lenses?

11.57. Explain the optical corrections in the following prescription for eyeglasses.

O.D.	$-4.00 + 1.50 \times 90$	$+ 2.3$
O.S.	-3.00	$+ 2.3$

11.58. Approximately how old is the person needing the glasses in Problem 11.57?

11.59. A contact lens made of material with refractive index 1.47 has a posterior radius of curvature set to match the anterior of the cornea (0.0078 m). Find the needed anterior radius of curvature to correct the vision of a:
(a) Myopic person needing −2.00 D correction.
(b) Hyperopic person needing +2.00 D correction.

11.60. A person wears glasses that provide −3.00 D of correction for myopia and wants to have eye surgery (by RK, PRK, or LASIK) to be able to see perfectly without glasses. The person has a cornea with anterior radius of curvature of 0.0078 m. What should the radius of curvature be after surgery? (Remember that −3.00 D refers to the Standard eye in air.)

11.61. A person with astigmatism has corrective lenses with +2 D spherical correction and +1 D cylindrical correction in what we will call the x direction. How could we rephrase spherical and cylindrical corrections for this same prescription if the cylindrical correction were that along the y direction (which is perpendicular to the x direction)?

11.62. We know that a person needs 4 D of accommodation to be able to see both near and far, but does that mean that someone with even more accommodation does not need corrective lenses? Consider as an example someone with a 22-mm-long eyeball in the Reduced eye approximation. Let us say that both persons A and B have 8 D of accommodation. The refractive power of person A can change from 58 to 66 D, while that of person B can change from 52 to 60 D.

11.63. The keratometer (Fig. 11.41) determines the radius of curvature of the anterior (outer) surface of the cornea by tracking the reflection from that surface. Keratometers convert this radius into a K reading, which is the refractive power of the air/cornea surface. However, keratometers sometimes assume slightly different refractive indices for the cornea, often 1.3375, but sometimes other values such as 1.336.
(a) Using $n = 1.3375$, determine the radius of curvature for K readings of 38, 42, 46, and 50.
(b) If the K reading is 44 and you are not sure which n was used, what is the uncertainty in the cornea radius of curvature?

11.64. The stage of keratoconus is characterized by the K reading, as measured by a keratometer (Fig. 11.41). It is mild for K readings $<$45 D, moderate up to 52 D, advanced up to 60 D, and severe above 60 D. What is the radius of curvature in each case? (The keratometer measurement is sensitive to the central region of the cornea and so these K readings are averaged over the apical region of the cornea.)

11.65. What happens if the contact lens base curve and the anterior surface of the cornea do not match?

11.66. A contact lens has a density of $0.9\,\mathrm{g/cm^3}$. It has a thickness as given in the chapter and a surface area that can be calculated from the lens diameters given in the chapter. (In calculating this area, assume the lens is flat.) Assume tears have the same surface tension as water.
(a) Estimate the forces needed to pry a typical soft and hard contacts lens loose off your eye.
(b) When you tilt your head down so the contact lenses on your eyes face down, they do not fall off. Why? (Give a numerical answer.)

Visual Perception and Vision in Animals

11.67. The normalized spectral luminous efficiency is 0.0040 at 420 nm, 0.060 at 460 nm, 0.323 at 500 nm, 1.00 at 555 nm, 0.6310 at 600 nm, 0.1070 at 650 nm, and 0.0041 at 700 nm. How many watts are needed to provide 1,000 lumens at each wavelength?

11.68. In a laboratory, a photometer measures $25\,\mu\mathrm{W}$ due to room lighting over its $0.5\,\mathrm{cm}^2$ area. Find the illuminance in the room in lux. Assume a spectrally-averaged luminous efficiency. Does this value make sense? Why?

11.69. A person with sensitive corneas is comfortable with normal room lighting but needs to wear fairly dark sunglasses – that transmit about 1/8 of the incident light – under normal conditions outside. Less dark sunglasses are not sufficient. Is this all consistent and does it make sense? Why?

11.70. (a) A filter that transmits a fraction T of incident light is said to have an optical density $\mathrm{OD} = -\lg(T)$. Express T in terms of the OD.
(b) Find the optical density of the sunglasses in Problem 11.69.

11.71. Light entering a pinhole camera with a pinhole aperture of diameter d forms an image on the back surface, a distance L_2 away, as in Fig. 11.61:

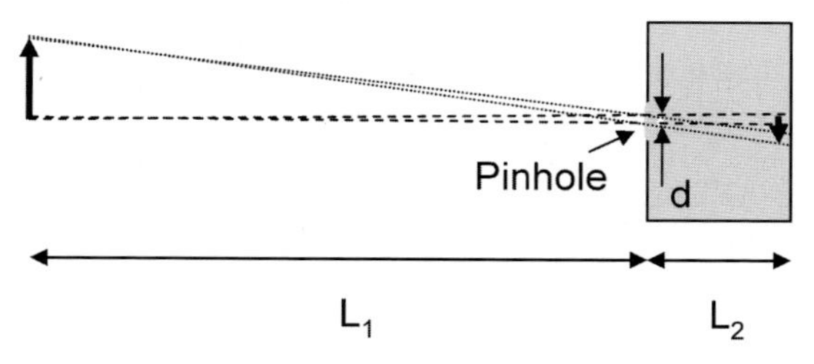

Fig. 11.61. Pinhole camera model of a pinhole eye. Imaging of an extended source is shown. For purposes of illustration a very large pinhole is shown. For Problem 11.71

(a) By using geometrical arguments, show that a point source a distance L_1 from the pinhole, forms an image of dimension $\Delta = [(L_1 + L_2)/L_1]d$, which is $\simeq d$ for the usual case of $L_1 \gg L_2$.
(b) What is the improvement in the resolution of the human eye over a pinhole eye with $d = 1\,\text{mm}$?
(c) Let us say d was made small enough so the resolution of the human and pinhole eye were the same. How much more light would be transmitted by the human lensing system? (If the pinhole were really that small, the transmitted light would diffract much and the imaging would, in fact, be very poor. Still, this illustrates that the human lensing system improves both resolution and light throughput over the pinhole analog.)

11.72. Let us model the eye as a lens with 17 mm focal length in air that forms a sharp image on the retina when it is 17 mm away and the source is at infinity (Standard eye model). In some animals accommodation occurs by changing the separation of the crystalline lens and retina, rather than changing the focal length of the crystalline lens. (Many fish can move their lenses using intraocular muscles.) How far would the crystalline lens need to move to form a sharp image on the retina when the object is only 30 cm away, and in which direction would the crystalline lens need to be moved?

12

Electrical and Magnetic Properties

Charge movement, electric fields, and voltages play essential roles in the body. The driving forces that induce such charge motion are complicated chemical and biological processes that are only partially understood. The interplay of the resulting charges and fields is physical in nature and is well understood.

We have addressed the importance of electricity in the body only briefly in previous chapters. In Chap. 3 we examined the electromyograms (EMGs) of muscle activity (Fig. 3.12), in Chap. 5 we saw that muscles are activated by electrical stimuli and the release of Ca^{2+} ions, and in Chap. 8 we learned that the polarization and depolarization of cell membranes in the heart provide the signals for electrocardiograms (EKGs, ECGs). We now discuss such electrical interactions in more depth as we focus on the electrical properties of the body, the propagation of electrical signals in the axons of nerves, and electrical potentials in the body (Table 12.1).

It is impossible to overemphasize the importance of this human "bioelectricity." The function of every cell depends on it. Every neuron in the brain, every neuron transmitting any information within the body, every neuron enabling skeletal, cardiac, and smooth muscles is yet another vital example. This chapter is largely a discussion of the physics of the motion of positive and negative ions in the blood and cells. We will be concerned with the motion of these ions across membrane boundaries, as in neurons, but not the underlying biology that controls these ion channels. Electric voltages measured at different places in the body describe electrical activity, as is seen in Table 12.1. We will emphasize the propagation of electrical signals in nerves and monitoring the EKG signals from the heart.

Electric and magnetic fields are closely coupled in many areas of physics; for example, electromagnetic waves (visible light, radio waves, X-rays, and so on) consist of electric and magnetic fields oscillating in phase. Magnetic fields appear when current flows. Although current flow is important in the body, the resulting magnetic fields appear to be relatively unimportant and we will address magnetism in the body only briefly in this chapter. For more details on the electrical and magnetic properties of the body see [566, 579, 581, 586, 594].

Table 12.1. Typical amplitude of bioelectric signals. (Using data from [567, 580])

bioelectric signal	typical amplitude
electrocardiogram (EKG/ECG, heart)	1 mV
electroencephalogram (EEG, brain waves)	10–100 μV
electromyogram (EMG, muscle)	300 μV
transmembrane potential	100 mV
electro-oculogram (EOG, eye)	500 μV

12.1 Review of Electrical Properties

We first review the various elements of electrostatics and current flow needed to understand electricity in the body, including the flow of an electrical pulse along an axon.

The electric field at a distance r caused by a point charge q is given by Coulomb's Law:

$$\mathbf{E} = \frac{kq\mathbf{r}}{r^3} = \frac{kq\mathsf{r}}{r^2}, \tag{12.1}$$

where the vector from the charge to the point is $\mathbf{r} = r\mathsf{r}$ and r is a unit vector from that charge to the point of interest, as illustrated in Fig. 12.1a,b. The constant $k = 8.99 \times 10^9$ N-m^2/C^2 for a charge in vacuum, where C stands for coulombs, and can also be expressed as $1/4\pi\epsilon_0$. In a medium of dielectric constant ϵ (where $\epsilon = 1$ in vacuum), $k = 1/4\pi\epsilon_0\epsilon$.

The potential of that charge is

$$V = \frac{kq}{r}. \tag{12.2}$$

and, as here, the potential is usually defined to be zero as r approaches infinity. The potential difference (or voltage) between two points "b" and "a" caused by a field is

$$\Delta V = V_\mathrm{b} - V_\mathrm{a} = -\int_{\mathbf{r}_a}^{\mathbf{r}_b} \mathbf{E} \cdot \mathrm{d}\mathbf{r}. \tag{12.3}$$

This can also be expressed as

$$\mathbf{E} = -\nabla V \tag{12.4}$$

or in one-dimension as

$$E = -\frac{\mathrm{d}V}{\mathrm{d}x}. \tag{12.5}$$

If there are two charges q and $-q$ in vacuum separated, say a distance $\mathbf{d}$ in the d unit vector direction (so the vector between them is $\mathbf{d}$), the electric field

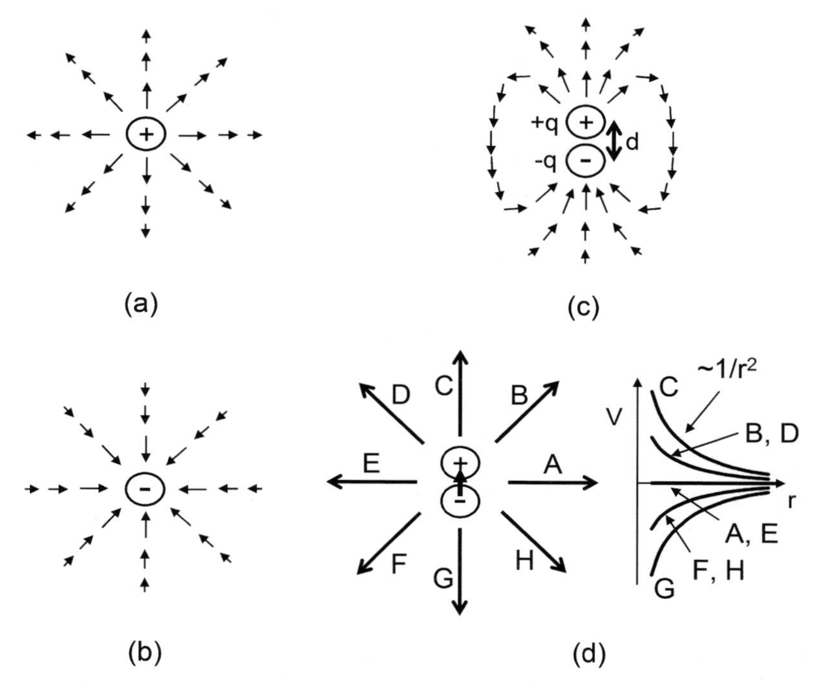

Fig. 12.1. The electric field vectors for (**a**) positive and (**b**) negative charges are shown, along with those for (**c**) a dipole of two charges $+q$ and $-q$, separated by a distance d, so the magnitude of the dipole moment is $P = qd$. The direction of the dipole moment is seen by the arrow within the dipole in (**d**). (d) also shows the potential along the different radial directions shown for this dipole, and the $1/r^2$ decrease in each of these voltages

is the vector sum of the contributions from (12.1) and the electric potential is still obtained using (12.3) (Fig. 12.1c,d). For $r \gg d$, the expression for the potential can be simplified to give

$$V = \frac{k\mathbf{P}.\mathbf{r}}{r^3}, \tag{12.6}$$

where $\mathbf{P} = q\mathsf{d}$ is the electric dipole moment vector, which has magnitude $P = qd$ and points in the d direction. If the angle between the dipole vector $\mathbf{P}$ and distance vector $\mathbf{r}$ is θ, then this equation becomes

$$V = \frac{kP\cos\theta}{r^2}. \tag{12.7}$$

Similarly, we can calculate the dipole moment for many charges separated by various distances. Evaluation of the fields caused by such electric dipoles is of particular value when there is no net charge in the collection of charges, as is true most everywhere in the body.

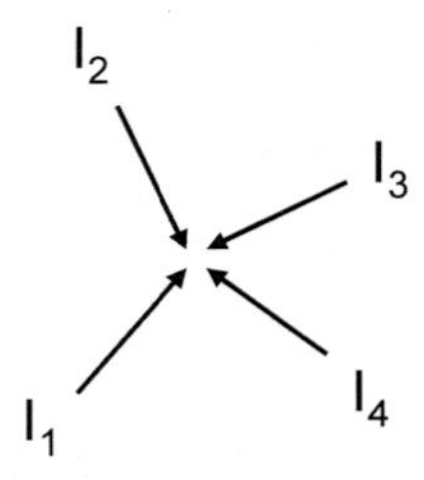

Fig. 12.2. Kirchhoff's 1st Law, showing that the algebraic sum of the current flows to a point must be zero (if charge is not accumulating or being depleted at that point). (Note that least one of the current flows must be negative, i.e., it must point outward)

Now let us consider a moving particle with charge q (in coulombs, C), the current, $I = \mathrm{d}q/\mathrm{d}t$, associated with such a charge or charges (which is the change in charge per unit time), and the associated current density, $J = I/A$ (which is the current flowing per unit area A). Charge is conserved, meaning that it is neither created nor lost. It also means that the vector sum of all currents entering a volume or a small volume element (such as a node) is zero in steady state (Fig. 12.2). This conservation of current (and charge) is known as Kirchhoff's 1st Law

$$\sum_n I_n = 0. \tag{12.8}$$

(The direction of current flow is important here, even thought the current is being expressed as a scalar.)

When a current flows along a material with *resistance* R (in ohms, Ω) (which we called R_{elect} in other chapters), there is a voltage drop V (in volts V) (which we called V_{elect} in other chapters) across the material given by Ohm's Law (Fig. 12.3a)

$$V = IR. \tag{12.9}$$

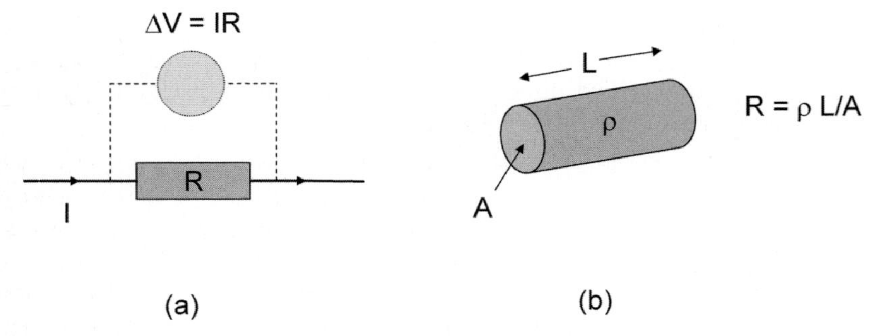

Fig. 12.3. (**a**) Ohm's Law and (**b**) evaluating resistance R from resistivity ρ

The resistance is an extensive property that depends on the intensive property *resistivity* ρ of the material, and the cross-sectional area A and length L of the structure (Fig. 12.3b)

$$R = \frac{\rho L}{A}. \tag{12.10}$$

For a cylinder with radius a, we have $A = \pi a^2$ and $R = \rho L/\pi a^2$. More generally, for a structure with uniform cross-section, the resistance R is proportional to length and we can define a resistance per unit length

$$r = \frac{R}{L} = \frac{\rho}{A}, \tag{12.11}$$

which equals $\rho/\pi a^2$ for conduction along a cylinder. The conductance G (units S (siemens), $1\,\mathrm{S} = 1\,\mathrm{mho} = 1/\mathrm{ohm} = 1/\Omega$) is $1/R$, the conductivity is $\sigma = 1/\rho$, and the conductance per unit area $g = G/A = 1/RA = 1/\rho L$. In the body, charged ions, such as Na^+, K^+, Ca^{2+}, Cl^-, and negatively-charged proteins, are the important carriers of charge. Electrons are the charge carriers in most man-made electronic circuits.

A voltage or potential difference V can also develop between two structures, one with a charge $+q$ and the other with charge $-q$, because of the electric fields that run from one to the other. This voltage is

$$V = \frac{q}{C}, \tag{12.12}$$

where C is the capacitance (in farads, F) of the system (called C_{elect} in other chapters). The capacitance C depends on the geometry of these two structures. For example, they could be two parallel plates or two concentric cylinders (Fig. 12.4), which is similar to the axon of a neuron.

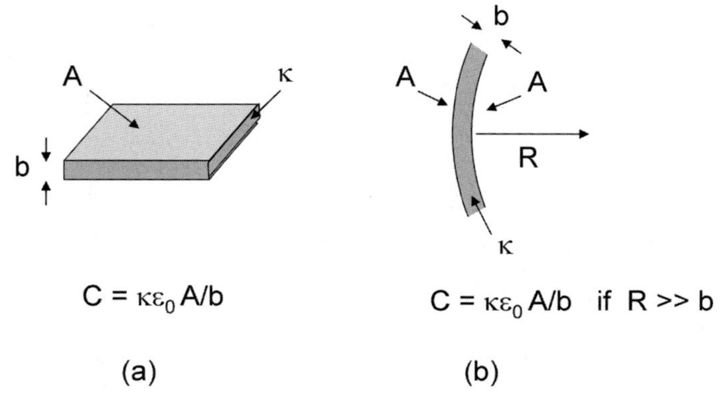

Fig. 12.4. Capacitance for (**a**) parallel plates and (**b**) cylindrical shells

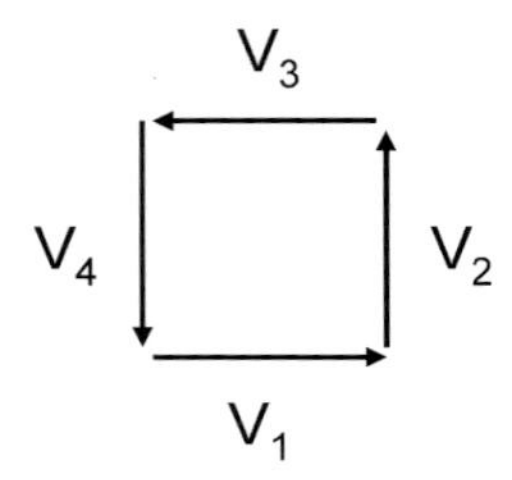

Fig. 12.5. Kirchhoff's 2nd Law, showing that the algebraic sum of the potential drops (voltages) along a closed loop is zero

For two parallel plates with area A separated a distance b by an insulator with dielectric constant κ, we see

$$C_{\text{parallel plates}} = \frac{\kappa \epsilon_0 A}{b}. \quad (12.13)$$

The charge density on each plate is $\sigma = q/A$.

The algebraic sum of all voltages along a closed loop circuit equals zero (Fig. 12.5). This is known as Kirchhoff's 2nd Law

$$\sum_n V_n = 0. \quad (12.14)$$

12.2 Electrical Properties of Body Tissues

12.2.1 Electrical Conduction through Blood and Tissues

When voltage is applied across a metal, a current flows because electrons move under the influence of an electric field. When a voltage is applied across a solution containing positive and negative ions, current flows because both ions move under the influence of the electric field. The conductivity σ of a solution is the sum of the contributions to the current flow for each ion. For low concentrations of these ions, this contribution is proportional to the concentration n_i for that ion, with a proportionality constant $\Lambda_{0,i}$, so

$$\sigma = \sum_i n_i \Lambda_{0,i}. \quad (12.15)$$

Table 12.2 gives $\Lambda_{0,i}$, the molar conductance at infinite dilution for several common ions, while Table 12.3 gives typical concentrations of common ions in the blood and in cells. The resistance of a path can be determined using $\rho = 1/\sigma$ and $R = \rho L/A$ (12.10).

As with many materials, body tissues have dielectric properties, but still have some conductivity, and therefore can be considered as leaky dielectrics. The resistivity of body tissues is shown in Table 12.4 and Fig. 12.6.

Table 12.2. The molar conductance at infinite dilution $\Lambda_{0,i}$ for different ions. (Using data from [596])

ion	$\Lambda_{0,i}$ (1/ohm-m-M)
H^+	34.9
OH^-	19.8
Na^+	5.0
Cl^-	7.6

Table 12.3. Ionic concentrations in blood and cell cytoplasm of unbound ions. (Using data from [597])

ion	blood concentration	cytoplasm concentration	ratio
Na^+	145 mM	12 mM	12:1
K^+	4 mM	140 mM	1:35
H^+	40 nM	100 nM	1:2.5
Mg^{2+}	1.5 mM	0.8 mM	1.9:1
Ca^{2+}	1.8 mM	100 nM	18:1
Cl^-	115 mM	4 mM	29:1
HCO_3^-	25 mM	10 mM	2.5:1

Table 12.4. Low frequency resistivity of some body tissues, in ohm-m (Ω-m). (Using data from [567, 573, 586])

tissue	resistivity
cerebrospinal fluid	0.650
blood plasma	0.7
whole blood	1.6 (Hct = 45%)
skeletal muscle	
– longitudinal	1.25–3.45
– transverse	6.75–18.0
liver	7
lung	
– inspired	17.0
– expired	8.0
neural tissue (as in brain)	
– gray matter	2.8
– white matter	6.8
fat	20
bone	>40
skin	
– wet	10^5
– dry	10^7

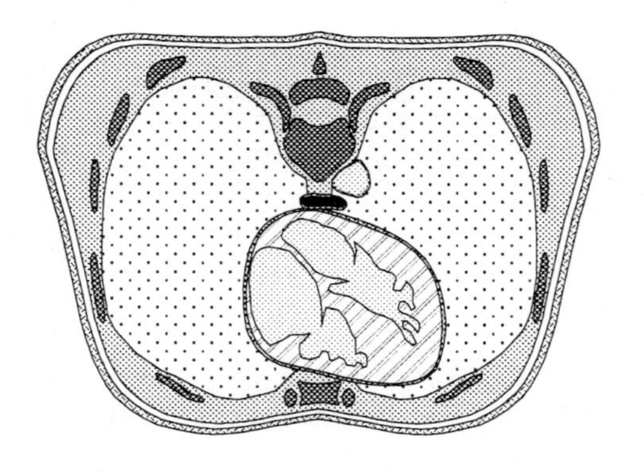

Tissue	Resistivity [Ω m]	
Blood	1.6	
Heart muscle	2.5	(parallel to fibers)
	5.6	(normal to fibers)
Skeletal muscle	1.9	(parallel to fibers)
	13.2	(normal to fibers)
Lungs	20	
Fat	25	
Bone	177	

Fig. 12.6. Cross-section of the thorax, with the electrical resistivity of six types of tissues. (From [586]. Used with permission)

12.3 Nerve Conduction

Figure 12.7 shows the structure of nerve cells or neurons with a nucleus, dendrites that receive information across synapses, an axon, and the axon terminals and synapses for signal transmission to other neurons. There are

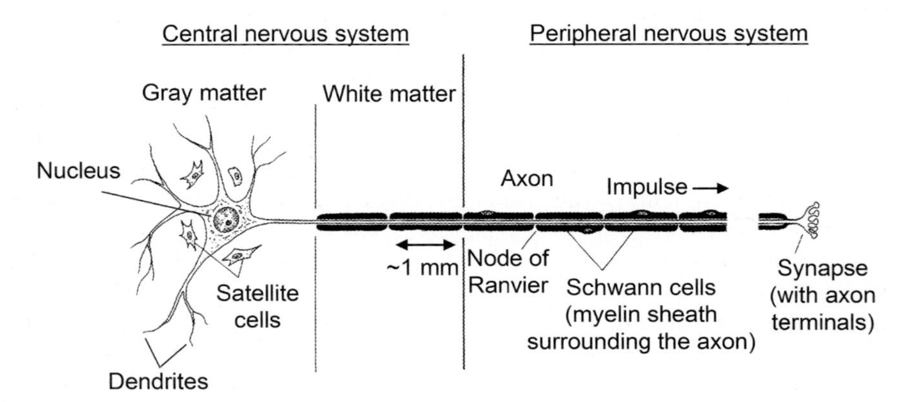

Fig. 12.7. Structure of a neuron. (From [592])

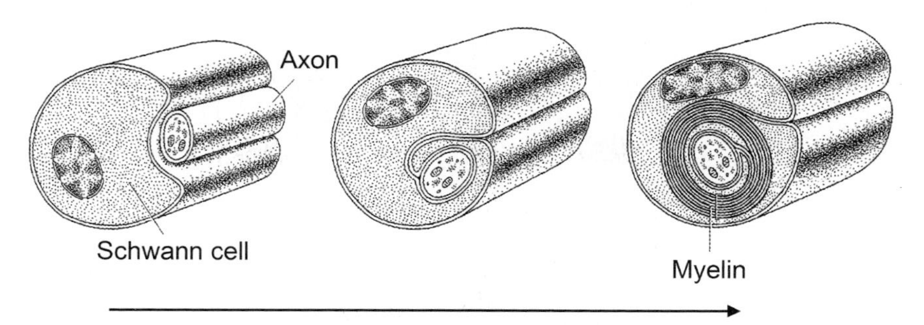

Fig. 12.8. The successive wrapping of Schwann cells about the axon of a neuron to form the myelin sheath of a myelinated nerve. (From [592])

many such neuron axons in a nerve. Unmyelinated axons have no sheath surrounding them. Myelinated axons have myelin sheaths in some regions, which are separated by nodes of Ranvier (ron-vee-ay'). These sheaths are formed by Schwann cells that are wrapped around the axon (Fig. 12.8), with successive wrapped cells separated at a node of Ranvier (Fig. 12.7). We will concentrate on how an electrical impulse travels along such axons.

Approximately 2/3 of the axon fibers in the body are unmyelinated. They have radii of 0.05–0.6 μm and a conduction speed of u (in m/s) $\approx 1.8\sqrt{a}$, where a is the radius of the axon (in μm). Myelinated fibers have outer radii of 0.5–10 μm and a conduction speed of u (in m/s) $\approx 12(a+b) \approx 17a$, where b is the myelin sheath thickness (in μm) (and $a+b$ is the total axon radius). The spacing between the nodes of Ranvier is $\approx 280a$.

Neurons whose axons travel from sensing areas to the spinal cord are called *afferent* neurons or input or sensory neurons. (They are "affected" by conditions that are sensed.) Neurons whose axons leave the ventral surface of the brain stem and the spinal cord to convey signals away from the central nervous system are *efferent neurons* or motor neurons, and these neurons exercise motor control. (They "effect" a change.) There are approximately 10 million afferent neurons, 100 billion neurons in the brain with 100 trillion synapses, and a half a million efferent neurons, so there are roughly 20 sensory neurons for every motor neuron and several thousand central processing neurons for every input or output neuron for processing. Bundles of these neuron axons are called nerves outside of the brain and tracts inside the brain. Details about the nervous system are given in [588].

There are approximately $1-2\times10^6$ optical nerves from the $1-2\times10^8$ rods and cones in our eyes, 20,000 nerves from the 30,000 hair cells in our ears, 2,000 nerves from the 10^7 smell cells in our noses, 2,000 nerves from the 10^8 taste sensing cells in our tongues, 10,000 nerves from the 500,000 touch-sensitive cells throughout our body, and many (but an uncertain number of) nerves from the 3×10^6 pain cells throughout our body.

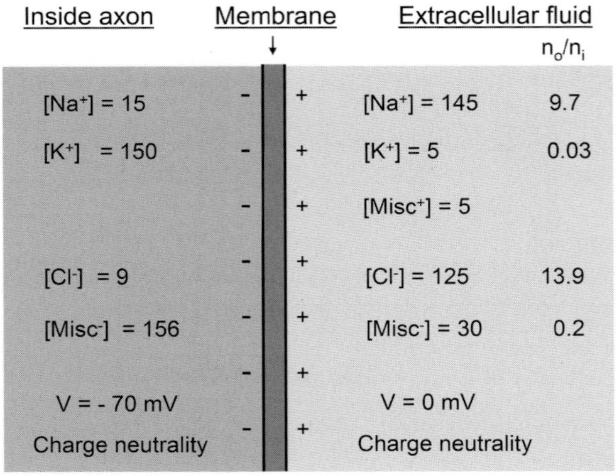

Inside axon Membrane Extracellular fluid

n_o/n_i

$[Na^+]$ = 15 - + $[Na^+]$ = 145 9.7

$[K^+]$ = 150 - + $[K^+]$ = 5 0.03

- + $[Misc^+]$ = 5

$[Cl^-]$ = 9 - + $[Cl^-]$ = 125 13.9

$[Misc^-]$ = 156 - + $[Misc^-]$ = 30 0.2

- +

V = - 70 mV V = 0 mV

Charge neutrality - + Charge neutrality

Fig. 12.9. Ion concentrations (in mmol/L) in a typical mammalian axon nerve cell (n_i) and in the extracellular fluid surrounding it (n_o), and their ratios (n_i/n_o). (Based on [581])

12.3.1 Cell Membranes and Ion Distributions

The cell membrane divides the intracellular and extracellular regions, in neurons and other cells. There are Na^+, K^+, Cl^-, negatively-charged proteins, and other charged species both in the neurons (intracellular) and in the extracellular medium. The concentrations of these ions are such that there is charge neutrality (i.e., an equal number of positive and negative charges) in both the intracellular and extracellular fluids. However, there are negative charges on the inside of the cell membrane and positive charges on the outside of this membrane that produce a *resting potential* of -70 mV (Fig. 12.9). This means that the intracellular medium is at -70 mV, when the *extracellular potential* is arbitrarily defined to be 0 V, as is the custom. Only potential differences are significant, so we are not limiting the analysis by fixing the extracellular potential. This resting potential is the usual potential difference when there is no unusual neural activity. This is known as the *polarized state.* (The propagation of an electrical signal would constitute this type of unusual activity.)

While there is charge neutrality both inside and outside the membrane, the concentrations of each ion are not equal inside and outside the cell, as we will see. The differences in ion concentrations inside and outside the cell membrane are due to a dynamic balance. When there are changes in the permeability of the cell membrane to different charged species, there are transient net charge imbalances that change the potential across the cell membrane. An increase in the membrane potential from -70 mV, such as to the -60 mV seen in Fig. 12.10, is known as *depolarization*, while a decrease from -70 mV to say -80 mV is called *hyperpolarization*. Depolarization is due to the net flow of

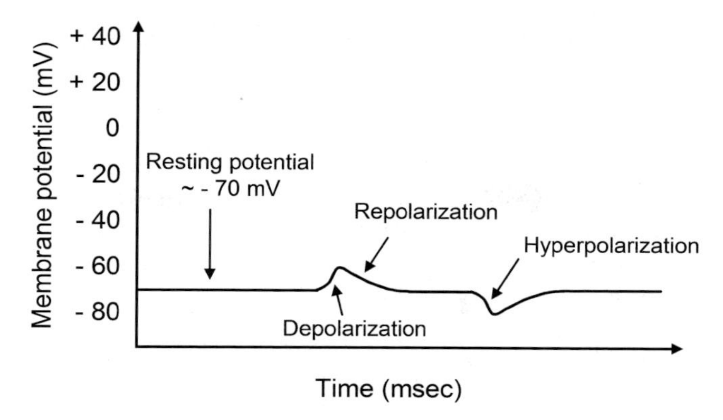

Fig. 12.10. The membrane resting potential of $-70\,\text{mV}$ (inside the membrane relative to the always fixed $0\,\text{mV}$ outside) – the polarized state, along with potential disturbances showing depolarization (voltage increases from the resting potential value), repolarization (returns to the resting potential), and hyperpolarization (decreases from the resting potential)

positive charges into the cell or negative charges to regions outside the cell. Hyperpolarization is due to the net flow of negative charges into the cell or positive charges to outside the cell. Such changes in ion permeability are often termed as changes in the *ion channel.*

Figure 12.9 also shows the concentrations of some of the important charged species inside and outside the cell under resting (i.e., polarized) conditions. We see that there are many more Na^+ outside (145 mmol/L) than inside (15 mmol/L) the cell, but many more K^+ inside (150 mmol/L) than outside (5 mmol/L). Including miscellaneous positive ions outside the cell, there are 165 mmol/L of positive ions both inside and outside the cell. Similarly, there are many more Cl^- outside (125 mmol/L) than inside (9 mmol/L) the cell, but many more miscellaneous negative ions (including proteins) inside (156 mmol/L) than outside (30 mmol/L). There are also 165 mmol/L of negative ions both inside and outside the cell.

There are several driving forces that determine the ionic concentrations, in general, and these intracellular and extracellular concentrations, in particular:

1. There is the natural tendency for concentrations to be uniform everywhere, so when there are concentration gradients across the cell membrane there are flows of these species from the regions of higher concentration to regions of lower concentrations, to equalize the intracellular and extracellular concentrations. This is described by Fick's First Law of Diffusion (7.51), $J_{\text{diff}} = -D_{\text{diff}}\, \mathrm{d}n/\mathrm{d}x$, where J_{diff} is the flux of ions in the x direction (the number of ions flowing across a unit area in a unit time), D_{diff} is the diffusion constant, n is the local concentration of ions, and $\mathrm{d}n/\mathrm{d}x$ is the local concentration gradient.

2. Because the potential is negative inside the cell, we would expect positive ions to enter the cell and be more dominant in the intracellular fluid than the extracellular fluid and for there to be such concentration gradients; this is true for K^+ but not for Na^+. Similarly, we expect negative ions to leave the cell because of the resting potential and be more dominant outside the cell than inside – and again for there to be concentration gradients; this is true for Cl^- but not for the negatively-charged proteins, which form the bulk of the miscellaneous negative ions.
   When charged species are in an electric field, they get accelerated and eventually attain a steady-state drift velocity, v_{drift}, because of collisions that act as a drag force. As shown in Problem 12.7, the drift velocity of a given ion is

$$v_{\text{drift}} = \mu E, \tag{12.16}$$

   where μ is called the mobility and E is the electric field. The flux of ions due to this electric field is

$$J_{\text{elect}} = n v_{\text{drift}} = n\mu E. \tag{12.17}$$

3. The cell membrane permeability and active processes cause the ion concentrations on either side of the membrane to deviate from the values expected from diffusion and the motion of charges in electric fields. The cell membranes are permeable to K^+ and Cl^-, which explains why they behave as expected. Proteins are never permeable to the cell membrane, which is why the concentration of negative-protein ions is unexpectedly high inside. The chemical mechanism called the Na^+ pump (or the Na^+-K^+ pump) actively transports $3Na^+$ from inside to outside the cell for every $2K^+$ it transports from outside to inside the cell; this keeps Na^+ outside the cell and K^+ inside.

The high Na^+ concentration outside the cell is the result of the Na^+ pump fighting against the driving electrical forces and the tendency to equalize concentrations (Fig. 12.11). The high K^+ concentration inside the cell is the result of the electric forces and the Na^+ pump fighting against the tendency to equalize concentrations. The high Cl^- concentration outside the cell is the result of the electrical forces fighting against the tendency to equalize concentrations. The concentration of negative protein ions is unexpectedly high inside because they are large and not permeable to the cell membrane.

Figure 12.12 depicts the directions of motion for charged and neutral molecules for either the random thermal motion in diffusion or the directed effect of an electric field. Figure 12.13 shows how a concentrated band of charged and neutral molecules changes due to either diffusion or an electric field.

Ionic Distributions (Advanced Topic)

What are the expected ionic distributions due to the membrane potential? First, let us consider the expected distributions for several steady state

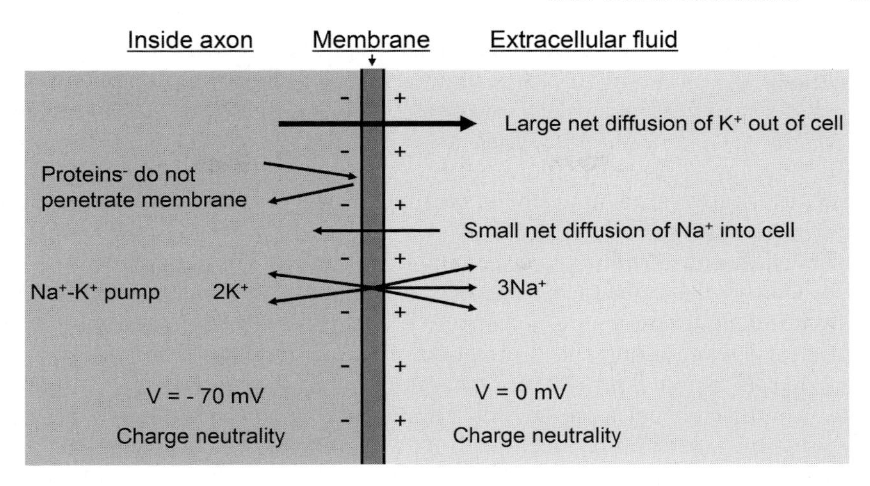

Fig. 12.11. Mechanisms for ion flow across a polarized cell membrane that determine the resting membrane potential

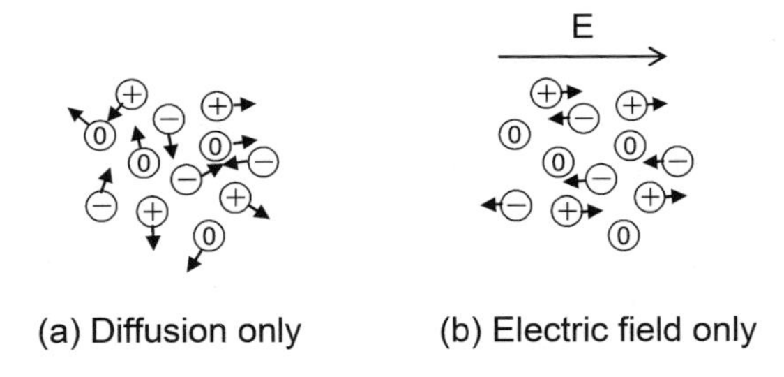

Fig. 12.12. The direction of motion for charged and neutral molecules due to (**a**) diffusion (at a given instant) and (**b**) an electric field

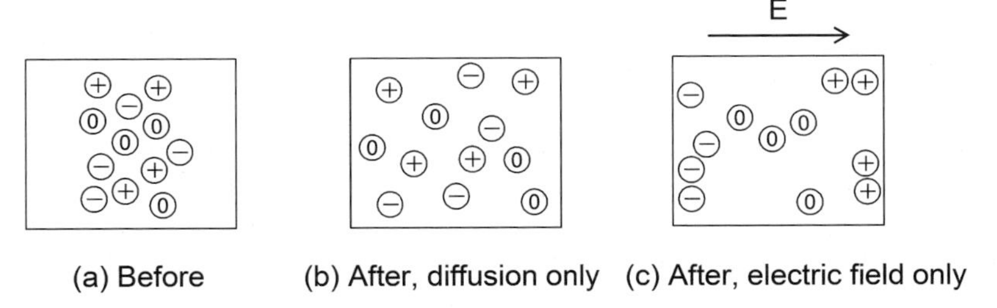

Fig. 12.13. An initial band of charged and neutral molecules (in (**a**)) changes very differently by the uniform thermal spreading in diffusion (in (**b**)) and the separation caused by an electric field (in (**c**))

conditions for a given ion. In steady state, the net flow of ions into any region is zero, so $J_{\mathrm{diff}} + J_{\mathrm{elect}} = 0$ and using (7.51) and (12.17) we see that

$$D_{\mathrm{diff}} \frac{\mathrm{d}n}{\mathrm{d}x} = n\mu E. \tag{12.18}$$

(In steady state n does not depend on time, so the partial derivative in (7.51) is not needed here.)

The diffusion coefficient, D_{diff}, and mobility, μ, are actually closely related. Consider a cylinder of cross-sectional area A and length $\mathrm{d}x$ along the x direction than contains a density n of ions of charge q. When an electric field E is applied along the x direction, the ions in the cylinder feel a force $(nq)(A\mathrm{d}x)E$, where nq is the total charge per unit volume and $A\mathrm{d}x$ is the volume. The mechanical force on this cylinder is due to the difference between the pressure $\times$ area on one side wall, $AP(x)$, and that on the other side wall, $AP(x + \mathrm{d}x) = A[P(x) + (\mathrm{d}P/\mathrm{d}x)\mathrm{d}x]$, or $-A(\mathrm{d}P/\mathrm{d}x)\mathrm{d}x$. The sum of these forces is zero in steady state, so $\mathrm{d}P/\mathrm{d}x = nqE$. Using the ideal gas law (7.2) $P = nk_{\mathrm{B}}T$ (which is an approximation here and where n is now the number of molecules per unit volume because k_{B} is used instead of the gas constant R), we see that $\mathrm{d}P/\mathrm{d}x = k_{\mathrm{B}}T(\mathrm{d}n/\mathrm{d}x)$ or $k_{\mathrm{B}}T(\mathrm{d}n/\mathrm{d}x) = nqE$. Comparing this to (12.18), gives the Einstein equation

$$\mu = \frac{qD_{\mathrm{diff}}}{k_{\mathrm{B}}T}, \tag{12.19}$$

a result we will use soon.

Now let us consider the charge current due to two ions, one of charge q (which we will say is >0), with density n_+ and mobility μ_+, and the other of charge $-q$, with density n_- and mobility μ_-. (We will now define the mobilities as being positive, so for this negative ion $v_{\mathrm{drift}} = -\mu_- E$.) If these are the only two ions, charge neutrality gives $n_+ = n_- = n$. The particle flux of each is determined by the concentration gradient of each and the motion of each in an electric field. The j_{flux} charge flux (or current density) is the ion charge $\times$ the ion flux. For the positive ion: $j_{\mathrm{flux},+} = q(J_{\mathrm{diff},+} + J_{\mathrm{elect},+}) = -qD_{\mathrm{diff},+}(\mathrm{d}n/\mathrm{d}x) + qn\mu_+ E$ and for the negative ion it is: $j_{\mathrm{flux},-} = -q(J_{\mathrm{diff},-} + J_{\mathrm{elect},-}) = qD_{\mathrm{diff},-}(\mathrm{d}n/\mathrm{d}x) + qn\mu_- E$, so the total current density is

$$j_{\mathrm{flux}} = -q(D_{\mathrm{diff},+} - D_{\mathrm{diff},-})\frac{\mathrm{d}n}{\mathrm{d}x} + qn(\mu_+ + \mu_-)E, \tag{12.20}$$

which is known as the *Nernst-Planck equation.* This can also be written as

$$j_{\mathrm{flux}} = qn(\mu_+ + \mu_-)\left(E - \frac{D_{\mathrm{diff},+} - D_{\mathrm{diff},-}}{\mu_+ + \mu_-}\frac{\mathrm{d}\ln n}{\mathrm{d}x}\right), \tag{12.21}$$

where we have expressed $(\mathrm{d}n/\mathrm{d}x)/n$ as $\mathrm{d}\ln n/\mathrm{d}x$. The prefactor on the right-hand side is the conductivity, $\sigma = qn(\mu_+ + \mu_-)$. (Also, the factor $\Lambda_{0,i}$ in (12.15) clearly equals $q_i\mu_i$.)

When there is no net current flow, we see

$$E = \frac{D_{\mathrm{diff},+} - D_{\mathrm{diff},-}}{\mu_+ + \mu_-} \frac{\mathrm{d}\ln n}{\mathrm{d}x}. \tag{12.22}$$

So, the voltage between two points, such as from the inside (with subscript i) of the membrane to the outside (with subscript o) is

$$\Delta V = V_\mathrm{i} - V_\mathrm{o} = -\int_{\mathrm{outside}}^{\mathrm{inside}} E\, \mathrm{d}x \tag{12.23}$$

$$= -\frac{D_{\mathrm{diff},+} - D_{\mathrm{diff},-}}{\mu_+ + \mu_-} \int_{\mathrm{outside}}^{\mathrm{inside}} \frac{\mathrm{d}\ln n}{\mathrm{d}x}\, \mathrm{d}x \tag{12.24}$$

$$= -\frac{D_{\mathrm{diff},+} - D_{\mathrm{diff},-}}{\mu_+ + \mu_-} \ln(n_\mathrm{i}/n_\mathrm{o}). \tag{12.25}$$

Using the Einstein relation, (12.19), we know that $D_{\mathrm{diff}} = \mu k_\mathrm{B} T/q$, and so

$$\Delta V = -\frac{k_\mathrm{B}T}{q} \frac{\mu_+ - \mu_-}{\mu_+ + \mu_-} \ln(n_\mathrm{i}/n_\mathrm{o}). \tag{12.26}$$

This is the *Nernst equation.*

Let us apply this to a membrane that is impermeable to negative ions, so $\mu_- = 0$ and

$$\Delta V = -\frac{k_\mathrm{B}T}{q} \ln(n_\mathrm{i}/n_\mathrm{o}). \tag{12.27}$$

Calling the charge $q = Ze$, where e is the magnitude of an elementary charge (electron or proton), we see that

$$\frac{n_\mathrm{i}}{n_\mathrm{o}} = \exp\left(-Ze(V_\mathrm{i} - V_\mathrm{o})/k_\mathrm{B}T\right). \tag{12.28}$$

This ratio is known as the Donnan ratio and this is known as *Donnan equilibrium.* The ion densities are considered constant within both the inside and outside regions.

This expression can also be derived by using the Maxwell–Boltzmann distribution, which gives the probability of a state being occupied, $P(E,T)$, if it has an energy E and is in thermal equilibrium with the environment at temperature T

$$P(E,T) = A \exp(-E/k_\mathrm{B}T). \tag{12.29}$$

The potential energy of the charge is $E = ZeV$. If a given species were in thermal equilibrium we would expect that its concentration n would be proportional to $\exp(-ZeV_{\mathrm{local}}/k_\mathrm{B}T)$, where V_{local} is the local potential, or more exactly

$$n = n_\infty \exp(-ZeV_{\mathrm{local}}/k_\mathrm{B}T), \tag{12.30}$$

where n_∞ is the concentration very far away, where the potential is zero. In particular, we would expect the ratio of the concentrations for each ion inside and outside the cell gives (12.28).

This Donnan ratio includes the physics of the first two driving forces explained earlier, as well as the physics of thermal equilibration. For the resting potential $V_\mathrm{i} - V_\mathrm{o} = -70\,\mathrm{mV}$ at $T = 310\,\mathrm{K}$ (core body temperature) and $Z = +1$, we expect $n_\mathrm{i}/n_\mathrm{o} = 13.7$ for this "Donnan" equilibrium; for Na^+ this ratio is $15/145 = 0.103$ and for K^+ it is $150/5 = 30$. For $Z = -1$ we expect $n_\mathrm{i}/n_\mathrm{o} = 1/13.7 = 0.073$; for Cl^- this ratio is $9/125 = 0.072$ and for miscellaneous singly negative charge ions it is $156/30 = 5.2$. There is relatively good agreement for K^+ and Cl^-, and great disagreement for Na^+ and the Misc.$^-$ for the reasons given earlier, such as the Na^+ pump for Na^+. (The agreement is not perfect for K^+ because the Na^+ pump brings K^+ into the cell.)

The theoretical *Nernst potential* V_Nernst is the potential that would lead to the observed concentration ratios

$$\left(\frac{n_\mathrm{i}}{n_\mathrm{o}}\right)_\mathrm{observed} = \exp(-ZeV_\mathrm{Nernst}/k_\mathrm{B}T). \tag{12.31}$$

For Na^+ it is 61 mV, for K^+ it is −91 mV, and for Cl^- it is −70 mV.

When the Nernst equation (12.27) is generalized to include the effects of many ions, such as Na^+, K^+, and Cl^-, and membrane permeability, the *Goldman Voltage equation* is obtained

$$\Delta V = -\frac{k_\mathrm{B}T}{q}\ln\frac{p_\mathrm{Na}n_\mathrm{Na,i} + p_\mathrm{K}n_\mathrm{K,i} + p_\mathrm{Cl}n_\mathrm{Cl,i}}{p_\mathrm{Na}n_\mathrm{Na,o} + p_\mathrm{K}n_\mathrm{K,o} + p_\mathrm{Cl}n_\mathrm{Cl,o}}, \tag{12.32}$$

with membrane permeabilities p (and with the subscripts i for inside and o for outside). For neurons and sensory cells the permeability for Cl^- is so small that it can often be neglected, and we find:

$$\Delta V = -\frac{k_\mathrm{B}T}{q}\ln\frac{p_\mathrm{Na}n_\mathrm{Na,i} + p_\mathrm{K}n_\mathrm{K,i}}{p_\mathrm{Na}n_\mathrm{Na,o} + p_\mathrm{K}n_\mathrm{K,o}} \tag{12.33}$$

or

$$\Delta V = -\frac{k_\mathrm{B}T}{q}\ln\frac{\alpha n_\mathrm{Na,i} + n_\mathrm{K,i}}{\alpha n_\mathrm{Na,o} + n_\mathrm{K,o}}, \tag{12.34}$$

with $\alpha = p_\mathrm{Na}/p_\mathrm{K}$. Using the earlier concentrations and $\alpha = 0.02$, this resting potential difference is −75 mV, which is closer to the real resting potential than $V_\mathrm{Nernst} = -91\,\mathrm{mV}$ for K^+.

Poisson–Boltzmann Equation (Advanced Topic)

So far we have determined the concentration ratios for a given potential. A more general problem, and one that is a bit beyond our scope, is to determine the potential V by using (12.1) and (12.3) and the densities of charges in the region. In other words, we also need to couple the potential with the distributions of ions.

By integrating the field over a surface $\mathbf{a}$, such as a sphere, around the charge q, Coulomb's Law (12.1) becomes Gauss' Law

$$\int \mathbf{E} \cdot \mathrm{d}\mathbf{a} = \frac{1}{\epsilon_0 \epsilon} \, q. \tag{12.35}$$

This can be converted into the differential form

$$\nabla \cdot \mathbf{E} = \frac{1}{\epsilon_0 \epsilon} \, \rho, \tag{12.36}$$

where ρ is the charge density

$$\rho = \sum_i Z_i e n_i. \tag{12.37}$$

In one-dimension, this form of Gauss' Law becomes

$$\frac{\mathrm{d}E}{\mathrm{d}x} = \frac{1}{\epsilon_0 \epsilon} \, \rho. \tag{12.38}$$

Using the relation between electric field and potential (12.4) and (12.5) these become Poisson's equation

$$\nabla^2 V = -\frac{1}{\epsilon_0 \epsilon} \, \rho, \tag{12.39}$$

which in one-dimension becomes:

$$\frac{\mathrm{d}^2 V}{\mathrm{d}x^2} = -\frac{1}{\epsilon_0 \epsilon} \, \rho. \tag{12.40}$$

Combining this with the Maxwell–Boltzmann relation (12.30) and with (12.37) gives the Poisson–Boltzmann equation:

$$\nabla^2 V = -\frac{1}{\epsilon_0 \epsilon} \sum_i Z_i e n_{i,0} \exp(-Z_i e V / k_{\mathrm{B}} T). \tag{12.41}$$

Without these free and mobile charges in solution, the potential from a charge Ze, such as an ion in solution, is given by (12.2), $V = Ze/4\pi\epsilon_0 \epsilon r$. These mobile charges partially *screen* or *shield* the potential due to this charge, as is seen by solving the Poisson–Boltzmann equation. When $Z_i e V / k_{\mathrm{B}} T \ll 1$, we can use $\exp(1 + x) \simeq 1 + x$ for $\mid x \mid \ll 1$ to approximate the exponential in (12.41) as $1 - Z_i e V / k_{\mathrm{B}} T$. This gives

$$\nabla^2 V = -\frac{1}{\epsilon_0 \epsilon} \sum_i Z_i e n_{i,0} + \frac{1}{\epsilon_0 \epsilon} \sum_i \frac{Z_i^2 e^2 n_{i,0} V}{k_{\mathrm{B}} T}. \tag{12.42}$$

In charge neutral regions the first term on the right-hand side sums to zero, leaving

$$\nabla^2 V = \frac{e^2}{\epsilon_0 \epsilon k_B T} \left(\sum_i Z_i^2 n_{i,0} \right) V \tag{12.43}$$

or

$$\nabla^2 V = \kappa^2 V, \tag{12.44}$$

where κ is the Debye–Huckel parameter given by

$$\kappa^2 = \frac{e^2}{\epsilon_0 \epsilon k_B T} \sum_i Z_i^2 n_{i,0} \tag{12.45}$$

This is solved in three-dimensions (see Problem 12.8 and Appendix C) to obtain the potential

$$V(r) = \frac{Ze}{4\pi\epsilon_0 \epsilon r} \exp(-\kappa r). \tag{12.46}$$

This means the charge is shielded beyond the Debye–Huckel length given by the radius $1/\kappa$.

12.3.2 Types of Cell Membrane Excitations

There are two qualitatively different types of axon excitations: graded potentials and action potentials.

Graded potentials (Fig. 12.14) are minor perturbations in the membrane potential due to the binding of neurotransmitters, the stimulation of sensory

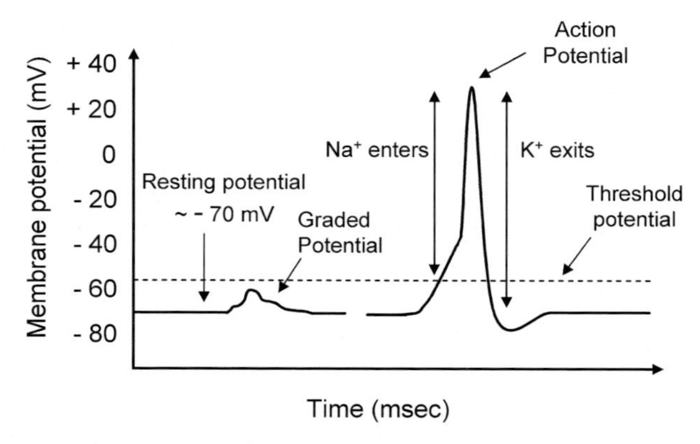

Fig. 12.14. The (subthreshold) graded potentials and (above threshold) action potentials

reception, or spontaneous ion leakage through the cell membrane. There is no threshold needed to stimulate a graded potential. They last for 5 ms to several min. Graded potentials can be either membrane depolarizations or hyperpolarizations. Successive graded potentials can add to one another. They propagate only short distances along the membrane before they decay.

Action potentials are qualitatively different from graded potentials in every way (Fig. 12.14). They initially have relatively large depolarizations by ~15–20 mV above the resting value of −70 mV to a threshold of about ∼ −55 mV. At this threshold potential the cell membrane opens up allowing Na^+ transport. The potential lasts for 1–5 ms, and it always involves depolarization of the membrane. Each action potential opens the cell membrane, and they do not add to one another. There is no decrease in potential along the entire length of the neuron cell axon, as this action potential leads to propagation of an electrical signal along the axon. We will analyze this quantitatively in Sect. 12.3.3.

Figure 12.14 shows the time sequence of the action potential at one point in the axon. After the threshold of ∼ −55 mV is reached, the voltage-gated Na^+ channels begin to open and Na^+ rushes into the cell due to the negative potential. There is an overshoot of positive ions inside the cell and the potential becomes positive, increasing to ~20 mV. This causes positive ions, such as K^+, to leave the cell and the potential decreases below the threshold potential to the resting potential (which is an overshoot). This electrical pulse travels along the axon. Figure 12.15 shows the depolarization and repolarization and the flow of ions for cardiac muscle. The local motion of ions near the membrane are shown in Fig. 12.16 during signal propagation.

12.3.3 Model of Electrical Conduction along an Axon

Neural axons can be treated as an electrical cable with passive parameters that characterize it per unit length, with one striking exception. The resistance of the fluids inside the axon, r_{i}, outside the axon, r_{o}, and of the axon membrane, r_{m}, can be characterized per unit length of the axon. The axon can also be characterized by its capacitance per unit length, c_{m}. The axon can then be modeled by the electrical cable in Fig. 12.17 with repeating units. So far, this description can explain only the decaying features of graded potentials. As we will see, the propagation of action potentials along the axon requires the additional current flow of ions across the axon membrane (see [569, 581, 582, 586]).

Properties of Neurons and Nerves

The parameters in Table 12.5 for unmyelinated and myelinated nerve axons will help us understand the electrical properties of the axon as we would any cable with a distributed resistance and capacitance. From Table 12.5, the resistivity for an unmyelinated nerve is typically $\rho_{\mathrm{i}} = 0.5$ ohm-m and the axon

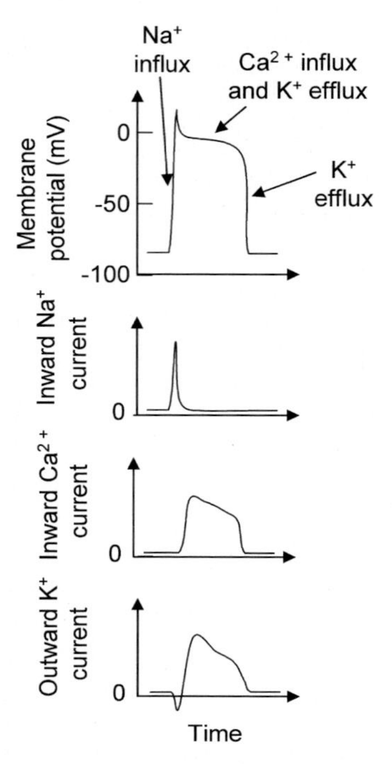

Fig. 12.15. The depolarization and repolarization of cardiac muscle, along with the flows of Na^+, Ca^{2+}, and K^+ ions. The inward flux of Na^+ and Ca^{2+} increases the potential and the outward flux of K^+ decreases it. (Based on [585])

radius is $a = 5 \times 10^{-6}$ m, and so the resistivity inside the axon per unit length along the axon is

$$r_\mathrm{i} = \frac{\rho_\mathrm{i}}{\pi a^2} = \frac{0.5\ \text{ohm-m}}{\pi(5 \times 10^{-6}\ \text{m})^2} = 6.4 \times 10^9\ \text{ohm/m} = 6.4 \times 10^3\ \text{ohm/}\mu\text{m}. \tag{12.47}$$

The resistivity of the membrane is $\rho_\mathrm{m} = 1.6 \times 10^7$ ohm-m, the membrane thickness is $b = 6 \times 10^{-9}$ m, and the cross-sectional area of the membrane normal to the axon axis is $A = 2\pi ab$. Therefore, the membrane resistivity per unit length along the axon is

$$r_\mathrm{m} = \frac{\rho_\mathrm{m}}{2\pi ab} = \frac{1.6 \times 10^7\ \text{ohm-m}}{2\pi(5 \times 10^{-6}\ \text{m})(6 \times 10^{-9}\ \text{m})} \tag{12.48}$$

$$= 8 \times 10^{19}\ \text{ohm/m} = 8 \times 10^{13}\ \text{ohm/}\mu\text{m}, \tag{12.49}$$

This resistivity is so high that for a given voltage drop along the axon, the current flow along the membrane is negligible compared to that in the fluid.

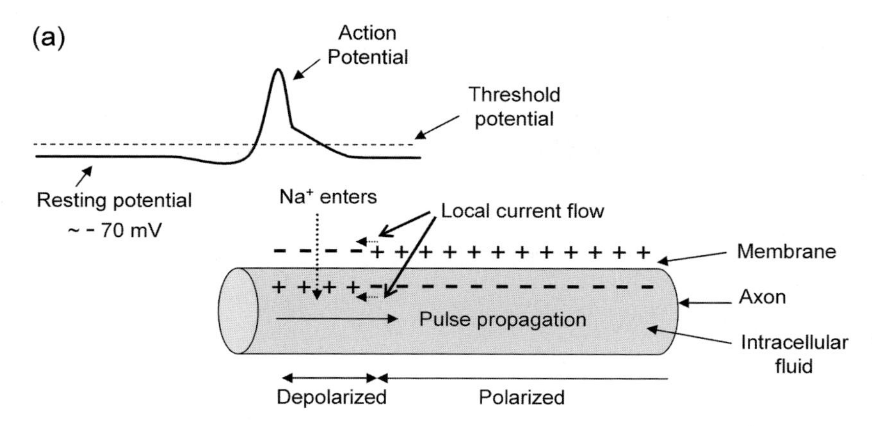

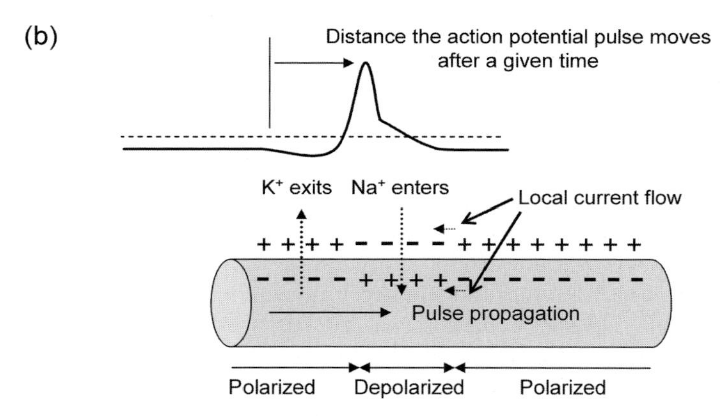

Fig. 12.16. The flow of ions across the membrane during action potential propagation (**a**) at a given time and (**b**) at a later time

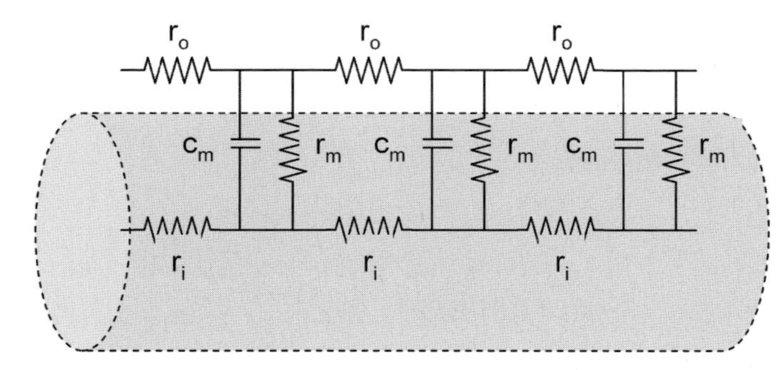

Fig. 12.17. Distributed circuit model of an axon, with resistance inside the axon r_i, membrane resistance r_m and capacitance c_m, and resistance outside the axon r_o, each per unit length

Table 12.5. Typical parameters for unmyelinated and myelinated nerves. (From [570, 571, 581])

		unmyelinated	myelinated
axon inner radius (m)	a	5×10^{-6}	5×10^{-6}
membrane/myelin thickness (m)	b	6×10^{-9}	2×10^{-6}
axoplasm resistivity (ohm-m)	ρ_{i}	1.1	1.1
membrane dielectric constant (s/ohm-m)	$\kappa\epsilon_0$	6.20×10^{-11}	6.20×10^{-11}
membrane/myelin resistivity (ohm-m)	ρ_{m}	10^7	10^7
resistance per unit length of fluid[a] (ohm/m)	r	6.37×10^9	6.37×10^9
conductivity/length axon membrane (mho/m)	g_{m}	1.25×10^4	3×10^{-7}
capacitance/length axon (F/m)	c_{m}	3×10^{-7}	8×10^{-10}

[a]Fluid both inside and outside the axon.

The transverse resistance across the membrane is $(\rho_{\mathrm{m}} b) A_{\mathrm{transverse}}$, so the conductance per unit area is

$$g_{\mathrm{m}} = \frac{1}{\rho_{\mathrm{m}} b}, \tag{12.50}$$

where conductance is the reciprocal of the resistance.

Because the axon radius a of an unmyelinated axon is much greater than the membrane thickness b, the cylindrical membrane can be unrolled along its length (much as in Fig. 8.23) and modeled very successfully as a plane parallel capacitor, with plate separation b and area $A = aL$, where L is the length of the axon unit. The material in the axon membrane has dielectric constant $\kappa = 7$, so with $\epsilon_0 = 8.85 \times 10^{-12}$ s/ohm-m, we see that $\kappa\epsilon_0 = 6.20 \times 10^{-11}$ s/ohm-m. From (12.13), the capacitance per unit length of an unmyelinated axon is

$$C_{\text{parallel plates,per length}} = C_{\text{parallel plates}}/L = \kappa\epsilon_0 a/b \tag{12.51}$$

$$= (6.20 \times 10^{-11}\text{s/ohm-m})(5 \times 10^{-6}\text{m})/6 \times 10^{-9}\text{m} \tag{12.52}$$

$$= 3 \times 10^{-7}\ \text{F/m} \tag{12.53}$$

and that per unit area is

$$c_{\text{parallel plates}} = C_{\text{parallel plates}}/La = \kappa\epsilon_0/b \tag{12.54}$$

$$= (6.20 \times 10^{-11}\text{s/ohm-m})/6 \times 10^{-9}\ \text{m} \tag{12.55}$$

$$= 0.01\ \text{F/m}^2. \tag{12.56}$$

Using (12.12), $q = CV$ and the charge density on the membrane walls is $\sigma = q/A = (C/A)/V$. For a -70 mV voltage drop, we see that $\sigma = (C/A)/V = (0.01\,\text{F/m}^2)(70\,\text{mV}) = 7 \times 10^{-4}\,\text{C/m}^2$. Because an elementary charge is

1.6×10^{-19} C, there are $(7\times10^{-4})(6.25\times10^{18}$ elementary charges$)/10^{12}$ μm^2 = 4.4×10^3 elementary charges/μm^2.

Does the flow of Na ions through the open ion channels appreciably affect the total number of such ions in the axon? This open channel corresponds to a voltage change from −70 to 30 mV or about 100 mV. Using the analysis of the previous paragraph, this corresponds to a change in charge of 6×10^3 elementary charges/μm^2. Consider a 1 μm long section of the axon. Its inner area is $2\pi(5$ μm$)(1$ μm$) = 31$ μm^2, so 2×10^5 Na^+ ions are transported into this volume, because they each have one elementary charge. Before the membrane opened there were 15 mmol/L of Na^+ ions inside the membrane, or $[(15 \times 6.02 \times 10^{20})/10^{15}$ μm$^3][\pi(5$ μm$)^2(1$ μm$)] = 7 \times 10^8$ Na^+ ions in this volume (and, similarly, 7×10^9 K^+ ions inside this volume). This means that this Na^+ ion transport increases the density by only about 0.03%. (Large changes in potentials are often caused by the transfer of very few charges!)

Model of Electrical Conduction in Axons (Advanced Topic)

Several things can happen when you apply a voltage to an axon of a neuron. There can be current flow of charged ions associated with the resistance in and about the axon; the voltage would drop with distance according to Ohm's Law and there would be dissipation of energy. There can be motion of charges to and from axon membranes and changes in the electric field energy stored between these charged surfaces, as characterized by their capacitance. There can also be changes in the transport of charges through these axon membranes.

Consider a cylindrical "pillbox" as shown in Fig. 12.18 of radius a and length δx, extending from x to $x + \delta x$ along the axon and with the curved cylinder surface within the cell membrane itself. The voltage at x is $V(x)$ and that at $x + \delta x$ is $V(x + \delta x)$. The current flowing (due to ions) within the axoplasm – i.e., the medium inside the axon – into this volume is $I_{\mathrm{i}}(x)$ and that leaving it is $I_{\mathrm{i}}(x + \delta x)$. There is a charge $+q$ on the outer membrane of the axon and a charge $-q$ on the inner membrane wall. A physical model for this is shown in Fig. 12.19.

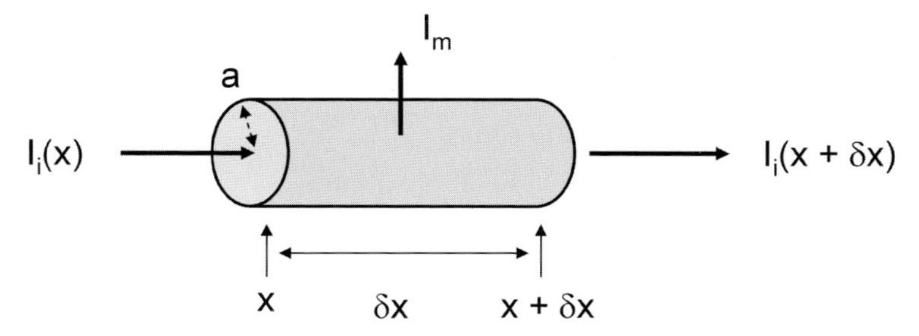

Fig. 12.18. Longitudinal current in an axon, with a "pillbox" for examining current flow, including the membrane current. (Based on [581])

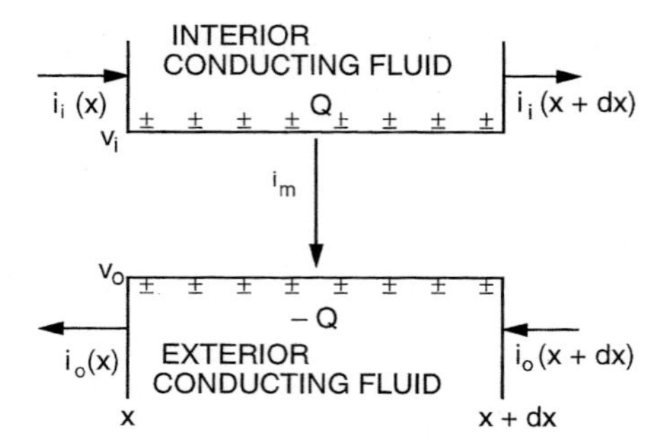

Fig. 12.19. A more physical model for the axon currents shown in Fig. 12.18. (From [581])

We will apply Kirchhoff's 1st Law (12.8) to this construct and sum all currents entering this pillbox. There are current flows inside the axon $I_{\mathrm{i}}(x)$ and $-I_{\mathrm{i}}(x+\delta x)$ entering the pillbox. There is also a current flow due to the flow of ions across the cell membrane I_{m}. We will say it is positive when it leaves the axon (Fig. 12.18), so $-I_{\mathrm{m}}$ enters it. The voltage across the cell membrane $V = q/C_{\mathrm{m}}$, where C_{m} is membrane capacitance. The time rate of change of the voltage is related to another current I_{c} associated with the change of charge on the cell membrane walls

$$\frac{\mathrm{d}V}{\mathrm{d}t} = \frac{\mathrm{d}q/\mathrm{d}t}{C_{\mathrm{m}}} = \frac{I_{\mathrm{c}}}{C_{\mathrm{m}}} \tag{12.57}$$

or

$$I_{\mathrm{c}} = C_{\mathrm{m}}\frac{\mathrm{d}V}{\mathrm{d}t}. \tag{12.58}$$

I_{c} flows to the outside, so $-I_{\mathrm{c}} = -C_{\mathrm{m}}(\mathrm{d}V/\mathrm{d}t)$ flows into the axon. Kirchhoff's 1st Law gives

$$I_{\mathrm{i}}(x) - I_{\mathrm{i}}(x+\delta x) - I_{\mathrm{m}} - C_{\mathrm{m}}\frac{\mathrm{d}V}{\mathrm{d}t} = 0 \tag{12.59}$$

or

$$I_{\mathrm{i}}(x) - I_{\mathrm{i}}(x+\delta x) - I_{\mathrm{m}} = C_{\mathrm{m}}\frac{\mathrm{d}V}{\mathrm{d}t}. \tag{12.60}$$

Using $I_{\mathrm{i}}(x+\delta x) \simeq I_{\mathrm{i}}(x) + (\mathrm{d}I_{\mathrm{i}}/\mathrm{d}x)\delta x$, we see that $I_{\mathrm{i}}(x) - I_{\mathrm{i}}(x+\delta x) \simeq -(\mathrm{d}I_{\mathrm{i}}/\mathrm{d}x)\delta x$ and this equation becomes

$$-\frac{\mathrm{d}I_{\mathrm{i}}}{\mathrm{d}x}\delta x - I_{\mathrm{m}} = C_{\mathrm{m}}\frac{\mathrm{d}V}{\mathrm{d}t}. \tag{12.61}$$

Using Ohm's Law, the voltage drop across the pillbox is

$$V(x) - V(x+\delta x) = I_{\mathrm{i}}(x)r_{\mathrm{i}}(\delta x), \tag{12.62}$$

where the resistance is $R_\mathrm{i} = r_\mathrm{i}\delta x$. Because $V(x+\delta x) \simeq V(x) + (\mathrm{d}V/\mathrm{d}x)\delta x$, we see that $V(x) - (V(x) + (\mathrm{d}V/\mathrm{d}x)\delta x) \simeq I_\mathrm{i}(x) r_\mathrm{i}(\delta x)$ or

$$I_\mathrm{i}(x) = -\frac{1}{r_\mathrm{i}}\frac{\mathrm{d}V}{\mathrm{d}x}. \tag{12.63}$$

Taking the first derivative of both sides gives $\mathrm{d}I_\mathrm{i}/\mathrm{d}x = -(1/r_\mathrm{i})\mathrm{d}^2V/\mathrm{d}x^2$, and (12.61) becomes

$$\frac{1}{r_\mathrm{i}}\frac{\mathrm{d}^2V}{\mathrm{d}x^2}(\delta x) - I_\mathrm{m} = C_\mathrm{m}\frac{\mathrm{d}V}{\mathrm{d}t}. \tag{12.64}$$

Dividing both sides by the membrane surface area is $(2\pi a)(\delta x)$ gives

$$\frac{1}{2\pi a r_\mathrm{i}}\frac{\mathrm{d}^2V}{\mathrm{d}x^2} - \frac{I_\mathrm{m}}{(2\pi a)\delta x} = \frac{C_\mathrm{m}}{2\pi a(\delta x)}\frac{\mathrm{d}V}{\mathrm{d}t}. \tag{12.65}$$

With the membrane current density (membrane current per unit area) defined as $J_\mathrm{m} = I_\mathrm{m}/(2\pi a(\delta x))$ and the membrane capacitance per unit area expressed as $c_\mathrm{m} = C_\mathrm{m}/(2\pi a(\delta x))$, this becomes

$$\frac{1}{2\pi a r_\mathrm{i}}\frac{\mathrm{d}^2V}{\mathrm{d}x^2} - J_\mathrm{m} = c_\mathrm{m}\frac{\mathrm{d}V}{\mathrm{d}t} \tag{12.66}$$

or

$$c_\mathrm{m}\frac{\partial V(x,t)}{\partial t} = -J_\mathrm{m} + \frac{1}{2\pi a r_\mathrm{i}}\frac{\partial^2 V(x,t)}{\partial x^2}. \tag{12.67}$$

This has now been expressed in terms of partial derivatives with respect to t and x, which means that the derivatives are taken with respect to t and x, respectively, treating x and t as constants. Also, the voltage is explicitly written as a function of x and t.

How do we treat active charge transport across the membrane? We model the membrane current as being $g_i(V - V_i)$ for each ion, with g_i the conductance per unit area and V_i a characteristic voltage being parameters for the specific ion. The total membrane current is

$$J_\mathrm{m} = \sum_i g_i(V - V_i) = g_\mathrm{Na}(V - V_\mathrm{Na}) + g_\mathrm{K}(V - V_\mathrm{K}) + g_\mathrm{L}(V - V_\mathrm{L}), \tag{12.68}$$

where we have included conduction by Na^+ and K^+ ions and by other ions (leakage, L). This *Hodgkin-Huxley model* is depicted in Fig. 12.20. So we find

$$c_\mathrm{m}\frac{\partial V(x,t)}{\partial t} = -\sum_i g_i(V(x,t) - V_i) + \frac{1}{2\pi a r_\mathrm{i}}\frac{\partial^2 V(x,t)}{\partial x^2} \tag{12.69}$$

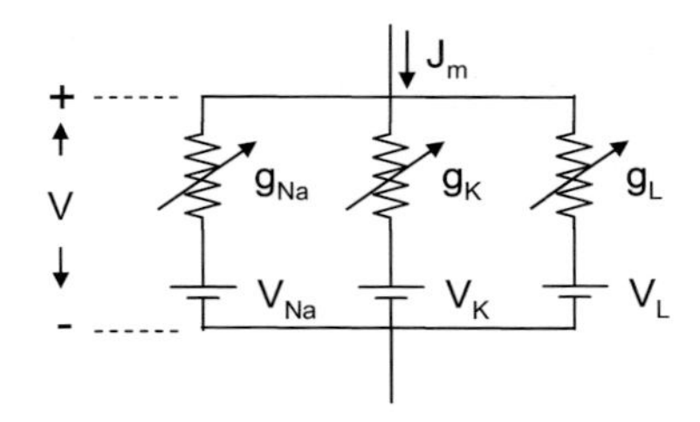

Fig. 12.20. Equivalent circuit of the Hodgkin-Huxley model of the membrane current, with variable resistors g_i and voltage sources V_i. (Based on [581])

or for only one ion

$$c_{\mathrm{m}} \frac{\partial V(x,t)}{\partial t} = -g_i(V(x,t) - V_i) + \frac{1}{2\pi a r_{\mathrm{i}}} \frac{\partial^2 V(x,t)}{\partial x^2}. \tag{12.70}$$

This last equation is known as the *Cable or Telegrapher's equation* because it also describes the propagation of electrical signals along long cables, such as submarine cables, as well as the propagation of such signals along axons in neurons. Remember that although the voltage disturbance propagates long distances, the charges move very little and in fact they move essentially only across the membrane wall, which is normal to the direction of wave propagation. (By the way, John Carew Eccles, Alan Lloyd Hodgkin, and Andrew Fielding Huxley shared the Nobel Prize in Physiology or Medicine in 1963 for their discoveries concerning the ionic mechanisms involved in excitation and inhibition in the peripheral and central portions of the nerve cell membrane, which are part of this Hodgkin-Huxley model.)

When this wave propagates at a speed u it travels as a pulse (Fig. 12.21) with unchanging shape that has constant $x - ut$. (See the discussion of sound wave propagation in Chap. 10.) It can then be shown that $\partial^2 V/\partial t^2 = u^2 \partial^2 V/\partial x^2$ or $\partial^2 V/\partial x^2 = (1/u^2)\partial^2 V/\partial t^2$ and so (12.70) can be written in terms of only derivatives with respect to time. Including the three ions, we see that:

$$\frac{1}{2\pi a r_{\mathrm{i}} u^2} \frac{\partial^2 V}{\partial t^2} - c_{\mathrm{m}} \frac{\partial V}{\partial t} = g_{\mathrm{Na}}(V - V_{\mathrm{Na}}) + g_{\mathrm{K}}(V - V_{\mathrm{K}}) + g_{\mathrm{L}}(V - V_{\mathrm{L}}) \tag{12.71}$$

with typical neuron properties given in Table 12.5. More details can be found in [581, 586].

Propagation Speed for Action Potentials

These equations must be solved numerically. Still we can gain some insight concerning the speed of these electrical signals along the axon by using an

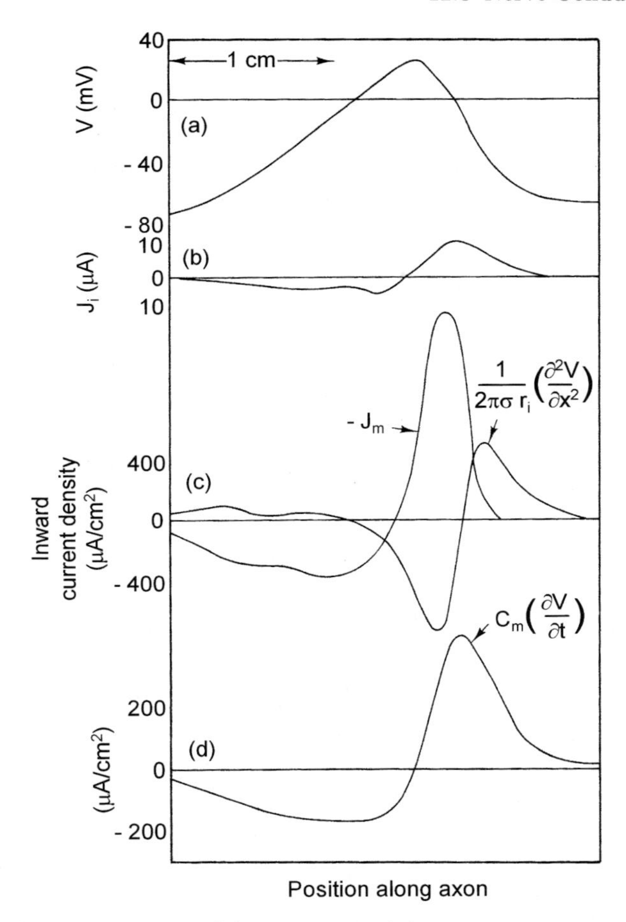

Fig. 12.21. A snapshot of the (**a**) voltage and (**b**) axon current of the pulse propagating along an axon; (**c**) current densities corresponding to the two terms on the right-hand side of (12.67); (**d**) current charging or discharging of the membrane. They are all calculated using (12.67). (From [581])

analytical method. Using (12.47), (12.50), and (12.54), we can rearrange (12.70) to give

$$\lambda^2 \frac{\partial^2 V(x,t)}{\partial x^2} - V(x,t) - \tau \frac{\partial V(x,t)}{\partial t} = -V_i, \tag{12.72}$$

where

$$\lambda = \sqrt{\frac{1}{2\pi a r_\mathrm{i} g_i}} = \sqrt{\frac{ab\rho_\mathrm{m}}{2\rho_\mathrm{i}}} \tag{12.73}$$

and

$$\tau = \frac{c_{\mathrm{m}}}{g_i} = \kappa\varepsilon_0\rho_{\mathrm{m}}. \tag{12.74}$$

We see that λ has units of distance and τ has units of time in (12.72), so it is not unreasonable to think that the conduction speed u is approximately

$$u \sim \frac{\lambda}{\tau} = \sqrt{\frac{ab}{2\rho_{\mathrm{i}}\rho_{\mathrm{m}}}\frac{1}{\kappa\varepsilon_0}}. \tag{12.75}$$

For an unmyelinated axon $b \approx 6\,\mathrm{nm}$ and so using the parameters in Table 12.5, we find

$$u_{\mathrm{unmyelinated}} \sim 0.27\sqrt{a}, \tag{12.76}$$

where u is in m/s and a is in μm. This is about 7× slower than observed, namely $1.8\sqrt{a}$.

For a myelinated axon, $b \approx 0.4a$, so $\lambda = 1,350a$ and

$$u_{\mathrm{myelinated}} \sim 2.2a, \tag{12.77}$$

which is again about 7× slower than observed, namely $17a$. This conduction model with Hodgkin-Huxley-type conduction across the membrane is not expected to be very accurate because, unlike that of the bare membrane, the conduction of the myelin sheath is independent of the voltage. Therefore, propagation occurs in the sheath region with this term and there is some decay until the signal reaches the next node of Ranvier. The signal is regenerated at this sheath-free membrane and then propagates until the next regeneration stage. If we instead assumed in the model of conduction in myelinated axons that the conduction speed is $u_{\mathrm{myelinated}} \sim D/\tau$, where $D \approx 280a$ is the distance between Ranvier nodes, the model conduction speed would be

$$u_{\mathrm{myelinated}} \sim 0.45a, \tag{12.78}$$

which is about 40× slower than observations.

The speed of nerve conduction can be measured by applying a stimulating voltage pulse at one place on the body and using electrodes to sense the time delay in the propagated pulse at another place on the body, as seen in Fig. 12.22 [568].

Passive Spreading

When voltages are below the threshold of $\sim -55\,\mathrm{mV}$, the graded potential–voltage disturbance decays along the axon and in time. We can use this model to understand this. This is equivalent to the distributed circuit model in Fig. 12.17.

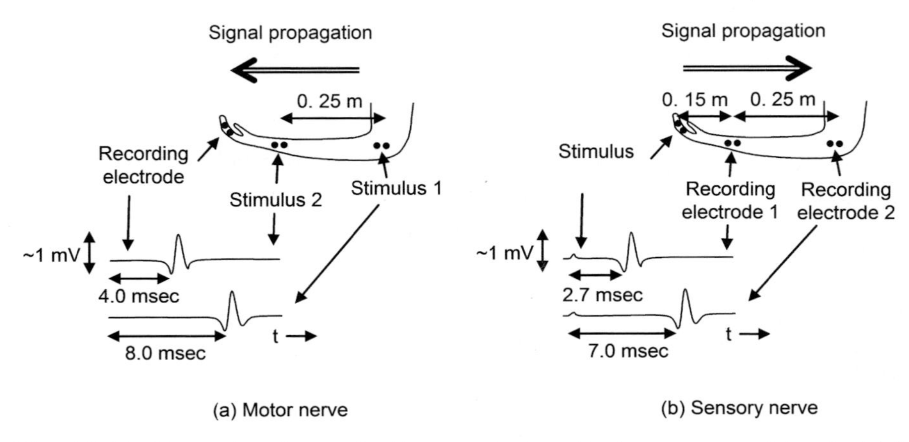

Fig. 12.22. Measuring the conduction speed along the lower arm and hand of (**a**) a motor nerve and (**b**) a sensory nerve, along with associated EMG signals. The conduction speed of the motor nerve in (a) is 62.5 m/s, and the conduction speed of the sensory nerve in (b) is 58.1 m/s (see Problem 12.18). (Based on [568])

Special Case: Only Resistance, No Capacitance, Infinitely Long Cable. If membrane capacitance is neglected in the model ($c_{\mathrm{m}} = 0$ and so $\tau = 0$), then (12.72) becomes

$$\lambda^2 \frac{\partial^2 V(x)}{\partial x^2} - V(x) = -V_i \tag{12.79}$$

and V does not depend on time. If at, say, $x = 0$, the voltage is held at $V = V_i + V_0$, the solution is

$$V(x) = V_i + V_0 \exp(-x/\lambda) \qquad \text{for } x > 0 \tag{12.80}$$
$$= V_i + V_0 \exp(+x/\lambda) \qquad \text{for } x < 0, \tag{12.81}$$

which can be proved by substitution. This means the subthreshold disturbance decays over a characteristic distance λ, as seen in Fig. 12.23.

Special Case: Only Resistance, No Capacitance, Cable of Finite Length. If the cable is semi-infinite or of finite length, the solution to (12.79) needs to be modified [589]. Such solutions are shown in Fig. 12.24 and are examined further in Problem 12.24 (for $V_i = 0$). Of particular importance for a cable of finite length is exactly how the axon is terminated at either end, i.e., the boundary conditions. Usually these boundary conditions are specified by giving the voltage V or the current flow (which is proportional to $\mathrm{d}V/\mathrm{d}x$) at the end of the axon cable.

Special Case: Only Resistance, No Capacitance, Infinitely Long Cable. If instead the axoplasm resistance is set equal to zero (such as by placing a wire

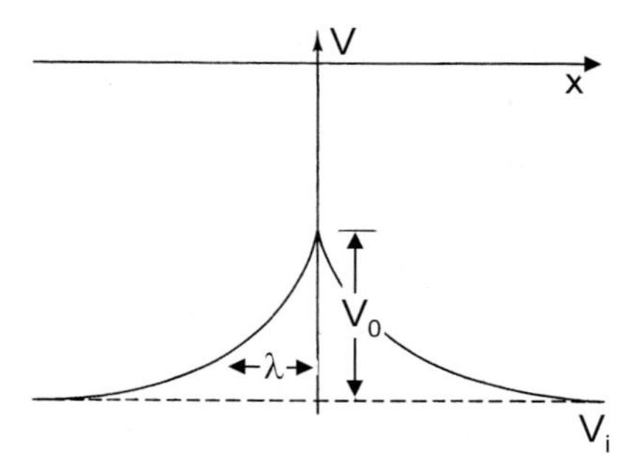

Fig. 12.23. With no axon membrane capacitance, a voltage disturbance decays over a characteristic distance λ. (From [581])

axially in the axon), then $\lambda^2(\partial^2 V(x,t)/\partial x^2) = 0$ and (12.72) becomes

$$\tau \frac{\partial V(x,t)}{\partial t} + V(x,t) = V_i. \tag{12.82}$$

(As in Fig. 12.17, transverse resistance is still possible.) If at, say, $t = 0$, the voltage were constrained to $V = V_i + V_0$ and the constraint were released, then for any x

$$V(t) = V_i + V_0 \exp(-t/\tau) \qquad \text{for } t > 0. \tag{12.83}$$

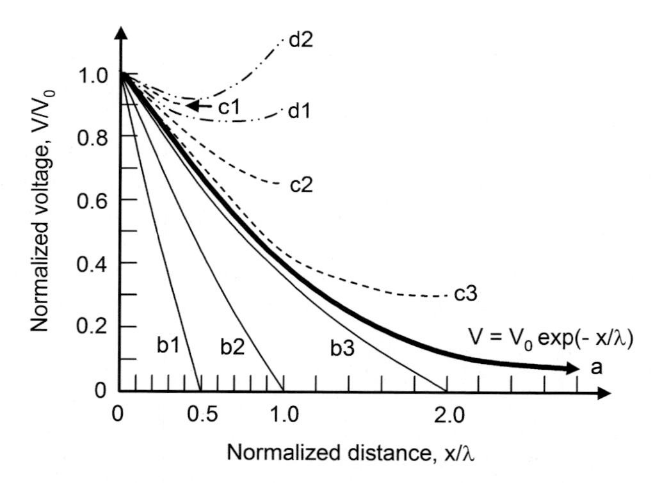

Fig. 12.24. Steady-state solutions to (12.79) for a (**a**) semi-infinite cable and (**b**)–(**e**) cables of finite length L with different boundary conditions at the end, with characteristic distance λ and $V = V_0$ at $x = 0$. For curves (b1)–(b3), $V(L) = 0$ (voltage clamped to zero), for $L = 0.5\lambda$, λ, and 2.0λ. For curves (c1)–(c3), $\mathrm{d}V/\mathrm{d}x = 0$ (current clamped to zero) at $x = L$, for $L = 0.5\lambda$, λ, and 2.0λ. For curves (d1) and (d2), the voltage is clamped to $0.9V_0$ and $1.1V_0$ at $x = L$, for $L = \lambda$. Also see Problem 12.24. (Based on [589])

Again, this can be proved by substitution. This means the subthreshold disturbance decays in a characteristic time τ.

General Case, Infinitely Long Cable. We can find a more general solution by substituting a trial solution

$$V(x,t) = V_i + w(x,t)\exp(-t/\tau), \tag{12.84}$$

into (12.72). This leads to:

$$\frac{\lambda^2}{\tau}\frac{\partial^2 w(x,t)}{\partial x^2} = \frac{\partial w(x,t)}{\partial t}. \tag{12.85}$$

This is the diffusion equation (Fick's Second Law of Diffusion, see (7.53); also see Appendix C). The disturbance w spreads in a gaussian-like manner over a distance λ in a time τ, approximately as

$$w(x,t) \propto \exp(-x^2/2D_{\text{diff}}t), \tag{12.86}$$

where the diffusion constant $D_{\text{diff}} = \lambda^2/\tau$. This assumes an initial voltage spike at $x = 0$. Using (12.84), the real voltage disturbance spreads as:

$$V(x,t) - V_i \propto \exp(-x^2/2D_{\text{diff}}t)\exp(-t/\tau), \tag{12.87}$$

which has an additional overall exponential decay in time with characteristic time τ.

12.4 Ion Channels, Hair Cells, Balance, Taste, and Smell

The previous section addressed the conduction of signals in an axon. Equally important is the actual generation of signals that are then conducted along an axon to the brain. We saw in the previous section that controlling the flow of ions across the cell membrane – by changes in the permeability of membranes to ions by the opening or closing of *ion channels* – is important in this conduction. It is also important in the generation of signals, as in sensing.

One interesting example is the excitation of hair cells, which is important in several parts of the body. Figure 12.25 shows that the "hair" in a hair cell is a hair bundle composed of an asymmetric series of 20–300 *microvilli*, which become successively larger in one direction. At the end of the bundle there is often one large cilium, which is called a *kinocilium*. When the hair bundle moves toward the kinocilium, the membrane potential depolarizes relative to the resting potential and when it moves away from it, the membrane hyperpolarizes, as is seen in Fig. 12.26. There is no change in membrane potential when the bundles moves perpendicular to the direction of increasingly large microvilli. One possible explanation for this depolarization is that Na^+ positive ion channels open when the hair bundle is displaced toward the kinocilium. The elastic response of the hair cell comes from the microvilli themselves, the elastic elements (*gating springs*) that pull on the ion channels, and the

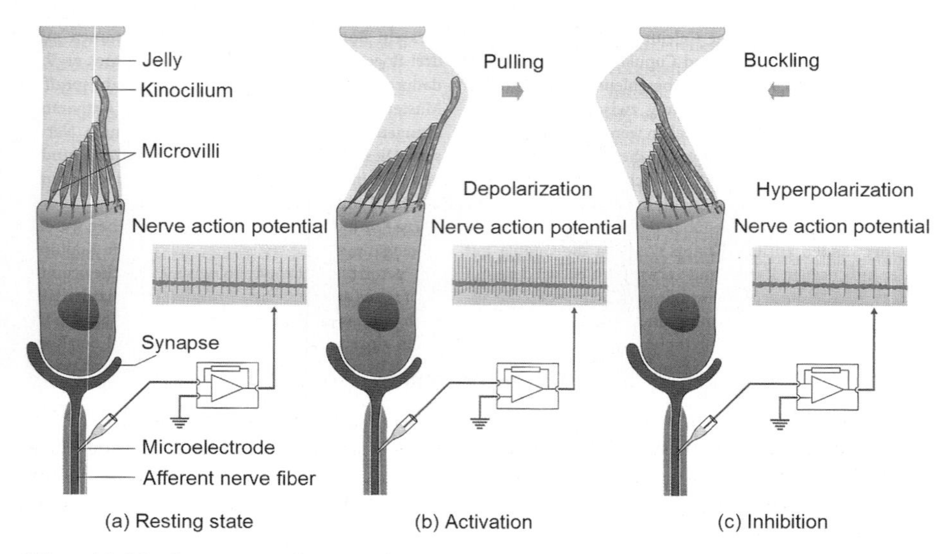

Fig. 12.25. In one mechanism for hair cell response, the hair bundle moves toward the kinocilium (hair with the bead) opening channels that are permeable to Na^+ (which is depolarization), as shown in (**b**). Resting activity is seen in (**a**) and hyperpolarization in (**c**). (From [593])

channels themselves. The response of hair to forces was discussed in Chap. 10 (text and Problem 10.55).

These hair cells are important in the ear, contributing both to the generation of auditory signals in the cochlea that travel to the brain to enable hearing and in the vestibular system in the ear that helps us maintain a sense

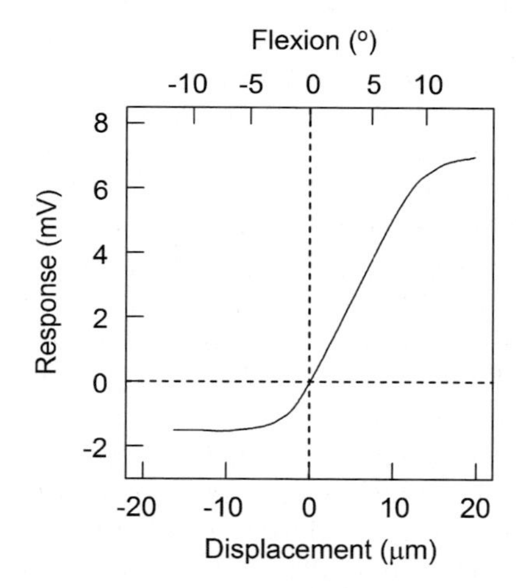

Fig. 12.26. Membrane potential vs. hair displacement (in position and angle). Positive displacements are toward the kinocilium. (Based on [574, 583])

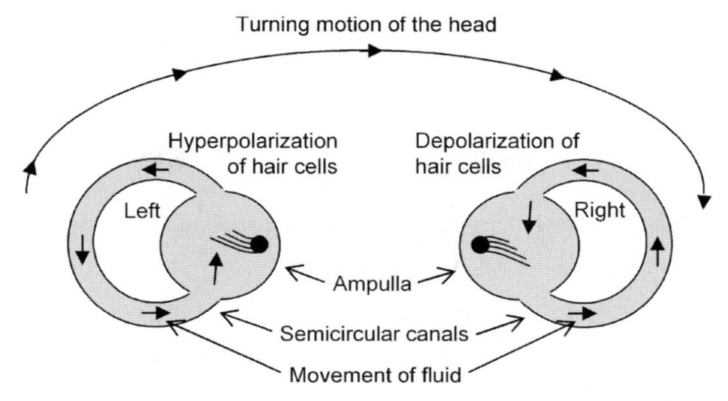

Fig. 12.27. The sense of balance is seen by examining the pair of horizontal semicircular canals, by looking at the head from above. When you turn your head clockwise there is counterclockwise motion of the cochlear fluid that depolarizes the hair cells in the semicircular canal in the right ear and hyperpolarizes them in the left ear. (Based on [574])

of balance. Each ear has three semicircular canals that are approximately orthogonal to each other, which provide us with a sense of balance through a sensing of the motion of fluid in them. Figure 12.27 shows how hair cells sense one such motion, that of turning your head to the right. This is clockwise looking from the top, as in the figure. The fluid in the two depicted horizontal semicircular canals lags behind this motion (Newton's First Law), and so it moves counterclockwise relative to the hair cells. This causes a depolarization of the hair cells in the right ear and a hyperpolarization of the hair cells in the left ear. These semicircular canals contain hair cells that are bathed in a fluid, the endolymph, which has high concentrations of K^+ and low concentrations of Na^+ and Ca^{2+}. Consequently, when the hair cells are stimulated, K^+ enters the cell through the channels during this depolarization. The hair cells in the cochlea are also bathed by this endolymph fluid in the scala media so the control of K^+ ion channels by the hair bundles is also important in hearing transduction (where transduction is the conversion of one kind of signal or stimulus into another by a cell, which in this case is the conversion of sound into an electrical signal). (The perilymph fluid in the scala vestibuli and the scala tympani is high in Na^+ and low in K^+, as are blood and cerebrospinal fluid.)

The importance of hair cells in the sense of touch (for hairy skin) was discussed in Chap. 2. The sense of touch by Merkel receptors, Meissner corpuscles, Ruffini cylinders, and Pacinian corpuscles in both hair-free and hairy skin arises from changes in the ion channels caused by applied pressure.

Taste bud sensors are found in clusters called taste buds on the tongue and other places in the oral cavity. There are several mechanisms that activate sensors of taste for sweet, sour, bitter, salty, and "umani," all involving the control of membrane ion channels. (Umani is the Japanese word for delicious. In this context it describes the taste of monosodium glutamate and other amino acids.) The conceptually simplest is that for saltiness, which is detected by a Na^+ channel that depolarizes the detector cell.

The olfactory receptor region in the nose has an area of $\sim$1–2 cm^2, with $\sim$12 million receptor cells. (There are $\sim$4 billion such cells in a German shepherd dog.) The sense of smell is activated by olfactory neurons, with the opening of ion channels. In many such neurons, this allows Na^+ to enter the cell during depolarization, which induces an increase in the firing of action potentials.

12.5 Electrical Properties of the Heart

The total charge of the heart is zero during the heart beat, but there are dynamic separations between positive and negative charges. These create an electric dipole that rotates as it becomes larger and then smaller in magnitude during each cardiac cycle. The electric potential at different places on the skin consequently changes with time during each cycle and this is what is sensed in an electrocardiogram (EKG or ECG). These potential differences are typically $\sim$30–500 μV. Usually 12-lead scalar EKG measurements are made, which give much information about the evolution of the cardiac dipole and sufficiently valuable information concerning potential abnormalities in the heart. Vector EKGs are taken less often; they can provide a more complete view of the evolution of the heart dipole during a heart beat. A typical EKG is shown in Fig. 12.28.

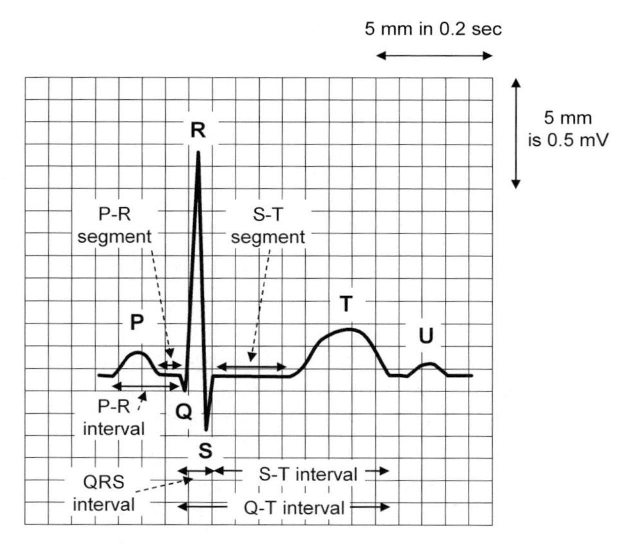

Fig. 12.28. A normal electrocardiogram (EKG/ECG), showing the P wave (atrial depolarization), QRS complex (ventricular depolarization), and T wave (ventricular repolarization) in a single cardiac cycle. Typically the scan proceeds with 25 mm/s and the signal strength is plotted as 10 mm/mV. (Based on [586])

From an electrical perspective, the heart can be described as an electric dipole whose magnitude and direction varies in a cyclic manner, repeating for each heart cycle. As for the axon described earlier in this chapter, the positions of charges in the cardiac muscle cells change during cell depolarization in muscle contraction and repolarization. This constitutes a change in the electric dipole moment of the individual cell. The electric fields due to all such heart muscle cells add to produce voltage variations in the body that are sensed by the EKG probes. The voltages vary with time indicating the depolarization (contraction) of the right and left atria (called the P wave), the depolarization (contraction) of the right and left ventricles during systole (the QRS complex) – the repolarization of the atria is masked by this, and the repolarization of the ventricles (the T wave). The time dependence can indicate normal or abnormal firing of the heart muscle, and this could, in principle, be determined from the voltage difference across two EKG probes. Analysis of the voltages across several pairs of electrodes provides important information that is used to spatially locate abnormalities in different parts of the heart muscle, such as after a heart attack. The EKG probes measure the electric potential (voltage) just below the skin. The resistance across the skin is not significant because the EKG probes are connected to the skin with a special contact jelly.

The difference in potential across the cell membrane of the cardiac muscle cell changes during the depolarization and subsequent polarization of atrial and ventricular heart muscles during each cycle, and this changes the electric potential near the heart. Because the tissues and blood of the body contain conductive ions, such changes in potential cause changes in currents and the net results affect the electric potential very far from the heart. As such, the cardiac muscle can be viewed as being placed in a volume conductor.

Why does this potential change with time, even for a single muscle cell? Let us follow the motion of charges during a cycle. The field across the cell membrane can be modeled locally as an electric dipole, with positive and negative charges, of equal magnitude, separated by a distance (Fig. 12.1). The electric field lines are shown for such a point dipole in Fig. 12.1.

A polarized cardiac muscle cell is a series of such dipoles as depicted in Fig. 12.29a, with about $-70\,\mathrm{mV}$ inside the cell relative to the outside, all around the cell. No potential (which is really a baseline potential) is seen at the electrode immediately to the right of the cell. As the depolarization wave propagates from left to right, the potential on the right increases and reaches a maximum when half the cell is depolarized, as in (c). As the depolarization wave arrives at the right end, this voltage decreases to zero as in (e). This is similar to the PQR wave in ventricular depolarization seen in the EKG in Fig. 12.28. This is what would occur if the potential across the membrane were zero after depolarization. Because it actually becomes slightly positive, the potential in (e) should dip slightly negative, as seen for the PQR wave. When the left side becomes repolarized, the potential becomes negative and a negative pulse develops, as in (f)–(h). This is similar to the S ventricular repolarization pulse in Fig. 12.28 except for its sign. Unlike that in Fig. 12.29h,

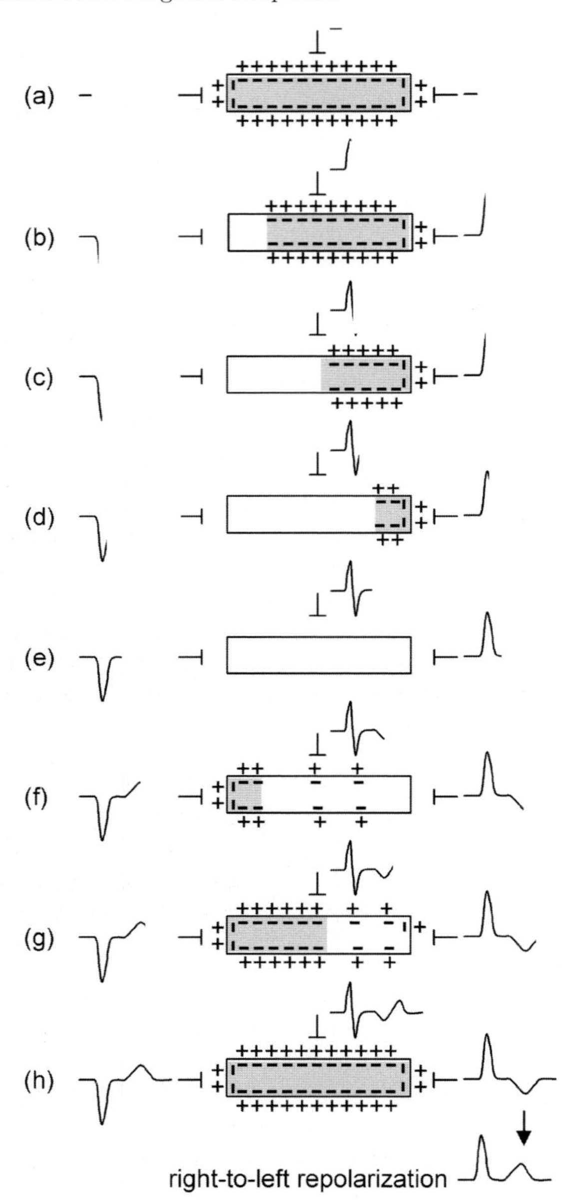

Fig. 12.29. Potential to the right of a strip of myocardium immersed in a volume conductor during (**a**)–(**e**) depolarization and (**f**)–(**h**) repolarization. The polarized section is gray and the depolarized section is white. For real cardiac muscle, the repolarization signal is positive, as is the depolarization signal (Fig. 12.28), because in the human heart repolarization proceeds in the direction opposite from depolarization, as shown on the bottom. (Based on [597])

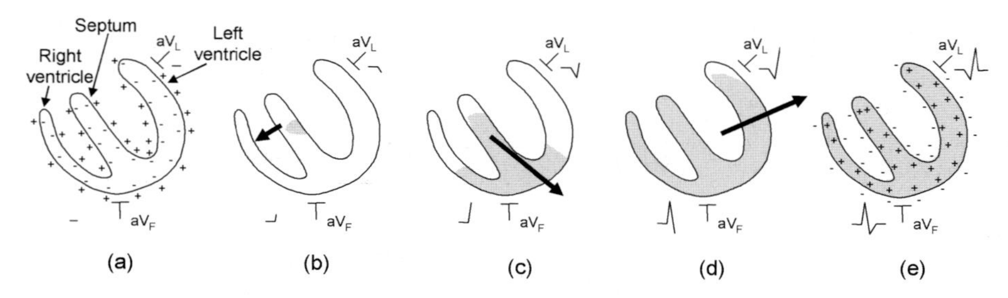

Fig. 12.30. Normal ventricular depolarization recorded by leads aV_L and aV_F, showing the change in magnitude and counterclockwise rotation of the projection of the cardiac dipole in the frontal plane. (Based on [584])

the second peak is positive because cardiac muscle repolarization proceeds in the direction opposite from depolarization, as for the lowermost trace. Furthermore, repolarization is a bit slower and more inhomogeneous than depolarization so the negative dip is broader (slower) and has a smaller magnitude (wider) than the first peak.

Figure 12.29 also shows that the potential is qualitatively different at the left and at the top of the cell during depolarization. This is why the EKG electrodes placed at different positions sense different signals (and can provide different information). Furthermore, the electrode placed on the left in Fig. 12.29 gives the negative of the signal of that placed on the right. (This makes sense. Why?)

Each of the four cycles of atrial and ventricular depolarization and atrial and ventricular repolarization does not occur simultaneously throughout the heart, and each is sensed by an EKG and can be analyzed separately. The evolution of the net cardiac electric dipole during ventricular depolarization (QRS cycle) is shown for a normal heart in Fig. 12.30.

Clearly the magnitude and direction of the dipole change greatly during each cycle. These are sensed by the exact placement of EKG electrodes, which can provide important details about cardiac function and malfunction. The location of the twelve leads for an EKG with a supine (lying down) person are described in Table 12.6. Six are on the ribs, and six others are on the arms and legs. Three of the latter have two leads (bipolar), one for monitoring and one for reference, while the other nine are single leads (unipolar). Figures 12.31 and 12.32 show the nine locations in the table where the 12 EKG leads are placed. Remember that voltage differences are being measured from one lead to another.

The earlier discussion shows that the EKG voltage is positive when the cardiac dipole points to the (positive side of the) EKG lead and negative when it points away from it. This is clear from Fig. 12.1d. Figure 12.33 shows the 12 EKG signals from a normal heart. Figure 12.30 shows the relation between the effective positioning of two of these bipolar leads, the evolution of the cardiac dipole, and the signal recorded by these leads.

Table 12.6. Position of electrodes in an EKG. See Figs. 12.31 and 12.32. (Using information from [575])

lead	electrode position
standard limb leads (bipolar)	
I	right arm and left arm
II	right arm and left leg
III	left arm and left leg
augmented leads (unipolar)	
aV_R	right arm
aV_L	left arm
aV_F	left leg
chest leads (unipolar)	
V_1	4th intercostal space, right side of sternum
V_2	4th intercostal space, left side of sternum
V_3	5th intercostal space, left side (between V_2 and V_4)
V_4	5th intercostal space, left side (midclavicular line)
V_5	5th intercostal space, left side (anterior axillary line)
V_6	5th intercostal space, left side (midaxillary line)

Three of the EKG electrodes are placed on the right and left arms and the left leg, and the voltages across the three pairs of these electrodes are monitored (along with the signals from the other probes), and are called I ($V_{\mathrm{I}} = V_{\text{left arm}} - V_{\text{right arm}}$), II ($V_{\mathrm{II}} = V_{\text{left leg}} - V_{\text{right arm}}$), and III

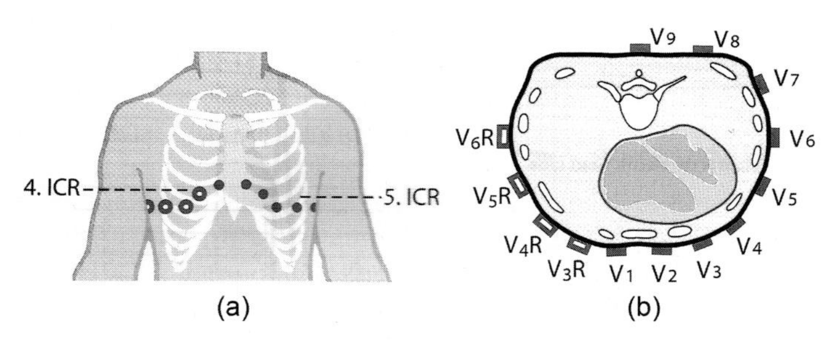

Fig. 12.31. Placement of the horizontal plane, precordial unipolar EKG electrodes. Only 6 of the 12 leads in (**b**), V_1–V_6, are used in usual EKGs (*solid circles*, in the region labeled by 5.ICR (5th intercostal space or region in the ribs) in (**a**)). The additional dorsal leads V_7–V_9 are specifically used to detect a posterior myocardial infarction. The additional right precordial leads, V_3R–V_6R (*open circles*, in the region labeled by 4.ICR (4th intercostal space or region) in (a)), are specifically used to detect a right ventricular myocardial infarction. (From [577])

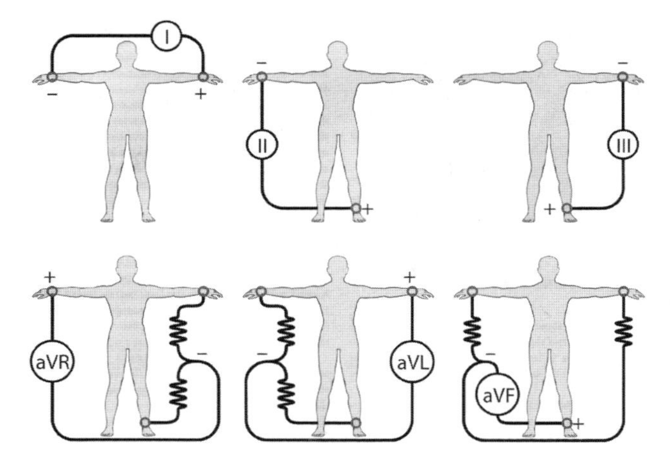

Fig. 12.32. Placement of the three unipolar and three bipolar front-plane limb leads. Sometimes an electrode is positioned on the right leg (not shown) to serve as an electrical ground. (From [577])

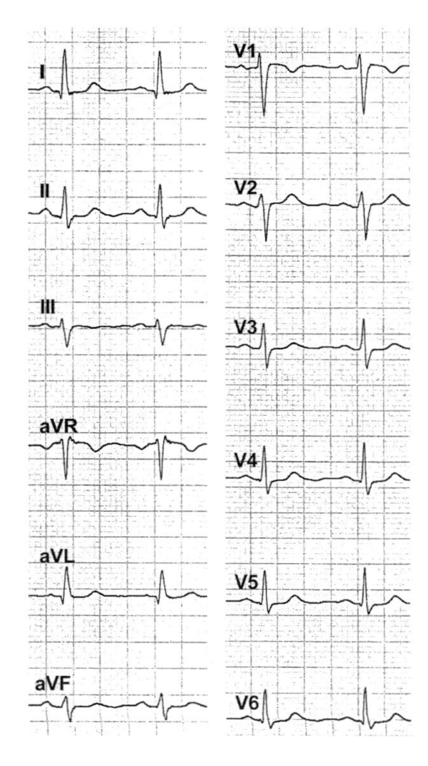

Fig. 12.33. Normal EKG patterns from the 12 electrodes. (From [577])

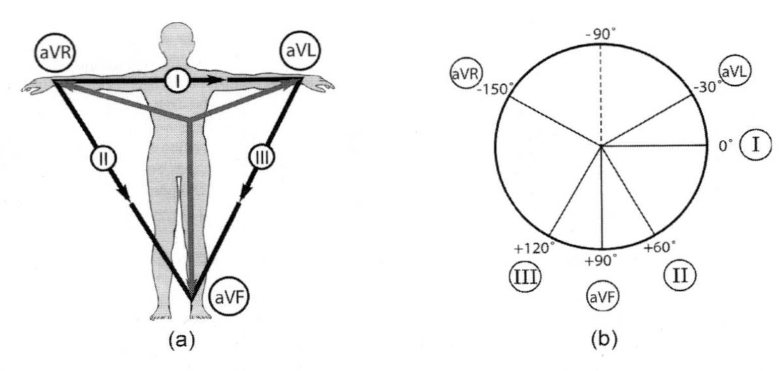

Fig. 12.34. (**a**) The three leads I, II, and III are arranged as Einhoven's triangle. (**b**) The effective directions for the six frontal leads shown in (a) are translated to form the triaxial reference system called Cabrera's circle. The signed, vector projection of the cardiac dipole onto these six directions gives the EKG signal for these six frontal plane leads. (From [577])

($V_{\mathrm{III}} = V_{\mathrm{left\ leg}} - V_{\mathrm{left\ arm}}$), as in Table 12.6. These three electrodes act as if they probe at the vertices of a triangle, which is usually called the *Einhoven's triangle*, as shown in Fig. 12.34a. Because the arms and legs do not have new sources of electric fields and the tissue in each is a conductor, the probes on the arms actually sense the same voltages as if they were instead placed on the respective shoulders and the probe on the leg has the same voltage as if it were placed on the bottom of the torso near the pubic area, and Einhoven's triangle is sometimes depicted for this smaller triangle. Using Kirchhoff's 2nd Law, (12.14), $V_{\mathrm{I}} + V_{\mathrm{III}} - V_{\mathrm{II}} = 0$ (the minus sign in front of the last potential indicates a different sign convention around the circuit than for the first two) or

$$V_{\mathrm{I}} + V_{\mathrm{III}} = V_{\mathrm{II}}. \tag{12.88}$$

Cabrera's circle in Fig. 12.34b shows the effective positioning of the six frontal plane leads, and this is used in Fig. 12.30.

Figure 12.34 shows how momentary cardiac dipoles in three-different directions cause momentary potential differences in these three electrode pairs. An appropriate sum of these three voltages, such as $V_{\mathrm{I}} + V_{\mathrm{III}} - V_{\mathrm{II}}$, serves as the electrical ground for measurements with each of the six chest probes. Also, note that the difference in the signals from the $\mathrm{aV_L}$ and $\mathrm{aV_F}$ leads in Fig. 12.30 should be similar to that from the III lead.

This EKG can provide important details about cardiac function and malfunction, and the location of the malfunction. This includes (1) the heart rate, (2) arrhythmia, (3) axis (giving the direction and magnitude of activity for atrial and ventricular contractions), (4) hypertrophy (which is an increase in the left or right ventricular muscle mass), (5) enlargement (which is an increase in the volume of the left or right atria chambers), and (6) infarction.

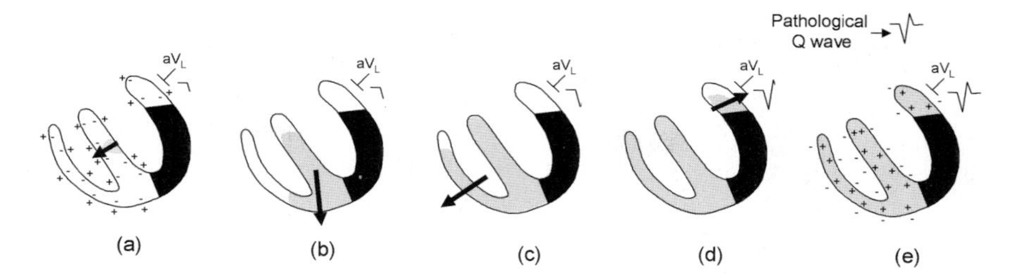

Fig. 12.35. Evolution of the cardiac dipole during ventricular depolarization after a lateral wall cardiac infarction, along with the EKG from lead aV$_L$. The larger-than-normal Q wave occurs because the site of the infarction (black) should be depolarizing and contributing a positive signal in (**c**). (Based on [584])

In arrhythmias, there can be a variable rhythm, a rhythm that is either too fast (tachycardia, >100 beats/min; but >250 beats/min – flutter or fibrillation – it can be life-threatening in the ventricles) or too slow (bradycardia, <60 beats per min, except it can be lower in trained athletes), and deviations from a 1:1 ratio of atrial and ventricular contractions, as described more in Chap. 8.

During a myocardial infarction part of the cardiac muscle is damaged and within a few hours these muscle cells usually die and then do not depolarize and repolarize. The absence of electrical signals from a given part of the left ventricle is seen in the EKG in Fig. 12.35. The dipole vector during depolarization points away from the black region of the infarction in step 3; if that black region were active it would then depolarize and the dipole would be pointing in the opposite direction. This is seen as an enhanced Q-wave. The formation of scarring in this damaged region can still be seen after recovery (Fig. 12.36).

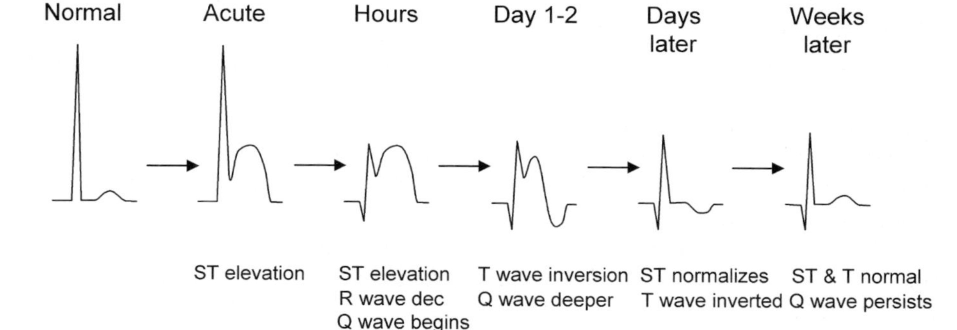

Fig. 12.36. EKG evolution during and after an acute Q-wave myocardial infarction. (Based on [584])

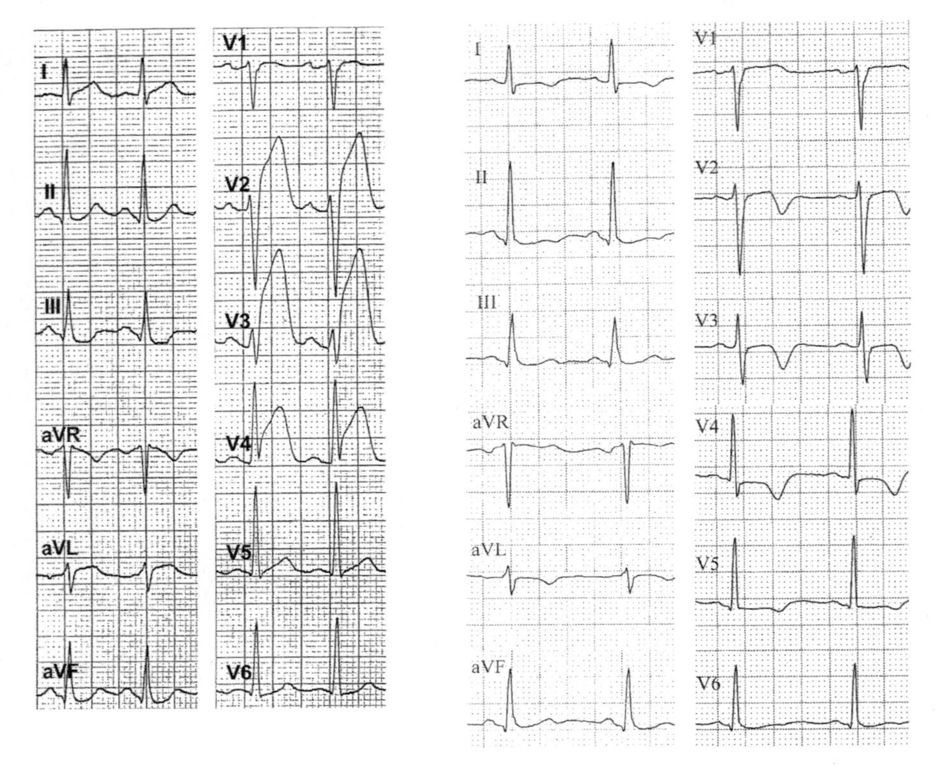

Fig. 12.37. EKG signals from the 12 leads (left) 2 h after an anteroseptal myocardial infarction an (right) 4 h later, after thrombolysis. (From [577])

In general, large Q waves in the I and aV_L traces indicate a lateral infarction, in the V_1, V_2, V_3, or V_4 traces an anterior infarction, and in the II, III, and aV_F traces an inferior infarction. A large R wave in the V_1 and V_2 traces indicates a posterior infarction.

Figure 12.37 shows all 12 EKG traces 2 h after an anteroseptal myocardial infarction (i.e., one with features of both anterior and (interventricular) septal myocardial infarctions) and 4 h after the infarction has been treated by thrombolysis (treatment to break up blood clots); there is no pathological Q wave in this case, but elevated ST waves at 2 h and a normal ST segment 4 h later, but negative T in some traces. (Compare both to the normal traces in Fig. 12.33.)

Information from these scalar measurements can be projected onto different body planes and provide information about the evolution of the cardiac dipole vectors in a process called vectorcardiography. Figure 12.38 shows the QRS vector evolution in a vectorcardiogram and its projections on the frontal and horizontal planes. For more on EKGs and diagnosis using EKGs see [572, 577, 584, 586, 587, 590, 591, 596, 597].

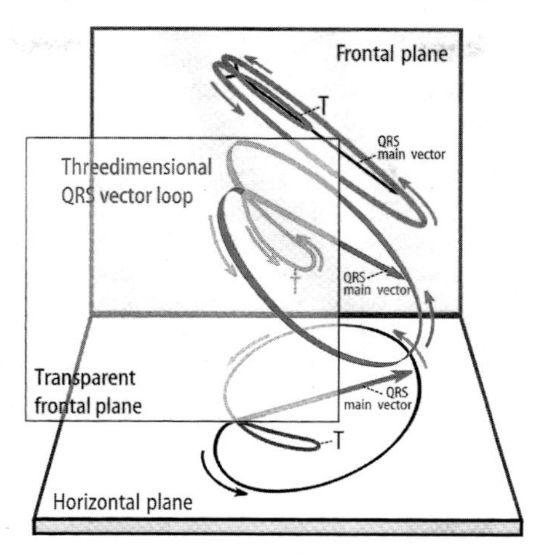

Fig. 12.38. A ventricular vectorcardiogram, showing the evolution of the QRS vector along its loop, along with its projections on the frontal and horizontal planes. The frontal projection gives the frontal plane EKG derived from the frontal plane leads, which need little correction; whereas the horizontal projection must be corrected to obtain the scalar EKG obtained directly from the precordial leads. (From [577])

12.6 Electrical Signals in the Brain

Electrical signals are also important in other parts of the body, as shown in Table 12.1, such as the electroencephalograms (EEGs) of brain waves in Fig. 12.39. In contrast to the very regular EKG patterns, the EEG signal is irregular, but it has identifiable rhythmic patterns: alpha waves (frequency of 8–13 Hz; awake, restful state), beta waves (14–25 Hz; alert wakefulness, extra

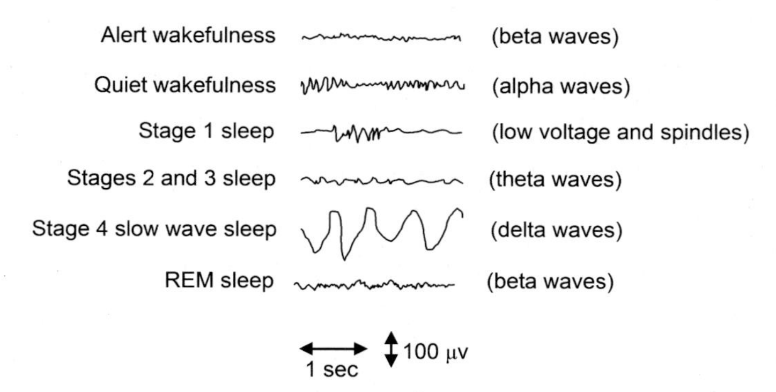

Fig. 12.39. Schematic of changes in brain waves during different stages of wakefulness and sleep. (Based on [580, 593])

Table 12.7. Effect of currents (in mA) on the human body (for about 1 s). (Using data from [595])

effect	DC	AC (60 Hz)
slight sensation at contact point	0.6	0.3
perception threshold	3.5	0.7
shock		
– not painful, no loss of muscular control	6	1.2
– painful, no loss of muscular control	41	6
– painful, let-go threshold	51	10.5
– painful, severe effects: muscular contractions, breathing difficulty	60	15
– possible ventricular fibrillation (loss of normal heart rhythm)	500	100

All values are approximate.

activation, tension), theta waves (4–7 Hz, mostly in children, also adults with emotional stress and with many brain disorders), and delta waves (<3.5 Hz; deep sleep) [586].

12.7 Effects of Electric Shock

External electrical currents running in the body can cause damage by interfering with normal bodily function – such as by preventing your otherwise operational skeletal and cardiac muscles from functioning normally – and by destroying tissues by thermal heating (Table 12.7). Muscles are controlled by a series of electrical impulses sent by the brain. External AC currents (60 Hz) above 10 mA or so override these signals and prevent you from exercising control over your muscles. You can barely control your muscles at 10 mA and barely "let go" of an object. At higher currents your muscles are under external control, possibly leading to breathing and circulatory difficulties. Ventricular fibrillation occurs from 100 mA to 4 A and paralysis occurs, along with severe burns (and death), over 4 A. For weak shocks, the sensation of shock varies as the 3.5 power of the applied 60 Hz voltage (Stevens' Law (1.6), Table 1.15), so the perception of electric shock is very superlinear with stimulus.

The skin is a very important barrier to current flow (I). The resistance (R) through dry skin is roughly 100,000–600,000 ohms and through wet skin it is only about 1,000 ohms. If the skin barrier is overcome, the resistance drops (so there is more current flow per unit voltage, as per Ohm's Law). Figure 12.40 shows that the internal body resistance is low, approximately 400–600 ohms from head to foot and 100 ohms from ear to ear. The amount of current that can flow in the body induced by a voltage source (V) is limited by two factors

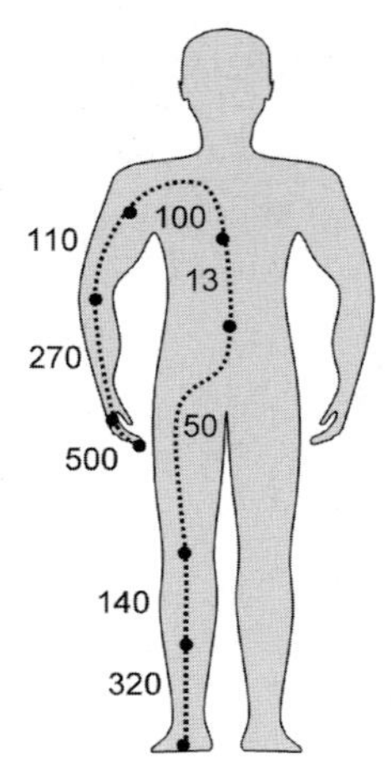

Fig. 12.40. Body segment resistance (in ohms), ignoring skin contribution. As shown, 500 ohms is the contribution from one finger. (Based on [579])

(1) Ohm's Law says the current will be $I = V/R$. (2) The current is sometimes limited by the voltage source itself.

Let us assume the skin barrier has been broken so the effective body resistance is about 500 ohms. (Please do not attempt this!!!) The 120 V AC from a wall outlet will produce a current of 240 mA, which is over twice that needed to cause death through ventricular fibrillation. Circuit breakers typically trip at 15 A, so this flow through the body will be uninterrupted by the circuit breaker. How about DC sources? The current induced by the often-used 9 V battery is 18 mA, which can cause a shock. (You can easily draw this current from such a battery.) The voltage across a car battery is 12 V with 400–600 A (cranking amps), so it can shock you even worse (Problem 12.5).

12.8 Magnetic Properties

The magnetic fields in the body are due to electric currents and are extremely weak. Typical magnetic fields in the body that can be measured are shown in Table 12.8, and are all much weaker than the 5×10^{-5} T (0.5 Gauss) magnetic field of the earth. (For comparison, the maximum human-made magnetic fields approach 100 T.)

12.8.1 Magnetic Field from an Axon

The Biot-Savart Law determines the magnetic field from currents. Consider a continuous current I flowing along the infinitely long z-axis. Using the Biot-Savart Law, one can show that a distance R away the magnetic field B has a

Table 12.8. Typical amplitude of biomagnetic signals. (Using data from [567])

biomagnetic signal	typical amplitude (pT)
magnetocardiogram (MCG)	50
fetal MCG	1–10
magnetoencephalogram	1
evoked fields	0.1
magnetomyogram	10
magneto-oculogram	10
Earth's field	50×10^6

magnitude

$$B = \frac{\mu_0 I}{2\pi R} \tag{12.89}$$

and is in the radial direction, according to the usual right hand rule.

This analysis does not exactly apply to signals along a neural axon. A voltage pulse traveling along an axon is a pulse and not a continuous current. It is in a medium that is fairly conductive. Also, there are several directions of the current flow, along the axon, transverse to the membrane, etc. Still, let us estimate the field strength just as the pulse passes by, and model it as a continuous current. The current along the axon is the most important. Using (12.63) $I_\mathrm{i}(x) = -(1/r_\mathrm{i})\mathrm{d}V/\mathrm{d}x$, we estimate the magnitude of this current to be

$$I \sim \frac{1}{r_\mathrm{i}} \frac{V}{\lambda}. \tag{12.90}$$

Therefore, the magnetic field magnitude 1 mm away from the axon is approximately

$$B \sim \frac{\mu_0 V}{2\pi R r_\mathrm{i} \lambda} \tag{12.91}$$

$$= \frac{(4\pi \times 10^{-7}\ \mathrm{T\text{-}m/A})(0.1\ \mathrm{V})}{2\pi(0.001\ \mathrm{m})(6.4 \times 10^9\ \mathrm{ohm/m})(3.8 \times 10^{-4}\ \mathrm{m})} = 8\ \mathrm{pT}, \tag{12.92}$$

with $\lambda = 3.8 \times 10^{-4}$ m. This value is consistent with the low values in Table 12.8. (This estimate is reasonable even though some of the assumptions are not perfect.)

12.8.2 Magnetic Sense

Humans (apparently) cannot sense magnetic fields, but magnetic fields do help several animals sense direction (as with a compass) and/or location due to the presence of 50-nm diameter magnetite (Fe_3O_4) particles in their bodies. (These particles are sometimes arranged in chains.) For example, this magnetic sense is very strong in pigeons, who have 10–20 nT sensitivity, and dolphins, who have $<2\,\mu$T sensitivity [565]. The magnetic sense mechanism may involve a torque that is induced on this particle system by the field and this in turn may induce a torque on intracellular filaments; this movement of the filaments triggers a sensory neuron. (Electric fields induce a torque on an electric dipole, as is easily seen by examining the Coulomb forces on the individual charges in the electric dipole. Similarly, magnetic fields induce a torque on a magnetic moment. This analogy is valid even though there are no magnetic charges.)

12.9 Electromagnetic Waves

Radio waves, microwaves, infrared radiation, visible light, ultraviolet light, X-rays, and gamma rays are all electromagnetic waves. Each propagates at the same speed of light c in vacuum. Each has a frequency of oscillation ν and wavelength λ related by $c = \lambda\nu$ (11.2). For each, this oscillation consists of electric and magnetic fields sinusoidally oscillating in phase in vacuum. They differ only in their frequency (and consequently wavelength), which increases (decreases) in going from one of these regimes to the next. We discussed visible light at length in Chap. 11.

The penetration of electromagnetic radiation through the body is sometimes of interest. The attenuation factor for such radiation plotted in Fig. 12.41 is α_{light} from Beer's Law, (10.18),

$$I(z) = I(z = 0)\exp(-\alpha_{\text{light}} z). \tag{12.93}$$

Attenuation consists of losses from absorption and scattering, and is clearly very dependent on frequency. This figure shows trends, but not all details. For example, the attenuation factor in the microwave is due to nonresonant processes. The body absorbs microwave radiation at 2.45 GHz used in microwave ovens much more strongly than indicated there because of the strong resonant absorption by water at this frequency. (This is why this frequency is used in microwave ovens.)

12.10 Summary

Electrical processes are essential to the operation of the body and have proved to be very important in medical diagnostics. Electrical conduction is important in most parts of the body. Models of the propagation of electrical signals in nerves can explain the physical basis of perhaps the most important mechanism of regulation in the body. Electrical processes are integral to cell operation, including to the physics of cell membranes. The electrical nature of the heart has led to the use of EKGs as a diagnostic that can be interpreted

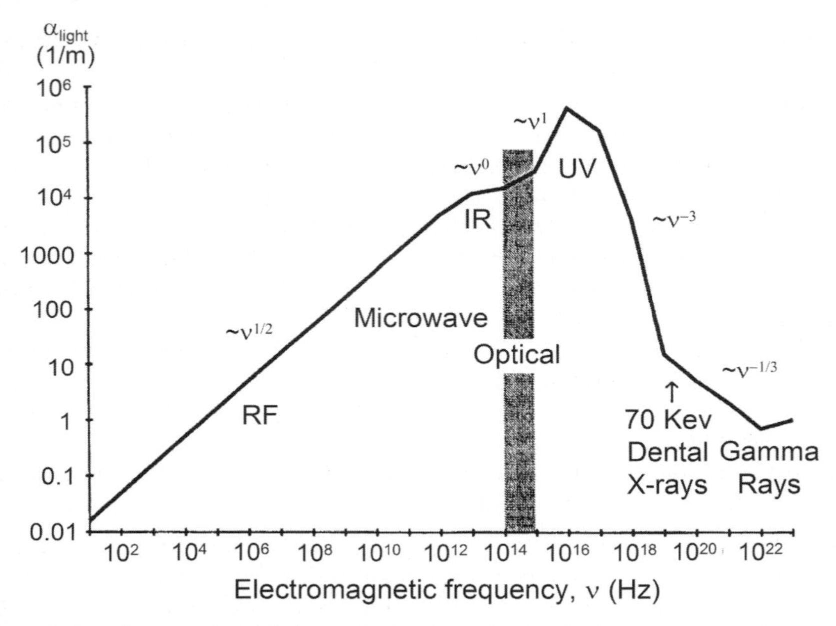

Fig. 12.41. Attenuation of electromagnetic radiation in human tissue, due to absorption and scattering. (From [576]. Courtesy of Robert A. Freitas Jr., Nanomedicine, Vol. 1 (1999), http://www.nanomedicine.com)

by using simple models of the dipole nature of the heart. Electrical signals in other parts of the body are also used in diagnostics. Naturally occurring magnetic signals are relatively less important in the body.

Problems

Conductance, Transmission, and Potentials

12.1. Use Table 12.4 to compare the conductivity in a cell and in blood.

12.2. Estimate the electrical resistance of the blood in a 50-cm long, 3-mm diameter artery.

12.3. There are two common relations for the resistivity of blood as a function of the hematocrit Hct: $\rho = 0.537\exp(0.025\text{Hct})$ and $\rho = 0.586(1 + 0.0125\text{Hct})/(1 - 0.01\text{Hct})$, which is called the Maxwell–Fricke equation [586]. How do they differ in the range of Hct from 10%–60%, and specifically at the normal value of 45%?

12.4. During an accident, 120 V AC from a wall socket connects your body to electrical ground, from hand to hand:

(a) If the resistance across the body is 500 ohms, what is the current flow?
(b) Is this dangerous?
(c) If the region from hand to hand can be modeled as a cylinder of constant diameter (equal to the diameter of the upper arm) and length (from finger tip to finger tip) of your own body, and all material is assumed to be uniform, estimate the electrical resistivity of the body tissue.
(d) How much power is dissipated in this section? (Calculate both the total power and the power per unit volume.) (Remember that the power dissipated is $P = IV$ and Ohm's Law is $V = IR$. Assume here and below that the power is the same as that for a DC voltage source.)
(d) What is the heat capacity of this cylindrical section? Assume the average specific heat of the body.
(e) Ignoring heat flow, how much would the temperature of this section increase per unit time?
(f) How long would it take to denature the proteins in this cylindrical section? (See the information provided in Chap. 13.)

12.5. Compare the amount of current that could be drawn from a car battery in an electrical shock to the maximum amount of current that could be drawn from it. What does this mean? Why is the shock worse than that from a 9-V battery?

12.6. What ranges of electromagnetic radiation can penetrate through your (a) eyelid, (b) finger, and (c) chest?

12.7. An ion of mass m and charge q moves at a speed v under the influence of an electric field E. It suffers a drag force that relaxes its speed with a characteristic time τ:
(a) Show that force balance on the charge gives: $m\mathrm{d}v/\mathrm{d}t = -mv/\tau + qE$.
(b) Show that in steady state, the ion moves at the drift velocity (which is really the drift speed here), $v_{\mathrm{drift}} = qE\tau/m = \mu E$, where $\mu = q\tau/m$ is called the mobility.

12.8. (advanced problem) Show that the potential of a charge Ze shielded by mobile charges is given by $V(r) = (Ze/4\pi\epsilon_0\epsilon r)\exp(-\kappa r)$, where $\kappa = \sqrt{(e^2/\epsilon_0\epsilon k_{\mathrm{B}}T)\sum_i Z_i^2 n_{i,0}}$. Do this by substituting this solution for $V(r)$ into the Poisson–Boltzmann equation (12.41), which can be expressed in three-dimensions as:

$$\frac{1}{r}\frac{\mathrm{d}^2(rV)}{\mathrm{d}r^2} = -\frac{1}{\epsilon_0\epsilon}\sum_i Z_i e n_{i,0}\exp(-Z_i eV/k_{\mathrm{B}}T) \tag{12.94}$$

for this spherically symmetric potential. Assume the region is electrically neutral and that $Z_i eV/k_{\mathrm{B}}T \ll 1$. (Hint: See Appendix C.)

12.9. (advanced problem) Repeat Problem (12.8), this time solving the Poisson–Boltzmann equation under the stated conditions in one-dimension. How does this solution differ from the spherically-symmetric three-dimensional solution?

12.10. In her famous 1973 rendition of the song "Killing Me Softly With His Song," Roberta Flack sang, "He sang as if he knew me in all my dark despair. And then he looked right through me as if I wasn't there." [578] If this were literally true, electromagnetic radiation would have to be able to be transmitted through her body. Over what wavelength ranges would that be possible? (Of course, there would also have to be a source of such radiation and his eyes would have to be sensitive to those wavelengths.) (By the way, this song was written by Norman Gimbel and Charles Fox for Lori Lieberman, who sang it in 1971, and it was sung in the 2001 movie, "About a Boy." A modified version was also released by the Fugees in 1996, however without the cited lyrics.)

Neuron Transmission and Membranes

12.11. Which of the four mechanisms involved in ion transport in an axon membrane shown in Fig. 12.11 contribute to the negative charge inside the cell and which to the positive charge?

12.12. The capacitance of a cylinder of length L, inner radius a, and outer radius $a + b$, with electrodes separated by material with dielectric constant κ, is $C_{\text{cylinder}} = 2\pi\kappa\epsilon_0 L/\ln(1+b/a)$. Show that unfolding the cylinder and treating it as a parallel plate capacitor is an excellent approximation for unmyelinated axons, but not for myelinated axons.

12.13. (a) Estimate the effective dielectric constant κ of myelin, by suitably weighting the averages of the dielectric constants of its components. These water, lipid, and polar components have $\kappa = 80$, 2.2, and 50, respectively, and have effective thicknesses t of 2.2, 4.2, and 10.8 nm, respectively, in the repeated 17.1 nm bilipid layered structure in the myelin [581]. These can be considered as capacitances in series, so

$$\kappa_{\text{eff}} = \frac{t_{\text{total}}}{t_{\text{water}}/\kappa_{\text{water}} + t_{\text{lipid}}/\kappa_{\text{lipid}} + t_{\text{polar}}/\kappa_{\text{polar}}}. \tag{12.95}$$

(b) How does this answer help explain why $\kappa = 7$ is reasonable for the axon membrane?

12.14. (advanced problem) Derive the relation in Problem 12.13(a).

12.15. Compare the numerical values of the graded potential decay length λ for typical unmyelinated and myelinated axons.

12.16. It is assumed that the spatial decay of the graded potential in myelinated axons is slow enough that there is little decay before the signal reaches the next node of Ranvier for regeneration. Is this assumption valid?

12.17. Some pain receptors transmit signals on myelinated axons in neurons with conduction speeds up to 30 m/s and others are transmitted on very slow unmyelinated axons with speeds of 2 m/s and lower. How long does it take such receptors on your finger tips to be transmitted to your brain?

12.18. Show that the conduction speeds as given in the caption of Fig. 12.22 are consistent with the EMGs given in the figure.

12.19. Determine the characteristic time τ for unmyelinated and myelinated axons.

12.20. Use substitution to show that (12.80) is the solution to (12.79).

12.21. Use substitution to show that (12.83) is the solution to (12.82).

12.22. Show that substituting (12.84) into (12.72) gives (12.85).

12.23. (advanced problem) Use substitution to show that each of the following is a solution to the voltage along an axon cable in steady state (12.79) for $V_i = 0$:

$$V(x) = A_1 \exp(x/\lambda) + A_2 \exp(-x/\lambda), \tag{12.96}$$

$$V(x) = B_1 \cosh(x/\lambda) + B_2 \sinh(x/\lambda), \tag{12.97}$$

$$V(x) = C_1 \cosh((x - L)/\lambda) + C_2 \sinh((x - L)/\lambda). \tag{12.98}$$

12.24. (advanced problem) Use substitution and evaluation at the boundaries to show that each of the following is a solution to the voltage along an axon cable (for $x \geq 0$) in steady state (12.79) as in Fig. 12.24, for $V_i = 0$ as the boundary condition $V(x = 0) = V_0$ and [589]:

(a) An semi-infinitely long cable (Fig. 12.24a):

$$V(x) = V_0 \exp(-x/\lambda), \tag{12.99}$$

(b) A cable of length L and the boundary condition that $V = 0$ at $x = L$ (which means the voltage is clamped at zero at the end of the cable, which is a *short-circuit boundary condition*) (Fig. 12.24b2):

$$V(x) = V_0 \frac{\sinh((L - x)/\lambda)}{\sinh(L/\lambda)}. \tag{12.100}$$

(c) A cable of length L and the boundary condition that $\mathrm{d}V/\mathrm{d}x = 0$ at $x = L$ (which means zero core current at the end of the cable, which is an *open-circuit boundary condition*) (Fig. 12.24c2):

$$V(x) = V_0 \frac{\cosh((L - x)/\lambda)}{\cosh(L/\lambda)}. \tag{12.101}$$

(d) A cable of length L and the boundary condition that $V = V_\mathrm{L}$ at $x = L$ (which means the voltage is clamped at $V = V_\mathrm{L}$ at the end of the cable)

(Fig. 12.24d):

$$V(x) = \frac{V_0 \sinh((L - x)/\lambda) + V_\mathrm{L} \sinh(x/\lambda)}{\sinh(L/\lambda)}. \tag{12.102}$$

12.25. Sketch V in (12.83) vs. t. (Label $t = \tau$.)

12.26. Sketch V in (12.87) vs. t and also vs. x. (Label $t = \tau$ and $x = \sqrt{D_\mathrm{diff} t}$, in the respective sketches.)

EKGs

12.27. By a series of diagrams similar to Fig. 12.29, show that the repolarization waves traveling to the left and right, respectively, produce signals that are the negative of each other.

12.28. Find the heart rate from the EKGs in Fig. 12.33. Each big box is 0.2 s wide.

12.29. Show that the integration of the cardiac dipole electric field gives a potential that is positive at a point the dipole points to, negative at a point the dipole points away from, and zero at a point where the dipole points in a transverse direction.

12.30. Assume that in (A)–(D) in Fig. 12.42 the cardiac dipole is initially zero, increases to the maximum dipole vector shown, and then decreases to zero, always in the direction shown. (This is not what normally happens. Why?) Match each dipole in (A)–(D) to the EKGs (a)–(d) for the EKG Type I lead shown.

12.31. Assume the same dipole dependencies as in Problem 12.30. Sketch the EKGs for EKG leads II and III for cases (A)–(D) in Fig. 12.42.

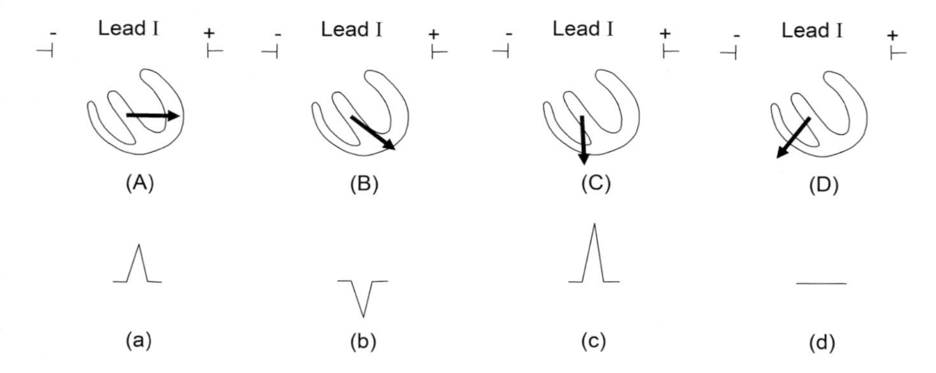

Fig. 12.42. Examples of cardiac dipoles and EKG lead I. (Based on [584].) For Problems 12.30–12.32

12.32. Explain why the sequence of cardiac dipole evolution is normally (D), (C), (B), (A) for the dipoles shown in Fig. 12.42.

12.33. (a) Use the normal sequence of cardiac dipole evolution to sketch the evolution of the EKG signal for EKG leads I, II, and III.
(b) The three lead potentials should always sum to zero. Confirm that your EKGs do so.

12.34. Compare the EKGs in Fig. 12.37 – taken at two times after a heart attack – with each other and then with the normal traces in Fig. 12.33.

13

Feedback and Control

We consciously control our thought processes to achieve physical goals, in part by using our senses to provide response for feedback to control our actions. For example, when we place an object on a table we control our actions by using our eyes to provide feedback on the relative positions of the object and the table as it approaches the table and our sense of touch to provide feedback as we place it on the table. Such feedback and control is also important in all manufacturing processes. Our bodies function amazingly well because of constant feedback and control processes.

One very essential feedback and control system is *homeostasis*, which technically is the stability of the chemical and physical conditions of the fluid surrounding the body cells. This extracellular fluid constitutes about one third of the total body fluid, and includes the blood plasma and interstitial fluid – the tissue fluid that is in the spaces between the cells. The other two thirds of the fluid is intracellular. The extracellular fluid is controlled to regulate (a) body temperature, (b) pressure in blood vessels, (c) oxygen and carbon dioxide concentrations, (d) pH, (e) the concentrations of ions, such as Na^+, K^+, and Ca^{2+}, (f) volume, (g) osmolality (water/dissolved particle ratio), and (h) the organic nutrient concentrations, such as glucose.

We have many other important control systems. Our eye irises open and close in response to light levels. The ciliary muscles control the focal length of the crystalline lenses in our eyes to focus on objects (accommodation, Fig. 11.22). The force we use to grasp objects depends on the normal force feedback we get from it (and if the object "gives" we apply less force). During exercise the intercostal and abdominal muscles that control breathing are regulated by the brain, which receives input from chemical receptors in the blood, mechanical and metabolic receptors in skeletal muscles, as well as from receptors in the lungs (Fig. 13.1). Similarly, during exercise the brain receives input from the heart and muscles and then sends signals to control the heart and blood vessels to increase cardiac output and blood pressure (Fig. 13.6). Body heat balance and temperature are also controlled during exercise (Fig. 13.4). Of course, these control systems are also essential during

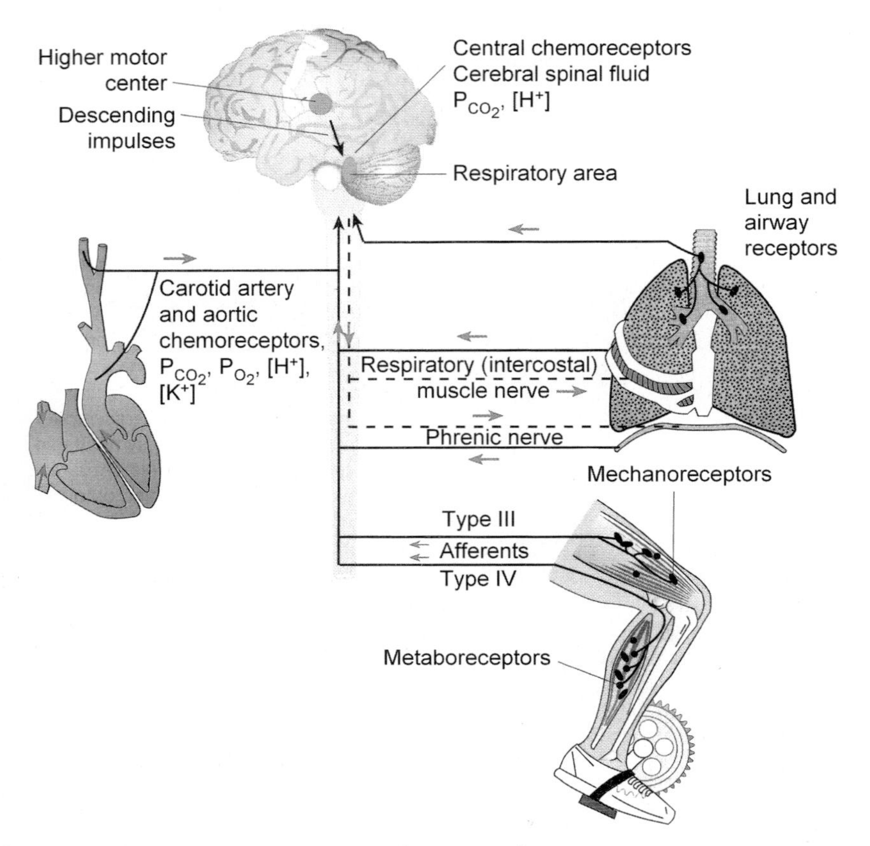

Fig. 13.1. Control of the respiratory (ventilation) system during exercise. The respiratory area in the brain gets signals from the brain motor region, central chemoreceptors, and other receptors, including carotid artery and aortic chemoreceptors, lung intercostal muscle and diaphragm receptors, and skeletal muscle mechanical and metabolic receptors. During exercise this results in increased neural activity to the intercostal muscles and diaphragm through the phrenic and intercostal nerves. This increases the rate and depth of breathing, which regulates arterial oxygen, carbon dioxide, and pH. (From [601]. Used with permission)

resting. For general discussions about such control systems and control in the body see [598, 601, 602, 604, 605, 607, 608].

13.1 Basics of Feedback and Control

Figure 13.2 is a block diagram depicting *feedback and control.* There is a sensor or receptor that measures a quantity (stimulus), such as blood pressure by baroreceptors. The signal from this receptor is transmitted to an integration

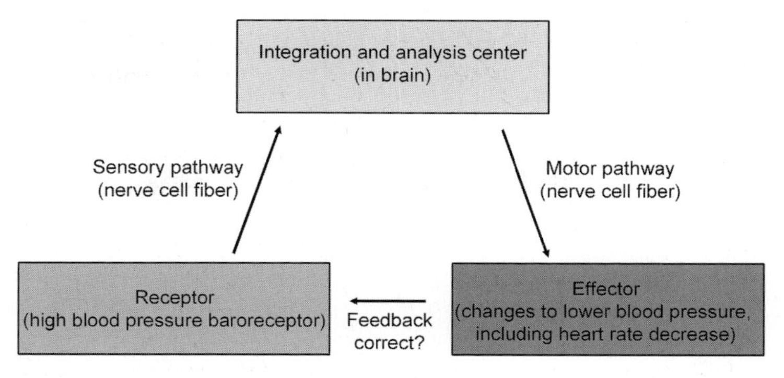

Fig. 13.2. Feedback and control in the body, with an example of controlling high blood pressure in parenthesis. (Based on [607])

center, such as the transmission of signals from sensory centers to the brain by nerves. The integration center induces an effect. This response is transmitted by nerves (motor pathways) to an effector control center. There is some effect or response of interest, such as changing the heart rate or stroke volume to change the blood pressure. The success of this control is determined by measuring the response, the new blood pressure, which provides feedback for the control.

Negative feedback reverses the direction of the change of a variation, to keep the measured parameter near the desired *set point*. This type of feedback is very common in the body. There is usually an operating range centered about this set point bounded by allowable values. Figure 13.3a shows that the effector is activated to correct the parameter when it wanders above the highest value allowed or below the lowest value allowed. Room thermostats usually operate in the same manner, with a several degree operating range about the set point; our bodies also have thermostats.

Positive feedback causes the effector to produce more of a change in the same direction that the parameter is already changing, as in Fig. 13.3b. This type of feedback is rare in the body. One example is suckling which leads to the production of more milk in mothers.

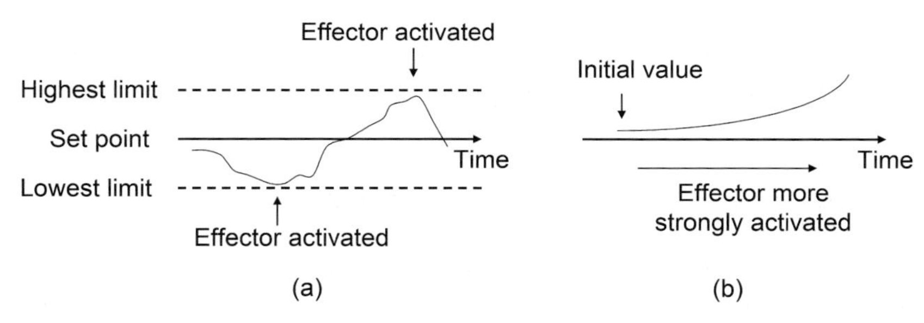

Fig. 13.3. (**a**) Negative and (**b**) positive feedback. (Based on [607])

13.1.1 Control Theory (Advanced Topic)

Feedback is needed to control a parameter P, such as blood pressure or temperature, to keep it near a set point P_0. The level of feedback is often proportional to the difference between the current parameter value and the desired set point [605]. The gain g is the proportionality factor and this determines the magnitude of the response. This response causes a change $\mathrm{d}P$ in time $\mathrm{d}t$ and so

$$\mathrm{d}P = -g(P - P_0)\mathrm{d}t. \tag{13.1}$$

This feedback is not instantaneous. It is delayed by a time τ_d due to delays associated with sensing, neuronal transmission to the brain, processing in the brain, neuronal transmission to the effectors, and how fast the effectors can cause a change. Consequently, a change at time t is due to feedback provided from sensors at time $t - \tau_\mathrm{d}$, and so (13.1) changes to

$$\mathrm{d}P(t) = -g(P(t - \tau_\mathrm{d}) - P_0)\mathrm{d}t, \tag{13.2}$$

or defining $p = P - P_0$

$$\frac{\mathrm{d}p(t)}{\mathrm{d}t} = -gp(t - \tau_\mathrm{d}). \tag{13.3}$$

The desired response is a monotonic decrease of the magnitude of p, a stable oscillation of p about 0 (and therefore of P about P_0), or some combination of the two. *What values of g and τ_d can produce this desired stable negative feedback?* Obviously g must be positive and it may seem that the larger the gain the better, but we will see that large values of the gain are not always desirable.

Let us try to solve (13.3) by substituting the possible solution $p(t) = p\exp(zt)$ into it. In general, z can be a complex number, with $z = x + \mathrm{i}y$. The imaginary term y gives oscillatory behavior about the set point, while $x < 0$ gives exponential decay to the set point and $x > 0$ gives an exponential increase away from the set point, and this last case is undesirable. Equation (13.3) then becomes

$$z = -g\exp(-z\tau_\mathrm{d}). \tag{13.4}$$

Reasonable solutions can exist for $x \leq 0$. Because $\exp(\mathrm{i}q) = \cos q + \mathrm{i}\sin q$, (13.4) becomes separate equations for the real and imaginary parts.

$$x = -g\exp(-x\tau_\mathrm{d})\cos y\tau_\mathrm{d}, \tag{13.5}$$

$$y = g\exp(-x\tau_\mathrm{d})\sin y\tau_\mathrm{d}. \tag{13.6}$$

These are transcendental equations that in general must be solved numerically. We will, however, be able to solve it analytically in special regimes.

The solution with $x = 0$ is one that oscillates about the set point. Using $x = 0$ in (13.5) leads to $0 = -g \cos y\tau_\mathrm{d}$, which gives $y = \pi/2\tau_\mathrm{d}$. Using $x = 0$ in (13.6) leads to $y = g \sin y\tau_\mathrm{d}$ and with $y = \pi/2\tau_\mathrm{d}$, this becomes $\pi/2\tau_\mathrm{d} = g \sin((\pi/2\tau_\mathrm{d})\tau_\mathrm{d}) = g$. This means

$$g\tau_\mathrm{d} = \frac{\pi}{2}, \tag{13.7}$$

and so if the delay time is τ_d, there is an oscillatory response when the gain $g = \pi/2\tau_\mathrm{d}$. This is an acceptable solution.

Now let us say that x has a small magnitude but is not necessarily 0. Equation (13.5) still gives $y \simeq \pi/2\tau_\mathrm{d}$ and so $\sin y\tau_\mathrm{d} \simeq 1$. Because $\exp(q) \approx 1 + q$ for $|\, q \,| \ll 1$, (13.6) becomes $y \simeq g(1 - x\tau_\mathrm{d}) \sin y\tau_\mathrm{d} \simeq g(1 - x\tau_\mathrm{d})$ and so

$$x \simeq \frac{1}{\tau_\mathrm{d}} \left(1 - \frac{\pi}{2g\tau_\mathrm{d}} \right). \tag{13.8}$$

For $g\tau_\mathrm{d} < \pi/2$, we see that $x < 0$ and the parameter P approaches the set point as an oscillation with a magnitude that decreases with time. For $g\tau_\mathrm{d} = \pi/2$, it is clear that $x = 0$ and the parameter P oscillates about the set point, as shown above. For $g\tau_\mathrm{d} > \pi/2$, we see that $x > 0$ and the parameter P oscillates about the set point with a magnitude that increases with time and this is unstable. Therefore, this stability criterion sets an upper limit to the possible gain for a given delay time, given by

$$g\tau_\mathrm{d} \leq \frac{\pi}{2}. \tag{13.9}$$

13.2 Regulation of the Body

13.2.1 Regulation of Temperature

The regulation of body temperature is closely tied to the discussion in Chap. 6 of the production of heat by the metabolism and the modes of heat loss from the body. The normal core body temperature is 37 ± 2°C (98.6 ± 3.6°F). Large increases and decreases mean trouble. Far above normal body temperatures, at 41°C (106°F) the central nervous system begins to deteriorate and convulsions occur. At 45°C (115°F), proteins denature, followed by death. For lower than normal temperatures, at 33°C (91°F), nervous functions are depressed to the point that consciousness is lost. At 30°C (86°F), the temperature regulation system fails. At 28°C (82°F), there is cardiac fibrillation, leading to death. The body is quite good at regulating core temperature, as is clear from Fig. 6.18, but it has its limits and can fail, as is clear from Figs. 6.19–6.21. Figure 13.4 illustrates how the receptors in the skin and in the core sense temperature and send information to the brain. Temperatures outside the range of normality lead to conscious acts and automatic activity by the body to correct the temperatures.

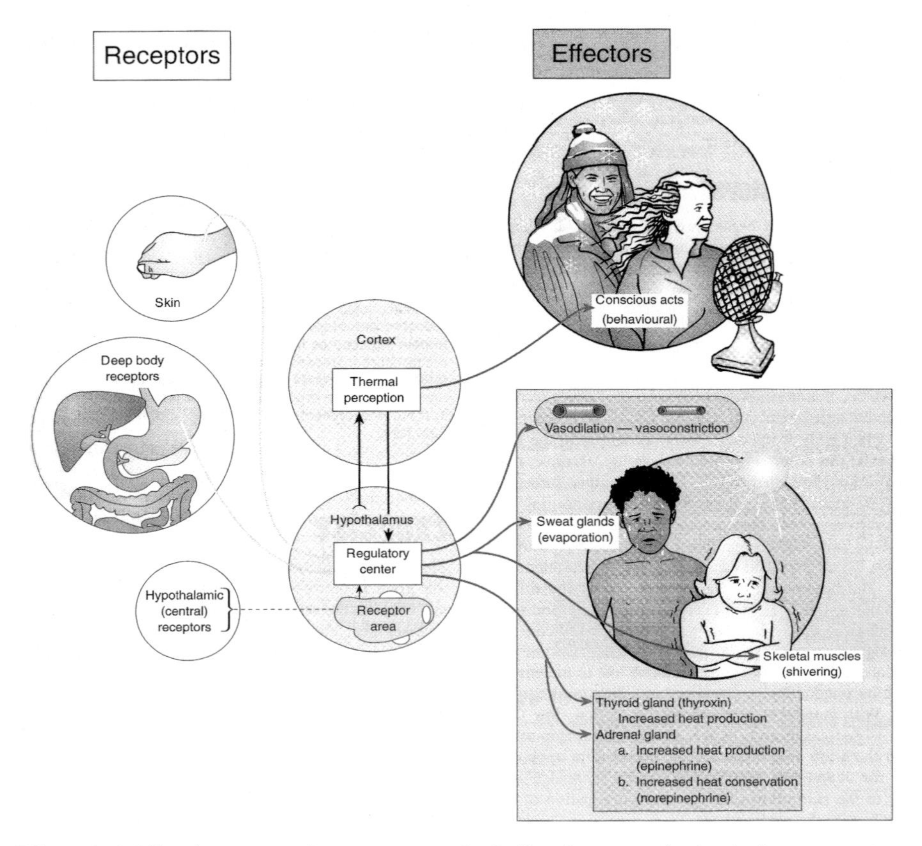

Fig. 13.4. The thermoregulatory system, including input to the brain from receptors and output from the brain to effectors. The hypothalamic center relays information to the effector organs to stimulate the heat production and conservation processes shown when the core temperature receptors differ from the 37°C set point. (From [601]. Used with permission)

Each person has a set point temperature, which varies by $\simeq \pm 0.3$°C (± 0.5°F) among people [600]. The mean set point for skin temperature is approximately 33°C (91.4°F). (In Chap. 6 we used a skin temperature of 34°C.) Skin temperatures above 34.5°C (94°F) cause active sweating, while those below 30°C (86°F) lead to the increased metabolic activity associated with shivering.

Recalling our discussion in Chap. 6, the conservation of energy implies a conservation of heat creation and loss

$$\left(\frac{\mathrm{d}Q}{\mathrm{d}t}\right)_{\text{net}} = \left(\frac{\mathrm{d}Q}{\mathrm{d}t}\right)_{\text{metabolism}} + \left(\frac{\mathrm{d}Q}{\mathrm{d}t}\right)_{\text{passive loss}} + \left(\frac{\mathrm{d}Q}{\mathrm{d}t}\right)_{\text{body controlled}} . \tag{13.10}$$

The left side is the net heating rate due to all modes of heat production and heat loss. The first term on the right-hand side is the heat production from metabolic activity and physical activity; it is >0. The second term is heat loss due to passive modes, such as from the skin, radiation, convection, etc., it is <0. The third term is that due to body-controlled heating (>0) and/or heat removal (<0) [598].

Our discussion in Chap. 6, as in (6.5) and (6.36), showed us how body temperature changes with a net flow of heat into or out of the body, according to

$$\left(\frac{\mathrm{d}Q}{\mathrm{d}t}\right)_{\mathrm{net}} = C\frac{\mathrm{d}T}{\mathrm{d}t} = (60\ \mathrm{kcal/^\circ C})\,\frac{\mathrm{d}T}{\mathrm{d}t}, \tag{13.11}$$

where C is the heat capacity = 60 kcal/°C for a 70 kg person. The heat capacity C = mass (= 70 kg) × specific heat c (= 0.83 kcal/°C-kg). We know that the minimum heating rate due to the metabolism is the BMR, so it can be written as

$$\left(\frac{\mathrm{d}Q}{\mathrm{d}t}\right)_{\mathrm{metabolism}} = f(\mathrm{BMR}), \tag{13.12}$$

where f is the activity factor – a multiplier between 1 and about 20, and BMR = 70 kcal/h.

If there is no loss of heat (and the last two terms of (13.10) are zero), we see that

$$C\frac{\mathrm{d}T}{\mathrm{d}t} = f(\mathrm{BMR}) = f\ 70\ \mathrm{kcal/h} \tag{13.13}$$

so

$$\frac{\mathrm{d}T}{\mathrm{d}t} = f\frac{\mathrm{BMR}}{C} = f\frac{70\ \mathrm{kcal/h}}{60\ \mathrm{kcal/^\circ C}} = 1.2f\ ^\circ\mathrm{C/h}. \tag{13.14}$$

This is similar to (6.36). Without adequate means of heat loss we would not survive heavy exercise for long. With $f = 20$ for heavy exercise, we see that $\mathrm{d}T/\mathrm{d}t = 24^\circ\mathrm{C/h}$.

Now let us consider what happens when there is passive loss of heat only, due to radiation and convection, with no control. From (13.10), the loss of heat due to radiation and convection is

$$\left(\frac{\mathrm{d}Q}{\mathrm{d}t}\right)_{\text{passive loss}} = -(Ah_{\mathrm{r}} + Ah_{\mathrm{c}})(T_{\mathrm{skin}} - T_{\mathrm{room}}) = -\lambda(T_{\mathrm{skin}} - T_{\mathrm{room}}), \tag{13.15}$$

where $\lambda = A(h_{\mathrm{r}} + h_{\mathrm{c}})$. For a person with a body area of 1.5 m^2, the radiation parameter $Ah_{\mathrm{r}} \sim 12$ kcal/h-°C. The convection parameter is $Ah_{\mathrm{c}} \sim 13$ kcal/h-°C for a nude person in the presence of a 9 m/s wind speed. This is much higher than typical values for clothed people. For a nude person in a 15 m/s breeze it is larger by a factor of 5 than the value used here.

Equation (13.15) now becomes

$$C\frac{\mathrm{d}T}{\mathrm{d}t} = f(\mathrm{BMR}) - \lambda(T_{\mathrm{body}} - T_{\mathrm{room}}) \tag{13.16}$$

or

$$(60\ \mathrm{kcal/^\circ C})\,\frac{\mathrm{d}T}{\mathrm{d}t} = f(70\,\mathrm{kcal/h}) - (25\ \mathrm{kcal/h\text{-}^\circ C})(T_{\mathrm{body}} - T_{\mathrm{room}}), \tag{13.17}$$

where we have called the body temperature T_{body} and are now ignoring the differences between the skin and core temperatures. These equations can be solved by substituting the form

$$T_{\mathrm{body}}(t) = \alpha + \beta\exp(-t/\tau_{\mathrm{r}}), \tag{13.18}$$

where τ_{r} is the characteristic response time. This substitution gives

$$C\beta(-1/\tau_{\mathrm{r}})\exp(-t/\tau_{\mathrm{r}}) = f(\mathrm{BMR}) - \lambda(\alpha + \beta\exp(-t/\tau_{\mathrm{r}}) - T_{\mathrm{room}}) \tag{13.19}$$

$$= f(\mathrm{BMR}) - \lambda(\alpha - T_{\mathrm{room}}) - \lambda\beta\exp(-t/\tau_{\mathrm{r}}). \tag{13.20}$$

Because this must be valid for all time t, we can equate the coefficients of the $\exp(-t/\tau_{\mathrm{r}})$ terms to get

$$C\beta(-1/\tau_{\mathrm{r}}) = -\lambda\beta, \tag{13.21}$$

$$\tau_{\mathrm{r}} = \frac{C}{\lambda} = \frac{60\ \mathrm{kcal/^\circ C}}{25\ \mathrm{kcal/h\text{-}^\circ C}} = 2.4\ \mathrm{h}, \tag{13.22}$$

and then equate the constant terms to get

$$f(\mathrm{BMR}) = \lambda(\alpha - T_{\mathrm{room}}), \tag{13.23}$$

or

$$\alpha = T_{\mathrm{room}} + f\frac{\mathrm{BMR}}{\lambda} = T_{\mathrm{room}} + f\frac{70\ \mathrm{kcal/h}}{25\ \mathrm{kcal/h\text{-}^\circ C}} = T_{\mathrm{room}} + 2.8f\ ^\circ\mathrm{C}. \tag{13.24}$$

At $t = 0$ we will call $T_{\mathrm{body}}(t = 0) = T_{\mathrm{body,initial}}$ (which is the initial condition), so,

$$T_{\mathrm{body}}(0) = T_{\mathrm{body,initial}} = \alpha + \beta \tag{13.25}$$

and therefore

$$\beta = T_{\mathrm{body,initial}} - \alpha = T_{\mathrm{body,initial}} - T_{\mathrm{room}} - 2.8f\ ^\circ\mathrm{C}. \tag{13.26}$$

The final solution is

$$T_{\text{body}}(t) = T_{\text{room}} + 2.8f\ ^\circ\text{C} + (T_{\text{body,initial}} - T_{\text{room}} - 2.8f\ ^\circ\text{C})\exp(-t/\tau_{\text{r}}). \tag{13.27}$$

This is an example of *open loop* (*or open cycle*) heating, because there is no feedback and control. The problems in this temperature regulation system are clear from (13.27). The final steady-state temperature (i.e., that for $t \gg \tau_{\text{r}} = 2.4\,\text{h}$) is

$$T_{\text{body,steady state}} = T_{\text{room}} + f\frac{\text{BMR}}{\lambda} = T_{\text{room}} + 2.8f\ ^\circ\text{C}. \tag{13.28}$$

If $T_{\text{room}} = 21^\circ\text{C}$ ($69.8^\circ\text{F} \simeq 70^\circ\text{F}$) and we were resting, so $f \sim 1$, our steady-state body temperature would be $\approx 24^\circ\text{C}$ (75°F), which is much lower than our desired body temperature. If we were exercising pretty hard, so $f = 10$, our steady-state body temperature would be $\approx 49^\circ\text{C}$ (120°F), which is too high. We could consciously change λ a bit (through h_{c}: by decreasing it by putting on heavy clothes in the first case or by increasing it by increasing the wind speed on our nude bodies in the second case), but this would help a bit. Ideally, we would want to fix

$$T_{\text{body,steady state}} - T_{\text{room}} = 37^\circ\text{C}\ (98.6^\circ\text{F}) - 21^\circ\text{C}\ (69.8^\circ\text{F}) = 16^\circ\text{C}\ (28.8^\circ\text{F}) \tag{13.29}$$

$$= f\frac{\text{BMR}}{\lambda} = 2.8f\ ^\circ\text{C}, \tag{13.30}$$

so, $f \sim 6$ with our current λ. We could regulate temperature by changing f or λ. This means to maintain our body temperature we would have to constantly engage in moderate activities, such as slow swimming (for the same λ). If we bundled up a bit (decreased λ), we could lower f a bit, to say $f = 4$, which would mean we would have to still engage in activities, such as moderately fast walking – even while we were asleep (such as constant sleep walking).

Another negative feature of the consequences of (13.27) is that it has a time constant $\tau_{\text{r}} = 2.4\,\text{h}$ which is much too long. If we decided to change our activity level (and therefore change f) it would take about 5 h for our body temperature to stabilize to the new steady-state temperature; this is unreasonable. These seemingly ridiculous results cannot be attributed to the assumptions we have made in the heat gain and loss terms we have used so far. We have neglected the last term in (13.10) which describes the feedback and control for *closed loop (or closed cycle)* temperature regulation. We need this to maintain body temperature within the normal range.

The last term in (13.10) $(\text{d}Q/\text{d}t)_{\text{body controlled}}$ describes the involuntary responses by the body to maintain a normal body temperature, which we can call the set point – the targeted temperature. When an increase in temperature is needed, this term is positive due to increased metabolic activity and motor

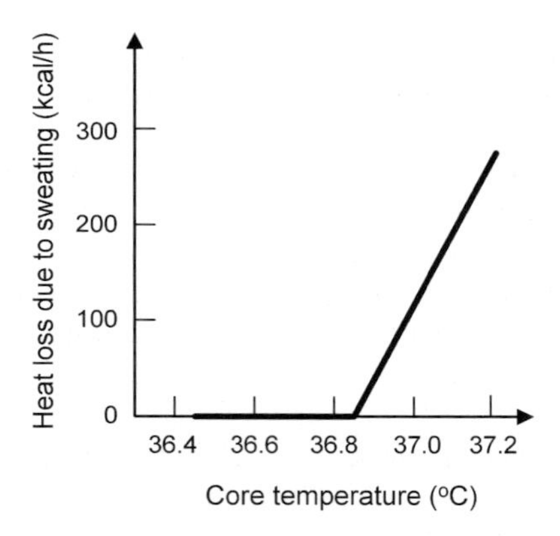

Fig. 13.5. Rate of heat loss due to sweating, for skin temperatures between 33 and 38°C. (Based on [598] and [599])

activity, such as shivering (up to the summit metabolic rate, Chap. 6); this heat goes directly to the core and muscles. When a decrease in temperature is needed this control term is negative, e.g., due to sweating. The body can lose 1 L/h water by sweating; if all of this sweat evaporates (and is not wiped off), then with the 540 kcal/L heat of evaporation of water, the heat loss through sweating can be 540 kcal/h. Through vasomotor control, this control term can be positive or negative. If blood is directed to vessels closer to the skin, cooling is more effective and the last term in (13.10) is negative. If blood is diverted to vessels farther from the skin and closer to the core, cooling is less effective and the control term is positive. One reason cooling is less effective is the countercurrent heat exchange mechanism described in Fig. 6.17.

Figure 13.5 shows the rate of heat loss due to sweating as a function of the body core temperature. This loss is zero when the core is cooler than normal and increases fairly linearly with temperature as $K_{\text{sweat}}(T_{\text{body}} - 36.85°\text{C})$ where $K_{\text{sweat}} \sim 750\,\text{kcal/h-}°\text{C}$, when the core is higher than normal (for skin temperature between 33 and 38°C). Now (13.10) becomes

$$C\frac{\mathrm{d}T}{\mathrm{d}t} = fP_0 - \lambda(T_{\text{body}} - T_{\text{room}}) - K_{\text{sweat}}(T_{\text{body}} - 36.85°\text{C}), \tag{13.31}$$

or

$$\begin{aligned}\left(60\ \text{kcal}/°\text{C}\right)\frac{\mathrm{d}T}{\mathrm{d}t} &= f(70\ \text{kcal/h}) - (25\ \text{kcal/h-}°\text{C})(T_{\text{body}} - T_{\text{room}}) \\ &\quad -(750\ \text{kcal/h-}°\text{C})(T_{\text{body}} - 36.85°\text{C}).\end{aligned} \tag{13.32}$$

Note that T_{body} in the passive heat loss term really refers to the skin temperature, while that in the control term is the core temperature. The solution to this equation has a form similar to (13.27) with the response time

$$\tau_{\text{r}} = \frac{C}{\lambda + K}. \tag{13.33}$$

Because $K/\lambda = 30$, the time constant is now smaller by a factor of 31, giving $2.4\,\mathrm{h}/31 \simeq 4.7\,\mathrm{min}$, which is much more reasonable.

Note that we have been able to see how the body can increase the gain to produce a response time that is reasonable on a physiological timescale. We have ignored the issue of the delay time τ_d, as discussed in the previous section. For this process at least $\tau_\mathrm{d} \ll \tau_\mathrm{r}$ and so the delay time is not important here.

Our bodies usually have a set point for the core temperature of about 37°C. Sometimes when we are ill, bacterial toxins called pyrogens increase this set point. When this set point increases, we feel cold until our voluntary and involuntary actions cause our temperature to increase to this new level. We can decrease heat loss by covering ourselves with blankets (which is voluntary) and decreases blood flow to the skin (involuntary). We can increase our metabolic rate by shivering (involuntary).

13.2.2 Control of Blood Pressure

The brain continuously receives input from the heart and muscles and then sends signals to control the heart and blood vessels to control blood pressure (Fig. 13.6). For example, the carotid sinus baroreceptors (pressure sensors in the carotid artery) fire a signal that is transmitted by the carotid sinus nerve that goes to the brain [606]. The frequency of this signal F is roughly proportional to how much the arterial pressure – as sensed by the baroreceptors p_b – exceeds a threshold pressure p_t, and this dependence has been modeled as

$$F + K_1 \frac{\mathrm{d}F}{\mathrm{d}t} = K_2(p_\mathrm{b} - p_\mathrm{t}) + K_3 \frac{\mathrm{d}p_\mathrm{b}}{\mathrm{d}t}, \tag{13.34}$$

where K_1, K_2, and K_3 are constants. Note the similarity of this equation to that describing the Kelvin model (4.68). The frequency is not directly proportional to the excess pressure because of the two derivative terms. The $\mathrm{d}p_\mathrm{b}/\mathrm{d}t$ term describes the observed overshoot in frequency. The $\mathrm{d}F/\mathrm{d}t$ term describes how the rates of receptor firing continue to change even after the pressure has stabilized.

In the overall feedback system there is a decrease in arterial pressure Δp_a observed in response to a change in the pressure in the baroreceptors Δp_b, which is given by

$$\Delta p_\mathrm{a} = -g \Delta p_\mathrm{b}, \tag{13.35}$$

where g is the gain factor. A more exact expression is

$$\Delta p_\mathrm{a} = -g\, \Delta p_\mathrm{b} - K_4 \frac{\mathrm{d}p_\mathrm{b}^+}{\mathrm{d}t}, \tag{13.36}$$

where the last term denotes the rate of change of pressure only when it is increasing. The open loop gain factor g_o has been measured to range from

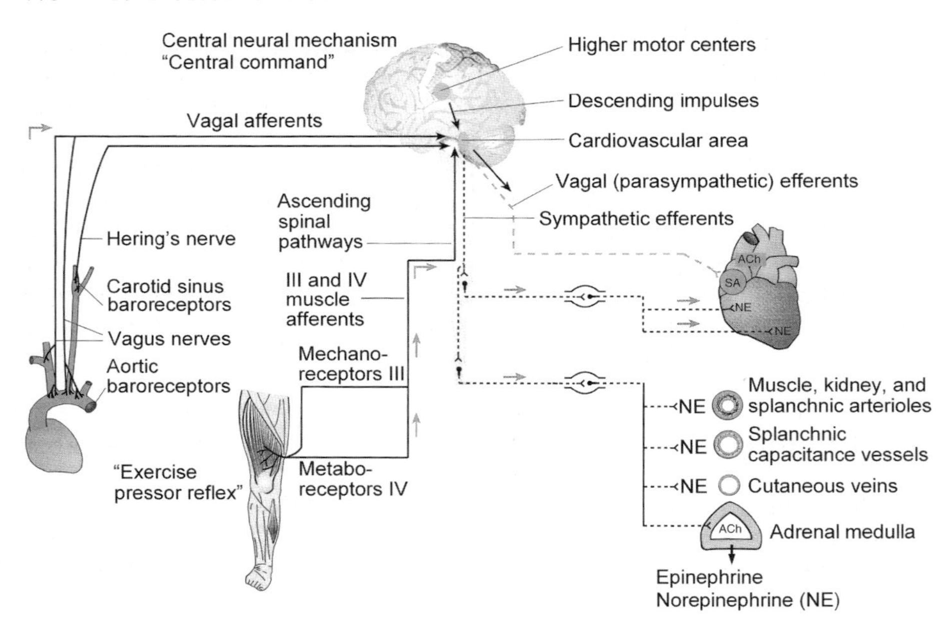

Fig. 13.6. Control of the cardiovascular system during exercise, showing control inputs on the left and outputs on the right. Nerve impulses from the motor region of the cerebrum in the brain, and the afferent input from the arterial baroreceptors and skeletal muscle receptors converge on the cardiovascular region of the medulla in the brain. During exercise this results in reduced parasympathetic activity to the heart and increased sympathetic activity to the heart, blood vessels, and the adrenal medulla. (The parasympathetic and sympathetic nervous systems are automatic regulation systems that, respectively, "relax" and "push" systems, such as organs, in the body.) The leads to an increase in blood pressure and cardiac output. (From [601]. Used with permission)

2 to 9 in dogs and humans in isolated systems with no feedback. The factor in a closed loop system with feedback g_c can be smaller; in particular the body makes it smaller during hemorrhaging.

13.2.3 Regulation During Exercise

Several functions and properties of the body are under active control during exercise. Arterial baroreceptors and skeletal muscle mechanoceptors are inputs to the cardiovascular area of the medulla in the brain, along with impulses from the motor region of the cerebrum, to regulate the cardiovascular system (Fig. 13.6). This results in sympathetic activity in the heart, blood vessels, and adrenal medulla, which leads to increases in blood pressure and cardiac output.

For this increased blood flow to be useful, there must be a concomitant increase in air intake and therefore control of the ventilation system is vital (Fig. 13.1). In addition to the signals used for cardiovascular control are signals from receptors that control the respiratory components, such as the intercostal muscles and diaphragm. This results in neural activity to increase the rate and depth of breathing, which regulates the arterial O_2, CO_2, and pH. (See [603] for a model of breathing.) Of course, during exercise the body must control temperature (Fig. 13.4).

During exercise the metabolism rate increases and the body must actively lose heat. This occurs through the temperature control system, outlined above and in Fig. 13.4.

13.3 Summary

Feedback and control are essential to every aspect of normal body function and regulation, and these processes can be modeled. Models of controlling body temperature by passive and active processes can be developed by using the models of heat formation and loss presented in Chap. 6.

Problems

Problem Type

13.1. Use the decision-making boxes as in Fig. 13.2 to devise a feedback and control system that redirects the flow of blood superficial or deep in the body for temperature regulation, as in the countercurrent mechanism described in Chap. 8.

13.2. Someone has a body temperature set point of 37.0°C. Her temperature feedback and control system turns on whenever her temperature strays by ±0.5°C from this set point, and returns it to the set point in 2 min. Unfortunately, when this feedback system is not on, her temperature drifts by +2°C/h. How often does her feedback system turn on?

13.3. What is the maximum gain for stable feedback and control if the delay is alternatively 1 s or 1 ms?

13.4. (a) Your 70 kg body stops generating heat (but somehow functions well anyway), while you are indoors under normal conditions. Estimate how long it will take for you to cool from 37 to 28°C (and death).
(b) As your temperature approaches 28°C in (a) your metabolism turns on again and your metabolic rate is your normal BMR (Chap. 6). Estimate how long it will take for your body temperature to reach 37°C again.
(c) Repeat (b) if your metabolic rate is at its summit value (Chap. 6).

13.5. Solve (13.33) for an arbitrary activity level.

Appendix A

Symbols and Units

Table A.1 gives many of the physical constants used in this text. The names of general variables are provided in Table A.2, along with their units. Table A.3 lists coefficients and parameters. Some more specific parameters and constants with acronyms, are provided in Table A.4. Most variables and parameters are defined locally in the chapters, sometimes a bit differently than in these tables when there is an overlap in the use of symbols. For example, in most of Chap. 8 the flow or vascular resistance is called R_{flow} (Table A.3) to avoid confusion with R used for radius (Table A.2). Elsewhere where there can be no confusion it is called R (and is locally defined as such).

Table A.1. Physical constants

parameter (variable)	value (in SI units)
Avogadro's number (N_A)	6.02×10^{23} (per mole)
Boltzmann constant (k_B)	1.381×10^{-23} J/K
Coulomb's Law constant ($k = 1/4\pi\epsilon_0$)	8.99×10^{9} N-m^2/C^2
electric permittivity (ϵ_0)	8.854×10^{-12} F/m
elementary charge (e)	1.602×10^{-19} C
gas constant ($R = N_A k_B$)	8.315 J/mole-K
gravitation constant (g)	9.8 m/s^2 = 32.2 ft/s^2
magnetic permeability (μ_0)	$4\pi \times 10^{-7}$ N/A^2
Planck's constant (h)	6.626×10^{-34} J-s
speed of light (c)	3.0×10^{8} m/s
Stefan-Boltzmann constant (σ)	5.67×10^{-8} W/m^2-K^4

Table A.2. General variables and units. Also see Fig. D.1

parameter (variable)	common units (Definition)
acceleration (a, $a_{\mathrm{decel}} = -a$)	$\mathrm{m/s^2}$
angular momentum (L)	$\mathrm{kg\text{-}m^2/s}$
area (A, in flow S)	$\mathrm{m^2}$
body height (H or H_{b})	m^a
body mass (m_{b})	1 kg = 1,000 g
body weight (W_{b})	N^b
charge (q)	coulombs (C)
charge, number of elementary charges (Z)	unitless, $q = Ze$
current (I_{elect}, I)	amps (A), 1 A = 1 C/s
current density (J_{elect}, J)	$\mathrm{A/m^2}$
density (mass) (ρ)	$1{,}000\,\mathrm{kg/m^3} = 1\,\mathrm{g/cm^3}$
density (number) (n)	$\#/\mathrm{m^3}$, $\#/\mathrm{cm^3}$
diameter (d, D)	1 m = 100 cm = 3.28 ft
dipole moment (P)	1 Debye (D) = 3.33610^{-30} C-m
distance (L), height (h, y)	m^a
electric field (E)	V/m
energy (E)	J^c
flux (particle) (J)	$\#/\mathrm{m^2\text{-}s}$
focal length (f)	m
force (F, M (for muscle))	N^b
frequency (in space) (k)	1/m
frequency (in time) (f, F, ν)	Hz, cycles per second (cps)
frequency (in time) (radial) (ω)	radians per second, $\omega = 2\pi f$
heat flow, amount (Q)	J^c
heat flow, rate ($\mathrm{d}Q/\mathrm{d}t$)	kcal/h, W^d
intensity (acoustic, optical) (I)	$\mathrm{W/m^2}$
Intensity (acoustic reference) (I_{ref})	$10^{-12}\ \mathrm{W/m^2}$
kinetic energy (KE)	J^c
loudness (L_{p}, L_{s})	phons, (10.64); sones, (10.65)
mass (m)	1 kg = 1,000 g
magnetic field (B)	1 T (T) = 10^4 gauss (G)
magnification (M)	unitless
mobility (μ)	$\mathrm{m^2/V\text{-}s}$
normal force (N)	N^b
osmotic pressure (Π)	Pa^e
potential energy (PE)	J^c
power (P_{power} or P, mechanical, metabolic)	W^d
pressure (P)	Pa^e
radiation flux (R)	$\mathrm{W/m^2}$
radius (radius of curvature) (r, R)	m^a
reaction force (R)	N^b
reflection coefficient (R_{refl}, R)	unitless
refractive power (P)	1/m = 1 D (Diopter)
speed, angular, rotational (Ω)	rad/s

(Cont.)

Table A.2. (*Continued*)

parameter (variable)	common units (Definition)
speed, velocity (v), flow (u, v)	m/s[f]
strain (ϵ)	unitless, mm/mm
stress (σ)	Pa[e]
temperature (T)	$T(\mathrm{K}) = T(^{\circ}\mathrm{C}) + 273^{\circ}$
tension (T)	N[b] (for force, as in Chap. 5)
tension (T) (surface tension)	N/m (force/length (7.4))
torque (τ) or moment (M)	N-m
transmission coefficient (T_{trans}, T)	unitless
volume (V or V_{flow})	$1\,\mathrm{L} = 1{,}000\,\mathrm{mL} = 1{,}000\,\mathrm{cm}^3$
volume flow rate (Q)	$1\,\mathrm{L/s} = 1{,}000\,\mathrm{mL/s} = 1{,}000\,\mathrm{cm}^3/\mathrm{s}$
voltage, potential difference (V_{elect}, V)	volts (V)
wavelength (λ)	m, $1\,\mathrm{nm} = 10^{-9}\,\mathrm{m}$
work (W)	J[c]
vergence (V)	1/m = 1 D (Diopter)

[a] 1 m = 100 cm = 3.28 ft, 1 mile = 5,280 ft.
[b] 1 N = 10^5 dynes = 0.225 lb, (Table 2.5).
[c] 1 J = 0.239 cal = 0.000948 BTU, 1 kcal = 4,184 J.
[d] 1 W = 0.86 kcal/h = 1/746 hp = 0.00134 hp (horsepower) (Table 6.1).
[e] 1 Pa = 1 N/m^2, 1 MPa = 1 N/mm^2 = 7,600 mmHg = 10,300 $\mathrm{cmH_2O}$ = 10 bar = 9.87 atm. Table 2.6.
[f] 1 m/s = 3.6 km/h = 3.28 f/s (fps, feet per second) = 2.24 mph (miles per hour, 1 mile = 5,280 ft).

Table A.3. General coefficients and parameters, and units. Also see Fig. D.1

parameter (variable)	common units (definition)
absorption coefficient (sound, light) (γ)	1/m
activity factor (f)	unitless
admittance ($Y = 1/Z = G + \mathrm{i}B$)	1 mho = 1/ohm
area moment of inertia (I_{A})	m^4, (4.38)
capacitance (C_{elect} or C); per unit length	farads (F) = C/V; F/m
capacitance per area (c)	$\mathrm{F/m}^2$
compliance (C_{flow} or C)	cm^3/bar, L/mmHg
conductance (electrical, G)	siemens, 1 S = 1/ohm
conductance per unit area (g)	1/ohm-m^2
conductivity (σ)	1/(ohm-m), $\sigma = 1/\rho$
dashpot constant (c)	N-s/m
dielectric constant (κ)	unitless
diffusion coefficient (D_{diff})	m^2/s, cm^2/s
distensibility (D_{flow})	1/Pa, (8.20)
drag coefficient (C_{D})	unitless
efficiency (ϵ)	$0 \le \epsilon \le 1$
emissivity (ϵ)	$0 \le \epsilon \le 1$, 1 for a black body

(*Cont.*)

Table A.3. (*Continued*)

parameter (variable)	common units (definition)
friction coefficient (static, kinetic) (μ_s, μ_k)	unitless
heat capacity (C)	kcal/°C, 1 MJ/K = 239 kcal/K
heat transfer coefficient ($h = K/d = 1/I$)	W/m^2-°C, $kcal/m^2$-h-°C
impedance ($Z = R + iX$)	ohm
index of refraction (n)	unitless
insulation ($I = 1/h = d/K$)	m^2-°C/W, m^2-h-°C/kcal
lift coefficient (D_{lift})	unitless
moment of inertia (I)	kg-m^2, (3.23), (3.24)
Poisson's ratio (υ)	unitless, (4.7)
radius of gyration (ρ)	m
reactance (X)	ohm
resistance (flow, vascular, R_{flow} or R)	mmHg-s/cm^3 [a]
resistance (electrical, R_{elect} or R)	ohm (Ω)
resistance (electrical; per unit length r)	ohm/m
resistivity (ρ)	ohm-m
scattering coefficient ($\alpha_{light\ scattering}$)	1/m
skin friction coefficient (C_{sf})	unitless
specific heat (c)	kcal/kg-°C[b]
specific heat ratio (γ)	unitless, $= c_p/c_v$
speed of sound (v_s)	m/s
spring constant (k)	N/m
stroke volume (V_{stroke})	1 L = 1,000 mL = 1,000 cm^3
surface tension (γ)	1 N/m = 1,000 dynes/cm
susceptance (B)	1 mho = 1/ohm
thermal conductivity (K)	W/m-K
total volume flow rate (Q_t)	1 L/s = 1,000 mL/s = 1,000 cm^3/s
viscosity coefficient (dynamic, absolute) (η)	Pa-s[c]
viscosity coefficient (kinematic) ($\upsilon = \eta/\rho$)	Pa-s/(kg/m^3)
Young's modulus (Y), elastic modulus (E)	Pa, 1 MPa = 1 N/mm^2

[a] 1 mmHg-s/cm^3 = 1 mmHg-s/mL = 1 PRU.
[b] per mass or volume, kcal/kg-°C, 1 MJ/m^3-K = 239 kcal/m^3-K.
[c] 1 Pa-s (Poiseuille, PI) = 1 (N/m^2)s = 1 kg/m-s = 10 poise (P) = 1,000 cP.

Table A.4. Acronyms, including those of parameters, and units

parameter (variable)	common units
adenosine triphosphate, diphosphate (ATP, ADP)	Fig. 6.3
basal metabolic rate (BMR)	kcal/h[a]
body mass index ($BMI = m_b/H^2$), Quételet's index (Q)	kg/m^2
center of mass (CM)	
chromatic aberration (CA)	
coefficient of restitution (COR)	unitless; (3.97)
electrocardiogram (EKG, ECG)	

(*Cont.*)

Table A.4. (*Continued*)

parameter (variable)	common units
electron transfer system (ETS)	
focal length, back, effective, front (BFL, EFL, FFL)	m
focal point, plane (first: F, **F**; second: F′, **F**′)	
forced expiratory volume (FEV)	L
functional residual capacity (FRC)	L
Gadd Severity Index (GSI)	s; (3.103)
Head Injury Criterion (HIC)	s; (3.105)
inspiratory, expiratory reserve volume (IRV, ERV)	L
intraocular pressure (IOP)	Pa, mmHg[b]
left atrium, ventricle (LA, LV)	
metabolic equivalent (MET)	unitless
metabolic rate (MR)	kcal/h[a]
near point, far point (NP, FP)	m
nodal point (first: N; second: N′)	
peripheral resistance unit (PRU)	mmHg-s/cm^3
phosphocreatine (PCr)	
physiological cross-sectional area (PCA)	$1\,cm^2 = 0.155\,in^2$
principal point, plane (first: P, **P**; second: P′, **P**′)	
residual volume (RV)	L
respiration exchange ratio (RER)	unitless, Table 6.2
Reynolds number (*Re*)	unitless; (7.11)
right atrium, ventricle (RA, RV)	
specific stature ($S = H/m_b^{1/3}$), Ponderal index	m/$kg^{1/3}$
spherical aberration (SA)	
Strouhal frequency, number (*St*)	unitless; (7.47)
tidal volume (TV)	L
total lung capacity (TLC)	L
total peripheral vascular resistance (TPVR)	mmHg-s/cm^3
transient ischemic attack (TIA)	
ultimate bending stress (UBS)	Pa[b]
ultimate compression stress (UCS)	Pa[b]
ultimate strain, ultimate percent elongation (UPE)	unitless
ultimate tensile stress (UTS)	Pa[b]
visual acuity (VA)	unitless
vital capacity, forced vital capacity (VC, FVC)	L

[a] 1 kcal/h = 1.162 W, 1 W = 0.86 kcal/h = 1/746 hp = 0.00134 hp (horsepower) (Table 6.1).
[b] 1 MPa = 1 N/mm^2, 1 Pa = 1 N/m^2, 1 MPa = 1 N/mm^2 = 7,600 mmHg = 10,300 cmH_2O = 10 bar = 9.87 atm. Table 2.6.

Appendix B

Locator of Major Anatomical and Anthropometric Information

This appendix cites the figures (Table B.1) and tables (Table B.2) that describe the main features of human anatomical and anthropometric information, which are used throughout this text.

Table B.1. Figures describing human anatomy and anthropometry

figure	content
1.1	Directions, orientations, and planes
1.2	Anatomy of the skeletal system
1.3	The knee synovial joint
1.8	Anterior and posterior view of several large skeletal muscles
1.9	Antagonistic motions allowed by synovial joints
1.10	More antagonistic motions allowed by synovial joints
1.14	Ocular muscles
1.15	Body segment lengths
1.16	Postures for opposing motions
2.7	Bones of the arm, anterior view
2.8	Bones of the arm, posterior view
2.14	Bones of the leg and hip, anterior view
2.15	Bones of the leg and hip, posterior view
2.33	The vertebral column (spine)
2.38	The intermediate layer of back muscles
2.49	Cross section of skin
3.2	Anterior and medial muscles of the thigh
3.3	Posterior thigh and gluteal region muscles
3.4	Lateral views of the right leg
5.26	Extensor muscles of the forearm
5.27	Flexor muscles of the forearm
8.1	Blood circulation system
8.2	Diagram of the heart
8.3	Major arteries in the body

(Cont.)

Table B.1. (*Continued*)

figure	content
8.4	Major veins in the body
9.1	Diagram of parts of the respiratory system
9.2	The relationship between the lung and heart
10.26	Diagram of the outer, middle, and inner ear
10.27	Diagram of the middle ear and uncoiled chochlea
11.1	Structure of the eye
12.7	Structure of a neuron

Table B.2. Tables describing human anatomy and anthropometry

table	content
1.1	Anatomical terms in anterior regions
1.2	Anatomical terms in posterior regions
1.5	The "Standard Man"
1.6	Body segment lengths
1.8	Masses and mass densities of body segments
1.8	Distance of the center of mass from either segment end
1.9	Radius of gyration of a body segment
1.11	Mass and volume of the organs
1.13	Allometric parameters for mammals
1.15	Steven's Law parameters
5.1	Percent PCA of muscles crossing the hip joint
5.2	Percent PCA of muscles crossing the knee joint
5.3	Percent PCA of muscles crossing the ankle joint
7.4	Approximate flow rates of the alimentary system
8.1	Normal resting values of blood pressures and volumes
8.2	Approximate quantification of individual vessels
8.3	Approximate quantification of total vessel systems
8.4	Tension in blood vessel walls.
9.1	Approximate quantification of the bronchial system

Appendix C

Differential Equations

The same form of simple differential equations is used to model very different problems throughout this text. They are presented here along with their solutions. The solutions can be checked by substituting them in the differential equation and showing that the equation is satisfied. This appendix is not meant to serve as a primer on differential equations or their solutions.

Solutions to first- and second-order differential equations, respectively, have one and two free parameters that are satisfied by the conditions of the problem. When the independent variable is time, t, these conditions are called *initial conditions*, so for the dependent variable $q(t)$, q is specified at a given time, such as at $t = 0$ for a first-order differential equation. For a second-order equation, both q and $\mathrm{d}q/\mathrm{d}t$ at $t = 0$ can be given. The dependent variable q can be a coordinate, such as x and angle θ or something else, such as force F. When the independent variable is a spatial coordinate, such as x, these conditions are called *boundary conditions.*

Unless otherwise specified, F and G are constants.

C.1 Simple First- and Second-Order Differential Equations

In these differential equations the derivatives of the dependent variable, q, depend on the independent variable t.

First-Order, Constant Driving

A variable $q(t)$ obeying

$$\frac{\mathrm{d}q}{\mathrm{d}t} = F \tag{C.1}$$

has solution

$$q(t) = Ft + q(t = 0), \tag{C.2}$$

where $q(0) = q(t = 0)$ is the initial condition.

This type of equation is used to describe the temperature rise of the body with metabolic heating and no heat loss (6.36).

Second-Order, Constant Driving

A variable $q(t)$ obeying

$$\frac{\mathrm{d}^2 q}{\mathrm{d}t^2} = F \tag{C.3}$$

has solution

$$q(t) = \frac{Ft^2}{2} + \frac{\mathrm{d}q(0)}{\mathrm{d}t}\, t + q(0), \tag{C.4}$$

where the initials conditions are q and $\mathrm{d}q/\mathrm{d}t$ evaluated at $t = 0$.

This type of equation is used in the model of ball throwing (3.76).

Second-Order, Increasing Driving

A variable $q(t)$ obeying

$$\frac{\mathrm{d}^2 q}{\mathrm{d}t^2} = F + Gt \tag{C.5}$$

has solution

$$q(t) = \frac{Ft^2}{2} + \frac{Gt^3}{6} + \frac{\mathrm{d}q(0)}{\mathrm{d}t}\, t + q(0), \tag{C.6}$$

where the initials conditions are q and $\mathrm{d}q/\mathrm{d}t$ evaluated at $t = 0$.

This type of equation is used in the model for bending a cantilever (4.44) with the position x as the independent variable.

First- and Second-Order, Increasing Driving

A variable $q(t)$ obeying

$$\frac{\mathrm{d}(t\ \mathrm{d}q/\mathrm{d}t)}{\mathrm{d}t} = Gt \tag{C.7}$$

or equivalently

$$t\,\frac{\mathrm{d}^2q}{\mathrm{d}t^2}+\frac{\mathrm{d}q}{\mathrm{d}t}=Gt \tag{C.8}$$

has solution

$$q(t)=q(0)+\frac{Gt^2}{4}, \tag{C.9}$$

where the initial condition for q is evaluated at $t = 0$, and q and $\mathrm{d}q/\mathrm{d}t$ are finite at $t = 0$.

This type of equation is used in determining the viscous flow in a tube (7.31).

C.2 Exponential Decay and Drag

In these differential equations the first derivative of the dependent variable, q, depends on q and in some cases on the dependent variable t.

First-Order, Proportional Drag, No Driving

A variable $q(t)$ obeying

$$\frac{\mathrm{d}q}{\mathrm{d}t}+\frac{q}{\tau}=0 \tag{C.10}$$

decays in time t as

$$q(t)=q(0)\exp\left(-t/\tau\right), \tag{C.11}$$

where $q(0)$ is the initial condition and τ has units of time (s) and is called a *time constant.* The value of q decays exponentially in time with this characteristic time constant. This can be due to a "frictional force" or *damping* with a rate $1/\tau$.

This type of equation is used in the mechanical models of non-Hookean materials ((4.22) with the strain ϵ as the independent variable) to describe the speed when there is Stokes-type drag that is proportional to speed (7.59) and in describing pulsatile flow in (8.94), (8.105), and Problem 8.49. This equation is equivalent to the first two terms of (C.30) describing position.

This type of equation is used in the viscoelastic mechanical models of materials with constant driving terms, including the Maxwell (4.52), Voigt (4.57), and Kelvin/standard linear (4.68) models, a model of muscles (5.9), the arterial pulse (Problem 8.49), and temperature regulation (13.18).

Equation (C.10) can also be phrased as

$$\frac{\mathrm{d}q}{\mathrm{d}t} + \gamma q = 0, \tag{C.12}$$

where the damping constant $\gamma = 1/\tau$ is defined. The solution (C.11) becomes

$$q(t) = q(0) \exp(-\gamma t). \tag{C.13}$$

First-Order, Proportional Drag, Constant Driving

A variation of (C.10),

$$\frac{\mathrm{d}q}{\mathrm{d}t} + \frac{q}{\tau} = F \tag{C.14}$$

with constant term F, has solution:

$$q(t) = (q(0) - F\tau) \exp(-t/\tau) + F\tau. \tag{C.15}$$

This type of equation is used in the viscoelastic mechanical models of materials with constant driving terms, including the Maxwell (4.52), Voigt (4.57), and Kelvin/standard linear (4.68) models, a model of muscles (5.9), the arterial pulse (Problem 8.49), and temperature regulation (13.18).

First-Order, Proportional Drag, Increasing Driving

A variation of (C.14) includes a driving term that varies linearly with the independent variable

$$\frac{\mathrm{d}q}{\mathrm{d}t} + \frac{q}{\tau} = F + Gt. \tag{C.16}$$

It has solution

$$q(t) = (F\tau - G\tau^2)(1 - \exp(-t/\tau)) + q(0) \exp(-t/\tau) + Gt\tau. \tag{C.17}$$

This type of equation is used in the Kelvin/standard linear viscoelastic mechanical model with a linearly increasing driving term (4.72).

First-Order, Proportional Drag, Arbitrary Temporal Driving

A variation of (C.12) and (C.14) includes a driving term that varies arbitrarily on the independent variable

$$\frac{\mathrm{d}q}{\mathrm{d}t} + \frac{q}{\tau} = F(t). \tag{C.18}$$

Substituting $q(t) = s(t)\exp(-t/\tau)$ into this gives

$$\frac{\mathrm{d}s}{\mathrm{d}t} = \exp(t/\tau)\, F(t), \tag{C.19}$$

so

$$s(t) = s(0) + \int_0^t \exp(t'/\tau) F(t')\mathrm{d}t' \tag{C.20}$$

and

$$q(t) = \exp(-(t/\tau)) \left(q(0) + \int_0^t \exp(t'/\tau) F(t')\mathrm{d}t' \right). \tag{C.21}$$

This type of equation is used for pulsatile flow (8.103).

First-Order, Higher-Order Drag, No Driving

A variable $q(t)$ obeying

$$\frac{\mathrm{d}q}{\mathrm{d}t} + Aq^n = 0 \tag{C.22}$$

varies as

$$q(t) = \left(q(0)^{1-n} + (n-1)At\right)^{1/(1-n)}, \tag{C.23}$$

for $n \neq 1$, where $q(0)$ is the initial condition. For $n = 1$, see (C.10) and (C.11).

For $n = -4$ this describes flow with resistance and compliance (8.26).

For $n = 2$ this describes the equation of motion for hydrodynamic drag where q is speed (7.64). Then

$$q(t) = \frac{q(0)}{1 + Aq(0)t}, \tag{C.24}$$

If $q = \mathrm{d}p/\mathrm{d}t$, where p would be the position for this type of drag, then

$$p(t) = p(0) + \frac{1}{A}\ln(1 + Aq(0)t). \tag{C.25}$$

C.3 Harmonic Oscillator

In these differential equations, the second derivative of the dependent variable, q, depends on q and in some cases on the dependent variable t.

Harmonic Oscillator: Undamped, Not Driven

A variable $q(t)$ obeying

$$\frac{d^2q}{dt^2} + \omega_0^2 q = 0 \tag{C.26}$$

oscillates as

$$q(t) = A\cos(\omega_0 t + \phi), \tag{C.27}$$

where A is the amplitude, ω_0 is the resonant frequency of this harmonic oscillator (with units rad/s), and ϕ is the phase. Alternatively, this solution can be expressed as

$$q(t) = B\cos(\omega_0 t) + C\sin(\omega_0 t), \tag{C.28}$$

where B and C are amplitudes. The frequency, f, is $\omega/2\pi$, and has units of Hz (Hertz) or cps (cycles per second), and so (C.27) would be

$$q(t) = A\cos(2\pi f_0 t + \phi). \tag{C.29}$$

This type of equation is used in the models of the harmonic motion of a mass on a spring (3.7), the simple (3.14) and complex pendulums (3.26), and Euler buckling (4.86).

Harmonic Oscillator: Damped, Not Driven

Adding damping to the harmonic oscillator equation (C.26) with damping constant $\gamma = 1/\tau$ gives

$$\frac{d^2q}{dt^2} + \gamma\frac{dq}{dt} + \omega_0^2 q = 0, \tag{C.30}$$

with solution

$$q(t) = A\exp(-\gamma t/2)\cos(\omega_0 t + \phi), \tag{C.31}$$

where A is the amplitude, ω_0 is the resonant frequency of this harmonic oscillator (with units rad/s), and ϕ is the phase. This solution is not exact, but is valid for $\omega_0 \gg \gamma$. This harmonic oscillation damps in a time $\sim 1/\gamma$, which corresponds to about $\omega_0/(2\pi\gamma)$ cycles; ω_0/γ is often called the quality factor Q of the system, as is discussed in the Chap. 10 discussion of acoustic resonances and more generally in Appendix D.

This type of equation is used in the models of harmonic oscillators, and simple and complex pendulums.

Harmonic Oscillator: Undamped, Driven

The equation

$$\frac{\mathrm{d}^2 q}{\mathrm{d}t^2} + \omega_0^2 q = F\cos(\omega t) \tag{C.32}$$

looks like the equation of motion for a simple harmonic oscillator of frequency ω_0 (C.26) with an extra term (the last one), which drives the oscillator with a "force" that oscillates at a frequency ω; ω can differ from the resonant frequency ω_0. The particular solution to this equation is

$$q(t) = \frac{F}{\omega_0^2 - \omega^2}\cos(\omega t), \tag{C.33}$$

to which the solution (C.27), $q(t) = A\cos(\omega_0 t + \phi)$ (of the homogeneous equation (C.26)) is added to set the initial conditions by the proper choice of A and ϕ. Without the driving term ($F = 0$), the solution is the usual harmonic solution (C.27).

This type of equation is used in the models of pulsatile blood flow (8.51) and the general models in Appendix D (D.2).

Harmonic Oscillator: Damped, Driven

If ω_0 were to approach ω, the response for the undamped, driven harmonic oscillator, (C.33), would approach infinity because of this resonance. There is always some damping that adds a term $\gamma\,\mathrm{d}q/\mathrm{d}t$ to (C.32) to give the new equation of motion

$$\frac{\mathrm{d}^2 q}{\mathrm{d}t^2} + \gamma\frac{\mathrm{d}q}{\mathrm{d}t} + \omega_0^2 q = F\cos(\omega t). \tag{C.34}$$

This has a particular and steady-state solution

$$q(t) = \frac{\left(\omega_0^2 - \omega^2\right)F}{\left(\omega_0^2 - \omega^2\right)^2 + (\gamma\omega)^2}\cos(\omega t). \tag{C.35}$$

The homogeneous solution (C.31), $q(t) = A\exp(-\gamma t/2)\cos(\omega_0 t + \phi)$ for $\omega_0 \gg \gamma$, is added to this to set the initial conditions by the proper choice of A and ϕ. Without the driving term ($F = 0$), the solution is the usual damped harmonic solution (C.31).

This type of equation is used in the models of pulsatile blood flow (8.53), acoustic impedance (10.21), and the general models in Appendix D (D.2).

C.4 Partial Differential Equations

Partial differential equations contain derivatives of more than one independent variable.

The Diffusion Equation

The diffusion equation (7.53) in one dimension (x) has the form

$$D\frac{\partial^2 q}{\partial x^2} = \frac{\partial q}{\partial t}. \tag{C.36}$$

The formal solution gives $q(x,t)$ from $q(x',t=0)$, the distribution for all x (called x') at an earlier time (defined as $t=0$). It is

$$q(x,t) = \frac{Q}{\sqrt{4\pi Dt}} \int_{-\infty}^{\infty} q(x',0) \exp\left(-(x-x')^2/4Dt\right) \mathrm{d}x', \tag{C.37}$$

where Q is the integral of q over all x at any time – which means that the total amount of the entity undergoing diffusion, such as the mass or number of particles, does not change during diffusion.

The importance of this diffusion is most simply seen when the initial distribution is gaussian and has an initial spread $\sigma(0)$,

$$q(x,0) = \frac{Q}{\sqrt{2\pi\sigma^2(0)}} \exp\left(-x^2/2\sigma^2(0)\right). \tag{C.38}$$

Then the solution becomes

$$q(x,t) = \frac{Q}{\sqrt{2\pi\sigma^2(t)}} \exp\left(-x^2/2\sigma^2(t)\right), \tag{C.39}$$

where

$$\sigma^2(t) = \sigma^2(0) + 2Dt. \tag{C.40}$$

If the initial spread is not gaussian, the solution is slightly different but approaches this for large x and/or large t. (Sometimes σ is defined a bit differently than it is here, as in (7.55).)

This type of equation is used in diffusion (7.53).

The integral over a gaussian probability curve,

$$\mathrm{erf}(x) = \frac{2}{\pi^{1/2}} \int_0^x \exp(-z^2)\mathrm{d}z, \tag{C.41}$$

is known as the error function. It increases from 0 to 1 as x increases from 0 to ∞. The error function is used in the statistics describing head injury as in Fig. 3.59.

The Poisson–Boltzmann Equation

The Poisson–Boltzmann Equation (12.44) is of the form

$$\nabla^2 q = \kappa^2 q. \tag{C.42}$$

where ∇^2 is the Laplacian. In one dimension $\nabla^2 q = \partial^2 q/\partial x^2$, while in Cartesian coordinates in three dimensions it is $\nabla^2 q = \partial^2 q/\partial x^2 + \partial^2 q/\partial y^2 + \partial^2 q/\partial z^2$. In three dimensions it can be expressed as $\nabla^2 q = (1/r)(\mathrm{d}^2(rq)/\mathrm{d}r^2)$ when there is no angular dependence (spherical symmetry), where r is the radial coordinate. Using this in (C.42) gives

$$\frac{1}{r}\frac{\mathrm{d}^2(rq)}{\mathrm{d}r^2} = \kappa^2 q. \tag{C.43}$$

Replacing $rq(r)$ by $p(r)$, this reduces to

$$\frac{\mathrm{d}^2 p}{\mathrm{d}r^2} = \kappa^2 p, . \tag{C.44}$$

with solution $p(r) = p(0)\exp(-\kappa r)$ valid for all r, and so

$$q(r) = q(0)\frac{\exp(-\kappa r)}{r}. \tag{C.45}$$

This is used to determine the potential of a charge in a neutral region with mobile charges, as in (12.44) and Problem 12.8.

Appendix D

Similar Model Systems

This appendix describes the models used throughout the text to describe mechanical, fluid flow, electrical, and acoustic systems. Figure D.1 shows the analog in the driving forces, currents, resistances, capacitance, and inductance in each of these models.

There are many examples of these models in the text. Chapter 4 covers the spring model of the elastic properties of materials (Fig. 4.3), the dashpot model of the viscous properties of materials (Fig. 4.48), the viscoelastic model of mechanical properties of materials, including the Maxwell (Fig. 4.52), Voigt (Fig. 4.57), and Kelvin/standard linear (Fig. 4.68) models. In Chap. 5, a mechanical model of muscles, with springs and dashpots (Fig. 5.9) is presented. Models of fluid flow are described in Chaps. 7 and 8. In fact, the Windkessel models of circulation in Chap. 8 are explicitly expressed in terms of electrical components (Fig. 8.57). The mechanical and flow model of breathing in Fig. 9.16b includes compliance, resistance, and inertance. Acoustic impedance (Fig. 10.19) and admittance (Fig. 10.20) are described in Chap. 10 (and in the problems in that chapter) in relation to mechanical analogs. In Chap. 10 there also are mechanical models of vibrations in the vocal tract (as vibrations in pipes and voice filtering theory) and vocal folds (vibrations in strings, (10.41), and mechanical model with mass, springs, and dashpots (Fig. 10.13) and the two-tube models of vowel formation (Fig. 10.25). In that chapter there is also a mechanical model of the outer and inner ears (Fig. 10.28), and the vibrations of the eardrum (Fig. 10.56), tapered, uncoiled cochlea (Fig. 10.38), and hair cells (Fig. 10.57). Axon nerve conduction in Chap. 12 involved a distributed model. (See Sect. D.1)

Equivalent mechanical, electrical, and acoustic models are shown in Fig. D.2 of a typical system. The electrical model is described by

$$V = L\frac{\mathrm{d}I}{\mathrm{d}t} + IR + \frac{q}{C}, \tag{D.1}$$

for a voltage V producing a current I; $I = \mathrm{d}q/\mathrm{d}t$ where q is the charge. (The subscripts specific for the electrical model are omitted for simplicity.)

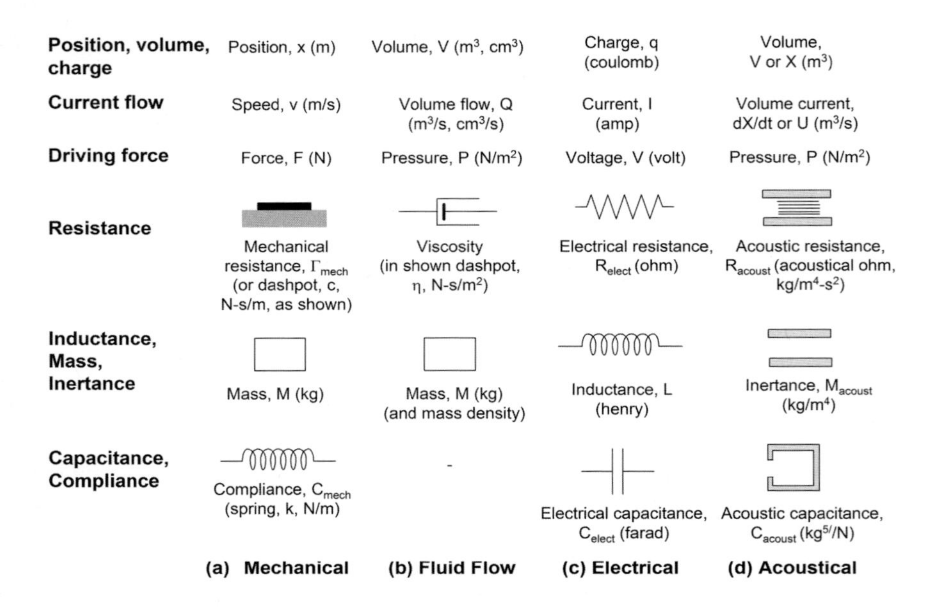

Fig. D.1. Model symbols are shown for **(a)** mechanical, **(b)** fluid flow, **(c)** electrical, and **(d)** acoustic models, along with the parameters and common units for each. The mechanical model is for linear (rectilinear) motion. Analogous parameters exists for the rotational mechanical model, such as for a pendulum. Viscosity is also important in mechanical models. Also see Table D.1 below. (Based on [609] and [610])

The other models are described similarly, simply by changing the parameters. Equation D.1 also be written as the second-order differential equation

$$V = L\frac{\mathrm{d}^2 q}{\mathrm{d}t^2} + R\frac{\mathrm{d}q}{\mathrm{d}t} + \frac{q}{C}, \tag{D.2}$$

as in (C.30) and (C.34).

The general solution for an oscillating voltage $V(t) = V_0 \exp(i\omega t)$ is

$$I(t) = \frac{V_0 \exp(i\omega t)}{R + i\omega L + 1/(i\omega C)} = \frac{V(t)}{Z}, \tag{D.3}$$

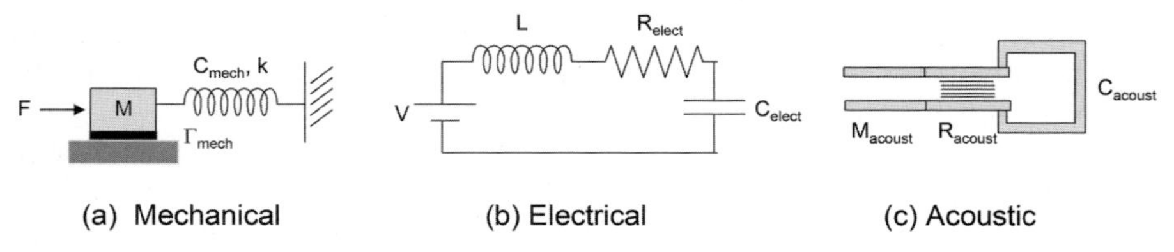

Fig. D.2. Equivalent **(a)** mechanical, **(b)** electrical, and **(c)** acoustic models. (Based on [609] and [610])

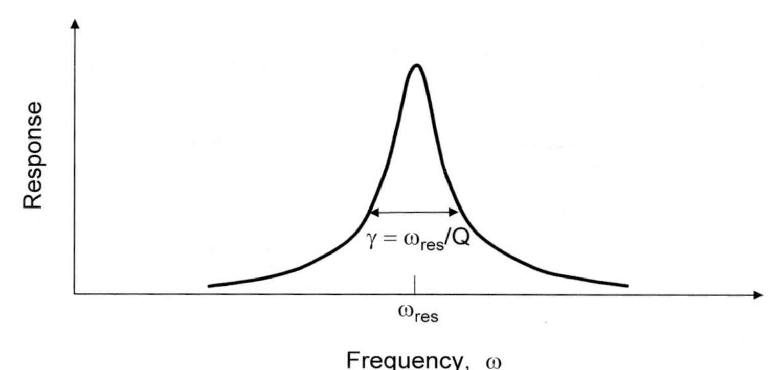

Fig. D.3. Resonant response

with complex impedance

$$Z = R + i\omega L + \frac{1}{i\omega C}. \tag{D.4}$$

The resonant frequency is seen in Fig. D.3 given by

$$\omega_{\text{res}} = \frac{1}{\sqrt{LC}} \tag{D.5}$$

in rad/s and $f_{\text{res}} = 1/(2\pi\sqrt{LC})$ in Hz or cps, and the quality factor Q is given

$$Q = \frac{\omega_{\text{res}} L}{R}. \tag{D.6}$$

Using the notation of (C.30) and (C.34)

$$Q = \frac{\omega_{\text{res}}}{\gamma}. \tag{D.7}$$

The full width of the resonance (between the points at half-maximum response) is ω_{res}/Q, as is illustrated in Fig. D.3. (This is actually the full width only for sharp resonances, for which $\omega_{\text{res}} \gg \gamma$ and $Q \gg 1$.) This full width is sometimes called the bandwidth (when expressed as f in Hz), Δf, and so an alternative definition of Q is $Q = f/\Delta f$; this is an equivalent definition in the low-loss, high Q limit.

After the excitation is turned off ((C.34) becoming (C.30)), the energy in the system exponentially decays to $1/e$ of the initial value in a time $t = 1/\gamma = Q/\omega_{\text{res}}$, where γ is the damping rate (as in (C.30)), which is R/L is here. This decay occurs in $Q/2\pi$ oscillation periods. This is consistent with the definition of Q as 2π(energy stored)/(energy dissipated per cycle).

Table D.1. Analog of blood flow and electrical circuits (with units)

blood circulation parameter	electrical parameter
volume, V_{flow} (m^3)	charge, q (C, coulomb)
blood flow rate, Q (m^3/s)	current, I (A, ampere)
pressure, ΔP (N/m^2)	voltage, V_{elect} (V, volt)
vascular resistance, R_{flow} (N-s/m^5)	resistance, R_{elect} (Ω, ohm)
inertance, L_{flow} (kg/m^4)	inductance, L_{elect} (H, henry)
compliance, C_{flow} (m^5/N-s)	capacitance, C_{elect} (F, farad)

D.1 Distributed vs. Lumped Models: Electrical Analogs of Blood Flow (Advanced Topic)

So far we have discussed lumped parameter models in this appendix. In (8.2) and (8.11) flow was analyzed with the vessel as a "lumped" parameter. We have also examined cases in this text in which the parameters are distributed per unit length, such as flow resistance per unit length for volumetric flow along an artery in Chap. 8 ((8.14) and (8.25)) and electrical resistance per unit length for current flow along an axon in Chap. 12 ((12.60) and (12.67)). These are "distributed" or "transmission-line" models.

A discretized version of the distributed electrical model is shown in Fig. 12.17. Let us say that each repeated section has (very short) length Δx. The changes in electrical voltage (the driving force) and current (the response) (Table D.1) along this length of an electrical cable are described by [376]

$$V_{\text{elect}}(x+\Delta x) - V_{\text{elect}}(x) = \frac{\partial V_{\text{elect}}}{\partial x}\,\Delta x = L_{\text{elect}}\frac{\partial I}{\partial t} + IR_{\text{elect}} \quad \text{(D.8)}$$

$$I(x+\Delta x) - I(x) = \frac{\partial I}{\partial x}\,\Delta x = C_{\text{elect}}\frac{\partial V_{\text{elect}}}{\partial t} + \frac{V_{\text{elect}}}{R_{\text{elect}}}. \quad \text{(D.9)}$$

V_{elect} and I are functions of x and t. The resistance, inductance, and capacitance are those for this length Δx, and can also vary with x. These equations can be obtained using Kirchhoff's Laws (the 2nd and 1st laws, respectively). They were derived and then combined in the discussion of electrical signals along nerves in Chapter 12 (Fig. 12.17) to give the telegraph equations. Part of the first equation is Ohm's Law: $\Delta V_{\text{elect}} = (\partial V_{\text{elect}}/\partial x)\Delta x = IR_{\text{elect}}$.

The analogous equations for blood flow along a vessel of length Δx are:

$$P(x+\Delta x) - P(x) = \frac{\partial P}{\partial x}\,\Delta x = L_{\text{flow}}\frac{\partial Q}{\partial t} + QR_{\text{flow}} \quad \text{(D.10)}$$

$$Q(x+\Delta x) - Q(x) = \frac{\partial Q}{\partial x}\,\Delta x = C_{\text{flow}}\frac{\partial P}{\partial t} + \frac{P}{R_{\text{flow}}} \quad \text{(D.11)}$$

where now the pressure is the driving force for the blood flow rate Q and R_{flow} is the vascular resistance. Without the inertance term, the first equation is just Poiseuille's Law (7.25): $\Delta P = (\partial P/\partial x)\Delta x = QR_{\text{elect}}$. The flow parameters are

$$R_{\text{flow}} = \frac{8\pi\eta L}{A^2} \tag{D.12}$$

$$C_{\text{flow}} = \frac{3LA(1+r/w)^2}{Y(1+2r/w)} \tag{D.13}$$

$$L_{\text{flow}} = \frac{\rho L}{A}, \tag{D.14}$$

where A is the cross-sectional area of the vessel, L is its length (which is Δx for the discretized model), r is its radius, w is its wall thickness, and ρ is the blood mass density. These equations are useful for tracking blood flow within vessels with both resistive and compliant properties.

All models of materials (and systems and processes) can be improved mathematically by adding more terms, such as in the mechanical model in Fig. 5.13. The bigger issues are whether the elements in such simple or more complex models correspond to the physical components of the material. Even if they do not, it is still important to learn if the model can be used to predict operation correctly when conditions are changed.

Appendix E

Biophysics of the Human Body

This appendix places the contents of this text within the field of *biophysics*.

Biophysics is hard to define well, as is illustrated by the many definitive, yet different definitions of biophysics provided in [611, 612, 613, 614, 615, 616, 617, 618, 619, 620, 621, 622]. Broadly speaking, biophysics is the applications of physics and physical principles to biology. In this context, virtually everything presented in this book is biophysics. However, this term is often used in the more restricted sense of the use of physics at a more molecular and cellular level. We will use this narrower context for the rest of this appendix, and in this restricted sense many topics covered here are still biophysics, but many areas in biophysics have not been covered. Yet another definition of biophysics is the study of biology using physical methods. This is distinguished from *biological physics*, which is the study of the physical properties of biology.

One topic in biophysics is the molecular structure of biological systems. This includes the electrostatics of ions in solutions (Chap. 12), the structure of biomacromolecules, such as proteins and areas such as protein folding, structure, and properties of interfaces between biological media such as cell membranes (surface tension in Chaps. 7 and 9, nerve cell membranes in Chap. 12), and ion channels in membranes (which is very briefly touched in nerve conduction in Chap. 12).

Statistical mechanics is the examination of systems composed of many similar objects or systems, each of which is well characterized. The whole ensemble of systems often is in thermal equilibrium and, after statistical mechanical analysis, can be treated by using thermodynamics. The treatment of ions in solution in Chap. 12 is the result of statistical mechanics. As stated in Chap. 5, the Hill force–velocity curve of muscles can be derived from statistical mechanical analysis of the many actin–myosin cross bridges (which was not done here). Many aspects of cell membranes and protein structure – such as folding – can be examined by using statistical mechanical methods.

Biophysics includes the bioenergetics of the photosynthesis process and the synthesis of ATP and its use. In Chap. 6 we examined the biophysics of energy usage in the human body. The movement of organisms, such as bacteria motion and muscular movement, are part of molecular and cellular biophysics; the microscopic basis of muscle operation was explored in Chap. 5. The electrochemical properties of cell membranes and nerve signals, as discussed in Chap. 12, have always been central topics in biophysics. Some include within biophysics the higher-level integration and combinations of molecular and cellular systems, such as memory, control of movement, visual integration, and consciousness and thinking.

The use of physical characterization to biological problems plays a central role in biophysics, such as the use of X-ray diffraction (XRD) to determine molecular structure, nuclear magnetic resonance (NMR) to study molecules in more natural environments than X-ray diffraction can be used, scanning tunneling microscopy (STM) to examine the atomic structure of surfaces, atomic force microscopy (AFM) to examine surfaces and to measure forces, and optical tweezers to manipulate molecules. Both AFM and optical tweezers have been instrumental in studying the fundamental interactions in muscles, such as individual actin–myosin cross bridges (Chap. 5).

Solutions to Selected Problems

Problems of Chapter 1

1.3 Medial.

1.13 Head.

1.25 (a) (partial answer) For lower legs 3.72–9.30 kg and 5.53–9.55 kg.

1.30 (b) 0.25 m and 0.50 m.

1.31 (partial answer) Surface area is 20.1 sq ft.

1.44 (a) 92.6 kg (204.2 lb) for Man A and 84.3 kg (185.9 lb) for Man B; (b) 0.4 kg (0.9 lb) for Man A and 8.7 kg (19.2 lb) for Man B.

1.49 (b) 21.1 and 27.0.

1.57 Bigger in cold climate.

Problems of Chapter 2

2.1 Third class lever.

2.2 Triceps brachii, second class lever.

2.3 (a) First; (b) second; (c) third class levers.

2.11 (b) $m_{\text{leg}}(x_{\text{extended leg}} + x_{\text{balancing leg}}) = (m_{\text{torso+head}} + 2m_{\text{arm}})x_{\text{upper body}}$.

2.13 (a) $-m_{\text{b}}x$.

2.14 $T_1 = W_1 = 223.6$ N, $T_2 = W_2 = 282.8$ N, $\alpha = 26.6°$.

2.17 (b) 310.6 N.

2.21 (b) $T = 131$ lb, $F = 208$ lb, angle of $\mathbf{F}$ is 29.8°.

2.33 (partial answer) −6 N-m for 20 cm deep.

2.34 (partial answer) −6.9 N-m for upright, −19.25 N-m for bent.

2.35 (partial answer) −21.25 N-m for bent over, with bent knees and the object far from her body.

Problems of Chapter 3

3.6 0.27, easy to achieve with cleated running shoes.

3.10 1.7127 kg-m^2.

3.26 2/3.

3.35 Yes, by −0.07 m, yes.

3.36 For (a) $0.254H$; (b) $0.150H$; (c) $0.077H$; (d) $0.150H$; (e) $0.254H$; 14 cm.

3.40 Yes, because how fast the body can take off at a given angle depends on the construction of the body's feet and legs.

3.64 44.0°, 31.1 m/s.

3.72 (partial answer) Elastic collisions are likely fatal for collision times $<$120 ms.

3.78 (a) 150 ft/s^2, $4.7g$, (b) 1,150 lb.

3.80 1.44, 1.06, 1.01, and 0.18 m.

3.81 0.50, 0.53, and 0.55.

3.88 48.0 mph before, 24.7 mph after.

3.89 40.5 oz and about 80 mph.

Problems of Chapter 4

4.2 (c) 480 MPa.

4.5 (a) 30 Pa, (b) 67 mm^2, (c) 1%.

4.6 (partial answer) 1.6 MPa for nails.

4.7 (partial answer) 0.0031 for nails.

4.8 (partial answer) 8,000 N/m^3 for nails.

4.9 (partial answer) 780 N/m^3 for nails.

4.15 (a) Tension.

4.20 (a) $\lambda = 2$, $\epsilon_{\text{small}} = 1$, $\epsilon_{\text{general}} = 3/2$.

4.21 (a) $\lambda = 2$, $\epsilon_{\text{small}} = 1$, $e = 3/8$.

4.22 (a) $\lambda = 2$, $\epsilon_{\text{small}} = 1$, $\ln \lambda = 0.69$.

4.23 (a) Yes, because the dashpot resistive force increases with speed; (b) Yes, because the spring supplies the needed restoring force to return it to its equilibrium position.

4.24 Length is 3.1 cm, dx/dt is 0.05 cm/s.

4.28 (a) $2\theta(t+1) - 2\theta(t-3)$, with all times in seconds.

4.33 It becomes the Voigt model, the Maxwell model, and a dashpot, respectively.

4.42 (a) 6,900 N.

Problems of Chapter 5

5.8 (a) 0.11, it is larger than the 0.09 listed in the table—but in linear theory it would be expected to be $\text{UTS}/Y = 0.22$, (b) $fY = 135\,\text{N/cm}^2$, (c) the diameter of the tendons is 0.073× that of the muscle, (d) $2{,}800\,\text{N/cm}^2 = 28\,\text{MPa}$, which is less than the 54 MPa UTS listed in Table 4.2, so it is less than it, even with linear theory.

5.10 (a) (partial answer) 2,770 W.

5.15 The muscles are fairly near their optimal lengths. However, there are significant changes in the lengths during bicycling, but less than the maximum expected for muscles are shown for several reasons: (a) The decreases in contracted muscle length are actually greater than those shown because tendon extension will lessen the decrease in the (plotted) total muscle/tendon length, (b) the bicycle is set to use muscles in their optimal state, both in muscle length and speed, so the muscles will not be much longer or shorter than their optimal length.

5.18 (a) (partial answer) $\sqrt{3}NF_{fiber}/2$.

5.23 (a) 40 s.

Problems of Chapter 6

6.1 (partial answer) 0.3°C.

6.4 (b) 144 BTU.

6.6 (a) (partial answer) 7.1 kcal/g.

6.14 (a) 261 g, compared to 260 g and 280 g; (b) 45%; (c) 18%.

6.17 90% of the fruit and 40% of the dried fruit is water plus non-metabolizable matter.

6.24 2,200 kcal.

6.36 (a) 139 kg; (b) This is much more than the body mass of 60 kg; (c) 1,400 cycles/day, 0.95 cycles/min.

6.30 (a) −12.6 kcal/mol; (b) −14.0 kcal/mol.

6.31 66–70%.

6.41 147 moles of ATP.

6.47 6.9.

6.56 (partial answer) 580 kcal/h for 50 kg college-age women, using 40 mL/kg-min and a calorific equivalent of 4.83 kcal/L O_2.

6.57 (a) 89.9 m/min, 1.50 m/s, 3.35 mph, very good agreement; (b) 259 kcal/hr.

6.61 (partial answer) (a) 1048 kcal/h and 3.6 L O_2/min for Stage VIII, (b) 35 kcal for Stage VIII.

6.64 (a) 305 J, (b) 36 kcal, (c) no, but it excludes the pushing for 5–10 s during each play, which obviously accounts for most of the energy expenditure (although relatively little of the work done).

6.67 12.2 kJ, 11.7 kcal.

6.74 Activity factor is 1.48 (assuming "self-care" walking and also cycling at 5.5 mph), MR is 1920 kcal/day (using 1300 kcal/day BMR).

6.75 (a) 4.9, (b) 4.2.

6.80 15 kg, 33 lb.

6.88 2.2 L/h.

6.95 (a) 70 kcal/h.

6.100 $T_{\mathrm{wc}} = 35.74 + 0.6215T - 35.75w^{0.16} + 0.4275Tw^{0.16}$.

6.101 (a) A and B; (b) C; (c) A.

6.106 The first term is 40 kcal/day, second term in 220 kcal/day. The second term is very significant.

Problems of Chapter 7

7.3 2,240 cm^3, 2.49 kg, 24.4 N; using a mass density of 1.11 g/cm^3.

7.7 No, his density is then 1.01 g/cm^3, which is above that of water, using a fat density of 0.8 g/cm^3.

7.18 (a) Q/N; (b) The diameter of the small tubes would then be $N^{1/4}D$, which is not possible because the diameter of the small tubes would then exceed that of the larger tubes.

7.25 (partial answer) 0.32 cm in a gas.

7.38 $Re = 0.001$, viscous/laminar.

Problems of Chapter 8

8.5 You should be concerned, but not about your blood pressure (which is really 120 mmHg/80 mmHg), but about the person who told you your blood pressure in absolute pressure instead of the standard gauge pressure.

8.12 (a) 121 mmHg, which is 32% higher than the 92 mmHg base case, (b) 115 mmHg, which is 25% higher.

8.15 (a),(b) They change by a factor of 1/8.

8.19 1 N-s/m^5 = 10^6 (N/m^2)/(cm^3/s) = 10^5 dyne-s/cm^5 = 1.32×10^8 PRU.

8.27 $u_1/4$.

8.34 98.6 mmHg, using a blood density of 1060 kg/m^3.

8.39 Type I skeletal muscle, because very fast response is not needed and endurance is essential.

8.43 0.128 L = 128 cm^3.

8.45 (partial answer) 3.6 cm inner radius.

8.53 Heart beat rate: 83/min for the man, 91/min for the woman, 161/min for the infant.

Problems of Chapter 9

9.6 (partial answer) 4,720 for $z = 0$ (turbulent), 0.18 for $z = 20$ (laminar).

9.16 0.0078 cmH_2O/(L/s), which is much smaller than the total resistance of ~2 cmH_2O/(L/s).

9.20 2.7 cmH_2O-s/L.

9.26 Larger in a mouse (0.005 L/kg-cm-H_2O) than in man (0.003 L/kg-cm-H_2O).

9.33 (partial answer) 570 L of O_2 (at 1 atmosphere oxygen pressure) are consumed per day.

9.37 (partial answer) 225 mmHg total, 42 mmHg oxygen.

Problems of Chapter 10

10.4 Respectively at 0, 20, and 25°C, v_{air} is 331, 343, and 346 m/s; ρ_{air} is 1.292, 1.204, and 1.184 kg/m^3; and Z_{air} is 428, 413, and 410 kg/m^2-s.

10.6 0.00002–2000 dyne/cm^2.

10.11 70 dB SPL.

10.14 (a) 100 dB SPL, 40 dB SPL.

10.20 (a) 0.27.

10.33 ~1,000 Hz, which makes sense since the voices of children are higher pitched than those of adults.

10.38 160 Hz, assuming a mass with the mass of the vocal folds is attached to a massless spring with the force constant of the vocal folds. This really requires the analysis of a freely oscillating massive spring, which shows that oscillation frequency is $\pi/2 \times$ this value [458].

10.40 Lower, because they have higher fundamental buzzing frequencies.

10.47 3×10^{-13} m.

10.49 It is 9×10^6 larger at 3,000 Hz than that at 1 Hz.

10.55 (a) Radii of 0.4 µm for 20 Hz to 400 µm for 20 kHz.

10.60 17,000 Hz would be best because the auditory sensitivity of older people is very low at this frequency relative to that of younger people, at 250 Hz and 1,000 Hz the auditory sensitivity is not that different for older and younger people, at 30,000 Hz humans have no auditory sensitivity.

10.61 About 200/s.

10.66 100–8,000 Hz, over 40 dB.

10.67 40–14,000 Hz, over 70 dB.

10.72 The former (60 dB SPL) is a bit louder than the latter (59 dB SPL).

Problems of Chapter 11

11.7 (a) (partial answer) 2.5% for the first surface, (b) no.

11.14 7.51 mm, −7.51 mm.

11.15 6.04 mm.

11.17 The retina.

11.22 More damage is done if you look in the direction of the beam because it will focus on the fovea. Damage to the fovea can hurt sharp vision, leaving you with fuzzy vision.

11.23 8.8 cm.

11.34 (partial answer) Refractive index is larger at 630 nm by 0.0008 (if everything else is the same).

11.42 58.62 D (smaller than before), 22.8 mm (longer), mostly due to the smaller refractive index of the crystalline lens.

11.48 −1.71 D, −1.67 D.

11.53 (c) Myopia. A correction of −2 D would lead to good vision. (The patient has 4 D of accommodation, which is sufficient with this correction.)

11.59 (b) 7.55 mm = 0.00755 m.

11.68 250 lux assuming 500 lumens/W (Fig. 11.49). It is consistent with the levels given in Table 11.5.

Problems of Chapter 12

12.2 110,000 ohms.

12.4 (a) 240 mA, (b) shock and possible ventricular fibrilation would result.

12.11 The large net diffusion of K^+ outside, the impermeability of the membrane to the proteins, which are negatively charged, and the Na^+ pump contribute to the cell being negative relative to the extracellular fluid for a resting axon. The small net diffusion of Na^+ into the cell adds slightly to the positivity inside the cell.

12.19 (partial answer) 6.20×10^{-4} s for unmyelinated axons.

12.28 ~80/min.

12.30 The dipole usually rotates also. A(c); B(a); C(d); D(b).

Problems of Chapter 13

13.2 Every 17 min.

13.3 (partial answer) 1.6/s for a 1 s delay.

References

Preface

1. B.K. Ahlborn: *Zoological Physics: Quantitative Models, Body Design, Actions and Physical Limitations in Animals* (Springer, Berlin Heidelberg New York 2004)
2. R. McNeill Alexander: *The Human Machine* (Columbia University Press, New York 1992)
3. G.B. Benedek, F.M.H. Villars: *Physics with Illustrative Examples from Medicine and Biology*, Three volumes, 2nd edn (AIP/Springer, New York/Berlin Heidelberg New York 2000)
4. B.H. Brown, R.H. Smallwood, D.C. Barber, P.V. Lawford, D.R. Hose: *Medical Physics and Biomedical Engineering* (Institute of Physics, Philadelphia 1999)
5. R.F. Burton: *Physiology by Numbers: An Encouragement to Quantitative Thinking* (Cambridge University Press, Cambridge 1994)
6. J.R. Cameron, J.G Skofronick, R. Grant: *Physics of the Body*, 2nd edn (Medical Physics, Madison, WI 1999)
7. E. Carson, C. Cobelli: *Modelling Methodology for Physiology and Medicine* (Academic, San Diego 2001)
8. R.M.J. Cotterill: *Biophysics: An Introduction* (Wiley, New York 2002)
9. P. Davidovits: *Physics in Biology and Medicine*, 2nd edn (Elsevier/Academic, San Diego 2001)
10. M.L. Foss, S.J. Keteyian: *Fox's Physiological Basis for Exercise and Sport*, 6th edn (McGraw-Hill, Boston 1998)
11. A.C. Guyton, J.E. Hall: *Human Physiology and Mechanisms of Disease*, 6th edn (Saunders, Philadelphia 1997)
12. F.R. Hallett, R.H. Stinson, P.A. Speight: *Physics for the Biological Sciences: A Topical Approach to Biophysical Concepts* (Methuen/Chapman & Hall, Toronto 1982)
13. J.L. Hargrove: *Dynamic Modeling in the Health Sciences* (Springer, Berlin Heidelberg New York 1998)
14. R.K. Hobbie: *Intermediate Physics for Medicine and Biology*, 3rd edn (AIP, New York 1997)

15. F.C. Hoppensteadt, C.S. Peskin: *Modeling and Simulation in Medicine and the Life Sciences*, 2nd edn (Springer, Berlin Heidelberg New York 2002) (First Edition from 1992 was called *Mathematics in Medicine and the Life Sciences*)
16. E.N. Marieb: *Human Anatomy & Physiology*, 5th edn (Benjamin Cummings, San Francisco 2001)
17. P. Maróti, L. Berkes, F. Tölgyesi: *Biophysics Problems: A Textbook with Answers* (Akadémiai, Budapest 1998)
18. H.J. Metcalf: *Topics in Classical Biophysics* (Prentice-Hall, Englewood Cliffs, NJ 1980)
19. P. Nelson: *Biological Physics: Energy, Information, Life* (Freeman, New York 2003)
20. M. Nordin, V.H. Frankel (eds.): *Basic Biomechanics of the Musculoskeletal System*, 3rd edn (Lippincott Williams & Wilkins, Philadelphia 2001)
21. P.E. Pack: *Anatomy and Physiology* (Cliff Notes, Lincoln, NE 1997)
22. H.D. Patton, A.F. Fuchs, B. Hille, A.M. Scher, R. Steiner (eds.): *Textbook of Physiology*, 21st edn (Saunders, Philadelphia 1989)
23. R.R. Seeley, T.D. Stephens, P. Tate: *Anatomy & Physiology*, 6th edn (McGraw-Hill, Boston 2003)
24. L. Sherwood: *Human Physiology: From Cells to Systems*, 5th edn (Thomson Brooks/Cole, Belmont, CA 2004)
25. D. Shier, J. Butler, R. Lewis, *Hole's Human Anatomy and Physiology*, 9th edn (McGraw-Hill, Boston 2002)
26. A. Stalheim-Smith, G.K. Fitch: *Understanding Human Anatomy and Physiology* (West, St. Paul, MN 1993)
27. G.J. Tortora, S.R. Grabowski: *Principles of Anatomy and Physiology*, 10th edn (Wiley, New York 2003)
28. J.A. Tuszynski, J.M. Dixon: *Biomedical Applications of Introductory Physics* (Wiley, New York 2002)
29. J.B. West (ed.): *Best and Taylor's Physiological Basis of Medical Practice*, 12th edn (Williams & Wilkins, Baltimore 1991)
30. J.H. Wilmore, D.L. Costill: *Physiology of Sport and Exercise*, 3rd edn (Human Kinetics, Champaign, IL 2004)

Chapter 1

31. B.K. Ahlborn: *Zoological Physics: Quantitative Models, Body Design, Actions and Physical Limitations in Animals* (Springer, Berlin Heidelberg New York 2004)
32. R.M. Alexander: *The Human Machine* (Columbia University Press, New York 1992)
33. J.T. Barter, I. Emanuel, B. Truett: *A Statistical Evaluation of Joint Range Data*, WADC-TR-57-311 (Wright-Patterson Air Force Base, Ohio 1957)
34. P.J. Berenson, W.G. Robertson: Temperature. In *Bioastronautics Data Book*, J.F. Parker Jr., V.R. West, eds. (NASA, Washington, DC 1973), Ch. 3, pp. 65–148
35. J.H. Brown, G.B. West (eds.): *Scaling in Biology* (Oxford University Press, Oxford 2000)

36. W.A. Calder III: *Size, Function and Life History* (Harvard University Press, Cambridge, MA 1984)
37. J.R. Cameron, J.G Skofronick, R. Grant: *Physics of the Body*, 2nd edn (Medical Physics, Madison, WI 1999)
38. D.B. Chaffin, G.B.J. Andersson: *Occupational Biomechanics* (Wiley, New York 1984)
39. D.B. Chaffin, G.B.J. Andersson, B.J. Martin: *Occupational Biomechanics*, 3rd edn (Wiley-Interscience, New York 1999)
40. C. Clauser, J.T. McConville: Aerospace Medical Research Laboratories, Wright-Patterson Air Force Base, 1964–1965
41. M.L. Foss, S.J. Keteyian: *Fox's Physiological Basis for Exercise and Sport*, 6th edn (McGraw-Hill, Boston 1998)
42. R.A. Freitas Jr.: *Nanomedicine, Volume I: Basic Capabilities* (Landes Bioscience, Austin 1999)
43. S.A. Gelfand: *Essentials of Audiology*, 2nd edn (Thieme, New York 2001)
44. E.N. Lightfoot: *Transport Phenomena and Living Systems* (Wiley, New York 1974)
45. T. McMahon: Size and shape in biology. Science **179**, 1201 (1973)
46. T.A. McMahon, J.T. Bonner: *On Life and Size* (Scientific American Books, New York 1983)
47. H.J. Metcalf: *Topics in Classical Biophysics* (Prentice-Hall, Englewood Cliffs, NJ 1980)
48. D.I. Miller, R.C. Nelson: *Biomechanics of Sport: A Research Approach* (Lea & Febiger, Philadelphia 1973)
49. K.L. Moore, A.M.R. Agur: *Essential Clinical Anatomy*, 2nd edn (Lippincott Williams & Wilkins, Philadelphia 2002)
50. J.F. Parker Jr., V.R. West (eds.): *Bioastronautics Data Book* (NASA, Washington, DC 1973)
51. F.A. Parker, D.A. Ekberg, D.J. Withey et al.: Atmospheric Selection and Control for Manned Space Stations (General Electric Co., Missile and Space Division, Valley Forge, PA), Presented at the International Symposium for Manned Space Stations, Munich, September 1965
52. S. Pinker: *How The Mind Works* (Norton, New York 1997)
53. J.A. Roebuck, K.H.E. Kroemer, W.G. Thomson: *Engineering Anthropometry Methods* (Wiley-Interscience, New York 1975)
54. T.H. Schiebler: *Anatomie*, 9th edn (Springer, Berlin Heidelberg New York 2005)
55. K. Schmidt-Nielsen: *Scaling Why is Animal Size So Important?* (Cambridge University Press, Cambridge 1984)
56. J.M. Smith: *Mathematical Ideas in Biology* (Cambridge University Press, Cambridge, 1971)
57. S.S. Stephens: To Honor Fechner and Repeal His Law. Science **133**, 80–86 (1961)
58. S.S. Stephens: The surprising simplicity of sensory metrics. Am. Psychol. **17**, 29–39 (1962)
59. B.N. Tillmann: *Atlas der Anatomie des Menschen*, (Springer, Berlin Heidelberg New York 2005)
60. D. Vaughn, T. Asbury: *General Ophthalmology*, 10th edn (Lange Medical, Los Altos, CA, 1983)

61. Webb Associates: *Anthropometic Source Book*, Vol. I, NASA Ref. 1024 (National Aeronautics and Space Administration), Chs. VI and VII (Laubach) (1978)
62. J.H. Wilmore, D.L. Costill: *Physiology of Sport and Exercise*, 3rd edn (Human Kinetics, Champaign, IL 2004)
63. D.A. Winter: *Biomechanics and Motor Control of Human Movement*, 3rd edn (Wiley, New York 2005)

Chapter 2

64. G.B.J. Andersson, R. Örtengren, A. Nachemson, G. Elfström: Lumbar disc pressure and myoelectric back activity during sitting. 1. Studies on an experimental chair. Scand. J. Rehabil. Med. **3**, 104–114 (1974)
65. G.B. Benedek, F.M.H. Villars: *Physics with Illustrative Examples from Medicine and Biology*, Three volumes, 2nd edn (AIP/Springer, New York/Berlin Heidelberg New York 2000)
66. B.H. Brown, R.H. Smallwood, D.C. Barber, P.V. Lawford, D.R. Hose: *Medical Physics and Biomedical Engineering* (Institute of Physics, Philadelphia 1999)
67. J.R. Cameron, J.G Skofronick, R. Grant: *Physics of the Body*, 2nd edn (Medical Physics, Madison, WI 1999)
68. A. Cappozzo: Compressive loads in the lumbar vertebral column during normal level walking. J. Orthop. Res. **1**, 292 (1984)
69. D.B. Chaffin, G.B.J. Andersson, B.J. Martin: *Occupational Biomechanics*, 3rd edn (Wiley-Interscience, New York 1999)
70. Y.C. Fung: *Biomechanics: Motion, Flow, Stress, and Growth* (Springer, Berlin Heidelberg New York 1990)
71. Y.C. Fung: *Biomechanics: Mechanical Properties of Living Tissues*, 2nd edn (Springer, Berlin Heidelberg New York 1993)
72. S. Garfield: *Teeth, Teeth, Teeth: A Thorough Treatise* (Simon & Schuster, New York 1969)
73. E.B. Goldstein: *Sensation and Perception*, 4th edn (Brooks/Cole, Pacific Grove 1996)
74. J.G. Hay: *The Biomechanics of Sports Techniques*, 4th edn (Prentice-Hall, Englewood Cliffs, NJ 1993)
75. B. Le Veau: *Williams and Lissner: Biomechanics of Human Motion*, 2nd edn (Saunders, Philadelphia 1977)
76. R.B. Martin, D.B. Burr, N.A. Sharkey: *Skeletal Tissue Mechanics* (Springer, Berlin Heidelberg New York 1998)
77. D.I. Miller, R.C. Nelson: *Biomechanics of Sport: A Research Approach* (Lea & Febiger, Philadelphia 1973)
78. K.L. Moore, A.M.R. Agur: *Essential Clinical Anatomy*, 2nd edn (Lippincott Williams & Wilkins, Philadelphia 2002)
79. K.L. Moore, A.M.R. Agur: *Essential Clinical Anatomy*, 3rd edn (Lippincott Williams & Wilkins, Philadelphia 2007)
80. A. Nachemson: Towards a better understanding of back pain: a review of the mechanics of the lumbar disk. Rheumatol. Rehabil. **14**, 129 (1975)
81. R. Nanda (ed.): *Biomechanics in Clinical Orthodontics*, (Saunders, Philadelphia 1997)

82. M. Nordin, V.H. Frankel (eds.): *Basic Biomechanics of the Musculoskeletal System*, 3rd edn (Lippincott Williams & Wilkins, Philadelphia 2001)
83. M. Nordin, S.S. Weiner, M. Lindh: Biomechanics of the lumbar spine. In *Basic Biomechanics of the Musculoskeletal System*, M. Nordin, V.H. Frankel, eds., 3rd edn (Lippincott Williams & Wilkins, Philadelphia 2001), Ch. 10, pp. 256–284
84. M. Nordin, V.H. Frankel: Biomechanics of the knee. In *Basic Biomechanics of the Musculoskeletal System*, 3rd edn, M. Nordin, V.H. Frankel, eds. (Lippincott Williams & Wilkins, Philadelphia 2001), Ch. 7, pp. 176–201
85. M. Nordin, V.H. Frankel: Biomechanics of the hip. In *Basic Biomechanics of the Musculoskeletal System*, 3rd edn, M. Nordin, V.H. Frankel, eds. (Lippincott Williams & Wilkins, Philadelphia 2001), Ch. 8, pp. 202–221
86. N. Ozkaya, M. Nordin: *Fundamentals of Biomechanics: Equilibrium, Motion, and Deformation*, 2nd edn (Springer, Berlin Heidelberg New York 1999)
87. W.R. Proffit, H.W. Fields Jr.: *Contemporary Orthodontics*, 3rd edn (Mosby, St. Louis 2000)
88. T.H. Schiebler: *Anatomie*, 9th edn (Springer, Berlin Heidelberg New York 2005)
89. R.F. Schmidt, F. Lang, G. Thews: *Physiologie des Menschen*, 29th edn (Springer, Berlin Heidelberg New York 2005)
90. F.G. Shaw, D.C. Scott: *Practical Exercises in Dental Mechanics*, 3rd edn (Henry Kimpton, London 1968)
91. C.J. Snijders: Engineering Approaches to Standing, Sitting, and Lying. In *Basic Biomechanics of the Musculoskeletal System*, M. Nordin, V.H. Frankel, eds., 3rd edn (Lippincott Williams & Wilkins, Philadelphia 2001), Ch. 17, pp. 420–436
92. H.J. Wilke, P. Neef, M. Caimi et al.: New in vivo measurements of pressures in the intervertbral disc in daily life. Spine **24**, 755 (1999)
93. B.N. Tillmann: *Atlas der Anatomie des Menschen*, (Springer, Berlin Heidelberg New York 2005)
94. M. Williams, H.B. Lissner: *Biomechanics of Human Motion* (Saunders, Philadelphia 1962)
95. D.A. Winter: *Biomechanics and Motor Control of Human Movement*, 3rd edn (Wiley, New York 2005)
96. H. Yamada: *Strength of Biological Materials* (Williams & Wilkins, Baltimore 1970)

Chapter 3

97. R.K. Adair: *The Physics of Baseball*, 3rd edn (HarperCollins, New York 2002)
98. R.K. Adair: The physics of baseball, Phys. Today 26–31 (May 1995)
99. R. McNeill Alexander: *The Human Machine* (Columbia University Press, New York 1992)
100. R. McNeill Alexander: Leg design and jumping technique for humans, other verterbrates and insects. Philos. Trans. R. Soc. **B347**, 235–248 (1995)
101. R. McNeill Alexander: Storage and release of elastic energy in the locomotor systems and the stretch-shortening cycle. In *Biomechanics and Biology of Movement*, B.M. Nigg, B.R. MacIntosh, J. Mester, eds. (Human Kinetics, Champaign, IL 2000), Ch. 2

102. T. Bahill, D.G. Baldwin: The rising fastball and other perceptual illusions of batters. In *Biomedical Engineering Principles in Sports*, G.K. Hung, J.M. Pallis, eds. (Kluwer/Plenum, New York 2004), Ch. 10, pp. 257–287
103. A.T. Bahill, W.J. Karnavas: Determining ideal baseball bat weights using muscle force–velocity relationships. Biol. Cybernet. **62**, 89–97 (1989)
104. A.E. Barr, S.I. Backus: Biomechanics of gait. In *Basic Biomechanics of the Musculoskeletal System*, 3rd edn, M. Nordin, V.H. Frankel, eds. (Lippincott Williams & Wilkins, Philadelphia 2001), Ch. 18, pp. 438–457
105. G.B. Benedek, F.M.H. Villars: *Physics with Illustrative Examples from Medicine and Biology*, Three volumes, 2nd edn (AIP/Springer, New York/Berlin Heidelberg New York 2000)
106. Braininjury.com
107. P.J. Brancazio: Swinging for the fences: the physics of baseball bats. New England Section of the American Physical Society meeting (October 1987)
108. H. Brody: The sweet spot of a baseball bat. Am. J. Phys. **54**, 640–643 (1986)
109. J.A. Buckwalter, T.A. Einhorn, S.R. Simon (eds.): *Orthopaedic Basic Science*, 2nd edn (American Academy of Orthopaedic Surgeons, Rosemont, IL 2000)
110. J.R. Cameron, J.G Skofronick, R. Grant: *Physics of the Body*, 2nd edn (Medical Physics, Madison, WI 1999)
111. P.R. Cavanagh (ed.): *Biomechanics of Distance Running* (Human Kinetics, Champaign, IL 1990)
112. P.R. Cavanagh, M.A. Lafortune: Ground reaction forces in distance running. J. Biomech. **13**, 397–406 (1980)
113. D.B. Chaffin, G.B.J. Andersson: *Occupational Biomechanics* (Wiley, New York 1984)
114. D.B. Chaffin, G.B.J. Andersson: *Occupational Biomechanics*, 2nd edn (Wiley, New York 1991)
115. D.B. Chaffin, G.B.J. Andersson, B.J. Martin: *Occupational Biomechanics*, 3rd edn (Wiley-Interscience, New York 1999)
116. T.M. Fraser: Human response to sustained linear acceleration. NASA SP-103 (NASA, Washington, DC 1966)
117. T.M. Fraser: Sustained linear acceleration. In *Bioastronautics Data Book*, J.F. Parker Jr., V.R. West, eds. (NASA, Washington, DC 1973), Ch. 4, pp. 149–190
118. T.M. Fraser: Rotary acceleration. In *Bioastronautics Data Book*, J.F. Parker Jr., V.R. West, eds. (NASA, Washington, DC 1973), Ch. 5, pp. 191–219
119. Y.C. Fung: *Biomechanics: Motion, Flow, Stress, and Growth* (Springer, Berlin Heidelberg New York 1990)
120. C.W. Gadd: Use of a weighted impulse criterion for estimating injury hazard. Tenth Stapp Car Crash Conference (Society of Automotive Engineers 1966), pp. 95–100
121. J.R. Gage: *Gait Analysis in Cerebral Palsy* (MacKeith, Oxford 1991)
122. T.J. Gay: *Football Physics: The Science of the Game* (Rodale, Emmaus, PA 2004)
123. Sir James Gray: *How Animals Move* (Pelican Books, Harmondsworth, England 1959)
124. E.S. Gurdjian, H.R. Lissner, F.R. Latimer, B.F. Haddad, J.E. Webster: Quantitative determination of acceleration and intracranial pressure in experimental head injury. Neurology **3**, 417–423 (1953)

125. E.S. Gurdjian, J.E. Webster, H.R. Lissner: Observations on the mechanism of brain concussion, contusion and laceration. Surg. Gynecol. Obstetr. **101**, 688–890 (1955)
126. J.G. Hay: Length changes of muscle-tendon units during athletic movements. In *Biomechanics and Biology of Movement*, B.M. Nigg, B.R. MacIntosh, J. Mester, eds. (Human Kinetics, Champaign, IL 2000), Ch. 3
127. J.G. Hay: *The Biomechanics of Sports Techniques*, 4th edn (Prentice-Hall, Englewood Cliffs, NJ 1993)
128. R.N. Hinrichs: Upper extremity function in distance running. In *Biomechanics of Distance Running*, P.R. Cavanagh, ed. (Human Kinetics, Champaign, IL 1990), Ch. 4, pp. 107–133
129. R.N. Hinrichs: Upper extremity function in running. II. Angular momentum considerations. Int. J. Sport Biomech. **3**, 258 (1987)
130. T.A. Hoyt, T.F. MacLaughlin, J.W. Kessler: Experimental pedestrian accident reconstructions – head impacts. Report no. DOT HS 807 288 (US National Technical Information Service, Springfield VA, 1988)
131. M.L. Hull, D.A. Hawkins: Analysis of work in multisegment movements: application to cycling. In *Multiple Muscle Systems*, J.M. Winters, S.L.Y. Woo, eds. (Springer, Berlin Heidelberg New York 1990), pp. 621–638
132. G.K. Hung, B. Johnson, A. Coppa: Aerodynamics and biomechanics of the free throw. In *Biomedical Engineering Principles in Sports*, G.K. Hung, J.M. Pallis, eds. (Kluwer/Plenum, New York 2004), Ch. 14, pp. 367–390
133. G.K. Hung, J.M. Pallis (eds.): *Biomedical Engineering Principles in Sports* (Kluwer/Plenum, New York 2004)
134. T.P. Jorgensen: *The Physics of Golf*, 2nd edn (AIP/Springer, New York/Berlin Heidelberg New York 1999)
135. A.I. King: Survey of the state of human biodynamic response. In *Aircraft Crashworthiness*, K. Saczalski, G.T. Singley III, W.D. Pilkey, R.L. Huston, eds. (University Press of Virgina, Charlottesville, VA 1975), pp. 83–120
136. P. Kirkpatrick: Batting the ball. Am. J. Phys. **31**, 606–613 (1963)
137. K. Laws: *The Physics of Dance*, (Schirmer Books, New York 1984)
138. K. Laws, C. Harvey, M. Swope: *Physics, Dance, and the Pas De Deux*, (Schirmer Books, New York 1994)
139. K. Laws, M. Swope: *Physics and the Art of Dance: Understanding Movement* (Oxford University Press, Oxford 2002)
140. J. Leavy: *Sandy Koufax: A Lefty's Legacy* (HarperCollins, New York 2002), pp. 4–5
141. R.S. Levine: Injury to the extremities. In *Accidental Injury: Biomechanics and Prevention*, A.M. Nahum, J.W. Melvin, eds., 2nd edn (Springer, Berlin Heidelberg New York 2002), Ch. 19, pp. 491–522
142. D. Lind, S.P. Sanders: *The Physics of Skiing: Skiing at the Triple Point* (AIP/Springer, New York/Berlin Heidelberg New York 1996)
143. H.R. Lissner, M. Lebow, F.G. Evans: Experimental studies on the relation between acceleration and intracranial pressure changes in man. Surg. Gynecol. Obstetr. **3**, 329–338 (1960)
144. P. Maróti, L. Berkes, F. Tölgyesi: *Biophysics Problems: A Textbook with Answers* (Akadémiai Kiadó, Budapest 1998)
145. I.S. McClay, M.J. Lake, P.R. Cavanagh: Muscle activity in running. In *Biomechanics of Distance Running*, P.R. Cavanagh, ed. (Human Kinetics, Champaign, IL 1990), Ch. 6, pp. 165–186

146. T.A. McMahon: *Muscles, Reflexes, and Locomotion* (Princeton University Press, Princeton, NJ 1984)
147. H.J. Mertz: Injury risk assessments based on dummy responses. In *Accidental Injury: Biomechanics and Prevention*, A.M. Nahum, J.W. Melvin, eds., 2nd edn (Springer, Berlin Heidelberg New York 2002), Ch. 5, pp. 89–102
148. H.J. Metcalf: *Topics in Classical Biophysics* (Prentice-Hall, Englewood Cliffs, NJ 1980)
149. D.I. Miller, R.C. Nelson: *Biomechanics of Sport: A Research Approach* (Lea & Febiger, Philadelphia 1973)
150. M.J. Milliron, P.R. Cavanagh: Sagittal plane kinematics of the lower extremity during distance running. In *Biomechanics of Distance Running*, P.R. Cavanagh, ed. (Human Kinetics, Champaign, IL 1990), Ch. 3, pp. 65–105
151. A.E. Minetti: The three modes of terrestrial locomotion. In *Biomechanics and Biology of Movement*, B.M. Nigg, B.R. MacIntosh, J. Mester, eds. (Human Kinetics, Champaign, IL 2000), Ch. 5
152. K.L. Moore, A.M.R. Agur: *Essential Clinical Anatomy*, 2nd edn (Lippincott Williams & Wilkins, Philadelphia 2002)
153. K.L. Moore, A.M.R. Agur: *Essential Clinical Anatomy*, 3rd edn (Lippincott Williams & Wilkins, Philadelphia 2007)
154. J.B. Morrison: The mechanics of the knee joint in relation to normal walking. J. Biomech. **3**, 51 (1970)
155. V.C. Mow, C.T. Hung: Biomechanics of articular cartilage. In *Basic Biomechanics of the Musculoskeletal System*, 3rd edn, M. Nordin, V.H. Frankel, eds. (Lippincott Williams & Wilkins, Philadelphia 2001), Ch. 3, pp. 60–97
156. A.M. Nahum, J.W. Melvin (eds.): *Accidental Injury: Biomechanics and Prevention*, 2nd edn (Springer, Berlin Heidelberg New York 2002)
157. A.M. Nathan: Characterizing the performance of baseball bats. Am. J. Phys. **71**, 134–143 (2003)
158. New York Times, Sunday, April 4, 2004, page 3, and the American Sport Medicine Institute
159. B.M. Nigg, B.R. MacIntosh, J. Mester (eds.): *Biomechanics and Biology of Movement* (Human Kinetics, Champaign, IL 2000)
160. M. Nordin, V.H. Frankel (eds.): *Basic Biomechanics of the Musculoskeletal System*, 3rd edn (Lippincott Williams & Wilkins, Philadelphia 2001)
161. E.L. Offenbacher: Physics and the vertical jump. Am. J. Phys. **38**, 829–836 (1970)
162. J. Perry: *Gait Aanlysis: Normal and Pathological Function* (SLACK, Thorofare, NJ 1992)
163. S. Pinker: *How The Mind Works* (Norton, New York 1997)
164. S. Plagenhoef: *Patterns of Human Motion – A Cinematographic Analysis* (Prentice-Hall, Englewood Cliffs, NJ 1971)
165. J. Rose, J.G. Gamble: *Human Walking*, 2nd edn (Lippincott Williams & Wilkins, Philadelphia 1994)
166. G.J. Sammarco, R.T. Hockenbury: Biomechanics of the foot and ankle. In *Basic Biomechanics of the Musculoskeletal System*, 3rd edn, M. Nordin, V.H. Frankel, eds. (Lippincott Williams & Wilkins, Philadelphia 2001), Ch. 9, pp. 222–255
167. R.A. Saul, J.F. Edlefson, K.L. Jarrett, J.R. Marous: Vehicle interactions with pedestrians. In *Accidental Injury: Biomechanics and Prevention*, A.M. Nahum,

J.W. Melvin, eds., 2nd edn (Springer, Berlin Heidelberg New York 2002), Ch. 22, pp. 588–617
168. J.M. Smith: *Mathematical Ideas in Biology* (Cambridge University Press, Cambridge 1971)
169. R.G. Snyder: Impact. In *Bioastronautics Data Book*, J.F. Parker Jr., V.R. West, eds. (NASA, Washington, DC 1973), Ch. 6, pp. 221–295
170. A. Stalheim-Smith, G.K. Fitch: *Understanding Human Anatomy and Physiology* (West, St. Paul, MN 1993)
171. D.J. Stefanyshyn, B.M. Nigg: *Work and Energy Influence by Athletic Equipment*, In *Biomechanics and Biology of Movement*, B.M. Nigg, B.R. MacIntosh, J. Mester, eds. (Human Kinetics, Champaign, IL, 2000)
172. J. Versace: A review of the severity index. Fifteenth Stapp Car Crash Conference (Society of Automotive Engineers 1971)
173. R.G. Watts, A.T. Bahill: Keep Your Eye on the Ball: Curve Balls, Knuckleballs, and Fallacies of Baseball (Freeman, New York 2000)
174. H.S. Weiss, R. Edelberg, P.V. Charland, J.I. Rosenbaum: The physiology of simple tumbing. Part 2. Human studies, WADC-TR-53-139 (Wright-Patterson Air Force Base, Ohio 1954)
175. K.R. Williams: A biomechanical evaluation of distance running efficiency, Unpublished PhD dissertation (Pennsylvania State University, University Park 1980), p. 157
176. N. Zheng, G.S. Flesig, S. Barrentine, J.R. Andrews: Biomechanics of pitching. In *Biomedical Engineering Principles in Sports*, G.K. Hung, J.M. Pallis, eds. (Kluwer/Plenum, New York 2004), Ch. 9, pp. 209–256

Chapter 4

177. H. Abé, K. Hayashi, M. Sato (eds.): *Data Book on Mechanical Properties of Living Cells, Tissues, and Organs* (Springer, Berlin Heidelberg New York 1996)
178. R. McNeill Alexander: *The Human Machine* (Columbia University Press, New York 1992)
179. S.W. Alpert, A. Ben-Yishay, K.J. Koval, J.D. Zuckerman: *Fractures and Dislocations: A Manual of Orthopaedic Trauma* (Lippincott-Raven, New York 1994)
180. D.D. Anderson, D.J. Adams, J.E. Hale: Mechanical effects of forces acting on bone, cartilage, ligaments, and tendons. In *Biomechanics and Biology of Movement*, B.M. Nigg, B.R. MacIntosh, J. Mester, eds. (Human Kinetics, Champaign, IL 2000), Ch. 16, pp. 283–306
181. T. Azuma, M. Hasegawa: Distensibility of the vein: from the architectural point of view. Biorheology **10**, 469–479 (1973)
182. G.B. Benedek, F.M.H. Villars: *Physics with Illustrative Examples from Medicine and Biology*, Three volumes, 2nd edn (AIP/Springer, New York/Berlin Heidelberg New York 2000)
183. J.A. Buckwalter, T.A. Einhorn, S.R. Simon (eds.): *Orthopaedic Basic Science*, 2nd edn (American Academy of Orthopaedic Surgeons, Rosemont, IL 2000)
184. A.H. Burstein, T.M. Wright: *Fundamentals of Orthopaedic Biomechanics* (Williams & Wilkins, Baltimore 1994)

185. W.D. Callister Jr.: *Materials Science and Engineering: An Introduction*, 5th edn (Wiley, New York 2000)
186. J.R. Cameron, J.G Skofronick, R. Grant: *Physics of the Body*, 2nd edn (Medical Physics, Madison, WI 1999)
187. D.R. Carter, W.E. Cater, D.M. Spengler, V.H. Frankel: Fatigue behavior of adult cortical bone: the influence of mean strain and strain range. Acta Orthop. Scand. **52**, 481–490 (1981)
188. D.B. Chaffin, G.B.J. Andersson: *Occupational Biomechanics* (Wiley, New York 1984)
189. J.D. Currey: *The Mechanical Adaptations of Bones* (Princeton University Press, Princeton 1984)
190. F.A. Duck: *Physical Properties of Tissue: A Comprehensive Reference Book* (Academic, London 1990)
191. J. Enderle, S. Blanchard, J. Bronzino: *Introduction to Biomedical Engineering* (Academic, San Diego 2000)
192. R.F. Fisher: The elastic constants of the human lens. J Physiol. **212**, 147–180 (1971)
193. R.F. Fisher: Presbyopia and the changes with age in the human crystalline lens. J Physiol. **228**, 765–779 (1973)
194. R.F. Fisher: The force of contraction of the human ciliary muscle during accommodation. J Physiol. **270**, 51–74 (1977)
195. W. Flügge: *Viscoelasticity*, 2nd edn (Springer, Berlin Heidelberg New York 1975)
196. V.H. Frankel, A.H. Burstein: *Orthopaedic Biomechanics* (Lea & Febiger, Philadelphia 1970)
197. V.H. Frankel, M. Nordin: Biomechanics of bone. In *Basic Biomechanics of the Musculoskeletal System*, 3rd edn, M. Nordin, V.H. Frankel, eds. (Lippincott Williams & Wilkins, Philadelphia 2001), Ch. 2, pp. 26–55
198. Y.C. Fung: *Biomechanics: Mechanical Properties of Living Tissues*, 1st edn (Springer, Berlin Heidelberg New York 1981)
199. Y.C. Fung: *Biomechanics: Mechanical Properties of Living Tissues*, 2nd edn (Springer, Berlin Heidelberg New York 1993)
200. M.A. Gomez, A.M. Nahum: Biomechanics of bone. In *Accidental Injury: Biomechanics and Prevention*, A.M. Nahum, J.W. Melvin, eds., 2nd edn (Springer, Berlin Heidelberg New York 2002), Ch. 10, pp. 206–227
201. R.W. Hertzberg: *Deformation and Fracture Mechanics of Engineering Materials*, 4th edn (Wiley, New York 1996)
202. M.H. Holmes, W.M. Lai, V.C. Mow: Compression effects on cartilage permeability. In *Tissue Nutrition and Viability*, A.R. Hargens, ed. (Springer, Berlin Heidelberg New York 1986), Ch. 4, pp. 73–100
203. J.D. Humphrey, S.L. Delange: *An Introduction to Biomechanics: Solids and Fluids, Analysis and Design* (Springer, Berlin Heidelberg New York 2004)
204. The Journal News, December 25, 2003
205. J. Kastelic, A. Galeski, E. Baer: The multi-composite structure of tendon. Connect. Tissue **6**, 11–23 (1978)
206. T.M. Keaveny, W.C. Hayes: Mechanical properties of cortical and trabecular bone. Bone **7**, 285–344 (1993)
207. J.C. Kennedy, R.J. Hawkins, R.B. Willis et al.: Tension studies of human knee ligaments: yield point, ultimate failure, and disruption of the cruciate and tibial colateral ligaments. J. Bone Joint Surg. **58A**, 350–355 (1976)

208. R.S. Levine: Injury to the extremities. In *Accidental Injury: Biomechanics and Prevention*, A.M. Nahum, J.W. Melvin, eds. 2nd edn (Springer, Berlin Heidelberg New York 2002), Ch. 19, pp. 491–522
209. A.B. Mann, M.E. Dickinson: Nanomechanics, chemistry and structure at the enamel surface. In *The Teeth and Their Environment*, R.M. Duckworth, ed. Monographs in Oral Science (Karger, Basel 2006), Vol. 19, Ch. 5, pp. 105–131
210. R.B. Martin, D.B. Burr, N.A. Sharkey: *Skeletal Tissue Mechanics* (Springer, Berlin Heidelberg New York 1998)
211. J.H. McElhaney: Dynamic response of bone and muscle tissue. J. Appl. Physiol. **21**, 1231–1236 (1966)
212. C. McGowan, *A Practical Guide to Vertebrate Mechanics* (Cambridge University Press, Cambridge 1999)
213. Merck Manual of Diagnosis and Therapy, on-line version
214. D.P. Miannay: *Fracture Mechanics* (Springer, Berlin Heidelberg New York 1998)
215. V.C. Mow, S.C. Kuei, W.M. Lai, C.G. Armstrong: Biphasic creep and stress relaxation of articular cartilage in compression: theory and experiment. J. Biomech. Eng. **102**, 73–84 (1980)
216. V.C. Mow, W.C. Hayes (eds.): *Basic Orthopaedic Biomechanics*, 2nd edn (Lippincott-Raven, New York 1997)
217. V.C. Mow, C.T. Hung: Biomechanics of articular cartilage. In *Basic Biomechanics of the Musculoskeletal System*, 3rd edn, M. Nordin, V.H. Frankel, eds. (Lippincott Williams & Wilkins, Philadelphia 2001), Ch. 3, pp. 60–97
218. A.M. Nahum, J.W. Melvin (eds.): *Accidental Injury: Biomechanics and Prevention*, 2nd edn (Springer, Berlin Heidelberg New York 2002)
219. J.A. Newman: Biomechanics of head trauma: head Protection. In *Accidental Injury: Biomechanics and Prevention*, 2nd edn, A.M. Nahum, J.W. Melvin, eds. (Springer, Berlin Heidelberg New York 2002), Ch. 14, pp. 303–323
220. B.M. Nigg, W. Herzog (eds.): *Biomechanics of the Musculo-Skeletal System*, 2nd edn (Wiley, New York 1999)
221. B.M. Nigg, B.R. MacIntosh, J. Mester (eds.): *Biomechanics and Biology of Movement* (Human Kinetics, Champaign, IL 2000)
222. M. Nordin, T. Lorenz, M. Campello: Biomechanics of tendons and ligaments. In *Basic Biomechanics of the Musculoskeletal System*, 3rd edn, M. Nordin, V.H. Frankel, eds. (Lippincott Williams & Wilkins, Philadelphia 2001), Ch. 4, pp. 102–120
223. M. Nordin, V.H. Frankel (eds.): *Basic Biomechanics of the Musculoskeletal System*, 3rd edn (Lippincott Williams & Wilkins, Philadelphia 2001)
224. F.R. Noyes: Functional properties of knee ligaments and alterations induced by immobilization. Clin. Orthop. **123**, 210–242 (1977)
225. P.E. Pack: *Anatomy and Physiology* (Cliff Notes, Lincoln, NE 1997)
226. W.D. Pilkey: *Formulas for Stress, Strain, and Structural Matrices*, 2nd edn (Wiley, Hoboken, NJ 2005)
227. E.L. Radin, I.L. Paul, M. Lowy: A comparison of the dynamic force transmitting properties of subchondral bone and articular cartilage. J. Bone Joint Surg. **52A**, 444–456 (1970)
228. B.D. Ratner, A.S. Hoffman, F.J. Schoen, J.E. Lemons (eds.): *Nanomedicine, Biomaterials Science: An Introduction to Materials in Medicine* (Academic, San Diego 1996)

229. G.J. Sammarco, A.H. Burstein, W.L. Davis, V.H. Frankel: The biomechanics of torsional fractures: the effect of loading on ultimate properties. J. Biomech. **4**, 113–117 (1971)
230. R.J. Sanford: *Principles of Fracture Mechanics* (Pearson, Upper Saddle River, NJ 2003)
231. T.H. Schiebler: *Anatomie*, 9th edn (Springer, Berlin Heidelberg New York 2005)
232. R. Skalak, S. Chien (eds.): *Handbook of Bioengineering* (McGraw-Hill, New York 1987)
233. B.N. Tillmann: *Atlas der Anatomie des Menschen*, (Springer, Berlin Heidelberg New York 2005)
234. J. Wolff (translated by P. Maquet, R. Furlong): *The Law of Bone Remodelling* (Springer, Berlin Heidelberg New York 1986)
235. S.L.-Y. Woo, J.M. Hollis, D.J. Adams, R.M. Lyon, S. Takai: Tensile properties of the human femur-anterior cruciate ligament-tibia complex: the effect of specimen age and orientation. Am. J. Sports Med. **19**, 217–225 (1991)
236. H. Yamada: *Strength of Biological Materials* (Williams & Wilkins, Baltimore 1970)

Chapter 5

237. H. Abé, K. Hayashi, M. Sato (eds.): *Data Book on Mechanical Properties of Living Cells, Tissues, and Organs* (Springer, Berlin Heidelberg New York 1996)
238. R. McNeill Alexander, A. Vernon: The dimensions of the knee and ankle muscles and the forces they exert. J. Hum. Mov. Stud. **1**, 115–123 (1975)
239. R. McNeill Alexander: *The Human Machine* (Columbia University Press, New York 1992)
240. X. Aubert, M.L. Rouquet, J. Van der Elst: The tension–length diagram of the frog's sartorius muscle. Arch. Int. Physiol. **59**, 239–241 (1951)
241. A.E. Barr, S.I. Backus: Biomechanics of gait. In *Basic Biomechanics of the Musculoskeletal System*, 3rd edn, M. Nordin, V.H. Frankel, eds. (Lippincott Williams & Wilkins, Philadelphia 2001), Ch. 18, pp. 438–457
242. S.M. Block, L.S.B. Goldstein, B.J. Schnapp: Bead movement by single kinesin molecules studied with optical tweezers. Nature **348**, 348–352 (1990)
243. D.B. Chaffin, G.B.J. Andersson: *Occupational Biomechanics* (Wiley, New York 1984)
244. D.B. Chaffin, G.B.J. Andersson, B.J. Martin: *Occupational Biomechanics*, 3rd edn (Wiley-Interscience, New York 1999)
245. R.M.J. Cotterill: *Biophysics: An Introduction* (Wiley, New York 2002)
246. J. Enderle, S. Blanchard, J. Bronzino: *Introduction to Biomedical Engineering* (Academic, San Diego 2000)
247. J.T. Finer, A.D. Mehta, J.A. Spudich: Characterization of single actin–myosin interactions. Biophys. J. **68**, 291s–297s (1995)
248. S.J. Fleck, W.J. Kraemer: *Designing Resistance Training Programs*, 2nd edn (Human Kinetics, Champaign, IL 1997)
249. L.E. Ford, A.F. Huxley, R.M. Simmons: Tension responses to sudden length change in stimulated frog muscle fibers near slack length. J. Physiol. **269**, 441–515 (1977)

250. M.L. Foss, S.J. Keteyian: *Fox's Physiological Basis for Exercise and Sport*, 6th edn (McGraw-Hill, Boston 1998)
251. E.L. Fox: *Sports Physiology*, 2nd edn (Saunders College, Philadelphia 1984)
252. R.A. Freitas Jr.: *Nanomedicine, Volume I: Basic Capabilities* (Landes Bioscience, Austin 1999)
253. S.C. Gandevia, R.M. Enoka, A.J. McComas, D.G. Stuart, C.K. Thomas: *Fatigue Neural and Muscular Mechanisms, Advances in Experimental Medicine and Biology*, Vol. 384 (Plenum, New York 1995)
254. A.C. Guyton, J.E. Hall: *Human Physiology and Mechanisms of Disease*, 6th edn (Saunders, Philadelphia 1997)
255. D.D. Hackney: The kinetic cycles of myosin, kinesin, and dynein. Annu. Rev. Physiol. **58**, 731–750 (1996)
256. W. Herzog: Muscle. In *Biomechanics of the Musculo-Skeletal System*, 2nd edn, B.M. Nigg, W. Herzog, eds. (Wiley, New York 1999) Sect. 2.7, pp. 148–188
257. W. Herzog, T.R. Leonard: Validation of optimization models that estimate forces exerted by synergistic muscles. J. Biomech. **24(S1)**, 31–39 (1991)
258. W. Herzog, V. Zatsiorsky, V. Prilutsky, T.R. Leonard: Variations in force–time histories of cat gastrocnemius, soleus, and plantaris muscles for consecutive walking steps. J. Exp. Biol. **191**, 19–36 (1994)
259. F.C. Hoppensteadt, C.S. Peskin: *Modeling and Simulation in Medicine and the Life Sciences*, 2nd edn (Springer, Berlin Heidelberg New York 2002) (First Edition from 1992 was called *Mathematics in Medicine and the Life Sciences*)
260. J. Howard: The movement of kinesin along microtubules. Annu. Rev. Physiol. **58**, 703–729 (1996)
261. J. Howard: *Mechanics of Motor Proteins and the Cytoskeleton* (Sinauer, Sunderland, MA 2001)
262. H.E. Huxley: Electron microscope studies on the structure of natural and synthetic protein filaments from striated muscle. J. Mol. Biol **7**, 281–308 (1963)
263. H.E. Huxley: The mechanism of muscle contraction. Science **164**, 1356–1366 (1969)
264. A.F. Huxley, R.M. Simmons: Proposed mechanism of force generation in striated muscle. Nature **233**, 533–538 (1971)
265. M. Ikai, T. Fukunaga: Calculation of muscle strength per unit cross-sectional area of human muscle by means of ultrasonic measurements. Int. Z. Angew. Physiol. **26**, 26–32 (1968)
266. J. Keener, J. Sneyd: *Mathematical Physiology* (Springer, Berlin Heidelberg New York 1998)
267. K. Kitamura, M. Tokunaga, A.H. Iwane, T. Yanagida: A single myosin head moves along an actin filament with regular steps of 5.3 nanometers. Nature **397**, 129–134 (1999)
268. W.J. Kraemer, K. Adams, E. Cafarelli, G.A. Dudley, C. Dooly, M. S. Feigenbaum, S.J. Fleck, B. Franklin, A.C. Fry, J.R. Hoffman, R. U. Newton, J. Potteiger, M.H. Stone, N.A. Ratamess, T. Triplett-McBride: ACSM position stand: progression models in resistance training for healthy adults. Med. Sci. Sports Exercise **34**, 364–380 (2002)
269. T. Lorenz, M. Campello, M.J. Pitman, L. Peterson: Biomechanics of skeletal muscle. In *Basic Biomechanics of the Musculoskeletal System*, 3rd edn M. Nordin, V.H. Frankel, eds. (Lippincott Williams & Wilkins, Philadelphia 2001), Ch. 6, pp. 148–174

270. B.R. MacIntosh, R.J. Holash: Power output and force–velocity properties of muscle. In *Biomechanics and Biology of Movement*, B.M. Nigg, B.R. MacIntosh, J. Mester, eds. (Human Kinetics, Champaign, IL 2000), Ch. 11
271. T.A. McMahon: *Muscles, Reflexes and Locomotion* (Princeton University Press, Princeton, NJ 1984)
272. A.J. McComas: *Skeletal Muscle: Form and Function* (Human Kinetics, Champaign, IL 1996)
273. H.J. Metcalf: *Topics in Classical Biophysics* (Prentice-Hall, Englewood Cliffs, NJ 1980)
274. K.L. Moore, A.M.R. Agur: *Essential Clinical Anatomy*, 2nd edn (Lippincott Williams & Wilkins, Philadelphia 2002)
275. K.L. Moore, A.M.R. Agur: *Essential Clinical Anatomy*, 3rd edn (Lippincott Williams & Wilkins, Philadelphia 2007)
276. B.M. Nigg, W. Herzog (eds.): *Biomechanics of the Musculo-Skeletal System*, 2nd edn (Wiley, New York 1999)
277. M. Nordin, V.H. Frankel (eds.): *Basic Biomechanics of the Musculoskeletal System*, 3rd edn (Lippincott Williams & Wilkins, Philadelphia 2001)
278. J.L. Palladino, A. Noordergraaf: Muscle contraction mechanics from ultrastructural dynamics. In *Analysis and Assessment of Cardiovascular Function*, G.M. Drzewiecki, J.K.-J. Li, eds. (Springer, Berlin Heidelberg New York 1998), Ch. 3, pp. 33–57
279. T.D. Pollard, S.K. Doberstein, H.G. Zot: Myosin-I. Annu. Rev. Physiol. **53**, 653–681 (1991)
280. T.H. Schiebler: *Anatomie*, 9th edn (Springer, Berlin Heidelberg New York 2005)
281. R.F. Schmidt, F. Lang, G. Thews: *Physiologie des Menschen*, 29th edn (Springer, Berlin Heidelberg New York 2005)
282. T.A. Schroer, M.P. Sheetz: Functions of microtubule-based motors. Annu. Rev. Physiol. **53**, 629–652 (1991)
283. J.M. Squire (ed.): *Molecular Mechanisms in Muscular Contraction* (CRC, Boca Raton, FL 1990)
284. A. Stalheim-Smith, G.K. Fitch: *Understanding Human Anatomy and Physiology* (West, St. Paul, MN 1993)
285. W. von Rohmert: Arbeitsmedizin, Sozial medizin, Arbeitshygiene **22**, 118 (1968)
286. A. Thorstensson, G. Grimby, J. Karlsson: Force–velocity relations and fiber composition in human knee extensor muscles. J. Appl. Physiol. **40**, 12–16 (1976)
287. R.D. Vale: The molecular motor toolbox for intracellular transport. Cell **112**, 467–480 (2003)
288. P.A. Wahr, J.M. Metzger: Peak power output is maintained in rabbit psoas and rat soleus singe muscle fibers when CTP replaces ATP. J. Appl. Physiol. **85**, 76–83 (1998)
289. T.L. Wickiewcz, R.R. Roy, P.L. Powel, V.R. Edgerton: Muscle architecture of the human lower limb. Clin. Orthop. Rel. Res. **179**, 275–283 (1983)
290. J.J. Widrick, J.G. Romatowski, M. Karhanek, R.H. Fitts: Contractile properties of rat, rhesus monkey, and human type I muscle fibers. Am. J. Physiol. Regul. Integr. Comp. Physiol. **272**, R34–R42 (1997)
291. J.H. Wilmore, D.L. Costill: *Physiology of Sport and Exercise*, 3rd edn (Human Kinetics, Champaign, IL 2004)

292. D.A. Winter: *Biomechanics and Motor Control of Human Movement*, 3rd edn (Wiley, New York 2005)

Chapter 6

293. B.K. Ahlborn: *Zoological Physics: Quantitative Models, Body Design, Actions and Physical Limitations in Animals* (Springer, Berlin Heidelberg New York 2004)
294. R. McNeill Alexander: *The Human Machine* (Columbia University Press, New York 1992)
295. W.O. Atwater, F.G. Benedict: *Experiments on the Metabolism of Matter and Energy in the Human Body*. Agricultural Bulletin No. 136 (U.S. Department of Agriculture, Washington, DC 1903)
296. G.B. Benedek, F.M.H. Villars: *Physics with Illustrative Examples from Medicine and Biology*, Three volumes, 2nd edn (AIP/Springer, New York/Berlin Heidelberg New York 2000)
297. P.J. Berenson, W.G. Robertson: Temperature. In *Bioastronautics Data Book*, J.F. Parker Jr., V.R. West, eds. (NASA, Washington, DC 1973), Ch. 3, pp. 65–148
298. K. Blaxter: *Energy Metabolism in Animals and Man* (Cambridge University Press, Cambridge 1989)
299. R.W. Bullard: Temperature regulation. In *Physiology*, 4th edn, E.E. Selkurt, ed. (Little Brown, Boston 1976), p. 689
300. J.R. Cameron, J.G Skofronick, R. Grant: *Physics of the Body*, 2nd edn (Medical Physics, Madison, WI 1999)
301. N.A. Campbell, J.B. Reece: *Biology*, 7th edn (Pearson-Benjamin Cummings, San Francisco 2005)
302. K. Cena, J.A. Clark: Thermal insulation of animal coats and human clothing. Phys. Med. Biol. **23**, 565–591 (1978)
303. J.L. Clapperton, J.P. Joyce, K.L. Blaxter: Estimates of the contribution of solar radiation to the thermal exchanges of sheep at a latitude of 55°N. J. Agri. Sci. (Cambridge) **64**, 37–49 (1965)
304. P.S. Dodds, D.H. Rothman, J.S. Weitz: Re-examination of the 3/4-law of metabolism. J. Theor. Biol. **209**, 9–27 (2001)
305. FAO/WHO/UNU: *Energy and Protein Requirements. Report of a Joint Experts Consultation*. WHO Technical Report Series No. 724 (World Health Organization, Geneva 1985)
306. M.L. Foss, S.J. Keteyian: *Fox's Physiological Basis for Exercise and Sport*, 6th edn (McGraw-Hill, Boston 1998)
307. E.L. Fox, D.K. Matthews: *Interval Training Conditioning for Sport and General Fitness* (Saunders, Philadelphia 1974)
308. R.A. Freitas Jr.: *Nanomedicine, Volume I: Basic Capabilities* (Landes Bioscience, Austin 1999)
309. J.S. Garrow: Resting metabolic rate as a determinant of energy expenditure in man. In *Substrate and Energy Metabolism in Man*, J.S. Garrow, D. Halliday, eds. (Libbey, London 1985), pp. 102–107
310. R.C. Garry, R. Passmore, G.M. Warnock, J.V.G.A. Durnin: Expenditure of energy and the consumption of food by miners and clerks, Fife, Scotland. Medical Research Council Special Report Series 289 (HMSO, London 1955)

311. S.E. Gebhardt, R.G. Thomas: *Nutritive Value of Foods*, U.S. Department of Agriculture, Agricultural Research Service, Home and Garden Bulletin 72 (2002)
312. J.F. Gillooly, J.H. Brown, G.B. West, V.M. Savage, E.L. Charnov: Effects of size and temperature on metabolic rates. Science **293**, 2248–2251 (2001)
313. C.V. Gisolfi, C.B. Wenger: Temperature regulation during excerise: old concepts, new ideas. Exercise Sport Sci. Rev. **12**, 339–372 (1984)
314. A.C. Guyton, J.E. Hall: *Human Physiology and Mechanisms of Disease*, 6th edn (Saunders, Philadelphia 1997)
315. W.V. Judy: Body temperature regulation. In *Physiology*, 5th edn, E.E. Selkurt, ed. (Little Brown, Boston 1984), Ch. 28
316. M.C.K. Khoo: *Physiological Control Systems: Analysis, Simulation, and Estimation* (IEEE, Piscataway, NJ 2000)
317. M. Kleiber: Body size and metabolism. Hilgardia **6**, 315–353 (1932)
318. M. Kleiber: Body size and metabolic rate. Physiol. Rev. **27**, 511–541 (1947)
319. M. Kleiber: *The Fire of Life: An Introduction to Animal Energetics* (Wiley, New York, 1961)
320. A.L. Lehninger: *Biochemisty: The Molecular Basis of Cell Structure and Function*, 2nd edn (Worth, New York 1975)
321. H. Lodish, A. Berk, P. Matsudaira, C.A. Kaiser, M. Krieger, M.P. Scott, S.L. Zipursky, J. Darnell: *Molecular Cell Biology*, 5th edn (Freeman, New York 2004)
322. A.G. Marshall: *Biophysical Chemistry* (Wiley, New York 1978)
323. T.A. McMahon: Size and shape in biology. Science **179**, 1201–1204 (1973)
324. T.A. McMahon: *Muscles, Reflexes, and Locomotion* (Princeton University Press, Princeton, NJ 1984)
325. H.J. Metcalf: *Topics in Classical Biophysics* (Prentice-Hall, Englewood Cliffs, NJ 1980)
326. J.L. Monteith: *Principles of Environmental Physics* (Edward Arnold, London 1973)
327. L.E. Mount: *Adaptation to the Thermal Environment* (Edward Arnold, London 1979)
328. L.E. Mount. In *Heat Loss from Animals and Man*, J.L. Monteith, L.E. Mount, eds. (Butterworths, London 1974)
329. D.L. Nelson, M.M. Cox: *Lehninger Principles of Biochemisty*, 4th edn (Worth, New York 2005)
330. R. Passmore, J.V.G.A. Durnin: *Energy, Work and Leisure* (Heinemann Educational Books, London 1967)
331. A. Pietrobelli, D.B. Allison, S. Heshka, M. Heo, Z.M. Wang, A. Bertkau, B. Laferrere, M. Rosenbaum, J.F. Aloia, F.X. Pi-Sunyer, S. B. Heymsfield: Sexual dimorphism in the energy content of weight change. Int. J. Obesity **26**, 1339–1348 (2002)
332. P.E. di Prampero: The energy cost of human locomotion on land and in water. Int. J. Sports Med. **7**, 55–72 (1986)
333. L.G.C.E. Pugh: The influence of wind resistance in running and walking and the efficiency of work against horizontal and vertical forces. J. Physiol.(London) **48**, 518–522 (1971)
334. R.M. Ross, A.S. Jackson: *Exercise Concepts, Calculations, and Computer Applications* (Benchmark, Carmel, IN 1990)

335. W.N. Schofield: Predicting basal metabolic rate, new standards and review of previous work. Hum. Nutr.: Clin. Nutr. **39C, Suppl. 1**, 5–41 (1985)
336. A. Stalheim-Smith, G.K. Fitch: *Understanding Human Anatomy and Physiology* (West, St. Paul, MN 1993)
337. R.G. Steadman: The assessment of sultriness. Part I. A temperature–humidity index based on human physiology and clothing science. J. Appl. Meteorol. **18**, 861–873 (1979)
338. R.B. Stull: *Meteorology for Scientists and Engineers*, 2nd edn (Brooks Cole, Pacific Grove, CA 1999)
339. W.C. Swinbank: Long wave radiation from clear skies. Q. J. R. Meterol. Soc. **89**, 339–351 (1963)
340. J.H. Wilmore, D.L. Costill: *Physiology of Sport and Exercise*, 3rd edn (Human Kinetics, Champaign, IL 2004)
341. D.A. Winter: *Biomechanics and Motor Control of Human Movement*, 3rd edn (Wiley, New York 2005)
342. M. Wishnofsky: Caloric equivalents of gained or lost weight. Am. J. Clin. Nutr. **6**, 542–546 (1958)

Chapter 7

343. B.K. Ahlborn: *Zoological Physics: Quantitative Models, Body Design, Actions and Physical Limitations in Animals* (Springer, Berlin Heidelberg New York 2004)
344. G.B. Benedek, F.M.H. Villars: *Physics with Illustrative Examples from Medicine and Biology*, Three volumes, 2nd edn (AIP/Springer, New York/Berlin Heidelberg New York 2000)
345. J.R. Cameron, J.G Skofronick, R. Grant: *Physics of the Body*, 2nd edn (Medical Physics, Madison, WI 1999)
346. C.G. Caro, T.J. Pedley, R.C. Schroter, W.A. Seed: *Mechanics of the Circulation* (Oxford University Press, Oxford 1978)
347. J.E. Counsilman, B.E. Counsilman: *The New Science of Swimming* (Prentice-Hall, Englewood Cliffs, NJ 1994)
348. J.D. Cutnell, K.W. Johnson: *Physics*, 3rd edn (Wiley, New York 1995)
349. G.M. Drzewiecki: Modeling of noninvasive arterial blood pressure methods. In *Analysis and Assessment of Cardiovascular Function*, G. Drzewiecki, J.J.-K. Li, eds. (Springer, Berlin Heidelberg New York 1998), Ch. 13, pp. 203–221
350. W. Flügge: *Viscoelasticity*, 2nd edn (Springer, Berlin Heidelberg New York 1975)
351. R.A. Freitas Jr.: *Nanomedicine, Volume I: Basic Capabilities* (Landes Bioscience, Austin 1999)
352. Y.C. Fung: *Biomechanics: Mechanical Properties of Living Tissues*, 1st edn (Springer, Berlin Heidelberg New York 1981)
353. Y.C. Fung: *Biomechanics: Motion, Flow, Stress, and Growth* (Springer, Berlin Heidelberg New York 1990)
354. Y.C. Fung: *Biomechanics: Mechanical Properties of Living Tissues*, 2nd edn (Springer, Berlin Heidelberg New York 1993)
355. H.L. Goldsmith, V.T. Turitto: Rheological aspectsof thrombosis and haemostastis: basic principles and applications. Thromb. Haemostasis **55**, 415–435 (1986)

356. A.C. Guyton, J.E. Hall: *Human Physiology and Mechanisms of Disease*, 6th edn (Saunders, Philadelphia 1997)
357. G.J. Hademenos, T.F. Massoud: *The Physics of Cerebrovascular Diseases* (Springer/AIP, Berlin Heidelberg New York/New York, 1998)
358. F.R. Hallett, R.H. Stinson, P.A. Speight: *Physics for the Biological Sciences: A Topical Approach to Biophysical Concepts* (Methuen/Chapman & Hall, Toronto 1982)
359. J.G. Hay: *The Biomechanics of Sports Techniques*, 4th edn (Prentice-Hall, Englewood Cliffs, NJ 1993)
360. R.K. Hobbie: *Intermediate Physics for Medicine and Biology*, 3rd edn (AIP, New York 1997)
361. J.D. Humphrey, S.L. Delange: *An Introduction to Biomechanics: Solids and Fluids, Analysis and Design* (Springer, Berlin Heidelberg New York 2004)
362. N.C. Korotkoff: On the subject of methods of determining blood pressure. Imperial Mil. Med. Acad. St. Petersburg **11**, 365–367 (1905)
363. B.D. Ratner, A.S. Hoffman, F.J. Schoen, J.E. Lemons (eds.): *Nanomedicine, Biomaterials Science: An Introduction to Materials in Medicine* (Academic, San Diego 1996)
364. R.F. Schmidt, F. Lang, G. Thews: *Physiologie des Menschen*, 29th edn (Springer, Berlin Heidelberg New York 2005)

Chapter 8

365. B.K. Ahlborn: *Zoological Physics: Quantitative Models, Body Design, Actions and Physical Limitations in Animals* (Springer, Berlin Heidelberg New York 2004)
366. H. Barcroft, A.C. Dornhorst: The blood flow through the human calf during rhythmic exercise. J. Physiol. **109**, 402–411 (1949)
367. G.B. Benedek, F.M.H. Villars: *Physics with Illustrative Examples from Medicine and Biology*, Three volumes, 2nd edn (AIP/Springer, New York/Berlin Heidelberg New York 2000)
368. R.M. Berne, M.N. Levy: *Cardiovascular Physiology*, 8th edn (Mosby, St. Louis 2001)
369. B.H. Brown, R.H. Smallwood, D.C. Barber, P.V. Lawford, D.R. Hose: *Medical Physics and Biomedical Engineering* (Institute of Physics, Philadelphia 1999)
370. R. Burattini: Identification and physiological interpretation of aortic impedance modeling. In *Modelling Methodology for Physiology and Medicine*, E. Carson, C. Cobelli, eds. (Academic, San Diego 2001), Ch. 8, pp. 213–252
371. J.R. Cameron, J.G Skofronick, R. Grant: *Physics of the Body*, 2nd edn (Medical Physics, Madison, WI 1999)
372. C.G. Caro, T.J. Pedley, R.C. Schroter, W.A. Seed: *The Mechanics of the Circulation* (Oxford University Press, Oxford 1978)
373. E. Carson, C. Cobelli: *Modelling Methodology for Physiology and Medicine* (Academic, San Diego 2001)
374. Y.-B. Chen, R.R. Liberthson, M.D. Freed: Congenital heart disease. In *Pathophysiology of Heart Disease: A Collaborative Project of Medical Students and Faculty*, 3rd edn, L.S. Lilly, ed. (Lippincott Williams & Wilkins, Philadelphia 2003), Ch. 16, pp. 347–370

375. S. Chien: Shear dependence of effective cell volume as a determinant of blood viscosity. Science **168**, 977–979 (1970)
376. U. Dinnar: *Cardiovascular Fluid Dynamics* (CRC, Boca Raton, FL 1981)
377. G.M. Drzewiecki: Modeling of noninvasive arterial blood pressure methods. In *Analysis and Assessment of Cardiovascular Function*, G. Drzewiecki, J.J.-K. Li, eds. (Springer, Berlin Heidelberg New York 1998), Ch. 13, pp. 203–221
378. G. Drzewiecki, J.J.-K. Li (eds.): *Analysis and Assessment of Cardiovascular Function* (Springer, Berlin Heidelberg New York 1998)
379. D. Dubin: *Rapid Interpretation of EKG's*, 6th edn (COVER, Tampa, FL 2000)
380. G.S.M. Dyer, M.A. Fifer: Heart failure. In *Pathophysiology of Heart Disease: A Collaborative Project of Medical Students and Faculty*, 3rd edn, L.S. Lilly, ed. (Lippincott Williams & Wilkins, Philadelphia 2003), Ch. 9, pp. 211–236
381. M.L. Foss, S.J. Keteyian: *Fox's Physiological Basis for Exercise and Sport*, 6th edn (McGraw-Hill, Boston 1998)
382. R.A. Freitas Jr.: *Nanomedicine, Volume I: Basic Capabilities* (Landes Bioscience, Austin 1999)
383. Y.C. Fung: *Biodynamics: Circulation* (Springer, Berlin Heidelberg New York 1984)
384. Y.C. Fung: *Biomechanics: Motion, Flow, Stress, and Growth* (Springer, Berlin Heidelberg New York 1990)
385. Y.C. Fung: *Biomechanics: Circulation*, 2nd edn (Springer, Berlin Heidelberg New York 1997)
386. M. Gertsch, *The ECG: A Two-Step Approach to Diagnosis* (Springer, Berlin Heidelberg New York 2004)
387. D.C. Giancoli: *Physics: Principles with Applications*, 4th edn (Prentice-Hall, Englewood Cliffs, NJ 1995)
388. J.F. Green: *Fundamental Cardiovascular and Pulmonary Physiology*, 2nd edn (Lea & Febiger, Phildelphia, 1987)
389. A.C. Guyton, C.E. Jones, T.G. Coleman: *Circulatory Physiology: Cardiac Output and its Regulation* (Saunders, Philadelphia 1973)
390. A.C. Guyton, J.E. Hall: *Human Physiology and Mechanisms of Disease*, 6th edn (Saunders, Philadelphia 1997)
391. G.J. Hademenos, T.F. Massoud: *The Physics of Cerebrovascular Diseases* (Springer/AIP Berlin Heidelberg New York/New York 1998)
392. G.J. Hademenos, T.F. Massoud, D.J. Valentino, G.R. Duckwiler, F. Vinuela: A mathematical model for the development and rupture of intracranial saccular aneurysms based on a biomathematical model. Neurol. Res. **16**, 376–384 (1994)
393. G.J. Hademenos: The physics of cerebral aneurysms. Phys. Today **48**, 24–30 (1995)
394. F.C. Hoppensteadt, C.S. Peskin: *Modeling and Simulation in Medicine and the Life Sciences*, 2nd edn (Springer, Berlin Heidelberg New York 2002) (First Edition from 1992 was called *Mathematics in Medicine and the Life Sciences*)
395. J.D. Humphrey: *Cardiovascular Solid Mechanics: Cells, Tissues, and Organs* (Springer, Berlin Heidelberg New York 2002)
396. J.D. Humphrey, S.L. Delange: *An Introduction to Biomechanics: Solids and Fluids, Analysis and Design* (Springer, Berlin Heidelberg New York 2004)
397. T. Karino, M. Motomiya, H.L. Goldsmith: Flow patterns in model and natural vessels. In *Biological and Synthetic Vascular Prostheses*, J.C. Stanley, ed.

Annals of the New York Academy of Science (Grune & Stratton, New York 1982)

398. T. Karino, H.L. Goldsmith, M. Motomiya, S. Mabuchi, Y. Sohara: Flow patterns in vessels of simple and complex geometries. In *Blood in Contact with Natural and Artifical Surfaces*, E.F. Leonard, V.T. Turrito, L. Vroman, eds. Annals of the New York Academy of Science, Vol. 516, pp. 422–441 (1987)
399. H.L. Klawans: *Toscanini's Fumble and Other Tales of Clinical Neurology* (Contemporary Books, Chicago 1988)
400. J.M. Kotchen, H.E. McKean, T.A. Kotchen: Blood pressure trends with aging. Hypertension **4, Suppl. 3**, 111–129 (1982)
401. L.S. Lilly: The electrocardiogram. In *Pathophysiology of Heart Disease: A Collaborative Project of Medical Students and Faculty*, 3rd edn, L.S. Lilly, ed. (Lippincott Williams & Wilkins, Philadelphia 2003), Ch. 4, pp. 75–110
402. L.S. Lilly (ed.): *Pathophysiology of Heart Disease: A Collaborative Project of Medical Students and Faculty*, 3rd edn (Lippincott Williams & Wilkins, Philadelphia 2003)
403. R. Malholtra, G.H. Williams, L.S. Lilly: Hypertension. In *Pathophysiology of Heart Disease: A Collaborative Project of Medical Students and Faculty*, 3rd edn, L.S. Lilly, ed. (Lippincott Williams & Wilkins, Philadelphia 2003), Ch. 13, pp. 289–310
404. J. Malmivuo, R. Plonsey: *Bioelectromagnetism* (Oxford University Press, New York 1995)
405. P. Maróti, L. Berkes, F. Tölgyesi: *Biophysics Problems: A Textbook with Answers* (Akadémiai Kiadó, Budapest 1998)
406. H.J. Metcalf: *Topics in Classical Biophysics* (Prentice-Hall, Englewood Cliffs, NJ 1980)
407. C.J. Mills, I.T. Gabe, J.H. Gault, D.T. Mason, J. Ross Jr., E. Braunwald, J.P. Shillingford: Pressure-flow relationships and vascular impedance in man. Cardiovasc. Res. **4**, 405–417 (1970)
408. K.L. Moore, A.M.R. Agur: *Essential Clinical Anatomy*, 2nd edn (Lippincott Williams & Wilkins, Philadelphia 2002)
409. M. Motomiya, T. Karino: Flow patterns in the human carotid artery bifurcation. Stroke **15**, 50–56 (1984)
410. H.D. Patton, A.F. Fuchs, B. Hille, A.M. Scher, R. Steiner (eds.): In *Textbook of Physiology*, 21st edn (Saunders, Philadelphia 1989)
411. R.M. Ross, A.S. Jackson: *Exercise Concepts, Calculations, and Computer Applications* (Benchmark, Carmel, IN 1990)
412. R. Rushmer: *Cardiovascular Dynamics* (Saunder, Philadelphia 1970), p. 196
413. K. Sagawa: Analysis of the CNS ischemic feed back regulation of the circulation. In *Physical Bases of Circulatory Transport: Regulation and Exchange*, E.B. Reeve, A.C. Guyton, eds. (Saunders, Philadelphia 1967)
414. A.M. Scher: Cardiovascular control. In *Textbook of Physiology*, 21st edn, H.D. Patton, A.F. Fuchs, B. Hille, A.M. Scher, R. Steiner, eds. (Saunders, Philadelphia 1989), Ch. 40, pp. 834–847
415. A.M. Scher: Cardiovascular control. In *Textbook of Physiology*, 21st edn, H.D. Patton, A.F. Fuchs, B. Hille, A.M. Scher, R. Steiner, eds. (Saunders, Philadelphia 1989), Ch. 51, pp. 972–990
416. T.H. Schiebler: *Anatomie*, 9th edn (Springer, Berlin Heidelberg New York 2005)

417. A. Stalheim-Smith, G.K. Fitch: *Understanding Human Anatomy and Physiology* (West, St. Paul, MN 1993)
418. B.N. Tillmann: *Atlas der Anatomie des Menschen*, (Springer, Berlin Heidelberg New York 2005)
419. J.B. West: *Respiratory Physiology: The Essentials*, 7th edn (Lippincott Williams & Wilkins, Philadelphia 2005)

Chapter 9

420. G.B. Benedek, F.M.H. Villars: *Physics with Illustrative Examples from Medicine and Biology*, Three volumes, 2nd edn (AIP/Springer, New York/Berlin Heidelberg New York 2000)
421. W.A. Briscoe, A.B. DuBois: The relationship between airway resistance, airway conductance and lung volume in subjects of different age and body size. J. Clin. Invest. **37**, 1279–1285 (1958)
422. J.R. Cameron, J.G Skofronick, R. Grant: *Physics of the Body*, 2nd edn (Medical Physics, Madison, WI 1999)
423. H.K. Chang, M. Paiva (eds.): *Respiratory Physiology: An Analytical Approach*, Lung Biology in Health and Disease, Vol. 40, C. Lenfant, exec. ed. (Dekker, New York 1989)
424. R.G. Crystal, J.B. West, P.J. Barnes, E.R, Weibel (eds.): *The Lung: Scientific Foundations*, 2nd edn (Lippincott-Raven, New York 1997)
425. M.L. Foss, S.J. Keteyian: *Fox's Physiological Basis for Exercise and Sport*, 6th edn (McGraw-Hill, Boston 1998)
426. R.A. Freitas Jr.: *Nanomedicine, Volume I: Basic Capabilities* (Landes Bioscience, Austin 1999)
427. Y.C. Fung: *Biodynamics: Circulation* (Springer, Berlin Heidelberg New York 1984)
428. Y.C. Fung: *Biomechanics: Circulation*, 2nd edn (Springer, Berlin Heidelberg New York 1997)
429. J.F. Green: *Fundamental Cardiovascular and Pulmonary Physiology*, 2nd edn (Lea & Febiger, Phildelphia, 1987)
430. J. Hildebrandt: Structural and mechanical aspects of respiration. In *Textbook of Physiology*, 21st edn, H.D. Patton, A.F. Fuchs, B. Hille, A.M. Scher, R. Steiner, eds. (Saunders, Philadelphia 1989), Ch. 52, pp. 991–1011
431. J. Milic-Emili, E. D'Angelo: Work of breathing. In *The Lung: Scientific Foundations*, 2nd edn, R.G. Crystal, J.B. West, P.J. Barnes, E.R, Weibel, eds. (Lippincott-Raven, New York 1997), Ch. 105, pp. 1437–1446
432. G. Nucci, C. Cobelli: Mathematical models of respiratory mechanics. In *Modelling Methodology for Physiology and Medicine*, E. Carson, C. Cobelli, eds. (Academic, San Diego, 2001), Ch. 10, pp. 279–304
433. T.J. Pedley, R.C. Schroter, M.F. Sudlow: The prediction of pressure drop and variation of resistance within the human bronchial airways. Respir. Physiol. **9**, 387–405 (1970)
434. T.J. Pedley, R.D. Kamm: Dynamics of gas-flow and pressure-flow relationships. In *The Lung: Scientific Foundations*, 2nd edn, R.G. Crystal, J.B. West, P.J. Barnes, E.R. Weibel, eds. (Lippincott-Raven, New York 1997), Ch. 99, pp. 1365–1381

435. S.E. Rees, S. Kjaergaard, S. Andreassen: Mathematical modelling of pulmonary gas exchange. In *Modelling Methodology for Physiology and Medicine*, E. Carson, C. Cobelli, eds. (Academic, San Diego, 2001), Ch. 9, pp. 253–278
436. T.H. Schiebler: *Anatomie*, 9th edn (Springer, Berlin Heidelberg New York 2005)
437. R.F. Schmidt, F. Lang, G. Thews: *Physiologie des Menschen*, 29th edn (Springer, Berlin Heidelberg New York 2005)
438. S. Singhal, R. Henderson, K. Horsfield, K. Harding, G. Cumming: Morphometry of the human pulmonary arterial tree. Circ. Res. **33**, 190–197 (1973)
439. A. Stalheim-Smith, G.K. Fitch: *Understanding Human Anatomy and Physiology* (West, St. Paul, MN 1993)
440. E.R. Weibel: *Morphometry of the Human Lung* (Academic, New York, 1963)
441. E.R. Weibel: Lung morphometry and models in respiratory physiology. In *Respiratory Physiology: An Analytical Approach*, Lung Biology in Heatlh and Disease, H.K. Chang, M. Paiva, eds., C. Lenfant, exec. ed. (Dekker, New York 1989), Vol. 40, Ch. 1, pp. 1–56
442. E.R. Weibel: Design of airways and blood vessels considered as branching tress. In *The Lung: Scientific Foundations*, 2nd edn, R.G. Crystal, J.B. West, P.J. Barnes, E.R, Weibel, eds. (Lippincott-Raven, New York 1997), Ch. 74, pp. 1061–1071
443. J.B. West: *Respiratory Physiology: The Essentials*, 7th edn (Lippincott Williams & Wilkins, Philadelphia 2005)
444. J.B. West: *Respiratory Pathophysiology: The Essentials*, 6th edn (Lippincott Williams & Wilkins, Philadelphia 2003)
445. J.H. Wilmore, D.L. Costill: *Physiology of Sport and Exercise*, 3rd edn (Human Kinetics, Champaign, IL 2004)

Chapter 10

446. B.K. Ahlborn: *Zoological Physics: Quantitative Models, Body Design, Actions and Physical Limitations in Animals* (Springer, Berlin Heidelberg New York 2004)
447. R. McNeill Alexander: *The Human Machine* (Columbia University Press, New York 1992)
448. J.B. Allen, S.T. Neely: Micromechanical models of the cochlea. Phys. Today 40–47 (July 1992)
449. C. Avendano, L. Deng, H. Hermansky, B. Gold: The analysis and representation of speech. In *Speech Processing in the Auditory System*, S. Greenberg, W.A. Ainsworth, A.N. Popper, R.R. Fay, eds. (Springer, Berlin Heidelberg New York 2004), Ch. 2, pp. 63–100
450. P. Avery, S. Ehrlich: *Teaching American English Pronounicaiton* (Oxford University Press, New York 1992)
451. M.J. Ball, J. Rahilly: *Phonetics: The Science of Speech* (Arnold, London 1999)
452. G. Bekesy: Zur Theorie des Horens bei der Schallauftiahme durch Knochenleitung. Ann. Phys. **13**, 111–136 (1932)
453. G. von Békésy: *Experiments in Hearing* (McGraw-Hill, New York 1960)
454. Bell Labs. Record **12**(6), 314 (1934)

455. J.R. Cameron, J.G Skofronick, R. Grant: *Physics of the Body*, 2nd edn (Medical Physics, Madison, WI 1999)
456. G. Clark: *Cochlear Implants: Fundamentals and Applications* (Springer, Berlin Heidelberg New York 2003)
457. E.A. Culler, J.D. Coakley, K. Lowy, N. Gross: A revised frequency-map of the guinea-pig cochlea. Am. J. Psychol. **56**, 475–500 (1943)
458. J.T. Cushing: The spring-mass system revisited. Am. J. Phys. **52**, 925–933 (1984)
459. P. Dallos, A.N. Popper, R.R. Fay (eds.): *The Cochlea* (Springer, Berlin Heidelberg New York 1996)
460. R. Daniloff, G. Schuckers, L. Feth: *The Physiology of Speech and Hearing: An Introduction* (Prentice-Hall, Englewood Cliffs, NJ 1980)
461. P.B. Denes, E.N. Pinson: *The Speech Chain: The Physics and Biology of Spoken Language*, 2nd edn (W.H. Freeman and Co., NY 1993)
462. R.L. Diehl, B. Lindblom: Explaining the structure of feature and phoneme inventories: the role of auditory distinctiveness. In *Speech Processing in the Auditory System*, S. Greenberg, W.A. Ainsworth, A.N. Popper, R.R. Fay, eds. (Springer, Berlin Heidelberg New York 2004), Ch. 3, pp. 101–162
463. G.L. Fain: *Sensory Transduction* (Sinauer, Sunderland, MA 2003)
464. G. Fant: *Acoustic Theory of Speech Production* (Mouton, The Hague 1970)
465. R. Fettiplace, C.M. Hackney: The sensory and motor roles of auditory hair cells. Nat. Rev. Neurosci. **7**, 19–29 (2006)
466. N.H. Fletcher: *Acoustic Systems in Biology* (Oxford University Press, New York 1992)
467. R.A. Freitas Jr.: *Nanomedicine, Volume I: Basic Capabilities* (Landes Bioscience, Austin 1999)
468. B.O. French: Appraisal of Apollo launch noise. Aerospace Med. **38**, 719–722 (1967)
469. S.A. Gelfand: *Hearing: An Introduction to Psychological and Physiological Acoustics* (Dekker, New York 1981)
470. S.A. Gelfand: *Essentials of Audiology*, 2nd edn (Thieme, New York 2001)
471. D.C. Giancoli: *Physics*, 4th edn (Prentice-Hall, Englewood Cliffs, NJ 1995)
472. E.B. Goldstein: *Sensation and Perception*, 4th edn (Brooks/Cole, Pacific Grove 1996)
473. D.M. Green: *An Introduction to Hearing* (Lawrence Erlbaum, Hillsdale, NJ 1976)
474. S. Greenberg, W.A. Ainsworth, A.N. Popper, R.R. Fay (eds.): *Speech Processing in the Auditory System* (Springer, Berlin Heidelberg New York 2004)
475. J. Gruber: Basic multifrequency tympanometry: the physical background, www.lymenet.de/symptoms/tympanom/basictym.htm
476. A.C. Guyton, J.E. Hall: *Human Physiology and Mechanisms of Disease*, 6th edn (Saunders, Philadelphia 1997)
477. C.M. Hackney, D.H. Furness: Mechanotransduction in vertebrate hair cells: structure and function of the stereociliary bundle. Am. J. Physiol. Cell. Physiol. **268**, 1–13 (1995)
478. W.M. Hartmann: *Signals, Sound, and Sensation* (Springer, Berlin Heidelberg New York 1998)
479. H.L. Hawkins, T.A. McMullen, A.N. Popper, R.R. Fay (eds.): *Auditory Computation* (Springer, Berlin Heidelberg New York 1996)

480. D.C. Hodge, G.R. Garinther: Noise and blast. In *Bioastronautics Data Book*, J.F. Parker Jr., V.R. West, eds. (NASA, Washington, DC 1973), Ch. 15, pp. 693–750
481. International Standards Organization (ISO): *Recommendation 226: Normal Equal Loundness Contours for Pure Tones and Normal Thresholds of Hearing Under Free-Field Listening Conditions* (ISO, Geneva 1961)
482. J.F. Kennedy Space Center. Results of AS-503 facilities and environmental measurements, Vol. 1. Technical Report 927, January 1969
483. C.P. Lebo, R.C. Reddell: The presbycusis component in occupational hearing loss. Laryngoscope **82**, 1399–1409 (1972)
484. L.S. Lilly: Heart sounds and murmurs. In *Pathophysiology of Heart Disease: A Collaborative Project of Medical Students and Faculty*, 3rd edn, L.S. Lilly, ed. (Lippincott Williams & Wilkins, Philadelphia 2003), Ch. 2, pp. 29–43
485. R.G. Loudon, R.L.H. Murphy: Lung sounds. In *The Lung: Scientific Foundations*, 2nd edn, R.G. Crystal, J.B. West, P.J. Barnes, E.R, Weibel, eds. (Lippincott-Raven, New York 1997), Ch. 100, pp. 1383–1391
486. T.J. Lynch III, V. Nedzelnitsky, W.T. Peake: Input impedance of the cochlea in cat. J. Acoust. Soc. Am. **72**, 108–130 (1982)
487. E.B. Magid, R.R. Coermann, G. H. Ziegenruecker: Human tolerance to whole body sinusoidal vibration. Aerospace Med. **31**, 921 (1960)
488. R.H. Margolis, L.L. Hunter: Acoustic immittance measurements. In *Audiology: Diagnosis*, R.J. Roeser, M. Valente, H. Hosford-Dunn, eds. (Thieme, New York, 2000), Ch. 17, pp. 381–423
489. W.N. McDicken: *Diagnostic Ultrasonics: Principles and Use of Instruments*, 3rd edn (Churchill Livingstone, Edinburgh 1991)
490. H.J. Metcalf: *Topics in Classical Biophysics* (Prentice-Hall, Englewood Cliffs, NJ 1980)
491. H.F. Olson: *Acoustical Engineering* (Van Nostrand, Princeton 1957)
492. H.F. Olson: *Music, Physics, and Engineering*, 2nd edn (Dover, New York 1967)
493. G.E. Peterson, H.L. Barney: Control methods used in a study of vowels. J. Acoust. Soc. Am. **24**, 175–184 (1952)
494. D.W. Robinson, R.S. Dadson: A re-determination of the equal loudness relations for pure waves. Br. J. Appl. Phys. **7**, 166–181 (1956)
495. D.W. Robinson, G.J. Sutton: Age effect in hearing: a comparative analysis of threshold data. Audiology **18**, 320–334 (1979)
496. R.J. Roeser, M. Valente, H. Hosford-Dunn (eds.): *Audiology: Diagnosis* (Thieme, New York 2000)
497. R.F. Schmidt, F. Lang, G. Thews: *Physiologie des Menschen*, 29th edn (Springer, Berlin Heidelberg New York 2005)
498. E.D. Schubert: *Hearing: Its Function and Dysfunction* (Springer, Berlin Heidelberg New York 1980)
499. H. Spoendlin: The afferent innervation of the cochlea. In *Electrical Activity of the Auditory Nervous System*, R.F. Naunton, C. Fernandez, eds. (Academic, London, 1978), pp. 21–41
500. A. Spoor: Presbycusis values in relation to noise-induced hearing loss. Int. Audiol. **6**, 48–57 (1967)
501. K.N. Stevens: *Acoustic Phonetics* (MIT, Cambridge, MA 1998)
502. S.S. Stevens, H. Davis: *Hearing: Its Psychology and Physiology* (Wiley, New York 1938), Reprinted by the Acoustical Society of America (1983)

503. S.S. Stevens, J. Volkman: The relation of pitch to frequency. Am. J. Psych. **53**, 329–353 (1940)
504. I.R. Titze: *Principles of Voice Production* (Prentice-Hall, Englewood Cliffs, NJ 1994)
505. J. Tonndorf, S.M. Khanna: The role of the tympanic membrane in middle ear transmission. Ann. Otol. Rhinol. Laryngol. **79**, 743–753 (1970)
506. S. Webb (ed.): *The Physics of Medical Imaging* (Adam Hilger, Bristol 1988)
507. D.A. Winter: *Biomechanics and Motor Control of Human Movement*, 3rd edn (Wiley, New York 2005)
508. A.B. Wolbarst: *The Physics of Radiology*, (Appleton & Lange, Norwalk, CT 1993)
509. W.A. Yost: *Fundamentals of Hearing: An Introduction*, 3rd edn (Academic, San Diego 1994)
510. E. Zwicker, H. Fastl: *Psychoacoustics: Facts and Models* (Springer, Berlin Heidelberg New York 1990)

Chapter 11

511. M. Bass (ed.): *Handbook of Optics, Volume III: Classical Optics, Vision, X-Ray Optics*, 2nd edn (McGraw-Hill, New York 2001)
512. R.E. Bedford, G. Wyszecki: Axial chromatic aberration of the human eye. J. Opt. Soc. Am. **47**, 564–565 (1957)
513. A.G. Bennett, R.B. Rabbetts: *Clinical Visual Optics*, 2nd edn (Butterworth, London 1989)
514. E.A. Boettner, J.R. Wolter: *Transmission of the ocular media*, Invest. Ophthalmol. Vis. Sci. **1**, 776–783 (1962)
515. P.K. Brown, G. Wald: Visual pigments in single rods and cones of the human retina. Science **144**, 45–51 (1964)
516. P. Buser, M. Imbert, translated by R.H. Kay: *Vision* (MIT, Cambridge, MA 1992)
517. F.W. Campbell, R.W. Gubisch: Optical quality of the human eye. J. Physiol. **186**, 558–578 (1966)
518. W.N. Charman: Optics of the eye. In *Handbook of Optics, Volume I: Fundamentals, Techniques and Design*, 2nd edn, M. Bass, ed. (McGraw-Hill, New York, 1995), Ch. 24
519. H.H. Emsley: *Visual Optics* (Hatton, London 1936)
520. G.L. Fain: *Sensory Transduction* (Sinauer, Sunderland, MA 2003)
521. R.F. Fisher: Presbyopia and the changes with age in the human crystalline lens. J Physiol. **228**, 765–779 (1973)
522. B. Fortner, T.E. Meyer: *Number by Colors: A Guide to Using Color to Understand Technical Data* (Springer/TELOS, Berlin Heidelberg New York/New York 1997)
523. E.B. Goldstein: *Sensation and Perception*, 4th edn (Brooks/Cole, Pacific Grove 1996)
524. P.R. Greene: Mechanical considerations in myopia. In *Refractive Anomalies: Research and Clinical Applications*, T. Grosvenor, M.C. Flom, eds. (Butterworth-Heinemann, Boston 1991), pp. 287–300

525. A. Gullstrand: The optical system of the eye. In *Physiological Optics*, H. von Helmholtz, ed. Vol. 1, pp. 350–358 (1909) Appendix II.3. English tranlsation: J.P.C. Southall, ed. (Optical Society of America, New York). Reprinted (Dover, 1962)
526. A.C. Guyton, J.E. Hall: *Human Physiology and Mechanisms of Disease*, 6th edn (Saunders, Philadelphia 1997)
527. W.M. Hart Jr. (ed.): *Adler's Physiology of the Eye: Clincal Application*, 9th edn (Mosby, St. Louis 1992)
528. J. Hartstein: *Questions and Answers on Contact Lens Practice*, 2nd edn (Mosby, St. Louis 1973)
529. E. Hecht: *Optics*, 4th edn (Addison Wesley, Reading, MA 2001)
530. D.H. Hubel: *Eye, Brain, and Vision* (Scientific American Library, New York 1987)
531. D.H. Hubel: Exploration of the primary visual cortex, 1955–78 (Nobel Lecture). Nature **299**, 515–524 (1982)
532. D.H. Hubel, T.N. Wiesel: Brain mechanisms of vision. Sci. Am. **241**, 130 (1979)
533. D.B. Judd, G. Wyszecki: *Color in Business, Science, and Industry*, 3rd edn (Wiley, New York, 1975)
534. M.V. Klein, T.E. Furtak: *Optics*, 2nd edn (Wiley, New York 1986)
535. Y. Le Grand, S.G. El Hage: *Physiological Optics* (Springer, Berlin Heidelberg New York 1980)
536. E.R. Lewis, Y.Y. Zeevi, F.S. Werblin: Scanning electron microscopy of vertebrate visual receptors. Brain Res. **15**, 559–562 (1969)
537. A. Lewis, L.V. Del Priore: The biophysics of visual photoreception. Phys. Today, 38–46 (January 1988)
538. W.F. Long, R. Garzia, J.L. Weaver: Assessment of refraction and refractive errors. In *Handbook of Optics, Volume III: Classical Optics, Vision, X-Ray Optics*, 2nd edn, M. Bass, ed. (McGraw-Hill, New York, 2001), Ch. 11
539. I.A. Mackie: *Medical Contact Lens Practice: A Systematic Approach* (Butterworth Heinemann, Oxford 1993)
540. W.B. Marks, W.H. Dobelle, E.F. MacNichol Jr.: Visual pigments of single primate cones. Science **143**, 1181–1183 (1964)
541. M.J. Mannis, K. Zadnik, C. Coral-Ghanem, N. Kara-Jose: *Contact Lenses in Ophthalmic Practice* (Springer, Berlin Heidelberg New York 2004)
542. J.T. McIlwain: *An Introduction to the Biology of Vision* (Cambridge University Press, Cambridge 1996)
543. D.D. Michaels: *Visual Optics and Refraction: A Clinical Approach*, 3rd edn (Mosby, St. Louis 1985)
544. P. Mouroulis, J. Macdonald: *Geometrical Optics and Optical Design* (Oxford, New York 1997)
545. National Eye Institute, National Institutes of Health. http://www.nei.nih.gov/photo/charts/index.asp
546. Y. Ohno: Radiometry and photometry review for vision optics. In *Handbook of Optics, Volume III: Classical Optics, Vision, X-Ray Optics*, 2nd edn, M. Bass, ed. (McGraw-Hill, New York, 2001), Ch. 14
547. N. Ohta, A.R. Robertson: *Colorimetery: Fundamentals and Applications* (Wiley, New York 2005)

548. J.M. Palmer: Radiometry and photometry: units and conversions. In *Handbook of Optics, Volume III: Classical Optics, Vision, X-Ray Optics*, 2nd edn, M. Bass, ed. (McGraw-Hill, New York, 2001), Ch. 7
549. M.S. Rea (ed.): *The IESNA Lighting Handbook*, 9th edn (Illuminating Engineering Society of North America, New York 2000)
550. P. Riordan-Eva, J.P. Whitcher: *Vaughan and Asbury's General Ophthalmology*, 16th edn (Lange Medical Books/McGraw-Hill, New York 2004)
551. R.F. Schmidt, F. Lang, G. Thews: *Physiologie des Menschen*, 29th edn (Springer, Berlin Heidelberg New York 2005)
552. S.H. Schwartz: *Visual Perception: A Clinical Orientation*, 2nd edn (Appleton and Lange, Stamford, CT 1999)
553. S.H. Schwartz: *Geometrical and Visual Optics: A Clinical Introduction* (McGraw-Hill, New York 2002)
554. D.H. Sliney, Ocular radiation hazards. In *Handbook of Optics, Volume III: Classical Optics, Vision, X-Ray Optics*, 2nd edn, M. Bass, ed. (McGraw-Hill, New York 2001), Ch. 15
555. W.J. Smith: *Modern Optical Engineering: The Design of Optical Systems*, 2nd edn (McGraw-Hill, New York 1990)
556. H.A. Stein, B.J. Slatt, R.M. Stein, M.I. Freeman: *Fitting Guide for Rigid and Soft Contact Lenses: A Practical Approach*, 4th edn (Mosby, St. Louis 2002)
557. L.N. Thibos, Y. Ming, Z. Xiaoxiao, A. Bradley: The chromatic eye: a new reduced-eye model of ocular chromatic aberration in humans. Appl. Opt. **31**, 3594–3600 (1992)
558. B.N. Tillmann: *Atlas der Anatomie des Menschen* (Springer, Berlin Heidelberg New York 2005)
559. D. Vaughan, T. Asbury: *General Ophthalmology*, 10th edn (Lange Medical, Los Altos, CA 1983)
560. B.A. Wandell: *Foundations of Vision*, (Sinauer, Sunderland, MA 1995)
561. G. Westheimer: The eye as an optical instrument. In *Handbook of Perception*, J. Thomas et al., eds. (Wiley, New York 1986), pp. 4.1–4.20
562. T.N. Wiesel: Postnatal development of the visual cortex and the influence of environment (Nobel Lecture). Nature **299**, 583–591 (1982)
563. D.R. Williams, D.H. Brainard, M.J. MacMahon, R. Navarro: Double-pass and interferometric measures of the optical quality of the eye. J. Opt. Soc. Am. A **11**, 3123–3135 (1994)
564. G. Wyszecki, W.S. Stiles: *Color Science: Concepts and Methods, Quantitative Data and Formulae*, 2nd edn (Wiley, New York 1982)

Chapter 12

565. B.K. Ahlborn: *Zoological Physics: Quantitative Models, Body Design, Actions and Physical Limitations in Animals* (Springer, Berlin Heidelberg New York 2004)
566. G.B. Benedek, F.M.H. Villars: *Physics with Illustrative Examples from Medicine and Biology*, Three volumes, 2nd edn (AIP/Springer, New York/Berlin Heidelberg New York 2000)
567. B.H. Brown, R.H. Smallwood, D.C. Barber, P.V. Lawford, D.R. Hose: *Medical Physics and Biomedical Engineering* (Institute of Physics, Philadelphia 1999)

568. J.R. Cameron, J.G. Skofronick, R. Grant: *Physics of the Body*, 2nd edn (Medical Physics, Madison, WI 1999)
569. R.M.J. Cotterill: *Biophysics: An Introduction* (Wiley, New York 2002)
570. P. Davidovits: *Physics in Biology and Medicine*, 2nd edn (Elsevier/Academic, San Diego 2001)
571. T.A. Delcha: *Physics in Medical Diagnosis* (Chapman & Hall, New York 1997)
572. D. Dubin: *Rapid Interpretation of EKG's*, 6th edn (COVER, Tampa, FL 2000)
573. F.A. Duck: *Physical Properties of Tissue: A Comprehensive Reference Book* (Academic, London 1990)
574. G.L. Fain: *Sensory Transduction* (Sinauer, Sunderland, MA, 2003)
575. M.L. Foss, S.J. Keteyian: *Fox's Physiological Basis for Exercise and Sport*, 6th edn (McGraw-Hill, Boston 1998)
576. R.A. Freitas Jr.: *Nanomedicine, Volume I: Basic Capabilities* (Landes Bioscience, Austin 1999)
577. M. Gertsch: *The ECG: A Two-Step Approach to Diagnosis* (Springer, Berlin Heidelberg New York 2004)
578. N. Gimbel, C. Fox: "Killing Me Softly (With His Song)", written by Norman Gimbel and Charles Fox, published by Fox-Gimbel (BMI) and Rodali Music (BMI)
579. S. Grimnes, Ø.G. Martinsen: *Bioimpedance and Bioelectricity Basics* (Academic, San Diego, 2000)
580. A.C. Guyton, J.E. Hall: *Human Physiology and Mechanisms of Disease*, 6th edn (Saunders, Philadelphia 1997)
581. R.K. Hobbie: *Intermediate Physics for Medicine and Biology*, 3rd edn (AIP, New York 1997)
582. F.C. Hoppensteadt, C.S. Peskin: *Modeling and Simulation in Medicine and the Life Sciences*, 2nd edn (Springer, Berlin Heidelberg New York 2002) (First Edition from 1992 was called *Mathematics in Medicine and the Life Sciences*)
583. A.J. Hudspeth, D.P. Corey: Sensitivity, polarity, and conductance change in the response of vertebrate hair cells to controlled mechanical stimuli. Proc. Natl Acad. Sci. USA **74**, 2407–2411 (1977)
584. L.S. Lilly: The Electrocardiogram. In *Pathophysiology of Heart Disease: A Collaborative Project of Medical Students and Faculty*, 3rd edn, L.S. Lilly, ed. (Lippincott Williams & Wilkins, Philadelphia 2003), Ch. 4, pp. 75–110
585. R. Malhotra, E.R. Edelman, L.S. Lilly: Basic cardiac structure and function. In *Pathophysiology of Heart Disease: A Collaborative Project of Medical Students and Faculty*, 3rd edn, L.S. Lilly, ed. (Lippincott Williams & Wilkins, Philadelphia 2003), Ch. 1, pp. 1–27
586. J. Malmivuo, R. Plonsey: *Bioelectromagnetism* (Oxford University Press, New York 1995)
587. H.J. Metcalf: *Topics in Classical Biophysics* (Prentice-Hall, Englewood Cliffs, NJ 1980)
588. K.L. Moore, A.M.R. Agur: *Essential Clinical Anatomy*, 3rd edn (Lippincott Williams & Wilkins, Philadelphia 2007)
589. W. Rall: Cable theory for dendritic neurons. In *Methods in Neuronal Modeling: From Synapses to Networks*, C. Koch, I. Segev, ed. (MIT, Cambridge, MA 1989), Ch. 2, pp. 9–62
590. R.M. Ross, A.S. Jackson: *Exercise Concepts, Calculations, and Computer Applications*, (Benchmark, Carmel, IN, 1990)

591. A.M. Scher: The electrocardiogram. In *Textbook of Physiology*, 21st edn, H.D. Patton, A.F. Fuchs, B. Hille, A.M. Scher, R. Steiner, ed. (Saunders, Philadelphia 1989), Ch. 38, pp. 796–819
592. T.H. Schiebler: *Anatomie*, 9th edn (Springer, Berlin Heidelberg New York 2005)
593. R.F. Schmidt, F. Lang, G. Thews: *Physiologie des Menschen*, 29th edn (Springer, Berlin Heidelberg New York 2005)
594. A. Stalheim-Smith, G.K. Fitch: *Understanding Human Anatomy and Physiology* (West, St. Paul, MN 1993)
595. Thomas Jefferson National Accelerator Facility (U.S.), Jefferson Lab Environment, Safety, Health, and Quality Manual, http://www.jlab.org/ehs/manual/EHSbook-429.html
596. J.A. Tuszynski, J.M. Dixon: *Biomedical Applications of Introductory Physics* (Wiley, New York 2002)
597. J.B. West (ed.): *Best and Taylor's Physiological Basis of Medical Practice*, 12th edn (Williams & Wilkins, Baltimore 1991)

Chapter 13

598. G.B. Benedek, F.M.H. Villars: *Physics with Illustrative Examples from Medicine and Biology*, Three volumes, 2nd edn (AIP/Springer, New York/Berlin Heidelberg New York 2000)
599. T.H. Benzinger: Heat regulation: homeostasis of central temperature in man. Physiol. Rev. **49**, 671–759 (1969)
600. P.J. Berenson, W.G. Robertson: Temperature. In *Bioastronautics Data Book*, J.F. Parker Jr., V.R. West, eds. (NASA, Washington, DC 1973), Ch. 3, pp. 65–148
601. M.L. Foss, S.J. Keteyian: *Fox's Physiological Basis for Exercise and Sport*, 6th edn (McGraw-Hill, Boston 1998)
602. R.K. Hobbie: *Intermediate Physics for Medicine and Biology*, 3rd edn (AIP, New York 1997)
603. J. Keener, J. Sneyd: *Mathematical Physiology* (Springer, Berlin Heidelberg New York 1998)
604. M.C.K. Khoo: *Physiological Control Systems: Analysis, Simulation, and Estimation* (IEEE, Piscataway, NJ 2000)
605. H.J. Metcalf: *Topics in Classical Biophysics* (Prentice-Hall, Englewood Cliffs, NJ 1980)
606. A.M. Scher: Cardiovascular control. In *Textbook of Physiology*, 21st edn, H.D. Patton, A.F. Fuchs, B. Hille, A.M. Scher, R. Steiner, ed. (Saunders, Philadelphia 1989), Ch. 51, pp. 972–990
607. A. Stalheim-Smith, G.K. Fitch: *Understanding Human Anatomy and Physiology* (West, St. Paul, MN 1993)
608. J.H. Wilmore, D.L. Costill: *Physiology of Sport and Exercise*, 3rd edn (Human Kinetics, Champaign, IL 2004)

Appendix D

609. H.F. Olson: *Acoustical Engineering* (Van Nostrand, Princeton 1957)

610. H.F. Olson: *Music, Physics, and Engineering*, 2nd edn (Dover, New York 1967)

Appendix E

611. R.M.J. Cotterill: *Biophysics: An Introduction* (Wiley, New York 2002)
612. M. Daune: *Molecular Biophysics: Structures in Motion* (Oxford University Press, Oxford 1999)
613. R. Glaser: *Biophysics* (Springer, Berlin Heidelberg New York 2000)
614. F.R. Hallett, R.H. Stinson, P.A. Speight: *Physics for the Biological Sciences: A Topical Approach to Biophysical Concepts* (Methuen/Chapman & Hall, Toronto 1982)
615. P. Maróti, L. Berkes, F. Tölgyesi: *Biophysics Problems: A Textbook with Answers* (Akadémiai Kiadó, Budapest 1998)
616. H.J. Metcalf: *Topics in Classical Biophysics* (Prentice-Hall, Englewood Cliffs, NJ 1980)
617. P. Nelson: *Biological Physics: Energy, Information, Life* (Freeman, New York 2003)
618. B. Nolting: *Methods in Modern Biophysics* (Springer, Berlin Heidelberg New York 2003)
619. R.B. Setlow, E.C. Pollard, *Molecular Biophysics* (Addison-Wesley, Reading, MA 1962)
620. M.P. Sheetz (ed.): *Laser Tweezers in Cell Biology*, Methods in Cell Biology, Vol. 55 (Academic, San Diego 1998)
621. R. Phillips, S.R. Quake: The biological frontier of physics. Phys. Today **59**(5), 38–43 (2006)
622. R.E. Goldstein, P.C. Nelson, T.R. Powers: Teaching biological physics. Phys. Today **58**(3), 46–51 (2005)

Index